BSCS Biology

BSCS Biology
An Ecological Approach

EIGHTH EDITION
BSCS Green Version

TEACHER'S EDITION

BSCS
Pikes Peak Research Park
5415 Mark Dabling Blvd.
Colorado Springs, CO 80918-3842

Revision Team
William J. Cairney, BSCS, Revision Coordinator
J. Frank Cassel, North Dakota State University
Patricia Cully, Educational and Editorial Consultant, Needham, Massachusetts
Janet Chatlain Girard, BSCS, Art Coordinator
Kenneth G. Rainis, Ward's Natural Science Establishment, Inc.
Gordon E. Uno, University of Oklahoma, Norman, Oklahoma
Linda K. Ward, BSCS, Executive Assistant

KENDALL/HUNT PUBLISHING COMPANY
4050 Westmark Drive, Dubuque, IA 52004

Contributors

Barbara Andrews, Colorado Springs, Colorado
Charles R. Barman, Indiana University at Kokomo
Don Chmielowiec, Ward's Natural Science Establishment, Inc.
Edward Drexler, Pius XI High School, Milwaukee, Wisconsin
Vernon L. Gilliland, Liberal Senior High School, Liberal, Kansas
Lynn Griffin, Cherry Creek High School, Englewood, Colorado
Donald P. Kelley, South Burlington, Vermont
Joseph D. McInerney, BSCS
George Nassis, Ward's Natural Science Establishment, Inc.
Larry N. Norton, Quantum Technology, Inc.
Chester Penk, Quantum Technology, Inc.
Barry Schwartz, Chatfield High School, Littleton, Colorado
John Settlage, Boston, Massachusetts
Carol Leth Stone, Alameda, California
Richard R. Tolman, Brigham Young University, Provo, Utah
David A. Zegers, Millersville University, Millersville, Pennsylvania

Safety Consultant

Kenneth G. Rainis, Ward's Natural Science
    Establishment, Inc.

BSCS Administrative Staff

Timothy Goldsmith, Chair, Board of Directors
Joseph D. McInerney, Director
Larry Satkowiak, Chief Financial Officer

Artists

Susan Bartel
Bill Border
Marjorie C. Leggitt
Audrey Penk
Brent Sauerhagen
Page Louis Thomas
Linn Trochim/Animart

Grateful acowledgement is made to the following individuals for permission to reprint from their work:

    D. Nelkin and L. Tancredi, *Dangerous Diagnostics: The Social Power of Biological Information* (New York: Basic Books, 1989).

    N. A. Holtzman, *Proceed with Caution: Predicting Genetic Risks in the Recombinant DNA Era* (Baltimore, MD: Johns Hopkins University Press, 1989).

Acknowledgements continue on page 771

Editorial, design, and production services provided by
PC&F, Hudson, New Hampshire

Teacher's Edition Contents

The Green Version Program

Introduction to the Teacher's Edition

The Teacher's Edition (TE) of *BSCS Biology: An Ecological Approach* includes all of the student material, as well as a variety of philosophical, instructional, and procedural aids especially prepared for Green Version teachers. It is designed to help you and your students gain maximum benefit from using the Green Version program and should greatly facilitate your work.

The Teacher's Edition consists of front matter and annotations in the student text. The first three features of the front matter explain the principles of the Green Version program and describe the program's components. The next seven features help you plan for the school year, order and prepare the necessary materials, and develop a laboratory safety program. Four short features then introduce theory-based teaching strategies that are effective for learning secondary biology as well as other subjects. Further information on these topics can be found in the references listed on page T33. The last two features offer suggestions for evaluating student work and keeping current in a rapidly changing field.

The bulk of the front matter is devoted to teaching strategies for each chapter. These strategies supply information for planning ahead, list major concepts and objectives, and offer specific strategies, references, and suggested answers to Applications and Problems at the end of each chapter. Preparation requirements and materials lists for a class of 30 are given for each investigation in the chapter.

Annotations in the Teacher's Edition include a list of major concepts and questions that will help you assess students' prior knowledge at the start of each chapter, and notes that provide supplementary information, suggest questions to raise with students and answer questions asked in the figure legends. Numbers beside the text indicate answers to the Concept Review questions. Each investigation is annotated with tips for conducting the investigation and answers to the questions in the procedures and discussions.

We welcome your comments about any aspect of this program. Please direct your correspondence to:

BSCS, GV8
Pikes Peak Research Park
5415 Mark Dabling Blvd.
Colorado Springs, CO 80918

Instructional Theory of the BSCS Green Version

One of the practices of the BSCS is to utilize the results of current research in learning, cognitive psychology, and science education in the development of life sciences curricula. Perhaps the most important theory in education today is the theory of **active learning.** Educational research indicates that students who are actively engaged in the learning process are the most successful learners. If a student is involved directly (i.e., physically, emotionally, and mentally) with the concepts or skills to be learned, he or she acquires a deeper understanding of the material that is retained longer than if the learning experience is passive. Thus, lectures, although superficially expedient, are not effective teaching strategies. The BSCS Green Version program offers a rich array of experienced-based instructional activities. Among these activities are laboratory investigations, projects, illustrations, interactive readings, and inquiry discussions. Because the program contains more activities than can be done in a single school year, you can select those that are most appropriate for your students.

Because it is student-centered, **inquiry** is central to active learning. A pragmatic expression of the principle of inquiry is, "Let the student figure out the concept, rather than telling it to them." Thus, most of the activities in the BSCS Green Version are inquiry-based and may appear to be indirect when compared with traditional instruction. As a rule, less material can be covered using inquiry-based instruction than using lecture-based instruction. You as the teacher must decide how to trade quantity for quality. It is the firm belief of BSCS that thoroughly developing selected concepts central to biology is preferable to covering many topics superficially.

A great deal of what is taught in secondary science traditionally is abstract, but the vast majority of adolescent learners are not yet abstract thinkers. Most ninth- and tenth-graders, however, are capable of understanding abstract concepts if the learning stage is appropriately set. Research in cognitive psychology supports the following instructional hypotheses.

1. **Learners benefit most from concrete learning experiences *prior* to trying to understand abstract science concepts.** Reading and lectures

should follow, rather than precede, an engaging activity, such as a laboratory investigation, that is related to the concepts students will study. All but the most motivated science students learn little from lectures and reading alone. A productive learning sequence might be (a) orientation, (b) hands-on investigation, (c) discussion and some lecture, and then (d) reading and working on problems.

2. **Learners must reconstruct knowledge as if it were entirely new to them.** Most knowledge, if it is to be applied, cannot be merely imparted. Learners must interact with and reconstruct concepts for themselves. Although interaction does not need to be with concrete objects always, there should be meaningful interaction between the learner and the knowledge to be acquired.

3. **Learners attempt to build onto their existing cognitive framework.** Having a conceptual framework and organizers that fit into the framework allows students to fit what they are learning into what they already know. When the connection is successful, the new knowledge is retrieved more easily, lasts longer, and is more meaningful when applied in other contexts.

The activities in this program utilize these three principles. Used collectively and in an appropriate sequence, they can provide a positive learning experience for your students.

The BSCS Philosophy—A Conceptual Framework for Biology

The BSCS was established to improve biological education. Among its early concerns was the formulation of goals for those who, under the BSCS aegis, would be developing new educational programs and for the teachers and students who would be using the new materials. Those goals, which still are valid, are to develop an understanding and appreciation of:

◆ the nature of scientific inquiry. Science is an open-ended, intellectual activity. What is known presently or believed is subject to change at any time.

◆ the limitations of science and of scientific methods.

◆ the diversity of life and the interrelations existing between organisms.

◆ the biological bases of problems in medicine, public health, agriculture, and conservation.

◆ the historical development of biological concepts and the relationship of these concepts to the society and technology of each age.

◆ the beauty and drama of the living world.

◆ the place of humans in nature. As living organisms, we have much in common with other organisms. We interact with all organisms in the biological systems of the earth, and we must share the earth with them.

BSCS Biology attempts to present biology as an experimental science, to demonstrate the status of biology in the twentieth century, and to illustrate its usefulness for students who will spend most of their lives in the twenty-first century. The BSCS staff believes that many teachers are not satisfied with teaching biology as a series of vocabulary terms or a taxonomic exercise in rote memory. Accordingly, BSCS materials move beyond the two levels of the whole organism and its organs and tissues. Students are exposed to seven levels of biological organization—from the molecular level to the level of the biosphere.

The Green Version focuses on the content of biology at the levels of organization of populations, communities, and the biosphere. In so doing, the Green Version aspires to give students insight into the biosphere—insight that will enrich their lives and their ability to become responsible citizens.

Woven through the text are 10 basic biological themes that form a framework for teaching the course.

1. **Evolution: change of living things through time.** The theory of evolution is basic to biology. The study of evolution permits biologists to make order out of the similarities and differences among living things. The theory of evolution, like other theories, is a body of interrelated data. As new data are obtained, interpretations may change, but that does not mean the basic theory is unsound.

2. **Diversity of type and unity of pattern in living things.** The diversity of living forms and their adaptation to widely differing environmental conditions are due to evolution and natural selection. Unity is represented by the similarities that exist in characteristics and in biochemical processes.

3. **The genetic continuity of life.** Life is a continuing stream of genetic information passed from generation to generation. The characteristics of the genetic mechanisms not only bring about duplication of living forms but also introduce variations through genetic recombinations and through errors of genetic replication that create new material for natural selection.

4. **The complementarity of organism and environment.** Organisms and the environment interact at

all levels of biological organization, from genes and cell organelles to populations and their ecosystems. There is a reciprocal relationship between organisms and their environment. Organisms, by their very existence, modify their environment and exploit it. In turn, they are limited by specific environmental conditions.

5. **The biological roots of behavior.** Behavior is an organism's reaction to its environment, both internal and external, that affects the organism's, and thus the species', ability to survive. It is the product of genes influenced by the environment—a biological process with a genetic basis that is shaped by evolution. The biology of organisms imposes limits on what they are capable of doing; their behavior is a reaction to stimuli from the external environment, such as gravity and light, as well as stimuli from the internal environment, such as hormones and enzymes. Adaptive behavior enhances an individual's probability of surviving to produce offspring; the number of descendants affects the survival of the species as a whole. Thus, the behavior of each member of a species helps to determine whether the species continues to reproduce and thrive.

6. **The complementarity of structure and function.** What an organ or organelle does depends on its structure. Conversely, function can be inferred from a given structure. Complementarity of structure and function apply to all levels of biological organization.

7. **Regulation and homeostasis: preservation of life in the face of change.** Through homeostasis and regulation, organisms have the capacity to adjust to change. Homeostasis—the maintenance of a stable internal environment—is made possible by regulatory feedback mechanisms that sense changes in the internal and external environments and respond to them.

8. **Science as inquiry.** An inquiry approach allows students to see the problems posed and the experiments performed, the data found, and the interpretation made that converts those data into scientific knowledge. It also includes a fair treatment of the doubts and incompleteness of science, thereby contradicting the notion that all is known and that science consists of unalterable fixed truths. Laboratory teaching allows student some insight into the world of biologists by reflecting the investigative, experimental approach of scientific enterprise, which includes the possibility that scientific knowledge can change with further inquiry. Students see how scientific data are derived and develop a conviction that the scientific process is valid.

9. **The history of biological concepts.** The investigation of the history of scientific concepts helps students acquire a realistic understanding of science and scientists. By understanding history, students can add to their knowledge of scientific processes and learn that scientific research does not invariably have a formal plan and clear purpose—that chance, intuition, and serendipity sometimes intervene. The history of scientific concepts underlies much of our changing technology, agriculture, medicine, and management of natural resources, as well as our changing body of scientific knowledge.

10. **Science and society.** The study of biology is most useful when students apply their understanding and skills to personal issues and to societal programs. By presenting biological concepts in the context of important social issues such as genetic engineering, pollution, and population growth, we help students to develop a deeper understanding of biology and a richer appreciation of the central role of science in society.

The BSCS position on biological education begins with a clear rationale for teaching and learning biology. The following is a summary of that position.

The BSCS . . . [is] concerned not only with improving the subject matter being presented under the title "biology" but also with the manner of presentation, the emphasis, and the focus. . . .

As the BSCS works on the high school biology program, we hope that biology—and indeed all science—will be presented as an unending search for meaning, rather than as a body of dogma. . . . Our main objective is to lead each student to conceive of biology as a science, and of the process[es] of science as reliable method[s] of gaining objective knowledge.

To a very great extent the key to this understanding lies in meaningful laboratory and field study that incorporates honest investigation of real scientific problems.

The aim of the BSCS is to place biological knowledges in its fullest modern perspective. If we are successful, students of the new biology should acquire an intellectual and esthetic appreciation not only for the complexities of living things and their interrelationships in nature but also for the ways in which new knowledge is gained and tested, old errors eliminated, and an ever closer approximation to truth attained.

(H. Bentley Glass, 1960)

The Green Version reflects this rationale and is constructed on the premises presented in the themes. The book is structured around two main threads into which the other content areas are woven—ecological relationships and evolution. It is aimed at the general student of biology, not just college-bound individuals. The intention of the writers and of the BSCS is to present students with a method for investigating and testing their own environment in a culture dominated by technology. As participants in a democratic society, these young adults will be responsible for maintaining their country, and to some extent the world, as a healthy and viable environment for themselves and future generations.

Components of the Program

The Student Text

The student text is a complete program for introductory biology. It consists of the following features:

- A table of contents that provides an overview of the organization of the book

- Section and chapter opening photographs with legends that encourage student interaction with the text

- Section introductions that include a list of the chapters and a brief overview to set the stage for the section

- Chapter introductions that preview the chapter and relate it to the other chapters

- Major headings that divide each chapter into parts that deal with a major topic

- Guidepost questions that arouse student interest and curiosity and help students identify important ideas

- Numbered minor headings that develop a specific idea within each major heading

- Concept Review questions at the end of each major headings that reexamine key ideas within the heading through a series of recall questions that allow students to check their understanding

- Illustrations that help to explain concepts and processes, emphasize specific points in the text, and portray some of the diversity and wonder of life. Figure legends relate to the text and often extend it.

- New terms central to understanding that are defined directly within the text, usually with examples, and printed in boldface type with pronunciations in parentheses

- Laboratory investigations that are integrated into each chapter to engage student interest through discovery or to verify and extend the material being studied. These investigations often allow students to form and test hypotheses, and in so far as possible, are inquiry oriented.

- Biology Today features that contain information on careers in biology and the latest research in biology and technology

- Chapter summaries that review the major concepts presented in each chapter

- Application questions that test students' ability to synthesize the knowledge they have gained from studying the chapter

- Problems that extend learning beyond the chapter and offer opportunities for creativity, research, or independent study

- Four appendices that further extend the text. Appendix 1 contains safety guidelines, general laboratory procedures, and information about the metric system. Appendix 2 includes investigations on laboratory safety and the use of microscopes. Appendix 3 contains tables and charts referenced in the chapters. Appendix 4, *A Catalog of Living Things,* provides a convenient means of identifying many organisms.

- A glossary that defines all the important terms used in the book, including the boldfaced terms defined in the text

- An index that acts as a guide to all the information in the book

The Student Study Guide

The *Student Study Guide* is a unique resource of the Green Version. As a resource book of learning skills activities, it contains a variety of activities that develop the basic concepts of the program and go beyond the student textbook. The *Student Study Guide* is designed to enhance students' abilities in three specific areas: communication skills, science skills, and general cognitive skills. Activities for each chapter use information related to chapter content to develop these skills. In the Green Version program, communication skills include the following:

- using the SQ3R method

- reading topic sentences

- skimming for main points

- identifying the main idea in a paragraph

- using two-column note-taking

- giving instructions (writing directions)

- outlining

- freewriting to begin discussion

- editing for syntax and clarity

- rewriting a given passage

Science skills include the following:

- predicting

- inferring

- classifying

- distinguishing data from opinion

- identifying variables

- interpreting and making graphs and tables

- observing

- describing properties and changes

- formulating, testing, and refuting hypotheses

- forming and using models

- interpreting data in charts and graphs

General cognition skills (mixed communication and science):

- developing concept maps

- defending a thesis with evidence

- using the library to extend text ideas

- refuting arguments

- focusing on schema

- making flash cards based on skimming a chapter

- metacognition: planning how to approach studying a chapter or book

- visualizing spatial orientation of 3-D figures

- analyzing new vocabulary (using prefixes and suffixes, using contextual clues, using vocabulary in writing)

- formulating questions and answers about the text

- generalizing text information to new instances; linking abstract and concrete ideas

The Teacher's Edition of the *Student Study Guide* contains suggested student responses. Used by itself, it is a source of many instructional ideas for the teacher

The Teacher's Resource Book (TRB)

The *Teacher's Resource Book* contains six sections: Software and media Resources that provide up-to-date technology support; Supplementary Topics that augment the text, encourage discussion, and offer additional genetic problems; Supplementary Investigations, including traditional and microcomputer-based labs; Readings from the Literature—reprints of selected articles from *Science* magazine; Invitations to Enquiry revised from *The Biology Teacher's Handbook;* and Blackline Masters of illustrations from the text and data tables for investigations.

The Computer Test Bank

The computer test bank is supported by a user-friendly program that includes graphics and a user's manual. The test items are mostly multiple choice, but essay questions have been added. All items have been revised for the eighth edition. Approximately half of the items are application/inquiry level questions that use students' higher level thinking skills. The Test Item File Manual for the computer test bank includes a printout of all test items that can be accessed by chapter, difficulty level, and biological concept.

Scheduling for the Year

Because the Green Version contains more than can be accomplished in one school year, you must decide which units, chapters, and investigations your students will study. There are several possibilities, depending on the students' past experiences in life science courses and the emphasis you wish to give your course. Certain portions of the text (for example Section One, most of Section Two and parts of Section Five) are considered essential, basic biology, and you should include as much of these as possible. Even within the sections you choose to emphasize, it may not be realistic to attempt all the investigations in a given chapter. Some investigations are more difficult than others and may be reserved for more able students.

Three schedules follow, with emphasis in ecology, physiology, or diversity. Many others are possible, but these three usually can be completed in any school with a mixed class of average to above-average students. None of the schedules purports to cover the field of biology but rather to give the students some basic science experiences using selected major concepts in biology. Approximately 75 percent of the material is common to all three proposed schedules.

You may wish to tackle a more ambitious schedule if your class consists of high-ability (honors) students. The schedules assume you plan to spend approximately 40–50 percent of class time on the investigations and the remaining time on discussion, evaluation, and other activities. A "day" is defined as a class period of 50 minutes in length, and a typical school year is assumed to have 160 full periods of actual instruction. (The time estimate for Chapter 1 includes all four investigations.)

Ecology Emphasis

Section One: The World of Life 38 days
Chapter 1   12 days
Chapter 2   8 days
Chapter 3   10 days
Chapter 4   8 days

Section Two: Continuity in the Biosphere 37 days
Chapter 5   10 days
Chapter 6   6 days
Chapter 7   3 days
Chapter 8   10 days
Chapter 9   8 days

Section Three: Diversity and Adaptation 20 days
Chapter 10   4 days
Chapter 11   3 days
Chapter 12   3 days
Chapter 13   5 days
Chapter 14   5 days

Section Four: Functioning Organisms 21 days
Chapter 15   4 days
Chapter 16   4 days
Chapter 17   4 days
Chapter 18   5 days
Chapter 19   4 days

Section Five: Patterns in the Biosphere 44 days
Chapter 20   8 days
Chapter 21   6 days
Chapter 22   12 days
Chapter 23   10 days
Chapter 24   8 days

Physiological and Cellular Emphasis

Section One: The World of Life 30 days
Chapter 1   9 days
Chapter 2   7 days
Chapter 3   6 days
Chapter 4   8 days

Section Two: Continuity in the Biosphere 48 days
Chapter 5   12 days
Chapter 6   12 days
Chapter 7   6 days
Chapter 8   12 days
Chapter 9   6 days

Section Three: Diversity and Adaptation 20 days
Chapter 10   4 days
Chapter 11   3 days
Chapter 12   3 days
Chapter 13   5 days
Chapter 14   5 days

Section Four: Functioning Organisms 42 days
Chapter 15   8 days
Chapter 16   8 days
Chapter 17   8 days
Chapter 18   9 days
Chapter 19   9 days

Section Five: Patterns in the Biosphere 20 days
Chapter 20   6 days
Chapter 21   2 days
Chapter 22   4 days
Chapter 23   4 days
Chapter 24   4 days

Diversity and Evolutionary Emphasis

Section One: The World of Life 29 days
Chapter 1   10 days
Chapter 2   7 days
Chapter 3   6 days
Chapter 4   6 days

Section Two: Continuity in the Biosphere 39 days
Chapter 5   10 days
Chapter 6   6 days
Chapter 7   3 days
Chapter 8   10 days
Chapter 9   10 days

Section Three: Diversity and Adaptation 43 days
Chapter 10   6 days
Chapter 11   7 days
Chapter 12   8 days
Chapter 13   10 days
Chapter 14   12 days

Keeping Current in Biology

One of the great challenges and sources of excitement in teaching biology is learning what is new in the field. World knowledge in biology doubles almost every seven years. This makes it impossible for even the research biologist to keep up in all areas of biology. Molecular biology is booming; our understanding of genetics is changing rapidly; classification schemes evolve as new data are found; ecology and environmental awareness appear to be having a rebirth; new discoveries in mathematics (such as chaos theory), chemistry, physics, astronomy, and earth sciences are having profound influences on biology.

How does a biology teacher stay current? Although it is not easy, staying relatively up-to-date is possible by selectively reading the literature and being active in professional organizations. If you do not already subscribe to *Discover,* we recommend you do so because it covers new findings in the sciences that are interesting reading both for adults and adolescents. If you keep past issues of *Discover* in your classroom, you may find that your students will take an interest in some of the articles. Numerous other journals are valuable, ranging from *Science* and *Nature,* which are rather technical, to *Scientific American,* with its useful summarizing articles, to more popular journals such as *Natural History, National Geographic, Popular Science, Science News,* and *World Watch.*

Consider joining a professional association for science teachers, particularly the National Association for Biology Teachers, which publishes the *American Biology Teacher,* or the National Science Teachers Association, which publishes *Science Teacher.* Both of these award-winning journals are dedicated to helping the science educator.

Investigations

Guided Inquiry for Students

Investigations are integral to the Green Version, and a great deal of class time should be devoted to them. Many teachers have their students spend approximately one-half of their actual class and homework time on the investigations. The 72 laboratory and field investigations include 9 microcomputer-based labs; a series of traditional experiments in which the students identify, control for, and manipulate experimental variables; and investigations in which students make observations about biological phenomena and attempt to conceptualize their observations. Some activities are simulations—paper-and-pencil or group discussion activities that are productive alternatives to lectures. Many of these are unique in that they use concrete objects to represent abstract biological concepts. Although these activities are simulations, they engage the students in firsthand experiences with biology. Most require inexpensive materials that can be used again or no materials at all. Active participation in all of the investigations helps students understand the nature and meaning of science and provides extensive experience in such scientific thinking skills as observing, hypothesizing, and inferring cause-and effect relationships between variables.

Each investigation uses a guided inquiry approach that requires students to think through all parts of the investigation yet provides enough guidance for them to carry out the investigation successfully. All the investigations contain an introduction that states the purpose of the activity and often asks the students to construct hypotheses; a materials list; a brief procedure (not lengthy, recipe like instructions) that frequently asks the students to collect, analyze, and draw inferences from data; and finally, discussion questions that lead the students from their observations to an understanding of the biological concepts. Questions may lead the students to go beyond their data and develop further hypotheses or to consider whether or not their observations can be generalized.

At the end of some investigations, a For Further Investigation section contains suggestions for individual students who have extra energy and drive. Some suggestions call for fairly simple extensions of the procedures in the investigation; some entail original thought and design. In most cases, specific directions are lacking, and students must design their own procedures.

A bound notebook is the most convenient way to handle the recording of data. Appendix 1 explains to students the use of a data book. You should encourage your students to regard their data books as primary, but permanent, records. As such, the books must meet the hazards of the laboratory table and will receive hurriedly made records.

Several variations are possible in almost every investigation. Some variations are noted in the preparation section; others may be dictated by classroom necessity. Staying as close to the printed form as your local situation permits allows you to establish more easily the connections between variations in procedure and variations in results. If you teach your students to follow procedures carefully, class data can be combined.

Ground rules can help make laboratory work meaningful to students. Some items to take into consideration when formulating ground rules are

- the location of work stations and the regulation of student mobility;

- a scheme for distributing and collecting materials;

- the principles of cooperative teamwork—leadership, acceptance of responsibility, and coordination of efforts; and

- the importance of discussion and methods of evaluating laboratory work.

If your school has more than one biology teacher, you may want to share preparatory tasks to save time and labor. Keep an organized stockroom or preparation room where all items have assigned places. Each investigation should be followed by a class discussion. If an investigation is observational, you can help students to relate their observations to the purpose of the investigation. If the investigation is experimental, illuminate the course of the reasoning from hypothesis through experimental design and data to conclusions.

A written report for each investigation is neither necessary nor feasible. Which investigations you select for written reports is a matter of personal choice, although it is more reasonable to require a written report for an experimental investigation rather than an observational one. In general, a written report should include a title; relevant data worked up from the data book; answers to questions in the investigation; and a brief conclusion that relates the results to the initial hypothesis. Prompt evaluation and return of the report to the students makes possible a worthwhile follow-up discussion.

Safety

Safety must be a major consideration in a biology laboratory. The safety information in this Teacher's Edition is not intended to be a complete guide but rather to help you organize your own laboratory safety program. Consult your school authorities and local and state regulations for further information. Keep informed of new safety data made available by government agencies, educational organizations, and other sources, and update your safety programs as necessary.

Before each laboratory experience, anticipate possible accidents and take steps to prevent them. Preventing an accident is, after all, the goal of a safety program. Base your conduct and expectations on the students' age, background, and intelligence. Do not expect them to behave as responsible adults.

Post laboratory rules in a conspicuous place in the laboratory. Insist on a safety contract between the student and the school. Devise your own or use the sample presented on the following page. For your convenience, this safety contract is reproduced as a blackline master in the Teacher's Resource Book.

Students should be instructed in techniques of laboratory safety and given the opportunity to demonstrate their knowledge of proper safety practices. When students learn what is expected of them, and when you show that you are safety conscious, they will be more likely to follow appropriate safety procedures. A list of basic safety considerations follows.

1. Have a thorough understanding of each investigation and the potential hazards of the materials, equipment, and procedures required.

2. Prior to conducting an investigation, be sure that all safety and personal protective equipment is present and in good working order. Before students begin an investigation, review specific safety rules and demonstrate proper procedures.

3. Never permit students to work in or be present in the laboratory without supervision. No unauthorized investigations should be conducted, and no unauthorized materials should be brought into the laboratory.

4. Lock the laboratory and storeroom when you are not present. Do not allow students to enter the storeroom at any time.

5. Mark locations of, and call students' attention to, eyewash stations, safety showers, and fire blankets in the laboratory and storeroom. Also mark locations of chemical spill kits, fire extinguishers (ABC tri-class), and first aid kits.

6. Post an evacuation diagram and procedure by each exit.

7. Provide for separate, labeled disposal containers for glass and sharp objects and separate, labeled disposal containers for individual waste chemical reagents.

8. For safety and economy, use small hot plates having an on/off switch and indicator light whenever possible. Do *not* use alcohol lamps.

9. Allow no food or beverages in the laboratory and no application of cosmetics. Guard against toxic exposure by providing adequate ventilation, reminding students not to ingest chemicals, and identifying plants or animals that may cause irritation or poisoning by contact or by bite. Caution students to keep their hands and fingers away from their faces and to wash their hands with soap and water before leaving the laboratory.

10. Know the location of the master shut-off for laboratory electrical circuits, gas, and water.

11. Notify those in authority of the existence or development of any hazard.

12. Remind students that any accident, no matter how trivial, must be reported directly to you.

Keep written records of events related to accidents.

Personal Protective Equipment

Whenever chemicals or laboratory equipment are used, everyone in the laboratory should wear safety goggles and laboratory aprons. Loose clothing, full blouses, ties, bows, etc., should be tucked in. Long hair should be tied back securely. If a chemical spill occurs on someone's clothing or soft cloth shoes, the individual should remove the article and wash the skin thoroughly with running water. Do not attempt to wash off a harmful chemical while the clothing is on the body. Use the safety shower in such cases. (See "Safety Shower," below, for details.) Contaminated shoes cannot be reused; contaminated clothing must be laundered separately before reuse.

When corrosives are used, students should wear both safety goggles and a face shield, as well as a laboratory apron and impervious gloves (nitrile rubber). A safety shower and eyewash station should be within a 30-second walking distance. Specifications for protective equipment follow:

Laboratory Safety Agreement

I _____ , agree to abide by the following laboratory safety regulations whenever performing a biology investigation. I will

1. use the science laboratory for authorized work only.

2. remove contact lenses and wear safety goggles when instructed to do so.

3. know the four hazard classes and control measures.

4. study the laboratory investigation before coming to the lab. (If in doubt about any procedure, I will ask the teacher.)

5. know how to use the safety equipment and know the location of the fire extinguisher, eyewash station, safety shower, and fire blanket.

6. in case of fire, alert the teacher and leave the laboratory.

7. carefully check for the presence of any ignition source (open flames, electric heating coils) before using flammable materials such as alcohol.

8. place broken glass and disposable materials in their designated containers.

9. report any incident, accident, injury, or unsafe procedure to the teacher at once.

10. never taste, touch, or smell any substance unless directed specifically by the teacher to do so.

11. handle chemicals carefully, check the label of every bottle or jar *before* removing the contents, and never return unused chemicals to reagent containers.

12. when heating a substance in a test tube, make sure that the mouth of the test tube points away from other people and away from myself.

13. use proper equipment to handle hot glassware.

14. tie back long hair, remove dangling jewelry, roll up loose sleeves, and tuck in loose clothing.

15. at the end of the lab, clean the work area, wash and store all materials and equipment, and turn off all water, gas, and electrical appliances.

16. wash my hands throughly with soap and water before leaving the laboratory.

_____   _____
Student's Signature   Parent's or Guardian's Signature

_____
Date

Lab apron—Gray or black rubber-coated cloth. Tyvek, or vinyl (nylon-coated), halter type are recommended when working with corrosives or solvents. Disposable polythene is recommended only to prevent physical contact with water-based reagents that are not, in themselves, corrosives or solvents.

Gloves—Nitrile or neoprene rubber is recommended when handling acids, caustics, or organic solvents. Polyethylene or natural latex gloves should be used only for protection against water-based reagents that are not corrosives or solvents.

Safety goggles—Clear, high-impact polystyrene; must meet ANSI Standard Z87.1

Face shield—Must meet ANSI Standard Z87.1; should be used in combination with safety goggles when working with corrosives, reactives, or solvents.

Contact lenses—Liquids can be drawn under a contact lens and into direct contact with the eyeball by capillary action. Therefore, wearing contact lenses for cosmetic reasons should be prohibited in the laboratory. Students who must wear contact lenses prescribed by a physician should wear eye-cup ANSI Z87.1 approved safety goggles. These are similar to the goggles sometimes worn when swimming underwater. If an accident occurs (despite the protection of safety goggles), the student should immediately remove the goggles and the contact lenses and flush the eyes, including under the eyelids, while moving the eyeball from side to side and up and down, at the eyewash station for at least 15 minutes. Meanwhile, call a physician.

Eyewash station—Must meet ANSI Standard Z358.1 and be within a 30-second walking distance from any spot in the room. The device must be capable of delivering a gentle but full flow of water to both eyes for at least 15 minutes. Portable liquid supply devices are not satisfactory and should not be used. A plumbed in fixture or a perforated spray head on the end of a hose attached to a plumbed-in outlet and designed for use as an eyewash fountain is suitable if it meets ANSI Standard Z358.1. Demonstrate the use of the eyewash station to your students. Follow the procedure described in "First Aid—Eyes," on the next page.

Safety shower—Must meet ANSI Standard Z358.1 and be within a 30-second walking distance from any spot in the room. Students should be instructed in the use of the safety shower in the event of a fire or chemical splash on clothing. Chemicals should be flushed off the bare skin for at least 15 minutes while under the safety shower. Help students to understand that contaminated clothing, shoes, wristwatches, etc., *must be removed while under the shower.* False modesty is a poor exchange for permanent injury. Call a doctor while the victim is still under the shower.

No safety shower is referred to in the cautionary statements that accompany each investigation because the quantities of chemicals used in this laboratory program are kept sufficiently small. A safety shower should be present in the laboratory, however, as a precaution against fires or chemical spills related to other laboratory procedures.

Understanding Chemical Hazards

General Information

Some degree of hazard or risk is associated with every chemical that you or your students will handle. Using chemicals safely means understanding the hazards and taking the appropriate measures to prevent harm. A hazardous chemical is any substance likely to cause injury if precautionary measures are not taken. The hazards presented by any chemical can be grouped into the following categories: flammables, poisons (toxins), corrosives, and reactives. A particular chemical may present more than one hazard.

When dealing with chemical hazards, you should be aware of and follow general safety procedures regarding storage, disposal, spills, and first aid, as listed below. Detailed information about specific chemical categories follows afterward.

Signal Words

The signal words *CAUTION, WARNING,* and *DANGER* are used in this program specifically to inform both the teacher and the student about the degree of risk or physical harm associated with a particular material or activity.

CAUTION

denotes a low level of risk associated with use.

WARNING

denotes a moderate level of risk associated with use.

DANGER

denotes a high level of risk associated with use.

For example, the signal word *DANGER* is used to denote a high potential of risk (corrosivity) when handling solid sodium hydroxide, but the signal word *CAUTION* is used when handling a $0.1M$ solution of this same material.

Chemical Hazard Labeling

ANY container used by students must be accurately labeled with the following information:

- the *name of the material* and its concentration (if in solution)

- the *names of individual components* and their respective concentrations (if a mixture)

- the appropriate *signal word*

- an affirmative *statement of the potential hazard* or hazards

- *precautionary measures* to be taken to avoid the hazard

- immediate *first aid measures*

For example, a stock 70% isopropyl alcohol solution should be labeled as follows:

70% Isopropyl Alcohol
WARNING: Flammable liquid

Avoid open flame, heat, or sparks.
Do not ingest. Avoid skin/eye contact.

Flush spills and splashes with water for 15 minutes; rinse mouth with water. Call your teacher.

Small student-use containers such as dropping bottles must be labeled with the name of the chemical and the caution statement. Reagent bottles must have complete labels as shown in the example.

Refer to the "Materials (per class of 30)" lists for each investigation in *Teaching Strategies by Chapter* as well as A Table of Recipes, beginning on page T28, for appropriate safety information when writing label warnings.

Storage

Specific information about chemical storage may be found in Management of Chemicals on page T18. In general, chemicals should be stored in a cool, dry place away from direct sunlight and local heat and segregated according to storage colors. Exceptions are noted in the Remarks column of A Table of Recipes.

Spills

Solids—sweep up material; avoid dusting; place in a suitable container; wash the area with water and discard water.

Liquids—check the pH with litmus or other indicator; if necessary, adjust pH to neutrality with small amounts of $1M$ acid or base (exceptions are stated in A Table of Recipes); wipe up with absorbent material and discard; wash spill area with water.

Disposal (for chemicals used in this program)

Recommended disposal procedures are found in A Table of Recipes for all prepared solutions that require special disposal steps. Keep in mind that these procedures may be preempted by state or local regulations. Consult the Material Safety Data Sheet (MSDS) for specific procedures for stock reagents before disposal.

Aqueous liquids—test the pH with litmus or other indicator; if necessary, adjust pH to neutrality by addition of small amounts of $1M$ acid, base, or other reagent as required. (Exceptions are stated in the table.) In all cases, dilute aqueous liquid waste material at least 1:20 with water and flush to a sanitary sewer (not a drain that leads to a septic tank).

Solids—dissolve small amounts of the material completely in water; dilute this volume 1:20 with water again; flush to a sanitary sewer. (Exceptions are stated in the table.)

Biological materials—Actively growing culture materials should be autoclaved or steam-sterilized in a pressure cooker at 15 psi for 15 minutes. Use autoclavable bags. If an autoclave or pressure cooker is unavailable, aseptically add (in a fume hood) just enough full-strength chlorine laundry bleach or 70% isopropyl alcohol to cover the growing surface. Cover the container, close the hood door, and allow at least 8 hours contact time before disposal. **WARNING: Alcohol is flammable. Extinguish all flames and avoid other ignition sources.** Dilute with water 1:20 and flush to a sanitary sewer. Autoclave or steam-sterilize contaminated objects or place them, in the hood, in covered pans or trays containing liquid chlorine laundry bleach. Allow 24 hours contact before diluting with water 1:20 and discarding the bleach to a sanitary sewer. Wash decontaminated objects with soap and water.

First Aid

Before using any chemical, read the label and the MSDS and follow the recommended procedures. In case of spills or splashes, carry out the following **immediate** first aid measures:

Eyes—Immediately flush eyes, including under eyelids, with flowing water for at least 15 minutes at an eyewash station. Roll eye from side to side and up and down while flushing. Call a physician.

Skin—Wash with flowing water for at least 15 minutes. Contact a physician if redness, blisters, continued irritation, or painful symptoms develop.

Clothing—Remove any contaminated clothing within 5 minutes and wash skin as above. (For concentrated chemicals used by the teacher, go to the safety shower immediately and remove clothing while under the shower.) Launder or decontaminate any article before wearing. Contaminated clothing includes shoes, belts, watches and watch straps, jewelry, etc. If laundering or decontamination is not possible, discard.

Inhalation—Remove to fresh air. Begin CPR if victim has stopped breathing. Get immediate medical attention.

Ingestion—For mouth contact, spit out, wash mouth with running water for at least 15 minutes. Contact a physician immediately.

Poisons (Toxins)

> **Protective Equipment: Gloves, safety goggles, lab apron, container for sharp objects, chemical fume hood, secured storage area, ventilation sufficient to keep breathing air concentrations well below TLV and/or PEL limits.**

Typically, toxic chemicals can injure the body through one or more exposure routes: inhalation, ingestion, injection, and absorption through intact skin or through a break in the skin.

There are two types of toxic effects: acute and chronic. An acute effect usually occurs on exposure or within a few hours following exposure. A chronic effect is noted only following repeated exposures or after a prolonged single exposure.

Important information about toxicity in the MSDS for each substance is provided by the supplier. (See the section titled "Health Hazard Data," or similar title, on the MSDS.)

Prevention/Control Measures

1. Treat all chemicals as potentially toxic. Use barriers, cleanliness, and avoidance when handling any chemical.

2. Wear eye protection.

3. Handle contaminated glass and metal carefully. Sharp objects are vehicles for injecting substances into the body.

4. Provide enough ventilation to keep vapor, mist, and dust concentrations well below the threshold limit value (TLV) or permissible exposure limits (PEL) as stated in the MSDS. Use a chemical fume hood if required.

5. Recognize symptoms of overexposure for each chemical used during an investigation. These usually described in the MSDS.

6. Become familiar with immediate first-aid measures for each chemical used during an investigation. See MSDS and label.

7. Be scrupulous in housekeeping and personal hygiene.

8. Never consume food or beverages or apply cosmetics in the laboratory.

9. Wash your hands thoroughly with soap and water before leaving the laboratory.

10. Post the phone number of the nearest Poison Control Center and consulting school physician on your telephone.

Flammables

> **Protective Equipment: Safety goggles, approved flammables storage cabinets, fire blanket, safety shower, fire extinguishers (ABC tri-class)**

Flammable substances are solids, liquids, or gases that will burn readily. The process of burning involves: fuel, oxidizer, ignition source. For burning to start, all three components must be present. To stop a fire or prevent it from starting, you must remove or make inaccessible at least one of those components.

Prevention/Control Measures

1. Store away from oxidizers.

2. Store only in approved containers in an approved flammable-liquid storage cabinet. Minimize the quantities available in the laboratory—usually 100 mL per bottle and 600 mL per room.

3. Remove ignition sources. Extinguish lighted burners. Check for and eliminate sources of ignition, such as sparks from static charge, friction, or electrical equipment, and hot objects such as hot plates or incandescent bulbs. Keep all ignition sources 30 feet away. If the ignition source is 6 feet above, a distance of 15 feet away is usually sufficient. (Flammable vapors are usually heavier than air and can travel long distances before being diluted below ignitable concentrations.)

4. Electrically bond and ground all metal containers before and during the dispensing of flammable liquids. Check with your local fire department for the correct procedure.

5. Ensure that ABC tri-class fire extinguishers are present in the laboratory and storeroom and that teachers have used these in at least one practice drill, supervised by a firefighting official, within the past year.

6. Drill students in exactly what should be done if clothes or hair catch fire. Practice "drop and roll" techniques. Be sure a safety shower is available and is in working order. A fire blanket should be available to cover a prone victim but should not be used to wrap smoldering or burning clothing, except in emergencies.

7. Conduct a fire inspection with members of the local fire department at least once a year. Practice fire drills at least annually.

8. Provide adequate ventilation so as to keep breathing air concentrations well below TLV and/ or PEL limits.

9. Prepare for spills by having absorbent, vapor reducing materials close at hand. (These are available commercially.) Plan to have enough absorbent material to handle the volume of flammables on hand.

Reactives

> **Protective Equipment: Safety goggles, lab apron, gloves, segregated storage location**

Reactives are chemical substances that undergo violent reactions, generating heat, light, flammable and non-flammable gases, and toxicants under certain ambient or induced (by mixing, shock, or disturbance) conditions. Categories of reactives include, but are not limited to, the following:

- Acid-sensitives—react with acids or acid fumes

- Water-sensitives—react with moisture

- Oxidizer—promote rapid burning or explosion in materials that can burn

- Unstable—spontaneously explode when handled, moved, exposed to sunlight, rapid temperature changes, etc.

Prevention/Control Measures

1. In storage, isolate compounds of a given hazard class **away** from other hazard classes. Consult color storage codes. Note that certain chemicals classed "white" or "red" should be stored separately—away from other chemicals with the same storage code color. Chemicals requiring special storage assignments are identified in A Table of Recipes beginning on page T8.

2. Protect from physical shock.

3. Provide a ready water source for dilution (except for water-sensitives).

4. Keep water away from water-sensitives.

5. Store in a cool, dry place away from sunlight and localized heat.

6. Familiarize yourself with any incompatibilities for *all* chemicals used or stored.

Corrosives

> **Protective Equipment: Safety goggles, lab apron, face shield, nitrile rubber gloves, safety shower, eyewash stations**

Corrosives are solids, liquids, or gases that, by direct chemical action, destroy body tissues. Irritants are a group of chemicals that cause less serious injury. Sensitizers are allergenic. Hence, injury may range from sensitization/irritation to actual physical destruction of body tissues. Categories of corrosives include the following:

- Corrosive—causes destruction and irreversible alterations in living tissue

- Irritant—causes reversible inflammation in living tissue

- Sensitizer—causes allergic reaction in normal tissue of a substantial number of individuals after more than one exposure

Prevention/Control Measures

1. Store corrosives below eye level. Keep containers closed.

2. Always wear safety goggles and lab apron. Also wear a face shield and gloves when handling any corrosive material above 1M.

3. Have at least one eyewash station and safety shower in close proximity. Be sure they are in working order.

4. Never wear cloth-covered, woven-leather, or open toed shoes in the laboratory.

5. Prepare for spills by having neutralizing kits readily available in sufficient quantity for the corrosives on hand.

6. Always wash hands with soap and water after working with corrosives.

Management of Chemicals

Storage

Never store chemicals alphabetically unless they have been segregated into color-coded storage areas (see below). Alphabetical storage greatly increases the risk of promoting a violent reaction. A list of storage suggestions follows.

1. Store chemicals in a cool, dry place, away from sunlight and rapid temperature changes.

2. Never store chemicals on the floor or above eye level.

3. Firmly secure all shelf assemblies to the wall.

4. All shelves should have antiroll lips.

5. Shelves should be permanently fixed, not adjustable.

6. Store flammables in a dedicated flammable-storage cabinet.

7. Store poisons in a locked, dedicated poison storage cabinet.

8. Store chemicals by color-code classification, in separate dedicated storage areas as described on the following page.

Many chemical and biological supply companies use color codes to designate general hazards and to facilitate handling and storage of chemicals. These codes are present on reagent labels, but each chemical or biological supplier uses a different color-code scheme. Many, but not all, of the supplier color codes *appear* to be based on single hazardous characteristics, such as flammability or corrosivity. In fact, they are based on multiple reactivity characteristics. Be sure you are familiar with the color codes for the chemicals used in this course as they are listed in A Table of Recipes, beginning on page T28. Use appropriate color dots (available at stationery stores) to label all chemical containers not color coded or with color codes inconsistent with the table.

In A Table of Recipes, storage recommendations for the chemicals used in the BSCS Green Version laboratory program are given in terms of colors. The instructions indicated by the colors are based on the scheme that follows. **Note that this color scheme applies only to the chemicals used in the program. Color coded labels from other sources may or may not fit this Green Version scheme. Chemicals coded a specific color according to the table cannot necessarily be stored safely with chemicals coded the same color by suppliers or by some other program.**

Red:

All reds in the table are flammables. Store only in a flammable-storage cabinet away from all other chemicals. Store all chemicals coded red in the table with each other, unless otherwise indicated in the Remarks column. Do not store any other chemicals in the flammable-storage cabinet.

Yellow:

All yellows in the table pose reactivity hazards and should be stored with the other chemicals coded yellow in the table and away from all other chemicals. See the Remarks column of the table for special instructions.

White:

White indicates a corrosive/contact hazard. Store away from all other chemicals. *Note that certain white-coded chemicals in this program cannot be stored with other white-coded chemicals.* See the Remarks column of the table for special instructions.

Blue:

All blues pose toxic/health hazards. Store in a secured poison-storage area and away from all other chemicals.

Green:

All other chemicals are coded green. Store these green coded chemicals together and away from all other chemicals. See the Remarks column of the table for special instructions.

When storing chemicals from the Green Version program, observe the following recommendations.

1. Store each color category in its own separate location, as far away as possible from other color categories.

2. Do not store chemicals from this program with chemicals of other programs without first identifying possible incompatibilities; then store only with compatible chemicals.

In addition, chemicals are given a hazard rating in the four categories listed above (health, flammability, reactivity, contact) on a scale of 0 to 4 as follows: 0 = minimal; 1 = slight; 2 = moderate; 3 = serious; 4 = extreme. In many cases, the numbers assigned in the table are different from NFPA numbers found on many reagent labels because numbers in the table pertain to laboratory hazards, whereas NFPA numbers provide useful information to fire fighters when fighting a fire.

Chemical Inventory

It is important to compile an inventory and location map of *all* chemicals and reagents in the school as well as to obtain a Material Safety Data Sheet for each. Together, these form a critical data base for protecting your own health and safety and those of your students. Have this information available in a central location in the science area and also give it to your local fire marshall or fire chief. Your inventory form should include the following categories: substance, protective equipment, storage color code, hazards, amount on hand, location.

Material Safety Data Sheets

The purpose of a Material Safety Data Sheet (MSDS) is to protect users and others from harm by supplying readily accessible information on hazards and precautionary measures. Typically, an MSDS is organized into sections, which include the following: manufacturer and material identification, hazards, physical data, fire and explosion data, reactivity data, health hazard information, spill and leak procedures, special information, exposure guidelines, and special handling precautions.

Under federal requirements, all manufacturers and suppliers of hazardous chemicals must provide MSDSs. Most biological supply houses include an MSDS with chemical at the time of shipment. To request an MSDS simply call or write the supplier, giving the product name and catalog number.

MSDSs should be kept on file and referred to *before* handling *any* chemical. The MSDSs also can be used to instruct students on chemical hazards and to evaluate spill and disposal procedures and incompatibilities with other chemicals or mixtures.

Emergency Procedures

What would you do if a student dropped a 1-L bottle of isopropyl alcohol or hydrochloric acid? Are you prepared? Could you have altered your handling and storage methods to prevent or lessen the severity of the incident? Plan now how to react effectively *before* you need to. Some planning tips include the following.

1. Post the phone numbers of your regional Poison Control Center, fire, police, and hospital on your telephone.

2. Practice fire and evacuation drills as well as what students must do in case of fire or chemical contact or exposure.

3. Ensure that all personal and other safety equipment is available and tested, if appropriate.

4. Compile an MSDS data base and inventory of all chemicals.

5. Prepare in advance for spill-control procedures.

6. Under no circumstances should students fight fires or handle spills.

7. Appoint a hazardous-material-response team of knowledgeable individuals who are prepared to handle spills or leaks. Agree beforehand who in the school has the ultimate decision-making authority for evaluating a hazardous-material incident. Know whom to call for help and when **not** to handle an incident yourself.

8. Be trained in first aid and basic life support (CPR) procedures. Have first-aid kits readily available.

9. Fully document *any* incident that occurs. Documentation is a critical tool in helping to identify areas of laboratory safety that need improvement.

Additional Safety Notes

1. Use only nontoxic marking pens. Many types of permanent marking pens release hazardous vapors.

2. Use nonmercury or digital thermometers in the laboratory. Mercury vapors from broken thermometers are poisonous. In the event a mercury thermometer is broken, or if mercury is spilled, collect all droplets and pools at once with a suction pump and aspirator bottle with a long capillary tube (commercially available). Cover fine (invisible) droplets in inaccessible cracks with calcium polysulfide and excess sulfur. Combine all contaminated mercury in a tightly stoppered bottle. Contact a registered and approved disposal agency. **NOTE: If the mercury in a small clinical thermometer were dispersed in a closed 30 m x 30 m x 4 m room, the TLV (threshold limit value) would be exceeded. Thermometer mercury spills are insidious and potentially dangerous.**

3. When treating a student who has a bleeding cut, protect yourself against the possibility of exposure to blood-borne diseases by wearing impervious gloves.

4. Electrical sockets in the laboratory must be protected with a GFI (ground fault interrupter) type of circuit breaker. Each electrical outlet in the lab must be 3-hole, and each set of 3-holes must be checked with a circuit tester in advance of any use to make certain that the wiring has been correctly connected to that set of 3-holes. (That is, there should be no "open ground," "open neutral," or "open hot" wiring, no "hot/ground reverse," and no "hot/neutral reverse" wiring. These wrong conditions are indicated by the circuit tester by various configurations of red and green lights. Circuit testers can be purchased at hardware and radio supply stores for about $5.00.) Do not use extension cords, except as instructed, as they might create tripping hazards that could result in falls.

5. All electrical equipment should have a 3-wire cord with an attached 3-prong plug.

Safety Information Resources and References

American Chemical Society Health and Safety Service, American Chemical Society, 1155 Sixteenth St., N.W., Washington, DC 20036, (202) 872-4511. This service refers inquiries to appropriate resources to help find answers to questions about health and safety.

Hazardous Materials Information Exchange (HMIX). HMIX can be accessed at no charge (other than the telephone call) by personal computer having a modem (300, 1200, or 2400 baud) with communication parameters set to no parity, 8 data bits, and 1 bit stop. Dial (312) 972-3275. The bulletin board is available 24 hours a day, seven days a week.

HMIX is sponsored by the Federal Emergency Management Agency and the U.S. Department of Transportation and serves as a reliable online data base, accessed through an electronic bulletin board. It provides information about instructional material and literature listings, hazardous materials and emergencies, and applicable laws and regulations.

Council Committee on Chemical Safety. "Safety in Academic Chemistry Laboratories." 4th ed. Washington, DC: American Chemical Society, 1990.

Gerlovich, J. A., et al. "School Science Safety: Secondary School." Batavia, IL: Flinn Scientific, Inc., 1984.

Lefevre, M. J. *The First Aid Manual for Chemical Accidents.* Stroudsburg, PA: Dowden, 1989. "Revised by Shirley A. Conibeau."

Pipitone, D., ed. *Safe Storage of Laboratory Chemicals.* NY: John Wiley, 1984.

Strauss, H., and M. Kaufman, eds. *Handbook for Chemical Technicians.* NY: McGraw-Hill, 1981.

Windholz, M., ed. *The Merck Index* 11th ed. Rahway, NJ: Merck, 1983.

Young, J. A., ed. *Improving Safety in the Chemical Laboratory:* A Practical Guide. NY: Wiley Interscience, 1987.

A Guide to Information Sources Related to the Safety and Management of Laboratory Wastes from Secondary Schools. NY: New York State Environmental Facilities Corp., 1985.

NIOSH Pocket Guide to Chemical Hazards. 5th ed. United States Department of Health and Human Services. DHEW (NIOSH) Publication No.78-210. Superintendent of Documents. Washington, DC: United States Government Printing Office, 1985.

Manual of Safety and Health Hazards in the School Science Laboratory. Cincinnati, OH: National Institute for Occupational Safety and Health, 1980. This publication is available through the Council of State Science Supervisors, Attention: Mr. Frank Kizer, Rt. 2, Box 637, Lancaster, VA 22503.

Prudent Practices for Handling Hazardous Chemicals in Laboratories. Committee on Hazardous Substances in the Laboratory. National Research Council. Washington, DC: National Academy Press, 1981.

Safety Awareness for Teachers. Rochester, NY: Ward's Natural Science Establishment, 1987. *Ward's MSDS Users Guide.* Rochester, NY: Ward's Natural Science Establishment, 1989.

Special Guidelines for the Biology Laboratory

Because biology involves the study of organisms, both living and preserved, certain safety considerations are unique to the biology laboratory

Safety Using Plants*

1. Become familiar with poisonous plants common to your area.

* (Adapted from *Heath Biology Laboratory Investigations* Lexington, MA: D. C Heath and Company, 1989.)

2. Have students observe the following rules:

   a. Do not eat any parts of plants intended for use in laboratory work.

   b. Do not rub sap or plant juice on the eyes, mucous membranes, skin, or an open wound.

   c. Do not inhale, or expose skin or eyes to, the smoke of any burning plant.

   d. Do not pick wildflowers or cultivated plants with which you are unfamiliar.

   e. After handling any plants, wash your hands thoroughly with soap and water before leaving the laboratory.

3. Do not work with plants that may have been sprayed with insecticides.

4. If any student exhibits signs of plant poisoning, such as headaches, dizziness, nausea, constriction of pupils, sweating, muscle tremor, or indications of convulsion, call the school nurse or a physician. The Poison Control Center may be able to offer suggestions for first aid.

The following is a partial list of potentially dangerous plants developed by the National Safety Council. Be sure to add to the list any dangerous plants specific to your area.

House and Garden Plants—autumn crocus, bleeding heart, castor bean (seeds), daffodil (bulbs), dieffenbachia, Dutchman's breeches (foliage, roots), elephant's ear (all parts), foxglove (leaves), hyacinth, iris (underground stems), larkspur (young plant, seeds), lily of the valley (leaves, flowers), mistletoe (berries), monkshood (fleshy roots), narcissus, oleander (leaves, branches), poinsettia (leaf), rhubarb (leaf blade), rosary pea, star-of Bethlehem (bulbs)

Trees, Shrubs, and Vines—azalea (all parts), black locust (bark, sprouts, foliage), cherries, wild and cultivated (twigs, foliage), daphne (berries), elderberry (shoots, leaves, bark), golden chain (capsules), jessamine (berries), lantana (green berries), laurel, oaks (foliage, acorns), poison ivy (leaves), poison oak (leaves), poison sumac (leaves), rhododendron, wisteria (seeds, pods), yew (berries, foliage)

Wildflowers—Jack-in-the-pulpit (all parts, especially roots), May apple (apple, foliage, roots), moonseed (berries), nightshade (all parts, especially the unripe berry), poison hemlock (all parts), thorn apple (all parts), water hemlock (all parts)

Pollen and Mold Spores

Handle pollen-producing plants and spore-producing fungi carefully so that pollen and spores are not spread throughout the classroom. Many people are allergic to either pollen, spores, or both.

Safety Using Microbes

1. Pathogenic bacteria, fungi, or protists are not appropriate investigational tools in the high-school laboratory and should never be used.

2. Demonstrate correct aseptic technique to students prior to conducting an investigation. Never transfer liquid media by mouth or mouth suction. Flame wire loops before and after transferring bacterial cultures.

3. Treat all microorganisms as pathogenic. Seal with tape plates containing bacterial cultures. Do not use blood agar plates, and never attempt to cultivate flora from a human or animal source.

4. Never allow students to clean up bacteriological spills. Keep on hand a spill kit containing 500 mL of chlorine laundry bleach, biohazard bags (autoclavable), forceps, and paper towels. In the event of a bacteriological spill, cover the area with a layer of paper towels. Wet the paper towels with the disinfectant solution; allow to stand for 15 to 20 minutes. Wearing gloves and using forceps, place the residue in the biohazard bag. If broken glass is present, place the bag in a suitably marked container.

5. Consult with the school nurse to screen students who may be receiving immunosuppressive drug therapy that could lower immune response. Such individuals are extraordinarily sensitive to potential infection from nonpathogenic microorganisms and should not participate in laboratory investigations involving microorganisms unless permitted to do so by a physician. Do not allow students with cuts, abrasions, or open sores to work with microorganisms.

6. Never discard microbe cultures without first sterilizing. Autoclave or steam-sterilize all used cultures and any materials that have come in contact with them at 15 psi for 15 minutes. If these devices are not available, flood or immerse these articles in either chlorine laundry bleach or 70% isopropyl alcohol for 30 minutes and then discard. While sterilizing the cultures, keep them covered and in a fume hood. **WARNING: Alcohol is flammable. Extinguish all flames and avoid other ignition sources.** Do not allow students to use a steam sterilizer or autoclave.

7. Wash the lab surface with a disinfectant solution before and after handling bacterial cultures.

Safety Using Preserved Materials

Biological supply firms use dilute formalin-based fixatives of varying concentrations (0.9 to 5 percent) for initially fixing zoological and botanical specimens. Usually, it is the practice to post-treat and ship specimens in holding fluids or preservatives that do not contain formalin. Ward's Natural Science Establishment provides specimens that are freeze-dried and rehydrated in a 10% isopropyl alcohol solution. In these specimens, no other hazardous chemical is present. Many suppliers provide fixed botanical materials in 50% glycerin.

Formaldehyde is a suspected carcinogen. Every effort has been made to eliminate the use of materials containing formaldehyde from the BSCS Green Version. Because your lab supplies may contain specimens fixed in formaldehyde, you should be aware of the following safety precautions. Be sure the formaldehyde concentration in the air is less than the permissible exposure level (PEL). (Currently the PEL is 1 ppm with an "action level" of 0.5 ppm.) To be sure that exposure levels do not exceed acceptable standards, you can measure the concentration of formaldehyde in the air. Your lab supplier can suggest appropriate measures and technical equipment.

The following personal protective safety equipment is mandated when handling preserved specimens or when in contact with preserving fluids: safety goggles; protective gloves (nitrile, polyethylene, latex, neoprene); and lab apron or smock.

To reduce free formaldehyde, prewash specimens in a container left ajar in running water for 1 to 4 hours to dilute the fixative. Formaldehyde also may be chemically bound to reduce off-gassing by immersing washed specimens in a 0.5–1.0% potassium bisulfate solution overnight.

The following safety practices are recommended when handling or dissecting any preserved material specimen.

1. **Never** dissect road kills or nonpreserved slaughterhouse material. Doing so increases the risk of infection.

2. Have students wear prescribed personal protective equipment (see above). Additional safe equipment includes eyewash station(s) within 30-second walk from any location in the lab.

3. Do not allow the preserving fluid to come in continuous contact with the skin. Follow supplier's recommendations for prolonged contact ingestion, or eye contact.

4. Conduct dissections in an area sufficiently ventilated to keep hazardous substances well below their PEL in the air.

Disposal of Specimens/Preserving Fluids

Neither preserved specimens nor preserving fluids are considered by the Environmental Protection Agency (EPA) to be a "hazardous waste" under the Resource and Recovery Act (RCRA), but local regulations may take precedence. Contact your supplier for recommended disposal procedures for the specific fluids provided.

Release of Biological Organisms

Nonindigenous species and certain microorganisms should not be reintroduced into local habitats. The responsible handling of life forms requires that you be informed of their specific habitat requirements or potential for negative impact on local fauna or flora. The acquisition of any exotic or nonindigenous life form requires preplanning to assure that it can be properly maintained throughout the entire year. See the table at the end of this section for specific information about life forms used in the Green Version.

Use the following guidelines in making a decision about the release of any organism into a local habitat.

1. Certain organisms, particularly insects, are specifically regulated, and you must have a permit to possess or release them. Check with your local office of the Animal and Plant Health Inspection Service (APHIS), U.S. Department of Agriculture, or contact your biological supplier. Examples of regulated organisms include cockroaches and termites.

2. Bacteria, fungi, yeasts, growth media, or materials that have been in contact with these organisms should never be discarded without prior sterilization. See Safety Using Microbes, on page T23. For certain microorganisms (*Agrobacterium, Erwinia,* and others), you will require a permit (in certain states) prior to shipment. Check with your biological supplier for restrictions before ordering.

3. Certain aquatic plants, particularly elodea (*Anacharis*), should not be introduced into local habitats. This plant is regulated as a pest in Canada and certain northern U.S. locales.

4. Ornamental plants should not, as a rule, be introduced into native habitats but instead should be kept indoors. Some states (California and others) have strict rules regarding procurement or introduction of plants containing root-bearing soils from outside the state that may contain plant damaging nematodes. Usually plants shipped from outside these states must pass an inspection or have a shipping permit. Check with your biological supplier before purchasing plant materials outside your state. Locally cultivated plants, as a rule, may be reintroduced by replanting if desired.

5. Nonindigenous macroinvertebrates and vertebrates should not be introduced into native habitats. In many cases these organisms will not survive local climates or may compete or otherwise interfere with local fauna and flora. Contact your local APHIS office or your biological supplier for specific information about whether a particular organism would be considered nonindigenous in *your* area. In extreme cases, when release or continued maintenance of an organism is impossible, the animal must be humanely destroyed. A recommended resource for the care and handling of invertebrate and vertebrate animals is F. B. Orlans, *Animal Care from Protozoa to Small Mammals* (Menlo Park, CA: Addison-Wesley, 1977). Biological supply companies usually provide information regarding proper euthanasia for animals supplied.

6. Most microinvertebrates and protists may be freely released in aquatic environments. Nematodes should never be introduced but should be destroyed by sterilization. Marine forms should be released only into marine habitats.

A double safety standard must be maintained when live animals are used in the laboratory for observation and experimentation. The humane treatment of the animals is one objective and the safety of the student is the other. The following National Association of Biology Teachers "Guidelines for the Use of Live Animals" was revised in January 1990. It is considered a comprehensive policy concerning the use of live animals in the instruction of biology. The student text contains general rules on which you may elaborate.

NABT Guidelines for the Use of Live Animals*

Living things are the subject of biology, and their direct study is an appropriate and necessary part of biology teaching. Textbook instruction alone cannot provide students with a basic understanding of life and life processes. The National Association of Biology

* © 1990 by the National Association of Biology Teachers, Reston, Virginia. Reprinted with permission.

Care and Release of Life Forms

Life Form	Care	Release
Bacteria (Eubacteria)	Slant or broth cultures may be stored for extended periods (up to six months) at refrigeration temperatures.	Discard only after autoclaving cultures or contaminated materials.
Butterflies/Moths	House in a perforated box covered with muslin netting. Keep caterpillars dry. Provide a continuous supply of fresh leaves. Use egg cartons as surface for pupation. Keep humidity at 60 percent.	May be freely released shortly after emergence.
Chaeopterus	Requires either an established 25-gallon marine aquarium or an equal amount of conditioned sea water. Acclimate the animals to the seawater. Aerate. Feed by introducing hatched brine shrimp. If care is exercised, the animals may be returned to their tubes and left in the aquarium following the conclusion of the investigation. Do not attempt to maintain these animals except in an established marine aquarium.	Release only in Florida gulf area. Otherwise, euthanize using MS-222 (tricane methanesulfonate).
Crickets/Grasshoppers/ Small Insects	House in plastic containers that are "sealable." Screen in air holes. Use egg cartons as "apartments." Keep temperature from 21–31° C (70–90° F), humidity at around 60 percent. Feed dried food pellets. Provide a number of watering devices (pieces of wet sponge or cotton in a plastic petri dish).	May be freely released.
Earthworms/Other Worms	Place commercial earthworm bedding or rich soil in a plastic washtub. Introduce worms. Allow approximately 150 cm³ per worm. Cover with sphagnum moss or leaves. Cover container with muslin. Long term storage is best under refrigeration.	May be freely released.
Elodea	Place in established freshwater aquarium. Provide a photoperiod of 16:8 using plant grow lights (fluorescent or incandescent). Light source should be placed approximately 16–18 inches over plants	Do not release.

Care and Release of Life Forms (continued)

Life Form	Care	Release
Frogs	Southern species (*R. belandearei*): Place up to five animals in a 10-gallon aquarium or similar container that is slightly tilted. Provide about 1–2 inches of water in lower end. Feed crickets 2–3 times/week. Keep tank covered to prevent escape. Place tank in a suitable location away from direct sunlight and excessive temperature. Northern species (*R. pipiens*): These animals may be stored at refrigeration temperatures for no longer than 5 days. They may be housed for longer periods as described above.	May be released into appropriate habitats.
Geraniums/Coleus	Pot in potting soil. Maintain a 16:8 photoperiod. Use of plant grow lights is recommended.	May be transplanted.
Hermit Crabs	Place in a plastic dishpan with 1–2 inches of dry aquarium gravel. Add two finger bowls, one containing dry pellet rat chow and the other filled with moderate-sized rocks submerged in water. Hermit crabs eat apples, will climb on a small branch, and need some cover under which to hide.	Release only in southern United States, subtropical climates.
Hydra	Store in jars at refrigeration temperature; change water (spring or pond) weekly. Feeding is not necessary unless animals are stored at room temperature; animals can be stored safely for up to 3 weeks at refrigeration temperatures. If stored at room temperature, feed weekly by introducing daphnia into jar.	May be freely released.
Gerbils	For up to three gerbils, provide a commercial enclosure or use a plastic washtub with screened cage top. Use a commercial water bottle introduced through the cage top. House at room temperature of 21–23°C (70–74° F). Use sawdust bedding changed at least weekly. May be fed commercial pellet food (rat/mice) or a mixture of seeds and cereals.	Not recommended. Usually pet stores will accept these animals.
Goldfish	15–25 gallon aquarium; allow 60 cm^3 of water for each fish. Use aeration. Maintain water at 20–24°C (68–75° F). Use of aquatic plants recommended. Goldfish are omnivorous; feed them commercially prepared food daily. Remove uneaten food.	Not recommended. Euthanize using MS-222 (tricane methanesulfonate).

Care and Release of Life Forms (continued)

Life Form	Care	Release
Guppies	10–25 gallon aquarium; allow 10 cm³ for each fish. Use aeration. Maintain water at 24–27° C (70–85° F). Use of aquatic plants recommended. Guppies are omnivorous; feed them commercially prepared food daily. Remove uneaten food.	Not recommended. Euthanize using MS-222 (tricane methanesulfonate).
Mice	For up to three mice, provide a commercial enclosure or use a plastic washtub with screen cage top. Use a commercial water bottle introduced through the cage top. House at room temperature of 21–23°C (70–74° F). Use sawdust bedding changed at least weekly. May be fed commercial pellet food (rat/mice) or a mixture of seeds and cereals.	Not recommended. Usually pet stores will accept these animals.
Microinvertebrates	Short-term holding: Unscrew jar caps and place in a cool area of lab away from direct sunlight. For long-term culture, place in jar containing one-half inch of pond mud that also contains sprigs of elodea. Place in an area that will receive southern light exposure but not direct sunlight. Gentle aeration may be applied.	May be freely released.
Planaria	Store in jars at refrigeration temperature, change water (spring or pond) weekly. Feeding is not necessary unless animals are stored at room temperature; animals can be stored for up to 3 weeks at refrigeration temperatures. If stored at room temperature, feed weekly by introducing small strips of fresh liver into jar.	May be freely released.
Protists	Short-term holding: Unscrew jar caps and place in a cool area of lab away from direct sunlight. Refer to culture instructions for long-term culture.	May be freely released.

Teachers recognizes the importance of research in understanding life processes and providing information on health, disease, medical care, and agriculture.

The abuse of any living organism for experimentation or any other purpose is intolerable in any segment of society. Because biology deals specifically with living things, professional biology educators must be especially cognizant of their responsibility to prevent the inhumane treatment of living organisms in the name of science and research. This responsibility should extend beyond the confines of the teacher's classroom to the rest of the school and community.

The National Association of Biology Teachers believes that students learn the value of living things, and the values of science, by the events they witness in the classroom. The care and concern for animals should be a paramount consideration when live animals are used in the classroom. Such teaching activities should develop in students and teachers a sense of respect and pleasure in studying the wonders of living things. NABT is committed to providing sound biological education and promoting humane attitudes toward animals. These guidelines should be followed when live animals are used in the classroom.

A. Biological experimentation should be consistent with a respect for life and all living things. Humane treatment and care of animals should be an integral part of any lesson that includes living animals.

B. Exercises and experiments with living things should be within the capabilities of the students involved. The biology teacher should be guided by the following conditions:

   1. The lab activity should not cause the loss of an animal's life. Bacteria, fungi, protozoans, and invertebrates should be used in activities that may require use of harmful substances or loss of an organism's life. These activities should be clearly supported by an educational rationale and should not be used when alternatives are available.

   2. A student's refusal to participate in an activity (e g., dissection or experiments involving live animals, particularly vertebrates) should be recognized and accommodated with alternative methods of learning. The teacher should work with the student to develop an alternative for obtaining the required knowledge or experience. The alternative activity should require the student to invest a comparable amount of time and effort.

C. Vertebrate animals can be used as experimental organisms in the following situations:

   1. Observations of normal living patterns of wild animals in their natural habitat or in zoological parks, gardens, or aquaria.

   2. Observations of normal living functions such as feeding, growth, reproduction, activity cycles, etc.

   3. Observations of biological phenomena among and between species such as communication, reproductive and life strategies behavior, inter-relationships of organisms, etc.

D. If live vertebrates are to be kept in the classroom the teacher should be aware of the following responsibilities:

   1. The school, under the biology teacher's leadership, should develop a plan on the procurement and ultimate disposition of animals. Animals should not be captured from or released into the wild without the approval of both a responsible wildlife expert and a public health official. Domestic animals and "classroom pets" should be purchased from licensed animal suppliers. They should be healthy and free of diseases that can be transmitted to humans or to other animals.

   2. Animals should be provided with sufficient space for normal behavior and postural requirements. Their environment should be free from undue stress such as noise, overcrowding, and disturbance caused by students.

   3. Appropriate care—including nutritious food, fresh water, clean housing, and adequate temperature and lighting for the species—should be provided daily, including weekends, holidays, and long school vacations.

   4. Teachers should be aware of any student allergies to animals.

   5. Students and teachers should immediately report to the school health nurse all scratches, bites, and other injuries, including allergies or illnesses.

   6. There should always be supervised care by a teacher competent in caring for animals.

E. Animal studies should always be carried out under the direct supervision of a biology teacher competent in animal care procedures. It is the responsibility of the teacher to ensure that the student has the necessary comprehension for the study. Students and teachers should comply with the following:

   1. Students should not be allowed to perform surgery on living vertebrate animals. Hence, procedures requiring the administration of anesthesia and euthanasia should not be done in the classroom.

   2. Experimental procedures on vertebrates should not use pathogenic microorganisms, ionizing radiation, carcinogens, drugs or chemicals at toxic levels, drugs known to produce adverse or teratogenic effects, pain-causing drugs, alcohol in any form, electric shock, exercise until exhaustion, or other distressing stimuli. No experimental procedures should be attempted that would subject vertebrate animals to pain or distinct discomfort, or interfere with their health in any way.

   3. Behavioral studies should use only positive reinforcement techniques.

   4. Egg embryos subjected to experimental manipulation should be destroyed 72 hours before normal hatching time.

5. Exceptional original research in the biological or medical sciences involving live vertebrate animals should be carried out under the direct supervision of an animal scientist e.g., an animal physiologist, or a veterinary or medical researcher, in an appropriate research facility. The research plan should be developed and approved by the animal scientist and reviewed by a humane society professional staff person prior to the start of the research. All professional standards of conduct should be applied as well as humane care and treatment, and concern for the safety of the animals involved in the project.

6. Students should not be allowed to take animals home to carry out experimental studies.

F. Science fair projects and displays should comply with the following:

1. The use of live animals in science fair projects shall be in accordance with the above guidelines. In addition, no live vertebrate animals shall be used in displays for science fair exhibitions.

2. No animal or animal products from recognized endangered species should be kept and displayed.

In addition to the NABT guides the following are recommended for student safety.

1. All mammals used in the biology laboratory should have been inoculated for rabies unless they were purchased from a reliable biological supply house or pet dealer.

2. Wild animals should never be brought into the laboratory.

3. Any student who is scratched or bitten by an animal should receive immediate attention by the school nurse or a physician.

Teaching with Green Version

Assessment

For most students, the reward for accomplishment is a "good grade," and they should not be deprived of this opportunity to achieve success. However, goals must be set so that each individual can achieve according to his or her interests and abilities. Do not use the same meterstick to measure each student. If you establish valid objectives that are consonant with the skills and abilities of your students, and if your tests measure these objectives, most of your students will be successful.

Tests can increase student comprehension, and they can help you identify concepts that should be retaught. Students should be aware that "missed ideas" will receive your attention. To reteach such ideas, select alternative ways of presenting them, and/or test with a different format, and give your students a second chance to learn. You also can use tests to evaluate your own performance and the adequacy of the materials you have used. If, for example, most students missed an item on a topic you thought had been well covered, reevaluate the materials or the teaching method used for that topic. Using student examinations to evaluate your own performance can help you improve your teaching.

Test results, however, should be viewed within the context of the total process of evaluation. Student performance and the attainment of objectives must be evaluated on more than the scores of examinations. Laboratory skills, the ability to handle ideas, contributions to class discussion, creativity, and other such parameters are difficult to quantify by simple paper-and-pencil examination techniques. A complete evaluation program should consider these parameters.

Although securing a broad data base for evaluating progress is difficult, it is an important task. The development of effective tests requires you to keep in mind the means and ends of the instructional program. Unfortunately, tests, more than sound objectives and effective lessons, often control what students will attempt to learn. Specific facts are the easiest things to teach, and their retention is easy to assess. Although facts are important, often they are only the small ideas that support the big ideas in science. You should emphasize these big ideas because they are the goals of science education.

To help establish a broader base from which to evaluate your students, ask yourself these questions frequently.

◆ Do my students read and comprehend suggested reading material?

◆ Do they notice related articles in newspapers or magazines?

◆ Do they mention related television programs?

◆ Do they anticipate laboratory sessions eagerly?

◆ Do they gain laboratory skills as the year progresses?

◆ Do they talk with me about science at times other than during scheduled class periods?

◆ Does their readiness to enter into class discussion increase as the year progresses?

◆ Does each individual get the opportunity to demonstrate his or her ability to inquire into problems?

◆ Does their competency in discussing substantive ideas improve with time?

◆ Can I find new ways to incorporate the spectrum of data into my evaluation of each student?

Tell students the criteria you will use in evaluating their performances, and explain how the criteria will be weighed. Consider demonstrations of logical thinking, willingness to examine alternative explanations, and the formulation of sound hypotheses as a basis for evaluation. The following suggestions may assist you in collecting evaluation data. In these activities, adjust your expectations to the experiences of your students and increase your expectations as the year progresses.

1. Duplicate a selected newspaper article (or ask students to bring one in) that presents a local view on issues related to the environment, biotechnology, bioethics, or general biology. Help students analyze the article and recognize any biases. Have them suggest ways to check the accuracy of the article, devise alternative solutions to the problem, or suggest public action that might alleviate the situation.

2. Divide the class into groups, and give each group the same discussion problems. Allow them time to read, discuss, and reach conclusions. Have one member of each group take notes on the remarks

and conclusions offered. Circulate among the discussion groups. Bring the class together and have the reporter from each group present the group's remarks and conclusions. Help the class to realize how different groups can reach different, but legitimate, points of view.

3. Show selected films with the sound turned off. Present the visual data in a way that elicits inquiry behavior and/or application of knowledge. The BSCS Classic Inquiries can be used this way. Groups of students can react to each question posed in the inquiry. (Classic Inquiries are available in video from Media Design Associates, Inc., P.O. Box 3189, Boulder, CO 80307.)

Assessing Laboratory Skills

To assess laboratory skills, plan simple, interesting investigations that can be done in a series of short steps. These should be related to concepts the class has studied. You may want to have all the necessary equipment ready or have students gather the equipment as part of the "test." Rate each student on how accurately and efficiently he or she accomplishes each task. During the year, structure these activities so you can measure the improvement in student competency with laboratory techniques. As a review, periodically offer a laboratory practical test in which each student moves from one lab task to the next and records answers to simple and clear requests at each station.

Investigative laboratory activities give students experience in generating knowledge directly from their observations. These activities demand a different type of evaluation. Provide students with opportunities to make conceptual statements based on observations of natural phenomena and to verify their interpretations through scientific procedures. This aspect of laboratory work also can be tested by a laboratory practicum, by paper-and-pencil tests, and by observing and questioning the students as they work in the laboratory and the field. You may find that students who do not do well on standard examinations can demonstrate their understanding of concepts and their logical insight by solving problems in the laboratory. An example of a laboratory practicum that contains elements common to most BSCS investigations may be found in the Teacher's Resource Book.

Controversial Issues in Biology

Progress in science and technology invites controversy almost as a matter of course. Whether the issue is research in astrophysics that affirms an ancient age for the universe, or the development of chorionic villus sampling, which allows first-trimester detection of certain birth defects, science and technology challenge traditional values and traditional views of the world. The incredible rate of progress in science and technology ensures that there will always be great disparity between what is possible and what people may find acceptable.

Although an introductory course in biology should not make controversial issues its central thesis, neither should it evade such issues when they arise naturally from the scientific content. Ours is a society rooted in science and technology; these disciplines derive from and help shape societal values. Indeed, many of the important decisions today's students will face as individuals and as members of the voting electorate will have their roots in science and technology. The proper use of gene therapy, the disposal of nuclear waste, and the management of water resources are but a few examples.

In BSCS *Green Version,* your students will confront two major categories of scientific controversy: debates within the scientific community and debates about the use of science and technology that extend into the community at large. Students must understand that debates between scientists are essential or there will be no science. Science is a dynamic, self-correcting enterprise that continually tests new information and ideas in the marketplace of open, even confrontational, debate. Debates within the scientific community do not demonstrate that the concepts under scrutiny are intellectually bankrupt; rather, they are indications of community health.

The current debate between those evolutionary biologists who believe in gradualism and those who believe in punctuated equilibria is a case in point, as is the dispute over whether *A. africanus* is ancestral to *A. afarensis* or vice versa. Neither debate calls into question the validity of evolution theory. The former is an argument about the pace of evolutionary change; the latter, a disagreement over the sequence in which our hominid ancestors diverged. Each debate has a healthy effect on evolutionary biology because the scientists involved must work harder and be more creative and insightful to establish the rigor of their arguments.

Although debates within the scientific community might or might not attract public attention, other issues that derive from biological progress are certain to do so. Your students still encounter a number of those issues in Green Version—for example, reproductive biology, conception, abortion, prenatal diagnosis, and genetic screening. These issues call into question the validity of many long-standing values and moral traditions. Controversy surrounding these issues does not mean that established values and traditions will necessarily be found wanting. It does mean that new knowledge and new techniques raise what once were intellectual abstractions to the level of hard, often painful, reality for individuals, families, and policy makers.

Students in biology who will spend most of their lives in the twenty-first century have a right to be exposed to societal issues that grow naturally out of science because it is within the societal context—not the scientific—that the average student will grapple with these problems. Informed citizenship requires that students understand the parameters of such issues, and your students will benefit from exposure to and analysis of these issues in your course.

We have not gone out of our way purposely to create controversy where none exists, but we have not avoided it. You will find in this teacher's edition suggestions for dealing with potentially controversial issues in the sections where they arise. We hope you find these suggestions helpful to your teaching program and hope also that you will share with the BSCS any comments you might have on these matters.

Dealing with Controversial Issues

Genetic screening and testing, gene sequencing, prenatal diagnosis, genetic counseling, abortion, and other topics are potentially controversial. How much controversy develops depends on many factors: the socioeconomic climate of the community, the religious preferences of your students, the degree of ethnic diversity in your classroom, and the value systems of your students and their parents. Most important, however, it will depend on how you handle these issues in the classroom. Suggestions for dealing with controversy are described below.*

1. Present as much information about the issue as possible.
   Frequently, the narrow and rigid viewpoint of students is due to their having little, or erroneous, information about an issue.

2. Allow all opinions or feelings to be expressed.
   It is not hard to determine when a student is saying something for its shock value, which is not constructive to the discussion. However, do not be a censor and forbid certain views because they are radical or shocking. Instead, observe whether other students have picked up on the inappropriate comment and ask them to respond.

3. Acknowledge each opinion in the same evenhanded manner.
   Open discussion and debate will be destroyed if the class senses you favor one group of ideas over

*Adapted from *Basic Genetics: A Human Approach,* 2nd ed. by BSCS. Published by Kendall/Hunt Publishing Company.

another. Your open attitude to all views will create an atmosphere in which students are comfortable to express their views and to contribute to discussions in a positive manner.

4. Create an open and nonjudgmental classroom environment.
   This type of atmosphere does not develop automatically. The development of a rapport between teacher and students and among the students must begin on the very first day of class.

5. Emphasize that everybody—you and the students— must be open to diverse views. We cannot make intelligent decisions if we close ourselves off from some viewpoints. Even if we cannot agree with or are offended by a viewpoint, it is important at least to hear it, so we know it exists and can consider it in our decision making.

6. Keep your personal views out of the discussion. Neutrality on the part of the teacher is the key to a successful discussion of controversial issues. Experts in education recommend that teachers withhold their personal opinions in classroom discussions. The position of the teacher carries with it an authority that might influence some students to accept the teacher's opinion without question—thus missing the point of the activity. There also is a danger that the discussion could develop into an indoctrination of a particular value position rather than an exploration of several positions. If your students ask what you think, respond with "My personal opinion is not important here. We want to consider your views." Make sure you consider alternative points of view, so that your students are able to define the relevant arguments and counter arguments. Allow students to freely express alternative points of view.

7. Create a sense of freedom in the classroom for the students.
   Freedom implies the opportunity to take advantage of openness in the classroom, producing positive results for all. There is a fine line between freedom and license. In general, freedom is a positive influence whereas license may have negative results.

8. Finally, your students should know that you respect them for who they are, not for what they might say during classroom discussions.
   If students feel they must respond in a certain manner to gain your approval, you do not have an open discussion. If they know that you are accepting of what they say because of who they are, discussion should be open and valuable.

Cooperative Learning Strategies

Cooperative learning is made up of a set of instructional methods that encourage or require students to work together on academic tasks. These methods may be as simple as having students sit together to discuss or to help one another with classroom assignments, or they may be as complex as each student having a mutually agreed-on task that is part of a larger project. Small heterogenous teams of three to five students should be used in cooperative learning situations. You should construct and assign the teams, which should stay together for a series of activities. Well-designed cooperative lessons plan for a collective goal, individual accountability, rules governing team interaction, monitoring, task processing, and team processing. Strategies for including these facets follow.

1. *Collective goal*   There are many ways to structure a goal for the team: Limit the team's resources so that all must share; "jigsaw" materials and information so that each member learns a part and then teaches it to the others; ask for one product from the team; give a single team grade; and give bonus points when every member of the team achieves success.

2. *Individual accountability*   Have students first work on their own and then bring their work to the team. Randomly select one student to represent the team. Assign specific jobs or roles to each student. Give bonus points if all team members do well individually.

3. *Team rules*   Clearly state the expectations for team behavior and then actively monitor the teams to assess the students' ability to follow the team rules.

4. *Monitoring*   Monitor the progress of the teams, intervening only when necessary to redirect the team. Interventions should take the form of questions that feed the problem back to the team.

5. *Task processing*   Decide how to bring closure to the lesson. It is important to summarize what has been accomplished, review concepts and skills, and reinforce the work done.

6. *Group processing*   Structure time in each cooperative lesson to process team effectiveness. Discuss how well teams followed the team rules, how each team solved problems that arose in the group, and what team members would do differently next time they worked in a team.

Laboratory activities appear to be particularly well suited to cooperative learning because of their specific task orientation. The more specific the group's task, the better. Students in a lab team can decide for themselves how to divide the labor, or you can specify the divisions. In any case, each individual should receive feedback from the remainder of the group on how he or she is doing. In assessing a cooperative lesson, pose questions that evaluate individual and team accountability. The total group score on an investigation is the sum of each individual's scores. Each student can receive both an individual score and a group score. If the groups frequently are reshuffled, the hard-working and successful student still will receive high scores in the course. In contrast to traditional instruction, there is peer pressure in cooperative learning for each member of the team to do his or her best.

Content Area Reading Skills

Content area reading instruction should teach reading and study skills that can help students understand the ideas and facts contained in the text.

Determine the prior experiences and knowledge of your students and link these to the assigned topic. Such links improve students' comprehension of the reading assignment because a reader's existing knowledge and experiences influence the content and form of new knowledge. Use the following suggestions to help organize instruction.

1. *Interactive legends*   Before beginning a section or chapter, encourage your students to study the section or chapter opening photograph. Each section and each chapter begins with a photograph of something biologically interesting. The legend accompanying the photo invites students to interact with the picture.

2. *Prereading activity*   This example is for Chapter 3, Communities and Ecosystems. Have students work in groups of three, with one spokesperson for each group. Write the word *community* on the chalkboard or overhead and ask the groups to define a community. Each group should reach consensus on a single definition. Students may need guidance on this at first.

    Give some examples of a community. Wait 5 or 10 minutes and then ask each spokesperson to state the group's definition and examples. Write each group's definition and examples on the chalkboard/overhead without comment. Ask the class to group the definitions that are similar and condense them into one, two, or three definitions. Relate each example to a specific definition.

Ask the students if they think these communities always have been the same and will remain unchanged. Ask each group to suggest one or two factors that might cause these communities to change. Again, record these factors on the chalkboard.

Assign the introduction through Section 3.2 as a reading. Have the students make a list of new information they have gained from their reading and a list of new vocabulary terms. Present the original definitions to the class after the reading and ask how they would modify their original ideas. Students usually are surprised and impressed by the knowledge they already possess and perceive the reading tasks as less formidable.

3. *Previewing the vocabulary* Students often are overwhelmed by the new terms in a biology text. Lacking skills for learning vocabulary, they may attempt to learn the terms solely from the context of the reading. Often the result is an imprecise understanding of the meaning of the terms and of the information in the text. Many students, especially those with marginal or poor reading skills, are intimidated by the sight of unfamiliar words. They tend to skip over phrases or even sentences that contain new terms and therefore lose the continuity of the material.

Do not review every word that might be troublesome. Select the terms you think are necessary for a full understanding of the reading assignment. The following suggestions, based on the introduction to Chapter 4, Matter and Energy in the Web of Life, provide a structure students can use in subsequent reading assignments.

a. *Contrast clue or specification of term* In the second sentence the terms *matter* and *energy* are contrasted in a specific example: "Using energy from the sun and absorbing matter from the surrounding soil and air, producers, such as green plants, make their own food."

b. *Summary clue* In the third sentence, a summary reinforces the definitions, ". . . must obtain their matter and energy from other organisms."

c. *Synonym or near-synonym* "This matter and energy (food) is made up of biological molecules—the molecules that are found in all living things."

d. *Structural analysis* Because many scientific terms have a Latin or Greek origin, it is helpful for students to learn a small number of prefixes, suffixes, and roots. The prefixes *mono-, di-,* and *poly-* frequently are used in biology. In Chapter 4, these prefixes are used with the root *peptide*. Other often-used roots include *carbo, hydra/hydro-,* and *nucleo-.*

e. *Prereading activity* (This works well in groups of three.) Prepare a worksheet that includes terms you think the students should have in their working vocabulary. Include columns for predicted meaning, confirmed meaning, and clues. Have them predict the meaning prior to reading, then confirm it from their reading, adding clues to help them remember the meaning. For example:

Word	Predicted Meaning	Confirmed Meaning	Clue Words
atoms	small pieces of something	made of protons and electrons	small particles
molecules	small pieces of matter, made of atoms	made of two or more atoms chemically bonded together	food

4. *Reviewing the vocabulary* Reinforcement exercises in the form of crossword puzzles, filling in blanks, word searches, and other games are valuable tools to encourage vocabulary retention. Computer programs are available that generate many of these activities, given the input of vocabulary terms.

Constructing Concept Maps

Success in learning depends on the motivation and effort of each individual student. No method or process alone can guarantee meaningful learning; the students themselves must make the effort. Concept mapping is a tool that can help students learn by building on what they already know.

Concept maps demonstrate meaningful relationships between concepts through the use of proposi-tions. A concept is a mental image, such as *plant, photosynthesis, solar energy*. A proposition is two or more concepts linked by words in a phrase or thought. The linking words show how the concepts are related. In developing a concept map, concepts and propositions are linked in a hierarchy, progressing from the more general and inclusive concepts at the top to the more specific at the bottom. The three concepts mentioned above could be linked in several ways. For example:

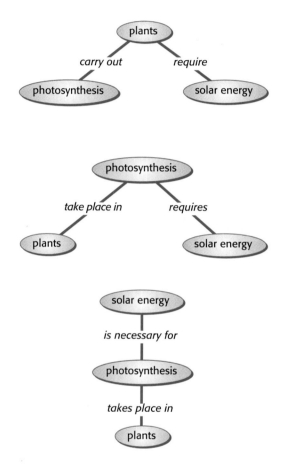

As much as possible, links should be functional rather than descriptive (for example, "plants *use* energy" rather than "plants *such as* trees"). Using these three concepts, it is possible to construct a useful concept map with the addition of a few related concepts, such as *leaves, chlorophyll, water, air, carbon dioxide, roots, soil,* and *vascular tissue.* For example:

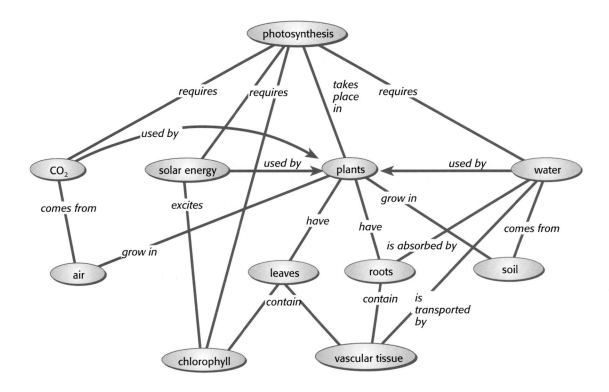

There is no single correct way to develop a concept map. Choice of concept words (usually nouns) and linking words depends on the information to be organized and the aim of the inquiry. Arrows, when used, indicate lateral or upward linkages.

You may need to lead and structure the development of the first few maps, but students quickly learn to construct maps with little or no guidance. First, ascertain students' backgrounds and help them realize that they do have prior knowledge to contribute and to build on. For example, before beginning study of Chapter 1, The Web of Life, place on the chalkboard or overhead the following:

Ask students how living things and nonliving things are related. Usually it is easiest for students to suggest specific examples of concepts, so you might expect numerous examples of living things such as cat, rabbit, frog, bird, grass, flower, and so on. Try to get students to consider the larger categories of plants and animals. Students' examples of nonliving things also are likely to be very specific, but it should be possible to lead them to the two major categories, matter and energy. These could be put together something like this:

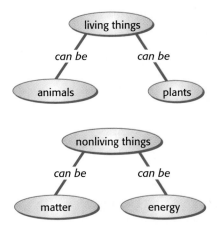

Now ask students how these concepts are linked. Try to elicit lateral relationships, such as that some animals eat plants, that matter has energy, and so on. To summarize, you could introduce the vocabulary term, *biosphere,* guiding students to an understanding of how much they already know about this term before they begin to study. The following is an example of how this concept map might look when finished.

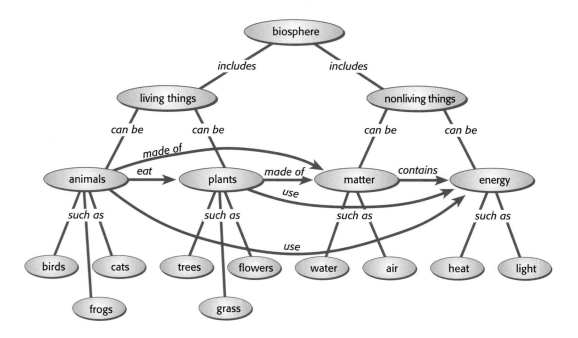

After establishing the collective background knowledge of the class, provide a backbone map on which they can structure their new knowledge as they begin to read the chapter. Write the following relationships on the chalkboard or overhead.

You also might establish that both plants and animals need energy and that it is found in their food.

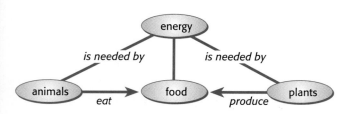

At this point, have the students complete the map, either in small groups or as a homework assignment, and hold a follow-up class discussion to compare the completed maps. Post maps around the room. The variety should make it clear that there are manyvalid ways to structure a map. The follow-up also permits you to be sure that all maps show correct relationships and to discuss why some relationships are not correct.

Make it clear that concept maps are meant to be rearranged and redrawn. As students work and learn, they will envisage new or different relationships. Explain that the first draft of a concept map almost certainly will have gaps or flaws and can be improved. If concept maps are a regular part of learning, students will become more skillful at the process.

As an introduction to a topic, a concept map helps students focus on the small number of key ideas required for a specific learning task. It also shows them what they already know about the material to be studied. Used again at the end of the topic, concept maps demonstrate graphically how much the students have learned. In the process of studying the unit, students can exchange views on why a particular linkage is good or valid, and they can learn to recognize missing links between concepts. In laboratory and field experiences, concept maps can help students to identify key concepts and relationships, which in turn will help them interpret their observations. Concept maps also can be used to plan written assignments. A simple map of the four of five most important concepts to be included in the assignment establishes a framework within which to begin writing.

Concept mapping works well in cooperative learning situations or in groups of three, ideally with students of varying abilities. Although learning is an individual responsibility, the cooperative nature of concept mapping enhances the chances of learning and encourages students to approach a new subject. Concept maps help both students and teacher to recognize the importance of prior information, whether correct or incorrect, to the acquisition of new knowledge.

The materials used to develop a concept map should allow for flexibility, negotiation, and changes, so that students can add, remove, and rearrange their concepts and linking words. Working in small groups or utilizing other cooperative learning strategies, students find that learning the meaning of a piece of information requires talking, exchanges, sharing, and sometimes compromise. Construction of concept maps acknowledges the contribution of all students.

References

Anderson, J. R. *Language, Memory, and Thought.* Hillsdale, NJ: Erlbaum, 1976.

Ausubel, David P. *The Psychology of Meaningful Learning.* NY: Grune and Stratton, 1963.

Ausubel, David P., Joseph D. Novak, and Helen Hanesian. *Educational Psychology, A Cognitive View.* NY: Holt, Rinehart and Winston, 1968.

Bloom, Benjamin S. *Taxonomy of Educational Objectives: The Classification of Educational Goals, Handbook 1: Cognitive Domain.* NY: McGraw-Hill, 1956.

Braus, J. Environmental Education. *BioScience* (June 1995) :S-45–S-51.

BSCS. *Biology Teacher's Handbook.* 3rd ed. NY: Macmillan Publishing Company, 1978.

Gagne, E. D. *The Cognitive Psychology of School Learning.* Boston: Little, Brown and Company, 1985

Gowin, D. Bob. "The Structure of Knowledge." (*Educational Theory*) (Fall 1970): 319–28.

"Philosophy of Science in Education." In *Encyclopedia of Educational Research,* edited by H. E. Mitzel, 5th ed., vol. 3. NY: Free Press, 1982.

Johnson, D. W. *Social Psychology of Education.* NY: Holt, Rinehart and Winston, 1970.

Johnson, D. W., and R. T. Johnson. *Learning Together and Alone: Cooperative, Competitive and Individualistic Learning.* Englewood Cliffs, NJ: Prentice-Hall, 1975 and 1987.

"Toward a Cooperative Effort: A Response to Slavin." *Educational Leadership* (April 1989): 80–81.

Lawson, A. E. "A Better Way to Teach Biology." *American Biology Teacher* (May 1988): 266–73.

Leonard, W. H. "What Research Says About Biology Laboratory Instruction." *American Biology Teacher* (May 1988): 303–06.

Loman, N. L., and R. E. Mayer. "Signaling Techniques that Increase the Understandability of Expository Prose." *Journal of Educational Psychology* (June 1983): 402–12.

Lorch, R. F., and E. D. Lorch. "Topic Structure Representation and Text Recall." *Journal of Educational Psychology* (April 1985): 137–48.

Malone, John, and John Dekkers. "The Concept Map as an Aid to Instruction in Science and Mathematics." *School Science and Mathematics* (March 1984): 220–32.

Mayer, R. E. "Structural Analysis of Science Prose: Can We Increase Problem-solving Performance?" In *Expository Prose,* edited by J. Black and B. Britton. Hillsdale, NJ: Erlbaum, 1984.

Mayer, R. E., et al. "Techniques that Help Readers Build Mental Models from Scientific Text: Definitions, Pretraining, and Signaling." *Journal of Educational Psychology* (December 1984): 1089–1105.

Meyer, B. J. F. "The Structure of Prose: Effects on Learning and Memory and Implications for Education Practice." In *Schooling and the Acquisition of Knowledge,* edited by R. C. Anderson, et al. Hillsdale, NJ: Erlbaum, 1977.

Novak, Joseph D. *A Theory of Education.* Ithaca, NY: Cornell University Press, 1977.

"An Alternative to Piagetian Psychology for Science and Mathematics Education." *Science Education* (October/December 1977): 453–77.

"Applying Psychology and Philosophy to the Improvement of Laboratory Teaching." *American Biology Teacher* (November 1979): 466-70.

"Learning Theory Applied to the Biology Classroom." *American Biology Teacher* (May 1980): 280–85.

Applying Learning Psychology and Philosophy of Science to Biology Teaching." *American Biology Teacher* (January 1981): 10–12. "A Need for Caution in the Use of Research Claims to Guide Biology Teaching." *American Biology Teacher* (October 1982): 393.

Novak, Joseph D., and D. Bob Gowin. *Learning How to Learn.* Cambridge: Cambridge University Press, 1984.

Pressley, Michael, and Joel R. Levin, eds. *Cognitive Strategy Research.* New York: Springer-Verlag, 1983

Rubin, A., and P. Timar. "Meaningful Learning in the School Laboratory." *American Biology Teacher* (November/December 1988): 477–82.

Santeusanio, R. P. *A Practical Approach to Content Reading.* Reading, MA: Addison-Wesley Publishing Company, 1983.

Schwab, Patricia N., and Charles R. Coble. "Reading, Thinking and Semantic Webbing." *The Science Teacher* (May 1985): 68–71.

Slavin, R. E. "Cooperative Learning." *Review of*

Educational Research (Summer 1980): 315–42.

Cooperative Learning. NY: Longman, 1983.

Stewart, James, Judith VanKirk, and Richard Rowell. "Concept Maps: A Tool for Use in Biology Teaching." *American Biology Teacher* (March 1979): 171–75.

Tierney, R.J., J. E. Readence, and E. K. Dishner. *Reading Strategies and Practices: Guide for Improving Instruction.* Newton, MA: Allyn and Bacon, 1980.

Trombulak, S. "Merging Inquiry-Based learning with Near-Peer Teaching." *BioScience* (June 1995): 412–416.

Vacca, R. T. *Content Area Reading.* Boston: Little, Brown and Company, 1981.

Teaching Strategies by Chapter

CHAPTER 1
THE WEB OF LIFE

Planning Ahead

The requirements of the first week are pressing, and ideally, planning and ordering should have been done in the spring. If not: (a) collect organisms or organism parts for Investigation 1.1 (see lab suggestions); (b) prepare growing plants to demonstrate how plants need light energy to live; one plant can be kept in the dark for several days (or weeks) and an identical plant can be left in light for comparison; you also can allow a plant to die in a closet; and (c) check the list of materials and equipment needed for the investigations you plan to do, and either order what you do not have or consider feasible substitutions.

Concepts

The Major Concepts listing at the beginning of each chapter is a short version that often represents only a portion of the important concepts of the chapter. The expanded concepts listing given here represents the heart of the chapter and should be the focus of instruction. There is necessarily a close correspondence between these concepts and the objectives of this program, both of which emphasize global biological themes. Thus, many of the concepts and objectives are developed in various sections of a chapter and cannot be keyed to specific parts. The concepts frequently include traditional biology content, various science process skills, and attitudes of awareness. Students should have enough experiences with these concepts to demonstrate an understanding of them rather than simply reiterate them. For these reasons, the objectives frequently are stated in terms of the concepts.

◆ All organisms must interact both with their environment and with other organisms to survive. An ecologist studies the interactions of organisms with each other and with their environment.

◆ Any event that affects one organism in an ecosystem indirectly affects all others. This is the ecological concept of interdependence.

◆ Organisms in an ecosystem can be grouped by the way they acquire and contribute food. Thus, producers, consumers, and decomposers all have unique niches in a food chain or web.

◆ All organisms must acquire matter and energy to live. Photosynthetic organisms can convert lightenergy to chemical energy stored in the structure of molecules. This chemical energy is made available to other organisms when they consume the photosynthesizers as food. Food also contains chemicals that serve as building blocks for body structure.

◆ Humans are a part of the web of life and can affect the biosphere in many ways.

◆ To answer questions about our environment, scientists use a variety of processes such as observing, collecting data, and drawing conclusions.

◆ Scientists frequently use controlled experiments that test specific hypotheses to determine cause-and-effect relationships between variables.

Objectives

Students should be able to:

◆ *demonstrate* observational skills.

◆ *diagram* two or more food chains.

◆ *distinguish* between a food chain and a food web.

◆ *identify* interrelationships among organisms observed in nature.

◆ *explain* simply the changing of light energy to chemical energy.

◆ *discuss* producers and consumers in energy and food.

◆ *describe* ways in which energy is lost from the system of living things.

◆ *name* several of the most abundant chemical elements found in living things.

◆ *distinguish* between the flow of energy and matter in living things.

◆ *discuss* the relationship between balance and change among living things.

◆ *construct* a statement or concept map that includes the principal attributes subsumed under the term biosphere.

◆ *describe* ways in which humans affect the state of the biosphere.

◆ *suggest* several current biologically based problems.

◆ *identity* the elements of scientific procedure in an investigatory situatuion.

◆ *construct* hypotheses appropriate to a given problem.

◆ *discuss* the values of measurement in scientific investigation.

◆ *record* data obtained from a given procedure.

◆ *differentiate* between data and interpretations or conclusion.

◆ *distinguish* between the terms *observation, hypothesis, experiment,* and *verification.*

Strategies

Chapter 1 lays some groundwork for the rest of the year. It covers the interdependence of organisms in the transfer of energy and in the cycling of matter, and the interaction of living systems with the physical environment. These ideas are never again stated so explicitly, but they persist throughout the course.

Because the use of microscopes has increased in junior high and middle schools, Chapter 1 does not include investigations teaching microscope skills. Nevertheless, some students may lack such skills or may need a review. Appendix 2 includes two investigations that teach microscope techniques. If necessary, use these investigations before beginning Chapter 2.

The four laboratory investigations are broad and introductory, and because they are not lengthy or complex to set up, your students should to all of them in order. Begin the year with Investigation 1.1 perhaps on the first or second day of class. Much of Investigation 1.2 can be done at home, although it is more productive if students can share their results with their classmates and have class discussion on the activity. Investigation 1.3 can be done in almost any field setting, including an urban park, during a class period. If it is not practical to take your students to the field during class time, they can do the investigation as an assignment, with guidance from you. Again, budget time for a class discussion of results and foreclosure. Investigation 1.4 helps students understand experimental design and should be done in class so you can help individuals and groups work through the concepts. Variables and controls are abstract concepts, but they are very important, and it is well worth the investment of two periods of class time to conduct and discuss this investigation thoroughly.

Short study assignments are preferable as you learn the reading capabilities of your students. Encourage students to respond to the interactive legends at the beginning of Section One and Chapter 1. After the discussion of Investigation 1.1, students should read Sections 1.1 and 1.2 and complete the Concept Review questions. Have a brief class discussion of Sections 1.1 and 1.2. After discussion have students do Investigation 1.2, preferably within the first week of school, followed by reading and discussion of Sections 1.3–1.5 and Concept Review questions. Get students outside the classroom during the second week with Investigation 1.3. Discuss the investigation and have students read Sections 1.6 and 1.7 and answer the Concept Review questions. A brief discussion of hypotheses and variables in conjunction with Section 1.7 should precede Investigation 1.4. Spend some class time discussing the Applications. Assign some of the Problems and the Biology Today, then discuss these in class as you complete the chapter.

Because whole organisms are the biological materials with which students have had the most experience, the course begins with whole living things. Have plenty of these in your classroom. Establish a terrarium in which caterpillars, frogs, grasshoppers, etc. can live. (See "Care and Release of Life Forms" in the Safety section for suggestions.) Whether your room is large or small, and whether or not you have the latest biological equipment, you can have at least a few living things that are conspicuous on the first day of school. With encouragement, students quickly will add to the array.

Starting the course with what is familiar can be dangerous, because students may mistake familiarity for understanding. You should, therefore, repeatedly emphasize the complexities of interaction behind the surface phenomena of the biosphere. You can do this best by maintaining a questioning attitude yourself. Questions in marginal notes and in figure legends in the text will help you to do this.

Throughout Section One, remember the trio of ideas that underlie it: (1) individuals as biological units, (2) verifiable observations as the foundation of biological concepts, and (3) the flow of energy as the core of ecosystem functions. This is an introductory chapter, and the depth of understanding students achieve at this point need not be great.

References

Bicak, L. J., and C. J. Bicak. "Scientific Method: Historical and Contemporary Perspective." *American Biology Teacher* (September 1988): 348–53.

Doubilet, D. "Journey to Aldabra." *National Geographic* (March 1995): 90–113.

"Energy." *Scientific American* (September 1990): entire issue.

"Managing Planet Earth." *Scientific American* (September 1989): entire issue.

Marshall, L. "The Terror Birds of South America." *Scientific American* (February 1994): 90–95.

Weinburg, S. "Life in the Universe." *Scientific American* (October 1994): 44–49.

Suggested Answers to Applications

1. The balance in a food web is shifted noticeably when there is a sudden increase in the number of one type of organism. A population of organisms cannot increase dramatically without an energy source. Such sudden increases in natural populations are usually only temporary.

2. If the supply of nutrients for producers were increased by such an occurrence as the dumping of sewage in the pond, then the producers might increase greatly in numbers, but only as long as as the enriched nutrient supply lasted. In turn, the consumers that feed on the producers might increase greatly in numbers, but only as long as the producers are in increased supply. Without an increase in producers, consumers are unlikely to increase, but if one type did increase, competition for food would probably reduce their numbers to far below the pre-increase number.

3. The food webs of each animal will vary, but the main idea should be that the fawn could become food for predators and, eventually, for humans. On the other hand, in Western society the puppy is unlikely to become food for another consumer.

4. Answers will vary. Some students may see the fawn as "better" because it may be used as food for humans.

5. Without light, the first organisms to die probably would be the plants. Animals, which depend on plants for their food, would die next. Decomposers would have plenty to decompose—for a while. Eventually, they too would have no further food and all would die, if they did not freeze first.

6. Because we get all of our food from plants, either directly as in salads or indirectly as in hamburgers, our bodies may be said to be rearranged plant molecules.

7. Minor disturbances such as a single broken strand of a spider web will leave the web functional. Similarly, a few organisms lost within a community should not disrupt the community entirely. However, the more strands and the more organisms lost, the greater the probability of complete disruption.

8. All living things require a source of energy. Only photosynthetic organisms can convert light energy into the chemical energy that can be used as food by consumers. (You may want to introduce other food chains such as those near deep sea vents in which heat-using microorganisms are the producers.)

9. Data are observations, measurements, or other types of information that do not change. Hypotheses are possible explanations of data.

Suggested Answers to Problems

1. Answers will vary.

2. The main point is that organisms are interrelated. A change that affects one organism can affect many others.

3. The main point is to design a low-weight but high efficiency system that recycles materials. Finding small producers with a high output would be the place to begin. Students can try to go on from there.

4. When planning meals, a school nutritionist evaluates foods according to their bulk (matter), calorie content (energy), and diversity of needed nutrients. The nutritionist is the school cafeteria's link to the plants and animals in people's food chains. Another link—to materials cycles—is in the disposal of garbage from the school cafeteria. Possibly the school system has a contract to supply the garbage to a stock farm or other agricultural business. If not, then the garbage is still decomposed in garbage dumps somewhere, and the materials reenter the materials, cycles.

5. Local newspapers and environmental groups can serve as resources for information on issues that are important to your community. You may want to refer to BSCS, *Investigating the Environment: Land Use* (Dubuque, IA: Kendall Hunt, 1984) to help your students organize a study.

6, 7. Students make "educated guesses" all the time. Help them to see that they arrive at these hypotheses through a process of elimination in which ridiculous suggestions are dismissed and logical explanations accepted. Relate this process to the scientific method and scientific investigation.

8. Your local museum may serve as a starting point for students to learn about native plants and animals of the past. Regional histories and naturalist studies will help as well.

Materials and Preparations for Investigations

INVESTIGATION 1.1 The Powers of Observation

Materials (class of 30, teams of 3)
10 10X hand lenses or stereomicroscopes
10 millimeter rulers
10 different sets of 4 specimens of one species

Preparations

Put four specimens of the same species in each set. Do not use chemically preserved specimens. If at all possible, use live organisms. Ward's Natural Science Establishment has prepared a special kit for this investigation (catalog #87M9100) that contains four specimens each of the following: mealworms, lichens, coleus, cacti, earthworms, planaria, goldfish, spirogyra, and newts. Allow two weeks for delivery.

In early fall, many live specimens are available in most areas. Collect organisms before the first week of school; try to provide variety, including representatives of as many kingdoms as possible. If it is appropriate to place the four specimens in the same container, label each specimen with a letter tag (A, B, C, and D). If the specimens are in separate containers, label the containers.

As students will be writing descriptions that are clear enough to distinguish one specimen from the other three in a set, the organisms should be similar in size, color, and shape. Try to select specimens of each type that demonstrate some differences but avoid specimens that have a readily distinguishable characteristic, such as a cricket missing a leg or a branch with far fewer (or more) leaves than the others in the group.

Include both macroscopic and microscopic specimens. With each set of organisms, supply a hand lens or stereomicroscope and a millimeter ruler.

Suggested organisms to be observed with the naked eye include: goldfish, guppies, crickets, grasshoppers, butterflies, moths, geraniums, coleuses, branches with leaves, cut flowers that stay fresh (such as carnations or sunflowers), cacti, earthworms, mice, gerbils, oranges, apples, mushrooms (store-bought), moldy bread, or fruit.

Suggested organisms to be observed with a hand lens or stereomicroscope include: hydra, planaria, small nonparasitic worms, mosses, algae, lichens, small insects.

INVESTIGATION 1.2 You and the Web of Life

Materials (class of 30, teams of 1)
30 data books
30 pencils

Preparations

You may need to make a list of foods that animals eat. For example, cattle eat alfalfa, grass, hay, corn, barley, oats; chickens eat corn, fish products, etc.

INVESTIGATION 1.3 Field Observation

Materials (per class of 30, teams of 1)
30 data books
30 pens

Preparations

The study can be performed in any setting, even an urban one. You may wish to stay near the school building or go to a local park. If you live in a rural area, choose a site close to the school that provides a diversity of habitat. Be sure to select an area that has some exposed ground, if only the soil in a tree planter or flower box. Students should have their own data books by now, but have extra pens, paper, and clipboards available in case students forget their own. Provide plant and animal keys for the identification of species likely to be present.

INVESTIGATION 1.4 How Do Flowers Attract Bees? A Study of Experimental Methods

Materials (per class of 30, teams of 3)
30 data books
30 pens

CHAPTER 2
POPULATIONS

Planning Ahead

Inventory your supplies and be sure you have enough for the number of students in your classes. Check your materials for Investigations 3.1 and 3.3, your supply of graph paper for Chapters 2, 3, and 4, and your supply of solutions for Chapter 4. Order AV materials for later chapters. Investigation 2.1 is the first opportunity to use the LEAP-System™. If you are unable to purchase this computer interfacing system, there are viable non-computer alternatives for seven of the nine investigations developed for it. Nevertheless, giving the students experiences in using state-of-the-art technology is an important part of their education, and you should make this a priority in your budget. Although the LEAP-System is a significant investment, it can be used extensively in other science courses as well.

Concepts

◆ Populations are groups of organisms of the same species that live in close enough proximity for potential interaction and interbreeding.

◆ Predictable factors that influence the size of a population are mortality, natality, immigration, and emigration.

◆ Population size is limited by environmental factors such as density, availability of nutrients, habitat availability, climate, and by such interactions between organisms as predation and parasitism.

◆ Carrying capacity is the maximum population size an environment can support over a sustained period. The carrying capacity is subject to change as the environment and organisms change.

◆ Population sizes fluctuate naturally through time. Environmental factors tend to keep populations from exceeding the carrying capacity, and survival mechanisms tend to keep populations from becoming extinct. This is an example of homeostatic mechanisms working at the population level.

◆ Populations tend to be restricted to certain types of environments. Factors responsible for the range of a population include the organisms' tolerance, dispersal mechanisms, geographic barriers, and some of the same factors that limit population size.

◆ Because humans can modify the environment through technology, there appear to be few natural barriers or limits to human population growth. Although humans can increase artificially the carrying capacity for our population, there remains an ultimate limit. Availability of resources such as clean water and energy will influence this upper limit, as will environmental contamination, famine, and disease.

◆ Global cooperation is necessary to solve environmental problems. The quality of life on the future planet Earth will depend on our ability to educate ourselves about wise use of the environment and on the degree to which all humans cooperate in using the earth as a resource.

Objectives
Students should be able to:

◆ *determine* the factors limiting the growth of a closed population.

◆ *discuss* difficulties in defining the term *individual*.

◆ *define* the term *population* according to type, place, and time.

◆ *solve* simple problems about the rate of population change.

◆ *predict* population size from given rates of the determiners.

◆ *solve* simple problems about population density.

◆ *construct* graphs from data of population changes.

◆ *interpret* graphs of population data.

◆ *explain* ways in which environmental factors affect population determiners.

◆ *plan* experiments to test hypotheses about changing populations.

◆ *interpret* data about population size.

◆ *describe* population fluctuation.

◆ *relate* carrying capacity to limitations on population growth.

◆ *compare* passive and active dispersal.

◆ *describe* barriers to population dispersal.

◆ *cite* three or more means of dispersal.

◆ *relate* these means of dispersal to effective barriers.

◆ *describe* the distribution of three or more species in terms of barriers to their dispersal.

◆ *explain* the relationship between the human population and the earth's carrying capacity.

◆ *apply* principles of change in population size to current statistics of the human population.

◆ *list* the limiting factors for the human population.

◆ *demonstrate* an awareness of personal water use.

◆ *relate* the causes of famine in Africa to population growth.

◆ *describe* several human-caused changes to the environment.

◆ *suggest* ways the global population can begin to manage the earth in a sustainable way.

Strategies
The concepts of individuals and populations form the foundation on which much of the remainder of the course rests. Section 2.1 introduces the relationship of individuals to populations. Again, encourage students to respond to the interactive legend at the beginning of the chapter. Sections 2.2–2.4 discuss the many factors that determine population sizes. Section 2.5 explains a basic population concept: fluctuation in size. Sections 2.6 and 2.7 develop factors determining the range of a given population. Section 2.9 uses the concept of carrying capacity to bring a focus to human populations, and Sections 2.10–2.12 discuss the factors and activities that affect the carrying capacity for humans on the earth. These are vital concepts if the human species is to solve its problems. The major difficulties students encounter in this chapter are abstract ideas, new vocabulary

(sometimes disguised as familiar words), and, worst of all in the view of many students, mathematics.

Mathematics teaching methods are essential to the study of population density and the interaction of rates. Work through several examples with the class and provide further practice with numerous problems. Some practice problems are supplied, but you will have to devise many of your own. Problems with a local flavor are best. Review these problems occasionally during the months after you have left Chapter 2.

Students should begin the chapter with the setup of Investigation 2.1. Although this is a major undertaking, it is one of the most important investigations in the course because it develops both population concepts and numerous process skills. Its preparation, requirements, and complexity are balanced by the fact that Investigations 2.2 and 2.3 are paper-and-pencil investigations that can be done during the 10-day data collection period for Investigation 2.1. Sections 2.1–2.7 can be assigned and discussed during the first few days of Investigation 2.1 data collection, followed by the assignment of the Concept Review questions at the end of Section 2.6. Students can do Investigation 2.2 during two other data collection days for Investigation 2.1. Then, assign and discuss Sections 2.8 and 2.9. Investigation 2.3 is best done in class so students can pool ideas. Follow it by assignment of Sections 2.10–2.12 and the Concept Review questions at the end of Section 2.12. The 14 Applications make an excellent review for the chapter test and can occupy an entire period. Let the faster students tackle some of the Problems at the end of the chapter.

References

Almanac and Book of Facts. New York: Newspaper Enterprise Association, published annually. Brown, L. R., W. U. Chandler, and S. Postel. "State of the Earth." *Natural History* (April 1985): 51–86.

Anderson, D. "Red Tides." *Scientific American* (August 1994): 62–68.

Blaustein, A. and D. Wake. "The Puzzle of Declining Amphibian Populations." *Scientific American* (April 1995): 52–57.

Bongaats, J. "Can the Growing Human Population Feed Itself?" *Scientific American* (March 1994): 36–42.

Kevfitz, N. "The Growing Human Population." *Scientific American* (September 1989): 118–27.

Southwick, C. H. *Global Ecology.* Sunderland, MA: Sinauer Association, 1985.

"State of the Species." *Science 86* (January 1986): 3 articles.

Statistical Abstract of the United States. Washington, DC: United States Department of Commerce, published annually.

Suggested Answers to Applications

1. The duck population cannot be calculated accurately unless you know all of the factors that influence its size, including the immigration and emigration rates of the birds.

2. Because each type of organism has different requirements, the carrying capacity of the environment in any particular area is different for each type. For example, an acre of pasture land can certainly support more grass plants than cattle. Organisms in the same trophic level may have similar limitations, however, because they compete for some of the same resources. Thus, the number of cattle that can be supported on a piece of land is much closer to the number of sheep than it is to the number of grass plants.

3. Humans are not invincible. They do have many limiting factors. Think about the small populations in the far north or in Antarctica. We can evade many of our limiting factors through our technology, but there are still limiting factors.

4. Food and water are two resources essential to living things. Other factors, however, prevent the continued growth of populations within a closed system; these include a buildup of toxic wastes produced by the increasing number of organisms themselves, the spread of diseases between individuals in close contact, and the increase in stress because of a lack of space, which can lead to a decrease in the reproductive rate.

5. Death rate, or mortality, can be a characteristic only of a population. Because each individual dies only once, an individual cannot have a death rate. The rate at which reproduction increases the population is the birthrate, or natality. Again, this is a characteristic of a population, not of an individual.

6. Biologists are uncertain about which factor—food, energy, or space—will ultimately limit the human population. Everything known about populations indicates that the number of people will be limited. The earth is finite and its carrying capacity is finite. The only question is how the limit will be reached. Will mortality through starvation, disease, and war rise to equal present-day natality? Or will natality be lowered to match today's low death rates?

7. The effects of a single individual seem insignificant compared with the size of the biosphere. It is difficult to realize that those effects, multiplied by more than five billion, can be catastrophic. Historically, humans have considered themselves

independent of the rest of the living world. Cooperation is not easy to establish: people resist changes in their life-styles.

8. By adding natality and immigration (43 per year) and subtracting the sum of mortality and emigration (38 per year), it becomes apparent that there was an addition of 5 box turtles per year, owing mostly to surplus natality. The population was increasing. Because emigration was higher than immigration, the area was supplying box turtles to other places, rather than vice versa.

9. The average annual change due to immigration and emigration was 5. In 10 years the box turtle population would equal 65.

10. Answers will vary, but students should stress the potential conflicts between biology, ethics, and morals. A useful reference is G. Hardin, *Naked Emperors, Essays of a Taboo Stalker* (San Francisco: William Kaufman, 1982).

11. Although the text does not list examples of how human activities are straining and destroying the balance in the biosphere, students should be able to supply them from newspapers and other sources. Automobile exhaust fumes and industrial discharges are the chief causes of changes in the atmosphere. Changes to the land and water of the biosphere result from many causes.

12. Ecologists are interested in changes in density; for this purpose, estimates from samplings are quite useful if they are made by a consistent method. Another rule for making valid estimates that some students might understand is that the extent of population fluctuations should be greater than the margin of error in the estimates.

Suggested Answers to Problems

1. Almanacs should include the changes in the United States population figures since 1790 and will indicate, under separate tables, the numbers of immigrants, births, and deaths. Your students should be able to calculate the total population change based on the changes in the individual rates of these contributing factors.

2. The !Kung bushmen of Africa have been studied extensively, and information about them and other hunter-gatherers may be found in such books as *Food, Energy and Society* by D. Pimentel and M. Pimentel (London: Edward Arnold, 1979).

3. *The Ecology of Invasions by Animals and Plants* by C. S. Elton (New York: Routledge, Chapman

and Hall, 1977) chronicles several major effects following dispersal events.

4. Answers will vary. The effect of increasing populations can be seen clearly in areas of rapid growth across the United States. Students should see that growth is not always good for the community, especially when resources and services become strained.

5. The curve may show accelerated growth (California, Florida), a decline (Arkansas), or a stable density.

6. Duckweed (*Lemna*) is obtainable from biological supply houses, but it can also be collected from ponds in most parts of the United States. In one greenhouse experiment under optimal conditions, duckweed plants were observed to double in number approximately every 4 days.

7. Students may take many different approaches. Among these are: enough food energy does not guarantee enough necessary nutrients; availability—countries where starvation occurred lacked food resources; problems of distribution; high standard of living of Western countries; position in the food web—whether grain was eaten directly or fed to animals for meat production. This question helps students relate what they have been studying in Chapters 1 and 2 to a real-world situation. They should begin to recognize the nonbiological factors that influence solutions to real world problems.

8. This problem often evokes much student interest. Careful consideration of each point made can result in some disciplined thought that effectively reviews many of the themes of this book.

9. Contraction of geographic ranges may be caused by (a) climate changes, (b) evolution of tolerances to a more specific or limited range, or (c) results of species interactions (e.g., predation, parasitism, and competition).

Materials and Preparations for Investigations

INVESTIGATION 2.1 Study of a Population

Materials (per class of 30, teams of 2)
30 pairs of safety goggles
30 16-mm × 150-mm screw-cap test tubes, each with 10 mL sterile broth medium
30 18-mm × 150-mm test tubes
30 cover slips

15 ruled microscope slides (2-mm × 2-mm squares)
15 1-mL graduated pipets
15 dropping pipets
Apple lle, Apple IIGSF Macintosh, or IBM computer
LEAP- System™
turbidity probe
30 density indicator strips (optional)
15 compound microscopes
15 test-tube racks
15 glass-marking pencils
15 metric rulers
60 sheets semilog graph paper
1 vial freeze-dried yeast (*Saccharomyces cerevisiae*)

Preparations

Appendix 2 includes two investigations that teach microscope techniques if needed before starting this investigation. Ruled slides, freeze-dried yeast, and screw-cap test tubes containing 10 mL sterile Sabouraud dextrose broth are available from Ward's Natural Science Establishment; allow one week for delivery. It is best to begin the experiment on a Monday; rehydrate the freeze-dried yeast according to the instructions in the package on Friday, or on the day before the experiment.

Students may require help on Day 0 to become familiar with the yeast organisms and with the counting techniques. Make sure they count individual cells and not clumps of cells, air bubbles, or extraneous materials. Suggestions under "For Further Investigation" provide opportunities to study the relationships between yeast and various limiting factors. If possible, have a small group of students work on each suggestion and present their results to the class.

If a LEAP-System is not available, students can use density indicator strips (Ward's catalog #15M1996) to determine turbidity. Instructions for use are included with the strips. Or, students can make direct cell counts each day.

The LEAP-System is a computer interfacing system. It consists of various probes that measure environmental conditions, such as temperature and pH, with hardware and software that translate these measurements into visual data, such as graphs and stripcharts. Study the LEAP-System Open Me First, Reference Manual, and BSCS Biology Lab Pac 2 to become familiar with how to operate the system, allocate disk space for saving data, connect the probes, boot the system, and bring up the menu on the computer screen. Be sure students know how to put the system on WAIT, take the system off WAIT, REINITIALIZE the system, put the system on SAVE, and QUIT the system for your computers; and that space is available on the disk(s) for each class so data can be saved and stripcharts run.

Calibrate the turbidity probe with a test tube of growth medium before use; directions are included in the LEAP-System BSCS Biology Lab Pac 2. Make certain the cap of the turbidity probe is securely in place before taking readings. Extraneous light may produce spurious data.

Students will use the turbidity probe each day to estimate the growth of the populations, and they will make sample counts on days 0, 1, 4, and 7 to help them relate the turbidity to actual organisms. Sample counts will be used in calculating predicted population size from the turbidity data. Have teams record their data each day on a class master table on the chalkboard or bulletin board. From the pooled data, students can determine a class average for each day of growth. They will need their individual data and the class data to construct growth-curve graphs.

Use of semilog graph paper is the best way for students to gain an understanding of exponential growth. This is a good opportunity to emphasize the dependence of science on math. You may want to enlist the aid of a math or physics teacher for a lesson on graphing with semilog paper.

A common source of error is uneven distribution of yeast cells in the medium. Instruct students to invert the test tubes several times before making counts or readings and to make the transfer to the slides quickly. It is not necessary to use sterile techniques when counting. Should a culture become contaminated, discuss that as another example of a limiting factor in population growth. Impress on the students the scientific necessity for uniformity of procedure.

INVESTIGATION 2.2 Population Growth

Materials (per class of 30, teams of 3)
10 pocket calculators
30 sheets of graph paper

Preparations

Students may need guidance in selecting scales for their graphs in Step 8. You may wish to prepare a sample for them. Increments of 200 years work well on the horizontal axis (AD 1 to AD 2000) and increments of 1000 million on the vertical axis (0 to 6000 million).

INVESTIGATION 2.3 Water—A Necessity of Life

Materials (per class of 30, teams of 1)
30 pocket calculators (optional)
30 sheets of graph paper

Preparations

You may wish to bring in various containers so students have an idea of how much a gallon is, how many glasses of water make a gallon, and so on.

CHAPTER 3
COMMUNITIES AND ECOSYSTEMS

Planning Ahead

Now is a good time to secure *Elodea*, prepared slides of frog blood, and the materials for the solutions and phenolphthalein agar used for the investigations in Chapter 5.

Concepts

♦ Ecosystems are made up of living (biotic) and nonliving (abiotic) components that interact continuously with each other. Changes in one almost always cause changes in the other.

♦ All organisms are interdependent. An organism's niche and habitat determine its unique position in a community. The community functions of one organism are interdependent with the functions of all other organisms.

♦ Energy interactions between organisms may be beneficial or harmful.

♦ Ecosystems are not discrete units. Their boundaries often overlap and change.

♦ Most organisms use up about 90 percent of the food energy they acquire, leaving only about 10 percent to contribute to a consuming organism. Therefore, as a given amount of energy passes through the ecosystem, the total amount of energy available to organisms eventually is depleted. The unavailable energy usually is lost as heat.

♦ Humans are subject to the same laws of nature as all other organisms. To ensure our own preservation and well-being, we must learn to live in harmony with our environment.

♦ Humans have a greater effect on communities and ecosystems than do other organisms, and our attempts to modify the environment for our own benefit usually are detrimental to the environment. These changes often result in harm to us as well as to all other organisms.

Objectives

Students should be able to:

♦ *distinguish* between communities and ecosystems.

♦ *list* abiotic factors that affect population sizes of given familiar organisms, including humans.

♦ *describe* the effects the activities of one or more types of organisms other than humans produce on abiotic factors.

♦ *discuss* pollution in terms of human effects on abiotic factors.

♦ *measure* temperature and relative humidity.

♦ *order* with respect to degree of complexity the biological systems designated as individual, population, and community ecosystems.

♦ *describe* a familiar ecosystem in such a way that the essential features of the ecosystem concept are included.

♦ *identify* several specified relationships within a familiar community.

♦ *describe* at least three types of ecological relationships and give examples of each.

♦ *explain* the terms *harmful, beneficial,* and *neutral* with respect to evaluation of ecological relationships.

♦ *distinguish* between *niche* and *habitat.*

♦ *describe* the effects of competition on ecosystem structure.

♦ *describe* the relationship between environmental factors and ecosystems.

♦ *explain* ecological productivity in relation to energy flow through ecosystems.

♦ *explain* the 10 percent rule of energy flow through ecosystems.

♦ *state* the apparent relationship between community diversity and community stability.

♦ *provide* examples of the ways in which humans have upset ecosystem stability.

♦ *suggest* ways in which humans can decrease the loss of biodiversity.

Strategies

The break between Chapters 2 and 3 is one of convenience. The sequence from individual through population and community to ecosystem overarches both chapters.

The investigations in Chapter 3 offer firsthand experiences with ecosystems. In addition to making use of these local outdoor and indoor experiences, try to broaden your students' knowledge of ecosystems with still and motion pictures, although you may want to delay some of these until Section 3.5. During both direct and vicarious experiences, call attention to ecological relationships, which are the essence of the ecosystem concept.

Use the recommended sequence of events for a chapter. First, do the investigation and assign the corresponding reading (including reacting to the interactive legend). Then, follow these events with a class discussion and the Concept Review questions. Repeat this sequence; include a reading and brief discussion of the Biology Today where appropriate. At the end of the chapter, have students do the Applications and selected

Problems if you so desire. Let the investigations drive all other activities, even though students just at the beginning of the learning process may not clearly grasp all concepts derived from the investigations.

Begin the chapter with Investigation 3.1, followed by reading and discussion of Sections 3.1–3.3. Investigation 3.2 focuses on interspecific competition. If you wish students to collect data from local fast-food restaurants, have them do so during study of Sections 3.2–3.4.

Investigation 3.3 extends for 10 days. The results may not be obtained until after the chapter has been finished; they can serve to focus on ecology in the midst of the chapter on chemistry. This should emphasize the lack of compartmentalization of the course. This investigation can be set up any time after the second reading assignment, Sections 3.5–3.7, is begun.

The last assignment, Sections 3.8–3.12, brings together the previous topics in the chapter and relates them to human ecology.

References

Colinvaux, P. A. "The Past and Future Amazon." *Scientific American* (May 1989): 102–09.

Jones, P. D., and T. M. Wigley. "Global Warming Trends." *Scientific American* (August 1990): 84–91.

Menard, H. W. *Islands.* New York: W H. Freeman,1986.

McClanahan, L. L., R. Ruibal and V. H. Shoemaker. "Frogs and Toads in Deserts." *Scientific American* (March 1994): 82–88.

Reganold, J. P., R. 1. Papendick, and J. F. Parr. "Sustainable Agriculture." *Scientific American* (June 1990): 112–21.

Repetto, R. "Deforestation in the Tropics." *Scientific American* (April 1990): 36–45.

Romme, W. H., and D. G. Despain. "The Yellowstone Fires." *Scientific American* (November 1989): 36–47.

Ruddiman, W. F., and J. E. Kutzbach. "Plateau Uplift and Climate Change." *Scientific American* (March 1991): 66–75.

Safina, C. "The World's Imperiled Fish." *Scientific American* (November 1995); 46–53.

Wayne, R. and J. Gittleman. "The Problematic Red Wolf." *Scientific American* (July 1995): 36–39.

White, R. M. "The Great Climate Debate." *Scientific American* (July 1990): 36–45.

Suggested Answers to Applications

1. The biosphere is too big to study as a whole. Studying a part of it—an ecosystem—and how each part interacts with another and with its abiotic environment helps us to understand the biosphere.

2. The community food web would be seriously disrupted. Many carnivores eat turtles; other carnivores eat the carnivores that eat turtles.

3. Intense competition for available herbivores would result. As the food supply was reduced, some carnivores might emigrate. Others would weaken and become susceptible to disease. Many might die.

4. Predators kill their prey; parasites *may* kill their hosts. The death of its host is a disadvantage and often a disaster for the parasite. When parasites kill, they usually kill indirectly.

5. A habitat tells *where* an organism lives; a niche tells *how* an organism lives. Organisms can live in the same place, yet live in different ways.

6. Every organism has many interactions with other organisms and is affected in many ways by the physical environment. Each of these interactions may be affected by changes in other organisms and in the environment. To catalog a complete set of interactions would be difficult, time consuming, and probably not worth the effort.

7. Competition occurs for scarce resources. Although temperature itself is never in limited supply, organisms may compete for areas where the temperature is in the right range for growth and reproduction.

8. The plants may have had a parasite-host relationship. If they had, the plant that died after the separation could not live without the sugar, nutrients, and/or water supplied by the host. When the parasite was no longer taking away its resources, the host grew much faster.

9. Probably the best plan would be (b), to eat the chickens first, then the corn. Feeding the chickens any corn would mean that much of the energy in the corn would be lost through the chickens' respiration and would not be available to you. You need to get rid of the animals that are your competitors for food.

10. The major advantage is that they are natural controls—no chemicals or other substances are added to the food chain. The predators may be locally available and inexpensive. They also might become well established and thus a permanent method of pest control. Disadvantages include difficulty in mass-producing the natural predators, establishing them successfully, protecting them from pesticides, and ensuring that they do not become pests themselves.

Suggested Answers to Problems

1. Ecologists must first define *harmful* in a way that allows them to collect data to prove or disprove hypotheses. They might define *harmful* to mean that an increase in one population brings about a decrease in another population. *Beneficial* then would mean that an increase in one population brings about an increase in another. *Neutral* would mean that a change in the size of one population has no effect on the size of another.

2. No single answer can be given to these questions. The important thing is for students to realize how much energy is imported into a city and how much the biotic community depends on people. Be sure the commensals of human beings—such as house mice, brown rats, and cockroaches—are not forgotten.

3. This problem can lead a student in many directions. It is a good preliminary to the ecological concept of infectious disease developed later. Chestnut blight is an example of a host-parasite relationship that has not evolved into a steady state. Immunity has not developed, and the parasite *Endothia parasitica* has decimated the population of the host organism. In the case of whooping cough, on the other hand, natural immunity to the parasite *Bordetella pertussis* has developed. Hence, there is low fatality to the host. Also, artificial immunization of the host means the disease is no longer a hazard to an entire population.

4. Students will be most likely to list food animals, pets, and houseplants first. Identifying and cataloging those organisms with which they have direct contact will be easier for them than working with indirect relationships. It also may be difficult for students to see how the physical environment affects the organism.

5. Growing a totally biocide-free food supply is probably impossible in the United States because of the mechanized agricultural system and the number of people this entire system must serve. The focus of this essay should be on what the risks are when we do or do not use biocides—the financial consequences and the potential health consequences.

6. Pictures of rare, exotic, endangered plants and animals, such as pandas, may be easier to find than pictures of local species. Nevertheless, it is important for your students to understand that the problem of endangered species can lie close to home. This assignment gives your artistic students a chance to express themselves.

7. An alien species may have no natural predators or may be a better competitor for scarce resources than a native population. As a result, the alien species may completely eliminate one of the native populations. The loss of one population can have numerous effects on other populations in the area. Kudzu and zebra mussels are examples.

Materials and Preparations for Investigations

INVESTIGATION 3.1 Abiotic Environment: A Comparative Study

Materials (per class of 30, teams of 6)

15 watches
15 metersticks
15 nonmercury thermometers (−10° C to +100° C)
15 nonmercury thermometers of same range, with cotton sleeves over the bulbs
5 screw-top jars containing 30–50 mL distilled water
15 pieces stiff cardboard
5 umbrellas or other shade devices
6 tables of relative humidity (in Appendix 3)

Preparations

Although you may delegate the procedure to a single team, replication is desirable; also, the study and interpretation of the data will be more meaningful for all if the whole class is involved in the procedure.

Differences in temperature between the three environments are more pronounced in early autumn than late autumn. For best results, the work should be done on a sunny day, vegetation should not have become dormant, and the sun should be fairly high in the sky. To minimize topographic differences, the three environments should be as near to each other as possible.

Mount the thermometers on a stick or pole. Except when making a reading, students should not stand near the thermometers. Body heat or shading can influence temperatures significantly. If bulb sleeves are used repeatedly, they should be soaked in distilled water. If you have several teams, gather data into a master chart on the chalkboard or overhead projector. A blackline master of Table 3.1 is included in the Teacher's Resource Book; you may want to provide each team with several copies.

INVESTIGATION 3.2 Competition

Materials (per class of 30, teams of 3)
none

Preparations

You can use the data in Table 3.2 or have your students collect their own entree data from FFRs in your area. If students collect their own data, it can be compared from year to year.

You may wish to introduce the concept of species at this time. Students can use petri dishes for making their diagrams. The Teacher's Resource Book includes a blackline master of the diagram. Be sure students make the labels as small as possible to allow for maximum space in each overlapping area. Students should label the species in a clockwise direction (see Figures T3.1–T3.3). Some hypotheses: Giveaways

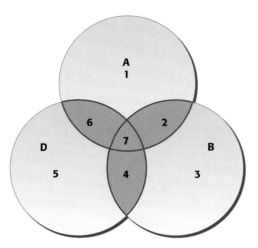

Figure T3.1

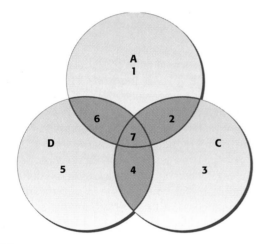

Figure T3.2

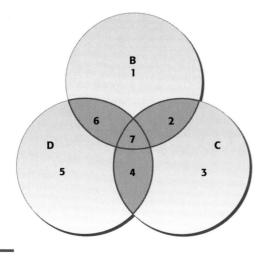

Figure T3.3

(glasses, comic books, free soft drinks); sales (buy one get one free, reduced price); better service (in and out in less than 60 seconds); drive-through service; higher-quality product; catchy advertising.

INVESTIGATION 3.3 Acid Rain and Seed Germination

Materials (per class of 30, teams of 2)
30 pairs of safety goggles
30 lab aprons
30 pairs of plastic gloves
15 petri dishes
15 glass-marking pencils
1 roll absorbent paper towels
15 pairs of scissors
15 transparent metric rulers
30 sheets of graph paper
colored chalk or felt-tip markers
300 mL each of water solutions ranging in pH from 2 to 7
rainwater
60 bean seeds

Preparations
Be sure students have hypothesized a pH they think will be most beneficial to seed germination. The data they collect will be used to evaluate their hypotheses.

Prepare the six pH solutions according to the directions in A Table of Recipes. Be sure students know which pH solution they have been assigned. Have some students use rainwater as a control. If rainwater is not available, substitute distilled water.

Use the chalkboard or a large piece of white construction paper on a visible wall or board for the class graph. You will need seven different colors of markers for the six pH solutions and the control. After about three or four average points have been plotted, have the students draw a curve connecting points of the same pH.

CHAPTER 4

MATTER AND ENERGY IN THE WEB OF LIFE

Planning Ahead
Purchase liver for Investigations 4.3 and 4.1 and fresh peroxide for Investigation 4.3. Order the two mating strains of yeast and the media for Investigation 7.1 and Chaeopteris for Investigation 7.2.

Concepts
◆   Biological activities are based on the chemical reactions of atoms and molecules, the building-blocks of living things. Atoms are made up of smaller particles. The movement of electrons from

one atom to another or the sharing of electrons between atoms determines the type of chemical reaction in which a given atom can participate.

◆ Molecules and/or ions can react with other molecules and/or ions, resulting in products different from either of the reactants. Organisms carry out synthesis reactions that store energy or build body parts and decomposition reactions that use energy or metabolize wastes.

◆ Energy is used to move matter and maintain order. It can take many forms, such as chemical (for example that stored in glucose or ATP), kinetic (motion), and heat. In most ecosystems, energy useful to organisms comes from the sun and is captured by the photosynthetic factory in plants.

◆ All living things contain the element carbon, which is found in biological molecules such as carbohydrates and lipids (energy storage), proteins (building materials), and nucleic acids (hereditary material).

◆ Chemical reactions in a cell are accelerated or inhibited by other chemicals called enzymes, which interact with the reactants.

◆ Ribonucleic and deoxyribonucleic acid are large, complex chemicals that control cell activities and inheritance in all living things.

◆ Carbon passes through many living and nonliving systems in the biosphere as it becomes part of atmospheric carbon dioxide, organic food materials (carbohydrates, lipids, proteins, etc.), organic wastes, or rocks (limestone). This is the carbon cycle.

Objectives

Students should be able to:

◆ *distinguish* between atoms and molecules.

◆ *demonstrate* the formation and breaking of bonds in a molecular model or chalkboard drawing.

◆ *classify* a given pH as basic, acidic, or neutral.

◆ *describe* the types of reactions that take place in living organisms.

◆ *explain,* using experimental evidence, the role of buffers in the maintenance of homeostasis.

◆ *define* the term *energy* and name several forms.

◆ *explain* how energy is used in living systems.

◆ *describe* what happens to a living system when the input of energy stops.

◆ *describe* in general terms the process of photosynthesis.

◆ *compare* the release of energy by burning and the release of energy in cells.

◆ *describe* the role of ATP in living systems and the importance of the ADP-ATP cycle.

◆ *name* food sources for the major biological molecules.

◆ *cite* examples of complex molecules with stored energy.

◆ *identify* carbohydrates, proteins, fats, and nucleic acids as the major groups of organic compounds and give examples of each.

◆ *name* the building units in carbohydrates, proteins, fats, and nucleic acids.

◆ *list* several characteristics and actions of enzymes.

◆ *interpret* experimental data showing the effects of environmental factors on enzyme actions.

◆ *explain* how nucleic acids control the activities of a cell.

◆ *list* the ways plants use carbon-containing sugars.

◆ *describe* the role of the carbon cycle in global warming.

◆ *diagram* the carbon cycle and relate it to the processes of photosynthesis and cellular respiration.

Strategies

This chapter offers a simple explanation of the chemical basis of life. Although most students study biology before chemistry, they need some understanding of chemistry to comprehend such topics as cellular respiration, photosynthesis, and the structure of nucleic acids. This chapter may be used both as background for those topics and as a reference throughout the year.

Living things are shown to be made of molecules that react to carry out all body functions. The role of energy in life also is emphasized.

The conceptual load in this chapter is comparatively heavy. Give short assignments, as follows: Sections 4.1–4.2, 4.3, 4.4–4.6, 4.7–4.9, 4.10–4.13, and 4.14. A short, upgraded quiz following each assignment may help students to check their own understanding and to review especially difficult parts of the chapter. Begin the chapter with Investigation 4.1 and use the remaining two investigations to break up the reading assignments and to illustrate the ideas in the previous sections. Investigation 4.3 is most essential. Students will enjoy Investigation 4.1, especially if the LEAP-System is used, although it works just as well using broad-range pH paper. If you omit any investigation, it probably should be 4.2.

References

Berner, R. A., and A. C. Lasaga. "Modeling the Geochemical Carbon Cycle." *Scientific American* (March 1989): 74–81.

Cech, T. "RNA as an Enzyme." *Scientific American* (November 1986): 64–75.

"Energy." *Scientific American* (September 1990): entire issue.

Ingelsrud, D. E. "How Living Things Obtain Energy." *American Biology Teacher* (February 1989): 89–97.

Mullis, K. B. "The Unusual Origin of the Polymerase Chain Reaction." *Scientific American* (April 1990): 56–65.

McPherson, A. "Macromolecular Crystals." *Scientific American* (March 1989): 62–73.

Olson, A.J., and D.S. Goodsell. "Visualizing Biological Molecules." *Scientific American* (November 1992): 76–81.

Orgel, L. "The Origin of Life on the Earth." *Scientific American* (October 1994): 76–83.

Rebek, J. "Synthetic Self-Replicating Molecules." *Scientific American* (July 1994) : 48–55.

Richard, F. M. "The Protein-Folding Problem." *Scientific American* (January 1991): 54–65.

Stryer, L. *Biochemstry*. 3rd ed. San Francisco: Freeman, 1988.

Weintraub, H. M. "Antisense RNA and DNA." *Scientific American* (January 1990): 4047.

Suggested Answers to Applications

1. Understanding the pathway of carbon can help clarify carbon transfers in an individual food chain or a complex food web.

2. Wheat enzymes put amino acid building units in sequences of characteristic of wheat physiology. Through digestion, wheat proteins are reduced to amino acids, and these then are synthesized into new proteins characteristic of human physiology. Because wheat proteins do not contain all the amino acids humans need, wheat alone is not a sufficient source of protein in the human diet.

3. Carbon dioxide in the atmosphere produces a greenhouse effect: radiant energy from the sun penetrates the atmosphere to reach the hydrosphere and lithosphere, but the radiation of heat from the earth is hindered. Higher concentrations of carbon dioxide during the Carboniferous might explain the warm conditions that apparently prevailed in most of the middle latitudes.

4. Humans contribute carbon dioxide directly as a result of cellular respiration and breathing and indirectly by burning fossil fuels and trees.

5. No. Plants can return carbon to the atmosphere through cellular respiration and then recapture it through the process of photosynthesis. Animals exhale carbon dioxide to the air, but the cycle would continue without them.

6. A sugar molecule is highly ordered or organized, especially as compared with the many molecules that go into the production of sugar. This order came about by the input of energy, which is released when the sugar molecule is broken down and can be used by the cell as a source of energy.

7. It is important for students to know how these structures relate to each other.

   a. Because there are numerous protons in each atom (except the hydrogen atom), you would expect to find more protons.

   b. Because nucleotides are subunits of DNA, there should be more nucleotides.

   c. Because humans are roughly 70% water, in any typical cell you might expect to find more water molecules.

   d. Because enzymes are made up of amino acids, you should find more amino acids. (Other types of proteins in the cell also are made up of amino acids.)

8. Lipids have more calories per gram than do proteins and carbohydrates. That is why they make convenient storage compounds in seeds, which provide food for the young plants. That also is why seeds are a major source of food for humans—think about all the legumes and grains we eat.

9. An increase of 5°C would cause melting of polar ice caps, raising sea levels and inundating coastal areas. These changes, in turn, would change global weather patterns, resulting in drought in central continental areas such as the midwestern United States. Food production would be reduced and desertification increased. An increase of 15°C would exacerbate these problems.

Suggested Answers to Problems

1. Answers will vary.

2. The increased use of fossil fuels has added vast amounts of carbon dioxide to the atmosphere. If the trend continues, it could bring about the

greenhouse effect, causing the average temperature of the earth to increase. This increase might melt the ice caps, raising the sea level.

3. In some fatty acids, double carbon bonds occur instead of bonds with hydrogen. Such a fatty acid contains less hydrogen than it contains without the double carbon bonds and is said to be "unsaturated" with hydrogen. Some fats contain many double carbon bonds in their fatty acids: they are polyunsaturated and are liquids or oils at room temperature.

4. Examples may include hair shampoos, medicines such as buffered aspirin, household cleansers that may be alkaline or acidic, soaps, beverages such as coffee and cola drinks, and foods such as vinegar, yogurt, and lemon juice.

5. This is the opportunity for you and your students to study creatures taken from science fiction. Based on what we know about life on earth, how accurate is the description of life forms on other planets? Does it even make sense to make comparisons with life on earth when the physical environment of other planets may be so different from ours?

6. It is difficult for most students to understand the relationships between DNA, RNA, and protein synthesis because they cannot actually see what happens within cells. Try to produce large, accurate models that can be used throughout the course to help students visualize these molecules and their roles in living things. You can make your whole room one cell with enormous organelles and macromolecules.

Materials and Preparations for Investigations

INVESTIGATION 4.1 Organisms and pH

Materials (per class of 30, teams of 4, 2 teams per LEAP-System)
30 pairs of safety goggles
30 lab aprons
8 50-mL beakers or small jars
8 50-mL graduated cylinders
24 colored pencils (8 of each color)
4 Apple lle, Apple IIGSE Macintosh, or IBM computers
4 LEAP-Systems™
printer
8 pH probes
8 pH meters or 8 rolls wide-range pH paper (optional)
8 forceps (optional)
tap water

8 dropping bottles 0.1M HCl (160 mL total) labeled
  CAUTION: Irritant
8 dropping bottles 0.1M NaOH (160 mL total) labeled
  CAUTION: Irritant
400 mL sodium phosphate pH 7 buffer solution
400 mL liver homogenate
400 mL potato homogenate
400 mL egg white (diluted 1 : 5 with water)
400 mL warm gelatin suspension, 2%

Preparations

Review the LEAP-System Open Me First, Reference Manual, and BSCS Biology Lab Pac 2 to become familiar with how to operate the system, allocate disk space for saving data, connect the probes, boot the system, and bring up the menu on the computer screen. Be sure students know how to put the system on WAIT, take the system off WAIT, REINITIALIZE the system, put the system on SAVE, and QUIT the system for your computers; and that space is available on the disk(s) for each class so data can be saved and stripcharts run. Check the calibration of the pH probes before use; directions are included in the LEAP-System BSCS Biology Lab Pac 2.

The standard LEAP-System includes specifications and probes sufficient to run each experiment for one team at a time. The system also can handle multiple teams concurrently. See the reference manual for detailed instructions about how to create new experiments or update existing ones. Using these procedures, you can generate your own multi-team experiments. Disks with completed multi-team experiments also are available for purchase. Additional probes may be required in some cases. When more than one team is using one LEAP-System, it may be difficult to differentiate all the lines on the graph. Under these circumstances, instruct students to read the numeric data displayed near the graph.

Table T4.1   Homogenates

Material	pH	Material	pH
Apples	2.9–3.33	Milk, cow	6.4–6.8
Beans	5.0–6.0	Oranges	3.0–4.0
Blood plasma, human	7.3–7.5	Peas	5.8–6.4
		Pickles, dill	3.2–3.5
Bread, white	5.0–6.0	Saliva, human	6.0–7.6
Carrots	4.9–5.2	Salmon	6.1–6.3
Corn	6.0–6.5	Sea water	8.0–8.4
Ginger ale	2.0–4.0	Shrimp	6.8–7.0
Grapefruit	3.0–3.3	Tomatoes	4.1–4.4
Lemons	2.2–2.5	Urine, human	4.8–8.4
Limes	1.8–2.0	Vinegar	2.4–3.4
Magnesia, milk of	10.5	Water, distilled	7.0
		Wines	2.8–3.8
Milk, human	6.6–7.6		

After they have completed the experiment, students can stripchart and list their individual team data. See the reference manual for details on stripcharting and listing data by lab team.

Before printing stripcharts, check "System Maintenance" in the LEAP-System Main Menu to be sure the printer specified corresponds to the one you are using. See the reference manual for details on how to specify the printer.

Directions for preparing the solutions are provided in A Table of Recipes. Assign a specific biological material to each team for Step 6. Liver and potato homogenates, egg white, and gelatin work well. A blackline master of Table 4.1 is included in the Teacher's Resource Book. Table T4.1 lists other common materials that can be used. Check to be sure students understand they are to repeat the same procedure followed in Steps 2–5.

Students may require help in preparing the graphs. The results vary with tap water, but the graphs in Figure T4.1 give typical results that should help students with Questions 5 and 6. You may need to help students make the connection between the phosphate-buffer model and the homeostatic mechanism for pH control.

INVESTIGATION 4.2 Compounds in Living Organisms

Materials (per class of 30, teams of 4)

30 pairs of safety goggles
30 lab aprons
30 pairs of plastic gloves
8 250-mL beakers
8 10-mL graduated cylinders

48 18-mm × 150-mm test tubes
8 test-tube clamps
4 hot plates
8 dropping bottles Benedict's solution (72 mL total) labeled
    CAUTION: Irritant
8 dropping bottles Biuret solution (48 mL total) labeled
    CAUTION: Irritant
8 dropping bottles indophenol solution (40 mL total) labeled
    CAUTION: Irritant
8 dropping bottles Lugol's iodine solution (24mL total)
    labeled ***WARNING: Poison if ingested: Irritant***
8 dropping bottles 1% silver nitrate solution (24 mL total)
    labeled ***CAUTION: Irritant***
8 screw-top jars 99% isopropyl alcohol (240 mL total)
    labeled ***WARNING: Flammable liquid***
96 5-cm × 5-cm squares of brown wrapping paper
30 mL for each test series of apple, egg white, liver, onion,
    orange, potato, or other food of your choice (3 foods
    for each team)

Demonstration

1/4 tsp butter or vegetable oil
5 mL 10% gelatin suspension
5 mL 2% NaCl solution
5 mL 10% starch suspension
5 mL 10% glucose solution
5 mL vitamin C solution (1% ascorbic acid)

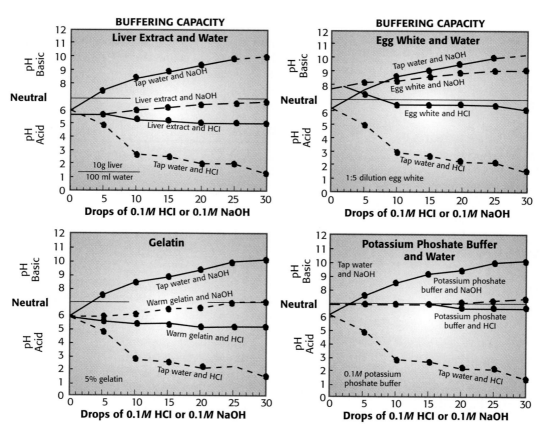

Figure T4.1

Preparations

Part A is written as a demonstration, which allows the students to concentrate on the results of a test rather than becoming involved in the test itself. Steps 5–10 describe the test procedures. Directions for preparing the solutions are provided in A Table of Recipes in the Teacher's Resource Book. In Part B students use the test results to determine the absence of the six compounds in foods. If time is limited, each team could test one food for the six compounds. Specific tests may be assigned to individual teams and the results recorded on the chalkboard. Ideally, each team should test as many different foods as possible. Frozen liver is preferable to fresh. For ease in handling the foods, grind them in a blender. Blackline masters of Tables 4.2 and 4.3 are included in the Teacher's Resource Book.

INVESTIGATION 4.3 Enzyme Activity

Materials (per class of 30, teams of 3)
30 pairs safety goggles
30 lab aprons
10 50-mL beakers
20 250-mL beakers
10 each 10-mL and 50-mL graduated cylinders
10 reaction chambers
60 18-mm × 1 50-mm test tubes
10 forceps
10 square or rectangular pans
10 test-tube racks
10 nonmercury thermometers
350 filter-paper disks
ice
water bath warmed to 37° C
40 mL each buffer solutions labeled *pH 5, pH 6, pH 7, pH 8*
150 mL catalase solution (liver homogenate)
1 L fresh 3% H_2O_2 labeled
    CAUTION: Reactive material

Preparations

The reaction chamber (see Figure 4.22) consists of a 25-mL square glass bottle, a one-hole rubber stopper, and a dropping pipet (Ward's catalog #17M0230). Insert the dropping pipet into the rubber stopper so that it points out of the chamber. Lubricate the stopper hole with glycerine or water to ease insertion, and protect your hands with cloth pads. Do not allow students to disassemble this equipment. The water containers can be dissecting pans, disposable aluminum baking pans, or plastic containers. They should be long enough (30 cm) to immerse a 50-mL graduated cylinder on its side and deep enough (7 cm) so that the reaction chamber and mouth of the graduated cylinder are covered with water.

It is essential for the 3% H_2O_2 to be fresh. Purchase it just before use. It is available at most pharmacies. Use a paper punch to prepare a large quantity of filter-paper disks.

Prepare stock catalase solution by grinding 5 g fresh (not frozen) liver in 100 mL distilled water in a blender. Directions for preparing buffers are provided in A Table of Recipes in the Teacher's Resource Book along with a completed blackline master of Table 4.4.

CHAPTER 5
CONTINUITY IN CELLS

Planning Ahead

Prepare the solutions for the first three investigations in Chapter 5. You may want to begin planning for the genetics unit and to order the yeast strains and media for Investigation 8.3. You may want students to collect samples from ponds, streams, or puddles in preparation for Investigation 12.1. Maintain the samples in the classroom until needed.

Concepts

◆ Cells are the basic structural and functional units of life. All organisms are made of one or more cells, and all organisms today develop from pre-existing cells.

◆ There are two basic types of cells—prokaryotic and eukaryotic. The differences between them separate all organisms into two large groups.

◆ Membranes organize eukaryotic cells into specialized compartments and organelles that allow for specialization and division of labor and result in increased efficiency.

◆ Although cells are relatively independent, they require energy and materials from the environment. Each cell type has unique mechanisms for acquiring energy and building materials. Water is the medium in which cells take in nutrients and give off wastes.

◆ Most eukaryotic cells are similar in size because of physical constraints. Cells must have a minimum size to carry out necessary processes. A maximum size is imposed by the decreasing surface-area-to-volume ratio as the cell size increases.

◆ Mitotic cell division is a small part of the cell cycle—the continuous sequence of events in the life of a cell. Most organisms grow in size by cell division, which results in two like cells. This process assures that each offspring cell will have the same organelles and genetic instructions as the parent cell.

Objectives

Students should be able to:

◆ *state* the two principal parts of the cell theory.

◆ *order* chronologically two or three principal events in the discovery of the cellularity of organisms.

◆ *explain* how microscopes have advanced knowledge of cells.

◆ *distinguish* between the two major cell types.

◆ *identify,* in material they have prepared, nuclei, plasma membranes, cytoplasm, cell walls, and chloroplasts.

◆ *describe* the functions of at least seven structures observable in cells.

◆ *distinguish* between plant and animal cells on the basis of their structures.

◆ *discuss* the relationship between energy and metabolism.

◆ *demonstrate* diffusion through a differentially permeable membrane.

◆ *explain* diffusion in terms of concentration gradients.

◆ *describe* evidence that supports the concept conveyed by the term *active transport.*

◆ *distinguish* between active and passive transport.

◆ *explain* the significance of the surface-area-to volume ratio in terms of movement of materials into a cell.

◆ *describe* the events in the cell cycle.

◆ *distinguish* between the different phases in the cell cycle.

◆ *distinguish* between cell division and mitosis.

◆ *order* a set of six or more models or pictures of cells in various stages of mitosis.

◆ *explain* the apparent significance of mitosis.

Strategies

Knowledge of cellular structure and function is essential to students' understanding of biology. The term *cell* is not new to most high-school students, and some students may have had experience with cells. Even though they may have observed cells previously and may have learned some generalizations about the cellularity of organisms, probably few students have grasped the full force of the cell theory.

Sections 5.5–5.7 deal with cell physiology and depend heavily on an understanding of cell structure and some acquaintance with molecular theory. The amount of understanding required for Chapter 5 is neither broad nor profound, but you should check on the backgrounds of your students and supply additional information if necessary.

The best summary of the work on mitosis probably is provided by a motion picture. Avoid films that combine mitosis and meiosis (or stop such a film before meiosis appears). Meiosis is not relevant to Chapter 5 and will only confuse the students. If they thoroughly understand mitosis now, it will be easier for them to understand meiosis when they encounter it in Chapter 6.

Let the activities drive this chapter and, if time allows, attempt all of them. Remember to have your students respond to the interactive legends at the beginning of Section Two and at the beginning of each chapter. Then, begin the chapter with Investigation 5.1. Sections 5.1–5.5, the first assignment, are a review of that investigation. Investigation 5.2 is an introduction to diffusion and osmosis. Investigation 5.3 is a visual demonstration of the effects of surface area/volume on diffusion rate. It is also another opportunity for students to attempt to quantify their results. Investigation 5.4 is a more traditional observation of mitosis on slides; it should be supplemented with a video or film illustrating the dynamics of this process.

References

Allen, R. D. "The Microtubule as an Intracellular Engineer." *Scientific American* (February 1987): 42–49.

Baringa, M. "Secrets of Secretion Revealed." *Science* (23 April 1993): 487–489.

Darnell, J. E., Jr., J. H. Lodish, and D. Baltimore. Molecular Cell Biology. New York: *Scientific American Books,* 1986.

Howells, M. R., J. Kirz, and W. Sayre. "X-Ray Microscopes." *Scientific American* (February 1991): 88–95.

Murray, A. Q., and M. W. Kirschner. "What Controls the Cell Cycle?" *Scientific American* (March 1991): 56–65.

McIntosh, J. R., and K. L. McDonald. "The Mitotic Spindle." *Scientific American* (October 1989): 48–57.

Pennisi, E. "A Room of Their Own; Finding the Place Where Immune Cells Process Undesirable Proteins." *Science News* (21 May 1994): 335.

Ptashne, M. "How Gene Activators Work." *Scientific American* (January 1989): 40–47.

Rasmussen, H. "The Cycling of Calcium as an Intracellular Messenger." *Scientific American* (October 1989): 66–73.

Stossel, T. "The Machinery of Cell Crawling." *Scientific American* (September 1994): 54–63.

Streit, W. and C. Kincaid-Colton. "The Brain's Immune System." *Scientific American* (November 1995): 54–61.

Todorov, I. N. "How Cells Maintain Stability." *Scientific American* (December 1990): 66–75.

Suggested Answers to Applications

1. Phase-contrast microscopy has made possible more effective examination of living, unstained cells. These observations can be compared with stained cells to determine whether staining changed the cells.

2. Using a typical compound microscope, students should be able to see the cell wall of plant cells, the location of the plasma membrane (especially when the cell is plasmolyzed), the nucleus, chromosomes during mitosis, chloroplasts, and a vacuole. All other cell structures are too small or too thin to see.

3. Seen under a microscope, cell structures such as chloroplast, cell wall, and centrosome would be evidence that the material is either plant or animal. Chemical tests for cellulose would confirm the identification.

4. A molecule first must pass through the dead cell wall in a plant cell. It then is screened at the plasma membrane. If it can pass through this, it then will pass through the cytosol until it reaches the vacuole. Here, it must pass through one more membrane before it gets inside the organelle.

5. The jellyfish would swell up if placed in fresh water, and the frog would dehydrate if placed in ocean water. Fish that live in both fresh and salt water maintain water balance in their tissues through several mechanisms. In salt water, they continuously drink, void little urine, and excrete salts actively through their gills. In fresh water, they take in only a little water by drinking, absorb salt through their gills, and excrete water as urine.

6. This question has to do with diffusion and osmosis. The water is, of course, seawater, which contains a lot of salt. When exposed to salt water, the wooden boards of the ship shrink as the water inside them diffuses out. If the mariner drinks seawater, he will become dehydrated and die, partly because of the diffusion of water out of his body cells in response to the exposure to salt water.

7. If the THC inhibits chromosome movement and if chromosome replication already has occurred, then one resulting cell might be expected to have double the normal number of chromosomes and might live. The other cell, containing no chromosomes, will die. (This is one way polyploid cells are produced.)

8. Amazingly enough, all the genetic information for an entire individual is found in the chromosomes of the zygote. This circumstance is called *totipotency* and is the basis for the process of cloning, in which a cell or group of cells taken from one individual is used to produce an identical individual.

9. If cancer is uncontrolled cell division, you might expect cancers to develop in those tissues that contain young cells likely to undergo mitosis. It is more difficult for older, specialized cells to divide, and, therefore, you might expect fewer occasions of cancer in such tissues.

Suggested Answers to Problems

1. Prokaryotic.

2. Answers will vary depending on the cells selected.

3. Nucleus—capital, library
   Mitochondria-power plants
   Golgi apparatus-factories, warehouses
   ER—highway and railway systems
   Lysosomes—waste treatment plants

4. Mitochondria, DNA, plasma membrane.

5. Answers will vary.

6. In division of a bacterium the single chromosome replicates, and each new cell receives one chromosome. In a paramecium the macronucleus merely pulls apart, but the micronucleus divides by mitosis.

7. In multicellular animals a wide variety of connective tissues hold cells together. All these have an extensive matrix of extracellular material containing fibers. Such extracellular material is manufactured by the connective tissue cells and holds these cells together. In turn, the connective tissue holds other tissues—such as muscle and nerve—in place. Plant cells are held together primarily by the adhesive qualities of cellulose cell walls.

Materials and Preparations for Investigations

INVESTIGATION 5.1 Observing Cells

Materials (per class of 30, teams of 2)

30 pairs of safety goggles
30 lab aprons

30 microscope slides
30 cover slips
15 compound microscopes
15 forceps
15 scalpels
15 dissecting needles with corks for the tips
paper towels
15 dropping bottles Lugol's iodine solution (375 mL total)
    labeled **_WARNING: Poison if ingested; Irritant_**
15 dropping bottles of 5% salt solution (375 mL total)
1 large onion
2 or 3 sprigs of elodea (*Anacharis*)
15 prepared slides of stained human cheek cells
15 prepared slides of frog blood

Preparations

Cut an onion into pieces like orange sections and keep the pieces under water in finger bowls. A student can remove a piece of the fleshy leaf from a section and bend it backward until it snaps. This usually leaves a ragged piece of epidermis. Remind students to use small pieces of material and discard the onion piece after they have removed the epidermis.

Elodea is easy to maintain in an aquarium if direct sunlight is excluded. To increase the likelihood of observing cyclosis in cells, illuminate for at least 12 hours before using. Directions for preparing the solutions are provided in A Table of Recipes in the Teacher's Resource Book.

Students should avoid using circles to frame their sketches because circles suggest the entire field of view, which would require sketching everything seen in the field. Only a few cells need to be sketched to show their arrangement. Cell drawings should be large enough to indicate structural details. Sketch an example on the chalkboard but use a type of cell the students will not observe.

INVESTIGATION 5.2 Diffusion Through a Membrane

Materials (per class of 30, teams of 3)
30 pairs of safety goggles
30 lab aprons
20 250-mL beakers
10 glass-marking pencils
20 15-cm pieces of dialysis tubing
40 10-cm pieces of string
150 mL soluble-starch suspension
150 mL concentrated glucose solution
10 dropping bottles Lugol's iodine solution (200 mL),
    labeled **_WARNING: Poison if ingested; Irritant_**
10 urine glucose test strips

Preparations

Directions for preparing the solutions are provided in A Table of Recipes. Dialysis tubing can be ordered from your biological supplier. It should be about 10 mm in diameter. The tubing comes in rolls with the sides flattened and pressed together. To open the tubing, cut it to the desired lengths and soak it in water for one or two minutes. After soaking, separate the sides by rubbing the cut ends between the thumb and forefinger. Test strips used by diabetics to test for sugar in urine are available from pharmacies and biological suppliers.

Two setups are used because iodine sometimes interferes with the use of urine glucose test strips. If you prefer, students can use Benedict's solution to test for glucose; they will need hot plates, water baths, and test tubes. If Benedict's solution is used, only one setup is necessary, and both starch and glucose solutions can be placed in one piece of tubing.

You may want to have a team of students demonstrate osmosis in a cell model. Fill a piece of dialysis tubing with a concentrated sucrose solution and tie it over the end of a 1-mL pipet, as shown in Figure T5.1. Suspend the tubing in water and hold the pipet vertical with a ring stand and clamp. Students can measure the rate of the liquid moving up the pipet. Add food coloring to the sucrose solution to make it more visible.

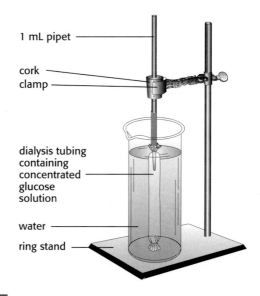

Figure T5.1

In order to draw any conclusions from this investigation, students must recognize the reactions between starch and iodine, glucose and urine glucose test strips, water and urine glucose test strips, and glucose and iodine. Before students begin their work, perform quick, silent demonstrations of any combinations that are unfamiliar to your students.

It is important that the results of this investigation be related to living things. Give special attention to the discussion questions. Glucose, starch, and water are common substances in living things; iodine is used merely as an indicator. The diffusion of glucose and water and the lack of starch diffusion can be linked to the storage of starch in plant cells, the need to digest starch, the possibility of feeding glucose by direct injection, and many other biological matters.

INVEST1GATION 5.3 Cell Size and Diffusion

Materials (per class of 30, teams of 2)
30 pairs of safety goggles
30 lab aprons
30 pairs of plastic gloves
15 250-mL beakers
15 millimeter rulers
15 plastic spoons
15 plastic table knives paper towels
2.25 L 0.1% NaOH labeled **CAUTION: Irritant**
1.5 L phenolphthalein agar

Preparations

The day before class, prepare enough phenolphthalein agar blocks to allow each team to have a block 3 cm × 3 cm × 6 cm. Directions for preparing the blocks and the solutions are provided in A Table of Recipes.

Caution students not to slice or make deep scratches in the surface of the cubes when using the spoons.

When slicing the cubes in half, the knife should be thoroughly dried between operations. Otherwise, NaOH will be smeared on the cut surface of each succeeding block, providing misleading results.

You may want to review with students the calculations for finding the surface area and volume of a cube.

$$\text{Surface area} = \text{length} \times \text{width} \times \text{number of sides}$$
$$\text{Volume} = \text{length} \times \text{width} \times \text{height}$$

INVESTIGATION 5.4 Mitosis and Cytokinesis

Materials (per class of 30, teams of 1)
30 compound microscopes
30 prepared slides of *Allium* root tip
30 prepared slides of *Ascaris* embryo cells

Preparations

Prepared slides may not have the impact that squashes of fresh root tips have, but if often is difficult to harvest root tips at the time of maximum mitosis. Your students will better understand the slides if you grow a few onions with roots. Whitefish slides can be substituted for *Ascaris,* but they are not as clear. If at all possible, have students observe a motion picture of mitosis so it can be seen as a continuous process.

If neither prepared slides nor onion root tips are available, students can model the process, using modeling clay (as in Investigation 6.1) or pipe cleaners for chromosomes and beads or short pieces of pipe cleaners for centromeres. Have them make three pairs of chromosomes, each pair a different length, and several large sketches of a plant cell on which to demonstrate the events of mitosis. Although it is not necessary to distinguish maternal and paternal chromosomes for mitosis, you may wish to anticipate meiosis by providing two colors for each pair of chromosomes.

CHAPTER 6
CONTINUITY THROUGH REPRODUCTION

Planning Ahead

Investigation 6.1 requires pipe cleaners and two colors of modeling clay. Having available a variety of live organisms that show some form of reproduction will be helpful for chapter discussions. Because Investigation 7.1 bridges Chapters 6 and 7 and continues for 8 days, you may want to have students begin it during the discussions of human reproduction in the latter part of Chapter 6.

If you do not purchase colored paper clips for Investigations 8.2 and 10.1, you need to begin preparing these now. The least expensive way is to use one can each of good quality gloss enamel spray paint in white black, red, and green, and 1000 jumbo or regular paper clips. Spread out 250 close together on newspaper, spray with one color, let dry, and spray the other side. Repeat with the other colors. Respraying may be necessary from year to year, and you can reuse the original spray cans if you discharge them upside down for about 10 seconds after each use to clean the paint out of the nozzle and prevent gumming. You can use the same clips for Investigation 8.2, but you need about twice as many as for Investigation 10.1, plus about 100 left unpainted. If you plan to do both labs, you will need at least 2100 paper clips.

Concepts

◆ Reproduction is a fundamental characteristic of life. Organisms must reproduce to pass on their genetic material and to continue their species.

◆ Asexual reproduction produces a new organism exactly like the parent. Although it occurs in all five kingdoms, it is most common among prokaryotes and protists and least common among animals.

◆ Sexual reproduction provides sources of genetic variation. The combination of gametes from each parent produces an organism that is a slight variation of the parents.

◆ Gametes produced by meiosis have only one of each chromosome pair found in a body cell. The gametes from each parent unite to form a new individual with a complete set of chromosomes.

◆ The combination of meiosis and fertilization maintains the chromosome number of a species generation after generation.

◆ Crossing-over between homologous chromosomes during meiosis can cause genetic variation.

◆ Human sperm, the male gametes, are produced continuously by the testes beginning at the onset of puberty.

- Human ova, the female gametes, are produced by the ovary each month from puberty to menopause.

- Menstruation is due to the sloughing off of a thin layer of cells in the human female uterus each month.

- The production of gametes and the female menstrual cycle are controlled by the body's nervous and hormonal systems and are closely tied to the developmental stage of the individual.

- Birth control can take place at several points before or during fetal development.

- Fertilization of an ovum requires only one sperm. Although a human ovum is viable for only about one day, human sperm can live in the female reproductive tract for about two days.

Objectives

Students should be able to:

- *describe* the events in a life cycle.

- *distinguish* the role of reproduction in the existence of a population from its role in the life of an individual.

- *explain* the advantage of sexual reproduction.

- *give* examples of organisms in which asexual reproduction occurs.

- *relate* the process of vegetative reproduction to the process of regeneration.

- *distinguish* between asexual and sexual methods of reproduction.

- *relate* the terms *male* and *female* to gametes.

- *explain* why meiosis is an essential process in a sexual reproductive cycle.

- *construct* a graphic representation of meiosis.

- *describe* the essential steps in meiosis.

- *compare* the outcome of meiosis with that of mitosis with respect to chromosome number.

- *compare* the process of gametogenesis in human males and females.

- *describe* the human female reproductive cycle (without necessarily naming the hormones).

- *explain* the significance of menstruation and menopause in the female life cycle.

- *relate* birth control by individuals to overpopulation and to family needs.

- *describe* methods of contraception and their advantages and disadvantages.

- *discuss* the process of fertilization in humans.

Strategies

This chapter continues to develop the ideas introduced in Chapter 5. Provide an abundance of living materials to illustrate reproductive processes. Strawberry plants, *Sansevieria, Kalanchoe,* and potatoes, all of which are easy to obtain, demonstrate vegetative reproduction. It is somewhat more difficult to get budding yeasts, fissioning paramecia, and budding hydra at the right time. For animal reproduction, the problem of timing is difficult to overcome, but guppies and pregnant rats or mice usually can be obtained. This chapter may be undertaken at a season when frogs exhibiting amplexus are available. Any fertilized eggs obtained from them can be used as a supplemental study of embryological development.

Have students read and discuss Sections 6.1–6.3, followed by Investigation 6.1. Section 6.4 reinforces the concepts and processes introduced in the investigation. Then, have students read the remainder of the chapter, followed by careful discussion.

References

Aral, S. O., and K. K. Holmes. "Sexually Transmitted Diseases in the AIDS Era." *Scientific American* (February 1991): 62–69.

Crews, D. "Animal Sexuality." *Scientific American* (January 1994): 108–114.

Corea, G. *The Mother Machine.* New York: Harper and Row, 1985.

Hart, S. "When *Wolbachia* Invades Insect Sex Lives Get a New Spin." (January 1995):4–6.

Lively, C. "Host-Parasite Coevolution and Sex." *BioScience* (February 1996): 107–114

Sagan, D., and L. Margulis. "The Riddle of Sex." *The Science Teacher* (March 1985): 16–22.

Singer, P., and D. Wells. *Making Babies: The Net Science and Ethics of Conception.* New York: Charles Scribner's Sons, 1985.

Suggested Answers to Applications

1. In prophase I of meiosis, matching pairs of chromosomes (homologous pairs) come together, but in prophase of mitosis, the matching pairs of chromosomes remain separate.

2. In mitosis, the matching chromosome pairs do not come close together, so there is no crossing-over.

3. This is a form of asexual reproduction because only one parent is involved.

4. The sexually reproducing species has a better chance because it will produce genetically varied offspring, some of which may survive the changing environment of the future.

5. A sperm carries a nucleus and almost no cytoplasm. Mitochondria in the middle section of the sperm provide power for the tail but do not penetrate the egg during fertilization. Thus, the only source of mitochondrial DNA is the mitochondria in the cytoplasm of the ovum.

6. This is most likely a gamete because it has an odd number of chromosomes, whereas most normal body cells contain an even number (2n).

Suggested Answers to Problems
6. Semen is composed of sperm and various fluids that come from the seminal vesicles, the prostate gland, and Cowper's glands. This fluid contains buffers that neutralize the normally acidic environment of the female reproductive tract.

7. In fraternal twinning (dizygotic), two eggs are released by the female and fertilized by the male. Identical twinning (monozygotic) results from the separation of the zygote into two masses at an early stage in development. Each mass then develops separately, producing two embryos.

Materials and Preparations for Investigations

INVESTIGATION 6.1 A Model of Meiosis

Materials (per class of 30, teams of 2)
1 lb red modeling clay
1 lb blue modeling clay
60 2-cm pieces of pipe cleaner
15 large pieces of paper

Preparations
You will need modeling clay or Play-Doh™ of two different colors (red and blue are suggested). The pipe cleaners (centromeres) can be pressed into the clay to hold the chromatids together. Pipe cleaners also can be used for the chromatids, but they are not as suitable as clay because they have to be cut and the pieces twisted together to show recombination.

The manipulation of the materials achieves the purpose of this investigation, but a number of points can be made in discussion. One worth considering (because of its bearing on Chapter 8) is the random distribution of paternal and maternal chromosomes during synapsis. There is likely to be considerable variation in results when chromosomes are moved to opposite poles. The consequent separation of paternal and maternal chromosomes may be pointed out to students.

CHAPTER 7
CONTINUITY THROUGH DEVELOPMENT

Planning Ahead
Check on your order for yeast strains and growth media for Investigation 8.3.

Concepts
◆ Major developmental changes occur in a multicellular individual, beginning with a fertilized egg and continuing until death. Development is physiological, anatomical, and behavioral.

◆ Embryological development occurs at the beginning of an individual's life and includes the division and movement of cells and the differentiation of cells into specialized tissues.

◆ During early embryological development, cells divide into smaller cells, move into the shape of a hollow ball, and eventually become a sphere with three distinct layers. These layers give rise to all the different tissues and organs of the body.

◆ Developmental changes are programmed genetically and follow characteristic stages for all individuals of a given species.

◆ Development is guided by chemical and neurological interactions between different cells, tissues, and organs.

◆ The "birth" of an individual varies widely among animals but usually signifies the animal's emergence from the egg or parent into the external environment.

◆ The more complex the animal, the greater is the developmental dependence of a new individual on its parent.

◆ Mammalian embryos are nourished by diffusion of substances from maternal blood through the placenta. Harmful chemicals in the body of the parent may enter the embryo by the same route.

◆ Occasional cells lose their ability to recognize and become the tissue from which they arose. This sometimes results in out-of-control tissue growth—cancer—that can interfere with the normal functioning of the surrounding tissues and organs.

◆ Some cancers appear to be governed by an interaction between specific cell genes and invasions of environmental factors such as viruses and radiation.

Objectives

Students should be able to:

◆ *describe* the stages in the life cycle of yeast and distinguish between the haploid and diploid stages.

◆ *describe* the early embryological development of animals in terms of three cell layers.

◆ *explain* what is meant by differentiation.

◆ *explain,* in terms of proteins and gene activity, how differentiated cells differ.

◆ *describe* the role of regulatory molecules in cell differentiation.

◆ *give* an example that demonstrates the importance of cell-to-cell interactions.

◆ *relate* the hard eggshell around a bird or reptile embryo to the control of evaporation.

◆ *explain* the importance of various extraembryonic membranes to the embryo.

◆ *describe* the development and functions of the placenta.

◆ *describe* several techniques used to help diagnose fetal abnormalities.

◆ *explain* how exposure of a pregnant woman to harmful agents can affect normal fetal development.

◆ *describe* the role of aging in development and in the life cycle of an individual cell.

◆ *compare* the growth of cancer cells with the normal cell growth.

◆ *describe* the roles of oncogenes and tumor-suppressor genes in cancer.

Strategies

This series of chapters dealing with continuity proceeds to the development of the zygote. The chapter establishes an integrated view of reproduction and development as basic processes in biology and provides background for later development of the concepts of heredity and evolution.

The chapter should help students to gain some scientific perspective on their own interests in reproductive processes. During the last parts of the chapter, some students, encouraged by the scientific atmosphere, may seek answers to personally perplexing questions. Be aware of your professional limitations when you deal with such questions.

Investigation 7.1 should begin immediately as it will take nearly half the class time for almost two weeks. This is an opportunity for students to engage in real laboratory science and is an important investigation. Students should start Part A on a Monday (or Tuesday at the latest) so they can complete Days 0, 1, and 2 with no more than a 24-hour interruption. Day 3 or Part B is best done on a Thursday or Friday so it can incubate over the weekend. The last of Part B and the first of Part C (Day 7) can be done the following Monday or Tuesday, followed by Day 8. Take a few minutes after the lab work each day to have the class discuss their procedures and observations. This sharing and summarizing is an important part of the scientific enterprise.

Intersperse Investigation 7.1 with discussions and reading assignments of Sections 7.1–7.3 of the text. Investigation 7.2 can be done in one period if the tasks are staggered, as follows: Teacher does Steps 1–6 for the first class about one hour before the class meets. The first class then does Steps 1–6 for the second class and Step 7 from what the teacher has set up. The second class does Steps 1–6 for the next class and Step 7 from the previous class, etc. This will avoid the 30 minute wait for each class, yet let all classes do all parts to of the investigation.

Follow Investigation 7.2 with a discussion the next day and then with readings and discussion of Sections in 7.4–7.8. Assign the Applications as homework for review and discuss these before the exam. Have the more eager students work on selected Problems.

References

DeRobertis, E. M., G. Oliver, and C. V. E. Wright "Homebox Genes and the Vertebrate Body Plan.2; *Scientific American* (July 1990): 46–53

Gillis, A.M. "Finding the Right Sperm for the Job." *BioScience* (September 1995): 525–526.

Kalil, R. E. "Synapse Formation in the Developing Brain." *Scientific American* (December 1989): 76–87.

Roush, W. "Sperm Protein Makes Its Mark Upon the Worm Embryo." *Science* (5 January 1996): 33.

Rusting, R.L. "Trends in Biology: Why Do We Age?" *Scientific American* (December 1992): 130–139.

Sacho, L. "Growth, Differentiation, and the Reversal of Malignancy." *Scientific American* (January 1986); 40–47.

Uvnas-Moberg, R. "The Gastrointestinal Tract in Growth and Reproduction." *Scientific American* (July 1989) 78–83.

Suggested Answers to Applications

1. The embryonic cell layers are the ectoderm, endoderm, and mesoderm, which give rise to the following structures:

a. ectoderm: brain, spinal cord, nerves, outer layer of skin, and skin derivatives (hair, nails, feathers, scales)

b. endoderm: lining of the alimentary canal, liver, lungs, and pancreas

c. mesoderm: muscles, skeleton, circulatory and excretory systems, inner layer of skin and gonads

2. Because almost all cells of an individual are genetically identical, it is possible to extract any cell or group of cells from one organism and use them to clone another, genetically identical, organism.

3. The blastula is developed through a series of mitotic divisions beginning with the zygote. For a while, the fertilized egg simply divides into smaller and smaller units. Although there are more cells, the total size remains the same.

4. The oak tree will not develop fingers because the regulatory molecules simply turn genes on and off. Since there are no "finger" genes in the oak tree to turn on, the regulatory molecules will have no effect.

5. Development is more than just an increase in the number of cells. Newly formed individual cells are smaller than at maturity and do not yet contain all of the internal workings of a mature cell, so single celled organisms do develop. On the other hand, single-celled organisms do not differentiate because they do not become substantially different in appearance or function from their parent cells.

6. This is an inefficient method of reproduction. Many of the eggs are lost or eaten. Many of the sperm are lost in the water. Thus, the males and females must produce enormous numbers of gametes to ensure that at least a few eggs are fertilized. Furthermore, in most cases, the adults do not tend the young; therefore, there is a tremendous loss of offspring.

7. In both, the developing embryo is housed until birth, nurtured in an aqueous environment, and protected from an external environment.

8. Cancer results from abnormal cell division and metabolism. Rapidly dividing cancer cells take the raw materials they use for division and growth from normal cells or organs. Because cell division is one of the basic components in development, cancer may be considered a developmental problem. A better understanding of developmental processes may help us learn to curb or prevent abnormal cell growths.

Suggested Answers to Problems

1. The phenomenon of regeneration occurs in varying degrees throughout the animal kingdom from sponges to man. Apparently the capacity for regeneration is limited to relatively unspecialized cells. In the course of development from zygote to adult, the potentiality of all cells becomes progressively restricted. Highly specialized cells, indeed almost all of the cells found in higher vertebrates, have almost lost the ability to regenerate easily.

2. A discussion of seeds and fruits is appropriate here. Students could dissect seeds, separating the embryo from its food source. Students could observe a progression of developing plant embryos made by beginning the germination of corn or bean seeds on successive days.

3. Experiments have been conducted in which a frog or toad blastula has been distorted with weighted glass plates that changed the position of cells and the pattern of cell division. Resulting organisms have structures in abnormal positions (for example, legs near head).

4. You also can focus on how certain factors operate to decrease life expectancy in humans.

5. Obtain a list of common household items with carcinogenic potential from a local member of an environmental group. Ask students what happens to these carcinogenic compounds when they are thrown away.

Materials and Preparations for Investigations

INVESTIGATION 7.1 The Yeast Life Cycle

Materials (per class of 30, teams of 2)

15 microscope slides
15 cover slips
15 dropping pipets (or 1 per table)
15 glass-marking pencils (or 1 per table)
15 compound microscopes
15 containers of clean flat toothpicks (or 1 per table)
15 bottles of water (or 1 per table)
90 self-closing plastic bags (or 1 per table per day)
45 growth medium agar plates (YED)
15 sporulation medium agar plates (YEKAC—"unknown")
7 biohazard bags
15 each agar slant cultures of **a**2 and *a*3 yeast strains

Preparations

Order freeze-dried yeast strains **a**2 HAR and *a*3 HBT from Ward's Natural Science Establishment (1-800-962-2660) within 3 to 4 weeks of use. Rehydrate the yeast and prepare agar slants 24 hours before the lab; directions and slant tubes are included in the kit. Bottles of prepared growth and sporulation media are supplied by Ward's; follow the directions to liquefy and pour into sterile petri dishes. Ward's will supply prepoured plates on request; allow two weeks for delivery. The plates can be stored at room temperature in a dark cupboard for as long as five months in tightly closed plastic bags. Growth medium plates should be made at least three days in advance; sporulation ("unknown") medium plates are best when at least one week old, and better after two or three weeks. Directions for preparing media are provided in A Table of Recipes in the Teacher's Resource Book if you prefer to make your own.

Use flat toothpicks from an unopened box. Cut a small hole in one corner of a new box so the students can shake out each toothpick far enough to grasp the pointed end as needed. The flat end is less likely to tear the agar.

Devote a few plates to practice so students can develop a light touch for streaking yeast on the agar surface without burying the toothpick. The students can pass the plates around to compare techniques.

You can shorten the experiment by preparing fresh cultures of the two strains and having the students start with Step 3. Good mating requires fresh cultures.

Always incubate the plates upside down so condensation does not drip on the agar surface. If there is condensation on the lid after refrigeration, keep the plates upside down while you remove the lid and wipe out the condensation with a sterile tissue.

If the students' plates are incubated at room temperature, the yeast can be allowed 2 days to grow at each step. In a 30° C incubator (optimal temperature for this yeast) they will grow faster, and each growth step will take only a day. Sporulation will require 4 days at either room temperature or at 30° C but is strongly inhibited by higher temperatures. The experiment can be held in the refrigerator at almost any stage. It is preferable to refrigerate after mating and zygote formation, not before. After sporulation, the cultures can be kept for more than a week at room temperature. Cultures then should be stored in the refrigerator for future use.

The sequence of events associated with mating, both in Step 6 and in Step 18, can be examined with greater time resolution by transferring different plates to the refrigerator at half-hour intervals during a 2-hour period, starting with plates that have been at room temperature or 30° C for 3 hours. Different teams can prepare and seal slides from different times and then take turns looking at each other's preparations. Sealing the edges of the cover slip with clear nail polish stops the streaming that often occurs and preserves the preparation for several days.

INVESTIGATION 7.2 Development in Polychaete Worms

Materials (per class of 30, teams of 2)
45 microscope slides
45 cover slips

15 dropping pipets
45 finger bowls
15 compound microscopes
15 forceps
15 pairs of dissecting scissors
30 dissecting needles with corks for tips
30 25-cm² pieces of cheesecloth
paper towels
1.5 L seawater
(for teacher prep)
40 L seawater
2 dissecting trays
1 male *Chaetopterus* (sufficient for 5 classes)
l female *Chaetopterus* (sufficient for 5 classes)

Preparations

Have 19 to 38 L of seawater prepared when the worms arrive. You can obtain a seawater mix from a local pet store or from a biological supply company. Follow the instructions enclosed with the worms to remove them from the tubes. Keep the males and females separate.

Fertilize a set of eggs from one parapodium the night before the investigation. Prepare a second set an hour before your first class. The students then will be able to observe swimming larvae and the initial stages of cleavage while they are waiting for fertilization to occur in their samples. This waiting time also can be used to transfer the recently fertilized eggs from previous classes to fresh seawater if that could not be completed in the previous period.

It helps to transfer all your cultures from the preceding 24 hours to a 250-mL beaker containing fresh seawater. This transfer will keep the larvae alive several days longer than the one day expectancy in the students' finger bowls. It may be helpful to transfer each class's embryos to separate culture dishes or beakers with fresh seawater, so that students can sample from the same container for additional observations after the first laboratory period.

One major difference between *Chaetopterus* eggs and amphibian or sea urchin eggs is that the *Chaetopterus* eggs have just started meiosis when removed from the parapodia. The first and second polar bodies are not cast off until fertilization has occurred. Students can see both polar bodies, providing a good opportunity to review their role in the formation of the egg. Table T7.1 shows the average times for some critical stages to occur. Temperatures above 26° C can kill the larvae. The times in the table represent averages, not absolutes.

CHAPTER 8
CONTINUITY THROUGH HEREDITY

Planning Ahead
Order seeds for Investigations 19.3 and 20.3 from a supply house or seed catalog.

Table T8.1 Average Development Times for Chaetopterus Stages

Stage	Time At 22–23° C	Time At 24–26° C
First polar body	14 min	11 min
Second polar body	28 min	18 min
"Pear" stage	42 min	36 min
Polar lobe	47 min	41 min
First cleavage	51 min	42 min
Second cleavage	71 min	59 min
Swimming trochophore larva	22–24 hr	8–20 hr

Source: From E. P. Costello, M. E. Davidson, Eggs, M. H. Fox, and C. Henley. *Methods for Obtaining and Handling Marine Eggs and Embryos* (Marine Biological Laboratory: Woods Hole, MA 1957), p. 247

Concepts

◆ Heredity is the transmission of genetic information from generation to generation.

◆ Genes, located on the cell's chromosomes, are the units of inherited information. The chemical make-up of genes tends to be stable from generation to generation.

◆ Inheritance occurs in regular patterns that can be predicted by the rules of probability.

◆ An allele is an alternative form of a gene at a given locus that can cause different expressions of a given trait. Alleles arise from mutations, chemical changes in a gene locus.

◆ Gregor Mendel identified basic principles that apply to the inheritance of simple traits.

◆ During meiosis, chromosome pairs separate into different gametes. As a result, each of the two alleles for a given trait will appear in a different gamete; this is the principle of segregation.

◆ Experimental work by many scientists was necessary to elucidate the mechanisms of inheritance. Outstanding among these were the experiments of Beadle and Tatum, Hershey and Chase, and Morgan and his group.

◆ In eukaryotes, genes are made of long, specific, base-pair segments of DNA. Genes may code for proteins or RNA or may serve regulatory functions.

◆ To make a protein, a gene is transcribed into mRNA in the nucleus. The mRNA moves to the ribosome and is translated into a specific sequence of amino acids that form a specific protein.

◆ The inheritance of alleles on one chromosome pair does not affect the inheritance of alleles on a different chromosome pair; this is the principle of independent assortment.

◆ Most inherited traits do not follow the patterns identified by Mendel. Other patterns include codominance, multiple alleles, multifactorial inheritance, nondisjunction, linkage, and X-linked traits.

◆ Many animals have a pair of chromosomes that control the sexual development of the individual. The sex chromosomes also contain genes for other traits for which the inheritance pattern is dependent on the sex of the individual.

◆ Risks and benefits must be weighed for the products of genetic engineering. Serious ethical, moral, environmental, and evolutionary questions need to be addressed regarding the safe use of this technology.

Objectives

Students should be able to:

◆ *discuss* the role of genes in heredity.

◆ *demonstrate* the laws of probability involved in the study of heredity.

◆ *describe* the ways in which Mendel's work differed from that of his predecessors.

◆ *relate* the terms *dominant* and *recessive* to breeding results.

◆ *distinguish* between genes and alleles.

◆ *distinguish* between an individual's phenotype and genotype.

◆ *identify* gene symbols indicating homozygous and heterozygous conditions.

◆ *solve* problems that involve the major principles of genetics, using appropriate symbols and diagrams.

◆ *relate* Mendel's observations to the chromosome theory of heredity.

◆ *describe* the work of Beadle and Tatum.

◆ *describe* the biochemical nature of a gene mutation.

◆ *explain*, using experimental evidence, how a mutant characteristic differs from other characteristics of a given individual.

◆ *recall* the steps by which it was determined that genetic information is transmitted by DNA.

◆ *explain* how the structure of DNA determines the gene properties of replication, mutation, and transmission of information to succeeding generations.

- *demonstrate*, using a model or diagram, the way in which DNA replicates.

- *contrast* the structure of a section of DNA with the mRNA formed from it.

- *relate* codons to the construction of polypeptide chains.

- *compare* the roles of mRNA and tRNA in the production of protein.

- *show*, using diagrams, how the 9: 3: 3: 1 ratio results from a Mendelian dihybrid cross.

- *cite* experimental and observational evidence for the idea that an organism's characteristics result from the interaction of its biological inheritance and its environment.

- *identify* evidence that requires modification of the idea of dominance.

- *construct* a diagram illustrating the inheritance of an X-linked trait.

- *explain* the genetic use of pedigrees.

- *name*, when given suitable case descriptions, various modes of inheritance, such as codominance, linkage, and recombination.

- *give* examples of genetic disorders.

- *identify* the results of nondisjunction in a karyotype.

- *identify* several types of chromosomal abnormalities.

- *explain* some practical applications of recombinant DNA technology.

- *describe* several ethical, legal, and policy issues that may arise from our increasing ability to screen for genetic disorders.

Strategies

Heredity is one of the most interesting and important areas of biology. As more is learned about genetic mechanisms at the molecular level, the emphasis on Mendelian principles will decrease. Mendelian rules are appropriate to begin a study of genetics, but they are far more the exception than the rule, especially for human inheritance. Be alert for signs of student confusion over the idea of recombinations arising from crossing-over and the material on enzymes. Reasoning from data is difficult for many students. The following topics deserve teacher explication: pedigrees, the Y chromosome, mapping genes, the reasoning in the *Neurospora* experiments, and the nature of the genetic code.

In chemistry and physics, solving problems is recognized as an important method by which students gain understanding. We have advised it in connection with population studies and do so again for Chapter 8. The Teacher's Resource Book includes a selection of genetics problems; most college genetics texts contain large numbers of problems. Select and adapt problems that are commensurate with your students' abilities. It is more desirable to supply many simple problems that illustrate a limited number of principles using different traits in different organisms than to cover a wide range of principles with a limited number of difficult problems. In working with genetics problems, it is more important for students to attempt to determine the patterns of dominance (if any) than to be told in advance what is dominant. Give students the observations and have them explain what is happening.

Investigation 8.1 can be started without introduction and can be followed by reading the introduction and Sections 8.1–8.4 and answering the Concept Review questions. Remind students to respond to the interactive legend at the beginning of the chapter, and reinforce this by letting them share their responses in class. Sections 8.5 and 8.6 should be read and discussed in class just before beginning Investigation 8.2. It is best not to rush students through this investigation; have them do Part C the next day if necessary. Discuss the entire investigation the next day, and then assign Section 8.7 and the Concept Review questions immediately following. The next five class days should be devoted to Investigation 8.3, an excellent, hands on activity that develops Mendelian principles. Assign the remainder of the chapter (Sections 8.8–8.14) and Concept Review questions during that week as home work. You may want to sprinkle in some of your favorite genetics problems during that week also. Finally, assign the Applications and discuss them prior to your exam.

References

Anderson, W. F. "Gene Therapy." *Scientific American* (September 1995): 124–128.

Barton, J. H. "Patenting Life." *Scientific American* (March 1991): 40–47.

BSCS. Advances in Genetic Technology. Lexington, MA: D. C. Health, 1989.

Capecci, M. R. "Targeted Gene Replacement." *Scientific American* (March 1994): 52–59.

Dewitt, P.E. "The Genetic Revolution." *Time* (17 January 1994): 46–53.

Holliday, R. "A Different Kind of Inheritance." *Scientific American* (June 1989): 60–73.

Lawn, R. M., and G. A. Vehar. "The Molecular Genetics of Hemophilia." *Scientific American* (March 1986): 48–57.

McInerney, J.D. "Genetics for High School Students and Undergraduate Non-majors." A Position Paper prepared for the National Science Foundation, February 1992.

Nathan, J. "The Genes for Color Vision." *Scientific American* (January 1989): 42–49.

Neufeld, P. J., and N. Colman. "When Science Takes the Witness Stand." *Scientific American* (May 1990): 46–53.

Ptashne, M. "How Gene Activators Work." *Scientific American* (January 1989): 40–47.

Ross, J. "The Turnover of Messenger RNA." *Scientific American* (April 1989): 48–55.

Spienza, C. "Parental Imprinting of Genes." *Scientific American* (October 1990): 52–61.

Verma, I. M. "Gene Therapy." *Scientific American* (November 1990): 68–84.

Weintraub, H. M. "Antisense RNA and DNA." *Scientific American* (January 1990): 40–7.

Welsh, M. and A. Smith. "Cystic Fibrosis." *Scientific American* (December 1995): 52–59.

Suggested Answers to Applications

1. There is a 50% chance that she carries the allele. Her parents are heterozygous, that is, *F/f* and *F/f*. She has a 50% chance of being *F/f* also but a 25% chance of being homozygous for the dominant trait, or *F/F*.

2. The probability is still 1½ % or 50%, no matter how many times "heads" or "tails" comes up in a row. The outcome of this single event is not based on previous events.

3. The nucleotides can be arranged in any order—it is the type and number of nucleotides in each different order that make up the genetic codes. There are more than enough ways to arrange the four nucleotides into combinations for all the different enzymes.

4. The nucleotide that is complementary to the missing DNA nucleotide would not be inserted into the mRNA. This would change all the succeeding codons, resulting in different amino acids being inserted into the protein and, therefore, in an abnormal protein.

5. The distribution of human multifactorial traits is continuous, as opposed to discontinuous (discrete) distribution of either/or traits.

6. Genes and combinations thereof determine the variations that are possible. The environment, however, determines which combinations survive and how they are expressed in the phenotype. An unfavorable trait or combination of traits probably will not survive. On the other hand, if the trait or traits do not adversely affect the individual, they probably will survive.

7. The principle of dominance explains why the peas were yellow or green, smooth or wrinkled. For example, the smooth form of the seed shape trait was dominant over the wrinkled form; therefore, in the first cross, all seeds were smooth. The principle of segregation explains how the combinations of alleles occur. Because the two alleles of a gene are segregated during gamete formation, three different combinations (*R/R, R/r, r/r*) are possible. The principle of independent assortment allows for the inheritance of certain traits independently of others. Both yellow smooth peas and yellow wrinkled peas are possible because the alleles controlling color are not linked to the alleles controlling skin type.

8. In the first meiotic division each chromosome pair separates independently of the other pairs. Thus, genes in one chromosome pair are assorted independently of genes in other chromosome pairs, giving rise to Mendel's ratios.

9. Because these plants are neither male nor female, they probably would lack sex chromosomes. Dioecious species, those with separate male and female plants, may have sex chromosomes.

10. If the genes are linked and if there are no crossing over, then fewer types of F_1 gametes will be produced and fewer types of F_2 offspring. With crossing-over, however, the number of types of offspring is increased.

11. On a strictly genetic basis, the father donates the Y chromosome to his sons, and the mother donates the X chromosome along with the potential to be bald. The father is not a good indicator, though the mother's brothers are.

12. DNA might be thought of as the basis on which different organisms are built. The structure of the individual's DNA determines its form and characteristics. The continuity of life in the biosphere depends on the DNA structure. If that structure is left alone, chances are the continuity will be undisturbed, but if the DNA structure in many organisms is substantially changed, the continuity of life in the biosphere could be disrupted.

Suggested Answers to Problems

1. Answers will vary.

2. There is a great deal of current information that students can use on this subject. The agent that causes AIDS is an RNA virus.

3. To obtain more information, contact the Crucifer Genetics Cooperative, Department of Plant Pathology, 1630 Linden Dr., University of Wisconsin, Madison, WI 53706.

4. For Down syndrome, the increase in risk with advancing maternal age has been known for many years. The risk does not change much up to the age of 29 but rises steeply beginning with the 35 to 39 age group. At age 45 or over, the risk for Down syndrome is 12 times as great as at age 29.

6. A three-dimensional or two-dimensional model of DNA, mRNA, or tRNA, and amino acids will help your students visualize the complicated process of protein synthesis beginning with DNA.

7. Genetic transformations of calves and other food producing animals have raised concerns about effects on the quality of meat and milk, and animal welfare groups are worried about harming animals and the environment. Several years ago, the introduction of ice-minus bacteria into an experimental garden in California was widely debated with much publicity. These are two examples that might serve to start a discussion of the problems and potential benefits of genetic engineering.

8. Traits and location of genes: (a) Rh blood group—chromosome 1; (b) ABO blood group—chromosome 9; (c) hemophilia—X chromosome; (d) MHC—chromosome 6; (e) Huntington disease chromosome—4.

9. Genetic engineers use reverse transcriptase, an enzyme that can synthesize DNA from eukaryotic mRNA. Because the introns already have been removed from the mRNA, the DNA made from it contains only spliced exons. It is this DNA that is inserted into the bacterium.

10. Information may be obtained from your local March of Dimes and from the National Huntington Disease Association, Inc., 1182 Broadway Suite 402, New York, NY 10001. For background and information about ethical issues, see BSCS, *Basic Genetics: A Human Approach,* 2nd ed. (Dubuque, IA: Kendall/Hunt, 1990).

Materials and Preparations for Investigations

INVESTIGATION 8.1 Probability

Materials (per class of 30, teams of 2)
30 pennies (15 shiny, 15 dull)
15 cardboard boxes

Preparations

Collect enough small boxes so that each team has one in which to toss the pennies. Students can supply their own pennies or you can provide them.

INVESTIGATION 8.2 DNA Replication and Transcription to RNA

Materials (per class of 30, teams of 2)
15 sets of paper clips: 40 black, 40 white, 32 red, 32 green, 5 silver per set (totals: 600 black, 600 white, 480 red, 480 green, 75 silver)
masking tape
pen
90 10-cm pieces of string (9 meters)

Preparations

Monitor the teams to make certain they are using the key properly and matching the base pairs correctly.

INVESTIGATION 8.3 A Dihybrid Cross

Materials (per class of 30, teams of 2) **PART C**
15 containers of clean flat toothpicks (or 1 per table)
15 glass-marking pencils
30 paper templates for streaking strains (or students can make these)
45 self-closing plastic bags labeled waste (or 1 per table per day)
3 biohazard bags
15 complete growth medium agar plates (YED)
15 minimal plus adenine growth medium agar plates (MV+AD)
6 YED plates with 24-hour cultures of yeast strains **a**1, **a**2, **a**3, **a**4, α*1*, α*2*, α*3*, and α*4*, with yeast growing in 1-cm streaks in the template (or 1 per lab table)

Preparations

Three to four weeks before use, order freeze-dried yeast strains **a**1 (HAO), **a**2 (HAR) **a**3 (HAT), **a**4 (HART), α1 (HBO), α2 (HBR), α3 (HBT), and α4 (HBRT) from Ward's Natural Science Establishment (1-800-962-2660). Rehydrate the yeast 24 hours before the lab; directions are included in the kit. Prepare one YED plate per lab table by making a 1-cm streak of each strain according to the template pattern. Bottles of prepared complete and minimal plus adenine media are supplied by Ward's; follow the directions to liquefy and pour into sterile petri dishes. Ward's will supply prepared plates on request; allow two weeks for delivery. The plates can be stored at room temperature in a dark cupboard for as long as five months in tightly closed plastic bags. Plates should be prepared at least three days in advance and maintained at room temperature. Discard any contaminated plates before inoculation. Directions for preparing media are provided in A Table of Recipes in the Teacher's Resource Book if you prefer to make your own.

Use flat toothpicks from an unopened box. Cutting a small hole in one corner of a new box will allow students to shake out each toothpick far enough to grasp the pointed end as needed. The flat end is less likely to tear the agar.

If your students did not do Investigation 7.1, devote a few plates to practice so they can develop a light touch for streaking yeast on the agar surface without burying the toothpick. The students can pass the plates around to compare techniques.

Always incubate the plates upside down to prevent condensation from dripping on the agar surface.

Blackline masters of Figures 8.15, 8.16, 8.17, and 8.18 are included in the Teacher's Resource Book.

CHAPTER 9
CONTINUITY THROUGH EVOLUTION

Planning Ahead

If you have not already done so, read the teacher's notes for Investigation 12.1 and decide whether to order the mixture or to collect organisms locally. If necessary, order the mixture, along with the organisms for Investigations 14.1 and 14.2.

Concepts

◆ Evolution, broadly defined, is change through time.

◆ Evolutionary theory is the major unifying theme in biology because it provides testable hypotheses that explain both similarities and differences among organisms.

◆ Unity and diversity are characteristics of all forms of life. Unity of pattern is demonstrated by the universal genetic code, the cell theory, structure and function, and reproductive and developmental processes. Diversity is evident in the many forms of life on earth.

◆ Variation is the genetic (often observable) differences within a species, whereas diversity is the differences between species (always observable).

◆ The fundamental reproductive unit in biology is the species—organisms that are structurally similar, have a common ancestral lineage, and can produce fertile offspring.

◆ Darwin's development of the theory of evolution by natural selection is a classic example of how scientific thought proceeds. His processes included asking questions, collecting an abundance of data, and drawing reasonable inferences from the data.

◆ Natural selection states that only those individuals best adapted to their environment survive to reproduce. Thus, the gene pool of the next generation comes only from the surviving parents.

◆ Genetic variation, resulting from mutation and recombination, is the raw material of evolution.

◆ The major sources of variation are mutation and genetic recombination resulting from meiosis and sexual reproduction.

◆ The major evolutionary forces affecting populations are mutation, migration, random change in allele frequencies, natural selection, and sexual selection.

◆ Speciation is a process in which characteristics of populations come to differ to such an extent that the individuals in one population can no longer interbreed with individuals of the other, thus dividing the gene pool into two. It most often occurs when a subset of the parent population, possessing a variable genetic makeup, becomes reproductively isolated from the original population.

◆ Major evolutionary changes probably occur relatively rapidly and during times of significant environmental changes in the habitat of a population.

◆ Many patterns occurring in evolutionary development provide clues to the varied relationships between organisms.

Objectives

Students should be able to:

◆ *construct* histograms from given data.

◆ *infer* evidence for evolution from examples of unity of pattern among living organisms.

◆ *describe* three ways the distinctness of a species is maintained.

◆ *explain* the concept of variation.

◆ *explain* how Darwin's observations while on the *Beagle* laid the groundwork for his development of the theory of evolution.

◆ *name* at least four types of evidence Darwin used to support the basic theory of evolution.

◆ *describe* the reasoning that led to the theory of natural selection.

◆ *explain* the mechanism of natural selection and give examples.

◆ *contrast* the Darwinian and Lamarckian theories.

◆ *explain*, in evolutionary terms, the significance of the changes in *Biston betularia* populations.

◆ *recognize* mutation as the basic source of change in hereditary characteristics.

◆ *identify* mutations and genetic recombinations resulting from meiosis and sexual reproduction as the major sources of variation.

◆ *describe* at least three evolutionary mechanisms that can affect a population.

◆ *relate* the biological definition of species to the requirement of reproductive isolation in speciation.

◆ *infer* the role of geographic isolation in permitting the development of genetic traits that might lead to reproductive isolation.

◆ *explain* the process of adaptive radiation.

◆ *discuss* the hypothesis of punctuated equilibria.

◆ *name* at least three types of evolutionary patterns and give examples.

◆ *explain* the concept of a population bottleneck and relate it to extinction.

Strategies

Chapter 9 illustrates the growth of an idea in the minds of scientists and depicts these scientists as humans. Few biologists fit the purpose as well as Darwin. From his youth, he was a man of many faults. After his one great adventure he led an outwardly dull and prosaic life. He was neither an amateur nor a professional by today's standards. He was remote from the universities and was a turgid writer—a nice antidote to the popular vision of the scientist as a superhuman figure in a white laboratory coat.

Laboratory work involving the manipulation of organisms and equipment is obviously impossible, considering the subject matter of Chapter 9. Manipulating data obtained by others and exercising thought two justifications for laboratory activity—are not only possible but necessary. The investigations in Chapter 9 require preparations as careful as any others in the course.

Investigation 9.1 vividly illustrates the fundamental concept of variation while further developing students' observational and mathematical skills. Unshelled peanuts work well, but counter their tendency to disappear by reminding students not to eat anything in the laboratory. Follow Investigation 9.1 with reading and discussion of Sections 9.1–9.4. Investigation 9.2 is a simulation of natural selection and develops Sections 9.5–9.7. Although somewhat tedious, Investigation 9.3 is recommended because it deals with a real evolutionary problem that was pursued by a contemporary biologist. Using actual field data, it gives students a feel for how research on evolution proceeds. Indeed, some of the questions arose during the original study, questions for which Dr. Stebbins's intensive fieldwork provided possible answers.

References

Alvarez, W., F. Asaro, and V. E. Courtillot. "What Causes Mass Extinction?" *Scientific American* (October 1990): 76–93.

Gould, S. J. "The Evolution of Life on the Earth." *Scientific American* (October 1994): 84–91.

Keating, J. F., O. B. Toon, and J. B. Pollack. "How Climate Evolved on the Terrestrial Planets." *Scientific American* (February 1988): 90–97.

Kerr, A. "Periodic Extinctions and Impacts Challenged." *Science* (22 March 1985): 1451–52.

Levington, J.S. "The Big Bang of Animal Evolution." *Scientific American* (November 1992): 84–91.

Lewin, R. "Punctuated Equilibrium Is Old Hat." *Science* (14 February 1986): 672–86.

Shipman, P. "Baffling Limb on the Family Tree." *Discover* (September 1986): 86–92.

Williams, M. B. "The Scientific Status of Evolutionary Theory." *American Biology Teacher* (April 1985): 205–10.

Suggested Answers to Applications

1. In domesticated species, humans select for many different characteristics, seeking varieties that can serve a particular economic function or those that are amusing or startling. We then curtail or even prevent crosses between these varieties. Natural selection, on the contrary, selects only the set of traits best suited to survival in a particular environment. Where environments differ within the geographical range of a species, some variation (subspecies) may occur.

2. Students may suspect, correctly, that polydactyly is not the only disorder caused by this dominant gene. The chief reason the phenotype is rare is that afflicted individuals often have more serious

disorders and do not live to pass on the gene. This suggests natural selection. Other polydactylys are caused by recessive genes that also may cause feeblemindedness and reduced fertility. Again the genes rarely are passed on. The *I* allele in the populations of some North American Indians apparently resulted from their origin as a small population that by chance was homozygous for this allele. Isolation preserved the allele.

3. In general, the less the chances for survival, the larger the clutch size. The important question is how many eggs survive to become reproductive adults. This number is undoubtedly lower as latitude becomes higher. Even if populations of adults at high latitudes were larger, this does not mean that they would replace the populations of lower latitudes. After all, the latter populations are probably already in balance with the carrying capacity of the land.

4. Mendel's work explains the genetic basis for inheritance of the variations on which natural selection acts.

5. Some medicines have been found to contribute to gene mutations or chromosome rearrangements. Others alleviate conditions that natural selection would have operated against. (Thus, there are more diabetics than formerly.) The trend has been toward human populations with more widespread and numerous abnormalities, yet medicine itself may end up dealing with some of these. For example, replacement of faulty genes in gametes may become possible in the future through gene splicing (or genetic engineering).

6. According to the biological definition, these populations may be part of the same species. Because they live in different habitats and look different from each other, however, they may be classified as separate species.

7. Genetic variation may be a problem for an individual because an individual with an unfavorable variation may die without reproducing. Such a death improves the gene pool of the species because it removes undesirable genes. Among genetically different offspring, those best adapted to the environment may live to reproduce, and some of the others may not. Without genetic variation, however, all members of a population would be equally adapted to the environment; and all might live, or die.

8. These are not equivalent terms, as many students believe. Evolution is change through time. Natural

selection is one of the mechanisms by which evolution occurs.

9. Yes, because genetic recombination would still supply different types of offspring. However, the rate of evolution would be reduced, and the long-term future of populations would be jeopardized.

10. Because factors that change the equilibrium of a population have a greater effect on small populations than on large ones, the rate of evolution should be higher in small populations. These same factors also increase the likelihood of extinction, with small populations facing a greater risk.

11. This phrase refers to the success of those individuals that reproduce the most. "Fitness" is determined by the number of offspring an individual produces. Those that are best adapted to their environment will reproduce the most. An organism can be weak, sneaky, and cowardly and still be successful at reproducing.

Suggested Answers to Problems

1. Genetic data alone would account for variety and changes in organisms through time and also would show relationships from which a theory of evolution could develop. Mutations change the genetic makeup and are inherited. Genetic engineering allows humans to manipulate genes and create new organisms. The theory, using molecular and genetic data, would resemble Darwin's. It would account for the origin of species and for broader categories of organisms. It would differ from Darwin's presentation by requiring less time for speciation (according to the theory of punctuated equilibria) and would not be based on morphology, embryology, or geographic distribution.

2. If species did not change and were unrelated to each other, there would be no way to classify them into various groups using common characteristics. Evidence pointing to relationships would be false, fossil evidence would be meaningless, and the adaptation of organisms to an environment would be impossible. Variations between populations of a species would exist only to a slight degree or not at all. Genetic data would be in error, and mutations and genetic engineering would be impossible. Biology would be harder to study because there would be no common elements among organisms. More than one million species would have to be studied separately, a much more difficult task than studying a few major phyla whose members have common attributes.

3. The population would have to be variable and be divided by an isolating mechanism. Thus, the original interbreeding population would become two populations with a barrier separating the two gene pools. In time, mutation and selection would make each genetically distinct. When that distinction became so great that the populations could no longer interbreed, two species would exist where one had existed before, even if the barrier were removed.

4. Guides to local flora will identify endemic plant species, and natural history guides should help identify endemic animals. The more different kinds of habitats in your community or the more isolated it is from other natural areas, the greater the probability of finding endemic species.

5. Refer to the many writings of Stephen Jay Gould on this topic.

6. Darwin studied and wrote about topics including coral reefs, orchids, climbing plants, domesticated plants and animals, earthworms, insectivorous plants, flower forms, and plant movements.

7. Have your students consult the chapter on life in the past as well as paleontology and anthropology texts.

Materials and Preparations for Investigations

INVESTIGATION 9.1 Variation in Size of Organisms

Materials (per class of 30, teams of 3)
10 15-cm metric rulers
20 sheets of graph paper
500 objects of one type (dried bean seeds, peanuts in the shell, leaves from the same type of tree)

Preparations
Almost any objects can be measured in this exercise. Items collected in the wild are better because they show more variation. Be certain the objects have sufficient measurable variation. Use your imagination to come up with a sample that is meaningful to your students. If you use dried bean seeds, purchase them in a supermarket, not from a seed store. Seeds for planting have been treated with fungicides that may be toxic.

INVESTIGATION 9.2 Natural Selection—A Simulation

Materials (per class of 30, teams of 6)
10 1-m × 2-m pieces of fabric, 5 each of 2 different colors and patterns
20 sheets of construction paper, 2 each of 10 different colors
16 self-closing plastic bags
5 small bowls
5 sets of colored pencils with colors similar to chip colors
25 sheets of graph paper

Preparations
Punch out quarter-inch paper chips from construction paper, 500 each of 10 different colors. Use a wide variety of colors such as red, orange, purple, green, blue, yellow, brown, gray, black, and white. To speed preparation, fold the paper to four thicknesses before you punch it out. Put chips of each color into separate plastic bags and shake well.

Remove 10 chips from each of the 10 bags to create populations of 100 each. Place chip populations in separate bags.

Choose fabric patterns that simulate natural environments, such as floral, leaf, or fruit prints. The patterns should have several colors and be of intricate design. Test the colored chips against the material to make sure some of them blend well. Select several designs, each with a different predominant color. This will make it possible to demonstrate the evolution of different adaptive color types from the same starting population.

Dim the lights if possible. Make sure all the participants stand with their backs toward the habitat so they do not prematurely locate any chips. At a signal, have the predators take turns picking out chips. Each student should have 15 turns (the quota for each predator depends on team size). The keeper should ensure that precisely 25 chips remain at the end of the selection process.

Other simulations are possible using these materials. Population growth and carrying capacity can be demonstrated by removing only 72 individuals during the first selection process and only 75 from each subsequent generation. Emphasize that each surviving individual asexually reproduces 3 offspring each generation and that the carrying capacity of the habitat is 100 individuals. Also emphasize that emigration and immigration do not occur, mutations do not occur, and neither the habitat nor the predators change during Part A of the investigation. At the end of three or four rounds, the population will far exceed the carrying capacity of the habitat. Ask students to extrapolate the population size at the end of 10 or 15 generations.

To stimulate mutation, have students add several new chips to an "adapted" population and continue the selection process. Selection for or against mutation can be demonstrated by having the mutated chips blend better or less well with the habitat than the chips of the original population.

By using two different sizes of paper chips or by marking half the chips with felt markers, you can select for two or more characteristics at the same time. You also can use thicker and thinner chips, or whole and half chips.

INVESTIGATION 9.3 A Step in Speciation

Materials (per class of 30, teams of 1)
30 outline maps of California (a blackline master of a full-size map of Figure 9.19 is in the Teacher's Resource Book.)

10 sets of colored pencils (or 1 set per table), 8
    colors each

Preparations

Several students can share each set of colored pencils. Avoid difficulties with pronunciation by referring to the subspecies by their numbers. Emphasize that correct key colors and accurate plotting are necessary for correct interpretations. Have students complete all of Part A. Then, discuss the questions as a class before the students begin Part B.

CHAPTER 10

ORDERING LIFE IN THE BIOSPHERE

Planning Ahead

Check your supply of agar plates for Investigations 11.1 and 11.2. Tell students they will need to supply household antibacterial agents for Investigation 11.2 (have a few of these on hand). You also will need to prepare cultures of *Micrococcus luteus* and *Escherichia coli* for these investigations. You will want to be knowledgeable enough about the current status of AIDS for Investigation 11.2 to be a meaningful experience for your students; see the references cited below. In addition to having consistently good articles relevant to the BSCS Green Version program, *Discover* magazine has regular articles on AIDS with up-to-date information. In planning Chapter 14, you may wish to ask a zoo curator or veterinarian to visit the class and bring along some live animals.

Concepts

◆ Living things can be classified into groups to clarify their relationships.

◆ The current biological classification system is based on similarities of characteristics between species, especially structural similarities, but also chemical similarities.

◆ Biological classification is hierarchical—taxa containing progressively larger numbers of species are more inclusive.

◆ Binomial nomenclature is a Latin-based system for universally identifying each species with its genus and species name.

◆ All organisms are either prokaryotes or eukaryotes. This fundamental dichotomy is based on profound differences in cell structure.

◆ Although all organisms currently are placed into one of five kingdoms, classifications change, and there is continuing debate about classification

strategies as we learn more about characteristics and origins of organisms.

◆ Because of the unique conditions that existed during the origin of life on earth, all present life may have come from a single ancestral species.

◆ The atmosphere of the early earth was different from that of today and probably contained methane and ammonia and lacked oxygen. Therefore, the events in the origin of life on the earth could not happen today.

◆ Life on the earth probably arose from chemical origins and first produced primitive heterotrophs.

◆ There is evidence that eukaryotic cells arose as symbiotic relationships among various prokaryotes.

◆ The organization of matter forms a continuum from subatomic particles to the biosphere. Each level exhibits greater complexity than the one before. The line between life and nonlife is not clear.

◆ Whether or not they are considered to be living, viruses evolved early in the earth's history and cannot function outside of a living cell.

Objectives

Students should be able to:

◆ *describe* the types of evidence used by taxonomists.

◆ *state* reasons for using a scheme of levels in biological classification.

◆ *identify* several types of homologies and give examples.

◆ *recognize* greater and lesser degrees of likeness among groups of organisms.

◆ *explain* the advantages of using a binomial system of naming organisms.

◆ *distinguish* between prokaryotic and eukaryotic cells.

◆ *describe* the distinguishing characteristics of the five kingdoms.

◆ *explain* why classification systems can change.

◆ *describe* the environmental conditions in which life might have originated.

◆ *explain* how biochemical experiments support speculations on the biochemical origin of life.

◆ *explain* the heterotroph hypothesis.

♦ *describe* several models of cell formation.

♦ *explain* how eukaryotic cells might have originated from prokaryotes.

♦ *relate* RNA as an enzyme to the evolution of life.

♦ *order* matter from most complex to least complex.

Strategies

To most tenth-grade students, the idea of classifying is well known. Therefore, you need to emphasize how and why biologists use categorization—particularly how they use it to express inferred relationships. Above all, remind students that the system of biological classification is a human construction and thus, artificial.

Continue to direct attention to structural adaptations and ecological relationships among major groups of organisms. This chapter permits you to develop two important aspects of humanistic science teaching.

1. Science is a seeking of new and better ways to order an ever-widening array of data, not the memorizing of a prescribed system into which data are to be fitted. Devote part of whatever time may be available for class discussion to problems involved in human attempts to impose order on the data of nature—as illustrated by difficulties of classification at the kingdom level, for example.

2. The historical background of present biological knowledge, briefly referred to in previous chapters, is developed more fully in this chapter. It should be a second focus of class discussion. A necessary corollary is the international nature of the scientific enterprise. Take every opportunity from this point forward to use biological names.

Investigation 10.1 links the earlier material on genetics to the modern methods of studying evolution. It should follow the first reading assignment, Sections 10.1–10.2. Encourage your students to share their responses to the interactive legends at the beginning of each chapter.

Investigation 10.2 is a transition from abstract ideas to the concrete details of the next part of the text. It can be a homework assignment, but if your students have reading difficulties, you might want to conduct it in class. It can be done at any time during the second reading and should be followed up with a class review/ discussion.

Ward's Natural Science Establishment offers several useful adjuncts to this chapter: a Five Kingdom Poster (catalog #33M5481), a classification lab based on the poster (catalog #32M2208), a set of color slides of the five kingdoms assembled by Margulis and Schwartz to complement their book Five Kingdoms

(catalog #170M0110), and a sound-filmstrip series (catalog #70M6441).

References

Badash, S. "The Age-of-the-Earth Debate." *Scientific American* (August 1989): 90–98.

Cairns-Smith, A. G. "The First Organisms." *Scientific American* (June 1985): 90–101.

Gulkis, S., P. M. Lubin, S. S. Meyer, and R. F. Silverberg. "The Cosmic Background Explorer." *Scientific American* (January 1990): 132–39.

Horgan, J. "In the Beginning." *Scientific American* (February 1991): 116–25.

Kemp, K. W. "Discussing Creation Science" *American Biology Teacher* (February 1988): 76–81.

LeGuenno, B. "Emerging Viruses." *Scientific American* (October 1995): 56–64.

Lewin, R. *The Bones of Contention.* New York: Touchstone, 1988.

Margulis, L., and K. V. Schwartz. *Five Kingdoms* 2d ed. New York: Freeman, 1988.

Novacek, M. "Where Do Rabbits and Kin Fit In?" *Nature* (25 January 1996): 299–300.

Schiebinger, L. "The Loves of The Plants." *Scientific American* (February 1996): 110–115.

Wissler, J. S., and T. R. Mertens. "Science and Creationism: An Annotated Bibliography of Recent Books." *American Biology Teacher* (November/ December 1986): 471–74.

Wright, C. G., and P. J. Bottino. "Mitochondrial DNA." *The Science Teacher* (April 1986): 27–31.

Suggested Answers to Applications

1. A person's first and last names identify one individual, just as a scientific name identifies one species. However, your name may be shared by others, whereas a scientific name is unique.

2. No, this would be a waste of time. First determine its major characteristics—prokaryotic or eukaryotic, heterotrophic or autotrophic. Then, determine what it resembles, using structural, cytological, and ecological characteristics. Once its close relatives are known, interbreeding experiments could determine relationships.

3. (a) Common; (b) from China; (c) white; (d) edible; (3) emetic; (f) three-petaled.

4. Tropical rain forests. Focus your discussion on the destruction of the tropical rain forests and the potential loss of food and industrial crops, plants and animals with medicinal value, and the effect on the climate and on native people living in these areas.

5. Viruses lack an independent metabolism, a characteristic often considered essential to the definition of a living thing.

Suggested Answers to Problems

1. The idea for Velcro is said to have come from the spiny cocklebur that sticks to people's clothes. This also would be a good time to illustrate the structural analogies between unrelated organisms.

2. *Gaia: An Atlas of Planet Management*, N. Myers, ed. (New York: Anchor Press,1984) contains much information about the earth and its environment

3. Gould writes about the panda's thumb, which is not, anatomically, a finger at all but, rather, is formed from an enlarged wrist bone. Gould discusses what this suggests in terms of "intelligent design."

4. A classroom herbarium of dried, pressed local plants can be developed. A collection of pictures of plants and animals also will help students learn to recognize them.

5. Students might start with the article by C. R. Woese, "Archaebacteria," *Scientific American* (June 1981): 98–122, although there is much additional, newer information on these organisms.

Materials and Preparations for Investigations

INVESTIGATION 10.1 DNA Sequences and Classification

Materials (per class of 30, teams of 4)
paper clips: 240 black, 240 white, 240 green, 240 red

Preparations
Poppit beads can be substituted for paper clips.

Point out to your students that this investigation only demonstrates how DNA hybridization can be used. The data accumulated when using this model should not be taken as actual support for one hypothesis or another. The questions raised in this investigation are still open to debate in the biological sciences. Students should not be led to believe that the data in this activity are definitive.

Students might be interested in the process by which cDNA is synthesized. It is roughly as follows: mRNA for the gene in question (in this case, hemoglobin) is isolated; the mRNA then is exposed in culture to activated nucleotides and the enzyme reverse transcriptase (reverse transcriptase reverses the normal process of transcription so that DNA is synthesized off an RNA template, rather than the other way around); the mRNA is dissolved, leaving the single strand of new DNA; that DNA is exposed to DNA polymerase, which synthesizes a new DNA strand that is complementary to the original gene for the protein in question. The 20-base sequence used in this investigation is not actually a part of the hemoglobin gene. Hemoglobin mRNA often is used in such studies, however, and is readily available in immature red blood cells.

INVESTIGATION 10.2 Levels of Classification

Materials (per class of 30, teams of 1)
none

Preparations
Blackline masters of the four forms of Table 10.4 are provided in the Teacher's Resource Book. Be sure that details do not obscure the basic point of the investigation: that the hierarchy of classification levels is an expression of degree of likeness. To be grouped together at the species level, organisms must be so much alike that they are capable of interbreeding.

CHAPTER 11
PROKARYOTES AND VIRUSES

Planning Ahead
If you plan to use glass petri dishes for the investigations in this chapter, be sure to have available a working pressure cooker or autoclave. Check your orders of live materials for the investigations in Chapters 12 and 13.

Concepts
♦ Prokaryotes (bacteria) are organisms that lack a nucleus and membrane-enclosed organelles. Their total biomass is very great, and they invade every imaginable place on earth and almost every niche.

♦ Prokaryotes come in almost all shapes and range in size from 0.2 to 20 μm. Although structurally simple, they are biochemically complex. They reproduce rapidly by fission; some reproduce sexually under the right environmental conditions.

♦ Archaebacteria are probably the most ancient of all organisms. They are adapted to extreme conditions and have significant biochemical differences from eubacteria.

♦ Eubacteria may be aerobic or anaerobic; photosynthetic, chemosynthetic, or heterotrophic; many are decomposers. Some form important symbiotic relationships, and a few are parasitic. Although some cause disease to other organisms, the vast majority are beneficial, even essential, to other forms of life.

♦ Disease results from an interrelationship between a pathogen and a host, and the severity of a particular disease depends on the virulence of the pathogen and the resistance of the host. Resistance, or immunity, may be acquired or inherited.

◆ Many disease-causing eubacteria harm the host by the release of metabolic products (toxins). Most eubacterial infections can be combated by antibiotics.

◆ Viruses are obligate parasites that act by injecting their DNA or RNA into a host cell, which in turn causes the production of more viral nucleic acid, thus interrupting the normal metabolism of the host cell. Some viruses alter the host's genetic makeup.

◆ Some human diseases caused by eubacteria and viruses are transmitted through direct skin or mucous-membrane contact. The means of such transmission is most often sexual.

◆ AIDS, which is caused by HIV, is a serious, worldwide human health problem. This virus attacks the helper T cells of the host's immune system, thus compromising the body's ability to combat other infections. Because HIV mutates so rapidly, the search for a vaccine is a major medical challenge.

Objectives

Students should be able to:

◆ *demonstrate* a method for collecting samples of prokaryotes from an environment.

◆ *use* effectively the appropriate equipment and instruments for handling and observing prokaryotes.

◆ *describe* the basic characteristics of prokaryotic cell structure.

◆ *list* characteristics that enable prokaryotes to live in almost every habitat in the biosphere.

◆ *explain* the differences between archaebacteria and eubacteria.

◆ *describe* the three types of archaebacteria and the environments in which they live.

◆ *explain* why, given their environmental requirements, archaebacteria may be the oldest living organisms.

◆ *distinguish* the three major groups of eubacteria.

◆ *discuss* several metabolic styles of eubacteria.

◆ *describe* the major events in the nitrogen cycle and the types of eubacteria involved.

◆ *distinguish* between infection and disease.

◆ *explain* the interaction between host and pathogen.

◆ *explain* the relationship between virulence of pathogen and resistance of host to the development of disease.

◆ *recognize* factors involved in the development of immunity.

◆ *explain* why, although pathogens are potentially worldwide in distribution, many infectious diseases have geographic limits.

◆ *describe* in ecological terms at least four infectious diseases.

◆ *relate* the means of disease prevention to the modes of transmission.

◆ *describe* the unique characteristics of viruses.

◆ *list* reasons for the increase in sexually transmitted diseases, despite antibiotic treatments for some of them.

◆ *present* arguments for and against widespread screening for AIDS.

◆ *describe* the effects of HIV and prospects for treatment of AIDS.

Strategies

Although disease is an important topic that occupies two-thirds of the chapter, it is important to emphasize the preponderance of beneficial effects of prokaryotes and their essential ecological roles—two topics often neglected. Classification of prokaryotes has been difficult because of their structural simplicity, but rapidly accumulating molecular data eventually will make a phylogenic classification possible. Archaebacteria and eubacteria are likely to be accorded kingdom status in the future; they are treated here as two major groups in the kingdom Prokaryote.

Chapter 11 lets students come to grips with the biological realities of concepts related to the spread and control of disease; apply these concepts in a local context. Use an infectious disease still important in your area to point up the concepts of transmission, symptoms, virulence, resistance, and epidemiology. The number of diseases described has been kept small. In class discussions, students may bring up many more, resulting in a morbid catalog. Try to make each disease a representative of a class of diseases.

This chapter contains scientific background for consideration of many human problems of social importance. If you have not already done so, establish bonds with other teachers in your school. They also may deal with topics of disease and medical care. The educational situation offers mutual reinforcements.

Laboratory work should occupy a large part of the class time devoted to Chapter 11. If you have not already set up Investigation 11.1, do so immediately to allow time for culture growth. When space and petri dishes are available, set up Investigation 11.2. Investigation 11.3 may be carried out at any convenient time. All students should participate in the discussion of this vital topic. The chapter reading assignments and class discussion can be sandwiched into the laboratory work in any convenient manner.

References

Aral, S. O., K. K. Holms. "Sexually Transmitted Diseases." *Scientific American* (February 1991): 62–69.

Balter, M. "Elusive HIV-Supressor Factors Found." *Science* (8 December 1995): 1560–1561.

Bardell, D. "The Changeable Nature of AIDS Virus." *American Biology Teacher* (February 1989): 79–83.

Blaser, M. "The Bacteria behind Ulcers." *Scientific American* (February 1996): 104–107.

Gallo, R. C. "The AIDS Virus." *Scientific American* (January 1987): 46–57.

Greene, W. "Aids and The Immune System." *Scientific American* (September 1993): 98–105.

Kartner, N., and V. Ling. "Multidrug Resistance in Cancer." *Scientific American* (March 1989): 44–53.

Margulis, L., D. Chase, and R. Guerreo. "Microbial Communities." *BioScience* (March 1986): 60–70.

Mills, J., and H. Masur. "AIDS-Related Infections." *Scientific American* (August 1990): 50–59.

Mitchison, A. "Will We Survive?" *Scientific American* (September 1993): 136–144.

Nowak, M. and P. J. McMichael. "How HIV Defeats The Immune System." *Scientific American* (August 1995): 58–65.

Oldstone, M. B. A. "Viral Alteration of Cell Function." *Scientific American* (August 1989): 42–49.

Suggested Answers to Applications

1. The very first organisms may have fed on the molecules of which they were formed. As they were broken apart, these organized structures would provide energy for the prokaryotes. Later in the evolution of these organisms, and before autotrophs appeared, some prokaryotes may have eaten each other.

2. In such an inhospitable habitat with no other living things, one might expect first to find an autotroph that could make its own food and live on the poor soil—for example, a cyanobacterium. After atomic tests in the desert, such organisms have been found to be the first colonizers.

3. The environment of the early earth was extremely different from today's—its habitats would be considered extreme by today's standards. Relatives of organisms that developed in such environments have been able to survive in similar conditions until today. This would be a good opportunity for you to review the origin of life on earth.

4. The text lists several important users of microorganisms, but you also can refer to any general microbiology text or such economic texts as: R. W. Schery, *Plants for Man* (New York: Prentice Hall, 1972).

5. The nitrogen cycle depends on the activities of several kinds of eubacteria—nitrogen-fixers, decomposers, nitrifiers, and denitrifiers. Only the nitrogen-fixers are able to utilize atmospheric nitrogen and convert it to a form that plants can use. (Animals, in turn, obtain their nitrogen from plants or other animals in the form or animal proteins.) Decomposers break down dead organic matter, making nitrogen available in the form of ammonia for nitrifiers, which convert the ammonia into nitrates usable by plants. Denitrifiers complete the cycle by converting remaining nitrates to nitrogen gas, which is returned to the air. The cycle could continue without plants and animals, but plants and animals could not survive without the activities of the eubacteria.

6. Vaccines are made from weakened or killed pathogens that stimulate production of memory cells and antibodies against that specific pathogen. Memory cells provide a rapid response to a subsequent exposure to the same pathogen. Once a person shows symptoms of the disease, however, a vaccine cannot combat the pathogen, which already has multiplied in the body.

7. Sexually transmitted diseases may be asymptomatic or produce only mild symptoms that are overlooked, allowing them to be spread unknowingly. In the case of AIDS, symptoms may not appear for many years. Widespread use of penicillin had led to the evolution of resistant strains of the eubacteria responsible for gonorrhea and syphilis. It often is difficult to trace all sexual contacts.

8. This is another example of an opportunistic disease that becomes lethal in times of stress when the body's immune system is weakened by starvation. With healthy children, diarrhea is an occasional, discomforting problem but usually not lethal.

9. Cancer, as a category of disease, may be environmental, infectious, *and* hereditary.

10. Viruses have two characteristics we usually associate with living things: the ability to reproduce and the ability to undergo changes in hereditary characteristics. Yet, what we call viral reproduction differs from reproduction of bacteria or other microbes. A virus cannot produce copies of itself unless it is within another organism.

11. Opportunistic infections attack AIDS patients because the immune system is not working properly. This deficiency gives a disease the opportunity to infect and kill, while in another individual with a properly working immune system the disease may never occur, or if it does, it may not be lethal.

12. Retroviruses such as HIV need the enzyme reverse transcriptase to make DNA copies of their RNA. Only by this means can they form the DNA that eventually controls the functioning of host cells and the production of more viruses.

Suggested Answers to Problems

1. In a typical cell, there are many copies of the same piece of RNA and perhaps only two of the DNA. Thus, it may be much easier to locate and study RNA. With modern techniques, however, it is possible to make many duplicates of a single gene, providing much material for genetic and phylogenetic studies. Refer students to modern molecular biology texts.

2. You might give your students J. D. Lennox, S. E. Lingenfelter, and D. L. Wance, "Archaebacterial Fuel Production: Methane from Biomass," *American Biology Teacher* (March 1983): 128-38. Another useful reference is G. T. Miller, Jr., *Living in the Environment,* 6th ed. (Belmont, CA: Wadsworth, 1990).

3. The text's illustration is greatly simplified and omits many of the organisms and processes involved in the complete cycle. Students should include these, and they may want to consider how to show the relative importance of each of the processes in the cycle.

4. Contact your local county extension agent affiliated with your state's agricultural college.

5. A good resource for human diseases is: *Control of Communicable Diseases in Man,* 12th ed. (1975). It is available from: American Public Health Association, 1015 Eighteenth St., NW, Washington, DC 20036. For plant diseases, a student might start with a general encyclopedia and then turn to textbooks of plant pathology or physiology. If a student is interested in animal diseases, he or she might ask a veterinarian for some references.

7. All the students' explanations should converge on the idea that the different influenzas are caused by different strains of viruses. Acquired immunity to one is not immunity to all. Individual resistance also varies.

8. Students should come up with their own answers where possible. Biochemists would analyze the contents of the victims' stomachs and bloodstreams for damaging chemical residues. They would also analyze the local water supply. Food inspectors would ask to see and examine samples of food the victims had eaten. Medical doctors would be engaged in many different tasks. Metallurgists would look for toxic meal substances in the victims, in the air-conditioning ducts of the buildings the victims occupied, and so on. Microbiologists would look for bacteria and viruses in the victims' blood and tissues. Pathologists would do careful autopsies, examining each part of every victim's body and studying slides of tissue samples. Toxicologists would work with the biochemists and try to find poisons of any chemical type, or of bacterial origin, in the victims' bodies.

9. Some infectious diseases in Third World countries include whooping cough, diphtheria, cholera, tuberculosis, and AIDS.

Materials and Preparations for Investigations

INVESTIGATION 11.1 Distribution of Microorganisms

Materials (per class of 30, teams of 3)
40 50-mL beakers
60 sterile swab applicators
10 glass-marking pencils
10 dispensers of transparent tape (or 1 per table)
10 stereomicroscopes or 10X hand lenses
10 self-closing plastic bags labeled waste (1 per table)
2 biohazard bags
40 sterile nutrient agar plates

Preparations

Student assistants who have learned the techniques of media formulation and sterilization can help prepare the plates. Bottles of sterile nutrient agar are available from Ward's Natural Science Establishment (catalog #88M1500). Follow the directions to liquefy and pour into sterile petri dishes, or order prepared plates (catalog #88M0905) from Ward's. Directions for preparing nutrient agar plates are included in A Table of Recipes in the Teacher's Resource Book.

Data from all the plates exposed at each location can be compared. Besides giving a total count of all colonies, the data may be classified into several categories—number of bacterial colonies, number of mold colonies, types of bacterial or mold colonies—as well as into number of colonies of each particular type.

Type of medium and incubation temperature influence the development of colonies. The medium used here is a general one on which many types of microorganisms grow.

INVESTIGATION 11.2 Control of Eubacteria

Materials (per class of 30, teams of 3)

30 pairs of safety goggles
180 construction-paper disks
10 glass-marking pencils
20 forceps
20 sterile swab applicators
10 dispensers of transparent tape (or 1 per table)
30 metric rulers
10 self-closing plastic bags labeled waste (1 per table)
20 sterile alcohol pads
3 biohazard bags
40 sterile nutrient agar plates
small containers of distilled water, antiseptics, and
     disinfectants
1 magazine each broad spectrum, gram-positive acting, and
     gram-negative acting antibiotic disks
1 vial each freeze-dried cultures of *Micrococcus luteus* and
     Echerichia coli

Preparations

Cut or punch small disks from construction or filter paper. Prepare agar plates as in Investigation 11.1 (see also A Table of Recipes in the Teacher's Resource Book), or order prepared plates from Ward's Natural Science Establishment. The day before the experiment, rehydrate the freeze-dried cultures of *Micrococcus luteus* (*Sarcina lutea*) and *E. coli* according to the instructions included with the cultures. Provide a broth culture of each eubacterium per lab table.

Ask students to bring small samples of disinfectants or antiseptics, but caution them to be careful in transporting them to class. Use small containers in which students can dip their disks. Provide three types of antibiotic disks—broad spectrum such as Chloromycetin, tetracycline, or neomycin; gram-positive acting, such as penicillin or bacitracin; and gram-negative acting, such as streptomycin or nalidixic acid.

Ward's supplies appropriate antibiotics as impregnated sterile disks in dispenser magazines (catalog #38M1607—broad spectrum; #38M4914—gram (+); #38M4916—gram (–)).

INVESTIGATION 11.3 Screening for AIDS

Materials (per class of 30, teams of 3)
none

Preparations

Prior to the small-group activity, you may want to respond to questions about our current understanding of AIDS.

The ELISA, which tests for HIV antibody to whole virus, is performed on *all* donor blood. Blood that repeatedly gives positive results with ELISA is tested using the Western blot. This test detects protein and can identify one or more of the nine different antibodies to the various antigens of HIV. In dealing with screening tests, students may encounter the terms *sensitivity* and *specificity*. Sensitivity is the ability of the test to identify *all* people who are infected; specificity is the ability of the test to identify *only* those who are infected. To stay current with the most recent information on AIDS, write and ask to be put on the mailing list of: HIV/AIDS Surveillance, Department of Health and Human Services, Public Health Service, Centers for Disease Control, Atlanta, GA 30333.

Discussion of screening tests performed on donated blood will help focus the activity. Screening for syphilis, hepatitis, or other communicable diseases may be used as examples of tests in which false positive and false negative results may occur. Students then should read the paragraphs in Part A.

When you are satisfied that students have enough information to handle the small-group discussion, divide the class into groups of 3 or 4 students. Designate each group as A or B and direct them to the appropriate discussion questions. During their discussions, move from group to group. Be available for any questions they may have or help them over hurdles that are hindering the discussions. When you are satisfied they have dealt with the discussion questions, conduct a class discussion and ask for the arguments they have developed for or against nationwide screening.

Issues such as AIDS are potentially controversial. How much controversy will develop depends on many factors: the socioeconomic climate of the community, religious preference of your students, and the value system of your students and their parents. Most important, it will depend on how you handle the issues in the classroom. A general method of dealing with controversy is described below.

1. Present as much information about the issue as possible. Often, the narrow and rigid viewpoint of students is a result of having too little or erroneous information.
2. Allow all opinions or feelings to be expressed. Do not censor radical or shocking views. On the other hand, if a student is saying something for shock value alone, point out the inappropriateness of the statement for the discussion.
3. Acknowledge each opinion equally and encourage a similar accepting attitude among the students. Do not

favor one viewpoint. Students should feel they have every right to say what they feel, as long as they are making a positive contribution to the discussion.

4. Create an open, nonhostile atmosphere in the classroom—not just for this discussion, but at all times.

5. Be careful to keep your personal values out of the discussion and be ready to assist students in defending differing points of view.

CHAPTER 12
EUKARYOTES: PROTISTS AND FUNGI

Planning Ahead

If you have not had much experience handling paramecia, try out the techniques in Investigation 12.1 before attempting to guide students. You will need fresh mushrooms from a supermarket for Investigation 12.3. Order the live animals for Investigation 14.1 unless you already have them in your classroom.

Concepts

◆ Protists comprise an extremely diverse kingdom of mostly microscopic organisms that are defined largely by exclusion. They are eukaryotes; they are not plants, animals, or fungi.

◆ Fungi comprise a kingdom of organisms that are heterotrophs and that absorb nutrients from the environment. Most of them decompose the bodies of other organisms and are important in energy and matter cycles.

◆ The kingdom Protista includes autotrophic and heterotrophic organisms. Many protists can switch from one form of nutrition to the other, depending on conditions.

◆ Algae are photosynthetic protists that are major producers in aquatic ecosystems. They include green algae, which may have been the ancestors of plants; diatoms, which have silica shells with intricate patterns; and brown and red algae, often called seaweeds, which have chlorophyll masked by other pigments.

◆ Protozoa are mostly unicellular, motile protists. They include flagellates, which move by whiplike structures; sarcodines, which move their shapeless bodies by flowing around their food; sporozoans, which parasitize animals; and ciliates which move rapidly about by short hairlike structures.

◆ Slime molds are multicellular, heterotrophic protists that are important decomposers of plant material.

◆ Fungi are widely known as mushrooms, mildew, mold, rusts, and yeasts. They decompose natural organic matter by secreting enzymes that digest the matter and then absorbing the resulting molecules into their bodies.

◆ Fungi can grow rapidly; some cause widespread damage to other organisms. They are resistant to very hostile environments and produce a variety of interesting metabolic products, some of which are beneficial and some harmful.

◆ Some fungi form specific and important symbiotic relationships with other organisms.

Objectives

Students should be able to:

◆ *state* the salient characteristics of the major groups of protists and fungi.

◆ *recognize* common examples of those groups.

◆ *recognize* several types of protists and cyanobacteria.

◆ *describe* the basic characteristics of the protists known as algae.

◆ *present* evidence for the hypothesis that plants evolved from green algae.

◆ *distinguish* between green algae, diatoms, red algae, and brown algae.

◆ *describe* how a paramecium carries out its basic life functions.

◆ *describe* animal-like and plantlike characteristics of an organism such as *Euglena*.

◆ *distinguish* between flagellates, sarcodines, sporozoans, and ciliates.

◆ *describe* the unique characteristics of slime molds.

◆ *explain* the origin of chloroplasts and mitochondria according to the endosymbiosis hypothesis.

◆ *describe* mechanisms of sexual and asexual reproduction in a fungus.

◆ *distinguish* between conjugating, sac, club, and imperfect fungi.

◆ *diagram* a food chain that includes fungi.

◆ *explain* how fungi, as decomposers, are essential in food webs.

◆ *describe* two important symbiotic relationships involving fungi.

Strategies

Students usually have had many experiences with animals, fewer with plants and fungi, and almost none with protists. Providing firsthand experience with protists is easy, and the core of this chapter is the laboratory experiences. Allow the major portion of time in this chapter for laboratory work.

If you plan to have students do Investigation 12.3 Part B, have them begin immediately as the fungi need 10 days to grow. Part A can be done at any time, with or without Part B, but Part B demonstrates sexual reproduction in fungi and is highly recommended. Investigation 12.1 can be done after a brief discussion of the introduction to the chapter. Then, have students read Sections 12.1–12.4. Investigation 12.2 requires students to infer function from structure and demonstrates the relationship between the two. This strongly inquiry-oriented investigation is well worth the investment of two or three class periods. Allow time for students to share their observations and interpretations in a class discussion. Then discuss protist diversity and assign the reading and Concept Review for Sections 12.5–12.9. Following this, have students do Investigation 12.3 Part A, discuss the fungi, read the remainder of the chapter, and finally complete Part B of Investigation 12.3. The Applications can be assigned as homework and completed with a review in class.

References

Ayers, J. C., J. O. Mundt, and W. E. Sandine. *Microbiology of Foods.* San Francisco: W. J. Freeman, 1980.

Baskin, Y. "Can Iron Supplementation Make The Equatorial Pacific Bloom?" *BioScience* (May 1995): 314–316.

Donelson, J. E., and M. J. Turner. "How the Trypanosome Changes Its Coat." *Scientific American* (February 1985): 44–51.

Foote, M. "Microbiology Gardens—A Close Look at Algae." *The Science Teacher* (May 1983): 38–43.

James, E. "Take a Dip! Culturing Algae Is Easy." *The Science Teacher* (May 1983): 44–47.

Lee, J. "Living Sands." *BioScience* (April 1995): 252–261.

Miller, A. "Clinical Opportunities for Plant and Soil Fungi." *BioScience* (November 1986): 56–59.

Sobieski, R. J. "Where Have All These Microbes Come From?" *The Science Teacher* (April 1984): 30–35.

Vida, G. "The Oldest Eukaryotic Cells." *Scientific American* (February 1984): 48–57.

Suggested Answers to Applications

1. Algae are more advanced because they possess nuclei and membrane-bounded organelles such as chloroplasts and mitochondria.

2. First, storing energy in the form of lipids or starch saves space—more energy can be packed into a cell in one of these chemical forms than in the form of sugar. Also, water and dissolved materials can pass into and out of the cells of aquatic organisms. If energy supplies were stored as sugar, which dissolves in water, they could be lost from the cell. Starch and other storage compounds do not dissolve easily in water and thus can be kept inside the cell.

3. In general, the answer is yes—green organisms usually are autotrophic, and nongreen ones usually are heterotrophic. There are, however, many brown, red, and yellow algae that still are perfectly good photosynthetic organisms. Euglenoids may be colorless heterotrophs under certain conditions but green autotrophs under other conditions.

4. The rate of contraction should slow down because water would tend to diffuse out of the paramecium into the salt water.

5. In the Blob's movement and method of engulfing its prey, it resembles the amoeba and the plasmodium of slime molds.

6. The enzymes fungi produce to break down cellulose also could break down their own cell walls if they were made up strictly of cellulose.

7. Fungi are heterotrophs, and the body of the fungus invades or comes in contact with its food source, which is on or in the ground. The body of the fungus is, therefore, rarely seen. The reproductive part of the fungus contains the spores, which must be exposed above ground to be dispersed by wind or some other agent.

8. Symbiotic relationship.

9. Lichens are formed by a close association between green algae or cyanobacteria and fungi. Although many of the producers can grow independently and can be recognized as species that live alone, the fungi usually are unlike any species that live alone.

10. The fungi enhance the growth of the plants by absorbing nutrients and water from a large area in the soil and transporting those substances to the roots.

Suggested Answers to Problems

1. Many manufactured foods with a smooth texture contain either carrageenan or algin. Examples include yogurt, ice cream, macaroni-and-cheese dinners, chocolate milk, and certain salad dressings.

2. A life-sized painting would be impressive because students could paint only a small part of the very large algae found along the coast. Unicellular algae and other microscopic organisms in the water could be drawn many times larger than actual size to illustrate their incredible diversity of shapes.

3. Your state or county health department may employ experts in water quality and sewage treatment and in the organisms found in the water and sewage.

4. The ovum contains mitochondria; the head of the sperm does not. The flagellum of the sperm has mitochondria (for energy production), but only the head of the sperm enters the ovum during fertilization.

6. Students can make spore prints of mushrooms, and the mushrooms can be preserved by drying. Lichens will remain dormant and retain their colors for years without special treatment.

Materials and Preparations for Investigations

INVESTIGATION 12.1 Variety Among Protists and Cyanobacteria

Materials (per class of 30, teams of 2)

30 microscope slides
30 cover slips
1 box of flat toothpicks
15 compound microscopes protist/cyanobacteria key
    (Appendix 3)
A Catalog of Living Things (Appendix 4)
5 squeeze bottles of Detain™
1 protist/cyanobacteria survey mixture samples from ponds,
    streams, or puddles

Preparations

Have students collect samples from ponds, streams, or puddles. (If necessary, they can be collected early in the fall and maintained in the classroom.) By collecting samples, students can observe the organisms' environments. Ward's Natural Science Establishment provides a survey mixture of 10 organisms prepared especially for this investigation (catalog #87M1525). Prepared slides also may be used but are not as exciting. The dichotomous key includes 52 protists and cyanobacteria most commonly present in freshwater samples, but it is helpful to provide additional keys such as T. L. Jahn, *How to Know the Protozoa* (Dubuque, IA: W. C. Brown, 1978), and G. W. Prescott, *How to Know the Fresh-Water Algae* (Dubuque, IA: W. C. Brown, 1978). Ward's *Dichotomous Key to Pond Protists* (catalog #74M1858) is a program for Apple computers that provides information about 75 protists. Line drawings of all organisms in the key are supplied as blackline masters in the Teacher's Resource Book, but it is better for students to work with the key before seeing the drawings.

INVESTIGATION 12.2 Life in a Single Cell

Materials (per class of 30, teams of 2)

30 18-mm × 150-mm test tubes
30 microscope slides
30 cover slips
15 dropping pipets
15 compound microscopes
15 rubber stoppers to fit test tubes
15 test-tube racks
5 incandescent light sources
15 forceps
1 roll of aluminum foil
1 box of flat toothpicks
5 squeeze dropping bottles of Detain™
5 squeeze dropping bottles of alizarin red stain
5 small bottles of India ink
5 squeeze dropping bottles of yeast-brilliant cresyl
    blue preparation
15 1-cm pieces of cotton thread wetted with 0.01M HCl
5 jars of paramecium culture

Preparations

Much of the success of this investigation depends on having rich cultures of paramecia. Begin preparations 3 or 4 days before the investigation. See A Table of Recipes in the Teacher's Resource Book for a simple culture medium. Prepare the yeast-brilliant cresyl blue mixture the day before its use; see A Table of Recipes. Alizarin red is a vital stain that does not kill paramecia. It stains nuclear material and the pellicle, making clear the three-dimensional character of the organism. If alizarin red is not available, you can use iodine or methylene blue, both of which kill the organism.

Although Detain™ forms a "skin" that helps prevent drying during microscopic observations, it still may be necessary for students to add water to their culture drops. Detain™ is a high-molecular-weight (>7 000 000), nontoxic polymer. Unlike methyl cellulose, it does not drug protists but does offer a physical barrier that slows their movement.

Some teachers find it preferable to make a new slide with fresh paramecia for each observation. Removing and then replacing the cover slip can be disruptive to the organisms in the culture.

INVESTIGATION 12.3 Growth of Fungi

Materials (per class of 30, teams of 3)
10 glass tumblers
10 microscope slides
10 cover slips
10 dropping pipets
10 compound microscopes
10 stereomicroscopes or 10X hand lenses
10 sheets of white paper
10 scalpels
10 flat toothpicks
30 sterile swab applicators
1 roll of masking tape
5 dispensers of transparent tape
10 metric rulers
10 glass-marking pencils
10 self-closing plastic bags labeled waste
        (1 per table)
biohazard bag
5 dropping bottles of 1:1 glycerine-water solution
        (50 mL total)
10 sterile potato-dextrose agar plates
1/2 lb fresh mushrooms slant culture of + strain of
        Rhizopus stolonifer
slant culture of − strain of *Rhizopus stolonifer*

Preparations
You may wish to purchase several varieties of mushrooms and have students compare the spore prints made by each. Allergies to mold spores may be a problem for some students. You can minimize the problem by having students replace the tumbler after removing the mushroom cap, or by preparing only one spore print (per mushroom variety) and dispensing the prints for examination. Spore prints can be made permanent by spraying with hair spray.

Slant cultures of + and − strains of *Rhizopus stolonifer* are available from Ward's Natural Science Establishment. They can be maintained in the refrigerator for several months. Bottles of sterile potato-dextrose agar are available from Ward's; follow the directions to liquefy and pour into sterile petri dishes. Or, order prepared plates from Ward's. Directions for preparing potato-dextrose agar plates are included in A Table of Recipes in the Teacher's Resource Book.

CHAPTER 13
EUKARYOTES: PLANTS

Planning Ahead
Try to have a sphygnomanometer available for Investigation 15.4.

Concepts
◆   Plants are multicellular, photosynthetic organisms with a life cycle that includes a distinct alternation of generations. They have invaded almost every possible habitat.

◆   Although their ancestors were aquatic, most plants are terrestrial. Adaptations that conserve water and allow plants to survive periods without water makes terrestrial existence possible. Nevertheless, water availability is the major limiting factor for the growth of land plants.

◆   Two groups evolved from multicellular algae: nonvascular and vascular plants.

◆   Bryophytes, such as mosses and liverworts, are relatively small organisms. Although they possess structures that appear to be roots, stems, and leaves, they are limited to moist areas because they lack vascular tissue and because their reproductive cycle requires water.

◆   Vascular plants have functional roots, stems, and leaves composed of vascular tissue that permits transport of water throughout the plant body. In addition, spores and/or pollen grains allow reproduction in dry environments.

◆   *Rhynia major,* a seedless vascular plant, represents the first land plants. Other seedless vascular plants include club mosses, horsetails, and ferns.

◆   The dominant organisms today in terms of biomass are seed-producing vascular plants such as conifers and flowering plants.

◆   Sophisticated adaptations for reproductive success have evolved in flowering plants, greatly increasing their habitat range and diversity.

◆   Humans are dependent on plants for food, clothing, housing, and medicines, as well as for oxygen. Our standard of life and success as a species necessitates the preservation of green plants throughout the biosphere.

Objectives
Students should be able to:

◆   *describe* the characteristics plants must have to survive on land.

◆   *contrast* the adaptations of vascular and nonvascular plants to terrestrial environments.

◆   *demonstrate* the ability to make and use a dichotomous key.

- *list* the adaptations of land plants for conserving water.

- *diagram* the life cycles of bryophytes and flowering plants.

- *contrast* reproduction in a moss and a flowering plant.

- *infer* from their life cycle how flowering plants are better adapted than mosses to a terrestrial existence.

- *compare* the relative importance of sporophyte and gametophyte generations in vascular and nonvascular plants.

- *contrast* bryophytes, horsetails, and ferns.

- *describe* the characteristics of conifers.

- *explain* the relationship between flowering plants and their pollinators and give an example.

- *describe* mechanisms in seed dispersal.

- *contrast* monocots and dicots.

Strategies

It is important to have abundant illustrative material on hand. Botanical gardens and conservatories are less numerous than zoos, but if there are any in your area, try to make use of them. Natural history museums often have excellent displays on plant adaptations and evolution. In assembling pictures, do not place too great an emphasis on flowers. Show whole plants as much as possible, and try to give balanced representation to all plant groups. Gauge the experience of your students. Many may be unacquainted with the whole plants from which come even familiar foods, such as potatoes, peanuts, squash, or beans.

Students should gain a panoramic view of plant diversity. Do not let taxonomic filmstrips and films limit their perspective to alternation of generations. Continue to emphasize the structural adaptations and ecological roles that are themes running throughout the book.

All investigations can be completed in one class period, but you should follow up each investigation with a half-period discussion. Begin the chapter with Investigation 13.1 and then assign the introduction and Sections 13.1 and 13.2. Follow Investigation 13.2 with a study of Sections 13.3–13.9, which can be a combination of reading, answering questions, and discussion. Investigation 13.3 should precede work on Sections 13.10 and 13.11.

References

Jones, A.M. "Surprising Signals in Plant Cells." *Science* (14 January 1994): 183–184.

Mangelsdorf, P. C. "The Origin of Corn." *Scientific American* (August 1986): 80–87.

Martinez del Rio, C. "Murder by Mistletoe." *Natural History* (February 1996): 64–70.

Raven, P. R. Evert 2nd S. Eichhorn. *Biology of Plants* New York: Worth, 1992.

Reganold, J. P., R. I. Papendick, and J. F. Parr. "Seed Dispersal by Ants." *Scientific American* (August 1990): 76–83.

Suggested Answers to Applications

1. Mosses may be considered less complex because they lack vascular tissue, cuticle, and reproductive structures that protect the embryo. Thus, they are confined to moist environments. Flowering plants are considered most complex because of the structures—especially the pollen tube and the seed that adapt them to dry land.

2. Yes. Students tend to think that one organism evolves into another organism through time and that once this happens the first organism disappears. Descendants of algae that evolved into land plants could still be around—it's just that some descendants branched off into the line of organisms that became land plants.

3. No, because the cuticle is a water-impermeable substance that would limit the amount of water absorbed by the root system.

4. The second question must be considered first. *Advantage* and *disadvantage* could be interpreted in terms of reproductive effectiveness. Because both spore beaters and seed bearers are abundant today and have been for long geological ages, it follows that the advantages and disadvantages have balanced out over time.

   Advantages: Seed-bearing plants—embryos in seeds have started to develop already; seeds contain a reserve food supply; fruit or seed coats may provide protection and mechanisms for dispersal. Spore-bearing plants—because they are small, spores can be produced in prodigious numbers; spores have low density and high surface-to-volume ratio, characteristics that favor dispersal by winds and air currents.

   Disadvantages: Seed-bearing plants—since seeds contain foods, they may be eaten by animals and digested unless protected by resistant coats; in many cases seeds are produced in relatively small numbers. Spore-bearing plants—food stored in a spore is very limited.

5. Ferns have swimming sperm and must have water for sexual reproduction. After a rainstorm, the

gametophytes are covered with water through which sperm can swim to the eggs. If the sporophyte appears only after a rainstorm, the ferns must live in a relatively dry area, which restricts sexual reproduction to certain wet times.

6. The only way for water to climb up a plant without vascular tissue is by capillary action, which is limited, thus restricting the height of a nonvascular plant.

7. Flowering palnts keep the spores on the parent plant which increases the probability that the spores will survive. Ferns release their spores to the environment as soon as they are mature. Many spores land in inappropriate places and die.

8. The energy that might have gone into producing sepals and petals can be used to produce the copious amounts of pollen formed by wind pollinated plants. In addition, the wind will not be impeded by small sepals and petals. Thus, pollen grains can be blown by the wind easily from one plant to another.

9. Those insects that fed at primitive flowers became diversified, which affected the evolution of plant species, which in turn affected insect evolution.

10. The plant parts that we use as food are those in which food is stored, partly because seeds, roots, and stems have the greatest nutritional value for us. Food storage in seeds represents an adaptive characteristic that increases reproductive effectiveness per seed. The food storage areas in plants also must provide a certain measure of protection. For example, food stored in underground parts of herbaceous plants is protected from animal depredation and trampling, and in certain climates, from damage by extremes of environmental temperatures or by desiccation.

Suggested Answers to Problems

1. These plants are xerophytes and have such adaptations as succulence, thick cuticles, deciduousness, and annual life cycles.

2. Students could begin their collection of fruits with Investigation 13.3. Because different plants will be reproducing throughout the growing season, students could collect different fruits from the same area at different times.

3. One of the better books on wood and its uses is the *Encyclopedia of Wood* (New York, NY: Facts on File, Inc., 1989).

4. Sculptures and paintings made by native peoples around the world often incorporate plants that were important to their lives. Students also should

investigate the incorrect belief that humans could determine the "intended" use of plants by looking at them. For instance, for headaches and related maladies, the walnut was considered important because of its resemblance to a human brain.

5. Many of today's fossil fuels come from plants that existed millions of years ago. Most of them looked very different from the plants familiar to students today, but several "living fossils" still exist.

6. Stems and leaves of aquatic plants have no cuticle; thus the plants absorb nutrients and gases directly from the water. The leaves are smaller and thinner than the leaves of land plants and lack supporting tissue. The thin, ribbonlike leaves provide a large surface area for absorption of carbon dioxide and diffused light. Aquatic plants usually have large chloroplasts. They lack stomates but have large air spaces in their leaves and stems that aid in buoyancy. Roots are usually smaller than those in terrestrial plants, have few or no branches, and no root hairs. Aquatic plants usually lack flowers and seeds and reproduce vegetatively.

Water is not a limiting factor for aquatic plants. Sunlight, however, can be a limiting factor because the light reaching the plant is diffused by the water.

7. This can be a challenging problem. Some characteristics that might be suggested are: (a) seeds readily carried (usually by wind) from one place to another because available places for growth are scattered; (b) ability to survive when little water is available in soils that are likely to be shallow and deficient in humus; (c) ability to grow with short periods of full sun because tall buildings shade many urban sites much of the day; and (d) ability to withstand such environmental chemicals as salt, carbon monoxide, and sulfur dioxide. All of these characteristics need to be considered by persons who wish to grow ornamental plants in the city.

9. Two good references are H. G. Baker, *Plants and Civilization,* 2nd ed. (Belmont, CA: Wadsworth, 1972) and R. W. Schery, *Plants for Man* (Englewood Cliffs, NJ: Prentice Hall, 1972).

Materials and Preparations for Investigations

INVESTIGATION 13.1 Increasingly Complex Characteristics

Materials (per class of 10, teams of 6)

5 stations, each with 10 labeled organisms of various kingdoms and divisions

5 compound microscopes
5 stereomicroscopes or 10X hand lenses
microscope slides and cover slips

Preparations

Students are to observe with as little help as possible, therefore, the organisms should show the characteristics needed for correct scoring. For example, conifer specimens should have seeds; flowering plant specimens should have fruits or flowers; mosses, ferns, and club mosses should bear spore cases. Some distinctions—as between shrub and tree—are difficult to exhibit in specimens small enough for the laboratory.

Try to provide a broad diversity of organisms. Fresh material is preferable to preserved. Most of the organisms can be collected in the autumn. Place a specimen of each type of organism at each station. If sufficient specimens are not available, devise a plan of rotation among stations. Suggestions: For compound microscope—green algae (*Spirogyra, Oedogonium, Ulothrix*), cyanobacteria (*Oscillatoria*), and yeasts; for hand lens or stereomicroscope—molds (*Rhizopus, Aspergillus*), liverworts (*Marchantia*), mosses (*Polytrichum, Mnium*), and lichens; for the naked eye—*Lycopodium*, ferns, pine, spruce, begonia, *Zebrina*, geranium, and firethorn (*Pyracantha*). The key allows inclusion of cyanobacteria. Ward's Natural Science Establishment has prepared a special kit for this investigation (catalog #87M9110) that includes the following organisms: *Gleocapsa, Oedogonium, Ulothrix, yeast, Aspergillus,* liverwort, moss, fern, begonia, and zebrina. Allow two weeks for delivery.

You can expedite the work by running through the scoring of a specimen (one not included in the investigation) with the students before they begin their own work. In the Teacher's Edition, the green names and numbers are examples of specimens for the 10 stations. The Teacher's Resource Book includes a blackline master of Table 13.1 with enough lines for 10 specimens.

The numbers in the table have been assigned so that when totaled they will yield low scores for organisms generally considered less complex and high scores for organisms generally considered more complex. Communicate to students the controversial nature of the scores. For example, a mycologist probably would object (justifiably) to the rather low score assigned to fungi.

INVESTIGATION 13.2 Reproductive Structures and Life Cycles

Materials (per class of 30, teams of 2)
30 microscope slides
30 cover slips
15 dissecting needles with corks for tips
15 scalpels
15 forceps
15 compound microscopes
15 stereomicroscopes or 10× hand lenses
modeling clay
prepared slide of filamentous stage of moss (whole mount)

15 prepared slides of moss male and female reproductive organs
10 mL of 15% sucrose solution
15 moss plants with sporophytes
fresh moss
3 stalks of gladiolus flowers
other simple flowers for comparison
15 fresh bean or pea pods

Preparations

The inquiry aspects of Part A depend on having available fresh moss plants and moss plants with sporophytes. Moss can be collected outdoors at almost any time of year or collected in the autumn and maintained in a terrarium. Even dry moss will revive and start growing in a few days. Moss plants with mature sporophytes are plentiful during spring and summer and can be preserved in a 50% glycerin solution. Preserved specimens of sporophytes and gametophytes are available from Ward's Natural Science Establishment. Ward's carries prepared slides of the filamentous stage, as well as a single slide (catalog #91 M4310) that includes the complete life cycle.

Gladiolus usually are available from local florists, especially if requested in advance. Lilies, tulips, snapdragons, pansies, poppies, *Tradescantia,* and *Vinca* also can be used.

Slides of germinated pollen are available from Ward's, but it is more exciting for students to germinate their own. Flowers of *Vinca,* obtainable from florists or nurseries, produce pollen that germinates in minutes. Many pollen grains grow in a 15% sucrose solution; finding the best concentration for particular flowers may require some experimentation. Students can use concavity slides, if available, although it may be difficult to provide sufficient moisture. Ward's carries a reusable silicone culture gum (catalog #37M9810) that is excellent for making the moist chambers.

Ward's also carries a special clone of *Tradescantia* (catalog #86M7305) used as a radiation detector, which blooms continuously, providing a ready supply of mature anthers. Order six to eight weeks before use to establish stable growth and bud development. Fresh beans or pea pods usually can be found in local supermarkets.

INVESTIGATION 13.3 Fruits and Seeds

Materials (per class of 30, teams of 3)
10 sets of fruits

Preparations

Gather fruits and seeds in the autumn and store them in your lab. Collect fruits during outdoor investigations, have students bring in fruits collected from natural (unrestricted) areas, and buy some exotic fruits at the supermarket. A good way to collect sticking fruits is to walk through a field with socks over your shoes. Include as wide a variety of dispersal mechanisms and fruit types as possible. Suggestions are listed below. Provide one seed/fruit of each type per team. To keep soft fruits intact for several years, spray them with Krylon™ or other preservative.

Dispersal mechanisms

Wind: dandelion, milkweed, any *Populus* species,
ash, elm, maple
Sticking: cocklebur, stickweed
Bird attractants: *Pyracantha,* berries, cherries
Laxative-containing: some mistletoe fruits

Fruit types

Simple (derived from a single ovary):
Pod: legume family—peanut, pea, bean, locust
Capsule: poppy, tulip, iris, lily, yucca
Silique: mustard family—candytuft, broccoli
Follicle: milkweed
Schizocarp: parsley family—carrot, celery, dill
Achene: composite family—sunflower, dandelion;
buttercup family
Grain: corn, wheat, barley, grasses
Samara: maple, ash, elm
Nut: walnut, chestnut, acorn
Drupe: olive, cherry, plum, almond, coconut
Berry: tomato, currant, grape
Pepo: cucumber, watermelon, pumpkin
Hesperidium: citrus
Pome: apple, pear

Compound (formed from several ovaries):
Aggregate fruit: strawberry, raspberry, blackberry
Multiple fruit: fig, pineapple

The Teacher's Resource Book includes a blackline master of a blank form like Table 13.2 for the keying activity in Part B.

CHAPTER 14

EUKARYOTES: ANIMALS

Planning Ahead

A model of a human torso will be a big help for Chapter 15. Begin to assemble the necessary equipment for Investigation 15.1.

Concepts

◆ Animals are a kingdom of multicellular, heterotrophic eukaryotes that reproduce primarily sexually and that have common evolutionary origins.

◆ Most animals, especially more complex and terrestrial animals, are motile and have advanced organ systems.

◆ Terrestrial animals have evolved special adaptations such as appendages, lungs, higher metabolic rates, and high motility.

◆ As newer animal types have evolved through time, there has been a trend toward increased structural and ecological complexity, size, diversity, and specialization.

◆ As a cell's size increases, its surface-area-to-volume ratio decreases. This results in a practical limit to the size a cell can achieve and still transport materials effectively. Therefore, increased size in animals is due to an increase in the number of cells rather than to increased cell size.

◆ Sponges and cnidarians are among the most ancient and least complex animals. They are all aquatic (nearly all marine), possess radial symmetry, and are mostly sessile.

◆ Flatworms have bilateral symmetry and contain some true organs and organ systems; many have evolved parasitic relationships with other organisms.

◆ Roundworms and mollusks have some well-developed organ systems, an internal body cavity, and complete digestive systems. Roundworms are extremely abundant and diverse.

◆ Annelids and arthropods have well-developed organ systems, segmented bodies, and are adapted for terrestrial habitats. The diversity and abundance of arthropods such as insects indicate evolutionary success.

◆ Chordates possess internal skeletons and especially well-developed nervous systems. Vertebrates appear to dominate the earth today and are capable of communication. Some mammals can learn and reason.

◆ Increasingly complex structural and behavioral adaptations that have evolved in the animal phyla enable them to carry out the life functions of digestion, gas exchange, transport, excretion, coordination, locomotion, and reproduction.

Objectives

Students should be able to:

◆ *describe* the characteristics of at least three different animals.

◆ *relate* bilateral and radial symmetry to particular life-styles.

◆ *compare* the advantages of aquatic and terrestrial environments for animals.

◆ *explain* the importance of appendages to terrestrial animals.

◆ *explain* why larger animals are made of more, rather than larger, cells.

◆ *recognize* a variety of structural adaptations in familiar animal species.

◆ *predict* the environment in which unfamiliar animals might suitably live, given a set of selected structural adaptations.

◆ *compare* the mechanisms of food-getting and digestion in a variety of animals.

◆ *explain* how some animals can exist without specialized organs for oxygen-carbon dioxide exchange.

◆ *identify* relationships between the process of oxygen-carbon dioxide exchange and water balance in both aquatic and terrestrial animals.

◆ *name* at least three means of transporting substances in animal bodies.

◆ *describe* the differences between closed and open circulatory systems.

◆ *name* at least three major types of substances regularly excreted by animals.

◆ *explain* how a terrestrial organism can have too much water.

◆ *explain* the role of a nervous system in coordination.

◆ *state* the advantages of endoskeletons and exoskeletons.

◆ *compare* reproductive adaptations of several animals.

Strategies

Diversity among animals is the obvious thread running through this chapter. Two strands in this thread are emphasized: ecological relationships and structural adaptations. Animal structure is correlated with the requirements of the animal way of life: the intake of materials, the disposal of excess and poisonous substances, the internal coordination of all metabolic activities, and the means of coping with the environment. To the greatest degree possible, allow students to observe diversity in living animals. In addition to displaying animals in your classroom, suggest that students visit a zoo or an aquarium. Use pictures for a still broader view. Many animal pictures are available on color slides or videodiscs. For bulletin boards, *National Geographic* and *Natural History* are particularly good sources. Students vary greatly in their acquaintance with animals; gauge your students' backgrounds and choose supplementary materials accordingly.

Begin the chapter with Investigation 14.1. It is of high interest to students because it allows them to study five diverse live animals. After students have read Sections 14.1 and 14.2, hold a class discussion of the reading, the students' responses to the interactive legend, and Investigation 14.1. Use many actual organisms in your class discussion of animal diversity. Then assign Sections 14.3–14.7 and the corresponding Concept Review questions. Assign Sections 14.8–14.9, then have students do Investigation 14.2, which teaches biology concepts and process skills (especially the handling of variables). Assignment of Sections 14.10–14.13, the summary, Applications, and selected Problems can follow.

References

Beecher, B. M. "The Birds of Paradise." *Scientific American* (December 1989): 116–25.

Funk, D. H. "The Mating of Tree Crickets." *Scientific American* (August 1989): 50–59.

Galdikus, B. M. "Living with the Great Orange Apes." *National Geographic* (June 1980): 830–53.

Jacobs, G. "Detection and Analysis of Air Currents by Crickets." *BioScience* (December 1995): 776–785.

Klowden, M. "Blood, Sex, and the Mosquito." *BioScience* (May 1995): 326–331.

Leggett, W. C., and K. T. Frank. "The Spawning of the Capelin." *Scientific American* (May 1990): 102–07.

O'Shea, T. "Manatees." *Scientific American* (July 1994): 66–72.

Pietsch, T. W., and D. B. Grobecker. "Frogfishes." *Scientific American* (June 1990): 96–103.

Rismiller, P. D., and R. S. Seymour. "The Echidna." *Scientific American* (February 1991): 96–103.

Sanderson, S. L., and R. Wassersug. "Suspension Feeding Vertebrates." *Scientific American* (March 1990): 96–102.

Seeley, T. D. "How Honeybees Find a Home." *Scientific American* (October 1982): 158–69.

Veit, P. G. "Gorilla Society." *Natural History* (March 1982): 48–59.

Suggested Answers to Applications

1. Plants are autotrophic—they make their own food. Since they don't have to seek out other organisms to keep themselves alive, the development of sense organs is not critical to their survival, and the coordination of these receptors in a head is not necessary.

2. Organs are specialized structures that control specific functions in the body of an animal. Thus, organs permit greater efficiency in the daily activities of acquiring food, eliminating waste, and avoiding predators.

3. Advantages: Exoskeletons provide support, protection against injury in collision, and protection against desiccation. An exoskeleton is inherently stronger than an endoskeleton, as a hollow cylinder is stronger than a solid cylinder made of the same amount of material. Disadvantages: Growth of an animal with an exoskeleton is discontinuous, and when the exoskeleton is shed, the animal is vulnerable to predation and desiccation. Such animals also are limited in size.

4. The caterpillar stage can be specialized for feeding and the butterfly stage for reproduction and dispersal. Also, the insect can make use of different habitats during its life.

5. Physical digestion increases the surface area of food, allowing the chemical part of digestion to take place more quickly. The lining of an animal's gut may have many folds, an adaptation that increases surface area and speeds the rate of food absorption. Because large animals do not have great enough body surface when compared to the total volume of the animal to allow for sufficient gas exchange, they must have organs that increase the surface area through which gas exchange can occur.

6. Chemical reactions generally speed up as temperatures increase and slow down as temperatures decrease. Constant body temperature allows birds and mammals to maintain high rates of metabolism and to be active at a wide range of environmental temperatures. Reptiles and other animals that cannot maintain a constant body temperature are less active in cool environmental temperatures.

7. Their skeletons, efficient gas exchange systems, systems for excreting nitrogenous wastes, internal fertilization, and shelled eggs or internal development of the embryos.

8. Answers depend on student selection.

9. Answers depend on student selection.

Suggested Answers to Problems

1. A survey of diets of people from other countries, especially those from Asia and those of hunting gathering tribes, will reveal the great diversity of organisms eaten—including insects, reptiles, worms, sea urchins, octopuses, jellyfish, and sea cucumbers, as well as many different kinds of mammals.

2. The popular press, as well as scientific journals, should have a number of articles about these infrequently seen insects.

3. Among many behavioral adaptations: Nocturnal activity and sleeping during the day; burrowing into the ground where temperatures are much lower than air temperatures. Among many physiological adaptations: Very dry waste products that conserve water; some organisms (for example, kangaroo rats) can survive without drinking water.

4. You may want to make up a data sheet on which your students can write about specific questions. Have them compare all of the primates at the zoo with other mammals and with humans. Or, have them compare all of the birds with each other.

5. This problem might be answered at various levels. The basic point is that different methods of successfully maintaining life functions depend on living in different environments. A sponge's methods would not be successful where a mountain goat's are and vice versa.

6. An excellent source of information on care given to the young of various species of animals and on number of offspring produced is P. L. Altman and D. S. Dittmer, eds. *Biology Data Book* 2nd ed. (Washington, DC: Federation of American Societies for Experimental Biology, 1974). This source also contains information on life spans. Other sources include biology textbooks and encyclopedias, but the information may be scattered. Students should be able to generalize that the greater the care of the young, the fewer the number of offspring.

Materials and Preparations for Investigations

INVESTIGATION 14.1 Diversity in Animals: A Comparative Study

Materials (per class of 30, teams of 6)
(See student materials list for organization of stations.)
9 Syracuse watch glasses
3 dropping pipets
1 aquarium, with water only
3 4-qt battery jars
9 stereomicroscopes or 10X hand lenses
9 compound microscopes
3 10X hand lenses
9 No. 2 paint brushes
12 sheets moistened paper towels
12 boxes of damp soil
3 large, shallow boxes or pans
10 to 20 pieces of lettuce and fruit
3 prepared slides of longitudinal sections of hydra

3 prepared slides of planaria cross sections
3 whole mounts of carbon-fed planaria
3 prepared slides of earthworm cross sections
5 to 10 live hydras
5 to 10 live planarians
5 to 10 live earthworms
3 live land hermit crabs
3 live frogs
1 daphnia culture, large (mature)

Preparations

The principal teaching problem is logistics. It can be difficult to assemble all the animals—in a healthy, active condition—simultaneously. Most of these species, however, are worth maintaining as permanent residents of your laboratory, thereby eliminating the problem of timing orders from suppliers. See "Care and Release of Life Forms" in Guidelines for Laboratory Safety and F. B. Orlans, *Animal Care from Protozoa to Small Mammals* (Menlo Park, CA: Addison-Wesley, 1977). Another useful book is M. P. Behringer, Techniques and Materials in Biology, 2nd ed. (Malabar, FL: Krieger Publishing, 1989). You may wish to add or substitute other animals, such as *Tenebrio,* crickets, snails, and daphnia. If you make substitutions, revise the specific directions.

These animals are best maintained in a refrigerator. Uncap the jars or containers before placing in the refrigerator. The earthworms and frogs may be released after the investigation. Be sure the frogs are released in a suitable habitat near wetlands. The hermit crabs should become a permanent part of your biology classroom.

For observations, two-gallon glass or clear plastic jars, such as those used for relish or mayonnaise, may be substituted for the battery jars. To keep the frogs from jumping out, provide cloth circles and rubber bands to use as lids.

Set up the five stations as far apart as your laboratory permits. Movable tables and peripheral facilities allow for the best arrangement; otherwise, adaptations can be made. Each group should be permitted about 10 minutes at each station. As the observation time occupies all of an ordinary class period, directions for observation must be studied thoroughly before the laboratory period, and everything must be in readiness at each station when the class arrives. Have the students prepare their data tables the day before, or distribute copies of the blackline master of Table 14.1 from the Teacher's Resource Book. Instruct students to tape the data table in their data books.

You may wish to omit some of the general questions or to add questions. Be sure that any added questions can be answered *from the material available for observation.* This is not the time to send students scurrying to reference books. Some directions for specific animals may have to be bypassed. You may need to demonstrate the proper method of capturing and holding a frog. Take care that students do not harm the animals. Try to keep one frog hungry by separating it from the others. Call the class together to demonstrate how frogs use their tongues to capture food.

INVESTIGATION 14.2 Temperature and Circulation

Materials (per class of 30, teams of 3)
10 50-mL beakers of tap water
10 500-mL beakers
10 nonmercury thermometers
10 10X hand lenses
10 clean, chemical-free dissecting trays
20 damp paper towels
5 L warm water ice
10 watches (or 1 clock) with second hands
10 large earthworms
1 box of damp soil

Preparations

Earthworms can be ordered from Ward's Natural Science Establishment or obtained at bait shops as "night crawlers." The worms can be recycled during the day, but they will be most responsive if used in alternate periods. Thus, 30 worms for an entire day of biology classes should be enough.

Because students will be working with living earthworms, the container used for observation must be clean and free of chemicals such as formalin. If your dissecting trays do not fit this description, substitute some other type of dish or pan. Check that students position their worms correctly. Caution students about the proper care and handling of laboratory animals. Do not allow them to play with the worms.

A hot water tap in the classroom probably will provide water of more than necessary warmth. Otherwise, provide water in 500-mL beakers on hot plates set on low (approximately 40° C). For convenience, provide ice and warm water at three or four stations around the room. Exact temperatures are not necessary. The important thing is that there is a range from very low to about 40° C. If the five temperatures at which data were collected are fairly close among the teams, data can be averaged.

CHAPTER 15

THE HUMAN ANIMAL: FOOD AND ENERGY

Planning Ahead

Gather any necessary models or charts for Chapter 16 or arrange to borrow them when needed. You will need narrow-range pH paper for Investigation 19.2 if you are using the alternate procedure, which does not require a computer. Test your local distilled water before and after blowing into it to determine which ranges to order.

Concepts

◆ Multicellular organisms have evolved digestive systems that are able to break food into particles small enough to enter the cell and then move food to organs that absorb the molecules into the bloodstream.

◆ Digestion in humans begins by mechanical processes such as chewing, mashing by the tongue, and churning in the stomach.

◆ Chemical digestion reduces carbohydrates, proteins, and fats to absorbable molecules by the actions of specific chemicals produced at designated sites along the digestive tract.

◆ Food is absorbed into the bloodstream from the small intestine; water is absorbed from the large intestine.

◆ Cellular respiration is necessary to provide the cell with an immediately available source of energy. The process uses oxygen to transfer energy from glucose to molecules of ATP. Carbon dioxide and water are produced as by-products of the process.

◆ In addition to its role in cellular respiration, the Krebs cycle plays a central role in cellular metabolism. Many of the intermediate reactions in the cycle provide building blocks for essential cell compounds.

◆ The human digestive system evolved to handle a high-fiber (unprocessed) diet consisting primarily of grains containing protein, carbohydrates, and unsaturated fats. The processing of foods removes both natural fibers and vitamins and tends to add unnatural amounts of refined sugars.

◆ The human body can process only the materials it takes in, so the quality of a diet can limit biochemical actions. In other words, "You are what you eat."

◆ There is a relationship between the intake of saturated fats, cholesterol levels, and cardiovascular disease. There also may be a relationship between fiber intake and cancer in the digestive tract.

◆ Carbohydrates and fats provide organisms with sources of energy. Fats also help to produce hormones. Proteins provide the building blocks for much of the body's structure.

◆ Nutritional disorders, although widespread among teenagers, can be treated or prevented altogether. Poor nutrition (rather than malnutrition) and obesity are among the nation's major health problems.

Objectives

Students should be able to:

◆ *identify* the principal digestive organs in a human.

◆ *distinguish* between ingestion and digestion.

◆ *relate* physical digestion to chemical digestion.

◆ *explain* the functions of digestive juices and enzymes.

◆ *describe* how and where the absorption of food and water occur.

◆ *demonstrate,* using experimental data, that foods contain energy.

◆ *list* the major functions and products of cellular respiration.

◆ *discuss* the roles of glucose and oxygen in cellular respiration.

◆ *diagram* the breakdown of the carbon chain of glucose during glycolysis.

◆ *describe* the role of the Krebs cycle in cellular respiration and in cellular metabolism.

◆ *explain* how ATP is formed in the electron transport system.

◆ *explain* the interactions between lactic acid, glycogen, and glucose in physical activity such as a marathon.

◆ *analyze* the nutritional value of a fast-food lunch.

◆ *compare* the proportions of fat, fiber, sugar, and salt in early and modern human diets.

◆ *assess* their risk for cardiovascular disease.

◆ *describe* the uses of lipids, carbohydrates, and proteins in the body.

◆ *calculate* the number of kcals needed for their own health.

◆ *explain* how to lower their risk of cardiovascular disease, including selection of appropriate foods.

◆ *distinguish* between healthy and obsessive attitudes toward thinness.

Strategies

This chapter focuses on humans, although ingestion, digestion, cellular respiration, and nutrition are common to all animals. In many school systems, human anatomy and some gross physiology are taught in earlier grades. Nevertheless, a good mannikin is an essential adjunct to the teaching of this chapter. Chapter 15 will stimulate questions about cardiovascular disease and other nutrition-related topics.

Investigations 15.1 and 15.2 offer opportunities to use the LEAP-System. Begin the chapter with a brief discussion of ingestion and digestion, assign Sections 15.1–15.4, and then have students do Investigation 15.1, which is a viable lab even if you do not have a LEAP-System. Students may need extra help with Sections 15.5–15.10, but do not get bogged down in the details of cellular respiration. Focus on the main concepts of step-by-step, enzyme-controlled energy release. Investigation 15.2 is an interactive simulation that teaches about homeostasis and feedback and relates energy use to running a marathon. Proceed to Section 15.11, followed by Investigation 15.3. Overweight students may be extremely self-conscious about weight. Do not insist that they are weighed. Obtain a sphygmomanometer for Part B if at all possible. Have students read (in class if time permits) and discuss the Biology Today on cardiovascular disease immediately after the investigation. Then, assign Sections 15.13–15.16 and the Concept Review questions, followed by the Applications.

References

Barman, C. R. "Nutrition Education: An Essential Ingredient of Biology Education." *American Biology Teacher* (September 1985): 333–39.

Cohen, L. A. "Diet and Cancer." *Scientific American* (November 1987): 42–49.

Department of Health and Human Services. Public Health Service. *FDA Consumer.* Washington, DC: Food and Drug Administration.

Eisenberg, M. S., et al. "Sudden Cardiac Death." *Scientific American* (May 1986): 37–43.

Hole, W. J. *Essentials of Human Anatomy and Physiology.* 3rd ed. Dubuque, IA: W. C. Brown, 1989.

Kokata, G. "How Important Is Dietary Calcium in Preventing Osteoporosis?" *Science* (1 August, 1986): 519.

Lawn, R.M. "Lipoproteins in Heart Disease." *Scientific American* (June 1992): 54–60.

National Geographic Society. *The Incredible Machine.* Washington, DC: National Geographic Society, 1986.

Taylor, M. F. "Simple Tools for Measuring Anaerobic and Aerobic Respiration." *American Biology Teacher* (October 1987): 439–41.

Weindrach R. "Caloric Restriction and Aging." *Scientific American* (January 1996): 46–52.

Suggested Answers to Applications

1. Chemical digestion provides raw materials for biosynthesis.

2. This is a good time to talk about form and function. The different types of human teeth play different roles in digestion. Sharp, pointed canines are ideal for tearing food. Incisors have a chisel shaped, sharp cutting edge and are good for biting. Bicuspids have two distinct edges and are good for grinding food. Molars are large and are good for grinding.

3. Conditions in the stomach are highly acidic, whereas those in the small intestine are close to neutral.

4. The small intestine, which is longer than the large intestine, has an immense surface area that is ideal for the absorption of food particles. The small intestine gets the food first and, being much longer, holds it for a longer time than the large intestine.

5. Water is absorbed in the large intestine.

6. In glycolysis and the Krebs cycle, carbon chains are broken, hydrogen atoms are removed, and a small amount of ATP is made as electrons give up energy. The electron transport system yields more energy.

7. The Krebs cycle provides the pathway for the final breakdown and energy release of all food molecules: carbohydrates, fats, and proteins. The compounds formed in the cycle provide carbon skeletons for all the major biosynthetic pathways.

8. Cellular respiration is greatly slowed at lower temperatures. When a child falls into cold water, the respiration rate is slowed, the requirement for ATP is less, and cells can survive for a longer than normal period without oxygen.

9. All living cells require a source of energy to keep them alive. When a person does not eat, there is no incoming source of energy. To obtain the required energy, the body will break down storage compounds. As this is done, the body loses weight.

10. Directly, or indirectly, all human food comes from plants (as in grass). The biological molecules in the grass plants or whatever food one eats are broken down and converted into the biological molecules of humans.

11. Complete proteins supply essential amino acids in the correct proportions.

12. Vitamins and essential elements act as cofactors or form parts of enzymes or carriers essential to cellular respiration and to biosynthesis reactions. If these reactions cannot take place at optimum rates health is impaired.

13. Humans lack the enzymes necessary to digest cellulose. Because fiber absorbs water, feces move more quickly and easily through the large intestine.

Suggested Answers to Problems

1. Cell processes are sometimes difficult to visualize but may be easier for students to understand after they take on roles of various molecules and enzymes. Your students should increase their understanding of the entire process as they figure out how to demonstrate the complex set of events in cellular respiration.

2. The dynamics of parent and sibling relationships may play a role. Low self-esteem is often a factor; anorexics may strive for the slim look emphasized in the media. Young people seem particularly susceptible at puberty or when contemplating a move. Any stress or life change can trigger anorexia or bulimia; for example, divorce, death, broken romance, or ridicule.

3. Milk contains the sugar lactose. Many of the people in the three continents mentioned cannot use the sugar because they have little of the intestinal enzyme lactase.

4. The principal functions that the artificial system would need would be movement of food through the system, digestion, and absorption.

5. This may be the first step in helping to lower the fat intake of students. If they are typical teenagers they will consume a high percentage of their calories in the fried foods, dairy products, and meats they eat. Today's packaging laws require nutritional information to be supplied on the package. Encourage students to maker use of this information.

6. It is important for students to understand the nature of valid, authenticated, experimental results. They need to develop a healthy skepticism about the claims in various advertisements for remarkable, amazing, miracle diet plans. Help them to unravel faulty or misleading information.

7. A few hunting-gathering tribes still exist around the world. Compare their diet with that of a modern American student. There is some evidence that human ancestors were scavengers of big game that had been killed by nonhuman predators. Get students to think about the difficulties in the hunting-gathering life style.

8. The incidence of diet-related health problems such as strokes and heart attacks, in Japan has increased dramatically over the last few years since the people there began to eat foods high in animal fats, such as ice cream and beef, coupled, of course, with high stress and other contributing factors.

Materials and Preparations for Investigations

INVESTIGATION 15.1 Food Energy

Materials (per class of 30, teams of 3, 2–3 teams per LEAP-System)

30 pairs of safety goggles
30 lab aprons
10 100-mL graduated cylinders
10 250-mL Erlenmeyer flasks
4 Apple lle, Apple IIGS, Macintosh, or IBM computers
4 LEAP-System™ (2 teams per system)
printer
10 thermistors
10 extension cables for thermistors
10 15-cm nonroll, nonmercury thermometers (optional)
10 balances
10 forceps
10 16-oz tin cans with cutout air and viewing holes
10 #20 corks with paper clip sample holder
1 box 2-1/4" kitchen matches (or fireplace matches)
10 3-cm² pieces of sandpaper
masking tape
10 20-cm × 30-cm pieces of extra heavy aluminum foil
20 pot holders
10 small containers of water
30 whole peanuts
30 walnut halves
duct tape, pliers, tin snips (for preparing tin cans)

Preparations

Review the LEAP-System Open Me First Reference Manual, and BSCS Biology Lab Pac 2 to become familiar with how to operate the system, allocate disk space for saving data, connect the probes, boot the system, and bring up the menu on the computer screen. Be sure students know how to put the system on WAIT, take the system off WAIT, REINITIALIZE the system, put the system on SAVE, and QUIT the system for your computers; and that space is available on the disk(s) for each class so data can be saved and stripcharts run.

The standard LEAP-System includes specifications and probes sufficient to run each experiment for one team at a time. The system also can handle multiple teams concurrently. See the reference manual for detailed instructions about how to create new experiments or update existing ones. Using these procedures, you can generate your own multi-team experiments. Disks with completed multi-team experiments also are available for purchase. Additional probes may be required in some cases.

When more than one team is using one LEAP-System, it may be difficult to differentiate all the lines on the graph. Under these circumstances, instruct students to read the numeric data displayed near the graph. After they have completed the experiment, students can stripchart and list their individual team data. See the reference manual for details on stripcharting and listing data by lab team.

Before printing stripcharts, check "System Maintenance" in the LEAP-System Main Menu to be sure the printer specified corresponds to the one you are using. See the reference manual for details on how to specify the printer.

Figure 15.5 shows the components of the calorimeter. To support the 250-mL flask, use a 1 6-oz tin can (vegetable can size). With a can/bottle opener, punch out four triangular openings on the top of the can and two on the side near the top. Use tin snips to cut out a viewing hole at the bottom large enough to view the burning sample (about 5 cm tall, 2.5 cm wide at the top, and 4 cm wide at the base). Cover the cut edges with duct tape. Mold a piece of aluminum foil around the cork to prevent it from catching fire. (If the cork is more than 2 cm high, cut off the excess so the viewing hole does not need to be too large.) Use a paper clip to fashion a sample holder such that the sample fits firmly between the wires. Insert the holder into the cork. For ease of igniting the sample, it should be held 1.5 cm above the cork surface.

Oil-roasted peanuts and walnuts burn better than the dry-roasted variety; the increased kcals are insignificant in this crude calorimeter. Snack foods such as potato chips, corn chips, cheese snacks, and so on also burn well.

The Teacher's Resource Book includes a blackline master of Table 15.2.

INVESTIGATION 15.2 The Marathon

Materials (per class of 30, teams of 3)
4 Apple lle, Apple IIGSS Macintosh, or IBM computers
4 LEAP-Systems™ (2 or 3 teams per system)
20 dials
printer (optional)

Preparations

Review the LEAP-System Open Me First, Reference Manual, and BSCS Biology Lab Pac 2 to familiarize yourself with how to operate the system, connect the probes, allocate disk space for saving data, boot the system, and bring up the menu on the computer screen. Be sure students know how to put the system on WAIT, take the system off WAIT, REINITIALIZE the system, put the system on SAVE, and QUIT the system for your computers; and that space is available on the disk(s) for each class so data can be saved and stripcharts run.

For a more thorough analysis of various strategies, have students save their data and run stripcharts. The marks on the stripchart indicate when changes in rate of travel and/or glucose were requested and can help students understand the relationship between those changes and their success or lack of success in completing the marathon.

Before printing stripcharts, check "System Maintenance" in the LEAP-System Main Menu to be sure the printer speci-

fied corresponds to the one you are using. See the reference manual for details on how to specify the printer.

The appearance of the screen on your computer may differ from the one illustrated in Figure 15.13. See the LEAP System BSCS Biology Lab Pac 2 for details.

Many factors influence a runner's performance during a marathon, only a few of which are considered in this simulation. Liver glycogen is the major energy reserve for aerobic respiration. As glycogen reserves diminish, and as anaerobic conditions prevail, lactic acid builds up in the muscles and in the blood. As glycogen reserves diminish and lactic acid levels build up, marathon runners "hit the wall," and running becomes a painful process. At this time, the body turns to a different energy reservoir, and free fatty acids increase in the blood. To keep the investigation manageable, other important variables, such as cardiac output, heart rate, breathing rate, and body fluid, are ignored.

Students can control two variables—rate of travel and blood glucose levels. The computer program is designed so the runner will crash if blood glucose levels drop below 50 mg/100 mL, if glycogen reserves drop below 10 mg/g liver tissue, or if lactic acid levels rise above 10 mM. These critical levels are reached by keeping the blood glucose levels too high and/or running the race too rapidly. An initial sprint that lasts too long also can contribute to an early crash.

In the class discussion of the homeostatic mechanisms involved, include aerobic and anaerobic pathways and their consequences as well as the different energy sources for metabolic activity. Stress the importance of cooperation among the partners controlling a runner and relate this cooperation to the homeostatic mechanisms that function constantly in the human body. Help the students construct "if/then" hypotheses about the best strategies to use. Some students may need help with phrasing hypotheses in this format. It is important for them to construct hypotheses about the race strategy rather than to continue to use trial and error. Record running times on the chalkboard to stimulate development of improved strategies, but check with each team to help them formulate hypotheses based on their strategies.

Hints for maximum performance: (Wait until the students have tried their strategies before giving these hints.)

1. Do not exceed 7 m/sec. Rates greater than 7 m/sec will deplete the glycogen/glucose levels and build up the lactic acid too quickly.

2. Do not allow the glycogen to drop below 10 mg/g. This can be accomplished by varying the rate of travel and the level of glucose. When the glycogen drops below 10 mg/g, the runner will crash in about 20 race minutes (20 seconds) as the glucose level is depleted and there is no glycogen to generate more glucose.

3. There is no benefit to keeping the glucose at less than 100 mg/100 mL as long as there is an adequate supply of glycogen. In fact, it is a disadvantage because glucose

less than 100 mg/100 mL will decrease the attained rate of travel, the runner will take longer to go the 42 kilometers, and more glucose will be used rather than less.

4. The glucose should be kept around 100 mg/100 mL until the glycogen has dropped below 10 mg/g. At that time, the inevitable crash can be held off by adjusting the requested glucose downward as the attained glucose falls. At this point, try not to request more glucose than can be attained.

5. As long as the glucose is 100 mg/100 mL or more, it is never advantageous to keep the requested rate of travel above the attained. From a physiological standpoint, the runner is straining to achieve more than the body can deliver, which is costly in glucose usage and buildup of lactic acid.

If, however, the requested glucose is held below 100 mg/100 mL as a strategy, it may be beneficial to raise the requested rate of travel above the attained. This will result in an increased attained rate of travel. It is a strategy worth pursuing, although it may take its toll in lactic acid buildup.

6. The greater the lactic acid level, the lower the attained rate of travel. The lactic acid can be lowered by lowering the rate of travel to about 2.1 km/sec. At this rate, the runner can blow off the CO_2 and decrease the lactic acid. Of course, this will slow down the runner (see 3).

7. The best strategy is to start the race at 7 km/sec for 8000 to 10 000 meters, then decrease to about 4.8 to 5.0 m/sec for the remainder of the race. At the same time, keep the glucose at 100 mg/100 mL until the glycogen drops below 10 mg/g and then follow the attained glucose down with the requested glucose.

A kick at the end also may be beneficial, but at that point, the lactic acid buildup and the low amount of glucose remaining may prevent the runner from making this kick.

INVESTIGATION 15.3 Assessing Risk of Cardiovascular Disease

Materials (per class of 30, teams of 2)
15 centimeter tape measures
bathroom scale
sphygmomanometer (optional)
15 calculators

Preparations
Some students may be self-conscious about having peers know their real weight. Allow students to omit any part of this activity that could cause them emotional discomfort.

Most fabric shops carry tape measures marked in both inches and centimeters. If you can obtain tape measures marked only in inches, instruct students to multiply all measurements by 2.5 to convert to centimeters. One sphygmomanometer is adequate for this activity; if unavailable, have your students omit item 6 on the self-check.

CHAPTER 16
THE HUMAN ANIMAL: MAINTENANCE OF INTERNAL ENVIRONMENT

Planning Ahead
Chapter 18 requires a variety of plants; arrange to buy or borrow these now. You will need leafy plants for Investigation 19.1. If you plan to grow them from seed plant them now.

Concepts
♦ Because the human is a large, multicellular organism, it cannot rely solely on diffusion for transportation of nutrients to and wastes from the cells; therefore, it depends on systems of organs working in concert.

♦ The human circulatory system contains a pump (heart), tubes that carry the blood from the heart to the tissues (arteries) and from the tissues to the heart (veins), and capillaries that bathe the tissues and connect arteries to veins.

♦ Valves in the heart and in the veins help blood to flow in only one direction.

♦ Human blood is a watery plasma in which cells are suspended and salts, plasma proteins, nutrients, and wastes are dissolved.

♦ Platelets and plasma proteins prevent blood loss by forming a coagulated network when exposed to air.

♦ The body's major defense mechanisms are mucous membranes, a variety of specialized white blood cells that engulf foreign particles and provide immunity, and the lymphatic system.

♦ Specific immune responses involve complex interactions among a variety of white blood cells.

♦ Problems with the immune response are responsible for a variety of illnesses.

♦ The human respiratory system supplies oxygen and removes carbon dioxide and heat from the tissues.

♦ Red blood cells carry oxygen to the tissues. Most carbon dioxide is carried away from the tissues in a dissolved state.

◆ The kidneys filter nitrogenous wastes and other undesirable materials from the blood. They also help to recover water and other useful materials.

◆ Human metabolic reactions take place most efficiently at 37° C. The body regulates its temperature by radiation, increased activity, or evaporative perspiration.

Objectives

Students should be able to:

◆ *show* experimentally the relationship between exercise and pulse rate.

◆ *demonstrate* with a chart or model the course of blood flow through a mammalian heart.

◆ *describe* the functions of arteries, veins, and capillaries.

◆ *identify,* from a brief statement of their characteristics, the principal components of mammalian blood.

◆ *describe* in general terms the process of blood clotting.

◆ *compare* and contrast the composition and functions of lymph and of blood.

◆ *contrast* specific and nonspecific immunity.

◆ *describe* the general process by which antigens are removed from the body by the immune system.

◆ *diagram* the events in vaccination and subsequent exposure to an antigen for the same disease.

◆ *explain* allergies as a disorder of the immune system.

◆ *predict* the likelihood of future problems for an Rh⁻ woman who has given birth to an R⁺ or an Rh⁻ baby.

◆ *identify* the organs of respiration and describe their functions.

◆ *explain* how oxygen and carbon dioxide enter or leave the body, move between the circulatory and respiratory systems, and are transported.

◆ *relate* the sites and processes of general respiration to those of cellular respiration.

◆ *describe* the function of a kidney as a homeostatic organ.

◆ *compare* blood and urine with regard to the various components and their proportions.

◆ *distinguish* between excretion, secretion, and elimination.

◆ *explain* the advantages of maintaining a constant internal temperature.

◆ *diagram* the thermostatic control of the body's temperature through evaporation from the skin and changes in the blood vessels.

Strategies

Students see how the circulatory, respiratory, and excretory systems work in coordination to keep the body's temperature constant, to distribute materials, and to rid the body of wastes and foreign matter. As in the previous chapter, you will find models and charts invaluable.

Investigations 16.1 and 16.3 are designed for use with the LEAP-System, although there are alternatives for both. Begin with Investigation 16.1; the main emphasis should be on the formation and testing of hypotheses. Then, assign Sections 16.1–16.4. A brief discussion of immunity should precede Investigation 16.2; then assign 16.5–16.7. Students usually enjoy activities such as Investigation 16.3 that relate to themselves. Break Sections 16.8–16.14 into two assignments depending on your schedule.

References

Atkinson, M. A., and N. K. Maclaren. "What Causes Diabetes?" *Scientific American* (July 1990): 62–71.

Bardell, D. "The Changeable Nature of AIDS." *American Biology Teacher* (February 1989): 79–83.

Brownlee, S. "The Cellular Battlefield." *U. S. News and World Report* (28 March 1994): 66–68.

Goldenberger, A. L., D. R. Rigney, and B. J. West. "Chaos and Fractals in Human Physiology." *Scientific American* (February 1990): 42–49.

Lichtenstein, L.M. "Allergy and the Immune System." *Scientific American* (September 1997): 16–24.

Mills, J., and H. Masur. "AIDS-Related Infections." *Scientific American* (August 1990): 50–59.

Newman, J. "How Breast Milk Protects Newborns." *Scientific American* (December 1995): 76–79.

Rennie, J. "The Body Against Itself." *Scientific American* (December 1990): 106–15.

Rosenberg, S. A., "Adoptive Immunotherapy for Cancer." *Scientific American* (May 1990): 62–69.

Smith, K. A. "Interleukin-2." *Scientific American* (March 1990): 50–57.

Storey, K. B., and J. M. Storey. "Frozen and Alive." *Scientific American* (December 1990): 92–97.

Young, E. D., and Z. A. Cohn. "How Killer Cells Kill." *Scientific American* (January 1988): 38–45.

Suggested Answers to Applications

1. Nostril, pharynx, trachea, bronchus, air sac, capillary; attaches to hemoglobin in blood. Once attached to hemoglobin, it travels through vein,

heart, aorta, artery, capillary, and diffuses into cell. Used in cellular respiration at mitochondrion.

2. Both flushing and blushing result from vasodilation affecting the capillaries in the skin.

3. The hamburger, fries, and soft drink contain carbohydrates, fats, proteins, and cellulose. Students need to consider the enzymes affecting (or failing to affect) these substances and the discussions in this and the previous chapter.

4. Excretion is the removal of metabolic wastes from cells and the blood; secretion is the production and release by a cell or a gland of any substance other than waste; and elimination is the removal of undigested matter from the body. Excretion— kidneys and lungs; secretion—kidneys and glands; elimination—digestive system.

5. A variety of diagrams can be constructed based on the information in the chapter.

6. Increases manyfold the amount of oxygen that can be transported.

7. Because the amount of oxygen decreases as elevation increases, the breathing rate must rise to offset the lower availability of oxygen. Gas transport becomes less efficient and, at extremely high altitudes, brain and body function are impaired.

8. Each of many different influenza viruses has a different protein coat. You may be immunized against one strain during one year, but that strain may not be the major disease-causing agent the following year, so you must be immunized against the new Virus.

9. In both cases antibodies are produced, but in allergies, these antibodies combine with specialized body cells rather than with foreign materials.

10. The best donors, in order, would be: (1) identical twins—with whom you have all genes in common (2) fraternal twin; (3) parent; and (4) spouse— with whom you have few genes in common.

11. People with type O blood, having neither A nor B antigens, were called universal donors, which meant they could give blood to anyone. People with AB blood, having neither A nor B antibodies in the plasma, were called universal recipients which meant they could receive blood from any donor. Now other factors in the blood must be checked (e.g., Rh factor) before a blood transfusion is considered.

Suggested Answers to Problems

1. The basic principle of the artificial kidney is to pass blood through minute channels bounded by a thin membrane. On the other side of the membrane is a dialyzing fluid (chemically similar to blood plasma) into which waste substances from the blood pass by diffusion.

   In the kidney, blood flows through a capillary network surrounding a nephron. All constituents of plasma except proteins move in both directions by diffusion or active transport.

   In the artificial kidney, blood flows between two thin sheets of cellophane. The dialyzing fluid is outside of the sheets. All constituents of blood except proteins diffuse in both directions, depending on concentration.

2. Artificially induced hypothermia slows the heart and greatly depresses body metabolism. During open-heart surgery it is possible to stop the heart for many minutes at a time. Cooling is not so great as to cause serious physiological effects.

3. All artificial heart transplant recipients have experienced strokes. The problem must be overcome. The artificial heart also must be lightweight and portable, maintain blood pressure, and not damage blood cells.

4. CPR should be applied only to a victim who is not breathing *and* who does not have a pulse. (The pulse should be checked for 5 to 10 seconds at the carotid artery.) If the victim has a pulse, then only mouth-to-mouth resuscitation is necessary. If there is respiratory arrest, the heart can continue to pump blood for several minutes, and existing stores of oxygen in the lungs and blood will continue to circulate to the brain and other vital organs. If resuscitation is begun in time, a rescuer can prevent cardiac arrest. When there is primary cardiac arrest, oxygen is not circulated, and the stored oxygen is depleted in a few seconds. That is why the chest compression of CPR is vital when no pulse can be detected.

   One of the first precautions is proper training in the techniques of CPR. If you come across a victim in need of CPR *and* you are properly trained, you should be careful about the following: (a) Check carefully for a pulse. Performing external chest compressions on a patient who has a pulse may result in serious medical complications. (b) For proper air ventilation and blood flow to the brain, be sure the victim's head is not placed higher than the chest. (c) Distention of the stomach may result if too much air is breathed into the patient or if the airway is partially or completely blocked. Such distention might promote vomiting or reduce the amount of air that can reach the lungs. (d) Be sure that your hand position is correct. Applying pressure too low on the chest may cause internal bleeding because the sternum may cut into the liver. (e) Fingers should not rest on the victim's ribs during compression. Pressure of the fingers on the ribs increases the likelihood that ribs may be fractured. (f) Compressions should be smooth, regular, and uninterrupted except for rescue breathing. Avoid sudden or jerking movements. Any jabs can increase the possibility of injury to the ribs and internal organs and might reduce the amount of blood that can be circulated by each compression. Information from "Standards for CPR and ECC," *Journal of the American Medical Association* (6 June, 1986).

5. Machinery such as the ultrasound machine is extremely expensive to operate; however, you may be able to borrow and show a videotape of a patient who shows abnormalities of the heart.

6. Pictures of diseased lungs often bring home the danger of smoking cigarettes and may be a powerful reminder to students of why they should stop smoking,

7. In this chapter, surface area is mentioned in relation to the human lung. In Chapter 15, it was mentioned in relation to the lining of the intestine. Surface area also plays an important role in root hairs of plants and internal membranes of organelles such as mitochondria.

8. Comparing the constituents of either blood or urine to the standard, normal level of each constituent can yield a variety of information about the health of an individual. Remind students about home pregnancy tests and checks for such diseases as diabetes.

9. Films such as the National Geographic video "The Incredible Human Machine" show thermographic photographs of the body.

10. Young people often have the illusion of immortality. Do not focus on the negative consequences of a person's choice, but do focus on the need for students to be fully informed about all of the positive and negative aspects of a situation *before* they make decisions.

Materials and Preparations for Investigations

INVESTIGATION 16.1 Exercise and Pulse Rate

Materials (per class of 30, teams of 2)
4 Apple IIe, Apple IIGS, Macintosh, or IBM computers
4 LEAP-Systems™
4 pulse probes
(Note: readings are so fast that teams can rotate on the computers.)
15 watches or 1 clock with second hand (optional)

Preparations

Review the LEAP-System Open Me First, Reference Manual and BSCS Biology Lab Pac 2 to become familiar with how to operate the system, connect the probes, boot the system, and bring up the menu on the computer screen. Be sure students know how to put the system on WAIT, take the system off WAIT, REINITIALIZE the system, and QUIT the system for your computers. Discuss with students what they are seeing on the screen when using the pulse probe. Data from the pulse probe cannot be saved on the disk, but the readings are so rapid that several students using one probe can complete their experiments in a class period. You may wish to have half the class use the LEAP-System and the other half use the wrist method and then rotate; data from the two methods can be compared. The logistics depend on the number of computers available.

INVESTIGATION 16.2 The Alpine Slide Mystery

Materials (per class of 30, teams of 3)
none

INVESTIGATION 16.3 Exercise and Carbon Dioxide Production

Materials (per class of 30, teams of 2)
15 100-mL graduated cylinders
30 250-mL flasks
4 LEAP-Systems™ (2 teams per system)
4 Apple IIe, Apple IIGS, Macintosh, or IBM computers
printer
16 pH probes
15 pH meters or wide-range pH paper (optional)
15 forceps (optional)
30 wrapped drinking straws
15 watches (or 1 clock) with second hand

Preparations

Review the LEAP-System Open Me First, Reference Manual, and BSCS Biology Lab Pac 2 to familiarize yourself with how to operate the system, allocate disk space for saving data, connect the probes, boot the system, and bring up the menu on the computer screen. Be sure students know how to put the system on WAIT, take the system off WAIT, REINITIALIZE the system, put the system on SAVE, and QUIT the system for your computers; and that space is available on the disk(s) for each class so data can be saved and stripcharts run. Check the calibration of the pH probes before use; directions are included in the LEAP-System BSCS Biology Lab Pac 2.

The standard LEAP-System includes specifications and probes sufficient to run each experiment for one team at a time. The system also can handle multiple teams concurrently. See the reference manual for detailed instructions about how to create new experiments or update existing ones. Using these procedures, you can generate your own multi-team experiments. Disks with completed multi-team experiments also are available for purchase. Additional probes may be required in some cases.

When more than one team is using one LEAP-System, it may be difficult to differentiate all the lines on the graph. Under these circumstances, instruct students to read the numeric data displayed near the graph. After they have completed the experiment, students can stripchart and list their individual team data. See the reference manual for details on stripcharting and listing data by lab team.

Before printing stripcharts, check "System Maintenance" in the LEAP-System Main Menu to be sure the printer specified corresponds to the one you are using. See the reference manual for details on how to specify the printer.

Regardless of how the investigation is conducted, the pH values and graphs can be combined and average values displayed. Try to take advantage of any spurious results to talk about experimental error. A blackline master of Table 16.2 is included in the Teacher's Resource Book.

For the vigorous exercise portion, you may want to have the students partially exhale each breath before exhaling into the water. During deep breathing, the carbon dioxide concentrations will be greatest deep in the lungs, and the oxygen concentrations will be greatest higher in the lungs. Partially exhaling before blowing into the flasks will give better results.

INVESTIGATION 16.4 The Kidney and Homeostasis

Materials (per class of 30, teams of 3)
none

Preparations

Before students begin Step 2, discuss the data in Table 16.4 so there is no question as to their meaning. It is important that students understand the numbers *are proportions of molecules* or *ions* and not actual *numbers of molecules.*

CHAPTER 17

THE HUMAN ANIMAL: COORDINATION

Planning Ahead

About six days before the discussion of roots in Chapter 18, place radish seeds on moistened black paper in a petri dish to use as a demonstration of root hairs. For Investigation 18.2, start grass seeds 10 days and radish seeds 6 days ahead of time. In addition to checking the materials you will need for Investigation 19.1, practice the technique yourself. Order the aquatic organisms for the investigations in Chapter 23. To increase students' awareness, you might want to ask them to begin collecting articles about local environmental problems for Chapter 24.

Concepts

◆ Muscles contract in response to electrochemical impulses from nerves.

◆ The energy for muscle contraction is generated by a state of equilibrium between ATP, creatine phosphate, and glucose.

◆ Body movement is accomplished by the contraction of muscles that are attached to bones by tendons and that span joints where bones are attached to other bones by ligaments.

◆ Muscles usually exist in opposing pairs because they can only contract, thus generating movement in only one direction.

◆ Cardiovascular fitness is crucial to one's long-term health and can be maintained by regular, moderate exercise.

◆ Nerve impulses travel through a neuron by changes in the electrical charges that are set up by changes in the concentrations of ions; the impulse continues across the synapse by chemical transmitters.

◆ Neurons vary in function. Sensory neurons receive impulses by means of specialized nerve endings (receptors) and transmit them to the central nervous system. Motor neurons carry im pulses from the central nervous system to muscles and glands. Interneurons transmit impulses from one neuron to another.

◆ The central nervous system (brain and spinal cord) coordinates nerve, muscle, and endocrine functions; the brain itself has highly specialized areas for complex voluntary and involuntary actions, including learning.

◆ The autonomic nervous system coordinates unconscious organ function and maintains the body's homeostasis.

◆ Endocrine glands secrete chemicals into the bloodstream that have specific regulatory effects on selected tissues and organs; their release is controlled by an intricate feedback mechanism, frequently requiring the interaction of several different chemicals.

◆ There appears to be an interaction between psychological and physiological stress; because of the long-term effects of stress on the body, stress management is an emerging field of health care.

◆ The use of psychoactive drugs is highly controlled both legally and medically because their effects are profound; they can be physiologically destructive.

Objectives

Students should be able to:

◆ *distinguish* between the three types of muscle.

◆ *describe* the structure of a muscle at a microscopic level.

◆ *diagram* the food-glucose-glycogen-ATP pathway in connection with muscular exertion.

◆ *state* three functions of skeletal systems.

◆ *explain,* in terms of muscles and bones, simple movements of their own bodies.

◆ *evaluate* their own physical conditions and decide what dietary and exercise measures are needed for reaching and maintaining fitness.

◆ *demonstrate* the relationship between number of receptors and sensitivity to stimuli.

◆ *identify* the parts of a neuron and show the direction in which a nerve impulse passes.

◆ *explain* the transmission of nerve impulses along neurons.

◆ *diagram* the main sensory and motor pathways involved in a reflex arc.

◆ *explain* in general terms the functions of the major parts of the brain.

◆ *explain* the role of rods, cones, and the brain in vision.

◆ *describe* how the autonomic or the central nervous system controls a given physiological activity.

◆ *explain* in general terms how hormones affect their target organs.

◆ *analyze* the effects of some interruption of an endocrine feedback loop.

◆ *describe* a homeostatic mechanism involving either internal or external regulation in a human.

◆ *explain* in general terms the hypothalamus pituitary-endocrine interaction.

◆ *distinguish* between physiological stress and stressors.

◆ *describe* the physiological reactions to stressors.

◆ *name* at least three actions that can reduce stress.

◆ *list* the effects of various drugs on the nervous system and on the actions it controls.

◆ *make* informed decisions about drug use.

Strategies

This chapter concludes the chapters on the human animal by showing how the nervous and endocrine system coordinate the body's functions. Stressors and drugs are discussed in the context of their effects on the brain, behavior, and physiology.

Investigations 17.1 and 17.2 are easy to set up and carry out. Students may have done these or similar investigations in earlier courses; if so, the investigations can be upgraded by making them more quantitative and attempting to control for variables such as sex, size, musculature, or physical condition of individuals. Investigation 17.3 can be done only with the LEAP-System, but some discussion of a hypothetical bicycle trip can be substituted.

Have students read the introduction and Section 17.1 before doing Investigation 17.1. Remind them to respond to the interactive legend at the beginning of the chapter. Follow the investigation with reading and discussion of Sections 17.2 and 17.3. Investigation 17.2 introduces the nervous system and should precede discussion of Sections 17.4–17.6 and the Biology Today. Then, assign Sections 17.7–17.9, have students do Investigation 17.3 if possible, and conclude with Sections 17.10 and 17.11.

References

Allen, W. "Animals and Their Models Do Their Locomotion." *BioScience* (June 1995): 381–383.

Cantin, M., and J. Genest. "The Heart as an Endocrine Gland." *Scientific American* (February 1986): 76–82.

Fine, A. "Transplantation in the Central Nervous System." *Scientific American* (August 1986): 52–61.

Grillner, S. "Neural Networks for Vertebrate Locomotion." *Scientific American* (January 1996): 64–69.

Raichle, M.E. "Visualizing the Mind." *Scientific American* (April 1994): 58–64.

Searle, J. R. "Is a Brain's Mind a Computer Program?" *Scientific American* (January 1990): 26–31.

Weiss, J. "Unconscious Mental Functioning." *Scientific American* (March 1990): 103–09.

Wootton, R. J. "The Mechanical Design of Insect Wings." *Scientific American* (November 1990):

Suggested Answers to Applications

1. Food must be digested and delivered to the cells. Normal digestion time ranges from four to seven hours. However, the liver stores glucose in the form of glycogen and under hormonal control releases glucose for muscle activity. Food eaten four hours before may not be adequately digested and, thus, would not yield much energy.

2. Muscles on the left side of the neck contract to bring the left ear to the left shoulder and then relax while muscles on the right side of the neck contract to straighten the head. The reverse occurs when the right ear is moved to the right shoulder.

3. Many organisms move without muscles. Plants have movement (but not motility) due to differential growth of cells in different parts of the plant. Many single-celled organisms move with the aid of cilia or flagella, neither of which involves muscles.

4. During exercise, muscle cells use oxygen faster than it can be delivered by the blood. In an environment with low oxygen levels, lactic acid, or lactate, builds up, which causes pain. Have students relate this question to the chapters on anaerobic respiration and aerobic conditioning.

5. The thermostat is a type of a negative feedback system in which, once the desired temperature has been reached, the thermostat will shut down the air conditioner, or heater, until the temperature becomes too high, or too low. This feedback system is similar to most body feedback systems, such as glucose regulation and hormonal feedback systems.

6. Color blindness may result from an inherited defect of a light-sensitive pigment in the cone cells of the retina and/or from an abnormality in or reduced number of cone cells themselves.

7. When a person lifts weights, the muscle tissue hypertrophies—gets larger due to an increase in the size, rather than the number, of its constituent cells.

8. Stimulus, receptor, sensory neuron, synapse, interneuron, synapse, motor neuron, effector (muscle), movement.

9. All of these beverages contain caffeine, which is known to trigger muscle twitching and heart palpitations in some people. Caffeine promotes the release of epinephrine and norepinephrine, hormones that stimulate cell activity.

10. The pituitary secretes hormones that control reproductive cycles and development of secondary sex characteristics. A pituitary tumor could either advance or delay the onset of maturity.

11. Students should be able to write a description based on Figure 17.17; some might describe feedback relationships involving the ovaries.

12. Students sometimes forget that adults face many of the same problems as they do—worrying about their performance in their occupation, their social interactions, the outcomes of political or sporting events, and other matters that contribute to stress. Adults, for their part, sometimes forget that students have any problems at all.

13. It has a calming effect and releases pent-up emotions that can trigger the fight-or-flight response.

14. Hobbies divert attention from pressures that produce stress.

15. The individual might sense colors, sounds, and sights more intensely but experience loss of balance, acuity, or ability to perform.

Suggested Answers to Problems

1. Students may come up with a variety of materials to use, but you might suggest using dowel rods for bones, rubber bands for muscles, and nuts and bolts for joints.

2. Biological supply houses sell skeletons of many organisms.

3. The times will decrease for both running events, but they should begin to level off—there is a limit to the speed at which both races can be run. Discuss the physical limitations that might prevent faster race times.

4. Answers will depend on student selection.

5. There should be several different groups in your community that address dependence on alcohol and its effects on the individual as well as on his or her family.

6. You might encourage a competition between teachers and students to see which group can get the greater percentage of smokers to stop smoking.

7. At the Menninger Foundation in Topeka, Kansas, scientists observed that a woman's migraine headaches disappeared when blood flow to her hands increased. In experiments, migraine sufferers were taught to increase the blood flow to parts of the body other than the head. Eighty percent learned to control their headaches.

8. There are two possibilities. The most probable is that cocaine caused the brain to send out mixed and random electrical impulses. These signals going to the heart caused it to beat irregularly and brought about ventricular fibrillation, a purpose-less twitching, inducing cardiac arrest. A second possibility is that the upset of brain signals caused constriction of the coronary arteries, resulting in cardiac arrest.

9. Various kinds of hormones have been used to help animals grow faster and larger. Human growth hormone recently has been used to help small children become taller; check periodicals index for more information.

Materials and Preparations for Investigations

INVESTIGATION 17.1 Muscles and Muscle Fatigue

Materials (per class of 30, teams of 2)
15 watches with a second hand or one clock
15 8-inch × 10-inch sheets of thin cardboard
15 pairs of scissors
30 1-m pieces of string or yarn
15 brass brads
roll of masking tape
15 metric rulers

Preparations

The cardboard backing of 8-inch × 10-inch note pads works well for the model; manila folders also may be substituted. Yarn may be used instead of string for the muscles. Inexpensive hand exercisers (available at sporting goods stores) or old tennis balls can be used in lieu of making a fist. If your classroom does not have a clock with a second hand, stop watches or students' wrist watches can be used to keep time. If you have a LEAP-System, you may wish to substitute the investigation "Exercise and Muscle Fatigue" from the Teacher's Resource Book for Part B. Students can observe muscle fatigue graphically on the computer screen.

INVESTIGATION 17.2 Sensory Receptors and Response to Stimuli

Materials (per class of 30, teams of 2)
15 centimeter rulers
15 probes or blunt dissecting needles
15 each red and blue colored pencils
15 metersticks

Preparations

There should be as little distraction as possible during this investigation. Students must work quietly and concentrate totally on each task. If you do not have probes, blunt the tips of dissecting needles by rubbing them several times across a concrete surface. Be certain the test instruments are blunt.

INVESTIGATION 17.3 A Bike Trip

Materials (per class of 30, teams of 7)
4 Apple lle, Apple IIGSF Macintosh, or IBM computers
4 LEAP-Systems™ (1 or 2 teams per system)
8 dials
printer (optional)

Preparations

Review the LEAP-System Open Me First, Reference Manual, and BSCS Biology Lab Pac 2 to familiarize yourself with how to operate the system, allocate disk space for saving data, connect the probes, boot the system, and bring up the menu on the computer screen. Be sure students know how to put the system on WAIT, take the system off WAIT, REINITIALIZE the system, put the system on SAVE, and QUIT the system for your computers; and that space is available on the disk(s) for each class so data can be saved and stripcharts run.

For a more thorough analysis of various strategies, have students save their data and run stripcharts. The marks on the stripchart indicate when snacks/drinks were taken and can help students understand the relationship between glucose and fluid levels and timing of snacks/drinks.

Before printing stripsharts, check "System Maintenance" in the LEAP-System Main Menu to be sure the printer specified corresponds to the one you are using. See the reference manual for details on how to specify the printer.

The appearance of the screen on your computer may differ from the one illustrated in Figure 17.15. See the LEAP System BSCS Biology Lab Pac 2 for details.

Tables 17.1 and 17.2 are available as blackline masters in the Teacher's Resource Book; you may wish to give students several copies.

This simulation focuses on only a few of the many factors that influence performance during a day-long bike trip. The dial controls the gear in which the bike is running, assuming a constant rate of pedaling. Boxes on the screen display the time of day, the temperature, and the gear being used, and bars display the grade percentage, glucose demand, glucose level, fluid demand, and fluid level. A rider will "crash" if glucose drops below 60 mg/100 mL or fluid level drops below 29 L. Glucose and fluid demand can be decreased by changing to a lower gear, thus slowing the rate of travel. (Although liver and muscle glycogen serve as a reservoir from which glucose can be withdrawn, the level of glycogen reserves is not displayed on the screen.)

Temperature also affects the bike trip, which begins early in the morning with cool temperatures that rise during the day, increasing fluid loss from the body. Dehydration can be a problem, especially during exertion on a hot day. Heat cramps can occur when fluid loss reaches 5% of body weight; fluid loss of 5–10% can produce heat exhaustion, and fluid loss of more than 10% can result in heat stroke. The simulation assumes a body weight of 55 kg, of which 60%, or 33 kg, is fluid.

Rider 1 is an insulin-dependent diabetic. Students can control the success of the trip by carefully selecting the insulin dose for rider 1, the breakfast for each rider, the amount and type of food and fluid packed in the saddlebags, and the times at which they stop for snacks. Riders must stop to eat and/or drink to replenish their reserves before glucose drops below 100 mg/100 mL or fluid levels drop below 31 L. The rate at which the reserves are replenished depends on what is eaten or drunk. Table T17.1 provides information about glucose absorbance. The rate at which the reserves are depleted depends on the rate of travel (controlled by the gears), the work done (the grade), the insulin/glucose relationship, and the temperature.

In the class discussion, focus on the relationship between insulin, exercise, and diet, the effects of different insulin doses on glucose level and fluid demand, and other physiological effects occurring during the bike trip. Help the students con-struct if/then hypotheses about the best strategies to use. Some students may need help with phrasing hypotheses in this format. It is important that they construct hypotheses about the trip strategy rather than continue to use trial and error.

Emphasize that this is a trip for the riders to enjoy, not a bike race, and that cooperation among team members is critical to success. This activity should make it clear that, with proper attention to diet and insulin dosage, insulin-dependent diabetics can lead normal lives and participate in virtually any physical activity.

Hints for the Bike Trip (Wait until the students have tried several strategies before sharing these hints.)

1. Fill the saddle bags, including at least 1 L water per rider. Sports drinks are not good for replenishing body fluid because their high glucose content delays absorption.

Table T17.1 Glucose Absorbance by Percentage and Time

	Type	Total Grams	Grams Glucose	% glucose absorbance		
				30 min	60 min	60+ min
For future meal/ drink						
Peanut butter & jelly sandwich	S	120	35	14	57	29
Ham & cheese sandwich	S	090	30	00	33	67
Tuna fish sandwich	S	090	30	33	67	00
Pizza slice	S	120	30	00	33	67
Hot dog in bun	S	090	30	00	33	67
Water	F	250	00	00	00	00
Sports drink	F	250	11	100	00	00
Orange juice	F	250	30	100	00	00
Milk	F	240	12	42	57	00
Diet soda pop	F	360	01	100	00	00
Soda pop	F	360	35	100	00	00
Vegetable juice [V8]	F	180	07	71	29	00
For future snack						
Hard-boiled egg	S	060	01	00	00	100
Granola bar	S	030	15	07	20	73
Chocolate bar	S	045	25	08	24	68
Potato chips	S	030	15	00	47	53
Carrot sticks	S	030	05	00	100	00
Apple	S	120	30	83	17	00
Orange	S	120	25	80	20	00
Peanuts	S	030	01	00	00	100
M & M's	S	045	30	07	27	66
Canned pudding	S	120	30	33	67	00
Cheese	S	030	01	00	00	100
Breakfast						
Bacon & eggs	S	000	60	00	00	100
Jelly doughnut	S	000	155	32	35	33
Captain Krunch	F	000	37	54	46	00
Cold pizza	S	000	30	00	33	67
Bagel and OJ	F	000	59	49	25	26

S = Solid, F = Fluid

2. For the diabetic, low insulin is best. The answer to Discussion Question 4 explains the repercussions of other settings.

3. All riders should eat breakfast or they will start out in the hole on glucose. Jelly doughnut is the best breakfast as far as glucose is concerned, cold pizza the worst.

4. Once on the road, riders must maintain glucose and fluid levels. The riders will crash at a glucose level of 60 mg/ 100 mL or a fluid level of 29 L. The program assumes a body weight of 55 kg, so 29 L represents a fluid loss of 7% of body weight.

   This is only half the story, however. When the glucose level drops below 90 mg/100 mL, the riders start feeling woozy and their strength starts to wane. Once the fluid level drops below 30.25 L (fluid loss equal to 5% of body weight), they get heat cramps. Either or both of these conditions cause a loss of efficiency, which keeps the riders out on the road longer, using more glucose and fluid. This starts a vicious cycle that is the first step to disaster.

5. Levels exceeding 250 mg/100 mL of glucose or 33 L of body fluid will be lost because the body is unable to utilize higher amounts.

6. To finish the trip and get home in time for dinner, the riders must keep up their speed by using tenth gear on downhill and level grades and first gear on grade 6 to conserve energy (glucose) and avoid excessive perspiration (fluid loss).

7. The secret of what to do when is contained in two bars on the screen—grade and glucose demand. When glucose demand begins to exceed 0.5 mg/100 mL/sec, riders must gear down. On the other hand, when going downhill (negative grade), riders should use tenth gear.

   Finally, riders should not allow glucose level to drop below l20 mg/100 mL or fluid level below 31 L. Maintaining these levels will retain rider efficiency.

CHAPTER 18
THE FLOWERING PLANT: FORM AND FUNCTION

Planning Ahead
Begin to collect the common materials needed for Investigation 21.1 now. If you have none available, make arrangements to borrow them.

Concepts

◆ The basic organs of a vascular plant are leaves, stems, and roots.

◆ The shape and cellular structure of the leaf are designed to maximize the opportunities for photosynthesis; this is an example of complementarity of structure and function.

◆ Water loss from the leaf by transpiration is controlled by the guard cells; the size of the opening between them is influenced by the concentration of water in the cells.

◆ Some leaves have functions other than or in addition to photosynthesis, such as prevention of water loss, protection from herbivores and other animals, and capture of nitrogen sources.

◆ Stems have a variety of functions, such as support, transport of materials, photosynthesis, bud production, and storage of food. Each of the different specialty cells in a stem has a characteristic structure that supports its unique function.

◆ Xylem cells provide the major support structure for a vascular plant and transport water and nutrients to the leaves; phloem cells transport carbohydrates throughout the plant.

◆ Sugars move from the leaves to nonphotosynthetic parts of the plant by a combination of active transport, diffusion, and flow from areas of higher to lower pressure.

◆ Water moves upward in a vascular plant by a combination of transpiration from the leaf, the cohesion of water molecules, and the lower concentration of water in the roots than in the soil.

◆ Roots can provide support, water and nutrient absorption, food storage, and anchorage.

◆ The increased surface area provided by root hairs maximizes the uptake of water by a vascular plant.

◆ The most important plant nutrients, such as nitrogen, phosphorus, and potassium, must be absorbed by active transport because they usually are more concentrated inside the plant cells than in the soil.

◆ Plant growth begins in the embryo when a seed germinates, and it continues throughout the life of the plant by means of meristematic tissues that allow for increase in length (primary growth) and in diameter (secondary growth).

Objectives

Students should be able to:

◆ *describe* briefly the functions of epidermis, mesophyll, veins, and stomates in a leaf.

◆ *recognize* substances normally taken in and substances normally lost through stomates.

◆ *demonstrate* the relationship between turgor and rigidity in plant tissues.

◆ *describe* the action of guard cells.

◆ *explain* transpiration in terrestrial plants.

◆ *name* plant groups that might be found in such habitats as marshes, deserts, or seawater.

◆ *identify* three principal functions of stems.

◆ *describe* briefly the functions of pith, cambium, xylem, and phloem in a stem.

◆ *explain* how sugars and water are transported in a plant.

◆ *name* three principal functions of normal roots.

◆ *describe* the functions of the root cap, root hairs, and the cortex of a root.

◆ *identify* which types of soil substances enter the root by diffusion and which by active transport.

◆ *describe* the formation of a young seedling from a seed.

◆ *demonstrate* the measurement of root growth.

◆ *point* out the location of the principal meristems on a plant or plant diagram.

◆ *distinguish* between primary and secondary growth.

Strategies

As you begin this chapter, display plants that exhibit special adaptations. Some suggestions: *Aloe* or *Agave* (with thick fleshy leaves that resist desiccation), *Kalanchoe* (plantlets on leaf margins), *Monstera* (perorated leaves), *Maranta* (leaves folding at night), *Mimosa pudica* (leaflets folding when touched), *Saxifraga sarmentosa* (plantlets on hanging runners), and of course, various cacti. All of these are houseplants.

A slide presentation of unusual plants is a good idea. The actual plants are better than pictures, but you can introduce a wider variety with slides. A session or two devoted to observing the microstructure of plant organs may be desirable, although the physiology of plants will be more important to students than morphological detail.

Have students read the chapter introduction and Section 18.1 just before Investigation 18.1, which takes most of three class periods. Intersperse your discussion of plant transport with reading of Sections 18.2, 18.6 and the Biology Today. Investigation 18.2 requires a full period and discussion. It may be followed by reading Sections 18.7 and 18.8. It is not essential to carry out all investigations in this chapter—18.3 and 18.4 require about four class periods and are nice activities but not as critical as the first two. You might want to select one of them and build your discussion of plant growth (Sections 18.9–18.11) around that experience.

References

Abelson, P. H. "Plant-Fungal Symbiosis." *Science* (16 August 1985): 617–18.

Gillis, A.M. "Using a Mousy, Little Flower to understand the Flamboyant Ones." *BioScience* (May 1995): 309–313

Jones, A.M. "Surprising Signals in Plant Cells." *Science* (14 January 1994): 183–184.

Mitton, J. and M. Grant. "Genetic Variation and the Natural History of Quaking Aspen." *BioScience* (January 1996): 25–31

Newhouse, J. R. "Chestnut Blight." *Scientific American* (July 1990): 106–11.

Rosenthal, G. A. "The Chemical Defenses of Higher Plants." *Scientific American* (January 1986): 94–99.

Schmipf, D. J., and S. D. Flint. "Light and Germination: Alternatives in Lamp, Filter, and Seeds." *American Biology Teacher* (October 1985): 410–11.

Young, J. A. "Tumbleweed." *Scientific American* (March 1991): 82–87.

Streit, W. and C. Kincaid-Colton. "The Brain's Immune System." *Scientific American* (November 1995): 54–61.

Suggested Answers to Applications

1. Leaves of water lilies float because of their highly developed spongy tissue. Their stems and long petioles are fragile and soft, lacking the abundance of fibrous tissue characteristic of terrestrial tracheophytes. This fragility correlates with the fact that the weight of the plant is largely supported by water. Since water is directly available to all plant parts, the roots are not as extended or possessed of as many root hairs as are the roots of land plants.

2. The roots' environment—the soil—is dense, often moist, and little light penetrates it. Support for the plant is firm, and water usually is absorbed rather than lost. The shoot is surrounded by air, which offers almost no support against gravity. Wind,

rain, sleet, or snow may batter it; the shoot must withstand all types of weather. It loses water to the atmosphere continually. The shoot has an abundant oxygen supply and a steady supply of carbon dioxide, and it may receive varying intensities of light.

3. Carbon dioxide and oxygen might be given off or taken in; water comes to the leaf from other parts of the plant by way of veins. Water diffuses from the mesophyll cells into the air, and the plant loses water. Respiration and transpiration are the forces involved.

4. Cultivation breaks soil particles apart, increasing the size of the air spaces and reducing the upward capillary movement of water. At the same time, it allows air and water to penetrate into the soil. However, in regions with well-spaced rains during the growing season, cultivation is practiced largely to reduce weeds. With the use of chemical weed killers, even this may be unnecessary.

5. Botanists distinguish between a root and a stem by the arrangement of their tissues and by how they originate in embryos within seeds. The easiest way to make the distinction is to look for buds. Buds are found on stems, but not on roots. A leaf is composed of different tissues, each of which performs some function in the life of the plant. The leaf shape usually is distinctive and constant enough to be useful in identification.

6. Roots hairs are thin-walled extensions of epidermal cells and greatly increase the surface area of the root through which water can be absorbed.

7. The base of a young tree branch is enclosed by the growing wood and becomes a knot. You can see annual growth rings in many knots.

8. During the night, plants are undergoing cellular respiration (not photosynthesis), just as the patient is. Thus, both organisms are using oxygen, but the plant is using it at a much lower, and not harmful, rate.

9. Grasses have meristems at the base of leaves, so that clipping them with a mower will not take away the growing point. This is also why grasses can regrow after being eaten by grazing animals.

10. Although the many root hairs may help to hold a plant in the ground, they are easily stripped off. Their enormous surface area provides a major avenue through which water and nutrients can be absorbed.

11. No. Active transport requires energy, and wood cells are nonliving.

12. Water diffuses into the vegetables' cells, causing them to have greater turgor pressure, which makes them crisp. Biting into the vegetables releases the turgor pressure with a "snap."

13. Both contain a young embryo that feeds on the stored material inside the container protecting the developing young offspring.

14. No. Because root hairs extend perpendicularly from the root, they would be sheared off as the root elongated.

15. No. A cuticle is a waxy, water-impervious coating that would inhibit the absorption of water—one of the prime functions of a root system.

Suggested Answers to Problems

1. Mammals, birds, and insects have definite limits to growth. Many other animals and all multicellular plants continue to grow throughout their lives— usually at a decreasing rate. Most kinds of animal cells retain the ability to duplicate even after they differentiate. In vascular plants, however, cells that have differentiated do not duplicate.

2. The attached end of the fence is the same height above the ground as it was 10 years ago. The tree grew in height, but only the apical meristems moved upward; all tissues to which the fence was attached remained where they were when formed.

3. Comparatively large amounts of water are lost through the apple leaves, but this occurs only during the season when living leaves are on the tree. Comparatively small amounts of water are lost through the pine needles, but to some extent this occurs throughout the year.

4. a. A wet year causes an annual ring to be wide; a dry year results in a much narrower ring. b. Drought or defoliation by insects may temporarily slow the growth, producing a band of small, thick-walled cells followed by a band of large, thin-walled cells produced during late-summer rains or after appearance of a new crop of leaves. This pattern produces "false" growth rings, suggesting two growing seasons. c. Especially in arid and semiarid climates, past climatic patterns are reflected in the patterns of growth rings in trees. By comparative study of the sequence of these rings in trees and aged wood, dates of environmental events in former periods can be determined. d. Indistinct rings are produced in the bark by cork cambium and by

growth of phloem from the principal cambium. But these rings are compressed by the outward growth of the wood. Eventually the outermost layers of dead cells, unable to increase in circumference, break, forming a scaly, ridged, or roughened bark.

5. The three types of cuts—radial, tangential, and cross-sectional—reveal completely different grains.

6. Leaves are a very good indicator of the natural environment—vegetation with relatively smaller leaves tends to grow in sunnier and drier habitats than does vegetation with relatively large leaves.

7. The color of tissues as seen in prepared slides is, of course, due to stains; thus students should feel free to color these tissues as they wish. Coloring plant tissues on duplicated enlargements also is a good way for your students to identify which tissue is which.

8. The key here is to use elastic material for the guard cells and to show differential thickening on the two sides of each guard cell.

9. Boxes can be made with plexiglass and wood. An alternative is to buy large plastic drinking glasses with slanting sides. Fill the glasses with dirt and place a germinated seed along the edge.

10. Narrow rings could mean a lack of rain or an insect infestation, among other possibilities. Make certain each core is plugged up so the sampled tree is not damaged. (Petroleum jelly works in most cases.)

Materials and Preparations for Investigations

INVESTIGATION 18.1 Water and Turgor Pressure

Materials
30 pairs of safety goggles
30 lab aprons
10 50-mL graduated cylinders
30 25-mm × 200-mm test tubes
10 15-cm metric rulers
10 balances
10 glass-marking pencils
10 cork borers (5- to 10-mm diameter)
10 scalpels
10 dissecting needles with corks for tips
10 test-tube racks
roll of aluminum foil or plastic wrap
paper towels
500 mL 10% sucrose solution
500 mL 20% sucrose solution
500 mL distilled water
10 white potatoes

Preparations
Students may bring their own potatoes, but better comparisons can be made if the potatoes are all of the same type. Each potato should be large enough to provide three cores, each 30 40 mm long and at least 6 mm in diameter. The three cores for each team should come from the same potato. Directions for preparing a sucrose solution are in A Table of Recipes in the Teacher's Resource Book.

The balances should measure mass accurately to 0.1 g, or better yet to 0.01 g. If the number of balances is limited, some teams can carry out the mass measuring procedure while other teams make the remaining measurements.

To prevent evaporation, use small pieces of foil or plastic wrap to cover the test tubes, or substitute stoppers. If the room is warm, store the test tubes with the potato cores in the refrigerator to prevent spoilage.

The Teacher's Resource Book includes a blackline master of Table 18.1.

INVESTIGATION 18.2 Leaves, Stems, and Roots: Structural Adaptations

Materials (per class of 30, teams of 2)
15 microscope slides
15 cover slips
15 dropping pipets
15 compound microscopes
15 stereomicroscopes
15 forceps
15 sets of colored pencils (red, blue, brown, green)
15 prepared cross sections of a leaf
15 prepared cross sections of an herbaceous dicot stem
15 prepared cross sections of a root
variety of mature plants
15 young bean plants radish seedlings germinating
    in a petri dish
grass seedlings germinating on water
bunch of carrots

Preparations
Sprinkle grass seeds over the top of water in a finger bowl about 10 days before you plan to do this investigation. About 6 days ahead of time, place 2 or 3 radish seeds on a piece of moistened black paper in a covered petri dish; prepare one petri dish for each lab table.

INVESTIGATION 18.3 Seeds and Seedlings

Materials (per class of 30, teams of 2)
30 pairs of safety goggles
30 lab aprons
15 10X hand lenses
5 scalpels
20 mL plain agar
620 mL starch agar

15 dropping bottles Lugol's iodine solution (100 mL total) labeled **WARNING: Poison if ingested, Irritant**

105 soaked corn grains

15 soaked bean seeds

15 each bean seeds germinated 1, 2, 3, and 10 days

1 demonstration petri dish with starch agar—dish A

1 demonstration petri dish with plain agar—dish B

15 petri dishes with starch agar and 3 each germinating corn grains—dish C

15 petri dishes with starch agar and 3 each boiled corn grains—dish D

Preparations

Start soaking the bean seeds one day before they are dissected in class. For Part B, plant bean seeds on a schedule so that 1-, 2-, 3-, and 10-day-old seedlings will be available at the same time for student observation. Soaking the seeds in water 24 hours before planting will help ensure seedling growth.

Prepare materials for Part C three days before the scheduled lab. Bottles of sterile starch agar are available from Ward's Natural Science Establishment (catalog # 88M8905), or order prepoured plates (catalog # 88M0927). Directions for preparing plain agar, starch agar, and Lugol's iodine solution are in A Table of Recipes. Pour a thin layer of sterile medium into the petri dishes. Test dish A (plain agar) and dish B (starch agar) with Lugol's iodine solution and set up as a demonstration.

At least two days before they are to be used, cut germinating corn grains lengthwise and parallel to the flat surfaces. Place the cut surfaces on the agar. To kill germinating corn seeds, boil them 15 to 30 minutes. Cut these grains as you did the germinating grains and put them on starch agar controls.

INVESTIGATION 18.4 Root Growth

Materials (per class of 30, teams of 3)

30 petri dishes

60 pieces of paper towel to fit petri dishes

10 glass-marking pencils

waterproof ink

10 scalpels

80 cardboard tags

10 15-cm metric rulers

30 flat toothpicks

distilled water

120 germinated corn seeds

Preparations

Seeds can be obtained easily at your local garden shop. Radish seeds may be substituted for corn. Begin germination three days before students need the seeds. To germinate the seeds, soak them for 6 to 12 hours in distilled water in a large beaker. Make sure the water covers the seeds during this time. Place paper towels on the bottom of one or more large pans or dishes. Scatter the seeds on the paper, cover with more towels, and dampen. Cover each dish with aluminum foil or plastic wrap so the paper will not dry out rapidly. (Light

is not necessary for germination.) Have students check their seedlings daily to make sure they are not drying out.

Some students may wish to plant the seedlings instead of discarding them. If materials are available, allow students to do this and to observe the growth patterns of the shoot and leaves of the plant.

CHAPTER 19

THE FLOWERING PLANT: MAINTENANCE AND COORDINATION

Planning Ahead

If you plan to have a local ecologist visit the class during the study of Chapter 22, call or write now to make the arrangements. If possible, order an aerial photo of your local area from the USGS, NASS, or other map source. This will add interest to Investigation 24.1, though it is not necessary for carrying out the activity.

Concepts

◆ Photosynthesis, the conversion of solar energy to chemical energy in carbohydrates, is one of the most significant of all biological processes to almost all other organisms in the biosphere.

◆ The shape of the plant leaf and the layered construction of leaf tissues and chloroplasts are adaptations for efficient photosynthesis.

◆ During the light reactions of photosynthesis, solar energy is absorbed by electrons in chlorophyll and other pigments, water is separated into oxygen, protons, and electrons, and solar energy is converted to chemical energy as ATP and NADPH are formed.

◆ During the Calvin cycle of photosynthesis, ATP and NADPH are used to combine hydrogen atoms with carbon dioxide, forming 3-carbon sugars. These sugars may be used to regenerate the molecules of the Calvin cycle, to enter the pathway of glycolysis and cellular respiration, or to form sucrose, starch, amino acids, lipids, or other compounds needed in the plant cell.

◆ Four environmental factors that significantly affect the rate of photosynthesis are light intensity, temperature, and the concentrations of carbon dioxide and oxygen. Their effects are combined and interactive.

◆ Photorespiration is the light-dependent uptake of oxygen and release of carbon dioxide that prevents incorporation of carbon dioxide in the Calvin cycle and results in the loss of previously incorporated carbon dioxide.

◆ C-4 photosynthesis is an adaptation to hot, dry conditions that overcomes the effects of photorespiration. Specialized plant anatomy and enzyme systems are involved.

◆ Plants produce hormones that interact with each other and with the environment to regulate growth and development.

◆ Tropisms are plant responses to environmental stimuli such as light, gravity, and touch, or to periodic environmental cues such as day/night and seasons.

Objectives

Students should be able to:

◆ *discuss* the role of chloroplasts in the photosynthetic process.

◆ *distinguish* between an absorption spectrum and an action spectrum.

◆ *describe* the major events of photosynthesis.

◆ *explain* the relationship between gas exchange and photosynthesis.

◆ *explain* the role of chlorophyll in the light reactions.

◆ *list* the products of the light reactions and the Calvin cycle and describe how they are used.

◆ *relate* the rate of photosynthesis to environmental effects, using experimental data.

◆ *describe* adaptations that reduce the effects of photorespiration.

◆ *describe* the effects of various plant hormones on plant growth.

◆ *analyze* a nursery's problem with plant growth and prescribe appropriate measures.

◆ *relate* a tropism to photoperiodism and plant growth.

◆ *describe* a method for bringing a plant into bloom for a certain holiday, using photoperiodism.

Strategies

The focus of this chapter is photosynthesis—the production of stored chemical energy in food. Students can see experimentally how environmental factors affect the rate of photosynthesis and can learn about adaptations that reduce some of those environmental effects. Plant hormones, interacting with each other and with the environment, regulate plant growth, and plant tropisms are responses to the environment.

Investigation 19.1 can be begun immediately. Because it takes three class periods including the closure discussion, assign reading and study of Sections 19.1–19.5 in small chunks while the students work in class. Investigation 19.2 uses the LEAP-System (or, alternatively, narrow-range pH paper) and also takes three periods. As before, assign Sections 19.6–19.9 while students are working on the investigation. The first two investigations are critical to development of the concepts of plant maintenance and coordination because they provide concrete experiences to which reading and discussion can be attached. Investigation 19.3 is less critical; it requires one period for setup, about five minutes on five consecutive days for data collection, and one-half period for a closure discussion. If you do Investigation 19.3, assign Sections 19.11–19.12 after the lab discussion; if not, assign these two sections during Investigation 19.2.

References

Bazzaz, F. A., and E. D. Fajer. "Plant Life in a CO_2-Rich World." *Scientific American* (January 1992): 68–74.

Eisen, Y., and R. Stavy. "Students' Understanding of Photosynthesis." *American Biology Teacher* (April 1988): 208–12.

Evans, L. M., R. Moore, and K. H. Hasenstein. "How Roots Respond to Gravity." *Scientific American* (December 1986): 112–19.

Frumhoff, P. "Conserving Wildlife in Tropical Forests Managed for Timber." *BioScience* (July/August 1995): 456–464.

Govindjee, and W. J. Coleman. "How Plants Make Oxygen." *Scientific American* (February 1990): 50–59.

Jones, J. B. "Growing Plants Hydroponically." *American Biology Teacher* (September 1985): 356–58.

Martin, F. L. "Radioactive CO_2 Fixation in Geranium Leaves." *American Biology Teacher* (October 1987): 433–35.

Moses, P. B., and N. H. Chua. "Light Switches for Plant Genes." *Scientific American* (April 1988): 88–93.

Steucek, G. L., and R.J. Hill. "Photosynthesis I: An Assay Utilizing Leaf Disks." *American Biology Teacher* (February 1985): 96–98.

Weinberg, C. J., and R. H. Williams. "Energy from the Sun." *Scientific American* (September 1990): 146–55.

Wolinsky, C. "Australian Wildflowers." *National Geographic* (January 1995): 68–89.

Youvan, M. "Molecular Mechanism of Photosynthesis." *Scientific American* (June 1987): 42–49.

Suggested Answers to Applications

1. Photosynthesis is an energy-storing reaction; respiration is an energy-releasing reaction. The products of photosynthesis are the raw materials for respiration, and vice versa. Figure 19.9 compares these processes.

2. In both cases, ATP is formed primarily by means of an electron transport system. In respiration, ATP also is formed as glucose is degraded to carbon dioxide in glycolysis and the Krebs cycle.

3. Photosynthesis uses carbon dioxide from the atmosphere and stores the carbon in the plant structure. Respiration releases carbon dioxide into the atmosphere, replenishing the supply.

4. The concentrations of auxins and cytokinins are adjusted to stimulate cell division and differentiation. (See the figure in Biology Today.)

5. Ripening fruit produces ethylene. Ethylene promotes the ripening of fruit. By packaging green fruit in this manner, any ethylene that is produced will remain near the fruit and help to ripen it faster.

6. None. Food is a combination of energy and nutrients. A plant may get its nutrients from the soil, but it makes its own food by putting together the nutrients from the soil and air with the energy from the sun.

7. Plants convert light energy into chemical energy during the process of photosynthesis, and some of this chemical energy is lost as heat energy during cellular respiration. Also, when an organism moves, chemical energy is converted into kinetic energy.

8. Chlorophyll is a necessary part of photosynthesis, but it is neither a reactant nor a product of the process, and it is changed only slightly and temporarily during the reactions.

9. Sunlight strikes a chlorophyll molecule, which absorbs some of the energy. This energy is used to help make one of two high-energy molecules—ATP or NADPH—which pass their energy on to the sugar molecule formed in the Calvin cycle.

10. The percentage of the total number of grass species having the C–4 pathway is highest in those regions with the highest temperatures during the growing season—in the southern United States.

11. Diminishing supplies of auxins, cytokinins, and gibberellins in the mature leaf lead to development of an abscission zone at the base of the petiole. Ethylene is produced in the abscission zone and promotes leaf drop. Environmental stimuli might include falling temperatures, decreasing day length, and water shortage.

12. Change in day length occurs at a regular pace. Changes in temperature are not predictable and may fluctuate wildly from day to day. Thus, if plants relied on changes in temperature they would be "fooled" by cool days in summer or a warm trend in winter.

13. Short waves such as gamma and X rays are too energetic and could damage the chlorophyll molecules. Longer wavelengths such as infrared radiation do not have enough energy to trigger the initial reactions in the chlorophyll molecules.

14. With increasing temperature, the rate of cellular respiration continues to rise past the optimum point for photosynthesis. Thus, at high temperatures, such as those in Texas, the starch is used up in respiration instead of being stored in the potatoes.

Suggested Answers to Problems

1. IAA and NAA are used to stimulate the seeding of fruit. NAA also is used to prevent fruit from falling prematurely, and to accelerate production of flowers and fruits. Gibberellins stimulate stem elongation, and 2,4-D kills unwanted plants.

2. This is a rapidly expanding field. Sources for information include, but are not limited to, botany and organic chemistry departments of universities, the U.S. Department of Labor, the American Institute of Biological Sciences, and research divisions of oil companies.

3. More rapid photosynthesis could indicate either a richer supply of CO_2 or a homeostatic adjustment to supply the plants' needs for O_2 for cell respiration. This relationship would not affect the composition of the earth's atmosphere, to which the plants are well adapted.

4. Plants are much more dependent on their environment than animals are because plants cannot move away from their environment.

5. Science fiction is a good way to interest some students in the study of biology. It is also a way to get students to consider biological concepts while they are reading books other than their text.

6. Light filters may be made out of colored cellophane; however, this material is "leaky" and may

let through a lot of different colors of light in addition to the desired color. Make certain that intensity is controlled. One design is to place pots in boxes with the tops removed and the opening covered with the filter.

7. There is growing debate about the aftereffects of this defoliant on the health of people exposed to it. There is much popular literature expressing evidence on both sides of the debate. This debate could be developed in class, or a veteran who is familiar with Agent Orange could speak to your class.

8. Using a root promoter, students could measure the rate of root development in cuttings taken from a single plant placed in different jars of water with different concentrations of the root promoter.

9. Yes.

10. Such charts are commercially available. When students complain about the amount of chemistry they have to learn, refer them to such biochemical charts outlining complete series of reactions and their chemical constituents.

Materials and Preparations for Investigations

INVESTIGATION 19.1 Gas Exchange and Photosynthesis

Materials
Part A (per class of 30, teams of 2)
15 microscope slides
15 cover slips
15 dropping pipets
15 compound microscopes
15 forceps
15 scalpels
fresh leaves, several types

Part B (per class of 30, teams of 5)
30 pairs of safety goggles
30 lab aprons
6 400-mL beakers
24 petri dish halves
6 forceps
6 pairs of scissors
6 sheets of white paper
paper towels
18 swab applicators
2-4 self-closing plastic bags labeled waste
6 500-mL beakers of water
petroleum jelly
6 screw-cap jars 99% isopropyl alcohol (360 mL total), labeled ***WARNING: Flammable liquid***

6 screw-cap jars Lugol's iodine solution (360 mL total), labeled ***WARNING: Poison if ingested; Irritant***
6 small corked bottles Histoclear (60 mL total), labeled ***WARNING: Combustible liquid***
9 potted plants

Preparations
For Part A, keep the leaves in water or plastic bags to prevent wilting. Epidermis is easy to remove from swedish ivy, geranium, and Zebrina. Part B requires small potted plants with abundant, small leaves. Begonia is excellent. Depending on the plants selected, two to four light-treated plants should be sufficient for the entire class, but each team will need one dark-treated plant. If the plants are placed in the light and dark on a Friday, Part A can be done on that day and Part B completed during the following week. Remind students to water their plants. Instead of cutting notches, students can use paper punches to mark their leaves. Double tipped swabs, available at supermarkets, make good applicators for the Histoclear. See A Table of Recipes for preparation of Lugol's iodine solution.

INVESTIGATION 19.2 Photosynthetic Rate

Materials (per class of 30, teams of 4)
8 2-L beakers
8 100-mL beakers or small jars
8 250-mL flasks
16 25-mm × 200-mm test tubes
16 1-mL pipets (optional)
16 wrapped drinking straws
8 lamps with 100-watt spotlight
4 Apple lle, Apple IIGSF Macintosh, or IBM computers
4 LEAP-Systems™ (teams per 8-input LEAP-System)
printer
16 pH probes
8 thermistors
8 rolls narrow-range pH paper (optional)
8 nonmercury thermometers (optional)
1 roll each red, blue, and green cellophane
2 L distilled water
8 L tap water at 25° C
2 L ice water (10° C)
16 15-cm sprigs of young elodea

Preparations
Young, healthy elodea is essential to gathering useful data. Maintain the elodea under a grow light so it is actively photosynthesizing when students begin the experiment. Use only sprigs with growing tips. Water temperature should be at least 25° C. A 2° C to 4° C increase in water temperature during the course of the experiment is not significant, but students should maintain the temperature as closely as possible. The fluorescent bulbs now available for lamps give satisfactory results and do not produce heat. It is helpful to fashion a shield from aluminum foil to direct the light.

Results are better with distilled water in the test tubes. If using pH paper, test the pH of your distilled water to determine which narrow-range pH paper to purchase. (Also test the decrease in pH when you blow into the water for 2 minutes.) Ward's Natural Science Establishment carries narrow range pH paper with internal comparison that shows a 0.2 or 0.3 change in pH. Students can use 1-mL pipets (with a finger) to transfer a drop from each test tube to the pH paper, or the pH paper can be folded and dipped into the test tube using forceps.

Review the LEAP-System Open Me First, Reference Manual, and BSCS Biology Lab Pac 2 to become familiar with how to operate the system, allocate disk space for saving data, connect the probes, boot the system, and bring up the menu on the computer screen. Be sure students know how to put the system on WAIT, take the system off WAIT, REINITIALIZE the system, put the system on SAVE, and QUIT the system for your computers; and that space is available on the disk(s) for each class so data can be saved and stripcharts run. Check the calibration of the pH probes before use; directions are included in the LEAP-System BSCS Biology Lab Pac 2.

The standard LEAP-System includes specifications and probes sufficient to run each experiment for one team at a time. The system also can handle multiple teams concurrently. See the reference manual for detailed instructions about how to create new experiments or update existing ones. Using these procedures, you can generate your own multi-team experiments. Disks with completed multi-team experiments also are available for purchase. Additional probes may be required in some cases.

When more than one team is using one LEAP-System, it may be difficult to differentiate all the lines on the graph. Under these circumstances, instruct students to read the numeric data displayed near the graph. After they have completed the experiment, students can stripchart and list their individual team data. See the reference manual for details on tripcharting and listing data by lab team.

Before printing stripcharts, check "System Maintenance" in the LEAP-System Main Menu to be sure the printer specified corresponds to the one you are using. See the reference manual for details on how to specify the printer.

If you have time cards in your computers, students can collect data over longer periods. If any students want to try the investigation with other aquatic plants, encourage them to do so.

The Teacher's Resource Book includes a blackline master of Table 19.1.

INVESTIGATION 19.3 Tropisms

Materials

Part A (per class of 30, teams of 4)
4 petri dishes
4 pairs of scissors
4 glass-marking pencils
1 pkg of nonabsorbent cotton
heavy blotting paper
4 rolls of transparent tape
2 oz of modeling clay
16 soaked corn grains

Part B
16 flowerpots, about 8 cm in diameter
16 cardboard boxes, at least 5 cm higher than the flowerpots
1 roll each red, blue, and clear cellophane
4 pairs of scissors
4 rolls of transparent tape
160 radish seeds
1 25-lb bag of potting soil

Preparations

Soak corn grains for two days prior to beginning the investigation. Use the largest petri dishes possible. Use nonabsorbent cotton to pack the grains. Paper towels or 9-cm filter paper can be used instead of blotting paper, but desk blotter paper forms a better support for seeds and cotton. (Use only white blotters. Colored blotters may contain dyes that are harmful to corn grains.) One standard desk blotter (19 inch × 24 inch) is sufficient for four classes and can be purchased from an office supply store.

Various substitutes can be used for flowerpots. Milk cartons with drainage holes poked in the bottoms work well. If space is a problem, four pots can share one box that has a suitably large window. Plain sand or fine vermiculite can be used unless you plan to grow the seedlings for some other purpose. Avoid clay. Because radish seeds germinate quickly, no more than a week is needed to obtain results. The radish seedlings need to grow for only two or three days after they break through the soil.

CHAPTER 20

BEHAVIOR, SELECTION, AND SURVIVAL

Planning Ahead

Survey your pictorial and projection materials for Chapter 22, which requires student experiences with various biomes. Except for the biome in which you live students can gain that experience only through still and motion pictures. Check your school and local libraries for additional materials. Other schools may be willing to share their resources.

Concepts

◆ Behavior is the sum of all the activities of an organism in response to its environment.

◆ Innate behavior is genetically driven and does not require any prior experiences.

◆ Learned behavior requires actual experiences and often includes a stimulus to elicit a specific behavior.

◆ Animals with developed nervous systems communicate with members of the same species in the interest of food-getting, reproduction, and defense.

- Communication may be visual, auditory, or chemical

- Most forms of behavior result in adaptive advantages for improved survival of the organism and of its species.

- Compared with random mating, sexual selection tends to promote reproductive success and results in a superior genetic mix in the next generation and in improved care for the young.

- Organisms have many mechanisms for physiologically adjusting to changes in their environment.

- The ability of an organism to tolerate varying conditions in the environment will determine the geographical distribution of that species.

- The major disruptive factor to all natural ecosystems is the expanding growth of Homo sapiens

Objectives
Students should be able to:

- *explain* the relationship between a stimulus and conditioning.

- *distinguish* between innate and learned behaviors and give examples of each.

- *explain* how innate behaviors can be modified by experience.

- *explain* how behavior is adaptive.

- *list* the different types of communication and give examples.

- *describe* the function of communication in a society.

- *explain* the functions of rituals and aggressive interaction.

- *describe* the function of territories and explain how they are maintained.

- *define* the concept of reproductive success in a population.

- *discuss* in general terms the effects of sexual selection.

- *discuss* the value of cooperative behavior.

- *demonstrate* the ability to design an effective field study of animal behavior.

- *describe* some behavioral and physiological responses of animals to environmental change.

- *describe* maximum and minimum tolerances of a number of familiar organisms to a variety of abiotic environmental factors.

- *explain* ways in which environmental factors interact to set limits to geographic ranges.

- *construct* hypotheses by which tolerances might be tested.

- *describe* ways in which human activities affect the behavior and survival of other organisms.

Strategies
This chapter integrates concepts in behavior, selection, and survival. As students are reading Sections 20.8–20.9, you may wish to have an ecologist from a local college or university visit the classroom to demonstrate some sampling techniques (such as radiotelemetry), or you could show some slides of ecologists at work in the field.

Begin with Investigation 20.1, which takes only one class period and requires only a bell and the student text. Reading and Concept Review questions for Sections 20.1–20.6 follow naturally. Investigation 20.2 is a classic ecological study that students enjoy and that develops process skills, especially observation and inference. Most of the work is done on the student's own time, but you should approve their plans in class before they begin their observations and allow several days of "homework" to complete this important investigation. Investigation 20.3 is a good experimental activity that requires collecting data for 10 days. It should be followed by Sections 20.8–20.9.

References
Handel, S. N., and A. J. Beattie. "Seed Dispersal by Ants." *Scientific American* (August 1990): 76–83.
Kirchner, W. H., and W. F. Towne. "The Sensory Basis of the Honeybee's Dance Language." *Scientific American* (June 1994): 74–80
Leggett, W. C., and K. T. Frank. "The Spawning of the Capelin." *Scientific American* (May 1990): 102–07.
Narins, P. "Frog Communication." *Scientific American* (August 1995): 78–83
Sapolsky, R. M. "Stress in the Wild." *Scientific American* (January 1990): 116–23.
Seachrist, L. "Sea turtles Master Migration with Magnetic Memories." *Science* (29 April 1994): 661–662.
Suga, N. "Biosonar and Neural Computation in Bats." *Scientific American* (June 1990): 60–71.
Wilkinson, G. S. "Food Sharing in Vampire Bats." *Scientific American* (February 1990): 76–83.

Suggested Answers to Applications
1. Predators usually attack those individuals that show some sign of weakness. In the long run, the prey

population actually may benefit by removal of the genes of weak individuals from the gene pool.

2. It makes little evolutionary sense for an individual to reduce its own reproductive success through cooperative behavior. What appears to be "unselfish" behavior, however, actually may benefit an individual and enhance its ability to survive. Helping closely related individuals also may help ensure that some common genes are passed on.

3. Answers will vary. Some scientists call tropisms "behaviors"; others do not. Students may say that tropisms are similar to innate behaviors, but that learned behaviors would not occur.

4. Humans can avoid the environment by erecting buildings that shelter them from cold, rain, wind, and heat. We create an environment that is constant and suited to our needs. Only when we step outside our artificial environment are we confronted with the "real" world.

5. Sometimes the learning experience occurs at such an early age, or the indications are so subtle, that learned behaviors may be mistaken for innate behaviors.

6. Student answers will vary. A commercial hatchery may have some information about times of imprinting.

7. Experimental design will vary but should include some mention of control groups and isolated individuals reintroduced to a related group.

8. In the absence of any predators, the deer population would increase and might exceed the carrying capacity of the land. Deer would overgraze the land and then might move into the developed area and destroy ornamental and fruit trees and shrubs.

9. In both cases, the body goes into a state of dormancy where the rate of cellular respiration is greatly reduced. Among many differences are the length of time spent in each state and the underlying reason for the behavior.

Suggested Answers to Problems

1. Public libraries and bookstores have many books on languages and other communication systems.

2. Pheromones are useful and economically important in the control of the gypsy moth, for example.

3. Your local library may circulate foreign language records. Also, an anthropologist, linguist, or faculty member of a language department at your local university or college may be able to help you

locate audio tapes. Languages convey thoughts and feelings in organized combinations and patterns, called syntax.

4. *Investigating the Human Environment: Land Use* is available from Kendall/Hunt Publishing Company, Dubuque, IA, and would be appropriate as a supplement to several sections of the text.

5. Koko the gorilla and Washoe the chimpanzee are two famous subjects of these efforts. These studies, while extremely interesting, have not been accepted by the entire scientific community. Have your students discuss the controversy.

6. The study of how humans perceive and use space is called proxemics. Proxemics deals with the proximity of people to one another in various places and situations. Ideas for proxemics studies, as well as other studies, are found in Julia G. Crane and Michael V. Angrosino, *Field Projects in Anthropology* (Prospect Heights, IL: Waveland Press, 1984).

7. Have the students discuss in class their extinct species and the role humans play in the extinction of modern species.

Materials and Preparations for Investigations

INVESTIGATION 20.1 Conditioning

Materials (per class of 30)
bell or other noisemaker

Preparations

Proper presentation is very important. The paragraphs should be read at a normal reading pace without undue emphasis on the boldfaced words. Sound the bell immediately before or simultaneously with the pronunciation of a boldfaced word. To perfect your timing, practice using the bell or buzzer while reading the paragraphs. Do not read too fast or hesitate before pronouncing the boldfaced words. Stand at the back of the room while reading; this will eliminate all stimuli except sound. If you wish, make a tape recording of both paragraphs so that the timing can be perfected and class disturbances will have no effect on your performance. The Teacher's Resource Book includes blackline masters of Tables 20.1 and 20.2.

INVESTIGATION 20.2 A Field Study of Animal Behavior

Materials (per class of 30, teams of 1)
30 data books
30 pens

Preparations

The student instructions should give students all the direction they need for a successful field study. Emphasize that no domesticated animals should be used. If you know of specific environments in your area where a variety of animals would be available for observation, recommend these environments to your students. Nature walks in preserves are excellent places for observations. Students may know of some good locations as well and can share these with the class.

You might want to discuss whether a city park is a natural habitat for an animal. Will a squirrel or a bird that lives there behave in the same manner as those that live away from populated areas? This question may start a good discussion of urban wildlife. Invite the students to compare the environments they have chosen to study, which ideally will represent a broad range of environments.

INVESTIGATION 20.3 Environmental Tolerance

Materials (per class of 30, teams of 5)

30 pairs of safety goggles
30 lab aprons
30 pairs of gloves
24 50-mL beakers
2 nonmercury thermometers
30 100-mm × 15-mm petri dishes
30 clear plastic bags with twist ties
2 shoe boxes with lids
100 100-mm diameter paper towel disks
100 100-mm × 15-mm cardboard strips
6 forceps
6 glass-marking pencils
refrigerator
incubator
2 L distilled water
1 L fungicide (household bleach, diluted 1 :5), labeled
　　　CAUTION: Irritant
3 packages radish seeds (300 seeds)
3 packages nonhybrid tomato seeds (300 seeds)
3 packages vetch seeds (300 seeds)
3 packages lettuce seeds (300 seeds)

Preparations

The plant species selected for this investigation offer a variety of responses to the conditions of the experiment. Other seeds may be substituted as follows: For radish—tobacco, corn; for vetch—celery, larkspur, columbine; for tomato—beet, pepper, carrot, sunflower, cotton; for lettuce—African violet, evening primrose, onion, fireweed, phacelia. Ward's Natural Science Establishment carries the recommended seeds (including light-sensitive lettuce) in packets containing approximately 100 seeds.

Household bleach is recommended as a fungicide because of its availability and low cost. Dilute 1 part bleach with 5 parts water.

Satisfactory substitutes for petri dishes include jar lids, aluminum pot-pie dishes, or cardboard milk carton bottoms. Cover them with plastic wrap. Test metal containers to ensure that they do not get too hot in the incubator.

The strips of cardboard used as dividers should be as wide as the containers are deep. Plastic petri dishes with four sections can be purchased from Ward's Natural Science Establishment (catalog # 1 8M7 108).

Thin plastic bags should be used to allow free diffusion of gases. The plastic bags used by laundries to protect clothing are satisfactory. Provide scissors if you wish students to prepare their own paper towel disks.

For the cold environment, use a refrigerator with a small high-intensity lamp and a 15W or 25W bulb to provide continuous light. Connect the lamp to an extension cord long enough to reach an outlet. Tape together the connection, which should be outside the refrigerator, and securely tape the extension cord to the side of the refrigerator so it cannot be pulled free. Test the refrigerator a few days before the experiment to make sure it will maintain a temperature of 10° C to 12° C.

If you have a commercial incubator, supply a continuous source of light with a high-intensity lamp, observing the same cautions as for the refrigerator. If the incubator door will not close when the cord is in place, remove the thermometer and run the lamp cord through the thermometer hole. As a safety measure, wrap electrical tape or duct tape around the cord at the point where it enters the incubator. Test the incubator a few days before the experiment to ensure that it does not overheat with the lamp on.

High-intensity lamps can become hot when left on for a long time. Just before each class, unplug both lamps from the wall socket(s) to allow the lamps to cool before students remove their petri dishes of seeds.

Have the students discuss their hypotheses before performing the investigation. It is important that they consciously try to test their hypotheses. Table 20.3 is available as a blackline master in the Teacher's Resource Book.

CHAPTER 21
ECOSYSTEMS OF THE PAST

Planning Ahead

Continue to assemble as many biome pictures as possible for use in Chapter 22. *National Geographic* and *Natural History* are good sources. You may want to make arrangements to visit a zoo sometime during the study of Chapter 22. Zoo personnel usually are very cooperative when asked to give "behind the scenes" tours for biology classes.

Concepts

◆　We know much about organisms and ecosystems of the past through reliable indirect sources such as fossils; anatomical, physiological, and ecological similarities; sedimentary layers; and radioactive dating.

◆ Radioactive dating is based on the assumption that a specific radioactive substance will decay at a constant rate. The ratio of decay products to the original element can be used to calculate the approximate age of an object.

◆ Because the half-life of different radioactive isotopes varies from seconds to billions of years, the most accurate radioactive dating is accomplished by carefully selecting the most appropriate element for the analysis.

◆ A fossil is some surviving part or revealing impression of an organism of the past. Because most fossils are incomplete, reconstruction of ancient organisms is interpretative.

◆ Fossils are tangible evidence for the existence of organisms in the past. From this evidence, paleontologists have been able to piece together a sketchy history of ecosystems on the earth.

◆ Because older fossils lie deeper in the ground than more recent fossils, a pattern of life forms through time is revealed that includes extinctions, the origin of new species, and a great deal of biological stability through long periods of time.

◆ The fossil record indicates that throughout the biological history of the earth, as environments changed, once-abundant types of organisms became extinct, and new types appeared.

◆ Because the earth's crust floats over hot magma, the crust tends to break, reform, and shift, building new continents and changing the positions of continents through time.

◆ Plate tectonics, a modern reformulation of the continental drift hypothesis, is a geologic theory that offers insights into the past distribution of species.

◆ The evolution of life on earth can be structured around some major events such as long periods of cooling and warming of the earth, the adaptation of life to land, periods of mass extinction, and emergence of important plant and animal groups such as dinosaurs, insects, flowering plants, and mammals.

◆ The earth is about 5 billion years old and has contained life for the past 3.5 billion years. Life invaded land about 400 million years ago; dinosaurs and flowering plants were dominant about 100 million years ago.

◆ Mammals have been widespread during the past 60 million years, and walking primates (hominids) appeared 4 million years ago.

◆ The genus *Homo* emerged in Africa and has evolved for about 2 million years, producing three progressively more advanced species: *Homo habilis, H. erectus,* and *H. sapiens.* Neanderthals, the earliest subspecies of *Homo sapiens,* became extinct about 30 000 years ago and may have evolved into modern man.

Objectives

Students should be able to:

◆ *relate* major biological and geological events to the geologic time scale.

◆ *explain* briefly how geologists date fossil-bearing rocks.

◆ *discuss* ways in which scientists' ideas about the past may change as their work continues.

◆ *discuss* in general terms the nature of the fossil record.

◆ *explain* the theory of plate tectonics.

◆ *explain* how knowledge of past distributions of organisms illuminates present discontinuous distributions.

◆ *explain* how knowledge of present organisms and environments is the basis for interpretation of the past.

◆ *name* two or more major groups of organisms characteristic of Cambrian, Devonian, Carboniferous, Triassic, and Cenozoic times.

◆ *explain* how the biota of a paleoecosystem indicates the environmental condition in the ecosystem.

◆ *infer* the chronological order indicated in the fossil record for the origin of bacteria, trilobites, fishes, amphibians, reptiles, and mammals.

◆ *identify* at least four characteristics in which modern humans differ from other primates, both modern and fossil.

◆ *identify* at least four characteristics common to all primates.

◆ *identify* in a skeleton characteristics associated with human upright posture.

◆ *relate* hominid evolutionary events to the geologic time scale.

◆ *describe* the generally accepted evolutionary lineage of humans.

◆ *describe* the evidence on which present understanding of hominid evolution rests.

◆ *give examples* of how archaeologists interpret evidence.

Strategies

You may want to supplement this chapter with photos of fossils and museum displays of ancient organisms. If you school is near good fossil-bearing strata, plan a field trip—perhaps a Saturday expedition of volunteers. Regardless of your school's location, have a collection of fossils in your laboratory. Donations from a fossil collector usually can be arranged. You also can order fossils from biological supply companies. Most cities have a museum with fossils and displays of paleoecosystems of your area. Plan to visit such a museum.

The illustrations in the textbook include restorations and fossils. The former have their use but, in general, the more vivid the portrayal of a scene from the geological past, the further the artist probably has departed from the strict fossil evidence. Attempt to distinguish between data about fossils and interpretations of such data.

Refer students frequently to Section Three and *A Catalog of Living Things* in Appendix 4. Extinct groups of organisms are not included there, but the task of placing extinct groups among modern organisms is instructive. Most of the groups in the higher levels of classification have long histories, so mention of them recurs in Chapter 21. Referring to Sections 21.2 and 21.3 also will give you an opportunity to show how the arrangement of taxonomic groups reflects the efforts of taxonomists to portray phylogeny—a matter discussed in Chapter 10.

The search for human fossils is one form of science that captures the interest of almost everyone; the best indication is the space that newspapers are willing to devote to the subject. Discussions of human fossils are almost always lively. Be alert for and discuss any recent human fossil discoveries.

A visit to a natural history museum, especially one having exhibits about paleontological methods, would be an excellent addition to this chapter. Whether or not that is possible, be sure to emphasize the careful work that goes into the study of fossils. Students should understand that the paleoecosystems described here are not flights of artistic fancy but the results of scientific methods.

Be sure to do Investigation 21.1, as it is a relatively concrete way of allowing students to visualize the immensity of the evolutionary time scale. Have students work on Sections 21.1–21.10 in the text during the two periods spent on Investigation 21.1. Investigations 21.2 and 21.3 require only paper and pencil and should precede Sections 21.11–21.13 and the Biology Today.

References

Alvarez, W., F. Asaro, and V. E. Coutillot. "What Causes Mass Extinction?" *Scientific American* (October 1990): 76–93.

Badash, S. "The Age-of-the-Earth Debate." *Scientific American* (August 1989): 90–98.

Broecker, W. S., and G. H. Denton. "What Drives Glacial Cycles?" *Scientific American* (January 1990): 48–56.

Buffetant, E., and R. Ingavat. "The Mesozoic Verte

Coppens, Y. "East Side Story: The Origin of Humankind." *Scientific American* (May 1994): 88–95.

Dalziwl, I. "Earth Before Pangea'" *Scientific American* (January 1995): 58–63.

Fischman, J. "Putting a New Spin on the Birth of Human Birth." *Science* (20 May 1994): 1082–1083.

Fulkerson, W., R. R. Judkin, and M. K. Sanghvi. "Energy from Fossil Fuels." *Scientific American* (September 1990): 128–35.

Grieve, R. A. F. "Impact Cratering on the Earth." *Scientific American* (April 1990): 66–73.

Johnston, A. C., and L. R. Kanter. "Earthquakes in Stable Continental Crusts." *Scientific American* (March 1990): 68–75.

Keating, J. F., O. B. Toon, and J. B. Pollack. "How Climate Evolved on the Terrestrial Planets." *Scientific American* (February 1988): 90–97.

Molleson, T., "The Eloquent Bones of Abu Hureyra." *Scientific American* (August 1994): 70–75.

Novacek, M., et al. "Fossils of the Flaming Cliffs." *Scientific American* (December 1994): 60–69.

Shreeve, J. "Lucy, Crucial Early Human Ancestor Finally Gets a Head." *Science* (1 April 1994): 34-–35.

Shreeve, J. "Sexing Fossils: A Boy Named Lucy." *Science* (24 November 1995): 1297–1298.

Stringer, C. B. "The Emergence of Modern Humans." *Scientific American* (December 1990): 98–105.

Turner, C. G. "Teeth and Prehistory in Asia." *Scientific American* (February 1989): 88–97.

Walker, A., R. E. Leakey, J. M. Harris, and F. H. Brown. "2.5 Myr. *Australopithecus boisei* from West of Lake Turkana, Kenya." *Nature* (7 August 1986): 517–21.

White, R. "Visual Thinking in the Ice Age." *Scientific American* (July 1989): 92–99.

Suggested Answers to Applications

1. Some of the many answers to this question deal with differences in the environment in and out of the water. The oceans are a relatively stable physical environment, and therefore speciation has not been as dramatic as on land. In addition, the flower

has become attractive to a myriad of pollinators, each pollinator with its own set of preferences. Thus, there has been selection for a wide array of flower types-and species of flowering plants.

2. There is a growing consensus that at least some dinosaurs were warm-blooded animals, with narrowly fluctuating body temperatures, unlike modern-day reptiles whose behavior is greatly affected by their environment. You may want to look at A. J. Desmond, The Hot-Blooded Dinosaurs: A Revolution in Paleontology (New York: Dial Press, Inc., 1976).

3. Weather patterns and climates of inland and, especially, coastal regions of the North American continent are affected by the Japan Current and the Gulf Stream. These currents affect air temperatures, humidity, and precipitation.

4. Such an illustration would quickly reveal the location of plates actively moving across the surface of our planet. Obtain a world map and a world almanac to locate a few active volcanoes.

5. A nuclear winter is the term used to describe one predicted scenario after a nuclear war. According to this scenario, the smoke from burning buildings and forests and the dirt kicked up by bombs would enter our atmosphere, block the sunlight, and create winterlike conditions. Some scientists have suggested that a large meteor hit the earth millions of years ago, creating, in effect, a "nuclear winter" that helped to wipe out dinosaurs.

6. Standing upright, twisting, and bending weaken the lower spine so that heavy lifting can dislodge a vertebra and cause severe back pain. Arches of the feet collapse under the weight of upright bodies, and hernias result when intestines bulge out through weakened abdominal wall. Varicose veins occur because upright posture hampers the flow of blood throughout the body.

7. Fossils are less likely to be formed there, partly because dead organisms will decompose rapidly. Even if they did form, they would be difficult to find. Consider this comment from a paleontologist from southern California upon transferring to an eastern university: "All that green stuff covers everything up."

8. You should; the high arches are adaptations for bipedal walking, even though the brain size is relatively small.

9. Answers will vary. It could be a gorilla.

Suggested Answers to Problems

1. The Geological Survey produces a series of topographical and geological maps of every state as well as brochures on the geology and fossil history of many areas throughout the country.

2. Anthropology texts should contain a lot of information about these cultural tools, and your local museum should house information about ancient people who lived in your area.

3. Discuss with your students the type of observations to make and how to make notes about the animals. If you quantify behaviors—for example, how many times an animal scratched itself or how much time was spent grooming—compare one animal with another.

4. A trip to your local museum should give your students a good idea of the quality of the background art and how packed it is with useful information. This is an opportunity for your more artistic students to shine.

5. There are several alternatives as to the exact arrangement of continents, including the one illustrated in this text. Make sure the map used to make the tracings does not distort the size and shape of the continents.

6. Among other things, students should discover that many of the habitats of our primate relatives are being destroyed by human activities.

7. Ask students to speculate on why some of the creatures found in these deposits became extinct and why some were successful.

8. Students should have no trouble constructing a food web from the information given.

9. This problem often evokes much student interest. If the points made in its statement are considered carefully, it can result in some disciplined thought that effectively reviews Chapters 20 and 21.

Materials and Preparations for Investigations

INVESTIGATION 21.1 Evolution and Time

Materials (per class of 30, teams of 4, 6, and 10)
3 geologic time scales
3 tables of major geological events
3 tables of major biological events
75 sheets of construction paper (3 colors, 25 sheets of each)
1 15-m tape measure
3 150-cm tape measures

3 15-cm metric rulers
1 permanent felt-tip red marker
1 permanent felt-tip black marker
15 dark-colored crayons or broad felt-tip markers
43 clothespins
100 m of rope clothesline
3 pocket calculators (student material)

Preparations

Use three easily distinguished colors of construction paper light enough so the labels will show up clearly. Use a different color for each team. Small metal binder clips may be used in lieu of clothespins. You may wish to use stakes or similar supports to help hold up the time line. Be sure students understand the difference between events that took place millions of years ago and events' that took place millions of years after the earth formed. The tables for the investigation are in Appendix 3 and also appear as blackline masters in the Teacher's Resource Book.

INVESTIGATION 21.2 Interpretation of Fossils

Materials (per class of 30, teams of 2)
15 protractors
15 metric rulers
30 sheets of graph paper

Preparations

The Teacher's Resource Book includes a blackline master of Table 21.3.

INVESTIGATION 21.3 Archaeological Interpretation

Materials (per class of 30, teams of 3)
paper and pens

CHAPTER 22
BIOMES AROUND THE WORLD

Planning Ahead

Chapter 23 requires as many pictorial supplements as Chapter 22. Plan bulletin boards about standing and flowing inland waters and marine ecosystems. Some enthusiastic students may want to help with this project. Order or collect organisms for the investigations in Chapter 23.

Concepts

◆ The major factor in determining world climatic patterns is the distribution of solar energy in any given region. Because warm air holds more moisture than cool air, air temperature greatly influences the amount of moisture in a given air mass.

◆ Seasons are caused by the 23° tilt of the earth's axis with respect to the path of the earth around the sun. As a result, the sun's rays that strike the earth are more concentrated in the Northern Hemisphere when it is tilted toward the sun (summer) than when it is tilted away from the sun (winter). The effects are opposite in the Southern Hemisphere, and there are essentially no seasons near the equator.

◆ A biome is a region of the world with characteristic climatic patterns and corresponding life forms. The two major climatic factors that determine the types of life forms present in a biome are temperature and precipitation. Other significant factors include proximity to large water masses and mountain ranges, and air/ocean currents.

◆ A climatogram displays the annual pattern of precipitation and temperature in a given location of the world; one can infer many dominant life forms from climatogram data.

◆ The major world biomes can be grouped as follows: equatorial or tropical (rain forest, tropical deciduous forest, desert, savanna); midlatitude or moderate (temperate deciduous forest, grassland, desert, chaparral); cool (taiga); and cold or polar (tundra) .

◆ World climates tend to get cooler and drier as one moves from the equator to the poles. Increased altitudes on a land mass reveal biome patterns similar to those of increased latitudes.

◆ The same basic biome can be present in several different regions of the world if the climates are similar, and when this occurs, life forms that have evolved independently can appear quite similar.

◆ Communities are subject to change due to climatic shifts or to disruptions such as fire, disease, and human influences. There often is a predictable succession of life forms replacing one another in a given community until a point of stability (climax) is reached that is determined by the climatic conditions.

◆ Humans have drastically altered many large communities in the world in attempts to improve the quality of their own lives. Human-caused trends such as deforestation, desertification, and acid rain must be reversed if the earth is to be able to support future generations of humans.

Objectives

Students should be able to:

◆ *distinguish* between biome and ecosystem.

◆ *describe* the effects of air circulation and ocean currents on climate.

◆ *recognize* the major biomes when characterized in climatograms.

◆ *name* the biome in which they live.

◆ *construct* climatograms, given data of monthly precipitation and temperature.

◆ *describe* the principal climatic characteristics of each biome and the biological characteristics associated with the climatic zones.

◆ *describe* the relationship of biome structure to latitude when moisture is adequate.

◆ *explain* how to conduct a well-designed field study.

◆ *discuss* the relationship of precipitation to evaporation in desert and forest biomes.

◆ *relate* the idea of climax to succession and to their own region.

◆ *describe* characteristic successional communities in their own region.

◆ *explain* the role of chance in colonization and subsequent successional stages.

◆ *explain* the effects of fire on succession in at least one community.

◆ *explain* the effects of humans on succession in ecosystems.

◆ *name* organisms that owe their present distributions to humans' intentional and unintentional actions.

◆ *describe* three or more examples of how humans have altered natural biomes.

Strategies

Emphasize the biome in which your school is located. The more your students understand their local biome, the more they will understand others. Arrange for a half- or full-day trip that will display the salient characteristics of the biome. If late-winter weather is prohibitive, the trip can be postponed, but this should be necessary only in the northern states. If a trip to observe characteristics of your biome is not feasible, portray it with a collection of color slides. Use aerial photographs, which can be obtained from agencies such as the U.S. Forest Service, U.S. Geological Survey, Soil Conservation Service, and NASA.

Most museums have displays of local communities. Zoos and botanical gardens are good places to see plants and animals from all over the world all year long. An excellent source of information about distant biomes and climates often is overlooked—some of your students have lived in other biomes and may have photographs and be eager to talk about their former surroundings.

A biome is a biotic expression of a climate. Therefore, students must have some knowledge of (1) the atmospheric factors that—when statistically summarized—constitute climate and (2) the astronomical and geophysical phenomena that determine the distribution of climates. As in the case of distant biota, an understanding of distant climates is best developed against a background of familiarity with the local climate.

If your students have poor backgrounds in earth science, you may need to spend some time demonstrating the ideas in Section 22.1 with a globe. Investigation 22.1 requires two class periods, including discussion, and should clarify the descriptions of the biomes in Sections 22.3–22.10. Investigation 22.2 is a classic ecology activity and should be the high point of the entire year. Allow three to five class periods, depending on how far you have to travel and the volume of biota available. Investigation 22.3 can be assigned as homework, but it is best for students to work together in groups of three.

References

Holloway, M. "Nuturing Nature." *Scientific American* (April 1994): 98–106.

Jones, P. D., and T. M. Wigley. "Global Warming Trends." *Scientific American* (August 1990): 84–91.

Kusler, J. A., W. J. Mitsch, and J. S. Larson. "Wetlands." *Scientific American* (January 1994); 64B–70.

"Managing Planet Earth." *Scientific American* (September 1989): several articles.

Newell, R. E., H. G. Reichle, and W. Seiler. "Carbon Monoxide and the Burning Earth." *Scientific American* (October 1989): 82–89.

Rennie, J. "Living Together." *Scientific American* (January 1992): 122–130.

Repetto, R. "Deforestation in the Tropics." *Scientific American* (April 1990): 36–45.

Romme, W. H., and D. G. Despain. "The Yellowstone Fires." *Scientific American* (November 1989): 36–47.

White, R. M. "The Great Climate Debate." *Scientific American* (July 1990): 36–45.

Suggested Answers for Applications

1. The earth is tilted with respect to its orbit around the sun. As a result, its rotation produces a change in seasons that is more pronounced in the midlatitudes than in the equatorial regions.

2. If you were to travel along the 38th parallel from Washington, DC, to San Francisco, you would expect to see the following biomes: From Washington, DC, to central Missouri is midlatitude deciduous forest, such as oak and hickory. Eastern Kansas is predominantly mixed-grass prairie, with a few deciduous trees limited to banks of streams and rivers. Continuing across Kansas, grasses become shorter and shorter. In eastern Colorado, the grassland biome consists of grama and buffalo grasses. A series of complex mountain biomes forms in the Colorado Rockies. For example, from Denver, 1609 m above sea level, to the top of Mount Evans, which is 4348 m above sea level, is a distance of about 56 km; traveling those 56 km is the equivalent of going north about 2736 km to the arctic tundra. Continuing west into Utah is a series of complex mountain biomes interspersed with canyons and desert. Nevada is mostly desert with scattered mountain ranges covered with pine and juniper. The Sierra Nevada of California are covered with pine, spruce, fir, and juniper along the eastern margin of the mountains. Descending the western portion of the mountains, we find broadleaf deciduous trees, part of the complex mountain biome system of the Sierra Nevada. The large central valley of California supports a variety of grasses and is similar to the other grasslands to the east. The low coastal range is covered with grasses and trees. Just north and south of San Francisco, there are groves of redwood and fir, part of the Northwest coastal forest biome. There are many changes due to human activity, such as farming, forestry, highways, cities, and, in general, increased population.

3. Grasslands are undergoing changes due to loss of topsoil, increased alkalinity, and desertification. Tropical forests are undergoing rapid, destructive changes because of increasing human population and the demand for forest products. Once the forest has been removed, high rainfall leaches what little nutrients there are, and the soil becomes unproductive. The rich, biological diversity of the tropics is lost, and the agricultural productivity has not been increased. Millions of species of plants and animals are lost as the destruction of tropical forests continues. Acid rain is damaging both deciduous and coniferous trees in Europe, Scandinavia, and North America. Extinction of plants and animals probably will increase as the amount of oxides of sulfur and nitrogen increase. Any human endeavor will in some way alter any ecosystem. The degree of change will depend on the type and duration of activity. You need only look around the area in which you live to see the changes.

4. Students' speculations may range widely. The important thing is to keep them consistent with known facts concerning biomes and succession. Briefly, then, we can say: (a) the farm would go back to grassland; (b) after a very long time deciduous trees would occupy the cracks between sidewalk fragments; (c) after a period as a freshwater pond, the swimming pool would fill with sediment and be occupied by humid coniferous forest; and (d) without constant watering the Nevada park would revert rather quickly to desert.

5. Some organisms that might be found on these lists are: apples, grapes, oranges, plums, cherries, melons, potatoes, tomatoes, beans, squashes, corn, barley, wheat, rice, sorghum, house mice, Norway rats, cockroaches, bedbugs, house sparrows, starlings, and pheasants. Whether or not each of these could survive in your locality without human help depends in part on the climate of your region. Some of them, such as corn and perhaps bedbugs, probably could not flourish anywhere without humans.

6. Tradition, in part. There are many more plants than there are animals, and plants are the dominant feature of the landscape. Also, plants support all other organisms of these biomes.

7. The temperature of the earth is affected much less by the relatively small changes in the distance of the earth from the sun during the year than by the angle and amount of light rays that do penetrate the earth's atmosphere.

8. Get students to discuss possible differences in precipitation and temperature that might result from changes in currents, and the associated effects of physical features such as mountains.

9. Successful agriculture results from a combination of good soil and the right climate. The growing season is too short in the tundra and northern coniferous forest; precipitation is too low in the desert; and soils are poor in the tropical rain forest. Soils, temperature, and precipitation, however, are excellent in the midwestern United States—part of the grassland biome.

10. There are many more species in the tropical rain forest than in the tundra, and the connections between them are much more complicated.

11. Higher in Mexico. As you travel south, the annual average temperature increases at the same elevation. Colder temperatures are restricted to higher and higher elevations southward, and thus the alpine tundra is found at higher elevations as well.

12. The large animal has a smaller surface-to-volume ratio than does the small animal, which can lose body heat over a proportionately greater amount of its surface.

13. There is debate about this, but without small fires, litter builds up that can fuel future crown fires. Fires also can open up new habitats for colonizing plants and add nutrients to the soil.

14. Hibernation is a response to cold in which metabolic processes drastically slow down, allowing the animal to survive long periods of cold by using stored body fat. Estivation is a response to heat and dryness and usually is much shorter in duration. They both allow the animals to survive unfavorable conditions.

Suggested Answers to Problems

1. Must be done locally.

2. Prairie dogs are herbivores, feeding mostly on grasses and other plants. They compete with cattle for the grasses, and as a result, the prairie dogs are being eliminated throughout their range. Expansion of suburban communities also contributes to the elimination of prairie dogs. Their elimination has caused the near extinction of the black-footed ferret, which preys exclusively on prairie dogs.

3. Any piece of heavy equipment that is driven over the tundra damages the small plants to the point where the tundra cannot recover for many generations. The equipment also leaks lubricating oil and fuel, which permanently destroy the delicate tundra plants.

  As the caribou encounter more and more man-made structures, they may be forced to change their former migration routes. Oil, gas, and mineral exploration will certainly be a factor in the Arctic for some time to come, and the caribou will most likely encounter some new obstacle that may interfere with their migration. Scientists are not sure what causes caribou to migrate north toward their calving grounds, and any change could prove disastrous.

4. The tiny Central American country of Costa Rica, which had a birth rate of about 3 percent, was clearing about 140 000 acres of forest annually. During the late 1960s, concerned about the loss of plants and animals, Costa Rica established 20 national parks and preserves representing 8 percent of nearly 39 000 km² of the total land mass.

Destruction of India's forests dates back to the beginning of British colonial rule, and the problem hasn't improved during the last 40 years. It is estimated that only 12 percent of India's forest remains today. The state of Uttar Pradesh once had a wide variety of wildlife, including bear, deer, and tigers, but the 1962 border dispute between India and China opened up the forest area to logging. Today, as a result of depletion of the forests, only a few monkeys remain. In 1976, the government set aside 1200 acres and forbade any logging for a 10-year period. Reforestation has been intensified, but it will take 200 years for replanted forests to make an appreciable improvement. The one associated problem that won't go away is that of growing populations.

  There are many examples of wildlife preservation programs throughout the world, but one of the best is in India. The Indian government, along with the World Wildlife Fund, has set aside 15 preserves, saving not only the tiger, but ecosystems of teak, bamboo, and mangrove swamps. The number of tigers has risen from 2000 in 1972 to over 3000; of these, over a third are in tiger preserves.

  A good reference is the March/April 1984 issue of *International Wildlife*. The article entitled "Wild in London" describes an urban ecologist's efforts in developing ecological settings in London.

5. Refer students to W. H. Romme and D. G. Despain, "The Yellowstone Fires," *Scientific American* (November 1989): 37–46; and F. Stanb, "A Legacy of Wildfires: The Yellowstone Experience," *Ward's Bulletin* (1990).

6. This problem will test your students' skills in using references as well as their ability to think ecologically. No one reference will serve for all parts; perhaps the best single one would be an encyclopedia.

  The ecological equivalents of these animals in most of the Australian and North American grasslands today are cattle; in the steppes of Asia— Old World bison, horses, wild asses; in the pampas of Argentina—guanaco, pampas deer; in the veldt of South Africa—gnu and other large antelopes; in the tundra—reindeer, caribou, and musk ox.

  In the desert of South Africa, the ecological equivalents of the cacti of North American deserts are the euphorbias, which have succulent stems, are thorny, and live in arid climates (see Figure 9.25). In the tropical forests of the Old World, ecological equivalents of the hummingbirds of the

New World tropical forests are the sunbirds, which are small, have long bills, and feed on nectar.

7. Students could debate the merits of such programs after the visits made by the groups' representatives.

8. Encourage students to take photographs of native plant and animal life wherever they travel. Photographs also can be collected from old newspapers and magazines.

9. Current periodical literature is filled with background information on the problems of the tropical rain forest and different suggested solutions to the problem.

10. While the vegetation may be similar in appearance from one part of the world to another, the species of plants and animals are distinctly different. *National Geographic*, encyclopedias, and college texts on ecology and biogeography should be helpful.

11. Literature also offers a wealth of descriptions of the ecosystems in which stories unfold. Work with the English instructors of your school to develop a reading list of books in which the natural environment plays a key role.

12. This exercise should yield information on the successional patterns within your community. Students could look at the succession of communities within nearby city blocks that have been abandoned in different years.

13. The *Land Use* module guides students through all the stages of investigation surrounding a land use problem.

Materials and Preparations for Investigations

INVESTIGATION 22.1 Climatograms

Materials (per class of 30, teams of 1)
30–90 sheets of graph paper or climatogram forms

Preparations

If the graphs are drawn on a variety of grids, comparisons are difficult. Furnish all students with graph paper of the same type and have them duplicate the grids shown in the climatograms in Appendix 3—12 blocks wide and 18 high. Or, supply copies of the climatogram blanks provided as blackline masters in the Teacher's Resource Book. On such a grid, the April rainfall datum from Moshi, Tanzania, extends two blocks above the top, but these blocks can be added by hand.

Obtain local climatic data from the nearest U.S. Weather Service Office. To convert precipitation data in inches to centimeters, multiply by 2.54. Assign a few students the task of conversion. They should check each other's work before releasing the data for class use.

For class discussion, prepare classroom-size charts or transparencies of the climatograms representing the 10 major biomes. Blackline masters of these climatograms are in the Teacher's Resource Book.

Be prepared to discuss the relationship between climatic factors and the biota. For example, the climatogram for tundra indicates that the average monthly temperatures are above freezing for only three months of the year. Furthermore, the precipitation is quite low all year. Thus, photosynthesis can occur only during a fraction of the year. The amplitude of the yearly cycle of monthly temperatures implies a high latitude. This, in turn, ensures a long daily period of sunlight during the season when temperatures are high. The low temperatures imply a low rate of evaporation; thus, conditions for plant growth during the brief summer season are not as unfavorable as they might seem at first glance. These food-production conditions suggest the possible presence of a large migratory summer population of consumers. Such reasoning should be applied to all climatograms.

INVESTIGATION 22.2 Studying a Piece of the Biosphere

Materials (per class of 30, teams of 6)
14 metal coat hangers, or pieces of stiff wire, for small quadrats
28 wooden stakes and string for large quadrats
5 metersticks or metric tapes
5 trowels or small-bladed garden shovels
5 white-enameled pans or large sheets of white paper
15 pairs of plastic gloves (if specimens are to be collected)
35 collection containers such as jars and plastic or paper bags
30 forceps
6 wire screens of approximately 1-mm mesh curved into a bowl shape (window screening)
30 clipboards and paper
*10 50-cm pieces of PVC pipe (1/2-1 inch in diameter)
*20 corks to fit pipe
*5 50-cm wooden dowels that fit inside pipe
20 sterile nutrient agar plates

Preparations

A field study is central to an ecological approach to biology and should be the highlight of the year. For many reasons, it is impossible to prescribe a single procedure for the investigation. Students have been given enough information to enable them to take an active part in the planning. Helpful resources are: G. W. Cox, *Laboratory Manual of General Ecology* (Dubuque, IA: Wm. C. Brown Publishers, 1985), and J. Brower, J. Zar, and C. von Ende, *Field and Laboratory Methods for General Ecology* (Dubuque, IA: Wm. C. Brown Publishers, 1989).

* available as a kit (catalog #36M1701) from Ward's Natural
   Science Establishment

Provide additional class time, as needed, for students to collect and construct field equipment. The latter might include making insect nets from clothes hangers and old nylons. You also might have students practice setting up quadrats especially if they will be using the 100-m² quadrat.

Other things to provide: binoculars for bird-watching, a knife to cut through sod if a pasture or lawn study is done nutrient agar plates (see a Table of Recipes for preparation or purchase in bottles or prepared plates from Ward's Natural Science Establishment) for streaking soil samples to show how many microorganisms live in the soil; identification keys of local flora and fauna.

If at all possible, assemble a labeled collection of the species most likely to be encountered. This may be difficult the first time you do the investigation, but your collection will grow and will help students identify their organisms. Although specific identification is not necessary in this study students like to have names for things. Such knowledge increases their sense of accomplishment and will help them learn more about the organisms they name. Do not let identification become the goal of the work.

Either before or after class discussion of the questions each student should write a report on the work as a whole. It should include the purpose of the investigation, a brief account of the methods used, a summary of the data, and (most important of all) his or her own detailed interpretation of those data—a description of relationships either observed or Inferred.

INVESTIGATION 22.3 Long Term Changes in an Ecosystem

Materials (per class of 30, teams of 3)
none
A blackline master of Table 22.4 is included in the
    Teacher's Resource Book.

INVESTIGATION 22.4 Effects of Fire on Biomes

Materials (per class of 30, teams of 3)
none

Preparations

This investigation can be assigned as homework and then discussed in class, or it can be worked through in class, using small teams, without previous assignment. You may find it advantageous to do some background reading on the effects of fire in different ecosystems, or you may wish to focus on the Yellowstone fires. *Ward's Bulletin,* "A Legacy of Wildfires The Yellowstone Experience" (catalog #32M0857), is a good resource, as is their set of transparencies (catalog #170M0140). Ward's also carries serotinous cones (catalog #86M8370)—which require intense heat before they can open and release their seeds—with instructions for investigation.

CHAPTER 23
AQUATIC ECOSYSTEMS

Planning Ahead

Prepare copies of the role cards and map for Investigation 24.2; both are provided as blackline masters in the Teacher's Resource Book. The three sets of role cards can be reused, but the maps will be consumed in each class. This chapter includes four investigations; recheck your materials for those you plan to do.

Concepts

◆ A pond is a small, shallow body of standing water that forms an ecosystem supported by light penetration to all depths and relatively constant temperatures throughout.

◆ A lake is a larger and more permanent body of water in which light penetration is limited, allowing photosynthesis only in the upper layers. Lake ecosystems often include large vertebrates.

◆ Larger bodies of water, such as lakes and oceans, exhibit temperature changes at different depths. Generally, as surface water cools, it becomes more dense than the lower water and is replaced by water below. This characteristic of water causes seasonal mixing of the levels of lakes and seas and greatly diversifies the life types present in these aquatic ecosystems.

◆ Terrestrial springs frequently give rise to small brooks, which in turn collect to form streams and, eventually, rivers. Usually, the smaller the body of running water, the more oxygenated it is.

◆ Smaller bodies of running water tend to run faster, thus collecting more oxygen, and are higher (and cooler), thus retaining more dissolved oxygen. These differences in oxygen retention are major factors in the types of organisms adapted to these different environments.

◆ Rivers eventually run into seas or oceans, in which abundant dissolved and suspended materials continually are being deposited. These ecosystems support a diverse array of organisms, many having special mechanisms to deal with an environment high in chemicals such as sodium chloride.

◆ Because oceans and seas are much larger than lakes and there is greater mixing of their waters, they tend to have more stable temperatures than freshwater ecosystems.

◆ Because most ocean producers (phytoplankton) occur only in the upper layers, most ecological activities and by far the bulk of ocean biomass exist there as well. Shallow, coastal areas are particularly productive because of upwelling, tides, and river runoff. Coral reefs are also productive because they form a protective barrier against the open ocean.

◆ With the exception of some chemosynthetic bacteria, all organisms found in deeper ocean levels are consumers.

◆ The intertidal zone is a difficult place to live because of its extremes in wetness and temperature and because its occupants are subjected to pounding surf.

◆ Because wetland ecosystems are highly productive and because they act as environmental filters of toxic materials, they are important to the ecological survival of the entire biosphere. Human attempts to dam large rivers, to fill in marshlands for buildings, and to dump wastes into marshlands have threatened the health of wetland environments. This will ultimately threaten the well-being of all organisms in the biosphere.

◆ Most human wastes eventually end up in the oceans. Because the oceans are basically tubs, these wastes collect rather than go away. Eventually the wastes will enter the water and food cycles and return to humans. Therefore, we must consider carefully both the quality and quantity of what we allow to run into our oceans.

Objectives

Students should be able to:

◆ *describe* in general terms the nature of aquatic ecosystems.

◆ *distinguish* between inland and ocean waters.

◆ *name* at least five major types of aquatic ecosystems.

◆ *describe* characteristic abiotic factors in the ecosystems named.

◆ *name* a number of organisms characteristic of the ecosystems named.

◆ *relate* pollution of a water source to the numbers and types of organisms present.

◆ *describe* the effect of thermoclines on lake productivity.

◆ *contrast* the long and short cycles of the water cycle.

◆ *describe* techniques used in a field study of a stream.

◆ *explain* how measurements of biomass in an ecosystem are dependent on productivity.

◆ *relate* salinity to the survival of selected organisms.

◆ *describe* at least three major ways in which human activity has physically changed inland water ecosystems.

◆ *describe* the process of eutrophication and its effects on organisms.

◆ *explain* the changing effects of sewage inflow along a length of river.

◆ *list* specific practices that reduce water pollution.

Strategies

Chapter 23 should come at a season that is convenient for fieldwork and when standing water is available even in areas having dry summers. A field trip can focus attention again on individual organisms and their adaptations, on species populations, and on the interaction of both with the environment. A study of human involvement with aquatic ecosystems continues the program's emphasis on human ecology and serves as a bridge to Chapter 24.

There is little additional technical vocabulary in Chapter 23. The discussion and the inquiry stimulating questions and problems revolve primarily around the special (from our terrestrial viewpoint) environmental factors of aquatic ecosystems and adaptational responses of organisms. One general ecological concept is developed: biomass.

Have students begin Investigation 23.1 immediately. This is an important lab both biologically and ecologically because the concept of decreased diversity is used to indicate increasing pollution. It requires three class periods, including closure discussion. Have students read and work on Concept Review questions for Sections 23.1–23.2 as homework during these days.

Investigation 23.2 is another classic ecology field study, this one of a stream. Make every effort to do this investigation, which students enjoy while developing process skills. Allow three periods if the stream is near the school or a half- to full-day field trip otherwise. Students can work on Sections 23.3–23.4 immediately after their stream study.

Investigation 23.3 is a laboratory study of the effects of salinity on organisms. It takes only one class period but requires a few minutes of orientation the day before and should be followed by discussion the next day. Students then should work on Sections 23.5–23.10. The Biology Today can be read and discussed at any time. Investigation 23.4 uses the LEAP-System (or an alternative) and requires one class period for setup followed by readings for about the next 10 class periods. The investigation interacts with Chapter 24. Students can work on Sections 23.11–23.13 and the Applications and Problems immediately after setting up the investigation.

References

Barrett, S. C. H. "Water Weed Invasion." *Scientific American* (October 1989): 90–97.

Childress, J. J., H. Felbeck, and G. N. Somero. "Symbiosis in the Deep Sea." *Scientific American* (May 1987): 114-21.

Culotta, E. "Is Marine Biodiversity at Risk?" *Science* (18 February 1994): 918-920.

Grassle, J. F. "Hydrothermal Vent Animals: Distribution and Biology." *Science* (23 August 1985): 713-16.

Horn, M. H., and R. N. Gibson. "Intertidal Fishes." *Scientific American* (January 1988): 64-71.

Jannash, H. W., and M. J. Mottl. "Geomicrobiology of Deep-Sea Hydrothermal Vents." *Science* (23 August 1985): 717-25.

McKay, K., J. Ryan, and J. Stauffer et al. "African Tilapaia in Lake Nicaragua." *BioScience* (June 1995): 406-411.

Suggested Answers to Applications

1. Photosynthesis by phytoplankton produces oxygen during the day. At night the concentration decreases as consumers use the oxygen in respiration.

2. Cloudiness, by reducing the intensity of radiant energy, decreases photosynthesis. Windy conditions, however, circulate phytoplankton, gases, and nutrients; this increases photosynthesis because more phytoplankton are exposed to sunlight, and the nutrient supply near the surface of the lake is augmented. A rough water surface also reduces the reflection of light.

3. Ice reduces diffusion of oxygen into lake water but also slows further cooling of the water.

4. Seeds of plants can be distributed from one pond to another by the wind or carried there on the feet of aquatic birds or on the fur of animals. This reflects the unclear boundary between the water ecosystem and the terrestrial one surrounding it.

5. The sources are air and photosynthetic organisms. Brooks have little of the latter, but their turbulence promotes solution of atmospheric oxygen. The more quietly a river flows, the less oxygen it obtains from the air—but the more photosynthetic organisms it is likely to contain. Wave action mixes air with water in lakes and ponds, and there is more wave action in lakes.

6. David Quammen's essay, "Stalking the Gentle Piranha," from his collection entitled *The Flight of the Iguana* (New York: Anchor Press, 1988), describes an intimate link between Amazon fish and trees: trees produce fruit which drops into the water; the fish eat and digest the fruit and disperse seeds away from the tree when they defecate the seeds.

7. Although many decomposers are anaerobes, decomposition in general proceeds more rapidly where oxygen is abundant. Therefore, it proceeds more rapidly, provided temperature is favorable, at the bottoms of shallow ponds.

8. Blue light penetrates farthest, and thus deep water is blue. The Red Sea has cyanobacteria which, during blooms, color parts of the sea red.

9. A desert is a place where few plants are found. Terrestrial deserts are controlled by the amount of precipitation falling in the area. Aquatic deserts result from a lack of nutrients. Although there is plenty of water and sunlight, there are not enough nutrients to support many producers.

10. Most such substances are not completely excreted by organisms. They accumulate and are passed on to any other organism that uses the first as food. Therefore, a bass, which is a consumer several steps higher on the food chain, accumulates more than a ciliate, which is low on the food chain.

11. When a pond with a favorable food supply is stocked with a single species of fish, such as bluegills, rapid reproduction results in a dense population. Because of intraspecific competition, individuals are stunted. If at least one prey/predator combination is included, the predator tends to consume those stunted offspring, and the average size of remaining fish increases.

12. The number of snails is irrelevant. The total weight divided by the number of square meters gives the biomass: 534 g/5 m^2 = 107 g/m^2.

13. This raises the whole question of landfills and the toxic combination of household chemicals that

results from their accumulation and subsequent seepage into water supplies. Getting students to discover how much waste they produce per year is a good way to get them to start reducing that amount.

14. The original quotation reflected an "out of sight out of mind" mentality. Unfortunately, the pollution problem was not solved. Today, so many people are producing wastes that there is not enough available water to dilute the toxic chemicals to an insignificant level, especially near urban areas.

Suggested Answers to Problems

1. Answers will vary.

2. Answers will vary.

3. Answers will vary.

4. El Nino is a current of warm water that piles up off the coast of South America, pushing the normally cold, nutrient-rich current below the surface so that the fish and birds can no longer feed. Changes in atmospheric pressure cause the normal easterly winds to reverse directions and push a large mass of warm water along the coast of Peru and Ecuador. Weather patterns are affected as far away as Australia and the central Pacific islands.

5. Answers will vary.

6. Answers will vary.

7. Answers will vary.

8. Various environmental groups should be able to supply information expressing their points of view, while the Army Corps of Engineers might provide an alternative viewpoint. There is much information about the impact of dams on downstream whitewater rapids, a focus of certain sports enthusiasts.

9. This also could be a good time to debate issues of hunting and fishing. What are the benefits? Are there any negative consequences?

10. Many of the effects may last for decades. The periodical literature should have ample information on both spills.

11. This might be a way to capture the interests of fishermen in your class and to show them that understanding the biology and ecology of game animals is necessary to the continuation of sport fishing.

12. The July 1990 issue of *National Geographic* deals with the Florida water problem.

13. Your local pet store may be able to help you begin such a project. Marine ecosystems are much more difficult to maintain than are freshwater aquaria, so before your students begin, have them investigate all of the problems they might encounter.

Materials and Preparations for Investigations

INVESTIGATION 23.1 Water Quality Assessment

Materials (per class of 30, teams of 2)
45 microscope slides
45 22-mm × 22-mm cover slips
15 dropping pipets
15 1-L beakers plankton net
15 compound microscopes
protist/cyanobacteria key (Appendix 3)
A Catalog of Living Things (Appendix 4)
15 2-cm² pieces of graph paper with 1-mm squares
1 box flat toothpicks
pocket calculators (optional)
5 squeeze bottles of Detain™
5 water samples with algae and cyanobacteria
5 each water samples—clean, intermediate, and polluted

Preparations

Ward's Water Quality Assessment Kit (catalog x 86M3056) contains water samples from three aquatic environments—clean water, moderately polluted water, and highly polluted water. The kit includes student worksheets and a picture key showing all the organisms in the kit. (Some of the organisms are not indicators by the Palmer method.) The investigation also can be completed by gathering your own water samples from a local source and identifying the algae and cyanobacteria present. The student worksheet appears as a blackline master in the Teacher's Resource Book; provide four copies for each student, or have students prepare worksheets in their data books, using the headings in Table 23.2. Most of the organisms are included in the protist/cyanobacteria key and *A Catalog of Living Things*; see the preparations for Investigation 12.1 for additional resources. If you purchased 2mm × 2-mm ruled microscope slides for Investigation 2.1, they can be used for this investigation. Have students refer to Step 6 of Investigation 2.1, and be sure they count organisms in the same total area.

Your major role in this investigation is to help students with identification and calculation problems. Table T23.1 lists the Palmer scores for the organisms in Ward's Water Quality Assessment Kit.

INVESTIGATION 23.2 A Field Study of a Stream

Materials (per class of 30, teams of 5)
12 10X hand lenses
1 field microscope (optional)

Table T23.1 Palmer Scores for Algae and Cyanobacteria in the Ward's Samples

Algal Genera	Water Sample 1	Water Sample 2	Water Sample 3	Palmer Score
Euglena			X	5
Oscillatoria		X	X	S
Chlamydomonas		X	X	4
Scenedesmus		X	X	4
Chlorella	X	X	X	3
Synedra	X	X		2
Phacus	X	X		2
Stigeoclonium	X	X		2
Closterium	X	X		1
Pandorina	X	X		1
Micrasterias	X			
Staurastrum	X			
Pediastrum	X			
Bulbochaete	X			

6 nonmercury thermometers
18 forceps
6 stopwatches or wrist watches with second hands
36 30-cm wooden or metal stakes
3 100-m measuring tapes
12 laminated black-and-white cards
30 pairs of thick cotton gloves
30 pairs of waders, hip boots, or old sneakers
6 1 m × 0.75-m screened frames
6 1-m × 0.75-m screenless frames plankton net (optional)
12 large trays filled with stream water keys for aquatic organisms
6 lemons

Preparations

Try to choose a section of the stream that has a variety of habitats for aquatic organisms—relatively fast- and slow moving water, sunny and shady parts, rocky and sandy bottoms. For the first part of the investigation, try to find a 50-m section of the stream of relatively uniform width. You may wish to allow the class to study the same part of the stream initially and then divide them into teams, having each team develop its own hypotheses about the type of organisms found and the factor(s) that may be related to them. In any case, in the third part of the study it is vital that students generate their own hypotheses based on their data and observations.

You may want to prepare forms for recording the cover students find within the frames on the streambed and the count of organisms they collect. Use pencils to record data; ink may be lost with water.

The mesh of regular window screen is too large to catch microscopic invertebrates and larval insects. Using two layers of screen oriented differently in the frame will reduce mesh size. A plankton net can be made from a nylon stocking with the top stretched around a bent wire hanger. Cut off the stocking at the ankle and attach a small jar to collect microscopic organisms. If possible, provide dip nets and insect nets to collect materials in and around the water. Have 3 × 5 unlined index cards and 3 × 5 pieces of black construction paper laminated together for observations with the hand lens. Ward's Natural Science Establishment carries an inexpensive 20X field microscope (catalog # 25M5502) that will greatly enhance observations of microscopic organisms. Alternatively, students can use screw-cap jars to collect samples (including bottom matter) from various locations. The jars should be uncapped in the laboratory and allowed to settle before examination.

Encourage students to make sketches of the stream, the surrounding vegetation, and the types of organisms they find. Most students will not have seen the larval stages of stream dwelling organisms before. Some are quite bizarre and can be a challenge for the good artist to draw. Students who do not wish to enter the water may be assigned tasks along the banks, including responsibility for drawing organisms.

Useful references include the following:

R. W. Merritt and K. W. Cummins, *An Introduction to the Aquatic Insects of North America* (Dubuque, IA: Kendall/ Hunt, 1978);

H. F. Chu, *How to Know the Immature Insects* (Dubuque, IA: Wm. C. Brown, 1949);

S. Eddy and J. C. Underhill, *How to Know the Freshwater Fishes*, 3rd ed. (Dubuque, IA: Wm. C. Brown, 1978);

J. F. Fitzpatrick, Jr., *How to Know the Freshwater Crustacea* (Dubuque, IA: Wm. C. Brown, 1983);

T. L. Jahn et al., *How to Know the Protozoa*, 2nd ed. (Dubuque, IA: Wm. C. Brown, 1978);

D. M. Lehmkuhl, *How to Know the Aquatic Insects* (Dubuque, IA: Wm. C. Brown, 1979);

G. W. Prescott, *How to Know the Aquatic Plants*, 2nd ed. (Dubuque, IA: Wm. C. Brown, 1980);

G. W Prescott, *How to Know the Freshwater Algae*, 3rd ed. (Dubuque, IA: Wm. C. Brown, 1978).

INVESTIGATION 23.3 Effects of Salinity on Aquatic Organisms

Materials (per class of 30, teams of 2)
30 microscope slides
30 cover slips
15 dropping pipets
15 compound microscopes
15 pieces of paper towel
aged tap water
15 dropping bottles of 5% NaCl (100 mL total)
5 sprigs of *Elodea*

60 samples of aquatic organisms,
15 each in aged tap water and in 1%, 3%, and
  5% NaCl solutions

Preparations

Devote some time to a prelaboratory discussion. Be sure each student has developed appropriate hypotheses and understands how to use them in the experimental procedure. Each team should test the effects of all concentrations of NaCl on a particular organism. Each type of organism should be tested by more than one team to allow for comparisons. Provide a basis for comparison by having all students observe the effect of a 5% salt solution on *Elodea* cells. The "shrinking" of the cytoplasm away from the wall is very dramatic and easily observed.

Many different aquatic organisms and concentrations of NaCl may be used. The following are suggested largely on the basis of availability: elodea, *Spirogyra* or other filamentous algae, *Vorticella, Paramecium, Euglena, Hydra, Artemia* (brine shrimp), *Daphnia* or cyclops, and rotifers. Ward's Natural Science Establishment has prepared a kit for this investigation (catalog # 87M9120) that includes cultures of paramecia, daphnia, and spirogyra.

If you have rich cultures of several organisms, add 2 or 3 full droppers of culture to small containers of the appropriate NaCl solutions at the beginning of each class period. Use aged tap water that has been allowed to stand overnight to drive off the chlorine. You may want to prepare a hay or lettuce infusion 3 to 4 weeks before this investigation and have the students use whatever species develop in it. *Artemia* eggs can be hatched a few days before the investigation.

INVESTIGATION 23.4 Eutrophication

Materials (per class of 30, teams of 3)
30 pairs of safety goggles
30 lab aprons
30 18-mm: 150-mm test tubes
30 18-mm: 150-mm test tubes
10 dropping pipets
4 Apple IIe, Apple IIGSF Macintosh, or IBM computers
4 LEAP-System™
4 turbidity probes
printer (optional)
30 density indicator strips (optional)
10 test-tube racks
10 glass-marking pencils
4 L aged tap water
4 L aged detergent water
4 L aged fertilizer water labeled ***CAUTION: Irritant***
euglena culture

Preparations

Review the LEAP-System Open Me First, Reference Manual, and BSCS Biology Lab Pac 2 to become familiar with how to operate the system, allocate disk space for saving data, connect the probes, boot the system, and bring up the menu on the computer screen. Be sure students know how to put the system on WAIT, take the system off WAIT, REINITIALIZE the system, put the system on SAVE, and QUIT the system for your computers; and that space is available on the disk(s) for each class so data can be saved and stripcharts run. Calibrate the turbidity probe with a test tube of aged tap water before use; directions are included in the LEAP-System BSCS Biology Lab Pac 2. Make certain the cap of the turbidity probe is securely in place before taking readings. Extraneous light may produce spurious data.

If your computer has a time card, you may wish to take hourly readings of several test tubes over a weekend and save the data to print as a stripchart. Four or more turbidity probes can be connected to one LEAP-System, depending on the number of inputs. The standard LEAP-System includes specifications and probes sufficient to run each experiment for one team at a time. The system also can handle multiple teams concurrently. See the reference manual for detailed instructions about how to create new experiments or update existing ones. Using these procedures, you can generate your own multi-team experiments. Disks with completed multi-team experiments also are available for purchase. Additional probes may be required in some cases.

Before printing stripcharts, check "System Maintenance" in the LEAP-System Main Menu to be sure the printer specified corresponds to the one you are using. See the reference manual for details on how to specify the printer.

If the LEAP-System is not available, students can use Insta-Rate™ density indicator strips (Ward's catalog # 15M1996), which consist of an adhesive panel containing five screened bars that is affixed to the culture tube. Using the shaded bars, students can observe the increasing density of the growing culture and graph the growth (bar number vs. time). Although the determination of cell density is subjective, it is adequate for quantifying relative differences in cell density among various environmental treatments. Instructions for use are included with the strips.

Purchase a small container of fertilizer or ask students to bring some from home. Try to use a brand in which the phosphate level (the second number in the formula) is higher than the nitrate (first number) or potash (third number) levels. A 16-24-16 formula seemed to work well during the field testing of this investigation. Age the water for 24 hours to remove any chlorine, then add a rounded tablespoon of fertilizer to one of the containers. Clearly label this container *fertilizer water*. Add about a tablespoon of detergent (laundry or dishwashing) to a second container labeled *detergent water*. Label the third container *tap water.*

The time required to achieve good results depends on the strength of the fertilizer or detergent, the intensity of the light, the temperature, and the organisms used. Euglena cultures produce a good "bloom" in one to two weeks. Cyanobacteria such as *Oscillatoria, Anabaena,* or *Cyanophora* may take longer—two to three weeks. Lower temperatures and light levels will lengthen the time required to get results. If you use euglena cultures, students should store their test tubes in the light, but not in direct sunlight. For optimum results, use fluorescent, incandescent, or grow lights placed 3242 cm away from the cultures, and provide a cycle of 16 hours light and 8 hours dark.

CHAPTER 24
MANAGING HUMAN-AFFECTED ECOSYSTEMS

Planning Ahead

Begin thinking and planning for next year. Review the beginning sections in the Teacher's Edition, particularly those you did not have time to digest completely. Refer to the materials and safety sections before you place your orders for next year.

Concepts

◆ The increasing ability of humans to use the environment for their own benefit has become a major factor in ultimately causing world ecological problems.

◆ Nonconservative agricultural and ranching practices are the most visible signs of human ability to modify the environment and are evidence of world ecological deterioration.

◆ As technology continues to be more sophisticated, the demand for resources increases and the waste products of technological processes increase, resulting in depletion of resources and the accumulation of wastes in the environment.

◆ The historical environmental degradation of the earth is closely related to human population growth. The human population is dangerously near or past the carrying capacity of the earth. If the human population, as we know it, is to survive, its overgrowth must be brought into check.

◆ Environmental decay is affecting the entire world and requires coordinated international efforts, even world laws, that will benefit all humans.

◆ Rather than use technology to "conquer" nature, we must use these tools to establish a coexistence with our ecosystem.

◆ The promising concept of sustainable agriculture proposes systematic crop rotation and the return of crop wastes and organic materials to the soil, so that very little leaves or is added to the system other than the needed agricultural products themselves.

◆ There is hope for the human race, but only if we can contain our population growth, frugally conserve energy, water, and other natural resources,

immediately employ worldwide sustainable agriculture practices, minimize and contain our personal and industrial wastes, recycle materials we do use, and resolve world conflicts without the heavy ecological costs of war.

◆ The hope for the future includes a personal commitment to ecologically sound practices. The commitment will be demonstrated by making decisions at local through worldwide levels based on what is ethically best for the future of all organisms on the earth.

Objectives

Students should be able to:

◆ *analyze* history textbooks for ecological information.

◆ *relate* cultural evolution to technological development.

◆ *distinguish* between the roles of ancient or primitive humans and the roles of modern humans in ecosystems.

◆ *state* two or more ecological consequences of the agricultural revolution.

◆ *explain* the effects of developing technology on the extension of the ecosystems of which human populations are a part.

◆ *evaluate* arguments for and against additions to the human-affected environment.

◆ *analyze* newspaper data about population, transportation, and pollution, and project future scenarios.

◆ *infer* future population growth given current population structures.

◆ *explain* why international cooperation is needed to solve environmental problems.

◆ *give* examples of how technology can solve some environmental problems.

◆ *discuss* the principles of bioremediation.

◆ *explain* the principles of sustainable agriculture and contrast it to current practices.

◆ *demonstrate* the ability to make informed decisions about land use.

◆ *explain* their attitudes toward at least two major current human problems.

◆ *demonstrate* a personal commitment to conservation of the earth's resources and to preserving the quality of the environment.

Strategies

Though the book has emphasized the human ecosystem throughout the year, it is the special focus of Chapter 24. The history of humans is considered, along with our species' increasing impact on the biosphere. This chapter deliberately introduces value-laden, controversial topics—not for the sake of sensationalism but because these topics inevitably will be a part of students' lives, and biological knowledge is needed for making decisions about them.

Topics considered in the chapter should by no means limit the discussion. Whenever possible, the information and concepts gained during the year should be woven into projections for the future. Above all, try to make all students feel their personal responsibility for applying their knowledge of science to decisions their generation will make.

Chapter 24 links biology to social studies. Some students may want to investigate ecological issues in United States or in world history as a special project. In addition to its other functions, Chapter 24 presents an opportunity to review almost all aspects of the course. The marginal notes call attention to many of the ideas and some of the vocabulary in earlier chapters, but you can augment these references.

The activities in Chapter 24 can be done in almost any order because there are so many interrelationships among the concepts. You are especially encouraged to share with your students the concepts listed above, as these should be among the most important they take away with them. Devote at least one or two weeks to this chapter, which is one of the most important in the text. Use large-scale ecological problems or current world conflicts (such as the Persian Gulf oil spill and war of 1991) to illustrate the importance of global cooperation in resolving major environmental issues and to point out the terrible environmental cost of war. Students need to understand that most issues are not one-sided and that to resolve them requires understanding of the perspectives of all involved, followed by a discussion of the problems by all concerned. Once your students become aware of the concepts behind world ecological problems, they will be able to participate actively in their solutions.

References

Barrett, S. C. H. "Waterweed Invasion." *Scientific American* (October 1989): 90–97.

Caldwell, J. C., and P. Caldwell. "High Fertility in Sub-Saharan Africa." *Scientific American* (May 1990): 118–25.

Crosson, P. R., and N. J. Rosenberg. "Strategies for Agriculture." *Scientific American* (September 1989): 128–35.

Davis, D., and H. L. Bradlow. "Can Environmental Estrogens Cause Breast Cancer?" *Scientific American* (October 1995): 166–172.

Frost, R. "The Industrial Ecology of the 21st Century." *Scientific American* (September 1995): 178–186.

Gray, C. L., Jr., and J. A. Alson. "The Case for Methanol." *Scientific American* (November 1989): 108–15.

Kates, R. "Sustaining Life on the Earth." *Scientific American* (October 1994): 114–122.

"Managing Planet Earth." *Scientific American* (September 1989): several articles.

Reganold, J. P., R. I. Papendick, and J. F. Parr. "Sustainable Agriculture." Scientific American (June 1990): 112–21.

Sapolsky, R. M. "Stress in the Wild." *Scientific American* (January 1990): 116–23.

Suggested Answers to Applications

1. One major problem is that hunting is a high-risk, low-return activity in which the males could be hurt while hunting and in which there is no guarantee they will come back with meat. A bigger problem is the lack of control over the food supply from one year to another—it is either feast or famine. In bad years, the population of hunter-gatherers may decline dramatically.

2. These people would hunt and gather food in one familiar territory. However, this food supply soon would be depleted, forcing the people to move to better hunting grounds.

3. Agriculture probably developed independently in many different parts of the world. If agriculture had developed in only one area and the idea of agriculture was spread around the world, then the people who invented it also would have spread the domesticated plants with which they were familiar. To the contrary, however, people around the world first domesticated plants found growing wild in their own native regions.

4. Compared with many animals, humans really have few advantages other than their intelligence. (They are not the strongest, fastest, or fiercest.) Much learning involves trial-and-error, and if the life of an individual depends on such learning there is a good chance that the individual may not survive. If, however, an individual survives the trial, and if there is a way to communicate the results (through a symbolic language), then information can be passed from one individual to another, thereby greatly increasing the chances of survival for the others.

5. The domesticated species were useful to humans— they could be eaten or used to make clothing or medicines. Plants must return a lot of product for the effort involved in growing them, and they must be able to survive in the fields that humans create. Animals must have the right temperament to live closely together and to be managed without causing humans harm.

6. There is no "correct" answer to this question. It is designed to elicit reactions from students and should be used as the basis for a debate. Have students develop a list of reasons supporting their point of view, then bring the class together. Of course, the United States can feed more people, but it is not just feeding people that students need to be concerned with. What about the resources used by each person in the United States?

7. Their manufacture and use require large amounts of metals, glass, and energy. They have made suburban sprawl possible and modern highways necessary. They have contributed to air pollution.

8. The sparsely inhabited areas have little water and poor topsoil. For these areas to support larger populations, food and water would have to be taken from other areas, which might have none to spare.

9. The nitrogen that is depleted during one year is replaced by the nitrogen-fixing legumes during the next year.

10. This problem deals with the "lifeboat ethic." If you try to save everyone by pulling them into the lifeboat (feeding all the people) you risk drowning *all* the people in the lifeboat (creating more problems in starving countries). Is it better to save a few people than to risk the lives of many? Feeding these children would allow them to survive, grow up, and have more children, which could create an even greater population problem. Help your students understand the consequences of their actions and deal with the ethics of this question.

11. The Hopi Indians of the southwestern United States were among the people who shared this philosophy.

12. Less corn would be needed because much energy is lost when corn reaches us by way of beef.

Suggested Answers to Problems

1. Students could begin by looking at plans of the space shuttle. Focus on what is essential to humans for survival, such as a source of oxygen and heat, and what is necessary for long-term living, such as privacy, adequate lighting, and space.

2. Jerusalem artichokes, sunflowers, cranberries, walnuts, and maple syrup make up a large proportion of the foods students might eat that came from what is now the United States. The list is short. An excellent resource is N. W. Simmonds, ed., *Evolution of Crop Plants* (London: Longman Press, 1976).

3. **a.** Sri Lanka has higher birth and death rates; therefore, there is a great preponderance of young people, and the life expectancy of an individual is low.

   b. The young and the old age groups have increased in size relative to the middle age group in the United States because of better control of childhood infectious diseases and the diseases of old age. The increase in young people also results from a high natality rate. This distribution of age groups in the population places an increasing economic burden on the middle age class, which is the principal supporting one.

   c. The birthrate for the population as a whole will decline, since the number of reproducing individuals declines.

   d. The rate of population increase will be greater in nation B, since the earlier age of delivery of the first child reduces the length of a generation. If the proportions of the population of childbearing age in both nations were known, the hypothesis would be on firmer ground.

4. Students should consult information on France where nuclear power supplies almost 25 percent of the total energy needs of the country. Information about all the environmental problems discussed in this chapter can be found in such books as W. H. Corson, ed., *The Global Ecology Handbook* (Boston: Beacon Press, 1990).

5. Students should realize they use resources from around the world. Certainly machines that use fossil fuels would be affected if there were a shortage of gasoline and oil. Students also should know that many petroleum-based products they use—including anything made with plastic— would be affected. The United States imports other nonfuel mineral materials, such as columbium and manganese, which are necessary for the production of aircraft engines (among other things).

6. One trend was toward the greater mechanization of farming. Another was toward the increased use of fossil-fuel-based products such as gasoline, pesticides, herbicides, and fertilizers. Students

also can research changes in lawn care. How prevalent were power mowers 50 years ago—or watered, fertilized, and manicured lawns?

7. A small book by the Earthworks Group, *50 Simple Things You Can Do to Save the Earth* (Berkeley, CA: Earthworks Press, 1989), is one place for your students to get started.

8. The extent to which students can do this is a fair measure of their grasp of the major aspect of this course.

9. Answers will vary.

Materials and Preparations for Investigations

INVESTIGATION 24.1 Views of the Earth from Afar

Materials (per class of 30, teams of l)
none

Preparations
You may wish to obtain infrared aerial photos of tropical rain forest areas from NASA's Jet Propulsion Lab for students to compare.

INVESTIGATION 24.2 Decisions About Land Use

Materials (per class of 30, teams of 10)
3 sets of role cards
10 maps

Preparations
Provide several enlarged copies of the map for each team. The role cards are included in Appendix 3; both role cards and an enlarged map are provided as blackline masters in the Teacher's Resource Book.

You may wish to have students read the background information on Winterville outside of class. Review the basics of reading and interpreting land forms from a topographic map. Be certain students can distinguish a steep slope from nearly level land.

APPENDIX TWO

SUPPLEMENTARY INVESTIGATIONS

Materials and Preparations for Investigations

INVESTIGATION A2.1 Chemical Safety

Materials (per class of 30, teams of 2)
30 pairs of safety goggles
30 lab aprons
30 pairs of plastic gloves

PART A
none

PART B
45 microscope slides
45 cover slips
15 dropping pipets
45 flat toothpicks
15 compound microscopes
*15 dropping bottles concentrated brilliant cresyl blue stain labeled **CAUTION: Irritant**
15 dropping bottles dilute brilliant cresyl blue stain labeled **CAUTION: Irritant**
15 squeeze bottles
Detain™
1 jar paramecium culture

PART C
30 glass dropping pipets
30 100-mL beakers
white paper
15 pairs of scissors
*1.5 L red food dye (25 mL concentrated)

PART D
15 100-mL glass beakers
15 spatulas
*1 box baking soda (sodium bicarbonate)
*1 pint distilled vinegar (acetic acid) labeled **CAUTION: Irritant**

Preparations
Ward's Natural Science Establishment has prepared a kit for this investigation (catalog # 36M3951) that includes the items marked*. The concentrated stain for Part B is a 2% alcoholic solution of brilliant cresyl blue; dilute 1: 9 in distilled water for the diluted preparation. To prepare the red food dye for Part C, add the contents of the bottle provided in the kit to 1.5 L of water.

Wear safety goggles, face shield, lab apron, and nitrile rubber gloves to set up the demonstration of corrosive solids. Do not allow students to come into contact with any of the corrosive solids used. All students should wear safety goggles, lab apron, and protective gloves.

Materials
*5 plates of plain agar numbered *1* through *5*
*plastic forceps
*sodium hydroxide crystals or pellets (plate 1) labeled **DANGER: Corrosive solid**
*potassium hydroxide crystals or pellets (plate 2) labeled **DANGER: Corrosive solid**
*phenol crystals (plate 3) labeled **DANGER Poison; Corrosive solid**
*sodium chloride—lumps of rock salt (plate 4)
*sugar cubes (plate 5)

Use plastic forceps to remove small quantities (1–3 pellets or small lumps) of each of the test solids and place them on top of the agar in separate, numbered petri dishes. Cover and seal the plates and record the start time of the experiment.

At the conclusion of this demonstration, add 1M HC1, drop by drop, to each of the plates containing corrosive bases until neutrality is reached according to litmus test. Discard plates only after this step.

INVESTIGATION A2.2 The Compound Microscope

Materials (per class of 30, teams of 2)
45 cover slips
45 microscope slides
15 dropping pipets
15 compound microscopes
15 pairs of scissors
15 transparent metric rulers
15 pieces of lens paper
1 section newspaper want ads

Preparations
Newspapers or the indexes of catalogs are a good source of lowercase letters in fine print. If the letters o, c, e, and r are in limited supply, you can substitute letters that demonstrate the same phenomena. The letter o shows the degree of magnification and detail, c shows the reversal feature, and e or r shows that images are inverted as well as reversed by the lenses of the microscope.

INVESTIGATION A2.3 The Compound Microscope: Biological

Materials (per class of 30, teams of 2)
30 pairs of safety goggles
30 lab aprons
15 microscope slides
15 cover slips
15 dropping pipets
15 100-mL beakers containing water
15 compound microscopes
15 pieces of lens paper
15 paper towels
15 dropping bottles of Lugol's iodine solution labeled
   WARNING: Poison if ingested; Irritant
25 mL yeast culture
15 3-mm cubes of white potato

Preparations
Cut white potatoes into approximately 3-mm cubes. Store in water in a small container; the pieces can be kept through several periods of the day. See A Table of Recipes for preparation of Lugol's iodine solution and yeast culture.

BSCS Biology

BSCS Biology
An Ecological Approach

EIGHTH EDITION
BSCS Green Version

BSCS
Pikes Peak Research Park
5415 Mark Dabling Blvd.
Colorado Springs, CO 80918-3842

Revision Team
William J. Cairney, BSCS, Revision Coordinator
J. Frank Cassel, North Dakota State University
Patricia Cully, Educational and Editorial Consultant, Needham, Massachusetts
Janet Chatlain Girard, BSCS, Art Coordinator
Kenneth G. Rainis, Ward's Natural Science Establishment, Inc.
Gordon E. Uno, University of Oklahoma, Norman, Oklahoma
Linda K. Ward, BSCS, Executive Assistant

KENDALL/HUNT PUBLISHING COMPANY
4050 Westmark Drive, Dubuque, IA 52004

Contributors

Barbara Andrews, Colorado Springs, Colorado
Charles R. Barman, Indiana University at Kokomo
Don Chmielowiec, Ward's Natural Science Establishment, Inc.
Edward Drexler, Pius XI High School, Milwaukee, Wisconsin
Vernon L. Gilliland, Liberal Senior High School, Liberal, Kansas
Lynn Griffin, Cherry Creek High School, Englewood, Colorado
Donald P. Kelley, South Burlington, Vermont
Joseph D. McInerney, BSCS
George Nassis, Ward's Natural Science Establishment, Inc.
Larry N. Norton, Quantum Technology, Inc.
Chester Penk, Quantum Technology, Inc.
Barry Schwartz, Chatfield High School, Littleton, Colorado
John Settlage, Boston, Massachusetts
Carol Leth Stone, Alameda, California
Richard R. Tolman, Brigham Young University, Provo, Utah
David A. Zegers, Millersville University, Millersville, Pennsylvania

Safety Consultant

Kenneth G. Rainis, Ward's Natural Science
    Establishment, Inc.

BSCS Administrative Staff

Timothy Goldsmith, Chair, Board of Directors
Joseph D. McInerney, Director
Larry Satkowiak, Chief Financial Officer

Artists

Susan Bartel
Bill Border
Marjorie C. Leggitt
Audrey Penk
Brent Sauerhagen
Page Louis Thomas
Linn Trochim/Animart

Grateful ackowledgement is made to the following individuals for permission to reprint from their work:

D. Nelkin and L. Tancredi, *Dangerous Diagnostics: The Social Power of Biological Information* (New York: Basic Books, 1989).

N. A. Holtzman, *Proceed with Caution: Predicting Genetic Risks in the Recombinant DNA Era* (Baltimore, MD: Johns Hopkins University Press, 1989).

Acknowledgements continue on page 771

Editorial, design, and production services provided by
PC&F, Hudson, New Hampshire

Content Reviewers

W. Carter Alexander, Armstrong Research Laboratory, San Antonio, Texas
Kenneth J. Andrew, Colorado College, Colorado Springs, Colorado
J. Frank Cassel, North Dakota State University, Fargo, North Dakota
Russell DeFusco, USAF Academy, Colorado Springs, Colorado
James S. Kent, USAF Academy, Colorado Springs, Colorado
J. Douglas Ripley, USAF Directorate of Environment, Washington, D.C.
Gordon E. Uno, University of Oklahoma, Norman, Oklahoma

Teacher Reviewers

Richard Benz, Wickliffe High School, Chardon, Ohio
Richard Bergholz, Montesano High School, Montesano, Washington
Christine Breuker, Western Reserve Academy, Hudson, Ohio
Richard E. Carman, Wadena Senior High School, Wadena, Minnesota
Stuart Caudill, Ashbrook High School, Gastonia, North Carolina
Russel F. Hansen, Western Reserve Academy, Hudson, Ohio
Ned Hatfield, Bonny Eagle High School, West Boxton, Maine
Gregory S. Hennemuth, Lake Region Union High School, Orleans, Vermont
Peggy Keeling, St. Paul Academy and Summit School, St. Paul, Minnesota
Sandra Kransi, Athens High School, Athens, Michigan
Mariellen Larson, Troy Athens High School, Troy, Michigan
Rex Miller, Roy J. Wasson High School, Colorado Springs, Colorado
Bob O'Hara, Cooper High School, New Hope, Minnesota
Patricia D. Schmitt, The Bishop's School, Encinitas, California
Gary Smith, Katella High School, Anaheim, California
Paula J. Thorpe, Woodland Park High School, Woodland Park, Colorado
Holly Williams, Storm King School, Cornwall-on-Hudson, New York
Sandra Williams, Serra High School, El Cajon, California
Sherry Yarema, Naperville Central High School, Naperville, Illinois

Acknowledgements

The following teachers contributed comments and suggestions for this edition:

Amy Adams, Washington Middle School, Green Bay, Wisconsin; Bruce Alexander, Chapel Hill High School, Pittsboro, North Carolina; Mary Ashton, Orange High School, Pepper Pike, Ohio; Shirley Ballack, Joliet Central High School, Joliet, Illinois; Bob Beaudoin, Franklin Middle School, Green Bay, Wisconsin; Barbara R. Beitch, Hamden Hall Country Day School, Hamden, Connecticut; William Benner, Joliet Township High School, Peotone, Illinois; Richard Benz, Wickliffe High School, Wickliffe, Ohio; Rick Berken, Preble High School, Green Bay, Wisconsin; Dave Birschbach, East High School, Central Office, Green Bay Wisconsin; Judy Bond, Lakeland High School, Lakeland, Florida; Luann R. Bridle, Forsyth Country Day School, Lewisville, North Carolina; Kirk Brill, Southeast Polk High School, Runnells, Iowa; Don Brown, Joliet West High School, Joliet, Illinois; Don Buntman, West High School, Green Bay, Wisconsin; Doug Cameron, St. Andrews, Tennessee; Stuart Caudill, Ashbrook High School, Gastonia, North Carolina; Lyle Chee, Mid Pacific Institute, Hawaii.

James Colman, Bow Basin Junior and Senior High School, Bow Basin, Wyoming; Gale Cook, Northeast High School, Lincoln, Nebraska; Kathleen L. Davis, Roosevelt High School, Wyandotte, Michigan; Ken Davis, Woodway High School, Spokane, Washington; W. C. Dicks, Northville High School, Novi,

Michigan; Mary Diedrich, West High School, Depere, Wisconsin; Lucille M. Dostie, Wiscasset High School, Wiscasset, Maine; Clarice Douoguih, Montclair Kimberley Academy, Montclair, New Jersey; Mark Eberhard, St. Clair High School, New Baltimore, Michigan; John L. Eckert, Hutchinson High School, Hutchinson, Kansas; Rod Epp, Hastings Senior High, Hastings, Nebraska; Jon Forney, Grant High School, Grant, Nebraska; Claudia R. Fowler, Louisiana State University Laboratory School, Baton Rouge, Louisiana; Patricia Foy, Charter Oke High School, Corina, California; Conradt Fredell, Clear Creek High School, Idaho Springs, Colorado; Martha Friedlander, Dexter, Michigan; Francis H. Gately Jr., Algonquin Regional High School, Northboro, Massachusetts; Vernon Gilliland, Liberal, Kansas.

Mary Ann Gillis, Franklin Middle School, Green Bay, Wisconsin; James Glock, West High School, Green Bay, Wisconsin; Dwane Grace, Celina Senior High School, Celina, Ohio; Mark E. Greene, Algonquin Regional High School, Northboro, Massachusetts; Howard Grimm, Upper Arlington High School, Arlington, Ohio; Eileen T. Grosso, Minooka High School, Minooka, Illinois; Warren Hagestuen, Park Center Senior High, Brooklyn Park, Minnesota; Thomas Hagewood, Schuyler Central High School, Schuyler, Nebraska; Russell Hanseter, Seymour High

School, Seymour, Wisconsin; Charles Hawkins, Franklin Middle School, Green Bay, Wisconsin; Philip Heck, Middle Township High School, Cape May Court House, New Jersey; Dale Hertel, Fargo South High School, Fargo, North Dakota; Holly Heverly, Lake Region Union High School, Orleans, Vermont; George Hrab, Montclair Kimberley Academy, Montclair, New Jersey; Louise Huey, Clemson, South Carolina; Stephanie Hunter, Richmond, Virginia; John E. Hutchins, Hanover High School, Hanover, New Hampshire; Debora Hutchison, Washington Middle School, Green Bay, Wisconsin; Robert Hutter, Southwest High School, Green Bay, Wisconsin; Leon Jacques, Edison Middle School, Green Bay, Wisconsin; P. R. Janda, Shelton High School, Shelton, Washington; Marion V. Jaskot, Scotch Plains-Fanwood High School, Scotch Plains, New Jersey; Ron Johnson, Sycamore High School, Sycamore, Illinois; Connie Jones, Enka High School, Enka, North Carolina; Judy Jones, Chapel Hill High School, Chapel Hill, North Carolina; Bob Keel, Sierra High School, Colorado Springs, Colorado; Tim Kennedy, Celina Senior High School, Celina, Ohio; Ed Kidder, Richmond, Virginia; Kenneth Kiesner, Edison Middle School, Green Bay, Wisconsin; Bruce Kilmer, Edison Middle School, Green Bay, Wisconsin; Michael L. Kimmel, Conneaut High School, Conneaut, Ohio; Dan Kirsch, Bay Port High School, Green Bay, Wisconsin; Lanny R. Kizer, Loup City High School, Loup City, Nebraska; Robert Klaper, Wheaton Central High School, Wheaton, Illinois; Michael Kobe, School City of Hammond Indiana, Hammond, Indiana; Evelyn F. Kolojychrek, Cranberry Area School District, Seneca, California; Robin Krause, Knob Noster High School, Knob Noster, Missouri; Steve Krings, Southwest High School, Green Bay, Wisconsin.

Laurel Krol, School City of Hammond Indiana, Hammond, Indiana; Beverly Lauring, St. Petersburg, Florida; Sandy Leotta, Casper, Wyoming; Stanley A. Likes, Pawnee High School, Pawnee, Illinois; Bill Lipp, Fargo South High School, Fargo, North Dakota; Jerry Loynachan, Wheaton Central High School, Wheaton, Illinois; Barbara Mack, Conway, South Carolina; Susan W. Majors, Weston High School, Weston, Massachusetts; Gary Manfready, Mulberry, Florida; James Martin, Wheaton North High School, Wheaton, Illinois; Nancy Martinez, Emery High School, Huntington, Texas; Larry M. Maurer, Concord High School, Wilmington, Delaware; Paul D. McIver, Englewood High School, Englewood, Colorado; Philip P. Merman, Fairfield High School, Fairfield, Connecticut; James H. Meyer, Cedar Falls High School, Cedar Falls, Iowa; Robert D. Miller, Parkland High School, Orofield, Pennsylvania; John Monsma, N. Michigan Christian High School, McBain, Michigan; Ted Munnezke, The Principia Upper School, St. Louis, Missouri; Ken Nicholson, Arlington Memorial High School, Arlington, Vermont; Susan Nuemann, Chapel Hill High School, Chapel Hill, North Carolina; Wayne Olm, Preble High School, Green Bay, Wisconsin; Jeff Olofson, St. Francis Community High School, St. Francis, Kansas; Albert A. Perez, Leilehua High School, Wahiawe, Hawaii; John Petrie, Milan High School, Milan, Michigan; Bob Pfister, Kaukauna High School, Kaukauna, Wisconsin; Randy Phillips, East High School, Green Bay, Wisconsin; Frank Pirman, Preble High School, Green Bay, Wisconsin; Herbert T. Potter, Rio Mesa High School, Oxnard, California; Leon Raether, West High School, Green Bay, Wisconsin; Cathy Rager, Bretwood Junior and Senior High School, Brentville, Pennsylvania;

Kathleen Ranwez, Moore Junior High, Arvada, Colorado; Fran Reinke, George Rogers Clark Junior and Senior High School, Hammond, Indiana.

Raybun Reynolds, Kathleen Senior High School, Winter Haven, Florida; Joan Richardson, Greenfield High School, Greenfield, Massachusetts; Sheila Richardson, Lakeview Academy, Atlanta, Georgia; Doug Rosendahl, Winona Senior High School, Winona, Minnesota; Robb Ross, Liberal High School, Liberal, Kansas; Pamela Roth, Caseville Public Schools, Caseville, Michigan; W. Bruce Roup, Holly Junior and Senior High School, Holly, Colorado; Roger Santille, Upper Arlington High School, Arlington, Ohio; Glenn Schlender, Southwest High School, Green Bay, Wisconsin; Gene Schmidt, Monroe High School, Fairbanks, Alaska; Robert Schoob, Bolingbrook High School, Bolingbrook, Illinois; Lisa Schweiner, Edison Middle School, Green Bay, Wisconsin; Calvin Scott, Illiara Christian High School, Lansing, Illinois; Sylvia Scyphers, Ariton, Alabama; Terri Seeley, George Walton Academy, Lawrenceville, Georgia; Tina Servais, Edison Middle School, Green Bay, Wisconsin; Jeanne Shaw, Liberty High School, Brentwood, California; Albert J. Shelley, Chesapeake High School, Baltimore, Maryland.

Forest H. Shoemaker, Stordley Lake High School, Broomfield, Colorado; Chris Smith, Bethel High School, Hampton, Virginia; Frances M. Smith, Easley, South Carolina; William G. Smith, Moorestown Friends School, Moorestown, New Jersey; Glenn M. Snyder, Wheat Ridge High School, Wheat Ridge, Colorado; Kris Spiegler, Hempstead High School, School, New York, New York; Grace Stephen, Estes Park High School, Estes Park, Colorado; Sarah Stewart, Quincy Senior High, Quincy, Illinois; Lois C. Stoner, Chestnut Ridge Senior High School, New Paris, Pennsylvania; Susan Straten, Montclair Kimberley Academy, Montclair, New Jersey; Edgar R. Stuhr, Lexington High School, Lexington, Massachusetts; Richard Summers, Willard High School, Willard, Missouri; Jean Paul Thibault, Cape Elizabeth High School, Cape Elizabeth, Maine; Brother Robert Thomas, De La Salle High School, Minneapolis, Minnesota; Tony Toston, Summit High School, Colorado; Judith A. Treharne, Ocean Township High School, Oakhurst, New Jersey; Frank Turner, McKinley High School, North Canton, Ohio; Lloyd A. Turner, San Fernando High School, San Fernando, California; Howie Usher, Mingus Union High School, Cottonwood, Arizona; Gene Vander Velden, East High School, Green Bay, Wisconsin.

Grandon Voorhis, Glen Rock High School, Glen Rock, New Jersey; Ruth B. Vredeveld, Girls Preparatory School, Chattanooga, Tennessee; Larry Wakeford, Chapel Hill High School, Carrboro, North Carolina; Randyll Warehime, Farrington High School, Honolulu, Hawaii; Mary Watson, Bartlesville Mid-High School, Bartlesville, Oklahoma; David Wehner, Roosevelt High School, Lansing, Michigan; Maryanne Weins, Marine City High School, Marine City, Michigan; Linden Welle, Fort Morgan High School, Fort Morgan, Colorado; Steve Wester, Odell High School, Odell, Nebraska; Joyce Whitney, Edison Middle School, Green Bay, Wisconsin; Michelle Will, Shoshone High School, Shoshone, Idaho; Jim Willard, Schenck High School, East Millinocket, Maine; Robert Williams, Canton High School, Plymouth, Michigan; Tom Younk, Preble High School, Central Office, Green Bay, Wisconsin.

Contents

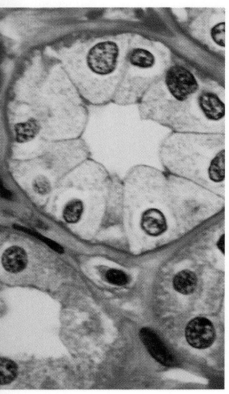

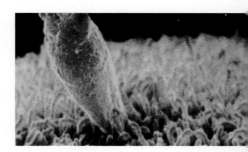

7 ◆ Continuity Through Development

8 ◆ Continuity Through Heredity

9 ◆ Continuity Through Evolution

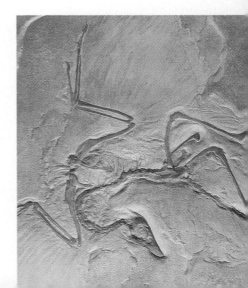

x   Contents

▬▬▬▬   Section Three **DIVERSITY AND ADAPTATION IN THE BIOSPHERE**

10 ◆ Ordering Life in the Biosphere

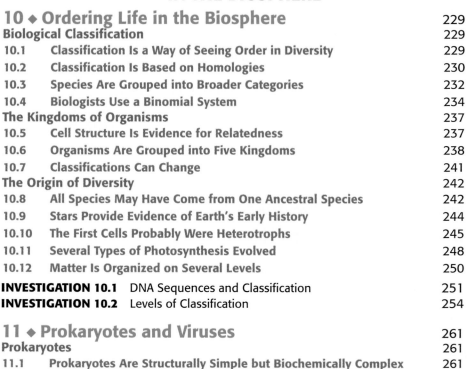

11 ◆ Prokaryotes and Viruses

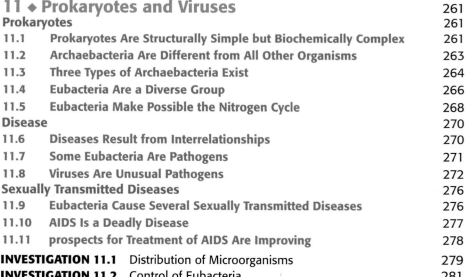

12 ◆ Eukaryotes: Protists and Fungi

13 ◆ Eukaryotes: Plants

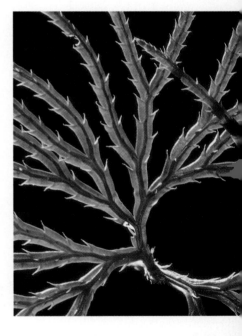

14 ◆ Eukaryotes: Animals

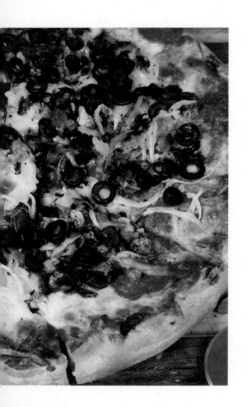

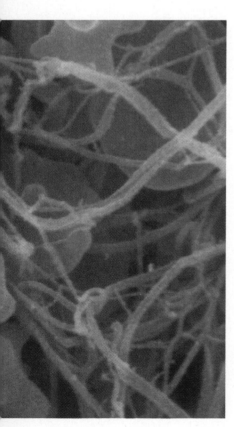

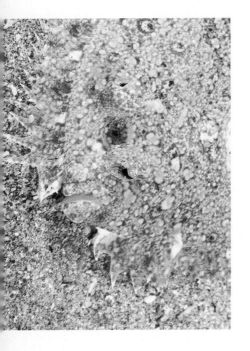

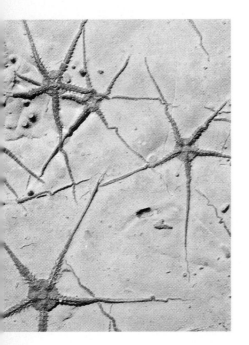

23 ◆ Aquatic Ecosystems

24 ◆ Managing Human-Affected Ecosystems

Biology Today Features

Pioneers Features

Foreword

This eighth edition of *BSCS Biology: An Ecological Approach* is the first edition of Green Version that will be used well into the twenty-first century, the beginning of a new millennium whose first decade will provide significant challenges to science education and to the quality of life on our ever-more-crowded planet. We have provided a variety of new features in this edition to help you and your students meet both challenges. These features address current research-based education and scientific concepts.

The publication of the *National Standards for Science Education* (NRC, 1995) and the *Benchmarks for Science Literacy* (AAAS, 1993) has encouraged all science teachers to examine their programs for the extent to which they meet the spirit of these documents. We have made a considerable effort in this edition to demonstrate the alignment of the textual materials and investigations with the standards and benchmarks. We have not had to change any of the central assumptions of Green Version to make it align with these documents because the standards and benchmarks are largely congruent with the approach to science education that BSCS has championed since 1958. The correlation of Green Version's content to national standards, therefore, is not based on a simple checklist of topics that appear in the program, but rather on an honest and rigorous analysis of the extent to which Green Version provides substantive opportunities for students to learn the concepts included in the standards and benchmarks.

This edition of Green Version provides other helpful tools to make your teaching of biology more efficient and effective. Examples include:

◆ an updated and expanded section on software and media resources and references, included in the Teacher's Resource Book.

◆ a new feature titled "Pioneers," which introduces students to pioneering ideas, people and technologies in biology.

◆ more multi-cultural examples of biological phenomena and more examples of female scientists in the features on "Pioneers" and "Biology Today."

◆ world-wide geographical examples in investigations and applications, and people and examples from various regions world wide in the Chapter 22, "Biomes Around the World."

The intent of these and other features, of course, is to help make an ecological view of biology more understandable and more interesting for your students. The world your students will shepherd into the next century will be filled with challenges and opportunities rooted in biology. The challenges range from the stewardship of declining biodiversity and the maintenance of environmental quality to the protection of individual privacy in an age of genetic testing and screening. The opportunities range from the practical, such as the increasing ability to improve the health status of much of the world's population, to the philosophic, such as the ability to enhance peaceful discourse among all of the world's diverse peoples as knowledge of the biological similarities that unite us comes to replace the fear and misunderstanding generated by cultural and ethnic differences.

We are hopeful that the conceptual basis of biology in Green Version, in conjunction with the implicit and explicit examples of the nature and methods of science, will provide your students with the knowledge and skills required to embrace the challenges and opportunities of the twenty-first century. As always, we welcome your feedback to help us determine whether we have succeeded.

Green Version Revision Team
William J. Cairney, Coordinator

April 1996

"How might pictures like this taken from space allow us to see the effects of humans on the earth?"

"This image was produced by the Jet Propulsion Laboratory's Oceanography Group using data from the TOPEX/Poseidon radar altimeter. TOPEX/Poseidon is a joint U.S./French satellite mission to study the earth from space. Scientists at NASA's Goddard Space Flight Center compiled the data for this planetary picture. NASA's long-term effort may reveal the impact of human activity on the earth. The color effects represent constantly changing features of ocean "climate.""

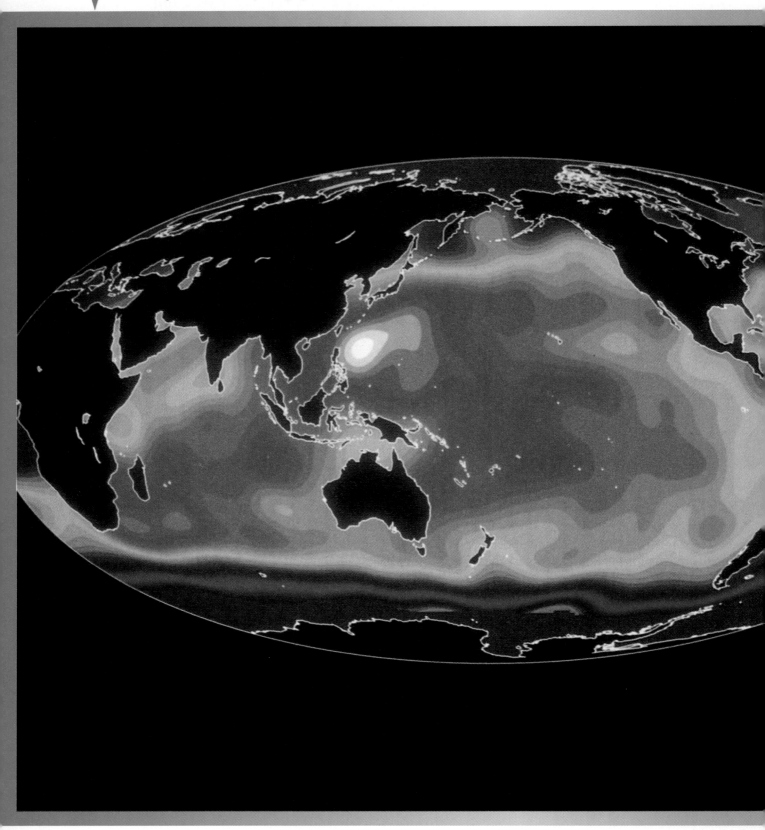

Section One

THE WORLD OF LIFE: THE BIOSPHERE

◆ Except for the most desolate and forbidding regions of the polar ice caps, all of the earth teems with life. The study of life is called biology, and it includes not only a study of the structure and function of individual organisms, but also a study of the relationships between organisms themselves and between organisms and their environment.

Sea horses, mushrooms, lilacs, mosquitoes, and parakeets are just a small fraction of the types of life found on earth. Of all the diverse forms of life, however, humans have the greatest effect on the earth and on other organisms. Section One examines the relationships between some familiar living things and begins to explain just how profound the effects of human activities can be. ◆

The Web of Life

As secure as they may appear to be, these red fox kits (Vulpes vulpes) are dependent on many things in their environment. Name three examples of such dependence in this photo.

They are dependent on plants, which provide food for their herbivore prey. The plants, in turn, are dependent on the nutrients, water, and sunlight in the environment. The foxes also are dependent on shelter and water.

◆ All living things are part of a complex network of interactions called the web of life. Much as the strands of silk in a spider's web bind one segment to another, feeding and other interactions bind all organisms to one another. Like other animals, humans are part of the web of life. We are reminded of that whenever our actions have visible effects on our surroundings.

Because of our numbers and our use of natural resources, we have the greatest impact of any living thing on the web of life. We can affect it unknowingly because many of the interrelationships are difficult to see or to understand. We may not notice that something has gone wrong until the problem is so large that we no longer can ignore it.

If the human population is to continue to thrive on earth, we must understand the interrelationships among all living things. If necessary we must modify our actions in order to decrease our impact on earth and the life it supports. When we have done this we can begin to solve such problems as global warming and acid rain. This chapter introduces the basic interrelationships among living things and provides a foundation on which to build solutions to some of the problems facing humankind. ◆

Each major heading in the chapter is accompanied by a guidepost in the form of a question. Direct students to look for answers to the question as they read. The class might construct a concept map of the biosphere to draw on knowledge students already have about biology.

MAJOR CONCEPTS

◆ All organisms interact both with their physical environment and with other organisms in the web of life.
◆ Any event that affects one organism in an ecosystem indirectly affects all others.
◆ All organisms must acquire energy and matter to live.
◆ Organisms in an ecosystem can be grouped by the way in which they acquire and contribute energy.
◆ Scientists use a variety of processes to answer questions about our environment.
◆ Controlled experiments help to determine cause and effect relationships among variables.

KNOWLEDGE CHECK

◆ How are organisms interrelated?
◆ What does interact mean?
◆ How do humans fit in the world?

Interactions Among Living Things

1.1 Organisms Interact with Organisms They Eat

Living things may affect one another in many ways. They may crowd each other, protect each other, poison each other, or give each other shelter. At the most basic and important level, however, living things consume one another. Feeding relationships are the major strands in the web of life.

Consider a brightly colored grasshopper sitting on a young plant and chewing a leaf. The grasshopper (see Figure 1.1a) is red, yellow, and black and very small. When the leaf is completely eaten, the grasshopper jumps toward another plant but lands on a sticky thread. The thread is just one of many that are carefully woven together into a large, shiny trap—a web. As the grasshopper struggles to free itself from the thread, more threads stick to it. With each movement, the grasshopper sends a vibration from thread to thread and finally to the maker of the trap. In the blink of an eye, a large brown and yellow spider (see Figure 1.lb) seizes the grasshopper and kills it with poison injected from the spider's fangs. Digestive fluid from the spider's mouth liquefies the grasshopper's body, and the spider sucks up the resulting broth. Later, the spider drops the empty body of the grasshopper on the ground and waits for another insect.

> **Guidepost**
>
> Why is it difficult to study an organism in isolation from its surroundings?

a

b

Figure 1.1   What activities are illustrated here?   **Feeding.**

Explain or demonstrate how if a plant is shut away from all light, it will die.

This story is repeated every day all around the world. Animals eat plants or other animals or both. Organisms are connected to each other in the web of life by their need for energy to live and to reproduce. Green plants, such as those in Figure 1.2, get energy directly from sunlight and use it to make their food. Animals that eat plants or other animals get energy indirectly. No matter where the energy comes from, without it, an organism dies. If the spider web is broken by a falling twig, it must be repaired, or the trap can no longer catch food for the spider. The spider's link to the supply of energy cannot be broken if the spider is to live. If grasshoppers eat all the leaves of the young plant, the plant can no longer make its own food, and it dies.

Many things can change the relationships between the plant, the grasshopper, and the spider. Caterpillars, such as the one shown in Figure 1.3, may eat the plant, which means less food for the grasshoppers.

Figure 1.2   What relationship is shown between these flowering plants and the sun? **The plants use sunlight to make food.**

Without enough rain, the plant may die before it can reproduce or provide food for an animal. The spider's web may trap flies and moths as well as grasshoppers. A bird may eat the grasshopper or even the spider or caterpillar. Just as the spider's web has many connected threads, in the web of life many organisms are connected to each other. The more closely we view the world, the more complex this web becomes and the more we become aware of the impact of humans on the web. The smallest change in the web can have a major effect on the connected organisms.

You have seen how plants and animals interact within the web of life. Nonliving things, such as temperature, sunlight, and rainfall, also affect the growth of plants and animals. The study of the living and nonliving parts of the environment and how they affect organisms is called **ecology**. Scientists who study ecology are called ecologists. Ecology is only one aspect of **biology,** the study of living things. In this textbook, you will study many areas of biology, but you will focus on ecology.

1

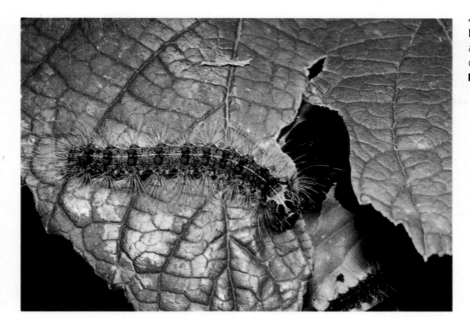

Figure 1.3 A gypsy moth caterpillar eating a leaf. How is the food supply of other organisms affected?
Decreased.

1.2 Plants, Animals, and Other Organisms Make up a Food Chain

Not far from the spider's web discussed in Section 1.1 is a raspberry bush. Underneath the bush, a rabbit (see Figure 1.4) finds shelter and a place to hide from animals that may kill it. The bush is an ideal place to hide because its thorns can tear clothes or dig into the flesh of larger animals, and its low-hanging red fruit provides the rabbit with food. A small bird feasts on the fruit near the top of the bush. Because rabbits usually do not eat raspberry leaves, the rabbit ventures out to look for grasses to eat once the berries are gone. Its movement is spotted by a hungry fox, which slinks forward and suddenly makes a leap for the rabbit. The rabbit looks up just in time, and a wild chase begins. This time, the rabbit reaches safety in another raspberry bush.

Give students an example of the ripple effect from their experience, such as how a change of socks from one color to another may require different shoes, different shoes may lead to a different color shirt, the new shirt may lead to a different pair of pants, etc.

Figure 1.4 Describe all the relationships you see among the plants and animals and their environment.

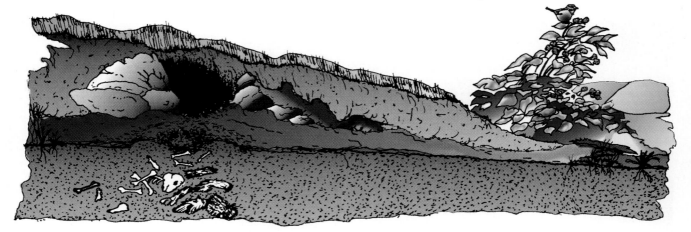

Figure 1.5   What role do decomposers play in the food web?
They decompose organic materials and assist in the cycling of chemicals.

The terms *producers* and *consumers* are
used in preference to *autotrophs* and
heterotrophs. The latter two terms are
introduced later.

Not far from the rabbit's bush is the fox's den. The fox had carried last
week's rabbit to the den and eaten most of it. What he did not eat, he buried.
Microorganisms—organisms too small to be seen with the unaided eye,
such as bacteria—began to break down the remains, causing them to decay.

Rabbits, raspberries, and the other nearby plants and animals play various   2
roles in the web of life. The green plants use light energy to make food.
Because they make their own food, they are called **producers**. Animals cannot
make their own food, so they eat plants or other animals or both. Organisms
that are unable to make their own food are called **consumers**. During the
process of decay, consumers that break down the bodies of dead plants and
animals are called **decomposers**. Bacteria and mushrooms, such as those   5
shown in Figure 1.6, are examples of decomposers. The producers, consumers,
and decomposers that live and interact in one area form a **community**.

The raspberry bush, the rabbit, and the fox can be connected in a **food
chain,** a pathway that tells what eats what. Several food chains are shown in
Figure 1.7. In Path a, the rabbit eats raspberries from the bush, and the fox

Figure 1.6   What role do mushrooms play
in a food web?   **They are decomposers.**

eats the rabbit. Path b illustrates a
shorter food chain—the bird eats the
raspberries. These two food chains
are connected by the raspberry bush.   3
Section 1.1 described a food chain in
which a spider ate a grasshopper that
ate a plant (Path c). If the bird that
eats raspberries also eats the spider,
two more food chains are connected
(Path d).

When all the food chains in a
community are connected to each
other, a **food web** is formed. Figure
1.7 does not include all the plants a
rabbit might eat, all the animals that
might eat a rabbit, or any decomposers. What would the web look
like if there were two spiders instead
of just one? Figure 1.8 shows a larger food web that includes the organisms discussed in this chapter. You
can see that a food web can be large
and complex.

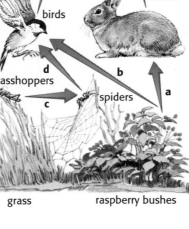

foxes

a

rabbits

birds

d   b

grasshoppers

c   spiders   a

grass   raspberry bushes

Figure 1.7   There is more than one food
chain here. How many can you find?

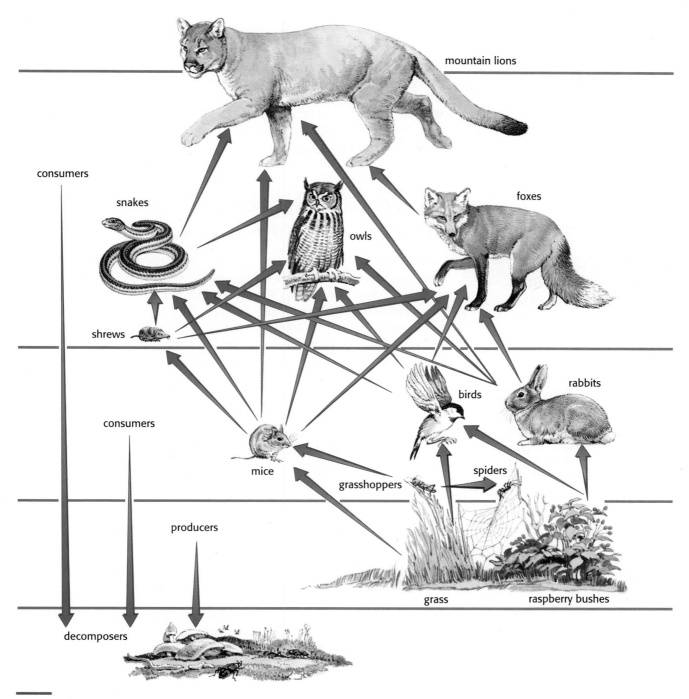

consumers

mountain lions

snakes

owls

foxes

shrews

consumers

mice

birds

rabbits

grasshoppers

spiders

producers

grass

raspberry bushes

decomposers

Figure 1.8   Can you find any more relationships?
Yes, for example, the bird may eat the grasshopper.

Food webs and food chains tend to keep living organisms in balance. The rabbits live off the green plants, and many other animals, including humans, live off the rabbits. This might appear to be hard on the rabbits, but rabbits produce many offspring in a short time. Imagine how many rabbits there would be if they reproduced without control. They soon would be so numerous that they would eat all the plants. Without the plants, the rabbits would starve. Foxes and other animals that eat rabbits may help to keep the rabbit population in balance. Disease or lack of food also may help to keep the rabbit population from growing too large. These controls, or checks, apply to all living organisms, including humans, and are just one part of the balance of nature. Investigation 1.2 may help you understand your place in the web of life.

CONCEPT REVIEW

1. What parts of the environment does an ecologist study?
2. How do producers differ from consumers?
3. How is a food chain related to a food web?
4. Explain how reproduction and death are part of the balance of nature.
5. In what way do decomposers differ from other consumers?

Matter and Energy—The Foundations of Life

1.3   All Biological Activity Requires Energy

Food chains and food webs are based on the flow of energy and matter from one organism to another organism. The details of this flow are developed throughout this course. Here, you will look at just the broad outline.

All of an organism's activities require energy. Imagine the marathon runner shown in Figure 1.9a trying to run a race without having eaten high energy foods. The activity does not have to be very great to require energy; even the movement of a tiny one-celled organism across a drop of water requires energy (see Figure 1.9b). Whenever you see any type of biological activity, you need to ask, "Where does the energy come from to support this activity?"

Notice that energy is not defined. Instead, familiar forms of energy are given. The classical definition "the ability to do work," which is an oversimplification, is used in Chapter 4.

Where do you get your energy? It may take some imagination to see energy in a hamburger and a pile of french fries. There is energy in this food, however. It is **chemical energy**. Chemical energy is found in the structure of the molecules that make up the meat and the potatoes. Other forms of energy include electrical, mechanical, heat, light, and nuclear energy. The most important form of energy for you is the chemical energy stored in the food you eat. You begin to release this energy as you digest your food. Most of the energy from your food is released within your cells in a complex series of chemical reactions. You use this energy to grow and to develop.

1.4   Photosynthesis Supplies Food Energy

Remember that you are part of many food chains and one giant food web. The hamburger that contains chemical energy came from a cow; cows eat only grasses and grains. A grass plant and a potato do not eat other organisms, so where do they get their energy?

Figure 1.9   Where do these organisms get their energy?   **Their food.**

a

b
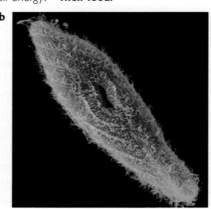

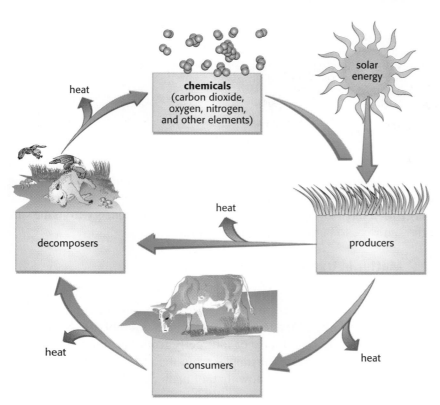

Figure 1.10 Why is it impossible for organisms to recycle energy?
Their activities change it into heat, which cannot be used to do work.

1   All green plants grow in light. In the process of **photosynthesis**, plants absorb light energy and convert it to the chemical energy found in sugars. The sugars formed by photosynthesis provide food for the plant. The plant then can use the energy in the sugars to grow and reproduce. Energy that is not used may be stored in the form of starch to be used at a 2 later time. The potato plant, for example, stores energy as starch. When you eat the potato, you benefit from the chemical energy stored in it.

     Photosynthesis is the basis for almost all of the food energy in the world. After grass makes its own food in photosynthesis, it uses some of that food to grow. Thus, some of the energy that is captured from sunlight is used before it reaches the cow, even if the cow eats the whole plant. Only certain bacteria are able to make their own food in other ways.

     Because no animal can make its own food, it must get its energy from plants or other animals. The cow eats the grasses and uses the chemical energy in them to grow, to produce milk, and to move across the pasture. The energy not used by the cow remains in the waste products dropped in the pasture. This energy is not completely lost to all organisms. Decomposers break down the cow dung and use the energy from it for their own growth and reproduction. Decomposers also get energy from the bodies of the cows and plants that die. This series of steps is referred to as the flow of energy through a system.

     The conversion of chemical energy to the energy used by living organisms is not efficient. Whenever an organism breaks down its food, some of the energy escapes as heat. If you touch your arm, for example, it feels warm. This is because some of the chemical energy stored in the fries and hamburger you ate is converted to heat energy. Your body uses the rest of the chemical energy to keep you alive and growing. Although some heat energy keeps the body warm, which allows biological activities to continue, most of it is lost to the air. No matter where energy comes from, all the 6 energy that enters a food web eventually is lost from the community in the form of heat (see Figure 1.10 above). No organism can use heat energy for

Students should be aware that plants can grow in visible light from any source, including lamps.

Ask students to name some decomposers, producers, and consumers other than those named in the reading.

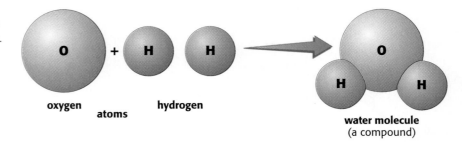

Figure 1.11 Water is a compound made up of two elements—hydrogen and oxygen.

oxygen
atoms

hydrogen

water molecule
(a compound)

growth. Therefore, energy must enter a community continually, beginning with photosynthesis, or the community will die. The source of chemical energy for almost all communities in the world is light energy from the sun. If the sun were to burn out, life on earth would cease.

1.5   Matter Is Used to Build Living Things

Unlike energy, which flows one way through a food web, matter does not leave the food web. Instead, it changes from one form to another within the web. This cycling of matter is important in the web of life. Most living things get their energy directly or indirectly from the sun, but all living things get their matter, or substance, from the earth and the air around them. You probably already know that all matter is made of minute particles called **atoms**. Matter that is made of a single type of atom forms an **6** **element**. Most elements, however, occur as **molecules** (MOL uh kyoolz), which are usually combinations of two or more atoms. A **compound** is matter that is made up of more than one type of atom chemically combined. Water, for example, is a compound made of atoms of two different elements—hydrogen and oxygen (see Figure 1.11 above).

Do not dwell on the difference between an atom and a compound. It is discussed again in Chapter 4.

Because living things are so different from nonliving things, scientists once believed that living matter contained unique elements. We now know this is not the case. Of the more than 100 different elements found on the earth, only about 30 are used in the makeup of organisms. Most of these elements, such as the hydrogen found in a water molecule and the carbon found in the carbon dioxide of the air, are very common. Figure 1.12 shows the proportions of elements in humans.

Plants and animals are made up of many different compounds, but the atoms used to make up these compounds occur all around you in the nonliving world. For example, plants use the simple compounds of carbon dioxide and water in photosynthesis, but only small amounts of chemical **5** energy are present in these compounds. During photosynthesis, the plants build complex compounds. Using light energy, they link together the **3** atoms from carbon dioxide and water to make sugars. Both energy and matter are stored in sugars. A plant can use these sugars as a source of

Figure 1.12 Elements present in the human body.

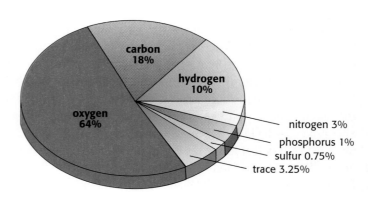

carbon
18%

hydrogen
10%

oxygen
64%

nitrogen 3%
phosphorus 1%
sulfur 0.75%
trace 3.25%

energy, as discussed in Section 1.4, or the plant can use the sugar molecules to make other molecules that it needs to build its body. To do this, a plant rearranges the atoms in the sugar molecules and adds new atoms. Sugars, therefore, are a food—a substance that an organism can break down to get energy for growth, body maintenance, and repair. Food also is matter that can be used to build the structure of the body.

When an animal eats a plant as food, both energy and matter are passed from one organism to another in a food web. Unlike the flow of energy in a food web, however, the flow of matter is not one way. Matter cycles within a community. Plants use carbon dioxide, water, and other substances during photosynthesis. These same substances are given off by organisms after they have used a plant as food. Other plants then can use these materials again to continue the process of photosynthesis.

In this manner, the same matter is used over and over again in a community. Matter travels in cycles from the nonliving environment into food webs and back to the nonliving environment. It then enters the food webs once again in photosynthesis. Figure 1.13 summarizes the relationship between matter and energy in the biosphere.

Pollution affects the cycling of matter. Pollutants can tie up materials that would ordinarily cycle between the living and nonliving parts of the world. If the pollutants can be changed to other materials, they can be made less harmful, and perhaps some of the matter can return to normal cycling.

energy
(as light)

Biosphere

matter as
simple
compounds

matter as
complex
compounds

energy
(mostly as heat)

Figure 1.13 Explain the relationships energy and matter have with a community. **Energy flows and matter cycles.**

You might use students' ideas to develop a diagram on the chalkboard that shows the idea of cycling. The carbon cycle is discussed in Chapter 4.

CONCEPT REVIEW

1. What is the source of energy for almost all living things?
2. How are light and chemical energy related in photosynthesis?
3. How does matter get from the soil and air to animals?
4. How is the flow of energy through a community different from that of matter?
5. In what way do simple compounds differ from complex compounds?
6. Explain why energy flows and chemicals cycle through food webs.

Studying the Living World

1.6 The Biosphere Is Home to All Living Things

The living world forms a thin layer around the nonliving world. This layer is called the **biosphere** (BY oh sfir). It includes all the organisms and the air, soil, and water surrounding them. The biosphere extends from the bottom of the oceans to the air above the earth.

Because you are a living organism, you are part of the biosphere. You have interrelationships with your family, friends, and teachers. Because you are part of the web of life, you also have relationships with many other organisms. Some relationships are obvious, such as those with the plants and animals you eat. However, you may not recognize the relationship that exists between you and animals that eat the same foods you eat. The relationship between you and a grasshopper that eats and damages lettuce is indirect, but any damage to your food plants can affect your food supply.

Humans have far-reaching effects on the biosphere. Data being collected today show that many human activities are straining and destroying the delicate balance of nature in the biosphere. In the spider's web, pulling

Guidepost

How do humans fit into the biosphere?

Figure 1.14 Why is an oil spill bad? **Answers will vary but should include thoughts about damaging wildlife, economic loss (fisheries), and clean-up cost.**

Give students examples of how their actions affect the biosphere, such as how building a house can change a small community. Give as much detail as you can.

a single thread affects the entire web. So, too, do your actions affect the **4** world. What you, your friends, and your family do affects the rest of the biosphere in many subtle ways.

The 1989 *Exxon Valdez* disaster is a dramatic example of the impact of humans on the biosphere. The tanker disgorged more than 10 million gallons of crude oil into Alaska's Prince William Sound (see Figure 1.14), and it will be many years before the wildlife in this area recovers. Yet, when compared with the far-reaching effects of excessive energy consumption and pollution around the world, the oil spill is a trivial occurrence. In 1989, for example, the population increased by about 87.5 million, further straining supplies of food and other resources. Deforestation and the burning of fossil fuels added at least 19 billion tons of carbon dioxide to the atmosphere, thus aggravating the global warming process. In addition 11.3 million hectares (28 million acres) of tropical forest were destroyed along with the organisms living there. The hole in the ozone layer over the Antarctic remained alarmingly large, and scientists report that another hole is developing over the Arctic, increasing ultraviolet radiation to the earth and the subsequent risk of skin cancers.

These incidents show that our planet is in trouble. If we do not take action, our planet one day may become unfit as a human habitat. People are gradually beginning to recognize and deal with these problems. Among the international initiatives currently underway is a world climate conference. It is hoped that the conference can lay the groundwork for a global limitation on the emission of greenhouse gases and, thus, the stabilization of the world's climate. Two other goals for the conference are the drafting of a biodiversity treaty, which would provide financial incentives for the protection of tropical forests, and the development of an agreement to phase out the production of chlorofluorocarbons (CFCs)—the major culprits in the depletion of the ozone layer. Other measures should be taken under consideration, including worldwide family planning to reduce population growth and new manufacturing techniques to reduce waste. At a more personal level, every individual can recycle discarded materials, insulate his or her home, and use mass transit and carpooling to cut energy use.

Because the United States has the technology to help other nations with environmental problems and because it consumes a large part of the world's resources, it has a major role to play in these areas.

1.7   A Hypothesis Is a Statement that Explains an Observation

Humans look at the biosphere from many different viewpoints, but each viewpoint reveals only a small part of the complex whole. In our working roles, we have limited points of view. An artist looks at living things for their beauty of color and form. A farmer looks for ways to manage the growth of living things. A biologist views the world scientifically.

As a science, biology does not deal with value judgments. Beliefs and value judgments differ from person to person. For example, some students looking at Figure 1.15 on page 16 might prefer the fawn, others the puppy. **2** These are value judgments. They are based on how you feel about puppies and fawns.

Biology is concerned with data—observations that do not differ from one person to another. The Steller's jay in Figure 1.15 is blue. This is an observation that can be agreed on by everyone with normal color vision. Biologists collect and organize data about organisms. They use the data

Ecologist

Tropical forests supply most of the oxygen that airbreathing organisms need for life. The destruction of these forests by burning and cutting releases a tremendous amount of carbon dioxide into the atmosphere. Increased carbon dioxide levels are one factor implicated in the predictions of global warming (see Chapter 4, Biology Today). Ariel Lugo is one of the many people who are studying the effects of tropical forest destruction. Dr. Lugo studies the role of tropical forest ecosystems for the United States Forest Service (a division of the Department of Agriculture) in Puerto Rico. Dr. Lugo was not always interested in ecology. In fact, he admits his passion originally was baseball. His family, however, encouraged him to study medicine. While at medical school, Dr. Lugo began to work with a prominent ecologist who was conducting radiation experiments in the rain forest. Dr. Lugo says, "After a summer of carrying cement blocks through the rain forest and after obtaining a D in embryology, I was sure that medicine was out and ecology was in—to the distress of my family."

Currently, Dr. Lugo is conducting research in three areas. First, he and a colleague, Dr. Sandra Brown, are analyzing the content of tropical forest soils that have been exposed to a variety of land use practices. These analyses allow them to determine how organic matter in the soil changes over time under different types of use. These studies will help to determine the role and importance of tropical forests in the global carbon balance.

A second area of research involves the study of forested wet lands. Forested wetlands are lowland ecosystems, such as marshes or swamps, that are saturated with fresh or salt water. Saline (salty) forested wetlands consist mostly of mangrove trees, which grow on tropical coastlines and can tolerate seawater. Freshwater forested wetlands include high-elevation palm forests. Dr. Lugo's interests lie in the study of the structure and function of these ecosystems and how they change through time.

The third area of research involves the dynamics of tropical tree plantations. Dr. Lugo is comparing plantations with natural areas of similar age to determine what elements of the natural area remain in the artificial plantation and how the chemical cycling in each area differs. Understanding these patterns is important because plantation agriculture is one way of rehabilitating damaged lands and restoring natural forest ecosystems in areas where humans have devastated the tropical forest.

Dr. Lugo has never questioned his decision to quit medical school and pursue a career as an ecologist. Being an ecologist has allowed him to meet many talented people, travel the world, and visit some of the most exciting locations on earth.

Figure 1.15   "Which is nicer, the fawn or the puppy?" "Which bird is blue?" What is the difference between these two questions?   **The first is a value judgment; the second deals with a verifiable observation—color.**

in various ways to gain more information. Science, however, is more than just the collection of data. It is a systematic way of looking at the world, of obtaining data, and of interpreting it. It is a continuous process of inquiry, the product of which is a body of knowledge. This body of knowledge is subject to change and revision as we acquire new information. Above all science is a human endeavor because people are involved in the process of inquiry.

The methods of obtaining scientific knowledge involve a series of steps that begin with observations of the living world. In the field with the spider and the rabbit, many flowers have insects buzzing around them. A biologist observes this field and asks the question, "What attracts bees to the bright red flowers in the field?" Library research may provide information about previous investigations of the question or suggest appropriate approaches to the problem. Reading and thinking about the question may produce thoughts such as, "Bees are attracted to this type of flower because of its red color." This thought is a **hypothesis** (hy POTH uh sis), a statement that explains an observation. A good hypothesis predicts a relationship between a cause and an effect. It can be tested by an experiment specifically designed to collect evidence that either will support or will not support the hypothesis. Evidence cannot *prove* a hypothesis, however, because it is always possible that new evidence may provide a better explanation. You can examine how hypotheses are tested in Investigation 1.4.

An understanding of science and of scientific methods is particularly important today as we face such problems as overpopulation, starvation, pollution, and the loss of forests. By using scientific methods to study our impact on the world around us, we can begin to understand how our actions affect the biosphere. Out of this understanding, we can begin to develop solutions that will minimize our impact and allow us to live in balance with the web of life.

CONCEPT REVIEW

1. What parts of the earth make up the biosphere?
2. How does a value judgment differ from data?
3. What is the relationship between hypotheses and observations?
4. How do humans fit in the biosphere? Why is understanding our role in the biosphere important?
5. How does space travel affect our concept of the biosphere?

INVESTIGATION 1.1   The Powers of Observation

You can explore the complex interrelationships among living things in many ways. In this investigation, you will make careful observations and then report and verify those observations.

Materials (per class)
10x hand lens or stereomicroscope   millimeter ruler
labeled specimens of organisms

Procedure

1. Located around the room are sets of organisms—living things—or parts of organisms. Each set contains four specimens. (A specimen is a sample individual or, in the case of large plants, a characteristic part, such as a leaf.) Each set is labeled with a group name and a number. Each specimen is labeled with a letter.

2. Work in teams of two to four. Each team will begin with a different set of specimens. You will have approximately 10 minutes to observe and to describe the set.

3. Select one person on your team to take notes. Observe the four specimens and make notes on differences you see among them. If appropriate, take measurements. Remember, the differences must be in the organisms, not in their containers.

4. When your team has decided on the differences between the four specimens, choose one. On a separate sheet of paper, write as complete a description as possible of the specimen. Do not use the specimen's letter on your description sheet. Write the group number and letter of the specimen described on another slip of paper and give it to your teacher.

5. When your teacher signals that time is up, place the description your group has written next to the set of specimens.

6. When your teacher tells you, move to the specimen set having the next highest number. (The team with the highest number will go to Set 1.) You will have several minutes to observe the set. Select one team member to take notes. Then, read the description next to the set and decide as a team which of the specimens it describes.

You may wish to have students read Investigation 1.1 ahead of time and write out the procedure in their own words. Give them the appropriate format for writing lab reports. Each student should be familiar with accepted record-keeping procedures and use a data book and pen to record data.

NOTE: Before beginning this or any other investigation, discuss the safety guidelines in Appendix 1 with your students. Be sure they are familiar with the laboratory rules, know where the rules are posted, and have returned a signed laboratory safety agreement.

This investigation allows students to sharpen their observational skills while working with a variety of organisms. Positioned at the beginning of the chapter, the investigation sets the stage for discussing the web of life. Teams of 2 to 4 students work well; 8 to 10 is the optimal number of sets of organisms.

Time: Approximately two 45-minute class periods or one 90-minute laboratory period. The time required for steps 1 through 5 is relatively short; steps 6 through 8 should take about 3 to 5 minutes per set of organisms. Students will need to refer to specimens and descriptions when they work through step 9.

7. In your data book, prepare a table with these headings:

Set Number	Specimen Fitting Description

Record the letter of the specimen in your table.

8. Repeat Steps 6 and 7 until you return to your starting point.

9. Your teacher will list the correct answers on the chalkboard. Check your table against this list. If your conclusions do not agree with the list, recheck the group of specimens. Did you miss anything? Was the description complete?

10. Wash your hands thoroughly before leaving the laboratory.

Discussion

1. Which was easier, writing a clear description or selecting the specimen another team had described? Why?

2. For each description, what information could be added to make it clearer? Does everyone in the class agree on what could be added? Why or why not?

3. For each description, what information could be removed and still leave the description clear? Does everyone agree on what information to remove? Why or why not?

4. Was there a set of specimens that you would have liked to describe? Why?

5. In what ways did reporting and verifying observations in this investigation increase your knowledge of these organisms?

INVESTIGATION 1.2   You and the Web of Life

How do you fit into the web of life? By relating the food you eat for one day to the plants and animals from which it came and to the other organisms with which those plants and animals interact, you can begin to form a picture of your role in the biosphere.

Materials (per person)

data book   pen

Procedure

1. In your data book, list all the foods you ate yesterday.

2. Separate these items into foods that came from animals and foods that came from plants. Remember, many foods are combinations of different foods. List the ingredients of each food separately. Then indicate whether each was from a plant or animal and the type of plant or animal. For example, if you had cake for lunch you should list: flour—plant, wheat; sugar—plant, sugar beets; eggs—animal, chicken. For every animal you have listed, list several foods that it eats. For example, if you had milk with your cake, list a cow and then list grass and corn as food the cow eats.

3. Across the bottom of a blank sheet of paper, write the names of all the plants you have mentioned. In a row above this row, list all plant-eating animals (herbivores) that eat any of the plants in the first row. From each plant, draw a line to every animal that eats it.

4. Above the herbivore line, enter the names of all animals (carnivores) from your list that eat other animals. From each herbivore, draw a line to the carnivores that eat that herbivore. The food web you have drawn shows some of the interrelationships between the plants and animals that provided your food for one day.

Discussion

1. Which, if any, of the items you listed as foods for animals are also foods that you could eat?

2. Did you include organisms that might compete with you for your food? Add as many as you can to your food web.

Discussion

This investigation helps students to understand, at a personal level, some of the interrelationships that are central to life on earth. The activity reinforces the discussion of food webs in Section 1.2 and sets the stage for the discussion of energy relationships in Section 1.3. Although it may be helpful for them to discuss the process in small groups of 3 or 4, students should prepare their food webs individually. Follow the activity with a class discussion that focuses on how humans have changed the environment to produce the food we eat, but how we still are linked to the global web of life.

Time: **Students should list their foods before the lab period. One period for preparation of the food webs; one-half period for general class discussion. Preparation of the food webs can be assigned as homework, although this early in the year the activity is likely to be more meaningful if done in class.**

Ask students to write down everything they eat for a day and bring the list to class for Investigation 1.2.

Discussion

1. **Answers will depend on individual food webs.**

2. **Beetles that damage grains; bacteria and fungi that cause decay; herbivores such as birds, insects, and rabbits are some possible competitors.**

3. What about the role of decomposers in your food web? Indicate with lines how they might be involved.

Going Further

Research the origins of several of the plants and animals that appear in your food web. What were the original ranges of those species? Where were they domesticated? Where are they raised now? What are the requirements of those plants and animals? How might they be affected by global climatic change? What environmental alterations are necessary for the production of those foods? In what other ways could these plants and animals be produced?

INVESTIGATION 1.3   Field Observation

Life is everywhere, although at times it may be difficult for you to see it. Some organisms are widespread, others found only in certain places. All organisms, however, need food, water, and protection to survive. In this investigation, you will sharpen your observation skills and try to explain the presence or absence of organisms in a study area.

Materials (per person)

data book   pen

Procedure

1. Look around you. What type of environment are you in? Is it next to a city street, in a park, or in the country? Being as specific as possible, record your answer in your data book.

2. What time of day is it? What are the weather conditions? Record your answers.

3. What type of life do you see? The type of life you observe may be flowering plants in a planter, people in cars, cows and grain in a field, or squirrels in trees. Record as many types of life as you can and be sure to note where you see them.

> **Do not handle any organism you observe, even if you are certain it is harmless. Unnecessary handling may be harmful to the organism and yourself.**

4. What type of birds do you see? Are they pigeons or some other bird? Where do you see them? Record your answers.

5. Find a patch of ground away from the other members of your class. Look at the area closely. Are there any plants growing there? Do you see any other organisms? How many? Record what you see.

6. What might explain the presence of these organisms?

7. What types of organisms common to your area, such as ducks or lizards, do you not see? Write down a few examples.

8. What might explain why some organisms are not present?

9. How would you determine if the time of day or the weather influences the types of organisms you see?

Discussion

1. Share your observations with the other members of the class. What organisms did they see? Did everyone see the same ones?

2. Where were the organisms found? Why do you think they were found there?

3. Do you think the same organisms could live in the downtown area of a large city? In the country? Why or why not?

4. What organisms were not found? Share your explanations of their absence.

Going Further

Use plant and animal keys to identify some of the organisms you observed.

---

3. **Decomposers should be included at each level of the food web.**

This investigation encourages students to observe and account for what they see. It is not a rigorous field study. Rather, it offers an early, meaningful field experience and prepares students for the discussion in Section 1.6. Students should work individually.
    Time: **One class period.**

Safety

Caution students not to handle any organisms, even if the organism is harmless.

Procedure

1. **Answers will depend on the site.**
2. **Confirm the time of day.**
3. **The larger types of observed organisms should be the same. Smaller types will vary.**
4. **In the city: pigeons, perhaps finches or sparrows. In the country: many types. Help students identify birds or describe them in detail for later identification. Locations will vary.**
5. **Answers will vary.**
6. **Explanations should include some mention of water, food, and protection.**
7. **Answers will depend on location. Make sure students do not make wild observations and that their answers are reasonable.**
8. **Explanations should include a connection between the environment and the requirements of organisms.**
9. **Answers will vary but should include mention of around-the-clock observation.**

Discussion

1. **Very unlikely.**
2. **Answers will vary. Students should understand that organisms usually are found where the necessities for their life exist.**
3. **Answers depend on the organisms.**
4. **Answers will vary. Time of day may or may not have an effect.**

This investigation builds on the discussion of hypotheses in Section 1.7 and provides students with practice in analyzing a problem and projecting the course of its investigation. The investigation is self-clued and requires no additional introduction. The procedural steps are illustrated in Figures 1.16, 1.17, and 1.18. Students will benefit most by working in teams of 3.

Time: Two periods—one to work through the investigation in small groups, and one for class discussion.

Procedure

1. **Place flowers cut from red paper in a field to see if they attract bees.**
2. **Yes, but the data are limited, and there may be other factors involved.**
3. **No; hypotheses cannot be proved, only disproved. If repeated experiments fail to disprove a hypothesis, the hypothesis is strengthened.**
4. **Change the position of the flowers in the experiment.**
5. **Color.**
6. **Size, shape, odor, size or shape of the petals, etc.**
7. **The paper flowers should differ only in color.**
8. **Refutes it.**
9. **No, because the hypothesis was refuted.**

Figure 1.16 *Observation:* Bees are attracted to Flower 1 but not to Flower 2. *Hypothesis:* If bees are attracted to Flower 1 by scent, then adding scent to a flower that does not attract bees should cause the bees to be attracted to that flower.

INVESTIGATION 1.4   How Do Flowers Attract Bees? A Study of Experimental Methods

In this investigation, you will learn about the experimental process by examining in detail several possible experiments.

Materials (per team of 3)

3 data books   3 pens

Procedure

PART A Controlling for Variables

1. Recall the hypothesis "Bees are attracted to this type of flower because of its red color." This hypothesis can be investigated if it is reworded as follows: If bees are attracted to the flower by its red color, then an artificial flower cut out of red paper may attract the same bees to it when placed in the field. The "if" part of the hypothesis refers to the cause—what is being investigated. The "then" part of the hypothesis refers to the effect—the outcome of the experiment. How would you investigate this hypothesis? Record your answer in your data book.

2. Suppose you carried out an experiment in which you cut out paper flowers of several colors (for example, red, yellow, green, and blue) and placed these in a field. Suppose also that a bee landed on the red paper flower but not on the other paper flowers. Does this observation support your hypothesis? Explain.

3. Your observations provide information, or data, to help answer your question or to test your hypothesis. A scientist uses data to draw conclusions about an experiment. Do the above data prove your hypothesis was true? Why or why not?

4. How could you test the hypothesis that bees landed on the red paper flower because it was the most convenient, not because it was red?

5. You know that flowers differ in many ways. These differences are called **variables**. Some of the variables for flowers in a field are size, shape, odor, color, and the size, position, or shape of the petals. A good experiment tries to control all variables except the one being studied—the experimental variable. What is the experimental variable in the flower experiment described in Step 2?

6. The colored paper flowers that are not red are **controls**. Controls are the parts of an experiment that make it possible to eliminate variables other than the one being studied. In this case, you are testing the variable of color, making certain it is the red color that attracts the bees. What are some variables that should be controlled for in this experiment?

7. You want to improve this experiment by making it more valid. Other than differing in color, what must these flowers be like?

8. Assume for a moment that no bees land on the red paper flowers. Does that support or refute the hypothesis in Step 1?

9. Usually, once a hypothesis has been refuted, it does not have to be tested again. It is thereafter considered false, assuming the experiment is a valid one. Given your answer to Question 8, in this case would you need to continue the experiment? Why?

PART B Analyzing an Experimental Design

10. Study the observation and hypothesis in Figure 1.16.

Flower 1   **Flower 2**

11. Examine Experiment 1, represented in Figure 1.17. Here the scent of a flower that attracts bees is added to a flower that does not usually attract bees. What can you tentatively conclude from this experiment?

Flower 1   **Flower 2**
                              scent added

Figure 1.17 **_Experiment 1_**: Scent of Flower 1 is added to Flower 2.

12. What does Experiment 1 lack?
13. Is scent the only variable?
14. What other variables are there between these two types of flowers?

11. Bees are attracted to the flowers by scent.
12. Experiment 1 lacks a control for variables such as flower shape and color.
13. No.
14. Flower color, shape, size, and the background in which the flower grows may be other variables.

Discussion

1. Variable aspects of its appearance, such as color, shape, size, arrangement of the petals and sepals.
2. Preexperimental observation: the bees do not visit Flower 2. The untreated Flower 2 is a control to make sure this observation continues to hold. Otherwise, the original observation was faulty.
3. The control flowers with unscented spray will test whether or not the spray solvent (apart from the scent)

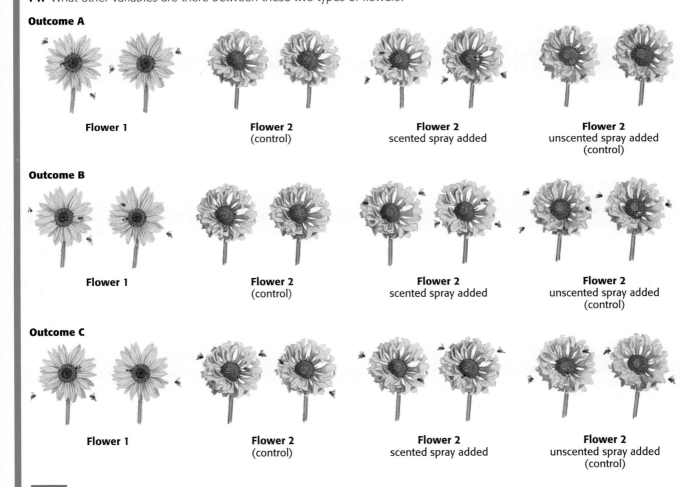

Outcome A

Flower 1   **Flower 2**   **Flower 2**   **Flower 2**
                  (control)   scented spray added   unscented spray added
                                                              (control)

Outcome B

Flower 1   **Flower 2**   **Flower 2**   **Flower 2**
                  (control)   scented spray added   unscented spray added
                                                              (control)

Outcome C

Flower 1   **Flower 2**   **Flower 2**   **Flower 2**
                  (control)   scented spray added   unscented spray added
                                                              (control)

Figure 1.18 **_Experiment 2_**: This is a redesign of Experiment 1 to consider other variables that might be responsible for the original observation. Flower 2 in this case is a different type from Flower 1, although both appear similar.

attracts bees. If so, the experiment must be redesigned using a different spray solvent that does not attract the bees.
4. Outcome A supports the experimental hypothesis that bees are attracted to Flower 1 by its scent. The scent also attracts them to Flower 2 in the presence of the scent.

15. Carefully study Experiment 2, represented in Figure 1.18 Be prepared to defend or criticize it. A, B, and C show different possible outcomes.

5. **Outcome B supports a hypothesis that scent is not the only factor attracting the bees. The unscented spray, however, is not a natural factor, as the untreated Flower 2 indicates. A new spray solvent must be selected that does not attract bees.**

6. **Outcome C appears to indicate that the preexperimental observation was wrong. Bees do visit the untreated Flower 2. The only alternative is a new hypothesis suggesting that a previously undetected variable must exist in Flower 2 and that variable has changed during the experiment.**

7. **Outcome B distinguished between Flower 2 in its natural state and with unscented spray added. The possibility of a difference in the bees' behavior supports the hypothesis that the spray solvent is another variable that must be controlled.**

8. **A further control could eliminate any differences in temperature between Flower and Flower 2 by keeping them at the same temperature. Still another could keep them at the same light intensity. A third could keep the soil in which they are grown at the same level of moisture. Still further controls could be devised. Whether their variables will affect the outcome is questionable. A significantly lower temperature for Flower 2 than for Flower 1 would be an important variable to eliminate or control, because bees must warm up to be able to fly at lower temperatures. Hence, if Flower 2 grows in the shade and Flower 1 in the sun, then light, temperature, and moisture could be important variables.**

9. **The larger the experimental population of bees, the less any variables among the bees themselves will influence the outcome of the experiment.**

10. **The individual flowers selected for this experiment may not be representative of all flowers of this type (species).**

11. **If the difference in numbers of bees visiting each type of flower is consistent, then some additional variable is probably at work. It is not uncommon for two or more variables to interact in influencing a behavioral pattern. The problem is to identify the additional variables and to confirm them by experiment.**

12. **Similar results with different types of flowers would allow one to generalize the conclusions to more than one type of flower and perhaps to most flowers.**

Discussion

1. Compare the results for Flower 2 in Experiments 1 and 2. What variables are being controlled in the design of Experiment 2?

2. In Experiment 2, the scent-producing substance from Flower 1 is dissolved in a liquid (water or another solvent harmless to flowers) to produce a spray. What is the purpose of the control flowers with nothing sprayed on them?

3. What is the purpose of the control flowers sprayed with unscented spray?

4. What hypothesis would account for Outcome A in Experiment 2?

5. What hypothesis would account for Outcome B?

6. What hypothesis would account for Outcome C?

7. How does Outcome B illustrate the difference between the two sets of control plants?

8. What other variables might it be helpful to control in Experiment 27?

9. Why would it be useful to try a similar experiment with more than one group of the same type of bees?

10. Why would it be useful to repeat the experiment with more of the same type of flowers?

11. How would you interpret the results if the numbers of bees visiting each type of flower differed noticeably?

12. How would results similar to Outcome A with different types of flowers strengthen your conclusions?

13. Experiment 2 is still an imperfect attempt to test the original hypothesis. Design an improved, but similar, experiment that might provide additional support for your conclusions. Consider the issues raised in Questions 9, 10, l l, and 12 above. Consider also what additional information you may need to accept or reject the hypothesis. Remember that you want to manipulate only one variable, so all other possible variables must be controlled. Describe your design and procedure in a short paragraph.

SUMMARY

No organism is isolated from other living things. All organisms are part of a large, complex web of life. Plants are producers. They use the energy from the sun and simple compounds from the soil and air to produce molecules of food that allow them to grow and reproduce. Because animals are consumers and cannot do this, they depend, directly or indirectly, on plants as their source of energy and matter. Humans also rely on plants for food. Humans, however, have a greater effect on the biosphere than other organisms because their activities often result in permanent changes in the environment. Using the techniques of observation, hypothesis formation, and experimentation, we are becoming more aware of how those changes

13. **Responses will vary. A good design should include the following:**

a. **Sampling considerations: a large enough sample of flowers and bees (perhaps of different types) to make valid generalizations for that population; a random sampling procedure or some other consideration of representativeness of both a flower and bee population**

b. **Controls for other variables, such as color, size, shape, height, and species of flower used and its natural habitat; habitat, type, behavior of bees** (do these bees normally come in contact with these flowers?) possible climatic and terrain variables; the time of day/night, season, and lengths of exposure of flowers to the bees

c. **Consideration of how the flowers, especially those with added scent, are distributed in a field (mixing the experimental flowers in a random pattern is important)**

d. **Some attention to the duration of the experiment; for example, adequate exposure time**

affect the lives of all organisms in the biosphere. In response to growing global problems, we are beginning to take responsibility for our actions and to develop plans that will allow us to manage our planet in a manner that will sustain it for generations.

APPLICATIONS

1. What might happen to the balance in the food web of a pond if the number of one type of organism suddenly increased greatly? How long do you think such an increase would last?
2. Consider the pond in Question 1. Is your prediction affected by the type of organism involved? Consider first a producer, then a consumer.
3. Look at the fawn and the puppy in Figure 1.15. Draw a food web for each animal. Explain the role each animal plays in a food web.
4. After drawing a food web for the fawn and the puppy in Question 3, has your value judgment about "which one is better" changed? Why or why not?
5. Suppose that the sun were to die out, leaving the earth in darkness. The earth would not freeze immediately, and not all living things would die at the same time. Which do you think would die first—plants, animals, or decomposers?

Explain your answer.
6. Someone once said, "You are what you eat." If this is true, some people would say that your body is really rearranged plant fruits, roots, stems, and leaves. Would you agree or disagree? Explain.
7. Suppose you found a spider web and broke one of the strands at the outer edge. Do you think the web would still function and insects still could be caught in it? What if you broke two, three, or more strands? Now, think about the web of life in your community. Suppose a natural disaster killed 10 percent of the birds in the community. Do you think the community would still function? What if 50 percent of the trees died?
8. Why do almost all food chains begin with photosynthetic organisms of some type?
9. What is the relationship between hypotheses and data?

PROBLEMS

1. Wildlife can be found everywhere, even in cities. Animals such as skunks, raccoons, and squirrels are common in cities. Birds, of course, are among the most common types of urban wildlife. Make a list of the plants and animals in your area and use it to build a food web.
2. Describe how building a road through the mountains might affect the organisms that live in the area. List as many organisms that might be affected as you can and try to show some of the relationships among the organisms. For example, cutting down trees affects the availability of nest sites for birds.
3. In making an extended journey into outer space, astronauts would have to take along a part of our biosphere. Design an efficient "package" of the biosphere for such a journey.
4. Many people have careers related to biology. How does the career of the nutritionist in your school cafeteria relate to energy flow and materials cycles in a food web?

5. Choose a current environmental issue in the United States or in your local community. After some research, present to your class the causes of the problem and the potential solutions.
6. Suppose that one morning your family car does not start. Which do you think might be the more likely reason: (a) the car is missing a hubcap, or (b) the battery is dead? Why does one of these reasons seem more likely than the other? How would you test your hypothesis?
7. Make a list of hypotheses you make in your daily life during a weekend. Explain why you made these hypotheses and not others.
8. What plants and animals lived in your area *before* any humans lived there? Construct a food web for that community.

Some of the problems throughout the text involve applications of text understanding and laboratory investigations. Others require further study.

Populations 2

King penguins on South Georgia Island, Antarctica. A population is a group of organisms of the same species that live and interact in the same place at the same time. What is the evidence of a population in this photo? What is the evidence that the penguins are the same species? Identify some of the variations among individuals. How can the penguins have these individual variations and still be a population?

The penguins look similar, and they appear to be interacting. The presence of young penguins suggests they are capable of interbreeding. There are differences in the patterns of black throat stripes and frontal pigmentation. As long as the penguins can interbreed freely and successfully, they belong to the same species, although they may have striking individual differences.

◆ A herd of mule deer moves slowly across a shrubby meadow at dusk. Nearby, a scrub jay lets out an angry squawk, and soon its voice is joined by several others. The deer suddenly stop, but the jays continue their chorus. A coyote quietly walks out of a clump of scrub oak. Because the coyote is not a threat to them, the deer do not run. Instead, they resume browsing on young twigs and leaves.

Although the participants may differ, scenes similar to this are repeated everywhere. Each individual in these situations lives with other individuals—some like itself, some different. Many times, as with the coyote, only one individual can be seen. The coyote, however, lives with many other organisms in the meadow, and when it mates, it spends at least some time with another coyote. Individuals of the same type that interbreed with one another, like the deer or the coyotes, make up a population. In the same way that the web of life connects individuals, it also connects populations. We, as humans, have a great effect on other populations because we have a great ability to change the environment to suit our needs. Suppose that an apartment complex were built in the meadow. How might the complex affect the deer, the scrub jays, and the coyotes? Our needs often conflict with the needs of other organisms. This conflict creates many serious problems that all populations, including our own, must face. We are, however, the only population capable of solving the problems we have created. This chapter discusses the nature of populations, how they are interrelated, and special characteristics of the human population. ◆

Individuals, Populations, and Environment

2.1 Populations Are Made Up of Individuals

In general, life processes occur in separate "packages." You are such a package—an individual. You carry on the activities of life within your body apart from the life processes that occur in the bodies of your parents, brothers, or sisters. Each cow in a herd, each corn plant in a field also is an individual.

A group of interbreeding individuals of the same type that live in a particular area is called a population. To define a population, you need to identify the type of individuals, the time, and the place. Thus, for example, you can refer to the population of pigeons in Cincinnati, Ohio, in 1989,

MAJOR CONCEPTS

- ◆ Populations are groups of organisms that interbreed.
- ◆ Predictable factors influence the size of populations.
- ◆ The carrying capacity of an environment tends to determine a population's maximum size.
- ◆ Population sizes tend to fluctuate naturally through time.
- ◆ Populations tend to be restricted to certain types of environments and to be affected by certain limiting factors.
- ◆ Some critical factors determine the carrying capacity for humans on earth.
- ◆ Global cooperation is necessary to solve some environmental problems.

KNOWLEDGE CHECK

- ◆ What is a population?
- ◆ How do populations affect one another?
- ◆ What are resources?
- ◆ What do organisms need to live?

> **Guidepost**
>
> How do populations change?

1

Figure 2.1 A population of ladybird beetles. How many differences can you see within this population? What is the range in number of spots present? What important characteristic of population does this illustrate? What is responsible for these differences? **Number, size, pattern, and percent coverage of spots. 0-11. Variation. Heredity.**

Emphasize the concept of unit area in defining a population. It may be a small area, such as the water droplets on moss, or a large area, such as a pond. Be sure students understand that the population must exchange genetic material. Just because individuals are of the same species does not place them in the same population.

the number of spring beauty flowers in an Oklahoma field in March, or the lady bird beetles shown in Figure 2.1. When biologists want to test hypotheses, they usually study populations. They do so because the individuals in a population usually vary. Looking at only a few individuals could be misleading. For example, a biologist who studied only a few human individuals who happened to have brilliant blue eyes might jump to the conclusion that all humans have brilliant blue eyes. It is important to look at enough individuals to be sure the conclusions and observations hold true for the population.

By studying populations, scientists often can see what is happening throughout a community. As one population changes in size, it may affect many other populations. The more we know about one population, the more we can predict about others.

2.2 Four Rates Determine Population Size

The size of a population changes through time. Suppose a biologist counted 700 ponderosa pines on a hill in Colorado in 1980 (see Figure 2.2). In 1990, when the biologist counted the trees again, there were only 500. In other words, there were 200 fewer trees in 1990 than in 1980, a decrease in the population of ponderosa pines. This change in population may be expressed as a rate—the amount of change divided by the amount of time for the change to take place. The rate is an average. In this example, the rate of change in the number of trees divided by the change in time may be expressed as: −200 trees/10 years = −20 trees per year. To the biologist, this means each year there were 20 fewer trees in the population. Keep in mind, however, that this rate is an average. It is unlikely the trees disappeared on such a regular schedule. All of the trees may have been lost in one year due to a fire, or the decrease may have been caused by selective cutting during several years.

One or two examples will not suffice unless your students are superior in mathematical background. Provide additional practice samples using simple examples with both positive and negative numbers.

What does the decrease of 200 pine trees in 10 years represent? Because pine trees cannot wander away, they must have died or have been cut down. In this situation, then, the decrease represents the death rate, or **mortality** (mor TAL ih tee) rate, of the pine population. The number of deaths in the pine population per unit of time is the mortality rate. Mortality is not the only change that can affect a population, however.

While some of the pines may have died, some young pine trees may have started to grow from seed. Death decreases a population; reproduction increases it. The rate at which reproduction increases the population is called the birthrate, or **natality** (nay TAL ih tee).

Organisms that can move have two other ways to bring about a change in population size. If you were studying the pigeon population in your city or town, you might discover that a certain number of pigeons flew into the city in one year and a certain number flew out. **Immigration** (im uh GRAY shun) occurs when one or more organisms move into an area where others of their type can be found. Immigration increases the population. **Emigration** (em uh GRAY shun) occurs when organisms leave the area. Emigration decreases the population. In any population that can move, then, natality and immigration increase the population, and mortality and emigration decrease the population. Thus, the size of any population is the result of the relationships among these rates.

Natality, mortality, immigration, and emigration rates apply to every population, including the human population. The sum of these rates makes up the growth rate of a population.

2.3   The Environment Limits Population Size

In nature, a population varies between some upper and lower limit, depending upon its natality, mortality, immigration, and emigration rates. Each population is like a swing moving back and forth between two points, or between a high and a low number. The lower limit is, of course, zero. At that point, the population no longer exists. What is the upper limit on population size? It depends on the **environment** (en VY run ment), everything that surrounds and affects an organism. The environment may slow the individual's growth, may kill the individual, or may stimulate the individual's growth and reproduction. In any case, the environment affects individuals and, thus, the ultimate size of a population.

The environment is made up of two parts: the living part and the non-living part of an organism's surroundings. The living (or recently living) part is called the environment. For you, the **biotic** environment includes your neighbors, houseplants, the dog, the fleas on the dog, and all the

2 organisms you eat. The nonliving part is called the **abiotic** (AY by OT ik) environment. The abiotic environment includes such things as living space, sunlight, soil, wind, and rain. Both biotic and abiotic factors affect the size of a population. Any biotic or abiotic factor that can affect the growth of a population, such as temperature, is called a **limiting factor**.

The climate is a group of abiotic limiting factors (weather factors) that affect all plants and animals. Weather factors include temperature, sunlight intensity, precipitation (rainfall, snowfall, fog), humidity (amount of moisture in the air), and wind. Although each factor may be measured alone, each factor affects the other, and together, they affect population size. The effect may be either direct or indirect.

Water is another important abiotic factor. All living organisms need water. Almost all chemical reactions needed to keep an organism alive take place in water, and indeed, water molecules are a part of many chemical reactions. Although a few organisms can survive by becoming inactive when there is no water, most organisms die. Because water is essential for all living organisms, it often is a limiting factor. Seedlings, such as those shown in Figure 2.3, grow and gradually establish themselves when they receive enough water. Because of other climatic factors,

4

Figure 2.2   How did the biologist count the number of trees on this hillside? **Probably estimated by counting the number in several sample areas and extrapolating the information.**

Using the term *biotic* to refer to things that are alive or were recently alive may be confusing to some students. A dead elm tree or a dead zebra being eaten by lions is still a biotic environmental component. A dinosaur that is fossilized or plants that have become petroleum can be considered abiotic.

Encourage students to give other examples of local weather effects. Note, however, that short-term weather effects primarily affect the population size of organisms with short life spans. Long-term changes, such as drought, can affect the population size of organisms with longer life spans.

Figure 2.3 A pine seedling. What environmental factors are important to the survival of this seedling? **Answers should include enough water, adequate sun, moderate temperatures.**

You may want to introduce density calculations at this time. The rate at which the density of a population changes is equal to the change in density divided by the change in time.

You may wish to explain the rabbit problem in Australia and relate the lack of natural predators to population size.

Ask students why, when considering carrying capacity, we must think of the whole earth for humans but not for house sparrows or pine trees. Pines (as a genus) and house sparrows are widely spread over the earth, but they are not as ubiquitous as people. Moreover, human activities do not affect only the parts of the biosphere humans inhabit but all parts—for example, the stratosphere and the oceans.

however, adequate rain and snow alone may not ensure enough water. Wind, for example, speeds up the rate of water evaporation. Low humidity in the desert also results in an increased evaporation rate. Thus, high rates of evaporation can affect the survival of certain organisms, even when precipitation is adequate. In some cases, a high rate of evaporation together with low rainfall may permit only plants with special features to survive.

Another abiotic limiting factor for populations is space. Although every individual needs living space, some organisms need less space than others. For example, individual corn plants grow well when they are planted closely together. A mountain lion, on the other hand, usually requires many square kilometers for its range. In general, organisms that move about, such as animals, need more space than stationary organisms, such as plants.

The amount of space needed by a single organism is related in part to a biotic factor—the availability of food energy. However, it also is affected by population density, the number of individuals in relation to the space the population occupies (see Figure 2.4). For example, in a classic experiment, illustrated in Figure 2.5, mice in cages were given more than enough food each day. As the mice reproduced, the density of the population increased, and the cages became very crowded. Some female mice stopped taking care of their nests and young. Mice continued to be born, but many newborn mice died from neglect. Eventually, mortality of the young mice reached nearly 100 percent. The high mortality of the young mice kept the population density from increasing further. **6**

2.4 Abiotic and Biotic Factors Work Together to Influence Population Size

Not only do abiotic and biotic factors in the environment influence one another, they also affect the size of a population. For example, a population is limited by the amount of matter and energy, or **resources**, available **3** to it. In any particular space, even under the best conditions, matter and energy are limited. In other words, resources are finite, or limited.

The organism's relationships with others around it can affect the availability of these resources. Predators, disease, and competition with other organisms for scarce resources may take their toll on a population. If an animal population is so large that it uses all the available water or food, it will decrease in size. Rabbits that exhaust the supply of grass and other foodstuffs in their area, for example, will die from starvation and disease. Usually, however, the size of the rabbit population is controlled by other populations in the environment. In this case, because of the coyotes that live in the same area, the rabbit population probably will not grow large enough to exhaust the food supply. Environmental stress brought about by abiotic factors, such as extreme cold or heat, also may limit population size.

The greatest number of individuals that a space can support indefinitely without degrading the environment is called its **carrying capacity**. Although a larger number of individuals may be supported temporarily in the space, **5** the environment will suffer. Ultimately, the carrying capacity of the environment is the most important measure in determining population size.

A given valley in the mountains, for example, has a carrying capacity of only so many deer. Suppose the deer population there increases. Gradually, the types of plants the deer eat become scarce. A few deer may emigrate and look for food in other places. Those that stay in the valley

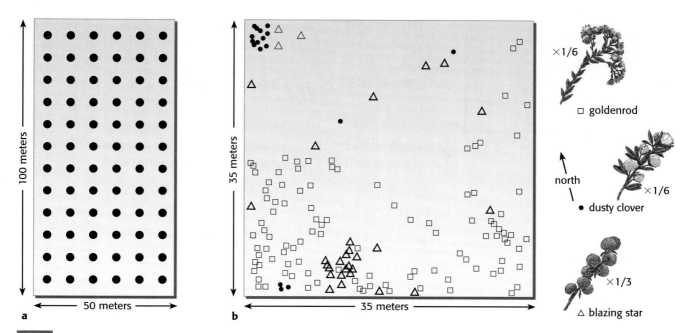

Figure 2.4 (a) Plan of an orchard. Each dot represents one tree. What is the density of trees in the orchard? (b) Under natural conditions, organisms are rarely distributed evenly. Calculate the density of dusty clover in the field as a whole and then only in the northwest corner. Compare.

become undernourished and weak. They are easily caught by mountain lions, killed by disease, or become too weak to search for food when the snow is deep and therefore starve. Because of this emigration and mortality, the deer population decreases, which means that fewer plants are eaten. As the plants that remain continue to grow, more food becomes available for the surviving deer. Deer stop emigrating, and a few hungry deer even move into the valley from the surrounding hills. All the deer now find plenty of food and become healthy. Mountain lions may still catch some of these healthy deer, but the other factors influencing deer mortality decrease. With less mortality and more immigration, the deer population again increases.

This deer population example illustrates **homeostasis** (hoh mee oh STAY sis), the tendency for a population to remain relatively stable in size.

Use a local example of a predator-prey relationship.

Contrast a lion's prey with that of a hunter. Ask students for their views on the impact of hunting on any animal population.

You may wish to compare this definition of homeostasis with the physiological one.

Figure 2.5 In this experiment, mice were provided with more than enough food (left). As a result (right), the population has grown dramatically. Predict what will happen if the mice experience a food shortage as well as overcrowding.

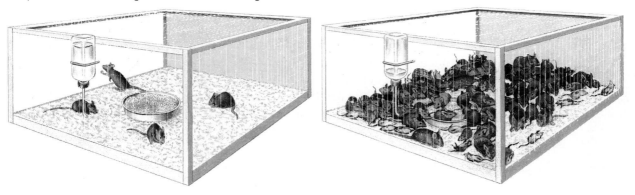

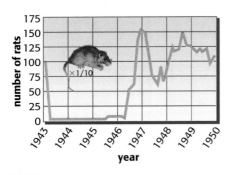

Figure 2.6 Calculate the changes in the rat population between 1946 and 1947, 1947 and 1948, 1948 and 1949, and 1949 and 1950. What might account for these changes?

The investigators estimate that their counting methods involved about 10% error. Of course, rats may have immigrated from surrounding blocks, but the investigators found that rats seldom crossed streets unless the rat population was quite dense.

Through time, the deer population remains between the upper and lower limits that are defined by mortality, natality, and population movements.

2.5   Population Density May Fluctuate

Any population has a built-in, characteristic growth rate—the rate at which the group would grow if food and space were unlimited and individuals bred freely. Environmental factors affect a population's growth **6** rate. The interaction of the population's with the environment determines the density of the surviving population.

If you measure the density of a population at intervals during a year you seldom find any two consecutive measurements the same. Density increases or decreases seasonally. Most natural populations are open populations, or populations in which individuals are free to emigrate or immigrate and in which the birth and death rates fluctuate.

The graph in Figure 2.6 is based on data collected while doing studies of Norway rats in Baltimore, Maryland. In 1942, the city's health department conducted a poisoning campaign that seemed to wipe out the rat population in the city block from which the data were collected. Because it is difficult to count individuals in natural populations, it is possible that a few rats survived or that a few immigrated to the block from elsewhere. In either case, a new rat population was present and increasing by early 1945.

Look at the line for the later years of the rat study. The population decreased after a peak density was reached, yet the decrease did not continue. Open populations usually increase again, just as the rat population did. After an open population peaks, it again decreases, and then may stabilize. Natural populations normally show fluctuations, or ups and downs. Variables in the environment, such as climate, available food, or the activities of natural enemies, are the causes of the fluctuations.

Sometimes population fluctuations are fairly regular, and the peaks on a graph are at approximately equal distances. For example, populations of lemmings (see Figure 2.7) often peak every three or four years. Many of the animals that live in the northern parts of Europe, Asia, and North America show similar population cycles. Although the data show very regular cycles when they are plotted on a graph, the reasons for the seemingly regular cycles are not well understood. A combination of purely chance events also can produce apparently regular cycles.

Although populations may change cyclically, some population changes are permanent. If a population becomes extinct, for example, the change is permanent. Any permanent change in a population is a change in the community to which the population belongs. Permanent changes in one population also may affect other populations of organisms in the same community.

2.6   Populations May Spread to Neighboring Areas

Environmental factors keep populations in a specific area between upper and lower limits, but what factors restrict the spread of populations into new areas? For example, although the climate and other abiotic conditions at the North Pole and the South Pole are similar, polar bears roam the ice **8** floes only in the Arctic, and penguins are found only in the Antarctic (see Figure 2.8). Why are there no penguins near the North Pole and no polar bears near the South Pole?

Apparently an organism's ability to tolerate some environmental factors, such as cold climates, does not entirely explain its geographic range, or

Figure 2.7 Lemmings are mouse like animals that live in the northern parts of North America, Europe, and Asia. What factors might affect the size of the lemming population? **Cold or warm weather, availability of food, predators.**

×1/2

Figure 2.8   How are the environments of the polar bear and the penguin similar? How are they different?   **Both are cold, icy, snowy, and have little or no vegetation; they differ in latitude.**

where it lives. Just because an organism *can* live in a particular place does not mean that it *does* live in that place. To understand the geographic ranges of penguins, polar bears, and other organisms, we need additional information.

All populations of living things have the tendency to spread from one place into others. This spreading of organisms is called **dispersal**. Have you ever noticed, for example, how a single dandelion in your yard can turn into 50 dandelions, seemingly overnight? Similarly, the European starling.

In the case of organisms that can move, dispersal may occur through flying, swimming, walking, running, crawling, or burrowing. For organisms that cannot move, dispersal is passive. In other words, these organisms are spread from one place to another by other agents (see Figure 2.9). Seeds, spores, and eggs can remain alive for long periods. During this time, they may be carried great distances by air or water currents. They even may travel in the mud on a bird's foot. Sometimes organisms that can move may be transported by other agents much farther than they can travel themselves. Floating ice may carry a polar bear hundreds of kilometers, and a spider may be blown long distances by the wind. Of course, dispersal alone does not change an organism's geographic range. Unless the organisms are able to survive and reproduce in the new location, their range has not changed.

Penguins do cross the equator, but only in the cold waters of the Humboldt Current along the west coast of South America.

2.7   Barriers Can Prevent Dispersal

Because every type of organism has some means of dispersal, you might expect that all species would be found wherever environmental conditions
7   were suitable. Actually, however, a broad geographic range is the exception, not the rule. What limits the dispersal of populations?

Physical barriers, such as mountains, prevent dispersal. For most terrestrial (land-dwelling) organisms, large bodies of water are effective barriers. On the other hand, for aquatic animals, land may be a barrier. Besides these physical barriers, there are ecological barriers. For organisms that live in the forest, grasslands and deserts may be barriers. Some barriers are behavioral. You might suppose that flying birds are found everywhere, but many birds that can fly great distances remain in very restricted regions. In these cases, distribution apparently is limited by the behavior of individuals

Figure 2.9   Dandelion seeds. Do dandelions emigrate or immigrate?

Figure 2.10 The distribution of tree and meadow pipits is limited by behavioral barriers.

Ask students to name tree species whose seeds are easily carried by wind. Examples include ashes, maples, and cottonwoods.

in selecting their living area. For example, the tree pipit (*Anthus trivialis*) and the meadow pipit (*A. pratensis*) live in similar areas, except that the tree pipit breeds only in areas having at least one tall tree. Both pipits nest on the ground, and they feed on the same types of organisms. Both have similar songs when they fly, but at the end of its song, the tree pipit perches in a tree, and the meadow pipit lands on the ground (see Figure 2.10). This difference in behavior limits the dispersal of the two species.

An increase in a population of organisms that can move encourages emigration to less populated areas that may have resources. Sometimes dispersal takes place rapidly as the population grows. If the barriers are great, however, dispersal may be slow even though the population is increasing rapidly.

For organisms or their seeds that undergo passive dispersal, the means of transportation help to determine the rate of spread. Some types of trees with seeds that are carried away and buried by squirrels are estimated to spread only about 1.6 km in 1000 years. By contrast, organisms that are caught in a tornado may be carried 30 km in a few hours.

How does this discussion of dispersal and barriers explain the absence of polar bears from the Antarctic? First, assume that polar bears originated in the Arctic and that the population, like many populations, tended to disperse. Although polar bears could tolerate the cold temperatures at the South Pole, the temperate environment they would have to travel through to reach the South Pole would be too hot. Therefore, it would act as a barrier to their dispersal. This wide barrier has existed for at least as long as polar bears have existed. Thus far no part of the population has been able to move across it successfully. Similar reasoning may be applied to lack of dispersal by penguins.

Limiting factors not only shape the survival of populations but also the environments in which they can live. If a population has the characteristics needed for survival in an area, the individuals in that population will grow and reproduce. If not, the individuals will be unable to produce enough offspring to maintain the population. Eventually, the population will no longer exist in that area.

CONCEPT REVIEW

1. What components must be identified in order to define a population?
2. How do limiting factors affect population growth?
3. How does the availability of resources affect the growth and size of a population?
4. In what ways is water important to living things?
5. Explain the concept of carrying capacity and relate it to limiting factors.
6. In what ways is the density of a population important?
7. What barriers might affect the dispersal of a population?
8. What factors interact to determine the geographic range of a population?

Guidepost

What are the characteristics of the human population, and how is this population affected by limiting factors?

Human Populations

2.8   Few Barriers Prevent Human Dispersal

You are one of more than 5.3 billion living humans on the earth today. In 1989, the human population grew by an estimated 87.5 million people—more people than live in California, New York, Texas, and Pennsylvania combined. Experts have estimated that at the current rate of growth, the human population on earth in the year 2025 will be double what it was in 1975. In other words, when you are about 50 years old, the world may be twice as crowded as it was when you were born. Think about what would happen if the student population in your school doubled, but the school's resources remained the same. You would have twice as many students, but the same number of teachers, classrooms, desks, and lunches. Could your school handle the increase? What would happen if the population doubled again?

Just like the student population in the crowded school, the human population keeps growing, but many resources do not. We survive because agricultural productivity has thus far kept up with our population increase. Why is the human population growing so rapidly? Look at Figure 2.11. Simply put, natality exceeds mortality. Our ability to control many diseases and provide a reliable food supply has resulted in a decrease in mortality for humans. In other words, we have temporarily disrupted homeostasis for the human population. More people produce more babies, and the human population continues to grow.

Although the geographic range of a population is determined by its environment, the dispersal of the population within that range is affected by various barriers. Polar bears are not found in the Antarctic because they cannot tolerate heat, and they would have to disperse through a temperate climate. Think about all the places on the earth where humans are found. What barriers affect the dispersal of the human population? If we did not modify our natural environment by using central heating and air conditioning, where would humans be found? Would we be so widespread if we did not have cars, planes, trains, and ships?

Technology has allowed the human population to occupy and modify every environment on earth and to move among these environments. Through technology, we have eliminated or reduced the incidence of some diseases that helped to control population growth in the past. Therefore, we now live longer on the average. Technology consumes many resources that are limited. Given a growing population and a finite supply of resources, what may happen?

2.9   Earth's Carrying Capacity Is Limited

Food, water, and available space are some of the limiting factors that can affect the size and growth rate of the human population. Like all areas, the

These and similar data can be obtained from *The World Almanac, Statistical Abstract of the United States* or other sources. Ask students how these population increases will affect state parks, beaches, and stadiums.

The growth of a population through greater survival of old (postreproductive) individuals will reduce the calculated birthrate, even though the actual number of births increases. Nevertheless, the ratio of birthrates/death rates per 1000 individuals determine whether a population increases or decreases.

Among these population crashes would be the Inca and Maya of South and Central America, Easter Island, and tribes of the Sahara. Reference books available to students would be helpful. Ask students to name some current populations faced with potential crashes, such as Somalia.

Point out to students that many parts of the earth are sparsely populated. Ask students why space still is a limiting factor.

Figure 2.11   Human population growth from 1 000 000 BC to the year 2000. What might account for the abrupt increase in numbers?

Beginning of agriculture and the Industrial Revolution.

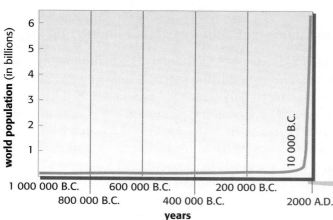

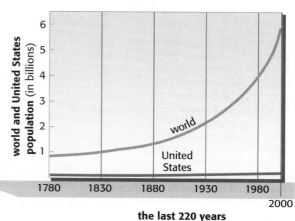

Figure 2.12   As the population increases, is land more valuable for growing food, housing people, or as habitat for other organisms? What might be responsible for this warning? **This is a value judgment. Water pollution.**

Ask students how much space a human needs. This is an open question and should stimulate student discussion. It provides an opportunity to relate rural isolation, urban density, suburban sprawl, and cultural differences. It might also be interpreted in terms of resources, particularly the amount of space required to supply food for humans. It also can mean personal space.

You may want to give some examples here. Water-breeding insects carry malaria (800 million infections per year) and yellow fever. Trachoma, leprosy, and conjunctivitis are spread by washing in polluted water. Typhoid, cholera, dysentery, and diarrhea are waterborne, and defective sanitation spreads intestinal worms.

Foods are complex organic compounds—carbohydrates, proteins, and lipids—that contain usable energy in their molecular structure. Nutrients are inorganic compounds, such as nitrogen, nitrates, phosphates, and potassium, that do not contain energy available for metabolism. The terms *food* and *nutrient* are used in different ways by animal and plant biologists. Here, food includes both energy containing compounds and those elements and compounds that do not contain energy but are essential for life.

earth as a whole has a carrying capacity. Scientists do not agree about the exact carrying capacity of the earth for humans. If food were the only factor, some people claim the earth could support 7 to 8 billion people. Other people think that not even the current population can be maintained for long. Even with these disagreements however, all agree that the human population cannot grow forever.

Humans need space for living. Although experiments on overcrowding, such as the one discussed in Section 2.3 involving mice, help to predict what may happen to other populations under similar conditions, it usually is difficult to apply the results directly to humans. Part of the space needed by humans for living must be used to grow food (see Figure 2.12). Unfortunately, the amount of land on which food can be grown decreases each year because other demands for space increase. People want more space for their houses and other buildings, but each house occupies about one-eighth of a hectare. Eight houses occupy one hectare, which is enough land to provide food for three people.

Water is another limiting factor for the human population, just as it is for other populations. Without an adequate supply of water, humans die. Furthermore, the water must be sanitary. Because many diseases are spread by water, a source of pure drinking water is vital to the health of humans and other animals. Unclean conditions only increase the risk of exposure to disease. In some parts of the world, the only water available for drinking contains disease-causing organisms and traces of human wastes. Toxic chemicals that have seeped into wells and reservoirs of drinking water also can cause illnesses. In some parts of the United States, toxic chemicals have threatened the purity of water supplies.

2.10   Uneven Distribution of Food May Limit Population Size

In many parts of the world today, the food supply is a major limiting factor for humans. The overall production of food in the world may be enough to feed all the people, but food resources are unevenly distributed. Sometimes, getting enough of the right type of nutrition is not easy. Not all people have equal access to the same quality food sources. Living organisms need a wide variety of foods for maintenance and growth. If adequate amounts of food are available, a population may grow. Excess amounts can be harmful. Eating too much of a particular fat, for example, may lead to heart disease. Scarcity, on the other hand, may limit the growth of a population. Many places in the world experience periodic times of **famine** (FAM in), a great shortage of food. This lack of food may slow growth or even cause death. In such places as Asia, Africa, and Latin America, the number of people who must be fed is increasing far more quickly than the amount of food that can be grown to feed them.

Food energy is usually measured in units called calories. A calorie is the amount of energy required to raise the temperature of one gram of water one degree Celsius. The calorie you probably are familiar with, however, is the food calorie, which is really a kilocalorie or kcal—1000 calories. Teenage females need 1200 to 3000 kcals per day, depending on how active they are. Teenage males need 2100 to 3900 kcals, and a football player may consume more than 6000 kcals.

Although life-styles have changed drastically in industrialized countries in the last 100 years and many people have become less active, they have continued to eat about the same number of kcals. In other parts of the world, however, some people are not able to obtain enough kcals to maintain normal activity levels. Without a constant and sufficient supply of kcals, activity levels drop, and eventually muscles are broken down by the body to supply cells with energy.

The number of kcals is not all that is measured when judging the nutritional value of food. Humans need certain foods that are not abundant in plants. A meal of only corn bread or white rice is not nutritious. Animal products may provide enough nourishment and energy, but animals are costly to produce and to buy. To provide a human with 3000 kcals of food that meets all nutritional requirements, a farmer must raise about 30 000 kcals in plant substance. In the United States, almost 90 percent of those 30 000 kcals is used to feed animals raised for meat. In other parts of the world, many people cannot afford to buy meat or raise animals for meat, so they lack both the kcals and the proper foods needed for good health. This problem is not limited to any one part of the earth. Overall, the amount of land that can be farmed is limited; therefore as the population increases, the amount of land available for farming decreases. Although a small amount of land can be very productive, it is costly to maintain.

You may wish to introduce students to the concept of the 10% rule in biomass pyramids and give some examples. These concepts are discussed in Chapter 3.

Most farmland in the United States, for example, is very productive, but this high productivity is maintained only by using huge amounts of fuel energy and fertilizers. Nonmechanized farming practices in tropical and warm climates may return 10 to 20 kcals in food per kcal of energy spent. In the United States, however, for each kcal of energy used to grow food, we get back only 0.1 kcal of energy. In other words, more energy is used to grow and transport food than is gotten back from the food itself. To continue to produce food in this way, we must use our supplies of oil and gas. Oil and gas, however, are nonrenewable resources—materials that never can be produced again. If we use all of these resources now, we will be unable to support our agricultural system in the future. In other words, we cannot keep growing food indefinitely with an energy-losing system.

2.11 Human Activities Change the Environment

Humans have modified the environment more, and in a shorter period of time, than has any other organism. For example, while the composition, temperature, and self-cleansing ability of the earth's atmosphere have varied since the formation of the planet, the pace of the changes in the past two centuries has quickened remarkably. These changes include "acid rain" and similar processes, urban smog, and a thinning of the protective

Figure 2.13   How does the removal of these trees affect the environment?   **Effects include increased runoff and flooding; leaching of soil nutrients; destruction of habitat that affects other organisms.**

ozone layer that shields the earth from harmful radiation. Some scientists ⁵ think the planet soon may begin to warm rapidly (see Chapter 4, Biology Today), causing drastic climatic changes. Certainly, some changes in the atmosphere can be attributed to natural causes, but the activities of humans account for many of the most rapid changes. These activities include the burning of fossil fuels (coal and petroleum) for energy, slash- ⁵ and-burn agricultural methods, and deforestation (see Figure 2.13).

Human activities also affect the water supply As the demand for water by the human population becomes greater than the supply, shortages will occur. Depletion of groundwater, for example, already is common in India, China, and the United States. Because deforestation removes the tree roots that hold soil in place and help retain water, deforestation leads not only to increased soil erosion but also to increased water runoff and flooding. Instead of entering the groundwater, rain runs into rivers.

2.12   Managing Earth Requires Global Cooperation

Our interactions with the environment are complicated by the sheer number of humans on earth. These numbers have increased the rate, scale, and complexity of those interactions. What once were local incidents of pollution now often involve several countries. For example, acid rain falls in most countries of Europe and North America. These days, preservation of the environment has many economic and political aspects because human development has led to environmental changes that affect the entire planet.

To plan effective management of these environmental changes, we must ask three questions: What type of planet do we want? What type of planet can we get? How do we get it? Although many individuals have begun to change their values, beliefs, and actions, it is as a global population that we are transforming the planet. By pooling our knowledge, coordinating our actions, and sharing the resources of the planet on a global scale we can manage the earth in a sustainable fashion. Some steps have been taken, such as the development of new irrigation techniques that save

The great Ogallala Aquifer, which stretches northward through the Midwest from western Texas to northern Nebraska, is an example of underground water that is being depleted. Because of low rainfall, the aquifer is replenished at a lower rate than it is used. Although all forms of irrigation do not necessarily deplete aquafers, water tables in the area are dropping, and the amount of irrigated land is decreasing.

Be certain that students understand we cannot *design* the earth, but to a certain extent, we can determine the changes that take place on it.

Biology Today

Famine in Africa

In late 1984, the world learned of a severe food shortage in Africa. The shortage was greatest in the Sahel, a swath of land cutting across the northern desert. By early 1985, the famine had killed 300 000 people in Ethiopia and another 200 000 in Mozambique. Officially more than 30 countries in Africa are listed as famine areas. In the most desperate of these, whole populations may be wiped out. How did this happen?

This worsening situation is caused by a number of interacting factors:

◆ Worldwide food distribution is uneven. If all the food produced in the world were divided equally, there would be enough to keep 6 billion people alive (2.5 billion on a diet typical of developed countries). Food is distributed unevenly because of differences in soil, climate, and income.

◆ Africa has the fastest population growth of any continent. Every three weeks an additional one million people are added to the population.

◆ A severe, 17-year drought has compounded the problem. Rainfall in the region can vary by as much as 40 percent from one year to the next. Prolonged droughts are not uncommon.

◆ Much of Africa has poor soil. Three-tenths of the land is covered by desert or is too sandy to grow crops. Poor farming practices have degraded marginal soils, and much of the land has been overgrazed and deforested. These conditions lead to severe topsoil erosion and desertification (the desert spreads out to cover degraded areas).

◆ Africa has poor food distribution systems. The lack of adequate, fast transportation hinders distribution. Often, perishable food rots before it reaches its destination.

◆ Foreign aid is used to raise cash crops on irrigated land rather than to improve dry-land farming, grow food for domestic use, conserve soil and water resources and replant trees.

In late 1984, a massive relief effort began, but it was stymied by corruption at the local level, mismanagement, and civil strife. Although the rains returned in 1985, hunger continues. The land has been so overworked and overgrazed that its carrying capacity has been permanently lowered.

Satellite images indicate that the land can recover some of its productivity. In good years of rain, the "green wave" of vegetation extends farther north into the Sahel than it does in dry years. Although high-quality vegetation will not return quickly to overgrazed areas, if controlled grazing programs are followed, much of the land may recover. New irrigation techniques help to conserve water, and planting several crops in the same field may increase the yield for all crops by reducing soil erosion, weed growth, and nutrient loss.

Experts agree that merely providing food for the hungry does not solve the problem. The only reasonable way to reduce hunger is to teach people how to produce food without degrading the environment. Organizations such as the Consultative Group on International Agricultural Research and the United Nations Environment Program are developing new crops and farming methods to increase self-sufficiency without harming the environment.

Vegetation is shown on the scale from little or none (tan) through heavy (purple). The year 1980 (upper map) had average rainfall. The year 1984 (lower map) was very dry.

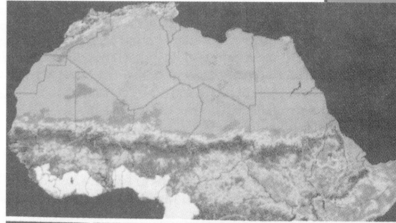

a

b

Figure 2.14   Old irrigation techniques (a) sprayed water over a large area. Some techniques (b) deliver water directly to the plants. How is this new technique beneficial? **Water goes directly to plant roots; less water evaporates.**

Emphasize that the number of humans on the earth is a major contributing factor. Even 50 years ago, the overall impact of activities such as manufacturing, clear cutting, and automobile use was less because there were fewer people.

water by sending it directly to the roots of plants (see Figure 2.14). A multinational agreement to cut in half the emission of some gases that harm the atmosphere also helps. Many other steps need to be taken, however, before the problems are under control.

CONCEPT REVIEW

1. What is the relationship between resources and the carrying capacity of a particular area?
2. Discuss the relationship between kcals and nutrition.
3. How have humans disrupted the homeostasis of our population?
4. Discuss the interaction of limiting factors for the human population. What is likely to happen in the future?
5. How have humans affected their environment?
6. If adequate amounts of food are produced overall, why do some countries experience famine?

In this investigation, students will obtain firsthand experience with population dynamics and acquire practice with experimental methods and the use of microscopes and microcomputers. Teams of 2 are preferable.
   Time: **Seven class periods—one period to set up and take initial counts and readings; thereafter, approximately 15 minutes per day (depending on class size) to take turbidity readings. Allow 15 additional minutes on days 1, 4, and 7 to make counts, and one period for class discussion of the results.**

INVESTIGATION 2.1   Study of a Population

Because they reproduce rapidly and are small, yeast organisms are useful for studying populations. You will observe a population of yeast cells growing in a broth medium in a test tube.

   This type of population is a closed population. In nature, open populations increase or decrease in size as organisms enter or leave them. In open populations, materials can cycle through the population. Conditions in a closed population are somewhat different, but you can estimate the rate of population growth more easily than in an open system. Using the Leap-System™ (a computer interfacing system) with a turbidity probe, you can measure the turbidity or cloudiness, of the medium to determine how the yeast population changes through time. By making cell counts, you can correlate changes in the number of yeast cells with the turbidity readings. Instructions for use of the LEAP–System are provided in the LEAP-System BSCS Biology Lab Pac 2.

   Read the investigation carefully and develop three hypotheses that might explain how a population changes during a period of time. Record these hypotheses in your data book and evaluate them with the data you collect during this investigation.

Table 2.1 Number of Yeast Cells

Count	Day 0		Day 1		Day 2		Day 3		Day 4		Day 5		Day 6		Day 7	
Tube	A	B	A	B	A	B	A	B	A	B	A	B	A	B	A	B
First sample																
Second sample																
Total																
Average																
Dilution																
Average X dilution																
Turbidity reading																
Estimate of count																

Materials (per team of 2)

2 pairs safety goggles
2 16-mm × 150-mm screw-cap test tubes, each with 10 mL sterile broth medium
2 18-mm × 150-mm test tubes cover slip
ruled microscope slide (2-mm × 2-mm squares)
1-mL graduated pipet dropping pipet

Apple lle, Apple IIGS, Macintosh, or IBM computer
LEAP-System™
turbidity probe
microscope
test-tube rack
metric ruler
glass-marking pencil
4 sheets of semilog graph paper

Put on your safety goggles.

Procedure

Day 0

1. In your data book, prepare a data table similar to Table 2.1.

2. Use the glass-marking pencil to label one screw-cap test tube A and the other B. Mark each test tube with your team symbol. Do not remove the caps.

3. Your teacher will transfer 0.1 mL of a yeast stock culture to test tube A. Gently invert the tube several times to distribute the cells evenly. Nothing will be added to test tube B. Slightly loosen the caps and store the two test tubes in the test-tube rack and in the location specified by your teacher. What is the purpose of test tube B?

4. Test tube A contains the start of a new population. Review your hypotheses regarding what you think might happen during the next week.

5. To determine how rapidly the yeast population grows, you will take population counts during the course of the experiment. Refer to Figure 2.15 for the appearance of yeast cells. Tighten the cap and gently invert test tube A several times to distribute the yeast. Then use a dropping pipet to transfer 1 drop from test tube A to the grid of the slide. Carefully place a clean cover slip over the drop, trying not to trap any air bubbles. Examine the slide with the high-power lens of a microscope. Note: Yeast cells are difficult to see if the light is too bright.

6. To count the number of individual yeast cells in the center square, position the upper left corner of the square under the high-power lens and count the cells in that field. Move the slide to the upper right corner and count those cells.

Safety

When a computer is used in the lab, the lab electrical sockets must be protected with a GFI (ground fault interrupter) circuit breaker. Each electrical outlet in the lab must be three-hole, and each set of three-holes must be checked with a circuit tester prior to use to make certain that the wiring has been correctly connected to that set. (That is, there should be no "open ground," "open neutral," "open hot," "hot/ground reverse," or "hot/ neutral reverse"

wiring. The circuit tester indicates these wrong conditions by various configurations of red and green lights. Circuit testers are widely available at hardware and radio supply stores for under $5.)

It is advisable for students to wear safety goggles when working with any liquid. Caution students to invert culture tubes gently to avoid foaming, which would result in inaccurate counts. Caps should be loosened slightly before storage in case of gas buildup. Remind them to wash each day.

Procedure

2. Test tube B is a control.

Figure 2.15 Yeast (X600). Note the buds still attached to the yeast cells. These will become new individuals.

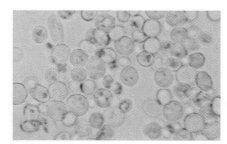

Figure 2.16   The LEAP-System with turbidity probe.

Continue clockwise to the lower right corner and then the lower left corner until you have counted all the yeast cells in the square. Make certain you are observing yeast cells and not other material. The cells often stick together, but count each cell in any clump separately. Buds also count as individuals.

7. Count at least 300 cells. If your count is less than 300 cells, count the cells in additional squares around the original square until you have counted 300 cells. Calculate the average number of cells per square by dividing the number of cells by the number of squares counted.

8. To obtain the number of cells per cubic centimeter (1 mL), multiply the number of cells by 2500. Each square is 2 mm × 2 mm; the cover slip is approximately 0.1 mm above the counting area; each square, therefore, has a volume of 2 mm × 2 mm × 0.1 mm = 0.4 mm³. There are 1000 mm³/cm³; therefore,

$$\frac{cells}{0.4mm} \times \frac{1000\ mm^3}{cm^3} = \frac{2500 \times cells}{cm^3}$$

To obtain the total number of cells in test tube A, multiply the final number by 10, because test tube A contains 10 mL of culture medium.

9. Have your partner count in the same way from a new sample. Record these counts in the data table and calculate the average of the two counts. Record this Day 0 population size in your data table and on the class master table.

10. Repeat Steps 5–9 for test tube B.

11. Use the turbidity probe (Figure 2.16) to take readings of both test tubes for about 10 seconds each. Remove the screw cap before placing the cap on the probe and replace after removing the test tube from the probe. Calculate the average of the readings (the computer takes one reading every second) and enter the averages under Day 0 in your data table and on the class master table. What do you predict will happen as the yeast organisms grow? Record your prediction in your data book.

12. Wash your hands thoroughly before leaving the laboratory.

Days 1-7

13. Invert test tube A several times to distribute the yeast organisms evenly. Using the turbidity probe, take readings for about 10 seconds. Record the average of the readings under Day 1 in your data table and on the class master table. Repeat the same procedure with test tube B.

14. Repeat the counting procedure in Steps 5–9, using one drop from test tube A and then one drop from test tube B. Record your data under Day 1 in your data table and on the class master table.

15. Based on the information you now have, you can start a graph that will help you estimate the yeast population size from the turbidity readings on the days you do not make counts. Draw x and y axis lines on a sheet of semilog graph paper. Label the y (vertical) axis *Number of Yeast Cells*. Label the x (horizontal) axis *Turbidity Readings*. The values on the y axis should be from 105 to 109. The values on the x axis should be from 0 to 100.

16. Make a mark on your graph at the point where the Day 0 turbidity reading and the yeast count for test tube A intersect. Label this point *Day 0*. Make a second mark on your graph at the point where the Day 1 turbidity reading and the yeast count intersect. Label this point *Day 1*. Use a ruler to draw a line connecting the two marks. Extend the line until it reaches both edges of the graph. This line now can be used to help you estimate your yeast counts on Days 2 and 3. Repeat for test tube B, using a dotted line.

17. **Students must multiply by 10 for the first dilution, by 20 for the second.**

17. Repeat Steps 13 and 16 each day (skip the weekend). Include Step 14 on Days 4 and 7. If your count is more than 300 or too many to count, make a dilution as follows. Use the 1-mL graduated pipet to transfer 0.9 mL of water to a clean test tube; label this test tube *D1*. Invert test tube A several times until

the organisms are evenly distributed, and immediately transfer 0.1 L to test tube D1. Invert test tube D1 several times to mix, and then use the dropping pipet to transfer one drop to the grid of the slide. Add a cover slip as in Step 5 and count the cells as in Step 6. If necessary, make another dilution by transferring 0.1 mL from test tube A to a fresh tube containing 1.9 mL of water; label this tube *D2*. Calculate the population size as in Step 8. What additional calculation must you make if you have made the first dilution? The second dilution? Record the calculated results under the appropriate day in the data table.

18. On Days 4 and 7, make an additional mark on your graph for each test tube at the points where your yeast counts and the turbidity readings intersect. If necessary, modify your lines after each mark is entered on the graph.

19. On Day 7, calculate the average of all the team readings from the class master table.

20. Each day, wash your hands thoroughly before leaving the laboratory.

Discussion

In this investigation, you estimated the population size in your test tube by counting the individuals in a sample. This method is called sampling. You also obtained an estimate based on the turbidity in the test tube. To increase the accuracy of your estimates and the computer readings, you took certain precautions. You inverted the test tube several times to distribute the organisms evenly. You used a counting procedure to estimate the population size on Days 0, 1, 4, and 7. The averaging procedure smoothed out chance differences between your counts. You also estimated the population size based on the turbidity readings obtained with the computer. On the class master table, you averaged the data obtained by all teams. That further smoothed out chance differences among the test tubes. The final count for each day's population is the average number of organisms predicted by the computer data.

1. On a fresh sheet of semilog graph paper, list the ages of the cultures (in days) on the horizontal axis. List the number of organisms per estimate on the vertical axis. Plot your own data for test tubes A and B and then use a different color to plot the averages of all teams for both test tubes.

2. On the basis of your discussion of this investigation, explain similarities and differences among the graph lines representing data from different teams.

3. Is there any general trend in the graph line representing the average data of all teams? If so, describe it.

4. Review the three hypotheses you developed before starting this investigation. Are any of them supported by the data that you have accumulated? Are any of them not supported? Explain.

5. Why was it necessary to make the yeast counts on Days 0, 1, 4, and 7?

6 Were there changes in turbidity in test tube B7? How can you account for those changes? How do those changes affect the accuracy of your readings for test tube A?

7. What limiting factors could have influenced the growth of the yeast population?

For Further Investigation

How does temperature affect the growth of a yeast population? Repeat the procedures, but incubate the cultures at a constant temperature 15° C above or below the average temperature at which the tubes were incubated before. Or, place one set of test tubes in a water bath at 30° C, and another set in a water bath at 50° C. On Day 2 of the experiment, place one set of test tubes in the freezer and another in the refrigerator. After one or two weeks, remove these test tubes, make counts (after the frozen tube thaws), and continue the experiment.

Discussion

2. There will be a wide range of numbers between counts of different teams. Fluctuations in the populations as measured by one team are likely to be so great that to detect any pattern of growth may be difficult. Averaging the team counts will result in a good growth curve. Sample turbidity readings, given in parentheses, and counts are as follows: Day 0: (2) 9.3×10^5; 1: (12) 1.5×10^8; 2 (18) 2.6×10^8; 3: (23) 3.5×10^8; 4: (25) 4.0×10^8; 7: (17) 2.2×10^7. (Note that dead, lysed cells will affect the turbidity readings, so counts may decrease more rapidly than readings as population size drops.) To obtain a curve that can be explained in terms of population theory is, of course, desirable and satisfying to students. Failure to obtain such a curve, however, must not be interpreted as failure of the investigation. Discussion Question 2 is the pivot on which the investigation turns. Whatever the results, they will provide material for a fruitful discussion of sources of error in an experimental procedure and of the need for teamwork in some kinds of scientific work. In any case, the hypotheses set up at the beginning of the investigation must be considered at the end.

3. There may be increases and decreases during the 7-day period, but if they concentrate on a general trend, students should be able to see a rapid increase, a leveling off, and then a decrease.

4. The three hypotheses that should be most obvious to students are (1) increase, (2) decrease, and (3) remain the same. Emphasize that the purpose of all the techniques being used and all the data being collected is to evaluate these hypotheses. Depending on what their graphs look like, it should not be too difficult to make a decision.

5. The yeast counts provide additional data on which to base an estimate of the population size.

6. Changes in the control tube most likely would be due to contamination (from unclean pipets or dust or fungal spores entering tubes when they are uncovered to take turbidity readings). Assuming similar contamination in the experimental tube, the control-tube reading should be subtracted from the experimental tube reading before graphs are made.

7. Food, space, and toxic waste products are limiting factors. If one or more cultures become contaminated, then immigration of a new population becomes a factor.

This investigation, which develops the concepts of carrying capacity and doubling tame, allows students to gain an understanding of exponential growth and how it affects population size as well as resources. The activity serves to summarize and reinforce concepts presented in Sections 2.3, 2.4, and 2.5. Students will benefit by working in teams of 3.

Time: **One period should be sufficient to perform the activities and discuss the questions.**

INVESTIGATION 2.2   Population Growth

Four rates—natality, mortality, immigration, and emigration—determine a population's size. A given area can support only a certain number of individuals on a long-term basis. That number is known as the carrying capacity. Populations grow at different rates; the *doubling time* is the number of years required for a particular population to double its size. In this investigation, you will compare the growth of two populations and examine the roles played by carrying capacity, doubling time, and the four rates that result in a population's size.

Materials (per team of 3)
pocket calculator   graph paper

PART A Reindeer Population
Procedure

1. In 1911, 25 reindeer—4 males and 21 females—were introduced onto St. Paul Island, one of the Pribilof Islands in the Bering Sea near Alaska. St. Paul Island is approximately 106 km² in size (41 square miles), and is more than 323 km (200 miles) from the mainland. On St. Paul Island there were no predators of the reindeer, and no hunting of the reindeer was allowed. Study the graph in Figure 2.17 and answer the following questions (steps 2 through 7) in your data book.

2. What was the size of the population at the beginning of the study? In 1920? What was the difference in the number of reindeer between 1911 and 1920? What was the average annual increase in the number of reindeer between 1911 and 1920?

3. What was the difference in population size between the years 1920 and 1930? What was the average annual increase in the number of reindeer between 1920 and 1930?

4. What was the average annual increase in the number of reindeer between 1930 and 1938?

5. During which of the three periods—1911–1920, 1920–1930, or 1930–1938— was the increase in the population of reindeer greatest?

6. What was the greatest number of reindeer found on St. Paul Island between 1910 and 1950? In what year did this occur?

7. In 1950, only eight reindeer were still alive. What is the average annual decrease in the number of reindeer between 1938 and 1950?

Part A Procedure

2. 25; 200; +175; 19.4
3. +200; +20
4. +200
5. 1930–1938
6. 2000; 1938
7. −166

Figure 2.17 Changes in the reindeer population on St. Paul Island between 1911 and 1950.

Discussion

1. **Probably not. St. Paul Island is more than 323 km (200 miles) from the mainland. Reindeer are strong swimmers, but the distance is too great for emigration or immigration to have a major effect.**
2. **Population growth is exponential. Exponential growth is characterized by doubling. A few doublings lead to enormous numbers. With a larger population base to start with in 1930, the growth between 1930 and 1938 would be great.**

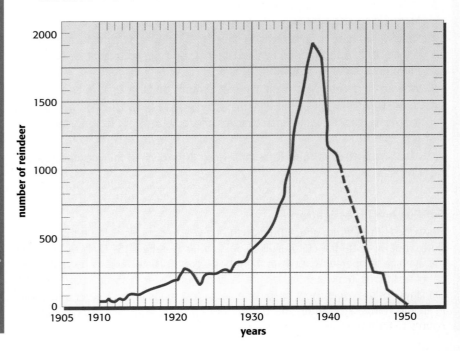

Discussion

1. Could emigration or immigration have played a major role in determining the size of the reindeer population? Explain your answer.

2. What might account for the tremendous increase in the population of reindeer between 1930 and 1938, as compared with the rate of growth during the first years the reindeer were on the island?

3. What effect might 2000 reindeer have on the island and its vegetation?

4. Consider all the factors an organism requires to live. What might have happened on the island to cause the change in population size between 1938 and 1950?

5. Beginning in 1911, in which time spans did the population double? How many years did it take each of those doublings to occur? What happened to the doubling time between 1911 and 1938?

6. If some of the eight reindeer that were still alive in 1950 were males and some females, what do you predict would happen to the population in the next few years? Why?

7. What evidence is there that the carrying capacity for reindeer on this island was exceeded?

8. What does this study tell you about unchecked population growth? What difference might hunters or predators have made?

3. **Overgrazing, death of plants, destruction of habitats, accumulation of wastes.**

4. **Overgrazing resulted in death of plants and insufficient food. Weakened by a lack of food, the reindeer were prey to disease, and the reproductive rate declined drastically.**

5. **Population doubled in 1912 (1 year); 1915 (3 years); 1920 (5 years); 1930 (10 years); 1934 (4 years); 1937 (3 years). Doubling time between 1911 and 1938 became longer until 1930 and then shorter.**

6. **The population might die out because it is too small to recover, or it might slowly increase. Rationales will vary.**

7. **The population crested after 1938.**

8. **Natural controls take effect and can have drastic results, such as the total population dying out. Predators and hunters might have controlled the population, preventing exponential growth and the destruction of the environment and thus maintaining the carrying capacity of the environment.**

PART B Human Population
Procedure

8. On a piece of graph paper, plot the growth of the human population using the data in Table 2.2.

9. Use your graph to determine the doubling times for the human population between AD 1 and 1990. How much time elapsed before the human population of AD 1 doubled the first time? the amount of time needed for the human population to double increasing or decreasing? What does that indicate about how fast the human population is growing? Record your answers in your data book.

10. Extend your graph to the year 2000. What do you estimate the human population will be in that year?

Part B Procedure

9. **1660 years; decreasing. The human population is growing exponentially.**

10. **About 5800 million.**

Table 2.2 Human Population Growth Between AD 1 and 1990

Date AD	Human Population (millions)	Date AD	Human Population (millions)
1	250	1920	1800
1000	280	1930	2070
1200	384	1940	2300
1500	427	1950	2500
1650	470	1960	3000
	(Black Death)	1980	4450
1750	694	1985	4850
1850	1100	1990	5300
1900	1600		

Source: American Association for the Advancement of Science, 1951

11. $\dfrac{(5300-4450) \times 100}{4450 \times 10} =$

$\dfrac{85\ 000}{44\ 500} = 1.91\ \%$

$\dfrac{70}{1.91} = 36.6$ years

$1990 + 37$ years $= 2027$

Discussion

1. **Both show exponential growth curves.**
2. **Food, crop land, grazing land, forests, water, air.**
3. **Earth is a finite environment—an island in space. Yes, because the earth has finite resources.**
4. **Decrease because of starvation or disease.**
5. **Birth control, abortion, infanticide, restriction on number of children parents may have. Student opinion.**
6. **South America, Central America, Africa, India: high birthrate, poor economy, insufficient food: Canada, United States: severe impact on resources.**
7. **Northern Europe: low birthrate, overall level of education.**
8. **Immigration, increasing water and air pollution, decreasing groundwater, increased soil erosion, decreased soil nutrients, loss of wildlife and wildlife habitat.**
9. **Water, living space, food supply, quality of life, justice.**

Although water is generally considered renewable students probably will be surprised at how much water they use in a day. The investigation reinforces the discussion in Section 2.9 and may encourage students to be more aware of the finiteness of resources. Students should work individually and as a class.
   ***Time:* One period should be sufficient to perform the activities and discuss the questions.**

Procedure

1. **Have students complete this step without looking further in the investigation. Emphasize that these estimates reflect direct water use only. Indirect water uses will increase the estimate greatly.**
2–8. **Depends on student data.**

11. Using the equations below, estimate the doubling time for the current population based on the rate of growth from 1980 to 1990. In what year will the present population double?

Rate of growth (in percent) $= \dfrac{(\text{population in 1990} - \text{population in 1980}) \times 100}{\text{population in 1980} \times \text{number of years}}$

Doubling time $= \dfrac{70}{\text{rate of growth}}$

Discussion

1. What similarities do you see between the graph of the reindeer population and your graph of the human population?
2. What are the three or four most important factors required to sustain a population?
3. In what ways is the earth as a whole similar to an island such as St. Paul? Does the earth have a carrying capacity? Explain your answer.
4. What might happen to the population of humans if the present growth rate continues?
5. What methods could be used to reduce the growth rate?
6. Cite a place in the world where population growth is a problem today. How is it a problem?
7. Cite a place in the world where population growth is not a problem today. Why is it not a problem?
8. Suggest several problems in your country that are related to the human population.
9. What are the most important three or four factors to think about with regard to the world population?

INVESTIGATION 2.3   Water—A Necessity of Life

In this investigation, you will calculate how much water you use on a daily basis and then calculate the water needs for your family, your class, your school, your town, and your state. You also will discover some of the ways in which you use water indirectly.

Materials (per person)
pocket calculator (optional)   graph paper

Procedure

1. In your data book, record how many gallons of water you think you use in an average day. Later, you will compare this *estimated* daily water use with your *calculated* daily water use.
2. As a group, list all the ways in which the members of your class use water on a day-to-day basis.
3. Using the data in Table 2.3, determine your *individual* water use per day for each activity your class listed. Include your share of general family uses such as dishwasher and washing machine. Then determine your individual *total* water use per day.
4. Compare the individual water use you calculated with the water use you estimated. Is your calculated figure higher or lower than your estimated figure?
5. Find out how many people are in your school, including teachers and students. Find out how many people live in your town or city and your state.
6. Calculate the total amount of water that is used each day by your family, your class, your school, your town, and your state, for each of the activities the group listed in Step 2.
7. Draw a bar graph to illustrate how much water is used by your class for each activity. Which activities require the most water?

Table 2.3 Domestic Uses of Water

Activity	Amount Used (gallons)
Automatic dishwasher	15
Brushing teeth	2–10
Cooking a meal	5–7
Flushing toilet	3.5–8
Getting a drink	0.25
House cleaning	7
Leaking faucet	25–50/day
Shaving	20 (2/min)
Showering	20–25 (5/min)
Tub bathing	25–35
Washing dishes	30 (8–10/meal)
Washing machine	24–50
Washing hands	2
Watering lawn	10/min (102/1000 m²)
Faucet and toilet leaks in New York City	= 757 million/day

Source: From *Living in the Environment: An Introduction to Environments Science,* 6th ed., by G. Tyler Miller, Jr. © by Wadsworth, Inc. Used by permission.

Discussion

1. Many water uses are not obvious to most people. Consider, for example, how much water is necessary to raise one calf until it is fully grown (see Table 2.4). Why do you think so much water is needed to raise a calf?

2. Make a list of the ways you use water indirectly—for example, in the production of the food you eat or the materials you use.

3. Compare your list with Table 2.5. How many of the indirect uses of water shown in the table did you list?

4. How could you reduce your indirect use of water?

5. What could you do to reduce your direct use of water?

Discussion

1. **About 65 percent of its body weight is water. There is a continual loss of water in urine, feces, respiration and evaporation. This loss must be replaced.**

2. **Student answers will vary. They may be interested in this data: It takes 650 gallons of water to make the steel for one bicycle and 200 gallons to make the rubber for one car tire.**

3. **Depends on student answers to Question 2.**

4. **Students have very little control over this, but being aware of indirect uses may lead to more and better conservation practices.**

5. ***Automatic dishwasher:* run only when full; brushing teeth: do not let water run; cooking a meal: do not let the water run; *flushing toilet:* put something (a brick) in tank to take up space so that the toilet will use**

Table 2.4 How to Make a Cow

Ingredients*			
1	80-lb calf	2500	lbs corn
8	acres grazing land	350	lbs soybeans, insecticides, herbicides, antibiotics, hormones
12 000	lbs forage	1.5	acres farmland
125	gals gasoline and various petroleum byproducts	1.2	million gals water, to be added regularly throughout
305	lbs fertilizer (170 lbs nitrogen, 45 lbs phosphorus, 90 lbs potassium)		

*These are the ingredients used to produce choice beef. Lesser amounts are required for today's leaner beef.

Take one 80-pound calf, allow to nurse and eat grass for 6 months, then wean. Over next 10 months, feed 12 000 pounds of forage. Use about 25 gallons of petroleum to make fertilizer to add to the 1.5 acres of land. Set aside rest of gasoline to power machinery, produce electricity, and pump water. Plant corn and soybeans; apply insecticides and herbicides. At 24 months, feed cow small amounts of crop and transfer to feedlot. Add antibiotics to prevent disease and hormones to speed up fattening. During next 4 months, feed remaining crop mixed with roughage. Recipe yields about 440 usable pounds of meat—1000 7-ounce servings. Option: Bake the 2500 pounds of corn and 350 pounds of soybeans into bread and casseroles—18 000 8-ounce servings.

Source: Excerpted from *Cousteau Almanac* by Jacques-Yves Cousteau. Copyright © 1980, 1981 by the Cousteau Society, Inc. Reprinted by permission of Bantams Doubleday, Dell Publishing Group, Inc.

less water; *getting a drink* do not let water run to get cold, use ice cubes or keep a pitcher of water in the refrigerator; *house cleaning*: put water in sink, tub, or bucket and clean entire room with same water; *leaking faucet*: fix it, or plug sink or tub to catch water for watering plants, cleaning, or any other uses you can think of; *shaving*: do not let the water run; *showering*: take shorter showers, turn water off when lathering hair and/or body; *tub bathing*: do not fill tub; washing dishes: do not let the water run, fill basin with hot water for rinsing, use the dishwasher if you have one; *washing hands*: do not let water run, put just enough in the sink to clean hands; *washing machine*: wash only full loads and use water-saver features if available; *watering lawn*: water in the AM when it is cool, do not let water run down sidewalks or into the street, use grass species that demand less water.

6. **Data from water departments would give this information**

Table 2.5 Indirect Uses of Water

Agricultural		Industrial	
Item	*Gallons used*	*Item*	*Gallons used*
1 loaf bread	150	1 kg aluminum	2200
1 kg corn	374	1 car	100 000
1 kg grain-fed beef	1760	1 kilowatt of electricity	80
1 kg cotton	4440	1 gallon gasoline	10
1 kg rice	1232	1 kg paper	220
		1 kg steel	25
		1 kq synthetic rubber	660

Source: From *living in the Environment An Introduction to Environments Science,* 6th ed., by G. Tyler Miller, Jr. © 1990 by Wadsworth, Inc. Used by permission

6. Is there any evidence that the water supply you use daily is decreasing in size or is being contaminated by pollutants? How could you go about obtaining this information?

For Further Investigation

1. Research actual water use for one of the items you listed in Question 2.
2. Examine your family utility bills for the past year and note the units of water used in each month. Each unit represents one cubic meter of water. If one gallon equals 0.004 cubic meters, calculate your family's average monthly water use in gallons. How does actual use compare with your calculations in this investigation?

SUMMARY

We are all part of a population. Like all other populations, the human population is affected by both biotic and abiotic factors. Whereas some of these factors help to increase a population's size, other factors cause it to decrease. Populations have species specific growth rates, but their growth is limited by environmental factors that may act as barriers to dispersal and affect the density of a population. Any given area has a carrying capacity; that is, it has resources to support only so many individuals. Once the density of the population exceeds the carrying capacity of the area, the necessities for life are in short supply, and the population size decreases. How long will it take before the carrying capacity of the earth for humans is reached? The human population is huge and growing. We affect other organisms because of the amount of resources we use and the amount of waste we produce. Through our use of resources, we change the environment around us. When these changes occur, other populations are affected. These changes occur because all populations are linked together in the biosphere. We are now recognizing how our actions affect the organisms around us and we are attempting to develop reasonable and workable management plans to reduce our impact on the environment and to extend the life of the human population and all other populations that share the earth.

APPLICATIONS

1. Suppose you know the size of a population of ducks on your local lake in 1990, and you know how many ducks will be hatched and will die during the next 10 years. Can you determine the duck population size at the end of 10 years? Why or why not?
2. Is the carrying capacity of an area the same for all types of organisms that live there? Why or why not?

3. Humans can live almost anywhere and our population has steadily increased through time. Does this mean that limiting factors do not affect humans? Explain your answer.
4. Suppose there is a population of some type of animal in a huge space, such as rabbits on a large ranch or fruit flies in a big building. If you gave these animals all the food and water they needed, would their populations continue to increase in

size forever? Explain your answer.

5. Why are the terms *mortality* and *natality* as they are defined in this text—not applicable to individuals?

6. Discuss the present relationship between the human population and the carrying capacity of the earth.

7. Why is it difficult for people to see the effects they have on the biosphere?

8. A team of biologists studied a population of box turtles in an Ohio woodlot for a period of 10 years. They determined that the natality averaged 40 per year, the mortality 30 per year, immigration 3 per year, and emigration 8 per year. Was the population increasing or decreasing? Was the area supplying box turtles to other places or vice versa?

9. In Question 8 what was the average annual change due to immigration and emigration? If the initial population was 15 turtles, what was the population at the end of 10 years?

10. Suppose a particular human population continues to reproduce and multiply, far outstripping the carrying capacity of the land. Food is in extremely short supply, water supplies are limited, and what water is available is polluted. Disease is everywhere. Discuss possible solutions to this problem and include the implications of each solution. For example, one solution is to supply food to the people, thus enabling them to live longer and raise their families. What are the implications of this solution? What solutions are ethically possible?

11. Name several ways that humans continue to change the biosphere and discuss the effects of these changes on the biosphere and other populations. What solutions can you identify?

12. It is seldom possible to count all the individuals in a natural population. How can a biologist study populations without such data?

PROBLEMS

1. Use a world almanac to determine how the population of the United States has changed through time. How much of the increase in population may be attributed to immigration? to natality? to mortality?

2. Suppose you had to collect all of the food you eat for yourself or your family. There are groups of people in the world—hunters and gatherers—who must do this on a daily basis. Find out about a group of these people—what they eat, how much territory they need to locate their food, and how many people live in each group.

3. Individuals of certain species have been dispersed to new areas with dramatic consequences to both the population and the area in which the species is introduced. Find out what happened to the European starling after it was introduced to the United States, to the pig after it was introduced to the Hawaiian Islands, the African bee after it was introduced to South America, or the prickly pear cactus after it was introduced to Australia.

4. You have just been elected mayor of your city. Unfortunately, the population of your city will double within the next 10 years. How are you going to handle all of the changes? What plans do you have to make to handle the changes?

5. Obtain the census data for your state. Make a graph of these data, beginning with the first census after your state entered the union. How does the form of this graph compare with that for the population of the United States as a whole? Try to explain any differences.

6. Obtain some duckweed, an aquatic plant that grows on the surface of ponds. Tap the mass of plants with your finger to separate a single individual. Place the single duckweed plant in pond or aquarium water in a petri dish. Make counts of the numbers of individuals at intervals of two to four days. Keep a record of dates and numbers. Construct a graph to show the growth of the duckweed population.

7. Grains, such as wheat and rice, are a basic food throughout the world. The 1985 worldwide production of grains was approximately 1.7 billion tons. This was enough grain to provide 4.6 billion people with enough food energy for normal activity for a year. Still, hundreds of thousands of people in Asia, Africa, and South America were malnourished, and thousands of others starved to death. How can you explain this? *The World Almanac* and books such as L. Brown et al., *State of the World* (New York: W. W. Norton, 1990) may give you some insights.

8. Suppose two populations of a protozoan are grown in an aquarium but are separated by a barrier. Each population reaches 1000, the maximum that can be supported. Then the barrier is removed. If population A is a killer group that attacks population B, what will happen to the two populations in terms of dispersal and eventual population size?

9. Explanations of the present geographic ranges of organisms may depend not only on expansion from former ranges but also on reduction of former ranges. What are some factors that might decrease a group's range?

Communities and Ecosystems

3

What does the ant appear to be doing with these aphids? Who is benefiting from this relationship? Is any organism harmed? Give two other examples of similar relationships in which neither party is harmed and one or both parties benefit.

The ant appears to be tending or herding the aphids. Both the ant and the aphids benefit—the aphids get protection and the ants get honeydew. Neither appears to be harmed. Other examples of symbiotic relationships are a termite and its intestinal protozoa and a grouper and a cleaner wrasse.

◆ In the web of life, boundaries mean little. Individuals and populations interact with one another; they affect and are affected by the nonliving parts of their environment. Under favorable conditions, populations may grow larger and spill into new areas. Under unfavorable conditions, populations may decline and leave space for other organisms to move in. Boundaries constantly shift and change. This chapter discusses interacting populations within specific boundaries. Establishing boundaries, even if they are artificial or temporary, makes it possible to study one section of the web of life at a time. ◆

Life in a Community

3.1 Many Interactions Are Indirect

All populations, including human populations, interact with one another in a complex web of relationships. The set of interacting populations present during one time in one place is called a community. In your community, there may be dogs, cats, trees, weeds, and humans that interact (see Figure 3.1). When you mow your lawn or your dog bites the mailman, for example, the interaction is very direct. Most of the time, however, the interactions are indirect, and people tend to ignore them. Suppose, for example, that you had vegetable soup for lunch and threw the can away. A garbage truck might haul the can to the dump, where it would fill with rainwater. A female mosquito might lay her eggs in the water, and her offspring, once mature, might fly from the dump and bite a person or another animal. Perhaps neither would have been bitten if you had not eaten the soup, but since you did, you have had an indirect relationship with the mosquitoes and their victims.

Every community is affected by the abiotic environment—sunlight, soil, wind, rain, and temperature. Together, a community and its physical environment make up an **ecosystem** (EE koh sis tum). Both the terms *community* and *ecosystem* can describe the same group of organisms, but the term ecosystem refers to the community and its abiotic environment. Although the biosphere is too big to study as a whole, it is possible to study a part of it—an ecosystem. In Investigation 3.1, you can study two of the abiotic factors in an ecosystem that affect organisms.

3.2 Many Populations Interact in the Florida River Community

An aquatic community lives in the rivers along the west coast of Florida. Among the largest individuals in this community are the river turtles

MAJOR CONCEPTS

◆ Ecosystems include nonliving and living components that interact.
◆ All organisms are interdependent.
◆ Energy interactions among organisms may be beneficial or harmful.
◆ Organisms can contribute to a consumer only a small fraction of the energy they acquire.
◆ Human attempts to modify the environment sometimes have been detrimental.
◆ Humans are subject to the same laws of nature as all other organisms.
◆ Humans can have a greater effect on communities and ecosystems than most other organisms.

KNOWLEDGE CHECK

◆ How do populations affect one another?
◆ What is a system?
◆ How do ecosystems change?

If a student has any idea about the term *system*, it is likely to be one that has static connotations. Emphasize the idea of

Guidepost
How do different populations in the same area affect each other?

interacting entities in systems generally and in ecosystems particularly.

In the usage adopted in this course, a community involves interspecific relationships, and a society involves interspecific relationships. For example, in a beehive there is a society, not a community, of bees.

Figure 3.1 How do these organisms interact? **Answers may vary.**

Figure 3.2 What interactions do you see? **Turtles eating grass, a snail on the grass and another turtle eating a snail.**

shown in Figure 3.2. The adult turtles in the community eat not only many of the plants that grow in the rivers but also the long, narrow blades of the tape grass, which is their favorite food.

Consumers that eat only plants are called **herbivores** (HER bih vorz). Unlike the adult turtles, young river turtles are **carnivores** (KAR nih vorz), **consumers** that eat other animals. The young turtles feast on snails, aquatic insects, and worms. Carnivores that eat both plants and animals are **omnivores** (OM nih vorz). Because the adult turtles are herbivores and young turtles are carnivores in this community, they have different roles in the food web.

Many other carnivorous animals are in the community. These animals eat turtle eggs and young river turtles (see Figure 3.3). The eggs have the highest mortality rate. Female turtles lay their eggs in holes they dig on land, but they do not guard their eggs. The eggs may be dug up and eaten

The description of the Florida river serves as a background from which examples of a community may be drawn. Use this community as a basis for comparison to stimulate discussion of communities your students may observe in their own neighborhoods.

The skunk and heron eat the eggs on the bank, then the turtle and gar eat the young hatchlings.

Figure 3.3 Four animals that eat young river turtles. Describe the direct and indirect relationships shown here.

skunk

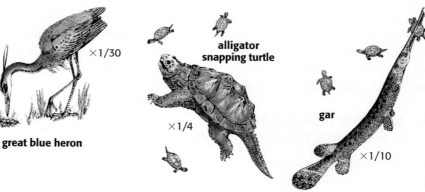

great blue heron   alligator snapping turtle   gar

by skunks, raccoons, or snakes. Unhatched turtles also are killed by molds that live in the soil and grow through the thin shells into the eggs. If the eggs survive, the young turtles that hatch still may be eaten by snakes or raccoons before they reach the river. Even if the young turtles reach the river, they are vulnerable to carnivores. Large fish, alligators, herons, and snapping turtles may eat them. For the young turtles that have survived the trip to the river, mats of floating plants, tangled tree roots, and sunken logs offer some protection from carnivores.

When the young turtles have grown larger, few organisms can kill them directly. Leeches may become attached to the turtles and suck their blood, but that usually does not kill the turtles. Turtles do die of disease, of accidents, and of old age. After they die, their bodies become food for the decomposers that return all the substances in the turtle's body to the abiotic world.

Each of the relationships described so far is a direct one between river turtles and other types of organisms. Like all organisms, however, river turtles have many indirect relationships with other organisms. For example, plants provide not only food and hiding places for the turtles but also oxygen for the fish that might eat young turtles. Snails eat tape grass and therefore compete with the adult river turtles for the same food supply. Another type of turtle, the musk turtle, eats nothing but snails. By reducing the number of snails that eat tape grass, musk turtles increase the supply of tape grass and thus have an indirect effect on the river turtle population. Because many of the carnivores that eat young river turtles also eat musk turtles, the more musk turtles there are in the river, the less likely it is that young river turtles will be eaten.

> Have students write the names of these organisms on paper and connect them with arrows leading from the eaten to the eater. Nonfood relationships may be shown by simple lines without arrowhead endings. Note that young and adult river turtles must be listed separately. The most successful diagram is one in which the organisms' names are arranged so that intersecting lines are least numerous.

3.3   A Niche Is the Role of an Organism

Adult river turtles and young river turtles have different functions in the community—different **niches** (NITCH ez). The niche of an organism is its role in the community: what it eats, what organisms eat it, and what indirect relationships it has with organisms. An organism's niche affects and is affected by its **habitat** (HAB ih tat)—the place where it lives. No two types of organisms occupy exactly the same niche within a community. For example, there are many different types of turtles living in a Florida river. They live in the same habitat, but their niches are different because each type lives in a manner distinct from the others and has different relationships with other members of the community. Understanding a community requires understanding all the relationships among all the organisms in all their niches. That is not an easy task, even for a small community.

> Have students attempt to describe niches or organisms that are familiar to them or have them match a set of niche descriptions with a set of organisms. Niches should include parasitism and other relationships.

> Let students know that it is possible for two different organisms to occupy the same habitat but have different roles in the community.

Consider the effect of adding just one more population, such as a human population, to the river community. Humans trap skunks, kill snakes, and catch fish that eat young river turtles. All of these activities indirectly reduce the mortality of young river turtles. Humans, however, also may dredge the rivers and thus kill the tape grass, which then will increase adult river turtle mortality.

3.4   Organisms May Benefit or Harm One Another

Many niches are easy to recognize. For example, algae, tape grass, and other green plants are producers. River turtles, snails, skunks, and fish are consumers. There are, however, other types of relationships that help to form the community's web of life. Each of these relationships involves at least

Figure 3.4   Competition does not always involve food. What might the bluebird and the starling be competing for? Who do you think wins?   **Nestholes; the starling.**

Ask students whether pond turtles compete with river turtles for tape grass. This question should emphasize the fact that competition is not on-off but a matter of degree. Competition with river turtles for tape grass approaches zero because pond turtles only appear in the rivers when tape grass is very abundant.

Students often are fascinated by the commensal and mutualistic relationships that have been discovered by naturalists. Refer them to T.C. Cheng, *Symbiosis: Organisms Living Together* (Indianapolis: Bobs–Merrill, 1970).

Some lichens consist of a cyanobacterium and a fungus.

Figure 3.5   How many different types of organisms do you see?

Most students will answer four or five—one for each color of lichen. Actually, there may be 10 or more—as many as 10 different algae and 10 different fungi.

two different organisms. In a **predator-prey** relationship, for example, one type of organism (the prey) is eaten by the other (the predator). In this relationship, one organism benefits from the relationship, but the other does not.

Another type of relationship, **competition**, benefits neither organism. Consider the Florida river community. Both snails and river turtles eat tape grass. The tape grass that is eaten by a turtle cannot be eaten by a snail. The grass eaten by a snail cannot be eaten by a turtle. Therefore, the presence of one of these herbivores would be potentially harmful to the other if there were insufficient tape grass to sustain them both. Organisms may compete for such things as food, space, sunlight, nutrients, or water. The competition is always for a scarce resource (see Figure 3.4). In the Florida river community, if both tape grass and other plants are scarce, competition for tape grass by herbivores is intense. If plants are abundant, competition for the tape grass is reduced.

When different organisms live in direct, physical contact with one another, the relationship is called **symbiosis** (sim by OH sis). For example, **lichens** (LY kenz) are leaflike, or crustlike structures that grow on rocks or the bark of trees, such as those shown in Figure 3.5. Lichens may be gray-green, black, yellow, or orange in color. Although a lichen appears **4** to be a single organism, it is actually two different organisms that live in close association with one another. In the most common lichens, one partner is an alga, a microscopic producer that makes food by photosynthesis. The other partner is a fungus, a consumer that gets its food from the alga but in return provides moisture for the alga. Their relationship benefits both organisms.

In many symbiotic relationships, both partners benefit, or at least no one is harmed. Figure 3.6 shows a remora and a shark. Using the suction disk on the top of its head, the remora attaches itself to the shark. The shark is not harmed, and the remora expends little energy because the shark pulls it around. In addition, the remora eats small bits of the shark's prey as they float by.

A special, and common, form of symbiosis is **parasitism** (PAIR uh sih tiz um). In parasitism, one organism (the parasite) lives on or in another organism, using it as a food source. The food source (the host) usually remains alive during the interaction. A leech, for example, clings to a turtle's skin and sucks its blood. Parasitic microorganisms in the turtle absorb **5** food directly from its blood. Like a leech, a tick on your skin sucks your blood. Plants also may have parasites. Some molds, microorganisms, and even other plants, such as mistletoe (see Figure 3.7), are parasites. Large microorganisms may have smaller parasitic microorganisms in them. Unlike a predator, which kills its prey outright, a parasite may kill its host indirectly by weakening it. Once it is weakened, the host becomes more susceptible to disease or an easier prey for predators.

Organisms may play different roles within their community at different times. For example, recall the difference in roles between young river turtles and adults. Or, an organism may play more than one role in a community. Snapping turtles prey on young river turtles, which makes them predators, but they are also **scavengers** (SKAV en jerz), organisms that eat the flesh of dead animals they did not kill. You also play more than one role in your community. When you eat a salad, you are a herbivore. When you eat both salads and hamburgers, you are an omnivore. Every community includes many different types of organisms and an individual can have many types of relationships.

Figure 3.6   What relationship do you see?

Figure 3.7   Mistletoe (yellow-green) growing on the branch of a ponderosa pine. What relationship is involved here? **Parasitism.**

Ask students what happens to tuberculosis microorganisms when a tubercular person dies. The question is intended to point out the idea that a parasite that kills its host usually is "unsuccessful."

CONCEPT REVIEW

1. Give an example of a direct and an indirect interaction between two organisms.
2. Distinguish between a community and an ecosystem.
3. How does an organism's niche differ from its habitat?
4. How does symbiosis differ from a predator-prey relationship?
5. In what ways are predation and parasitism alike? How do they differ?

Ecosystem Structure

3.5   Boundaries of Ecosystems Overlap and Change

What are the boundaries of an ecosystem? You might think that the boundary in the Florida river example is easy to define—the edge of the water (see Figure 3.8). Turtles, however, crawl onto the river banks to lay their eggs. Herons get most of their food from the river, but they nest in tall trees near the water. Frogs are caught and eaten by raccoons on land. Although some organisms such as fish spend all of their lives in the river, other organisms in the river do not.

Guidepost

How are biotic and abiotic environmental factors related to each other in an ecosystem?

Figure 3.8   A Florida river. What is the boundary of this ecosystem?

Figure 3.9 What abiotic factors might produce the effect shown here? **Principal factor is exposure. In the northern hemisphere, a south-facing slope (right), because of its angle of exposure to solar rays, receives more solar radiation per day and a greater total radiation per year than does a north-facing slope. Therefore, south-facing slopes are hotter and drier than north-facing slopes. In a climate where precipitation is barely enough for tree growth, these factors cause the striking difference between north and south-facing slopes.**

This reference is to the deep-sea vent communities in the Pacific Ocean and the Gulf of Mexico. In these communities, the producers are bacteria that use sulfur compounds welling up from vents in the earth as a source of energy.

When abiotic factors are considered, the question of boundaries becomes complex. Energy for the entire ecosystem comes from the sun, which is a great distance from the river's shoreline. Rain falls from the sky and may carry soil from the riverbanks to the river. The wind may blow seeds into the river from communities that are many kilometers away. Thus, the boundaries of this one ecosystem can expand to include many ecosystems.

The Florida river is not a special case. All ecosystems are connected to others around them. A river ecosystem is linked to a forest; a forest ecosystem is linked to the grassland, and so on. In fact, all ecosystems on the earth are connected to one another to form the biosphere. Because they are connected, a change in one ecosystem may affect many others.

Because of the web of relationships in a community, when one population is affected by abiotic factors, many others in the same community are affected at the same time. If the temperature drops to abnormal lows in Florida, river turtle eggs and young river turtles may be killed while they are still on land. With fewer river turtles, more tape grass will grow, but there will be less food for snapping turtles. If a heavy rainfall washes mud into the river, light will not penetrate as deeply in the water, and fewer plants will be able to photosynthesize. That, in turn, will mean less oxygen and less food for herbivores. Thus, each environmental change has many effects. One change causes many others, as shown in Figure 3.9. In Investigation 3.3, you will examine the effects of one environmental change on seed germination.

3.6   Most Communities Have More Producers Than Consumers

One of the most important abiotic factors that affects relationships in a community is energy. As discussed in Chapter 1, organisms in an ecosystem are tied together by the flow of energy and matter from one organism to another. The food chain that exists when a herbivore eats a plant and a carnivore eats a herbivore depends on the energy entering the community in the form of sunlight. Without the sun, there would be no green plants, no herbivores, and no carnivores. There are a few ecosystems that get their energy from another source. These unusual communities are described in Chapter 23.

The size of a community, therefore, is limited by the amount of energy entering it through its producers. The total amount of chemical energy stored by photosynthesis is the gross primary productivity of the community. Much of that energy is used by the producers to grow and to maintain themselves. The remaining energy, which is available to the consumers as food, is the net primary productivity of the community. One way to measure the net primary productivity of a community is to find the mass of all producers. Mass is a measure of the amount of matter in an object. Because much of the mass of living organisms is water, the producers first must be dried for a truer estimate of their mass. For example, the energy found in a fresh apricot and a dried apricot is the same, though the mass of the dried fruit is much less.

The greater the net primary productivity of a community, the greater the amount of food available to consumers. When net primary productivity is high, more consumers can live in the community. No matter how great the productivity, however, there is always some limit—the carrying capacity—to the number of organisms that can be found in any community.

Like a pyramid, which decreases in size from bottom to top, a community usually has many producers at its base, fewer herbivores in the middle, and even fewer carnivores at the top. Usually, the total mass of all the producers is much greater than that of the consumers.

Without more food producers than food consumers in a community, there will not be enough food to go around, and consumers will die. Because plants use for their own activities much of the energy they pro-
3 duce, a large number of plants is needed to support a few herbivores. Energy that has been used to keep a plant alive cannot be taken in by a herbivore. Because herbivores also use some of the energy to keep themselves alive, not all of the energy they take in is available to carnivores.

The loss of energy from a food chain may be represented by an **energy pyramid** (see Figure 3.10), in which producers form the foundation. In
4 general, the energy at each level of the pyramid is one-tenth that at the level below it. This concept often is referred to as the 10 percent rule because only 10 percent of the energy at any given level is available for use by organisms in the next higher level.

Because of this steplike decline of available energy in a food chain, the number of consumers supported at the top level is substantially smaller than the number of consumers and producers supported at the lower levels. Only about one one-thousandth of the chemical energy used by producers can flow all the way through a three-consumer food chain to a top predator. For example, in a small valley there may be millions of individual producers, such as grasses, but only a few hundred herbivores, such as rabbits. In this same valley, you may have to search carefully to find just one or two carnivores, such as foxes.

The concept of an energy pyramid has serious implications for the human population. Eating meat is an inefficient way to obtain the energy originally captured by plants. Recall the recipe for making a cow in Investigation 2.3. You can obtain far more kcals by eating a given amount of grain yourself than by eating the chicken or beef supported by the same amount of grain.

You may wish to introduce the terms *pyramid of numbers* and *pyramid of biomass*.

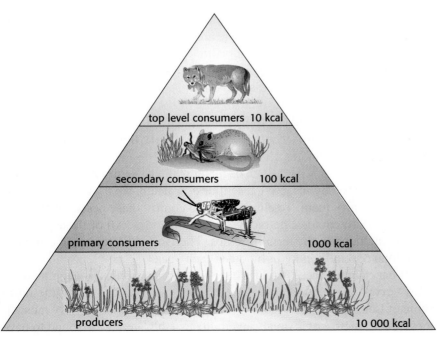

top level consumers 10 kcal

secondary consumers   100 kcal

primary consumers   1000 kcal

producers   10 000 kcal

Figure 3.10   An idealized energy pyramid. How much energy is lost at each level? **90% of the level below.**

Ecosystem Stability and Human Influences

3.7   Humans Lower the Stability of Ecosystems

> ### Guidepost
> How do the complex activities of humans affect ecosystems?

Ecosystems tend to be stable through time because of the great number of different types of organisms and interconnecting relationships found in the ecosystems. Usually, the greater the number of types of organisms and, thus, the greater the number of links in the food web, the more stable the community is thought to be. A large community is like a web with many threads. If only one of the threads is broken, the web as a whole may still function. If a disease kills many of the rabbits in a community, the foxes can eat mice and squirrels until rabbits reproduce or immigrate into the community. Even though many rabbits have died, the rest of the ecosystem remains intact. It may be changed, but it is not destroyed. This is an example of homeostasis, discussed in Chapter 2.

Human activities, however, can change ecosystems greatly. Humans tend to disrupt homeostasis. In doing this, we can have positive effects on other members of the community, or we can have dramatic negative effects. Farmers, for example, eliminate many members of a food web when they raise crops to meet human needs (see Figure 3.11). To feed the human population, it is necessary to plant large fields of producers. To use machinery efficiently, a single crop is planted in a large field. One cornfield may be 16 hectares, contain 76 000 plants, and yield 7258 m³ of corn. Although a wild field of corn would provide energy for a great number of birds, insects, and animals that eat birds and insects, these consumers compete with humans for the crop. Frequently, they are eliminated by humans, as are other plants that compete with the corn for light, water, and nutrients. With such a large population of plants, however, plant diseases can spread easily. The disease might be a parasite of the corn. If the farmer is to grow crops economically, he or she must control the food web and reduce the pests and diseases. Often, such control involves widespread spraying of chemicals on crops, which may have unplanned side effects.

Sometimes humans change ecosystems in the quest for health. Because parasitic diseases, such as malaria and African sleeping sickness, which are carried by insects, cause illness or death, we try to eliminate the insects that carry them. To eliminate or to control these pests, we use poisons called insecticides, herbicides, and fungicides. All of these products may be included in the term **biocides.** The United States alone produces more than 636 million kg of biocides each year.

What effect do these biocides have on the relationships in a community? DDT serves as a typical example. At one time, marshes along the north shore of Long Island, New York, were sprayed with DDT to control mosquitoes. Later, microscopic organisms in the marsh waters were found to have about 0.04 parts per million (ppm) of DDT in their cells (see Figure 3.12).

Figure 3.11 What members of the food web might be eliminated by crop dusting?

Primary consumers, decomposers.

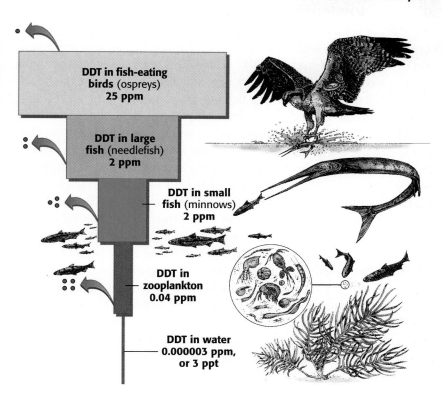

DDT in fish-eating
birds (ospreys)
25 ppm

DDT in large
fish (needlefish)
2 ppm

DDT in small
fish (minnows)
2 ppm

DDT in
zooplankton
0.04 ppm

DDT in water
0.000003 ppm,
or 3 ppt

Figure 3.12 What happens to the concentration of DDT as it travels up the food chain?
It increases.

While this is a low level of poison, no one had expected to find any poison in the organisms. The minnows, clams, and snails that ate these organisms had levels of DDT more than 10 times higher— between 0.5 and 0.9 ppm. The eels, flukes, and billfish that ate the snails and small fish had levels of DDT ranging from 1.3 to 2.0 ppm. The ospreys, herons, and gulls that ate the eels and minnows had levels of DDT between 10 and 25 ppm. Thus, the concentration of DDT in the tissues of the organisms in this food chain increased almost 10 million times from the amount in the seawater and nearly 625 times from the producer level at the bottom of the pyramid to the consumer level at the top.

Because each consumer ate a large number of producers, DDT became concentrated at upper levels in a food chain. Although the body of each producer can be broken down and used by the consumer, the DDT in the producer's body was not broken down. Each of the producers, then, contained a little DDT, and these small amounts built up to a large amount within the consumer. The greater the concentration of DDT in an organism, the greater the damage. Ospreys and pelicans with high levels of DDT accumulated in their bodies, for example, could not produce normal eggs. The egg shells were so thin that the eggs were crushed by the nesting parent. At one time, some of the osprey and pelican populations were nearly exterminated because of the effects of DDT. No one expected that the DDT used to control mosquitoes could affect so many other organisms far from the area where the poison was sprayed. Fortunately, humans are capable of learning and responding. Since the use of DDT has been banned in this country, the affected populations have recovered.

DDT also has been used to combat the mosquitoes that carry the microorganisms causing malaria. A person with malaria experiences successive bouts of chills, followed by high fever. In some cases, the person may die. At one time, malaria was a major health problem on the island of Borneo in Indonesia. To help the people of Borneo, workers from the World Health Organization sprayed remote villages and nearby areas with

Ask students why other countries still use DDT. Answers might include economic need and lack of a winter with killing frosts.

DDT. Most of the mosquitoes in the sprayed area died, but other organisms also were affected. Flies and cockroaches, the favorite food of the lizards found in Borneo, died from the DDT. The lizards gorged themselves on the DDT poisoned insects, and they, too, began to die. Local cats ate the infected lizards and died. After the cats died, the rat population grew unchecked.

The rats in the villages carried a disease that affected humans. Now, the people on the island no longer had to worry about malaria, but they began to die from the disease carried by the rats. To restore the balance in the community, cats were brought into the country and parachuted into remote villages to eat the rats (Figure 3.13). This is another example of the unplanned and long-lasting effects that we, as humans, may have on other organisms and on ourselves because of interactions in communities and of how we can respond with restorative measures.

Some pest populations include individuals that are resistant to a biocide. These individuals vary in their ability to tolerate, detoxify, or avoid the poison. When most of the pests are killed by a biocide, these few individuals may survive and reproduce. They pass their biocide resistance to their offspring. Then, when people try to kill these offspring with the same biocide, the poison has little effect. To be effective, the strength or the amount of the spray must be increased, which then affects other organisms. In some cases, no amount of spray will control the pests. In these cases, a different pesticide may or may not be effective. Today, one can find many resistant disease-causing and disease-carrying organisms in places where biocides have been used often. DDT-resistant mosquitoes, herbicide-resistant weeds, and antibiotic-resistant bacteria are not uncommon. Although the use of DDT has been banned in the United States, U.S. chemical companies make DDT and export it to other nations that still use this biocide.

3.8   Humans Threaten the Diversity of Organisms

At the same time that human activities are raising the number of biocide resistant organisms, many potentially important organisms are disappearing, resulting in a decrease in **biodiversity**—the number of different types of organisms that live on earth. Ecosystems with a large number of different types of organisms are usually quite stable. As our human population grows and we expand our activities, we occupy more land and therefore destroy the habitats of many organisms. Often this damage extends to beneficial and desirable organisms (see Figure 3.14). The smog created

Figure 3.13   Borneo rat patrol.

Figure 3.14   What might be killing the forest?   **Smog.**

by automobiles and industry is killing many types of trees over a wide area of southern California. The needles of ponderosa pines, for example, gradually turn brown, and the tops of palm trees have only small tufts of fronds. When this happens, photosynthesis is drastically reduced, and the plants die.

The Everglades National Park in southern Florida depends on a slowly moving sheet of water that flows from north to south. Drainage ditches built at the northern edge of the Everglades have decreased the flow of water over the entire area. As a result, many alligator holes (see Figure 3.15) have dried up. These holes helped to contain fires in the Everglades. Now, destructive fires are more frequent in this national park.

Plants and animals in heavily populated areas are not the only ones threatened. Tropical rain forests, such as the one shown in Figure 3.16, are the most diverse ecosystems on the earth. They provide habitats for many different **species**, groups of organisms that only can reproduce successfully with others of the same type. Unfortunately, deforestation for lumber, grazing and crop land, and other uses is destroying countless species every day. Two-thirds of the world's species are located in the tropics and subtropics. If the current rate of habitat destruction in these areas continues, nearly half of the earth's species of plants, animals, and microorganisms will become **extinct**—gone forever—or severely threatened, during the next 25 years. Hundreds of species of plants and animals, including the whooping crane and some rare pitcher plants (see Figure 3.17), are threatened with extinction today. Again, fortunately, the whooping crane population is recovering through great management effort. Although extinction is a natural process, the process has been greatly accelerated due to human activity.

Why does it matter if whooping cranes and pitcher plants become extinct? One argument against human-caused extinction comes from genetics. If all the crop plants in a field are genetically similar and one individual gets a disease, all the plants may die. As long as wild populations exist, a vast pool of genetic characteristics remains available. These characteristics may be used to prevent widespread death among the genetically similar individuals of our crops and domesticated animals (see Chapter 9, Biology Today). The extinction of each wild population erases genetic material that could mean healthy crops and animals. Once extinction occurs, the genetic material is gone forever.

The current number of described species is approximately 1.4 million; the absolute number of living species, described and undescribed, may be as high 30 million. Most experts agree the absolute number will exceed at least 5 million.

Figure 3.16 Tropical rain forests are the most diverse ecosystems on the earth.

Figure 3.17 What difference does it make if whooping cranes (a) and pitcher plants (b) become extinct?

a

×1/20

b

Ask students to think of ways humans are threatening, either directly or indirectly, other organisms in your area.

Figure 3.18 Madagascar periwinkle.

Figure 3.19 Rocky Mountain National Park in Colorado

A second argument against human-caused extinction is related to the instability of simplified ecosystems. Think of a field of corn as a simplified ecosystem. If something should happen to the corn, the whole ecosystem would collapse. To prevent such a collapse in a natural ecosystem, a community must have a wide diversity of species. The fewer the species in a community, the easier it is for homeostasis to be disrupted.

A third argument comes from research on plants. The island of Madagascar, off the east coast of Africa, is the only known habitat of the Madagascar periwinkle (see Figure 3.18). This plant produces two chemicals not produced by other plants—vincristine (vin KRIS teen) and vinblastine (vin BLAS teen). Both of these chemicals are used to help combat Hodgkin's disease, a leukemia like disease that affects thousands of people each year. As the human population on Madagascar grew, the habitat for the periwinkle shrank almost to the point of making the periwinkle extinct. Fortunately, botanists collected and grew some of these plants before they were gone forever. The medicines made from the Madagascar periwinkle are worth millions of dollars each year and help many people with Hodgkin's disease to live longer. These medicines never would have been known if the plant had become extinct.

3.9   Humans Also Can Help Conserve Species

We can take many measures to decrease the loss of species diversity. One such action is the establishment of wilderness areas, such as the one shown in Figure 3.19. Only foot and horse travel is allowed in these areas. In some areas, camping may be restricted to designated areas. Some wilderness areas require a permit for use. What good are these places if more people cannot use them? Wilderness areas act as one type of habitat preservation. In these areas, the effect of human activities on local organisms is reduced. In this way, an "island" of unspoiled habitat may help to ensure the safety of some species. Many countries now have park or conservation systems similar to those in the United States.

Biology Today

Naturalist/Interpreter

Dave Marshall is a naturalist/interpreter for the Bear Creek Nature Center in Colorado. He grew up in Colorado Springs, close to the Rocky Mountains. His father was an outdoors man who took Dave on many fishing trips to the nearby streams and lakes. This, along with his father's interest in birds, sparked Dave's curiosity about nature, and he decided to pursue a career related to natural history. Originally, he had planned to study forest management and ecology. After discovering a program in environmental interpretation in a college catalog, however, he decided that this new career "seemed to fit," and he went on to obtain degrees in environmental education and in resource and recreation management.

One of Dave's major duties is to conduct natural history programs for elementary school students at the Bear Creek Nature Center. He takes the children on short walks on the trails, shows them the exhibits at the center, introduces them to a display animal, such as a turtle or a garter snake, and gives a short talk on the specific program. Dave is in charge of the mammal natural history records, the insect collection, and the Bear Creek Nature Center's library. Dave also leads nature hikes to various types of natural areas.

Dave's favorite part of being a naturalist/interpreter is leading programs and meeting the public. He particularly enjoys working with children, and he feels a special satisfaction when he sees their curiosity and imagination stimulated by his program. Dave feels that being a good communicator is the most important quality for a naturalist/interpreter. By communicating what you know, you stimulate people's curiosity about the natural world.

Other duties of a naturalist/interpreter include designing and constructing exhibits, preparing brochures and educational materials, providing information for visitors, doing natural history research, identifying plants and animals, caring for small animals, and training volunteers. Because naturalist/ interpreters sometimes lead hikes into the back country, they must be in good physical health.

Many people interested in natural history become trained volunteers, or *docents.* Docents assist in all the duties of the naturalist/interpreter and often present programs and lead walks themselves. Volunteers also may become involved in programs with senior citizens and scouts. Docents with special skills or interests, such as photography and art, often are able to use these skills. No special education is required to become a volunteer— just a strong interest in natural history and a desire to share this knowledge with others.

Positions such as naturalist volunteers and naturalist/interpreters can be very rewarding because they provide a service to continue to enhance the public's environmental awareness.

The preservation of existing habitats may not be enough, however. Not all types of habitat are represented in preservation systems. The size of the areas set aside may not be adequate to stop the extinction of some species, particularly those requiring a large range or those sensitive to small climatic changes. Furthermore, many of the conservation areas do not have enough professional, well-trained, and well-equipped staff to manage them. What else can be done to slow the loss of species? Many countries have plans for replanting deforested areas, actively managing wildlife populations, and instituting nondestructive farming practices. Brazil, for example, recently developed a conservation system that encompasses almost 15 million hectares. These areas include good examples of many of Brazil's major ecosystems. China recently has replanted 30 million hectares with trees.

Many times, the loss of wild populations is due to habitat destruction, economic and cultural pressures, like those that have led to the over hunting of the African elephant. Sometimes overall population numbers can be increased by using such technologies as the establishment of breeding populations in captivity, the reintroduction of populations into the wild, and the relocation of populations from one "island" of habitat to another. For example, the Andean condor has been successfully reintroduced, many types of monkeys have been moved to new areas where their populations are less threatened, sandhill cranes have been used to incubate the eggs of whooping cranes, and the eggs of giant pandas have been artificially fertilized. In many cases, zoos and botanic gardens have established breeding populations of many different plants and animals. The need for these technologies may well increase in the years to come as the biosphere changes from a system of wild ecosystems to a system of managed nature preserves and zoos. These preserves may become the final refuge for many species.

CONCEPT REVIEW

1. How does the number of types of organisms in an ecosystem relate to the stability of that ecosystem?
2. What can happen to the concentration of some biocides in the members of a food chain?
3. Explain how humans can influence ecosystems by changing the environment.
4. Why is maintaining biodiversity important?

This investigation demonstrates the effects of abiotic factors in an ecosystem as students observe differences both among and within environments. Teams of 6 with each team divided into three pairs work best.

INVESTIGATION 3.1   Abiotic Environment: A Comparative Study

Two important abiotic factors are temperature and relative humidity. Relative humidity is the percentage of water vapor *actually* in the air at any given temperature compared with the amount of water vapor that the air *could* hold at that temperature. Organisms usually lose water faster in an atmosphere with low relative humidity than in an atmosphere with high relative humidity.

Read through the procedure and then develop hypotheses that describe expected differences in the three environments and at the four heights.

Materials (per team of 6)

3 watches
3 metersticks
3 nonmercury thermometers
   (–l0°C to + 100°C)
3 nonmercury thermometers of same
   range with cotton sleeves over
   the bulbs

3 screw-top jars containing 30–50 mL
distilled water
3 pieces of stiff cardboard
3 umbrellas or other shade devices
table of relative humidities
(see Appendix 3)

Time: One class period to make the measurements, calculate the relative humidities, and answer the questions; one–half period for class discussion.

Safety

Use only alcohol–filled thermometer to avoid any mercury hazard.

Procedure

1. In your data book, prepare a table like Table 3.l, shown below.

2. Divide your team into three pairs of students. Each pair will measure the temperature in one of three types of environment: a dense cover of vegetation (a woods, a thicket, or a mass of shrubbery); a single layer of herbaceous vegetation (a meadow or a lawn, preferably not cut close to the ground); no vegetation (bare ground or a tennis court). The environments should be located as close to each other as possible. In each team pair, one person will read both thermometers and the other will fan the thermometers with the stiff cardboard and record the data in the data table. Before starting, the people on the team who will record the data should synchronize their watches and agree on the time at which each measurement will be made. Schedule at least 8 minutes between readings. Write these times on the first line of the data table.

3. Take four sets of measurements in each environment: at ground level, at 30 cm above the ground, at 90 cm above, and at 150 cm above.

4. Take readings on both thermometers at the same time. Position the thermometers at least 5 minutes before the readings are taken. For example, if the first reading is to be taken at 1:30 PM, both thermometers should be in the first position at 1:25 PM. Use the umbrellas to shield the thermometers from the direct rays of the sun. To take a wet-bulb reading, soak the sleeve of the thermometer in water and fan it vigorously for at least 2 minutes; then read the thermometer. Plan for about 8 minutes between readings to allow for positioning and the 5-minute wait.

5. Using both your dry-bulb and wet-bulb thermometer readings and the relative humidity table, determine the amount of water vapor in the air compared with the amount the air could hold at that temperature. (The necessary calculations were made when the table was constructed.)

Discussion

1. At ground level, which environment is coolest and most humid?

2. At ground level, which environment is warmest and least humid?

3. How do these two environments compare in temperature and humidity at higher levels above the ground?

4. At which level above the ground are all three environments most alike in temperature and humidity?

Discussion

1-8. **These depend on the results obtained. On a sunny day, bare ground is usually the warmest and driest habitat at all levels; vegetation-covered habitats are cooler and moister below the top of the vegetation than above it. In general, these measurements illustrate the modifying effect of vegetation, the conversion of radiant energy to heat on contact with the soil, and, finally, the idea of microclimate variations.**

Table 3.1 Temperature Readings

Location	Height			
	0 cm	*30 cm*	*90 cm*	*150 cm*
Time				
Dry-bulb temperature				
Wet-bulb temperature				
Relative humidity				

5. How does the greatest temperature difference *within* the same environment compare with the greatest temperature difference *among* the environments?

6. What differences among the three environments might account for differences in temperatures and relative humidities?

7. How do these differences show the interaction of biotic and abiotic factors in an ecosystem?

8. You have been examining the differences among environments. Now turn to differences within an environment. How does the temperature in each environment vary with respect to elevations? Is the variation the same for each environment? If not, in which is the variation greatest? How does the humidity in each environment vary with respect to elevation? Is the variation the same for each environment?

9. In weather forecasts, temperatures predicted for the center of a city often differ from those predicted for the suburbs. Relate this fact to the situations you have been observing.

10. What differences in temperature and humidity would be experienced by a beetle crawling on the ground in a meadow and a gnat hovering at 1.5 m above the meadow?

11. Whereas we may say the beetle and the gnat are in the same environment, small differences within that environment are often important to the existence of some organisms. We therefore can distinguish small environments within larger ones on the basis of measurements such as those you have made in this investigation. Would it be useful to measure such differences if you were studying the ecological relationships among cows in a meadow? Explain.

9. Compared with concrete and asphalt, vegetation is a moderator of temperature. Other factors, such as different concentrations of artificially heated buildings, also are important.

10, 11. Microclimates are important to an organism whose size does not exceed the extent of one distinguishable microclimate (beetle and gnat) but of little importance to an organism whose size encompasses the whole range of the microclimate variation (cow).

INVESTIGATION 3.2   Competition

Organisms need certain resources to live. For survival and reproduction, they may need food, water, light, breeding space, mates, and other substances. If these resources are in short supply, the organisms may compete for them. Sometimes the competition between two groups of organisms is so intense that one group loses completely and no longer competes. The winner in such contest may be able to use the resource or may prevent the loser from using it. In these situations, the losing group of organisms usually moves to another place, where the competition is less intense or where they are better competitors, or the group dies out.

To study how competition works, you will compare it to something familiar. Here the groups are fast food restaurants (FFRs), and the resource in short supply is the consumer's money. Although the resource may be in short supply, obviously FFRs do coexist, often in close proximity to one another.

 Before beginning this investigation, develop several hypotheses to answer the question: "How might these FFRs avoid going out of business?" Another way of asking the same question is: "What might individual FFRs do to increase their own chances of survival?" Record your hypotheses in your data book.

This investigation demonstrates how competition can lead to specialization, how generalists and specialists can coexist, and how an ecological perspective helps us understand our world. The activity prepares students for the discussion in Section 3.4. Students may benefit by interacting in groups of 3.

Time: **One class period if students read the investigation the day before.**

Procedure

3. Be sure students understand the directions for each of the three comparisons.

Group A-B-D

1. roast beef, turkey, reg. cheeseb, hotdog
2. fish sand, breakfast, reg hamb
4. chick pcs, child's meal, Lg hamb, Lg cheeseb, XL hamb
5. baked potato, chili, XL cheeseb
6. chick sand, salad bar, bacon hamb

Materials (per person)
none

Procedure
PART A   **Fast Food Restaurant Competition**
1. Study the entrees from four nationally franchised FFRs shown in Table 3.2. The categories of entrees are general because each FFR has its own set of unique names for many entrees. Therefore, different types of hamburgers, for example, were grouped on the basis of size (all quarter-pound hamburgers were categorized as *large,* and anything smaller was labeled *regular*).

Table 3.2 Entrees of Four Nationally Franchised Fast Food Restaurants

Entree	A	B	C	D	Entree	A	B	C	D
Roast beef sandwich	X				Large cheeseburger	X	X		X
Turkey sandwich	X				Extra-large hamburger	X	X		X
Chicken pieces		X	X	X	Extra-large cheeseburger	X			X
Chicken sandwich	X		X	X	Salad bar	X		X	X
Fish sandwich	X	X	X		Baked potato				X
Breakfast	X	X	X		Chili				X
Children's meal	X	X	X		Hot dog	X			
Regular hamburger	X	X	X		Ham-and-cheese sandwich			X	
Regular cheeseburger	X		X		Bacon hamburger	X		X	X
Large hamburger	X	X	X						

Group A–C–D

1. roast beef, turkey, hotdog
2. fish sand, breakfast, reg hamb, reg cheesb
3. ham & cheese sand
4. chick pcs, child's meal, Lg hamb, Lg cheeseb, XL cheesb, XL hamb
5. baked potato, chili
7. bacon hamb, chick sand, salad bar

Group B–C–D

2. fish sand, breakfast, reghamb
3. ham & cheese sand, reg, cheesb
4. chick sand, XL cheesb, salad bar, bacon hamb
5. baked potato, chili
7. Lg hamb, Lg cheesseb, XL hamb, chicks pcs, child's meal

2. Figure 3.20 shows three overlapping circles that contain different numbered areas. Draw three of these diagrams for use in comparing similar FFRs or use the diagrams provided by your teacher. Label the circles in the first diagram *Group A*, *Group B*, and *Group D*. Label the circles in the second diagram *Group A*, *Group C*, and G*roup D.* Label the circles in the third diagram *Group B*, *Group C*, and *Group D*. Number the areas of each set of circles as shown in Figure 3.20. Each set of circles represents a neighborhood with three FFRs in it.

3. Write the name of the entree that fits each category in the proper numbered area on each diagram. For example, one entree unique to Group A is a roast beef sandwich. Write *roast beef sandwich* in Area 1 of the Group circle of the A-B-D set. Complete the list of entrees for all three sets of FFR species and use the information to answer Questions 4 and 5.

4. How is coexistence possible for the FFRs in each neighborhood?

5. Do your data support the following hypothesis, "Entree specialization explains coexistence"?

PART B   Competition Among Birds

6. In the short-grass prairie of the Great Plains, yellow-headed and red-winged blackbirds live together in the same marshes. These two groups have basically

4. Note how few items are held in common. Each FFR (except B) has some unique items, demonstrating specialization as a result of competition. Students should conclude that there is a high degree of competition among the three and that they probably would not be in close proximity to one another unless there was a large resource (human population) available.
5. Yes. We infer that this specialization is the result of competition.
7. In direct competition, red-winged black birds are inferior to yellow-headed blackbirds.
8. Red-winged blackbirds are able to nest in areas too dry for yellow-headed blackbirds.
9. Coexistence is possible because the specialist B is the better competitor.

Figure 3.20   Diagram for comparison of FFR entrees

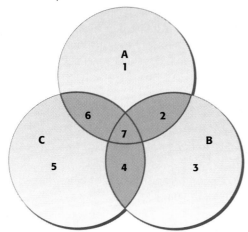

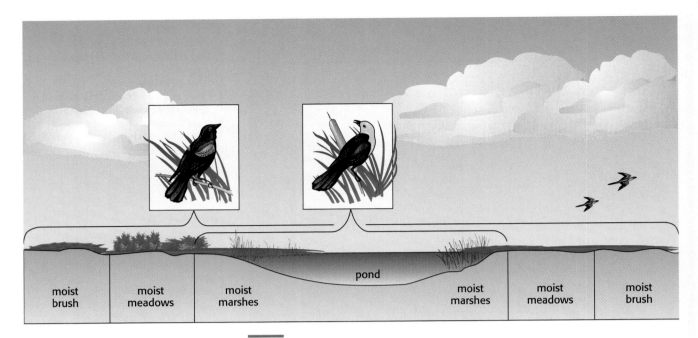

Figure 3.21   Overlapping habitats of red-winged and yellow-headed blackbirds.

C is more of a generalist, offering six items that B does not. Therefore, C is not driven to extinction even though B is the better competitor.

Discussion

1. **It is unlikely that E would be successful. If B and E are equal competitors, B should persist simply because it was there first. Chances are slim that such a group of fish could establish itself in the lake. An exception might be if the new fish were more aggressive competitors than any of the fish groups already established in the lake.**
2. **Variables include nesting places and food, the location of nests, the means of finding food, and the types of food eaten.**
3. **A highly successful competitor often has a habitat that overlaps that of a less successful competitor.**
4. **The most successful competitor persists because it is superior. The less successful group also persists if it can use resources in some way different from that of the most successful competitor.**
5. **Some examples of specialization of services in the commercial endeavors: clothing stores (low vs. high quality, large vs. petite sizes, sportswear vs. formal wear, men's vs. women's wear); auto repair (self- vs. fullservice stations; muffler vs. radiator repair, tune-up only vs. general repair)**

the same niche. The only difference is that red-winged blackbirds are generalists. They can nest in a variety of habitats ranging from moist, brushy areas to moist meadows. They do, however, prefer to nest in extremely moist marshes at the edge of the open water of ponds and lakes. Yellow-headed blackbirds are specialists. They require the extremely wet part of the marsh, next to open water, to nest successfully. Because of their larger size and more aggressive nature, yellow-headed blackbirds will displace red-winged blackbirds in open wet marsh. Thus, the yellow-headed blackbird has a niche totally included within the niche of the red-winged blackbird, as shown in Figure 3.21. Use this information to answer Questions 7 through 9.

7. How do the competitive abilities of red-winged blackbirds compare to those of yellow-headed blackbirds?
8. Why are all red-winged blackbirds not displaced by yellow-headed blackbirds?
9. Recall your information from Part A. Compare FFRs B and C. Using what you have learned about blackbirds, how is coexistence possible between these two FFRs?

Discussion

1. A new FFR chain, opens a restaurant one block from the established restaurant B. If E offers exactly the same menu and service as B, what do you think will happen, and why? What does this suggest to you about the chances of a particular group of fish that swims up a river into a lake and attempts to establish itself?
2. Five similar groups of small, insectivorous birds live in the same spruce forest and apparently coexist. What variables might you investigate to explain this?
3. Competitive exclusion does not always occur among competing groups. The groups coexist?
4. How can two groups of organisms coexist when one group has a niche that overlaps the niche of the other?
5. How can the concepts of competition, specialization, and overlapping niches be applied to aspects of urban life?

INVESTIGATION 3.3   Acid Rain and Seed Germination

Acid rain is something of a misnomer because natural rainwater is acidic and has a pH of about 5.6. (pH is a measure of acidity; the lower the number, the greater the acidity. A pH of 7 is neutral.) Acid rain, therefore, refers to rain with a pH lower than 5.6. Acid rain is produced when sulfur and nitrogen compounds are released into the atmosphere where they combine with water to form sulfuric and nitric acids. Sulfur compounds may come from such natural sources as decomposing organic matter, volcanos, and geysers. The environmental problem known as acid rain, however, does not arise from natural sources. It is caused primarily by fossil fuel combustion. When coal, oil, and gas are burned, large amounts of sulfur and nitrogen are released as gases and combine with water to make the rain more acid than usual. Acid rain has many effects on an ecosystem. In this investigation, you will examine just one of them.

 Read through the procedure and develop a hypothesis about the ideal pH for bean seed growth.

To demonstrate the effects of abiotic factors on a community, students will investigate the effects of acid rain on seed germination by conducting an experiment with bean seeds under varying pH conditions. It is best for students to work individually, but teams of 2 can be used.

Time: Two weeks: one-half class period to organize groups and set up the activity, a few minutes at the beginning of each period to water and measure seed growth and to record data from individual and class groups on days 1–11, and one period to summarize and discuss results.

Materials (per team of 2)

2 pairs of safety goggles
2 lab aprons
2 pairs of plastic gloves
petri dish
glass-marking pencil
absorbent paper towels
scissors

transparent metric ruler
graph paper
colored chalk or magic markers
water solutions ranging in pH from
  2 to 7
rainwater
4 bean seeds

Put on your safety goggles, lab apron, and gloves.

SAFETY

Safety
pH values below 4 are irritants; label the bottles of water solutions accordingly.

Procedure
Day 0

1. In your data book, prepare a table in which to record seed lengths and averages for the duration of the investigation.
2. Cut four discs the size of the petri dish from the absorbent paper towel.
3. Dampen the discs with the water assigned to you. Record the ph of the water.

 Solutions of pH 4 or less are irritants. Avoid skin/eye contact; do not ingest. Should a spill or splash occur, call your teacher immediately; flush the area with water for 15 minutes; rinse mouth with water.

WARNING

---

4. Place two of the paper discs on the bottom of the petri dish.
5. Use the transparent ruler to measure the lengths of your four seeds in millimeters. Determine the average length and record under Day O in your data table.
6. Sketch the shapes of the seeds and note their color.
7. Arrange the seeds in the petri dish and cover with the two remaining discs. Make sure the discs are still moist. If not, add more of your assigned pH solution.
8. Place the lid on the petri dish and label with your team name.
9. Wash your hands thoroughly before leaving the laboratory.

Days 1–10

10. Remove the lid from the petri dish and lift off the discs covering the four seeds.
11. Measure the length of the seeds in millimeters, average the lengths, and record in your data table under the appropriate day.
12. Sketch the shapes of the seeds and note their color.
13. Cover the seeds with the paper discs. If necessary, moisten the paper with the assigned pH solution and replace the lid.
14. On a piece of graph paper, set up a graph labeled *Age in Days* on the horizontal axis and *Length of Seeds (mm)* on the vertical axis.
15. Plot the average length of your seeds for the two measurements (Day 0 and Day 1) you have made.
16. Repeat Steps 10–15 each day for the duration of the investigation. If the seeds begin to germinate during this time, include the length of any growth you observe in your measurements.
17. Combine your data with those of other students using the same pH solution and determine the average. Record the average on the class graph, using the color assigned to your particular pH.
18. Each day, wash your hands thoroughly before leaving the laboratory.

Discussion

1. Study the data on the completed class graph. What appears to be the optimal pH solution for successful bean seed germination?
2. What appears to be the worst pH solution for successful bean seed germination?
3. What pH do you think the rainwater has, based on the data gathered? Determine its pH.
4. What impact on local crops might an increased rain acidity have?
5. Do you think there is reason for concern about the rain in your area?

Discussion

1–3. Depends on class data.
4. **The seeds watered with the lower pH solutions should have a slower growth rate than the others. From this information, it can be assumed that rainwater with high acidity will have similar effects on crop growth: slow growth rate and a subsequent drop in food production (assuming that soil conditions do not exert any buffering effect).**
5. **That depends on the rainwater data and how it compares to the data from the various pH solutions. The pH of rainwater is variable depending upon amounts of sulfur and nitrogen compounds entering the atmosphere. A normal acid rain today might be a high acid rain next week.**

SUMMARY

Some of the relationships in a community are direct, but most are indirect. A single organism may have many types of relationships with other organisms, and these relationships define the organism's niche. Relationships include predator/prey, symbiosis and competition. The organism's habitat, on the other hand, is the sum of the biotic and abiotic environment in which it lives. The niches of organisms are distinct, but the boundaries between ecosystems are not clear because energy and matter pass from one ecosystem to another. In most ecosystems, producers greatly outnumber consumers, and there is stability through time. Humans can affect the populations of producers or consumers by what they do. Sometimes, as when DDT was sprayed, the results are many, surprising, and undesirable. In some cases, human activity can lead to the extinction of another species, sometimes many species. This loss of biodiversity may have negative consequences for humans in the future. Humans can use strategies, such as preserving natural areas and reintroducing populations, to help slow down the rate of species diversity loss.

APPLICATIONS

1. How does the concept of ecosystems help us to understand the biosphere?
2. What would happen if all the turtles were killed within the Florida river community?
3. Predict what would happen to a community if the number of carnivores present suddenly were doubled.
4. A predator affects the population of its prey, but a parasite may not have the same effect on its host population. Why?
5. Explain the relationship between habitats and niches.
6. Why is it difficult to describe an organism's niche accurately?
7. Most organisms can survive only within a narrow temperature range. Does this mean that temperature is an environmental factor for which organisms compete? Explain your answer.
8. Suppose there are two different types of plants growing together in one pot. When the two plants are separated and planted in different pots, one plant dies and the other plant starts to grow much larger. What type of relationship may these plants have had?
9. Suppose you were trapped on a desert island with 12 chickens and a bushel of corn as your only food. You will not be rescued for a month, which is not a long enough time to grow more corn. Which would you do to survive: (a) eat the corn first, then the chickens; (b) eat the chickens first, then the corn; (c) feed the corn to the chickens, then eat the eggs laid by the chickens? Give reasons for your selection.
10. Although pesticides may be beneficial to farmers, they may have serious side effects on ecosystems and nonpest organisms, including humans. Many people advocate using natural predators to control pest populations. What are some advantages to this solution? What are some drawbacks?

PROBLEMS

1. How might an ecologist find out whether the effect of one type of organism on another is beneficial harmful, or neutral?
2. Study a community in your city or town. Gather information to answer these questions: (a) How does the community get its energy? (b) Which organisms would be present and which organisms would be absent if humans were not part of the community?
3. Investigate these two parasite-host relationships: (a) chestnut tree and chestnut blight; (b) whooping cough and humans. How are the hosts affected, and how have humans tried to control the parasites?
4. Describe your niche. Build lists of organisms that you affect and that affect you. Describe what each interaction is, and whether it is direct or indirect.
5. Do you think that biocides should be sprayed on your food or not? Write an essay describing the hazards of biocide use, and the potential benefits and problems associated with growing food entirely without biocides.
6. What does "endangered species" mean? You can obtain a list of the endangered plants and animals in your state from the state's Wildlife or Game and Fish Department or the Extension Service. Collect or create drawings, photos, or other illustrations of endangered species to display in your classroom. What can you do to keep these species from vanishing?
7. Two major wildlife problems are human activities, which can lead to or speed up the extinction of species, and the introduction of alien species into an ecosystem. How can the introduction of an alien species, such as the water hyacinth or the Norway rat, upset the balance in an ecosystem? Research an alien species and prepare a report for the class.

Matter and Energy in the Web of Life

What happened to this area about three weeks before this photo was taken? Nearly all the energy stored in the organisms previously present was consumed. Give three specific examples of what remains in this environment that these seedlings are using.

They probably are using moisture (water), chemicals such as iron and nitrates, and the ground itself as an anchor.

◆ All living things are tied together by their need for matter and energy. Using energy from the sun and absorbing matter from the surrounding soil and air, producers, such as green plants, make their own food. Consumers, on the other hand, must obtain their matter and energy from other organisms. Humans, for instance, eat plants and animals to get their matter and energy. This matter and energy (food) is made up of biological molecules—the molecules that are found in all living things. We eat the molecules of plants and animals and then rearrange them to make our own molecules. This chapter examines these important molecules and investigates some of the characteristics of matter and energy. ◆

MAJOR CONCEPTS

- ◆ Biological activities are based on the chemical reactions of atoms and molecules, the building blocks of living things.
- ◆ Energy is used to move matter and to maintain order. It can take many forms.
- ◆ All living things contain the element carbon.
- ◆ Carbon cycles through living and non-living systems.
- ◆ Enzymes promote chemical reactions.
- ◆ Nucleic acids control cell activities and inheritance in all living things.

KNOWLEDGE CHECK

- ◆ What is energy, and where does it come from?
- ◆ What is a chemical reaction?
- ◆ What are enzymes?

Matter and Energy

4.1 Matter Is Made of Atoms

Biological molecules are **organic compounds**—compounds built of carbon combined with other elements. Like all other compounds, organic compounds can be broken down to the elements from which they were formed. For example, molecules of sugar, an organic compound, can be broken down to the elements carbon, hydrogen, and oxygen. Elements cannot be broken down to other types of substances. Carbon, iron, nitrogen, gold, silver, calcium, and chlorine are some elements with which you probably are familiar. Four elements—carbon, oxygen, hydrogen, and nitrogen—are common to all living systems.

Elements are composed of atoms, which are made of even smaller particles. Each atom has a core, or **nucleus** (NOO klee us), that contains positively charged particles called **protons** (PROH tahnz) and uncharged particles called **neutrons** (NOO trahnz). Rapidly orbiting the nucleus are one or more **electrons** (ee LEK trahnz), which are negatively charged particles. Because the number of electrons equals the number of protons, an atom is electrically neutral.

The numbers of particles in an atom determine what element is formed. For example, a hydrogen atom is made of one proton and one electron. (Hydrogen is the only element that does not have neutrons in its atoms.) An atom of carbon contains six protons, six neutrons, and six electrons. Oxygen atoms are composed of eight protons, eight neutrons, and eight electrons. Figure 4.1 shows models of these atoms.

> ### Guidepost
> How are matter and energy related in the biosphere?

Refer students to the periodic table of the elements in Appendix 3. Help them understand the information the periodic table provides.

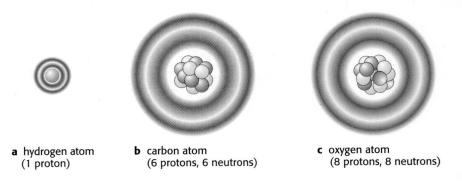

Figure 4.1 Simplified models of (a) hydrogen, (b) carbon, and (c) oxygen. Red circles represent protons, yellow circles represent neutrons. The models show electrons as a cloud of negative charge around the nucleus. The number of electrons in the outer level determines chemical activity.

a hydrogen atom (1 proton)   **b** carbon atom (6 protons, 6 neutrons)   **c** oxygen atom (8 protons, 8 neutrons)

Reactions between atoms depend on the number of electrons the atoms have. Sometimes a reaction involves an electron moving from one atom to another. For example, when atoms of sodium and atoms of chlorine react to form table salt, each sodium atom gives up an electron to a chlorine atom, as shown in Figure 4.2. As a result, the number of protons and the number of electrons in the sodium atom are no longer equal. **2** Because the sodium atom has 11 protons but only 10 electrons, it has a positive charge. The chlorine atom, which captured the electron from the sodium atom, now has one more electron that it had originally. Consequently, it has a negative charge. A charged particle that has either a negative or positive charge is called an **ion** (EYE on). The chlorine atom has become an ion with a negative charge—a chloride ion. The sodium atom has become an ion with a positive charge. The positively charged sodium ions and the negatively charged chloride ions are attracted to each other and come together, forming sodium chloride, or table salt.

Often when atoms react, they do not gain or lose electrons. Instead they share electrons. For example, in a molecule of water, one oxygen atom shares electrons with two hydrogen atoms (see Figure 4.3). Mole- **2** cules of carbon dioxide, hydrogen gas, and oxygen gas also are formed by shared electrons. The attractions that hold atoms or ions together are called **chemical bonds.**

Explain that a chemical bond is very specific and involves electrons. If conditions are right and interacting atoms/molecules are present, then a chemical bond is formed.

4.2   Chemical Reactions Are Essential to Life

There are two basic types of chemical reactions in living cells. When sodium and chlorine combine to form table salt, or oxygen and hydrogen combine to form water, compounds are built up. Reactions that lead to the buildup of compounds are called **synthesis** (SIN thih sis) reactions. Compounds also may be broken down. When this happens, the reaction is

Figure 4.2 Sodium and chlorine can react to form the salt sodium chloride (NaCl). By losing one electron, sodium becomes a positive ion, and by gaining one electron, chlorine becomes a negative ion, chloride.

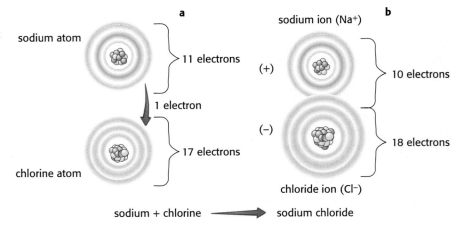

a

sodium atom

11 electrons

1 electron

chlorine atom

17 electrons

b

sodium ion (Na⁺)

(+)

10 electrons

(−)

18 electrons

chloride ion (Cl⁻)

sodium + chlorine ⟶ sodium chloride

known as a **decomposition** (de kom poh ZISH un) reaction. The digestion of foods involves decomposition reactions.

For chemical reactions to take place, the reacting substances must come in contact with each other. This happens most easily when the substances are in solution, that is, dissolved in water. When table salt dissolves in water, the sodium and chloride ions separate from each other, but they remain as ions in solution. Nonionic compounds, such as water, also can be converted into ions through a process called **ionization** (see Figure 4.4). Stated in simple terms, during ionization, water molecules separate into hydrogen ions (H⁺) and hydroxide ions (OH⁻). A hydrogen ion is a single proton—a hydrogen atom that has lost its only electron. The missing electron is held by the hydroxide ion, which consists of an oxygen atom, a hydrogen atom, and the electron the hydrogen atom has lost.

Although water is a common compound, it has unique properties, such as ionization, that make it essential for life. Only about one in 10 million molecules of water forms ions, but all life processes depend on this small percentage. Hydrogen and hydroxide ions are involved in most of the reactions that occur in organisms. If more hydrogen ions than hydroxide ions remain in solution after a reaction, the solution is said to be acidic. If more hydroxide ions than hydrogen ions remain, the solution is said to be basic, or alkaline. The relative levels of hydrogen and hydroxide ions are very important to organisms because of their effects on chemical reactions.

The hydrogen ion level of a solution is described by a range of numbers known as the **pH scale** (see Figure 4.5). The scale runs from 0 to 14. A solution that has the same number of hydrogen and hydroxide ions is
3 magnesia neutral and has a pH of 7. Pure water has a pH of 7. As the hydrogen ion level rises, the solution becomes more acidic, and the pH drops. Thus, a bleach solution with a pH of 2 is highly acidic. Solutions with a pH above 7 are basic. They have relatively low levels of hydrogen ions and high levels of hydroxide ions. Figure 4.6 shows the pH of several common substances.

Organisms have an internal pH that must remain fairly stable for chemical reactions to occur. Environmental factors may affect that stability in a variety of ways. In Investigation 4.1 you can discover how internal pH is regulated.

Even when they are dissolved in water, most atoms and molecules react extremely slowly, if at all. Not enough of them come into contact with each other. Certain substances, however, promote chemical reactions by increasing the chances of contact. These substances are called **catalysts** (KAT uh lists).
5 Catalysts are present only in small amounts, and although they participate in

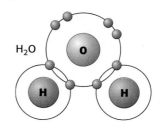

Figure 4.3   In a molecule of water, the oxygen atom forms an electron sharing bond with each hydrogen atom.

Reacting substances may be dissolved in organic solvents, but water is the most common solvent.

Figure 4.4   Water molecules ionize into hydrogen and hydroxide ions.

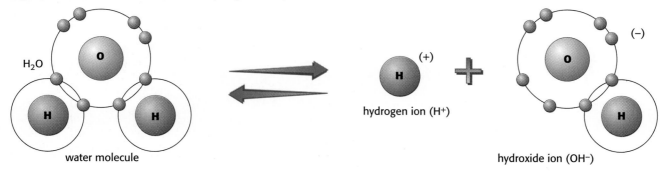

water molecule   hydrogen ion (H⁺)   hydroxide ion (OH⁻)

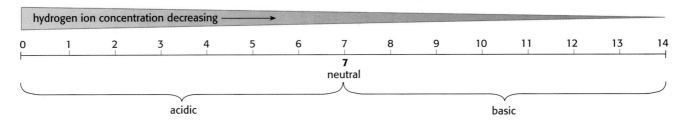

hydrogen ion concentration decreasing ⟶

| 0 | 1 | 2 | 3 | 4 | 5 | 6 | 7 | 8 | 9 | 10 | 11 | 12 | 13 | 14 |

7
neutral

acidic   basic

Figure 4.5 A pH scale. When is a substance acid? When is it basic?

Energy can be classified as either potential or kinetic. Potential energy is associated with position or arrangement of parts; kinetic energy with motion. Before work can be done, energy must become kinetic, as in light, heat, electrical, mechanical, sound, and nuclear energy. Water behind a dam and chemical-bond energy are forms of potential energy.

This discussion deals with entropy and the second law of thermodynamics. It is kept simple purposely.

Figure 4.6 The pH of several common substances.

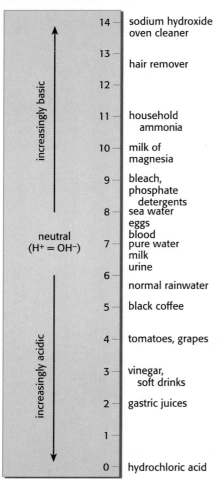

increasingly basic	14	sodium hydroxide oven cleaner
	13	hair remover
	12	
	11	household ammonia
	10	milk of magnesia
	9	bleach, phosphate detergents
	8	sea water
		eggs
neutral ($H^+ = OH^-$)	7	blood pure water milk
		urine
	6	normal rainwater
	5	black coffee
increasingly acidic	4	tomatoes, grapes
	3	vinegar, soft drinks
	2	gastric juices
	1	
	0	hydrochloric acid

the reactions, they themselves are not changed or used up in the reactions. Catalysts make it possible for reactions to occur at rates high enough to sustain life. The specialized and highly specific catalysts present in organisms are called **enzymes** (EN zymz). These are discussed in Section 4.11.

4.3   Energy Makes Work and Order Possible

Chemical reactions usually involve energy. In general, synthesis reactions require an input of energy. Decomposition reactions usually release energy. What is energy, and how is it used in living systems?

Understanding matter is fairly easy. You can see it, you can touch it, and you can weigh it. Understanding energy is more difficult. In a general way, energy can be defined as the ability to do work or to cause change. It is work to move an arm, play tennis, jump the high jump, heat a house, or build a skyscraper (see Figure 4.7). Growing a leaf or a wing also can be considered work because energy is used in these processes. In a cell, energy is used to move substances and to build complex molecules. This, too, is work.

Energy also is required to make and maintain order. Living things are extremely complex. Their atoms and molecules are arranged in highly organized systems. High levels of organization, however, can be unstable. If left to themselves, all systems tend to become simple and random, or disorganized. Only by a continual input of energy can organization be maintained.

Think about your room. When it is clean and neat, it is an organized system. How long does it stay that way? The tendency is for your room to become disorganized with your books and clothes spread out in a random fashion (see Figure 4.8). To get your room organized again requires energy (work).

Figure 4.7   Is this person doing work?   **Yes.**

Figure 4.8 The tendency toward increasing disorder of a system and its surroundings.

Students may suggest that a room can be ordered, locked up, and then remain organized indefinitely. Given enough time, however, even under these conditions, disorganization will occur.

Like your clean room, a living organism such as a frog is an organized system. The frog maintains its internal organization by eating flies and other insects that contain energy. This energy keeps the frog alive and allows it to grow and reproduce. If the frog does not get enough food, it dies. As soil decomposers break down the frog's body, it becomes disor-
6 ganized. The decomposers live by using the matter and energy remaining in the frog's body cells.

Many life processes tend to bring about an organized state with a minimum of randomness. That requires energy. Where does the energy come from?

Ask students to identify the source of energy and raw materials required to keep them in a steady state. Anything students eat may be considered a source of energy. However, debates may arise about "junk" food. These are discussed in Chapter 15.

CONCEPT REVIEW

1. How are the protons, neutrons, and electrons within an atom related to one another?
2. How are chemical bonds formed?
3. In terms of pH values, what is the difference between neutral, acidic, and alkaline solutions?
4. How is matter different from energy?
5. What is a catalyst?
6. Why do all organisms need energy?

Energy for Life

4.4   **Photosynthesis Is the Source of Your Energy**

Biological activity, like all other types of activity, requires energy. Consumer organisms get their energy from the food they eat, but where do the producers get their energy? Usually, their energy comes from the sun. Because no organism can use light energy directly from the sun as a source of food energy, the energy must be converted to chemical energy.

In the process of photosynthesis, green plants (and some other organisms) convert light energy from the sun into chemical energy that is stored in complex food molecules. That chemical energy then can be used either by
1 the plants themselves or by organisms that eat the plants. Because animals cannot make their own food, most animals depend on plants for their source of energy as well as matter. Directly or indirectly, therefore, photosynthesis is the source of energy for biological activity in almost all organisms on earth.

Guidepost
What is the source of energy for living organisms and how is it used?

An exception is the deep sea vents in which sulfur compounds are the source of energy. Even here, the chemosynthetic bacteria at the base of the food chain are indirectly dependent on plants and photosynthesis as their source of oxygen.

Figure 4.9 The energy in sunlight is converted to chemical energy during photosynthesis.

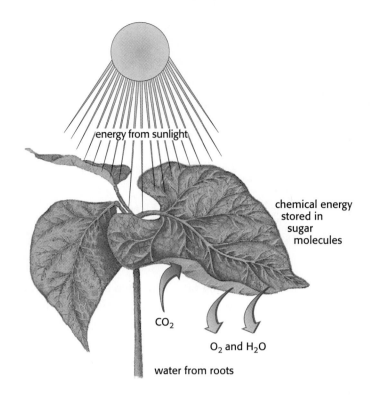

energy from sunlight

chemical energy
stored in
sugar
molecules

CO_2

O_2 and H_2O

water from roots

Plants, animals, and most protists use essentially the same chemicals and processes to obtain energy from their stored fuel. Stress this biochemical unity of living things.

You may find students saying that animals use respiration but that plants use photosynthesis to obtain energy. Through photosynthesis, plants build up food reserves. These food reserves then are broken down by respiration to release energy for the plant.

Any chlorophyll-containing organism can use light energy to make its own food. Much of the photosynthesis on the earth is carried out by algae and cyanobacteria. The discussion here is limited to green plants for simplicity.

The "lost" heat of metabolism provides cells with an internal environment that is usually warmer than the external environment. In general, this allows the cells to carry on metabolic reactions at a greater rate. To some extent this is true of any organism. Animals that regulate their body temperature have the means to conserve metabolic heat. At the same time, these animals must have the means to dissipate metabolic heat when the environmental temperature is high.

The first step in photosynthesis is the absorption of light energy by a green plant. The energy is absorbed primarily by **chlorophyll** (KLOR uh fil), a green pigment that gives plants their color. A plant also absorbs carbon dioxide molecules from the air and water molecules from the soil. Light energy, carbon dioxide, and water are the raw materials used to make sugars. The light energy absorbed by the plant is used to break down water molecules into hydrogen and oxygen. The hydrogen is combined with carbon dioxide to form sugar molecules. Thus, some of the light energy absorbed by the plant is stored in the sugar molecules as chemical energy. The oxygen is released into the air as oxygen gas. Figure 4.9 highlights these events.

When sugars are formed, several small molecules are linked together by chemical bonds. The energy used to form the sugars is stored in the structure of the molecules. When sugar molecules are broken down in a cell, the energy stored in them is released. That energy is used by the cell to do cellular work.

4.5 Energy Is Released as Food Is Broken Down

How is the energy stored in the structure of the molecules released? The major energy-releasing process is **cellular respiration** (SEL yoo ler res pih RAY shun), a series of chemical reactions that occur in all living cells. During these reactions, the sugars made in photosynthesis are broken down, and energy is released.

The way in which different organisms release energy from food is remarkably similar. Foods are compounds of carbon, hydrogen, oxygen, and often other elements—building blocks for the cell. Chemically, foods are similar to substances such as wood, coal, and oil, which can be used as fuels because they contain chemical energy. During the chemical reactions that occur in burning, fuels are reduced to simpler compounds. The chemical energy from the fuel is released in the form of heat and light, as shown in Figure 4.10a.

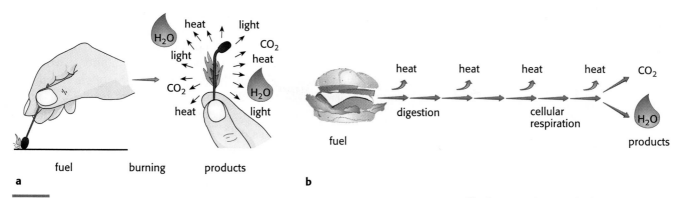

Figure 4.10   A comparison of energy released during (a) burning and during (b) cellular respiration. How are they the same? How are they different?

The chemical energy in food is also released by chemical reactions. The chemical reactions in a cell, however, are quite different from those in a fire. When fuels burn, a large amount of energy is released in a short time. The sudden release of energy produces high temperatures—high enough to provide heat for cooking. Cells, however, would be destroyed by such high temperatures. Energy-releasing chemical reactions occur in cells at low temperatures.

In cells, the same total amount of energy is released as in burning, but it is released gradually, in many small steps that are controlled by enzymes (see Figure 4.10b). The cell uses the energy in food that is released during respiration to move substances and to carry out other cell work. Carbon dioxide and water molecules are the by-products that are formed as the food molecules are broken down. Notice that carbon dioxide and water are the same molecules the plant uses to make sugars in photosynthesis. Figure 4.11 summarizes the relationships between photosynthesis and cellular respiration.

Alike because they use fuel and produce heat; different because burning releases heat and light all at once, whereas cellular respiration releases heat in small steps.

In living systems, energy (except for heat) is never released; rather it is transferred to another molecule. Some energy is lost in every reaction as heat.

Distinguish between breathing and respiration. Breathing is the process that moves respiratory gases to and from respiratory organs. Respiration is the exchange of O_2 and CO_2 by extension, the chemical process that involves this exchange in cells is known as cellular respiration.

Fats are another energy-rich food used in respiration. Even amino acids from proteins are broken down directly occasionally when there is an excess or when they are needed as food for a body suffering from malnutrition.

Figure 4.11   Photosynthesis and cellular respiration are two important links in an ecosystem.

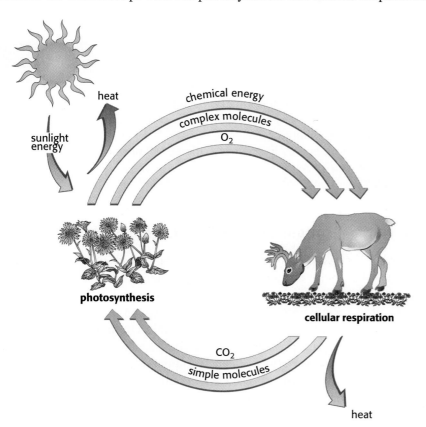

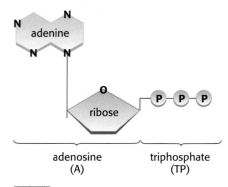

Figure 4.12   Where is the energy stored in ATP?   **The structure of the molecule.**

The phrase "high-energy phosphate bond" is an oversimplification that is not accurate. An understanding of the complexity of energy storage in ATP requires a knowledge of thermodynamics. For students, the important ideas are those presented here.

Figure 4.13   Explain the ATP-ADP cycle.

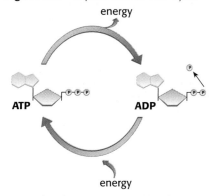

ATP molecules are continually rebuilt from ADP molecules, phosphates, and chemical energy.

Guidepost

How are carbon-containing molecules important to living things?

4.6   Energy Is Used in Small Packets

If you had one $100 bill, you might find it difficult to buy small things such as a hamburger, a pack of notebook paper, or a comb. It would be much easier if you had 100 $1 bills. So, too, with the energy in a cell. The large amounts of energy in food molecules are converted into "small change." This "small change" is the chemical energy stored in compounds such as **ATP**, adenosine triphosphate (uh DEN oh seen try FOS fayt).

ATP is the most important of several energy-transfer compounds that are found in all organisms. These compounds are involved in all the energy processes of living cells. The energy released during respiration is temporarily transferred to molecules of ATP. As you can see in Figure 4.12, each ATP molecule is made up of a main section (A) to which are attached three identical groups of atoms called phosphates (TP).

As food molecules are broken down to simpler compounds, a great deal of energy is released. The energy is used to make ATP molecules. Energy is stored in ATP until it is released by reactions that remove the third phosphate group from ATP. This energy is used to help the cell do its work. The work may be to move a muscle, to send a nerve impulse, to grow, or to form new compounds. Thus, ATP is a carrier of chemical energy in the cell.

Each ATP molecule releases a bit of energy whenever a phosphate group is broken off or transferred to another molecule. The molecule that remains, which has only two phosphate groups, is called **ADP**, adenosine diphosphate (uh DEN oh seen dy FOS fayt). Just as you cannot keep spending money from your pocket without eventually putting more money in, a cell cannot continue to "spend" its ATP without rebuilding it. As you can see in Figure 4.13, there is a continual ADP-ATP cycle. To make ATP molecules, an ADP molecule and a phosphate group are needed as well as chemical energy to combine the two. This energy comes from the breakdown of food molecules.

CONCEPT REVIEW

1. In what way is photosynthesis important for all living organisms?
2. What are the two products of photosynthesis?
3. How are the reactions of photosynthesis and cellular respiration similar? How are they different?
4. Is there more energy in one molecule of ATP or one molecule of sugar? How do you know?
5. Explain the relationship between ATP and ADP.

4.7   Carbon Is Found in All Living Things

Although organisms are composed of many different chemical elements, carbon is the central element for all living systems. Carbon atoms can join together to form chains or rings, as shown in Figure 4.14. Furthermore, carbon atoms can combine with hydrogen, oxygen, nitrogen, sulfur, and phosphorus to form a vast number of organic compounds. In fact, the atoms of the elements present in organic compounds can be arranged in so many ways that the variety of organic compounds is almost limitless. This variety ensures the uniqueness of each organism.

Organic compounds are the essential building blocks for organisms and are also their major source of chemical energy. Four basic types of carbon-containing molecules are found in all organisms. **Carbohydrates** (kar boh HY drayts) and **lipids** (LIP idz) are important energy-storing compounds. They also form part of the cell structure. The sugars produced in photosynthesis and used in respiration are carbohydrates. Starch is a large carbohydrate molecule made by joining many individual sugar molecules. Oils and fats are examples of lipids. **Proteins** (PROH teenz) function as enzymes and form part of the cell structure. Muscles ("meat") are composed largely of protein. **Nucleic** (noo KLEE ik) **acids** are the hereditary, or genetic, material for all organisms. They also coordinate the activities of the cell.

Because all living organisms contain the same types of biological molecules, you can get the molecules you need for life from another organism. In fact, that is the only way you can get the molecules you need. To live, you must eat a plant or animal and rearrange its molecules and atoms into your own molecules and atoms.

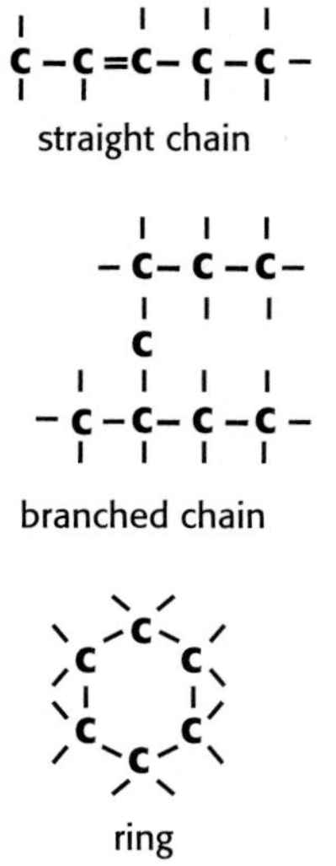

4.8 Carbohydrates Are Used for Energy, Storage, and Building

Carbohydrates contain only the elements carbon, hydrogen, and oxygen. Hydrogen and oxygen are present in the same proportions as in water. The subunits, or building blocks, of carbohydrates are single sugars, such as glucose and fructose (see Figure 4.15a), which contain no more than seven

2 carbon atoms in each molecule. Glucose molecules can be changed into other biological molecules within the cell. They are a major source of

3 energy for most organisms. Two single sugars may bond together to form a double sugar, as shown in Figure 4.15b. The most familiar double sugar is sucrose, commonly called cane, beet, or table sugar. Sucrose is formed by a chemical reaction that combines a glucose and a fructose molecule.

Figure 4.14 Carbon atoms can bond in several ways. The unconnected lines protruding from some of the carbon atoms show that any one of a number of elements can bond with the carbon in these positions.

Make a paper tetrahedron to explain the unique structural conformation of the carbon atom and its bonding pattern.

Molecular models are useful throughout the teaching of this section. Kits are available from supply houses, or models can be constructed from toothpicks and jelly beans.

Not all sugars are sweet; most are not as sweet as sucrose. When eaten by humans, however, they all are degraded by the human digestive system to their respective single sugars.

Figure 4.15 Single sugars (a) can combine to form double sugars (b). Starch, glycogen, and cellulose (c) are formed by linking together many glucose units.

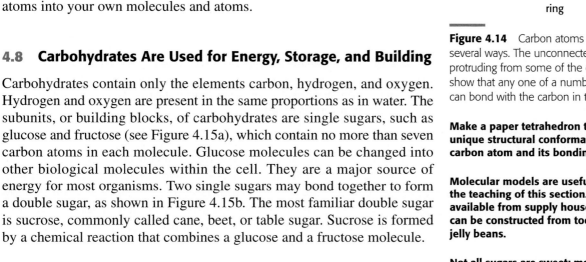

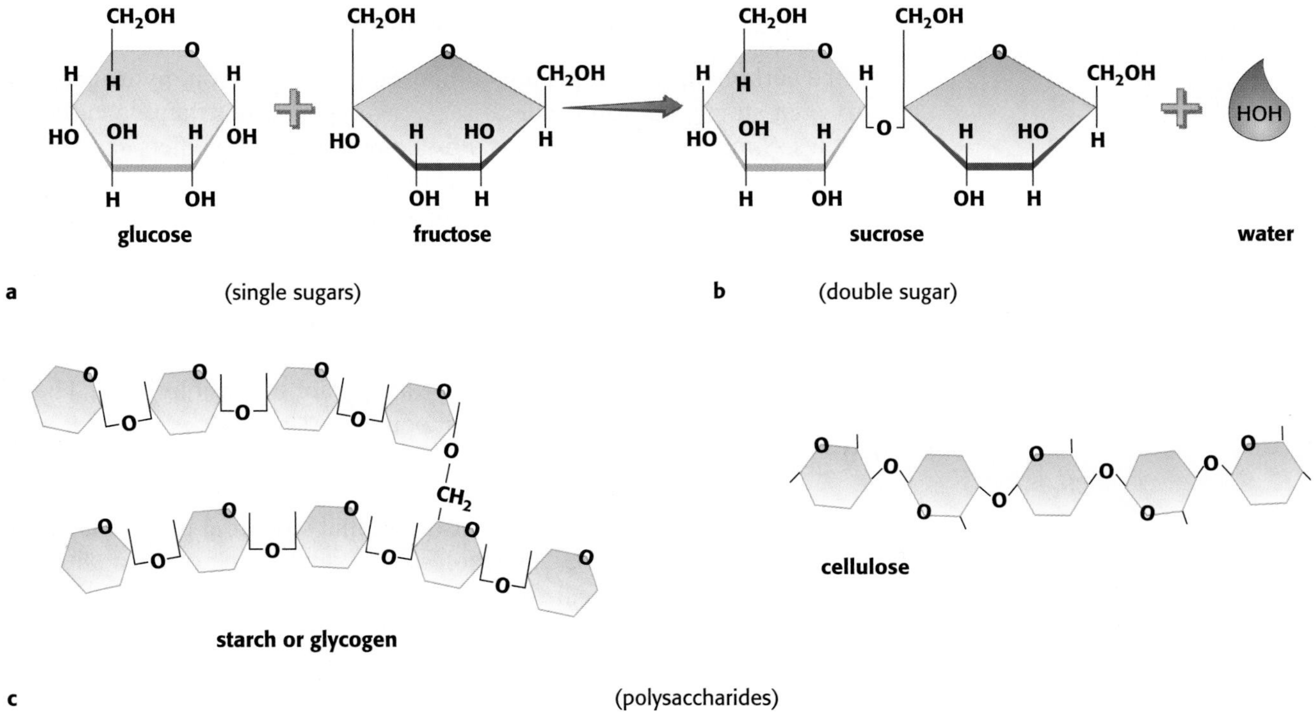

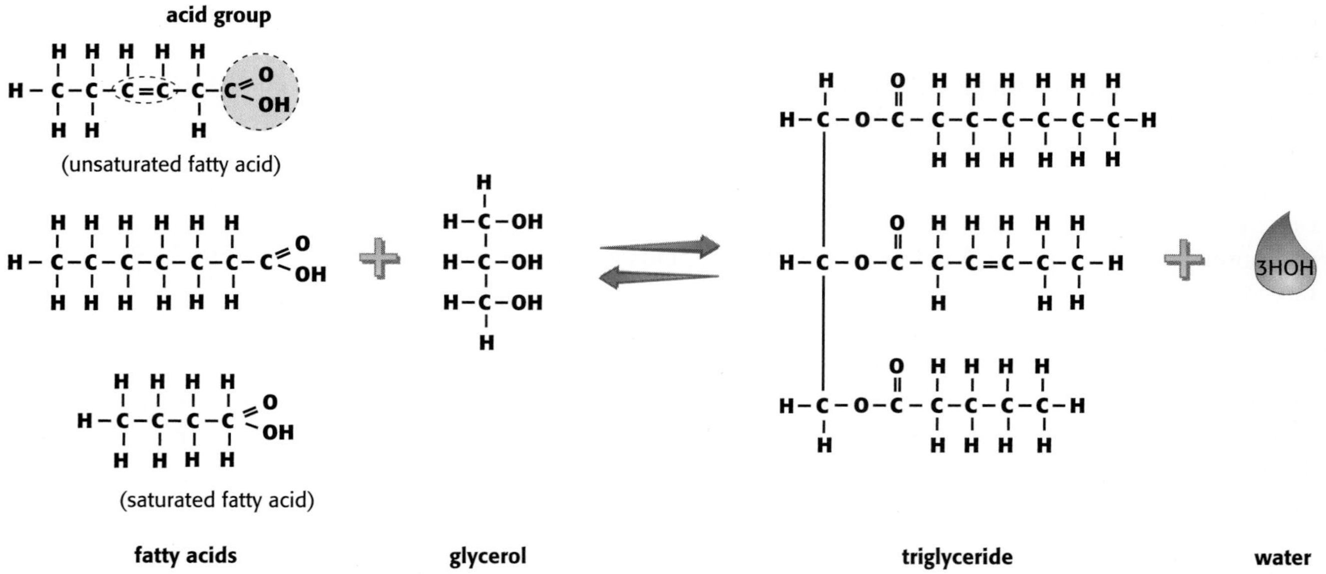

acid group

(unsaturated fatty acid)

(saturated fatty acid)

fatty acids

glycerol

triglyceride

water

Figure 4.16 A fat molecule consists of fatty acids joined to a glycerol molecule. To form a fat, one molecule of glycerol combines with three molecules of fatty acids. The fatty acids in one fat may be alike or different. Examine the diagram. What is a by-product of this reaction? **Three molecules of water.**

In synthesis reactions, many glucose molecules may bond together to build complex carbohydrates such as starch and **cellulose** (SEL yoo lohs), represented in Figure 4.15c. Starch is an energy-storage compound in many plants and an important food source for humans. (See Chapter 15.) Cellulose, a major part of wood and cotton fibers, gives the cell walls surrounding plant cells their rigidity. In the human liver and muscles, carbohydrates are stored as glycogen, also called animal starch. Molecules of starch, cellulose, and glycogen consist of thousands of molecules of glucose units and have no fixed size.

Mass is an impediment to locomotion, so a low-mass fuel is an advantage to a locomoting organism.

Fats that are in the liquid state at room temperature (about 20° C) are called *oils*.

4.9 Lipids Are Efficient Energy Storage Compounds

Like carbohydrates, lipids are composed only of carbon, hydrogen, and oxygen atoms, but lipids contain fewer oxygen atoms than do carbohydrates. Simple fats are the lipids most common in human diets and bodies. These fats are made of two types of lipid building blocks: glycerol, which **2** is a single sugar, and fatty acids, which are chains of carbon and hydrogen with an acid grouping on one end (see Figure 4.16). Both building blocks can be formed in cells from glucose.

Both carbohydrates and lipids are important energy-storage compounds in organisms. A gram of fat, however, contains more than twice as much chemical energy as a gram of carbohydrate. There- **3** fore, fats are better storage compounds. As animals prepare for winter when food is scarce, they eat large amounts of food. Much of this food energy is converted into fat, and the fat levels in their bodies **4** increase dramatically. Lipids are also essential structural parts of all cells. Lipids include plant waxes and **cholesterol** (koh LES ter ol), in addition to simple fats. The chemical structure of cholesterol is illustrated in Figure 4.17. Although cholesterol is needed by the body, excess cholesterol in the blood has been linked to heart disease.

Figure 4.17 A diagram representing the structure of cholesterol.

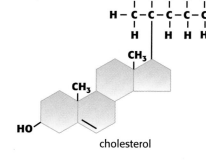

cholesterol

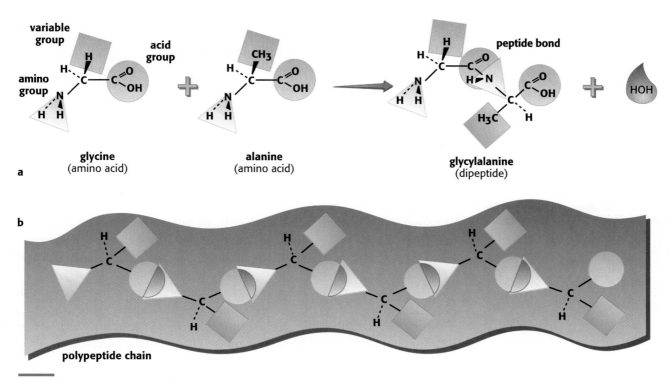

a

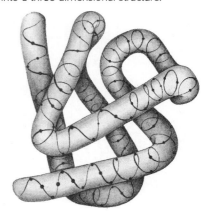

Figure 4.18 (a) The formation of a dipeptide from two amino acids. Polypeptide chains (b) are formed from long strings of amino acids.

4.10 Muscles, Enzymes, and Many Cell Parts Are Made of Protein

Protein molecules form part of the structure of every cell. Proteins also form enzymes and muscle tissue. Usually, protein molecules contain thousands of atoms—sometimes tens of thousands. The subunit of a protein molecule is an **amino** (uh MEEN oh) **acid,** shown on the left side of Figure 4.18a. Amino acids always contain at least four types of atoms: carbon, hydrogen, oxygen, and nitrogen. There are two amino acids that also contain sulfur. Twenty different types of amino acids can be found in a protein molecule. Green plants can synthesize all of these amino acids from simple materials. Animals cannot. They must get some amino acids ready-made in the food they eat. Unfortunately, not every type of food contains all the needed amino acids. Therefore, animals need a balanced diet of protein sources. Without it, protein-deficiency diseases may occur.

To synthesize a protein, amino acids must be linked together. First, two amino acids are linked together to form a dipeptide (DY PEP tyd), as shown in Figure 4.18a. A long chain of amino acids is a **polypeptide** (POL ee PEP tyd), illustrated in Figure 4.18b. Some proteins are made of only one polypeptide, but most are made of two or more polypeptides bonded together. Polypeptide chains are coiled and folded into complex three-dimensional structures to form proteins such as myoglobin, shown in Figure 4.19. The structure of the protein determines how it will function.

Thousands of different proteins can be made from various combinations of the 20 different types of amino acids. This variety is important in a living organism because each type of chemical reaction is controlled by a different type of enzyme, which is formed from proteins. The production of many different enzymes makes possible thousands of chemical reactions.

Essential amino acids are those that cannot be made in human cells and must be obtained in foods. They are essential in human nutrition, just as vitamins are. Many animals can change one type of amino acid to another. Human cells can change about 10.

Figure 4.19 The polypeptide chain in a myoglobin molecule is coiled and folded into a three-dimensional structure.

4.11   Enzymes Catalyze Cell Reactions

Most enzymes are large, complex proteins made by the cell. Like other catalysts, enzymes promote reactions but are not used up in the reactions. They allow chemical reactions to take place at the temperature of the cell. Enzymes are needed in only small amounts, because one enzyme molecule can complete the same reaction thousands of times in a single minute.

As shown in Figure 4.20, the specific reaction catalyzed by an enzyme depends on the molecular structure and shape of a small area of the enzyme known as the **active site**, which can attract and hold only specific molecules. An enzyme must bind closely to the molecules—the **substrates** (SUB strayts)—on which it acts. Sometimes the enzyme changes shape slightly after the substrate binds, creating the necessary fit. Because only a few substrate molecules are enough alike in structure and shape to fit the active site, each enzyme can catalyze only specific chemical reactions.

To act as a catalyst, an enzyme must take part temporarily in a chemical reaction. The reacting molecules combine with the active site of an enzyme, forming an **enzyme-substrate complex**. The enzyme aligns the reacting 5 molecules precisely and permits chemical changes to be completed rapidly at low temperatures. Once the reactions are complete, the new molecules break away, leaving the enzymes as they were before the reaction.

Enzymes catalyze both synthesis and decomposition reactions. Usually a different enzyme catalyzes each reaction. In a synthesis reaction, two or more small molecules combine with the enzyme. The enzyme provides the proper alignment, which enables these small mole- 7 cules to join into one large molecule. In a decomposition reaction, the substrate combines with the enzyme and is split into two or more smaller molecules. Figure 4.21 shows examples of both types of reactions.

Two aspects of enzyme activity are very important to cells. Enzyme reactions are faster at high temperatures but only to a certain point. At higher temperatures, the enzymes may begin to lose their shape. Because fit is so important for proper enzyme action, enzymes that lose their shape no longer function. Enzyme activity also varies with the pH of the solution. Thus, the temperature and the pH must be at the right levels for enzymes to act effectively, as you observe in Investigation 4.3.

The change in shape is somewhat like a glove conforming to fit the hand.

Figure 4.20   (a) A substrate binds to the enzyme at the active site. (b) Sometimes the active site changes shape after the substrate binds, bringing about the necessary fit. The binding produces an enzyme-substrate complex.

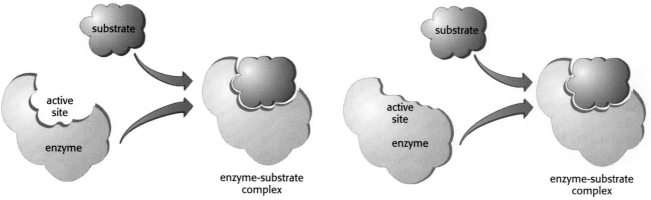

a

b

4.12   Nucleic Acids Control the Activities of the Cell

Two types of nucleic acids are present in all cells and are vital to cell function: **RNA**, ribonucleic (ry boh noo KLEE ik) acid, and **DNA**, deoxyribonucleic (dee OK sih ry boh noo KLEE ik) acid. An organism may be made up of billions of cells, but each living cell has its own DNA and RNA. Information stored in DNA controls all cell activities and determines the genetic, or hereditary, characteristics of the cell and the organism. RNA is required for the synthesis of proteins, including enzymes.

Both DNA and RNA are made up of individual subunits called **nucleotides** (NOO klee oh tydz). Each nucleotide, in turn, is made up of three small molecules linked together: a phosphate group, a five-carbon sugar, and a nitrogen base (see Figure 4.22 on page 84). The nitrogen base may be one of five different types. Each type is made up of carbon, hydrogen, oxygen, and nitrogen atoms. RNA nucleotides contain a single sugar called ribose (RY bohs). DNA, shown in Figure 4.23 on page 84, contains a slightly different sugar called deoxyribose (dee OK sih RY bohs), which has one less oxygen atom than ribose. DNA molecules are made of two strands with thousands of linked nucleotides. The two strands are attached in a specific way and coiled to form a double helix. RNA molecules are usually smaller and may be singlestranded.

Figure 4.21   In synthesis, two or more substrate molecules join at the active site forming one larger molecule. In decomposition, the substrate combines with the enzyme and is split into two or more smaller molecules.

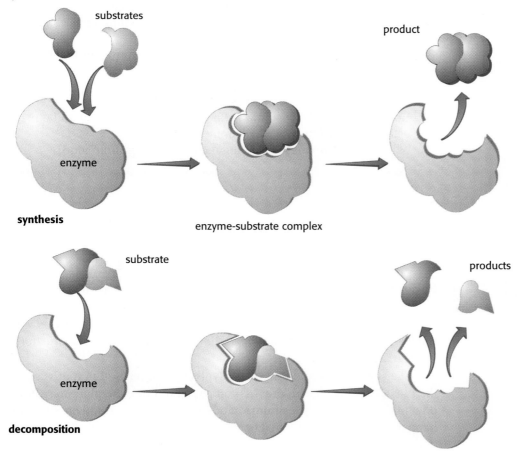

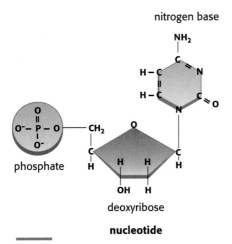

nitrogen base

phosphate

deoxyribose

nucleotide

Figure 4.22   The parts of a nucleotide.

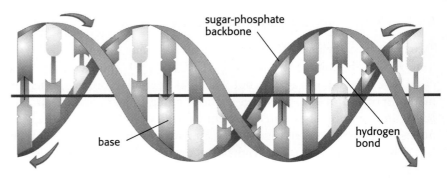

sugar-phosphate backbone

base

hydrogen bond

Figure 4.23   The structure of DNA. Alternating sugar and phosphate molecules connect each chain of nucleotides. Two chains are held together by specific pairing of the bases and coiled around a central axis to form a double helix. The axis is shown to help visualize the correct twisting. The two strands run in opposite directions, and there are 10 nucleotides per turn of the helix. Only a small portion of a DNA molecule nitrogen base is shown.

Explain that ribose and deoxyribose are 5 carbon, nonsweet sugars.

It is helpful to have models of DNA available. They also can be made from Tinker Toys and cardboard.

It is the information stored in DNA that determines the sequence of amino acids in proteins. This information also plays a major role in controlling when each protein is made. By controlling the synthesis of enzymes necessary for chemical reactions in the cell, DNA controls the activities of the cell. All the cells in an individual have the same unique genetic information in their DNA. That unique pattern of DNA results from the combination of genetic information provided by the individual's parents during sexual reproduction.

CONCEPT REVIEW

1. Why is the element carbon so important to living things?
2. What are the subunits of carbohydrates? of fats? of proteins? of nucleic acids?
3. Name one important function of each of the biological molecules named in Question 2.
4. Why is fat a better storage compound than starch?
5. How do enzymes work to catalyze a chemical reaction?
6. How is it possible for so many different types of proteins to be made with only 20 different types of amino acids?
7. What role do enzymes play in the release of energy from food?

Life is Based on Carbon

4.13   Plants Make and Use Carbon-Containing Sugars

Guidepost

What is the source of matter for living things, and what happens to the matter?

Recall from Section 4.4 that plants take up carbon atoms in the form of carbon dioxide from the air. During photosynthesis, they use the energy in sunlight to make sugars from carbon dioxide and water. In this way, the energy from sunlight and the carbon from carbon dioxide are stored in the sugars.

The sugars created during photosynthesis can be used in four ways by the plant, as shown in Figure 4.24 on page 86. First, the plant may break down the sugar molecules immediately to release the stored energy. This happens during respiration. The energy that is released from the sugars during respiration may be used by plant cells to continue the activities of

The Carbon Cycle and Global Warming

Because carbon is present in many of the gases that compose the earth's atmosphere, it is a major determinant of climate. Although some carbon dioxide is released by the respiration of organisms, larger amounts are released when fossil fuels are burned. Since the Industrial Revolution the atmospheric content of carbon dioxide has increased by about 25 percent.

As early as the nineteenth century, experts recognized that carbon dioxide in the atmosphere gives rise to a "greenhouse" effect. Think how a glass greenhouse allows sunlight to stream in but keeps heat from escaping. Similarly, carbon dioxide and other "greenhouse gases," such as methane, ozone, and nitrous oxide, allow sunlight to reach the earth but trap and reflect the heat. If there were no greenhouse gases, the earth would be cold and devoid of life. Too great a concentration of such gases, on the other hand, would cause the temperature to rise.

Human activity interrupts the carbon cycle in two ways. First, the burning of carbon-containing fuels adds carbon dioxide to the atmosphere. Second, the destruction of rain forests and other vegetation leaves less plant life to absorb carbon dioxide. When unharvested plant matter decays and when roots in the exposed soil react with oxygen, still more carbon dioxide enters the atmosphere.

Some scientists expect the increased levels of carbon dioxide, other gases, and water vapor to raise the earth's temperature from 1° C to 6° C in the next 100 years. Such an increase could affect global climate, food-producing patterns, and world population. Suppose, because of the rise in temperature, a population can no longer grow enough food to feed itself. If famine results, the population may migrate into a new area, compounding the problems of the population there and perhaps creating new ones.

Most experts agree on what the greenhouse effect is and how it works, but they disagree on the extent of the effects. Critics contend the computer models used to predict the greenhouse effect are so weak they do not account for the 0.5° C warming that has occurred in the last 100 years. Many experts feel that ignorance of the interactions between the oceans and the atmosphere limits their ability to predict effects. A new study, however, indicates that increased evaporation of the earth's oceans due to higher temperatures will amplify the greenhouse effect.

On the other hand, in February 1990, a federal government study indicated that the temperature in the southeastern United States actually has fallen 1° C during the past 30 years. Although the report does not disprove the occurrence of global warming, it should stimulate the debate.

1. Sunlight penetrating the atmosphere warms the earth's surface.

2. The earth's surface radiates heat to the atmosphere.

3. Greenhouse gases and water vapor absorb some and re-radiate part of them toward the earth.

4. When greenhouse gases build up in the atmosphere, more heat is trapped near the earth's surface.

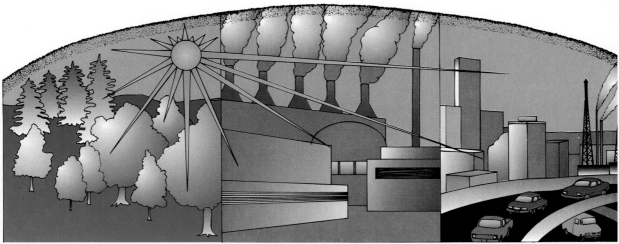

For the most part, cellulose is an undigestible molecule. Its function as fiber in human digestion is discussed in Chapter 15. Termites are able to digest wood because flagellated protozoa in their gut provide cellulose.

life. Second, a plant may use the sugar molecules for growth. In this case, many sugar molecules are joined together to make the building materials **1** necessary for more cells. Cellulose, for example, is one of these materials. Third, the plant may store sugars for future use. Starch is an important storage compound found in many parts of a plant. When the plant needs energy, starch is first broken down to individual sugar molecules. During respiration, the sugars are broken down to release energy. Fourth, sugar molecules may be converted into the other biological molecules needed for life.

4.14   Carbon Cycles Within an Ecosystem

As a plant grows, its body becomes larger. If the plant is eaten, the carbon in the plant is passed from producer to consumer. For the consumer to use the food, it must break down the plant body. As this happens, the carbon containing molecules are broken apart, releasing both carbon atoms and energy. Much of the energy released is used for the activities of the consumer. Some of the carbon from the body of the plant is added to the body of the consumer. The rest of the carbon is exhaled into the air as carbon dioxide. For example, you take in carbon in all the foods you eat. You **2** return carbon dioxide to the air every time you exhale. A plant also returns carbon dioxide to the air when it uses its own sugars as a source of energy. When another plant takes in this carbon dioxide during photosynthesis, the cycle of carbon through the community is complete.

Carbon dioxide also is returned to the air by decomposers. When producers or consumers die, decomposers begin their work. As its source of energy, a decomposer uses the energy locked in the bodies of dead organisms, and it uses the carbon from the bodies to build its own body. Carbon that is not used is returned to the air as carbon dioxide. Eventually, almost all the carbon that is taken in by plants during photosynthesis is returned to the air by the activity of decomposers.

Hundreds of millions of years ago, many energy-rich plant bodies were buried before decomposers could get to them. When that happened, the bodies slowly changed during long periods of time. They became a source of fuels—coal, oil, and natural gas. When these fuels are burned,

Figure 4.24   Plants use the sugars made during photosynthesis in a variety of ways.

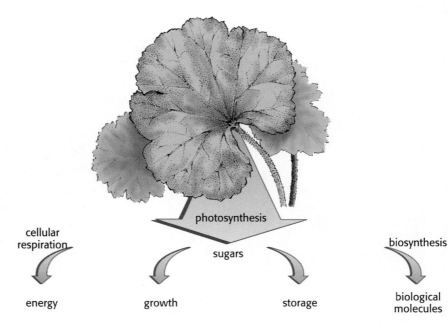

cellular respiration

photosynthesis

sugars

biosynthesis

energy   growth   storage   biological molecules

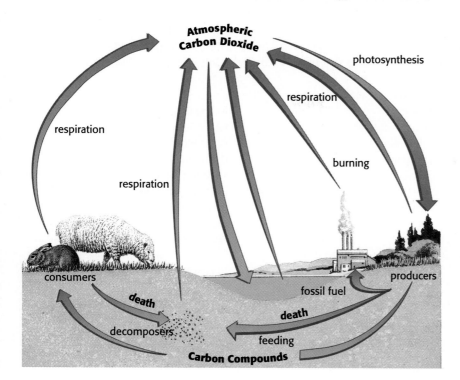

Figure 4.25   Where would you place humans in the carbon cycle?
Consumers, respiration, death, burning

energy is released, and the carbon in the fuels is returned to the air as carbon dioxide. You can see that even the energy obtained from fuels is a result of photosynthesis. The process in which carbon is passed from one organism to another, then to the abiotic community, and finally back to the plants is called the **carbon cycle** (see Figure 4.25). Some of the other cycles found within ecosystems—the water cycle, the sulfur cycle, and the nitrogen cycle—are described in later chapters.

CONCEPT REVIEW

1. Of the four ways a plant uses the sugars it makes, which is the only way that does not add material to the plant?
2. How are producers, consumers, and decomposers involved in the carbon cycle?
3. What is the Greenhouse effect and how does it work?

INVESTIGATION 4.1   Organisms and pH

Chapter 2 discussed the tendency for populations to maintain a relative stability, or homeostasis. Individual organisms and cells must maintain an internal homeostasis, but many factors can affect that stability—for example, the relative concentrations of hydrogen ions (H^+) and hydroxide ions (OH^-). The biochemical activities of living tissues frequently affect pH, yet life depends on maintaining a pH range that is normal for each tissue or system. Using a pH probe and the LEAP-System™, or a pH meter or wide-range pH paper if the LEAP-System™ is not available, you can compare the responses of several materials to the addition of an acid and a base. Instructions for use of the LEAP-System™ are provided in the LEAP-System™ BSCS Biology Lab Pac 2.

This investigation demonstrates two of the BSCS themes: the complementarily of organisms and the environment on a cellular and functional level and homeostasis in terms of internal regulation of pH. Using biological materials, students can observe the maintenance of homeostasis (pH) in spite of changes in the external

Figure 4.26 The LEAP-System™ with pH probe.

environment. A buffered system is introduced as a model of how pH might be regulated; the questions elicit student understandings. Teams of 4 will work if using the LEAP–System™ or pH meters, (two teams per LEAP–System™); teams of 2 are preferable if using pH paper.

Time: Two class periods—one to perform the experiment and one for discussion of the results. Allow extra time if this is the students' first experience with the LEAP–System™.

Students often think enzymes act only in digestion. Emphasize that without enzymes, chemical reactions in a cell would not take place or would be too slow to sustain life.

Safety

See the safety remarks for Investigation 2.1 when working with the LEAP-System™ equipment. If the investigation spans more than one day, remind students at the break point to wash their hands before leaving and to observe safety precautions on the day class resumes.

Warn students not to ingest any materials.

Before you begin, study the investigation and develop a hypothesis that answers the question "How do organisms survive and function despite metabolic activities that tend to shift pH toward either acidic or basic ends of the scale?"

Materials (per team of 4)

4 lab aprons
4 pairs of safety goggles
50-mL beaker or small jar
50-mL graduated cylinder
3 colored pencils
Apple lle, Apple IIGSX or IBM computer
LEAP-System ™
printer
pH probe
pH meter or wide-range pH paper
(optional)

forceps (optional)
tap water
0.1*M* HCl in dropping bottle
0.1*M* NaOH in dropping bottle
sodium phosphate pH 7 buffer solution
liver homogenate
potato homogenate
egg white (diluted 1:5 with water)
warm gelatin suspension, 2% solution

Put on your safety goggles and lab apron.

Procedure

1. In your data book, prepare a table similar to Table 4.1.
2. Pour 25 mL of tap water into a 50-mL beaker.
3. Record the initial pH by using a pH probe with the LEAP-System™ (Figure 4.26), a pH meter, or by dipping small strips of pH paper into the water and comparing the color change to a standard color chart.
4. Add 0.1*M* HCl a drop at a time. Gently swirl the mixture after each drop. Determine the pH after 5 drops have been added. Repeat this procedure until 30 drops have been used. Record the pH measurements in your table.

◆ **0.1 *M* HCl is an irritant and may destroy clothing. Avoid skin/eye contact; do not ingest. Should a splash or spill occur, call your teacher immediately; flush the area with water for 15 minutes; rinse mouth with water.**

---

5. Rinse the beaker thoroughly and pour into it another 25 mL of tap water. Record the initial pH of the water and add 0.1*M* NaOH drop by drop, recording the pH probe changes in exactly the same way as for the 0.1*M* HCl.

Table 4.1 Testing pH

| Solution Tested | Tests with 0.1*M* HCl | | | | | | | Tests with 0.1*M* NaOH | | | | | | |
| | *pH after addition of* | | | | | | | *pH after addition of* | | | | | | |
	0	5	10	15	20	25	30 drops	0	5	10	15	20	25	30 drops
Tap water														
Liver														
Potato														
Egg white														
Gelatin														
Buffer														

>
> **0.1***M* **NaOH is an irritant and may destroy clothing. Avoid skin/eye contact; do not ingest. Should a spill or splash occur, call your teacher immediately; flush the area with water for 15 minutes; rinse mouth with water.**

6. Using the biological material assigned by your teacher, repeat Steps 2–5. Record the data in your table.

7 Test the buffer solution (a nonliving chemical solution) using the same method outlined in Steps 2–5. Record the data in your table.

8. Wash your hands thoroughly before leaving the laboratory.

Discussion

1. Summarize the effects of HCl and NaOH on tap water.

2. What was the total pH change for the 30 drops of HCl added to the biological material for the 30 drops of NaOH added? How do these data compare with the changes in tap water?

3. Run strip charts to obtain individual graphs of the data. In your data book, prepare a simple graph of pH versus the number of drops of acid and base solutions added to tap water. Plot two lines—a solid line for changes with acid and a dashed line for changes with base. Using different colored solid and dashed lines, add the results for your biological material. Compare your graph to the graphs of teams who used a different biological material. What patterns do the graphs indicate for biological materials?

4. How do biological materials respond to changes in pH?

5. Use different colored solid and dashed lines to plot the reaction of the buffer solution on the same graph. How does the buffer system respond to the HCl and NaOH?

6. Is the pH response of the buffer system more like that of water or of the biological material?

7. How does the reaction of the buffer solution serve as a model for the response of biological materials to pH changes?

8. Would buffers aid or hinder the maintenance of homeostasis within a living cell in a changing environment?

9. What does the model suggest about a mechanism for regulating pH in an organism?

Discussion

1. HCl decreases the pH about 0.5 for each 5 drops; NaOH increases the pH but levels off slightly above 10.5.
2. Answers will vary but usually less than 2 for both acid and base. Changes are not as great as with tap water.
3. Small changes occur with each 5 drops, usually around 0.1.
4. Biological materials have the capacity to control the extent of pH changes as HCl and NaOH are added.
5. There is little change with the buffer solution.
6. Biological materials.
7. The buffer solution shows the capacity to limit the extent of pH changes in the presence of added acid or base. In this respect, it is similar to the situation with the biological materials.
8. Buffers would aid greatly in maintaining a stable internal environment.
9. Organisms might use buffer systems to help regulate the pH of their systems.

BIOLOGICAL MOLECULES

INVESTIGATION 4.2   Compounds in Living Organisms

The compounds your body needs for energy and building materials are carbohydrates, proteins, fats, vitamins, and other nutrients. These compounds are present in the plants and animals you use as food. In this investigation, you will observe tests for specific compounds and then use those tests to determine which compounds are found in ordinary foods.

Materials (per team of 4)

4 pairs of safety goggles	Biuret solution in dropping bottle
4 lab aprons	indophenol solution in dropping bottle
4 pairs of plastic gloves	Lugol's iodine solution in dropping bottle
250-mL beaker	1% silver nitrate solution in dropping bottle
10-mL graduated cylinder	
6 18-mm × 150-mm test tubes	isopropyl alcohol (99%) in screw-top jar
test-tube clamp	brown wrapping paper
hot plate	apple, egg white, liver, onion, orange,
Benedict's solution in dropping bottle	potato, or other foods of your choice

This investigation gives students an opportunity to examine some common foods for the presence of several compounds important in nutrition and makes more meaningful the discussion of those compounds in later sections. Teams of 4 work well.

Time: One class period is sufficient if you do the reagent tests as a demonstration.

If you prefer to have the students conduct the tests in Part A and discover the results for themselves, the investigation requires two class periods.

Demonstration

1/4 tsp butter or vegetable oil
5 mL 10% gelatin suspension
5 mL 2% NaCl solution
5 mL 10% starch solution
5 mL 10% glucose solution
5 mL vitamin C solution (1% ascorbic acid)

Safety

Remind students that alcohol is flammable and that its vapors can explode. Extinguish all flames before any alcohol is used. It is a good idea to keep the hot plates in an area separate from the area where the tests are performed. Restrict the amount of alcohol to 600 mL total in the laboratory at any one time. Store in bottles no larger than 100 mL capacity. Use screw-top jars for storing isopropyl alcohol during the investigation.

Results of Reagent Tests

1. The Biuret test gives a pink-to-purple reaction in the presence of protein. Biuret powder (caramyluea) can be used to demonstrate this classic color reaction.

2. Benedict's solution shows a positive test for simple sugars with a color change that ranges from green to yellow or orange.
3. The iodine test is positive for starch if there is a color change to blue-black.
4. Blue indophenol turns colorless in the presence of vitamin C. Disregard the intermediate pink stage.
5. Silver nitrate forms a white precipitate when added to a solution of sodium chloride.
6. Fat makes a translucent greasy spot on the paper.

Table 4.2 Reagent Tests of Known Food Substances

Food Substance	Reagent Test	Results
Gelatin	Biuret solution	
Glucose	Benedict's solution	
Starch	Lugol's iodine solution	
Vitamin C	Indophenol solution (0.1%)	
Sodium chloride	Silver nitrate solution (1 %)	
Butter or vegetable oil	Brown paper	

Put on your safety goggles, lab apron, and gloves. Tie back long hair.

Procedure

PART A TEST DEMONSTRATION

1. In your data book, prepare a table similar to Table 4.2.
2. Scientists use special chemical solutions, or reagents, to detect the presence of certain compounds. Observe the six reagent tests your teacher performs. In your table, describe the results of each test.

PART B COMPOUNDS IN FOODS

3. In your data book, prepare a table similar to Table 4.3. Then, record the presence (+) or absence (–) of each substance in the foods you test.

 The reagents you will use in this procedure may be corrosive, poisonous, and/or irritants, and they may destroy clothing. Avoid skin and eye contact; do not ingest. Should a splash or spill occur, call your teacher immediately; flush the area with water for 15 minutes; rinse mouth with water.

4. Predict what substances you will find in each sample your teacher assigns to you. Then, test the samples as your teacher demonstrated or as described in Steps 5–10. Record the result of each test in your data book, using a + or –.

5. Protein test: Place 5 mL of the assigned food in a test tube. Add 10 drops of Biuret solution.

Table 4.3 Analysis of Compounds in Common Foods

Substance		Protein	Glucose	Starch	Vitamin C	Chloride	Lipid
Egg	Prediction						
	Test results						
Potato	Prediction						
	Test results						
Etc.	Prediction						
	Test results						

6. Glucose test: Add 3 mL of Benedict's solution to 5 mL of the assigned food. Place the test tube in a beaker of boiling water and heat for five minutes.

> **Use test-tube clamps to hold hot test tubes. Boiling water will scald, causing second-degree burns. Do not touch the beaker or allow boiling water to contact your skin. Avoid vigorous boiling. Should a burn occur, call your teacher immediately; place burned area under cold running water.**

7. Starch test: Add 5 drops of Lugol's iodine solution to 5 mL of the assigned food.
8. Vitamin C test: Add 8 drops of indophenol to 5 mL of the assigned food.
9. Chloride test: Add 5 drops of silver nitrate solution to 5 mL of the assigned food.
10. Fat test: Rub the assigned food on a piece of brown wrapping paper. Hold the paper up to the light. When food contains only a small amount of fat, it may not be detected by this method. If no fat has been detected, place the assigned food in 10 mL of a fat solvent such as isopropyl alcohol (99%). Allow the food to dissolve in the solvent for about five minutes. Then, pour the solvent on brown paper. The spot should dry in about ten minutes. Check the paper.

> **Isopropyl alcohol is flammable; its vapors may explode. Keep away from heat and sparks. Extinguish any open flames in the area.**

11. Wash your hands thoroughly before leaving the laboratory.

Discussion

1. How did your predictions compare with the test results?
2. Which of your predictions was totally correct?
3. Which foods contained all the compounds for which you tested?
4. On the basis of your tests, which food could be used as a source of protein? glucose? starch? vitamin C? fat?
5. How might the original colors of the test materials affect the results?

Discussion

1–4. **Predictions depend on student's observations and the variety of material tested.**
5. **A blue result may give a green color on a yellow surface. Color of a material such as tomato may mask the test result or give a falsely positive reading.**

INVESTIGATION 4.3   Enzyme Activity

In this investigation, you will study several factors that affect the activity of enzymes. The enzyme you will use is catalase, which is present in most cells and found in high concentrations in liver and blood cells. You will use liver homogenate as the source of catalase. Catalase promotes the decomposition of hydrogen peroxide (H_2O_2) in the following reaction:

$$2H_2O_2 \xrightarrow{\text{catalase}} 2H_2O + O_2$$

Hydrogen peroxide is formed as a by-product of chemical reactions in cells. It is toxic and would kill cells if not immediately removed or broken down.

Materials (per team of 3)

3 pairs of safety goggles	test-tube rack
3 lab aprons	nonmercury thermometer
50-mL beaker	filter-paper disks
2 250-mL beakers	buffer solutions: pH 5, pH 6, pH 7, pH 8
10-mL and 50-mL graduated cylinders	catalase solution
reaction chamber	fresh 3% H_2O_2
6 18-mm × 150-mm test tubes	ice
forceps	water bath at 37° C
square or rectangular pan	

This investigation allows students to compare enzyme activity under a variety of conditions and to experience some of the factors that affect enzyme regulated reactions. Performed before the discussion in Section 4.11, the activity is one of discovery. It also provides practice in graphing experimental results. Teams of 2 to 4 work well.
   Time: Allow two to three class periods—one periods for Parts A and B, one period for Parts C and D, and one period for discussion. You can shorten the time by dividing activities B, C, and D among lab teams, but all teams should perform Part A.

Table 4.4 Catalase Activity Under Various Conditions

Experiment	mL O_2 Evolved/30 sec																			
Reading	1	2	3	4	5	6	7	8	9	10	11	12	13	14	15	16	17	18	19	20
(Part A) Full concentration																				
(Part B) 3/4 concentration																				
Etc.																				

Put on your safety goggles and lab apron. Tie back long hair.

Procedure

In all experiments, make certain that your reaction chamber is scrupulously clean. Catalase is a potent enzyme, and if the chamber is not washed thoroughly, enough will adhere to the sides to make subsequent tests inaccurate. Measure all substances carefully. Results depend on comparisons between experiments, so the amounts measured must be equal or your comparisons will be valueless. Before you do the experiment, read through the instructions completely. Make sure that you have all the required materials on hand, that you understand the sequence of steps, and that each member of your team knows his or her assigned function.

PART A The Time Course of Enzyme Activity

1. Prepare a table in your data book similar to Table 4.4.
2. Obtain a small amount of stock catalase solution in a 50-mL beaker.
3. Obtain a reaction chamber and a number of filter-paper discs.
4. Place 4 catalase-soaked filter-paper discs high on *one* interior sidewall of the reaction chamber. (They will stick to the sidewall.) Prepare a disc for use in the reaction chamber by holding it by its edge with a pair of forceps and dipping it into the stock catalase solution for a few seconds. Drain excess solution from the disc by holding it against the side of the beaker before you transfer it to the reaction chamber.
5. Stand the reaction chamber upright and carefully add 10 mL of 3% hydrogen peroxide (H_2O_2) solution. *Do not allow the peroxide to touch the filter-paper discs.*

 H_2O_2 is a reactive material. Avoid skin/eye contact; do not ingest. Should a splash or spill occur, call your teacher immediately; flush the area with water for 15 minutes; rinse mouth with water.

6. Tightly stopper the chamber.
7. Fill a pan almost full with water.
8. Lay the 50-mL graduated cylinder on its side in the pan so that it fills with water completely. If any air bubbles are present, carefully work these out by tilting the cylinder slightly. Turn the cylinder upside down into an upright position, keeping its mouth underwater at all times.
9. Making certain the side with the discs is at the top, carefully place the reaction chamber and its contents on its side in the pan of water.

10. Move the graduated cylinder into a position so that its mouth comes to lie directly over the tip of the dropping pipet extending from the reaction chamber, as shown in Figure 4.27. One member of the team should hold it in this position for the duration of the experiment.

11. Rotate the reaction chamber 180° on its side so that the hydrogen peroxide solution comes into contact with the soaked discs.

12. Measure the gas levels in the graduated cylinder at 30-second intervals for 10 minutes. Record the levels in your data table.

PART B The Effect of Enzyme Concentration on Enzyme Activity

13. Test 3/4, 1/2, and 1/4 concentrations of enzyme solution, using the procedure for Part A with the following changes:

   a. 3/4 concentration: Use 3 catalase-soaked discs instead of 4.

   b. 1/2 concentration: Use 2 catalase-soaked discs as well as a 10-mL graduated cylinder.

   c. 1/4 concentration: Use 1 catalase-soaked disc and a 10-mL graduated cylinder.

14. Record all data in your data table.

PART C The Effect of Temperature on Enzyme Activity

15. Add 10 mL of 3% H_2O_2 to each of two test tubes. Place one test tube in a beaker of ice water and the other in a beaker with water maintained at 37° C.

16. When the temperature of the chilled H_2O_2 reaches approximately 10° C, repeat Part A with the following changes:

   a. In Step 5, use 10 mL of chilled 3% H_2O_2.

   b. In Step 7, add ice to the pan to chill the water to approximately 10° C.

17. When the temperature of the warmed H_2O_2 reaches approximately 37° C, repeat Part A with the following changes:

   a. In Step 5, use 10 mL of warmed 3% H_2O_2.

   b. In Step 7, fill the pan with water warmed to approximately 37° C.

18. Record the data in your data table.

PART D The Effect of pH on Enzyme Activity

19. Label 4 test tubes *pH 5, pH 6, pH 7,* and *pH 8,* respectively. Add to each of these 8 mL of 3% H_2O_2.

20. Add 4 mL of pH 5 buffer solution to the *pH 5* test tube, shaking well to ensure mixing. Do the same for the other buffer solutions.

21. Repeat Part A for each pH value, substituting the buffered 3% H_2O_2 solutions.

22. Record the results in your data table.

23. Wash your hands thoroughly before leaving the laboratory.

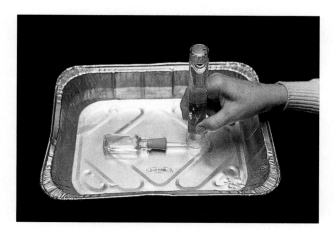

Figure 4.27 Apparatus for measuring O_2 production in a reaction between catalase and hydrogen peroxide.

Discussion

1. No. Graphs should show that enzyme action remains constant for the duration of experiment.
2. Enzyme activities generally directly proportional to enzyme concentration.
3. Lower temperatures slow down enzyme action; higher temperatures speed it up. Ask students what would happen if they increased the temperature to 45°C or higher.
4. Students should observe that there is an optimum pH for enzyme activity and that activity is slower at other pH values. The optimum pH for catalase is pH 7 to 73.
5. Students should recall from Investigation 4.1 that a buffer is a substance that prevents large changes in pH. The other parts of the experiment might have worked better had a buffer at the optimum pH been used, but the liver extract also contained buffers.
6. Results from the experiments should show that optimum concentration, temperature, and pH are necessary for effective enzyme action. Students may need some help in realizing that different enzymes have different optimum conditions.
7. Answers will vary but should include use of reaction chambers with and without catalase-soaked discs.
8. Enzymes are proteins; they can remain active unless they are denatured.

Discussion

1. In your data book, plot the data of Part A on a graph. Label the horizontal axis *Time (sec)*, and label the vertical axis *mL O_2 Evolved*. Does the action of catalase change through time?
2. Plot the data of Part B on the grid used for Part A and label the concentrations on the graph. Based on these data, how does enzyme activity vary with concentration?
3. Copy the graph for Part A and plot the data from Part C on it. Based on these data, how does temperature affect enzyme action?
4. Plot the results of all four runs of Part D on a third graph. How does pH affect enzyme action?
5. What is a buffer? Would Parts A, B, and C have been different if buffers had been used in them too? If so, how?
6. Summarize the general conditions necessary for effective enzyme action. Are these conditions the same for each enzyme? Why?
7. How would you design an experiment to show how much faster H_2O_2 decomposes in the presence of catalase than it does without the enzyme?
8. Inasmuch as the liver from which you obtained catalase was dead, how does it happen that the enzyme is still active?

SUMMARY

Energy is needed to do work and to make and maintain order. Without energy, highly organized systems such as living things could not exist. The need for energy is continuous, and therefore, organisms must obtain and use energy throughout their lives. Organisms also must obtain matter to build up their bodies. Without matter, no organism could grow. Plants make sugars from simple molecules in the process of photosynthesis. They convert these sugars into other biological molecules, including proteins, lipids, and nucleic acids, all of which also contain carbon. These biological molecules are built up and broken down in many different chemical reactions. All chemical reactions are catalyzed by enzymes, and many of them produce ATP as a temporary energy-transfer molecule.

APPLICATIONS

1. How is knowing the carbon cycle useful to biologists in making a study of food chains or food webs?
2. The proteins in the cells of a wheat plant are different from the proteins in your cells. How can the differences be explained? What must happen when you use wheat as a nutrient for the formation of your proteins?
3. Many botanists believe that the concentration of carbon dioxide in the air was much greater during the Carboniferous period, when most of the large coal deposits were being formed, than it is at present. What might be the basis for their belief?
4. In what ways do you contribute to the carbon cycle?
5. Are animals needed for carbon to cycle within a community? Explain.

6. Does a sugar molecule have order and organization? How does this question relate to the use of sugar as a source of energy by living things?
7. In terms of calories, why do you think seeds (small structures containing young plants) are high in fat content? Examples are the peanut, corn, and sunflower seeds from which we get cooking oils. Why are such seeds good sources of food for humans?
8. Apply what you have learned about carbon dioxide and the carbon cycle to global warming. Suppose the temperature of the earth increased 5° C. What effects might this have? What if the temperature increased 15° C?

PROBLEMS

1. Make two lists: (1) all the ways you use energy in a day, and (2) all the different types of energy you use in a day (e.g. light, heat, mechanical, chemical).

2. One hundred years ago, the carbon dioxide in the atmosphere was measured at 0.0283 percent. Today the level is 0.0330 percent. What factors during the last 100 years may have contributed to this increase? What might be some possible future consequences if this trend continues?

3. Many fats and cooking oils are called "polyunsaturated." Find out what this means in chemical terms.

4. Each day, you come in contact with many basic, acidic, and neutral solutions. You may have seen products that are labeled "pH-balanced" or "buffered." What does this mean? Make a list of such products or foods and explain how they affect you. Do you use these products or eat these foods because of their acidity or alkalinity, or doesn't it make any difference what their pH is?

5. *The Andromeda Strain*, by Michael Crichton, is a novel about a life form that comes to earth from another galaxy. Read it, or a similar science-fiction novel, and compare the extraterrestrial life form to life forms on earth. What is its chemical composition and structure? Is it based on carbon? Does it require energy?

6. Build three-dimensional models of DNA, RNA, and an enzyme, that can be hung from your classroom ceiling. After reading Chapter 5, create cell parts to hang from the ceiling and turn your classroom into a model of a large cell.

A cheetah and its young. What evidence is there for continuity in the biosphere? What factors and forces are responsible for this continuity?

The two young look very much like their parent, suggesting that continuity is preserved in the inheritance of material carried from generation to generation.

Section Two

CONTINUITY IN THE BIOSPHERE

◆ Individuals within a population come and go, but the population itself can exist for a long time. Fossil evidence indicates the biosphere itself has endured for more than 3.5 billion years. Thus, there is a continuity in the biosphere, but there also is change. The fossil record reveals that change usually has occurred slowly and systematically—that homeostasis exists between organisms and their environment. How continuity is maintained in populations and how populations change over time are considered in Section Two. ◆

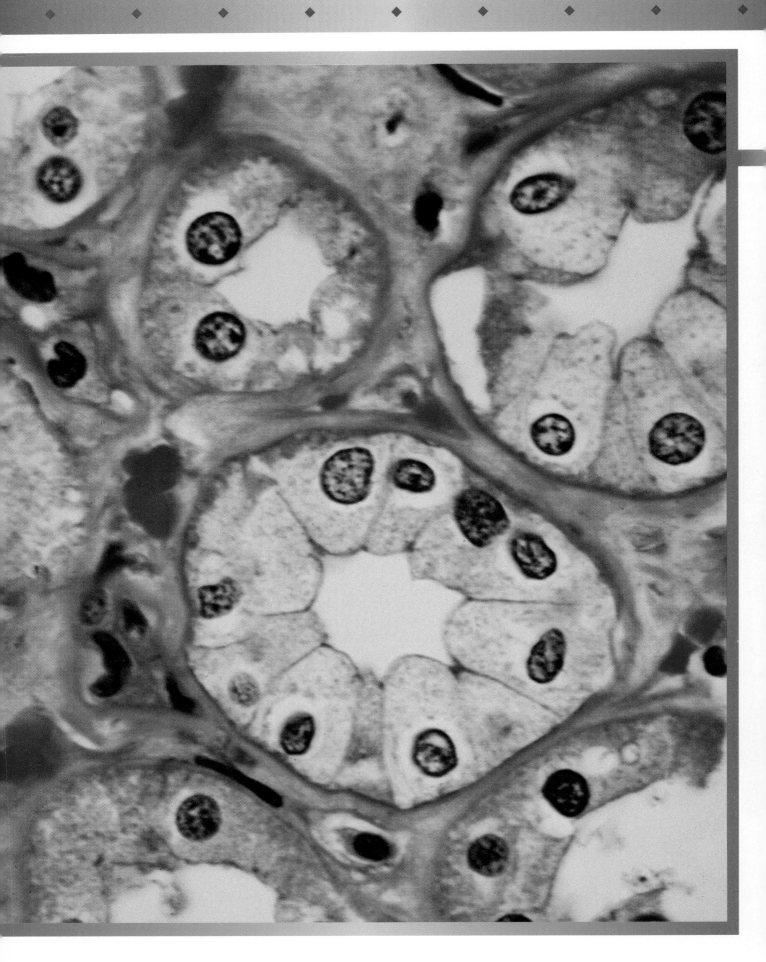

Continuity in Cells 5

These tubule cells from a human kidney are enlarged approximately 250 times and are about 50 µm in diameter. What is the large pigmented spot near the center of each cell? Estimate the distance across this spot. How many darkly pigmented dots are inside these spots? What are these dots, and how large are they?

The nucleus is approximately ¼ of the cell's diameter or about 12µm. The three or more nuclei are each about ⅙ the diameter of the nucleus, or approximately 2µm each.

◆ In learning how the world of life is organized, you have looked at large living systems—populations, communities, ecosystems, and the entire biosphere. To learn more about these complex systems, you have read about some of the important biological molecules that make up organisms.

Biological molecules are arranged not randomly but in organized patterns. Amino acids, for example, are linked together to make proteins, and single sugars are linked together to form complex carbohydrates. Proteins, sugars, and various other compounds are arranged in patterns that make up the membranes and various other structures that are found within cells. In turn, cells are organized into more complex arrangements called tissues. Tissues are organized into organs, organs into body systems, and body systems into complete organisms.

Each of these levels of organization enables the organism to solve its problems of staying alive and of passing on its distinctive nature to its offspring. Even the simplest organisms have various parts that allow them to take in food, to move, to detect the environment, and to reproduce. Biologists study all these parts to learn how they function, how they interact, and how they permit life to continue. This chapter is about cells, the basic units of life—how they are constructed, how they work, and how they reproduce. ◆

Cell Structure

5.1 Cells Are the Units of Life

All the organisms scientists have examined are made of tiny, highly active units called cells. Some organisms are unicellular. Each one consists of only a single cell that carries out all the activities of the organism. Others are multicellular and may be composed of trillions of cells. As illustrated in Figure 5.1, there are many different types of cells. In multicellular organisms, many of the cells are specialized for particular functions. Because cells are found in all organisms, scientists consider them to be the basic units of life.

The idea that cells are the basic units of life began to take shape in the 1700s and early 1800s. This idea, which now is known as the cell

MAJOR CONCEPTS
- ◆ Cells are the basic structural and functional units of life.
- ◆ There are two types of cells: prokaryotic and eukaryotic.
- ◆ Membranes organize eukaryotic cells into specialized compartments that allow for a division of labor.
- ◆ Multicellular organisms have tissues made of specialized cells.
- ◆ Cells exchange energy and materials with the environment.
- ◆ Enzymes control chemical reactions such as those active in photosynthesis, reproduction, metabolism, and growth.
- ◆ Physical constraints limit cell size.
- ◆ Mitotic cell division is a small part of the cell cycle, which is a continuous sequence of events in the life of a cell.

KNOWLEDGE CHECK
- ◆ What are the two basic types of cells?
- ◆ What is the nucleus, and what is its importance to a cell?
- ◆ What is the importance of the plasma membrane?
- ◆ What is the cell cycle?
- ◆ How is a cell's genetic material passed on to offspring?

Guidepost
What features do cells have in common?

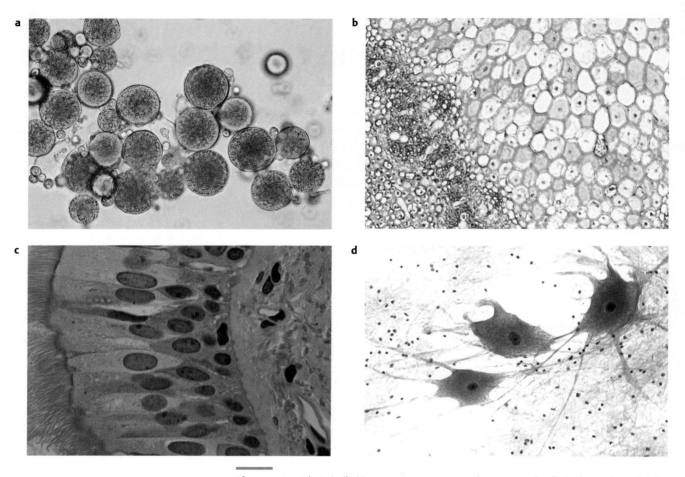

Figure 5.1 What similarities can you see among the groups of cells in these photos? (a) Unicellular green alga, (b) cross-section of a young maple stem, (c) cells from the trachea, (d) human nerve cells.

theory, has undergone modifications through the years but can be stated generally as follows:

1. Cells, or products made by cells, are the units of structure and function in all organisms.
2. All cells come from preexisting cells.

Like other theories in science, the cell theory is a model that explains many observations and predicts the outcomes of future observations. Recall that a hypothesis is a testable statement that explains an observation. A theory, on the other hand, is a model that explains a great many observations and predicts the outcomes of future observations. Theories that serve as models in science, such as the cell theory, have been supported repeatedly by data resulting from testable hypotheses. In addition, almost no data opposing these theories have been found. Any theory, however, no matter how well established it has become, can be changed or discarded if new data do not support the model. All the theories that serve as models in science have been modified as new research has been performed and new information is found.

5.2   Biologists Use Microscopes to Study Cells

Biologists began to study cell structure and function in detail as the cell theory developed. To do so, they used different tools and techniques. The light microscope, based on microscopes invented hundreds of years ago,

Make certain students understand the scientific use of the term *theory* and do not confuse it with the lay definition of "a guess." You may want to review the definition of *hypothesis* in Section 1.7 to ensure that students can distinguish between a hypothesis and a theory. Avoid the thinking that once a hypothesis gains enough support it is "elevated" to a theory. A hypothesis is a potential answer to a single question and can be tested; a theory is an organizing model based on data from many tested hypotheses.

Biology Today

The Cell Theory

Much of the biological progress that occurred in the seventeenth century depended on the development of glass lenses. By 1650, the art of grinding and polishing pieces of glass into lenses had so developed that it became possible to build good microscopes. Anton van Leeuwenhoek (1632–1723), a Dutch civil servant, was a master lens maker. Leeuwenhoek placed his lenses into simple microscopes. Using these microscopes, Leeuwenhoek discovered an amazing, invisible world. For 50 years, he described and made careful drawings of bacteria, detailed structures in small insects, and even sperm cells from humans, dogs, frogs, and insects.

In the late 1600s, an Englishman named Robert Hooke also made microscopes and used them to examine small objects. When Hooke examined the cork layer of bark from an oak tree, he observed rows of compartments. The compartments reminded him of the small cells in which monks lived in medieval monasteries. For that reason, he called the compartments cells.

Scientists gradually began to realize that cells are the fundamental units of living organisms. In 1838, two German biologists, Matthias Schleiden, a plant biologist (botanist), and Theodor Schwann, an animal biologist (zoologist), proposed the cell theory. According to the cell theory, all organisms consist of cells and cell products, and one can understand how living creatures are built and how they function if cells can be understood. Until the development of the cell theory, the emphasis had been on the cell walls. In looking at animal cells, however,

Schwann and others had been unable to find the boxlike cells seen in plants. Schwann interpreted his observations in a new way, emphasizing what was inside the "box," rather than the box itself. Further studies showed that certain structures are common to plant and animal cells as well as those of microorganisms. Once the contents of cells began to be studied, ideas about the origins of organisms began to change.

For hundreds of years, people had thought that organisms could come from nonliving matter, an idea known as spontaneous generation. Several different studies disproved the idea. In 1855, Rudolph Virchow, a physician and biologist, proposed that all cells produce more cells through time, but his idea was not accepted by everyone. In 1864, Louis Pasteur, a French scientist working with yeast cells, demonstrated that microorganisms cannot arise from completely nonliving matter. By the 1880s, the work of French and German scientists showed how cells divide and produce more cells. Organisms, therefore, come from existing organisms, and their cells arise from existing cells.

Today, the cell theory is summarized in two main ideas: (1) cells are the units of structure and function in living organisms, and (2) all new cells come from cells that already exist. New technologies have made possible extremely detailed studies of cell structure and function, and although there are conflicting ideas about how cells first came to be, the cell theory is still one of the major unifying themes of biology.

a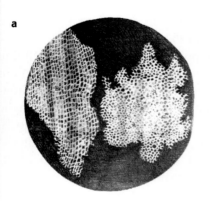

b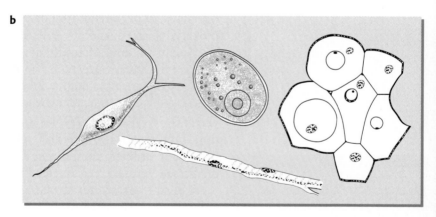

(a) Cork cells as drawn by Hooke, and (b) animal cells as drawn by Schwann.

The development of the cell theory illustrates how increased understanding frequently depends on improved technology. Use of dyes and of phase-contrast and electron microscopes has revealed increasingly detailed information about cells.

You may want to discuss the two types of electron microscopes and display appropriate micrographs. Point out the main difference in the two types. The transmission electron microscope (TEM) sends electrons through thinly sliced specimens, revealing internal structures. It cannot be used to observe living organisms. The scanning electron microscope (SEM) bounces electrons off the surface of an otherwise undisturbed, sometimes living, specimen.

Alert your students that many new terms are introduced in the discussion of cell physiology. Familiarity with these new terms is important at this time as well as in future discussions.

Figure 5.2 Cells from a human cheek lining as seen through (a) an ordinary light microscope (X400) and (b) phase contrast microscope (X1000).

a

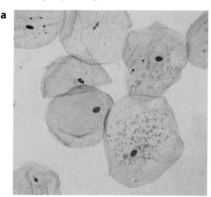

b

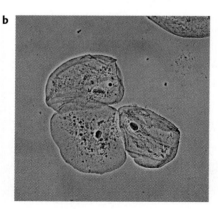

still is used in laboratories for many basic tasks. With a light microscope, light waves pass through a small organism, or thin slices of a larger organism, and the structures are magnified through the lens system.

Tiny organisms, such as those found in a drop of pond water, can be studied in the living state. Colonies of bacteria and many other organisms used in studies of inheritance can be identified or counted using a low 3 power light microscope. Light microscopes with fine quality lenses can magnify structures up to 1500 times.

Because most cells are nearly transparent, dyes often are used to stain structures within a cell. The structures then can be studied through a light microscope. Dyes, however, may kill or distort cells. To study the smallest 2 structures and events in living cells, biologists must use other methods and instruments, such as the phase-contrast microscope.

The phase-contrast microscope is a special type of light microscope that increases the differences in light and dark areas, thus making cell structures more visible. This instrument makes it possible to study living cells and events that occur within them, such as cell division. Figure 5.2 shows the same cells as seen under an ordinary light microscope and as seen through a phase-contrast microscope. 3

Electron microscopes show cell parts at very high magnifications. They use an electron beam rather than a light beam to illuminate the object. Modern transmission electron microscopes can enlarge cell structures as much as one million times and provide photographs that show remarkable detail (see Figures 5.7 through 5.11). Scanning electron microscopes allow detailed observations of the surface of biological objects, such as the coleus leaf in Figure 5.3.

5.3   Cells Are of Two Basic Types

Profound differences in the structure of their cells separate all known living organisms into two groups: **prokaryotes** (pro KAIR ee ohtz) and **eukaryotes** (yoo KAIR ee ohtz). Cells of prokaryotes—bacteria—do not have a membrane enclosing their DNA. Cells of eukaryotes—plants, animals, and other familiar organisms—usually have at least one membrane-enclosed structure, the **nucleus,** which contains DNA.

Besides a nucleus, eukaryotic cells also have other membrane-enclosed structures, or **organelles** (or guh NELS), that prokaryote lack. Compare the prokaryotic and eukaryotic cells shown in Figure 5.4.

Scientists think that prokaryote represent the earliest type of living 4 cell, yet they are everywhere—in the soil, the air, and the water, and in or on every organism, including humans. They also inhabit extreme environments such as salt flats, hot springs, and thermal vents located deep in the ocean floor. Most prokaryotes are microscopic in size, averaging 1 to 10 µm (micrometers) in length. Up to 35 000 could fit side by side on the head of an ordinary thumbtack, and even the point of a pin can hold hundreds (see Figure 5.5). Chapter 11 describes prokaryote in more detail.

Cells of eukaryotes are larger (10 to 100 µm) and structurally more complex than those of prokaryote. The remainder of this chapter deals with the structure and function of eukaryotic cells.

5.4   Membranes Organize Eukaryotic Cells

Each type of cell found in a multicellular organism has a specific role in making life possible for the organism in a changing environment. In a large animal such as a human, there are at least 200 different types of

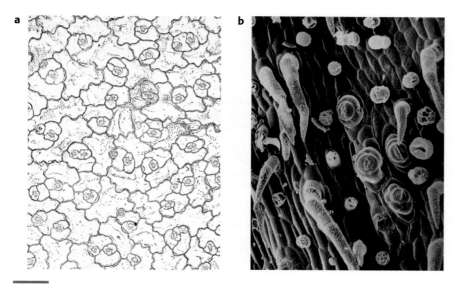

Figure 5.3 The surface of a coleus leaf, as photographed through (a) a light microscope (X25) and through (b) a scanning electron microscope (X161 000).

cells. The many types of cells have a similar basic pattern. Depending on their roles, however, they may differ in some details. Figure 5.6 shows the basic structures in most plant and animal cells. Refer to this illustration as you read the descriptions of these structures.

Every cell has an outer membrane known as the **plasma membrane.** The plasma membrane encloses the cell's contents. It is a thin but active structure that controls the passage of materials into and out of a cell. The

5 plasma membrane is made of two thin layers of lipid molecules arranged in a definite pattern (see Figure 5.6a). Protein molecules on and within the lipid layers perform special functions, such as helping molecules to move in and out of the cell and receiving chemical messages. These proteins and their functions vary in different types of cells.

The cell contents outside the nucleus are known as the **cytoplasm** (SYT oh plaz um). Extending throughout the cytoplasm is a complex network of

An adult human is composed of about 60 billion (6×10^{12}) cells.

The plasma membrane is only 4 to 5 nanometers (nm) thick and, therefore, too thin to study with a light microscope. Early researchers hypothesized its presence, but its existence was confirmed only with the advent of the electron microscope.

Figure 5.4 Photomicrograph of (a) eukaryotic plant cells in a pine needle and (b) prokaryotic cell. Notice generally more complex structure in the eukaryotic cell.

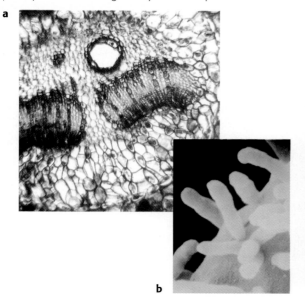

Figure 5.5 Prokaryotes are extremely small. Many prokaryotes are visible in this scanning electron micrograph of a finger print ridge magnified 47,000 times.

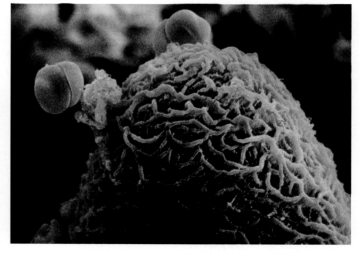

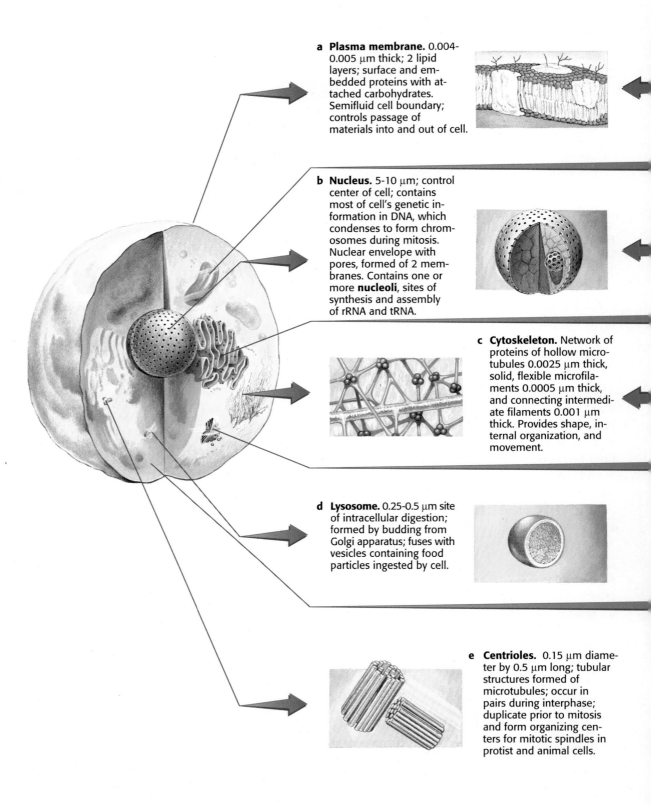

a Plasma membrane. 0.004-0.005 μm thick; 2 lipid layers; surface and embedded proteins with attached carbohydrates. Semifluid cell boundary; controls passage of materials into and out of cell.

b Nucleus. 5-10 μm; control center of cell; contains most of cell's genetic information in DNA, which condenses to form chromosomes during mitosis. Nuclear envelope with pores, formed of 2 membranes. Contains one or more **nucleoli**, sites of synthesis and assembly of rRNA and tRNA.

c Cytoskeleton. Network of proteins of hollow microtubules 0.0025 μm thick, solid, flexible microfilaments 0.0005 μm thick, and connecting intermediate filaments 0.001 μm thick. Provides shape, internal organization, and movement.

d Lysosome. 0.25-0.5 μm site of intracellular digestion; formed by budding from Golgi apparatus; fuses with vesicles containing food particles ingested by cell.

e Centrioles. 0.15 μm diameter by 0.5 μm long; tubular structures formed of microtubules; occur in pairs during interphase; duplicate prior to mitosis and form organizing centers for mitotic spindles in protist and animal cells.

Figure 5.6   Generalized animal (left) and plant (right) cells, with enlargements of the major organelles. Some of the terms used here are not explained in this chapter. They are defined in later chapters.

f Cell wall. 0.1-10 μm thick; formed by living plant cells; made of cellulose fibers embedded in a matrix of protein and poly-saccharides; provides ri-gidity to plant cells and allows for development of turgor pressure.

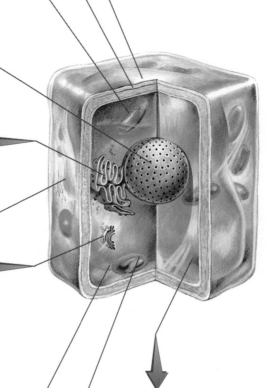

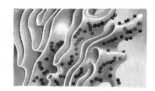

g Mitochondrion. 2-10 μm long by 0.5-1 μm thick; enclosed in double mem-brane; inner membrane much folded; most re-actions of cellular respiration occur in mitochondrion; contains small amounts of DNA and RNA; may be several hundred per cell.

h Endoplasmic reticulum (ER). 0.005 μm diameter; tubular membrane system that compartmentalizes the cytosol; plays a central role in biosynthesis re-actions. Rough ER is studded with **ribosomes**, the site of protein syn-thesis; smooth ER lacks ribosomes.

i Golgi apparatus. 1 μm diameter; system of flat-tened sacs that modifies, sorts, and packages macromolecules in vesicles for secretion or for delivery to other organelles.

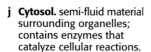

j Cytosol. semi-fluid material surrounding organelles; contains enzymes that catalyze cellular reactions.

k Chloroplast. 5 μm long by 0.5-1 μm thick; enclosed by double membrane; third membrane system forms thylakoids in which light-absorbing pigments are embedded. All reactions of photosynthesis occur in chloroplasts.

l Vacuole. Variable size; large vesicle enclosed in single membrane; may occupy more than 50% of volume in plant cells; contains water and diges-tive enzymes; stores nutri-ents and waste products.

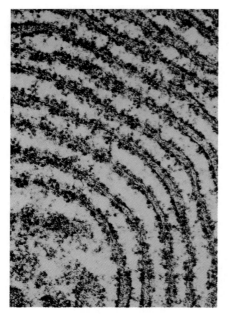

Figure 5.7 Transmission electron micrograph of rough endoplasmic reticulum in a liver cell (X40 800). What are the round structures along the membranes?

Chloroplasts and mitochondria also contain genes that direct synthesis of some of their proteins. In fact, the presence of genes acts as evidence for the bacterial origin of these organelles.

membranes that form flattened sheets, sacs, and tubes. Figures 5.6h and 5.7 show a portion of this large network, which is called the **endoplasmic reticulum** (en doh PLAZ mik reh TIK yoo lum), or ER. The ER specializes in the synthesis and transport of lipids and membrane proteins. Still other membranes surround the organelles, which often have their own internal membranes and may be very complex.

Membranes organized as stacks of flattened sacs form the **Golgi** (GOHLjee) **apparatus** (Figures 5.6i and 5.8). This organelle helps to package cell products for export from the cell. Materials synthesized in the ER are then transferred to tiny saclike structures, or vesicles, that are formed by the membranes of the Golgi apparatus. The vesicles move to the plasma membrane and release their products to the outside of the cell.

Clearly, membranes serve an important purpose in cells. The eukaryotic cell may be viewed as a highly organized system divided into distinct parts by membranes. Within each part, a special job is done. Molecules pass back and forth across the membranes within a cell and through the membrane that separates the cell from its outside environment.

5.5   Eukaryotic Cells Contain Various Organelles

In many cells, the most obvious organelle is the nucleus (Figures 5.6b and 5.9), a rounded body surrounded by a double membrane called the **nuclear envelope.** The nucleus contains most of the **genes,** the units of hereditary information that control the basic functions of the cell. For this reason, the nucleus is considered the control center of the cell. Genes are composed of DNA nucleotides. These nucleotides contain information that tells the cell how to function and directs the synthesis of protein molecules. Long strands of nucleotides wrapped around protein molecules form **chromosomes** (KROH moh sohmz), which become visible only during nuclear division.

The powerhouses of the cell are the **mitochondria** (my toh KON dree uh; singular, mitochondrion). They not only help to release energy from food molecules but also help to synthesize building blocks of larger molecules. The chemical reactions that occur in mitochondria provide new supplies of ATP (see Section 4.6), the energy molecules of the cell. The

Figure 5.8 Transmission electron micrograph of a Golgi apparatus (X21 930). Part of the nucleus can be seen at lower left. Note the vesicles at the right.

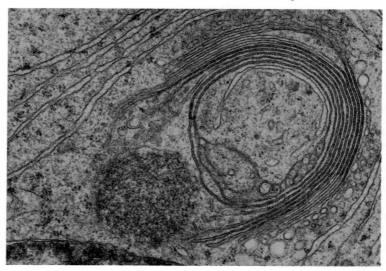

6 ATP then is used for all types of cell activities. Mitochondria are especially abundant in the parts of an organism that have high energy requirements, such as a muscle in an animal or a growing root tip of a plant. Although they may have different shapes in different cells, all mitochondria consist of two layers of membranes—an outer membrane and a folded inner membrane—as shown in Figure 5.6g.

Along the membranes of the ER may be found tiny bodies called **ribosomes** (RY boh sohmz). Proteins are manufactured on the ribosomes, which are visible in the electron micrograph in Figure 5.7.

Two other types of organelles are common in animal cells. **Lysosomes** (LY soh zohmz), shown in Figure 5.6d, are vesicles containing a variety of digestive enzymes that help break down large molecules and worn-out cell parts. **Centrioles** (SEN tree ohlz) are paired cylindrical structures (Figure 5.6c) that play an important role in cell division.

Many plant cells have **chloroplasts** (KLOR oh plasts). These organelles contain chlorophyll, a pigment that is essential for capturing the energy in sunlight. The chloroplasts (Figure 5.6k), which are the sites of photosynthesis, have a complex arrangement of membranes. The energy captured from the sun and stored in the structure of biological molecules during photosynthesis is released later in the mitochondria. Most plant cells also contain one to several **vacuoles** (VAK yoo ohlz), fluid-filled structures that are surrounded by single membranes (Figure 5.6l). Vacuoles contain enzymes, store nutrients and waste products, and regulate the water content of cells. They may occupy as much as 95 percent of an older cell's volume.

All the cytoplasmic membranes and organelles are bathed in a gelatin like substance, the cytosol (SYT oh sol), which is illustrated in Figure 5.6j. Many of the chemical reactions of a cell take place in the cytosol. A complex network of ultrafine protein filaments within the cytosol forms an internal skeleton, the **cytoskeleton** (Figure 5.6c). Constructed of hollow microtubules and solid rods that can be assembled and disassembled rapidly, the cytoskeleton gives the cell its distinct shape and its capacity for movement. The cytoskeleton also may provide internal organization and a framework for the attachment of various enzymes and organelles within the cytosol.

Some types of cells have still other types of organelles, and not all cell structures lie within the plasma membrane. In plant cells and in some other cells, there is a cell wall (see Figure 5.6f) outside the plasma membrane. The cell wall is made of cellulose and other materials that the cell secretes and exports through the Golgi apparatus. The cell wall provides strength and protection. Often when the cell dies, the cell wall remains behind. When Robert Hooke looked at thin slices of cork under his microscope (see Biology Today), he actually was observing the empty cell walls in the bark of a cork oak. For early biologists, the presence of cell walls made plant cells easier to see than animal cells. Most of a large tree's trunk, for example, consists of nonliving cell walls. The living cells of the tree are found in a thin band just underneath the bark of the trunk.

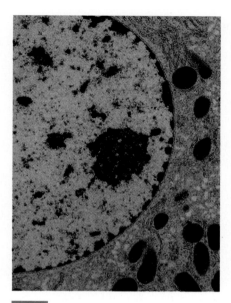

Figure 5.9   Transmission electron micrograph of a mouse liver cell nucleus (X51 300). Note the double membrane that makes up the nuclear envelope. The large dark round body is called the nucleolus.

CONCEPT REVIEW

1. What are the main ideas of the cell theory?
2. How did the use of stains advance the study of cells?
3. Compare the advantages of the light, phase-contrast, and electron microscopes.

4. Describe the major differences between prokaryotic and eukaryotic cells.
5. Describe the structure and function of the plasma membrane.
6. Discuss the functions of three cell organelles.
7. How has the meaning of the word cell changed since the time of Robert Hooke?

Cell Functions

5.6 Cell Activities Require Energy

All the different parts of a cell are important to the life of the cell. Each part has a particular role to play, but it is not always possible to tell what that role is just by looking at the part. If all the cogs, wheels, springs, and bearings of a watch were spread on a table, you probably could not understand exactly how the watch worked just by looking at the parts.

Biologists are in much the same situation. They can describe the various parts of a cell and even separate them, but they cannot always figure out how the parts all fit together or how they enable the cell to work. That is the challenge of modern biology. In trying to find out, it usually is more useful to study living cells than nonliving ones.

Scientists, for example, cannot see cells use energy. Instead, it is inferred from what they observe happening in living cells. All cells are able to release energy from complex molecules. The energy in a sugar molecule is released by the mitochondria in small steps. After a cell has taken in energy-rich molecules that can be used as fuel, the molecules are channeled into pathways in the mitochondria where the energy is released. The more active a cell is, the more mitochondria it is likely to contain.

Energy makes it possible for chemical reactions to occur in cells. The **1** sum of all the chemical reactions in a cell or organism is known as its **metabolism** (meh TAB oh liz um). As a part of a cell's or organism's metabolism, biologically important molecules are processed, energy is transferred to ATP, and waste materials are released. Excess carbon atoms, **2** which are released as carbon dioxide (CO_2) are a part of these waste materials. In animals, waste materials include extra nitrogen atoms that can be released in nitrogen-containing molecules such as ammonia (NH_3). These wastes must be removed because they can become toxic to the cell.

Not all energy is used immediately for metabolism. Some cells, for example the cells in green plants that contain chloroplasts, store energy in complex molecules such as sugars. The sugar molecules then can be transformed into other types of molecules such as amino acids, lipids, or nucleotides. These molecules are used to maintain the life of the cell, for growth, and to help build new living substances.

5.7 Substances Enter and Leave Cells by Diffusion

Atoms, molecules, and small particles are in constant motion. Although they may move in any direction, in general, molecules tend to move from an area where they are more concentrated to an area where they are less concentrated, until their concentration is the same everywhere (Figure 5.10). This movement is called **diffusion** (dih FYOO zhun). The concentrations of molecules at various points between the high and the **3** low areas form a gradient, which is known as the concentration gradient.

Place a drop of oil of cloves or oil of peppermint in one corner of the classroom and elicit student reactions from various parts of the room.

Figure 5.10 Diffusion is the movement of molecules from an area of high concentration to an area of low concentration until the concentration is the same everywhere. In (a), a crystal of potassium permanganate is dropped into a glass of water. The molecules diffuse through the water (b) until they are evenly distributed throughout (c).

a

b

c

Because molecules diffuse from areas of higher concentration to areas of lower concentration, they are said to move down the concentration gradient. Many substances can move in and out of cells by diffusion alone. Once a substance has passed into a cell through the plasma membrane, it can continue to diffuse throughout the cytosol down its concentration gradient. Thus, diffusion aids in the even distribution of materials within a cell.

Water moves in and out of cells and diffuses down its concentration gradient in the same manner as other substances. The diffusion of water through membranes is called **osmosis** (os MOH sis). For example, suppose a living cell is placed in a salty solution. If the concentration of water molecules is lower in the solution than it is in the cell, water diffuses out of the cell (from an area of higher concentration in the cell to an area of lower concentration), and the cell shrinks (see Figure 5.11a).

If the concentration of water molecules is greater outside a cell, water diffuses inward, and the cell swells (Figure 5.11b). When an animal cell is placed in pure water, osmosis usually causes it to burst. A plant cell swells but does not burst because of the rigid cell walls. (This swelling enables well-watered plants to stand upright and explains why many plants wilt and droop when their cells have lost water.) Figure 5.12 shows how animal and plant cells react to changes in water concentrations as a result of osmesis.

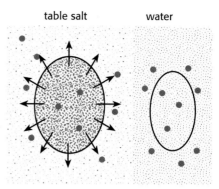

table salt   water

a   if concentration of water is higher inside the cell, water diffuses out of the cell

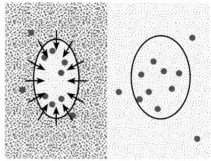

b   if concentration of water is higher outside the cell, water diffuses into the cell

Figure 5.11   A diagram of osmosis. When water concentration is higher inside the cell (a), water moves out of the cell, and the cell shrinks. When water concentration is higher outside the cell (b), water moves into the cell, and the cell swells. Does this diagram help you understand why supermarkets spray their produce with fresh water?

Figure 5.12   An animal cell (red blood cell) and a plant cell in different solutions. An isosmotic solution contains the same concentrations of dissolved material and water as the cell. a. if concentration of water higher inside the cell, water diffuses out of the cell b. if concentration of water is higher outside the cell, water diffuses into the cell

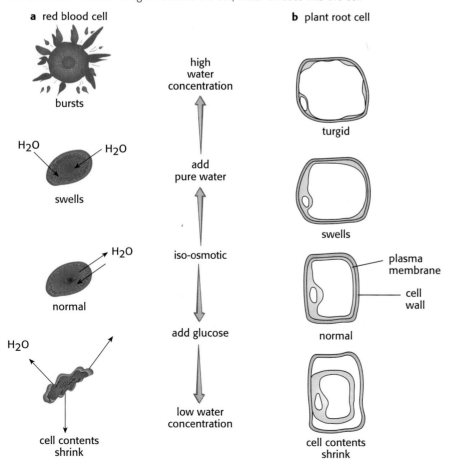

a red blood cell

bursts

H_2O H_2O

swells

H_2O

normal

H_2O

cell contents shrink

high water concentration

add pure water

iso-osmotic

add glucose

low water concentration

b plant root cell

turgid

swells

plasma membrane

cell wall

normal

cell contents shrink

Students may have difficulty understanding the idea of *water concentration*. Point out that a 5% salt solution is 95% water.

You can demonstrate osmosis by placing lettuce leaves in salt water and fresh water. Another example students may be familiar with is the Venus's flytrap. A change in water concentration causes the leaf to spring shut.

5.8   Cells Move Substances in a Variety of Ways

The cell wall, if it is present, allows free diffusion of most substances. The plasma membrane, however, is differentially permeable. This means that only certain molecules can diffuse through it freely. A small glucose molecule, for example, may pass through the membrane easily, but a **5** large protein molecule cannot. Membranes also may be impermeable to particular substances, which cannot pass through at all. Some substances are too large to pass through a membrane. Others have an electrical charge that makes diffusion impossible because the plasma membrane also is electrically charged. These substances enter or are released from a cell in other ways.

Figure 5.13 illustrates two additional methods by which cells transport substances across membranes. Both methods require special transport proteins in the membrane. In **passive transport,** no energy is expended by the cell. Transport proteins move substances down their concentration gradient. Each protein can transport only a specific substance. Cells transport many essential ions by this means. In **active transport**, energy in the form of ATP is used to move substances through the transport proteins. By means of active transport, substances can move **6** across a membrane *against* concentration gradient. For example, the soil around a plant may contain low amounts of elements necessary for the plant's growth. Ordinary diffusion would not provide the plant with enough of these elements. By means of active transport and other processes, however, the plant's root cells can accumulate these elements in relatively high amounts. The elements then can be transported to all parts of the plant.

In another transport mechanism, the plasma membrane may fold inward, forming a tiny "pocket" that fills with fluid and particles from the cell's surroundings (see Figure 5.14a). The membrane then closes over the pocket, forming a vesicle that is released into the cell. In this manner, large molecules and particles from the surroundings can be incorporated into the cell. This activity can occur in the reverse direction to expel materials from the cell, as shown in Figure 5.14b.

Figure 5.13   In passive transport (a), substances move through membrane proteins down their concentration gradient. Active transport (b) requires the use of ATP energy as well as membrane proteins. Active transport can move substances against their concentration gradient as well as with their concentration gradient.

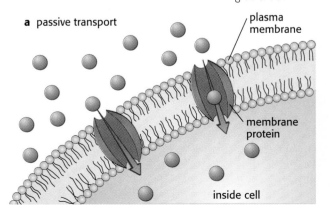

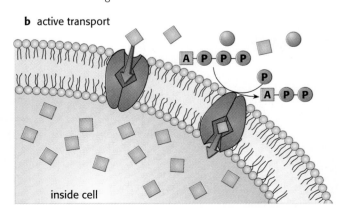

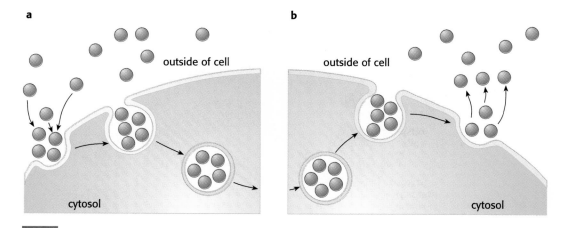

Figure 5.14   The transport of masses of material into a cell and out of a cell. These processes require energy.

CONCEPT REVIEW

1. What activities are involved in metabolism?
2. Why do cells constantly take in and get rid of substances?
3. What is a concentration gradient? How does it affect diffusion?
4. How does osmosis affect an animal cell? How does it affect a plant cell?
5. Why are membranes said to be differentially permeable?
6. Distinguish between passive and active transport. How is each important to a cell or organelle?

Cell Reproduction

5.9   The Cell's Life Is a Cycle

Although diffusion supplies cells with needed materials and helps to remove waste products, it takes a long time for substances to diffuse great distances. Therefore if a cell is to function well, it cannot be too large. Usually, when a cell reaches a certain size, it begins a series of changes that permit it to divide into two cells.

The life of a eukaryotic cell consists of a continuous sequence of events called the **cell cycle** (see Figure 5.15 on page 112). The cell cycle starts with the formation of a new cell and continues until the cell itself has divided into two new cells. Each new cell then begins the cell cycle anew. The dramatic and visible events of cell division are only a brief portion of the cell cycle. A cell spends most of its life between divisions in a stage known as **interphase** (IN ter fayz). Although a cell may *appear* to be inactive in interphase, chemically it is very active indeed. During interphase, all normal metabolic activities take place and, in addition to these activities, intricate preparations for division occur in a precise sequence.

New cells usually contain the same structures as their parent cells. That means the chromosomes, the organelles in the cytosol, and the plasma membrane must be duplicated before a cell divides to form two new cells. If these structures were not duplicated, the offspring cells would be incomplete and unable to survive. Preparations for division take place in

Guidepost

Why is it difficult to study an organism in isolation from its surroundings?

This discussion applies only to cells that divide. Most plant cells (leaf, stem, and root) and some animal cells (nerve and muscle) never divide; they remain in an extended G$_1$ phase often called G$_0$. These cells are quite active and serve important functions in the organisms.

Figure 5.15 The cell cycle includes cell division and interphase. Most of a cell's life is spent in the various phases of interphase.

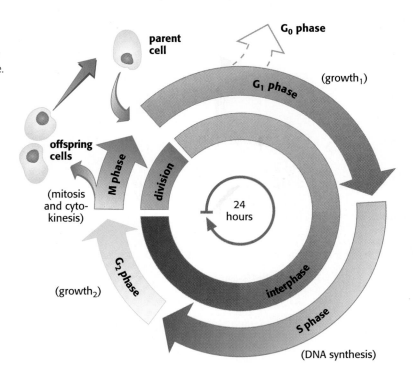

the nucleus and the cytosol in distinct phases. The events of the cell cycle occur in fixed order, and each successive step depends on a preceding one.

Just after division, the cell enters the G_1 (Growth 1) phase. During this phase, the cell grows and is metabolically active. New proteins and RNA are made, and other molecules needed for new cells are gathered. Organelles may be duplicated. All these activities, which take place in the cytosol, require energy, raw materials, and enzymes.

Next, the cell enters the S (Synthesis) phase. During this phase, the DNA content of the cell doubles. In the doubling process, the genetic information is copied exactly. This exact copying, or replication, of the **3** DNA ensures that a complete and identical set of genes will be available for each new offspring cell formed during cell division.

Think of this genetic information, which is replicated in each cell generation, as a program that the cell can call on as needed. In a multicellular organism, not all the genetic information is needed or used by every cell, but all of it is replicated in all cells. In unicellular organisms, each new cell generation needs all the information.

When the S phase is complete, the cell enters G_2 (Growth 2), another phase of growth and metabolism. The cell again synthesizes protein, RNA, and other large molecules but often in smaller amounts than in the **2** G_1 phase. A cell in G_2 contains twice as much DNA in its nucleus as a cell in G_1. During G_2, the cell also synthesizes a substance that triggers the beginning of cell division.

During the M (Mitosis) phase, which occurs after G_2, the nucleus reproduces through a series of events called **mitosis** (my TOH sis; from the Greek *mitos* meaning thread). During this phase, the chromosomes, which usually are stretched out in a threadlike network, become visible with the use of a light microscope. Toward the end of mitosis, the cytosol, with its organelles, also divides. This part of cell division is called **cytokinesis** (syt oh kih NEE sis; from the Greek *cyto,* meaning cell, and *kinesis,* meaning movement).

Thus during its life, a cell passes through two main stages: interphase and cell division. In interphase (G_1, S, G_2), cells carry out normal

So much attention is given the process of nuclear division that its name, mitosis, often is incorrectly applied to the whole process of cell division.

metabolic activities and preparations for division. During the preparatory phases, the chromosomes, organelles, and cytosol are very active. After a complex series of events, cell division results in two offspring cells. These two new cells then enter interphase again and resume their metabolic activities, including preparations for division.

5.10   The Events of Mitosis Occur as a Continuous Process

Although mitosis is a continuous process, it is convenient to freeze the action at intervals and to describe the process as a series of stages. This is similar to looking at a single frame in a motion picture film. The photographs in Figure 5.16 show the stages of mitosis in plant cells. Refer to the photographs as you read the discussion that follows.

Because mitosis is a continuous process, it can be illustrated adequately only by motion pictures. Make every effort to obtain such a film. The BSCS inquiry film, *Mitosis*, is suggested.

It may help students remember the names of mitotic stages if they know that *pro* = first; *meta* = change; *ana* = up; and *telo* = end.

Figure 5.16   Stages of mitosis in the blood lily, *Haemanthus:* (a) interphase, (b) prophase, (c) late prophase, showing disintergration of the nuclear envelope, (d) early metaphase, (e) metaphase, (f) early anaphase, (g) late anaphase, (h) early telophase, (i) late telophase showing formation of the cell plate.

a
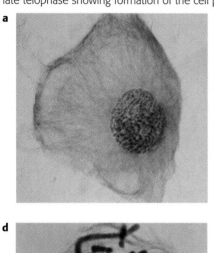

b
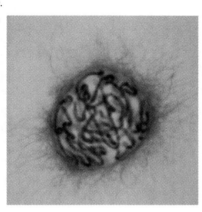

c
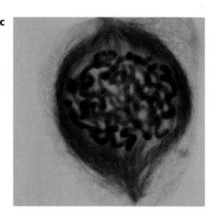

d
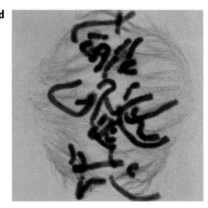

e

f
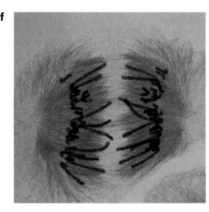

g
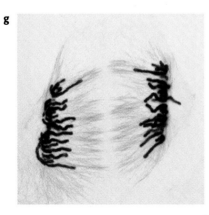

h
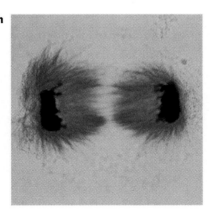

i
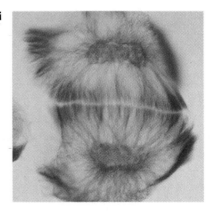

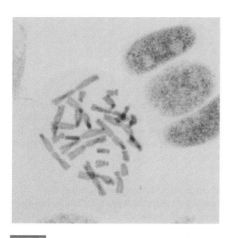

Figure 5.17 Chromosomes in a root tip cell of the hyacinth, *Hyacinthus orientalis*. What indicates that replication already has taken place?

During the first stage of mitosis, prophase (see Figure 5.16b, c), the chromosome strands slowly begin to coil like tiny springs. As they coil, they become shorter and thicker. The nuclear envelope begins to disappear. At this time, the chromosomes barely are visible under a microscope. Careful examination shows that each chromosome is double (see Figure 5.17) having replicated during the S (Synthesis) phase. A framework of fibers, known as the spindle apparatus (shown in Figure 5.16d), forms within the cytoskeleton of the cell. The spindle apparatus determines the direction of the cell's division. Some of the spindle fibers attach themselves to the chromosomes.   **4**

During the next stage of mitosis, metaphase (Figure 5.16d, e), the chromosomes move to the center of the cell, perpendicular to the spindle. As anaphase begins (Figure 5.16f, g), the chromosome strands separate. Pulled by the spindle fibers in opposite directions toward the two poles of the dividing cell, one strand of each replicated chromosome moves toward either pole.

Now the final events of mitosis occur. During telophase (Figure 5.16h, i), the chromosomes gather at opposite ends of the cell, and a new nuclear envelope forms around each group of chromosomes. Thus, two new nuclei are formed. Each nucleus contains a set of chromosomes identical to those   **6** of the parent cell.

Meanwhile, the other organelles, which also duplicated during interphase, segregate. The cytoplasm divides into two parts, more or less equal in size. In plant cells, a cell plate (Figure 5.16i) begins to form across the middle of the cell, dividing the cell in two. This is the developing cell wall.

Mitotic cell division in animal cells is similar to cell division in plants.   **5** Animal cells, however, unlike plant cells, contain a pair of centrioles Figure 5.7e that duplicate as mitosis begins. One pair gradually moves toward the opposite side of the cell from the other pair, as shown in Figure 5.18. The pairs of centrioles act as the poles of the cell. Later in prophase as the nuclear envelope breaks down, the spindle apparatus begins to form between the pairs of centrioles. No cell plate forms in animal cells. Instead, the cell constricts (see Figure 5.19) across the middle as cytokinesis begins. A new plasma membrane forms across the constricted portion of the cell. Table 5.1 summarizes the events in the different phases of the cell cycle.

Figure 5.18 Before mitosis begins, the centrioles duplicate and move to opposite poles of the cell.

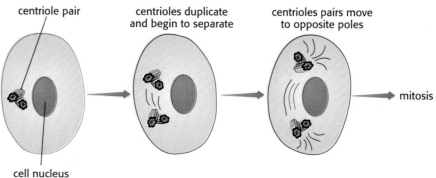

Table 5.1 Events in the Cell Cycle

Phase		Events	Cell Part Affected
Interphase	G₁ (Growth 1)	Cell makes new proteins, RNA, and other molecules. Organelles may be duplicated. Energy, materials, and enzymes used.	Cytoplasm
	S (Synthesis)	Cell duplicates its DNA. 2 complete, identical sets result. In animals, centrioles duplicate.	Nucleus Cytoplasm
	G₂ (Growth 2)	Similar to G₁, but molecules often produced in smaller amounts. Cell synthesizes substance that triggers the beginning of mitosis.	Cytoplasm
Mitosis and Cytokinesis	Prophase	Chromosomes condense, nuclear envelope disappears. Spindle forms, spindle fibers attach to chromosomes.	Nucleus Cytoplasm
	Metaphase	Chromosomes move to center of cell.	Nucleus
	Anaphase	Chromosome strands pulled toward poles.	Nucleus Cytoplasm
	Telophase	Chromosomes gather at poles, nuclear envelopes form. Division of cytoplasm, cell plate formed (in plants). Cytokinesis, new plasma membrane.	Cytoplasm New nuclei Cytoplasm

5.11 Cells Become Specialized in Development

Complex organisms, such as yourself, develop from the division of a single cell (see Chapter 7). During development, a human may form about 200 different types of cells, each type having a specialized function. As cells become specialized, they take on specific shapes, and their divisions may become more distinctive. For example, different types of cells may require different amounts of time to complete a cell cycle. Some cells divide about every eight hours. Others divide only once every several months. Certain types of cells—such as muscle and nerve cells—almost never divide in adults. In plants, cell division is confined to specialized cells in the tips of stems, branches, and roots, and in bark.

Almost all normal animal cells undergo a certain limited number of divisions—usually no more than 50. There appears to be some type of brake on cell division, although scientists do not know why. Some cells seem to lose control, however, and divide rapidly and abnormally, often at the expense of the organism. These are cancer cells, shown in Figure 5.20. Cancer cells grow and divide uncontrollably. If their growth cannot be controlled or the cells are not removed in time, they may cause the death of the entire organism. In many ways, cancer cells act as rejuvenated, undifferentiated cells. If we can learn why and how cells differentiate, a

Figure 5.19 A scanning electron micrograph of a dividing animal cell (X42).

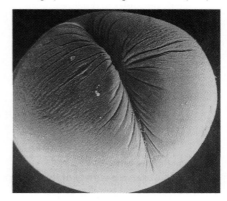

Students have no difficulty recognizing the distinct differences among cells in various tissues and concluding that differentiation is a reality. They are less likely to see that the characteristics of a cell are to be expected in its offspring cells. Help your students recognize the paradox.

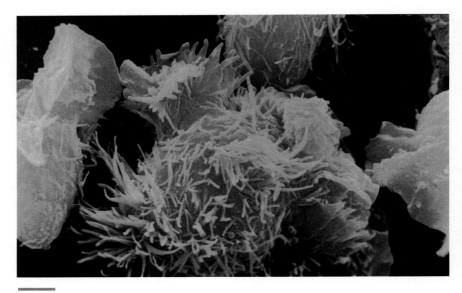

Figure 5.20   Scanning electron micrograph of dividing cancer cells (X6000).

way may be found to stop the cancer cell's process of undifferentiated growth and reproduction.

Because cells live for a limited time, life on the earth can continue only through repeated cell division. The complex events of cell division ensure accurate reproduction of chromosomes and genes. Although accidents do happen, most cells receive complete and correct genetic information. When the processes of cell reproduction are interrupted or interfered with, abnormal cell activities (such as those in cancer cells) result, or the death of the cell occurs.

CONCEPT REVIEW

1. In what way might diffusion limit the size of cells?
2. Describe the activities in each phase of the cell cycle.
3. What does the term replication mean?
4. What is the role of the spindle apparatus in mitosis?
5. In your own words, describe the process of mitosis.
6. What is the biological importance of cell division?

This investigation allows students to observe a variety of living cells and to become familiar with basic cell structures. If students perform the investigation before reading the text material, the activity becomes one of discovery rather than verification. Students can work individually or in teams of 2.

Time: **One period to make the observations and sketches, one-half period for discussion.**

INVESTIGATION 5.1   Observing Cells

Many types of cells have been described and photographed. In this investigation, you will examine some of them.

Materials (per team of 2)

2 pairs of safety goggles	paper towel
2 lab aprons	Lugol's iodine solution in dropping bottle
2 microscope slides	5% salt solution in dropping bottle
2 cover slips	onion
compound microscope	elodea leaf (*Anacharis*)
fine-pointed forceps	prepared slide of stained human cheek
scalpel	cells
dissecting needle	prepared slide of frog blood

<table>
<tr><td>

Put on your safety goggles and lab apron.

</td><td> </td></tr>
</table>

Procedure

1. Separate one layer from an onion quarter and hold it so that the concave surface faces you. Snap it backwards (see Figure 5.21) to separate the transparent, paper thin layer of cells from the outer curve of the scale.

2. Use forceps to peel off a small section of the thin layer and lay it flat on a microscope slide. Discard the rest of the onion piece. Trim the piece with a scalpel if necessary and smooth any wrinkles with a dissecting needle. Add 1 or 2 drops of water and a cover slip.

<table>
<tr><td>

   Scalpel blades and needles are sharp; handle with care.

</td><td></td></tr>
</table>

3. Using your microscope, examine the slide under low power. Look for cell boundaries. Sketch a small part of the field of view to show the shapes and arrangement of the cells.

4. Place a drop of Lugol's iodine solution at one edge of the cover slip. Touch the tip of a piece of paper towel to the opposite edge of the cover slip, as shown in Figure 5.21b. The paper acts as a wick, pulling the Lugol's solution across the slide and into contact with the cells. Record any changes that occur.

Procedure

4. Addition of Lugol's iodine solution should make the nucleus and nucleoli clearer, although they frequently can be observed without any stain.

<table>
<tr><td>

   Lugol's iodine solution is poisonous if ingested, irritating to skin and eyes, and can stain clothing. Should a spill or splash occur, call your teacher immediately; flush the area with water for 15 minutes; rinse mouth with water.

</td><td></td></tr>
</table>

5. Switch to high power and sketch a single cell. Include as much detail as you can see. Save your sketch for reference in Section 5.3 and Section 5.4.

Figure 5.21   Removing the thin layer of cells from a section of onion, and (b) adding salt solution under a cover slip.

a

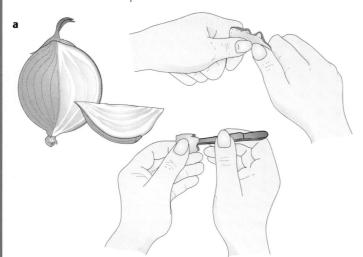

b

7. **From previous microscope use, students should have some concept of depth of focus. If not, illustrate on a chalkboard.**

8. **Students often consider *cyclosis* as evidence of life. (The term is not used in the text. You may wish to introduce it.)**

10. **The salt solution should cause the cytoplasm to draw away from the cell wall. The boundary of the cytoplasm, the plasma membrane, is not visible. Students may need help interpreting their observation.**

11. **Students should note the irregular cell boundary and absence of chloroplast.**

12. **Students may see white blood cells. Provide clarification if necessary.**

Discussion

1. **Most students will be able to identify the cell wall, nucleus, and cytoplasm; some will recognize the chloroplast. With help, they usually can identify the plasma membrane and nucleoli in the onion cells.**

2. **Remind students of limitations in magnification and resolving power. Adjustment of light intensity and special staining procedures may be required to reveal cell structures.**

3. **The rigid cellulose-containing cell walls of most plant cells usually are associated with definite cell boundaries. (During the post laboratory discussion, use either projected prepared slides or photographs to increase your students' awareness of cell variability. Invite students to identify cells shown as either animal or plant cells.)**

4. **The cell wall.**

5. **The solution made some cell parts darker and easier to see. Biologists use stains to make different structures more visible.**

6. Using forceps, remove a young leaf from the tip of an elodea plant. Place it upside down on a clean slide. Add a drop of water and a cover slip.

7. Observe the leaf under low power. By slowly turning the fine adjustment back and forth, count and record the number of cell layers in the leaf.

8. Switch to high power. Select an "average" cell and focus on it. Is there any evidence that the cell is living? If so, what is the evidence?

9. Sketch the leaf cell, including as much detail as you can see. Label any parts you can identify. Keep this sketch for later reference.

10. Place a drop of 5% salt solution at one edge of the cover slip and draw it under the cover slip as you did in Step 4. Observe the cell for any changes and describe what you see. What, if any, structures are evident now that you were not able to see before?

11. Examine a prepared slide of stained human cheek cells. Find several cells separated from others and sketch one or two of them. Include as much detail as you see and label any parts you can identify. What, if any, differences from the onion and elodea cells do you see?

12. Under low power, examine a prepared slide of frog blood. Find an area where the cells are not too crowded and switch to high power. Sketch one or two cells and label any parts you can identify.

13. Wash your hands thoroughly before leaving the laboratory.

Discussion

1. Construct a table in your data book In the first column, list all the types of cells you observed. Head the other columns with the names of cell parts that you identified. Review your sketches and notes. For each type of cell examined, place an X beneath the name of each cell structure observed.

2. Does the lack of an X indicate that the structure was not present? Why or why not?

3. On the basis of your observations, which kind of cell seems to have less-rounded shapes? Using your observations, which kind of cell has more clearly defined boundaries?

4. What structure may be involved in determining a cell's shape?

5. How did the Lugol's iodine solution help? Why do biologists use different stains to study cells?

Laboratory experience with diffusion and osmosis supports the text material about the movement of materials into and out of cells. Terms such as *differential permeable, concentration, gradient osmosis*, and *diffusion* become more real and meaningful, and the concepts learned are useful in many other chapters and investigations. Teams of 2 or 3 are best for this investigation.

   ***Time:* One class period for setup and initial tests, a few minutes for observation the next day, and at least one-half period for discussion of the observations.**

INVESTIGATION 5.2   Diffusion Through a Membrane

How do things get in and out of cells? Using dialysis tubing, you can create a model of a plasma membrane to begin to answer this question.

Materials (per team of 3)

3 pairs of safety goggles	4 10-cm pieces of string
3 lab aprons	15 mL soluble-starch suspension
2 250-mL beakers	15 mL glucose solution
glass-marking pencil	Lugol's iodine solution in dropping bottle
2 15-cm pieces of dialysis tubing	urine glucose test strips

Put on your safety goggles and lab apron.

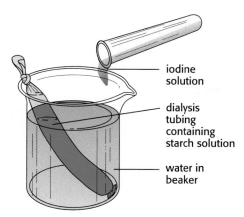

Figure 5.22   Setup for diffusion

- iodine solution
- dialysis tubing containing starch solution
- water in beaker

Procedure

1. Twist one end of a piece of dialysis tubing. Fold the twisted end over and tie it tightly with a piece of string. Prepare the other piece the same way.
2. Pour soluble-starch solution to within 5 cm of the top of one piece of tubing. Twist and tie the end as in Step 1. Rinse the tubing under running water to remove any starch from the outside.
3. Place the tubing in a beaker of water labeled A (see Figure 5.22). Add enough Lugol's iodine solution to give the water a distinct yellowish color.

 Lugol's iodine solution is poisonous if ingested, irritating to skin and eyes, and can stain clothing. Should a spill or splash occur, call your teacher immediately; flush the area with water for 15 minutes; rinse mouth with water.

4. Repeat Step 2 with the second piece of dialysis tubing, using glucose solution instead of the soluble starch. Place this tubing in a beaker of water labeled B.
5. Allow the pieces of tubing to stand for about 20 minutes. Dip a urine glucose test strip into the water in beaker B. Record the color on the strip.
6. Observe the tubing in beaker A. Record any changes, including color, that you see in either the tubing or the water in the beaker.
7. Let beakers A and B stand overnight. Record any changes observed the next day.
8. Wash your hands thoroughly before leaving the laboratory.

Discussion

1. On the basis of the chemical test for starch, what must have happened to the iodine molecules in beaker A?
2. On the basis of the chemical test for glucose, what must have happened to the glucose molecules in beaker B?
3. From the evidence obtained after the beakers stood overnight, what other substances passed through the membrane in beaker B?
4. Which substance did not pass through a membrane? How do you know the substance did not?
5. Physicists can show that all the molecules of a given substance are about the same size but that the molecules of different substances are different in size. Measurements show that iodine molecules and water molecules are very small, glucose molecules are considerably larger, and starch molecules are very large. On the basis of this information, suggest a hypothesis to account for your observations.
6. What assumption did you make about the structure of the dialysis tubing?

Discussion

1. The iodine molecules diffused into the tubing.
2. The glucose molecules diffused into the water.
3. The turgidity of the glucose tubing after 24 hours not only indicates the diffusion of water but also demonstrates osmotic pressure. (Turgor is perhaps the most important biological effect of osmotic pressure, but you may wish to demonstrate how osmotic pressure can support a column of liquid against the force of gravity by using the apparatus shown in Figure T5.1.)
4. If starch passed through the membrane, then the water in beaker A should be blue. (After 24 hours the water in beaker A is usually clear, because most of the iodine has moved into the tubing, where it has bound to the starch molecules.)
5. The simplest hypothesis is that the tubing contains submicroscopic pores large enough to allow molecules of iodine, water, and glucose but not starch to pass through.
6. The dialysis tubing is similar in structure to a plasma membrane.

Figure T5.1   Demonstration setup.

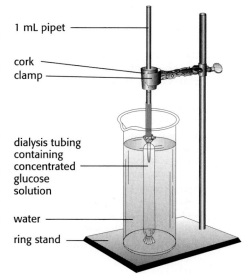

- 1 mL pipet
- cork
- clamp
- dialysis tubing containing concentrated glucose solution
- water
- ring stand

This investigation uses cell models to illustrate the relationship between the surface area of a cell and its volume, a concept important in many biological phenomena, including absorption, excretion, diffusion, and cell size. Students watch a substance diffusing into agar block "cells" and turning the "cells" pink as the process proceeds. Teams of 2 are best.
Time: One class period.

Safety

Students should wear gloves when handling the agar cubes that have been in the beaker of NaOH

Discussion

1. The phenolphthalein in the agar turns pink when the NaOH combines with it; therefore, you can tell how far the NaOH has diffused by the development of color. The NaOH diffuses about the same distance into each side of each block in a given time. Small blocks turn pink sooner, not because of a difference in rate but because of the difference in size.
2. Agar cubes in order from largest to smallest: 3cm, 2cm, 1cm. Agar cubes in order of surface-area-to-volume ratio: 1 cm, 2 cm, 3 cm. The order is reversed.
3. Surface area of cube 0.01 cm on a side = $0.01 \times 0.01 \times 6 = 0.0006$ cm². Volume of a cube 0.01 cm on a side = $0.01 \times 0.01 \times 0.01 = 0.000\,001$ cm³.

Ratio of surface area to volume =

$$\frac{0.0006}{0.000\,001} = \frac{600}{1} \text{ or } 600:1$$

INVESTIGATION 5.3   Cell Size and Diffusion

Why are cells so small? Why do they stop growing after reaching a certain size? What causes cells to stop growing and then to divide? One way to investigate these questions is to build a model. A model is often a small copy of something large. Here you will build a large model of something small.

 After reading the procedure, develop a hypothesis that predicts which cell model will do the best job of moving materials in and out of the cell.

Materials (per team of 2)

2 pairs of safety goggles
2 lab aprons
2 pairs of plastic gloves
250-mL beaker
millimeter ruler
plastic spoon

plastic table knife
paper towel
150 mL 0.1% NaOH
3-cm × 3-cm × 6-cm block of
    phenolphthalein agar

Put on your safety goggles, lab apron, and plastic gloves.

Procedure

1. Using the plastic knife, cut the agar block into three cubes—one 3 cm, one 2 cm, and one l cm on a side.
2. Place the cubes in the beaker and cover them with 0.1% NaOH (sodium hydroxide). Record the time. Use the plastic spoon to turn the cubes frequently for the next 10 minutes.

I   **NaOH is an irritant and can destroy clothing. Avoid skin/eye contact; do not ingest. Should a splash or spill occur, call your teacher immediately; flush the area with water for 15 minutes; rinse mouth with water.**

---

3. Prepare a data table like Table 5.2 and do the calculations necessary to complete it. The surface-area-to-volume ratio is calculated as follows:

Ratio of surface area to volume = $\dfrac{\text{surface area}}{\text{volume}}$

This ratio also may be written "surface area : volume." The ratio should be expressed in its simplest form (for example, 3 : 1 rather than 24 : 8).

4. Wear gloves and use the plastic spoon to remove the agar cubes from the NaOH after 10 minutes. Blot them dry. Do not handle the cubes until they are dry. Use the plastic knife to slice each cube in half. Record your observations of the sliced surface. Measure the depth of diffusion of the NaOH in each of the cubes.

Table 5.2 Comparison of Agar Cubes

Cube Dimension	Surface Area (cm²)	Volume (cm³)	Simplest Ratio
3 cm			
2 cm			
1 cm			
0.01 cm			

Discussion

1. What evidence is there that NaOH diffuses into an agar cube? What evidence is there that the rate of diffusion is about the same for each cube? Explain.

2. List the agar cubes in order of size, from largest to smallest. List them in order of the ratios of surface area to volume from the largest to the smallest ratio. How do the lists compare?

3. Calculate the surface-area-to-volume ratio for a cell model that is 0.01 cm on a side.

4. Which has the greater surface area, a cube 3 cm on a side or a cube 0.01 cm on a side? Which has the greater surface area *in proportion* to its volume?

5. What predictions can you make about the cell model in Question 3?

6. What happens to the surface-area-to-volume ratio of cubes as the cubes increase in size?

7. Most cells and microorganisms measure less than 0.01 cm on a side. What is the relationship between rate of diffusion and cell size?

8. Propose a hypothesis to explain one reason why large organisms have developed from *more* cells rather than *larger* cells.

9. In what other ways can models be used?

4. **A 3-cm cube has a greater surface area than a 0.01-cm cube. A 0.01-cm cube has a greater surface area in *proportion* to its volume.**

5. **High surface area to volume would make this cell efficient.**

6. **The surface-area-to-volume ratio decreases.**

7. **The rate of diffusion remains the same, but because cells are so small, diffusion occurs quickly.**

8. **Small cells enable large organisms to keep the balance between diffusion and active transport efficient. (The hypothesis can be tested but not with school laboratory equipment.)**

9. **Models can be used to look for general principles in simplified systems and to study processes and structures too small to be seen or too difficult to study in living organisms.**

INVESTIGATION 5.4   Mitosis and Cytokinesis

The growth of an organism is the irreversible increase in the number and size of cells. In other words, a plant or animal grows when it produces new cells that increase in size. When an organism grows, cell parts must be made for each new cell formed. For growth to occur, mitosis and cytokinesis must take place. Not all parts of a plant or an animal grow, and so not all cells of a plant or animal need to carry out mitosis and cytokinesis.

An onion placed in water sprouts slender white roots. This growth occurs partly by repeated duplication of cells. You might expect to see cells undergoing mitosis in a root. Because mitosis is only a small segment of the cell cycle, however, root cells are not always dividing. It is easiest, therefore, to examine the stages of mitosis in stained, prepared slides of root tips that have been sliced very thin longitudinally.

Materials (per person)
compound microscope
prepared slide of *Allium* (onion) root tip
prepared slide of *Ascaris* embryo cells

Procedure

1. Scan the entire length of the *Allium* root tip slide. Study each section of the root tip under low power and then under high power. In which region of the slide do you find many small, tightly packed cells? Are these cells old or new cells? Can you recognize any cells that are undergoing mitosis? How can you tell? Do you see any cells undergoing cytokinesis?

2. The strings, or bars, that appear in the nucleus are chromosomes. Examine these structures in several cells. What happens to the chromosomes during mitosis?

3. Mitosis is a continuous process in which the contents of dividing cells change shape and position. In the region in which you think mitosis is occurring, sketch at least five complete cells that look different from each other. How do the size, shape, and contents of cells undergoing mitosis compare with those of cells in other parts of the root?

4. The continuous process of mitosis often is divided into several stages for ease of study. Compare your sketches with the photographs of a dividing cell shown on the following page and try to identify the stages in your sketches.

This investigation precedes the discussion of mitosis so that students can engage in discovery and appreciate the difficulty of working out the proper sequence of events in mitosis and cytokinesis. The following concepts should be most emphasized. First, although there are minor differences between mitosis in plants and mitosis in animals, the process is similar in all eukaryotic cells. Second, two complete, identical sets of chromosomes move to opposite ends of the parent cell. Third, if cytokinesis follows mitosis, the cytoplasmic material divides. Fourth, the two new cells are like the parent cell; they contain all the nuclear material necessary to have complete biological instructions as in the parent cell.

If sufficient microscopes and slides are available, it is desirable for each student to work alone.
Time: One class period.

Procedure

1. **Comparing cell size in embryonic tissue with cell size in older tissue farther away from the root tip develops the idea that root growth depends on: (1) the increase in number of cells and (2) their enlargement after cell division. Most small cells are younger than larger, differentiated cells. Cells undergoing mitosis should have visible chromosomes, and these cells should be in the tip where new cells are being produced. Help students to distinguish between mitosis, which involves nuclei, and cytokinesis, which involves cell formation.**

2. **The chromosomes appear to be dividing into two groups. This can be seen easily in a motion picture but not in the static prepared slide, although students will be able to see two groups of chromosomes in anaphase and telophase.**

3. **A discussion of changes in shape and size will lead into a consideration of differentiation. Emphasize that the drawings should be large enough to show all observed detail.**

4. **Have your students match their sketches with the different stages of mitosis. Emphasize the process, however, not just the names of the stages.**

5. **The prepared slide represents a set of cells in the root tip caught in the act of mitosis. The longer the time it takes for a phase of mitosis to take place, the greater the chance that one of the cells in this phase will be found. The number of cells found in interphase will be the greatest by far. This indicates that interphase takes the longest time to complete.**

6. **Differences include the absence of a cell plate, the presence of centrioles, and constriction of the cytosol.**

Discussion

1. **The chromosomes must have doubled.**

2. **Chromosomes are made up of thousands of nucleotides linked together in a precise order. When the chromosomes duplicate all of the nucleotides must be matched up exactly, which takes a long time.**

3. **Similarities include: (a) two complete and identical sets of chromosomes move to the opposite ends of the parent cell; (b) cytokinesis occurs; (c) the two offspring cells are replicas of the parent cell; and (d) the two new nuclei contain chromosomes that are replicas of the parent cell nucleus. Differences are the absence of a cell plate in animal cells and the presence of centrioles and pinching in of the cytoplasm. Ask students why plant cells do not pinch in during division. An explanation could include the presence of the rigid plant cell wall.**

5. Compare the length of time for each of the different mitotic stages. To do this, count the number of cells within your field of view that are in each of the 5 stages. As a class, determine the average number of cells found in each of the different stages. What does this suggest about the length of time for each stage? Explain your answer.

6. Examine an *Ascaris* slide and look for stages similar to those you saw in the onion root tip. What differences do you see? Sketch several cells.

Discussion

1. Mitosis produces two nuclei from one nucleus. The number of chromosomes in each new nucleus is the same as that in the nucleus from which they were formed. What does this suggest must happen to the number of chromosomes in the nucleus *before* it divides?

2. In which stage do you think the chromosomes are duplicated? Why does it take so long for this to occur?

3. Based on your observations, describe the differences and similarities of mitosis and cytokinesis in plant and animal cells.

For Further Investigation

Suppose you suspect that the frequency of mitosis in onion roots varies with the time of day. How would you go about getting data to confirm or refute your suspicion?

SUMMARY

Cells are the basic units of organisms, and all cells arise from preexisting cells. Different types of microscopes make possible the detailed study of cell structures.

A plasma membrane surrounds each cell and helps to control the passage of materials in and out of the cytoplasm. Materials may move into a cell by diffusion, passive transport, or by energy-requiring active transport. In eukaryotic cells, internal membranes form a variety of structures and enclose organelles, converting the cell into several parts. Each part has a specialized role in the life of a cell. The chemical reactions of a cell—its metabolism—take place in the organelles and in the cytosol, the gelatin like portion of the cytoplasm.

The continuous sequence of events in the life of a cell is known as the cell cycle. Cells that divide spend most of their life in interphase, the time between divisions. During this phase, cells are metabolizing and preparing for division. Not all cells divide, however. Most plant cells and the nerve and muscle cells in animals remain in a growth phase of the cell cycle.

The cell nucleus contains most of the genes, which control the basic functions of a cell. The genes in the nucleus are located in the chromosomes, each of which replicates itself in the S phase of the cell cycle. During mitosis and cytokinesis, the doubled chromosomes separate, the cytoplasm divides, and two new cells are formed from one existing cell. Thus, each cell receives genetic instructions from its parent cell. In the development of complex organisms, many different types of cells are formed. Each type of cell possesses a specific structure and function. Continued life depends on cell division.

APPLICATIONS

1. During the late nineteenth century, most of the knowledge of detailed cell structure was gained by studying dead cells. Some biologists of that time objected to many of the conclusions that

were drawn from such observations. They argued that the processes of killing, staining, and mounting cells could cause a cell's structure to appear different from what it was in a living cell. What types of evidence are available today to meet at least some of these objections?

2. Which cell parts can you see using the highest magnification of your microscope? What does this indicate about the sizes of these parts compared with all the other cell parts you cannot see?

3. While working in a police laboratory, you are given a tiny sample of material and asked to identify it as either plant or animal matter. How could you decide which it is?

4. Trace the path of a molecule as it enters and moves through a plant cell and eventually enters a vacuole. Through which structures must it pass?

5. On the basis of your understanding of osmosis, describe what would happen to (a) a marine jellyfish placed in a freshwater stream and (b) a frog placed in ocean water. Some fish (for example, shad and striped bass) annually swim from the ocean into freshwater rivers and back. How are they able to do this?

6. Here is a quotation from *The Rime of the Ancient Mariner*, a poem about a sailor.

    Water, water, everywhere,
    And all the boards did shrink;
    Water, water, everywhere,
    Nor any drop to drink.

    Why did the boards of the ship shrink, and why couldn't the mariner drink any water?

7. THC is a chemical that stops chromosomes from separating during anaphase of mitosis. What do you think the nuclei of the two resulting cells might look like if a cell were sprayed with THC during mitosis?

8. Almost every living thing, including you, starts off as a single cell. When you grow, however, you have cells that are extremely different from each other. Consider the difference between your brain cells, skin cells, and stomach cells—all of which were formed by the genetic information in your chromosomes. What does this suggest about the genetic information found in that first cell?

9. In which type of cells is a cancer more likely to begin—a young cell or a highly specialized cell?

PROBLEMS

1. Find out what the oldest known cells on earth looked like. Do they show more resemblance to a modern-day prokaryotic cell or eukaryotic cell?

2. Using a microscope or photomicrographs in books, examine various types of cells from multicellular organisms. Discuss the relationships between the structural forms of the different cells and their functions.

3. A cell can be compared to a small country. A cell has a membrane at its outer edge that screens what comes into the cell. Likewise, a small country may have border guards at its edges to prevent the entry of certain people or objects. Make a list of cell parts. Then, try to think of parts of the country that may have similar functions.

4. What parts and processes do all cells have in common? Consult a cell biology text to determine which organelles and cell structures are found in plant, animal, bacterial, and fungal cells. What does this suggest about the similarity of related functions in all of these cells?

5. There are many plant cells that humans use for things other than food. Consult a book on economic botany to see what these uses are.

6. This chapter describes mitosis in cells that have single, well-defined nuclei. Investigate what is known about what happens to nuclear material during division in (a) a cell that lacks a nucleus (a bacterium, for example), and (b) a cell with more than one nucleus (a paramecium, for example).

7. It may be possible for you to attach a 35-mm camera onto a microscope. Take a series of photographs of cells in different stages of mitosis. Enlarge them and make a display for your entire class. If a camera setup is not available, make large drawings of each stage. For a more difficult assignment, try making all of the drawings in three dimensions.

8. In unicellular organisms, cells usually separate shortly after division. In some unicellular organisms, however, they remain attached and form colonies. In multicellular organisms, cells remain attached, but more strongly in some cases than in others. Investigate the ways in which the cells are held together.

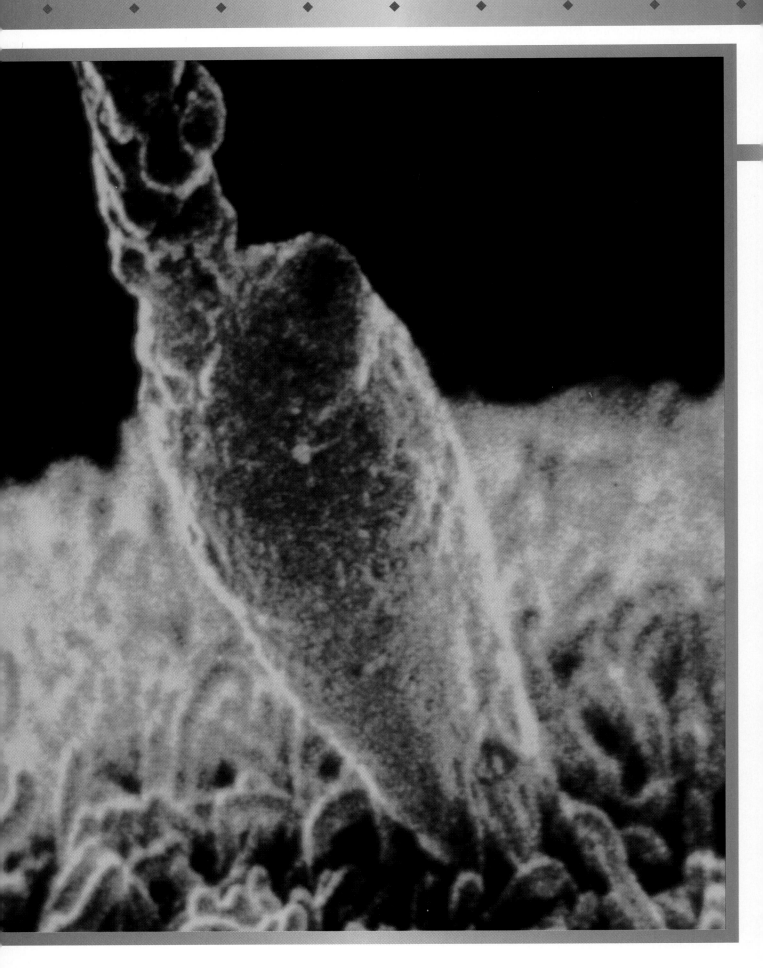

Continuity Through Reproduction

6

This is a human sperm attempting to penetrate an egg. What is the evolutionary advantage of having contributions from both parents during sexual reproduction?

It provides genetic variation in the next generation.

◆ Chapter 5 describes how cells reproduce by dividing in two. When that happens, the "parent" cell disappears, but its life continues in its offspring. The process of reproduction for entire organisms, however, usually is not that simple. When entire organisms are produced, the parent usually lives on as an individual and in many cases cares for the offspring.

Producing offspring is not necessary for individual survival. For example, a plant that does not get enough water during a particular year may survive but be unable to flower or produce seeds. The life of the individual plant, therefore, does not depend on its production of seeds every year. If the entire species is to survive, however, individuals of the species must produce new members. Because individuals die, the species will disappear if new individuals are not formed. The formation of new individuals that are like their parents is called reproduction.

All the organisms alive today arose from preexisting organisms. Although organisms have many different methods of reproduction, every method leads to a marvelous event—the beginning of a new life. This chapter discusses some of these methods. ◆

Reproduction

6.1 Reproduction Is Essential for the Continuation of Life

A new organism grows, develops, and eventually becomes large enough and mature enough to reproduce. The time when an organism becomes able to reproduce varies among species. Some organisms, such as bacteria, can reproduce after 20 minutes or so of life; other organisms may require months or years to become mature.

For a unicellular organism, reproduction occurs only once. Multicellular organisms, in contrast, may continue to reproduce for days, months, or years. These reproductive periods may form a large part of their life cycle. A life cycle is all the events that occur between the beginning of one generation and the beginning of the next. Because life cycles are continuous, they often are shown in diagrams as circles (see Figure 6.1).

Some organisms, such as humans, outlive their reproductive periods. Other organisms continue to reproduce until they die. The giant tortoises of the Galapagos Islands may live as long as 200 years, but they cannot

> **Guidepost**
>
> Why is reproduction important?

The ability to reproduce is a fundamental, unique characteristic of living organisms. Throughout this chapter, stress the genetic variability that results from sexual reproduction, as well as the ways new combinations of genes help to ensure the survival of a species.

Ask students what problems arise with our increasing ability to keep people alive.

Figure 6.1   In the life cycle of the leopard frog (*Rana pipiens*), the animal changes rapidly (in the course of several weeks) from an embryo within an egg, to a tailed tadpole with structures and behaviors suited for life in the water, to a tailless adult frog suited to life on land.

Point out to students that some of the Galapagos tortoises alive today may have seen Darwin when he visited the islands.

reproduce until they are about 40 years old. Thereafter, they may continue to reproduce as long as they live.

Even though some organisms live for a long time, no organism lives forever. Although the life span of an organism is part of its genetic program, not all organisms live out their genetic life spans. Some are eaten or die of accidents, infections, or other diseases. The secret of biological success for a species lies in its members' ability to reproduce and leave offspring before death occurs. When organisms do reproduce successfully, their species continues. A species becomes extinct if its members fail to reproduce or if they fail to leave enough offspring to allow for deaths as a result of accidents and disease. Successful reproduction ensures the continued existence of a species.

6.2   Reproduction May Be Sexual or Asexual

In complex organisms, reproduction generally requires two parents and two different special cells, one from each parent. When these cells are united, a new individual begins. This is known as **sexual reproduction**. Humans, frog, worms, and flowering plants reproduce sexually. In sexual reproduction, only certain types of cells can unite to form a new individual. These special cells are called **gametes** (GAM eets). The other cells that make up the individual are known as body cells.

Sexual reproduction affects the behavior of organisms, their genetic makeup, and their evolution. The nuclei of gametes, like those of all cells, contain chromosomes. Each chromosome in a gamete carries some genes from the parent organism that produced the gamete. When parents reproduce, they pass genetic information on to their offspring, and the new individuals then develop in response to these genetic instructions. Because it combines genetic material from two different individuals, sexual reproduction increases the amount of genetic variation in a population.

In **asexual reproduction,** new individuals originate from a single parent. Either the parent divides into two (or more) individuals, or new individuals arise as buds from the parent's body (see Figure 6.2). No gametes are needed for asexual reproduction. In plants, asexual reproduction sometimes is called vegetative reproduction.

Some animals and plants can reproduce both sexually and asexually. For example, the potato is a flowering plant that can use sexual reproduction to produce seeds, but it also can reproduce asexually. The little "eyes" on potatoes are actually buds—groups of cells that can undergo rapid mitosis and develop into new plants. When potatoes are not used, the buds

2 sprout and begin to grow, as shown in Figure 6.3a.

A sea star is another organism that can reproduce sexually or asexually. To obtain food, sea stars invade oyster beds in coastal waters, open the

3 oyster shells with their powerful arms, and then feast on the oysters inside. Scuba divers used to collect the sea stars to save the oyster beds. In an effort to destroy the sea stars, the divers would cut off the arms and throw the pieces back into the water. A sea star arm, however, can regenerate an entire sea star if part of the central body is attached to it (see Figure 6.3b). In their efforts to eliminate sea stars from the oyster beds, the divers actually increased the sea star population.

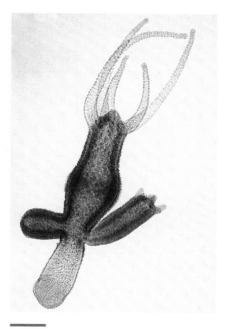

Figure 6.2 A budding hydra.

Because an asexual offspring is essentially a genetic copy of its parent, asexual reproduction in complex organisms is regarded as an evolutionary dead end. In microorganisms, however, asexual reproduction does not inhibit change because enormous numbers of mutant individuals can arise rapidly following a favorable mutation (e.g., antibiotic resistance in bacteria).

Contrast sexual and asexual reproduction in terms of adaptability of a species to a changing environment. In general, asexual reproduction threatens survival because it results in uniformity, whereas sexual reproduction guarantees some variability.

CONCEPT REVIEW

1. What is the main difference between sexual and asexual reproduction?
2. What do regeneration and vegetative reproduction have in common?
3. How can having two different types of reproduction help some species survive?
4. From the viewpoint of the individual, how does reproduction differ from other life processes, such as digestion and cellular respiration?

Figure 6.3   Asexual reproduction in a plant and an animal. The sprout on the potato (a) grew from a bud that can produce an entire new potato plant. The sea star can regenerate a new body from one arm (b).

a

b

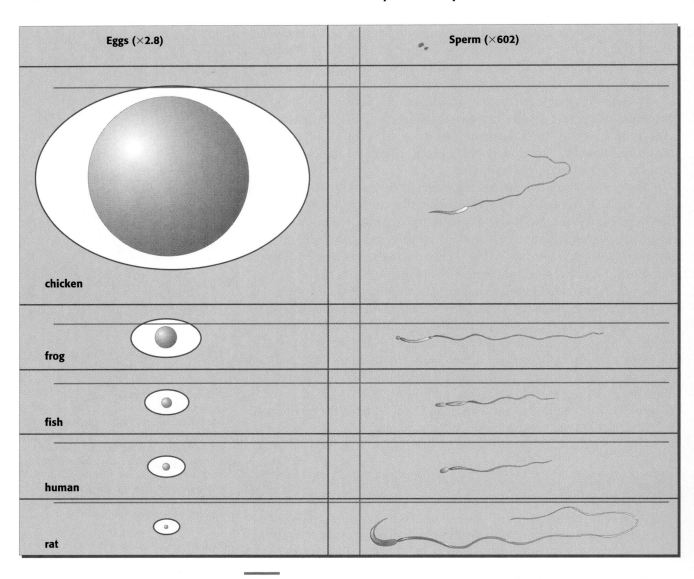

Eggs (×2.8)	Sperm (×602)
chicken	
frog	
fish	
human	
rat	

Figure 6.4   Eggs and sperm of several organisms. Note that the sperm are enlarged much more than the eggs, and direct comparison between the two is not possible. In the chicken, frog, and fish, the ova are surrounded by other materials (shown in outline).

Guidepost

How do gametes differ from other cells?

Figure 6.5   A scanning electron micrograph of human sperm showing the head and the middle piece that contains the mitochondria (×19 500). The mitochondria provide power for the tail.

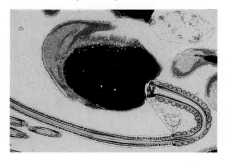

Formation of Reproductive Cells

6.3   Gametes Are Reproductive Cells

In most types of sexual reproduction, each parent produces a different type of gamete. For example, one parent may produce large, stationary gametes, and the other parent may produce smaller gametes that can move. In some organisms, the reproductive cells of both parents look similar. Gametes usually differ in appearance from other cells in an organism, and they contain fewer chromosomes.

Males produce gametes called **sperm cells,** which usually are quite small. (The word *sperm* can be used to refer to one or many sperm cells.) Females produce egg cells, or **ova** (singular, ovum). An ovum may be large. Sometimes it is thousands of times larger than a sperm cell. Figure 6.4 compares the ova and sperm of several organisms.

A sperm cell consists of little more than a nucleus in the head, a tail that can move the cell about, and an energy generator—its mitochondria (see Figure 6.5). An ovum, on the other hand, contains, in addition to its nucleus, other organelles, such as mitochondria and ribosomes, that are important for cell function. In many cases, an ovum also contains a reserve supply of food for the new organism. Biologists use the differences in size and function of gametes to define gender. An organism that produces ova is called female. An organism that produces sperm is called male. Some organisms, however, actually change their sex during their lifetime. Such an ability can be extremely useful. An individual can be whatever sex is more productive at a given moment, thus helping to ensure the continuation of the species. For example, the blueheaded wrasse shown in Figure 6.6 lives in tropical reefs and eats parasites off the scales of other fish. It usually lives in groups of one male and five or six females. The male, largest of all, mates with all the females. The females have a social hierarchy based on size. When a female of high status dies, each female of lower status moves up a notch. When the male dies, however, the largest female fish becomes male. Within one hour after the male's death, the dominant female starts to act like a male. Within two weeks, the fish begins to produce sperm.

Figure 6.6   A dominant male blueheaded wrasse (*Thalassoma bifaciaum*). When this male dies, the largest female will assume his role.

6.4   The Chromosome Number of a Species Remains Constant

The cells of an organism contain the number of chromosomes that are typical of its species. For example, the body cells of corn plants have 20 chromosomes, the body cells of the common fruit fly, which is used in genetic research, have 8 chromosomes, and the body cells of humans have 46 chromosomes. Each chromosome in a body cell has a partner of the same length and appearance. In other words, the chromosomes occur in matching pairs, and for the most part, each pair has genes that code for the same traits. One member of each chromosome pair originally came from the sperm and the other from the ovum that produced the organism.

A new individual begins at **fertilization**, when the nuclei of two gametes—one gamete from each parent—unite and produce a **zygote** (ZY goht). Gametes have only half the number of chromosomes that body cells have—*one* chromosome from each matching pair. Because gametes contain only one chromosome from each matching parental pair, they are said to be **haploid** (HAP loyd) cells. Haploid cells are represented by the symbol *n*. Cells with both chromosomes of each pair are called **diploid** (DIP loyd) and are represented by the symbol 2*n*. The union of two haploid gametes restores the chromosomes to the diploid number, as you can see in Figure 6.7. In this illustration, the diploid number is four—two pairs of chromosomes. For humans, normal haploid gametes contain 23 chromosomes. The diploid body cells each contain 46 chromosomes.

Gametes become haploid through a special type of cell division known as **meiosis** (my OH sis). In sexual reproduction, meiosis and fertilization are complementary processes: meiosis produces the haploid (*n*) chromosome number in gametes, and fertilization restores the diploid (2*n*) chromosome number in the zygote.

6.5   Gametes Are Formed by Meiosis

The number of chromosomes is reduced from diploid to haploid when gametes are formed. This reduction takes place during meiosis, which requires two cell divisions in succession. The stages of meiosis are similar to those of mitosis, described in Section 5.9.

The term *egg* probably should be used to describe an ovum plus other substances, as is the case with a bird's egg. Frequently the term is used ambiguously.

The attempt here is to get students to think about maleness and femaleness in a fundamental way.

Here and in later discussions, emphasize that the new individual is unique as a result of the combination of genetic material from each unique parent.

If the study of genetics is to be meaningful, students need to understand the chromosome reduction at meiosis and the increase at fertilization. Discuss each step thoroughly.

Figure 6.7   Gametes unite in the process of fertilization, restoring the diploid number of chromosomes. What number of chromosomes in the new individual come from each parent? **Half the total number of chromosomes come from each parent.**

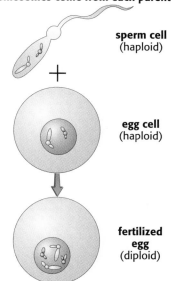

sperm cell
(haploid)

+

egg cell
(haploid)

fertilized
egg
(diploid)

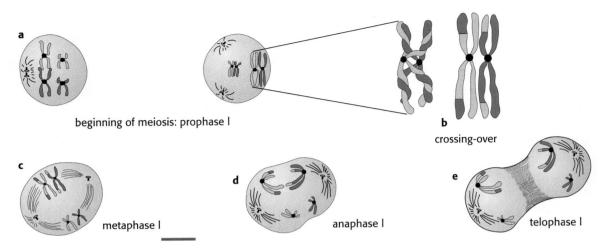

beginning of meiosis: prophase I

b
crossing-over

c
metaphase I

d
anaphase I

e
telophase I

Figure 6.8 The stages of meiosis—first meiotic division.

Mitosis and meiosis are similar in both name and process. Challenge the students to invent some mnemonics to help alleviate the confusion that inevitably occurs. Ask students to model both processes, using colored pipe cleaners or modeling clay as chromosomes. The chromosomes can be cut to different lengths so that both their colors and their lengths will key them for continuous identification during the processes. Both chromosomes of a homologous pair should be the same length. If students are to identify all the possible combinations that can occur in meiosis, the chromosomes of a pair should be different colors.

Be certain students understand that segments of a chromosome contributed by one parent (in the last generation) are exchanged with segments of the partner chromosome contributed by the *other* parent. An understanding of this concept will help students comprehend how new combinations of genes occur.

During the S phase of the cell cycle preceding meiosis, the chromosomes replicate and produce twin structures. Each structure is called a **chromatid** (KROH muh tid). A specialized area known as the **centromere** (SEN troh meer) connects the twin chromatids, which contain identical DNA. Meiosis consists of one replication of chromosomes followed by two divisions. The first meiotic division is called I; the second is called II. Once meiosis begins in a cell, no more DNA is made—the DNA already present is distributed to the gametes during the two divisions. Figure 6.8a-h illustrates the events of meiosis in a cell with three pairs of chromosomes. Refer to the illustration as you read the description that follows.

As meiosis begins (a), each chromosome is a double structure made of two chromatids joined at the centromere. (The number of centromeres equals the number of chromosomes.) In prophase I, a unique event occurs. Remember that the sperm contributes one chromosome (paternal) to every pair of chromosomes and the ovum contributes the other (maternal). After the chromosomes replicate, the matching pairs come close together and may twist around each other (b). In this position, parts of a chromatid of one chromosome may break off and exchange places with the identical parts of a chromatid of the other matching chromosome. This process is called **crossing-over,** because chromatid segments are exchanged between the two chromosomes of a pair, which results in new combinations of genes.

Toward the end of prophase I, the chromosome pairs move to the equator of the cell. Metaphase I (c) marks their arrival in the center of the cell, where

Figure 6.8 The stages of meiosis (continued) second meiotic division.

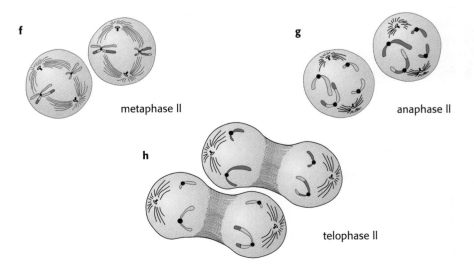

f

metaphase II

g

anaphase II

h

telophase II

they line up in pairs. In anaphase I (d), the matching chromosomes of each pair separate and begin to move toward opposite poles of the spindle. Each chromosome still consists of two chromatids attached at the centromere.

The matching chromosomes, one from each pair, continue to move in opposite directions. As they approach the poles in telophase I (e), cytokinesis occurs, resulting in two cells. As telophase I ends, the matching chromosomes are in different cells for the first time. Each new cell now has only half as many chromosomes as a body cell and, therefore, contains half the parent cell's total genetic information. Because the amount of genetic information per cell has been reduced by one-half, the first meiotic division often is called *reduction division.*

Meiosis continues with a short prophase II in which no additional crossing over occurs. In metaphase II (f), the chromosomes again move to the equator in each of the two cells. This time, however, there are no matching chromosome pairs. The chromosomes line up in single file down the center of each cell, just as in mitosis.

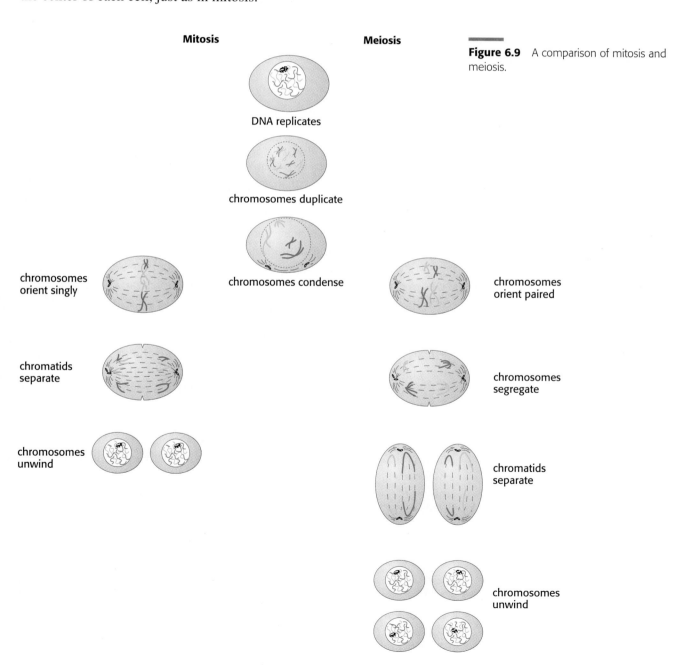

Figure 6.9 A comparison of mitosis and meiosis.

Mitosis

DNA replicates

chromosomes duplicate

chromosomes condense

chromosomes orient singly

chromatids separate

chromosomes unwind

Meiosis

chromosomes orient paired

chromosomes segregate

chromatids separate

chromosomes unwind

Figure 6.10 In human males, meiotic division usually results in four equal sized sperm, each with 23 chromosomes. How does the formation of an egg differ from the formation of sperm? **Two meiotic divisions form four sperm; two meiotic divisions in egg formation form only one ovum. The other 23 chromosomes are given off in polar bodies.**

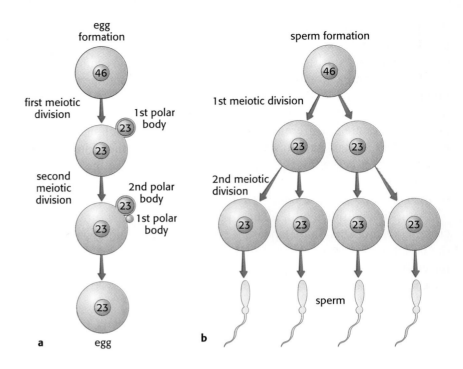

In anaphase II (g), the centromeres divide, and the two chromatids of each chromosome separate and move toward opposite poles. The chromatids (now called chromosomes) gather at the poles and are enclosed by new nuclear envelopes in telophase II (h). Cytokinesis occurs again, resulting in four haploid cells, each containing *n* single chromosomes. You can become more familiar with the events of meiosis by modeling them in Investigation 6.1.

Although the similarities between mitosis and meiosis are striking, their differences are more important. See Figure 6.9 on page 131.

Male animals, including human males, usually produce four equal-sized sperm from each original cell. Further cytoplasmic changes occur as each sperm develops a tail and the DNA becomes concentrated in the head of the sperm cell. The mitochondria packed into the middle part of the sperm provide energy for the tail, which will move the sperm through the female reproductive tract. When the cytoplasmic changes are complete, the cells are fully formed sperm.

Females usually produce only one ovum from each original cell that begins meiosis. In the first meiotic division, one complete set of chromosomes is cast off in a small cell called a polar body, which consists of little but the chromosomes. Most of the cytoplasm remains in the larger cell. In the second meiotic division, another polar body with a haploid set of chromosomes is cast off. In most species of animals, the two polar bodies disintegrate. The remaining cell, which becomes the ovum, contains a haploid set of chromosomes and cytoplasm. Figure 6.10 compares sperm and egg formation.

CONCEPT REVIEW

1. Compare the two types of gametes.
2. Explain the biological distinction between male and female.
3. How do gametes differ from other cells of the body?

4. What is the significance of the chromosome number in gametes?
5. How does fertilization affect the chromosome number of a cell?
6. How is a zygote formed?
7. Compare and contrast the stages of meiosis and mitosis.

The Human Reproductive System

6.6   Gamete Development Differs in Males and Females

In humans and other animals, meiosis and gamete formation take place in special reproductive organs called **testes** (singular, testis) in males and **ovaries** (singular, ovary) in females. In early development, these reproductive organs look alike.

When the developing organism, the **embryo**, is about eight weeks old, these reproductive organs become clearly male or female. Like the ovaries, the testes develop in the abdominal cavity of an embryo. Before birth, however, the two testes move downward and to the outside of the body, where they are housed in a pouch called the scrotum, which is located just below the penis (see Figure 6.11a).

Each testis is made up of thousands of tiny tubules, shown in Figure 6.11b, in which meiosis occurs. Figure 6.11c shows a cross section of one tubule. The diploid cells that will undergo meiosis are located around the outer edge of each tubule. As meiosis proceeds, the new cells are formed toward the center of the tubule. Cells in the various stages of meiosis are shown in Figure 6.11c. The last stages of meiosis occur near the center of

> **Guidepost**
>
> How does reproduction occur in humans?

In mammals, production of live sperm seems to depend on maintaining the testes (in humans, testicles) at a temperature lower than that of the body. In many mammals, the testes descend into the scrotum only during the breeding season.

Figure 6.11 (a) The human male reproductive system. (b) Each testis is composed of packed coils of tubules in which sperm develop. (c) Cross Section through a tubule, ×100. Around the edge of the central canal are the dark heads of the sperm, with their tails extending into the canal.

a

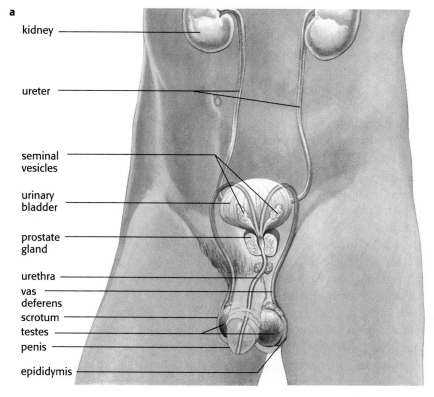

kidney

ureter

seminal
vesicles

urinary
bladder

prostate
gland

urethra

vas
deferens

scrotum

testes

penis

epididymis

b

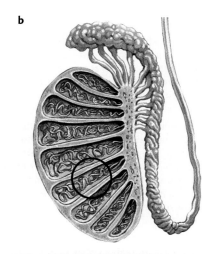

c

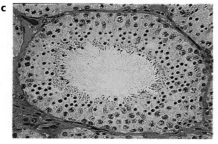

the canal, where the dark heads of mature sperm can be seen with their long tails extending into the center of the canal.

The sperm move out of these canals into a special collecting duct near the testis. This curved duct, which is about 50 cm long, can hold billions of sperm. The sperm remain there until they are released in sexual activity, or until they are reabsorbed by the body. Mature human males can produce huge numbers of sperm and may release from 200 million to 500 million at one time.

In humans, meiosis and the formation of sperm cells take about nine weeks. Remember, however, that the development of sperm is a continuous process, and only mature sperm cells are absorbed or released. As some sperm are absorbed or released, others have matured and moved from the tubules in the testis to the collecting duct.

Gametes are not produced until puberty, or sexual maturity, in males, but in females, meiosis begins in the embryo. Millions of potential ova form in the ovaries of the female embryo. Certain diploid cells of the newly formed ovary begin meiosis, but the first meiotic division soon is interrupted. The potential ova stop developing in prophase I and remain in suspended division for many years.

Meiosis in the ovary does not begin again until puberty, when a girl enters adolescence. At this time, about once a month, one cell resumes meiosis. That potential ovum continues into the second meiotic division, and the process stops once more. Later, the cell is released from the surface of the ovary, a process called ovulation (ohv yoo LAY shun). In humans, one ovum usually is released each month, although sometimes two (or more) may be released.

Only about 400 to 500 ova are released during an entire lifetime, and the reproductive span of a female may extend from puberty (which usually begins between ages 9 and 14) until age 45 or 50. Thus, some of the partly mature ova may take as long as half a century to complete meiosis.

6.7   Hormones Control Reproductive Cycles

Both the onset of puberty and the lessening of gamete formation in later life are controlled by changes in the body chemicals that influence the reproductive organs. Many different organs in the body produce these chemicals, called **hormones**. Traveling in the blood to all parts of the body, hormones act as messengers that influence other organs. A hormone from one organ may affect another organ, which can respond in striking ways. The first organ, in turn, can be influenced by hormones from still other organs. This complex set of interactions, a network of stimuli and responses, is called a **feedback** system. This is one way the various parts of the body communicate with each other.

Hormones produced in the brain stimulate the testes and ovaries to secrete distinctive sex hormones. Although the embryo's sex is determined by the sex chromosome at the time of fertilization, the development of an embryo into a male or a female baby is controlled by the sex hormones. Sex hormones also help bring about some of the noticeable indicators of approaching sexual maturity—changes in voice and body proportions, growth of hair, and increased interest in sex. Although hormones are essential for sexual reproduction in both males and females, the hormone interactions in females often are more complex.

In females, the nervous system, several glands and organs, and a variety of hormones interact to regulate a monthly ovum-releasing cycle,

The importance of the interaction of hormones in the various stages of reproduction, especially in the female, cannot be overstated. You may wish to introduce this topic with the subject of ovulation.

Women who take fertility drugs often release a number of ova—resulting in the fertilization of several eggs and the birth of more than a normal number of offspring, sometimes five or six.

The number of ova produced is related to the probability of young being born and reaching maturity. Ask students to account for the enormous number of ova produced by some fish compared with the relatively small number produced by most mammals.

Point out that the hormonal regulation of the reproductive cycles is so similar in mammals that most knowledge we have of human cycles was derived from studies of laboratory animals such as the white rat.

Biology Today

Birth Control

The earth's population is increasing rapidly. In an effort at control, some overpopulated countries are trying to limit the number of children born. To do so, they rely on birth control, or contraception.

Although abstinence (refraining from sexual intercourse) is the only *sure* way to prevent pregnancy, contraception can prevent fertilization of an egg or disrupt development of the zygote. It does so by preventing (a) development of gametes, (b) physical contact between two gametes, or (c) implantation of a fertilized egg.

Many women prevent the development of gametes by taking birth control pills containing estrogen and a synthetic hormone called progestin. The combination of the two hormones inhibits the development of egg follicles in the ovaries. The pill is about 99 percent effective, but in some women it may have serious side effects and must be prescribed by a doctor. To date, studies on the side effects have been inconclusive. A similar method of birth control is a hormonal contraceptive implanted under the skin of the upper arm that can prevent pregnancy for up to five years. Several other countries have injectable contraceptives that last for one to two months. Not all these options are available in the United States.

Contact between gametes can be prevented by interfering with the movement of ova or sperm cells. A membranous sheath, or condom, can be placed over the penis prior to intercourse, thus preventing sperm from entering the body of the female. Condoms also prevent the spread of AIDS and other sexually transmitted diseases. Another method uses a flexible cap, or diaphragm, over the cervix, the entrance to the uterus, to prevent sperm from entering.

Surgical procedures, or sterilization, also can prevent the movement of ova or sperm. In males, the duct that carries sperm to the penis may be tied and cut. This simple procedure, called a vasectomy (vuh SEK tuh mee), is widely used in some countries. In females, the oviducts may be tied and cut. This method is called tubal ligation (TOO buhl ly GAY shun). Both methods generally are considered irreversible.

If pregnancy does occur, it can be terminated involuntarily or voluntarily. Many embryos are aborted naturally. If a major chromosomal or genetic abnormality is present, the body may abort the embryo during its first few days of development.

Many abortions are undertaken because contraceptives failed. Experts blame contraceptive failure for about 2 million unwanted pregnancies a year in the United States. Many of these pregnancies are allowed to continue, but every year about 0.8 million of them are ended by abortion. That represents about half of the 1.5 million abortions occurring annually. The debate surrounding planned abortions is complex and emotionally charged. Because there are no clear-cut answers, the debate is likely to continue for a long time. Many experts feel that the United States lags many years behind other countries in contraceptive development and that new, easier-to-use contraceptives might help reduce our abortion rate, which is one of the highest in the industrialized world.

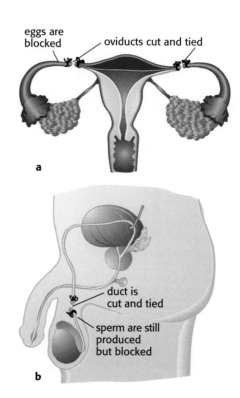

Tubal ligation (a) and vasectomy (b) are permanent methods of preventing pregnancy.

known as the **menstrual cycle** (from the Latin *mensis,* meaning month). During this cycle, usually a single ovum is discharged from one of the ovaries, which are located low in the abdominal cavity (see Figure 6.12). The ovum enters one of two tubular structures called oviducts. The oviducts act as passageways through which the ovum moves into the **uterus,** a muscular, pear shaped organ. If the ovum is fertilized by a sperm, the fertilized egg develops in the uterus until the time of birth. At that time, strong muscular contractions push the baby out of the uterus through the vagina.

During the menstrual cycle, the inner lining of the uterus builds up in preparation for a fertilized egg. If the egg is not fertilized, the lining, which develops a rich blood supply, disintegrates and passes from the uterus through the vagina to the outside of the body. Discussions of the menstrual cycle usually take as their starting point the first day of menstrual flow (menstruation, or "period").

The hormonal interactions that regulate the menstrual cycle are shown in Figure 6.13, opposite. Refer to the diagram as you read the descriptions that follow. In the menstrual cycle, the **hypothalamus,** a part of the brain shown at (a), acts in much the same way that a thermostat acts in a house. A thermostat set at 20°C does not keep the room at exactly 20°C. The temperature fluctuates somewhat as the furnace is turned on and off. Just as the thermostat monitors and adjusts temperature, the hypothalamus monitors and helps to adjust the level of hormones in the body. The hormonal fluctuations serve to coordinate the release and fertilization of an egg with the preparations for a receptive environment in the uterus.

At the start of the menstrual flow, two of the hormones secreted into the bloodstream by the ovaries are at low levels. These low levels cause the hypothalamus to secrete a releasing hormone that stimulates a gland,

Figure 6.12 The human female reproductive system.

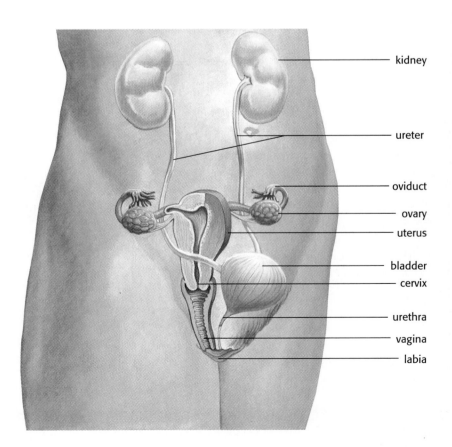

kidney

ureter

oviduct

ovary

uterus

bladder

cervix

urethra

vagina

labia

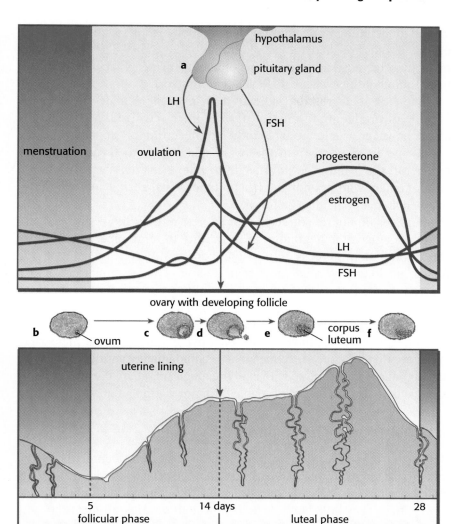

Figure 6.13 The menstrual cycle. Interactions among the nervous system and several organs, glands, and hormones regulate the cycle.

known as the **pituitary** (pih TOO ih ter ee), just underneath the brain. As a result, the pituitary secretes two hormones that act on the ovaries. These hormones are called follicle-stimulating hormone (FSH) and luteinizing hormone (LH). FSH causes an ovum to begin maturing inside a fluid-filled sac, or follicle, on the surface of the ovary (b).

The two pituitary hormones also stimulate the maturing follicle (c) to secrete the hormone **estrogen** (ES troh jen). Estrogen affects the uterus, causing an increase in the blood supply and a thickening of the lining. The effects of estrogen and the two pituitary hormones build for about 14 days. Just before day 14 a sudden increase in LH causes the follicle to burst and release the ovum. This process is known as ovulation (d).

The ruptured follicle is converted to the corpus luteum (KOR pus LOOT ee um). The corpus luteum (e) releases estrogen and another hormone, **progesterone** (proh JES tuh rohn). Progesterone and estrogen act together to stimulate additional buildup of the lining of the uterus.

In the brain, meanwhile, the feedback system enables the hypothalamus to detect the increased levels of estrogen and progesterone in the bloodstream and to slow down secretion of its releasing hormone. This change causes the pituitary to secrete less FSH and LH. If the ovum is not fertilized as it travels down the oviduct toward the uterus, the corpus luteum starts to degenerate (f), and the levels of progesterone and estrogen drop. The drop in estrogen and progesterone levels decreases the blood supply to the lining of the uterus. Menstruation begins as the blood-filled uterine tissues are shed and passed through the vagina. The onset of the menstrual flow marks the beginning of another cycle. Notice the approximate timing of the events in the cycle in Figure 6.13.

The first menstruation indicates that a young female has become capable of producing ova and of having them fertilized. Menstruation alone, however, is an unreliable indicator of fertility. Remember that an ovum first is released from the ovary and begins to travel down the oviduct. If the ovum is *not* fertilized, then the menstrual cycle begins. A young female who has not had her first menstrual period still may have released an ovum that can be fertilized.

Females do not remain fertile for the rest of their lives. Just as menstruation signals the beginning of the ability to reproduce, **menopause** (MEN oh pahz), the cessation of the menstrual cycle, indicates that a female no longer is capable of reproducing. The hormonal effects of menopause are just as profound as those of the menstrual cycle and vary widely from individual to individual. Menopause is a normal part of the life cycle of a human female and usually occurs between the ages of 45 and 50. Males do not have a hormonal cycle similar to the menstrual cycle, nor do they experience menopause, although the number of sperm a male produces may decline with age. If enough sperm are present, however, a male of advanced age is capable of impregnating a female.

6.8   Only One Sperm Fertilizes the Ovum

Before fertilization can take place, sperm cells must enter the body through the vagina. This usually occurs through sexual intercourse. As the male becomes sexually excited, the penis fills with blood and becomes rigid. The sperm cells begin to move through a series of ducts into the penis. After further stimulation, the male reproductive system responds by **4** expelling 3 to 4 mL of **semen** (SEE men). In addition to hundreds of millions of sperm, this whitish fluid contains the secretions of various male glands. These glandular fluids provide food and lubrication for the sperm and counteract the acidity of urine, which may be harmful to sperm. (Urine and semen pass through a common duct in the penis.) Figure 6.11 shows reproductive structures in males. After sperm cells are released into the vagina, they swim in all directions. Many of them swim up along the moist linings of the female reproductive tract, enter the uterus, and swim into the two oviducts.

The released ovum's trip to the nearest oviduct may take two days. The ovum, however, can be fertilized successfully for only 24 hours after its release. Human sperm, on the other hand, may live for 48 hours after **3** they are deposited. If sexual intercourse occurs during the 24 hours after an egg has been released, or if live sperm are present from previous intercourse at the time the egg is released, the ovum may meet many sperm cells, and one sperm may penetrate the outer membranes of the ovum.

The membranes of the ovum immediately react to fertilization and act as a barrier to the entry of other sperm cells. Only one sperm cell fertilizes each ovum, and only the head of the sperm, which contains the genetic material, enters the egg. If only one sperm fertilizes the ovum, why are so many sperm released in semen? Sperm secrete an enzyme that helps to break down the layer of cells that surrounds an ovum. Apparently, many sperm are required to provide enough enzyme.

The events of fertilization and implantation institute a new sequence of controls, causing cessation of the menstrual cycle and maintenance of pregnancy. If the opportunity arises, this is a good place to mention the hormonal basis of some birth control procedures.

You may wish to discuss the reproductive patterns of mammals with estrus cycles. The bleeding that occurs with dogs and cattle while in estrus is caused by red blood cells passing through the vaginal wall, not by sloughing off of the uterine lining.

Most aquatic vertebrates have no need for a penis, as the sperm and ova are released into the water, where fertilization occurs. Only reptiles and mammals have a true penis, and in some mammals, there is a bone in the penis to assist in the rigidity necessary for successful intercourse.

Unlike the diploid cell that enters meiosis and forms four haploid cells in males, in females each diploid cell that enters meiosis eventually forms only one haploid ovum. Virtually all the cytoplasm of the original cell is preserved in the ovum, where it provides the material and energy needed for the early cell divisions of development. After the embryo is attached to the inner layer of the uterus, it receives nourishment from its mother.

CONCEPT REVIEW

1. Describe the male and female reproductive systems.
2. Describe what happens in the menstrual cycle.
3. If an ovum can be fertilized only during a 24-hour period, how can pregnancy occur even when a female does not engage in intercourse during the period?
4. Trace the movement of sperm from the time they are released to the time they encounter an ovum.

INVESTIGATION 6.1   A Model of Meiosis

Many biological events are easier to understand when they are explained by models. In this investigation, you will use a model to duplicate the events of meiosis.

The evolutionary importance of retaining most of the cytoplasm in the ovum is discussed in Chapter 7.

Materials (per team of 2)
modeling clay, red and blue
4 2-cm pieces of pipe cleaner
large piece of paper

This hands-on model of the events of meiosis serves to reinforce the discussion in Section 6.5 and gives students a real grasp of the process. Teams of 2 allow each student to form one pair of homologous chromosomes.
Time: **One class period.**

PART A Basic Meiosis
Procedure

1. Use the clay to form two blue and two red chromatids, each 6 cm long and about as thick as a pencil.
2. Form four chromatids, two of each color, 10 cm long.
3. Place the pairs of similar chromatids side by side. Use pipe cleaners to represent centromeres. Press a piece of pipe cleaner across the centers of the two red, 6-cm chromatids. This represents a chromosome that has replicated itself at the start of meiosis. Do the same for the other three replicated chromosomes (see Figure 6.14).
4. On a sheet of paper, draw a spindle large enough to contain the chromosomes you have made. Assume that the spindle and chromatids have been formed, and the nuclear membrane has disappeared.
5. Pair the two 6-cm chromosomes so that the centromeres touch. Pair the two 10cm chromosomes. Assume that the red chromosome of each pair was derived from the female parent (maternal). Its matching chromosome, the blue one, came from the male parent (paternal).
6. Arrange the two chromosome pairs along the equator (middle) of the spindle so that the red chromosomes are on one side and the blue on the other.
7. Holding on to the centromeres, pull the chromosomes of each matching pair toward opposite poles of the spindle. Once the chromosomes have been moved to the two poles, the first meiotic division is completed.

Figure 6.14   Chromosome models.

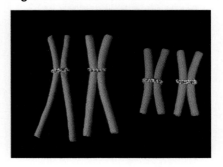

8. Draw two more spindles on the paper. These new spindles should be centered on each of the poles of the first meiotic division. Both spindles should be perpendicular to the first spindle. Your model is now ready for the second division of meiosis.

9. Place the chromosomes from each pole along the equator of each of the two new spindles. Unfasten the centromere of each chromosome. Grasp each chromatid at the point where the centromere had been attached. Pull the chromatids to opposite poles of their spindles. Draw a circle around each group of chromosomes that you have.

Discussion PART A

1. **One; four cells.**
2. **Four; two. Meiosis results in each cell containing half the number of chromosomes that were found in the original cell.**
3. **The resulting cells are called gametes or sperm or ova.**
4. **Two should contain only red chromosomes; two should contain only blue chromosomes.**

Discussion PART B

1. **These gametes should contain one red and one blue chromosome.**
2. **This increases the genetic variation found in the gametes, which leads to greater genetic variation within the population.**
3. **The number of possible meiotic outcomes is equal to 2^n (not considering crossing over), where $n =$ number of chromosome pairs.**

Discussion PART C

1. **Answers will vary depending on how students align the chromosomes on the spindles.**
2. **Crossing over increases the number of types of gametes that are produced, which increases the genetic variation in the population.**
3. **Genes (alleles) are being exchanged.**
4. **Because the model is large and can be manipulated, the events are easy to observe. The action can be stopped at any stage, reversed, or repeated. This could not be done with living material.**
5. **Answers will vary.**
6. **This model is not the real thing. A model is a form of analogy and has, in some degree, all the advantages and disadvantages attached to the verbal form. A simulated series of events cannot anticipate the variations in a living system, and a model takes some liberties and shortcuts to show aspects of a process. There is a danger of misunderstanding at every point at which a model differs from the biological reality—for example, size, color, materials, and metabolic activity.**

Discussion

1. How many cells were there at the start of meiosis? How many cells are formed at the end of meiosis?
2. How many chromosomes were in the cell at the beginning of meiosis? How many chromosomes were in each of the cells formed at the end of meiosis?
3. At the end of meiosis, what are the cells called?
4. How many of your cells at the end of meiosis had only red chromosomes in them? How many had only blue chromosomes in them?

PART B Effects of Chromosome Position
Procedure

10. A real cell is three-dimensional. Although the red chromosomes (from the female) may be on one side and the blue (from the male) may be on the other when they line up at the equator, there is an equal chance that one red and one blue chromosome will be on the same side. Attach the chromatids as they were at the beginning of the investigation. Go back to Step 6 and arrange the chromosomes so that one blue and one red chromosome are on each side. Now complete meiosis.

Discussion

1. How do these gametes compare with those you made earlier?
2. What difference does this make in terms of genetic variation in the offspring?
3. How many different types of gametes could be made if there were three sets of chromosomes instead of just two?

PART C Effects of Crossing Over
Procedure

11. Reassemble your chromosomes. To show crossing-over (Figure 6.8b), exchange a small part of the clay from a chromatid making up one chromosome with an equal part from a chromatid of its matching pair. The colors make the exchange visible throughout the rest of the investigation.

12. Place your chromosome pairs along the equator of the spindle as in Step 6, and complete meiosis.

Discussion

1. How many different types of gametes did you form? Did you form any different ones from those formed by others in your class?
2. In general, how do you think crossing over affects the number of different types of gametes that are formed?
3. In crossing over, what actually is exchanged between the chromatids?
4. What are some of the advantages of using a model to visualize a process?
5. How did this model improve your understanding of the process of meiosis?
6. What are some disadvantages of this model?

SUMMARY

Reproduction is essential for the continuity of life on the earth. Organisms usually reproduce either sexually or asexually. Sexual reproduction increases genetic variability in a population because genetic material is provided by two different parents.

Sexual reproduction requires two parents. Each parent contributes a gamete to the new individual, forming a zygote. The gametes contain only half the genetic information of the parent. The reduction of genetic material is accomplished by meiosis. Genetic material is exchanged and reduced by half in two suc-cessive cell divisions. In males, a single diploid cell gives rise to four haploid sperm, but in females, a diploid cell results in only one ovum. When the gametes combine in fertilization, the chromosome number is replenished.

The menstrual cycle is regulated by a complex feedback system involving hormones produced by the brain and other organs. During the cycle, the lining of the uterus builds up in preparation for receiving a fertilized egg. If fertilization of the ovum does not occur, the blood-rich lining of the uterus disintegrates and passes from the body.

APPLICATIONS

1. Explain how the behavior of chromosomes during prophase I in meiosis differs from chromosome behavior during prophase in mitosis.
2. Why is crossing over possible in meiosis but difficult in mitosis?
3. Bacteria reproduce by splitting into two different cells. Do you think this is sexual reproduction or asexual reproduction? Explain.
4. Suppose there are two similar plant species one that reproduces only by sexual reproduction, and one that reproduces only by asexual reproduction. Which species has a better chance of surviving over the next one million years, and why?
5. Mitochondria have their own DNA, which is circular. This DNA differs from the DNA found in the nucleus. Studies show that mitochondrial DNA is derived from the female parent, not the male parent. Use your knowledge of meiosis and fertilization to explain this phenomenon.
6. Suppose you find a cell that contains 15 chromosomes. Is this cell a gamete or a body cell? How do you know?

PROBLEMS

1. Most nursery managers use asexual reproduction to propagate some of their plants. Invite a local gardener or nursery manager into your class to demonstrate techniques of cloning, grafting, or vegetative propagation.
2. What effect does the environment have on the development of genetically identical organisms? Take several cuttings from a single plant, making sure that the cuttings come from parts of the plant that are of similar age. After the cuttings have developed their own roots, plant them in identical pots, but expose them to different environmental factors. Observe their growth.
3. Many crop plants have twice the number of chromosomes as their ancestors. These plants are called polyploids individuals that have more than the normal number of sets of chromosomes. Investigate the evolution of several crop plants and report to the class on the importance of polyploidy in their development. One useful reference is N. W. Simmonds, ed., *Evolution of Crop Plants* (New York, Longman Inc., 1979).
4. Is human reproduction lacking in seasonality? Record by months the birthdays of members of your biology class and as many other classes as possible. Present the data in the form of a bar graph. What do the data indicate?
5. There is some indication that the age at which human females reach sexual maturity, or puberty, is declining. Collect data on this phenomenon and on its possible causes.
6. The environment of the female reproductive tract is slightly acidic, which can be harmful to sperm cells. Investigate the components of semen and the ways in which they help counteract the acidity.
7. There are two types of twins: fraternal (dizygotic) and identical (monozygotic). Explore the cause of twinning in each case.
8. Make a model of the female hormonal interactions that regulate the menstrual cycle. Indicate in your model when pregnancy can occur, and what will happen to the hormonal levels if pregnancy does occur.

Continuity Through Development

7

These young Cooper's Hawks (*Accipiter cooperi*) do not appear similar to the adult bird. What biological processes must occur from now until the time when they will resemble their parent?

Growth, continued differentiation, and the development of tissues.

◆ Frogs croaking in a marsh, cats crying in the alley, insects chirping outside your window, all are the sounds of the living world. Reproduction is a fundamental part of the living world, but how does the new life created by this process come about?

In most complex organisms, the fertilization of an egg is the first in a series of intricate steps. The fertilized egg is still a single cell, yet within that one cell is all the genetic information needed to form an organism that has millions or billions of cells. Many of these cells differ greatly in both structure and function. How can one set of genetic instructions determine the fate of so many different types of cells? This chapter describes the series of events that occur after fertilization and result in a multicellular organism. ◆

Development

7.1 A Zygote Gives Rise to Many Cells

1 At the moment of fertilization, when the haploid (*n*) set of chromosomes in the sperm nucleus joins with the haploid (*n*) set in the ovum nucleus, the diploid (2*n*) number of chromosomes is restored. The zygote that results is ready to begin **development**, a series of events that converts a single cell into a fully formed individual. Three basic activities result in embryonic development: cell division, cell movement, and cell **differentiation**—the process by which new cells specialize and become different in appearance and function from their parent cells.

Soon after fertilization, the zygote divides into two cells by mitosis. Each of these two cells divides again to form four cells, then eight cells, and so on. No new cytoplasm is added between the divisions. Soon the tiny embryo is made of dozens of cells. In animals, there is no growth for a while—the cells continue to divide, forming new cells smaller than the parent cells.

These early cell divisions are called **cleavage** (KLEE vaj) because the original mass of the egg is cleaved, or divided, into smaller and smaller cells. Cleavage converts the zygote into a **blastula** (BLAS choo luh), a hollow ball of several hundred cells. Figure 7.1 shows some of the stages of cleavage.

Once the blastula has formed, both individual cells and entire sheets of cells begin to change position in an ordered and precise manner. Many
2 cells flow from the surface to the interior of the blastula, so that a new

MAJOR CONCEPTS

◆ Development is a process that encompasses the entire lifetime of the individual, including the production of offspring and aging.

◆ For any species, development has characteristic stages that are genetically programmed.

◆ Specialized cells arise by differentiation from unspecialized cells.

◆ Organs arise from three cell layers in the developing embryo.

◆ Many embryos develop in close association with a parent.

◆ Uncontrolled cell division is detrimental to an organism.

KNOWLEDGE CHECK

◆ What is development?
◆ How does development differ among animals?
◆ What controls the specialization of cells into tissues?
◆ How are certain stages in development similar among differing animals?

Guidepost

How does a zygote become a multicellular, complex organism?

Changes in cell shape also play an important role in development. Differentiation is an intriguing and little-understood process. Urge your students to investigate this topic further.

Cleavage in frogs is temperature dependent. At room temperature, *Xenopus* eggs complete cleavage within about 12 hours and reach the gastrula stage after about 18 hours. In humans, cleavage takes place slowly as the egg travels down the oviduct (four to five days). The blastula stage normally is reached as the embryo enters the uterus.

143

Figure 7.1 Stages of cleavage in the frog: (a) fertilized egg, (b) 2-cell stage, (c) 4-cell stage, (d) 8-cell stage, (e) early blastula, and (f) late blastula.

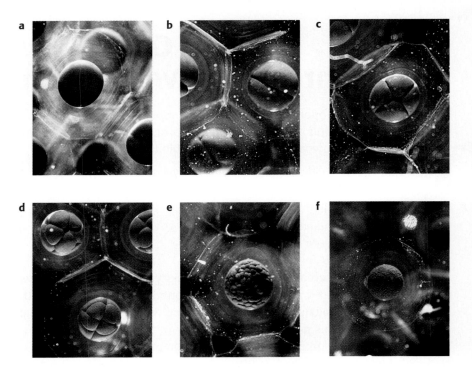

The endoderm is responsible for the gill arches found in the early embryonic states of vertebrates. In fish, these develop into respiratory organs. In land vertebrates, they aid in the development of part of the tonsils, inner and middle ear, parathyroid, and thymus.

layer of cells lines the original blastula. As a result of these cell movements, the blastula becomes a **gastrula** (GAS truh luh). Figure 7.2a illustrates this process, which is known as gastrulation. The gastrula has three distinct cell layers—the original surface layer, or ectoderm (EK toh derm); the new inner layer, or endoderm (EN doh derm), that formed from cells flowing into the blastula; and a third layer, the mesoderm (MEZ oh derm), that forms between those two layers. Figure 7.4b shows these three cell layers, which give rise, either singly or in combination, to all the different tissues of the body. Investigation 7.2 illustrates these developmental stages in a marine worm.

Figure 7.2 Development of a frog gastrula (a). Cross section of a vertebrate embryo, showing the locations of the three primary tissue layers (b).

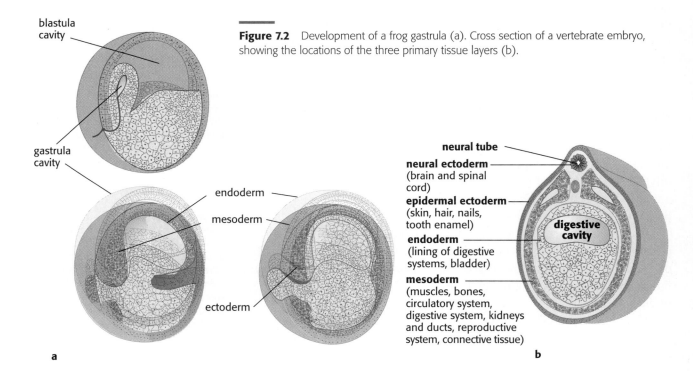

7.2   Cell Differentiation Is a Major Part of Development

How does a single cell—the zygote—develop muscle tissue, blood tissue, skin tissue, nerve tissue, and all the other tissues necessary to the functioning of a complete organism? Each of these tissues is made of specific types of cells—muscle cells, blood cells, skin cells, nerve cells—and each of these cell types not only looks unlike the others but also produces different proteins. Red blood cells, for example, produce the protein hemoglobin, whereas intestinal cells produce digestive enzymes. Because genes control the production of proteins, scientists once wondered if each cell type contained a different set of genes.

Experiments with amphibian eggs, however, have shown that almost every type of cell in an organism contains a set of genes identical to the set in the zygote. Study the experiment shown in Figure 7.3. In this experiment, an unfertilized egg was exposed to ultraviolet light to destroy its nucleus. Then, the nucleus was withdrawn from a skin cell that had undergone differentiation. When the nucleus was injected into the unfertilized egg, the egg was able to develop into a normal tadpole. This could not have happened unless a complete set of genes was present in the differentiated cell nucleus.

If all cells in an organism contain identical, complete sets of genes, then there must be differences in the genes' activity. As cells begin to specialize, different sets of genes are activated in different cells for varying lengths of time. Thus, after an initial period of activity, one set of genes may stop its activity while another continues. The specific genes and the sequence depend on factors in the genes' environment.

Study again the experiment shown in Figure 7.3. Note that when the nucleus of the differentiated cell was injected into the egg, the gene action changed to resemble the action of a normal, undifferentiated egg nucleus. Genes that were inactive in the differentiated cell became active again, allowing a normal tadpole to develop. In some way, the cytoplasm of the egg (the environment of the genes) had modified the pattern of gene action in the cell nucleus.

Recall that during cleavage, the zygote divides into smaller and smaller cells, each containing a portion of the egg's original cytoplasm. Experimental evidence suggests that the cytoplasm in an egg is not

Figure 7.3   Diagram of an experiment showing that the nucleus of a differentiated cell contains the same set of genes as the zygote.

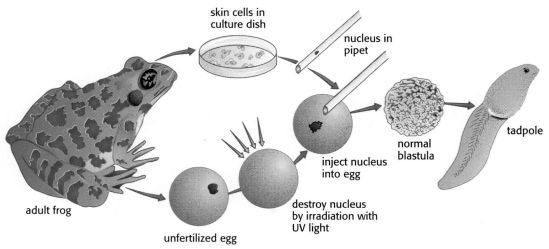

skin cells in culture dish

nucleus in pipet

inject nucleus into egg

normal blastula

tadpole

adult frog

destroy nucleus by irradiation with UV light

unfertilized egg

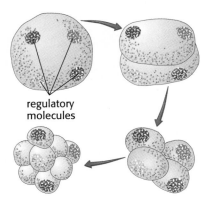

regulatory
molecules

Figure 7.4 Regulatory molecules are not uniformly distributed in the cytoplasm of an egg. As the egg divides, different molecules may be incorporated in the offspring cells, resulting in the activation of different genes.

Chapter 6 suggests that the production during meiosis, of one large ovum containing an abundance of cytoplasm may have a selective advantage. Ask students to relate that idea to this concept of nonuniform cytoplasm.

One of the most amazing aspects of nervous system development is the formation and growth of nerve axons. A single axon may be 1 m or more in length, but it originated as a single cell and elongated to a precise, predetermined destination.

Development relies on the correct sequencing of events and programming of cells. Hands start as paddle-shaped, but cells die and a normal hand results. The details of sequencing and preprogramming are unknown.

uniform. Instead, the various molecules that regulate gene activity are located in different portions of the cytoplasm. As cleavage proceeds, the newly formed cells receive only specific portions of the egg's original cytoplasm. As a result, each cell receives different regulatory molecules (see Figure 7.4). These regulatory molecules activate specific genes in each cell, leading to the production of specific proteins. Thus, although the cells all have the same genetic information, they specialize because different genes are activated and then different proteins are produced. Differences in the cytoplasm arising during cleavage, therefore, can play an important role in the differentiation of cells.

7.3   Interactions Between Cells Influence Differentiation

Interactions between different cell types also can play an important role in 3
differentiation. During the development of the eye, for example, the cells of the embryonic brain divide and expand until they contact the skin ecto- 4
derm. The skin ectoderm cells respond to this stimulus by thickening, pushing inward, and pinching off a rounded mass. Genes for specialized lens proteins are activated, causing the mass to become transparent and begin functioning as a lens. An extension of the brain forms a cup around the lens that becomes the retina (REH tin uh)—the part of the eye that is sensitive to light. The lens cells then stimulate other skin ectoderm cells to develop into the cornea (KOR nee uh), the transparent covering of the eye (see Figure 7.5). Cells from the brain and the ectoderm must interact to begin lens formation. In experiments in which a cellophane barrier was inserted between the brain and the ectoderm, the lens failed to develop.

How can one tissue or cell type influence the differentiation of other 5
cells? Regulatory molecules, in the form of chemical signals, appear to be responsible. Figure 7.6 suggests a general model for this process. A cell transmits a chemical signal to a second cell, thus activating genes that cause the second cell to differentiate into a specialized cell type. The second cell, in turn, may transmit a chemical signal to a third cell, and so on. Researchers have found, for example, that undifferentiated cells from the ectoderm do not develop into nerve cells unless a special chemical (fibronectin) is present.

Although the nature of most of these regulators is not known, various hormones can affect development dramatically. For example, the conversion of a tadpole to a frog is caused by a hormone from the thyroid gland, and the egg-larva-pupa-adult development in insects is regulated by a group of interacting hormones. Such interactions are a common feature of differentiation in many organisms.

Figure 7.5 Neural ectoderm (an outpocketing of the brain) induces formation of the lens of the eye in the skin ectoderm.

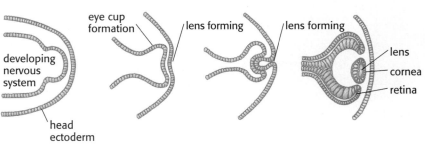

eye cup
formation
developing
nervous
system
head
ectoderm
lens forming
lens forming
lens
cornea
retina

The first organ system to develop in an embryo is the nervous system, which triggers the development of other systems. The ectoderm at the top of the embryo begins to form two long, parallel folds of cells, clearly shown in Figure 7.7. These folds enlarge and become prominent ridges that border a groove. Eventually the ridges join to become a hollow tube, called the neural tube, and other ectodermal cells close over it, as shown in Figure 7.8. The neural tube runs the length of the embryo and eventually gives rise to the brain and spinal cord.

Occasionally the neural tube fails to close completely, and part of the nervous system remains exposed. This condition, called spina bifida (SPY nuh BIH fih duh), often results in abnormalities in the functioning of the nervous system or even in the death of the child. Tests performed before birth can indicate whether a baby will be born with an exposed spinal cord.

The differentiation of cells illustrates the importance of environmental influences, even at the molecular level. Genes are influenced by their cytoplasmic surroundings, and cells in turn are influenced by chemical regulators in their surroundings. In ever-expanding circles, from DNA to whole organism, each structure is affected by its environment.

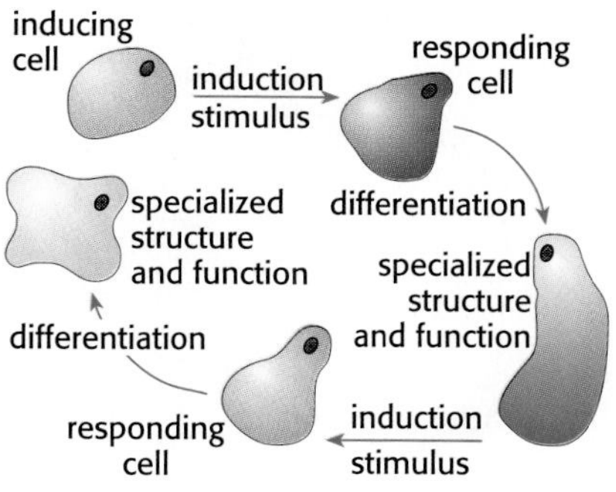

Figure 7.6   A model of cell-to-cell interactions. Chemical signals transmitted from one embryonic cell to another initiate differentiation into specialized types of cells. These signals may occur at different stages of development, bringing about increasing specialization.

CONCEPT REVIEW

1. What is the role of fertilization in triggering development?
2. Explain the difference between a gastrula and a blastula.
3. How can cells affect one another?
4. What cell layers contribute to the formation of the eye?
5. How do regulatory molecules affect cell differentiation?
6. What structures arise from the neural tube?

Figure 7.7   The neural groove in a frog embryo.

Animal Development

7.4   Some Animals Develop Independently of Their Parents

Like all cells, fertilized eggs need water to develop normally. Most female fish, as well as semi-aquatic female animals such as frogs and salamanders, release their eggs in water. In these cases, the eggs may be fertilized already, or the male may fertilize them after they have been released.

Guidepost

What are the relationship between animal embryos and their mothers?

You may wish to introduce the terms *external* and *internal fertilization.*

Figure 7.8   Development of the neural tube in a vertebrate embryo.

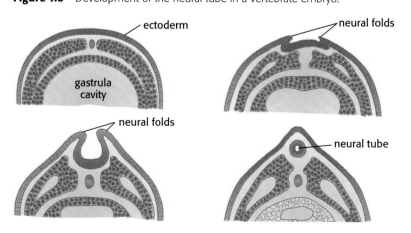

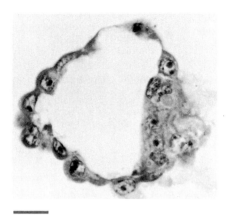

Figure 7.9 Section through a human blastocyst. The embryo develops from the inner cell mass, the thicker part on the left.

At implantation, the blastocyst is about six days old and is composed of about 100 cells.

You may wish to introduce the terms *allantois* **and** *yolk sac* **here. The allantois and yolk sac become the umbilical cord.**

Although its diameter is only about 18 cm, because of the proliferation of villi, the effective absorbing area of the placenta at term is about 13m². This is 50 times more than the area of the newborn's skin.

In fact, the need for water even affects mating. When salamanders or frogs mate, for example, the male awaits the release of hundreds or thousands of eggs by the female. The male then sheds large numbers of sperm directly over the eggs or nearby. Some of the sperm encounter eggs and fertilize them, but most of the sperm are lost in the water. Although this process may seem inefficient, frogs and salamanders have reproduced this way for millions of years and their survival demonstrates the success of this method.

Within the egg, young frogs develop as far as the tadpole stage. By then they can feed, and they have a tail and can swim. At this point, they hatch and become independent. Further development depends on their ability to find food for themselves and to avoid becoming food for predators.

Because reptiles and birds usually do not lay their eggs in the water, their eggs must be able to retain water for the developing embryos. Usually these eggs have a leathery or hard shell that retards evaporation. Reptiles, even reptiles that spend most of their lives in water, often bury their eggs to protect them from predators and from the hot sun.

7.5   Mammalian Embryos Develop Within the Mother

For a short time, a mammalian embryo depends on the small amount of material stored in the ovum. After the blastula becomes implanted in the wall of the uterus, it begins to absorb nutrients brought to the uterus by the mother's blood. The blastula (now called a blastocyst) has a thick mass of cells on one side. As shown in Figure 7.9, this side continues to develop as the embryo. Other cells in the mass develop into embryonic membranes that are important for the developing embryo. One membrane, the **chorion** (KOR ee on), extends fingerlike projections, or villi (VIL eye), into the inner lining of the uterus. The villi are supplied with tiny blood vessels from the embryo's developing circulatory system, as shown in Figure 7.10. The chorionic villi and the uterine lining form the **placenta** (pluh SEN tuh), through which the embryo receives nourishment from its mother and eliminates wastes. The embryo is connected to the placenta by a flexible cord, known as the **umbilical** (um BIL ih kul) **cord.** Gases and other molecules can diffuse from the blood of the mother through the placenta into blood vessels in the umbilical cord and from there into the circulatory system of

Figure 7.10 Cut-away view of a developing human fetus in the uterus. The illustration also shows embryonic membranes and the placental connection. Part of the placenta is enlarged to show the circulation of the mother and fetus. Note that the two circulatory systems do not mix.

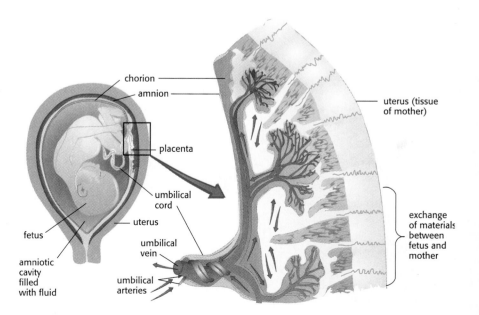

Biology Today

Progress in Prenatal Diagnosis

Today it is possible to diagnose many genetic diseases and congenital defects while the fetus is still in the mother's uterus. One procedure uses high frequency sound waves, or ultrasound, to determine the overall condition of the fetus and to confirm its age. Because it can be performed without the withdrawal of any fluid or tissue from the uterus or the fetus, ultrasound is called a noninvasive procedure. This test usually is performed during the fourth month of pregnancy. At this time, major developmental abnormalities, such as extra or misplaced limbs, can be detected, and the sex of the fetus can be determined.

Often ultrasound is used in conjunction with more invasive tests. Amniocentesis, which came into prominence in the 1970s, allows physicians to check for chromosomal abnormalities and certain enzyme disorders. For this test, the doctor first uses ultrasound to determine the position of the fetus and the location of the placenta. Next, a long thin needle is inserted through the mother's abdomen into her uterus. About 10 mL of amniotic fluid is withdrawn, along with cells that have been sloughed off from the fetus. These cells are cultured for several weeks and then analyzed. The amniotic fluid is checked for increased levels of a substance called alphafetoprotein (AFP). High levels of AFP are found in such disorders as spina bifida. Amniocentesis cannot be performed until about the sixteenth week of pregnancy because there is not enough room in the amniotic sac to allow the safe withdrawal of fluid. Amniocentesis is quite accurate and safe. The risk of spontaneous abortion after the procedure is less than 0.5 percent (1 in 200).

Chorionic villi sampling (CVS) was first introduced in the early 1980s. CVS can be performed as early as the ninth week of pregnancy. A catheter is inserted through the vagina into the uterus and directed by ultrasound to a region called the *chorion frondosum*, where there are many chorionic villi. Some villi are withdrawn by suction through the catheter. As the cells in the villi are growing and dividing, they can be analyzed quickly. Results from chromosome studies are available in a few days and results from enzyme studies in about one week. Because no amniotic fluid is withdrawn in this procedure, AFP analysis cannot be performed; otherwise, diagnoses from CVS are as accurate as those from amniocentesis. CVS carries a higher risk of spontaneous abortion—about 1 to 2 percent—than amniocentesis does, but it allows earlier diagnosis of many defects.

Prenatal diagnosis is recommended routinely for women 35 years of age and older because the risk of chromosomal disorders increases with advanced maternal age. Prenatal diagnosis also is recommended for women who already have a child with a genetic disorder or who have a history of a genetic disorder in the family of either parent. All detectable chromosomal disorders and more than 200 single gene disorders now are diagnosable using amniocentesis and CVS.

Chorionic villi sampling

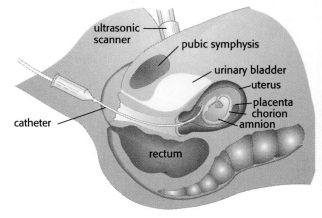

the embryo. Carbon dioxide and other wastes diffuse from the villi into the blood of the mother. The mother's lungs and kidneys dispose of the wastes. During these exchanges, the blood of the embryo and the blood of the mother normally do not mix. About this time, another embryonic membrane, the **amnion,** encloses the body of the embryo and fills with fluid. The embryo then develops in an aqueous environment.

With proper nourishment, the embryo grows, and its cells develop into distinctive types. Similar types of cells are found together in tissues. Tissues perform different functions for the body; muscular tissue contracts, for example, and glandular tissue secretes certain substances. When several tissues are grouped together in a special arrangement for a particular function, an organ is formed. The stomach and heart, for example, are organs. Finally, several organs may be related in general function. These organs form an organ system. The digestive system consists of a whole series of organs that carry food from the mouth through all the steps of digestion and absorption and eliminate the undigested remains.

Often the amniotic fluid contains cells that are cast off from the embryo during its growth. Physicians who wish to test the embryo for biochemical or chromosomal abnormalities without touching the embryo use these cells. A sample of amniotic fluid is removed, and the cells are grown in the laboratory. This process, called amniocentesis (AM nee oh sen TEE sus), is used to test for missing chromosomes, extra chromosomes, and some genetic disorders of a biochemical nature (see Figure 7.11).

A human develops in its mother's uterus for about 40 weeks. Figure 7.12 shows several stages of development. After 12 weeks, the embryo is called a **fetus.** At the completion of development (and sometimes earlier), the mother's body undergoes a complex series of hormonal interactions that result in birth. Strong muscular contractions push the fetus from the uterus through the vagina, which usually is able to expand sufficiently to allow the head and body to pass through. After birth, the placenta (afterbirth) is expelled. The vagina returns to normal size, and the uterus shrinks back to the size and shape of a pear.

Figure 7.11   Prenatal diagnosis of genetic disorders by amniocentesis. Fetal cells and amniotic fluid are withdrawn and cultured for later chromosomal and biochemical studies.

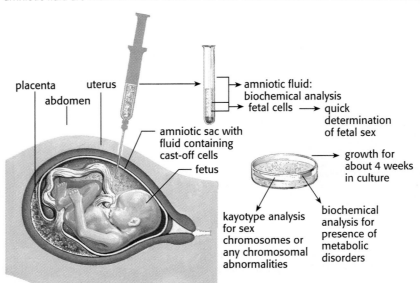

Occasionally a tear occurs in the placenta between the two circulatory systems, resulting in direct contact between the blood of the mother and the fetus. This is important if the blood types differ, particularly in the case of different Rh types. A student might want to investigate this situation and report to the class.

The sex of the fetus also can be determined through amniocentesis.

At the time the uterine muscle layers begin to contract and relax, another hormone, relaxin, is secreted by the corpus luteum. This hormone causes the cartilage between the two pubic bones that arch over the vagina to soften and stretch.

You may wish to discuss caesarean sections, what they are and why they are performed.

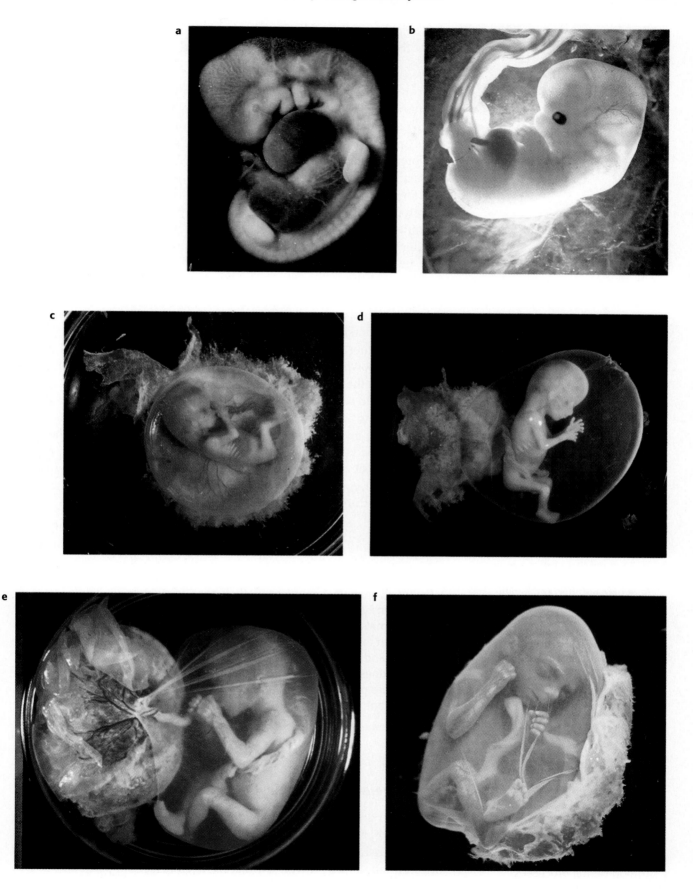

Figure 7.12 Early development of a human embryo: (a)
4 weeks, (b) 6 weeks, (c) 8 weeks, (d) 12 weeks, (e) 14
weeks, (f) 16 weeks.

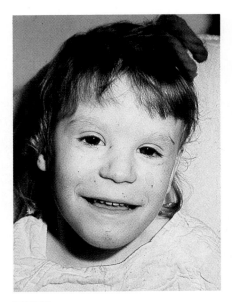

Figure 7.13 Four-year-old child with fetal alcohol syndrome. Note the short nose and the smooth, long, narrow upper lip as well as the narrow eye openings.

Stress the importance of proper prenatal care both for the mother and for the developing fetus. The placenta never is impervious to drugs. It is safe to say that the fetus is affected by drugs at least as much as the mother is. Addictions of the mother will be addictions of the newborn infant.

Have students further investigate the effects of smoking, caffeine, alcohol, and other drugs on the developing fetus. Point out that proper nutrition also is important. Even in the United States where nutrition and health care generally are adequate, between 250 000 and 500 000 infants are born each year with some type of mental or physical handicap. Many of these handicaps could be avoided with proper prenatal care.

7.6 Embryos Are Affected by Substances in the Mother's Blood

Physicians divide the total time needed for development into three periods, or trimesters. By the end of the first trimester (about 12 weeks), most organs are formed, and the skeleton is clearly visible. The last trimester is a time of rapid growth and the maturation of organs and systems as the fetus approaches the time of birth. Because so much cell differentiation and organ formation occur during the first 12 weeks, embryos in the first trimester are especially sensitive to harmful materials.

The growing embryo shares materials from the mother's blood and therefore can be affected by foods or drugs taken by the mother. For example, if the mother drinks alcoholic beverages, the alcohol in the mother's blood diffuses into the blood system of the fetus, and the fetus may be physically and mentally affected. Some children are born with fetal alcohol syndrome. These children have certain abnormal facial features (see Figure 7.13) and may be mentally retarded. Most doctors recommend that women abstain from alcoholic beverages while they are pregnant or nursing.

Many other substances, including marijuana, cocaine, and heroin, have been shown to cause abnormal fetal development. Smoking marijuana may harm the smoker's fetus, and a mother who smokes cigarettes risks having a child whose birth weight is significantly less than normal, thus threatening its health and development. If a pregnant woman uses a drug such as cocaine or heroin, the baby may be born addicted to the drug. Even caffeine is suspected of damaging the fetus. As you might expect, substances that can harm the mother are likely to harm the fetus. In addition, however, even substances that are harmless or beneficial to the mother may he harmful to the fetus.

CONCEPT REVIEW

1. Are all animals dependent on their parents? Explain.
2. Explain how substances pass from the mother to the embryo in mammals.
3. Explain the relationships between the amnion, the chorion, the umbilical cord, and the placenta.
4. How does the fetus get rid of waste products?
5. Why are proper nutrition and healthful behavior important for a pregnant woman?

Cancer

7.7 Development Usually Is Well Controlled

Guidepost

How is cancer related to normal developmental processes?

The entire drama of development depends on a certain number of cell divisions. Fertilization starts these divisions, and the resulting increase in cell number is accompanied by cell movement, changes in cell shape, and differentiation and growth. Ultimately, each cell has a particular task to perform in the development and life of the new individual. Aging is a normal part of development. At the end of the aging process, an organism dies. This process is built into the genetic program and is rather precise for each species.

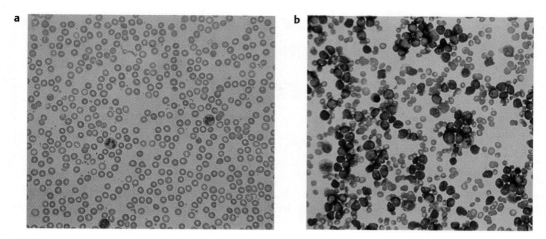

Figure 7.14   In cancer, abnormal numbers of certain cells are produced. These photomicrographs of stained blood cells show normal human blood (a) and blood from a person with leukemia (b). Note the large number of white blood cells in (b).

1   When an animal reaches adult size, most cell divisions often cease. Although some cells can continue to divide once the organism is fully grown, they do so only to replace worn-out or lost cells. For example, blood cells have a limited life span and are replaced regularly by new cells. Cells lining the digestive system also wear away and are replaced continuously. Outer skin cells are lost and are replaced from layers of cells below. Nervous system cells, on the other hand, rarely divide after differentiation. Although generally long-lived, these cells are not replaced. Normal life functions mainly require cell replacement once adult size is reached.

Most cells are able to divide only a certain number of times. One mechanism that controls cell division is called contact inhibition. Contact inhibition causes a cell that is touching another to stop dividing. In cancer, this control over cell division seems to be lost, and cells begin to divide rapidly (see Figure 7.14). These cells have lost their sensitivity to the growth regulation brought about by contact with other cells.

7.8   Cancer Cells Divide Without Limit

Just as organisms have certain life spans, so do the cells that make up the organisms. On rare occasions, however, the genetic controls over normal
3   cell division are altered. Sometimes the cell not only loses control over its division but also loses control over its position in a specific tissue. Normally, cells are able to recognize each other by particular proteins on the surface of each cell type. Cells with the same surface proteins bind together to form tissues and organs. When the genes responsible for those proteins malfunction, the cells may lose their identity and invade surrounding tissues. These cells form a growing mass called a tumor. If the tumor remains in one place near its origin, it usually is benign. Sometimes, small groups of these dividing cells may move to other parts
2   of the body and establish new tumors. Cell masses that continue to divide and to spread (or metastasize) at the expense of other tissues are called malignant tumors.

For a cell to become malignant, certain genes that regulate cell func-
4   tions must be present. These genes are called **oncogenes** (AHN coh jeenz). Oncogenes first were identified because of their presence in certain viruses that cause cancers in animals. More than 30 oncogenes have been

Growing tumors exert increasing pressure on surrounding cells, interfering with cellular functions. Cancerous cells do not necessarily grow at phenomenal rates, but they have lost the mechanism that directs them to stop.

Discovery of Breast Cancer Gene

In 1990, Mary-Clair King, a biologist at the University of California at Berkeley, made a key discovery. In her research on human chromosome 17, King discovered that certain genetic markers on the chromosome were inherited in families where the incidence of breast cancer was very high and where the disease occurred at an early age. Late in 1994, building on King's observations, an international research team led by scientists from the University of Utah, Myriad Genetics from Salt Lake City, the National Institute of Environmental Health Sciences, McGill University, and Eli Lilly and Company, confirmed the existence of a mutated gene that puts some women at very high risk of breast cancer. The gene was named BRCA1. Researchers pinpointed the location of BRCA1 on chromosome 17 and mapped its location in relation to other known disease genes on the chromosome. A woman with a defective BRCA1 gene may develop cancer very early in life, even in her 20s. Her risk of cancer is 60 percent by age 50.

Although the inheritance of the defective gene is responsible for only about 5 percent of all cases of breast cancer, the discovery of the gene and mapping its location were important. Scientists are still working out how this gene actually causes breast cancer. One idea is that the BRCA1 gene could code for the production of a particular protein that serves as a molecular switch in breast cell nuclei. Normal BRCA1 genes are thought to initiate a chemical signal that keeps cells from growing out of control. Generally, a woman would inherit two good copies of the BRCA1 gene. If one of the copies she inherits is damaged, the other good copy of the gene would protect her. If, however, a woman with a damaged gene somehow has the second copy damaged during her lifetime, the normal chemical would no longer be present. Breast cells then would grow out of control.

Because BRCA1 is very long and complex gene subject to several different kinds of mutations, a reliable test for the gene could take some time to develop. For families where women have a high incidence of breast cancer, however, devising a clinical test to detect defective BRCA1 is very important. Also, gaining a greater understanding of how both normal and damaged BRCA1 genes work may provide additional clues for understanding other forms of the disease.

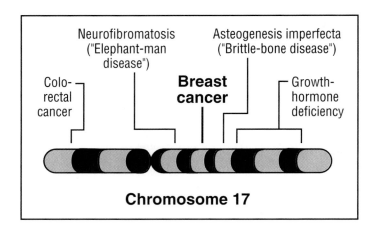

recognized. Oncogenes also are found in normal cells but in small amounts. Thus, oncogenes are considered normal cell-regulating genes that have gone awry.

At least three genes, called tumor-suppressors or anticancer genes, have been implicated in cancer. When these genes function normally, they make proteins that keep a cell's growth in check. If these genes become damaged, however, they are no longer able to produce those proteins, and cells become malignant. Researchers now think that these two types of cancer genes can work together—an oncogene turns dividing cells malignant after the tumor-suppressing gene has been damaged.

Scientists in many labs are working on new cancer drugs based on these tumor-suppressor proteins. The growth-controlling proteins would be made by healthy versions of the genes. These proteins then would be administered to patients to slow or reverse the growth of tumors. Eventually, it may be possible to replace patients' defective genes with healthy genes.

CONCEPT REVIEW

1. For what reasons do cells divide in adults?
2. Explain the differences between benign and malignant tumors.
3. Give a brief definition of cancer.
4. Compare and contrast the actions of oncogenes and tumor-suppressing genes.

INVESTIGATION 7.1   The Yeast Life Cycle

Baker's yeast (*Saccharomyces cerevisiae*) is a unicellular organism that reproduces both sexually and asexually. Because the cells have a characteristic shape at each stage, all the major stages of the life cycle can be seen under the microscope.

Yeast cells may be either haploid or diploid. Haploid cells occur in two mating types ("sexes"): mating-type **a** and mating-type α (alpha). When **a** and α cells come in contact, they secrete hormonelike substances called mating pheromones (FAYR uh mohnz), which cause haploid cells of the opposite mating type to develop into gametes. The **a** and α gametes pair and then fuse, forming a diploid zygote. The fusion of **a** and α gametes is similar to fertilization in animals, except that both parents contribute cytoplasm and nuclei. Yeast zygotes reproduce asexually by budding. When cultured on a solid growth medium, a yeast zygote may grow into a visible colony that contains up to 100 million cells.

Diploid yeast cells do not mate, but in times of stress, such as during a period of unbalanced food supply, the diploid cells may sporulate, or form spores. The spores remain together, looking much like ball bearings, in a transparent saclike structure called an ascus (ASK us).

In this investigation, you will start with two haploid yeast strains of opposite mating types, mate them in order to form a diploid strain, and try to complete an entire yeast life cycle.

Using yeast, students can observe all the stages in a sexual life cycle. The investigation connects Chapters 6 and 7 and serves as a bridge leading from the discussion of reproduction to the consideration of development and differentiation. For example, the production of four spores in an ascus is the visible result of meiosis, and the development of a gamete from a haploid cell in response to the mating pheromone is a visible example of differentiation. The procedure introduce simple microbiologic techniques and provide experience in preparing samples and observing them under the microscope. The diagrams make it easy to recognize the different types of cells. Teams of 2 work best.

Time: One or 2 weeks, depending on the incubation temperature. Plan to devote most of a class period to each of the days described.

Materials (per team of 2)

microscope slide
cover slip
dropping pipet
glass-marking pencil
compound microscope
container of clean, flat toothpicks
bottle of water

6 self-closing plastic bags labeled
    waste (1 for each day)
3 agar growth medium plates (YED)
1 "unknown" agar medium plate
agar slant cultures of **a** and α yeast
    strains

Figure 7.15 Procedure for making a mating mixture from **a** and α yeast strains (a), for subculturing the mating mixture (b), and for inoculating the "unknown" medium plate (c).

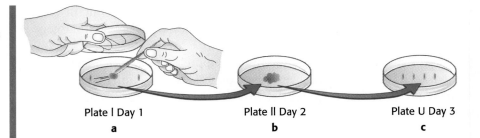

Plate I Day 1   Plate II Day 2   Plate U Day 3
a   **b**   **c**

SAFETY

Seal plates contaminated with mold or bacteria with masking or electrical tape if you wish students to examine them. Before discarding culture plates, disinfect them with household bleach or alcohol or by autoclaving. See "Safety Using Microbes" in *Guidelines for Laboratory Safety* **in the Teacher's Edition. Because these are** *Saccharomyces cerevisiae***, nonpathogenic laboratory strains of yeast, uncontaminated plates require no special treatment. It is good practice, however, to disinfect all cultures. Students should discard all used toothpicks in designated waste bags. Label self-closing plastic bags waste and tape one bag to each lab table. Collect these bags at the end of each class and place in a biohazard bag. Students also should discard used culture plates in biohazard bags. Autoclave all biohazard bags before discarding.**

Procedure

6. **Cream colored. (Cream color is the dominant form of the color trait. Students will study this trait in Investigation 8.3 and can be referred to this observation.) Under the influence of each other's pheromones, the mating-type cells should be developing into gametes. Because fusing gametes stick together at their smaller ends, the zygote is shaped like a peanut. The diploid zygote forms its first bud at the middle, so a budded**

Procedure
PART A Sexual Reproduction: Haploid to Diploid
Day 0

1. Prepare fresh cultures of the **a** and α yeast strains from the agar slants. Colonies of mating-type **a** are red; those of mating-type α are cream colored. Touch the flat end of a clean toothpick to the **a** strain, then gently drag it across the surface of an agar growth medium plate to make a streak about 1 cm long and 1 cm from the edge (see Figure 7. 15). Discard the toothpick in the self-closing plastic waste bag, being careful not to touch anything with it. Use a glass-marking pencil to label the bottom of the plate near the streak with an a. Use a new, clean toothpick to put a streak of the α strain near the opposite side of the agar on the same plate, and label it α. *Use a new, clean toothpick from the container for each thing you do. Be careful not to touch the ends of the toothpicks to anything except yeast or the sterile agar. Discard used toothpicks in the plastic waste bag. Keep the lid on the plate except when transferring yeast.* Label this plate *1* and add the date and your names. Incubate upside down for 1 day, or 2 days if your room is very cool.

2. Wash your hands thoroughly before leaving the laboratory.

Day 1

3. Use a microscope to examine some yeast cells of either mating type. To prepare a slide, touch a toothpick to the streak of either the **a** strain or the α strain in plate I, mix it in a small drop of water on the slide, and place a cover slip over the drop. Discard the toothpick in the waste bag. Examine the cells with the high power lens. In your data book, sketch the cells.

4. Use a clean toothpick to transfer a small amount of the **a** strain from the streak in plate I to the middle of the agar. Use another clean toothpick to transfer an equal amount of the α strain to the same place. Being careful not to tear the agar surface, thoroughly mix the two dots of yeast to make a mating mixture. Discard used toothpicks in the waste bag. Invert and incubate plate I at room temperature for 3 or 4 hours, then refrigerate until the next lab period. (Or, refrigerate immediately, and then incubate at room temperature for 3 or 4 hours before the next step.)

5. Wash your hands thoroughly before leaving the laboratory.

Day 2

6. Remove plate 1 from the refrigerator. What color is the mating mixture colony? Use the microscope to examine the mating mixture, as in Step 3. Discard used toothpicks in the waste bag. Sketch what you see and compare it with your earlier drawings. Describe any differences in the types of cells you see. When haploid cells of opposite mating types (**a** and α) are mixed, they develop into pear-shaped, haploid gametes. Do you see any gametes? A diploid zygote is formed when two gametes fuse. Growing diploid cells are slightly larger and more oval than haploid cells. Do you see any evidence that gametes may be fusing into zygotes? Compared to Day I, are there more or fewer diploid cells?

7. Make a subculture by transferring some of the mating mixture with a clean toothpick to a fresh agar growth medium plate. Discard used toothpicks in

the waste bag. Label this plate *11*. Invert and incubate at least overnight but not more than 2 nights.

8. Wash your hands thoroughly before leaving the laboratory.

Day 3

9. Use the microscope to examine the freshly grown subculture in plate 11. Discard used toothpicks in the waste bag. What types of cells are present? Sketch each type. If any of the types seen in Step 6 have disappeared, explain what happened to them.

PART B The Effect of the Environment on the Life Cycle
(Day 3, continued)

10. On a plate of "unknown" medium, make several thick streaks of the freshly grown subculture. Discard used toothpicks in the waste bag. Label this plate *U*. Invert and incubate at room temperature at least 4 days.

11. Wash your hands thoroughly before leaving the laboratory.

Day 7

12. Use the microscope to examine yeast from plate *U*. Discard used toothpicks in the waste bag. You may need to use the fine adjustment on the microscope to distinguish cells at different levels. What cell types are present now that were not present before? Sketch these cell types, and compare them with the cell types you saw at other stages.

13. Refer to the introduction. How do you think the "unknown" medium differs from the growth medium? If the cells in the sacs are most frequently found in groups of four, do you think they were formed by meiosis or mitosis? Explain. Are the cells in the sacs haploid or diploid? Explain. What part of the life cycle seen on Day 1 do these cells most resemble?

PART C Closing the Cycle: The Next Generation
(Day 7, continued)

14. Transfer some yeast from plate U to a fresh agar growth medium plate. Discard used toothpicks in the waste bag. Label this plate *111*, invert and incubate at room temperature for about 5 hours, and then refrigerate until the next lab period. (Or, refrigerate immediately and then incubate at room temperature for 5 hours before the next step.)

15. Wash your hands thoroughly before leaving the laboratory.

Day 8

16. Use the microscope to examine the growth from plate III. Discard used toothpicks in the waste bag. What life-cycle stages are present? Sketch the cells and compare them with the stages you observed before. What evidence is there that a new life cycle has started?

17. Discard used culture plates as directed by your teacher, and wash your hands throughly before leaving the laboratory.

Discussion

1. In the three parts of this investigation, you have observed the major events of a sexual life cycle. You could readily observe these events in yeast because it is a unicellular organism. In plants and animals, including humans, similar cellular events occur, but they are harder to see. Although the changes were too slow for you to see occurring, your sketches provide a record of the sequences. Think of them as "pauses" in a tape of a continually changing process. Notice in particular how the cycle repeats. Draw diagrams on one page, showing the different shapes of cells you observed in the order in which they appeared. Indicate where you first saw each type and when it disappeared if it did.

zygote resembles a fat clover leaf. It is difficult to discern the difference in size, but the more oval shape of diploids is apparent.

7. Growing a fresh culture from the mating mixture, at optimal nutrition and temperature conditions, is important for good sporulation. Transfer these cells to sporulation ("unknown") medium within 2 days without refrigeration.

9. There should be no gametes ("pears") or zygotes ("peanuts" or "clover leaves") in the culture, only growing haploid and diploid cells

12. The diploid cells should have sporulated on the YEKAC medium, so students should see clusters of four cells. Asci may be tetrahedral (in the shape of a pyramid), oval (all four lying flat), or (rarely) linear. Two- and three-spored asci are common. (The spores probably are still haploid; some do not mature for one reason or another. Cells that have not sporulated will appear distinctly unhealthy; many will be dead. Caution students that no organelles except a large vacuole can be seen in unstained yeast cells.)

13. Because sporulation occurs in response to food imbalance, the medium probably does not contain all necessary food sources for the yeast. (Carbon and nitrogen sources are most critical.) Meiosis, because it produces four haploid spores from a single diploid cell. Haploid, because the spores were formed by meiosis. These haploid spores resemble haploid gametes in terms of chromosome number per cell.

16. The cells should look much like those in the mating mixture in step 6 because each ascus contains two mating-type a and two mating-type α spores, which may mate soon after germination. There should be budded and unbudded haploid cells, zygotes, and budded zygotes, but no intact asci. The presence of the zygotes ("peanuts" or "clover leaves") indicates another life cycle has been started.

Discussion

1, 2. The students should be able to reconstruct the main sequence of events in the life cycle from their drawings by referring to the diagram. Comparison of the drawings made by different teams is useful. Have a few students pool the results of all the teams, then prepare a large-scale life-cycle diagram. Provide room for labeling the stages with the corresponding events in the human life cycle.

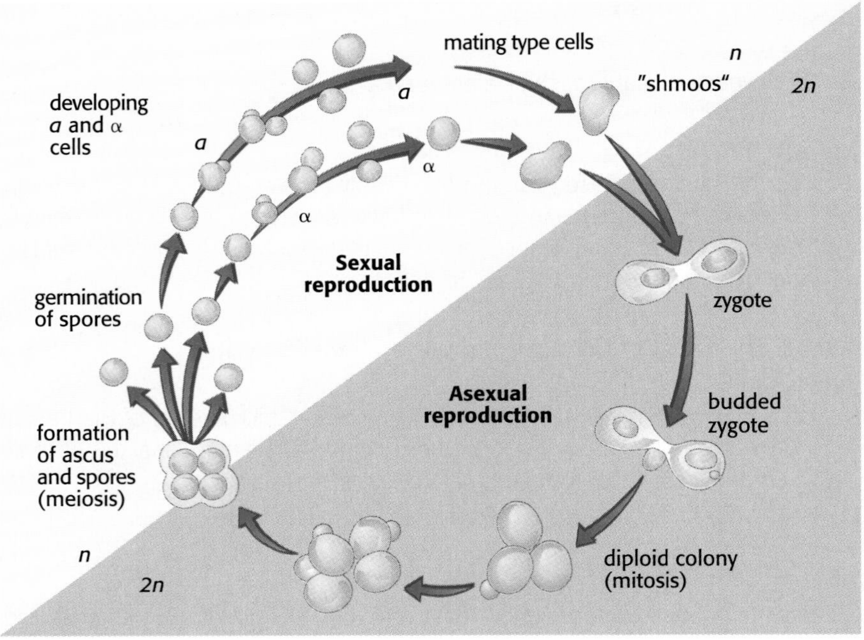

Figure 7.16 Yeast life cycle, showing both asexual and sexual reproduction.

3. **Spores, gametes, and all budding cells prior to zygote formation are haploid. Zygotes and budded zygotes are diploid. After zygotes have formed, but before the diploids have sporulated—steps 9 and 11—most cells are diploid because they outgrow the haploids.**

4. **Fusion and sporulation, respectively.**

5. **The two gametes contribute cytoplasm and nuclei equally to the zygote. Also, there are no readily observable secondary sexual characteristics. (There are, however, some subtle asymmetries in the roles of the two mating types at the biochemical level.)**

6. **In a cyclic process there is no beginning or end. (in the context of evolution, however, one can argue that the earliest form that was uniquely "yeast" was the beginning of the first cycle. Since mutation and selection operate more effectively on haploid forms, an asexually reproducing haploid form seems the most likely progenitor.)**

7. **Because a number of zygotes will form in this culture, there should be enough diploids for significant sporulation to occur. The frequency however, could be quite variable.**

8. **See Table 7.1.**

2. Compare your sketches with the diagram in Figure 7.16, and try to identify each of the forms you saw.

3. For each of the cell forms you observed, indicate whether it was haploid or diploid.

4. Mark the points in your diagram where cells changed from haploid to diploid and from diploid to haploid.

5. Why do you think the two different mating types are not called "female" and male"?

6. Can you think of a good argument for calling any particular point in the cycle the "beginning" or the "end"? Why or why not?

7. What would you expect to happen if you allowed the yeast cells in Step 14 to grow for another day and then put them on the "unknown" medium again?

8. Table 7.1 summarizes the similarities between stages in the yeast life cycle and the events in sexual reproduction in animals. For each change you observed in the yeast life cycle, indicate the step in the human sexual reproduction cycle that is most similar.

For Further Investigation

The mating mixture in Step 9 (Day 3) should contain mostly diploid cells, which cannot mate. Use the procedure of Step 4 to test mating ability by mixing samples with each of the **a** and α strains. What would you expect to see?

Table 7.1 Comparison of the Yeast Life Cycle and Sexual Reproduction in Animals

Yeast	Animals
Mating pheromones	Sex hormones
Mating type **a** and α gametes	Gametes (ova and sperm)
Fusion	Fertilization
Zygote	Zygote
Asexually reproducing diploid cells	Diploid body cells
Meiosis and formation of spores	Meiosis and formation of gametes

INVESTIGATION 7.2 Development in Polychaete Worms

The marine polychaete worm, *Chaetopterus variopedatus*, lives in a leathery, U-shaped tube in the sand near the low tide level. Figure 7.17 shows the major features of the worm. The sperm and eggs are visible inside the parapodia (leglike extensions from the body) near the posterior end of the worm. The sperm appear ivory-white and give the male parapodia a smooth, white appearance. The ovaries are in yellow coils, and the eggs inside them give the female parapodia a grainy appearance.

In this investigation, you will observe the worm, remove eggs and sperm, fertilize the eggs, and observe the development of the embryos. The initial stages of development you will observe appear to be the same for nearly all sexually reproducing organisms.

Materials (per team of 2)

microscope slides	dissecting scissors
cover slips	dissecting needle
dropping pipet	cheesecloth
finger bowls	paper towels
microscope	seawater
forceps	fresh water

Procedure

Your teacher will give you specific instructions on obtaining the eggs and sperm.

1. Get a small double-thickness piece of cheesecloth, rinse it in fresh water, and then rinse it in seawater. Place the wet cheesecloth in the bottom of a finger bowl.

2. Using forceps and dissecting scissors, remove 1 parapodium with eggs from the female worm and place it on the wet cheesecloth in the finger bowl. Use the pipet to remove any eggs that spilled from the parapodium and to drop them onto the cheesecloth. Cut open the parapodium and use the dissecting needle and forceps to release the eggs.

> **Dissecting needles are sharp, handle with care; replace cork on tip after use.**

3. Lift the cheesecloth and pour seawater into the finger bowl to a depth of 1 cm. Holding the cheesecloth by the four corners, gently lift and lower it in the water, and move it around slowly. The movement should allow most of the eggs to filter through the cheesecloth and drop into the seawater while sand and other debris stay on the cheesecloth. The eggs are tiny yellow or yellow-orange dots in the seawater. Note the time in your data book. Discard the cheesecloth in the container designated by your teacher.

4. You must wait 15 minutes from the time you put the eggs in seawater before you add sperm. *Do not add the sperm before 15 minutes have passed.* While you are waiting, add 10 mL of fresh seawater to another finger bowl. Remove a parapodium with sperm from the male worm and quickly place the parapodium into the finger bowl with the 10 mL of seawater. With a dropping pipet, pick up any sperm that were released in the bowl with the worm. (The sperm will appear as a small cloud in the water.) Add these sperm to your finger bowl.

5. Place one drop of the sperm and seawater mixture on a clean microscope slide and add a cover slip. Examine the sperm with low power of a microscope. They will appear as small, moving dots. Switch to high power and

Compared with *Xenopus* or sea urchins, *Chaetopterus variopedatus* (an annelid worm) has many advantages, and some disadvantages, in demonstrating developmental stages. The major advantages are reduced cost and the fact that hormone injections are not required. The major disadvantage is the small size of the embryos and the resultant need to use a compound microscope to see them. *Because these organisms are only available from May through December from Ward's Natural Science Establishment, you need to schedule this investigation carefully.*
Teams of 2 work best.

Time: If you have a double lab period, students will be able to complete all of the procedures in one day. If you have a 40-60 minute lab period, some staggering of tasks will be necessary.

Safety

Caution students to exercise care in handling dissecting tools. They should be stored in a tray or similar container when not in use. Place corks on the tips of dissecting needles and remind students to replace them after use.

Figure 7.17 Polychaete worm in a glass tube (top) and the worm's natural tube (bottom). The head is to the right, the parapodia (the fingerlike extensions) at the left.

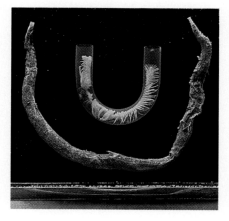

observe the sperm. If you see no moving sperm, take another sample from the 10 mL of seawater. If moving sperm still are not evident, remove another parapodium with sperm and repeat Steps 4 and 5.

6. After the eggs have been in the seawater for 15 minutes, add one drop of the sperm and seawater mixture to them. The eggs should all be fertilized within 30 minutes. After 30 minutes, remove the eggs from the finger bowl with the sperm and place them in another finger bowl with fresh seawater. The development of the embryos will take place in this bowl.

7. While you are waiting for the 30 minutes to pass, obtain fertilized eggs from bowls designated by your teacher and examine them under the microscope. Look for embryos at the 4-, 8-, and 16-cell stages and note any other developmental stages. Your teacher may also direct you to transfer fertilized eggs prepared by other classes to bowls of fresh seawater.

8. Wash your hands thoroughly before leaving the laboratory.

Discussion

1. Examine the illustration of a human sperm in Figure 6.4. How do the polychaete worm sperm compare with the picture of a human sperm? Describe any similarities or differences.

2. Describe the development process you observed with the polychaete worm embryos.

3. The polychaete worm larvae usually develop within 24 hours after fertilization. Based on what you observed and on your knowledge of cleavage and cell division, approximately how many cell divisions have occurred between the 1-cell stage and the swimming larvae?

4. Describe any similarities between the development of the polychaete worm embryos and the development of human embryos.

DISCUSSION

1. **They are nearly identical.**
2. **This will vary but should include cleavage and a progression of cell divisions through to the larval stage.**
3. **Accept any reasonable response—probable somewhere around 30-50 cell divisions.**
4. **Both go through cleavage beginning with the zygote; the initial stages are virtually identical.**

SUMMARY

Development is the series of events that begins when a zygote starts to divide and grow. Cells that are similar in early divisions gradually specialize to perform certain tasks. This differentiation appears to result in part from the action of regulator molecules located in different areas of the zygote cytoplasm. As the zygote divides, each cell receives duplicate genetic material but different regulators. These regulators influence the genes that direct cell specialization. Cell-to-cell interactions can determine when these regulators "switch on" specific genes, which in turn direct the production of specific proteins.

Development is aided by cell movements and tissue foldings. In animals, the blastula becomes a gastrula—an embryo that has three distinct layers of cells, each of which gives rise to specific tissues. One or more tissues may make up an organ, and groups of related organs make up organ systems.

Early animal development takes place in an aqueous environment. In mammals, the fetus develops within the mother and is surrounded by a fluid filled amnion. The fetus gets food and other substances from the mother's body systems, and so it is affected by substances in the mother's blood, such as alcohol or other drugs. Developing organisms continue to change throughout life.

These changes include aging, which eventually ends the life of the individual. The life span of an organism apparently is a part of the genetic program determined at fertilization.

Normally, development is under strict control, and cells in mature organisms rarely divide. When cells begin to divide without limit and invade other tissues, cancerous growth results. Oncogenes and tumor-suppressing genes are implicated in controlling cell division.

APPLICATIONS

1. Differentiation leads to specialized cells, tissues, and organs. Initially, in the embryo, it involves formation of specific cell layers. What are these embryonic layers, and what structures in the mature animal are formed from each of them?

2. Cloning is the process of producing several genetically identical individuals from a single organism. This practice is relatively common in research involving plants. How is cloning aided by the fact that almost all cells within an organism contain identical sets of genes?

3. If growth is defined as an increase in the size and number of cells, explain why there is no real growth in the developing organism through the blastula stage.

4. Suppose you take the regulatory molecules that control the production of fingers in humans and inject them into an oak tree. Will the oak tree develop fingers? Explain.

5. Do you think the word *development* applies to single-celled organisms? Do you think the word *differentiation* applies?

6. Water is necessary for the development of animal embryos. Frogs, fish, and many other organisms lay their eggs directly in the water. What are some of the disadvantages of such a reproductive system?

7. In what ways is a bird or reptile egg like the amnion of a human?

8. In what ways is the medical problem of cancer related to an understanding of developmental processes?

PROBLEMS

1. Generally, the ability to regenerate missing tissues or organs decreases during embryonic development as specialization increases. Investigate why this is so.

2. Find out about the growth and development of plant embryos. How are such embryos different from, and how are they similar to, the embryos of animals?

3. Using a frog blastula, design a hypothetical experiment that might show the importance of cell position in the blastula for normal tadpole development.

4. Why do cells, and people, grow old? Investigate several hypotheses biologists have made about aging, and identify the strong and weak points of each.

5. Carcinogens are cancer-causing agents. Make a list of potential carcinogens found in and around your home or school. What can be done to reduce your exposure to them, and what ways can you think of to reduce their use?

Continuity Through Heredity

8

How closely does this young Koala bear resemble its mother? How would you explain this phenomenon? Aside from contributing genetic information, how else might parents influence their offspring? What might the mother Koala be teaching her offspring here?

The offspring and the mother look similar because of the inheritance of genetic information. Parents contribute not only genes but experience as well. Many times offspring are taught necessary or beneficial behaviors by their parents. Here, the mother probably is teaching food gathering to her offspring.

◆ Bob and Mary were the proud parents of a new baby girl they named Lisa. Several weeks after Lisa came home from the hospital, Mary and her pediatrician noticed that Lisa was not gaining weight as fast as she should. In fact, she was losing weight. After running some tests, the doctor told Bob and Mary that Lisa probably had cystic fibrosis.

Children with cystic fibrosis secrete unusually thick mucus, which accumulates and damages the lungs, digestive tract, and other organs (see Figure 8.1). Untreated, children with this disorder normally die by the age of 4 or 5 from liver damage, pneumonia, or a secondary infection. With appropriate medication and treatment, however, they may live to adulthood.

Bob and Mary were crushed by this tragedy. They also were angry and perplexed. Although cystic fibrosis is the most common genetic disorder among Caucasians, the disorder had not been known to occur in either of their families. How was it possible for them to have a child with cystic fibrosis?

The genetic counselor in the hospital told them the defective gene had been present in both families for many generations. A defective gene is an altered form of a gene that is essential for normal development. The appearance of the disorder was a matter of chance.

How can an individual carry a defective gene without also having the altered trait? What is the chemical basis of genes that gives them the ability to carry information from parent to offspring and to direct the biochemistry of the organism?

This chapter discusses genes and inheritance. It describes the major events that have led to our understanding of how genetic continuity is achieved. ◆

The Dual Roles of Genetic Material

8.1 Genes Determine Biological Potential

Heredity is the transmission of genetic information from one generation to another. The basic unit of information passed along is a gene, which also controls the basic functions of the cell (see Chapter 5). Geneticists study heredity and gene action together, because the visible effects of gene action in individuals and populations provide clues to the genes that are present in an organism or in a population and how they work.

The information in genes is stored in the sequence of nucleotide bases that make up DNA. In response to environmental influences, this information, a type of molecular code, directs all the cell processes involved in the

MAJOR CONCEPTS

- ◆ Genes are the units of inherited information.
- ◆ Inheritance occurs in regular patterns that can be predicted by the rules of probability.
- ◆ Genes code for the production of specific proteins.
- ◆ The chemical makeup of genes tends to be stable.

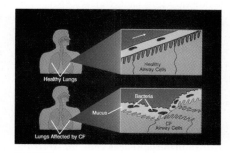

Figure 8.1 People with cystic fibrosis produce a thick mucus that damages the airways of the lungs.

- ◆ Chromosomes and genes are made of DNA.
- ◆ Genetic variation, arising from mutation and recombination, is the raw material of evolution.
- ◆ Risks and benefits must be weighed for the products of genetic engineering technology.

KNOWLEDGE CHECK

- ◆ What is inheritance?
- ◆ How is DNA involved in inheritance?
- ◆ What are genes?
- ◆ What are dominant and recessive traits?
- ◆ What is genetic engineering?

Guidepost
How was the genetic riddle solved?

The study of genetics has shown that all living organisms use the same information storage, transfer, and translation system. That similarity provides an explanation for the stability of life and its possible descent from a common ancestral form.

163

Figure 8.2 Gregor Mendel (1822–1884). Experimenting with peas in his monastery garden, Mendel developed the fundamental principles of heredity that became the foundation of modern genetics.

It is sometimes possible to predict the most probable outcome in any randomly ordered event, but the expected outcome may not match the actual results. Emphasize that a prediction based on observed data is just that—a prediction— and not a guarantee of a particular result.

Not all DNA sequences code for structural proteins or enzymes. Some sequences code for tRNA and rRNA, others code for regulatory proteins, and as much as 90 percent of the DNA in some organisms consists of introns, repeated sequences, and other noncoding sequences of unknown function.

In a sense, the primary function of an organism is to reproduce genes; the organism is merely a vehicle to transfer genes from one generation to its descendants. Ask your students how they feel about this idea.

Section 13.4 includes a detailed discussion of reproduction in flowering plants. The intent here is to provide only sufficient information to allow students to understand Mendel's experiments.

The scientific study of heredity did not begin with Mendel, but he made fundamental changes in biological thinking.

development and function of the organism, including both the reproduction of genes and the effects of the genes themselves. The genes passed along through reproduction provide the continuity between generations that is essential for the continuation of a species. They also provide instructions for the structure, function, and development of an organism during each generation.

How can we predict whether a particular hereditary trait will appear in an offspring of two organisms? When you predict something, you make a statement that something will or will not happen with a certain amount of confidence. In the study of heredity, this type of prediction is expressed in terms of **probability,** an application of mathematics that predicts the chances that a certain event will occur. Geneticists use probability to predict the outcomes of matings. You can develop the rules of probability for yourself by doing Investigation 8.1.

8.2 Mendel's Work Led to the Concept of the Gene

Today we know that each gene brings about the synthesis of a particular protein, such as an enzyme, a muscle protein, or a protein pigment that may affect skin color. The instructions for the synthesis of a particular protein are stored in the form of specific sequences of DNA nucleotides, which make up the genes. The genes are organized into chromosomes. Through the processes of meiosis and fertilization, chromosomes are passed on to each new generation.

Although these ideas may sound simple, it has taken centuries to assemble the information. Most of our knowledge about genetics has been gained in just the last 100 years. Biologists have joined together various hypotheses and experimental results, as if they were working on a jigsaw puzzle, to give us a broad picture of what happens in inheritance.

The first part of the puzzle was put together in a monastery in Austria. There, in the 1860's, Gregor Mendel (see Figure 8.2), a monk who was trained as a mathematician and natural scientist, set forth the basic principles of heredity. Mendel was interested in the inheritance of animal and plant features, or traits, and he began an eight-year series of experiments with ordinary garden peas. Garden pea flowers contain both male and female reproductive parts, as shown in Figure 8.3. Although under natural conditions the pea plant usually self-pollinates, it is possible to collect pollen grains from flowers of one pea plant and transfer them to the flowers of another plant. This technique, known as cross-pollination, results in seeds that are the offspring of two plants, not just one.

Mendel's work, which we now call Mendelian genetics, provided the basis for the modern study of heredity and variation. His experiments were unique in four important ways. First, he concentrated on one trait at a

Figure 8.3 Self-fertilization in a pea flower. The petals of the pea flower (a) completely enclose the reproductive organs. As a result, the pollen from the anthers falls on the stigma of the same flower (b). Pollen tubes grow down through the female reproductive organ to the ovules (immature seeds) in the ovary. The ovules develop into seeds, and the ovary wall develops into the pea pod (c).

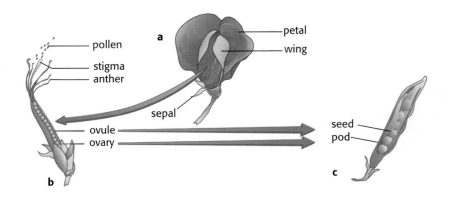

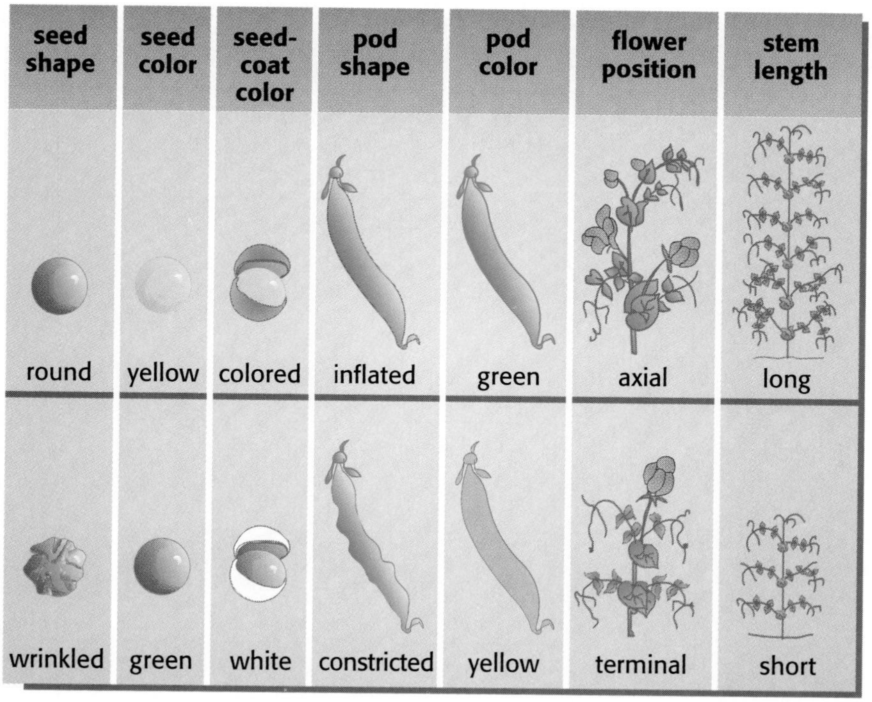

seed shape	seed color	seed-coat color	pod shape	pod color	flower position	stem length
round	yellow	colored	inflated	green	axial	long
wrinkled	green	white	constricted	yellow	terminal	short

Figure 8.4 The seven traits of garden peas Mendel studied.

2 time. Second, he used large numbers of organisms to minimize the influence of chance on his data. Third, he combined the results of many identical experiments. Fourth, he used the rules of probability to analyze his results. By using these methods, Mendel was able to recognize distinctive patterns of inheritance.

To begin his work, Mendel collected various strains of peas and tested each strain to make certain it was genetically pure, or pure breeding. Pure breeding plants produce offspring that are identical to themselves generation after generation. Mendel worked with strains that showed one of two different forms of the same trait. For example, plants were either tall or short or produced either green or yellow seed colors. By having distinct and contrasting forms of one trait, Mendel could follow the differences in the offspring. The seven traits shown in Figure 8.4 are the ones Mendel worked with in his experiments.

In his first experiments, Mendel crossed pure-breeding plants that grew from round seeds with pure-breeding plants from wrinkled seeds. Would the offspring produce round seeds, wrinkled seeds, or something in between? Mendel found that *all* the plants from this cross produced round seeds. The wrinkled form of the trait seemed to have disappeared. Today we call the parents of such a cross the P_1 (for parental) generation. The offspring are called the first filial, or F_1, generation.

Mendel then allowed the F_1 seeds to grow into plants and to self pollinate. The resulting seeds were planted and gave rise to the F_2 (second filial) generation. Of these F_2 plants approximately ¾ produced round seeds and ¼ produced wrinkled seeds. The wrinkled seeds that seemed to have disappeared in the F_1 generation reappeared in the F_2 generation. Mendel called the round seeds in all the F_1 plants the **dominant** form of the trait. He called wrinkled seeds that had reappeared in the F_2 generation the **recessive** form of the trait.

Mendel repeated this same two-generation cross for six other traits. Table 8.1 shows the data from the F_2 generations in all his experiments. He then calculated the ratio of dominant to recessive forms for each trait. In

Mendel conducted most of his studies between 1856 and 1864, using thousands of plants.

Remind students that a pea seed is the beginning of the next generation. The seed characteristics of the parent plants are those of the seeds from which they grew.

This observation is known as Mendel's *principle of dominance*. Note, however, that alleles are not dominant or recessive—only their effects on a trait are. At the molecular level, both alleles are expressed, although the effects of one allele (for the dominant form) may mask those of the other (for the recessive form).

Table 8.1 should be a focal point of discussion. Students should know what a ratio is and become aware of the significance of the data. On the basis of later experimental work, some biologists have declared that Mendel must have fudged his figures. More charitable writers have suggested that Mendel intended his figures to be illustrative only. We cannot know which is true. Mendel's logical reasoning and experiments, however, certainly led him to inclusive principles that have stood the test of time.

Table 8.1 Results from Mendel's Experiments

P_1 Cross	F_1 Plants	F_2 Plants	Actual Ratio
1. round × wrinkled seeds	all round	5474 round 1850 wrinkled 7324 total	2.96 : 1
2. yellow × green seeds	all yellow	6022 yellow 2001 green 8023 total	3.01 : 1
3. colored × white seed coats	all colored	705 colored 224 white 929 total	3.15 : 1
4. inflated × constricted pods	all inflated	882 inflated 299 constricted 1181 total	2.95 : 1
5. green × yellow pods	all green	428 green 152 yellow 580 total	2.82 : 1
6. axial × terminal flowers	all axial	651 axial 207 terminal 858 total	3.14 : 1
7. long × short stems	all long	787 long 277 short 1064 total	2.84 : 1

Supplement this discussion with an abundance of simple problems involving other organisms as well as Mendel's peas. Sample problems are provided in *Supplementary Topics* in the Teacher's Resource Book.

each case, the dominant form appeared in about ¾ of the F_2 plants, and the recessive form appeared in about ¼ of the F_2 plants. All the experiments showed the same 3 : 1 ratio no matter what trait was tested. What was the meaning of this result?

8.3   Mendel Identified the Unit of Heredity

Mendel proposed that each pure-breeding plant had two identical copies of a "factor" for a particular trait. He did not know where the "factors" were located in cells, but he hypothesized that only one copy of a factor went into each sperm or egg cell when gametes were formed. If a parent were pure-breeding "round," for example, all its gametes would have the "round seed factor." The same would be true for the "wrinkled-seed factor." Thus, the offspring of a round and wrinkled cross would receive a round-seed factor from one parent and a wrinkled-seed factor from the other parent. Mendel called this separation of factors the **principle of segregation.**

Today, we call Mendel's "factors" genes, and we call the alternative forms of a gene **alleles** (al LEELZ). The allele for the dominant form is commonly represented by a capital letter (in this cross, *R* for round seeds), and the allele for the recessive form is represented by the same letter in lower case (*r* for wrinkled seeds).

Study Figure 8.5 to help you understand Mendel's results. The symbols in the diagram represent the alleles in three generations of pea plants. The alleles form the genetic makeup, or **genotype** (JEE noh typ), of the plants. The genotype of an individual is responsible for its **phenotype**

The *R/r* notation is a shorthand way of indicating that and *R* and *r* are from different parents.

(FEE noh typ)—its appearance or observable traits. Note that both the *R/R* and *R/r* genotypes produce the same phenotype—round seeds.

In a pure-breeding plant that produces only round seeds, both alleles are the same in every cell: *R/R*. In the same way, pure-breeding plants that produce only wrinkled seeds also have two alleles that are the same, in this case *r/r* pure-breeding plants in which both alleles are alike are called

4 **homozygous** (HOH moh ZY gus; from the Greek *homo*, meaning alike, and *zygo*, meaning a pair). When the chromosome pairs containing the alleles separate during meiosis, each gamete receives only one allele. A plant homozygous for a particular trait can produce only one type of gamete (probability = 1, or 100%).

When gametes carrying an *R* allele unite with gametes carrying an *r* allele, all the offspring have the combination *R/r* (probability = 1). The new organisms are **heterozygous** (HET er oh ZY gus; from the Greek *hetero* meaning different) with respect to the alleles for round and wrinkled seeds. Heterozygous offspring are called **hybrids**. In his experimental

4 crosses, Mendel used *R/R* plants as either the male parent or the female parent. No matter which parent contributed which type of gamete, all F_1 plants produced round seeds. Apparently only one *R* allele is needed to direct the plant to produce round seeds, which are the dominant form of the trait for seed shape.

Both the male and female parts of *R/r* heterozygous F_1 plants produce two types of gametes: ½ carry *R* and ½ carry *r*. If the F_1 flowers self pollinate, fertilization is random. Any sperm nucleus can fertilize any egg nucleus, regardless of its genotype. Figure 8.5 shows how the probabilities are calculated for an *R/r* and *R/r* cross that gives rise to an F_2 generation.

By combining the ideas of probability and random mating as Mendel did, the mathematical regularity in these crosses becomes clear. If more than one trait is involved in the cross, the explanation is more complicated, but the same basic principles apply.

Mendel's experiments defined the basic unit of inheritance, but he could provide no information about its physical or chemical nature.

Figure 8.5   One of Mendel's crosses using round and wrinkled peas. Note that the alleles of a gene segregate during gamete formation. What is the probability of *R/r* in the F_2?

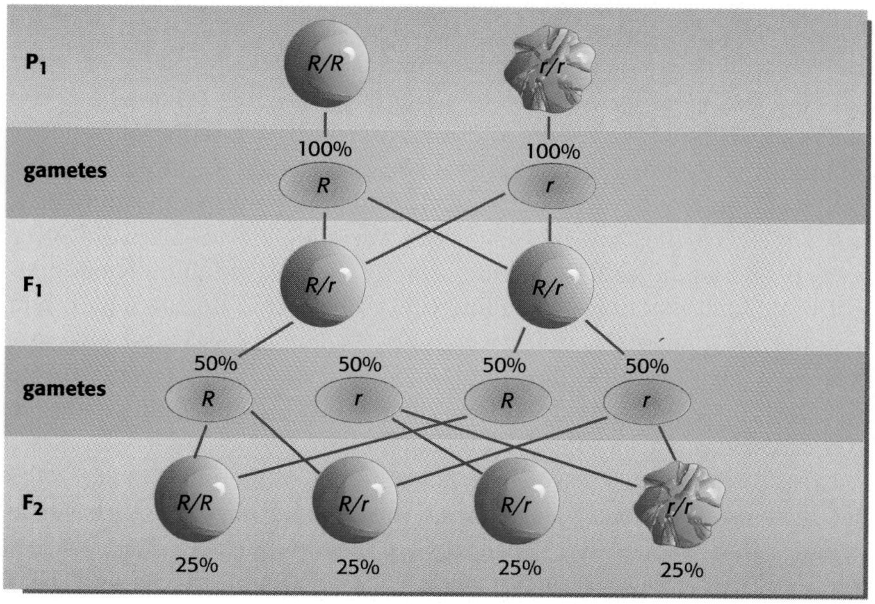

Work through Figure 8.5 with students to be sure they understand how to arrive at the probabilities shown.

Figure 8.6 Procedure used by Beadle and Tatum. In this case the *Neurospora* spore has lost the ability to synthesize substance c.

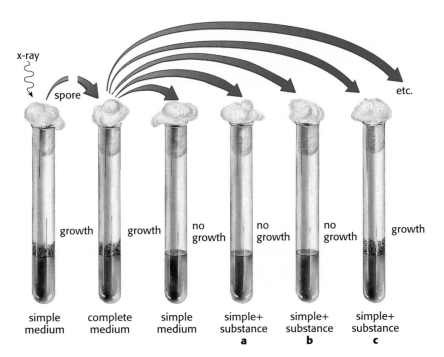

Unable to see such a unit in his studies, he inferred its existence from his experiments. The answer to the riddle of what genes are and how they work could not be known until nearly 100 years later, when techniques for studying biological molecules were developed.

8.4   Genes Direct Biosynthesis

Our understanding of the chemical basis of heredity has come primarily from studies of organisms less complex than garden peas. A giant step toward understanding how genes function came from studying a pink bread mold called *Neurospora crassa* (noo ROS poh ruh CRAS uh). *Neurospora crassa* reproduces both asexually by means of spores and sexually.

Normally, the asexual spores develop into mold that can grow on a nutrient-deficient, or minimal, medium because the mold itself produces most of the nutrients it needs. In the 1940's, George W. Beadle and Edward L. Tatum used X rays to produce heritable changes in the mold. Changes that can be passed on to offspring are called **mutations** (myoo TAY shunz). Beadle and Tatum found that the spores of some mutated molds could no longer grow on a minimal medium. For many of these mutants, however, the addition of a single vitamin or amino acid to the medium permitted the spores to grow. Apparently the radiation-induced mutations prevented the synthesis of some substance the mold normally produced. Although each mutant usually required only a single nutrient, it was not always the same nutrient, and some mutants lacked nearly every vitamin and amino acid the mold normally synthesized. If enzymes control biological reactions, including biosyntheses, as Beadle and Tatum hypothesized, then the defective molds lacked an essential enzyme. Adding a supplement to the simple medium would compensate for the lack of this enzyme and allow the mold to grow. Figure 8.6 illustrates the experimental procedure Beadle and Tatum used.

Beadle and Tatum's work included studies of the inheritance of the deficiencies they induced. It was from these studies that biologists began to understand what genes do. *Neurospora crassa,* like most molds, is able to reproduce sexually. It occurs in two mating types (sexes) that can be **6** crossed to yield offspring in much the same way plants reproduce. Beadle and Tatum found the nutritional deficiencies in the molds were inherited in the same way as the traits Mendel studied. In other words, the X rays produced mutant alleles of normal genes. Other experiments by Beadle and Tatum demonstrated that the forms of the traits produced by the mutant alleles were recessive to the forms produced by the normal alleles.

Eventually, each missing substance in the growth medium was linked to a particular missing enzyme in the mold. Each missing enzyme was the result of a single gene mutation. Beadle and Tatum proposed that each gene specifies the synthesis of one enzyme; they called this the one-gene one-enzyme hypothesis.

We now know that genes control not only the synthesis of enzymes but also the synthesis of all other proteins. Somehow a gene dictates the way amino acids are combined to form large protein molecules. In other words, genes code for protein synthesis. What in the genes controls the coding? The answer to that question would fill in the next major portion of the genetics picture.

> Because there is not a stable diploid phase in the *Neurospora crassa* life cycle, they employed a more indirect, but entirely valid, procedure of testing for dominance in molds that contain two different types of haploid nuclei.

CONCEPT REVIEW

1. What are the chances that a family with three children will have three daughters? After a family has two daughters, what is the probability that their next child will be a daughter?
2. How was mathematics important in Mendel's explanation of his results?
3. Distinguish between the terms *gene* and *allele.*
4. Distinguish between *genotype* and *phenotype*; between *homozygous* and *heterozygous.*
5. Explain how the principle of segregation applies to the movement of chromosomes in meiosis.
6. How do Beadle and Tatum's experiments provide a chemical explanation for dominance and recessiveness?

Chromosomes and Genes

8.5   Genes Are on Chromosomes

The foundation for Beadle and Tatum's experiments had been established earlier in the century when other biologists had found physical evidence of genes—Mendel's "factors." By 1902, the connection between the activity of chromosomes during meiosis and fertilization and the actions of genes in transmitting hereditary information had become clear. Experiments with egg and sperm cells had shown that the gametes make equal genetic contributions to a new organism. Two biologists—Theodore Bavaria and W. S. Sutton—had reasoned independently that if the contributions of sperm and eggs are the same, the genes must be located in corresponding parts of the two types of gametes. They noted that the nuclei of the two gametes are similar even though their cytoplasm differs greatly. Both men concluded that genes are located in the nucleus.

In the 1930's, various studies using the fruit fly *Drosophila melanogaster* (droh SOF il uh mel an oh GAS ter), shown in Figure 8.7, provided experimental support for the conclusions of Bavaria and Sutton. Work with *Drosophila* clearly demonstrated the direct relationship that exists between genetic and chromosomal events and firmly established the

Guidepost
What are the relationships between chromosome structure and gene function?

Figure 8.7   An adult fruit fly, *Drosophila melanogaster.*

Figure 8.8 The experiments of Alfred Hershey and Martha Chase showed that DNA carries the hereditary instructions of viruses. In (a), phages (viruses that attack bacteria) with protein coats labeled with an isotope of sulfur (³⁵S) were allowed to infect bacterial cells. In (b), phages with their DNA cores labeled with an isotope of phosphorus (³²P) were allowed to infect bacterial cells. Later, the bacteria and the phage particles that grew in the cells were tested for radioactivity. The presence of radioactivity inside only the bacteria that were infected with ³²P-labeled phage showed clearly that only the DNA of the phage entered the bacterial cell.

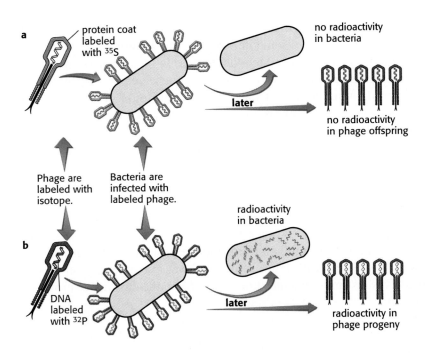

The advantages of using *Drosophila* in genetic research are many. The flies can be grown in bottles and fed on rotten fruit or yeast; the life cycle is completed in two weeks; females can lay hundreds of eggs in only a few days. That means nearly 30 generations, consisting of thousands of flies, can occur in one year.

chromosome theory of heredity, which states that genes are small particles located on the chromosomes.

In recent years, geneticists have learned that each chromosome contains a continuous sequence of DNA, which is associated with specialized proteins that help regulate gene activity. In other words, the genes are integral parts of the chromosomes.

Although the chromosome theory of heredity was established early in the twentieth century, the location of the genes within the chromosomes was less certain. Biologists knew that chromosomes contain both protein and DNA. They did not know, however, which one made up the genes and directed protein synthesis.

In 1952, Alfred Hershey and Martha Chase performed a series of experiments using viruses that attack bacteria. The viruses that Hershey and Chase used have a core of DNA (or sometimes RNA), surrounded by a protein coat. When Hershey and Chase labeled the protein coat with a radioactive marker in order to test for its presence, the bacteria did not show the marker (see Figure 8.8a). When they labeled the viral DNA with a radioactive marker, however, they found the labeled DNA had entered bacterial cells (see Figure 8.8b). Once inside the cell, the viral DNA took over the cell's machinery and materials and made new virus particles. The new viruses were similar to the original ones and had newly made protein coats. If the viral DNA was labeled with a radioactive marker, the offspring were radioactive also. It became clear to Hershey and Chase that DNA, not the protein coat surrounding the DNA, directed the synthesis of protein and transmitted the virus features from one generation to the next. These and many other experiments convinced scientists that DNA is the genetic material in chromosomes.

Piece by piece, experiment by experiment, biologists were learning how genes are inherited and what they do. These studies set the stage for identifying the molecular structure of the gene and its dual work in heredity and biochemical function.

8.6   Genes Are Long Chains of DNA Nucleotides

Once it was known that genes are composed of DNA and code for specific proteins, molecular biologists could begin to answer more specific questions about the chemical basis of genetics. What is the molecular structure of genes? How do they coordinate their dual roles of transmitting information to the next generation and programming the structure and function of the organism? What is the role of RNA, the other nucleic acid?

In 1953, J. D. Watson and F.H.C. Crick proposed a structure for DNA that explained how DNA could pass information from generation to generation. They hypothesized that a DNA molecule is a long, twisted, double stranded structure. Each strand consists of a chain of nucleotides, and each nucleotide consists of three smaller units: a sugar, a phosphate group, and a nitrogen-containing base. There are four types of nucleotides in DNA, and each one has a different base: adenine, thymine, cytosine, or guanine. The sugars and phosphates form the sides of the strands and join the nucleotides together. Each base from one strand pairs with a base from the other strand. In this way, thousands of base pairs join the two strands together, as shown in Figure 4.25.

According to Watson and Crick, the two strands are complementary because the molecular shape of each base can pair only with a particular complementary base. Adenine (A) can pair only with thymine (T), and cytosine (C) can pair only with guanine (G). These A—T and C—G base pairs occur along the entire length of a DNA molecule. The chemical bonds that hold the bases together are weak, however. When they are broken, two separate, complementary strands result. In DNA replication, each old strand serves as a pattern for the formation of a new, complementary strand. The result is two identical double strands of DNA, each of them exactly like the original double-stranded molecule. Figure 8.9 shows this process, which you can practice in Investigation 8.2.

The molecular DNA model developed by Watson and Crick explains several essential properties of genes: (1) genes replicate (exactly duplicate) during cell division; (2) genes mutate occasionally; and (3) genetic instructions are passed from generation to generation through meiosis and fertilization.

Continuity in the Biosphere

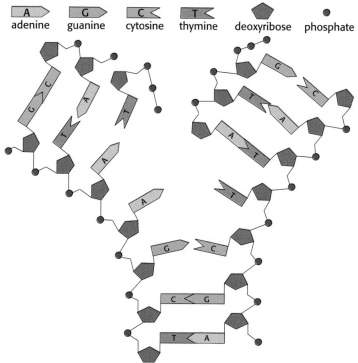

A adenine   G guanine   C cytosine   T thymine   deoxyribose   phosphate

Help students understand that to qualify as genetic material, a substance must have the following characteristics: it must reproduce accurately generation after generation, it must mutate infrequently, and it must be able to express its information.

Watson and Crick worked with data collected by M.H.F. Wilkins, Rosalind Franklin, and E. Chargaff. Watson, Crick, and Wilkins shared a Nobel Prize in 1962 for the model of DNA structure.

The story of the discovery of DNA and of the race won by Watson and Crick to discover its structure is dramatic as well as educational. Urge interested students to read Watson's entertaining personal account, *The Double Helix*.

Figure 8.9 Replication of DNA. The strands come apart at the bonds between the nucleotides. New nucleotides, which temporarily bear extra phosphates, are added one by one. Eventually two new DNA molecules are produced.

Still, biologists did not know how the genetic instructions, coded in sequences of nucleotides in nuclear DNA, are translated into the structure and function of proteins in the cytoplasm. What, they wondered, was the link between instructions for a protein and the construction of the protein in the ribosomes found in the cytoplasm.

8.7   DNA Makes RNA and RNA Makes Protein

The molecule that carries the information from the chromosomes in the nucleus to the ribosomes in the cytoplasm is RNA. Like DNA, RNA consists of chains of nucleotides. In each RNA nucleotide, however, the sugar is ribose instead of deoxyribose, and thymine is replaced by a similar nitrogen base called uracil. RNA is synthesized by copying one of the complementary DNA strands.

Three different types of RNA occur in cells. One type—ribosomal RNA (rRNA)—along with a number of proteins, makes up the ribosomes. Another type is messenger RNA, or mRNA. It carries the DNA message from the nucleus to the ribosomes. The third type, transfer RNA (tRNA), carries amino acids to the ribosomes. These amino acids are added to a growing chain of amino acids.

As long chains of amino acids are completed, proteins are formed. The 20 different types of amino acids found in proteins can be connected in any order. The structure and function of each protein is determined by the amino acids that are present and by the order in which the amino acids are attached. DNA, then, provides information about the precise sequence of amino acids in the protein chain. How can the DNA instructions guarantee the sequence?

For a moment, imagine the bases in DNA (A, T, C, and G) are like the letters in a sentence written in a secret code (see Figure 8.10). Somehow the code must specify all the arrangements of 20 different amino acids. Does

In prokaryotes, the translation of mRNA into protein can begin even before synthesis (transcription) is complete. In eukaryotes, however, RNA must pass through the nuclear membrane into the cytoplasm before translation can begin.

Figure 8.10 The DNA genetic code is written in mRNA. Each triplet is a codon for an amino acid. For example, UGG codes for the amino acid tryptophan. The codon AUG signals start, and UAA, UAG, and UGA signal stop. Several amino acids have more than one codon.

First letter	Second letter				Third letter
	U	**C**	**A**	**G**	
U	phenylalanine	serine	tyrosine	cysteine	**U**
	phenylalanine	serine	tyrosine	cysteine	**C**
	leucine	serine	stop	stop	**A**
	leucine	serine	stop	tryptophan	**G**
C	leucine	proline	histidine	arginine	**U**
	leucine	proline	histidine	arginine	**C**
	leucine	proline	glutamine	arginine	**A**
	leucine	proline	glutamine	arginine	**G**
A	isoleucine	threonine	asparagine	serine	**U**
	isoleucine	threonine	asparagine	serine	**C**
	isoleucine	threonine	lysine	arginine	**A**
	(start) methionine	threonine	lysine	arginine	**G**
G	valine	alanine	aspartate	glycine	**U**
	valine	alanine	aspartate	glycine	**C**
	valine	alanine	glutamate	glycine	**A**
	valine	alanine	glutamate	glycine	**G**

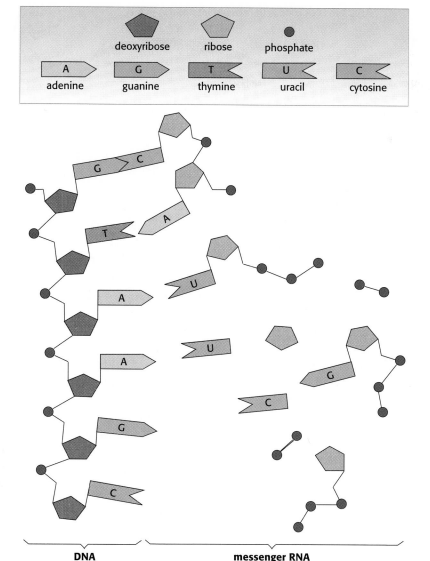

deoxyribose   ribose   phosphate

A adenine   G guanine   T thymine   U uracil   C cytosine

DNA   messenger RNA

Figure 8.11   Formation of part of a strand of mRNA along one strand of a DNA molecule.

each base in a DNA strand code for one amino acid? No, because the four bases could code for only four amino acids. A code of 2 bases could account for 16 amino acids—still not enough. A sequence of 3 bases, however, could provide 64 combinations of the 4 bases, more than enough to code for 20 amino acids. For this reason, scientists assumed the bases are arranged in codons, or "words," of three and that each codon stands for one amino acid. Experiments have supported this assumption. It took several years of work to determine which codon, or triplet of bases, codes for each amino acid. Most of the 64 triplets code for some amino acid; several amino acids are specified by more than one codon, as shown in Figure 8.10.

The steps by which genes coordinate molecular activity are now clear. DNA codes for protein structure. The coded instructions are transcribed when RNA is synthesized along one of the two DNA strands (see Figure 8.11). In its turn, mRNA carries these instructions from the nucleus to the ribosomes. In the ribosomes, a triplet of mRNA nucleotides specifies which tRNA will bring in a particular amino acid. At the ribosome, the amino acids are attached, one at a time, to the end of the growing protein chain. The sequence of amino acids determines how the chain can fold into a three-dimensional structure. At the end of the process, a new protein molecule has been formed. Figure 8.12 illustrates these steps.

4

The essential features of molecular genetics important for students are: (1) the self replicating ability of DNA; (2) the ability of DNA to direct, by way of mRNA, the synthesis of proteins; (3) the use of a code so that there is a point-by-point correspondence between the DNA and the protein molecules; and (4) the complete determination of the structure of a protein by its sequence. Point out that energy is required to synthesize proteins—to attach amino acids to tRNA and to join amino acids in the elongating polypeptide chain.

Figure 8.12 How DNA determines the formation of a protein.

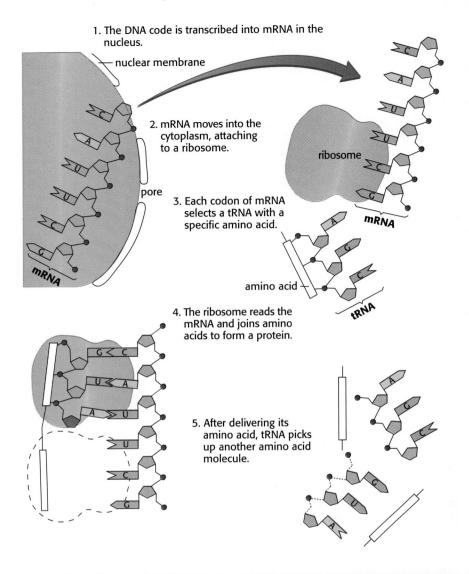

1. The DNA code is transcribed into mRNA in the nucleus.

nuclear membrane

2. mRNA moves into the cytoplasm, attaching to a ribosome.

pore

ribosome

mRNA

3. Each codon of mRNA selects a tRNA with a specific amino acid.

amino acid

tRNA

4. The ribosome reads the mRNA and joins amino acids to form a protein.

5. After delivering its amino acid, tRNA picks up another amino acid molecule.

mRNA

CONCEPT REVIEW

1. Why would the knowledge that genes are made of DNA rather than protein be helpful to biochemists trying to isolate and study genes?
2. Sketch or describe the structure of the DNA molecule.
3. Describe the evidence for the chromosome theory of heredity.
4. How is the genetic information coded in DNA translated into protein?

Patterns of Inheritance

8.8 Dihybrid Crosses Produce a Distinctive Pattern

Guidepost

How are different patterns of inheritance explained?

What has been described thus far about nucleic acids and chromosomes explains Mendel's results with crossing plants that differed in one trait. Each trait depended on the production of certain proteins, which were specified by certain genes. Mendel, however, also crossed plants that differed in two traits. For example, he crossed plants that were pure breeding for round seed shape and yellow seed color with plants that were pure breeding for wrinkled seed shape and green seed color. This type of cross is called a dihybrid cross because the offspring that result are heterozygous for two (*di*) different traits.

Figure 8.13 illustrates the four phenotypes Mendel saw in the F_2 generation of a dihybrid cross. Note that the phenotypes do not occur with equal frequency. Instead, the ratio is $\frac{9}{16}$ to $\frac{3}{16}$ to $\frac{3}{16}$ to $\frac{1}{16}$—9 : 3 : 3 : 1 ratio. How can these results be explained? Remember that during meiosis
1 the chromosome pairs separate independently. When gametes were formed in the F_1 generation, the genes for seed shape and color separated independently, resulting in four different combinations of the alleles of these genes. The phenotypes in the F_2 offspring can be predicted on the basis of their alleles. These results laid the foundation for Mendel's **principle of independent assortment:** Alleles for one trait segregate independently of alleles for other traits during gamete formation.

This modification of a Punnett square provides another approach to understanding a dihybrid cross. It is graphic but contains the danger (as does Figure 8.5) that students may count the gametes and peas instead of interpreting them as ratios. Note that the two characteristics are both of the seed. Difficulty may arise if one attempts to combine seed and mature plant traits, such as seed color and vine length.

Emphasize that *r* can combine with either *Y* or *y*.

Although Mendel used garden peas in his experiments, his findings are applicable to humans. For example, think of the little girl with cystic fibrosis discussed in the introduction. Medical geneticists now know that cystic fibrosis is a recessive disorder. What hypothesis about the biochemical basis of the disease does this knowledge suggest?

The disease is a consequence of the absence or a defect in the activity of some protein.

Assume that *F* represents the allele for the normal condition and *f* represents the allele for cystic fibrosis. What are the genotypes of Bob, Mary, and Lisa? What can the genetic counselor tell this couple about the probability that their next child will have cystic fibrosis?

Mary and Bob: *F/f*; Lisa: *f/f*. Probability for the next child to have cystic fibrosis: 25%.

8.9 Gene Expression Can Affect Patterns of Inheritance

Mendel worked with traits that showed clear dominant and recessive inheri-
2 tance. Complete dominance however, does not always occur. Sometimes the phenotype of hybrid organisms is different from the phenotype of either homozygous parent. When both alleles contribute to the phenotype of a heterozygote, the alleles are said to show codominance. The flower color of the morning glories illustrated in Figure 8.14 is an example of codominance.

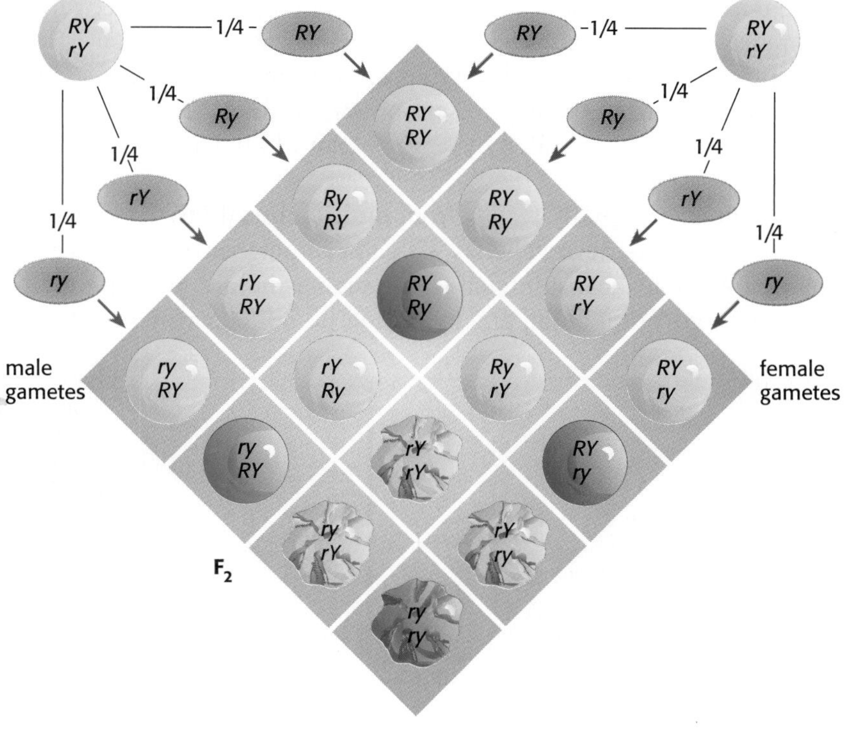

Figure 8.13 A dihybrid cross that illustrates the principle of independent assortment. What phenotypes are present in the F_2 generation, and in what ratios? These results demonstrate that the traits for seed shape and seed color assorted separately. Note that there are 12 round to every 4 wrinkled seeds, and 12 yellow to green seeds. Each trait independently shows the 3 : l ratio typical of a monohybrid. *R* = round seed; *r* = wrinkled seed; *Y* = yellow seed; *y* = green seed.

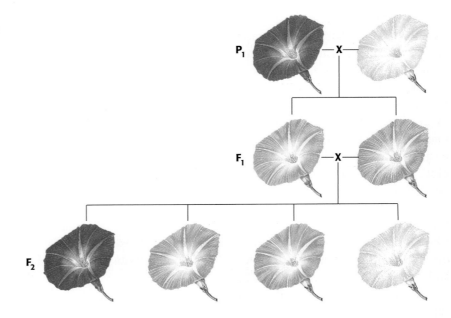

Figure 8.14 Inheritance of flower color in morning glories, an example of codominance.

Sometimes there are more than two alleles for a trait. ABO blood types in humans are determined by three alleles: I^A, I^B, and I. This system of human blood types illustrates the existence of **multiple alleles,** genes that have more than two allelic forms, and offers another example of **3** codominance. These blood types determine whether or not blood from one person can be transfused safely to another person. Allele I^A causes the formation of blood factor A. Allele I^B causes the formation of factor B. These two alleles show codominance; when a person has the genotype $I^A I^B$, both types of factors are produced. Allele I does not cause either factor to form. Table 8.2 shows the genotypes that are responsible for the various phenotypes. Note that an individual has only two of the alleles.

Many traits exhibit more than two phenotypes. These traits vary across a broad range. Suppose you plotted the heights of all the tenth graders in your school on a graph (see Figure 8.15). Between the shortest and the tallest person, there would be many other people who cover the whole range of height. This type of distribution is called continuous. Traits that exhibit more than two phenotypes are different from traits such as cystic fibrosis, which is either present or absent. Such "either/or" traits, which generally are controlled by a single pair of alleles, are called discontinuous, or discrete.

Continuous variability results from **multifactorial inheritance**—the interaction of at least several genes with a large number of possible environmental variables. Most human traits, such as height, weight, intelligence, hair color, skin color, and eye color, do not occur in an either/or **4**

Table 8.2 ABO Blood Types

Genotype	Blood (Phenotype)
$I^A I^A$ or $I^A i$	A
$I^B I^B$ or $I^B i$	B
$I^A I^B$	AB
ii	O

Multifactorial inheritance combines genetic and environmental factors. In polygenic inheritance, multiple genes are involved but no environmental influence is identified.

Figure 8.15 Continuous and discontinuous variations. Most human traits are determined by multiple genes and are influenced by the environment. Some traits (usually controlled by single genes) exist in only two forms. Histogram *a* shows the distribution of the heights of students in a high-school biology class. Histogram *b* shows the distribution of sexes in the same class.

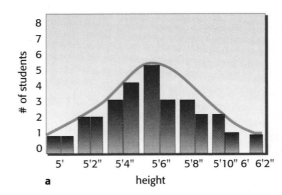

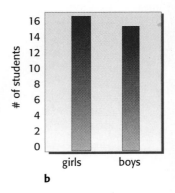

condition. Instead, these traits vary from person to person to person. Each trait is controlled by more than one gene, and each gene contributes to the final appearance of the trait. For example, the production of the skin pigment melanin is thought to be controlled by four major genes, and not one of these genes demonstrates simple dominance. The independent assortment of the alleles for melanin production results in a continuous variation in skin color among humans. Skin color, however, also is affected by the amount of exposure to sunlight. Some disorders, such as cleft lip and spina bifida, also may be caused by multiple factors. They are caused by a number of genes interacting with certain environmental factors present in the mother's uterus during pregnancy.

Although these examples do not display the same ratios as Mendel's simple crosses, they still are considered to result from Mendelian inheritance. That is, they can be explained in terms of genes residing on chromosomes transmitted through meiosis during sexual reproduction. Not all genetic information is transmitted this way. In many eukaryotic organisms, some of the genes are associated with cytoplasmic organelles, such as chloroplasts and mitochondria. This genetic information follows a pattern of inheritance unrelated to meiosis. In prokaryotic organisms, which lack a complete sexual cycle and do not undergo meiosis, there are a great variety of non-Mendelian patterns of inheritance.

8.10   X-Linked Traits Show a Modified Pattern of Inheritance

Remember that genes are integral parts of chromosomes, which occur in matching pairs. When chromosomes are stained, the stains concentrate in specific regions and thus create a characteristic banding pattern. The two chromosomes in each pair show the same banding pattern, and their pattern is different from the patterns of other chromosome pairs. Figure 8.16 illustrates how two similar-sized chromosomes can be distinguished on the basis of their banding patterns. Once the chromosomes are stained, they can be photographed through a microscope. When the photograph is enlarged, individual chromosomes may be cut out and arranged in order by size and shape to form a karyotype similar to the one shown in Figure 8.17. With some experience, counting and identifying chromosomes becomes a relatively simple procedure, and any unusual, missing, or extra chromosomes can be detected fairly quickly.

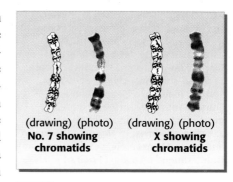

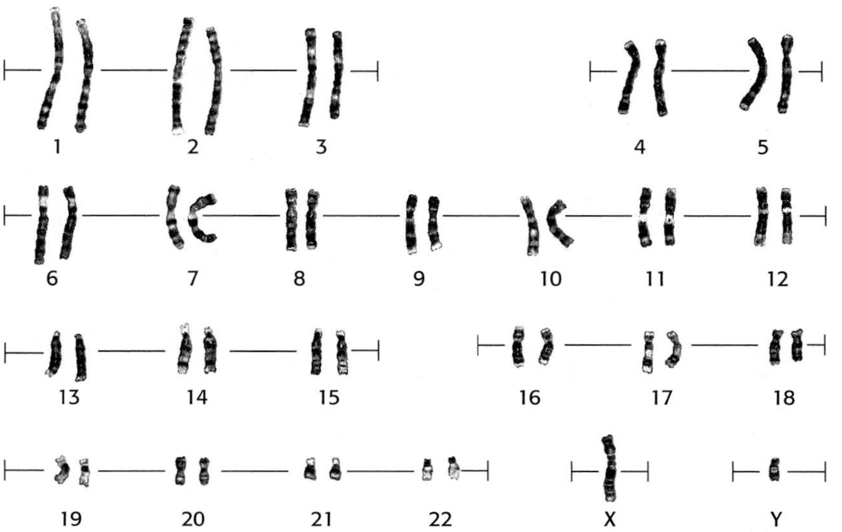

(drawing) (photo)   (drawing) (photo)
**No. 7 showing   X showing
chromatids   chromatids**

Figure 8.16 Human chromosomes 7 and X. The banding patterns evident in the photographs are shown in the diagrams. These two chromosomes, which have almost the same length and centromere location, can be distinguished by their banding patterns.

Figure 8.17 A karyotype is a display of human chromosomes arranged as homologous pairs. It is prepared by cutting out individual chromosomes from a photograph and matching them, pair by pair. Is this the karyotype of a male or a female?   **A male.**

Not all organisms have an X-Y system of sex determination. Some insects have an X-O system: females have two X chromosomes, males have one, and there is no Y chromosome. Birds, some fish, and some insects have a Z-W system: males are ZZ and females ZW. Some plants have an X-Y system, but most plants and some animals have no sex chromosomes. In *Neurospora crassa* and yeast, for example, sex (mating type) is determined by a single pair of alleles.

Nearly all mutations are potentially harmful to a species because they replace genes that have evolved and effectively served the organism through many generations.

Using karyotypes, biologists have found that humans have 23 pairs of chromosomes—22 pairs that are the same in both sexes (known as autosomes), and one pair that is different in males and females. This pair determines the sex of an organism, and the chromosomes that make up the pair are called the sex chromosomes. In humans and other mammals, the sex chromosomes are referred to as X and Y Chromosomes. Females have two X chromosomes in each body cell; males have one Y chromosome. Each ovum produced by a female, therefore, contains an X chromosome. Males, on the other hand, produce two types of sperm—half with an X chromosome and half with a Y chromosome. Thus during fertilization, it is the sperm that determines the sex of the offspring. If an ovum is fertilized by an X-bearing sperm, the zygote develops into a female. If the sperm contains a Y chromosome, the offspring is a male. Because each chromosome contains many genes, and each gene is capable of affecting some part of development, complete chromosomes are essential for normal development and sex determination.

While studying *Drosophila* at Columbia University in the early 1900s, Thomas Hunt Morgan noted some mutant flies with striking differences. Some mutants had white eyes instead of red; others had short wings instead of long; and still others had yellow or black bodies instead of gray ones. Morgan and his colleagues observed hundreds of mutations in fruit flies.

They carefully tested each mutation by mating mutant with nonmutant flies. In most cases, the mutations were inherited according to Mendelian patterns. One exception, however, was the first mutation Morgan had observed—a male with white eyes.

When Morgan crossed the white-eyed male with normal red-eyed females, the F$_1$ generation contained only red-eyed flies, supporting his assumption that the white-eye form of the eye color trait is recessive. The F$_2$ generation had the expected 3 : 1 ratio of red-eyed flies to white-eyed flies, as shown in Figure 8.18a. Morgan noticed, however, that only males showed the white-eye form. Although not all males were white-eyed, this form seemed to be linked to a fly's sex.

Figure 8.18   X-linked inheritance. When Morgan crossed a white-eyed male with a normal red-eyed female, all the F$_1$ flies (a) had red eyes. The F$_2$ flies showed the expected 3 : 1 ratio of red to white eyes, but only male flies showed the recessive trait. A reciprocal cross (b). All male offspring have white eyes because they inherited their x chromosomes from their mother. Compare this cross with the F$_1$ results of the first cross. The ratio of F$_2$ flies also is shown.

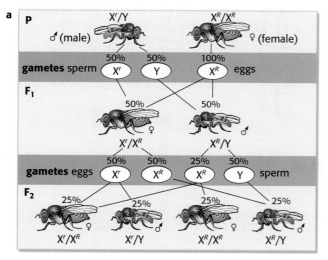

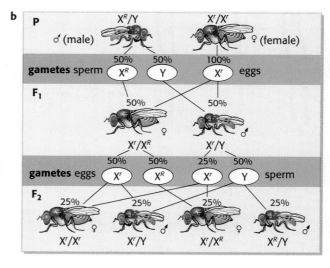

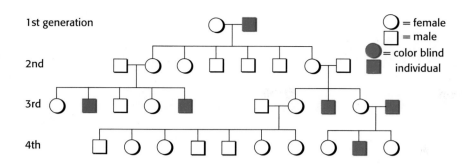

1st generation

2nd

3rd

4th

= female
= male
= color blind individual

Figure 8.19   A pedigree for red-green color blindness. How does the pattern of inheritance compare with the pattern for white-eye trait in *Drosophila*?

Morgan continued breeding his flies and eventually obtained some white-eyed females. He then crossed a white-eyed female with a red-eyed male and studied the inheritance of white eyes. The F$_1$ offspring of this cross were not all red-eyed, however. Only F$_1$ females showed the dominant red-eye form. All the males had white eyes. This cross is shown in Figure 8.18b.

6   Morgan hypothesized that the gene for eye color is on the X chromosome and there is no corresponding gene on the Y chromosome. Thus, a single allele for the recessive form, if on a male's X chromosome, could result in a white-eyed male. White eye color was the first example of an **X-linked trait,** a trait whose gene is on only the X chromosome.

Since then many other X-linked traits have been found in *Drosophila*, and more than 200 have been identified in humans. One well-known X-linked human trait is hemophilia, a disorder in which blood does not clot properly. Others include red-green color blindness and Duchenne muscular dystrophy, a crippling, muscle-wasting disease that results in premature death. The pattern of inheritance of these X-linked traits is the same as that of white eyes in *Drosophila*. Study the pedigree shown in Figure 8.19 and try to determine the genotype of each individual on the chart Charts illustrating family histories of traits are one of the major tools available for studies of inheritance in humans.

8.11   Abnormal Chromosomes Can Affect Patterns of Inheritance

Some types of birth defects are caused by abnormal numbers or types of chromosomes. These chromosomal abnormalities are present from birth and often from the moment of fertilization. One such condition is Down syndrome. (A syndrome is a group of symptoms associated with a particular disease or abnormality.) Individuals with this condition have distinctive features of the eyes, mouth, hands, and sometimes internal organs. All have retarded mental development, though the degree of retardation is highly variable.

Individuals with Down syndrome have 47 chromosomes in each cell instead of 46. The extra chromosome is the tiny number 21, visible in the karyotype shown in Figure 8.20. Down syndrome results from trisomy-21, which means the number-21 chromosome is present in triplicate.

Abnormal numbers of sex chromosomes cause several other syndromes. For example, individuals having only one X chromosome and no Y chromosome (a 45,X karyotype) usually are short, underdeveloped, and sterile females. This condition is known as Turner syndrome. Females who have a 47,XXX karyotype often have limited fertility and may have slight intellectual impairment. A 47,XXY karyotype results in Klinefelter syndrome. These males often are tall and sexually underdeveloped.

Not all afflicted individuals necessarily exhibit all the symptoms.

Down syndrome can be diagnosed prenatally by amniocentesis.

Figure 8.20 Karyotype in Down syndrome (a), a child with Down syndrome (b).

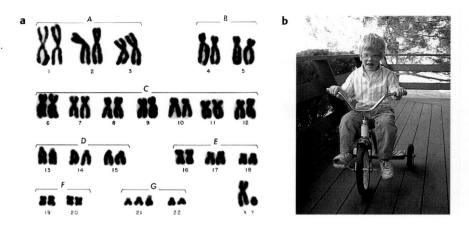

These abnormalities arise when nondisjunction occurs—when chromosome pairs do not separate in meiosis. Nondisjunction results in the formation of abnormal gametes, as shown in Figure 8.21. Some sperm or egg cells get extra chromosomes, and some have too few chromosomes. These gametes, when fused with normal gametes, usually result in abnormal development.

Apparently there must be at least two of each type of chromosome in every cell for survival of the embryo. A striking exception is the X chromosome—45,X individuals can survive a missing chromosome. Extra sex chromosomes (X or Y) usually permit a fetus to develop, although the development may be abnormal. Extra autosomes, however, except for trisomy-21, rarely permit development to continue. Examination of cells from spontaneously aborted fetuses indicates that most have abnormal chromosome numbers.

Why do most trisomies of autosomes result in death of the fetus, whereas trisomies of sex chromosomes do not? A possible answer was provided by the British geneticist Mary Lyon in 1961. She suggested that early in the development of a normal female, one X chromosome in each body cell is inactivated. Lyon proposed X-inactivation as a possible explanation for a darkly staining mass, called a Barr body, that normally **7**

Figure 8.21 Nondisjunction can occur during the formation of either sperm or ova. What combinations of sex chromosomes would result from fertilization between an abnormal sperm and an abnormal egg?

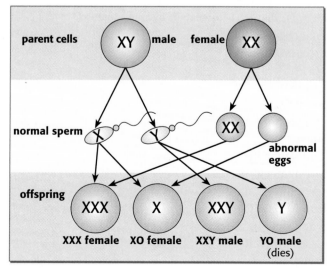

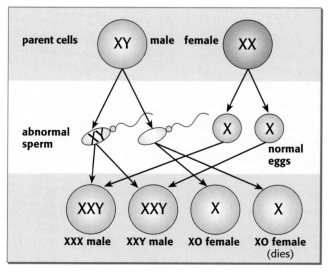

appears in the cell nuclei of females but not of males (see Figure 8.22). Females with three X chromosomes (47,XXX) have two Barr bodies, suggesting that all but one X chromosome in a cell are inactivated. One reason trisomies of sex chromosomes may not be as disruptive as autosomal trisomies is that the extra X chromosomes are inactivated and condensed into Barr bodies. Apparently a few genes remain active in the condensed X chromosomes, resulting in the abnormalities associated with extra X chromosomes.

Sometimes abnormalities result from altered chromosome structure. For example, a piece of chromosome may be missing (deletion), or a section may be reversed (inversion). The missing piece of one chromosome may attach to its pairing partner, producing a duplication, or it may attach to an unrelated chromosome, resulting in a translocation. Figure 8.23 shows diagrams of these alterations in chromosome structure. It is clear that the proper number and types of active chromosomes, as well as the proper structure, are essential for normal development. The consequences of abnormal chromosome numbers apparently reflect a delicate balance among the many gene products that participate cooperatively in the total development and function of an organism.

8.12   Many Genes Are on Each Chromosome

Because there are far more genes than chromosomes, many genes must be located on each chromosome. Genes that are close together on the same replicated chromosome are said to be linked, which suggests that they are inherited together. For example, in humans the gene for the ABO blood group is on chromosome 9. An oncogene that may be involved in certain types of cancer also is on chromosome 9, close to the ABO gene. A geneticist studying those traits would find little or no evidence of independent assortment. Why?

Linked genes do not always remain together. Recall from the discussion in Section 6.5 that chromosome pairs often exchange fragments during the first meiotic division (Figure 8.24). That exchange gives rise to recom-
8 binations that increase the variability resulting from sexual reproduction.

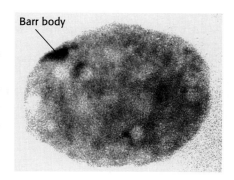

Figure 8.22   The darkly staining Barr body in this nucleus is the condensed x chromosome. This cell is not dividing, so the chromosomes are not visible as discrete bodies.

Ask students how many Barr bodies they would expect to see in cells of a person with Turner syndrome. None, because there is only one x chromosome.

Stress that crossing-over produces new combinations that may be transmitted together. This is important in discussing evolution.

Figure 8.24   Crossing-over frequently occurs between chromosome pairs during the early stages of meiosis.

Figure 8.23   If the piece of a chromosome carrying the genes B and C breaks off, four types of chromosomal abnormalities can result: deletion, duplication, inversion, or translocation to another chromosome.

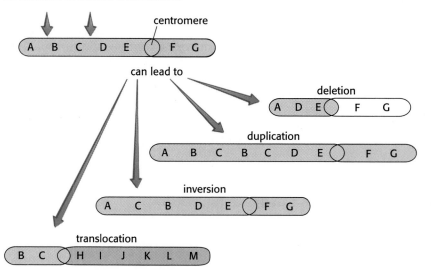

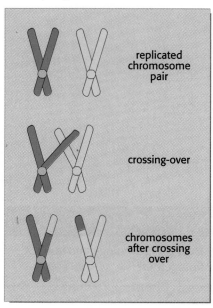

Figure 8.25 A few of the genes that have been mapped to the human X chromosome and to autosome 9.

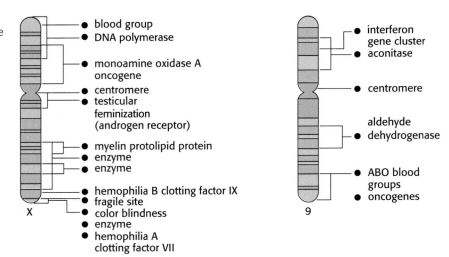

X
- blood group
- DNA polymerase
- monoamine oxidase A oncogene
- centromere
- testicular feminization (androgen receptor)
- myelin protolipid protein
- enzyme
- enzyme
- hemophilia B clotting factor IX
- fragile site
- color blindness
- enzyme
- hemophilia A clotting factor VII

9
- interferon gene cluster
- aconitase
- centromere
- aldehyde dehydrogenase
- ABO blood groups
- oncogenes

Genes on the same chromosome generally are inherited together. Crossing-over is unlikely between genes that are close together.

The farther apart genes are on a chromosome, the more potential breakage sites exist between them. Suppose that breakage sites occur at 1μm intervals; then two genes located 8μm apart are more likely to separate by crossing-over than two genes that are only 3μm apart.

The farther apart two genes are on a chromosome, the more likely a break will occur between them. If two genes are far enough apart on a chromosome, the breaks between the genes may be so frequent that the principle of independent assortment applies to their inheritance. In other words, these genes exhibit no linkage. Some of the genes in Mendel's experiments sorted independently for this reason.

The frequency with which two linked traits become separated provides a way to determine the relative distance between the two genes for those traits on the chromosome. With this information, geneticists can construct accurate maps showing the sequence of genes on chromosomes. Genetic maps now exist for chromosomes of humans (see Figure 8.25) and many other organisms.

CONCEPT REVIEW

1. How does the movement of chromosomes during meiosis explain the principle of independent assortment?
2. How does codominance account for the presence of more than two phenotypes of a trait?
3. What are multiple alleles? Can a person have more than two alleles for a single gene? Explain.
4. Describe how multifactorial inheritance explains continuous variability in a trait such as height.
5. How can karyotypes contribute to genetic research?
6. How does the inheritance of X-linked traits differ from that of other traits?
7. Why are the effects of genotypes with extra X chromosomes apparently less severe than those of other trisomies?
8. How does crossing-over affect linkage?
9. Describe how linkage can be used to prepare chromosome maps.

Genetics and Technology

8.13   New Tools Aid Genetics Research

New technology provides the tools for new discoveries that, in turn, lead to additional technology. The relationship between science and technology is illustrated best by two of the most important innovations in molecular

INFORMATICS-MAPPING AND SEQUENCING THE HUMAN GENOME

Informatics, the use of highly complex data systems and new electronic techniques to sort and analyze these systems has opened up a wide range of new research opportunities. Increasingly, the electronic management of information is becoming a central, indispensable feature of science, as research produces ever more data that must be made accessible to the scientific community. The accurate storage and rapid retrieval of scientific data are nowhere more critical than in the Human Genome Project (HGP), whose intent is to map and sequence the estimated 100,000 genes, containing approximately 3 billion nucleotides of DNA, a complete record of which ultimately will reside in electronic databases.

The HGP has two major objectives. The first is to develop detailed maps of the human genome and the genomes of several other well-studied organisms including bacterium, yeast, nematode, fruit fly, mouse, and a rapidly growing plant with a small genome, *Arabidopsis thliana*. The second objective is to determine the complete base sequences of these genomes. Because a written record of the human genome sequences would require the equivalent of 200 telephone books of 1,000 pages each, one important part of the project is the development of systems for electronic database storage and management to handle the tremendous amount of information the HGP will generate.

The HGP has the potential to increase our understanding of human variation, development, gene regulation, and patterns and products of human change through time. It will advance the practice of medicine by opening doors to new ways of diagnosing and treating inherited disorders. The HGP also will increase the demand for individuals trained in genetics, medical genetics, genetic counseling, and other health fields. The development of new technologies, especially those to help manage the enormous amounts of data the HGP will generate, will have applications far beyond the project itself.

Although progress in the development of electronic databases has been of great advantage to the HGP, it also has raised some important questions about the applications of such information. As researchers uncover more and more associations between particular segments of the human genome and complex human traits, there will be more concern over how these data will be used in areas such as heath care, employment, and insurance.

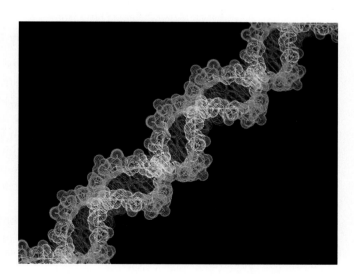

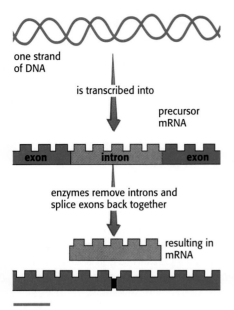

one strand
of DNA

is transcribed into

precursor
mRNA

| exon | intron | exon |

enzymes remove introns and
splice exons back together

resulting in
mRNA

Figure 8.26 Formation of eukaryote mRNA.

Students might understand an analogy to the product bar code, with the white spaces being introns.

Most prokaryotes do not have introns, and they are found only rarely in many of the lower eukaryotes. Split genes seem to be the rule only in some eukaryotes.

biology: the ability to determine the exact base sequence of DNA, and the ability to transfer DNA from the chromosomes of one organism to those of another.

Studies of the base sequence of genes have allowed evolutionary biologists to compare DNA from different organisms. Through these comparisons, biologists have been able to assess more carefully the degree to which those organisms are similar and thus possibly related.

These types of studies have led to other discoveries. In the late 1970's for example, biologists learned that many eukaryotic genes are not continuous. That is, the sequences that code for a protein are separated by noncoding intervening sequences, called introns (IN tronz). After the gene is transcribed, the introns are cut from RNA strand while it is still in the nucleus (see Figure 8.26). The remaining pieces—called exons (EKS onz) are spliced together into the mRNA that leaves the nucleus and directs the synthesis of protein on the ribosomes. While biologists had considered their relationship between DNA base sequences and the sequence of amino acids in the protein as simple, this discovery changed their ideas. The function and evolutionary significance of introns still are unknown. **2**

As a result of new techniques available from genetic engineering, the number of genes mapped to human chromosomes is increasing by at least 25 percent per year. This information is valuable in detecting carriers of genetic disorders (see Biology Today). A major project is under way to map and sequence the human genome—the entire complement of human genetic material in a haploid cell. The human genome consists of 3 billion base pairs of DNA. The goal of the human genome project is to identify the exact type and position of each of those DNA base pairs. This information will increase our understanding of the origins, nature, and variability of human traits. It also could open doors to new ways of diagnosing and treating such inherited disorders as cystic fibrosis. One of the expected benefits of the human genome project is a better understanding of the importance of introns as well as other DNA sequences with functions that presently are undetermined. **3**

8.14   Proteins Can Be Manufactured by Genetic Engineering

The genetic code is almost universal—the same genetic code functions in bacteria and in humans. Can the genetic instructions from one type of organism be interpreted by another type of organism?

The answer is yes. The protein-making machinery of a cell reads DNA instructions even when they come from another type of organism. By using special types of enzymes, DNA from one organism can be cut into small pieces and spliced into the DNA of another organism. The resulting DNA is called **recombinant DNA,** because DNA from different organisms has been recombined.

Molecular biologists first used these techniques to study how genes are turned on and off in response to hormones and other physiological and environmental factors. Knowledge of these mechanisms has provided new insights into cancer, AIDS, and many other diseases.

Other applications of transplanted genes also are widespread. For example, individuals who have diabetes produce too little insulin, a protein made by the pancreas. For years, diabetics have depended on insulin extracted from the pancreases of hogs and cattle. Although the animal insulin molecules are similar to human insulin, they are different enough to cause problems for some diabetics.

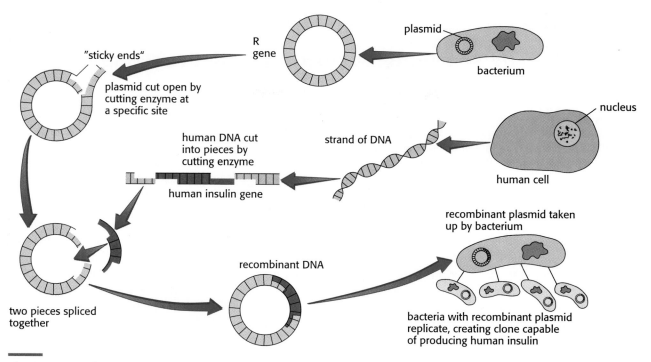

Figure 8.27 Techniques of recombinant DNA. To splice a human gene (in this case, the one for insulin) into a plasmid, scientists take the plasmid out of an *Escherichia coli* bacterium, open the plasmid at a specific site with a cutting enzyme, and splice in insulin-making human DNA. The resulting hybrid plasmid can be taken up by another *E. coli* bacterium, where it replicates together with the bacterium, making it capable of producing large quantities of insulin.

Today, diabetics can use a molecule that is exactly like the molecule their own pancreases should be making. Human insulin is manufactured in large amounts from recombinant DNA. The technique, represented in Figure 8.27, uses plasmids—small circles of DNA outside the chromosome of bacteria. A fragment of DNA coding for human insulin is inserted into a plasmid, and a bacterial cell takes in the plasmid from a culture
4 medium. The recombinant DNA replicates along with the DNA of the bacterial cell. This process is known as gene cloning. It is analogous to conventional cloning—the reproduction and growth of genetically identical cells or organisms by asexual reproduction. The new cells containing a cloned gene can produce whatever product is coded for by the recombinant DNA, in this case, human insulin. Cloning makes it possible to isolate quantities of a gene for chemical analysis and DNA sequencing.

Medicine is only one area using practical applications of genetic engineering. Genetic engineers also have produced a strain of bacterium that prevents the formation of frost on plants to temperatures as low as −6°C. Calves have been born from fertilized eggs that have had genes to speed
5 growth and make cattle leaner inserted into them. Genetic engineering has been used to improve other food-producing animals as well.

Genetic engineering also has been applied to genetic screening—including prenatal diagnosis—for disorders such as cystic fibrosis, Huntington disease, and muscular dystrophy. In 1989, Lap-Chee Tsui and John R. Riordan of the Hospital for Sick Children of Toronto and Francis Collins of the University of Michigan Medical School announced the identification and cloning of the cystic fibrosis gene. This unusually large gene consists of 250 000 base pairs on chromosome 7. Most of the gene is composed of introns that separate 24 exons. These exons code for a

Genetic Screening: Prevention with Problems

The isolation of the gene for cystic fibrosis was a major breakthrough in biomedicine. There is new hope now that research will produce a cure for the disorder. The ability to detect the CF gene through DNA analysis also makes prevention of the disease through genetic screening possible. DNA analysis could lead to the detection of carriers (heterozygotes) in the general population. Once recognized, carriers could be advised of the risk of having a child with CF and could choose alternative methods such as adoption, artificial insemination, or *in vitro* fertilization and embryo transfer. Any gametes used in these methods could be screened for the CF gene. DNA analysis also permits the detection of the disorder in developing fetuses (homozygotes). The parents of a fetus diagnosed with CF could choose to abort the pregnancy or to carry it to term.

While the ability to screen a population for a variety of genetic disorders allows for early detection and prevention, it also raises many difficult questions about ethics and public policy. According to Dorothy Nelkin and Laurence Tancredi, sociologists who study the social impact of medicine, DNA analysis in genetic screening "has shifted the focus of the health-care system from looking for *actual* disease in individuals to looking for *tendencies* to *develop disease*." As a result, there is a danger that people who have no symptoms of disease may be labeled as handicapped and may be discriminated against by health or life insurance companies. Potential employers, for example, may refuse to hire such individuals because they do not want to risk increased costs for medical benefits and do not want to spend time and money training an employee whose life expectancy might be limited.

Neil A. Holtzman, a pediatrician/geneticist at John Hopkins, claims the problems raised by DNA analysis are intensified because the tests may not always predict with certainty whether a person will develop a particular disor-der. This is especially true for multifactorial disorders such as heart disease, cancer, and schizophrenia, which result from a complex interaction of genes and environmental conditions. Furthermore, says Holtzman, there are too few laboratories that can perform DNA analysis with precision and not enough trained genetic counselors to interpret the results.

Holtzman, Nelkin, and Tancredi recommend the development of new regulations to control how genetic screening is done and what is done with the results. Without regulation and greater public education about genetics, there is danger that some individuals may lose some control of their lives. Carried to extremes, some people might be forbidden to marry, to have children, or to do certain kinds of work.

The ethical, legal, and policy issues raised by DNA analysis and genetic screening must be weighted against the potential benefits to society. Such value conflicts often accompany advances in science and technology and affect individuals, families, and society at large.

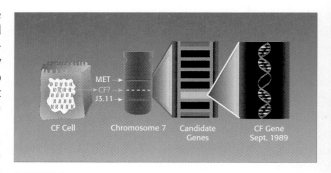

Method of identifying the CF gene.

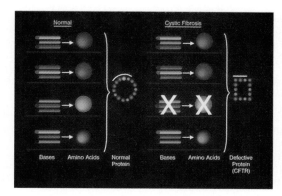

Figure 8.28 Loss of one codon in the gene for CF results in the detection of one amino acid, causing production of a defective protein.

protein made up of 1480 amino acids. The protein's shape suggests it extends through the plasma membrane and thus may have a role in membrane transport. Approximately 75 percent of the mutant genes that cause cystic fibrosis have lost a single codon, resulting in the deletion of a single amino acid from the corresponding protein (see Figure 8.28). Now scientists must determine the function of the protein and how it malfunctions in cystic fibrosis. The search for the remaining 25 percent of the mutations that alter this gene also is under way, but these mutations may be more difficult to find because the gene is so large that many abnormalities can occur.

Genetic engineering provides a good example of the relationship between science and technology. Most of the organisms engineered by geneticists are used in research to demonstrate the organization and function of genetic material, the molecular web that connects all living things. The same techniques and knowledge used to engineer organisms offer potential solutions to many practical and societal problems.

CONCEPT REVIEW

1. How do science and technology interact?
2. How are introns and exons involved in production of mRNA?
3. What is meant by the *human genome project*, and what are some potential benefits?
4. How can bacteria make human insulin molecules?
5. What is genetic engineering, and what are some of its applications?

INVESTIGATION 8.1   Probability

The probability of a chance event can be calculated mathematically using the following formula:

$$\text{Probability} = \frac{\text{number of events of choice}}{\text{number of possible events}}$$

What is the probability that you will draw a spade from a shuffled deck of cards like that shown in Figure 8.29? There are 52 cards in the deck (52 possible events). Of these, 13 cards are spades (13 events of choice). Therefore, the probability of choosing a spade from this deck is ¹³⁄₅₂ (or ¼ or 0.25 or 25%). To determine the probability that you will draw the ace of diamonds, you again have 52 possible events, but this time there is only one event of choice. The probability is ¹⁄₅₂ or 2%. In this investigation, you will determine the probability for the results of a coin toss.

The ideas in this investigation are essential for understanding genetics, particularly probability, which is not discussed in the text. It is important for students to realize that science deals largely with probabilities rather than with certainties. The principles of probability are at work in the disintegration of radioactive atomic nuclei and the collisions of molecules in gases, as well as in the distribution of genes from one generation to the next. Teams of 2 work well.
***Time:* One class period.**

Figure 8.29 The chance of drawing one ace is ⅟₁₃. What are the chances of drawing two, three, or four aces?

Discussion

1. **Five heads in 10 tosses. The observed number may be different.**
2-5. **Depends on student data. For example, if the first 10 tosses yield 4 heads and 6 tails, deviation is**

$$\frac{1+1}{10} = \frac{2}{10} = 0.5$$

6. **Increasing the number of tosses decreases the average percentage deviation. Point out therelationship between this conclusion about size of sample and the practice, in several past investigations, of combining team data.**
7. **Two columns: H/H and Dull H/Shiny T.**
8. **Depends on student data.**
9. **Two columns: H/H and Dull T/Shiny H.**
10. **Depends on student data.**
11. **Only one (H/H); note that the only use of T/T is to calculate the total.**
12. **Depends on student data.**
13. **It should be closest to the product.**
14. **The probability that two independent random events will occur simultaneously is the product of their individual probabilities. (Be sure students understand this principle as a generalization.)**

Materials (per team of 2)
2 pennies (1 shiny, 1 dull)   cardboard box

Procedure

1. Work in teams of two. One person will be student A and the other will be student B.
2. Student A will prepare a score sheet with two columns—one labeled *H* (heads), the other *T* (tails). Student B will toss a penny 10 times. Toss it into a cardboard box to prevent the coin from rolling away.
3. Student A will use a slash mark (/) to indicate the result of each toss. Tally the tosses in the appropriate column on the score sheet. After 10 tosses, draw a line across the two columns and pass the sheet to student B. Student A then will make 10 tosses, and student B will tally the results.
4. Continue reversing roles until the results of 100 (10 series of 10) tosses have been tallied.
5. Prepare a score sheet with four columns labeled *H/H, Dull H/Shiny T, Dull T/Shiny H,* and *T/T* (H = heads; T = tails). Obtain 2 pennies—1 dull and 1 shiny. Toss both pennies together 20 times, while your partner tallies each result in the appropriate column of the score sheet.
6. Reverse roles once so that you have a total of 40 tosses.

Discussion

1. How many heads are probable in a series of 10 tosses? How many did you actually observe in the first 10 tosses?
2. Deviation is a measure of the difference between the expected and observed results. It is not the difference itself. It is the ratio of the sum of the differences between expected and observed results to the total number of observations. Thus:

$$\text{deviation} = \frac{\begin{array}{c}\text{difference between heads expected and heads observed} + \\ \text{difference between tails expected and tails observed}\end{array}}{\text{number of tosses}}$$

Calculate the deviation for each of the 10 sets of tosses.
3. Calculate the deviation for your team's total (100 tosses).
4. Add the data of all teams in your class. Calculate the class deviation.
5. If your school has more than one biology class, combine the data of all classes. Calculate the deviation for all classes.
6. How does increasing the number of tosses affect the average size of the deviation? These results demonstrate an important principle of probability. State what it is.
7. On the chalkboard, record the data for tossing two pennies together. Add each column of the chart. In how many columns do data concerning heads of a dull penny appear?
8. In what fraction of the total number of tosses did heads of dull pennies occur?
9. In how many columns do data concerning heads of a shiny penny occur?
10. In what fraction of the total number of tosses did heads of the shiny pennies occur?
11. In how many columns do heads of both dull and shiny pennies appear?
12. In what fraction of the total number of tosses did heads of both pennies appear at the same time?
13. To which of the following is this fraction closest: the sum, the difference, or the product of the two fractions for heads of one penny at a time?
14. Your answer suggests a second important principle of probability that concerns the relationship between the probabilities of separate events and the probability of a combination of events. State this relationship.

15. When you toss two coins together, there are only three possibilities—H/H, T/T, or H/T. One of these three combinations will occur 100% of the time. The rules of probability predict that H/H and T/T each will occur 25% of the time. What is the expected probability for the combination of heads on one coin and tails on the other?

16. When you toss a dull penny and a shiny penny together, what is the probability that heads will occur on the dull penny? What is the probability that tails will occur on the shiny penny? Calculate the probability that the dull penny will be "heads" *and* the shiny penny will be "tails" if you toss the two pennies together. Compare this answer to the answer in Question 15. How do you account for the different answers? Are there other ways than Dull H/Shiny T to get the H/T combination?

17. How many different ways can you get the H/T combination on two coins tossed together? What is the probability of each of those different ways occurring? Is the probability of getting heads and tails in any combination of pennies closest to the sum, the difference, or the product of the probabilities for getting heads and tails in each of the different ways?

18. Your answer suggests a third important principle of probability that concerns the relationship between (a) the probability of either one of two mutually exclusive events occurring and (b) the individual probabilities of those events. State this relationship.

15. 100% − (25% + 25%) = 50%
16. 50%; 50%; 50% × 50% = 25%. Yes, you can get heads on the shiny penny and tails on the dull penny.
17. There are two different ways, and the probability of each is ¼. Thus, the probability of getting any H/T combination is the sum of the two probabilities, or ½.
18. The sum rule is: The probability of either one of two mutually exclusive events occurring is the sum of their individual probabilities.

INVESTIGATION 8.2 DNA Replication and Transcription to RNA

DNA replication is the process by which exact copies are made of the DNA found in prokaryotes and in the chromosomes of eukaryotes. During replication, the genetic code contained within a sequence of nucleotide bases in DNA is preserved. Transcription is the process by which the message in DNA is transferred to RNA. How do replication and transcription take place? In this investigation, you will observe some of the basic steps in these processes.

This investigation gives students hands-on-experience with DNA models to help them understand the process described in the chapter. Teams of 2 work best for this investigation.
Time: **One class period.**

Materials (per team of 2)
paper clips: 40 black, 40 white, 32 red,   pen
     32 green, 5 silver   string
masking tape

Procedure
PART A Building the DNA Molecule

1. Use the following key for nucleotide bases and the colored paper clips to build a double-stranded segment of DNA.

black = adenine (A)   white = thymine (T)
green = guanine (G)   red = cytosine (c)

2. Construct the first DNA strand by linking the colored paper clips together to represent the following sequence of nucleotide bases:
           A A A G G T C T C C T C T A A T T G G T C T C C T T A G G T C T C C T T

3. Attach a piece of masking tape to the AAA end of the strand, and label the strand with the Roman numeral *I* by marking the masking tape.

4. Now construct the complementary strand of DNA that would pair with strand I. Remember that thymine (T) pairs only with adenine (A), and guanine (G) pairs only with cytosine (C).

5. Attach a piece of masking tape to the TTT end of the complementary strand and mark a Roman numeral *II* on the masking tape.

6. Place strand *II* beside strand I and check to make certain you have constructed the proper sequence of nucleotide bases in strand II. Green paper clips should

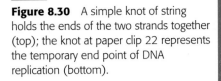

Figure 8.30 A simple knot of string holds the ends of the two strands together (top); the knot at paper clip 22 represents the temporary end point of DNA replication (bottom).

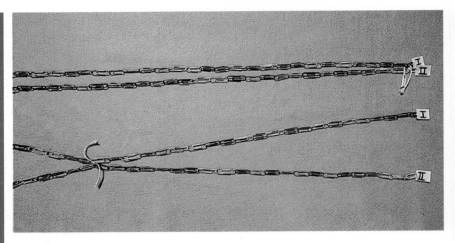

be opposite red paper clips, and black paper clips should be opposite white paper clips. Make any correction required in order to have the proper sequence of base pairs.

7. Using a short piece of string, tie a simple knot to join the first paper clips in strand I and strand II, as shown in Figure 8.30.

8. Repeat Step 7 for the paper clips on the opposite ends of strands I and II.

PART B DNA Replication

9. Place the double-stranded DNA molecule in a straight line on your table, with the AAA end of strand I on your right. (Do not untie the string at the end.)

10. Beginning at the AAA end of strand I, count 22 paper clips from right to left. Tie the 22d pair of paper clips together, as shown in Figure 8.30.

11. Untie the string holding the two strands together at the AAA end of strand I. Separate strand I and strand II so they form a Y.

12. DNA replication on strand I begins at the AAA end and proceeds toward the fork of the Y (the point at which the nucleotide bases are joined at the 22d base pair). Construct the new complementary strand for strand I beginning at the AAA end and working toward the fork of strands I and II. When you have finished, tie another knot at the AAA end of strand I to join it to its new complementary strand.

13. Replication of strand II begins at the fork of the Y and moves outward toward the TTT end. Build the complementary DNA strand for strand II, proceeding from left to right, and then tie the two strands together at the TTT end.

14. Now untie the knots that join strands II and at the 22d base pair and at the right hand end of the original double strand.

15. Continue the replication of strand I from right to left until you complete the new complementary strand. Tie another knot to join strand I and its new complementary strand at the left-hand end of the molecule.

16. Continue the replication of strand II from the left-hand end to the right. When you reach the 22d base pair, join the two segments of the newly formed strand (the complement of strand II).

17. Tie another knot in the left-hand end of strand II to join it to its new complementary strand.

PART C mRNA Formation

18. Repeat Steps 9-11, using one of the two double-stranded DNA molecules you just made. This represents one chromosome. Assume this chromosome has only two genes on it, one 21 and one 15 nucleotides long.

19. RNA is produced by transcription from strand I. The rules for forming mRNA are the same as for DNA except that uracil (silver) is used in place of thymine (white). Beginning at the AAA end of strand I, construct an mRNA molecule for the first gene on strand I.

20. Remove the new mRNA and rejoin the original DNA double strand with a simple knot.

Discussion

1. Compare the two new double-stranded molecules of DNA you made. How are they similar to the original DNA molecule containing strands I and II? How are they different?

2. Describe the differences in the ways strands I and II are replicated.

3. Describe how DNA replication makes it possible to produce two identical cells from one parent.

4. How many nucleotides are in the mRNA molecule? How many amino acids does this mRNA molecule code for?

5. How does the number of nucleotides in the two genes of this model compare with the number of nucleotides in your genes?

6. Would it make any difference in the type of protein formed if strand II were used as a blueprint for the mRNA instead of strand I?

Discussion

1. **Both should be identical. There should be a new strand I with the original strand II and a new strand II with the original strand I.**

2. **Strand I proceeds continuously from left to right; strand II proceeds in segments in the opposite direction.**

3. **Replication produces two identical sets of DNA so that the DNA in both cells is exactly the same.**

4. **There are 21 bases in the mRNA, one for each base in the gene. Each three adjoining bases is a codon for one amino acid; thus this mRNA codes for seven amino acids.**

5. **The genes in this model have few nucleotides compared with real genes.**

6. **The bases of strand II are complementary to those of strand I, and so mRNA formed on strand II would not be the same as mRNA formed on strand I, and the resulting protein would be different.**

INVESTIGATION 8.3   A Dihybrid Cross

Sexually reproducing organisms have haploid and diploid stages. In flowering plants and in animals, only the gametes are haploid, and traits (the phenotype) can be observed only in the diploid stage. When Mendel made dihybrid crosses to study the inheritance of two different traits, such as seed shape and seed color, he could observe the traits only in the diploid cells of the parents and their offspring. He had to use probability to calculate the most likely genotypes of the gametes.

In the yeast *Saccharomyces cerevisiae*, however, phenotypes of some traits can be seen in the haploid-cell colonies. In Investigation 7.1, for example, you could observe the color trait, red or cream, in both haploid and diploid stages of the yeast life cycle. Not all yeast traits are visible, however, so yeast geneticists use the methods Beadle and Tatum devised for studying mutations. That is, they isolate mutants that cannot make some essential substance. The inability of the mutant and the ability of the normal strain to make the substance define two different forms of a trait.

In this yeast dihybrid cross, you will follow two forms of each of two traits: red versus cream color, and tryptophan-dependent (requires this amino acid to grow) versus tryptophan-independent (does not require this amino acid). In Investigation 7.1, a red strain of yeast of one mating type is crossed with a cream-colored strain of the other type. The diploid strain is cream colored. If there is a single gene for this trait, then there must be one allele that determines cream color in the haploid strain and another allele that determines red color. In the diploid strain, then, there must be one of each of these alleles. Which form of the color trait is dominant?

By observing the color of a colony that a haploid strain forms, you can predict with certainty which allele it carries. In the case of a cream-colored diploid, however, you cannot be sure of the genotype. It could carry either one allele for cream color and one for red, or two alleles for cream color. The trait defined by tryptophan-dependence or independence works in much in the same way. A tryptophan-dependent haploid must carry the allele for tryptophan-dependence (a defective form of the gene) and a tryptophan-independent haploid must carry the allele for tryptophan-independence (the functional form of the gene), but a tryptophan-independent diploid could be either homozygous or heterozygous for the functional allele.

The experimenter using yeast has full control over the life cycle. Guesswork about the genotypes of the gametes that mate to form a particular hybrid is eliminated. In Parts A and B the students work through the method geneticists use to predict the types of gametes expected from a particular diploid and the probabilities of occurrence. This technique is particularly instructive because it is based on the mechanics of chromosome segregation in meiosis. Part C demonstrates visually that some aspects of the process are totally deterministic. The 9: 3: 3: 1 segregation ratio is not subject to chance if the gametes are as predicted. With higher organisms, the element of chance at this step comes from the random sampling of offspring. Teams of 3 or 4 can use cooperative learning strategies for Parts A and B; teams of 2 are best for Part C.

Time: Five class periods: one period for Parts A and B (best done in conjunction with the discussion of the text material on Mendel's experiments), and four periods for Part C.

Safety

Before discarding culture plates, disinfect them with household bleach or 70% isopropyl alcohol or by autoclaving. See "Safety Using Microbes" in *Guidelines for Laboratory Safety* in the Teacher's Edition. Because these are *Saccharomyces cerevisiae*, nonpathogenic laboratory strains of yeast, uncontaminated plates require no

Table 8.3 Growth of Yeast on Two Media

Haploid	Diploid	On COMP medium	On MIN medium
RT	R/R T/T R/r T/T R/r T/t	growth, cream	growth, cream
Rt	R/R t/t R/r t/t	growth, cream	no growth, (cream?)
rT	r/r T/T r/r T/t	growth, red	growth, cream
rt	r/r t/t	growth, red	no growth, (red?)

special treatment It is good practice, however, to disinfect all cultures. Students should discard all used toothpicks in designated waste bags. Label self-closing plastic bags waste and tape one bag to each lab table. Collect these bags at the end of each class and place in a biohazard bag. Students also should discard used culture plates in biohazard bags. Autoclave all biohazard bags before discarding.

Using symbols for the traits, the allele for the dominant cream form is R and the recessive red form is r; the normal, tryptophan-independent form of that gene is T and the tryptophan-dependent form is t. Whereas the color trait is visible, determining tryptophan dependence or independence requires a simple growth test.

Table 8.3 shows how all possible combinations of these two traits can be determined by testing the yeast on two kinds of growth medium: a nutritionally complete medium (COMP), and a medium lacking tryptophan (MIN).

Materials (per team of 2)
Part C
container of clean flat toothpicks
glass-marking pencil
2 paper templates for streaking strains
3 self-closing plastic bags labeled
   waste (1 for each day)
biohazard bag
complete growth medium agar plate
(COMP)

minimal + adenine growth medium
   agar plate (MIN)
COMP plate with 24-hr. cultures of
   yeast stains **a**1, **a**2, **a**3, **a**4, α1, α2,
   α3, and α4

Procedure PART A

1. See Figure T8.1.
2. From the *r/r/T/T* strain all the gametes should be *rT* and from *R/R t/t* all should be *Rt*.
3. The only diploid that can be formed is *R/r T/t*. It is heterozygous, or hybrid, for two pairs of alleles.

Procedure
PART A Predicting the Gametes of the Parents
The diploid parents of this dihybrid cross would have been *r/r T/T* and *R/R/t/t*. To predict the F_2 offspring, it is necessary to predict the different types and relative numbers of gametes that can be produced. The simplest way to make this prediction is to diagram how the chromosomes separate at meiosis. The diagram in Figure 8.31a shows the gametes predicted from the pure-breeding diploid red parent (*r/r T/T*).

Two pairs of chromosomes are represented, one carrying the gene for color (in this case red, *r*) and the other carrying the gene for the ability to make tryptophan (in this case tryptophan-independent, *T*). At the first meiotic division (M_1), chromosome pairs segregate (separate). At the second meiotic division (M_1), the two chromatids of each chromosome segregate into the nuclei of the gametes. In this case, segregation can occur in only one way because only one allele of each gene is represented. All the gametes from a pure-breeding diploid are the same (*rT*). The probability of this parent producing a gamete with the genotype of *rT* is $^1/_1$, or 100%.

1. In your data book, copy and complete the diagram shown below in Figure 8.31b (or use the diagram your teacher provides) for the cream, tryptophan-dependent parent (*R/R t/t*).

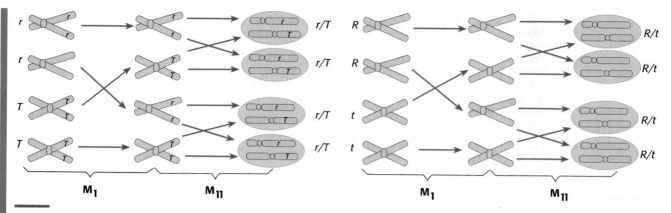

Figure 8.31 Probabilities of gametes that could be formed by yeast organisms that are homozygous for two traits.

2. Use the symbols given to describe for each parent strain the genotypes and phenotypes of all the possible gametes and the relative probabilities of their occurrence.

3. The haploid gametes produced from these two pure-breeding parents mate to form the diploid zygotes of the F₁ generation. Use the symbols to describe the genotype and phenotype of the diploid zygotes that could be formed from the fusion of these gametes. At this point your diagrams should show that they all will have the dihybrid genotype *R/r T/t*.

PART B Predicting the Gametes of the F₁ Diploids

Because there are two equally probable ways that the alleles of the two genes can separate at *M₁*, the chromosome diagram for the segregation of the *R, r, T,* and alleles is more complicated. These are shown in Figures 8.32a and 8.32b. Since these two patterns of segregation are equally probable, the two diagrams together illustrate the relative numbers of all possible genotypes.

4. Copy the diagrams in Figure 8.32 in your data book, or tape in the diagrams your teacher provides. Fill in the missing symbols. Each chromatid should have a symbol.

5. What is the total number of gametes represented in the diagram?

6. How many different genotypes are represented? Give their symbols.

7. How many times is each genotype represented?

8. What is the probability of occurrence of each genotype among the total number of gametes shown? Compare this prediction with the gametes in Figure 8.13.

9. Draw a checkerboard diagram for all the possible crosses among these gametes. This diagram should illustrate the 16 different combinations predicted. Instead of male and female gametes, use mating-type **a** and mating-type α haploid strains. In each square of the diagram construct the diploid genotype

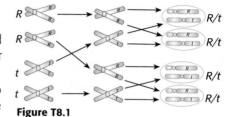

Figure T8.1

PART B

4. See Figures T8.2a and b.
5. 8.
6. 4: RT, Rt, ,rT, rt.
7. 2.
8. ⅛, or 25%.
9. The diagram should be the same as Figure 8.19 with the appropriate symbols for the yeast traits. It should predict 9 : 3 : 3 : 1 ratios.

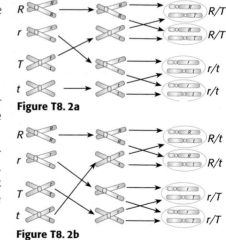

Figure T8. 2a

Figure T8. 2b

Figure 8.32 Gametes that could be formed by yeast organisms heterozygous for two traits.

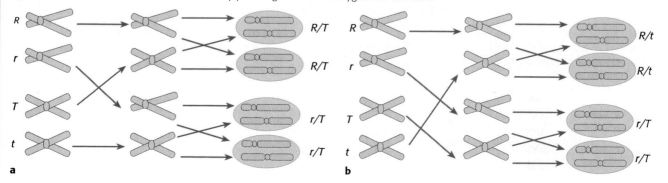

Mating-type a strains

	a1 R T	a2 R t	a3 r T	a4 r t
α1 R T				
α2 R t				
α3 r T				
α4 r t				

Mating-type α strains

Figure 8.33 Template for dihybrid yeast cross.

PART C

10. **Blackline masters of the templates are provided in the Teacher's Resource Book. Reduce or enlarge as necessary to fit your petri dishes. Or, make copies of Figure 8.33.**

15. **Encourage the students to make light streaks. Heavy streaks may lead to ambiguous results or false positives on growth tests because the yeast may contain enough nutrients to allow some growth even on the MIN medium. With light streaks, the growth is insignificant.**

16. **This step can be two days.**

21. **See Figure T8.3. (Crm = cream color on COMP, Red = red color on COMP, + = growth on MIN, and- = no growth on MIN.**

22. **9 are cream and grow, 3 are red and grow, 3 are cream and don't grow, and 1 is red and doesn't grow.**

23. **The results should confirm the predictions.**

that would result from the fusion of the corresponding gametes. How many different genotypes are there? How many different phenotypes? Does this diagram predict a 9 : 3 : 3: 1 ratio? What specific phenotypes would be represented in these ratios?

PART C Testing the Predicted Phenotypes

Day 0

10. Make two templates for setting up crosses. Copy the pattern shown in Figure 8.33 onto a piece of paper twice, so that each fits the bottom of a petri plate (or use the templates your teacher provides).

11. Tape one template to the bottom of a plate of complete growth medium (COMP) so that you can read it through the agar.

12. Transfer a small sample of each strain onto the agar directly over its corresponding label. To do this, touch the flat end of a clean toothpick to strain **a**1 on the plate your teacher provides. Then, gently drag the toothpick on the COMP agar plate to make a streak about one cm long. Discard the toothpick in the self-closing waste bag, being careful not to touch anything with it. Repeat this procedure for all eight strains, using a new toothpick for each transfer. *Be careful not to touch the ends of the toothpicks to anything except yeast or the sterile agar. Discard used toothpicks in the self-closing plastic waste bag. Keep the lid on the plate at all times except when transferring yeast.*

13. Invert and incubate at room temperature.

14. Wash your hands thoroughly before leaving the laboratory.

Day 1

15. On the same COMP plate, make a mating mixture for each of the mating-type **a** strains with each of the mating-type α strains. To do this, use the flat end of a clean toothpick to transfer a dot of the **a**1 strain of freshly grown cells to each of the boxes below it on the template. Discard the toothpick in the waste bag. Repeat this procedure for **a**2, **a**3, and **a**4, using a new toothpick for each strain. Using the same procedure, transfer a dot of the freshly grown cells of each α strain to each of the boxes to the right of the strain on the template. Place these dots side-by-side, but not touching, as shown in the first box in Figure 8.33. Use a clean toothpick to mix each pair of spots together. Be sure to use a clean toothpick each time you mix pairs together. Discard all toothpicks in the waste bag.

16. Invert and incubate at room temperature.

17. Wash your hands thoroughly before leaving the laboratory.

Day 2

18. Test each mating mixture and parent strain for its ability to grow on MIN agar. To keep track of the tests, tape a copy of the template to the bottom of the MIN plate. Then, use the flat end of a clean toothpick to transfer a small amount of each strain and mixture from the COMP plate to the corresponding position on the MIN plate. Be sure to use a clean toothpick each time you change strains and mixtures. Discard all toothpicks in the waste bag.

19. Invert and incubate both plates until the next day.

20. Wash your hands thoroughly before leaving the laboratory.

Day 3

21. Copy the score sheet in Figure 8.34 in your data book (or tape in the copy your teacher provides). Record the color and growth phenotypes of each parent haploid strain and F_2 diploid on the score sheet for both plates.

22. Tabulate the different phenotypes observed among the F_2 diploids and the number of times each one occurred among the 16 crosses.

23. Compare the F_2 phenotypes with the predictions you made in Part A. Explain how your results either support or contradict your predictions.

24. Discard all plates in the biohazard bag.

25. Wash your hands thoroughly before leaving the laboratory.

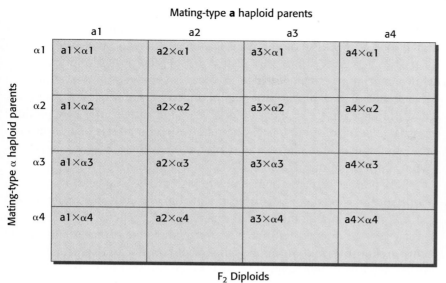

Mating-type **a** haploid parents

	a1	a2	a3	a4
α1	a1×α1	a2×α1	a3×α1	a4×α1
α2	a1×α2	a2×α2	a3×α2	a4×α2
α3	a1×α3	a2×α3	a3×α3	a4×α3
α4	a1×α4	a2×α4	a3×α4	a4×α4

Mating-type α haploid parents

F₂ Diploids

Figure 8.34 Score sheet for dihybrid yeast cross.

Mating-type a strains

	a1 R T	a2 R t	a3 r T	a4 r t
α1 R T	crm +	crm +	crm +	crm +
α2 R t	crm +	crm −	crm +	crm −
α3 r T	crm +	crm +	Red +	Red +
α4 r t	crm +	crm −	Red +	Red −

Mating-type α strains

Figure T8.3

Discussion

Because yeast exhibit most of the same traits in the haploid stage (gametes) and the diploid stage, you knew the precise genotypes of the gametes (haploid strains) that you mated to produce the F₂ diploids. This removed the element of chance at this step. In Part A, however, you had to deal with the role of chance to predict the numbers and genotypes of the gametes from the F₁ diploids. Throughout the entire process, beginning with the parental cross (P) and going through to the F₂ offspring, there are some steps at which chance plays a role, so the results can only be expressed as a probability. In other steps, chance is not a factor, so you can predict the outcome exactly.

1. List the steps in which chance is a factor.
2. List the steps in which chance is not a factor.
3. Explain why the outcome of some steps involves chance, whereas the outcome of others does not.

Discussion

1. **Chance is a factor at the first meiotic division in the dihybrid in determining whether R and T or R and t segregate together.**
2. **Chance is not a factor in the segregation of the parental strains at either meiotic division, or in the second division in the dihybrid.**
3. **Chance plays a role whenever there is more than one possible outcome of a step. In the present situation, this occurs when heterozygous alleles segregate at the first meiotic division.**

SUMMARY

Genetic information is passed to the next generation by heredity. The genes provide a set of instructions for this information in the sequence of DNA bases. A triplet of bases codes for one amino acid. RNA copied from the DNA transmits genetic information to the ribosomes, where amino acids are connected in chains to form distinctive proteins. The sequence of DNA bases specifies the sequence of amino acids in a protein.

Because genes are located on chromosomes, patterns of inheritance are explained by chromosome behavior in meiosis and fertilization. Sex determination in many organisms depends on X and Y chromosomes.

Genes exist in different forms called alleles. Normally, only two alleles for the same trait are present in an individual—one on each member of a chromosome pair. In meiosis, these paired alleles segregate into different gametes. Random fertilization brings alleles together in ratios that can be predicted by the rules of probability. Mendel observed the behavior of traits passed through successive generations. His observations were consistent with the behavior of alleles in gamete formation and fertilization. The principles of dominance, segregation, and independent assortment are the basis of Mendelian genetics.

Other patterns of inheritance are variations of Mendelian genetics. Alleles for some traits show codominance, and the inheritance of other traits, such as the ABO blood types, is controlled by multiple alleles. Both of these hereditary patterns result in several possible phenotypes. Multifactorial inheritance, which involves several genes interacting with environmental variables, results in continuous variability. X-linked traits and differences in the number or structure of chromosomes also produce unusual patterns of inheritance. The relatively mild effects of sex chromosome trisomies are explained by the hypothesis that all X chromosomes except for one are inactivated early in embryonic development.

Genes on the same chromosome are linked. The frequency of crossing over between linked genes is used to develop chromosome maps, which are helpful in understanding and screening for genetic disorders. The human genome project will increase our knowledge of genetic disorders and provide information about human variability and the function of introns and other DNA sequences that do not code for proteins.

Using recombinant DNA techniques, biologists can splice DNA from one species into the DNA from another species. Although the benefits from genetic engineering technology already are great and will increase, society is faced with many ethical and legal problems raised by applications of this technology.

APPLICATIONS

1. Cystic fibrosis is a common genetic disorder among Caucasians. The disorder is recessive; the gene is not X-linked. A young woman has a brother who has cystic fibrosis. The women visits a genetic counselor to find out whether she carries the allele for cystic fibrosis. Before performing any tests, what can the counselor tell her about the probability that she carries the allele? (Hint: What are the genotypes of the woman's parents?

2. Suppose you were flipping a coin, and the first nine times it came up "heads." What is the probability that it will be "heads" on the next and the flip?

3. For each characteristic you have, there is at least one special enzyme that controls the chemical reaction responsible for that characteristic. This is true for all living organisms,which means there must be millions of different types of enzymes. How is it possible that only four types of nucleotieds can produce the codes for all these different enzymes?

4. Suppose one of your chromosomes was damaged so that just one nucleotide of one gene was missing. Explain how the mRNA formed along this gene and the protein formed along the mRNA would be affected.

5. Most human traits are multifactorial. How does the distribution of such traits differ from either/or traits?

6. Barton Childs, a medical geneticist at Johns Hopkins School of Medicine, has said the following about the relationship between genes and the environment: "Genes propose, environments dispose." What does this statement mean?

7. How do Mendel's various principles explain his results?

8. How does the behavior of chromsomoes during the first meiotic division explain the ratios Mendel observed?

9. Most flowering plants have both male and female reproductive structures on the same individual. Would you expect these plants to have sex or expect these plants to have sex chromosomes? Explain.

10. In a dihybrid cross, how will the type of F_2 offspring differ if the two genes are on the same chromosome (linked together) rather than on separate chromosomes?

11. Baldness is an X-linked trait. If you are a male, can you tell by looking at your father whether or not you will become bald?

12. Genetic engineering involves the ability to transfer genetic material from one organism to another, even from one species to another. What does it tell you about the continuity of life in the biosphere?

PROBLEMS

1. Investigate two discontinuous human traits. Determine what is known about the genetics of those traits. Indicate how the environment influences the expression of those traits.

2. Some viruses contain RNA but no DNA. Find out how these viruses take control of human cells and reproduce other viruses.

3. "Wisconsin fast plants" are a species of mustard plant that grows from seed and that flowers within two weeks. Obtain some seeds and conduct crossing experiments with some of the many varieties.

4. Explore the relationship between advancing maternal age and the occurrence of chromosomal disorders such as Down syndrome.

5. There are a great many opportunities for employment in the field of genetics. Investigate the educational requirements for a particular job, such as a genetic counselor. Find out about the day-to-day tasks of a person employed in such a position.

6. Build a three-dimensional model of DNA.

7. Debate the merits and possible negative aspects of genetic engineering.

8. Many human genes have been mapped to specific locations on individual chromosomes. Using a genetics textbook investigate a current human gene map and locate the genes for the following traits: (a) Rh blood group; (b) ABO blood group; (c) hemophilia; (d) major histocompatibility complex (MHC) (these genes control the acceptance or rejection of tissues received in transplant operation); and (e) Huntington disease.

9. Human DNA can be inserted into a bacterium, and the bacterium will express the human genes. Because eukaryotic genes have introns and prokaryotic genes do not, bacteria do not have the enzymes necessary to cut the introns out of human DNA. Investigate how genetic engineers get around this problem. (Hint: The process involves an enzyme called reverse transcriptase.)

10. It is now possible to determine whether an individual carries the gene that causes Huntington disease. Find out about the disease and the method of detection. If you were at risk to develop the disease, would you want to know if you carried the fatal gene? What are the pros and cons about knowing?

a

b

Continuity Through Evolution

The organism in (a) died more than 150 million years ago, and its remains were preserved in rock. What does it look like to you? What familiar group of modern organisms does it most closely resemble? What does it have in common with these modern organisms? How does it differ? How would you account for these differences?

The organism shown in (b) is a modern bird, the anhinga (*Anhinga anhinga*). How do you know it is a bird? How does it resemble the organism in (a)? How does it differ? How are these two organisms, one extinct and the other very much alive, examples of *unity* of pattern and *diversity* of type? Discuss your responses with a classmate.

Archaeopteryx is the oldest bird known from the fossil record. It shares two characteristics with modern birds: feathers (the hallmark of birds) and a wishbone. Some authorities, however, believe that *Archaeopteryx* could not fly because fossils indicate it lacked a breastbone. *Archaeopteryx* also has many reptilelike characteristics, such as teeth, claws on its wings, and a bony tail. Evolution accounts for the differences between the two organisms. The anhinga was chosen because it visually resembles fossil *Archaeopteryx*. It is, however, no more closely related to *Archaeopteryx* than is any other modern bird. The similarities and differences that students find between the two organisms illustrate how organisms may share patterns because of common ancestry yet be very distinctive because of forces such as mutation, selection, and chance.

◆ All things are affected by time. Organisms are born, grow old, and die over periods of time. Plants sprout, grow, and bloom in some way before they die back with the changing seasons. The transmission of genetic material happens through time.

For an individual organism, 100 years is a long time, but changes in species require much longer periods of time than that. Each species alive today is the tip of a branching tree that extends far back in time to an ancient earth inhabited by gradually changing populations of organisms. How did today's species come to be? Evolution—change through time—is the biological process that links all species, no matter how they differ.

Evolution is the source of biological diversity and the one process that explains the unity in life. This chapter examines the evidence for ways in which evolution occurs. ◆

MAJOR CONCEPTS

◆ Evolution occurs in populations or species.

◆ Organisms similar in appearance and able to produce fertile offspring make up a species.

◆ New species arise because of reproductive isolation, selection, and the combination of variations within a population.

◆ Variations within a population or species arise from many sources.

◆ Only organisms that survive to reproductive age pass their genes to the next generation.

◆ Evolution may occur gradually or as spurts of change between periods of stability.

KNOWLEDGE CHECK

◆ What is evolution and how does it occur?

◆ Do individuals evolve?

◆ Is evolution taking place at the present time?

Diversity, Variation, and Evolution

9.1 Living Organisms Are Both Similar and Varied

The same types of cells and molecules can form different organisms. A daffodil is obviously different from an oak tree, and a horse does not look much like a pig. Even though these organisms look quite different, they share many similarities. Every type of organism also must have some way of obtaining energy, reproducing, and exchanging substances with the environment. Because there are a limited number of ways in which these processes can be accomplished organisms that appear quite different may have similar ways of accomplishing the processes.

Biologists have observed a unity of pattern in the structures and functions of different organisms. The most striking example of this unity of

> **Guidepost**
>
> What does diversity of type and unity of pattern mean?

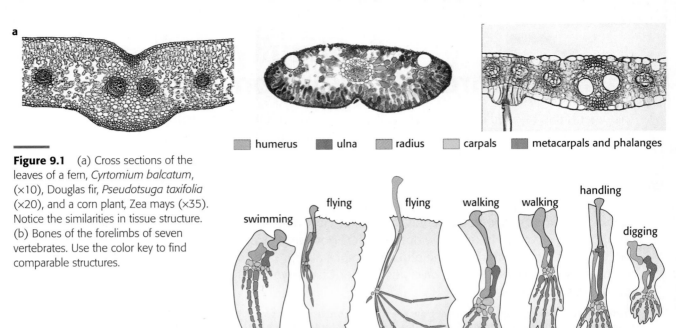

humerus   ulna   radius   carpals   metacarpals and phalanges

Figure 9.1 (a) Cross sections of the leaves of a fern, *Cyrtomium balcatum*, (×10), Douglas fir, *Pseudotsuga taxifolia* (×20), and a corn plant, Zea mays (×35). Notice the similarities in tissue structure. (b) Bones of the forelimbs of seven vertebrates. Use the color key to find comparable structures.

Four lines of evidence have been used as support for evolution: the number of species, biogeography, the fossil record, and homology. Use photos and charts to supplement the presentation of evidence in this chapter.

Be certain that the students understand the importance of this structural and functional similarity. Ask them for examples of how research with animals and other organisms has helped human lives (such as the treatment of leukemia).

pattern is the genetic code (the triplet sequence of bases in DNA that specifies which amino acids are incorporated into a protein) discussed in Chapter 8. There apparently is only one basic genetic code for all organisms, including humans. The cell theory discussed in Chapter 5 describes a unity of pattern at the cellular level.

This unity of pattern also can be seen in the structure of various organisms. Vertebrates, animals having backbones (a column of vertebrae), provide a good example of structural unity, as you can see in Figure 9.1. Even the most diverse types of vertebrates, such as whales and bats, have related limb patterns. Organisms also may share many functional features, such as ways of using food and eliminating wastes. Because of these similarities, the results of research on one group of organisms often are applicable to other groups, including humans. The nerve cells in a squid, for example, function much like the nerve cells in a human. Therefore, scientists can study squid nerve cells to gain an understanding of how human nerve cells function.

Reproductive and developmental processes also have a unity of pattern. The life cycles of animals with different sexes contain the same sequence of events meiosis produces gametes, fertilization forms zygotes, and mitosis eventually produces adults. As Figure 9.2 shows, for several vertebrates, the early stages of embryologic development are remarkably alike. These similarities do not mean that a human passes through fish, amphibian, or reptile stages during development. Rather, the similarities show that the same fundamental processes occur in the development of many different structures found in vertebrates.

Even with these similarities however, amazing diversity occurs in the types of organisms on the earth. Humans, squirrels, cows, frogs, spiders, **5** trees, orchids, mushrooms and bacteria are only a few of the different forms of life. There are millions of different organisms each with unique structures, behaviors and ways of life. How can you account for both great unity and great diversity in life?

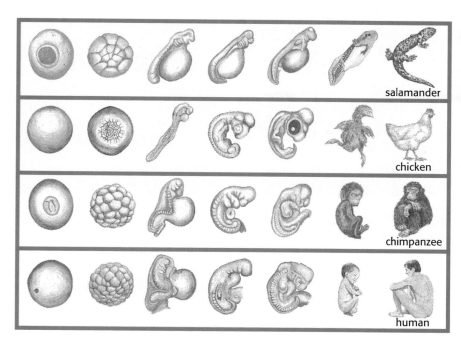

Figure 9.2 Embryologic development of some vertebrates. Zygotes are shown on the left, adults on the right, and comparable developmental stages are shown in between. Drawings are not to the same scale.

9.2   All Organisms Are Grouped into Species

2 Although every organism is unique, some have more characteristics in common than others. Organisms that are so closely similar that they can mate and produce fertile offspring are grouped in the same species.

The distinctness of a species is maintained in three basic ways. First behavioral, geographical, or physical differences may prevent one species from mating with another. The inability of one group to interbreed successfully with any other groups is known as **reproductive isolation**. Clearly, an elephant and a crab cannot mate and produce offspring. Alaskan brown bears and polar bears, however, have mated successfully in zoos, but no such cross has been discovered in the wild. Because brown bears live in forests and polar bears live on snowfields and ice floes, they rarely, if ever, encounter one another in nature. They are separate species.

Second, mating may result in offspring, but the offspring may not live.
3 If the eggs of a bullfrog are fertilized by a leopard frog, they develop for a short time, but soon die.

Third, organisms of two species may mate and produce live offspring, but the offspring are sterile, unable to reproduce. Mules, the offspring of a female horse and a male donkey, almost always are sterile (see Figure 9.3). Exceptions such as the Indian paintbrush shown in Figure 9.4, often

This section expands on the general definition of species in Section 3.8.

Organisms may not develop or may be sterile because of chromosome incompatibility during meiosis.

Krause was able to conceive because her extra chromosome was shed in a polar body.

Figure 9.3 Krause (center), a mule, with her two colts. Technically a mule is a cross between a male donkey and a female horse; normally mules are sterile but sometimes strange things happen. In 1984, Krause gave birth to a colt. Geneticists determined the number of chromosomes in both Krause and the new colt. The tests showed that Krause and her offspring were mules. Horses have 64 chromosomes, donkeys 62, and mules 63. In 1987, Krause gave birth to another colt. Examples such as these illustrate the difficulty in defining a species.

Figure 9.4   Indian paintbrush: *Castilleja rhexifolia* (a) and *C. sulphurea* (b), two different species, can form a fertile hybrid (c).

The species concept is useful but not very exact. The diversity of organisms is of such an overlapping nature that no exact definition holds for them all. Plants, especially, defy the usual breeding restrictions. The criterion of interbreeding is useless for organisms that are completely asexual in their reproduction.

Dogs vary as much as any other species of domestic animal. Obtain pictures of other animals, such as swine, cattle, horses, and chickens to show variation.

occur in plant species which can form fertile hybrids with other species. Generally however the term *species* refers to a group of actually or potentially interbreeding populations that are reproductively isolated from other such groups.

Sometimes, organisms in the same species may not look like one another. For example, not all dogs look alike. There are many intermediate breeds between a Great Dane and a beagle (see Figure 9.5). Middle-sized dogs can interbreed easily, but extremely different dogs, such as a Great Dane and a Pekingese, may have difficulty interbreeding because of physical differences. They are grouped within the same species because they can mate with intermediate breeds.

Differences among individuals of a species are called **variations.** Sometimes, as Investigation 9.1 demonstrates, members of the same species   **5** may look very similar but are quite different. The grass blades in a lawn, for example, may look identical. Closer examination, however, reveals that some blades are long, others are short, some broad, others thin. Variations occur among individuals of all sexually reproducing populations. Variations may be small or major. Humans, for example, do not all look alike. Many populations vary from one end of their geographic range to the other. For

Figure 9.5   Five breeds of dogs. Beginning with the Great Dane, breeds are represented by every other dog. Mongrel offspring are intermediate between the breeds. What types of variations can you see?   **Size, hair, length, limb length, and tail form are a few possible answers.**

great dane   collie

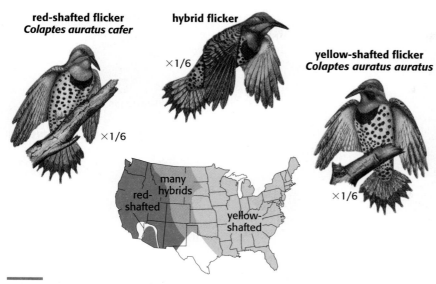

red-shafted flicker
Colaptes auratus cafer

hybrid flicker

×1/6

yellow-shafted flicker
Colaptes auratus auratus

×1/6

many hybrids

red-shafted

yellow-shafted

×1/6

Figure 9.6 The red-shafted flicker and the yellow-shafted flicker have large geographic ranges. They once were classified as separate species. Because they can and do interbreed, forming fertile hybrids, they now are thought of as one species.

Figure 9.7 Snow geese come in two color forms all white and mainly dark. Colonies often contain birds of both color forms, but geese usually pair with others of the same color.

example, the flickers shown in Figure 9.6 belong to the same species even though the populations on the West Coast are red-feathered and those on the East Coast are yellow-feathered. The hybrid individuals are coral or orange feathered. Some populations consist of two or more distinct varieties of individuals, such as the snow geese and blue geese shown in Figure 9.7.

You may wish to discuss clinal variations, such as those in response to cold, altitude, or latitude.

9.3 Darwin Observed Variation Among Organisms

How does such diversity and variation come about? In the nineteenth century, Charles Darwin offered one answer to this question. In 1831, at the age of 22, Darwin signed on as a naturalist with the H.M.S. *Beagle* for a five year voyage around the world (see Figure 9.8). During this trip, Darwin studied the natural history and biology of South America and various islands in the Pacific.

The Galapagos (guh LAH puh gohs) Islands, off the coast of Ecuador, particularly interested Darwin. As he explored these islands, Darwin

Figure 9.8 Young Charles Darwin (left) sailed around the world on the *Beagle*. The route taken (right) included stops in South America and the Galapagos Islands. During the trip, Darwin found evidence that would later support his theory of evolution by natural selection.

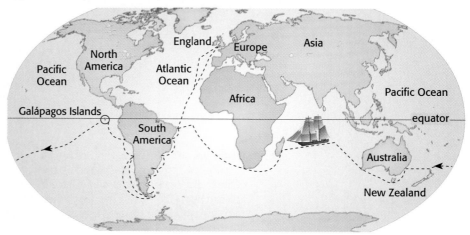

England   Europe   Asia

North America

Atlantic Ocean

Pacific Ocean

Africa

Pacific Ocean

Galápagos Islands

equator

South America

Australia

New Zealand

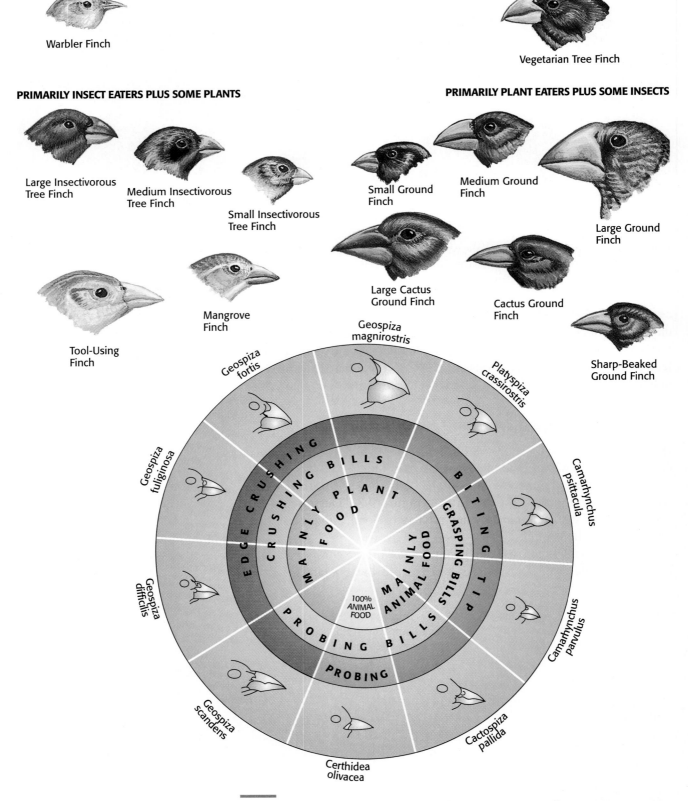

INSECT EATER

Warbler Finch

PLANT EATER

Vegetarian Tree Finch

PRIMARILY INSECT EATERS PLUS SOME PLANTS

Large Insectivorous Tree Finch

Medium Insectivorous Tree Finch

Small Insectivorous Tree Finch

Tool-Using Finch

Mangrove Finch

PRIMARILY PLANT EATERS PLUS SOME INSECTS

Small Ground Finch

Medium Ground Finch

Large Ground Finch

Large Cactus Ground Finch

Cactus Ground Finch

Sharp-Beaked Ground Finch

Geospiza fortis

Geospiza magnirostris

Platyspiza crassirostris

Geospiza fuliginosa

Camarhynchus psittacula

Geospiza difficilis

Camarhynchus parvulus

Geospiza scandens

Cactospiza pallida

Certhidea olivacea

EDGE CRUSHING

CRUSHING BILLS

MAINLY PLANT FOOD

100% ANIMAL FOOD

MAINLY ANIMAL FOOD

BITING TIP

GRASPING BILLS

PROBING BILLS

PROBING

Figure 9.9 The Galapagos finches were able to adapt to many different food sources because few species competed with them. Note the variations in the sizes and shapes of the beaks. The six species on the left are tree species feeding on various insects or on fruit and buds. The six dark species on the right feed on seeds of different sizes or on cacti. The finch at the top right uses cactus spines as tools to extract insects from under the bark of trees.

encountered a fascinating assortment of plants and animals. He collected many biological specimens, including marine and land iguanas and various types of birds.

One day, an island resident told Darwin that he could tell from which island any of the tortoises came by the size and shape of their shells and the length of their legs and necks. Although this observation meant little to Darwin at the time, years later it led him to wonder whether each of the islands was somehow producing its own forms of these tortoises and other creatures.

4   When Darwin finally returned to England, he continued studying the specimens he had collected during the long voyage. His collection of Galapagos finches was examined by specialists who determined that the specimens represented 13 species, differing primarily in the size and shape of their beaks (see Figure 9.9). Darwin surmised that these finches must originally have come from the South American mainland, but why, he wondered, were these island birds so different from those finches to be found on the mainland? Also, why did the assortment of finches differ so much from one island to the next?

During his studies, Darwin came across an essay by Reverend Thomas Malthus (1766-1834) that warned about the dangers of human overpopulation. Malthus pointed out that populations tend to increase and that if humans continued to reproduce at the same current rate, they eventually would outstrip their food supply. Darwin applied this idea to his own work and concluded that most species have a high reproductive potential but not all individuals reproduce. Populations, he argued, are kept in check in part because some organisms fail to reproduce. If this was true, then what was it that affected reproductive output, and did it affect all individuals in the same way?

> There were 14 species of finches, but Darwin collected only 13. One, the sharp-beaked ground finch, is now extinct. Darwin was sloppy about recording the islands from which he collected his specimens. Only after the voyage were the specimens sorted out by John Gould, who pointed out the differences in beaks. Darwin, in fact, had not even recognized them all as finches. The 13 species remaining today actually belong to 4 different genera. Darwin largely ignored the differences in tortoises because he thought they were an introduced species. After his collections were analyzed by Gould, Hooker, and others, Darwin was able to make his important insights. Frank Sulloway provides an excellent discussion of Darwin's work in *Biological Journal of the Linnean Society,* 21(1984): 29-59.

CONCEPT REVIEW

1. What does "unity of pattern" mean in evolution?
2. What is a species?
3. How do species maintain their distinctness?
4. How was Darwin influenced by the variety of organisms he observed on the Galapagos Islands?
5. Distinguish between variation and diversity.

Evolution and Natural Selection

9.4   Darwin Identified Selection as a Force in Evolution

1   Darwin knew that breeders could accentuate desirable characteristics by carefully selecting animals for mating. For example, by mating only the offspring of the greatest milk producers, breeders could develop high-yield dairy cattle. The roosters in Figure 9.10 also are the results of choosing animals for breeding. This process is known as **artificial selection.**

Darwin thought that some process similar to artificial selection might affect organisms in their natural environment. He suggested that in nature, the animals that reproduced were somehow selected by the environment. Individuals with characteristics that helped them survive in their environment could produce more offspring than individuals that did not have such

1   characteristics. He called this process natural selection. Scientists now

> **Guidepost**
>
> How does evolutionary change occur?

define **natural selection** as the process by which those characteristics that permit survival and reproduction are continued and eventually replace less advantageous characteristics. The characteristics that enable some members of a species or population to survive and reproduce are called **adaptations**. Adaptations are part of the variations in the group. **2**

If natural selection acts on variations among members of a population, what accounts for these variations? Darwin proposed that variations appear randomly and without design. If a new variation allows its bearer to produce more offspring, it will spread through future generations.

Darwin was not the only nineteenth-century scientist trying to explain how species change and variations are passed on. In France, Jean Baptiste Lamarck suggested that one species could give rise to another. He could not explain changes, but he surmised that some force causes an organism to generate new structures or organs to meet its biological needs. Once formed, he argued, these structures continue to develop through use, and the development attained by the parents is inherited by the offspring.

Today we know that the characteristics an organism develops during its life, called acquired characteristics, are not passed on to the offspring.

9.5   Darwin's Theory Changed Biology

Darwin developed his theory during a time when the prevailing scientific opinions were quite different. He lacked conclusive proof for his ideas. Although the field of genetics might have offered support, little was known about it then. Not having definite proof and expecting a storm of controversy, Darwin was reluctant to present his ideas. He did discuss them, however, with biologists who were close friends and they urged him to publish his theory.

The Teacher's Resource Book contains an expanded discussion of Lamarck's ideas of the passing on of acquired characteristics.

Note that evolution by natural selection is a theory, but that the scientific use of theory differs from the common use of the term. A scientific theory is not a guess. Rather, a theory provides a sound conceptual framework for its discipline and can withstand substantial testing. It also has explanatory and predictive values.

Darwin's work was extensive and extremely well thought-out. Point out that science often works in this manner—many different people may develop similar ideas. The best-developed and supported ideas usually receive the most credit. Wallace became famous for his work on biogeographical distribution.

Figure 9.10 An example of artificial selection. The red jungle fowl, a bird of Southeast Asia, is thought to be the species from which the many breeds of domestic chickens have been developed. Which of the breeds shown retains the largest number of the wild bird's traits? **The leghorn looks the most similar (except for the color).**

Figure 9.11 The basis for selection in the peppered moth (*Biston betularia*). Dark and light forms on a tree with light-colored lichens (a). The two forms of the moth on a tree blackened with soot (b). Which moth in each photo is most likely to be eaten by predators? **The dark moth on the light tree trunk; the light moth on the dark tree trunk.**

In 1858, Darwin received an unfinished paper from Alfred Russel Wallace, a young man who was working in Indonesia. Wallace's paper was, in essence, a sketchy outline of the principles of natural selection. Darwin had planned to set forth his findings at a scientific meeting, but now he insisted that Wallace disclose his ideas and share the credit. Both papers were read at the meeting, but Darwin's was presented first because he had extensive supporting evidence from his travels and observations. He subsequently published his ideas in a book entitled *On the Origin of Species*, in 1859.

As Darwin had feared, his theory of evolution by natural selection met with protest and opposition. His data came from his years of collecting on the *Beagle* journey, and his ideas were developed carefully. Nevertheless, the theory required the support and defense of some the best scientists of his time.

Theories usually are modified as our knowledge increases. This process of continuous, minor modification is an example of how science works. Occasionally, as in the case of Darwin's theory, a major theory is proposed that alters the way scientists look at their data. After that, the process of modification resumes in the light of the new theory.

Darwin's theory of natural selection has been supported by the findings of thousands of other biologists. A few "natural experiments" even have demonstrated evolution occurring. One example concerns the peppered moths found in the British countryside. In the 1700s, many trees in Britain were covered with pale gray-green lichens. Light-colored moths, like those shown in Figure 9.11, could be found on these lichens. Darker moths were rare, and they usually were eaten by birds because they were so visible against the light-colored trees. Later in the century, as the Industrial Revolution progressed, coal-burning factories spewed out clouds of black smoke. As soot darkened the tree bark and covered the lichens, the light colored moths became easier to see and darker moths became less visible. Birds began to eat the conspicuous light-colored moths, and their numbers dwindled. By the late 1840s about 98 percent of the moths were dark. Today, as pollution controls curb the production of

The complete title of Darwin's work was *On the Origin of Species by Means of Natural Selection or the Preservation of Favoured Races in the Struggle for Life.* The first printing of 1250 copies sold out the first day.

The *BSCS* inquiry video *The Peppered Moth: A Population Study* describes this natural experiment. The video is most effective if students have not read ahead, but it is useful in any case.

Figure 9.12 Natural selection explains the increase of penicillin resistant bacteria.

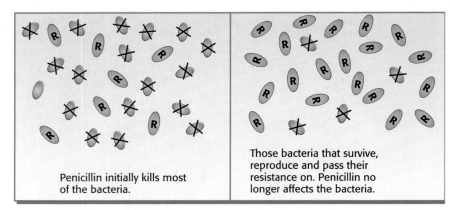

Penicillin initially kills most of the bacteria.

Those bacteria that survive, reproduce and pass their resistance on. Penicillin no longer affects the bacteria.

Figure 9.12 Natural selection explains the increase of penicillin resistant bacteria.

soot, more light-colored moths can be found. This change in color is the result of a change in the frequency of alleles for color in moths.

Other natural experiments have shown the same process of natural selection in bacteria (see Figure 9.12). About 30 years ago, penicillin killed most types of bacteria, but now many strains are unaffected by this antibiotic. This has come about because a few bacteria always had a genetic resistance to the penicillin. As nonresistant individuals were killed, the penicillin-resistant strains were left to reproduce. Therefore, there were a greater number of resistant individuals in the next generations. To avoid having this happen again, antibiotics are used more carefully today often on a rotating basis.

Extensive field work and experimentation continue to support Darwin's theory of natural selection as one evolutionary force. Mechanisms other than natural selection also have been proposed and have supporting evidence. Thus, now the question is not *whether* evolution occurs, but *how* it occurs, and how rapidly it occurs.

Darwin's book started a debate that never has abated completely. Because of its effect on scientific and religious thought, it can be considered one of the most influential intellectual achievements of all time.

CONCEPT REVIEW

1. Compare and contrast artificial selection with natural selection.
2. What are adaptations and how are they important to the survival of an individual and a population or species?
3. What is meant by evolution through natural selection?

Evolution and Genetics

9.6 Evidence from Genetics Supports Natural Selection

Guidepost

How do genetic mechanisms influence evolution?

In this century, although they are the framework for much of today's research, Darwin's original ideas have been modified and expanded greatly. Evolution may be defined as change in a species or population through time, but what type of change? Mendel's discoveries, which led to the chromosome theory of heredity, showed the way in which variations could be inherited and maintained in a population. Darwin clearly understood that only hereditary variations could have any meaning in the evolutionary process. Variations that are not passed on do not lead to changes in a species or a population. Hereditary variations, however, are products of an organism's genes.

There are at least two major sources of the variability required for evolution. One of these sources is mutations and another is genetic recombinations.

Mutations, which are discussed in Chapter 8, are changes in DNA molecules. The rate at which mutations occur is not known precisely. Some scientists maintain that the average mutation rate in humans is about one mutation per 100 000 gametes per generation. Others contend that the figure is lower about one mutation per one million gametes per generation. The mutation rate is difficult to determine because many mutations cannot be detected. A mutation in a gene for the recessive form of a given trait, for example, will not be exhibited in a heterozygous individual. Furthermore, there may be many mutations in introns—the pieces of DNA that are not translated into protein. At this time, in fact, biologists do not know if mutations in introns have any evolutionary significance. As DNA analysis improves, the ability to detect mutations also will improve, and a more accurate estimate of mutation rates will be possible. Whatever the actual mutation rate may be, however, a large body of evidence indicates that it has operated in large populations through long periods of time, producing enough variation for the process of natural selection to operate.

Genetic recombinations that result from meiosis and sexual reproduction increase the genetic variability in a given population by producing different combinations of genes (see Figure 9.13). A human, for example, possesses 23 pairs of chromosomes. Even if each chromosome pair contained only one pair of heterozygous alleles, independent assortment would produce more than 8 million different sorts of gametes. Thus, more than 60 trillion (6×10^{15}) genetic recombinations are possible in one child. Considering that there are a great many heterozygous pairs of alleles, the actual number of possible recombinations is enormous. This process of genetic recombination contributes to the **gene pool**—the sum of all the alleles of a population.

The rediscovery of Mendel's work at the turn of the century was a blow to Darwinian evolution, because the emphasis was on stability rather than change. De Vries's work (not described in this text) on macromutations in evening primroses demonstrated great genetic changes rather than the minor changes necessary for Darwinian evolution. Only more recently (since the 1930s) have geneticists found that most genetic changes are minor, supporting Darwin's conception of evolution.

The important points for students to understand are, first, that evolution acts in *populations*, not individuals per se. Therefore, a mutation rate per individual is not important. Second, evolution acts on mutations in populations *through time*.

The number of possible meiotic outcomes is equal to 2^n, where N = the number of chromosome pairs. In humans, N = 23, so the number of possible combinations of chromosomes in a gamete is 2^{23}, or 8 388 608.

Figure 9.13   Mutation and crossing over are the major sources of genetic variation. Mutations (a) give rise to new alleles, such as when *A* mutates to *a*. Before mutation, only the combination A*b* was possible. Crossing-over (b) provides new combinations. Before meiosis, only the combinations A*b* and a*B* occur. Crossing-over between *A* and *b* results in the recombinants A*B* and a*b* in the gametes.

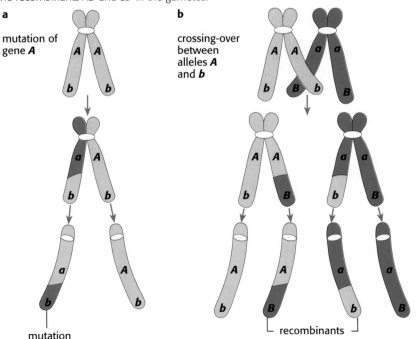

a mutation of gene **A**

b crossing-over between alleles **A** and **b**

mutation

recombinants

In any species, the variety of possible hereditary types is almost limitless because of genetic recombination. While not all that frequent, new mutations continue to occur and form new genotypes that produce new phenotypes. These phenotypes are subjected to natural selection, which eliminates certain gene combinations from the population's gene pool. Natural selection, however, is not the only evolutionary mechanism at work.

9.7   Several Evolutionary Mechanisms Can Affect Populations

In a plant population made up entirely of homozygous red-flowered plants, the gene pool for flower-color alleles consists of only the allele for red flowers—a frequency of 100 percent. In that homozygous population, the frequency of alleles for white flowers is zero. On the other hand, if the population consists entirely of *heterozygous* red-flowered plants, then the allele frequencies are 50 percent red-flower alleles and 50 percent white-flower alleles.

Study the cross shown in Figure 9.14, and determine the allele frequencies for each generation. In general, the frequency of alleles in a population's gene pool remains relatively stable, particularly if the population is large and well adapted to its environment. This type of population is said to be in equilibrium.

A number of factors, however, can change the frequency of alleles in a population. The first factor is mutation. A mutation results in the introduction of a new allele and causes an immediate, but small, shift in equilibrium. Populations of organisms (bacteria, for example) that produce a large number of offspring within a short time are influenced more by mutations than populations of organisms that produce a few offspring over a relatively long time span.

The second factor is migration, the movement of organisms into or out of the population and therefore the gene pool. If organisms entering or leaving the population have a genotype different from that of the rest of the population, a change in equilibrium occurs. The movement of genes into or out of a population because of migration is called gene flow.

The third factor is a *random* change in allele frequencies that results in a population with distinct characteristics. This random change, which usually occurs in small populations, is called **genetic drift.** The Dunker population of Germany (see Figure 9.15) is a good example of genetic drift. The Dunker population arose from a small group of people that had a higher frequency of a particular blood type allele (and a lower frequency of a second allele) than the surrounding population. Succeeding generations retained this frequency differential because of religious beliefs against marriage outside the group. When some of the group

The Hardy-Weinberg principle requires a diploid, sexual organism, nonoverlapping generations, random mating, and a large population.

An excellent example of a nonadaptive trait being established in a small population (a rare allele being over represented) is the combination of dwarfism and polydactylism which occurs among the Old Order Amish of Lancaster, PA. Sixty-one cases of this rare deformity have been reported in this population since the 1770s, nearly as many as have occurred worldwide. The Amish now number about 17 000, but the order was begun by only a few couples and has kept itself virtually isolated since then. By chance, one of the founding individuals carried the allele responsible for this defect.

Figure 9.14   In this cross of two heterozygous red-flower plants, what are the allele frequencies in the parental generation? In the offspring generation? **50% red, 50% white, in both generations.**

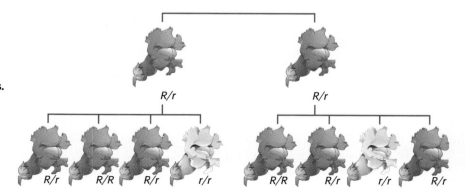

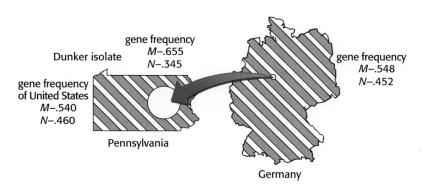

gene frequency
Dunker isolate   M–.655
                    N–.345

gene frequency
of United States   gene frequency
   M–.540   M–.548
   N–.460   N–.452

Pennsylvania

Germany

Figure 9.15 Genetic drift in *MN* blood groups. A Dunker group that had a relatively high frequency of the allele *M* immigrated to Pennsylvania. Descendants of these Dunkers continue to exhibit a higher frequency of the alleles for M and a lower frequency of the alleles *N* than others in Germany or the rest of the United States

emigrated to the United States, the differential frequency increased and then was maintained by the same marital bias.

2   A random change in allele frequencies may be caused by other factors as well. If a population is small, a storm or an attack by predators could kill a large proportion of the population, and the survivors may have a different allele frequency than that found in the original group.

The fourth factor is artificial or natural selection. These types of selection determine which individuals in a population will reproduce and pass on their genes. If there are strong selection pressures on the population, the equilibrium of the gene pool will change rapidly. Weak selection pressures will change the population more slowly.

The fifth factor that can influence the equilibrium of a gene pool is the nonrandom mating of individuals within the population. If certain individuals in a population show a preference for mating with individuals that have a particular phenotype, the mating pattern ceases to be random. Such nonrandom mating is known as **sexual selection** (see Figure 9.7 on page 203). Sexual selection influences the allele frequencies of a population in future generations. If these allele frequencies lead to reproductive isolation, the population may become a new species. Investigation 9.3 considers some of the factors that can lead to **speciation**, the development of new species.

9.8 Isolation Is Needed for Speciation

Reproductive isolation maintains the distinctness of a species. It also is necessary for the development of new species. The evolution of a new species through time, known as speciation, is shown in Figure 9.16. Speciation usually does not occur within a population in which individuals are free to interbreed. If a population separates into two small groups that remain isolated, however, changes between the two groups through
3   time may become so great that interbreeding is impossible even if the groups meet. When the two groups can no longer interbreed, speciation has occurred. For example, the Kaibab and Abert squirrels of the northern and southern rims of the Grand Canyon probably once belonged to the same species but diverged when they became separated by the canyon (see Figure 9.17).

The tendency of a population to disperse into new areas, which was discussed in Chapter 2, also may lead to speciation. As the parent population grows, offspring populations move off into new areas that offer different opportunities and different problems. Each population then gradually adapts to its new environment or dies out. In time, these adaptations render the populations so genotypically different that they do not

You must introduce the concepts of behavioral, ecological, geographic, mechanical, and seasonal isolation. The important point for students to remember, however, is that reproductive isolation—the inability of one group to mate and produce offspring successfully with any other groups for whatever reason—is the major criterion for determining a species.

Figure 9.16 A simple model of how two populations of a species may become isolated and evolve to produce two species. They may be considered separate species when they become so different that they can no longer interbreed even though little or no physical barrier remains.

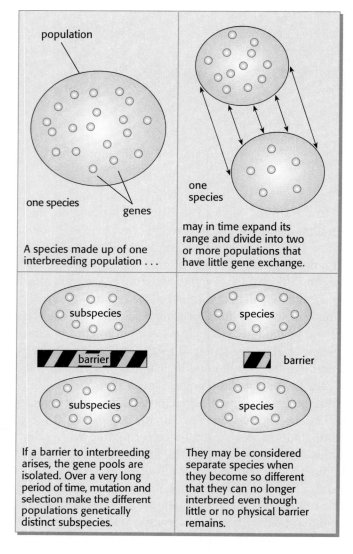

population

one species

genes

A species made up of one interbreeding population . . .

one species

may in time expand its range and divide into two or more populations that have little gene exchange.

subspecies

barrier

subspecies

If a barrier to interbreeding arises, the gene pools are isolated. Over a very long period of time, mutation and selection make the different populations genetically distinct subspecies.

species

barrier

species

They may be considered separate species when they become so different that they can no longer interbreed even though little or no physical barrier remains.

The adaptation of the organisms to the new resources and environment is the major point of adaptive radiation.

interbreed even when they inhabit the same area. They have become separate species. This process of dispersal, adaptation, and subsequent speciation is called **adaptive radiation** (Figure 9.18). The Galapagos finches are thought to be descendants of a mainland finch that dispersed to the islands and underwent adaptive radiation.

The development of species may not always take a long time. Most experts agree that the different organisms represented in the fossil record changed little for millions of years. Occasionally, however, many new

Figure 9.17 The Kaibab (a) and Abert (b) squirrels are related. When they became isolated on opposite rims of the Grand Canyon, they began to develop different characteristics. How many can you see?

Answers may include tail color, belly color and back color.

a

b

The Cheetah On the Way to Extinction?

Plant and animal extinction has many causes. Humans often cause changes in the environment that reduce or eliminate food, living space, and breeding grounds, thus modifying or destroying habitats. In the case of the cheetah, which the Convention of International Trade in Endangered Species has listed as threatened by extinction, other factors may be more important. Estimates of the cheetah population vary from 1000 to 25 000 animals. Recent studies have shown that a "population bottleneck" has limited the population. At some time in the cheetah's recent evolutionary history, an infectious disease or a natural catastrophe must have reduced its population. This in turn caused a reduction in the variety of alleles in its gene pool.

To test the hypothesis that a population bottleneck had reduced the genetic variation in the cheetah and was an important factor in the decline in the cheetah population, scientists used their knowledge of transplant and skin graft rejections in humans. Frequently, an organ transplant or a skin graft is rejected because the donated organ or skin is recognized as foreign by the immune system of the recipient. The greater the genetic variation between donor and recipient the greater the possibility of rejection. In genetically similar individuals, the chances of rejection are greatly reduced.

The researchers exchanged skin grafts between six pairs of unrelated cheetahs. They also grafted patches from one part to another of each cheetah. As a second control, the researchers grafted skin patches from another species, the domestic cat, onto each cheetah; the bottom photo on the right shows one of these grafts after 12 days. The grafts between unrelated cheetahs (on the left) and from one part to another of the same cheetah (in the middle) are healing. There was no rejection of the graft between unrelated cheetahs or between parts of the same cheetah. The graft from the domestic cat, however, shows acute rejection.

The rejection of the grafts from the domestic cat indicated that the cheetah's immune system was responsive to genetically different transplanted tissues. Therefore, if there were a normal variation of alleles in the gene pool, one would expect grafts between unrelated cheetahs to be rejected after 7 to 13 days. The nonrejection of these grafts supported the hypothesis that the population bottleneck reduced the allelic variation in the gene pool of the cheetah.

The cheetah population is an example of a genetically uniform species that could be wiped out by a deadly infectious disease. Should such a disaster happen, evolution of the cheetah species would stop, and it would become one of the many animal species that are known only through books and museums.

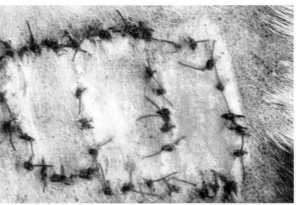

Young cheetah (top); skin grafts (bottom).

Figure 9.18 Adaptive radiation in mammals. As the first mammals dispersed to new areas, they adapted to the new environmental conditions. Gradually, these adaptations produced different species and, eventually, all the different mammals alive today. Representative animals from the orders included are (1) pangolin, (2) squirrel, (3) bat, (4) lion, (5) buffalo, (6) horse, (7) elephant, (8) armadillo, (9) rabbit, (10) monkey, (11) mole, (12) whale, (13) aardvark, (14) manatee, (15) kangaroo, (16) mastodon, (17) litoptern, (18) creodont, (19) *Brotothenum*, (20) *Uintathenum*.

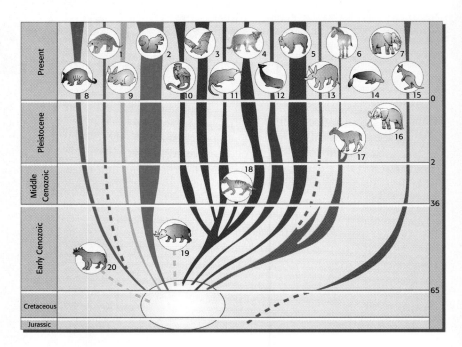

organisms, not represented in the earlier fossil record, appear. These periods of long stability, followed by the sudden appearance of new forms, are called **punctuated equilibria**. According to the punctuated equilibria hypothesis, major genetic changes may have caused new species to evolve in relatively few steps (see Figure 9.19). Although biologists debate whether the development of most new species has been relatively quick or a process of gradual change, even the punctuated equilibria model requires thousands of years for the formation of new species.

Figure 9.19 Two models for the rate of speciation. The traditional model (a) contends that new species branch off the parent species as small populations. Species descended from a common ancestor become more and more unlike as they acquire new adaptations. According to the model of punctuated equilibria (b), a new species changes the most at the time it branches from a parent species and changes little for the rest of its existence.

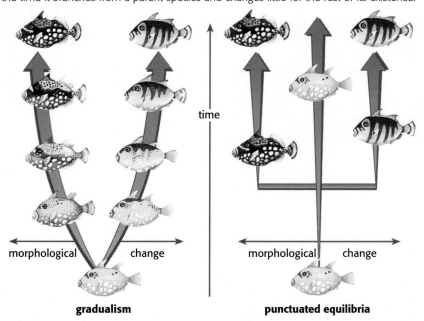

Figure 9.20   An ostrich (a) and rhea (b). What similarities and differences can you see? **Answers will vary.**

Debates over the mechanisms of evolution are good examples of the refinement of scientific theories. New information leads to new questions and hypotheses. Each new hypothesis then must be tested to determine its validity.

9.9 Various Evolutionary Patterns Occur

A variety of evolutionary patterns can be found in the living world. Some of these patterns occur between groups of closely related organisms. Divergence occurs when organisms within a species become so different that they can no longer interbreed. Even after divergence occurs, however,
5 closely related groups still can evolve in the same direction. When this happens, it is known as the process of parallel evolution. For example, the African ostrich and the South American rhea, shown in Figure 9.20, probably are descended from a fairly recent common ancestor. Even though the two groups of birds have diverged, they have evolved in a similar manner because they were subjected to the same type of environmental pressures.

Other patterns of evolution occur between more distantly related
4 groups of organisms. In convergence, different groups have developed the same adaptations because they live in the same type of environment. Compare the spiny, leafless euphorbia (yoo FOHR bee uh) of the African desert with a cactus from the Mexican desert. Their body forms are similar, but their flower structure shows that they are members of two different families (see Figure 9.21).

5 In coevolution, two unrelated groups become uniquely adapted to one another. One of the most common examples of coevolution is the adaptation that has developed between certain flowers and their pollinators. One such relationship is that of the yucca plant and the pronuba moth of the southwestern United States, shown in Figure 9.22. When the yucca flowers open, several small pronuba moths fly into the flowers and mate there. Each female moth collects pollen from that flower and rolls it into a ball. She then flies to another yucca flower, bores into its ovary, and deposits her fertilized eggs there. Afterward, she pollinates the flower with the ball of pollen. Later, the moth eggs hatch into larvae that feed on some, but not all, of the yucca seeds that are maturing in the ovary of the yucca blossom.

Both the yucca and the pronuba moth benefit from this relationship, and neither can complete its life cycle without the other. Other insects that are attracted to the yucca blossom may derive benefit for themselves but are unlikely to pollinate the flowers.

Similar in appearance; spines grown in rows. Differences include the shape of the spines and the type of flowers.

Figure 9.21   Convergence in plants. Cactus is on the top, euphorbia on the bottom. How are these plants similar? How are they different?

CONCEPT REVIEW

1. How does genetics help explain evolution?
2. Explain how different factors can affect a population's gene pool.
3. Explain how isolation is necessary for the formation of a new species.
4. Compare and contrast the evolutionary processes of convergence and divergence.
5. Explain the difference between parallel evolution and coevolution.

Figure 9.22   A yucca, *Yucca glauca*, and a Pronuba moth carrying a pollen ball on a yucca flower.

INVESTIGATION 9.1 Variation in Size of Organisms

Within a population of organisms, small or large variations make each individual different from the others. Variations among familiar organisms are fairly obvious. For example, you are very aware of differences among humans, but all bluebirds may appear alike to you. In this investigation, you will study individual variations in size among members of a sample population, draw graphs of the data, and interpret them.

This investigation acquaints students with the concept of variation within a population and with some of the mathematical treatments of data. The investigation lends itself to small group learning; teams of 3 are recommended. *Time:* **One class period.**

Materials (per team of 3)
metric ruler
graph paper
50 objects of one type—dried bean
    seeds, peanuts in the shell, leaves
    from the same type of tree

PART A Line Graph
Procedure

1. Choose one object and measure its length in millimeters. How "typical" do you think this object is in terms of its length? Guess how many millimeters larger the longest object will be. Then, guess how much smaller the smallest object will be. Record the object's length and your two guesses in your data book.

2. Chose two team members to measure the objects. The third member will record the data. Using a metric ruler, measure the length of the other 49 objects to the nearest millimeter. (If an object measures 11.5 mm, for example, round up to 12.0 mm.) Record the measurements in your data book with your first measurement.

3. Construct a grid. Label the *x* (horizontal) axis *Length*. The size of your largest object will determine the length of this axis. Divide the axis into 10 or 20 equal parts and label each mark (*10 mm, 20 mm,* etc.). Label the *y* (vertical) axis *Number of Objects*. Divide this axis into 20 equal parts and label each mark (*1, 2, 3,* etc.).

4. Draw a line graph on the grid. Group the objects according to their length. Then, count the number of objects in each group and mark this number on the graph. For example, suppose there were 3 objects that were 20 mm long. Find the 20 mm mark on the *x* axis and trace an imaginary line up from this point. Trace an imaginary line from the third mark along your *y* axis across the graph. At the point these lines meet, draw a large dot. Continue to construct the entire graph.

5. Connect all of the dots with one continuous line.

Discussion

1. Is there variance within your population? What is the shape of your line? Does it resemble those of your classmates?

2. What would be the shape of the line if you pooled all of the data in the class?

3. How close were your guesses about the largest and smallest objects? Do you think that using a small sample is a good way to predict the characteristics of a large population? Explain.

PART B Frequency Distributions and Histogram
Procedure

6. Using your data from Part A, construct a frequency distribution table as follows:

   a. Determine the range of length in the sample by finding the difference between the largest and smallest objects.

   b. Divide the range into 8 to 10 intervals, selecting a convenient size. For example, if the range of the sample is 23 (the smallest, 16 mm; the

PART A Procedure

1. **Many students will pick a "typical" object that will fall somewhere in the middle of the range, but some will choose large or small objects. Few students will pick the object with the average length.**

3. **To help your students understand graphing, put an example on the chalkboard. Graph paper is not essential but will make constructing graphs simpler.**

Discussion

1, 2. **There should be variation in all of the graphs. If every graph is in millimeters, then the wider the graph, the more variation is expressed in the population. Characteristics with continuous variation, such as the size of seeds, will show a bell-shaped curve if enough measurements are made.**

3. **Probably not. An individual rarely is a good representative of a population with variation. Although most individuals will be similar in size or shape, there will be noticeable, measurable differences. Make certain students understand that variation extends to internal structures and processes as well as external morphology.**

Table 9.1

x	f
16–18	2
19–21	7
22–24	4
25–27	9
28–30	14
etc.	etc.

PART B Procedure

8, 9, 10. The average probably will be slightly different for each group. Depending on the number of modes, the data may be unimodal, bimodal, or polymodal. For clarity, treat the data as unimodal. Mean, mode, and median are measures of central tendency—the tendency of data to congregate around a central point. The students should understand that these measures are used by biologists and provide different ways of using and interpreting data. Each is a valuable measure in demonstrating different points.

Discussion

1. **Answers will vary from group to group.**
2. **Although size differences may be only slight, they could make a profound difference in the life of an organism. For example, small seeds of many plant species tend to germinate faster than larger seeds. Small seeds will be dispersed farther from the parent plant than larger ones and will be harder to find by predators.**
3. **This is a subjective question. The difference in size may be more significant to some students than to others.**
4. **The size of many objects will vary with age. Also, objects from different individuals probably will show greater variation than objects from a single individual.**
5. **In this example, it will be easier for your students to make the histogram than the line graph. For data that are continuous, a line graph is most appropriate. For data that are discontinuous, a histogram is more appropriate.**

This investigation illustrates how natural selection operates and allows students to experience the interrelationship between coloration and habitat. It also emphasizes the relationship between predators and their prey. Teams of 6 are recommended.

Time: **One class period.**

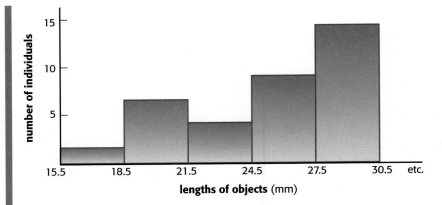

Figure 9.23   A histogram

largest, 39 mm), it could be divided into 8 intervals of 3 mm each, as follows: 16–18, 19–21, 22–24, 25–27, 28–30, 31–33, 34–36, and 37–39.
  c. Assemble your data in a frequency distribution table with the intervals listed in the x column and the number of individuals falling into each interval (frequency) listed in the F column, as shown in Table 9.1.
7. Construct a histogram similar to Figure 9.23 from your frequency distribution table. A histogram is simply a bar graph with the intervals on the horizontal axis and the frequency on the vertical axis.
8. Calculate the mean, or average, of your data. The mean is the sum of all measurements divided by the number of individuals you measured.
9. What is the mode, or high point, on the histogram? The value of the mode usually is given as the midpoint of the interval having the highest bar. If the high point falls in the 16 to 18 range for example, choose 17 as the mode. In Figure 9. 23, the mode is 29.
10. Find the median, which is the value for the middle of the sample, when the values (lengths) are arranged in order. If a series of measurements is 2, 2, 3, 4, 7, 8, 9, 9, and 11, the median would be 7.

Discussion

1. Look at the data and your histogram. What is the difference in length between the longest and the shortest objects in your sample?
2. Given the overall size of the objects, do you think this difference is important? What might be the advantage or disadvantage of being smaller or larger than average?
3. Would you have noticed the differences if you had not measured the objects?
4. Do you think there would be any size difference if the objects were of different ages or from different plants?
5. When do you think a histogram is a better, or more accurate, way to present data than a line graph?

INVESTIGATION 9.2   Natural Selection—A Simulation

Biologists consider natural selection to be the chief mechanism of evolutionary change and the process responsible for the diversity of life on earth. This investigation illustrates one way in which natural selection operates.

Materials (per team of 6)
2 l-m × 2-m pieces of fabric with different colors and patterns
100 paper chips of assorted colors in plastic bag

small bowl
1 set of colored pencils with colors similar to chip colors
5 sheets of graph paper

PART A Natural Selection
Procedure

1. Spread out the fabric "habitat" on a table top. Examine the paper chips and record the colors of the chips.

2. Appoint one team member as the keeper of the bag of paper chips. All other members are predators, whose prey are chips. The keeper will keep track of the number of turns each predator takes and the number of prey remaining.

3. Turn your backs to the table and allow the keeper to spread the chips uniformly over the fabric, making sure no chips stick together.

4. Imagine yourselves as predators, the paper chips as your prey, and the fabric background as your habitat. One at a time, turn around and select a paper chip using only your eyes to locate it. Do not use your hands to feel the chips. When you have selected a chip, place it in the bowl and turn around. Continue taking turns until only 25 paper chips remain on the fabric and the keeper signals you to stop.

5. Carefully shake the fabric to remove the survivors.

6. Group the survivors according to color by placing chips of the same color together. Arrange them in a row. Record the number of each color that survived.

7. Assume each survivor produces three offspring. Using your teacher's reserve supply, place three chips underneath each survivor.

8. Mix the survivors and their offspring thoroughly and distribute them as in Step 3. (Do not use the chips that were "eaten.")

9. Repeat the entire process of selection (Steps 3–7) four more times. Be sure to record the total number of chips of each color at the start of each round.

Discussion

1. Prepare a histogram for each generation (five histograms) using colored pencils that match as closely as possible the chip colors. Study the histograms of each generation. Was one color chip represented more than others in the first generation of survivors? Why do you think these chips "survived"? What, if any, change occurred between the first and second generation? The first and fifth generation?

2. Compare the original and the last survivor population. Which, if any, color from the original population is not represented in the survivor population? Why?

3. Examine the colors of chips in the fifth generation and the fabric habitat. How do the colors of the survivors relate to the colors of the habitat?

4. How are your results related to the process of natural selection?

5. Assuming no new individuals migrate into the habitat, how will the population change with time? How are your results related to evolution?

PART B Variation and Changes in the Environment
Procedure

10. Using a piece of fabric (habitat) with colors different from the first, begin a new series of predator-prey interactions with the 100 paper chips used in Part A.

11. Repeat Steps 3–8 five times.

Discussion

1. An organism is adapted to its environment if it can grow and reproduce in it. In Part A, which color of chip was "best adapted" to the habitat? Which color of chip was "least adapted"?

2. In Part B, which color of chip was "best adapted" to the new habitat? Which color was "least adapted"? How do these compare to the chips from Part A?

3. Changing colors of habitat is similar to a change in the environment. Name as many types of changes in the environment as you can.

Discussion

1. **The chip best blending into the background will have the most survivors. You can relate this to animals having camouflage. The change between the first and the fifth generations is much more dramatic than that between the first and the second.**

2. **Individuals that stand out from the habitat are easily seen by predators, captured, eaten, and their genes removed from the gene pool.**

3. **The colors of the survivors blend closely with the colors of the habitat.**

4. **These results mimic natural selection, where the best adapted individuals reproduce the most and pass their genes on to the next generation.**

5. **Eventually, the best adapted individuals will make up the majority within the population and the least adapted will become extinct.**

Discussion

1, 2. **This answer is similar to 1 above; however, students should relate adaptation to reproduction. Also, students should see that with a change in the habitat, a different color of chip is better adapted to the new environment, even if it survived poorly in the first habitat.**

3. **Seasonal changes and changes from one year to the next, such as dry or wet years. Changes through geological time, such as ice ages, also can affect the success of living things.**

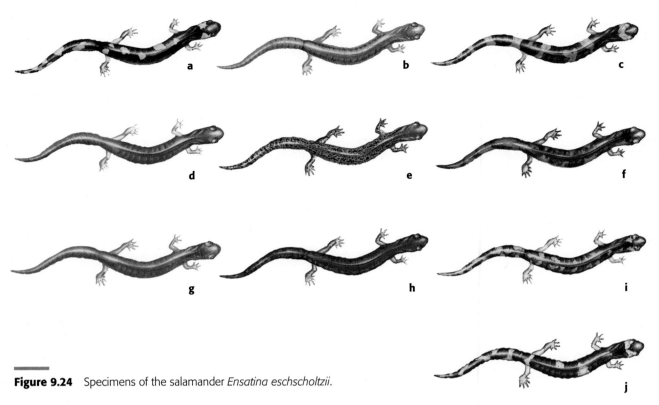

Figure 9.24   Specimens of the salamander *Ensatina eschscholtzii*.

4, 5. Variation increases the likelihood that at least a few members of a population will be able to survive even if the environment changes. If all the offspring were identical, then most would survive in a constant environment, but would die when they could no longer tolerate part of their environment.

4. How are species with genetic variation in their offspring able to survive in environments that change through time?

5. What do you think would happen to a population if it produced only offspring that were all alike?

For Further Investigation

If new individuals of different colors immigrate into your population, what will be the effect on the population, assuming that habitat and predators remain the same?

This investigation demonstrates how speciation occurs and should emphasize for students that a scientist's real work begins when she or he starts to organize and analyze data. Dr. Stebbins's research was published in *University of California Publications in Zoology* 48 (1949): 377-526. Students should plot data individually but use small groups to consider the discussion question
    Time: **One class period.**

INVESTIGATION 9.3   A Step in Speciation

The small salamanders of the genus *Ensatina* are strictly terrestrial. They even lay their eggs on land. Nevertheless, these salamanders need a moist environment and do not thrive in arid regions. In California, *Ensatina eschscholtzii* has been studied by R. C. Stebbins at the University of California (Berkeley). This investigation is based on his work.

Materials (per team of 1)
outline map of California
8 different colored pencils

PART A COLLECTION AREAS
Procedure

1. Imagine that you are working with Stebbins's salamander specimens, some of which are pictured in Figure 9.24. In the following list, salamanders are identified by subspecies. (A subspecies is a geographically restricted population that differs consistently from other populations of the same species.) The parentheses after each subspecies contain a number and a color. The number is the total of individuals Stebbins had available for his study. The color is for you to use in designating the subspecies. Following this is a list of collection areas. They are indicated by a code that fits the map of California in Figure 9.25. For example, 32/R means that one or more *E. e. croceator* specimens were collected near the intersection of Line 32 and Line R.

a. *E. e. croceator* (15; brown): 32/R, 32/S, 30/T, 31/T

b. *E. e. eschscholtzii* (203; red): 30/M, 32/O, 34/S, 35/V, 36/W, 35/Z, 38/Y, 40/Z

c. *E. e. klauberi* (48; blue): 36/Z, 38/a, 40/a, 39/a

d. *E. e. oregonensis* (373; purple): 9/B, 7/E, 6/E, 13/C, 10/C, 7/D, 15/D

e. *E. e. picta* (230; yellow): 2/B, 2/C, 3/C, 4/C

f. *E. e. platensis* (120; green): 8/J, 10/J, 11/M, 13/M, 15/M, 15/O, 17/M, 15/P, 20/Q, 24/S, 21/R, 25/T, 26/U

g. *E. e. xanthoptica* (271; orange): 17/G, 17/F, 19/H, 19/O, 20/I, 20/J, 21/I

2. Plot each collection area by making a small X mark on an outline map that has a grid like the one in Figure 9.25. Use the colors indicated for each subspecies population to make a distribution map of *Ensatina eschscholtzii* in California.

Discussion

1. Is the species uniformly distributed? Use your knowledge of the species' eco-logical requirements to offer an explanation of its distribution. Are there any other factors that might affect distribution?

2. Consider the physiography of California in Figure 9.25. Does the species seem more characteristic of mountain areas or of valley areas?

3. Do you expect any pattern in distribution of subspecies? Why or why not?

4. Examine the salamanders in Figure 9.24. Note that some subspecies have yel-low or orange spots and bands on black bodies. Some have fairly plain, brown-orange bodies. One has small orange spots on a black background. There are other differences as well. For example, some of them have white feet. Now refer to your distribution map. Does there appear to be any order to

Discussion

1. **No. In those areas where salaman-ders do not occur, there probably are specific limiting factors, such as arid or semiarid conditions. Historical factors, such as human habitation, also can affect the distribution.**

2. **Salamanders occur in the mountain-ous regions, except in mountainous desert ranges.**

3. **The fact that subspecies are geo-graphical variations within a species suggests there is an order to the dis-tribution of these subspecies. Because of ancestral relationships, adjacent species should be more like each other than widely separated subspecies. Exceptions can be explained. Stebbins postulated that subspecies 5, *E. e. picta*, is closest to the ancestral form and that subspeci-ation has taken place southward along the coastal and the inland mountains. This would explain the difference in pattern between coastal and inland mountain forms and the similarity among coastal forms and among island forms. At the end of the investigation, you may want to discuss with your students the most likely center of origin of the species, including a consideration of gene flow isolation mechanisms.**

Figure 9.25 Map of California, with the grid to be used in plotting distributional data.

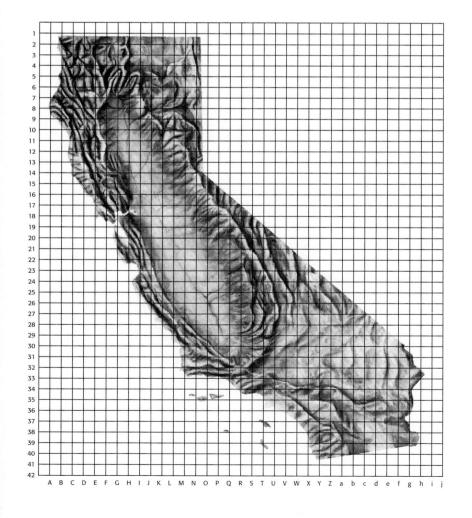

4. The spotted forms tend to be in the inland Sierra Nevada and the unspotted forms along the coast, except in southern California, where spotted and unspotted forms occur together.
5. Populations of *E. e. eschscholtzii* and *E. e. klauberi* occur in the same area in southwestern California.

Discussion

1. They represent genetic intergrades (hybrids) between subspecies.
2. The drawing should be an intergrade between *E. e. eschscholtzii* and *E. e. xanthopica*. Making such a drawing is a difficult but challenging assignment, requiring a bit of imagination. There are clines from north to south between the two adjacent subspecies (*E. e. xanthopica* in the north and *E. e. eschscholtzii* in the south). In general, from north to south the eyelids tend to become lighter and the dark pigmentation of body tends to disappear or become restricted to dots. To be more specific, toward the south:
   a. Lemon-yellow eye patch disappears.
   b. Orange coloration of ventral surface becomes restricted to underside of limbs and tail or is lost (sketches will not show this, of course).
   c. Yellow dots on sides of back appear.
   d. Small black dots on sides of back appear.
   e. Tip of tail becomes lighter.
   This question should encourage students to look very carefully at the drawings.
3. These two are geographically isolated by several hundred miles, with another subspecies occurring between them.
4. To see if there were any intergrade specimens and to see if the two subspecies populations occupied the same region.
5. The two subspecies are intermixed with no intergrades. (See the references listed in For Further Investigation.)
6. Population *E. e. klauberi*. This is the only form without intergrades with other forms. Students may wonder if further collecting would turn up intergrades. A provocative discussion might involve what the intergrades would be between and why. Incidentally, many biologists formerly concluded that *E. e. klauberi* was a separate species.
7. Between *E. e. klauberi* and *croceator*. Note that the intergrade

the way these color patterns occur in California? For example, do the spotted forms occur only along the coast? Do spotted forms occur in the north and unspotted ones in the south?
5. Subspecies *E. e. eschscholtzii* and *E. e. klauberi* are different from each other. What relationship is there between their distributions?

PART B Additional Collections
Procedure

3. You may wonder if there are salamanders in some areas for which you have no records. You also may wonder if there might be additional subspecies for which you have no specimens. A biologist faced with these questions would leave the laboratory and go into the field to collect more specimens. Imagine that you have done so and returned with the following data:
   b. *E. e. eschscholtzii* (l 6; red): 36/Z, 41 /Z, 33/M, 34/W, 34/U
   c. *E e. klauberi* (23; blue): 40/b, 40/Z, 36/a
   h. Unidentified population 8 (44; black and green): 4/l, 5/H, 7/H, 7/F, 6/J, 9/F
   i. Unidentified population 9 (13; black and red): 28/T, 27/T, 26/T, 28/S, 29/T
   k. Unidentified population 11 (131; black and blue): 23/J, 24/K, 24/l, 29/M, 25/J, 25/l
   l. Unidentified population 12 (31; black and yellow): 6/C, 7/C, 6/B
4. Mark with a 0 the following places that were searched for *Ensatina* without success: 11/l, 14/l, 17/K, 19/K, 22/N, 26/Q, 5/M, 32/U, 32/A, 35/F. Specimens of populations 8 and 9 are shown in Figure 9.24. There are no illustrations for populations 11 and 12.

Discussion

1. According to Stebbins, the unidentified populations are not additional subspecies. What, then, is the probable genetic relationship of populations 8, 9, and 11 to the subspecies plotted on the map?
2. On this basis, describe (or make a colored drawing of) the appearance you would expect specimens of population 11 to have.
3. Why is it unlikely that you would ever find individuals combining characteristics of *E. e. picta* and *E. e.xanthoptca*?
4. Look at the distribution of the original collections of *E. e. eschscholtzii* and *E. e. klauben*. What reasons were there for trying to collect additional specimens from extreme southwestern California?
5. How do the results of the additional collections differ from the results in other places where two different populations approach each other?
6. Bear in mind the biological definition of a species and also the appearance and distribution of the named populations of *Ensatina*. Which one of these populations could be considered a species separate from *E. e. eschscholtzii*? (This population was indeed once considered by biologists to be a separate species.)
7. Now imagine that, while examining salamanders in another collection, you find specimen j from population 10 shown in Figure 9.24. Compare its characteristics, especially the spotting pattern, with those of the named populations. Also consider the distribution of these populations. Between which two is this specimen most likely a hybrid? On your map, draw a line along which you might expect to collect other specimens like this one.
8. In a brief paragraph, explain why Stebbins concluded that there is only one species of *Ensatina* in California.

9. Suppose volcanic activity in northern California should become violent and completely destroy all the salamanders in that region. How would this event affect the species *Ensatina*?

For Further Investigation
What accounts for the one record of *E. e. xanthoptica* in the Sierras, with the rest of the subspecies occurring along the coast?

SUMMARY

Although all organisms are similar in some basic ways, they also exhibit a remarkable amount of diversity. A unity in pattern is evident from the genetic code common to all organisms and from similar developmental stages and structures. The process of evolution explains how these similarities and differences arise.

Charles Darwin developed a theory of evolution through natural selection, which explains how adaptations to particular environments allow organisms to survive and produce offspring. Figure 9.26 depicts the facts and inferences that Darwin used to develop his explanation. Our current knowledge of genetic mechanisms, such as mutation and recombination, explains how these adaptive variations are inherited by some of the offspring.

Populations are the units of evolution. Populations are affected by mutation and recombination, migration, genetic drift, natural and artificial

specimen has spots that tend to form bands. Individuals of *E. e. croceator* have definite spots; those of *E. e. klauberi* are banded. Note also that the two spots on the head almost form a band. The line is a clockwise arc from, roughly, 32/U to 35/Z (from the most southerly *E. e. croceator* around the east side of *E. e. eschscholtzii* to the most northerly *E. e. klauberi*.) These intergrades were collected at 33/Y and 35/Y.

8. The map shows intergrades between all the subspecies except the two that exist together in southern California. It would seem likely that subspeciation has taken place from a common ancestor in the north (closely related to *E. e. picta*), down the two separated mountain chains. The two subspecies that exist together without intergrading must have become so different from each other that they are reproductively isolated. You can see why these two were thought to be separate species before all the information on the intergrades was available. These species "rings" and "loops" may be a nuisance to a biologist who merely wants to classify objects, but they are exciting to a student of evolution.

9. It would break the chain of interbreeding subspecies because *E. e. eschscholtzii* and *E. e. klauberi* at the southern end of the range do not interbreed. This would produce two reproductively isolated populations—two species.

For Further Investigation
There are several possible explanations. The specimen may have been introduced accidentally by humans, or irrigation in this region may have allowed the specimen to move across the broad valley. More intensive collecting in the valley might show more specimens in the inland mountains, with perhaps a connection between the coastal and inland

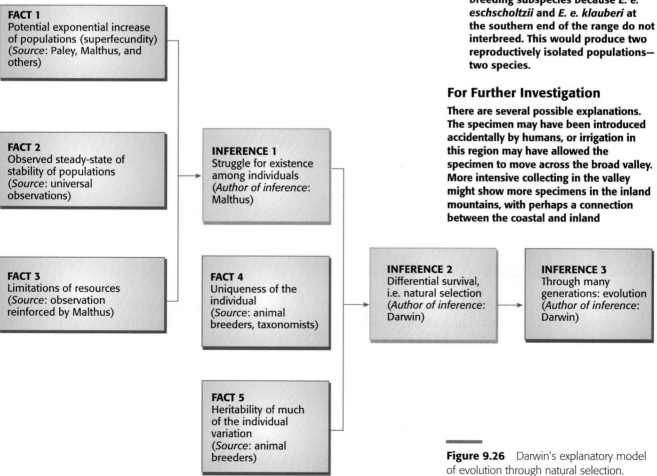

Figure 9.26 Darwin's explanatory model of evolution through natural selection.

populations in the not too distant past. In actuality, more than just this one individual were located in the inland mountains, in this same general region. Consult D. B. Wake and K. P. Yanev, "Geographic Variation in Allozymes in a 'Ring' Species. the Plethodontid Salamander *Ensatina eschscholtzii* of Western North America," *Evolution* 40 (1986): 702715, and D. B. Wake, et al., "Intraspecific Sympatry in a 'Ring' Species, the Plethodontid Salamander *Ensatina eschscholtzii* in Southern California," *Evolution* 40 (1986): 866-868.

selection, and nonrandom mating. When populations of a species become isolated from each other, changes can accumulate until they become separate species.

Although biologists do not completely agree about the mechanisms of evolution, the overwhelming majority agree that diversity of type and unity of pattern are best explained by evolution. Evolution pervades all aspects of biology and provides a foundation for understanding many phenomena.

 APPLICATIONS

1. How do you explain that the variability in a domesticated species is greater than in the same species, or a similar one, in the wild? (For example: dogs vs. wolves; chickens vs. red jungle fowl.)
2. Polydactyly (more than the normal number of fingers or toes) is caused by several different genes. One type of polydactyly is a dominant form of the trait, yet the phenotype is rare in humans. Type O blood is the recessive form of the blood type trait, yet in some populations of North American Indians, as many as 97 percent of the individuals may have type O blood. Explain these two situations.
3. In many species of birds, populations that live at higher altitudes lay more eggs per clutch than do those that live at lower altitudes. Would you expect the former gradually to replace the latter? Why or why not?
4. How does Mendel's work, of which Darwin was unaware, provide support for Darwin's theory of evolution by natural selection?
5. What effects may modern medicine have on the future evolution of humans? Distinguish carefully between the data and your interpretations of them.

6. Suppose there are two different populations of flowering plants living in fields next to one another. The plants of the two populations look different, but where the populations meet they form fertile hybrid offspring. Do you think these different populations belong to the same species?
7. Genetic variation is important for the longterm survival of a population, but how might it be considered a problem for individuals?
8. What is the difference between natural selection and evolution?
9. If there were no mutation, do you think that evolution still would occur?
10. Is the rate of evolution likely to be higher or lower in a small population compared with a large population? How about the likelihood of extinction?
11. Many people believe, incorrectly, that the phrase "survival of the fittest" means that "only the strong survive." What does "survival of the fittest" really mean?

 PROBLEMS

1. Suppose no data existed to support the theory of evolution until the twentieth century. Using only data compiled since 1900, how would scientists develop the theory of evolution today? In what ways would it resemble Darwin's presentation? How would it differ?
2. Suppose species of organisms were not related to one another and did not change through time. How would our view of biology change? Would it be easier or harder to study?

3. Using an imaginary population of organisms, detail how it could evolve into two distinct species. Describe the environment of the population and the factors that led to its separation into two species.
4. Endemic species are those that are restricted to certain areas. Identify some of the endemic plant and animal species of your community or state and try to learn what might have allowed them to become endemics.

5. What is the evidence for the punctuated equilibria theory of evolution? Present the evidence to you class.

6. Charles Darwin was a remarkably versatile scientist and naturalist. While most associate his name with natural selection and evolution, he was a prolific writer on a number of diverse subjects. Find a book other than *Origin of Species*, read it, and write a report on it.

7. Time is an important factor in the evolution of organisms. Make a time scale and place the major events in the evolution of life on earth on this scale. First, get a very long piece of butcher paper that covers the entire length of the longest wall of your classroom. Divide the scale into five equal parts each of these parts is equal to 1 billion years of time. Divide each of these sections into 10 equal parts of 100 000 000 years. Now indicate on the time scale major events beginning with the formation of the earth some 4.6 billion years ago and ending with the first fossil evidence of humans.

A scene from Organ Pipe Cactus National Monument, Arizona. How many different species of flowering plants can you count? Check your estimate with a classmate and try to resolve any differences you may have. What important biological phenomenon is illustrated?

At least four species of flowering plants are visible. There also are several other desert plants with less visible flowers. The phenomenon is diversity.

Section Three

DIVERSITY AND ADAPTATION IN THE BIOSPHERE

◆ Biologists have described an incredible diversity and number of organisms, and more are discovered every year, primarily in the tropical rain forests. Why are there so many types of living things? In what ways are all living things related? What adaptations have allowed them to survive? Section Three describes some representative forms of organisms and examines how each is uniquely adapted to its environment. It also explains how biologists name organisms and how they keep track of the different forms and bring order to the diversity of life. ◆

Ordering Life In the Biosphere

10

When you observe an organism, it is not always apparent what type of organism it is. What is the organism in this photo? Although you may think you see coral, if you look closer, you will see a tassled scorpionfish *(Scorpaenopsis oxycephala)*. What possible advantages would there be for a fish to look like this?

The tassled scorpionfish blends so much with its surroundings that it is hardly discernible. Such adaptations provide protection from predators and invisibility from prey.

◆ How many different types of organisms can you recognize? A hundred? A thousand? Biologists have described more than 1.4 million different species of organisms. Millions more have never been named or completely described. Some biologists estimate that we share this planet with as many as 10 million species, and that number does not include organisms that are extinct.

Evolution has produced overwhelming diversity among organisms. In fact, evolution has produced diversity in every aspect of life—in size, from the smallest microorganism to the largest whale; in habitat, from the deepest ocean bottom to the top of the highest mountain; and in way of life, from growing on a rooted tree to flying high above land or water. To make sense of the relationships that exist among organisms, we must have some method of arranging the organisms in a logical and meaningful way. For example, it is possible to classify birds according to the color of their feathers: all blue-colored birds in one group, gray birds in another group, and black birds in a third group. This type of system could be useful, but it has limitations. The type of information it conveys about an organism it does not take into consideration where the organism lives, what it eats, or how it may be related to other organisms. The classification system most widely used today considers how organisms are related and groups them on the basis of what is known of their evolutionary history. This chapter discusses how biologists classify organisms and examines some current ideas about the origins of this great diversity of life. ◆

MAJOR CONCEPTS

- ◆ Life is classified according to similarities in characteristics.
- ◆ The groups in a classification system become progressively more inclusive.
- ◆ Binomial nomenclature is a system that provides each species with a unique name.
- ◆ Currently all organisms are grouped in one of five kingdoms.
- ◆ The atmosphere of early earth was made up of water, methane, ammonia, and carbon dioxide primarily.
- ◆ The first life probably arose from simple chemicals.
- ◆ The first organisms to exist on earth may have been heterotrophs.

KNOWLEDGE CHECK

- ◆ Why do biologists classify organisms?
- ◆ How are organisms classified?
- ◆ How is a classification system meaningful?
- ◆ Does the classification of an organism ever change?

Biological Classification

10.1 Classification Is a Way of Seeing Order in Diversity

The diversity of the living world is awesome. Even the molecules and structures that make up the great variety of living organisms are diverse. From earliest times, however, humans have searched for order in the diversity of nature. They have tried to find ways to distinguish the many types of living things (see Figure 10.1). From these efforts has grown the science of classifying organisms, or **taxonomy** (tak SAHN uh mee).

The basic grouping used in biological classification today is the
1 species. Species differ from one another in at least one characteristic, and they usually do not interbreed freely with other species. Section 9.2

Guidepost
Why do biologists classify organisms, and how do they do it?

Taxonomy is merely the placing of organisms in a specific category. Systematics is the branch of biology that attempts to reconstruct phylogenetic histories. Systematics includes taxonomy but extends the study to include evolutionary relationships.

229

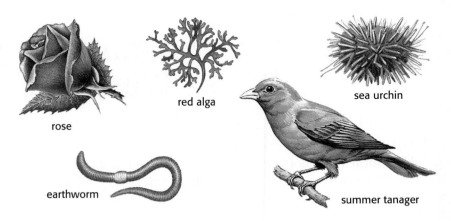

Figure 10.1 One similarity among these organisms is obvious—their color. How else are they alike?

rose

red alga

sea urchin

earthworm

summer tanager

discusses the nature of species and the problems with defining a species. Because some groups of organisms, bacteria and many eukaryotes, for example, reproduce asexually, they cannot be classified according to their interbreeding abilities. Biologists agree, in general, on the types of organisms they classify as species, but the criteria they use when classifying these organisms may differ.

10.2 Classification Is Based on Homologies

To classify organisms, specialists in taxonomy use a variety of characteristics. These characteristics include structure, function, biochemistry, behavior, nutrition, embryonic development, genetic systems, cellular and molecular makeup, evolutionary histories, and ecological interactions. The more constant a characteristic is, the more valuable it is in determining classification. For example, characteristics such as the size or color of organisms may vary greatly, but structures such as skeletal form, reproductive parts, and internal anatomy are less variable. Structure, therefore, generally provides a consistent and useful basis for classifying organisms.

Similarities of structure that indicate *related ancestry* are especially important in classification. For example, fish, amphibians, reptiles, birds, and mammals all share the same limb pattern, as discussed in Chapter 9. The limbs have the same relationship to the body, and they develop in the same way in the young. These types of relationships are called structural **homologies** (hoh MOL uh jeez).

Similarities between body substances (blood, for example) and between molecules (proteins, DNA, or RNA) are called biochemical homologies. Biochemical homologies have become increasingly important in determining relationships. Sometimes comparisons of DNA or RNA sequences have changed our understanding of the relationship between organisms. The greater the similarities in DNA sequences, the more closely related two organisms are thought to be. It also is possible to compare the nucleic acids or the resulting sequences of amino acids. For example, cytochrome *c* is a protein that plays an important role in respiration in animal and plant cells. Between species, the differences in cytochrome *c* are differences in amino acid sequence. If two species are closely related in their evolutionary history, there will be few amino acid differences between their comparable proteins.

Biochemical techniques have helped to clarify some classification problems. For example, for many years experts could not agree on the classification of the giant panda (see Figure 10.2). Some experts grouped

pandas with bears, others grouped them with raccoons. New techniques for studying the homologies in DNA and chromosomes, however, have led to a greater understanding of the evolutionary relationships between bears, raccoons, and pandas. Now, the giant panda is classified with the bears, and the evidence suggests that the raccoon and bear families diverged from a common ancestor between 35 and 40 million years ago.

In Investigation 10.1, you can use DNA sequences to determine relatedness. Sometimes studies of DNA sequences offer surprising results. For example, traditional studies of structural characteristics led scientists to think that the relationship between chimpanzees and gorillas was closer than that relationship between chimps and humans. DNA studies, however, have indicated that the relationship between chimpanzees and humans is closer than that between chimps and gorillas (Figure 10.3).

Still, structural characteristics remain particularly important in classification schemes for several reasons. First, biochemical techniques cannot be used on most **fossils**—the remnants or traces of organisms from past geologic ages. In some fossils, such as fossil bones, the living material has been replaced by minerals. Second, because these organisms no longer exist, scientists only can guess about their behavior. Using preserved specimens or geological records, however, a taxonomist can observe structural characteristics and make guesses about an organism's behavior, based on structural evidence. Sometimes, an organism has left behind tracks, such as the ones shown in Figure 10.4. The organism's behavior must be inferred from the pattern of tracks it leaves. Because organisms of the past may have been ancestors of living organisms, knowledge of past organisms can help to determine the relationships between living organisms.

Figure 10.2   Giant panda, *Ailuropoda melanoleuca.*

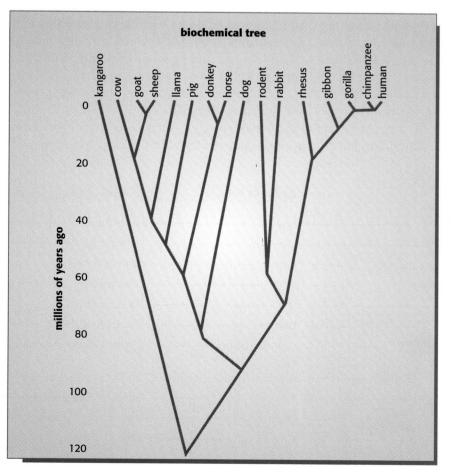

Figure 10.3   Biochemical trees represent the degree of relatedness of organisms based on analysis of biochemical data, such as amino acid sequences in proteins or base sequences in DNA. In this tree, goats and sheep are closely related while kangaroos are fairly distant relatives of all other organisms shown.

Although macroscopic organisms are classified primarily on the basis of structural characteristics, behavioral and biochemical data increasingly are being used in taxonomic determinations. Biochemical data are important in plant species determination, as well as in some groups of lichens. Biochemical data also are used in determining species of bacteria.

Figure 10.4 What type of organism may have made these trace fossils?
A burrowing worm.

This is a good time to call on students' prior knowledge of mammals.

10.3   Species Are Grouped into Broader Categories

When relatively few plants and animals were known, there was little need to group them into larger, more inclusive units that reflected common characteristics. Once thousands of plants and animals are known, however, larger groupings become necessary in order to distinguish specific organisms within this bewildering array. In a universal system of classification, organisms of different species are grouped into larger, more general categories based on homologies. For example, dogs, coyotes, and wolves are considered separate species, but they are similar in many ways. Species that share many similar characteristics are grouped into the same **genus** (JEE nus; plural *genera*). The genus for these doglike animals is *Canis*.

In much the same way, similar genera are grouped together in a **family.** Although foxes share many similarities with members of the genus *Canis*, there are also many differences. For example, foxes usually have a lighter build and are more catlike than members of the genus *Canis*. Therefore, taxonomists place foxes in a separate genus *(Vulpes)*, yet because of the similarities that exist, include them in the same family as dogs, coyotes, and wolves—the family Canidae (see Figure 10.5).

Weasels also resemble dogs and wolves in some ways, but they are less like them than are foxes. Taxonomists express this difference by placing weasels in a separate family, Mustelidae. Bears, which eat meat and fruit, have many structural differences from weasels or foxes. Taxonomists place bears in still another family, Ursidae. These three families (Canidae, Mustelidae, Ursidae) are all meat-eaters. Therefore, they are grouped with other families of meat-eaters into the same **order,** Carnivora.

Because monkeys appear to have little in common with wolves, weasels, and bears, they are placed in a different order, Primates. Monkeys (and many other organisms), however, do have some characteristics in common with wolves, weasels, and bears. These characteristics include hair and the production of milk to feed their young. Therefore, monkeys, wolves, weasels, and bears, as well as other animals that have hair and produce milk to feed their young, are put together in the next larger grouping, the **class**—in this case, the class Mammalia.

Continuing with this method of grouping, taxonomists combine classes containing birds, frogs, fish, and snakes with the class Mammalia to

Figure 10.6 Classification of several animals.

Common Name	Scientific Name
Human	*Homo sapiens*
Lion	*Panthera leo*
Coyote	*Canis latrans*
Wolf	*Canis lupus*
Dog	*Canis familiaris*
Gopher	*Thomomyus bottae*
Ground squirrel	*Citellus tridecemlineatus*
American robin	*Turdus migratorius*
European robin	*Erithacus rubecula*
Gopher turtle	*Gopherus polyphemus*
Green Frog	*Rana clamitans*
Bullfrog	*Rana catesbeiana*

bear ×1/56

wolf ×1/42

coyote ×1/47

fox ×1/28

weasel ×1/16

Figure 10.5 Some animals in the order Carnivora. Which two look least alike? In what ways? **To most students, the wolf and the coyote look alike. If some students think that either resembles the fox more, call attention to the forelimbs of the fox. They are more delicately constructed, and the whole animal is relatively lighter. Some foxes, such as the gray fox, are somewhat arboreal. Most of the structural differences between *Canis* and *Vulpes* that are important to taxonomists are not visible in the drawings.**

form the **phylum** (FY lum) Chordata. Animals in this phylum all have a primitive spine, or notochord. (Botanists group classes of plants as **divisions** instead of as phyla.) Finally, chordates, snails, butterflies, and thousands of other organisms are grouped into the **kingdom** Animalia. This kingdom contains all the living things we think of as animals.

Notice how the characteristics the organisms have in common become more general at each succeeding level from species to kingdom. At the species level the individuals are so much alike they can interbreed. At the kingdom level only a few characteristics are shared among all the individuals. Figure 10.6 shows the classification of twelve organisms.

By placing the dog family, the bear family, and the weasel family in the same order, taxonomists imply that all these animals long ago descended from a common ancestral group. The farther up the list of levels, the more distant the relationships become. When taxonomists place a dog and a goldfish in the same phylum but in different classes, they imply a very distant evolutionary relationship. Each level in the uni-

The term *phylum* indicates evolutionary relationships. Many plant systematists do not think enough is known about plant evolutionary relationships to justify using this term, so they use *division* instead.

Although the technical form of each group name is used, in class discussions you are urged to employ standard English whenever possible. Thus, say *chordates, arthropods, mammals,* even *canids*. Students should know the technical forms exist, but they do not need to burden themselves with exotic spellings and pronunciations.

Characteristics of several animal phyla are presented in Chapter 14 and in more detail in Appendix 4, *A Catalog of Living Things*. If it seems important to consider here what characteristics place a dog and a goldfish in the same phylum but in different classes, refer to the appendix.

Figure 10.6 Continued

Genus	Family	Order	Class	Phylum	Kingdom
Homo	Hominidae	Primates			
Panthera	Felidae				
Canis	Canidae	Carnivora	Mammalia		
Thomomys	Geomyidae			Chordata	Animalia
Citellus	Sciuridae	Rodentia			
Turdus	Turdidae	Passeriformes	Aves		
Erithacus					
Gopherus	Testudinidae	Chelonia	Reptilia		
Rana	Ranidae	Salientia	Amphibia		

Figure 10.7 A carnation

Linnaeus's *Species Plantarum* (1753) was the first book in which binomial nomenclature was consistently used. Animal nomenclature dates from the 10th edition of Linnaeus's *Systema Naturae*, 1758. Nomenclatures of some special groups that Linnaeus did not treat fully, such as fungi and bacteria, date from later works of several authors.

versal classification system deals with three things: characteristics that pertain to each group; organisms that belong to each group; and varying degrees of evolutionary relatedness among the group.

A *taxonomic classification is not necessarily permanent.* It is the result of the interpretations of data made by individuals developing the system of classification. Although it is possible to prove that cats and eagles and alligators have claws, whether these three organisms should be grouped together is subject to debate. Biologists do not totally agree about where many organisms fit into the classification scheme, especially at the species level. The more evidence taxonomists obtain, the more complex the relationships of organisms appear to be. Taxonomists also differ in how they interpret the evidence—different inferences may be drawn from the same set of observations. For these reasons, there are several schemes **3** of classification, all of which are designed within the same framework of levels. Investigation 10.2 acquaints you with some of the characteristics taxonomists use to classify organisms.

10.4 Biologists Use a Binomial System

Probably the first plants that were named were useful to humans as spices, medicines, and foods or for religious ceremonies. These useful plants were not given elaborate or formal names. In fact, the common names for many organisms are still important today. If, for example, you ask for some *Lactuca sativa* in a supermarket, you are not likely to get what you want. A request for lettuce works much better. Why, then, do biologists need special names?

Different languages usually have different names for the same organisms. For example, a potato is *papa* in Spanish and *Kartoffeln* in German. To compound the problem, the same word in a language may refer to different organisms. In Florida, *gopher* refers to a turtle; in Kansas, to a rodent. Biological names, then, are necessary for scientific exactness. There is no other single, agreed on set of names available for all organisms.

During the Middle Ages, scholars looking for a universal naming system attempted to fit the plant and animal names used by the ancient Greeks to the plants and animals of the rest of Europe. In order to recognize the differences in the plants and animals of England, Germany, and other northern lands from those of Greece, they attached new adjectives to the old name of a similar plant or animal. As new places and new organisms were discovered, the names became more complicated. By the eighteenth century, the name used for the carnation plant shown in Figure 10.7 was *dianthus floribus solitar squamis calycinis subovatis brevissimis corollis crenatis*, which meant "the pink [a general name for the carnation] with solitary flowers, the scales of the calyx somewhat egg-shaped and very short, the petals scalloped."

In 1753, a young Swedish botanist named Carolus Linnaeus (lih NAY us) developed the solution to this problem of scientific naming. Linnaeus used a system of **binomial nomenclature** (by NOH mee ul NOH men klay chur), or two-word naming, to identify each species. The first word in each name indicates the genus. Thus, all species of pinks are in the genus *Dianthus*. The second word in each name is the species name, and it indicates a group of similar individuals. For the common carnation, Linnaeus picked the species name *caryophyllus*. Neither the genus name nor the species name is, by itself, the scientific name. The scientific name consists

Reproductive Physiologist

Plants and animals are disappearing today at such an unprecedented rate that some researchers estimate more than one million species may become extinct by the end of this century. The primary cause of this decay in diversity is the destruction of habitats due to the expansion of the human population and its demand for space and goods. In some cases, zoos may be a species' last refuge.

Dr. Betsy Dresser has dedicated her life to preserving endangered wildlife. Currently she is the director of the Cincinnati Zoo's Center for Reproduction of Endangered Wildlife (CREW). CREW is known internationally for developing techniques that allow interspecies embryo transfer—the transfer of an embryo from an endangered species to a surrogate parent of a different, but closely related, species. For example, a domestic cat acting as a surrogate mother carried to term an Indian desert cat embryo. CREW's goals are the preservation and propagation of endangered species through the use of embryo transfer techniques and *in vitro* fertilization.

From an early age Dr. Dresser was fascinated with wildlife. After graduating from high school, she took a job in animal research. Although the job gave her hands-on experience with animals, it was not enough. She became a volunteer at the zoo and helped teach junior-high-school biology classes. Eventually she attended college where her interests in wildlife and biology soon centered on reproduction in exotic animals.

Dr. Dresser has worked with cats, sheep, mice, rhinos, and antelope. The common antelope she works with is the eland, and the exotic species is the bongo. Only about 60 bongos exist in captivity. To increase their numbers, seven-day-old bongo embryos were removed nonsurgically from a bongo at the Los Angeles Zoo and transferred into surrogates at the Cincinnati Zoo that same day. Twin bongo calves were born one week apart, one to a surrogate eland mother and one to a surrogate bongo mother.

The techniques Dr. Dresser uses require a great deal of skill. Once the sperm or embryos have been collected, they are frozen until needed. The water in the cells is replaced with glycerol, which, unlike ice, does not pierce the membranes and rupture the cells. The sample is placed in a plastic straw and the ends are heat sealed. The sample then is cooled to below freezing and plunged into a liquid hydrogen bath. Primarily because of Dr. Dresser's hard work and dedication, CREW has received funding to build a new center equipped with the latest technology. At the new center, new storage methods will be devised, and the technology will be expanded to the embryos of other species.

Dr. Dresser's long-term goal is to make herself obsolete by creating a world where her line of work is no longer needed. "It won't happen in my lifetime, I know that. The animals are dying so fast, and there is so much to be done and so much we don't know. All the technology in the world isn't the answer to conservation problems. CREW is just one of the tools."

Figure 10.8 A short section on poisonous mushrooms from a Chinese BSCS biology book. Note the scientific name.

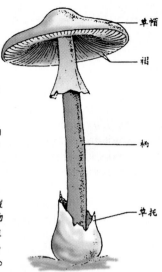

有些蕈菌含有毒性化合物，稱爲蕈菌素，若誤食後會引起呼吸及循環的失常，對人類是有害的。蕈類約有 70 種，其中以蕈菌 *Amanita verna* 最毒（圖 12—6），它有深白的子實體，雖僅食用少許的蕈帽，也必在一日之內致死，所以俗稱 "死神蕈"，真是名符其實。通常這類毒蕈，柄的基部有一 "杯狀物"（圖 12—6），這杯狀物常深藏土中不易發現，毒蕈與可食蕈的區別，普通傳述有如下的幾種認見："銀匙與 毒蕈共煮則呈黑色"；"可食的蕈帽易於剝落，毒 蕈則不然"；"若昆蟲或其他動物吃過的蕈，人亦 可食之" 等等，其實皆不可信。惟一可靠的選擇方 法是由市場購買人工栽培的食蕈。如從野外採來的 ，就應先經專家檢定後方可食用。

圖 12—6 蕈菌 *Amanita verna* 圖示蕈托
杯狀物位在柄的基部（只露地下）。這是許多毒蕈類的特徵。蕈帽下放射狀型褶是蕈的特徵。

of both genus and species names—*Dianthus caryophyllus*. Scientific names always are italicized (or underlined), and the first letter of the genus name always is capitalized.

Linnaeus established two rules that have made his system succeed. **4** The first rule is that the genus name can never be used for another group. *Dianthus,* for example, can be used only to describe pinks. The second rule is that the species name cannot be used for any other species within the same genus. For example, *caryophyllus* can never be used for any other species of *Dianthus*. The species name may be used with some other genus, but because a genus name cannot be repeated and because the scientific name of an organism always has two words, no other species can be called *Dianthus caryophyllus*.

Although Latin and Greek word roots frequently are used, the scientific name of an organism may come from any language, or it may be invented. For example, *Tsuga* (the hemlocks) comes from Japanese, and *Washingtonia* (a genus of palms) obviously does not come from Latin. All names are given Latin endings, however, and the names always will appear the same. Thus, in a Russian or Chinese biology book, biological names are printed in the same form as they appear in this book, though the rest of the printing is different (see Figure 10.8).

CONCEPT REVIEW

1. What is the basis for biological classification? What advantages does it have?
2. How does the number of characteristics shared by all members of a classification level change as you progress from species to kingdom?
3. Why can equally experienced taxonomists disagree about the classification of a particular species?
4. What are the basic rules of the binomial nomenclature system, and how is the system useful in science?

The Kingdoms of Organisms

10.5   Cell Structure Is Evidence for Relatedness

As the universal classification system progresses from species to kingdom, each category contains more organisms. The more organisms or species in a category, the more difficult it is to find homologies among them. At the same time, the more variations there are within a group, the more difficult it becomes to find definite characteristics that distinguish one group from other groups at the same taxonomic level.

As discussed in Chapter 5, all organisms are made up of cells—the basic units of structure and function—and cell products. These cells contain clues about the relationships between organisms. Based on their cell structure, all living organisms are either prokaryotes or eukaryotes (see Figure 10.9).

2   Prokaryotes (bacteria) are thought to represent the earliest type of living cell. They lack membrane-enclosed internal structures. This means they do not have a nucleus or cytoplasmic organelles such as mitochondria or chloroplasts. Their cellular material does not appear to move. Their rigid cell walls are formed from protein like chains and sugars. Their genetic material consists of a single circular thread of DNA, although some also contain several plasmids. Mitosis does not occur. Prokaryotes move by glid-
1   ing or by means of long, hairlike projections called **flagella** (fluh JEL uh).

Eukaryotic cells, which usually are larger than those of prokaryotes, have a membrane-enclosed nucleus and other membrane-enclosed organelles (see Figure 10.9). The cellular material often appears to move, or stream, within the cell. If cell walls are present, they usually are composed of cellulose or another carbohydrate, chitin (KYT in). The DNA of eukaryotes is organized into chromosomes, and there are no plasmids. Cell division includes mitosis and cytokinesis. Table 10.1 summarizes the major differences between prokaryotes and eukaryotes.

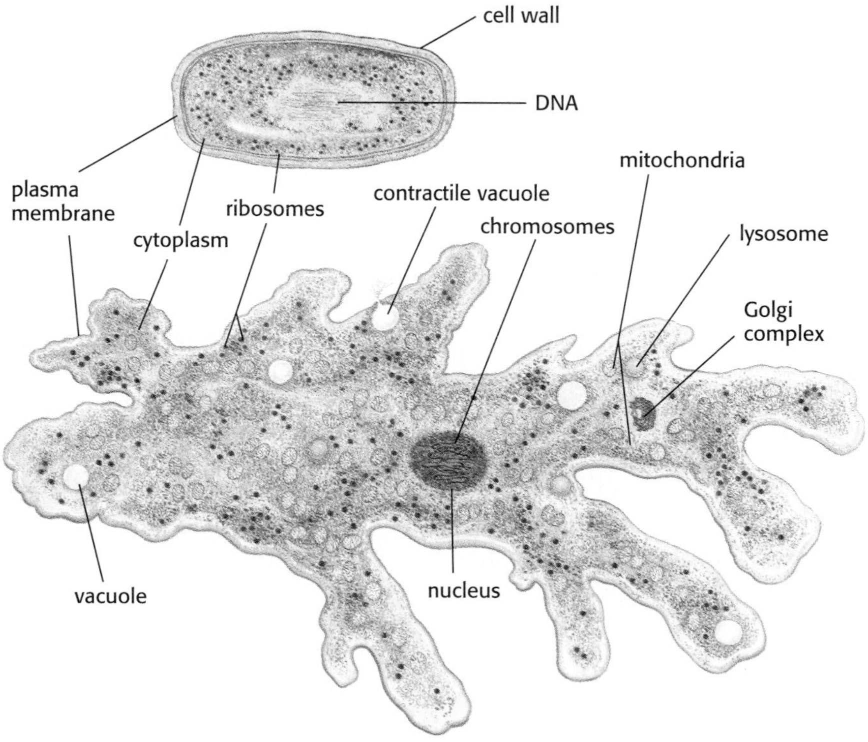

Guidepost

What characteristics determine how organisms are grouped into kingdoms?

Figure 10.9   A prokaryotic cell (top) compared with a eukaryotic cell (bottom). Note the greater complexity of the eukaryotic cell (an amoeba) and the presence of many membrane enclosed organelles.

Table 10.1 Some Differences Between Prokaryotic and Eukaryotic Cells

Eukaryotes Have	Prokaryotes Have
nucleus	no nucleus
membrane-enclosed organelles	no membrane-enclosed organelles
chromosomes in pairs	single chromosome
streaming in the cytoplasm	no streaming in the cytoplasm
cell division by mitosis	cell division without mitosis
complex, tubular flagella	solid flagella
larger ribosomes	smaller ribosomes
complex cytoskeleton	simple cytoskeleton
cellulose in cell walls	no cellulose in cell walls

10.6   Organisms Are Grouped into Five Kingdoms

Because organisms may look very unalike at the kingdom level of classification, biologists consider several questions about the evolutionary history of an organism before classifying it. Is it a prokaryote or a eukaryote? Is it a producer, or **autotroph** (AWT oh trohf)? Is it a consumer, or **heterotroph** (HET eh roh trohf)? Does it reproduce asexually or sexually, and does it develop from an embryo? Finally, biologists consider the general structure and function of the organism. None of these questions, however, resolves all the problems of classification, and the systems continually change as new knowledge emerges. Today, most biologists favor the five kingdom classification system shown in Figure 10.10.

At present, all prokaryotes are grouped in one kingdom called the **Prokaryotae** (proh kayr ee OH tee). Prokaryotes display a greater variety of chemical and functional patterns than do eukaryotes. Like green eukaryotes, many prokaryotes produce food through photosynthesis, but the prokaryotes use a greater variety of substances as raw materials. The energy source of prokaryotes differs, also. Some prokaryotes use energy obtained from inorganic chemicals (chemicals that do not contain carbon chains or rings) rather than light to produce food, and many prokaryotes are heterotrophs. Although most prokaryotes are unicellular, there are some prokaryotes that are multicellular. Reproduction usually is accomplished through asexual cell division. When sexual reproduction is used, the parents do not contribute equal amounts of genetic material. Figure 10.11 shows several prokaryotes.

Currently, the eukaryotes are divided into four kingdoms. Most autotrophic, multicellular eukaryotes that produce their own food through photosynthesis belong to the kingdom **Plantae** (PLAN tee). This kingdom includes all organisms developing from an embryo that lacks a blastula stage. Plants have cellulose-containing cell walls; their cells contain chloroplasts; they store food as starch; and most reproduce sexually. The bulk of the world's food and much of its oxygen are derived from plants. Figure 10.12 shows a variety of plants.

All organisms developing from an embryo that has a blastula stage are members of the kingdom **Animalia** (an ih MAYL yuh). All animals are heterotrophic and multicellular. Ranging in size from microscopic forms to giant whales, animals are the most diverse in form of all the kingdoms

The use of kingdom Prokaryotae, instead of Monera, conforms with the definitive work for the identification and classification of bacteria. See N. R. Krieg and J. G. Holt, eds. *Bergey's Manual of Systematic Bacteriology*, 9th ed. (Baltimore: Williams and Wilkins, 1984), vol. 1.

Ask students how the question of prokaryote versus eukaryote brings the ancestry of organisms into biological classification. Have the students describe homologies among prokaryotic cells, then homologies among eukaryotic cells. Stress that the variety they see in eukaryotes in the world is not as great at the cellular level as the biochemical variety in molecular structure and ways of life among the prokaryotes.

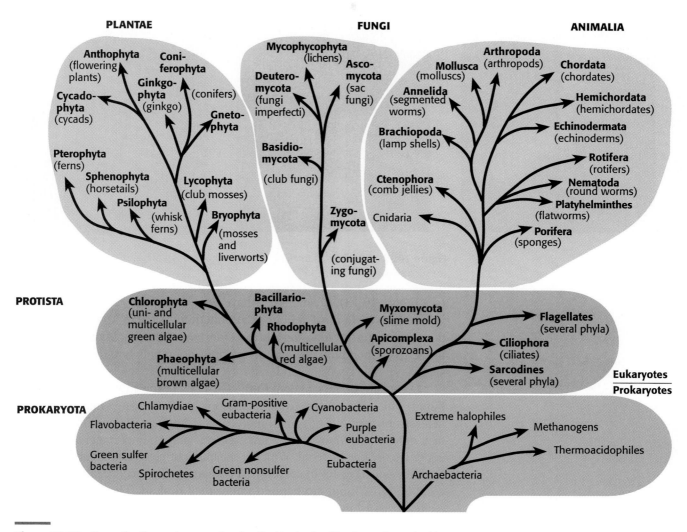

PLANTAE

Anthophyta
(flowering plants)

Cycado-phyta
(cycads)

Ginkgo-phyta
(ginkgo)

Coni-ferophyta
(conifers)

Gneto-phyta

Pterophyta
(ferns)

Sphenophyta
(horsetails)

Psilophyta
(whisk ferns)

Lycophyta
(club mosses)

Bryophyta
(mosses and liverworts)

FUNGI

Mycophycophyta
(lichens)

Deutero-mycota
(fungi imperfecti)

Asco-mycota
(sac fungi)

Basidio-mycota
(club fungi)

Zygo-mycota
(conjugating fungi)

ANIMALIA

Arthropoda
(arthropods)

Mollusca
(molluscs)

Annelida
(segmented worms)

Brachiopoda
(lamp shells)

Ctenophora
(comb jellies)

Cnidaria

Chordata
(chordates)

Hemichordata
(hemichordates)

Echinodermata
(echinoderms)

Rotifera
(rotifers)

Nematoda
(round worms)

Platyhelminthes
(flatworms)

Porifera
(sponges)

PROTISTA

Chlorophyta
(uni- and multicellular green algae)

Bacillario-phyta

Rhodophyta
(multicellular red algae)

Phaeophyta
(multicellular brown algae)

Myxomycota
(slime mold)

Apicomplexa
(sporozoans)

Flagellates
(several phyla)

Ciliophora
(ciliates)

Sarcodines
(several phyla)

Eukaryotes
Prokaryotes

PROKARYOTA

Chlamydiae

Flavobacteria

Green sulfer bacteria

Spirochetes

Gram-positive eubacteria

Green nonsulfer bacteria

Cyanobacteria

Purple eubacteria

Eubacteria

Extreme halophiles

Archaebacteria

Methanogens

Thermoacidophiles

Figure 10.10 Currently all organisms can be classified in the five kingdoms shown in this diagram. However, increased understanding of the prokaryotes may lead to that group being divided into more kingdoms.

Figure 10.11 Representative prokaryotes include (a) cyanobacteria, *Anabaena circinalis* (b) purple photosynthetic bacteria, *Rhodospirillum rubrum* (stained); and (c) nitrogen fixing bacteria.

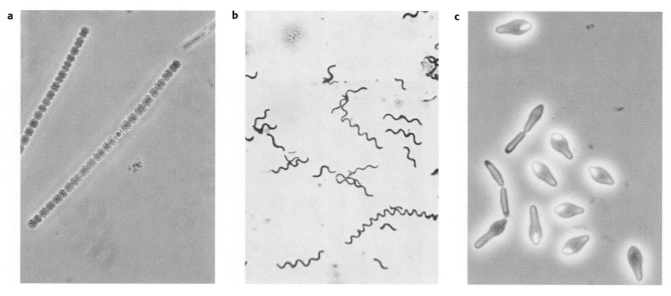

a

b

c

Figure 10.12   Members of the kingdom Plante (a) moss, (b) sword and wood ferns (c) sunflower

(see Figure 10.13). Most animals have senses and nervous systems that aid **4** in locomotion. In most animals, reproduction is sexual. Members of kingdom Animalia include vertebrates (animals with backbones), such as birds, frogs, and humans; and invertebrates (animals without backbones), such as sponges, worms, and insects.

Organisms of the kingdom **Fungi** (FUN jy) develop directly from spores. Fungi, which vary in size from microscopic species to large mushrooms, are heterotrophs that absorb small molecules from their surroundings through their outer walls. Many are decomposers and play an important role in breaking down organic material. With the exception of yeasts, most fungi are multicellular and have chitinous cell walls. Reproduction may be either sexual or asexual; most fungi are haploid rather than diploid. The kingdom Fungi includes yeasts, molds, bracket fungi, and mushrooms (see Figure 10.14).

All the remaining eukaryotes are grouped in the kingdom **Protista** (pro TIST uh). This kingdom includes algae (plant-like protists), protozoa (animal-like protists), slime molds (fungus-like protists), and other organisms. Many protists are microscopic and unicellular, but others are large multi- **4** cellular organisms that may reach 10 m or more in length. Protists also vary in cellular organization, methods of reproduction, and life-styles (see Figure 10.15). They may be producers, consumers, or decomposers; some even switch from one form of nutrition to another, depending on conditions in their environments. Despite their differences, these organisms probably

Figure 10.13   Animals exhibit great diversity of type. Shown here are (a) mountain goat with young (b) butterfly and (c) sea slug.

Figure 10.14 Typical fungi include (a) collard stinkhorn, *Dictyophora;* (b) jelly fungus, *Dacrymyces palmatus* and (c) hat thrower fungus, *Pilobolus.*

are more closely related to each other than they are to members of the other kingdoms. The classification of organisms representing each kingdom is shown in Figure 10.16 on the next page.

10.7 Classifications Can Change

The more information we obtain, the more complex the relationships between organisms appear to be. New knowledge often requires changes in classification, especially at the kingdom level. Any classification system should be exclusive as well as inclusive. That is, the characteristics used should allow us to form groups of similar organisms that are different from other groups. For example, members of the kingdom Prokaryotae are prokaryotes: all other organisms are eukaryotes. Plants are autotrophic; animals and fungi are heterotrophic. These characteristics serve to *include* similar organisms in one kingdom and *exclude* them from others.

> **This is a good example of how science works in general. New data require new interpretations of existing knowledge. You may wish to reinforce this idea here.**

At one time it appeared that all organisms could be classified in two kingdoms—animals and plants. As improved microscopes increased our knowledge of microscopic organisms, scientists realized that many did not fit well into either kingdom, and taxonomists added other kingdoms for the prokaryotes and protists. Because the fungi share few characteristics with the other groups, a separate kingdom was proposed for them.

Figure 10.15 Examples of Protista: (a) single-celled alga, *Clamydomonas,* (b) protozoa, *Vorticella,* and (c) slime mold (Slime molds are capable of movement when in the "plasmodium" phase.)

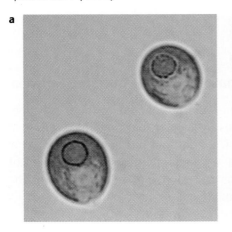
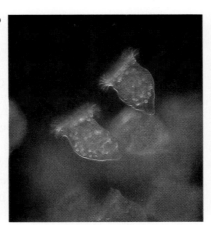
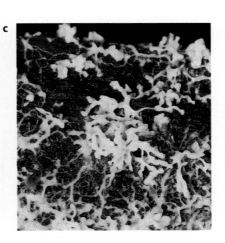

Figure 10.16 Classification of an organism from each kingdom.

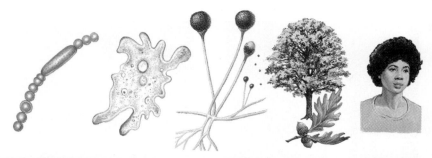

	Anabaena cyanobacteria	Amoeba	Rhizopus bread mold	Quercus alba white oak	Homo sapiens human
kingdom	Prokaryotae	Protista	Fungi	Plantae	Animalia
phylum/division	Cyanobacteria	Sarcodina	Zygomycota	Anthophyta	Chordata
class	Eubacteria	Lobosa	Phycomycetes	Dicotyledoneae	Mammalia
order	Oscillatoriales	Amoebina	Mucorales	Fagales	Primates
family	Nostocaceae	Amoebidae	Mucoraceae	Fagaceae	Hominidae
genus	Anabaena	Amoeba	Rhizopus	Quercus	Homo
species	circinalis	proteus	stolonifer	alba	sapiens

The five-kingdom system of classification presented here is not accepted by all biologists. Some people place multicellular algae with the plants and heterotrophic protists with the animals. Other specialists divide the bacteria into two different kingdoms. Further studies of bacteria may result in the addition of new kingdoms.

Species, genera, families, orders, classes, and kingdoms do not exist in nature; only individual organisms exist. Classification systems are simply a means to think more easily about the great diversity of the world. The changing systems of classification reflect how science works—it is 5 open to modification on the basis of new data. Appendix 4, *A Catalog of Living Things,* shows only a few examples of the diversity of life.

CONCEPT REVIEW

1. Compare and contrast eukaryotic and prokaryotic cells.
2. What is the most fundamental characteristic that separates organisms?
3. What characteristics do biologists use to sort organisms into kingdoms?
4. Describe the distinguishing characteristics of each kingdom.
5. Why do classification systems change?

The Origin of Diversity

10.8   All Species May Have Come from One Ancestral Species

Guidepost

How did life originate and evolve into the many forms found today?

Although organisms arise only from other organisms of the same type, individual organisms vary. The process of natural selection, then, may lead to an increase in the number of species. For example, all of the Galapagos finches probably are descended from the same ancestors. If many types of organisms evolved from a few, the farther into the past scientists look, the fewer species they should find. Ultimately, it is possible that all of today's species came from one ancestral species.

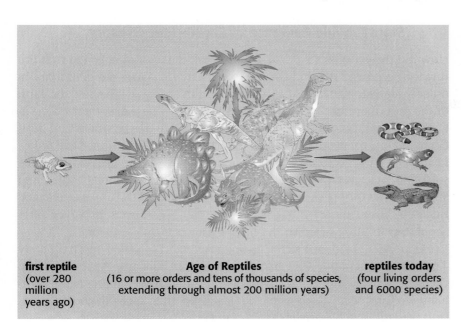

first reptile
(over 280
million
years ago)

Age of Reptiles
(16 or more orders and tens of thousands of species,
extending through almost 200 million years)

reptiles today
(four living orders
and 6000 species)

Figure 10.17 The age of reptiles produced tens of thousands of species, from small lizards to flying reptiles and giant dinosaurs. Turtles, alligators and crocodiles, lizards, and snakes are the only surviving orders of reptiles today.

Studies of fossils indicate that many forms have evolved from a few ancestral forms. As scientists have studied fossil evidence, they have found that one group of organisms would flourish, then die out—perhaps because of changes in the environment or because of competition from other groups. As one group faded, another group of organisms would increase in number and diversity. During the age of reptiles, for example, there was a greater variety of reptile species than there is today (see Figure 10.17) but there were only a few mammal species and almost no birds. When the number of reptiles declined, the number and types of mammals and birds increased.

This pattern appears in each age in the fossil record, and it appears in the history of each group of organisms. Throughout the history of life, new species have evolved and older species have become extinct. Again and again, more species have evolved from a smaller number of species that preceded them.

Given this pattern, ask students what "age" we are in today.

How far back can scientists trace life? Rocks in South Africa and western Australia have yielded the oldest known microfossils—microscopic prokaryotes. Because geologists have estimated that the rocks holding these fossils are between 3.0 and 3.5 billion years old, the fossils themselves must be that old. The Australian fossils were found associated with dome like structures called stromatolites (stroh MAT uh lyts). See Figure 10.18a, b. The domes, about 30 cm high by 1.5 m across, are constructed of many wafer-thin layers of rock. So far, scientists have been unable to identify any fossils older than these.

By reference to maps of early earth, students can see that fossils of some of the earliest life may be preserved in Antarctica also. Antarctica, however, drifted to the South Pole and has become covered with ice, making fossil expeditions there difficult. Other ancient rocks are in the rock shield around Hudson Bay in Canada.

a
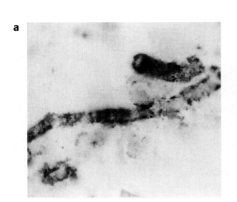

b

Figure 10.18 A 3.5 billion-year-old prokaryote fossil (a) in a stromatolite from western Australia. Modern living stromatolites (b) in Shark Bay, Australia, are built by cyanobacteria.

10.9   Stars Provide Evidence of Earth's Early History

How did these organisms evolve? How can the evolution of the first life forms be studied? It is impossible to recreate the earth in a laboratory in order to study how it was formed and how life appeared. To study the earth of the past, therefore, scientists make inferences based on their observations of stars. Many stars develop from revolving clouds of gas and dust. Giant clouds, often with trailing arms of gases, are visible in space today. From their observations, scientists infer that the sun condensed from the solar system's early gas cloud. They further infer that the cloud had trailing arms of gases, some of which may have condensed to form the planets. Because these are only inferences based on observations, however, little about the evolution of the earth or other planets is unquestioned.

Geological evidence suggests that after it first formed approximately 4.6 billion years ago, the earth grew very hot. Because heat speeds up chemical activity, atoms must have been constantly combining and recombining to form many types of molecules. Even after the earth cooled greatly, it was still much warmer than it is today.

Geological evidence also indicates that the early atmosphere of the earth was composed of volcanic gases, such as methane (CH_4) and carbon dioxide (CO_2). It also included ammonia (NH_3) instead of present-day nitrogen (N_2). Some hydrogen (H_2) and a great deal of water vapor (H_2O) also were present, but free oxygen (O_2) was absent.

With the energy of heat and lightning, the gases of the early atmosphere may have combined to form substances such as amino acids, which are found in all organisms today. Through millions of years, organic compounds may have accumulated in the warm oceans, lakes, and pools, forming an "organic soup." Along the shores, in tide pools, or on particles of clay, these simple compounds may have combined to become more complex molecules. Finally, these more complex molecules may have united somehow to form a simple type of reproducing "living thing." Figure 10.19 summarizes these ideas.

In 1952, to test these speculations, Stanley Miller and Harold Urey at the University of Chicago decided to find out what would happen if a simulated primitive atmosphere were exposed to an energy source. Using apparatus like

Oxygen was present combined in compounds in the early atmosphere, but students should understand that little or no free oxygen gas as O_2 molecules (or O_3 ozone), was present. Free oxygen appeared only after the evolution of oxygen-producing photosynthesis.

Figure 10.19   A sequence of events that might have led to the first life.

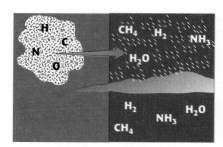
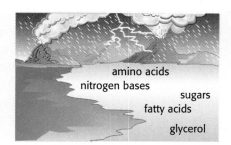
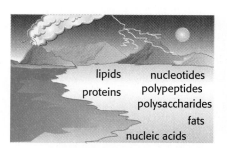
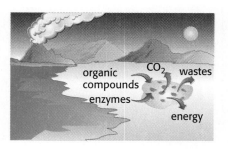

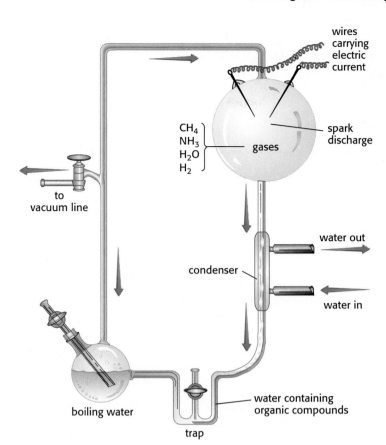

Figure 10.20 A drawing of Miller's apparatus in which conditions thought to exist in the primitive atmosphere were reproduced in the laboratory. Does this experiment prove how life originated on earth?

the one shown in Figure 10.20, they passed electric sparks (simulating lightning) through ammonia, methane, water, and hydrogen. When the resulting substances were analyzed later, simple amino acids had been produced.

Other scientists have repeated this experiment. Some investigators have used ultraviolet light instead of electric sparks and obtained the same type of results. Since Miller and Urey's first experiments, researchers have synthesized many other types of organic molecules, including nucleotides and carbohydrates. Although these experiments suggest a way in which life might have originated, it is still a long way from complex molecules to even the simplest of known organisms.

There are more questions about the origin of life than there are answers, but it is important to remember that the conditions under which
3 life originated are not the same as conditions today. If life on earth originated in the way that Miller and Urey have suggested, why has it not originated this way again and again during the history of earth? If organic compounds cannot be spontaneously generated today, then where did they originally come from? Why are conditions on earth different today? What has happened to change them? How might the first organisms have arisen?

The rationale for using ultraviolet radiation lies in the probability that it penetrated the primitive atmosphere much more than it does today.

Different combinations of starting molecules and different environments have yielded different organic molecules. For example, extreme pressure such as that caused by the movement of ice at the poles can cause the information of simple amino acids from constituent chemicals.

You may wish to introduce the idea of spontaneous generation.

10.10   The First Cells Probably Were Heterotrophs

The first life forms must have had two essential characteristics of life: first, a means of self-maintenance (maintaining their organization and order); and second, a means of self-replication (reproducing themselves). An early microorganism could not have been immortal. Sooner or later it would have been killed by storms or waves. Hence, for life to evolve successfully, it had to be able to reproduce. There must have been some way to store coded information about how to make new copies of itself.

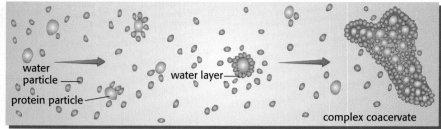

water particle
protein particle
water layer
complex coacervate

Figure 10.22 This drawing shows how a complex coacervate can form when a water layer surrounds a cluster of protein molecules.

Figure 10.21 Repeating structures in kaolinite clays. The first chemical systems to self-replicate may have been clays. Minerals in the clay may have formed a crystalline pattern that served as a blueprint for assembling more minerals into layers of similar patterns.

This account of protein like clusters of molecules may be revised considerably as research continues. The discovery that RNA can code hereditary instructions as well as catalyze chemical reactions may indicate that RNA was present along with protein-like materials in the first precells.

Figure 10.23 Microspheres are formed when dry mixtures of amino acids are heated and then placed in water. The dark material, including the boundary, is proteinoid. The individual microspheres in this picture are about 2μm in diameter.

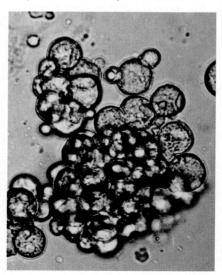

How could these abilities have developed? Scientists have two major hypotheses; one focuses on self-replication and the other on self-maintenance. According to the first hypothesis, the first life forms were not enclosed in a membrane but consisted of molecules or crystals that could self-replicate. Following this hypothesis, minerals dissolved in clay could have crystallized to form layers with a definite, repeating structure (see it; Figure 10.21). These layered clays could have acted as "crystal genes" able to store and transmit information in much the same way as DNA.

The second hypothesis suggests that the first life forms were cell-like structures enclosed within a membrane. Through time, some amino acids in the "organic soup" may have formed polypeptides and proteins. Other simple organic molecules also may have formed larger, more complex **4** moleclues. Eventually, some of the larger molecules combined into clusters and the clusters merged to form a primitive cell.

Scientists have proposed different models for these precells. A Russian scientist, A. I. Oparin, suggested that precells might have been like coacervates (koh AS er vayts). Coacervates are clusters of proteins or proteinlike substances held together in small droplets within a surrounding liquid, shown in Figure 10.22. Sidney Fox of the University of Miami has proposed a second model. He thinks precells were more like microspheres—cooling droplets from a hot water solution of polypeptides. Each microsphere may have formed its own double-layered boundary as it cooled (see Figure 10.23). Carl Woese at the University of Illinois has presented a third model for the formation of the first cells. According to his model, the early atmosphere had a high level of carbon dioxide that produced a greenhouse effect. Meteorites that hit the primitive earth provided large quantities of dust that were spread by the wind. Water vapor may have condensed around these dust particles, forming droplets. Each droplet, warmed by the atmosphere, then acted as a primitive cell in which chemical reactions occurred, allowing life to begin. All three models of cell formation have proponents and opponents. If a membrane enclosed the first life form, any one, or none, of the three models could account for the first cell.

What did the first living things use for food? It is tempting to suggest that they made their own food, but food-making is a complex process. The first cells probably could not have synthesized all the organic compounds **5** they needed. If evolution requires the fewest possible steps, the first living things would have been heterotrophs that used the supply of naturally occurring organic compounds for food. This is known as the **heterotroph hypothesis.** Heterotrophs would have required fewer evolutionary steps to develop than would have more complex autotrophs. Unless evolution took

RNA as an Enzyme

Like all other sciences, biology is an incomplete body of knowledge. New information is added constantly to the existing store of knowledge. A dramatic example of the revision of a long-held assumption in biology occurred in the early and mid-1980s, when a young University of Colorado biochemist began to question the role of RNA.

RNA is one of the two nucleic acids that store genetic information. For many years, biologists believed that information molecules and enzymes were mutually exclusive. That is, an enzyme could not serve as an information molecule, and an information molecule could not serve as an enzyme.

In 1981 and 1982, Thomas R. Cech and his colleagues, while working with the protist *Tetrahymena thermophila*, discovered that RNA can indeed act as an enzyme. Like all other eukaryotes, *T. thermophila* contains portions of DNA that are not translated into messenger RNA or protein. These segments, called introns, are cut out of the precursor mRNA that is made from the DNA in the nucleus. The remaining segments of mRNA, the exons, then are spliced together to form a shorter piece of mRNA. This mRNA is released from the nucleus into the cytoplasm, where it directs the synthesis of protein on the ribosomes.

The removal of introns and splicing together of exons requires energy and enzymes. Cech and his colleagues, Paula J. Grabowski and Arthur J. Zaug conducted splicing experiments. Assuming extracts from *T. thermophila* nuclei would supply the necessary enzymes, they used unspliced RNA with extracts and unspliced RNA without extracts. They found that a particular *Tetrahymena* intron was removed even without the extract from *T. thermophila* nuclei.

Thinking that a nuclear protein might be bound tightly to the RNA, the researchers synthesized artificial *Tetrahymena* RNA, using recombinant DNA techniques. As Cech explained in a 1986 *Scientific American* article, "the resulting RNA had never been near a cell and therefore could not be contaminated by splicing enzymes." Yet, when Cech and his team tested the newly synthesized RNA, the intron was still removed. The only possible conclusion was that the RNA itself acted as an enzyme.

In 1985, Zaug added another startling piece of information. He showed that *Tetrahymena's* self-spliced intron could organize short pieces of RNA into longer segments. In other words, RNA not only can act as a splicing enzyme but also can organize its own replication in certain situations. Because a system for the replication of information is required for the origin of life, this discovery has implications for evolution theory. For this work, Cech shared the 1989 Nobel Prize in chemistry with Sidney Altman of Yale university, who conducted similar, yet independent, research.

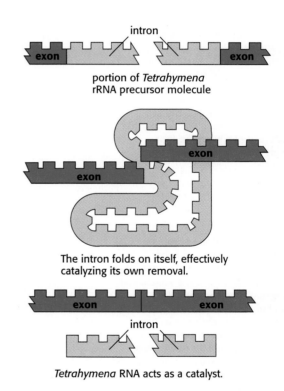

intron

portion of *Tetrahymena*
rRNA precursor molecule

The intron folds on itself, effectively catalyzing its own removal.

Tetrahymena RNA acts as a catalyst.

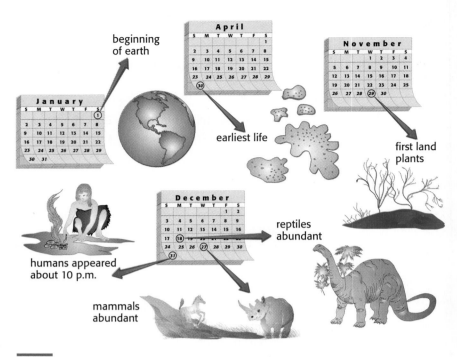

Figure 10.24 Moving clockwise, in this calendar 4.6 billion years of the earth's history are compressed into one year. Each day on the calendar is equal to almost 13 million years on earth.

Many biologists at first objected to this hypothesis, because most heterotrophs today require free oxygen. However, if free oxygen had been present in significant amounts in the early atmosphere, it would have destroyed naturally occurring organic compounds. Some heterotrophs can live without free oxygen even today, and so the problem of oxygen was solved by assuming that such heterotrophs were the first organisms living in an oxygen-poor atmosphere.

If heterotrophs developed first, as biologists believe, they used the available supply of organic compounds relatively inefficiently in the absence of free oxygen. Photosynthesis then evolved, but the first photosynthetic processes probably did not give off oxygen as a by-product. However, all photosynthesis produces new, energy-rich compounds. The first photosynthetic organisms probably were somewhat like anaerobic photosynthetic bacteria of today: they neither required nor gave off free oxygen.

a great many complex steps almost simultaneously, the heterotroph hypothesis seems the most reasonable explanation of the first cells. Thus, the earliest life had appeared and was well established by the time the earth was only one third of its present age (see Figure 10.24).

10.11   Several Types of Photosynthesis Evolved

Continuing the heterotroph hypothesis, as organic compounds were used for food by the primitive cells and as environmental conditions changed, the supply of compounds diminished. Under these conditions, some of the primitive cells must have begun to use visible light from the sun as a source for making food.

The first organisms able to use sunlight may have acted partly like heterotrophs, taking in organic compounds, and partly like autotrophs, making other organic compounds with energy from sunlight. Certain present-day bacteria have this ability. Gradually the ability to live entirely by means of photosynthesis evolved, although the photosynthesis did not produce free oxygen. Many types of bacteria living today use light energy to synthesize organic compounds from hydrogen sulfide or other compounds. These types of bacteria are **anaerobic** (an eh ROH bik)—able to live in environments in which oxygen is absent or present only in low concentrations. At some point, some primitive cells developed a form of photosynthesis that used water and released oxygen. Today, this type of photosynthesis is found in all autotrophic eukaryotic cells as well as in some modern photosynthetic bacteria.

Although this explanation may account for the evolution of autotrophs from heterotrophs, it does not clarify the evolutionary connection between prokaryotes and eukaryotes. Lynn Margulis of the University of Massachusetts has hypothesized that the mitochondria and chloroplasts found in eukaryotic cells originated as free-living prokaryotes (Figure 10.25). Margulis proposes that small, specialized prokaryotes may have formed close associations with larger anaerobic prokaryotes. The larger cells could have taken in the smaller ones as food-stuffs. Instead of being digested, however, the smaller prokaryotic cells continued their specialized activities in the host cell. The resulting increase in energy would have been beneficial to the host cell. Those prokaryotes that used oxygen to produce ATP could have been the ancestors of mitochondria in eukaryotic cells. Photosynthetic prokaryotes could have been the ancestors of chloroplasts. A large body of evidence supports this hypothesis, which is widely accepted.

With the advent of oxygen-producing photosynthesis, a great change took place in the atmosphere of the earth. Free oxygen began to accumulate in significant amounts, and **aerobic,** or oxygen-requiring, life became possible. Because oxygen molecules react readily, some of them probably combined to form ozone (O_3). As a layer of ozone developed in the upper atmosphere, most of the ultraviolet rays from the sun were blocked from the earth. Without that energy source, the synthesis of organic compounds in the seas was no longer possible. Heterotrophic cells became dependent on autotrophic cells for their source of food and energy.

Both chloroplasts and mitochondria have their own ribosomes, which resemble those of modern prokaryotes, and their own DNA. The DNA of mitochondria is a single, circular strand like the DNA of modern prokaryotes. Mitochondria are produced in cells only by other mitochondria. Chloroplasts are produced only by other chloroplasts. The reproduction of both organelles resembles that of free-living prokaryotic organisms. In addition, mutualistic relationships that exist today indicate that such associations are not difficult to establish or to maintain. Many modern organisms have bacteria, cyanobacteria, or algae living inside their own cells.

Figure 10.25   Mitochondria, chloroplasts, and cilia may have originated as free-living cells.

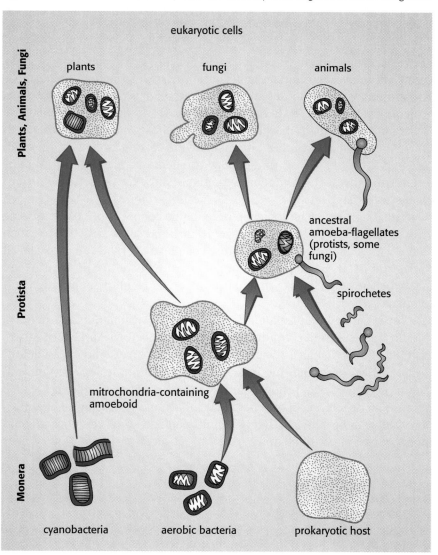

The Gaia hypothesis has been modified since it was first proposed. In light of the prospect of global climatic change, it has had a profound effect on the way some scientists study the earth and the questions they ask. There is some new evidence that one-celled organisms are at least partially responsible for regulating the temperature of the earth.

Life has greatly altered, and continues to alter, the atmosphere of the earth. James Lovelock, a British scientist, has developed a hypothesis to explain the changes in earth's atmosphere. This hypothesis, known as the Gaia (GY uh) hypothesis, suggests that the earth, including all of its abiotic and biotic components, consists of a huge, self-regulating system. The sum of all the living organisms on the earth controls various aspects of the atmosphere, oceans, and land. Is this system responsible for the advent of the current oxygen-rich atmosphere and the changes from the earlier, more hydrogen-rich atmosphere? If evidence for the Gaia hypothesis can be found, then all organisms on earth, including humans, would be functioning components of a larger system. While some data from various disciplines seem to support the hypothesis, it is not well accepted and makes for heated scientific debates.

10.12   Matter Is Organized on Several Levels

At what point did the clusters of proteins discussed in Section 10.10 become alive? It is difficult to give a simple answer. Obviously, a cow is alive and a stone is not. We can say that an amoeba is living and a coacervate is not. Difficulties arise, however, when biologists try to set up an exact classification system for living versus nonliving things.

Sometimes nonliving things appear lifelike. If a salt crystal is added to a concentrated salt solution, for example, it will grow and start the formation of other crystals. There is a difference, however, between this type of growth and the growth of organisms. The salt crystal can grow only by taking from the environment material of the same composition as itself. The primitive cells proposed in the heterotroph hypothesis also may have grown by taking materials like themselves from the environment. The **7** environment eventually became depleted of those materials, however. In the new environment, other organisms were favored. Because they can take in and use materials from the environment that are significantly different from themselves, modern organisms show few superficial resemblances to crystal growths.

Viruses provide a biological puzzle in this connection. Like salt, some viruses can be crystallized (Figure 10.26) and even stored. Yet when placed inside a living cell, these viruses can take over the cell and cause it to destroy itself, reproducing more viruses. Outside a cell, viruses are inactive. They cannot take in and use materials from the nonliving environment. Are viruses alive?

The organization of matter, from the smallest subatomic particles to the biosphere, forms a continuum like the one shown in Figure 10.27. Each level is dependent, to some degree, on the levels above and below. Where do we draw the line between living and nonliving? At present, there is no way to decide how complex a system must be before it is called living. The difference between a cow and a stone is obvious. The difference between the simplest living organism and the most complicated nonliving system is not so obvious.

Figure 10.26 This crystal of polio virus, is magnified 110,000 times.

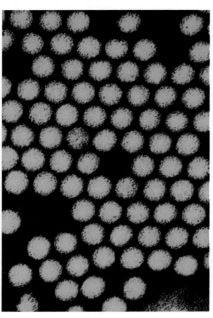

CONCEPT REVIEW

1. What evidence do we have that many species evolved from a few species?

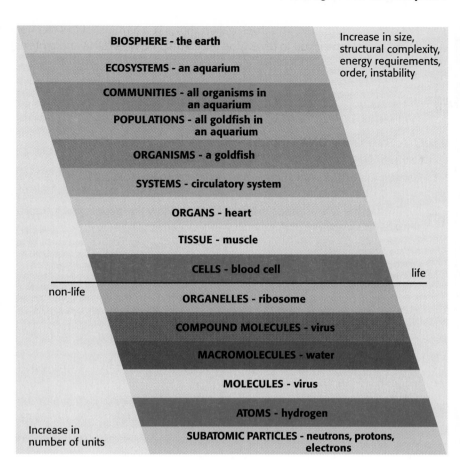

BIOSPHERE - the earth

ECOSYSTEMS - an aquarium

COMMUNITIES - all organisms in
an aquarium

POPULATIONS - all goldfish in
an aquarium

ORGANISMS - a goldfish

SYSTEMS - circulatory system

ORGANS - heart

TISSUE - muscle

CELLS - blood cell

ORGANELLES - ribosome

COMPOUND MOLECULES - virus

MACROMOLECULES - water

MOLECULES - virus

ATOMS - hydrogen

SUBATOMIC PARTICLES - neutrons, protons,
electrons

Increase in size, structural complexity, energy requirements, order, instability

non-life

life

Increase in number of units

Figure 10.27 The organization of matter on earth forms a continuum. The least complex is at the bottom; matter then increases in complexity to the top of the chart.

2. Compare environmental conditions on the earth during its early history with conditions today.
3. How can experiments such as those of Miller and Urey provide evidence for the origin of life?
4. How do scientists think primitive cells might have formed?
5. Why do scientists think it likely that the first cells were heterotrophs rather than autotrophs?
6. What changes did oxygen-producing photosynthesis cause on earth?
7. Why is it so difficult to draw a line between life and nonlife?

INVESTIGATION 10.1   DNA Sequences and Classification

Until the mid-1970s, taxonomists usually classified organisms by comparing observable structures in a given organism with those of another organism. For example, a taxonomist might compare the structure of forelimbs in mammals. In recent years, taxonomists also have been able to compare the structure of certain proteins in different organisms.

New techniques now allow biologists to compare the DNA that codes for certain proteins. A hypothesis known as the molecular clock hypothesis uses the comparison of DNA sequences to make predictions about the relatedness of the organisms from which the DNA was taken. According to the molecular clock hypothesis, any changes that occur in the genetic materials of isolated populations are due to mutations in the DNA that are passed from one generation to the next. The rate at which the mutations accumulate can be measured. On the basis of this information, the approximate point when two species diverged from a common

This investigation allows students to use a model of DNA hybridization to investigate the degree of relatedness of three organisms (chimpanzee, human, and gorilla) and to explore how a tool such as DNA hybridization can be used to study questions in evolutionary biology. Teams of 4 work well.
Time: One class period.

ancestor can be determined. This investigation shows you a model of this technology. It may be helpful for you to review the relationship between DNA and proteins discussed in Chapter 8.

Materials (per team of 4)
paper clips: 50 black, 50 white, 50 green, 50 red

PART A Comparing DNA Strands
Procedure

1. Consult Table 10.2. Which organisms share the most groupings? On the basis of the data in the table, which organisms are least closely related?

2. Synthesize the DNA strands indicated below. Allow the paper clips to represent the four bases of DNA according to the following key:

   black = adenine (A)   green = guanine (G)
   white = thymine (T)   red = cytosine (C)

   Hook together paper clips of the appropriate colors. Stretch the string of paper clips out on the lab table with position 1 on the left. Tape a small piece of paper next to the strand and label it as indicated. Each strand will represent a small section (20 bases) of a gene that codes for the protein hemoglobin.

3. Team member I will synthesize the following piece of DNA by hooking together paper clips in this sequence:

   pos. 1   pos. 20
   A G G C A T A A A C C A A C C G A T T A

   Label the strand *human DNA*. This represents a small section (20 bases) of the gene that codes for the protein hemoglobin in humans.

4. Team member 2 will synthesize the following strand:

   pos. 1   pos. 20
   T C C G G G G A A G G T T G G C T A A T

Part A Procedure

1. **Human and gorilla share the most groupings. Katydid is least closely related to the others in the table.**

Table 10.2   Examples of Classification of Animals

	Human	Gorilla	Southern Leopard Frog	Katydid
Phylum	Chordata	Chordata	Chordata	Arthropoda
Subphylum	Vertebrata	Vertebrata	Vertebrata	
Class	Mammalia	Mammalia	Amphibia	Insecta
Subclass	Eutheria	Eutheria		
Order	Primates	Primates	Salientia	Orthoptera
Suborder	Anthropoidea	Anthropoidea		
Family	Hominidae	Pongidae	Ranidae	Tettigoniidae
Subfamily			Ranidae	
Genus	*Homo*	*Gorilla*	*Rana*	*Scudderia*
Species	*Homo sapiens*	*Gorilla gorilla*	*Rana pipiens*	*Scudderia furcata*
Subspecies			*Rana pipiens sphenocephala*	*Scudderia furcata furcata*

Source: From C P. Hickman, L S. Roberts, and F M. Hickman, *Integrated Principles of Zoology,* 7th ed. (St. Louis Times Mirror/Mosey College Publishing, 1984), p. l48.

Label this strand *chimpanzee cDNA*. cDNA stands for complementary DNA. cDNA is a single strand of DNA that will match up with its partner strand. Remember, the bases in DNA are complementary. That is, adenine (A) always pairs with thymine (T), and cytosine (C) always pairs with guanine (G). This cDNA was made from the gene that codes for chimpanzee hemoglobin.

5. Team member 3 will synthesize the following piece of cDNA:

    pos. 1   pos. 20
    T C C G G G G A A G G T T G G T C C G G

Label this strand *gorilla* cDNA. This cDNA strand was made from the gene that codes for gorilla hemoglobin.

6. Team member 4 will synthesize the following piece of DNA:

    pos. 1   pos. 20
    A G G C C G G C T C C A A C C A G G C C

Label this strand *hypothetical common ancestor DNA*. This strand will be used in the second part of the investigation.

7. Compare the sequences of the human DNA and the chimpanzee cDNA. Match the human DNA and the chimpanzee cDNA base by base (paper clip by paper clip). Remember, black (adenine) always must pair with white (thymine), and green (guanine) always must pair with red (cytosine). Where the bases are complementary (that is, where the colors match correctly), allow the clips to remain touching, as shown in Figure 10.28. Where the bases are not complementary, separate the clips slightly to form a loop, as shown in Figure 10.28. Count the number of loops. Also count the total number of bases that do not match. Copy Table 10.3 in your data book and record the data in the appropriate columns.

8. Repeat Step 7, using the gorilla cDNA and the human DNA. Enter the data in the appropriate columns in the table in your data book.

Discussion

1. Based on the data you have collected for this one protein, is the gorilla gene or the chimpanzee gene more similar to the human gene?

2. Again, based on the data you have collected for this one protein, does the gorilla or the chimpanzee seem more closely related to humans?

3. Do the data prove that your answer to Question 2 is correct?

PART B An Evolutionary Puzzle
Procedure

Scientists have determined that mutations in DNA occur at a regular rate. They use this rate to predict how long ago in evolutionary history two organisms began to separate from a common ancestor. In this part of the investigation, you will use your paper-clip model to provide data in support of one of two hypotheses about a common ancestor for humans, chimpanzees, and gorillas.

9. Read the following information about a current debate among scientists who study human evolution:

    Most scientists agree that humans, gorillas, and chimpanzees shared a common ancestor at one time in evolutionary history. However, one group thinks the fossil record shows that gorillas, chimpanzees, and humans split

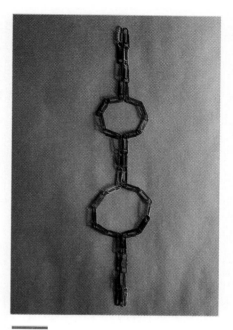

Figure 10.28   Paper-clip DNA sequences. Position 1 is at the top.

7–8. Chimpanzee cDNA hybridized to human DNA: 1 loop; 5 base differences. Gorilla cDNA hybridized to human DNA: 2 loops; 9 base differences.

Discussion

1. Based on the data collected, the chimpanzee gene is more similar to the human gene than is the gorilla gene.

2. Based on the data collected, the chimpanzee seems more closely related to humans than does the gorilla.

3. The data do not *prove* anything; they simply lend support to the hypothesis that the chimpanzee is more closely related to humans than is the gorilla.

Table 10.3 Human DNA

Hybridized to:	Chimpanzee cDNA	Gorilla cDNA
Number of loops		
Number of differences		

Figure 10.29 Hypotheses of primate divergence.

Part B Procedure

11. **Common ancestor DNA hybridized to: human cDNA: 2 loops; 10 base differences; chimpanzee cDNA: 2 loops; 8 base differences; gorilla cDNA: 1 loop; 3 base differences.**

Discussion

1. **Gorilla DNA is most similar to common ancestor DNA.**
2. **Human cDNA and chimpanzee cDNA give the most similar patterns of matching and looping when hybridized to the common ancestor DNA.**
3. **The data support the second model, which shows two separate divergences, rather than one three-way split.**
4. **The findings using this model do not *prove* the validity of the hypothesis, but they do provide some direction for additional research, such as comparing the genes for other proteins, or expanding the search for fossils that can shed more light on the likely pattern of divergence.**

This investigation allows students to practice using structural characteristics to group organisms into several classification levels. Students may work alone or in small groups of 3.
Time: One class period.

from one common ancestor at the same time. Their model for this split is shown in Figure 10.29a.

A second group thinks the fossil record shows there were two splits. In the first split, gorillas split from the common ancestor. Humans and chimpanzees then shared another common ancestor for perhaps 2 million years. They then split again and evolved into their present states. The model for this pattern of splitting is shown in Figure 10.29b.

10. Use your DNA model and the DNA sequences from Part A to investigate this debate. First, use your human DNA as a guide to synthesize a human cDNA. Do that step as a team. Make certain that the human cDNA is complementary to the human DNA strand.

11. The hypothetical common ancestor DNA synthesized in Part A is DNA for hemoglobin extracted from a hypothetical common ancestor. Now, match all three samples of cDNA (gorilla, human, and chimpanzee) with the common ancestor DNA, one sample at a time. Again, allow the paper clips to touch where the bases match correctly. Form loops where the bases do not match. Copy Table 10.4 in your data book and record your data.

Discussion

1. Which cDNA is most similar to the common ancestor DNA?
2. Which cDNAs give the most similar patterns of matching and looping when matched to the common ancestor DNA?
3. Which model in the evolutionary debate described above do your data support?
4. Do your findings prove that this model is the correct one? Explain.

INVESTIGATION 10.2   Levels of Classification

You will investigate some of the structural characteristics that taxonomists use in separating animal groups at different classification levels. Because you will use the observations of other people (recorded as drawings), your conclusions can be no more valid than those drawings.

Materials (per person)
none

Table 10.4 Hybridization Data: Common Ancestor

Common Ancestor DNA to:	Human cDNA	Chimpanzee cDNA	Gorilla cDNA
Number of loops			
Number of differences			

Table 10.5   Structural Characteristics

| Characteristics | Animals | | |
	1.	2.	3.
a.			
b.			
c.			
d.			
e.			
Classification level _____			

Procedure

1. In your data book, prepare four forms similar to Table 10.5, or tape into your data book the forms your teacher provides. Label the forms *Table A, Table B, Table C,* and *Table D.*
2. For Table A, use the information in Figure 10.30. In the spaces under *Animals*, write *human, chimpanzee,* and *gorilla*. In the spaces under *Characteristics*, copy the italicized key words in each of the following questions. These words should remind you of the full questions when you review the table.
   a. How does the *length of the arms* of the animal compare with the length of its legs?
   b. Are the *canine teeth large*, or are they *small* as compared with other teeth of the same organisms?
   c. How many *incisor* teeth are present in the upper jaw?

Procedure

In some cases *explicit* information concerning the group in which a specific animal is classified is not in A *Catalog of Living Things* in Appendix 4. Specific difficulties that may arise are given below.

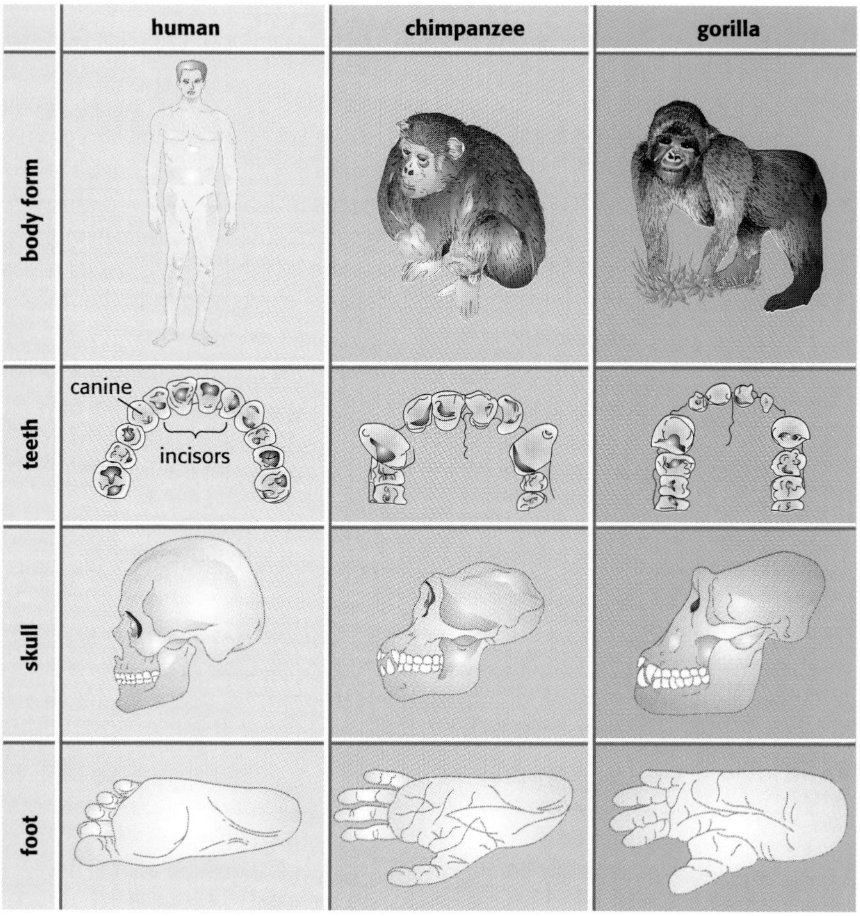

Figure 10.30

Figure 10.31

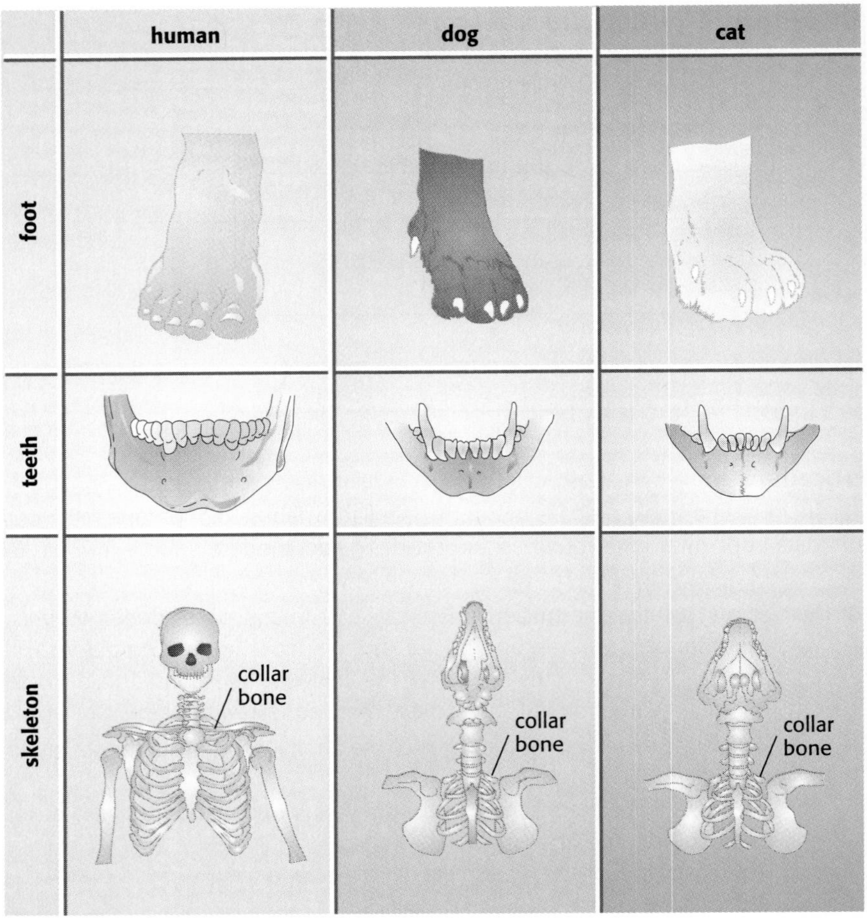

	human	dog	cat
foot			
teeth			
skeleton	collar bone	collar bone	collar bone

2. (Figure 10.30):
   e. If the past four items have led students to expect the chimpanzee and gorilla to be differentiated from humans on all counts, this item may be puzzling. The number of incisors is the same for all three species. Placement of chimpanzee and gorilla in the family Pongidae depends on a student's interpretation of the word *ape*. Although a chimpanzee is illustrated in Appendix 4, it is not explicitly identified as an ape.

4. (Figure 10.31):
   a. No picture is required for this question.
   c, d. Students should remember canines and incisors from Figure 10.30. Although no dog is shown under the order Carnivora, the necessary information is included in Figure 10.5. Neither is a cat (genus *Felis*) shown, but lions are sufficiently catlike that most students will, by inference, locate the order of the cat.

6. (Figure 10.32):
   a. Your laboratory should contain a specimen of a frog for students to examine.
   b. No picture is required for this question.

   d. Is the *brain* case of the skull *large*, or is it *small* as compared with the overall size of the skull?
   e. Is there an *opposable first toe on the foot?* (An opposable toe is one that can be pressed against all the others, just as your thumb can press against your other fingers.)

3. After studying Figure 10.30, fill in all the spaces in Table A with your answers for each animal. Then write *Family* in the space following *Classification level.* Refer to Appendix 4 to find the family into which each of these organisms has been placed. Write this information in the spaces at the bottom of the table.

4. For Table B, use the information in Figure 10.31. Under *Animals*, write *human, dog,* and *cat.* Under *Characteristics,* copy the italicized words in each of these questions:
   a. How many paired *appendages* (arms and legs) does the animal have?
   b. Are *nails* or *claws* present on the toes of the foot?
   c. How does the size of the *canine teeth* compare with that of other teeth in the lower jaw?
   d. How many *incisor teeth* are present in the lower jaw?
   e. How does the size of the *collarbone* compare with that of the other organisms?

5. After studying Figure 10.31, fill in the spaces in Table B with your answers for each animal. Write *Order* in the blank space following *Classification level.* From Appendix 4, select the order into which each of these organisms has been placed. Enter this information in the table.

6. For Table C, use the information in Figure 10.32 and the following questions:
   a. What type of *body covering* (hair, feathers, scales, none) does the animal have?
   b. How many paired *appendages* (arms and legs) does the animal have?
   c. Do the *ears project from* the surface of *the head?*

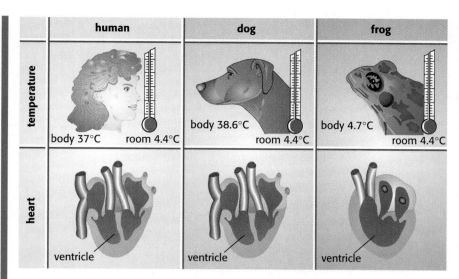

Figure 10.32

d.   Is the *body temperature* similar to the temperature of the environment or not?

e.   How many *ventricles* are *in the heart?*

7.   The classification level for this table is *Class*. Determine the class for each organism in Figure 10.32 and write it in the table.

8.   For Table D, use the information in Figure 10.33 and the following questions:

a.   What type of *skeleton* (internal or external) does the animal have?

b.   Is the *position* of the *nerve cord* along the back or along the belly?

c.   Compared with the rest of the nervous system, is the *brain* large or small?

d.   Are paired *appendages* present, or are they absent?

e.   Are there *grooves behind* the *head region* of the very young organism?

9.   Write *Phylum* in the space following *Classification level* and add the name of the phylum into which each animal in Figure 10.33 is placed.

8. (Figure 10.32):

c.   Some students may say a bird's brain is small, but point out that the task here is to make comparisons; compared with the brain of a crayfish, a bird's brain is large, even considering relative body size.

d.   This is another case in which attention is called to a point that does not make a distinction; all have paired appendages, though some students may call attention to a difference in numbers.

e.   All students recognize the embryonic similarity between human and bird. Some may, nevertheless. complain that they cannot definitely answer the question with respect to the crayfish. They are right, of course.

Figure 10.33

	human	bird	crayfish
skeleton			
nervous system			
young organism			

Discussion

1, 2. **Students may try to make these questions more complicated than they are, attempting to give reasons for the classification by repeating the structural features noted in the tables instead of merely citing the classification levels. This results from misreading. The question is not concerned with the evidence itself but with the way the evidence is expressed in the hierarchy of classification levels.**

3. **Species C and D are more similar than species A and B. On the basis of the information given, it is impossible to make any other general statement.**

Discussion

1. There are more structural similarities between chimpanzees and gorillas than between chimpanzees and humans. How does the classification system you used express this fact? Focus on the levels in the classification system into which these organisms are placed together.

2. How does the classification system you used express the following?
   a. There are more structural similarities between dogs and cats than between dogs and humans.
   b. There are more structural similarities between humans and dogs than between humans and frogs.
   c. There are more structural similarities between humans and birds than between humans and crayfish.
   d. There are more structural similarities between humans and chimpanzees than between humans and dogs.

3. You are told that species A and B are classified in the same kingdom but different phyla. You also are told that species C and D are classified in the same phylum but different classes. What general statement can you make about similarities among species A, B, C, and D?

SUMMARY

Classification systems enable biologists to study the great variety of organisms that exist in the biosphere. Taxonomists have developed a system that names organisms and indicates their relationships. That system is accepted throughout the world and allows biologists from different places to understand one another when discussing a particular organism. Classification systems may be based on many characteristics, but the most important ones are structural and biochemical homologies. Currently all organisms are classified in five kingdoms—Prokaryotae, Protista, Fungi, Plantae, and Animalia. Classification systems will continue to change as biologists learn more about the natural world.

The theory of evolution led to many predictions of what the first life was like. Based on those predictions, scientists have studied the evolution of the earth itself. A variety of experiments has provided evidence that life may have originated in a series of steps beginning with the formation of simple organic molecules. Later, those molecules may have assembled in a variety of aggregates, some of which had the ability to reproduce themselves. Those first cells probably were heterotrophs that used the organic molecules around them as sources of energy. Later, some cells developed the ability to use light energy. Fossil evidence supports the view of very simple beginnings; the oldest fossils resemble the prokaryotes of today.

APPLICATIONS

1. Compare your name to the scientific name of a plant or animal. How are they similar or different? Is your first name more like a genus or species name? What about your last name?

2. Suppose you joined an expedition into a tropical rain forest and found an organism that was previously unknown to modern science. How would you go about determining its closet relative? Do you think you should start with breeding experiments using different organisms to determine its relationship with them?

3. The species portion of an organism's scientific name usually helps to describe some characteristic of that organism. Find out what each of these frequently used words means: (a) *vulgaris*; (b) *chinensis;* (c) *alba*; (d) *edulis;* (e) *vomitorius*; (f) *tripetalus*. Can you find plants with these words in their names? How appropriate are their names?

4. What part of the earth has the greatest variety of plants and animals? What is happening to this area, and what effect will this have on the rest of the earth?

5. What characteristics do viruses lack that would suggest they are not living organisms?

PROBLEMS

1. Morphological analogues are structures having similar functions but different origins. For example, the wing of an airplane and the wing of a bird are analogous but certainly have different origins. List the ideas that humans have borrowed from plants and animals to use in building structures and making materials.

2. What would the atmosphere of earth be like if there were no life on it? Look at the Gaia hypothesis proposed by James Lovelock.

3. Write a report on *The Panda's Thumb* by Stephen Jay Gould. What does his first essay illustrate about evolution? About homologies?

4. You can increase your awareness of the different plants and animals in your area by consulting lists of local flora, field guides, and books on natural history. Set a goal for yourself of becoming familiar with several species of plants and animals previously unknown to you by the end of the year. Find out their names, where they grow, and how they live.

5. Some scientists place the archaebacteria in separate kingdom. Read about these organisms, and determine what characteristics might warrant their separation.

Prokaryotes and Viruses

11

These rocks are part of a beach on Prince William Sound, Alaska, the site of the *Exxon Valdez* oil spill. What might cause the rocks on one side of the photo inset to be clean, whereas the rocks on the other side still are covered with oil?

Researchers treated several test plots with oil-eating bacteria and fertilizers. Using such biological methods to clean up oil, toxic wastes, and other hazards is called bioremediation. The Biology Today feature in Chapter 24 discusses bioremediation in more depth.

◆ Members of the kingdom Prokaryotae—commonly known as bacteria—are the most ancient organisms on the earth. Prokaryotes are microscopic organisms that lack a nucleus and other membrane-enclosed organelles. Found in every available habitat, they affect the living world dramatically. They may be producers, consumers, or decomposers. Because they play essential roles in chemical cycles such as the nitrogen cycle, life on the earth would be totally different without them. Without prokaryotes, there would be few decomposers to break down the bodies of dead plants and animals, the oxygen supply would be decreased, and large oxygen-using animals such as humans would find life difficult. Although some prokaryotes can cause human diseases such as strep throat, rheumatic fever, and skin infections, most are beneficial. They help digest food, provide certain vitamins, and are used in manufacturing such foods as cheese and yogurt.

This chapter introduces a few of the many members of the kingdom Prokaryotae and examines their important roles in the biosphere. The chapter also discusses viruses—microscopic agents of disease that are not truly living but that dramatically affect the lives of many different organisms. ◆

MAJOR CONCEPTS

- ◆ Prokaryotes lack nuclei and membrane-enclosed organelles.
- ◆ Bacteria occupy every imaginable niche.
- ◆ Archaebacteria may be the most ancient of all organisms.
- ◆ Although some prokaryotes cause disease, the vast majority are beneficial and essential.
- ◆ Some diseases caused by bacteria and viruses are transmitted mainly by sexual contact.
- ◆ Viruses have profound effects on organisms.
- ◆ AIDS, caused by HIV, is a worldwide human health problem.

KNOWLEDGE CHECK

- ◆ What are prokaryotes and how do they differ from other organisms?
- ◆ What are the two major groups of prokaryotes and how do they differ?
- ◆ In what ways are prokaryotes important in the biosphere?
- ◆ How do viruses differ from living organisms?

Prokaryotes

11.1 Prokaryotes Are Structurally Simple but Biochemically Complex

Prokaryotes are extremely small—a spoonful of garden soil may contain 10 billion of them, and the total number in your mouth is greater than the number of humans who ever have lived on the earth. Prokaryotes cover the skin, line the nose and mouth, live in the gums and between the teeth, and inhabit the digestive tract of humans.

Most prokaryotes are one-celled organisms. The cell may be shaped like a straight rod, a sphere, a spiral, a filament, or a short, curved rod, as shown in Figure 11.1. Some prokaryotes join together in clusters and chains, forming multicellular structures. Just as the shapes of prokaryotes may differ, their reactions to certain stains and the size, shape, and color of their colonies also differ. Many prokaryotes move by means of flagella, but others glide over surfaces. Spiral-shaped prokaryotes move by using a corkscrew motion that involves flagella-like structures between the layers of their cell walls. Although all prokaryotes are microscopic, some are smaller than others. Rickettsia are among the smallest known cells, rang-

Guidepost
What are prokaryotes, and what are their roles in the biosphere?

Prokaryotic flagella are structurally different from eukaryotic flagella. They are not enclosed in the plasma membrane and do not contain fibrils. A prokaryotic flagellum has about the same diameter as a fibril. Some parkaryotes glide without visible structures.

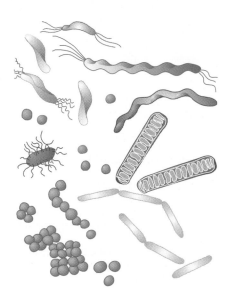

Figure 11.1 Prokaryotes may be round, rod-shaped, spiraled, or filamentous. Many have flagella, others form colonies.

Oil-soaked beaches around Prince William Sound were treated with a fertilizer developed to stimulate the growth of naturally occurring oil-degrading eubacteria. Clean-up results surpassed expectations, and elevated levels of the eubacteria persisted for five months.

The ecological roles of prokaryotes often are overlooked and should be emphasized.

See Elso S. Barghoom, "The Oldest Fossils," *Scientific American* **(May 1971):70.**

Figure 11.2 The structure of a typical prokaryotic cell.

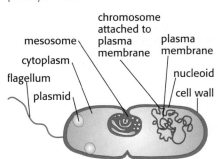

ing from 0.3 to 0.7µm in diameter. These tiny organisms, however, have been responsible for many human deaths from diseases such as Rocky Mountain spotted fever, in which the rickettsia are transmitted by ticks.

Most prokaryotic cells have rigid cell walls made of lipids, carbohydrates, and protein, but no cellulose. Figure 11.2 is a diagram of the structures of a prokaryote. Inside the cell wall is a plasma membrane that encloses the cell contents. An infolding of the plasma membrane forms the mesosome (MEZ oh sohm). The mesosome may aid in secretion, the movement of particles out of the bacterium. It also may have some role in copying the chromosome before cell division. Prokaryotes have one main chromosome, which is made of a continuous, circular molecule of double-stranded DNA. Because there is only one chromosome, the genes are not present in pairs as they are in eukaryotes. The chromosome is attached to the plasma membrane and located in an area of the cell known as the nucleoid (NOO klee oyd). In addition to one chromosome, prokaryotes usually contain one or more smaller circular DNA molecules known as plasmids. Each plasmid consists of a few genes. As discussed in Section 8.14, plasmids are important in genetic engineering and gene cloning.

Although many of the metabolic processes of prokaryotes are similar to those of eukaryotes, others are unique. As a group, prokaryotes can digest almost anything—even cellulose and petroleum. Prokaryotes are the major decomposers in most ecosystems. Because they break down complex organic compounds into simple inorganic materials used by plants, prokaryotes are essential to all food webs. They also transform inorganic materials into complex organic compounds and serve as food for protists and other microorganisms. Many prokaryotes are photosynthetic; still others are the source of antibiotics.

The fossil record of prokaryotes goes back 3.5 billion years (see Figure 11.3), which leads scientists to think they probably were the first living organisms. By comparison, records of animals date back about 700 million years and records of land plants, 470 million years. Some biologists think that about 2 billion years ago certain photosynthetic prokaryotes started a metabolic revolution that led to the increase of oxygen concentration in the atmosphere of the earth from less than 1 percent to the present-day 20 percent. Humans and other animals could not have evolved without this increase in oxygen.

Prokaryotes, which are among the most hardy organisms known, can survive extremely low temperatures, even freezing, for many years. Some

Figure 11.3 Fossil prokaryote *Eoastrion* (a), this section through Gunflint chert. These fossils are about 2 billion years old. Modern prokaryotes (b) very similar to *Eoastrion,* from a microbial mat. Organism (c) shown in (b), magnified 100 times.

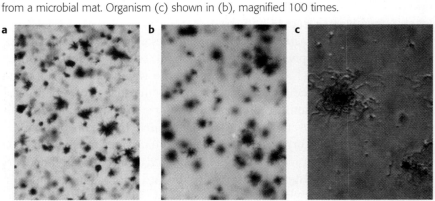

species live in boiling hot springs and others in hot acids. They can survive at high pressure in great oceanic depths and at low pressure high in the atmosphere. They can tolerate total drying by forming endospores (EN doh sporz), complex thick-walled structures that contain DNA (see Figure 11.4). They are among the first organisms to inhabit new environments such as burned or volcanic areas, and they help transform these environments into areas in which other organisms can survive.

All prokaryotes can reproduce asexually. A form of sexual reproduction known as conjugation may occur, but the genetic contribution of the parents is not equal. In conjugation, one prokaryote transfers a portion of its DNA through a tube into another, as shown in Figure 11.5. Genetic material also can be transferred between prokaryotes by viruses.

More than 10 000 species of prokaryotes have been described, but the vast majority have not been identified. Studies in the molecular biology of these organisms indicate there are two fundamentally different groups of prokaryotes: the **archaebacteria** (AR kee bak TIR ee uh; *archae* meaning ancient) and the **eubacteria** (YOO bak TIR ee uh; *eu* meaning true).

11.2   Archaebacteria Are Different from All Other Organisms

Although they share some characteristics with the eubacteria, archaebacteria differ in basic ways from all other forms of life. Their biochemistry and their adaptation to extreme environments suggest they were among the earliest organisms on the earth. The first archaebacteria may have lived 3.5 billion years ago.

The cell walls of archaebacteria lack a large molecule known as peptidoglycan (PEP tid oh GLY kan), which is made of sugar subunits cross linked by peptides. This molecule is present in the cell walls of all walled eubacteria. Plasma membranes of archaebacteria contain lipids that are not like those found in the membranes of either eubacteria or eukaryotes. The transfer RNA found in archaebacteria also is different from that found in eubacteria and eukaryotes. Ribosomes in archaebacteria are unique in shape, and the nucleotide sequence of the ribosomal RNA is different from that of eubacteria.

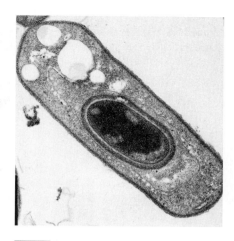

Figure 11.4   Some prokaryotes can survive harsh conditions by producing a thick-walled endospore (the dark body in the middle of this cell). Transmission electron micrograph (X100 000).

Two good references on Archaebacteria are C. R. Woese, "Archaebacteria," *Scientific American* (June 1981):98–122; and R. H. Evans, "Archaebacteria: A New Primary Kingdom for Our Classrooms," *The American Biology Teacher* (March 1983):139–43.

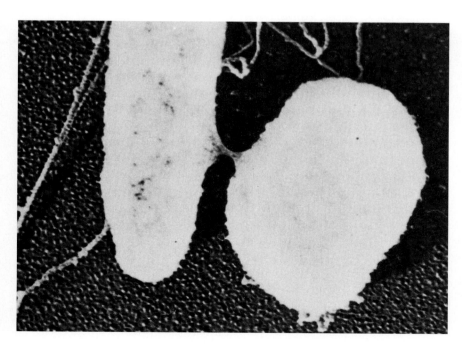

Figure 11.5   Visible evidence for the exchange of cell materials between prokaryotes. This electron micrograph reveals a tube between prokaryotes of two different strains. A plasmid can move through the tube.

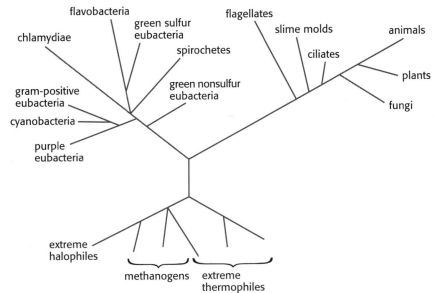

For example, the close relationship between humans and chimpanzees is based on a 98.4% similarity in their DNA.

Invite students to speculate on the evolutionary implications of the similarity in rRNA. In general, it is another piece of evidence for a common origin of life and life processes.

All organisms have ribosomal RNA (rRNA), and the study of an organism's rRNA can reveal its evolutionary history. The sequence of the nucleic acid bases within rRNA changes little through time. By comparing the rRNA of two different organisms, a scientist can estimate the relationship of the organisms to each other. If the sequence of rRNA bases is very similar, the organisms are thought to be closely related. The greater the differences in base sequence, the less closely related the organisms are thought to be. Figure 11.6 shows relationships among archaebacteria, eubacteria, and eukaryotes based on rRNA analysis.

11.3   Three Types of Archaebacteria Exist

There are three distinct types of archaebacteria. One type, the thermoacidophiles (THER moh a SID uh fylz), lives in hot, acidic environments in which temperatures average 70° to 90°C and the water may have a pH of 2 or less (see Figure 11.7). These environments include hot sulfur springs and smoldering piles of coal-mining debris. Despite the low external pH, the organisms maintain an internal pH close to neutral (pH 7).

Figure 11.7 Thermoacidophilic archaebacteria. *Pyrodictium occultum* (a) can grow at temperatures up to 100°C. *Pyrococcus furiosus* (b) grows at temperatures up to 103° C. *Acidianus infernus* (c) grows at temperatures up to 95°C and at pH 1.0 (the pH of concentrated hydrochloric acid). In all three transmission micrographs, the bar = 10μm.

a

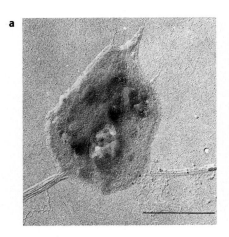

b

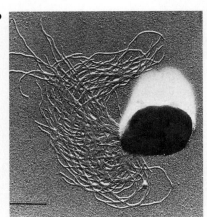

c

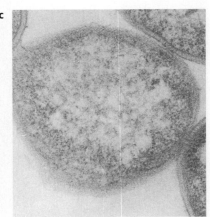

Another type, the extreme halophiles (HAL uh fylz), shown in Figure11.8, requires a high concentration of salt to survive. Halophiles can grow in salt brine. In their presence, salted fish may become discolored and spoiled. The extreme halophiles are found in salty habitats along ocean borders and in salty inland waters such as the Great Salt Lake and the Dead Sea. Although some are photosynthetic, they do not contain chlorophyll. Instead, they absorb light energy through the pigment bacterial rhodopsin (roh DOP sin), which is similar to a visual pigment found in human eyes.

The third type, the methanogens (meh THAN uh jenz), shown in Figure 11.9, is the best studied and most widely distributed of the archaebacteria. Because methanogens are killed by oxygen, they live only in anaerobic conditions. They produce methane (CH_4) gas from hydrogen and carbon dioxide. Methanogens often are found with the eubacteria that decompose organic matter and release hydrogen gas. They are common in stagnant water, sewage treatment plants, the ocean bottom, and hot springs. They play an important role in the carbon cycle by removing organic compounds from the sediments and releasing methane to the atmosphere, where it reacts with oxygen to form water and carbon dioxide. In this way, they increase the carbon dioxide level of the earth's atmosphere. Methanogens also live in the digestive tracts of animals, including cattle and humans.

Cows, goats, deer, sheep, and antelope, unlike other herbivores, have rumens (ROO mens), special enlargements of the digestive tract. Within the rumen, protozoans, eubacteria, and archaebacteria live in an unusual ecosystem. The protozoa and eubacteria contain enzymes that break down the cellulose in the grass and other plant material a cow or goat eats. The methanogens use the carbon dioxide and hydrogen released by those organisms as their source of food and produce methane gas.

Methane gas can be used as a fuel if enough is produced and captured. Some scientists have suggested using methanogens to convert the growing amount of garbage, sewage, agricultural waste, and manure to methane gas. Organic waste would be placed in a methane generator—an airtight, anaerobic container—along with methanogens and various eubacteria, which would grow and reproduce. Methanogens produce large amounts of

Figure 11.8   Halobacteria require a high concentration of salt to survive. This organism is growing beside a salt crystal (the large rectangle).

Ask students to speculate about the effect of methanogens on global warming.

Students may be familiar with ruminants as cud-chewers.

See J. E. Lennox, S. E. Lingenfelter, D. L. Wance, "Archaebacterial Fuel Production: Methane from Biomass," *The American Biology Teacher* (March 1983):128–38.

Figure 11.9   Two methanogens *Methanobacterium ruminatum* (a), from cow rumen. The bacterium has nearly finished dividing. *Methanospinilum hungatii* (b). Transmission electron micrographs; bar = 1μm.

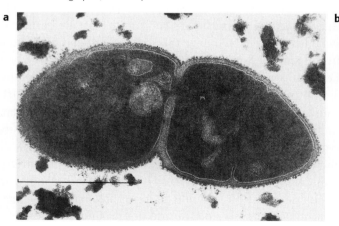

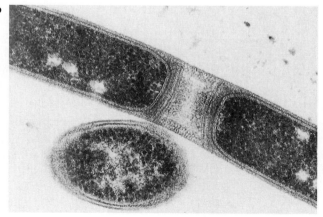

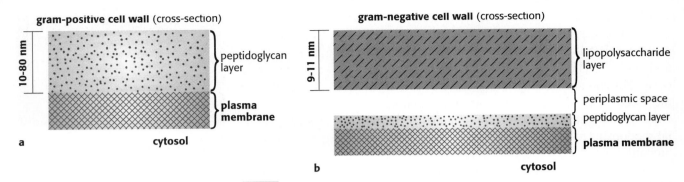

gram-positive cell wall (cross-section)

10–80 nm

} peptidoglycan layer

} **plasma membrane**

a   cytosol

gram-negative cell wall (cross-section)

9–11 nm

} lipopolysaccharide layer

} periplasmic space

} peptidoglycan layer

} **plasma membrane**

b   cytosol

Figure 11.10 Diagrammatic representation of cell wall cross sections of a gram-positive (a) and a gram-negative (b) eubacterium showing the differences in complexity.

methane gas, which can be collected and used to heat houses. Although presently rather unpredictable, improved generators may help solve two problems—waste and garbage buildup, and the energy crisis.   **4**

11.4 Eubacteria Are a Diverse Group

The eubacteria may be aerobic or anaerobic; photosynthetic, or chemosynthetic; or thermophilic (heat loving). In fact, their most prominent characteristic is their metabolic diversity. Autotrophs and heterotrophs are intermixed within the various eubacterial phyla. Although structure and function do not provide much information about their evolutionary relationships, a growing body of ribosomal RNA analysis is adding information rapidly. Until more data are assembled, however, classification of   **5** eubacteria at the phylum level remains uncertain.

Three major groups can be defined by their cell walls. Some lack a rigid cell wall; some have a thick peptidoglycan wall without an outer lipoprotein layer; and some have a thin peptidoglycan wall that has an outer lipoprotein layer. Figure 11.10 shows diagrams of the two wall types. The thick-walled types are known as gram-positive, because of their reaction to Gram's stain (a widely used bacterial stain); the thin-walled types are gram negative.

Eubacteria without cell walls are surrounded by a triple-layered membrane composed of lipids. These eubacteria are resistant to penicillin, because penicillin works by inhibiting the growth of the cell wall. Commonly known as mycoplasmas (my koh PLAZ muhs), these eubacteria are probably the smallest (0.2–0.3 μm) organisms capable of independent growth. They are of special evolutionary interest because of their extremely simple cell structure. Although certain species of mycoplasmas (see Figure 11.11) cause diseases such as pneumonia in humans and cattle, most are harmless.

Gram-positive eubacteria, which have thick peptidoglycan cell walls, are widespread in soil and air. Some gram-positive eubacteria produce lactic acid (a 3-carbon compound) in their metabolism. These are used in the preparation of food products such as sauerkraut, buttermilk, and yogurt. Others are the source of antibiotics, such as streptomycin, tetracycline, and erthyromycin. Still others form endospores that enable them to survive unfavorable growth conditions.

The remaining eubacteria are gram-negative and have a wide variety of metabolic styles. Although photosynthesis is widespread among the gram negative eubacteria, many photosynthetic forms are anaerobic—they do not use water in photosynthesis and thus do not produce oxygen gas. Based on the pigments present in their cells, these producers may be

Figure 11.11 *Mycoplasma pneumoniae* lives in human cells and causes a type of pneumonia.

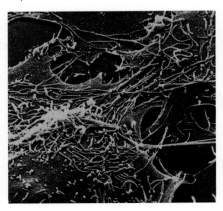

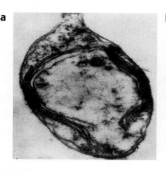
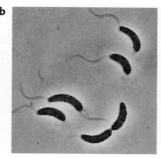
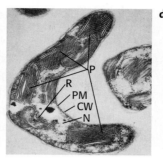
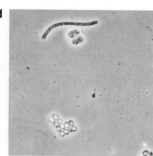

Figure 11.12 Anaerobic photosynthetic eubacteria. Transmission electron micrograph of *Rhodomicrobium vanielli* (a), a purple nonsulfur eubacterium. Note the photosynthetic membranes around the periphery of the cell; bar = .05 µm. *Rhodospirillum* (b), another purple nonsulfur eubacterium, can live also as a heterotroph. Transmission electron micrograph of *Ectothiorhodospira mobilis (c)*. P is the photosynthetic membrane system, R indicates ribosomes, PM is the plasma membrane, CW is the cell wall, and N is the nucleoid. Three species of purple sulfur bacteria: *Thiopedia, Thiospirillum,* and *Chromatium* (d).

called green eubacteria and purple eubacteria (see Figure 11.12). Many use hydrogen sulfide (H_2S) and release sulfur in photosynthesis; others use hydrogen gas (H_2) or compounds such as lactic acid. Like plants, however, they synthesize organic compounds from carbon dioxide and convert light energy to chemical energy.

The cyanobacteria (SY an oh bak TIR ee uh), or blue-green eubacteria, shown in Figure 11.13, are a photosynthetic group with structures and functions similar to those of algae and plants. Cyanobacteria inhabit a wide variety of environments, including hot springs and under the ice of antarctic lakes. They are responsible for the "algal" blooms on freshwater streams and lakes that have been polluted with phosphates. Some of the cyanobacteria grow in colonies that can be seen with the naked eye, although individual cells are microscopic.

Cyanobacteria contain chlorophyll associated with their plasma membranes, but the membranes are not organized into chloroplasts as they are in eukaryotic cells. Their photosynthesis, like that of plants and algae, uses water and results in the release of oxygen gas. Cyanobacteria are thought
6 to be primarily responsible for the start of the oxygen revolution that changed the earth's atmosphere, making possible the evolution of large oxygen-using organisms.

There are thousands of species of cyanobacteria, and thousands more existed in the ancient past. Remains of these ancient communities are found in fossil stromatolites. Stromatolites still are being produced in the salt flats along the Persian Gulf, the Bahamas, western Australia, and the west coast of Mexico.

Cyanobacteria were previously known as blue-green algae and were considered part of the plant kingdom.

For more information about prokaryotes, see L. Margulis and K. V. Schwartz, *Five Kingdoms*, 2nd ed. (San Francisco: Freeman, 1988).

Figure 11.13 Cyanobacteria are abundant in all aquatic ecosystems. *Oscillatoria* (a) (X100); *Gleocapsa* (b) (X250), *Fischeriella* (c) (X150).

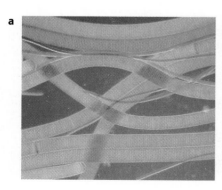
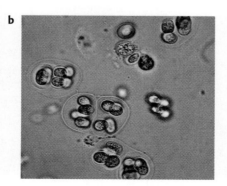
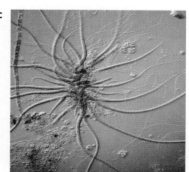

11.5   Eubacteria Make Possible the Nitrogen Cycle

Although cyanobacteria make their own food using energy from the sun and carbon from carbon dioxide, most eubacteria get their energy from foods produced by other organisms, either living or dead. If the food source is living, the eubacteria are called parasites. If the food source is dead, they are called decomposers. Some eubacteria neither photosynthesize nor use organic molecules as their source of energy. Instead, they obtain energy from inorganic molecules in the environment. These eubacteria are called chemosynthetic (KEE moh syn THEH tik). Both chemosynthetic and decomposer eubacteria play important roles in the nitrogen cycle.

Nitrogen is essential to all living things. In fact, it is found in two of the four types of biological molecules needed for life—proteins and nucleic acids. Just as carbon cycles through a community and back into the physical environment from which it came, nitrogen also cycles from the biotic to the abiotic parts of the ecosystem. Study the nitrogen cycle shown in Figure 11.14 as you read the remainder of this section.

Although nitrogen gas (N_2) makes up about 78 percent of the atmosphere, neither plants nor animals can use that form of nitrogen. Instead, all consumers get nitrogen-containing compounds in the things they eat, which, of course, can be traced to producers. Producers get their nitrogen bearing compounds from the soil (or water) in which they grow. Nitrogen gas dissolved in water is present in the soil. Also living in the soil or in the roots of certain plants are nitrogen-fixing eubacteria. These chemosynthetic eubacteria can "fix" nitrogen, converting it to a form they or other organisms can use. Nitrogen fixers convert nitrogen gas to ammonia (NH_3), which can be used to synthesize proteins and other nitrogen-containing compounds.   **7**

Figure 11.14   The nitrogen cycle.

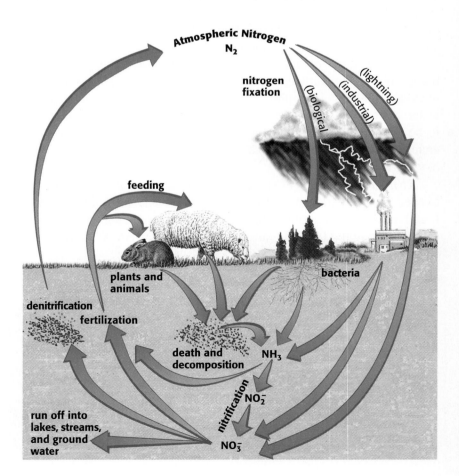

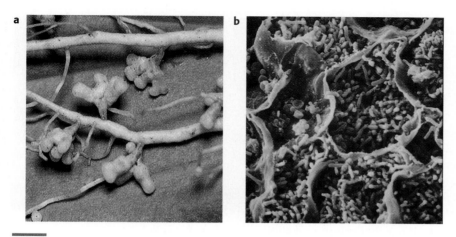

Figure 11.15   Soybean roots with nodules formed by a species of *Rhizobium* (a). *Rhizobium meliloti* nodules on sweet clover (b). Scanning electron micrograph (X640).

Farmers discovered long ago that soils in which clover, peas, beans, alfalfa, or other members of the legume family had been grown produced better crops than did other soils, although they did not know why. Today, scientists know the crops grow better because of an increase in the nitrates in the soil. The increased nitrate level in the soil is the result of nitrogen fixation and chemical reactions following fixation. The plants, however, do not fix the nitrogen. Eubacteria that live in the roots of these and many other legumes are responsible for nitrogen fixation. The eubacteria induce the plant to form swellings on its roots called nodules (NOD yoolz), shown in Figure 11.15. Under favorable conditions, eubacteria in these nodules can fix as much as 22.5 g of nitrogen per square meter every year. Some free-living eubacteria and cyanobacteria also have the ability to fix nitrogen. The ammonia produced by nitrogen-fixing eubacteria can be used by the legume plants, by eubacteria, or by other plants.

Decomposer eubacteria in the soil break down the complex organic compounds in dead plants and animals. They use the energy in these compounds and convert the compounds to simpler substances, including the gas ammonia. Some of the ammonia escapes into the atmosphere, but much of it dissolves in soil water, where it reacts chemically with hydrogen ions to form ammonium ions (NH_4^+). In the form of ammonium ions, nitrogen may be absorbed by the roots of plants. It then is built into living material again by the plants.

Other chemical reactions also may occur in the soil. Two groups of chemosynthetic nitrifying (NY trih fy ing) eubacteria (Figure 11.16) continue the nitrogen cycle. One group changes the ammonium ions (NH_4^+) to nitrite ions (NO_2^-). The other group changes the nitrite ions to nitrate ions (NO_3^-). Plants can absorb the nitrates from the soil and can use them to build proteins.

Nitrifying eubacteria operate only under aerobic environmental conditions. Oxygen from the air spaces found between soil particles dissolves in soil water. Sometimes, all the spaces fill up with water, leaving no room for air. When that happens, the soil environment becomes anaerobic, and nitrifying eubacteria cannot carry on their activities. Denitrifying (DEE ny trih fy ing) eubacteria, however, thrive in anaerobic environments. They change any remaining nitrates to nitrogen gas. The nitrogen gas gradually escapes into the atmosphere, and the cycle of nitrogen is complete.

Some farmers add ammonia (NH_3) directly to their irrigation water to provide nitrogen for plants, while others may apply nitrate salts to the soil. The constant use of these chemical fertilizers influences the soil ecosystem changing many of the populations of microorganisms that normally are involved in the nitrogen cycle.

Figure 11.16 Nitrifying eubacteria. *Nitrosomonas* (a) converts ammonia (NH_3) to nitrite (NO_2). *Nitrobacter winogradskyi* (b) converts nitrite to nitrate (NO_3). Transmission electron micrographs, (a) X121 300; (b) X213 400.

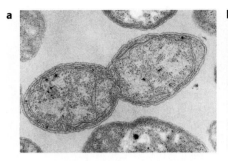
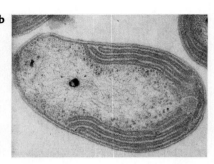

CONCEPT REVIEW

1. What characteristics distinguish prokaryotic cells?
2. In what ways are prokaryotes important in the biosphere?
3. In what ways do archaebacteria differ from eubacteria?
4. How can methanogens help with the energy crisis?
5. What characteristics do microbiologists use to distinguish eubacteria?
6. How might cyanobacteria have contributed to the evolution of animals?
7. How does nitrogen fixation take place?
8. How is nitrogen returned from organisms to the atmosphere?

Students are eager to recount symptoms and to testify to miraculous cures. To cut off *all* of this stifles interest, but you can use it to lead into discussions of principles. Emphasize the biological universality of disease by including animal and plant diseases.

Guidepost

What is the relationship between a pathogen and a disease?

Although the discussions of disease use the expression cause, emphasize that infectious disease involves an interrelationship. An entire group of organisms may be infected by a pathogen, but only some individuals may develop symptoms of disease. It can be argued that the cause of the disease in infected hosts with symptoms is their lack of resistance rather than the presence of a pathogen, which also is present in those with no symptoms. This interrelationship is a general—and important—feature of biological problems: A, B, and C may be necessary for a given outcome, but A, B, or C alone or in paired combination is not sufficient for the outcome to occur. This may be a difficult concept for students to grasp, yet the kind of thinking required— the balancing of two oppositely changing variables—is an important skill. Students may relate to this concept if reminded about the increased possibility of illness when fad dieting or not getting adequate rest.

The response of the immune system to a second infection by a pathogen may occur in 3 days, instead of 7 to 10 as with a first infection.

Disease

11.6   Diseases Result from Interrelationships

Disease is a condition that interferes with an organism's ability to perform a vital function. An infectious disease involves a disease-causing agent, or **pathogen.** The disease results from the interaction between an infected organism, or **host,** and the pathogen. The pathogen infects the host. If the pathogen harms the host, symptoms of disease may appear.   2

How serious a disease becomes depends on the characteristics of both the host and the pathogen. The ability to cause disease is called **virulence** (VIR yuh LENTS). The ability of an infected host to cope with a pathogen is called **resistance.** A pathogen with high virulence may cause death in a   1 host with low resistance. A host with high resistance, however, may show only mild symptoms of the same disease. Sometimes, a moderately virulent pathogen may produce serious illness. During a famine, for example, more people die from disease than from starvation. This is because a poorly nourished host may have much less resistance than a well-fed host.

Although genetics does play a role in the ability of an organism to resist disease, much of an individual's resistance to disease is acquired during the lifetime of the individual, rather than inherited. Resistance, whether acquired or inherited, is called **immunity.** When a pathogen   3 infects a human host, the host produces proteins called **antibodies** that can help destroy the pathogen. In addition to antibodies, certain cells of the body's defense, or immune system, have a chemical memory. (The immune system is discussed in Chapter 16.) If the host survives the initial infection, these cells retain the ability to produce the same antibodies. If the same pathogen infects the host a second time, the host's body can act against it more rapidly, preventing symptoms from developing or reducing their severity.

Each type of antibody is effective against only the particular pathogen with a that brought about its production, or pathogens that are very similar.

For example, a person can have chicken pox only once, because antibodies produced during the first infection usually prevent symptoms from developing after a second infection. These antibodies are not effective against measles, however. Because there are many types of pathogens that cause colds, immunity to one type does not confer immunity to all the others.

Many diseases can be prevented by **vaccines.** A vaccine contains only enough of the killed or weakened disease-causing agent to stimulate production of antibodies by the immune system, usually without producing
3 symptoms of the disease. Memory cells then act to keep later infections by the same pathogen from causing disease. Vaccines first were used by Edward Jenner, an English physician (see Figure 11.17).

Disease may result from a variety of causes. The pathogens involved in infectious diseases may be viruses or eubacteria or other organisms. For example, athlete's foot, ringworm, potato blight, and corn smut are caused by fungi. Protists cause malaria, African sleeping sickness, and amoebic dysentery. Many worms and insects cause diseases of plants and animals, and insects themselves are important vectors, or carriers, of disease. Other diseases, such as scurvy, are the result of diet deficiencies; still others are a result of advancing age; and some, such as asthma, are brought on by reactions to substances or pollutants in the environment. Finally, some disorders, such as cystic fibrosis or Huntington disease, are hereditary.

11.7   Some Eubacteria Are Pathogens

Although the vast majority of prokaryotes are beneficial or at least harmless to other organisms, some eubacteria are pathogenic. At the present time, however, no archaebacteria are known to cause disease.
4   Eubacterial diseases may be spread through water, food, or air. Many plant diseases are caused by eubacteria, and almost all types of plants are susceptible to one or more types of eubacterial disease. Fire blight, shown in Figure 11.18, is a common eubacterial disease that can destroy fruit trees. In Florida in 1984, an outbreak of a eubacterial disease called citrus canker led to the destruction of more than 4 million citrus seedlings in four months in an effort to halt the spread of the disease. Citrus canker is caused by one of more than 100 distinct varieties of the gram-negative eubacterium *Xanthomonas campestris.*Other varieties of this eubacterium cause diseases with similar symptoms in beans, cabbages, peaches, and other plants. Citrus canker now threatens the existence of the Florida citrus industry—an annual $2.5-billion business.

Eubacteria also cause human diseases such as cholera, leprosy, tetanus, eubacterial pneumonia, whooping cough, and diphtheria. Many eubacteria multiply in the human digestive tract and leave the body in the feces. If
4 untreated feces enter the water supply, as discussed in Chapter 3, the eubacteria may enter another organism when it drinks the water. Waterborne diseases such as typhoid kill 6 million children each year world-wide. Most cases of food poisoning are due to eubacteria. In some cases, the eubacteria produce poisonous substances, or toxins, that are released in the food. These toxins can cause nausea, vomiting, and diarrhea within a few hours of ingestion. In the case of botulism, even a small amount of the toxin can be fatal. In other cases, eubacteria ingested with food multiply in the intestine, causing symptoms even several days after ingestion.
4   Many eubacteria are airborne. Whenever you open your mouth and exhale, you release droplets of moisture, each of which may contain one or two eubacteria. With a sneeze, you release many thousands of droplets

Figure 11.17   The English physician Edward Jenner demonstrated in 1796 that by inoculating patients with a vaccine made of cowpox viruses he could protect them from the more serious disease of smallpox.

Ask students why archaebacteria do not cause disease. The extreme environments in which archaebacteria live exclude most organisms, thus removing the opportunity for archaebacteria to cause disease.

The most common type of food poisoning is caused by *Staphylococcus aureus* and involves foods such as cream-filled pastry, puddings, creamy salad dressings, poultry, meat and meat products, and egg and meat salads. Refrigeration keeps these foods relatively safe, because *Staphylococcus* cannot grow at low temperatures. Spores of *Clostridium botulinum* often contaminate raw foods before harvest or slaughter. If the foods are processed so the spores are killed, no problem arises. Adequate cooking (80°C for 10 minutes) destroys the toxin. *Salmonella* food infection is not, strictly speaking, food poisoning, because the organism must grow in the intestine before symptoms appear.

October 1990 reports suggest that some strains of streptococcus may cause symptoms similar to toxic shock syndrome.

Figure 11.18 Fire blight is a disease of apples, pears, and related trees and in beans, cabbages, peaches, and related trees and plants. It is caused by the bacterium *Erwinia amylovora*.

at great speed. Droplets from a sneeze, visible in Figure 11.19, have been clocked at 200 miles an hour. Droplets from a cough travel at about half that speed. Each time you sneeze, you can expel from 10 000 to 100 000 individual eubacteria. The eubacteria travel through the air without being killed and may infect another person standing close to you.

Toxic shock syndrome, which is characterized by a sharp drop in blood pressure followed by fever, vomiting, and diarrhea, is caused by certain strains of the eubacterium *Staphylococcus aureus.* Although about 85 percent of the cases of toxic shock syndrome reported in the United States have occurred in menstruating women who use tampons, both men and women can contract the disease. Researchers have found that two fibers contained in super-absorbent tampons absorb strong concentrations of magnesium from the surrounding blood and tissue. These high magnesium concentrations enhance the production of toxins by the eubacteria. As the eubacteria multiply, they produce more toxins, which are released into the body, killing cells and causing other symptoms of the disease. Most people have developed a resistance to the toxin by the age of 20.

Even dental caries, or tooth decay, is caused by eubacteria. The decay begins on the surfaces of the teeth in a film known as dental plaque. This film, which consists of large eubacteria in a complex sugar matrix (shown in Figure 11.20), builds up on unbrushed teeth and in crevices not reached by a toothbrush. Diets high in certain sugars especially sucrose (table sugar), are especially harmful because eubacteria break down the sugars to lactic acid, which causes the loss of calcium from the teeth. Enzymes released by eubacteria then begin to break down the tooth enamel, allowing other eubacteria to penetrate the teeth. Fluoride makes the teeth more resistant to decay because it retards the loss of calcium.

11.8   Viruses Are Unusual Pathogens

What do AIDS, hepatitis, measles, mumps, influenza, colds, and polio have in common? These diseases all are caused by viruses such as those shown in Figure 11.21. A virus is an infectious agent that contains a nucleic acid (either DNA or RNA) and a protein coat. Viruses are so small that they can be seen only with an electron microscope. In fact, viruses can pass through most bacteriological filters. Viruses may play one of two roles when they enter a host cell. As agents of disease, viruses enter the host cells, disrupt their normal functioning, and sometimes kill them. As

Figure 11.20 *Streptococcus mutans* is a lactic acid bacterium associated with tooth decay. (X8000)

Figure 11.19 Droplets from a sneeze.

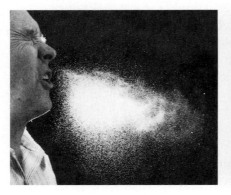

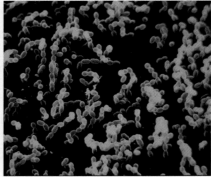

a
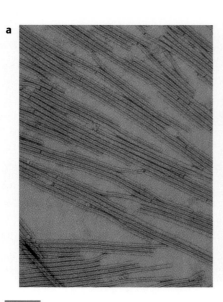

b

c
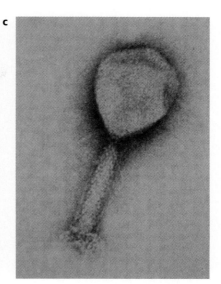

Figure 11.21   All viruses are pathogenic. Tobacco mosaic virus (a) causes a disease of plants (X185 000); adenovirus (b) causes respiratory illnesses in humans (X11 000); T4 bacteriophage (c) (X240 000).

agents of heredity, viruses can enter cells and cause permanent, inheritable changes. Often, the role the virus plays depends on the host cell and environmental conditions.

Viruses differ from living things in several important ways. They are able to produce copies of themselves only inside a living organism. Outside the host cell, viruses neither reproduce, feed, nor grow. They have no metabolism of their own. They do not take in and use energy. They do not have cell parts. Some can be crystallized and survive for years in that state. Only when they enter the appropriate host cells can they resume reproduction. Viruses infect plants, animals, fungi, and Prokaryotes (see

6 Figure 11.22). The virus first attaches to the host cell by means of its protein coat or membrane and then injects its DNA or RNA into the host cell. The viral nucleic acid takes over the machinery of the host cell, com-

7 manding it to make more viral protein and viral nucleic acid. The proteins and nucleic acids then are assembled into new virus particles, and the infected cell ruptures, releasing hundreds of newly made viruses. Each new virus has the ability to infect a single new host cell. Figures 11.23 and 11.24 on the following page show these events for a virus that has entered a eubacterium.

Usually, genetic information is stored in DNA, transcribed into mRNA, and then translated into protein. In most viruses, the viral genetic information actually is stored as mRNA. Certain RNA viruses, known as retroviruses, must go through an additional step before they can repro-

8 duce. Retroviruses must make copies of DNA from their RNA, the reverse of the normal flow of stored information in biological systems. Like other viruses, a retrovirus binds to the surface of a host cell and injects its RNA into the cell. Reverse transcriptase, a special enzyme associated with the

7 RNA, allows the retrovirus to make a complementary DNA copy of its made RNA (see Figure 11.25). The newly formed viral DNA then can be integrated into the host cell DNA, replicated at the same time, and transmitted to offspring cells. When the cell produces RNA from its own DNA, it also produces viral RNA, which becomes the source of new viral particles, thus continuing the infection.

Viruses are examples of natural gene-splicers. If the viral DNA is integrated into the host DNA instead of producing new viruses, a genetically modified cell results. When the host cell reproduces, the viral DNA is replicate along with the host DNA. Thus the cell and its descendants may be permanently changed.

Figure 11.22   TYMV (turnip yellow mosaic virus) in a young cabbage plant.

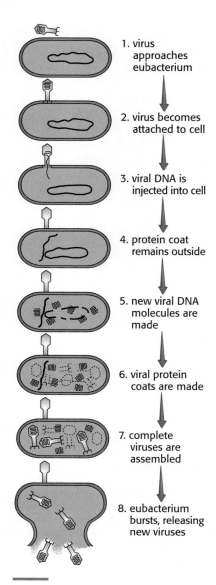

1. virus approaches eubacterium

2. virus becomes attached to cell

3. viral DNA is injected into cell

4. protein coat remains outside

5. new viral DNA molecules are made

6. viral protein coats are made

7. complete viruses are assembled

8. eubacterium bursts, releasing new viruses

Figure 11.23   A virus attacking a single eubacterium.

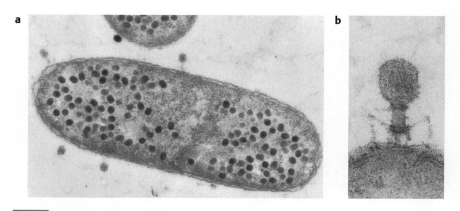

Figure 11.24   The eubacterium *Escherichia coli* (a) with T4 phage viruses attached to the surface and inside (X50 000). T4 phage (b) attached to the eubacterium after it has injected its DNA (X200 000).

Retroviruses are known to cause cancer in some animals and have been associated with certain types of cancers in humans. Human oncogenes have similarities to genes in these retroviruses—both can result in cancerous growth when they are somehow disturbed. The virus that causes AIDS is a retrovirus that attacks and kills certain cells of the immune system. Once these cells are killed, the immune system is unable to perform its normal function of defending the organism against disease.

CONCEPT REVIEW

1. Describe the interaction between host and pathogen that may result in disease.
2. Distinguish between infection and disease.
3. What is immunity? How can a vaccine bring about the development of immunity?
4. Describe, using examples, the major ways by which eubacterial diseases can be spread.
5. What is the role of eubacteria in dental caries?
6. What is the structure of a virus and how does it infect a host cell?
7. How does a virus direct the formation of other virus particles?
8. How does a retrovirus differ from a virus?

Figure 11.25   Retroviruses use the enzyme reverse transcriptase to make a DNA copy of their RNA. The viral DNA then is incorporated into the host cell DNA, and the host cell directs production of new viruses.

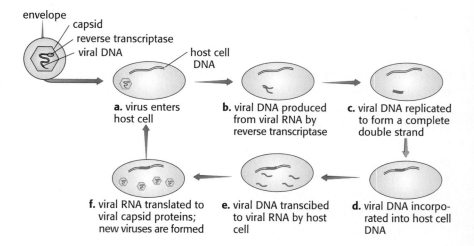

envelope
capsid
reverse transcriptase
viral DNA
host cell DNA

a. virus enters host cell

b. viral DNA produced from viral RNA by reverse transcriptase

c. viral DNA replicated to form a complete double strand

d. viral DNA incorporated into host cell DNA

e. viral DNA transcibed to viral RNA by host cell

f. viral RNA translated to viral capsid proteins; new viruses are formed

Biology Today

Slow Infections of the Brain

Since 1976, 22 cases of a rare and fatal brain infection have been diagnosed in a sparsely populated area of Czechoslovakia. This disease, identified as Creutzfeldt-Jakob disease, or CJD, kills its victims within seven months after the first symptoms appear. Normally, CJD occurs in one person per million every year throughout the world. The incidence in Czechoslovakia is several hundred times that.

CJD resembles kuru, a fatal infection that was common among Papua, New Guinean tribes who ate and handled the brains of their dead during mourning rituals. Kuru victims first lose their coordination and their ability to walk and then become insane. Now that the rituals have ended, the number of kuru victims has decreased. CJD also is similar to "mad cow disease" (bovine spongiform encephalopathy, or BSE), which has caused 15 000 cattle to be destroyed in Great Britain since 1986. Most investigators believe that CJD, kuru, and mad cow disease are unconventional slow infections that are variants of scrapie. Scrapie affects the central nervous system of sheep and goats. Animals infected with scrapie also lose their coordination and become unable to walk. Scrapie and scrapielike diseases are called slow infections because of the long incubation period—from several months to several years.

At least 25 cases of CJD in Western countries can be traced to technological misfortune. Injections of growth hormone from human pituitary glands are responsible for 14 cases. Because of the long incubation period of the disease, more cases caused by technological misfortune probably will appear. Three people who received skin grafts of the tissue that lines the inside of the skull developed CJD. The use of unsterile instruments and electrodes during neurosurgery has produced six cases. One case was transmitted during a cornea transplant and another by root-canal surgery.

The nature of the agent that causes CJD and other slow infections of the brain is controversial. The victims exhibit no immune response, and the infection is resistant to treatments that normally kill infectious agents. Even ultraviolet radiation, which normally destroys nucleic acids, appears to have no effect on the agent.

Some researchers believe the cause is a prion, a deviant piece of pure protein. If this abnormal protein can somehow catalyze its own formation from a healthy version of the protein present in all brains, it may not need any nucleic acid. Other researchers believe the protein-only hypothesis is implausible. Analyses of experiments dealing with radiation and heat lead some researchers to conclude the causative agent probably is a small virus. The most common view, however, is an intermediate one: The agent is a protein with a tiny amount of nucleic acid. The amount is too small to code for more protein but large enough to carry a scrap of genetic information. One researcher has found a fragment of mitochondrial DNA in scrapie proteins extracted from infected brains, but its role is unclear. Until researchers finally can identify the cause of scrapie and related diseases, treatment will remain impossible.

Slow Infections of the Brain

Infection	Affected Organisms
Scrapie	Sheep, goats
Bovine spongiform encephalopathy—mad cow disease	Cattle
Transmissible mink encephalopathy	Mink
Creutzfeldt-Jakob disease Kuru Gerstmann-Straussler syndrome	Humans
Chronic wasting disease with spongiform encephalopathy	Captive Rocky Mountain elk and mule deer
Spongiform encephalopathy	Nyala Gemsbok Domestic cat

Sexually Transmitted Diseases

11.9   Eubacteria Cause Several Sexually Transmitted Diseases

Oral contraceptives cause the body to mimic the pregnant state. One of the results is a lack of glycogen production in the vagina and a consequent rise in vaginal pH. The lactic acid bacteria normally present fail to develop under these conditions, and *N. gonorrhea* is able to colonize more easily than in an acidic vagina containing lactic acid eubacteria. This is another example of interaction in an ecosystem.

A number of diseases are transmitted by sexual contact with infected persons. These diseases include gonorrhea, syphilis, chlamydia, herpes, and AIDS (*A*cquired *I*mmune *D*eficiency *S*yndrome). Gonorrhea, syphilis, and chlamydia are caused by eubacteria; herpes and AIDS are caused by viruses.

Gonorrhea is one of the most widespread human diseases. It is caused by the eubacterium shown in Figure 11.26, which thrives in the human body but can be killed by drying, sunlight, or ultraviolet light. An infected person can transmit the eubacteria to his or her sexual partners through the moist membranes of the sex organs. In females, symptoms of the infection may be only a mild irritation of the vagina, which often goes unnoticed. In males, however, the organism causes a painful infection of the urethra that results in burning during urination and sometimes the discharge of pus. Gonorrhea also can cause eye infections in adults and in newborns of mothers with the disease. Most gonorrhea responds well to treatment with antibiotics, but the number of infections reported each year continues to increase for many reasons. First, because the symptoms can be mild in females, one female unknowingly can infect many men. Second, oral contraceptives, which are widely used today, alter the pH of the vagina and allow the eubacteria to grow more readily. Third, for unknown reasons, an individual does not develop immunity to the eubacteria, so repeated infections are possible. And fourth, because gonorrhea usually is treated with large doses of penicillin, penicillin-resistant strains are becoming more common. Left untreated, gonorrhea can result in sterility.

Less common but more serious is syphilis, which is caused by the spiral eubacterium shown in Figure 11.27. Syphilis can be fatal. The incubation period—the time between infection by the eubacterium and the appearance of symptoms—for syphilis is about three weeks. The first symptom is a hard, painful ulcer called a chancre (SHAN ker), which appears at the site of infection. The chancre disappears in about one to four weeks, and the infected person may experience no other symptoms for two to four months even if the disease is left untreated. After this time, the disease enters the second stage, which is marked by a skin rash and infections in various organs. The disease then may become latent, or hidden. Although the disease may remain latent for the remainder of an individual's life, in many people the disease enters a third stage. This stage can produce severe nervous system or circulatory system damage, insanity, and even death. Syphilis is treated with penicillin, but as with gonorrhea, many penicillin resistant strains have developed, and recent reports indicate an increase in the number of cases.

These infections are caused by the eubacterium *Chlamydia trachomatis*, which is difficult to culture.

More common than either gonorrhea or syphilis are chlamydial infections, which can be controlled easily with tetracycline. In females, chlamydial infections usually affect the cervical canal and the vagina and may cause pelvic inflammation. The symptoms may go unrecognized, but if not treated, they may result in sterility. Infected females may pass the disease to their children at the time of birth, resulting in eye and lung infections. Chlamydial infections in males cause irritation of the urethra and produce a discharge from the penis. Because chlamydial infections are difficult to diagnose, they are the most prevalent sexually transmitted disease today.

Figure 11.26 The eubacterium that causes gonorrhea, *Neisseria gonorrhea* (X1000).

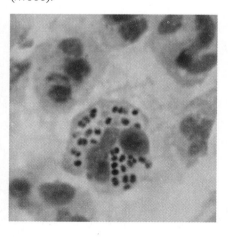

11.10   AIDS Is a Deadly Disease

4 The virus responsible for AIDS was identified in 1983 and named *H*uman *I*mmunodeficiency *V*irus, or HIV. HIV primarily infects those cells of the immune system known as helper T cells, but it also may infect brain or nerve cells. Helper T cells, brain cells, and nerve cells all have a particular cell-surface molecule called CD4 that serves as the site of attachment for the virus. Inside the host cell, the viral RNA is copied into DNA, which may be integrated into the host cell genome. Each infected cell then can serve as a factory for the production of HIV. When the cell eventually ruptures, hundreds of viruses, each able to infect another cell, are released, as shown in Figure 11.28. Reproduction of HIV in helper T cells leads to a gradual decline in the number of helper T cells, weakened function of the immune system, and increased susceptibility to opportunistic infections and cancers. Opportunistic infections are infections a normal, fully func-

5 tioning immune system would be able to repel. Although HIV can damage organs directly, it is the opportunistic infections that account for up to 90 percent of deaths from AIDS.

The incubation period for HIV infection is two weeks to three months. If infection has occurred, HIV antibodies can be detected in the blood at that time. The incubation period for AIDS is much longer. An individual infected with HIV may not develop AIDS for several months or even as long as 12 years. The variation in the AIDS incubation time is related to the length of time the virus remains inactive in the cells of the immune system. Symptoms vary widely, depending on the type and number of opportunistic diseases that occur, and often alternate with periods of relatively good health.

AIDS, the most severe stage of HIV infection, usually leads to death within two years of diagnosis. Diagnosis is based on the presence of certain infections or cancers in a person who tests positive for HIV antibod-

4 ies. Among those conditions are a variety of infections caused by fungi, protozoans, eubacteria, and other viruses seldom seen in the population (or only in much milder form), and a type of skin cancer known as Kaposi's sarcoma. Besides Kaposi's sarcoma, the most common opportunistic infection associated with AIDS is a pneumonia caused by a protozoan parasite, *Pneumocystis carinii.*. Tuberculosis, which is caused by a eubacterium, also is prevalent among HIV-infected individuals.

HIV usually is introduced into the body through sexual contact, although infection also can occur through contaminated blood. AIDS can occur in both homosexuals and heterosexuals, and heterosexual transmission is becoming more common. Abstaining from sexual activity is one way to reduce the risk of infection. Limiting sexual activity to one noninfected individual and using condoms are two other methods that reduce the risk of acquiring AIDS from sexual contact.

Many individuals with AIDS are drug addicts who use intravenous injections and have shared hypodermic needles with infected people. Some AIDS patients are hemophiliacs and other recipients of blood transfusions that contained HIV. Since 1985, however, all donated blood in the United States has been tested for HIV. Because the virus can cross the placenta and infection can occur at birth or through the mother's milk, an increasing number of AIDS cases are appearing in babies born to HIV-infected mothers. Although there is no evidence that AIDS can be transmitted through casual contact such as shaking hands or swimming in the same pool, several health care workers exposed to HIV by way of accidental cuts or injections have contracted AIDS.

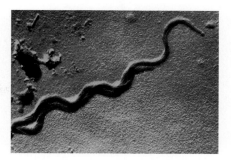

Figure 11.27   A scanning electron micrograph of *Treponema pallidum,* a spirochete that causes syphilis (X48 000).

A small number of biologists challenge the assumption that HIV is the causative agent in AIDS. You may wish to share this information with your students and discuss the role of multiple competing hypotheses as central to the progress of science.

New studies indicate *P. carinii* may be a fungus.

Figure 11.28   The several hundred small particles are HIV releasing themselves from a helper *T* cell, which they infect and ultimately destroy. Scanning electron micrograph. (X124 000).

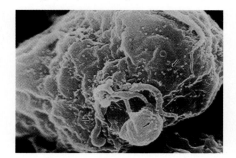

11.11   Prospects for Treatment of AIDS Are Improving

The efforts to understand AIDS have greatly increased our knowledge of how the immune system works and hold promise for better treatment in the near future. Although it was once thought impossible to vaccinate against AIDS because the protein coat of the virus can change, thus appearing to the immune system as a different virus, prospects for a vaccine are improving. **6** A study of pregnant women infected with HIV found HIV antibodies present in women who had uninfected babies, but not in women whose babies were infected. These results indicate that antibodies can protect against infection. A number of vaccines are being tested and others are in various stages of research and development. Of the trial vaccines under study, some protect chimpanzees from infection and some stimulate the production of antibodies in AIDS patients as well as in healthy people.

Researchers are trying to design specific drugs to block the action of HIV. For example, in test-tube studies, when CD4 (the cell-surface molecule to which HIV attaches) is introduced into the bloodstream, it binds with HIV. Thus HIV is no longer free to bind to CD4 on the cells. Experiments are under way with CD4 molecules attached to toxins that may kill the virus (see Figure 11.29). Three-dimensional computer models are being used to screen other drugs for the same binding capability. AZT, the first drug used against AIDS, interferes with HIV's ability to reproduce. Cytokines—molecules that act as immune-system regulators—have been able to turn off production of HIV in laboratory studies and to delay the onset of AIDS in people infected with HIV. These and other drugs under investigation have serious side effects, however, and many researchers think combinations of drugs will prove the most effective therapy. Although initial studies of various combinations are promising, particularly in treating and preventing recurrence of opportunistic infections, much remains unknown. **6**

There is, however, new hope for treating Kaposi's sarcoma. Recent evidence indicates that HIV stimulates tumor development by causing the production of a growth-stimulating factor on which the sarcoma is dependent. These findings may lead to new treatments with drugs that block the action or synthesis of the growth factors, thus stopping or slowing the tumor growth.

Figure 11.29   Linking a toxin to an AIDS virus molecule that recognizes glycoproteins antigen on an infected cell's surface should only destroy sick cells.

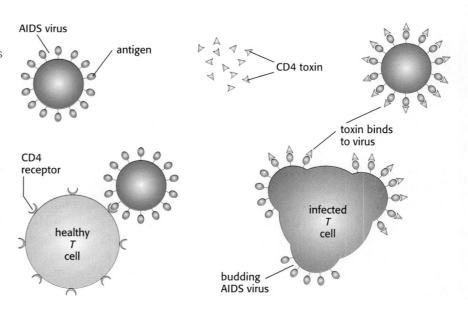

CONCEPT REVIEW

1. What are some of the reasons for the increase in gonorrhea despite the effectiveness of penicillin in treating it?
2. Describe the symptoms of each stage of syphilis.
3. What characteristics of chlamydial infections increase the spread of this disease?
4. Distinguish between HIV and AIDS.
5. What is meant by opportunistic infections?
6. Describe some of the problems involved in developing treatments for HIV infection and AIDS.

INVESTIGATION 11.1   Distribution of Microorganisms

Just where and in what abundance are prokaryotes and other microorganisms found? In this investigation, you will discover some of the answers to these questions.

Materials (per team of 3)

4 50-mL beakers
6 sterile swab applicators
glass-marking pencil
transparent tape

stereomicroscope or 10x hand lens
self-closing plastic bag labeled *waste*
4 sterile nutrient agar plates

This investigation provides a good introduction to the chapter. When it is completed, students should have an understanding of how widespread bacteria and other microorganisms are.

Teams of 3 allow each student to expose a plate.

Time: **Five days: one class period to inoculate the plates, about 15 minutes each day for observations, and about one-half period on the final day for class discussion.**

Procedure

1. Do not open the nutrient agar plates. Label the bottom of the four plates with your team symbol and the date. Number the plates from 1 to 4.
2. Do nothing to plate 1.
3. Uncover plate 2 and expose it to the air for 15 to 20 minutes. Each team should select a different area of the laboratory or the school building, including quiet corners and traffic areas. In your data book, note the exact location. Cover the plate after exposure.
4. Using plate 3, draw a line across the middle of the bottom of the plate. Label one side *Clean* and the other *Dirty*. Carefully remove a sterile swab applicator from its paper wrapper and rub the swab over part of your lab table. Without completely removing it, carefully lift the cover of the plate, as shown in Figure 11.30. Slip in the swab and *gently* rub it over the agar on the side labeled *Dirty*. Be sure the swab does not cut the agar surface or touch the clean side. Replace the used swab in its paper wrapper and discard in the self-closing plastic waste bag.

Safety

Check with the school nurse to identify any students using immunosuppressive drugs. These students should be excused from laboratory activities involving culturing of microorganisms. Pathogenic organisms can be picked up and cultivated on a simple nutrient agar. Do not allow students to cough into plates, culture any body secretions, or swab any body surfaces. After exposure, all plates must be taped closed, and observations must be made without opening the plates. Make certain that students wash up each day. Students should discard all used swabs in designated waste bags. Label self-closing plastic bags waste and tape a bag to each lab table. Collect these bags at the end of each class, place in a biohazard bag, and autoclave before discarding. Collect all used plates in a biohazard bag; autoclave before discarding or before opening plates for washing. Disposable plates are ideal for use in this experiment. See "Safety Using Microbes" in *Guidelines for Laboratory Safety.*

Figure 11.30   Inoculating a plate.

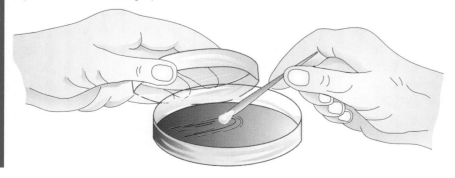

Discussion

1. **Plate 1 is the control. There should be no growth if it was sterilized carefully.**

3-5. **Airborne microorganisms transferred by contact easily contaminate surfaces, so bacteria are present on practically all surfaces. Microorganisms from an infected person can grow on a surface and then be picked up on the hands of another person. They can be transferred to the mouth or an open cut.**

6. **By cleaning surfaces, by maintaining personal habits of cleanliness (such as washing your hands or covering your mouth when sneezing or coughing), not spitting on the ground and keeping fingers and pencils out of your mouth. There should be fewer microorganisms on the clean side than on the dirty side of plate 3.**

7. **Sterile conditions are those resulting when all microorganisms have been killed. Clean conditions only reduce their number. (You might want to discuss the procedure** **both patient and physician must undergo before surgery to increase the likelihood of sterile conditions.)**

8. **Depends on student data.**

5. Use soap, or detergent, and water to wash carefully a small area of your lab table. Rub the clean area with a second swab, and then rub the swab over the side labeled *Clean*. Discard the swab in the waste bag as in Step 4. (Instead of wiping the swab over your lab table, you can compare freshly washed laboratory glassware with glassware that has been on the shelf several days.)

6. Divide plate 4 into equal sections. Allow one section to act as a control. Collect leaves, bark, pebbles, soil, or other materials. Using a clean beaker each time, mix each material with a small amount of sterile water, dip a fresh swab into the mixture, and rub over the agar surface of one section of the plate. As a control, dip a swab in plain sterile water and rub over one of the sections. Discard all swabs in the waste bag. Label the sections according to the material used.

7. Tape all plates closed and invert to prevent the water droplets that may condense inside the cover from dripping onto the agar. Incubate at room temperature for 3 or 4 days. In your data book, predict which plates will have the greatest number of microorganisms, which will have the fewest, and why.

8. Wash your hands thoroughly before leaving the laboratory.

9. Observe all plates daily, *without opening them*, and record your observations. A stereomicroscope or 10X hand lens may help in your observations.

> **Do not remove the cover of any plate. Harmful microorganisms can be picked up and cultivated even on this simple nutrient agar.**

10. On the fourth day, sketch the different types of colonies and write a description of each. There may be mold colonies present; they are fuzzy and larger than bacterial colonies. Record the number of each type of colony per plate.

11. Discard all plates in the biohazard bag designated by your teacher.

12. Each day, wash your hands thoroughly before leaving the laboratory.

Discussion

1. What are the results in plate 1? What was the purpose of plate 1?
2. What are the results in plates 2 through 4?
3. How do you account for the presence or absence of microorganisms on the examined surfaces?
4. Coughing or sneezing may spread droplets from mouth and nose to a distance of 3 m or more. The water in these droplets evaporates rapidly, leaving microscopic bits of dry matter that contain bacteria. Microorganisms from many other sources also may be carried on dust particles. How can you use this information to help interpret the class data?
5. Are microorganisms transmitted by touch alone?
6. In your daily living, how can you protect yourself and others from contamination by microorganisms?
7. Were any microorganisms found on the clean side of plate 3? What is the difference between clean and sterile conditions?
8. Were your predictions about the growth of microorganisms correct?

For Further Investigation

Does the type of medium used in the plates affect the count obtained at any one location? Does the temperature at which the plates are incubated affect the count? Design and carry out experiments that test hypotheses based on these questions.

INVESTIGATION 11.2   Control of Eubacteria

How effective are antibiotics, antiseptics, and disinfectants in controlling the growth of eubacteria? In this investigation, you can test a variety of substances with two common eubacteria.

 After reading the procedure, construct hypotheses that predict how the substances you select will affect the eubacteria.

Materials (per team of 3)

3 pairs of safety goggles
18 construction-paper disks
glass-marking pencil
stereomicroscope or 10X hand lens
2 pairs of forceps
2 sterile swab applicators
transparent tape
metric ruler

self-closing plastic bag labeled *waste*
sterile alcohol pad
4 sterile nutrient agar plates
small containers of distilled water, antiseptics, and disinfectants
6 antibiotic disks
broth cultures of *Micrococcus luteus* and *Escherichia coli*

Put on your safety goggles.

Procedure

1. Using a glass-marking pencil, label the bottoms of the four plates *A, B, C,* and *D*. Divide each plate into six equal, pie-shaped sections. Label the sections *1* through *6*. Label plates B and D *Control*.

2. Carefully remove a sterile swab applicator from its paper wrapper and use the swab to streak the entire surface of plates A and B with *Micrococcus luteus*, as shown in Figure 11.31. Replace the used swab in its paper wrapper and discard in the self-closing plastic waste bag.

3. In the same manner, streak the entire surface of plates C and D with *Escherichia coli*. Discard the swab in the waste bag as in Step 2.

4. Use forceps to place one plain disk each near the outside edge of sections 1, 2, and 3 in plates B and D. Be sure the disk adheres to the agar surface but does not break through it.

5. Using forceps, pick up a clean disk and dip it in sterile distilled water. Gently tap the disk against the side of the container to remove excess water and place it near the outside edge of section 4 in plate B. Repeat until you have placed a disk in sections 4, 5, and 6 of plates B and D. Why do you dip these disks in distilled water before placing them on the agar?

6. Your teacher will provide disks that have been soaked in three different antibiotics, *I, II,* and *III*. Place one disk of antibiotic I near the outside edge of section 1 in plate A and one disk near the outside edge of section 1 in plate C. In the same way, place disks of antibiotic II in section 2 of plates A and C, and disks of antibiotic III in section 3 of plates A and C.

7. Select any three of the antiseptic and disinfectant solutions your teacher provides. Using forceps, dip a clean disk into the first solution. Place one disk each in section 4 of plates A and C. Clean the forceps with a sterile alcohol pad before taking another disk.

8. Repeat Step 7 with two other solutions, placing them in sections 5 and 6 of plates A and C.

9. Cover and tape all four plates. Invert them and incubate at 37°C.

10. In your data book, record the substance into which you dipped each disk and the section in which you placed it.

11. Wash your hands thoroughly before leaving the laboratory.

 In this investigation, students can compare the effects of antiseptics, disinfectants, and antibiotics on two species of bacteria. Teams of 3 work well.
Time: Two class periods—one to inoculate the plates and one for observations and discussion.

Safety

Check with the school nurse to identify any students using immunosuppressive drugs. These students should be excused from laboratory activities involving culturing of microorganisms. Although the eubacteria being used are ordinarily harmless, observe sterile techniques throughout the investigation and impress this procedure on your students. Treat all eubacteria as potential pathogens. After inoculation, **all plates must be taped closed, and observations must be made without opening the plates. Make certain that students wash up each day. Students should discard all used swabs in designated waste bags. Label self-closing plastic bags waste and tape a bag to each lab table. Collect these bags at the end of each class, place in a biohazard bag, and autoclave before discarding. Collect all used plates in a biohazard bag; autoclave before discarding or before opening plates for washing. Disposable plates are ideal for use in this experiment. See "Safety Using Microbes" in *Guidelines for Laboratory Safety.***

Procedure

5. **Disks are dipped in water to provide a better control for the disks dipped in antiseptic and disinfectant solutions.**

Discussion

1. **The clear areas around the disks with chemicals indicate inhibited growth of eubacteria. The larger the zone, the greater the efficacy of the material being tested.**

2. **The control plates should show that the disks are not the inhibiting factor.**

Figure 11.31   Two patterns for streaking culture plates.

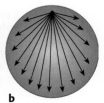

a   b

3. Eubacteria are not equally sensitive to all antibiotics. Penicillin is effective against gram-positive eubacteria such as *M. luteus*, but not as effective against *E coli*, which is gram-negative. Broad spectrum antibiotics are effective against certain eubacteria of both types, and nalidixic acid is particularly active against *E. coli*. Antibiotics work by inhibiting cell wall synthesis (penicillin), protein synthesis (neomycin), or eubacterial DNA replication (nalidixic acid).

4, 5. Depends on the substances used.

6. Any antibiotic that destroys intestinal eubacteria causes some disturbances, and in some cases the eubacteria must be replaced. The more serious problem is that in discriminate use promotes development of resistant strains.

7. Antibiotics may be taken internally or applied externally and inhibit the growth of specific microorganisms. Some antibiotics are highly toxic and must be used with great caution. Disinfectants are used on nonliving objects and destroy a wide range of microorganisms. Depending on their strength, many disinfectants would cause tissue damage if used externally or internally. Antiseptics also destroy a wide range of microorganisms but are not harmful to human tissue.

12. After one or two days, observe all plates, *without opening them*, and record your observations. A stereomicroscope or 10X hand lens may help your observations. Use a metric ruler to measure the width of any clear areas around the disks. Record your measurements.

> ⚠ **Do not remove the cover of any plate. Harmful microorganisms can contaminate the plates when you inoculate them.**

13. Discard all used plates in the biohazard bag designated by your teacher.
14. Wash your hands thoroughly before leaving the laboratory.

Discussion

1. What do the clear areas indicate? What do differences in the widths of the clear areas indicate?
2. What evidence do you have that the inhibition of these eubacteria is due to the chemicals on the disks and not the disks themselves?
3. Which antibiotic was most effective against *Micrococcus luteus*? Against *Escherichia coli*? What might account for the difference in effectiveness?
4. Which of the two species of eubacteria is more sensitive to all the chemicals?
5. Which antiseptic or disinfectant would you use to control *M. luteus*? To control *E. coli*?
6. *E. coli* is a normally harmless eubacterium that lives in the human intestines. Does the reaction of *E. coli* to antibiotics suggest that antibiotics should be used only when necessary? Explain.
7. How does an antibiotic differ from a disinfectant? From an antiseptic?

For Further Investigation

1. Test different concentrations of the same antiseptic or disinfectant.
2. Test the effects of heavy metals such as copper, lead, or zinc.
3. Check the antiseptics and disinfectants you used to determine the chemicals they contain. Use a reference such as *The Merck Index*, 11th ed. (Rahway, NJ: Merck, 1989) to determine which chemicals are likely to be the active ingredients.

Many of the issues that surround screening for AIDS apply to genetic screening as well and are likely to become of greater concern as our ability to screen for diseases in the general population increases. The aim of this investigation is to provide students with tools by which to assess such issues. Teams of 3 or 4 for the discussion work best.
    Time: One class period.

Part A Discussion

(Possible answers to student discussion questions.)
  1. Group A
     a. Screening protects the nation's blood supply from contamination.
     b. Screening protects the hemophiliacs from HIV antibody in blood products they need during bleeding episodes.

INVESTIGATION 11.3   Screening for AIDS

The public policy issues surrounding the detection, control, and treatment of AIDS are complex and open to debate. This investigation examines several of these issues.

Materials (per team of 3 or 4)
none

PART A Screening Donated Blood and Plasma for HIV Antibody

Procedure

1. For this investigation, you will be working in small discussion groups of 3 or 4 students. Read the following paragraphs before forming your small groups.

    HIV infection has been associated with several well-defined high-risk groups, including recipients of blood and blood products. Screening blood for contamination by the causative agent, HIV, has become one of the highest public health priorities and currently is in place nationwide.

    When your body is infected by a virus, it produces antibodies that help destroy the virus. The blood-screening test currently used on all blood detects HIV antibody, not the virus itself. Any blood that tests positive is tested again, using a different, more specific test. If both tests are positive, the person has been infected with HIV. That does not necessarily mean that the person has AIDS or even that he or she will develop AIDS at a later date. It does mean, however, that these possibilities exist, and that the donated blood from that person should not be transfused into other individuals.

Unfortunately, the current tests sometimes produce *false positive* results. False positive results indicate that a person *has* HIV antibody when in fact the person *does not* have the antibody in his or her blood. False positives are rare, but they do occur. Adverse publicity and personal trauma may result for a donor who receives a false positive report from a blood donation. Fear of AIDS and concern with the screening process could decrease the pool of volunteer blood donors.

For the person who is truly positive for HIV antibody, notification and elimination from the donor pool is, indeed, appropriate. However, for donors falsely labeled as antibody positive, these results may do needless harm to their social and personal relationships.

2. If your small group is designated group A, your task is to provide arguments in favor of the nationwide screening program for HIV antibody in blood banks. If you are group B, your task is to provide arguments against the nationwide screening program for HIV antibody in blood banks.

Discussion

1. Group A: During your discussion to develop arguments *for* the nationwide screening program, you might wish to consider the following questions:
    a. Of what value is the screening program for the general public?
    b. How do hemophiliacs, who may require frequent blood-cell products, benefit from the program?
    c. Of what value is the screening to blood recipients in general?
    d. Of what value is screening to an apparently healthy donor infected with HIV?
    e. What rights do blood donors have?
    f. What rights do blood recipients have?

2. Group B: During your discussion to develop arguments against the nationwide screening program, you might wish to consider the following questions:
    a. What problems might arise from screening for HIV and getting a false negative test, results that indicate a person *has not* been infected with HIV when in fact he or she *has* been infected?
    b. What problems might arise from a false positive result? Consider employment and medical insurance, for example.
    c. How might a false positive test for HIV antibody affect the life of that individual? Consider interpersonal relationships, friends, and family.
    d. In the use of screening tests, there is concern both for the public at large and for the individual. What types of conflicts might arise as a result of these concerns? How might society reach some compromise?

PART B Coping with a Positive Result

Procedure

3. Remain in your groups and read the following scenario:

Jean and Roger were married for three years before they decided to begin a family. When Jean's doctor determined that she was pregnant, she ordered an ultrasound to confirm the gestational age of the fetus. The ultrasound revealed that Jean was not carrying one child, but twins. The delivery was uncomplicated and both children, one girl and one boy, were quite healthy.

Two months later, Roger's boss was admitted to the hospital for surgery, and the employees decided to donate blood to the blood bank to help defray the costs of the surgery. All donated blood was screened for HIV antibody. Roger's test was positive, and the hospital performed the second screening test. That test also was positive, indicating that Roger had been infected with HIV.

c. **Screening helps ensure that the blood received does not contain HIV.**
d. **Possible identification of an antibody positive donor without symptoms of AIDS can result in changes in the behavior of that individual, which might prevent spread of the disease.**
e. **Donors deserve privacy and protection so that a positive result for HIV antibody does not lead to a personal disaster such as the loss of one's job due to the ignorance of an employer.**
f. **Recipients have a right to expect the blood they receive have been screened for HIV and other pathogens.**

2. **Group B**
    a. **This person might be HIV positive. If a unit of his or her blood or plasma is used in a transfusion, there is a risk of transmitting HIV infection.**
    b. **The individual may face discrimination in terms of medical or life insurance as well as in employment. There is a social stigma attached to a person so identified. This person might not be infected with HIV, and he or she might not even have been exposed to it.**
    c. **The individual's professional, social, and family life may be placed in a state of turmoil. Personal fear and depression might result for the individual who is falsely identified.**
    d. **When the public health is at stake, concerns for the welfare of the general population usually have superseded concerns for the individual. The conflict raises many complex questions about the rights of the individual and the duty to protect the public from harm.**

Summary

Blood recipients have a right to expect that all blood they receive has been tested by the most technologically advanced methods in order to minimize the risk of transmission of infectious disease agents. At the same time, donors have a right to expect consideration and protection so that their life-giving gift does not lead to a personal disaster. How can society reach a middle ground?

The specific issue of HIV screening may be resolved in time. Students should see screening as an example of the general problem of wholesale public health measures. No screening test will be perfect, whether it is for HIV, cystic fibrosis, or predisposition to heart disease. False positives and false negatives will occur.

Part B Discussion

1. **Problems include informing his wife and determining if his wife and children have been exposed to HIV. AIDS may or may not develop.**

2. **Jean and the children will have to be tested. If the results are positive, then the entire family will face the possibility that one or more family members may develop AIDS. Further planning and preparation for events depend on test results. Have students examine what some of this planning entails, such as wills, provisions for the care of the children should they outlive their parents, or coping with the possible loss of both children.**

3. **Student responses will vary, but answers should be based on scientific knowledge of AIDS transmission, not only emotion.**

4. **If both Roger and Jean test positive, then either person could be responsible for infecting the other. (Male–female transmission is more effective than female–transmission, however.) In this case, they both should be concerned about any sexual relationships they have had within the past several years. All sexual partners during this time should be alerted. Students should consider the implications of not alerting these people.**

5. **Answers will vary.**

6. **Answers will vary.**

Discussion

1. What problems does the positive result pose for Roger?
2. What might this result mean for his wife and children?
3. Do Roger's coworkers and his boss have a right to know about the test results? Why or why not?
4. Using what you know about HIV and AIDS, should Roger and Jean be concerned about any relationships they may have had before they were married? What should they do?
5. Assume that one child tests positive for HIV antibody and the other child tests negative. What rights does each child have?
6. If the children live to attend school, what rights do the other children have?

SUMMARY

Prokaryotes are the most ancient, most widespread, and most metabolically diverse of any organisms. They also are the most important ecologically. They are major decomposers, indispensable members of all food webs, and essential to the cycling of matter. They include two distinctly different groups—archaebacteria and eubacteria. Both groups lack a nucleus and other membrane-enclosed organelles. Archaebacteria have distinctive cell walls and other characteristics that distinguish them from eubacteria. Apparently they are as closely related to the eukaryotes as to the eubacteria. A large percentage of archaebacteria inhabit extreme environments like those that may have been typical on the early earth—temperatures near or above boiling, brine ponds saturated with salt, waters with very high or very low pH, and the extreme pressures of the ocean depths. Among the archaebacteria are the methanogens, which can produce methane gas from organic matter such as sewage, garbage, agricultural waste, manure, and other waste materials.

Among the eubacteria are many photosynthetic types, some of which are anaerobic, using substances other than water as a source of hydrogen. The cyanobacteria are aerobic photosynthesizes that most likely were responsible for the initial buildup of oxygen gas in the early atmosphere. Numerous types of eubacteria are involved in the nitrogen cycle.

Disease results from the interaction between a host and a pathogen. Pathogenic eubacteria are responsible for many infectious diseases, including whooping cough, typhoid, eubacterial pneumonia, syphilis, and numerous plant diseases. Viruses, which have a simple structure of DNA or RNA with a protein coat, also cause many plant and animal diseases, including tobacco mosaic, colds, influenza, chickenpox, and AIDS. Although AIDS continues to be a deadly disease, prospects for treatment are improving.

APPLICATIONS

1. If the first organisms on earth were heterotrophic prokaryotes, from where did they get their food if there were no other living things?

2. After an atomic blast that wipes out all living things in a local area, which type of microorganism might you expect to colonize the blast site first? Why?

3. The adaptations of archaebacteria to extreme modern-day environments suggest they were among the earliest organisms on earth. Explain this statement.

4. In what way are microorganisms economically important to humans? In other words, how do we use microorganisms for our own good?

5. How does the nitrogen cycle illustrate the interdependence of organisms in the biosphere?

6. Will a vaccine for a particular disease help an individual if he or she already has the disease? Explain your answer.

7. What characteristics of sexually transmitted diseases make them particularly difficult to combat?

8. During famines, diarrhea is one of the major killers, especially of infants. Explain why this might be.

9. How does cancer fit into the classification of diseases?

10. What distinguishes viruses from living organisms?

11. HIV does not kill people—the cause of death usually is an opportunistic infection. Explain what is meant by an opportunistic infection and why these occur in AIDS patients.

12. Why might a cure for AIDS be found if scientists could block the activity of reverse transcriptase?

 ## PROBLEMS

1. Both DNA and RNA are used to study the evolutionary relationships of organisms to each other. Compare the techniques. What are the advantages of each?

2. Design a methane generator for a home that uses food waste as an energy source. How would you control the smell of rotting food and prevent contamination? How would you collect the methane gas without allowing it to build up to dangerous levels? How would you provide the appropriate environment for the methanogens?

3. Draw a model of the nitrogen cycle based on the information in your text—try not to refer to your book's illustration. You also may want to refer to the textual material on the nitrogen cycle in a general ecology book, but do not look at the illustrations. After you have drawn your model, compare yours to the illustrations in any book you used. How are the illustrations similar or different? Which do you think is better?

4. Start a small experimental compost pile at your school. What kind of materials should you add to your compost? Which materials decompose and which do not? How fast do materials change and how do they change? What environmental conditions favor the development of a compost pile?

5. Investigate an infectious disease of a plant or animal using the following outline of topics: history, symptoms, pathogen, vector (where applicable), treatment, epidemiology. Suggested diseases: anthrax, bacterial meningitis, Dutch elm disease, filariasis, tsutsugamushi fever, hoof-and-mouth disease, black stem rust of wheat, brucellosis, fire blight of pears, or schistosomiasis.

6. Using the same method as in problem 5, study a noninfectious disease.

7. In recent years epidemics of virus-caused influenza have occurred and spread over many parts of the world. Yet vaccination against influenza viruses is common. Why are the vaccines ineffective against new epidemics? Use your understanding of pathogens, resistance on the part of both host and pathogen, and immunity to answer the question.

8. Scientists are aware that some infectious diseases have not been identified. First cases or infrequent cases may go unrecognized. The first epidemic draws attention to the disease. This was the case with the human disease now called Legionnaire's disease. It is named after the 1976 American Legion convention in Philadelphia at which an epidemic occurred. A number of people died. Consider such an epidemic, its victims, and their autopsies (studies of blood and other body parts after death). If you were attempting to identify the cause of the disease, how would you use people in each of the following careers: biochemists, food inspectors, medical doctors, metallurgists, microbiologists (virologists and bacteriologists), pathologists, and toxicologists? If necessary, do library research on the nature of these professions.

9. In developed countries, infectious diseases are no longer the leading causes of death as they are in many parts of the Third World. Choose one of the most common infectious diseases in the Third World and report on the causative agent, the frequency of the disease, and attempts to control it.

10. In the early 1980s, the World Health Organization reported that smallpox had been eliminated worldwide. Report on the campaign to eradicate this viral disease.

Eukaryotes: Protists and Fungi

This is a red raspberry slime mold. What does this organism look like on first observation? Share your idea with a classmate, and ask for the classmate's impressions.

The red raspberry slime mold is a fungus (*Tubifera ferreginosa*). Students may think it resembles a plant or an organism from a marine environment. It looks different from fleshy fungi such as mushrooms and morels with which students might be more familiar.

◆ Protists and fungi are eukaryotes. They possess organized nuclei and membrane enclosed organelles. Evidence suggests that some of these organelles originated from free-living prokaryotic ancestors.

The kingdom Protista includes a diverse group of organisms that have little in common. Its members are eukaryotes, but not plants, animals, or fungi. Protists may be unicellular, colonial, or multicellular. They may be consumers or producers, and some switch from one form of nutrition to the other, depending on conditions. All protists carry out aerobic cellular respiration in mitochondria. Most have flagella at some stage of their life cycle. Producer protists carry out photosynthesis in chloroplasts.

Organisms in the kingdom Fungi are unicellular or multicellular heterotrophs that absorb molecules from their surroundings. Many fungi are decomposers and play an important role in food chains by making nutrients available to plants. This chapter describes several species of protists and fungi and examines their roles in the larger community. ◆

MAJOR CONCEPTS

- ◆ Protists make up a diverse "catch-all" kingdom of organisms.
- ◆ Algae are photosynthetic protists that are major producers in the biosphere.
- ◆ Protozoa are heterotrophic protists that are singled celled and motile.
- ◆ Slime molds are multicellular heterotrophic protist and important decomposers.
- ◆ Fungi comprise a kingdom of multicellular heterotrophic organisms that absorb nutrients from their environment.
- ◆ Fungi are major decomposers in the biosphere.

KNOWLEDGE CHECK

- ◆ What distinguishes protists from other eukaryotes?
- ◆ What ecological roles do protists play?
- ◆ How do fungi differ from other eukaryotes?
- ◆ What major ecological roles do fungi play?

Autotrophic Protists

12.1 Algae Are Photosynthetic Protists

Algae, perhaps the most familiar protists, are aquatic and have relatively simple structures. They can be found just about anywhere you find tiny droplets of water—in the air, on tree trunks and branches, in the bottoms of streams, on soil particles, and on rocks at the seashore. They may be unicellular, colonial, or multicellular (see Figure 12.1). Multicellular species may form either long filaments or thin plates of similar cells or plant like structures that have division of labor.

Although all algae are photosynthetic, not all of them appear green. Both unicellular and colonial algae float near the surface of bodies of water, using energy from sunlight and carbon dioxide from the water to carry out photosynthesis. Multicellular algae, on the other hand, are adapted to living along shores and in shallow water, where nutrients are generally abundant but living conditions are difficult. These algae are subject to periodic drying and to variations in temperature, light, and other factors. As a result, different groups of algae have become adapted to particular niches.

The term *algae* is one of convenience. It is no longer used in formal classification because in recent years biologists have found differences in the biochemical characteristics of the various groups. Algae vary greatly in the compounds that make up their cell walls, in their food storage com-

Guidepost
What characteristics distinguish the various groups of algae?

A variety of freshwater algae can be maintained in classroom aquariums. Ask students to scrape the sides of an aquarium to collect and examine under a microscope the diatoms usually found there. Examples of marine algae mounted on herbarium sheets are available from Ward's.

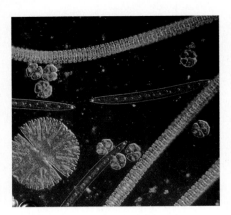

Figure 12.1 Mixed algae: *Micrasterias,* large two-lobed alga; *Cosmarium,* small two-lobed alga; *Closterium,* oblong alga with conspicuous vacuoles; and *Dismidium* filamentous alga (X100).

pounds, and in the pigments present in their cells. Like the cell walls of plants, the cell walls of most algae contain cellulose, but they may contain a variety of other compounds as well. Some algae store food as starch, but others store it as carbohydrates that are unique to algae.

In addition to chlorophyll, algae may contain various other pigments. Often, these pigments mask the bright green of the chlorophyll and produce the characteristic colors for which some algae are named. The variety of pigments suggests that the various groups may have evolved from symbiotic relationships with different photosynthetic prokaryotes.

12.2 Green Algae May Have Been the Ancestors of Plants

The majority of green algae (phylum Chlorophyta; kloh ROF it uh) live in fresh water. You can find them in your aquarium at home or on the shady side of trees. Many green algae are unicellular and microscopic, but they may be so abundant that they color the water of ponds and lakes green or form huge, floating mats. Some marine species form large, multicellular seaweeds. Figure 12.2 shows a few members of this large group.

Green algae reproduce both asexually and sexually. Some sexually reproducing species produce flagellated gametes that look identical (see Figure 12.3). Other species produce flagellated sperm and larger, nonmotile ova that are retained in the parent algae. Many of the multicellular green algae have complex life cycles, similar to those in plants, during which they produce both gametes and spores.

Biologists think plants originated from green algae. Several types of evidence support this hypothesis. Green algae, but not other algae, contain the same combination of photosynthetic pigments as plants do. Both store reserve foods as starch, and both have cellulose-containing cell walls. Like plants, several species of green algae have developed true multicellularity with some division of labor. Other species show intermediate steps between unicellularity and multicellularity. Finally, many species of green algae have patterns of reproduction and life cycles in common with plants.

12.3 Diatoms Are Important Producers

Diatoms (phylum Bacillariophyta; buh SIL er ee OFF ih tuh) are some of the most intricately patterned creatures on earth, as shown in Figure 12.4. Their cell walls are formed by a double shell, the two halves of which fit together like a box. The shells are made of organic materials impregnated with silica. Electron microscope studies have shown that the patterns on

Figure 12.2 Green algae; *Acetabularia,* a large one-celled alga (X1); (b) water net, *Hydrodictyon reticulatum,* a filamentous form (X100); and (c) sea lettuce, *Ulva,* a multicellular alga (X1).

a
b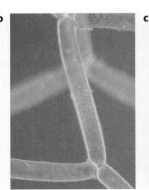
c

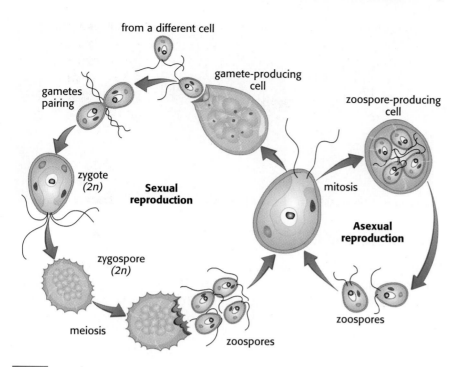

Figure 12.3 Life cycle of *Chlamydomonas*, a unicellular green alga, showing both sexual and asexual reproduction. Notice that the zygote is the only diploid stage in the life cycle.

2 the shells are created by the pores that connect the inside of the cell with the outside environment. Diatoms are microscopic, occurring as single cells, simple filaments, or colonies; some move by gliding. Abundant in both fresh and marine waters, they are the principal producers in many food webs and provide much of the world's oxygen. Diatoms also grow in damp places on land or even on moist flowerpots in greenhouses. Usually brownish in color, diatoms contain various yellow pigments that mask their chlorophyll. They store food in the form of oils, and they reproduce either asexually by cell division, or else sexually by producing gametes. The zygote resulting from fertilization grows to the full size for the particular species of diatom and undergoes cell division. The offspring cells then produce new silica containing shells.

Silica shells remain intact for many years, even after the organisms inside have died. Great masses of shells that have slowly accumulated through many years (diatoms date back to the Cretaceous period, 144 million years ago) now are mined as diatomaceous earth. That material is used many ways—for example, as a filter (both for beer and for swimming pools), as an abrasive in silver polish, and as a brightener for the paint used to mark highways.

Figure 12.4 Diatoms: (a) Strained mixture photographed through a light microscope (X200); and (b) scanning electron micrograph showing the double silicon shell and the pattern of pores (X4000).

Figure 12.5   (a) A brown alga, *Macrocystis pynfera*, showing air bladders that help keep it afloat; (b) a brown alga, *Postelsia*, showing hold fast, stipe and blade; (c) a coralline red alga, *Bossiella*.

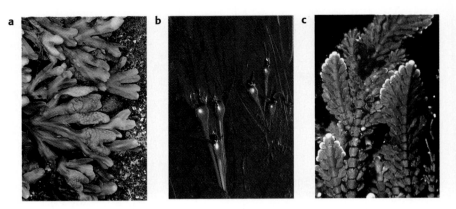

12.4   Brown and Red Algae Are Multicellular

The major seaweeds of the world are the multicellular brown and red algae. The brown algae (phylum Phaeophyta; fay OFF it uh) inhabit rocky shores in the cooler regions. Kelps, for example, form extensive offshore beds, where they are the primary producers for many communities of animals and microorganisms. Brown algae have cellulose in their cell walls. They store their food as a unique carbohydrate or as oil, rather than as starch, which is present in most other algae. Other pigments usually mask the chlorophyll in the brown algae.

> Brown algae store food as the compounds mannitol (an alcohol) and laminarin (a polysaccharide).

Brown algae (Figure 12.5a, b) may be very large, sometimes reaching as much as 100 m in length. Some are unspecialized, but others are differentiated into areas for anchorage (hold fast), support (stipe), and photosynthesis (blade), as seen in Figure 12.5b. Some kelps have conducting cells in the center of the stipe that resemble the sugar conducting cells of plant stems.

Most seaweeds of warm oceans of the world are red algae (phylum Rhodophyta; roh DOF it uh). They generally grow attached to rocks or other algae (Figure 12.5c). Because they absorb blue light, which penetrates farther into the water than other light wavelengths, red algae can grow at greater depths in the ocean than any other algae. Biologists think the chloroplasts of red algae may have evolved from ancestral cyanobacteria because the two groups contain similar unique pigments. The cell walls of red algae contain an inner layer of cellulose and an outer layer of other carbohydrates such as agar, used for culturing bacteria in the laboratory.   4

> Call attention here to the use of chemical characteristics in classification of this and other algae.

In many parts of the world, humans eat red and brown algae as part of their regular diet. Although algae do not provide many kcals, they do contain some vitamins and trace elements necessary to human health. Red and brown algae produce compounds used in the manufacture of food and goods that require a smooth texture. When you eat marshmallows, ice cream, or certain cheeses, and when you use hand lotion, lipstick, or paint, you probably are using products derived from algae.   3

CONCEPT REVIEW

1. What characteristics of green algae make them probable ancestors of all plants?
2. What makes diatomaceous earth useful, and how is it used?
3. How are brown and red algae important to humans?
4. Which algae group can live in deeper water than any other? Why?

12.5   Flagellates May Be Consumers or Producers

1 Protozoa are animal-like protists that are mostly heterotrophic and unicellular. Some are heterotrophic and others are autotrophic. Most flagellates are unicellular organisms that move by means of flagella. Some have only one flagellum; others have thousands. Flagellates are abundant in damp soil, fresh water, and the ocean. Most species reproduce asexually by dividing in two, but a few reproduce sexually by fusion of identical-appearing gametes. Among the several phyla of flagellates, some contain chlorophyll and synthesize food when light is present. Others are heterotrophic and capture smaller microorganisms, which they digest internally. Some flagellates can be either consumers or producers, depending on conditions.

Euglenoids are a small group of flagellates similar to the organisms in the genus *Euglena*, which are commonly found in fresh water (see Figure 12.6a). Many species contain chloroplasts and therefore manufacture their own food. The nonphotosynthetic forms absorb dissolved organic substances or ingest living prey. Most species orient to light by means of a light sensitive eyespot. Euglenoids lack rigid cell walls but have supporting strips of protein inside their plasma membranes.

Dinoflagellates, in contrast, are marine flagellates armored in stiff cellulose walls. Two beating flagella cause the cells to spin like tops as they move through the water (Figure 12.6b). Many dinoflagellates are red in color, and some produce powerful toxins that affect the human nervous system. If these reproduce in great numbers, they cause what are known as "red tides." During red tides, the toxins may become concentrated in food

2 chains. Clams and other organisms that eat dinoflagellates may become poisonous to humans at that time.

Many flagellates live in symbiotic relationships with other organisms. For example, the parasitic flagellate *Trypanosoma gambiense* (Figure 12.6c), causes African sleeping sickness in humans. Symbiotic flagellates live in the digestive tract of termites, where they break down the wood eaten by the insects. Without the flagellates the termites would starve to death because termites cannot digest cellulose any more than humans can.

Ask students what advantage the flagellates may gain. They probably benefit from the protective environment of the termite gut and the supply of minced wood particles.

Figure 12.6   Flagellates: (a) *Euglena* (x150); (b) a dinoflagellate, *Gymnodinium* (X12 000); (c) *Trypanosoma gambiense,* one of the parasites that causes African sleeping sickness in humans.

a
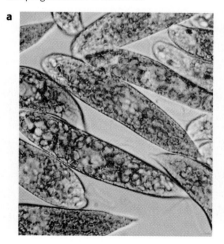

b
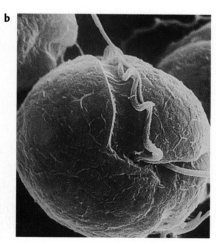

c
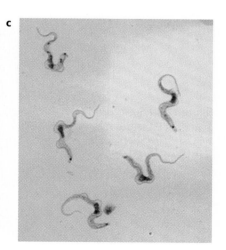

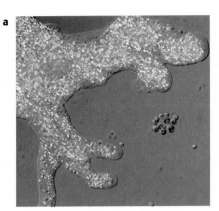

Figure 12.7 An amoeba (a) is a sarcodine that surrounds its prey with pseudopods and engulfs them (X180). Foraminiferans (b) build shells of calcium carbonate (X200).

If you have a microscope with an oil immersion lens, slides of human blood with one of the stages of malaria can be used as a demonstration.

About 100 million to 200 million cases of malaria are estimated to occur worldwide each year. The death rate is at least 10%.

12.6 Many Sarcodines Use Pseudopods

Amoebas and their relatives are protozoan protists called sarcodines. The sarcodines are thought to have evolved from flagellates, and many develop flagella during some stage of their life cycle. They reproduce asexually by cell division. Some amoebas live in ponds, puddles, and damp soil, whereas others live within larger organisms in symbiotic relationships that may be parasitic. The organism that causes amoebic dysentery in humans is one example of a sarcodine parasite.

The cytoplasm of a sarcodine is a granular, grayish or transparent mass that has no definite shape. In contrast to the fast-moving flagellates, sarcodines ooze or float about. They move and feed by means of pseudopods **3** (SOO doh podz), finger like extensions of cytoplasm. The cytoplasm continuously flows into one or several pseudopods, and the organism flows in the direction of whichever pseudopod is longest. Pseudopods also surround and engulf bacteria, protists, algae, or other aquatic organisms that the sarcodines use as food, as shown in Figure 12.7.

Most shell-bearing sarcodines live in the sea. Radiolarians, for example, have an outer shell of silica, a sandlike substance. Many long, stiff pseudopods radiate from a radiolarian's shell. The foraminiferan (FOR am ih NIF er un) shown in Figure 12.7b builds a shell of calcium carbonate which is the same material clam shells are made of. During millions of years, great numbers of foraminiferan shells have accumulated at the bottom of seas and gradually have solidified into rock. Chalky deposits, such as the white cliffs of Dover in England, result from these accumulations of shells.

12.7 Sporozoans Are All Parasites of Animals

Sporozoans (phylum Apicomplexa; ap ih kum PLEK suh) are spore-forming protozoa that are parasites of animals. Unlike the sarcodines or flagellates, these protozoa lack any means of self-propulsion. Their complex life cycles involve more than one host. Sporozoans reproduce sexually, and **4** their zygotes form thick-walled cysts that can be passed to new hosts through infected blood. Inside the cysts, the zygotes divide rapidly, forming large numbers of tiny spores. When the cysts break open, each one of the spores can infect a cell within a new host. **6**

In the human bloodstream, sporozoans like those shown in Figure 12.8 cause serious and even fatal diseases such as malaria. In the past, malaria has been controlled by drugs that inactivate the parasitic sporozoan and pesticides that kill the mosquitoes that carry infected blood from one host to another. Unfortunately, both sporozoans and mosquitoes have developed resistance to the chemicals used to control them. Today, malaria remains a principal cause of human death in the world.

Figure 12.8 Sporozoans: (a) the organism that causes malaria, *Plasmodiium vivax*, showing several different stages of the life cycle in a blood smear; and (b) *Gregarina,* a gregarine (X75).

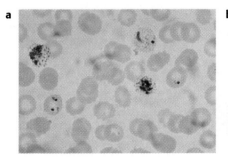

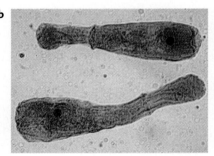

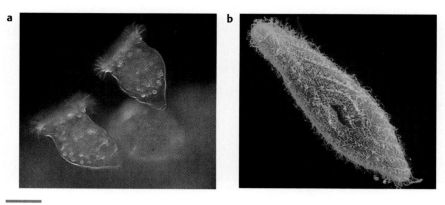

Figure 12.9 Ciliates: (a) *Vorticella campanula* (X125); (b) cilia of *Paramecium caudatum.* Scanning electron micrograph (X1500).

12.8   Ciliates Have Two Types of Nuclei

The most specialized and complex of the protozoan protists are ciliates (phylum Ciliophora; sil ee OFF er uh). Most are unicellular, free-living organisms found in marine or freshwater habitats. They have definite, semi-rigid shapes, as shown in Figure 12.9a. Active swimmers such as *Paramecium* have distinct anterior and posterior ends. All ciliates move by means of **cilia** (SIL ee uh)—short, whiplike extensions that beat in rhythm, driving the organism through the water (see Figure 12.9b). In some species, the beating of cilia causes a current that sweeps food particles into the organism. In others, bundles or sheets of cilia function as mouths, paddles, teeth, or feet.

Ciliates have two types of nuclei, macronuclei and micronuclei. The macronuclei are required for growth, reproduction, and cellular functions. Macronuclei do not divide by mitosis. Instead, when the cells divide, the macronuclei are distributed between the offspring. Micronuclei are exchanged by two ciliates during conjugation (see Figure 12.10).

12.9   Slime Molds Resemble Both Fungi and Protozoa

Slime molds (Myxomycota; mik soh my KOH tuh) are fungus like heterotrophs that usually grow among damp, decaying plant material. If you search through such material soon after a heavy rain, you may find a slime-mold plasmodium (plaz MOHD ee um)—a glistening sheet or net with no definite form. The plasmodium may be more than one meter in width and brightly colored. As it moves from one place to another, the plasmodium engulfs bacteria, spores, and small bits of dead organic matter.

As long as there is food and water, a plasmodium continues to grow. When food becomes depleted, the plasmodium may clump into mounds and form **sporangia** (spoh RAN jee uh)—stalked structures that produce reproductive cells or spores (see Figure 12.11a). When a sporangium breaks open, a large number of spores are released and carried away by wind, insects, or raindrops. The spores can survive unfavorable environmental conditions, but whenever they land in suitable spots with enough moisture, they can germinate. Each spore gives rise to a tiny organism with one or two flagella. After swimming about for a short time, these organisms lose their flagella and fuse in pairs. By feeding and growing, each pair develops into a new plasmodium. Figure 12.11b shows the life cycle of a slime mold.

5

7

6

Students might find it interesting to calculate the swimming speed of a ciliate. Ask them what the speed in meters per hour would be if a ciliate requires one second to swim across the low-power field of a microscope. The diameter of the low-power field (100X) of a compound microscope is about 1.4 mm. There are 3600 seconds in an hour. 3600 x 1.4 = 5040 mm = 5.04 m. The rate of movement is approximately 5 m per hour.

Here the word *plasmodium*, uncapitalized, is used to refer in a general way to slime molds. Make the distinction between the genus *Plasmodium* and this general reference.

Figure 12.10 Conjugating *Paramecium multimicronucleatum* showing the macronuclei (X100).

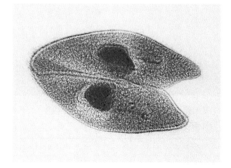

a

b

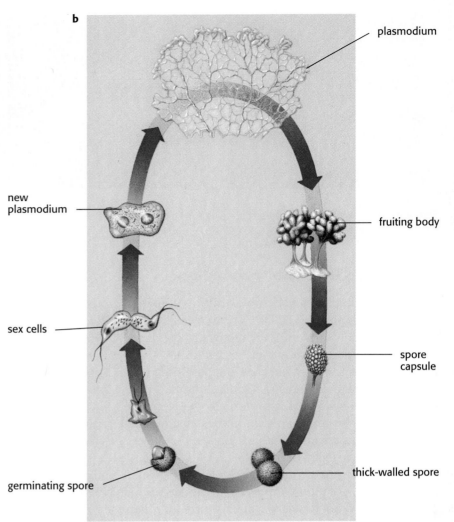

plasmodium

new plasmodium

fruiting body

sex cells

spore capsule

thick-walled spore

germinating spore

Figure 12.11 (a) Slime mold sporangium, *Hemitrichia clovata* (X60). The plasmodium stage of *Physarum polycephalum* also is visible. (b) Life cycle of a slime mold. Which part of the cycle is best adapted to a dry environment? Which parts of the cycle would require abundant moisture?
The spores are best adapted to a dry environment, and the sex cells, plasmodium, and fruting body stages would require more moisture.

CONCEPT REVIEW

1. Are flagellates algae or protozoa? Explain.
2. What roles do flagellates play in the biosphere?
3. Describe how a sarcodine moves and obtains food.
4. How do sporozoans differ from other protozoa?
5. In what way are ciliates considered complex?
6. How do spores of slime molds and sporozoans differ?
7. Which characteristics of slime molds are animal-like and which are fungus like?

Fungi

12.10   Fungi Are Important Decomposers

Fungi are heterotrophic eukaryotes. Most forms are non-motile. They include mushrooms, molds, mildews, rusts, smuts, and many other, less familiar, organisms. Many fungi cause diseases, especially plant diseases, but many more form essential symbiotic relationships with plants. Others are used for baking breads, making alcoholic beverages, and producing drugs. Fungi reproduce by means of spores that germinate and grow slender tubes called **hyphae** (HY fee). A large, tangled mass of hyphae called

Guidepost

What are the characteristics of the fungi?

Biology Today

Models of Eukaryotic Origins

Two of the major ideas about the origin of eukaryotic cells are known as the *autogenesis hypothesis* and the *endosymbiosis hypothesis*. According to the autogenesis model, eukaryotic cells evolved through the specialization of internal membranes that originally were part of the plasma membrane of a prokaryote. The internal membrane system, including the nuclear envelope, Golgi apparatus, endoplasmic reticulum, and other organelles enclosed by a single membrane may have resulted from the plasma membrane folding in on itself. Mitochondria and chloroplasts may have developed their double membranes through more complex foldings.

According to the endosymbiosis model, the ancestors of eukaryotic cells were symbiotic prokaryotic cells, with certain species living within larger prokaryotes. One modern example of endosymbiosis is found in the digestive cells of *Hydra viridis*, which contain algal cells of the genus *Chlorella*. These digestive cells recognize the appropriate algae and eject any others. Accepted algae are moved to the base of the digestive cell where they reproduce by mitosis until the normal algal population is reached. Thereafter the algae reproduce when the host cell reproduces. The presence of the algae enables the hydra to survive when food is scarce.

Many biologists think mitochondria originated as symbiotic aerobic bacteria. The bacteria that became mitochondria were engulfed by ancestral eukaryotic cells. Before these cells acquired the bacteria, they lacked the enzymes that would allow them to use oxygen in respiration. The engulfed bacteria, however, were able to perform these reactions, and they became the inner part of mitochondria. All the mitochondria within a eukaryotic cell arise from the division of the existing mitochondria, just as new bacteria are produced through cell division. Mitochondria divide by simple fission, splitting into two, and they apparently replicate and divide their circular DNA molecule as do bacteria. Mitochondrial replication, however, is impossible without a cell nucleus because most of the genes required for mitochondrial division are located in the nucleus and translated into proteins by ribosomes in the cell's cytosol. Moreover, mitochondria cannot be grown in a cell-free culture.

During the time that mitochondria have existed as endosymbionts in eukaryotic cells, most of their genes have been transferred to the chromosomes of the host cell. Mitochondria, however, still have genes of their own contained in a circular molecule of DNA, similar to that of bacteria. This mitochondrial DNA contains several genes that produce some of the proteins essential for the functioning of mitochondria in aerobic cellular respiration.

Symbiotic events similar to those proposed for the origin of mitochondria could explain the origin of chloroplasts, which are characteristic of photosynthetic eukaryotes. Some ancestral eukaryotic cells may have ingested photosynthetic bacteria, which evolved into chloroplasts. The incorporation of photosynthetic bacteria would have conferred one great advantage on the cell: It then could manufacture its own food. Like mitochondria, chloroplasts have their own circular DNA, but many of the genes that code for necessary chloroplast components are located in the nucleus.

It is possible that the evolution of eukaryotic cells required both autogenesis and endosymbiosis. Endosymbiosis may be responsible for mitochondria and chloroplasts, and autogenesis may be responsible for other organelles.

A hydra, *Chlorahyda virdiscima,* with buds.

a **mycelium** (my SEE lee um), shown in Figure 12.12, makes up the body of a fungus. Each fungal spore is capable of producing an entire mycelium. A mycelium will grow on virtually anything from which the fungus can obtain food. It grows especially well in moist soil that contains large amounts of dead plant and animal matter. Sometimes growth can be very rapid: A single fungus can produce a kilometer of hyphae in just 24 hours. The cell walls of fungi contain a carbohydrate known as **chitin** (KYT in). In some cell walls, the chitin is combined with cellulose, the same material found in plant cell walls. Chitinous walls are hard and resistant to water loss, which allows fungi to live in some extreme environments.

Along with bacteria, fungi are the most important decomposers in the biosphere. Digestion begins outside a fungus. It secretes digestive enzymes that break down food into small organic molecules. After the food is broken down, the fungus absorbs the small molecules. In this way, fungi decompose dead organisms.

Fungi reproduce both asexually and sexually. In asexual reproduction, the reproductive part of the fungus may appear above ground. A mushroom is an example. It is formed of tightly packed hyphae and may produce a million or more spores. Most spores are surrounded by tough, resistant walls that enable them to survive drying and extreme temperatures. Spores usually are dispersed by wind and air currents. Because their small size keeps them suspended in air for a long time, fungal spores are widely distributed. Fungi also may reproduce sexually by conjugation. In these fungi, the hyphae of different mating types come together and fuse. Because the hyphae of most fungi look very much alike, classification of fungi is based on differences among their sexual reproductive structures. The next sections describe some of the major divisions of fungi. (Phyla are referred to as divisions in fungi and plants.)

12.11   Conjugating Fungi Can Destroy Foods

The common black bread mold, *Rhizopus stolonifer*, is probably the most familiar example of conjugating fungi (division Zygomycota; ZY goh my KOH tuh). Without preservatives or refrigeration, bread will develop a black, fuzzy growth like that in Figure 12.13 in a few days. The growth begins as a spore that germinates on the surface of the bread, producing hyphae. Then, some of the hyphae push up into the air and form sporangia. Mature sporangia are black, giving the fungus its color. The sporangia eventually break open, releasing many spores, each of which can germinate and spread the mold.

None of the common names applied to fungi, such as mold, mildew, or mushrooms, have any taxonomic significance.

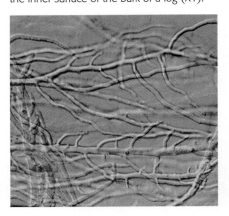

Figure 12.12 Fungal mycelium growing on the inner surface of the bark of a log (X1).

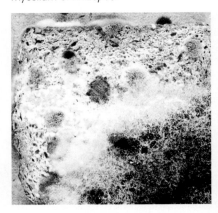

Figure 12.13 Bread covered with mycelium of *Rhizopus.*

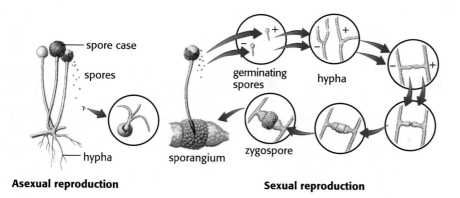

Asexual reproduction   Sexual reproduction

Figure 12.14   Asexual and sexual reproduction in *Rhizopus.*

Sexual reproduction in *Rhizopus* occurs when nuclei from the hyphae of two different mating types fuse (see Figure 12.14). The resulting zygotes develop into thick-walled zygospores (ZY goh sporz), structures that give this division its name. Zygospores can survive periods of extreme temperature or dryness. When they eventually germinate, they give rise to sporangia that release spores and begin the cycle again.

4 Black bread mold can appear on many foods, including fruits such as grapes, plums, and strawberries. There also are parasitic species in this division, some of which attack crop plants. Most conjugating fungi, however, live in the soil and decompose dead plant and animal material.

12.12   Sac Fungi Have a Special Reproductive Structure

The sac fungi (division Ascomycota; AS koh my KOH tuh) include yeasts, powdery mildews, and many common blue-green molds, as well as
4 morels and truffles. Some sac fungi cause diseases such as chestnut blight and Dutch elm disease, both of which have destroyed many trees in the United States. Sac fungi are distinguished from other fungi by a microscopic sexual reproductive structure called an ascus (AS kus), which gives this division its name. Figure 12.15 shows several examples of sac fungi.

When sac fungi reproduce sexually, two hyphae conjugate and form an ascus, illustrated in Figure 12.16. Inside the ascus, nuclei from the hyphae fuse and become ascospores. The ascospores eventually are released from the ascus and may travel long distances on wind currents. Each ascospore can germinate and produce hyphae. In asexual reproduction, an individual hypha segments into huge numbers of spores that are dispersed by wind, water, or animals.

Figure 12.15   Sac fungi: (a) leaves covered with a powdery mildew; (b) a morel, *Morchella augusticeps;* and (c) red cup fungus, *Peziza.*

fruiting body
of sac
fungus

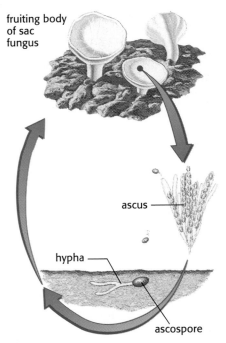

ascus —

hypha —

ascospore

Figure 12.16 Sexual reproduction in a sac fungus.

Yeasts are single-celled ascomycetes. They are small, oval cells (see Figure 2.15) that reproduce asexually by budding and sexually by the production of ascospores. Yeasts are essential to the production of breads and alcoholic beverages, such as beer and wine. Under anaerobic conditions, yeasts convert the sugars found in bread dough or fruit juices into carbon dioxide and ethyl alcohol. The carbon dioxide bubbles cause bread dough to rise and beer to foam. When bread dough is baked, the alcohol evaporates. **4**

12.13 The Reproductive Structures of Some Club Fungi Are Edible

Of all the fungi described here, club fungi (division Basidiomycota, buh SID ee oh my KOH tuh) are probably the most familiar. The mushrooms you buy in a grocery store are the sexual reproductive structures of club fungi. The division also includes smuts, rusts, jelly fungi, puffballs, and stinkhorns (see Figure 12.17). **4**

Most species of mushrooms, both edible and poisonous, are club fungi. A mushroom is a spore-producing body composed of masses of tightly packed hyphae. Under a mushroom cap, you can see structures called gills radiating out from the center. Between these gills many spores are produced on tiny clubs, or basidia (buh SID ee uh), which give this division its name (see Figure 12.18).

A toadstool is a poisonous mushroom. One of the most poisonous is the "death cap," *Amanita*. One bite of certain species of *Amanita* can be fatal. Some species of toxic mushrooms have hallucinogenic effects and sometimes are eaten deliberately. Other club fungi are plant parasites, such as the organisms causing wheat rust and corn smut.

12.14 Imperfect Fungi Have No Known Sexual Reproductive Structures

The classification of fungi is based principally on their sexual reproductive structures. Some fungi are unclassifiable because they seldom, if ever, show reproductive structures. Biologists have investigated certain fungi for years without discovering sexually produced spores. Some of these fungi, however, do resemble species whose sexual reproductive structures are known, and they can be classified accordingly. For example, if a fungus produces a large number of asexual spores, it can be classified on the basis of similarities in sporangia. Once the sexual structures of such fungi are discovered, they are reclassified on that basis. **5**

Figure 12.17 Club fungi: (a) puffballs, *Lycoperdon perlatum;* (b) bracket fungus from an Australian rain forest; (c) the poisonous *Amanita muscaria.*

a

b
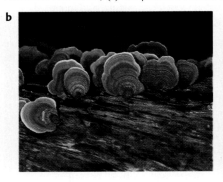

c

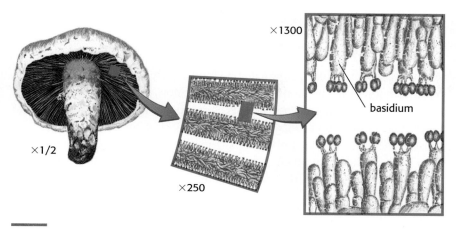

basidium

It is difficult even for mycologists to tell the poisonous from the edible mushrooms, but certain key characteristics are used to classify poisonous mushrooms. *Amanita phalloides*, one of the most deadly, is recognized by its white spores, a ring below the cap, a red yellow cap, and the torn remains of a veil on the top of the cap. Warn students not to pick or eat any wild mushrooms without advice from an expert.

Figure 12.18   Reproductive structures of a field mushroom.

If no sexual reproductive structures ever have been discovered for a fungus, it is called "imperfect." For convenience only, taxonomists place imperfect fungi in a separate group. Figure 12.19 shows two examples. Some imperfect fungi are important in making certain cheeses and antibiotics. For example, *Penicillium roquefortii* is responsible for the distinctive flavor and appearance of Roquefort cheese, *P. camembertii* is used to manufacture Camembert cheese, and *P. notatum* produces the drug penicillin. Other imperfect fungi are parasites of crop plants and animals. For example, athlete's foot is caused by an organism that for years was thought to be an imperfect fungus. Recently, sexual spores have been discovered, showing athlete's foot fungus (and a number of related skin disease fungi) to be a sac fungus.

Penicillium sometimes is listed as a sac fungus, but this classification is not approved by mycologists. No perfect forms are known for the majority of the species included in the "form genus" *Penicillium*, so they should be called imperfect fungi.

A good student report on Fleming's discovery of penicillin can do three things at this time: (1) relieve steady attention to taxonomy, (2) remind students of the role of serendipity in scientific research—with due attention to the principle that chance favors the prepared mind—and (3) remind students of the complexities of ecological relationships.

CONCEPT REVIEW

1. How are hyphae related to a mycelium?
2. What adaptations to a terrestrial environment do fungi show?
3. What characteristics are used to distinguish fungi from each other?
4. How are fungi important to humans?
5. What is meant by the term *imperfect fungi?*

a
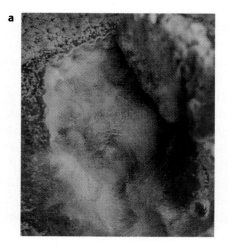

b
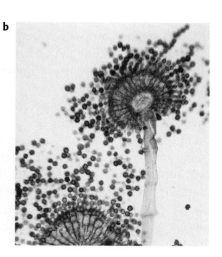

Figure 12.19   Imperfect fungi: (a) a species of *Penicillum* growing on an orange; and (b) asexual reproductive structures (conidiospores) of *Aspergillus* (X90).

Fungi in a Community

12.15   Fungi Support Many Food Chains

A soil ecosystem is like a city in that its food supply comes from the outside. Most soil organisms are consumers and live below the surface of the soil. In soil, food made during photosynthesis comes down into plant roots from the leaves and stems. Plant roots can serve as food for some consumers, but for most of them, the food supply comes from the remains of other organisms. Therefore, decomposers such as fungi and bacteria are very important in soil communities. Figure 12.20 shows a decomposer food chain.

A dead root in the soil still contains a large amount of starch; likewise, a dead animal still contains fats and proteins. Beetles and other small animals living on or just below the soil surface use these substances as food. Other substances, such as cellulose in plant bodies and chitin in insect bodies, can be broken down only by microorganisms and fungi. Decomposers that use cellulose and chitin as food leave simple organic substances as waste products. The wastes still contain energy that other fungi and bacteria can use. These fungi and bacteria, in turn, leave simple inorganic substances in the soil, which still other decomposers can use. Thus, one decomposer depends on another for its food supply.

This type of food chain is like an assembly line in reverse. Instead of building up step-by-step from simple to complex compounds, the decomposer food chain breaks down complex organic substances. At the end of the chain, only inorganic substances such as carbon dioxide, nitrates, ammonia, and water remain. Plants then can assemble these inorganic materials into organic compounds and begin the cycle again. **1**

Some soil organisms produce substances that harm other organisms. These substances may accumulate in the soil around the organism that formed them, reducing the growth of competing organisms. We use these substances as antibiotics to combat bacterial infections in animals. Aureomycin, for example, is derived from a bacterium, and penicillin is derived from a fungus.

Fungi even can act as predators (see Figure 12.21). Several species of soil fungi form hyphae with tough branches that curl into loops. The loops **2** from adjacent hyphae interlock, forming a network and producing a sticky fluid. Small worms living in the soil, called nematodes, are caught in the loops and held fast despite violent struggles. Other fungal hyphae then invade the bodies of the captive nematodes and digest them.

12.16   Mycorrhizal Fungi Help Many Plants to Grow

If you trace the networks of fungal hyphae down through loose soil, you will find some leading to plant roots. The relationship between the roots and the fungi, which causes slight modifications of the roots, is known as **mycorrhizae** (my koh RY zee).

Mycorrihizae are a symbiotic relationship. The fungi either form a sheath around the root or penetrate the root tissue. Some plants, such as orchids and pines, do not grow, or else grow poorly, if their mycorrhizal fungi are not present. The mycorrhizae increase the surface area of the root through outward extensions (Figure 12.22). The fungi absorb soil **3** nutrients and secrete an acid that makes the nutrients readily usable by the plant. Thus, the fungi transfer nutrients to the plant, and photosynthetic products from the plant nourish the fungi. Mycorrhizae also absorb water and protect the plant against certain pathogens in the soil.

**Decomposer
food chain**

1. mustard-yellow
   polypore (shelf
   fungus)
2. oyster
   mushroom
3. amanita
4. yellow morel
5. snail
6. sow bug
7. centipede
8. wood roach
9. springtails
10. mite
11. ant
12. carrion beetle
13. soil fungi
14. soil
    protozoans
15. earthworm
16. inorganic
    compounds
17. soil bacteria

Figure 12.20   A decomposer food chain. Identify each organism and the role it plays in
the process of decomposition.

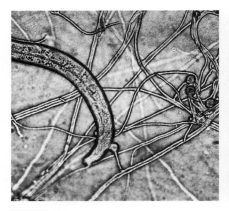

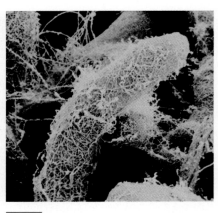

Figure 12.21 Predacious fungi capture small roundworms present in the soil *Dactylella drechsei* traps the worms with small adhesive knobs (X400).

Figure 12.22 A scanning electron micrograph of mycorrhizae between a *Boletus* mushroom and an aspen *Populus tremuloides* (X5O).

Bring rocks and other objects with lichens into the classroom for student observation.

When lichens are accorded a taxonomic grouping, they usually are considered a separate class of fungi. Often, however, they are divided between the sac fungi and the club fungi. A student might argue that because they are so different from other fungi, they should be given division rank.

Mycorrhizae occur in all plant families. They are found in some of the oldest plant fossils, dating from the Devonian period. Mycorrhizae even may have made it possible for plants to adapt to a terrestrial existence. Scientists speculate that the fungi may have transported nutrients to the roots of primitive plants and prevented them from drying out long enough for the plants to establish themselves on land.

12.17 Lichens Are Symbiotic Pairs of Organisms

You can find lichens on the bark of pine trees, on tombstones in New Hampshire, buried under arctic snow, or on rocks in Arizona. The shape and color of lichens are so characteristic that for several hundred years biologists described them as if they were single organisms. Actually, a lichen is made up of a fungus and either an alga or a cyanobacterium in a symbiotic relationship. More than 25 000 "species" of lichens have been described. They are named according to the fungus part of the lichen.

A lichen has the structural framework of fungal hyphae. Many small producers, usually green algae or cyanobacteria, live in the branches of the hyphae. The producers can live alone, but the lichen fungi do not grow well when separated from their partners. Many lichen fungi are classified as sac fungi, but they are unlike any of the sac fungi that do live alone.

Lichens secrete acidic compounds that break down rocks. Thus, they often are the first organisms to colonize bare rock, bare soil, or ice (see Figure 12.23). Lichens can survive in these extremely dry areas because **4** they have the ability to dry out completely and become inactive, or dormant. When dormant, lichens can tolerate temperature extremes that would kill other organisms, even hardy ones. However, they do not grow until they become wet again. In rain, fog, or dew, a lichen may absorb up to 30 times its own weight in water; then the producer in the lichen quickly begins to photosynthesize. Growth continues until the lichen dries out again. Because lichen growth occurs in spurts, the average growth rate is extremely low—sometimes just 0.1 mm per year. Thus some small lichens may be several hundred years old.

One common lichen, reindeer moss (Figure 12.24), covers great areas of the Arctic. Although it is called a moss, it is actually a lichen that reindeer and caribou eat. Reindeer moss, like all lichens, absorbs some nutrients from the soil or rock and from rainfall, then concentrates the nutrients

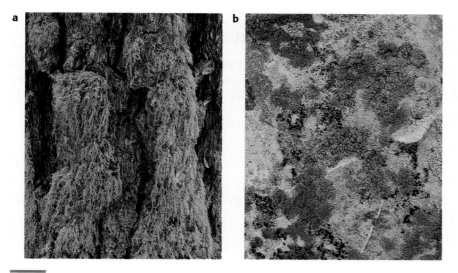

Figure 12.23 Fruitcose lichens (a) on a tree trunk, crustose lichens (b) colonizing on rock.

in its cells. Toxic compounds also may be absorbed and concentrated in the lichen bodies. If the air is extremely polluted, the lichens will die, thus providing an indication of how polluted the air is in a given area. Some
5 reindeer moss has accumulated radioactive materials from aboveground atomic bomb testing in the Arctic. Reindeer and caribou that have eaten the lichens have taken in the radioactive materials. When Eskimos have eaten these animals, they also have absorbed the radioactive materials. Thus materials may be passed along unexpectedly through a food chain to a consumer at the top of the chain.

Be certain students know the difference between moss and lichens. They often use moss to describe a lichen.

CONCEPT REVIEW

1. Compare and contrast a food chain involving only decomposers with one that includes plants and animals.
2. How are some fungi predators?
3. Describe mycorrhizae. What makes this relationship symbiotic?
4. What characteristics make lichens good pioneer organisms?
5. Why do lichens make good indicators of air quality?

Figure 12.24 Reindeer moss, a type of shrubby lichen.

This investigation allows students to become familiar with some common freshwater protists and cyanobacteria and to use a dichotomous key to identify them. Teams of 2 are satisfactory, but allow students to work individually if you have enough microscopes.
Time: **One class period.**

Discussion

1. **Answers will depend on the organism selected. Let students use their imaginations, but they should be able to justify their choices of the environment and the other organisms they have described.**
2. **Using the key, students should be able to summarize characteristics such as "bright green, chloroplasts, rigid cell wall" for chlorophyta, "moves by cilia" for ciliates, and so on.**

INVESTIGATION 12.1 Variety Among Protists and Cyanobacteria

Many protists live in fresh water as do many cyanobacteria In this investigation, you will use a dichotomous key to make a series of choices between alternative characteristics and identify a number of common protists and cyanobacteria.

Materials (per team of 2)

2 microscope slides
2 cover slips
flat toothpick
compound microscope
protist/cyanobacteria key

A Catalog of Living Things
Detain™
protist/cyanobacteria survey mixture
pond, stream, or puddle samples

Procedure

1. Place a drop of Detain™ (protist slowing agent) on a clean slide. Add a drop of either the survey mixture or a sample and use a toothpick to mix the two drops. Place a cover slip on the mixture.
2. Use the low power (100x) of a microscope to examine the mixture. For more detailed observations, switch to high power (400x).
3. Use the protist/cyanobacteria key in Appendix 3, *A Catalog of Living Things* (Appendix 4), and other references to identify as many organisms as possible.
4. Wash your hands thoroughly before leaving the laboratory.

Discussion

1. Choose one of the specimens you observed. Write a short paragraph describing the organism's environment, other organisms that might live in the same environment, and their possible interactions with the protist you observed.
2. Using the key's descriptions, summarize the distinguishing characteristics of cyanobacteria and the major groups of protists included in the key.

Using *Paramecium,* this investigation highlights the relationship between structure and function. It also develops students' observation skills as well as laboratory skills. Do not expect 100% association between structure and function on the part of students. Be satisfied with one for some students and expect a few more from others. Teams of 2 are recommended.
Time: **About three class periods for students to make the observations. You may wish to save time by dividing the tasks among teams. Part E can be set up as a demonstration.**

Safety

Although yeast-brilliant cresyl blue and 0.01*M* HCl are irritants, safety goggles are not necessary if you provide the stain in squeeze dropping bottles and place pieces of HCl-wetted thread in petri dishes. Remind students to wash up each day before leaving the laboratory.

INVESTIGATION 12.2   Life in a Single Cell

The relationship between structure and function in plants and animals is always significant but not always obvious. Careful and detailed observations of living organisms, sometimes over long periods of time, will reveal which structures are performing the life functions. In this investigation, you will observe structures of the protozoan *Paramecium* and form hypotheses about their possible functions.

Materials (per team of 2)

2 18-mm X 1 50-mm test tubes
2 microscope slides
2 cover slips
dropping pipet
compound microscope
rubber stopper to fit test tube
test-tube rack
incandescent light source
forceps

aluminum foil
6 flat toothpicks
Detain™ in squeeze dropping bottle
alizarin red stain in squeeze dropping bottle
India ink in small bottle
yeast-brilliant cresyl blue preparation in squeeze dropping bottle
cotton thread wetted with 0.1*M* HCl
paramecium culture

PART A Structure and Function
Procedure

1. Place a drop of Detain™ in the center of a clean slide. Add a drop of paramecium culture and use a toothpick to mix the two drops together. Place a cover slip on the mixture.

2. Use the low power (100X) of a microscope to examine the slide. Locate at least one paramecium.

3. In your data book, make a simple drawing of the paramecium that you have seen through the microscope. Include in your sketch only the parts of the organism you observe.

4. What structure performs what function? Label the structures in your drawing according to the following scheme. (You may not have observed all the structures listed below.)

a. Structure(s) associated with movement.

b. Structure(s) associated with food intake.

c. Structure(s) associated with digestion.

d. Structure(s) associated with elimination of undigested food.

e. Structure(s) associated with oxygen and carbon dioxide exchange.

f. Structure(s) associated with excretion of fluid wastes.

g. Structure(s) associated with removal of excess water.

It is not necessary at this time to know the names of the parts you have labeled.

5. Carefully raise the cover slip and add a drop of alizarin red stain to your slide. Mix with a toothpick and gently replace the cover slip.

6. Use low power to examine the slide. Can you see structures in the paramecium that you did not see before? Add these structures to your diagram, using a colored pencil to distinguish them from your first observations.

7. Use the list in Step 4 to develop hypotheses about the function of any newly visible structures.

Discussion

1. What difference do you observe in the paramecium after adding the alizarin red stain?

2. What structures could you see before adding the alizarin red? What structures could you see afterwards?

3. What functions of a living organism are not represented in your seven hypotheses?

4. Now look at Figure 12.25. What clues to functions do you find in the names of the structures?

5. With this new information, change any of the structure-function relationships you think need changing.

6. What should you do to determine if your hypotheses are good?

PART B Movement
Procedure

8. Prepare a fresh slide as in Step 1.

9. When the paramecia begin to slow down, switch to high power (400X) to observe movement in detail. Which end of the paramecium is usually in front as the organism moves?

10. The front end of an organism is referred to as anterior, the back end as posterior. Using the diagram you made in Part A, label what you believe to be the anterior and posterior ends. Can you identify a top (*dorsal*) and a bottom (*ventral*) side?

11. Diagram the motion of a paramecium as it moves in one direction. Use arrows to show the directions of body motions.

12. Carefully raise the cover slip and use the tip of a toothpick to add a small amount of India ink to the paramecium culture on the slide. Mix well; the culture should be light gray. Gently replace the cover slip.

13. Use high power to observe the surface of one of the organisms. The hairlike structures you see are cilia. Notice carefully what happens to the India ink particles that touch an organism. Adjust the light until you can see individual cilia.

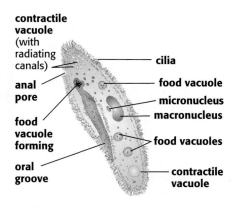

contractile vacuole (with radiating canals)

anal pore

food vacuole forming

oral groove

cilia

food vacuole

micronucleus

macronucleus

food vacuoles

contractile vacuole

Figure 12.25   Structures in a paramecium.

Part A Procedure

2. Cilia are best observed by adjusting the diaphragm on the microscope to reduce the amount of light.

Discussion

1. If alizarin red was used, the nuclear material and pellicle become visible. Iodine and methylene blue kill the organism.

2. All structures that can be seen before adding the stain usually are more visible after its addition. (If a contractile vacuole was contracted completely when iodine or methylene blue was added, it would not be seen after staining.)

3. Coordination (nervous or endocrine), reproduction, reception of stimuli, secrebon, chemical synthesis, growth.

4, 5. Depends on students' observations.

6. Test them with experiments.

Part B Procedure

9. The more rounded end.

10. The anterior end of the paramecium is more rounded; the posterior end is more pointed. Because the organism rotates like a forward-passed football, there is no dorsal or ventral side.

11. The organism rotates as it moves forward or backward.

Discussion

1. They act like tiny oars.

2. They beat strongly and swiftly in one direction, more slowly and weakly in the opposite direction.

3. Students should conclude that a paramecium moves by means of cilia.

Discussion

1. What appears to be the function of the cilia?
2. What does the pattern of movement of the India ink particles tell you about the activities of the cilia?
3. On the basis of your observations in Part B, review and revise (if necessary) the hypothesis you formed in Part A about how a paramecium moves from one place to another.

Part C Procedure

14. **Yeast cells in food vacuoles will change color in about 3 minutes. Observations must be made rapidly. Suggest to the students that they observe yeast cells in a food vacuole being formed at the base of the oral groove and observe what happens in and to this vacuole as it moves throughout the paramecium. Blue. They are ingested and enter through the oral groove. The oral groove and cilia. At the base of the oral groove.**
15. **Color changes from blue to red. The pH of the food vacuole changes (becomes more acidic).**
16. **They move away from the oral groove and seem to float through the cytoplasm.**
17. **At the anal pore, not far from where the food vacuole was formed.**

Discussion

2. **Food is swept into and down the oral groove by cilia. A small food vacuole forms at the base of the oral groove. The vacuole containing food is set free in the cytoplasm. In this vacuole, digestion proceeds. Assimilation occurs during the movement of the food vacuole. The undigested food is discarded to the outside through the anal pore.**

Part D Procedure

19. **Diffusion through the plasma membrane.**
20. **Leaving. Depends on students' hypotheses.**
21. **Two. One anterior, one posterior. Stay in one place. Clear. Varies greatly with conditions but averages about three per minute.**

PART C Ingestion, Digestion, and Elimination
Procedure

14. Prepare a fresh slide as in Step 1, but add a drop of yeast-brilliant cresyl blue preparation. Observe what happens to the yeast cells. Notice their color when they are outside the paramecium. Then, find some that have been ingested and observe their color. Yeast cells that have been ingested will be in small, round structures called food vacuoles. What color are the yeast cells when they are outside the paramecium? What happens to the yeast cells that are encountered by the paramecium? What parts of the paramecium are involved first in the ingestion of the yeast cells? Where are the food vacuoles formed in the paramecium?
15. Carefully watch a newly formed food vacuole and note any changes in it. Within 2 or 3 minutes you should begin to notice a color change in the food vacuoles. Describe the color change. How do you account for it?
16. Continue to observe the food vacuoles. By means of circles representing food vacuoles and arrows representing their movement, diagram the path a food vacuole takes from its formation to its disappearance. How do the food vacuoles move through the organism?
17. Watch carefully to see how undigested food in a food vacuole is eliminated from the paramecium. From what part of the body does this happen?

Discussion

1. Refer to the hypothesis you formed in Part A regarding the structures in the paramecium responsible for food intake, digestion, and the discharge of undigested food. Were you correct?
2. On the basis of your observations in Part C, summarize how you now think a paramecium ingests and digests food and eliminates wastes.

PART D Exchange of Materials
Procedure

18. Like most living things, a paramecium needs a constant supply of oxygen and must eliminate carbon dioxide, liquid wastes, and excess water. To discover the structures associated with these functions, prepare a slide as in Step 1 and observe under high power.
19. Considering the environment of the paramecium, through what structure would you expect oxygen and carbon dioxide to move? How do you think this movement occurs?
20. Concentrate your observation on either end of the paramecium. Locate a structure that contracts and then seems to disappear. It becomes visible again and repeats this process. The canals leading into this structure give it a star-shaped appearance. The structure is called a contractile vacuole. Is fluid entering or leaving the vacuole when it contracts? If you identified any contractile vacuoles in Part A, what hypothesis did you form about their function?
21. How many contractile vacuoles are there in a paramecium? Where are they located? Add them to your drawing. Do the contractile vacuoles move through

the cell as food vacuoles do, or do they stay in one place? Is the fluid within the enlarged vacuole clear or granular? How many contractions does the vacuole make in one minute?

Discussion

1. Students who investigated the question, "What is the function of a contractile vacuole?" made the following observation: Urea (an end product of nitrogen metabolism) has not been found in any great amount in the fluid of a contractile vacuole. What effect does that observation have on a hypothesis that the contractile vacuole is an excretory organ used for eliminating nitrogenous wastes?

2. Two further observations were: (a) If vacuoles of freshwater paramecia do not function, the body swells; and (b) Injection of distilled water into the organism increases the rate of contraction of the vacuole. Considering these observations, what probably is the function of a contractile vacuole in the paramecium?

3. List the ways in which water might enter a paramecium.

4. The students also observed that water intake along with food in food vacuoles is only a fraction of the amount of water that is expelled by the contractile vacuole. What would you suggest is the major method of water entry?

5. Develop a hypothesis that accounts for the function of a contractile vacuole.

PART E Behavior
Procedure

22. Fill two test tubes 1/4 full with paramecium culture. Place one test tube upright in a test-tube rack so that it receives uniform light from all directions.

23. Stopper the second test tube. Wrap the top half of the test tube with aluminum foil. Lay the test tube on its side and evenly illuminate the bottom half.

24. Now observe the upright test tube. In a rich paramecium culture, a concentration of the organism looks like a white cloud. Examine the tube for any concentration. What evidence do you find that the paramecia are or are not moving toward one part of the tube? In this controlled situation, is the tendency of organisms to concentrate in one part of the tube a response to gravity? Why do you think so? What other factors might account for this concentration? What additional investigations would you devise to check all variables, including gravity?

25. Observe the partially covered test tube placed on its side. Has the paramecium culture concentrated at one end of the test tube? Is there a response to light? How do you know? Why could this investigation not have been done with the test tube in an upright position? How might this response to light act as an advantage for a paramecium?

26. Place a drop of paramecium culture on a slide. Use forceps to place a small piece of thread wetted with 0.01M HCl across the middle of the drop. Add a cover slip. Using low power, focus on the piece of thread. In what way does the paramecium respond to the acid?

 HCl is an irritant; do not allow the thread to touch your skin. Should contact occur, call your teacher immediately; flush the area with water.

27. Each day, wash your hands thoroughly before leaving the laboratory.

Discussion

1. Summarize your observations of a paramecium's behavior.

Discussion

1. Because so little urea was found, it probably is not solely an excretory organ.
2. A structure to rid the organism of excess water.
3. Through the plasma membrane or in food vacuoles.
4. The plasma membrane must be the main passage for water into the cell.
5. The contractile vacuole controls the amount of water in the paramecium.

Part E Procedure

24. A white cloud at the upper end of the tube is evidence that the paramecia are moving toward that end of the tube. Possibly, because the paramecia move away from gravity. Oxygen, light, food, pH are possible factors. Oxygen could be ruled out if the tubes were completely filled with water and tightly corked. Light could be eliminated or controlled. Food could be evenly distributed throughout the tube. pH could be tested.

25. There should be a white cloud at the uncovered, lighted end. Apparently, although the response could be confused with the response to gravity. If the test tube were upright, a response to gravity might override a response to light. The food it consumes would more readily be found in the light.

26. Negatively.

Discussion

1. The response of *Paramecium* to gravity and acid is negative; to light, positive.

This investigation allows students to observe the fruiting body and spores of the mushroom, and sexual reproduction in the common bread mold, *Rhizopus stolonifer*. Teams of 3 are recommended.

   Time: 10 days: one class period for Day 1 of Parts A and B, one period for Day 2, 10 minutes each for Days 3-9, and one period for observations and discussion on Day 10.

Safety

Remind students never to eat anything in the laboratory. This rule applies especially to mushrooms, which never should be eaten unless the source is known. Use only mushrooms you have purchased from a grocery store; do not allow students to collect their own. Students should discard all used swabs in designated waste bags. Label self-closing plastic bags waste and tape a bag to each lab table. Collect these bags at the end of each class, place in a biohazard bag, and autoclave before discarding. Collect all used plates in a biohazard bag; autoclave before discarding or before opening plates for washing. See "Safety Using Microbes" in *Guidelines for Laboratory Safety*, in the Teacher's Edition. Remind students to wash their hands thoroughly each day before leaving the laboratory.

Procedure Part A

6. **The students should observe mushroom spores on the paper. Students should infer that the spores came from the gill structures inside the cap of the mushroom.**

8. **There probably will be several hundred spores, depending on the size of the swab used. Students should extrapolate to estimate the total number of spores in the entire mushroom. The figures should be in the hundreds of thousands.**

INVESTIGATION 12.3   Growth of Fungi

Mushrooms and molds are common fungi. In previous investigations some of your cultures may have been contaminated by molds, which may have given you the idea that molds are everywhere. If this is true, what features of molds make them so plentiful?

Materials (per team of 3)

glass tumbler	masking tape
microscope slide	transparent tape
cover slip	15-cm metric ruler
dropping pipet	glass-marking pencil
compound microscope	self-closing plastic bags labeled waste
stereomicroscope or 10X hand lens	glycerine-water solution (I : I)
sheet of white paper	sterile potato-dextrose agar plate
scalpel	fresh mushrooms
flat toothpick	+ strain of *Rhizopus stolonifer*
3 sterile swab applicators	– strain of *Rhizopus stolonifer*

PART A Observing Mushrooms
Procedure
Day 1

1. Examine a fresh mushroom. The mushroom itself is the reproductive body of this particular fungus. It consists of a stalk and an umbrellalike cap.

> **Do not eat the mushrooms.**

2. Use the scalpel to cut the cap from the stalk.

> **Scalpels are sharp; handle with care.**

   Look at the underside of the cap and find the gills. Place the cap, gill side down, on a sheet of white paper.

3. Label a piece of masking tape with your team name and place it on the bottom of the tumbler. Cover the cap with the tumbler to protect it from air currents.

4. Set the cap, paper, and tumbler in a place where they will not be disturbed.

5. Wash your hands thoroughly before leaving the laboratory.

Day 2

6. Remove the tumbler and lift the cap straight up from the paper. Replace the tumbler. What do you observe on the paper? What is the relationship between what you see on the paper and the structure of the mushroom cap?

7. Remove a sterile swab from its wrapper. Tilt the tumbler and use the swab to transfer a small amount of this material to a slide. Add a drop of glycerine-water and a cover slip. Replace the swab in its wrapper and discard in the waste bag.

8. Examine the material under both low and high power of the compound microscope. Approximately how many particles are present in your sample? Using this number, estimate the total number of such particles coming from the mushroom.

9. Wash your hands thoroughly before leaving the laboratory.

PART B Growing *Rhizopus*

Day 1

10. Draw a line across the middle of the bottom of a potato-dextrose agar plate. Label one side + and the other −.

11. Carefully remove a sterile swab applicator from its paper wrapper and transfer a few spores from the + culture of *Rhizopus stolonifer* to the + section of your plate. Make a 1-cm streak about 1 cm from the outside edge. Be careful not to dig into the surface of the medium. Replace the used swab in its paper wrapper and discard it in the waste bag.

12. Using a fresh sterile swab, repeat Step 11 with the − strain and place the spores in the − section of your plate.

13. Cover the plate and tape it closed. Invert and incubate at room temperature for 10 days.

14. Wash your hands thoroughly before leaving the laboratory.

Days 2–10

15. Observe the plate daily using a stereomicroscope or a 10X hand lens. As the mold starts to grow on the plate, you can see that it is composed of many branching threads called hyphae. Hyphae perform various functions. Some support the sporangia, and others anchor the mold to the agar and absorb water and nutrients.

16. Measure the growth of the mold daily and record the data in your data book. What difference is there in the growth rate of the two strains?

17. As the two strains grow toward the center of the plate, observe them carefully for changes, particularly where hyphae from the two strains have touched. The enlarged cells near the tips of the hyphae are sex cells. How do the sex cells of the two strains differ from each other? What occurs when the pair of cells touch one another?

18. Note carefully the zygospore formed by fusion of the two sex cells. How does it seem to be adapted to a land environment?

19. Each day, wash your hands thoroughly before leaving the laboratory.

Day 10

20. Unseal and partially open your plate. Use a flat toothpick to transfer a bit of material from the mature culture to a glass slide. Reseal the plate. Add a drop of glycerine water solution and a cover slip.

21. Observe under the low power of a compound microscope. Where are the sporangia produced? How might the position of the sporangia affect the distribution of spores? Does each sporangium produce a large or a small number of spores? Why is the number of spores produced important in the reproduction of *Rhizopus*?

22. Using high power, observe a few spores. How is their structure adaptive to a land environment?

23. Discard your plate in a biohazard bag as instructed by your teacher.

24. Wash your hands thoroughly before leaving the laboratory.

Discussion

1. Why is the world not covered with mushrooms? How numerous would mushrooms be if they produced only a small fraction of the number of spores you calculated for one mushroom in Part A?

2. Would the process you observed in Part B occur if both sides of the plate had + or both had − strains of *Rhizopus* on them? Explain.

3. How does the number of spores produced by *Rhizopus* compare with the number you calculated for the mushroom in Part A?

4. Which type of reproduction, sexual or asexual, appears likely to produce the greatest number of molds?

5. Which type of reproduction would be most advantageous to the mold in adjusting to changing environmental conditions?

Procedure Part B

16. There should be no difference in the growth rate of the two strains.

17. The sex cells are essentially the same, although one may be slightly smaller than the other. A bridge is formed between the two hyphae of the opposite strains. Refer students to Figure 12.15.

18. The zygospore has a very thick, tough coat.

21. Sporangia are produced at the tips of hyphae or by a germinating zygospore. The aerial location of the sporangia places them in a position to release their spores into the air currents. Large. The large number of spores improves the chances that a few will germinate and grow into new molds.

22. Spores have thick walls that make them resistant to drying.

Discussion

1. Few spores encounter proper growing conditions. The number would be smaller because of the reduced chance of locating the proper environment for growth.

2. No. The + and - strains are opposite mating types required for sexual reproduction.

3. Depends on student observations, but they should report large numbers of spores for both *Rhizopus* and the mushroom.

4. Asexual reproduction by spores produces the largest number of molds. Many more spores are produced by asexual means than zygospores by sexual ones.

5. Sexual reproduction would be most advantageous because it produces greater variability and thus a better chance for adjusting to a varied environment.

For Further Investigation

To avoid indiscriminate spread of mold spores, instruct students to grow molds in sealed containers or plastic bags and to make all observations and measurements without opening the containers.

For Further Investigation

With your classmates, conduct a mold-growing race. Select a medium, a source of spores, and an environment for growth. Keep a daily record of the mold growths. At the end of the race, present your procedures and results to the class.

SUMMARY

Protists and fungi are two kingdoms of eukaryotes. Algae are autotrophic protists and may be distinguished from each other by their dominant pigments—red, brown, yellow, or green. Heterotrophic protists include protozoa and slime molds. Most protozoa can be distinguished by their method of movement. Some do not move at all, whereas others move by using cilia, pseudopods, or flagella. Some protozoa are parasites that cause diseases of humans and other animals. Slime molds are multicellular, heterotrophic protists that exhibit animal-like and funguslike characteristics.

Fungi are all heterotrophic and can be distinguished from each other by their sexual reproductive structures. Whereas some fungi are edible, others destroy foods, and still others are parasites. In a community, fungi play a major role as decomposers. Fungi also may be part of symbiotic relationships with other living organisms, as they are in mycorrhizae and in lichens.

APPLICATIONS

1. What characteristics of eukaryotic algae make them more complex than the cyanobacteria discussed in Chapter 11?

2. For an aquatic organism, what are the advantages of storing energy reserves as lipids, starch, or some other carbohydrate, and *not* as sugar?

3. Is color a good characteristic to use to distinguish between autotrophic and heterotrophic organisms? Explain.

4. Suppose you moved a freshwater paramecium into salt water. What do you think would happen to the rate of contraction of the contractile vacuole? Explain.

5. *The Blob* was a movie about an alien organism that lands on earth and consumes all living things in its path. Which organism in this chapter does the Blob resemble and in what ways?

6. Many fungi are decomposers of plant tissues. Considering this, what is the significance of the fact that fungi have cell walls made up mostly of chitin?

7. The bodies of many fungi are located underground, but the reproductive structures are above ground. What is the advantage of this arrangement?

8. In which of the ecological relationships between organisms described in Chapter 3 are fungi involved?

9. How do the characteristics of lichens cause a taxonomic problem?

10. Mycorrhizal fungi use sugars from the plants to which they are attached. Why are these fungi not considered to be parasites of the plants?

PROBLEMS

1. Start reading the labels of the foods you eat. Make a list of those foods that include algae or algal products. Look for carrageenan or algin, two algal products commonly used.

2. Draw or paint a picture depicting an ecosystem in which algae play an important role. Show the algae and the organisms that depend on these producers. You might consider a freshwater pond or lake with different types of algae and fish or a coastal ecosystem that includes sea otters, brown algae, and sea urchins. Try to depict through your illustrations the relationships between the organisms in the ecosystem.

3. Algae are good indicators of water quality—many species are found only in standing, polluted waters. Ask a local phycologist (a person who studies algae) or a local expert on water quality to visit your class and speak on the types of organisms normally found in your water reservoirs, the problems they can cause, and how the problems are resolved.

4. Mitochondria have their own circular DNA that is different from the DNA found in the nucleus of the cell. Studies in humans have shown that mitochondrial DNA is of maternal origin, that is, it is inherited from the mother only. How might you explain this? (Hint: refer to the discussion of fertilization in Chapter 6.)

5. Investigate how fungi and bacteria are used to help fight disease—especially in the production of antibiotics.

6. Make a collection of mushrooms and lichens from areas where collecting is permitted. *Do not eat any of them.* Describe their size, shape, and color; if possible, note to what each was attached. What role do you think each played in the community?

7. How many of the groups of organisms discussed in this chapter can be found in your locality? Take into consideration wild and cultivated, indoor and outdoor, and aquatic and terrestrial organisms.

Eukaryotes: Plants
13

This is a curious-looking plant. What type of plant is it? See if you can locate a plant like it using *A Catalog of Living Things* (Appendix 4).

This is *Lycopodium tristachyum*, a club moss.

◆ Several hundred million years ago, the land surface of the earth was barren. Life existed only in the water. One-celled algae colored the surface waters of the oceans shades of green. Some algae joined together into multicellular organisms and evolved a degree of cell specialization. These organisms probably were the ancestors of modern plants (see Chapter 12).

This chapter examines several of the characteristics that enabled plants to invade the land and describes a few examples of modern plants. Today, plants adapted to life on land are found in most habitats, from prairies to tundra and in the forests and deserts of the world. They come in the widest possible range of sizes. The largest organisms in the world are the giant redwoods in California, such as those shown in Figure 13.1. Some of these trees are more than 100 m tall and 7 m in diameter. Between a simple multicellular green alga in the ancient sea and a giant California redwood tree with billions of cells, the difference in cellular coordination is tremendous. The evolution of such a complex land organism from a relatively unspecialized multicellular ancestor did not occur overnight. It took millions of years. ◆

The Evolution of Land Plants

13.1 Two Major Groups of Land Plants Evolved

Plants are multicellular, photosynthetic organisms adapted primarily for life on land. The evolution of multicellular green algae that had some division of labor among their cells produced many adaptations that allowed plants to live in this environment. The oldest known plant fossils are simple, branched structures that had several important adaptations.

Because the climate on land may vary from hot to freezing and from wet to dry, water is the major limiting factor for any organism. On land, the only reliable source of water is underground where it is too dark for 1 photosynthesis. What possible advantages of life on land could overcome such a major disadvantage?

Without the presence of other organisms, plants that were able to tolerate the dry conditions that exist on land would have had little competition for food, living space, and the other necessities of life. Another advantage would have been the abundance of carbon dioxide and oxygen present in the air. A third advantage for these plants would have been increased light levels, because the air does not absorb as much light as water does. Finally, there would have been plenty of space, and nutrients would have been readily available in the soil.

A multicellular alga tossed onto the shore would have had a better chance of surviving than would a one-celled alga. The outer cells of the

MAJOR CONCEPTS

◆ Plants are multicellular, photosynthetic organisms having alternation of generations.
◆ Many adaptations that permit success on land have evolved in plants.
◆ Two plant groups apparently evolved from multicellular algae: bryophytes and vascular plants.
◆ Conifers and flowering plants are the most widespread organisms on the earth.
◆ Pollen and seeds have increased the reproductive success of the complex plants.
◆ The quality of human life depends on the success of plants.

KNOWLEDGE CHECK

◆ In what ways do plants differ from one another?
◆ What adaptations do plants need for life on land?
◆ What is a bryophyte?
◆ How do plants reproduce?

Guidepost
What adaptations allowed plants to colonize land?

According to the *Guinness Book of World Records* (1983), the tallest tree in the world is a Coast Redwood, *Sequoia sempervirens,* in Redwood Creek Grove, Redwood National Park, CA. In 1970, it was estimated to be 111.6 m tall.

Figure 13.1 The giant redwoods *Sequoiadendron giganteum*, are the largest organisms in the world.

organism might have protected the inner ones from drying out rapidly, and the inner cells might have been efficient at photosynthesis. Other cells of the same organism might have specialized in collecting water or nutrients from the environment. The resulting division of labor would have enabled the alga to exploit the resources in its new environment. A specialized, multicellular alga, somewhat like the modern *Chara* in Figure 13.2, might have been able to live and reproduce on land if ocean spray or tides kept it moist.

Two plant groups apparently evolved from relatively complex multicellular green algae. The more complex group, which includes fossils of the oldest land plants, has many adaptations to life on land, including **vascular tissue**—cells joined into tubes that transport water and nutrients throughout the body of the plant. These plants are called vascular plants. **2** The less complex group, the **bryophytes** (BRY oh fyts), are not aquatic, yet they possess few adaptations to life on land. This group includes the true mosses, hornworts, and liverworts. Bryophytes are called nonvascular plants because they do not have vascular tissue.

Vascular plants can be divided into two groups: those that produce seeds, and seedless plants such as ferns that reproduce with spores. Vascular plants that produce seeds can be divided further into two groups: those that produce naked seeds, such as pine trees, and the flowering plants, which produce seeds enclosed in a fruit. Figure 13.3 represents an evolutionary history of plants.

13.2 Vascular Plants Have Adaptations That Conserve Water and Permit Gas Exchange

The success of plants on land depends largely on their ability to absorb and hold water. Structures that enable them to do so include roots, vascular tissue, and an outer covering that retards water loss.

Vascular plants such as ferns, conifers (cone-bearing plants), and flowering plants have well-developed root systems that penetrate into many parts of the soil. An extensive root system provides an efficient way to collect water and nutrients from the soil and bring them to the main body of the plant. Rooted plants offer a good example of division of labor—the cells in the root collect water and nutrients, and the aboveground cells absorb sunlight and produce food through photosynthesis.

Figure 13.2 The multicellular green alga *Chara*. The fossil record of *Chara*-like algae extends back about 400 million years. Why is *Chara* considered to be a model of the ancestors of land plants? **Because of its structure and reproductive patterns.**

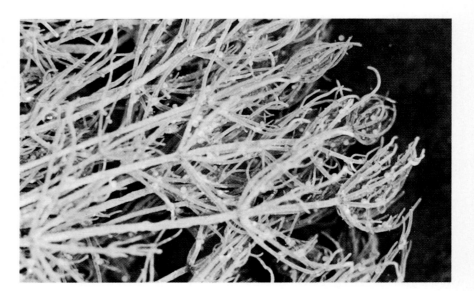

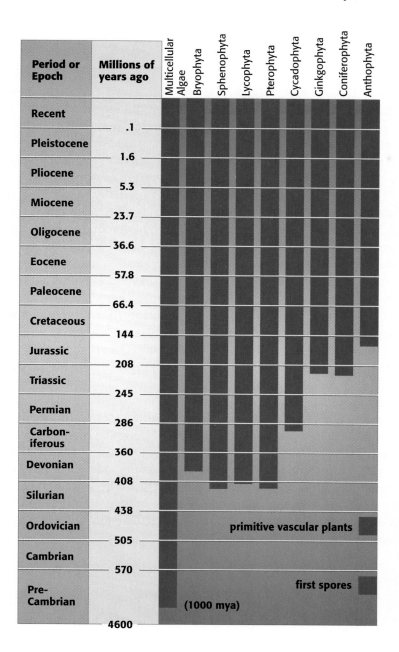

Period or Epoch	Millions of years ago	Multicellular Algae	Bryophyta	Sphenophyta	Lycophyta	Pterophyta	Cycadophyta	Ginkgophyta	Coniferophyta	Anthophyta
Recent										
	.1									
Pleistocene										
	1.6									
Pliocene										
	5.3									
Miocene										
	23.7									
Oligocene										
	36.6									
Eocene										
	57.8									
Paleocene										
	66.4									
Cretaceous										
	144									
Jurassic										
	208									
Triassic										
	245									
Permian										
	286									
Carbon-iferous										
	360									
Devonian										
	408									
Silurian										
	438									
Ordovician			primitive vascular plants							
	505									
Cambrian										
	570									
Pre-Cambrian		first spores (1000 mya)								
	4600									

Figure 13.3 The evolutionary history of plants. What evidence was used to construct this history? **Fossil evidence.**

Figure 13.4 Plants such as this *Sedum adolphic* have very thick cuticles. How are thick cuticles adaptive? **They allow the plants to survive in dry climates.**

Tall plants require cells that can transport water from the roots to the leaves and cells that can support an upright body. Because vascular tissue serves both these functions, vascular plants can grow taller than bryophytes and thus capture more sunlight. Bryophytes never grow very tall because they lack an efficient water transport system.

In bryophytes, water can evaporate from the entire surface, so a moss plant, for example, dries out quickly. Vascular plants produce a waxy covering, or cuticle, that covers the plant body above ground, reducing the amount of water that can evaporate from its surface. The **cuticle** is often thick on the leaves of plants living in dry places, such as the sedum in Figure 13.4. The covering does not prevent gas exchange with the environment because vascular plants have slit like openings, or **stomates** (STOH maytz), in the surface of their leaves. Stomates (Figure 13.5), found in the oldest fossil plants, permit carbon dioxide and oxygen to enter or leave the plant. Roots that absorb water, vascular tissue that supports the plant and conducts water, a cuticle that prevents evaporation, and stomates that permit gas exchange are characteristics that enable plants to live on land.

The cuticle is so hydrophobic that most plant sprays, such as biocides, contain a detergent to reduce surface tension of the water in the solution, allowing it to spread.

Figure 13.5 Stomates on the surface of a leaf.

As sperm enter the female gametophyte, they move down the neck to the egg. Neck cells secrete fluid, and the sperm move toward the area of greater fluid concentration.

The sporophyte usually contains chlorophyll until the spores are produced, after which it uses food from the gametophyte.

13.3   Bryophytes Require Water for Reproduction

Reproductive adaptations also enabled plants to survive on land. Bryophytes are restricted to moist areas because their sperm are flagellated, as are those of animals and algae. For sexual reproduction to occur, the plants must be bathed in water so the sperm can swim to the egg. Thus, bryophytes reproduce sexually only where water sprays them or after they are wet with dew or rain. Like all sexually reproducing plants, bryophytes have a life cycle in which a haploid *(n)* phase alternates with a diploid *(2n)* phase. This type of life cycle is called **alternation of generations.**

The carpet of moss shown in Figure 13.6a actually is many individual plants. These small plants are haploid, and each is called a **gametophyte** (guh MEET oh fyt). As the name implies, gametophytes produce gametes, usually in special structures near the tips of the plants. In wet conditions, sperm are produced in male gametophytes. The sperm swim in a film of water to the egg   **4** cell in the female gametophyte, and a diploid zygote is produced. In some species, both eggs and sperm are produced on the same gametophyte.

The zygote divides by mitosis and develops into a diploid embryo. Eventually, the embryo grows out of the female gametophyte into a stalk like structure, the **sporophyte** (SPOR oh fyt). The sporophyte is visible above the small haploid individuals shown in Figure 13.6b. Meiosis occurs within the capsule at the end of the sporophyte, and haploid spores are formed. The capsule helps to protect the spores. Once the spores are released, they are carried to wherever wind or water transports them. Most spores fall onto unfavorable habitats and die. If a spore reaches a favorable environment—usually a moist soil surface—its wall can burst open. The cell within begins to divide by mitosis, producing long green threads that resemble the filaments of many aquatic algae. The gametophyte moss plant develops from these threads. When a spore germinates and grows into another gametophyte plant, the life cycle diagrammed in Figure 13.7 is complete.

13.4   Flowering Plants Have Special Reproductive Adaptations

Unlike bryophytes and seedless vascular plants, which require water to be present for fertilization, seed plants produce special structures called **pollen**

Figure 13.6 Mosses (a) thrive in moist conditions such as streambanks. Moss sporophytes (b), *Polytrichium,* growing from the female gametophyte plants.

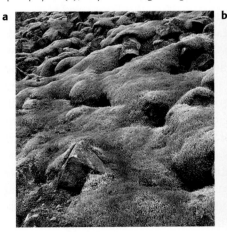

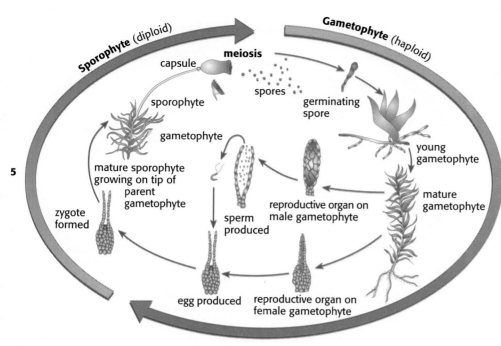

Sporophyte (diploid)

Gametophyte (haploid)

capsule

meiosis

sporophyte

spores

germinating
spore

gametophyte

young
gametophyte

mature sporophyte
growing on tip of
parent
gametophyte

reproductive organ on
male gametophyte

mature
gametophyte

zygote
formed

sperm
produced

egg produced

reproductive organ on
female gametophyte

5

Many bryophytes also are capable
of vegetative reproduction. Any immature cell of either the sporophyte
or gametophyte can produce a
new gametophyte. Also, some can
produce an asexual reproductive
body, the gemma, which gives
rise to a new gametophyte.

Figure 13.7 The life cycle of a moss. Which generation is predominant?
The haploid generation.

grains (Figure 13.8), in which the sperm develop. The pollen grains may be
blown by wind or may be accidentally carried by animals from one plant to
another. This efficient means of transferring sperm to the egg under dry conditions is most highly developed in the flowering plants. Brightly colored, or
scented flowers attract a variety of animals, such as hummingbirds, bats and
insects (mainly) and contain nectar that they drink. As the animal drinks, it
picks up pollen from one flower and carries it to another flower while searching for more nectar. Because seed plants are not restricted to moist conditions, sexual reproduction can occur whenever the sperm and egg are fully
developed. Like bryophytes, flowering plants have a life cycle with alternation of generations, but the gametophyte generation of flowering plants is
smaller and protected, at least for a while, by the sporophyte.

Flowering plants are considered to be the most complex of the vascular plants. Their reproductive structures are found in flowers. A flower is
actually a short branch bearing groups of specialized leaves. Some of
these leaves may resemble ordinary leaves in many ways, but others are so
different in structure that it is hard to think of them as leaves at all. If you
closely examine a flower such as the buttercup illustrated in Figure 13.9,
you will see a number of green, leaflike structures called sepals (SEE
pulz) on the underside of the flower. Before the bud opens, the sepals
cover and protect the other parts of the flower. The most conspicuous parts
of a flower are the colorful petals. Although petals are often leaflike in
shape, they are not usually green.

Just inside the circle of petals of a typical flower is a ring of male reproductive structures, the **stamens** (STAY menz). In the center of the flower
is the **carpel** (KAR puhl), the female reproductive organ. Although most
plants have both male and female organs within the same flower, a few
plants produce flowers with only female parts or only male parts. Stamens
usually have an enlarged tip, the anther, whereas the carpel tip (the stigma)
is more pointed. Despite their shape, however, both stamens and carpels are
thought to be modified leaves that have been adapted for reproductive roles.

Figure 13.8 Scanning electron
micrograph of pollen grains from
dandelion (X2500).

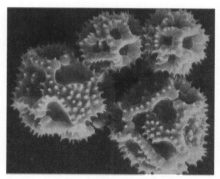

Figure 13.9 Diagram of the flower
structure of a buttercup.

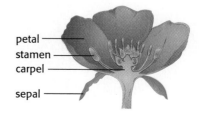

petal
stamen
carpel
sepal

Figure 13.10 The life cycle of a flowering plant. The parts are drawn to different scales. Which generation predominates? **The diploid, or sporophyte, generation.**

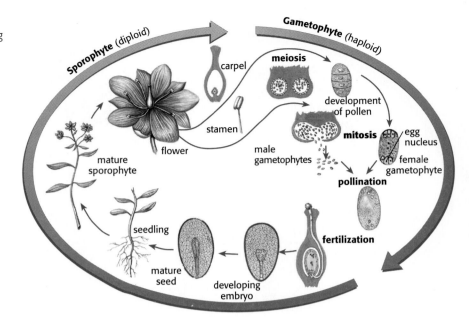

Figure 13.10 shows the relationship between reproductive structures and the life cycle of one type of flowering plant. At the base of the carpel is an enlarged portion, the ovary, that contains one or more small structures called **ovules** (OHV yoolz). Meiosis occurs in a special cell in each ovule, resulting in the formation of four haploid cells, the female spores. These spores do not separate from the sporophyte as they do in mosses. **6** Instead, three of the spores disintegrate. The fourth spore divides three times by mitosis, forming eight nuclei. The nuclei, with their surrounding cytoplasm, form seven cells, one of which is the egg cell. (One of the cells contains two nuclei, the polar nuclei.) Figure 13.11a shows these seven cells, which constitute the female gametophyte.

In the stamens, cells in the anthers undergo meiosis, each giving rise to four haploid cells, the male spores. Each spore contains one haploid nucleus that divides by mitosis, forming two nuclei. A spore wall thickens around each nucleus, forming a pollen grain. A single stamen may contain thousands of pollen grains. Each mature pollen grain is a single celled male gametophyte containing three nuclei (see Figure 13.11b).

Pollination is the transfer of pollen from the stamens to the carpel, either within a flower, between flowers of the same plant, or between plants of the same species. The sticky stigma at the top of the carpel can trap pollen grains carried to it by wind, water, or a visiting animal. Only pollen grains from flowers of the same species are useful in fertilization.

The hard wall of the pollen grain protects the haploid cell until it lands on a stigma. Once there, a thin finger of tissue, the **pollen tube,** grows from the grain into the carpel. Within the pollen tube, one nucleus, the tube nucleus, leads the way. The other nucleus divides to form two sperm nuclei. The pollen tube grows down the carpel, transporting the sperm nuclei to the ovule. Fertilization occurs when one sperm nucleus unites with the egg, forming a zygote. This diploid cell gives rise to the embryo. The other sperm nucleus unites with the polar nuclei, leading to the formation of **endosperm** (EN doh sperm), a mass of food-storing cells that **7** will later nourish the developing embryo. The endosperm is triploid—it has three sets of chromosomes, one set from the sperm and two from the polar nuclei. Figure 13.12 summarizes the stages of fertilization and embryo development in a flowering plant.

Figure 13.11 Female gametophyte of a lily (a). Mature ovule ready to be fertilized; the egg nucleus is one of those on the right. A pollen grain at the two-nucleate stage. (b). A pollen grain is the male gametophyte of a seed plant.

a
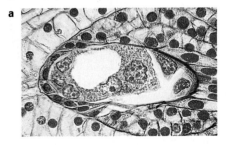

b
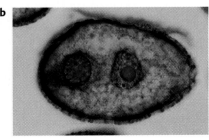

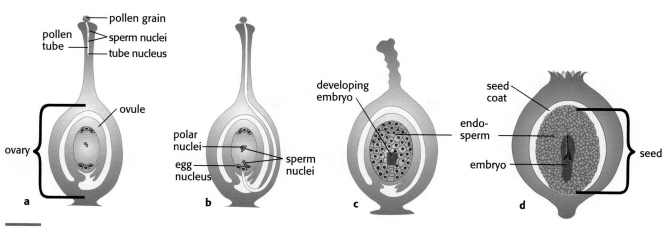

Figure 13.12   Pollination, fertilization, and development of the embryo, endosperm, and seed in a flowering plant.

The life cycle of a flowering plant is similar to that of a moss in two ways. First, meiosis occurs just before spore formation. Second, there is alternation of generations between the sporophyte and gametophyte portions of the life cycle. There are, however, several differences. First, abundant surface water is not necessary for fertilization in flowering plants. Second, the gametophytes are smaller than the sporophytes. Third, the gametophyte that produces an egg and the spore that produces the female gametophyte do not separate from the sporophyte plant. Thus, these structures are better protected from the environment than are their counterparts in an embryonic moss. Finally, the embryo (young sporophyte) grows for a short time, then becomes dormant. The embryo and its endosperm are surrounded by a protective coat formed from ovule tissues. This package is a **seed,** diagrammed in Figure 13.13. The seed protects the young sporophyte, which remains dormant until environmental conditions are suitable for germination. Moss embryos, on the other hand, cannot tolerate dry conditions. Also, moss spores contain no embryos and only a small amount of food.

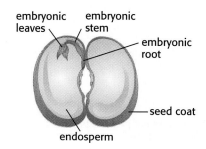

Figure 13.13   A bean seed, showing the embryo and endosperm.

Figure 13.14   Evolution and specialization in sporophytes and gametophytes.

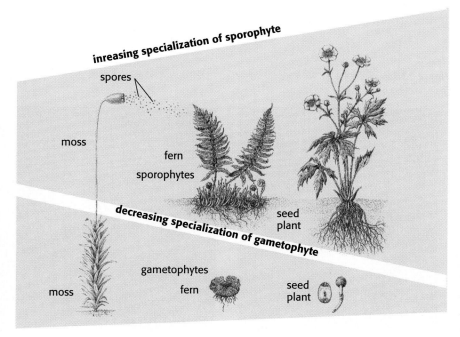

Land plants show a trend toward increasing specialization of the sporophyte and decreasing specialization of the gametophyte, as shown in Figure 13.14. Seed plants have succeeded because of their adaptations to the changing environment on land. Cuticle, vascular tissue, root systems, sperm carried in pollen, spores protected on sporophytes, and embryos protected in seeds makes flowering plants well suited to life on land.

CONCEPT REVIEW

1. Describe the limiting factors and advantages of life on land for the first land plants.
2. What types of adaptations were necessary for plants to live on land, and why were they important?
3. Briefly distinguish between the different types of land plants.
4. Compare and contrast gametophytes and sporophytes.
5. Compare and contrast the life cycles of bryophytes and flowering plants.
6. Explain the roles of the reproductive structures of flowering plants.
7. Describe the origin of the embryo, the endosperm, and the seed.

Bryophytes and Seedless Vascular Plants

13.5 Bryophytes Have No Roots, Stems, or Leaves

Bryophytes are pioneer plants on newly exposed substrates, such as bare rock.

Guidepost

How do bryophytes and seedless vascular plants differ in structure and habitat?

They build up organic materials and produce a water-and nutrient-reclaiming substrate that may be colonized by vascular plants.

Because many of these plants are less familiar to students than are seed plants, it is important to have on hand many specimens, preferably living. Collect living bryophytes at any time of the year and establish a terrarium.

Most bryophytes (division Bryophyta; bry OFF ih tuh) are relatively small. Few of them exceed 20 cm in height. Although they have structures resembling stems and leaves, these terms are not used in describing bryophytes because they lack the vascular tissue of other land plants. There are three classes of bryophytes: true mosses, the largest class; liverworts; and hornworts. Figure 13.15 shows several bryophytes.

True mosses often grow in clumps or small clusters in rock crevices and on the shady side of trees. An individual moss plant from such a clump is simply an upright, green, stemlike stalk with threadlike structures called **rhizoids** (RY zoidz), which perform the function of roots by aiding in absorption and helping to hold the plant in place. Many flat, green,

Figure 13.15 Bryophytes: (a) moss, *Polytrichium;* (b) liverwort, *Marchantia polymorpha;* (c) hornwort, *Anthoceros punctatus.*

a

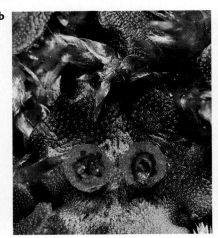
b

c

leaflike structures are attached spirally along the stalk. Water and nutrients are absorbed throughout the body of the bryophyte, so most grow in fairly damp places, and a few grow in water. Under dry conditions many mosses become dormant. When dormant, the life processes slow down, and the plant appears dead. Normal activities resume when the plant comes in contact with water. Because many bryophytes can photosynthesize in limited light, they often are found on the ground in forest ecosystems where other plants cannot grow.

One important group of mosses is found in boggy places in the cold and temperate parts of the world. These plants, from the genus *Sphagnum*, form peat bogs—small lakes and ponds completely filled with living and dead mosses. *Sphagnum* produces a very acidic condition in the water that keeps decomposers such as bacteria and fungi from growing, thus allowing these plants to build up through time, layer on layer. People in Ireland and other countries cut blocks of peat, dry it, and use it for fuel or to build small enclosures. Dry peat absorbs water quickly and holds the water well, characteristics that make peat attractive to gardeners, who add it to their soil to lower the pH and to increase the water-holding capacity.

Liverworts and hornworts grow in very moist areas, such as on the banks of streams where water spray keeps the soil wet. Some liverworts have flattened, lobed bodies, and other species may look "leafy," although they do not have true leaves. Liverworts have distinct top and bottom surfaces, with numerous rhizoids projecting from the bottom surface into the soil. The life cycle of liverworts is much like that of true mosses, and they can reproduce both sexually and asexually.

Hornworts look much like liverworts but are distinguished by their sporophytes, which are elongated capsules that grow like horns from the matlike gametophyte. Each photosynthetic cell of a hornwort contains a single large chloroplast, in contrast to the many smaller chloroplasts typical of the cells of most plants.

Although bryophytes are less complex than vascular plants, there is no evidence that they are the ancestors of vascular plants. The fossil record for bryophytes is small. The earliest bryophyte fossils are about 350 million years old. By that time, vascular plants were already established on land. The first fossils of vascular plants appear 50 to 100 million years *earlier* in the fossil record.

13.6 Club Mosses and Horsetails Are Seedless Vascular Plants

The roots, stems, and leaves of vascular plants make up the vascular system through which water, sugar, and dissolved nutrients move from one place to another in the plant.

Rhynia major (RY nyuh MAY jer), shown in Figure 13.16, best represents the oldest land plants. *Rhynia* had an underground stem that probably served to anchor the plant and to absorb water. From this underground stem grew upright branched stems that had stomates. At the tips of the stems were sporangia, which split open to release thick-walled spores. The closest living relative of *Rhynia* is the whisk fern, *Psilotum* (sy LOH tum), shown in Figure 13.17.

The club mosses (division Lycophyta; ly KOF ih tuh) are evergreen plants that seldom grow more than 40 cm tall. Although the word *moss* is a part of their name, they are not true mosses. Their branching, horizontal stomate stems grow on the surface of the soil or just below it. The most noticeable part of a club moss plant is an upright branch growing from one of these

In the succession of a peat bog, the *Sphagnum* forms a floating mat that is a suitable habitat for other plants.

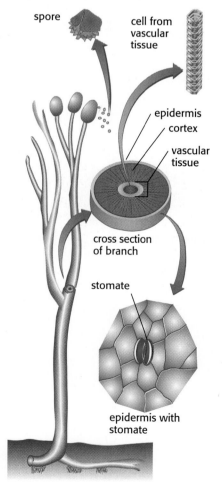

Figure 13.16   A reconstruction of the extinct plant *Rhynia*. It had no leaves or roots and stood about 30 cm high. Sporangia were produced at the tips of upper branches. The presence of stomates indicates that photosynthesis occurred in the branches. The cells of vascular tissue (upper right) were oblong and sometimes hollow. Thick-walled spores were produced in fours in the sporangia, an indication that the spores were produced by meiosis.

spore

cell from vascular tissue

epidermis

cortex

vascular tissue

cross section of branch

stomate

epidermis with stomate

Figure 13.17 The whisk fern, *Psilotum*. Note the sporangia along the stems.

Psilotum reproduces vegetatively by gemmae on stem tips that fall to the ground and grow directly into new plants.

stems (see the chapter opening photo). Club mosses reproduce by spores, which are produced on modified, specialized leaves. In many species, these leaves form club-shaped cones at the tips of short, upright stems. The name *club moss* is derived from this feature. Club mosses are rather common in the eastern and northwestern United States and often are used to make Christmas wreaths. They rarely grow in the dry states of the Southwest.

Horsetails (division Sphenophyta; sfen OFF ih tuh) have hollow, jointed, upright branches that grow from horizontal underground stems. Their small, scalelike leaves grow in a circle around each stem joint. Spores are produced in conelike structures at the tips of some of the upright branch- **4** es, such as those shown in Figure 13.18. In middle latitudes, horsetails rarely reach a height of two meters, but in the American tropics one species may grow several meters tall. They are found in moist places, such as along streams. Horsetails are harsh to the touch; their tissues contain silica, a compound present in sand. Because American Indians and the pioneers scrubbed pots and pans with them, they are commonly called scouring rushes.

Relatives of the club mosses and horsetails can be traced back about 430 million years. During the Coal Age, about 300 million years ago, great

Figure 13.18 Spores of *Equisetum* are produced in these conelike structures (a). The vegetative stage of *Equisetum* (b).

Paleontologist

Dr. Gurat Vermeij (pronounced Ver May) is a paleontologist and a professor at the University of California at Davis. As a paleontologist, he studies life forms from the ancient past through the examination of fossils. Dr. Vermeij specializes in the natural history of mollusks and their relationships with ancient predators. According to his research, mollusks have adapted to the presence of ever more efficient predators by developing progressively heavier and more protective shells. Dr Vermeij feels that this reaction to natural enemies provides a better explanation of how mollusks evolved than the usual explanations based on climatic changes and other large-scale physical factors.

Dr. Vermeij was born in the Netherlands and raised in New Jersy. His proximity to the Atlantic seaboard allowed him to pursue his early interest in shell collecting. He has developed this interest through world-wide travel and extensive field study. He has traveled widely in the South Pacific, Central and North America and Africa. His research on predatory patterns on mollusks has involved investigating shell geometry, breakage patterns, scarring, hole patterns, and evidence of repair following encounters with their marine enemies.

A graduate of Princeton and Yale Universities, Dr. Vermeij holds a particular distinction among paleontologists. He has been completely blind since age 3 and has never actually seen a single organism he has studied. Instead, he uses his fingers to feel all of the subtle features of the ancient marine creatures that he investigates. His sense of touch is so refined that he often is able to identify mollusks to subsecies based on the most minor of changes in the shapes, textures, and thickness of shells. Dr. Vermeij works closely with his wife and professional colleague, Dr. Edith Zipser, also a professor at the University of California at Davis. Together they have collected shells from their extensive travels and formulated their ideas on the evolutionary patterns of mollusks. Dr. Vermeij has risen above his physical restrictions and opposition from skeptical peers to be a premier level paleontologist and editor of *Evolution,* the field's most prestigious journal.

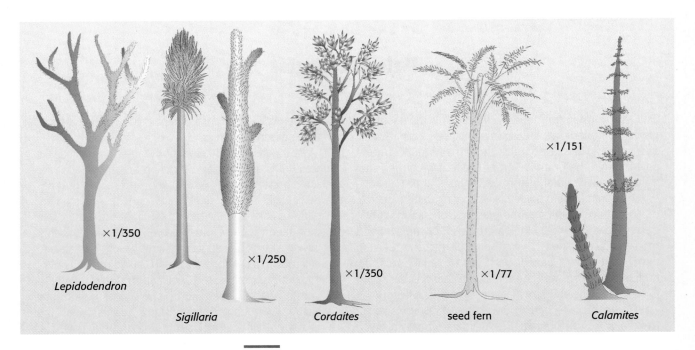

×1/350

×1/250

×1/350

×1/77

×1/151

Lepidodendron

Sigillaria

Cordaites

seed fern

Calamites

Figure 13.19   Some trees of the coal-age forests: *Lepidodendron* and *Sigillaria* were club mosses. *Cordaites* were primitive cone-bearing plants. The seed ferns have no living species. *Calamites* were horsetails.

Fern spores are easy to grow. Spread fern fronds with spore cases on white paper. Leave them undisturbed. The spores will be expelled as the fronds dry out. Dry sporangia also burst when placed in a drop of glycerol; this may be observed under low power of the microscope. Spores can be stored for several years in capped vials. To germinate and grow the gametophytes, float a few spores on a solution of commercial, indoor plant food in water about 1 cm deep in a petri dish. Cover and leave in indirect sunlight or under a grow lamp. Allow about 6 weeks for the gametophytes to form fully and 12 weeks to see the sporophytes growing out of them.

Figure 13.20   Sporangia on the underside of a licorice fern leaf, *Polypodium glycyrrhiza.*

parts of North America were covered by shallow swamps and seas. The warm and wet environment allowed plants to grow year round. Under these conditions giant relatives of today's club mosses, horsetails, and ferns, as well as seed-producing plants, covered the land. Some of these plants were more than 20 m tall (Figure 13.19). As they died, their large stems were covered with mud and soil before they completely decayed. A tremendous number of plants from the Coal Age were compressed over long periods of time and under high temperature and great pressure. Eventually, they became fossil fuels, mainly coal and some natural gas.

13.7   Fern Leaves Grow from Underground Stems

Ferns (division Pterophyta; ter OFF ih tuh), like the club mosses and horsetails, reproduce by spores. At certain times of the year, small brown spots develop on the undersides of fern leaves, as shown in Figure 13.20. Each spot consists of a cluster of sporangia. Each sporangium produces a large number of spores, which are almost microscopic in size.

When a sporangium is mature, it opens and the spores are thrown out   **4** into the air. Spores are very light and can be carried for incredible distances by wind. If a spore falls in a suitably moist place, it germinates and develops rapidly into a thin, green, heart-shaped plant that is rarely over one centimeter in diameter (see Figure 13.21a). This small gametophyte plant, which is completely different from the familiar fern with its large leaves, is seldom noticed in the woods. The gametophyte produces the sperm or eggs, or in some cases, both. Like the moss sperm, the fern sperm must swim in a film of water to fertilize the egg. The zygote pro-   **3** duced by that fertilization eventually grows into the conspicuous spore-bearing fern plant that most people recognize. This large plant is the sporophyte generation of the fern, seen in Figure 13.21b.

The ferns native to most of the United States are shade-dwelling plants with underground roots and stems. From these stems, roots grow down-

Figure 13.21   Fern gametophyte (a) and sporophyte (b).

ward and new sets of upright leaves appear above ground each spring. In Hawaii and elsewhere in the tropics, many species of ferns have stems that grow upright. These tree ferns may reach a height of 20 m with leaves 5 m in length, as Figure 13.22 illustrates. Most fern species are found in the tropics, but many can be found in forest ecosystems around the world.

CONCEPT REVIEW

1. How do bryophytes differ from vascular plants?
2. How does reproduction in seedless vascular plants differ from that in bryophytes?
3. What events must occur for a fern sporophyte to be produced from a gametophyte?
4. How is the reproductive strategy of ferns and horsetails adaptive?

A student who would like to prepare some herbarium specimens might well start with ferns. They are easy to collect and press. Further, they do not suffer much of the color loss that often is disappointing when pressing flowering plants.

Seed Plants

13.8   Many Conifers Are Evergreens

Humans have used plants for food, clothing, shelter, and medicines. The conifers (division Coniferophyta) are woody vascular plants with seeds borne in cones. Conifers provide all of the paper pulp and most of the lumber used in home construction and furniture. They include pines, firs, spruces, and junipers, among many other species. Figure 13.23 shows several conifers and related plants.

Almost all conifers are trees or shrubs, and all are at least somewhat woody. Many have leaves that are like needles or scales, such as those in Figure 13.24, and most of these plants are evergreen. An evergreen tree or shrub appears green throughout the year because it always maintains most of its leaves. A few leaves die at different times of the year and drop to the ground. Although the number of conifer species is small, the number of individual conifers is enormous.

Many common conifers are well adapted to life in dry habitats. For example, although pine trees may grow where there is much snow, the snow is really frozen water and is not available for growth. In the spring, 1 much of the snow evaporates, and the melted snow may run off into streams before it soaks into the soil. The leaves of pines are well adapted for growth in dry places. The long, narrow needles reduce the amount of water lost by evaporation. In addition, a pine needle often is covered by a thick, waxy cuticle that further reduces water loss.

Guidepost

How are seed plants adapted to life on land?

A relative of conifers, the maidenhair tree, or *Ginkgo biloba*, often is called a "living fossil." Derived from a wild population indigenous to china and Japan, it has remained basically unchanged since the late Paleozoic age.

Figure 13.22   Tree fern on the island of Sumatra.

Figure 13.23   Conifers and related plants: (a) Ponderosa pine, *Pinus ponderosa;* (b) cycad, *Dioon edule;* and (c) maidenhair tree, *Ginkgo biloba.* Seeds of these plants are not enclosed in tissues as are those of flowering plants.

Inquisitive students may present you with embarrassing specimens: the "fruits" of yew and ginkgo, both of which may occur in the autumn around many schools. A little examination shows that a yew seed, though deep in pulp, is not completely within it as, for example, is the seed of a peach. Although a ginkgo seed is completely encased in pulp, the pulp is part of the seed itself, not a structure developed from an ovary as are the pulp and stone around a peach seed. Therefore, ginkgoes and yews are said to bear naked seeds, not fruits.

Fresh pine branches can be collected at any time during the year. Only in the spring, however, will the branches have both the pollen and seed cones. The pollen-bearing cone consists of an axis bearing spirally arranged microsporophylls corresponding to the stamens of a flowering plant. The seed-bearing cone is what normally is thought as the pine cone. Old seed cones are available almost any time of the year. Preserved young seed and pollen cones can be obtained from Ward's. Pollen producing cones collected in the spring can be kept for later use by storage in plastic bags in a freezer.

Conifers reproduce by seeds that are attached to the upper surface of cone scales. A seed developing in a cone may be protected by the scales, somewhat as a small coin may be concealed between the pages of a book. **2** If two scales are separated slightly, you can see the seed between them, as shown in Figure 13.25. Thus the seeds in cones are not completely covered as they are in the fruits of flowering plants.

Conifer spores are of two types and are produced in different cones, illustrated in Figure 13.26. Typically, pollen develops from spores in the small male cone, and pollination occurs in the spring when the pollen is blown onto a female cone. The larger, more familiar female cones contain the ovules. Pollen grains, the male gametophytes, lodge in a sticky substance secreted by the ovule. Once lodged, the grains develop pollen tubes within which the sperm are formed. Within the ovule, the female gametophytes develop and produce eggs. Fertilization occurs approximately a year after pollination, and the seed requires an additional year to mature. Table 13.1 summarizes these events.

Figure 13.24   Many conifer leaves are needlelike or scalelike. (a) Single needles of Douglas fir, *Pseudotsuga menziesii;* (b) scalelike leaves of juniper, *Juniperus chinensis "pfitzerii."*

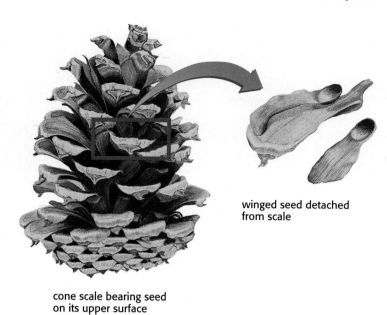

cone scale bearing seed
on its upper surface

winged seed detached
from scale

Figure 13.25   Pine cone and seed.

Table 13.1 Seed Development in a Pine

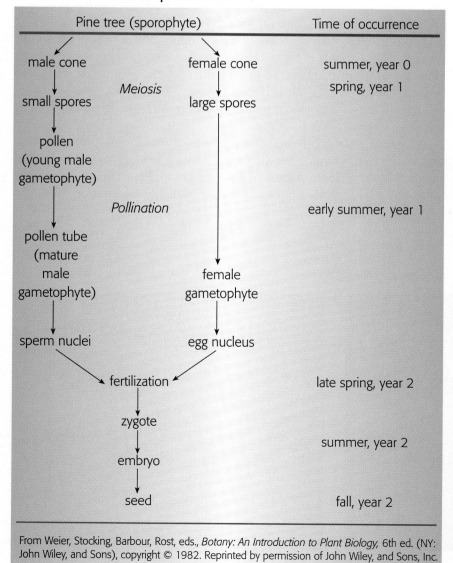

Pine tree (sporophyte)		Time of occurrence
male cone	female cone	summer, year 0
Meiosis		spring, year 1
small spores	large spores	
pollen (young male gametophyte)		
Pollination		early summer, year 1
pollen tube (mature male gametophyte)	female gametophyte	
sperm nuclei	egg nucleus	
fertilization		late spring, year 2
zygote		
embryo		summer, year 2
seed		fall, year 2

From Weier, Stocking, Barbour, Rost, eds., *Botany: An Introduction to Plant Biology*, 6th ed. (NY:
John Wiley, and Sons), copyright © 1982. Reprinted by permission of John Wiley, and Sons, Inc.

Figure 13.26   Male cone (a); female
cone (b) of the pinon pine, *Pinus edulis*.

a

b

Figure 13.27 A few examples of the diversity of flowers. (a) Indian paintbrush; (b) water lily; (c) elephant-head; and (d) bougainvilla.

13.9 Many Flowering Plants Have Special Pollinators

Many botanists support the recently proposed renaming of the division Anthophyta to Magnoliophyta, in keeping with the practice of naming a division for a representative group.

Try to have an assortment of flowers on hand for students to examine. Collect and press flowers in season or obtain several varieties from a florist.

Flowers are the distinguishing feature of the most successful division in the plant kingdom, the Anthophyta (an THOF ih tuh). Although we commonly appreciate flowers for their beauty, their major role is in reproduction. When a flower opens, it reveals the reproductive structures, the stamens and carpels. Insects, birds, bats, or other animals that visit the flower may pick up pollen from the anther and carry it to the next flower they visit. When pollen is transferred from a flower of one plant to a flower of another, the process is called cross-pollination. The main pollinators of flowers are insects. Pollen also may be transported from flower to flower by wind. Some flowers, however, seldom or never open. In this case, the pollen falls on the stigma of the same flower. These flowers are self-pollinating. Flower and vegetable gardeners say they "self." Most plants, however, have devices that prevent self-pollination.

The sepals and petals are not directly involved in seed formation, so a flower can function without them. In fact, in a few plants a flower may consist of only a single stamen or a single carpel. Petals and their adaptations, however, usually play a major role in flower pollination. Much of

Figure 13.28 Insect-pollinated flowers such as the aster in (a) usually are brightly colored. Wind-pollinated flowers often lack petals and sepals and produce an abundance of pollen. Shown in (b) is the male catkin of a willow, *Salix.* Note the many stamens.

the diversity among flowering plant species lies in their flowers (see Figure 13.27). This diversity usually is related to the way pollination occurs. If pollen is transferred from stamen to carpel by insects, the petals of the flower often are large and brightly colored, as in Figure 13.28a. The petals often have small glands that produce a sugar solution called nectar. **3** These adaptations attract pollinating insects. On the other hand, flowers in which pollen is transferred by wind usually have small sepals and petals or none at all. These flowers often are located high on the plant and produce an abundance of pollen (Figure 13.28b). Their carpels commonly have large, long, or feathery structures at the tips, which are covered with a sticky fluid. These adaptations increase the likelihood that some pollen will stick to the carpels.

The great variety of flowers has come about, in part, by the coevolution of flowers and pollination agents. For example, hummingbirds need lots of nectar to supply their energy needs. Although they do not have a good sense of smell, they can see the color red very well. Flowers pollinated by hummingbirds usually are well adapted to their pollinators, as you see in Figure 13.29. The columbine flowers shown in the illustration are red, have little or no scent, and produce copious amounts of nectar. The nectar is found at the bottom of a long tube formed by the red petals, a shape that makes it difficult for other organisms to rob the flower of nectar.

Hummingbirds, however, have long beaks that can probe the flower and reach the nectar. The stamens stick out in such a position that the hummingbird's head is dusted with pollen when it visits these flowers. When the hummingbird flies to another flower of the same species, the tip of the carpel is in a perfect position to have pollen from the hummingbird's head scraped onto it. Thus, the flower is pollinated while the hummingbird drinks nectar. Interactions of this type, in which both organisms become uniquely adapted to each other, have helped shape the way flowering plants have evolved.

13.10 Flowering Plants Produce Fruits with Seeds

After pollination and fertilization, seeds begin to develop. The carpel, often with other parts of the flower, develops into a protective fruit around the seed, as shown in Figure 13.30. There may be many seeds in a fruit. Each seed began its development when an egg cell in one ovule was fertilized by a sperm from one pollen grain. In flowering plants, then, an embryo protected within a seed, and seeds are protected within a fruit.

Part of the embryo in the seed consists of one or two modified leaves called **cotyledons** (kot ih LEE dunz). Another part is a beginning of a root. Each seed also contains a supply of food that is used when the embryo starts to grow. The food may be stored in the endosperm, or it may be stored in the embryo itself, usually in the cotyledon.

Beans and peas are examples of seeds. They are enclosed in protective pods, which are just one of many types of fruits. Each bean or pea contain a small embryo and a supply of stored food for its early development. The entire pea or bean can act as food for humans. Peas and beans can be germinated easily, and each embryo gives rise to a plant that, in turn, gives rise to new flowers and fruits. Apples and oranges are examples of fruit that contain a number of seeds. Under natural conditions, these fruits eventually decay, or are eaten by animals leaving their seeds behind to germinate and give rise to the next generation.

Ask students what type of ecological relationship exists between a plant and an animal in such cases. The relationship appears to be symbiotic.

Most fruits require seeds for normal development. If an apple has developing seeds on only one side, only that side will develop. Strawberries will not develop normally without seeds. The formation of fruit lacking seeds is called parthenocarpic fruit development. The fruit develops from many immature ovules. Examples are bananas, some melons, figs, and pineapples.

Have some examples of seeds on hand for students to examine. In addition to the old standbys, corn (a fruit) and beans, use almonds, avocados, pinons, and others.

spurs of flower

Figure 13.29. The columbine and hummingbird are uniquely adapted to each other.

You may wish to introduce the differences between annuals and perennials.

Ask students what are the oldest known trees. Redwoods may exceed the age of 2500 years. Some bristlecone pines (*Pinus arktata*) near the timberline in the White Mountains of California are even older, up to 4600 years.

Fruits show as much diversity as flowers, as indicated in Figure 13.31. This diversity is related to the method of dispersal of the fruits and their seeds—that is, how they are scattered from the parent plant. In Investigation 13.3 you will observe some of the structures that aid in seed dispersal. In many cases, part of the carpel becomes thick and fleshy, as in the fruits of peach, plum, and tomato. Fleshy fruits, often red in color, may be eaten by birds or mammals. The seeds in many such fruits have thick coats that permit them to pass through an animal's digestive system unharmed. They are dropped later at some distance from the parent plant. Many fruits are not fleshy but have other adaptations that aid in scattering **4** their seeds. These fruits may have spines that catch on the fur of an animal that brushes up against the plant. The fruit is carried from the plant and later falls off or is brushed off by the animal. Many fruits and seeds are lightweight and have special winglike projections that help them to be carried away from the plant by wind. An entire plant, such as the tumbleweed, can be broken off near the ground and blown about by the wind. As the tumbleweed bounces about, it drops its fruits (and the seeds within) all along its path.

13.11   Flowering Plants May Appear Very Different

There is great diversity in the size of flowering plants and in the life span of their shoots—the parts that appear above ground. Many flowering plants are trees. A tree bears leaves well above the ground where they are likely to receive more light than do the leaves of shorter plants. Because of their size trees can store large reserves of food in trunks and roots and can survive through bad years. Trees have relatively long life spans. A tree species flowers may survive even if a year's seed crop is destroyed. Most species of flowering plants are not trees, however. Some, such as roses and

Figure 13.30   Stages in the development of tomato fruit from flowers.

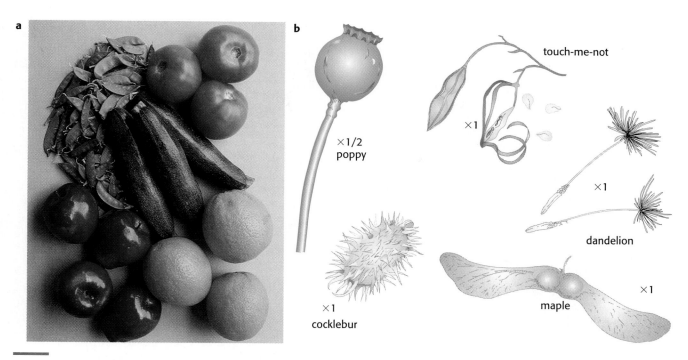

Figure 13.31 Each edible fruit is a mature ovary containing the seeds of the plant (a). Many fruits have structural adaptations for seed dispersal (b). Can you describe them? **Dandelion parachutes and maple wings carry on wind; poppy and touch-me-not scatter when touched; cocklebur attaches to animal fur and breaks when brushed off.**

raspberries, are woody shrubs. Others, such as ivy, grapes, and hundreds of tropical species, are woody vines that grow on rocks, walls, or other plants. Most, however, are neither trees, shrubs, nor vines. Instead, they are nonwoody, or herbaceous (her BAY shus), plants such as those in Figure 13.32.

Flowering plants are divided into two large classes—the **monocots** (MON oh kotz) and the **dicots** (DY kotz). In monocots, (monocotyledons), the embryo contains a single cotyledon. The monocots include grasses and grain-producing plants such as wheat, rice, and corn—the chief food plants of the world. The pasture grasses that feed cattle, another source of human food, are also monocots. Without monocots, the human population never could have reached its present state.

The seeds of the dicots (dicotyledons) have two seed leaves. This class is larger than the monocot class. Most fruits and vegetables, such as carrots, lettuce, apples, and grapes, are dicots. In addition, the so-called hardwoods used in furniture, flooring, hockey sticks, and baseball bats come from dicot trees. Almost all shade trees are dicots, also. Figure 13.33 illustrates the major differences between monocots and dicots.

In middle latitudes, the distinction between herbaceous and woody plants is fairly clear; herbaceous plants die back to the ground in winter. In the tropics, however, the distinction is not clear at all: the treelike banana plant is essentially a large herbaceous plant.

Ask students what monocot plants other than grains are important human food. Examples include bananas, pineapples, onions, yams, sugar cane and taro.

Figure 13.32 Herbaceous plants include marigolds, common garden flowers.

CONCEPT REVIEW

1. How are conifers adapted to dry conditions?
2. How do conifers differ from flowering plants?
3. What role does flower color play in the reproduction of flowering plants?
4. How are fruits related to seed dispersal?
5. In what ways do monocots and dicots differ?

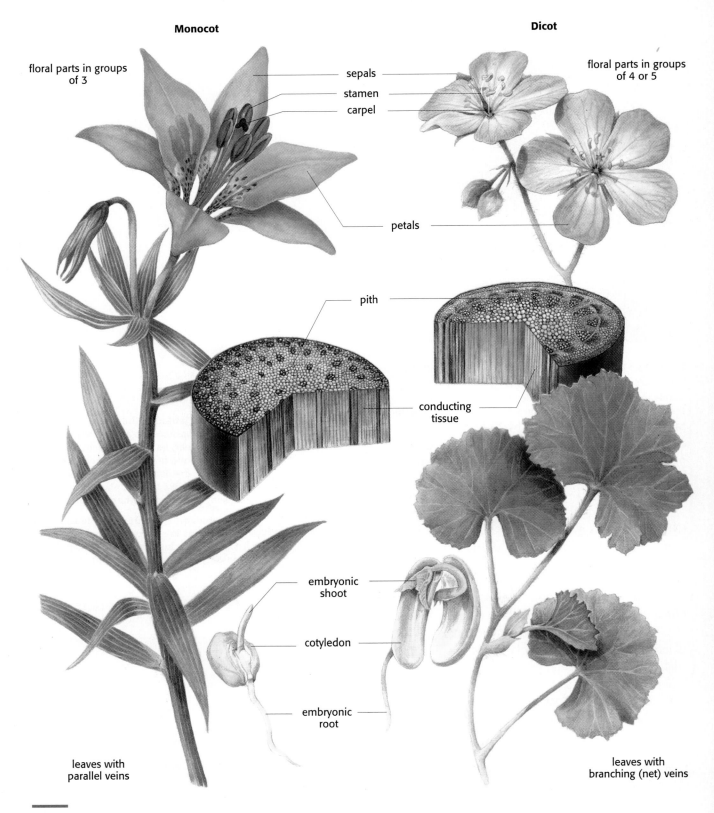

Monocot

floral parts in groups of 3

Dicot

floral parts in groups of 4 or 5

sepals

stamen

carpel

petals

pith

conducting tissue

embryonic shoot

cotyledon

embryonic root

leaves with parallel veins

leaves with branching (net) veins

Figure 13.33 Comparison of monocot and dicot characteristics.

Marjorie Leggit is a freelance biological illustrator. Biological illustrators illustrate reports, journal articles, and books prepared by researchers. Many illustrators work in natural history museums, painting and constructing the background scenery for different types of displays. These artists may recreate Alaskan ice floes to show off a group of polar bears or a South American hillside for a display of llamas.

Marjorie's family is composed of writers, sculptors, painters, and woodcarvers, so an interest in art comes naturally to her. Her interest in science, Marj thinks, may have developed because she could do so much drawing in science classes.

Although a biological illustration course in college sparked her interest, Marjorie continued to study fine arts. After studying for a year in Europe, however, she realized she missed science and the technical art that goes with it. She returned to college and designed her own program of independent studies that led to a degree in scientific art. Her program included an internship at the Denver Museum of Natural History in Denver, Colorado. While searching for a full-time job after graduation, Marj illustrated a professor's botanical field guide. Her persistence was rewarded eventually with a position at the Field Museum of Natural History in Chicago, Illinois.

Marj later returned to Denver to be married. Reestablishing her career there proved to be difficult because opportunities for biological illustration were far fewer than in very large cities where more research is conducted. For seven years she worked at art-related jobs, including geological drafting and computer-aided graphics. Although she learned a whole new field of art, Marj was not satisfied with the direction of her career.

To keep in touch with the field of biological illustration, she contacted the Guild of Natural Science Illustrators (GNSI) in Washington, DC. Knowing Colorado had nothing similar, she founded the Colorado Chapter of GNSI and made friends with a group of local illustrators.

In 1986, Marj went into business for herself as a biological illustrator and graphic artist. She contacted old and new local clients as well as publishers throughout the country. She spent many hours researching the possibilities for assignments and marketing her skills.

Marj did many of the illustrations used in this textbook, including Figure 13.33. To ensure accuracy in her work, Marj does a great deal of research and uses many references and photos. She consults extensively with authors when planning an illustration. It is essential that she understand the author's intent, what the illustration is meant to portray, and how it relates to the text and to other illustrations. Proper interpretation depends on good communication between author and artist. Marj strongly believes that the hard work involved in creating illustrations has its rewards. At times the hours are long, the deadlines close to impossible, and the challenges seemingly insurmountable. However, she is doing exactly what she wants to be doing—full-time biological illustration.

This investigation broadens students' firsthand experience with a variety of organisms as they consider evolutionary relationships and retention or change of ancestral characteristics. Because the investigation precedes most of the chapter, students can observe certain traits of plants before reading about them. Suggest they use the chapter and *A Catalog of Living Things* (Appendix 4) as references. Teams of 6 are workable.
    ***Time:* One class period.**

INVESTIGATION 13.1   Increasingly Complex Characteristics

Biologists sometimes use pairs of terms such as *less complex* and *more complex* when discussing diversity among organisms. A species that has changed little from its ancestors is said to be less complex than organisms that differ greatly from their ancestors. Conversely, a species that has few of the characteristics of its ancestors is said to be more complex. After studying many types of evidence in the fossil record and in living organisms, scientists have reached fairly general agreement about which characteristics have been in existence for a long time and which are more recent. Table 13.3 is based on such studies. Because there may be many degrees of complexity, the terms *less complex* and *more complex* are not absolute, but they are useful for making the types of comparisons you will make in this investigation.

Materials (per team of 6)

10 labeled specimens of organisms of various kingdoms and divisions
compound microscope

stereomicroscope or 10X hand lens
microscope slides
cover slips

Procedure

1. In your data book, prepare or tape in a table similar to Table 13.2, with enough lines for all 10 specimens.

Table 13.2

Name of Organism	Numerical Values of Choices Made	Total Complexity Score	Rank
1.			
2.			
etc.			

2. Determine the complexity score for each of the labeled specimens. Start at the left of Table 13.3. Arrows from the starting point lead to two descriptions. Choose the one that fits the organism you are scoring. More complex organisms such as plants are represented at each station as well as some less complex organisms from other kingdoms.

3. Proceed across Table 13.3 by following the arrows and choosing in each column the description that best fits each organism. Continue as far as the arrows go.

4. With each description there is a number. As you proceed, record the numbers of your chosen descriptions in the second column of your data table. The complexity score for the organism is the sum of all the numbers appearing after the descriptions you used in working through Table 13.3. The more alike two organisms are, the more alike their scores will be. The greater the difference between two organisms, the greater will be the difference in their scores. More complex organisms, such as plants, will have high scores (maximum 26), and less complex organisms, such as prokaryotes, will have low scores (minimum 3).

5. When you have the complexity score for each of the organisms, give the organism with the lowest score a rank of 1 and the organism with the highest score a rank of 10. Then rank the rest of the organisms according to their scores. Record the rankings in the column at the right side of your data table.

Discussion

1, 2. The greater the differences in scores between characteristics at any one dichotomy, the more important the difference was considered to be by the maker of the table.

3, 4. This entails listing the characteristics common to organisms that scored low and high, respectively.

5. Not only would roots help anchor a plant in the ground, they would provide great surface area through which the plant could absorb water. Stems with vascular tissue that transports water to all parts of the plant would allow a plant to grow tall, so it could obtain more light for growth. Seeds protect the

Table 13.3 Key for Determining the Complexity Score for an Organism

BEGIN HERE

- Plant kingdom. Has structures that look like roots, stems, or leaves. (4)
  - No true roots, stems, or leaves. May bear hairlike rhizoids instead of roots. Not more than 10 cm tall. (4)
    - Leaflike structures with midribs; spaced equally around stalk. (5) → **moss 7**
    - Leaflike structures without midribs, in two rows. (4) → **liverwort**
  - True roots, stems, or leaves. Usually with vascular tissue. Usually more than 10 cm tall. (4)
    - No seeds. Reproduction by spores. (6) → **ferns 8**
    - Produce seeds. (5)
      - Flowers present. (5)
        - Flower parts in threes. Leaves parallel-veined. (4) → **monocots**
          - trees or shrubs (1)
            - trees (1)
            - shrubs (2)
          - woody vines or herbs (2)
            - woody vines (1)
            - herbs (2) **10**
        - Flower parts in fours or fives. Leaves net-veined. (4) → **dicots**
          - trees or shrubs (1)
            - trees (1)
            - shrubs (2)
          - woody vines or herbs (2)
            - woody vines (1)
            - herbs (2)
      - No flowers. Seeds borne in cones. Leaves usually needlelike. (5) → **conifers 9**
        - trees (1)
        - shrubs (2)
- Not in plant kingdom. Does not have structures that look like roots, stems, or leaves. (1)
  - Not green. Usually white, gray, brown, or yellow. (1)
    - All individual parts microscopic in size. Occurring singly or in chains. (1) → **yeast 1**
    - Some or all individual parts not microscopic in size. (2)
      - Umbrellalike, shelflike, or spherical. Usually from 1 to 15 cm high. (3) → **basidiomycetes 3**
      - Hairlike, slender, usually fuzzy or powdery. Usually less than 1 cm high. (2) → **mold 2**
  - Green, blue-green, or gray green. (2)
    - Body crustlike and flat or upright and branched. (3)
      - Bright green. Generally on damp soil. (5)
      - Gray-green. Generally on rocks or tree trunks. (4) → **lichen 4**
    - Body not flat or leaflike. Usually aquatic. (2)
      - Blue-green. Prokaryotic. Chlorophyll diffused throughout cells, not in distinct structures. (3) → **5**
      - Grass-green. Eukaryotic. Chlorophyll in definite structures only. (4) → **6**

young embryo from various hazards, including drying out, until the proper environmental conditions occur for germination and growth. Flowers attract insects and other pollinators that pick up pollen grains containing the sperm of the plant. The sperm then are transported from one plant to the egg of another with out the need for water—an adaptation to land not present in plants with swimming sperm, except for the most primitive gymnosperms, such as the cycade and ginkgo.

This investigation develops the BSCS themes of complementarity of structure and function, diversity of type and unity of pattern, complementarity of organism and environment, the basis of the genetic continuity of life, and growth and development in the individual. Students will study alternation of generations and the diversity of organs in plants that are adapted for sexual reproduction. Teams of 2 are workable.
Time: Two class periods.

Safety

Provide small corks for the dissecting needle tip, and store the needles and scalpels in a tray when not in use.

Procedure Part A

2. **Spores.** The wind shakes the capsule and the spores are shed. They are carried or distributed by air currents. Spores have fairly thick walls composed of waterproof material that keeps them from drying out.

Discussion

1. Basing your conclusions on the way the complexity score key was designed, list some of the most important differences among the organisms you observed.

2. What are some of the less important differences?

3. Using the information in the table, list the characteristics you would expect to find in one of the less complex organisms.

4. Do the same for one of the more complex plants.

5. Evidence exists that today's land plants evolved from water-dwelling ancestors. Plants that live on land are in constant danger of drying out. Suggest how each of the following characteristics found in a more complex plant would help the plant live on land: (1) roots; (2) stems that contain vascular tissues; (3) seeds; and (4) flowers.

INVESTIGATION 13.2   Reproductive Structures and Life Cycles

Although their reproductive organs differ as do the environments in which they live and reproduce, the basic principles of sexual reproduction are the same in a moss, a flower, a bee, and a human. In this INVESTIGATION you will learn how the structures of a moss and a flower serve reproductive functions in their respective environments.

Mosses form mats on logs and on the forest floor, growing best in damp, shaded environments. Sporophytes grow out of the tops of the gametophytes and often look like hairs growing out of the mat of moss. Mosses cannot reproduce sexually unless they are wet. Flowering plants, on the other hand, are found in many different environments and climates. They need water to live but not to reproduce.

Materials (per team of 2)

2 microscope slides	prepared slide of moss male and
2 cover slips	female reproductive organs
dissecting needle	15% sucrose solution
scalpel	moss plant with sporophyte
forceps	fresh moss
compound microscope	gladiolus flower
stereomicroscope or 10X hand lens	other simple flowers for comparison
modeling clay	fresh bean or pea pods
prepared slide of filamentous stage	
of moss	

Procedure

PART A Moss

1. Examine a moss plant with sporophyte attached. The sporophyte consists of a smooth stalk terminated by a little capsule. Separate the two generations by pulling the sporophyte stalk out of the leafy shoot of the gametophyte.

2. Using a dissecting needle, break open the capsule of the sporophyte into a drop of water on a slide.

  Needles are sharp. Handle with care.

Add a cover slip and examine under the low power of a compound microscope. What are the structures you observe? How are these structures distributed in nature? How are they adapted for life on land?

3. Most moss spores germinate on damp soil and produce a filamentous stage that looks like a branching green alga. Examine a prepared slide of this stage.

4. The filamentous stage gives rise to the leafy shoot of the gametophyte. Using forceps, carefully remove a leafy shoot from the fresh moss. How does this shoot obtain water and nutrients for growth?

5. The reproductive organs of the gametophyte are at the upper end of the leafy shoot. Examine a prepared slide of these organs under the low power of a compound microscope. The male sex organs are saclike structures that produce large numbers of sperm cells. The female sex organs are flask-shaped and have long, twisted necks. An egg is formed within the base of the female organ. How does a sperm reach the egg? Would you expect to find moss plants growing where there was little or no water? Explain. The union of the egg and sperm results in a cell called the zygote. Where is the zygote formed? What grows from the zygote?

PART B Flowers

6. Examine the outside parts of a gladiolus flower. The outermost whorl of floral parts may be green and leaflike. These green sepals protected the flower bud when it was young. In some flowers, such as lilies, the sepals look like an outer whorl of petals. Petals are usually large and colored and lie just inside the sepals. Both sepals and petals are attached to the enlarged end of a branch. These parts of the flower are not directly involved in sexual reproduction. What functions might petals have?

7. Strip away the sepals and petals to examine the reproductive structures. Around a central stalklike body are 5 to 10 delicate stalks, each ending in a small sac, or anther. These are the male reproductive organs, or stamens. Thousands of pollen grains are produced in the anther. The number of stamens varies according to the type of flower. How many stamens are present in the flower you are using? How is pollen carried from the anthers to the female part of the flower?

8. If the anthers are mature, shake some of the pollen into a drop of 15% sucrose solution on a clean slide. Add a cover slip and examine with the low power of a compound microscope. What is the appearance of the pollen? How is the pollen adapted for dispersal?

9. Make another pollen preparation on a clean cover slip. Use modeling clay to make a 5-mm-high chamber, slightly smaller than the cover slip, on a clean slide. Add a small drop of water to the chamber and invert the pollen preparation over it. Examine after 15 minutes and again at the end of the lab period. What, if any, changes have occurred? (If no changes have occurred, store the slide in a covered petri dish containing a piece of cotton moistened with water, and examine it the next day.)

10. The central stalk surrounded by the stamens is the female reproductive organ, or carpel. It is composed of an enlarged basal part, the ovary, above which is an elongated part, the style, ending in a stigma. How is the stigma adapted to trap the pollen grains and to provide a place for them to grow?

11. Use a scalpel to cut the ovary lengthwise.

 Scalpels are sharp; handle with care.

Using a hand lens or stereomicroscope, look at the cut surface. How many ovules can you see? Each ovule contains one egg. To what stage of the moss life cycle is the ovule comparable? How close to the egg can the pollen grain get? If the pollen grain cannot get to the egg, how do the sperm produced by the pollen reach it? To what stage of the moss life cycle is a pollen grain comparable?

4. **The leafy shoot of a moss plant has rhizoids, which have a function similar to roots and root hairs.**

5. **The sperm swim to the female organ through water. Because the plants must be wet for fertilization to occur, mosses cannot reproduce in a dry area. In the base of the female reproductive organ. An embryo.**

Part B

6. **The petals identify the flower for a particular species of insects and advertise the presence of nectar.**

7. **In lily and gladiolus, the number is six; it varies with the flower used. Pollen may fall from the anther to the stigma or be carried by insects, birds, or wind. In some aquatic plants, water is the agent for pollination.**

8. **Shape depends on the pollen examined. The thick walls prevent drying during dispersal.**

9. **If the pollen germinates, a pollen tube should be visible after some time.**

10. **The stigma is sticky and moist.**

11. **Students must count the little white structures inside the ovary of the flower. The female gametophyte is contained within the ovule. The pollen grain itself gets no closer to the egg than the surface of the stigma. There must be a pathway (the pollen tube) for sperm cells to follow to the egg. The pollen grain is a male gametophyte.**

12. **The ripened ovary—a fruit. The ovule. A pea or bean seedling. An embryo.**

12. The union of egg and sperm causes extensive changes in the female reproductive parts. Fertilization of the egg stimulates the growth of the ovary and the enclosed ovules. Carefully examine a fresh bean or pea pod. Open the pod to find the seeds. What part of the female reproductive apparatus is the pod of a bean or pea? What is the origin of a seed? If you plant ripe bean or pea seeds and water them, to what will they give rise? What can you conclude develops within a seed as a result of fertilization?

13. If time permits, examine other types of flowers. Compare the numbers of various parts and the ways the parts are arranged with respect to each other.

14. Wash your hands thoroughly before leaving the laboratory.

Discussion

1. In alternation of generations in a moss, which is the predominant, independent generation? Which is the less conspicuous generation?

2. Compare the life cycle of a moss (with alternation of generations) with your life cycle (with no alternation of generations).

3. Would you expect the most variation in flowering plants or in those reproducing by asexual means? Explain.

4. Compare and contrast the life cycle of a moss with that of a flowering plant.

5. Do flowering plants represent more or less adaptation to a land environment than mosses? Explain.

Discussion

1. The gamethophyte. The sporophyte.
2. In the human life cycle, there is no stage corresponding to the vegetative gametophyte.
3. Flowering plants have greater genetic variability; asexual reproduction results in offspring exactly like the parents.
4. In a moss, the gametophyte is the dominant generation, and there is a small but visible sporophyte. In a flowering plant, the sporophyte is dominant, and the gametophyte is microscopic.
5. Flowering plants are more highly developed for a land existence because they do not need water during the time of fertilization. Thus, they can withstand drier conditions than the mosses.

This investigation continues the study of the evolution of flowering plants and allows students to develop their own dichotomous keys based on the fruits used in the investigation. Students should understand fruits and seeds as survival mechanisms important in the evolution of plants, particularly because two-thirds of all plant species form seeds. Teams of 3 allow for small-group interaction.
    Time: One class period for the activity: one-half period for follow-up discussion.

INVESTIGATION 13.3   Fruits and Seeds

The survival of plants depends on their ability to reproduce. In seed producers, the most complex plants, reproductive ability is enhanced by mechanisms that protect and disperse the seeds (and fruits) so the seeds do not compete for nutrients, light, and water, with the parent plants. Dispersal increases the likelihood that some seeds will not be eaten. In flowering plants, seeds are protected by the tissues of the mature ovary, or fruit. (Conifers, which lack an ovary, produce naked seeds on cone scales.)

In this investigation, you will try to determine how seeds of various fruits are dispersed. You also will use the fruits to make a dichotomous key, which can help you distinguish between objects by focusing on their similarities and differences.

Materials (per team of 3)
set of fruits

Safety

Remind students not to eat anything in the laboratory. If you ask students to collect fruits in the field, caution them to collect only those with which they are familiar.

Procedure

1. The maple, ash, poplar, milkweeds, and other fruits with wings, hairs, and similar appendages, or those fruits that are lightweight, should take flight.
2. Those seeds/fruits that hook onto clothing or hair will be dispersed by adhering to animal fur.
3. Bright colors of many fruits serve as an attractant to birds and other animals.

Procedure
PART A Seed Dispersal

1. Blow gently at the fruits. What happens? Why? What parts of the fruits are important in allowing them to be wind-dispersed?

2. Gently rest the sleeve of your blouse or shirt on the fruits and then lift up your arm. Which fruits were lifted? What parts of these fruits are important in allowing them to be dispersed by animals?

3. Some fruits attract birds and other animals that eat the fruits but do not damage the seeds inside. What characteristics of the fruits might serve as attractants?

4. Some fruits contain a chemical that acts as a laxative. How might this function in seed dispersal? How might the contents of bird droppings assist in survival of the new seedlings?

PART B Dichotomous Key

5. Design a dichotomous key using the fruits from Part A. A dichotomous key should separate all the available fruits into individual categories. Assemble the fruits and review the characteristics you observed in Part A.

6. To form a dichotomous key, use characteristics that some fruits have and some do not, rather than characteristics that all share. For example, *fruits with spines* versus *fruits without spines* might be a good characteristic to use in building your key. On the other hand, *fruits that contain seeds* would not be a good characteristic because almost all fruits contain seeds.

7. To divide the fruits into two groups, choose one major characteristic not shared by all the fruits. Separate your fruits into two groups—one group that possesses the characteristic and another group that does not. In your data book, prepare a table similar to Table 13.2 but with the lines only, or tape in the table your teacher provides. Write the characteristic on the table as shown below:

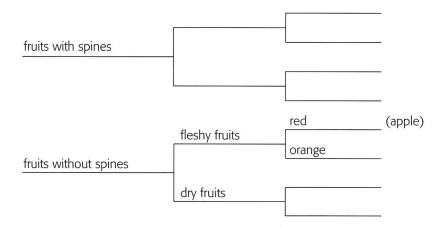

8. Now focus on all the fruits in just one of your two groups. Choose another characteristic that separates those fruits into two groups. Write this characteristic in the second column of the key.

9. Continue to select characteristics and write them on the key until you have produced a key that separates all the fruits into individual categories. In other words, your key should split the fruits into smaller and smaller groups until each fruit is in a group by itself. Once this is done, place the name of the fruit (maple, ash, etc.) next to the appropriate line on the right side of the key.

10. Exchange keys with another team. See if you can follow their key and they can follow yours.

11. Wash your hands thoroughly before leaving the laboratory.

Discussion

1. What are some of the ways seeds/fruits are dispersed?

2. Describe how a dispersal mechanism that relies on the presence of other organisms might develop in a plant species.

3. What would happen to the distribution of plants that produce cockleburs if they lived on an island having no animals?

4. Explain how poplar trees might come to inhabit an island in the middle of a large lake.

5. What is the purpose of a dichotomous key?

6. What characteristics did you use to develop your dichotomous key? What characteristics did your class as a whole use?

7. Could you develop a dichotomous key for organisms that looked identical to each other? Explain.

4. Those chemicals that act as a laxative guarantee a rapid movement of the fruit through a bird's digestive system. Other fruits require long periods of time in the digestive system in order to germinate. The droppings provide an available source of fertilizer.

5. Students may use any characteristics they choose to develop their dichotomous key, but they need to understand the ultimate goal of the key—to help identify objects by separating them into individual categories through a series of alternative choices.

6. Any characteristic could be used, but those that are easy to determine will be the best. For instance, presence of spines is a good characteristic to use, whereas number of hairs per square centimeter, while possibly useful, is tedious to determine.

Discussion

1. Seeds/fruits are dispersed by wind, animals, and water. Some stick to fur, feathers, or clothing. Others are enclosed in fruits that are eaten; the seeds are released undigested in the feces.

2. The animal dispersal mechanisms could develop if some seeds were picked up and then transported by animals to other areas where suitable growing conditions exist. Those seeds with this mechanism would continue to be dispersed from the parent plants as long as the animal was attracted to the fruits.

3. The plants probably would be limited to small areas where competition for resources was keen.

4. The poplar fruits could be blown by winds great distances across water.

5. A dichotomous key is useful in helping to identify unknown organisms or structures. It is a series of alternatives based on characteristics of the objects being studied, each alternative reducing the number of possibilities.

6. Depends on student responses.

7. Yes, but characteristics used would have to relate to the physiology or behavior of the organisms, rather than to their morphology.

SUMMARY

The ancestors of plants were probably multicellular algae. A multicellular organism has many advantages over a single cell, including size, division of labor, and ability to conserve water. Other adaptations that permit plants to absorb and hold water are roots, vascular tissue, and the cuticle. All plants have a life cycle that alternates between two different generations, a gametophyte and a sporophyte. In flowering plants, however, the egg cell, the gametophyte that produces the egg, and the spore that produces the gametophyte all remain protected in the flower. Mosses are nonvascular plants that lack roots, stems, and leaves. They possess some adaptations to life on land, but they still require a moist environment. Club mosses, horsetails, and ferns are seedless vascular plants, and many of their ancestors produced today's fossil fuels. Conifers and flowering plants are seed producing vascular plants, but flowering plants have a fruit that encloses their seeds. The great diversity of flowers and fruits is the result of the coevolution of plants and their agents of pollination or dispersal.

APPLICATIONS

1. Which organisms in this chapter would you consider to be less complex? Which would you consider to be more complex? Explain your answer.
2. Could descendants of the first land plants still be in existence today? Explain.
3. Would you expect to find a cuticle around the tissues of a plant's roots? Why or why not?
4. What advantages and disadvantages does a seed-producing plant have compared with one that produces only spores? How do you interpret the words *advantage* and *disadvantage* here?
5. Suppose you found a population of ferns in which the young sporophytes appeared only after a rainstorm. How would you explain this occurrence?

6. Why are nonvascular plants short?
7. Considering only spores, what reproductive advantage do flowering plants have over ferns?
8. What advantage do wind-pollinated plants gain by having flowers with very small sepals and petals?
9. The fossil record reveals a parallel increase in the diversity of flowering plants and the diversity of certain insect groups. How might you explain this?
10. Explain the following statement: The plant parts that furnish the greatest amount of human food are either seeds, roots, or underground stems.

PROBLEMS

1. Desert plants are land plants adapted to extremely dry conditions. What adaptations allow them to live with little water? Make a collection of desert plants or find out about them from a plant ecology text.
2. Fruits come in many sizes and shapes and with diverse appendages. Collect different fruits and try to identify their plants.
3. Bring to the classroom different pieces of furniture, tools, sculptures, or other objects made from wood. What qualities of the wood make it useful

for each object? Can you identify which tree provided the wood? Are some woods better than others for certain tasks?
4. Throughout history, plants have played religious and cultural roles in different societies. Many of these roles have been captured in paintings, sculptures, and tapestries. Find artwork that contains a plant or plants as an important part of the production. What does the plant signify?

5. Find or make some drawings of plants that were important during the time of the dinosaurs. Hang the drawings in your classroom. What happened to these plants? Did any of their descendants survive until today?

6. In what ways do plants that live entirely in water differ in structure, reproduction, and limiting factors from plants that live entirely on land?

7. Observe the plants that grow without human help in a city. Try to discover the characteristics that enable them to live successfully in an urban development.

8. How many of the different types of plants discussed in this chapter can be found in your locality? Consider wild and cultivated plants, indoor and outdoor plants, aquatic and terrestrial plants.

9. Choose some cultivated plants. Investigate the history of their domestication. Examples: wheat, apple, potato, cotton, corn, cabbage, sugarcane.

Eukaryotes: Animals

14

Members of the animal kingdom sometimes exhibit play behaviors. These grizzly bears, *Ursus arctos,* were photographed in the McNeil River, Alaska.

◆ The animal kingdom presents an enormous diversity of organisms, adapted to a wide variety of habitats and life-styles. Although they display many structural and functional differences, all animals share certain characteristics. All are essentially multicellular in organization and heterotrophic. Most are capable of locomotion for some part of their lives, and most are able to respond rather quickly to changes in their environments.

Animals show evidence for a common evolutionary origin. Thus, in spite of the great diversity among animals, their basic features are evident in even the most ancestral forms. Given similar habitats, similar responses to common problems may have occurred, even in unrelated groups. This chapter explores some of the problems animal groups encounter and some of the adaptations that have evolved to solve these problems. ◆

MAJOR CONCEPTS

- ◆ Animals are multicellular, heterotrophic, and show evidence for a common evolutionary origin.
- ◆ Most animals are motile and can reproduce sexually.
- ◆ Terrestrial animals show special adaptations to the land environment.
- ◆ There are physical reasons for a limit in the size of cells.
- ◆ Animals have consistently adapted to the demands of their environments.

KNOWLEDGE CHECK

- ◆ How has selection affected the structure of animals?
- ◆ What general types of structures and functions do all animals have?

The Animal Way of Life

14.1   Animals Adapt to the Demands of Their Environments

Animals are heterotrophic. Therefore, they usually must seek food. As a result, adaptations that allow animals to be **motile** (MOH tuhl), or capable of movement from place to place, result in the greatest survival success. Motility is easier if the organism is elongated in the direction of movement. Fish, for example, are streamlined in a manner that minimizes water resistance as they swim. Motility is helped further if the sensory organs—organs that detect food, light, and other stimuli—are concentrated in the end that meets the environment first, usually the head. An organism can move more easily if it has a balanced body, with neither side heavier or larger than the other.

Animals have the type of body plan that is best suited to their lifestyle (see Figure 14.1). For motile animals, the term **bilateral symmetry**

> ### Guidepost
>
> What are the problems of animal life, and what are some of the solutions?

Figure 14.1 Body symmetry.

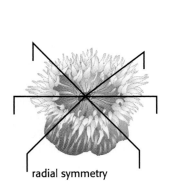

radial symmetry

bilateral symmetry

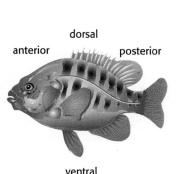

anterior   dorsal   posterior

ventral

343

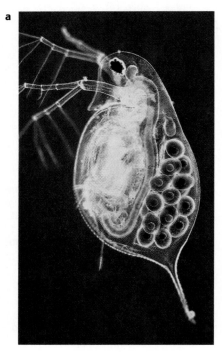

Figure 14.2 Animals vary greatly in size. The microscopic *Daphnia* (a) contains perhaps a few hundred cells (X40); the elephant (b) millions.

(by LAT er ul SIM eh tree) is used to describe the body plan. In bilateral symmetry, the right side is an approximate mirror image of the left side. Motile animals have an **anterior,** or head end, a **posterior,** or tail end, and **dorsal** (top) and **ventral** (bottom) surfaces.

As active movement requires coordination, successive adaptations have produced an area in which sensory organs and nerve cells are gathered—a head—which usually is found in the anterior end. Structures that help a motile animal capture prey, such as claws and teeth, usually are located toward this end. Digestive, excretory, and reproductive structures usually are located toward the animal's posterior.

Organisms that are **sessile** (SES il), or nonmotile, often have a body plan best described by the term **radial** (RAYD ee ul) **symmetry.** Their body parts radiate from a center, much as spokes radiate from the hub of a bicycle wheel. Radial symmetry allows sessile or drifting **aquatic** (uh KWAH tik) (water-dwelling) animals to interact with the environment from all sides. Sessile organisms, however, must deal with certain limitations. They must depend on food that floats by them, or they must create water currents to bring food to them. They cannot actively seek mates and, therefore, must rely on the chance meeting of eggs and sperm released into the water.

Every environment imposes special demands on the organisms living there. Sea water is the most uniform and least stressful environment for animal life. Oxygen is usually adequate, temperatures and salt content are fairly constant, and there is little danger that the organism will dry up. In contrast, the salt and oxygen contents of fresh water vary greatly.

Organisms that live in water have special adaptations. Gills, for example, allow the organisms to make use of the oxygen found in water, and circulatory systems (systems that transport materials) help to maintain the organisms' water and salt balance. On land, oxygen is plentiful, but the organisms that live there must protect themselves from the dangers of drying up—dangers that are increased greatly because air temperatures fluctuate daily and seasonally. Air does not provide the same buoyancy as water, so large **terrestrial** (ter ES tree uhl), or land-dwelling, animals require good supportive structures. On the other hand, there is less resistance to movement in air than in water. **Appendages** (uh PEN dij ez), such as arms and legs, may especially help with land locomotion.

14.2 As Size Increases, Complexity Increases

As shown in Figure 14.2, organisms exhibit an enormous range in size. Through natural history, animal groups have shown a trend toward increasing size, a result of their being multicellular. Although larger animals may be able to obtain more food than smaller ones and may get away from predators faster, increased size imposes special problems. Larger animals often require an increased division of labor among their body parts, which has, in some cases led to greater complexity in the organism. Consider how many different parts your body is made of and all their different functions.

No matter what the body size, individual cells are remarkably consistent in size among all organisms. Large animals have *more* cells than small animals but not larger cells. The presence of more cells can lead to increased specialization of tissues, which in turn requires more coordination. Through diffusion, small animals, for example, can exchange materials more easily with their environment, but within large animals diffusion is less efficient in moving materials. These animals may require specialized circulatory systems to obtain oxygen, remove wastes, and transport substances. These systems must be finely coordinated. In general, the larger the animal, the more complex are the controls necessary to coordinate its activities.

All animals face common problems of feeding, self-maintenance, and reproduction. In members of the same phylum, the solutions to these problems often are similar. The next section examines eight animal phyla, ranging from the least complex to the most complex. It describes the special features and modifications that improve the adaptation of each phylum to its environment. (Further descriptions of animal phyla are in Appendix 4, *A Catalog of Living Things.*)

The descriptions of the eight phyla are purposely brief. Emphasis is on whatever new structures that phylum has evolved to better perform life functions.

CONCEPT REVIEW

1. Compare and contrast bilateral and radial symmetry.
2. Compare the advantages and disadvantages of aquatic and terrestrial environments for animals.
3. Describe how increased size leads to increased complexity.
4. How does an animal's structure help it survive in different environments? Give some examples.

Diversity and Adaptation in Animals

14.3   Sponges and Cnidarians Are the Least Complex Animals

Sponges (phylum Porifera; poh RIH fer uh) are an ancient group with an extensive fossil record that dates back to the early Precambrian period, about 550 million years ago (see Figure 14.3). The several thousand species living today are all aquatic, occupying a variety of marine and freshwater habitats. They show considerable diversity of form, and many are brightly colored, as shown in Figure 14.4.

Sponges are sessile animals without any apparent symmetry. The name Porifera, pore-bearer, aptly describes the body of a sponge, which consists of a bag pierced by many pores and canals. The cells are loosely organized and do not form true tissues or organs. If a sponge were forced through a sieve, its cells would break apart and then flow back together again, reforming the sponge on the other side. The pores and canals function as a water-filtering system. Water containing oxygen and microscopic food is pulled through pores in a current created by the beating flagella of specialized cells that line the inner body. Water containing carbon dioxide and other cellular wastes flows back out of the body cavity through an opening called the mouth. The body wall is supported by interlocking particles of a tough material that form a type of skeleton. Figure 14.4c shows the sponge body plan.

Cnidarians (nih DAR ee unz), members of the phylum Cnidaria, are aquatic animals usually found in shallow, warm marine habitats, although some are found in freshwater and cold marine habitats as well. The fossil record of the cnidarians, like that of the sponges, goes back more than 600 million years. Cnidarians include corals, jellyfishes, sea anemones, and freshwater hydras (see Figure 14.5). Many of these organisms harbor symbiotic algae within their cells, which allow them to thrive in nutrient-poor waters. The algae produce some sugars useful to cnidarians, and the cnidarians

Figure 14.3   The appearance of animals throughout the earth's natural history.

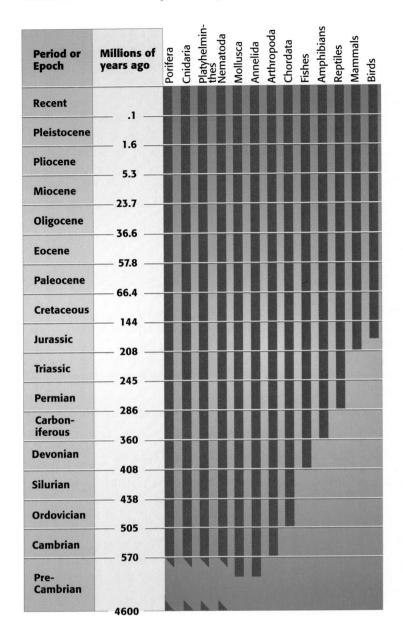

Period or Epoch	Millions of years ago
Recent	
	.1
Pleistocene	
	1.6
Pliocene	
	5.3
Miocene	
	23.7
Oligocene	
	36.6
Eocene	
	57.8
Paleocene	
	66.4
Cretaceous	
	144
Jurassic	
	208
Triassic	
	245
Permian	
	286
Carbon-iferous	
	360
Devonian	
	408
Silurian	
	438
Ordovician	
	505
Cambrian	
	570
Pre-Cambrian	
	4600

Figure 14.4   Sponges: (a) brown volcano sponge, *Hemectyon ferox;* (b) purple vase sponge, *Dasychalina cyathina;* (c) body plan of a sponge.

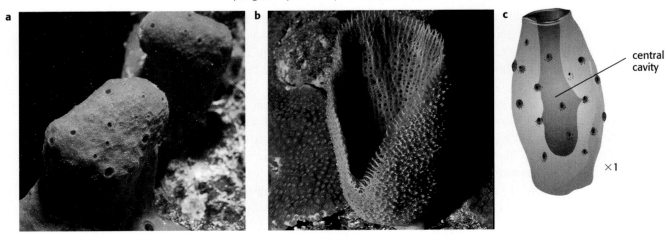

central cavity

×1

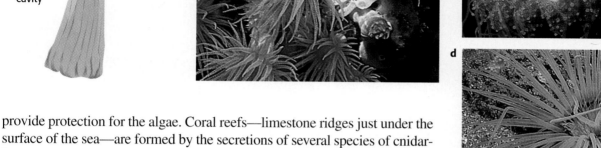

provide protection for the algae. Coral reefs—limestone ridges just under the surface of the sea—are formed by the secretions of several species of cnidarians and are home to the greatest diversity of animal life in the water. Corals have produced coral island groups such as the Bahamas and Bermuda.

Cnidarians are radially symmetrical and, thus, are adapted to a sessile existence, although some are motile. They have two basic body forms, tubular and bowl-shaped. Reef-building corals and freshwater hydras have tubular bodies that usually are attached to some surface. Free swimming or drifting jellyfishes have inverted, bowl-shaped bodies that have dangling tentacles.

The body of either type of cnidarian is little more than a thin-walled bag with an outer and inner tissue layer, as shown in Figure 14.5a. Sandwiched between the layers is a jellylike material that contains a network of nerve cells and contractile fibers. Digestion occurs in the central cavity of the bag; there are no internal organs. Food is taken in and undigested particles are released through the same opening. The opening is surrounded by tentacles containing specialized cells that sting prey on contact. These cells give the phylum its name, Cnidaria (nettles).

Figure 14.5   Cnidarians: (a) cnidarian body plan; (b) night-blooming orange cup corals, *Tubastraea coccinea;* (c) jellyfish, from the Red Sea; (d) burrowing sea anemone, *Pachycerianthus.*

Sessile cnidarians that live attached to each other-such as the corals—are fastened at the ends opposite their cavity openings, and their tentacles stretch upward. Other cnidarians—such as jellyfishes—swim or float in the water. Their cavity openings are directed downward, and their tentacles hang beneath.

14.4   Flatworms Have True Organs

Flatworms (phylum Platyhelminthes; plah tih hel MIN theez) represent the next evolutionary step in complexity. They are the least complex of the bilaterally symmetrical animals that have organs, organ systems, and a distinct head at the anterior end. They also have greater organ specialization and division of labor than are found in either sponges or cnidarians.

Flatworms (Figure 14.6a) are adapted to a great variety of habitats. Many, including tapeworms and flukes, are parasitic, and some are parasites of humans. Free-living forms, such as planarians, may be marine or may live in fresh water or moist soil.

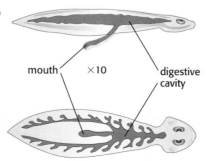

Figure 14.6   Flatworms: (a) marine flat worm, *Alleena mexicana;* (b) flatworm body plan. Note the much-branched digestive cavity.

a
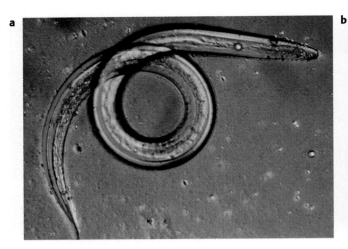

b
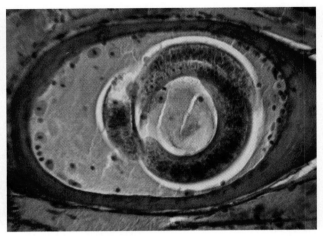

Figure 14.7   Nematodes *Necator americanus* DIC X65: (a) in garden soil; (b) *Trichinella spiralis,* a parasite, larvae in muscle (X400).

Flatworms lack circulatory systems. The typical flatworm digestive system is diagrammed in Figure 14.6b. It is a branched sac with only one opening, which serves to take in food and to eliminate wastes. The branches bring **2** food directly to the cells or within diffusible distance of them. Parasitic forms often have no digestive system. Instead, they absorb digested food directly from the host organism. Oxygen and carbon dioxide easily diffuse in and out of all cells because of the flat body plan.

14.5   Roundworms and Mollusks Have a Body Cavity

Roundworms (phylum Nematoda; nee muh TOH duh) are slender cylindrical animals that are present in abundance in almost every type of habitat (see Figure 14.7). There may be one million roundworms (also called nematodes) in a shovelful of good garden soil. Many are efficient predators, and some are parasites that are able to infect almost all types of plants and animals, often causing serious disease.

Roundworms have a complete digestive system with two openings. This adaptation permits continuous processing of food. The digestive system is usually tubelike and is suspended in a fluid-filled body cavity within the body wall. The internal body cavity represents an evolutionary advance in **2** animal body plans, and the resulting tube-within-a-tube shown in Figure 14.8 is the basic body plan of animals classified in all the more complex phyla.

An internal body cavity is also typical of mollusks (phylum Mollusca; muh LUS kuh), which include snails and slugs, clams, oysters, and scallops, as well as squid and octopuses (see Figure 14.9a, b). Most mollusks live in a variety of aquatic habitats, from the shallows along the shores to the ocean depths. Their diversity of form, feeding methods, and life-styles results from their adaptation to these very different habitats.

This might be a good place to describe the human digestive system as a complex tube.

Figure 14.8   Roundworm body plan, a tube within a tube.

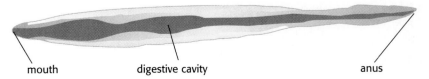

mouth   digestive cavity   anus

a
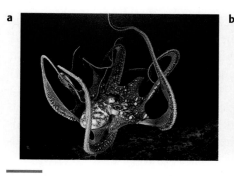

b

c
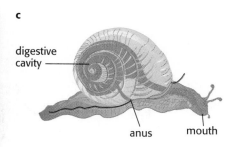

digestive cavity

anus   mouth

Figure 14.9   Mollusks: (a) Day octopus, *Octopus cyanea,* from Hawaii; (b) rock scallop, *Hinnites giganteus;* (c) molluscan body plan.

Mollusks have soft bodies in which the ventral wall is modified into a head-foot region adapted for feeding, sensory reception, and locomotion, as shown in Figure 14.9c. They have a circulatory system and structures specialized for exchanging gases with the environment. Their most distinctive characteristic is a mantle, a modified body wall that forms a cavity and encloses the internal organs. In many mollusks, the mantle secretes substances that create a shell. For these mollusks, the mantle forms the lining of the shell. Many mollusks protect themselves by withdrawing the head or foot into the cavity that the shell surrounds. In aquatic forms, muscular pumping of the mantle maintains a current of water that brings in food and oxygen and carries out wastes. The jet propulsion of squid and octopuses also is accomplished in this way.

14.6   Annelids and Arthropods Are Segmented

The basic adaptive feature of annelids (AN ih lidz; phylum Annelida; an NEL ih duh), or segmented worms, is their body segmentation, which shows clearly in Figure 14.10a. The annelid body consists of a series of ringlike compartments, each resembling the next. Many internal body organs are repeated through several compartments. The fluid-filled body cavity also is segmented. The contraction of muscles surrounding each segment allows these animals to burrow and swim.

Ask students how the lack of appendages in earthworms might favor burrowing. For an animal living in a narrow burrow not much wider than itself, movement would be impeded by appendages extending out from the animal's body. Locomotion in a confined space is made efficient by friction between the body surface and the burrow walls, in addition to the squirming motions of burrowing animals.

Figure 14.10   Annelids: (a) night crawler, *Lumbricus;* (b) marine annelid, *Sabella.*

a

b

mouth

digestive cavity

anus

Figure 14.11 Arthropod body plan. Note the appendages.

Two types of spiders that live throughout much of the United States are poisonous to humans: the black widow and the brown recluse. Chigger mites are annoying in much of the East. Wood ticks carry the rickettsias of Rocky Mountain spotted fever, particularly (paradoxically) in the Northeast.

Lyme disease, named for the Connecticut village of Olde Lyme, is spreading, although it occurs primarily in the Northeast. It is caused by spirochetes that live in wild animals and may be transmitted to humans by deer ticks. Because the symptoms are so general, many people suffering from Lyme disease never seek medical treatment. If the disease goes untreated, it may lead to chronic arthritis.

Although terrestrial annelids such as earthworms are most familiar to us, most annelids are aquatic. Marine annelids are abundant, diverse, and, important in food chains. They live under rocks or in coral crevices, burrow in sand and mud, or secrete tubes from which they extend feathery tentacles for feeding (see Figure 14.10b). Scientists have found worm burrows made during the Precambrian period. Leeches, another type of aquatic annelid, are adapted to bloodsucking and have supporting adaptations such as cutting jaws, suckers, and the production of substances that inhibit blood clotting.

Three-quarters of all known species of animals are arthropods (phylum Arthropoda; ar THROP uh duh), the most numerous animals on earth. The arthropods include crustaceans and insects as well as centipedes, millipedes, and arachnids. They are found in every type of habitat, have a wide variety of feeding habits, and are uniquely able to adapt to changing conditions. In fact, their adaptability has contributed to their great evolutionary success.

Arthropods have segmented bodies, usually with three major divisions: a head, a thorax (chest), and an abdomen. Many of the segments have pairs of jointed appendages, an adaptation that is efficient for locomotion. Appendages often are modified for specialized functions such as food getting, sensory reception, swimming, or flying. Figure 14.11 shows the basic arthropod body plan.

There are three major groups of living arthropods, the chelicerates (chih LIS er ayts) the crustaceans, and the uniramia (yoo nih RAY me uh). Chelicerates include horseshoe crabs, spiders, harvestmen, scorpions, and ticks and mites. Figure 14.12 shows examples of chelicerates. Unlike the separate head and thorax of many insects, the head and thorax of chelicerates are joined. They have four pairs of legs and two pairs of chelicerae— appendages adapted for feeding. Spiders are the most familiar chelicerates. Most spiders make webs, and all have poison glands that are used to immobilize their prey. Only a few, such as the brown recluse and black widow, are dangerous to humans. Many ticks and mites transmit pathogens that cause several serious diseases, including Rocky Mountain spotted fever and Lyme disease, which is characterized by fevers, chills, muscle aches, and pains.

Crustaceans are mostly aquatic, and many of their appendages are adapted for swimming. They breathe by means of gills and have three distinct body parts with two pairs of antennae or "feelers" (see Figure 14.13). Lobsters, crabs, and shrimp are among the larger and more familiar members of this group, but most crustaceans are small, even microscopic. Tiny

Figure 14.12 Chelicerates: (a) horseshoe crab, *Limulus polyphemus;* (b) tick, *Dermacenter;* (c) nursery web spider, *Pisaurina mira.*

a

b

c

a

b

Figure 14.13 Crustaceans: (a) spiny lobster, *Panulirus argus,* from the Florida Keys; (b) sow bug, *Oniscus.*

crustaceans such as krill are the basic food supply for saltwater fishes and whales. A number of crustaceans have adapted to a life on land. Sow bugs and pill bugs, which live in moist places such as under rocks, are common terrestrial animals.

Uniramia include the millipedes, centipedes, and insects. Insects (Figure 14.14) are the most numerous and diverse arthropods. There are more than one million species of insects. There are about one billion individual insects for every human on earth. Insects have three body parts, one pair of antennae on the head, and three pairs of legs on the thorax. The abdomen usually is visibly segmented. Most adult insects have one or two pairs of wings attached to the thorax. Many insects undergo metamorphosis (met uh MOR phuh sis), a change in body form and function that occurs during development. Moths and butterflies are examples of insects that undergo **metamorphosis.** Figure 14.15 illustrates the life cycle of a moth. In its immature, or larval (LAR vuhl), stage as a caterpillar, it feeds continually and may do great damage to trees, crops, and ornamental plants. Many adult insects live only long enough to mate and lay eggs.

Figure 14.14 Insects: (a) ant *Componotus* sp. from new South Wales, Australia; (b) dragonfly on wood lilly; (c) giant water bug, male insect carrying eggs on back.

a

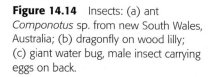

b

c

14.7   Chordates Have Internal Skeletons

At some time in their lives, all chordates (phylum Chordata; kohr DAH tuh) possess three characteristics: (1) paired gill slits; (2) a dorsal, tubular

Figure 14.15 The life cycle of the Cecropia moth, or American silkworm. An egg develops into a small caterpillar in 10 days. The caterpillar, or larva, eats leaves, molts 4 times, and grows up to 10 cm in length in several weeks. It then spins a cocoon around itself, and the pupa develops within the cocoon. The adult moth emerges in the spring.

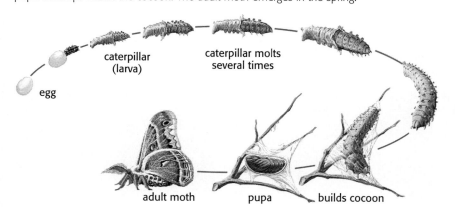

egg

caterpillar (larva)

caterpillar molts several times

adult moth   pupa   builds cocoon

Figure 14.16   Cartilaginous fishes: (a) whitetip reef shark, *Treaenodon obesus;* (b) stingray, *Dasyatis americana.*

nerve cord; and (3) a **notochord** (NOH toh kord)—a flexible, rodlike structure that extends the length of the body. In the vertebrate class, the notochord is replaced by a backbone made up of bony or cartilaginous pieces called vertebrae (VER teh bray) that surround the nerve cord. Most vertebrates have paired appendages such as legs, wings, and fins. They also possess a well developed brain that provides good coordination. Because chordates are highly adaptable, they have succeeded in occupying most types of habitats. Fish, amphibians, reptiles, birds, and mammals are all members of phylum Chordata.

Cartilaginous fishes, a class that includes sharks and rays, have an internal skeleton made up entirely of **cartilage**—a tough, elastic connective tissue that is stiff enough to provide support but is more flexible than bone. The rays shown in Figure 14.16, for example, have oddly flattened and flexible bodies that are adapted to gliding along on the sea bottom.

Bony fishes have skeletons made of bone, and most have an outer covering of scales. Almost all fishes obtain their oxygen through covered gills, although lungs that carry out gas exchange have evolved in air-breathing lungfishes. There is great diversity among bony fishes. In fact, of all the vertebrate classes, this one has the most species, occupying almost all the waters of the earth (see Figure 14.17).

Amphibians include frogs, toads, and salamanders, shown in Figure 14.18. Most live on land as adults but return to water to reproduce. Amphibians lay their eggs in water, where the young develop before returning to land. The limbs of amphibians, which are poorly adapted for walking, may have evolved from the lobed fins of the lungfishes. This group is of special importance in biology because of its evolutionary position between water-and land-dwelling animals.

Another class of fishes is the agnathans. These fishes lack paired fins. Their cartilaginous skeletons are poorly developed, and notochords are present in adults as well as in embryos. The best-known agnathans are the lampreys. A lamprey has no jaws. It feeds by attaching itself to a fish, rasping a hole in the body, and sucking out the body fluids of the victim.

Figure 14.17   Bony fishes (a) copper rockfish, *Sebastes caurinus;* (b) schoolmaster fish, *Lutjanus apodus.*

Figure 14.18   Amphibians: (a) phantasmal dart poison frog, *Epipedobates tricolor;* (b) marbled salamander, *Ambystoma opacum.*

a

b

Figure 14.19   Reptiles: (a) spectacled caiman; (b) smooth green snake, *Opheodrys vernalis.*

Reptiles (Figure 14.19) have completely adapted to terrestrial life, although some spend at least part of their lives in water. Their embryos develop in eggs with leathery shells that are laid on land. Reptiles have an outer covering of scales, which prevents excessive water loss, and they breathe by means of lungs all their lives. Some also can exchange gases through their digestive systems. Dinosaurs, which were early reptiles, are thought to be the ancestors of birds.

All birds (Figure 14.20) have feathers, and all animals with feathers are birds, although some feathers are highly modified. No other class of animals is so easy to characterize. All birds have wings, and although not all can fly, those that can have the greatest motility of any animal. In addition, all birds hatch from eggs that have hard shells. Unlike most other chordates, birds have a nearly constant body temperature, an adaptation that enables them to maintain a high rate of metabolism at all times.

Except for fishes, no group of vertebrates shows as much diversity of color as birds. Some students might like to study this phenomenon.

Top to bottom: whippoorwill, heron, pelican, and golden eagle. The whippoorwill has claws adapted for perching; the pelican has webbed feet for swimming; the heron has feet for perching on or around the shore; the eagle has claws for grasping prey.

Figure 14.20   Birds (a) mountain bluebird; (b) pelican; (c) diversity of types and unity of pattern. Although species of birds vary greatly, they share characteristics that group them together. How do these differences adapt each bird to its life style?

c

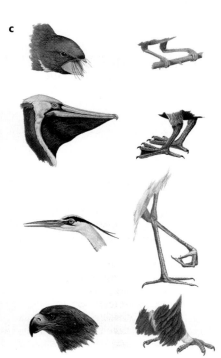

a

b

a

b

c

Figure 14.21   (a) mountain lion; (b) humpback whale; (c) porcupines. What similarities can you see? **Answers will vary.**

Bird feathers show many structural adaptations. Feathers form the soft down of ducks, the long, beautiful plumes of ostriches, and the waterproof coat of penguins. Bird wings also show many structural adaptations. The wings of penguins are not used for flying, but rather for swimming. Albatrosses, which have long, slim wings, spend almost all their lives gliding on air currents. Modification of other structures, such as feet and beaks, also has been important to the ability of birds to adapt to many types of ecosystems. Figure 14.20c illustrates several adaptations.

From the tiny shrew, which measures about 15 cm, to the giant blue whale, which is 30 m long and the largest animal that has ever lived, all mammals share two characteristics. First, all have hair, although it sometimes is not very evident. Even whales have hair at birth, though they lose it soon after. Second, all species of mammals feed their young with milk, a fluid secreted from mammary glands in the skin. This adaptation allows a long period of development under parental care and increases the chances that the young will survive. Like birds, mammals can maintain a constant body temperature regardless of the environmental temperature.

Although they have many characteristics in common, mammals show great diversity in size, structure, and color, some of which is evident in Figure 14.21. Diverse modifications for motility and for eating have adapted mammals to function in a variety of specific ecological niches. Modifications for reproduction permit mammals to be grouped into three categories: monotremes, which lay eggs; marsupials, which have pouches in which the young develop; and placentals, which have special internal structures that nourish the young until birth.

Students usually think of dinosaurs as being the largest animals. Not so.

You may wish to address the misconceptions many students have about the terms *warm-blooded* and *cold-blooded.*

Most students should be able to mention many structural adaptations in mammals—beginning with themselves.

6

6

CONCEPT REVIEW

1. What characteristics do sponges and cnidarians share, and how are they different?
2. What adaptations of evolutionary significance are shown by flatworms? by roundworms? by annelids?
3. What is the distinctive structure found in mollusks and what functions does it perform?
4. Describe characteristics that distinguish a spider, a crustacean, and an insect.
5. What characteristics distinguish chordates from other phyla?
6. What structural adaptations distinguish the vertebrate classes?

Biology Today

Marsupials and Monotremes

All animals of the class Mammalia have hair—at least at some point in their lives—and produce milk for their young. They also maintain a constant internal temperature. On the basis of reproductive modifications, the animals in this class can be divided into three groups: the monotremes, the marsupials, and the placental mammals.

Monotremes lay eggs as birds do, but fertilization occurs internally, as it does in other mammals. In contrast to marsupial and placental mammals, monotremes have not developed a great diversity of species. Today, only two members of this group survive: the platypus and the echidna, or spiny anteater. One possible explanation for this scarcity of species is that the early monotremes were displaced by the more rapidly evolving marsupials.

Because monotremes are not well represented in the fossil record, most information about them is derived from the two living genera. These animals resemble reptiles in their skeletal structure and in their method of reproduction. The eggs of the duck-billed platypus, for example, are fertilized in the oviduct. As the eggs continue down the oviduct, gland secretions add albumin and a thin, leathery shell. The platypus lays its eggs in a burrow and incubates them for about 12 days. Thus, monotremes have no period of pregnancy. The developing embryo is nourished entirely by nutrients stored in the egg. Because it has a bill, a newly hatched young platypus cannot suck milk from its mother—instead, it must lick milk from the mother's fur.

Marsupials are mammals, such as opossums, kangaroos, wombats, and koalas, that give birth to immature young, then carry them in protective pouches. There is more diversity among marsupials than there is among monotremes. The koala, for example, is specialized for life in the treetops. Its long arms, curved claws, and hands with a viselike grip permit the koala to climb very smooth tree trunks. Unfortunately, the destruction of forests, the killing of thousands of koalas for their fur, and now epidemic venereal disease among them have reduced their numbers to a dangerously low level.

The kangaroo family contains a great many species, but all kangaroos have elongated hind feet and a long, heavy tail that serves as a balance for jumping or as an extra leg for standing or walking. The young of the largest marsupial, the red kangaroo, are only 1.9 cm long at birth but can grow to over 1.65 m and can weigh as much as 81 kg. The first pregnancy of the season lasts 33 days. When the young "joey" is born, it crawls unaided to its mother's pouch, where it attaches to a nipple. Although the mother immediately becomes pregnant again, the presence of the joey in the pouch arrests the development of the new embryo in the uterus. When the joey leaves the pouch, the embryo resumes development and is born about a month later. Though the developmental pattern among other marsupials may differ somewhat, in all marsupials, the young are born after a very short pregnancy.

A platypus, *Ornithorhynchus anatinus*. Unlike the bill of a duck, the bill of the platypus is flexible and leatherlike. Males also have sharp venom spurs on their hind legs.

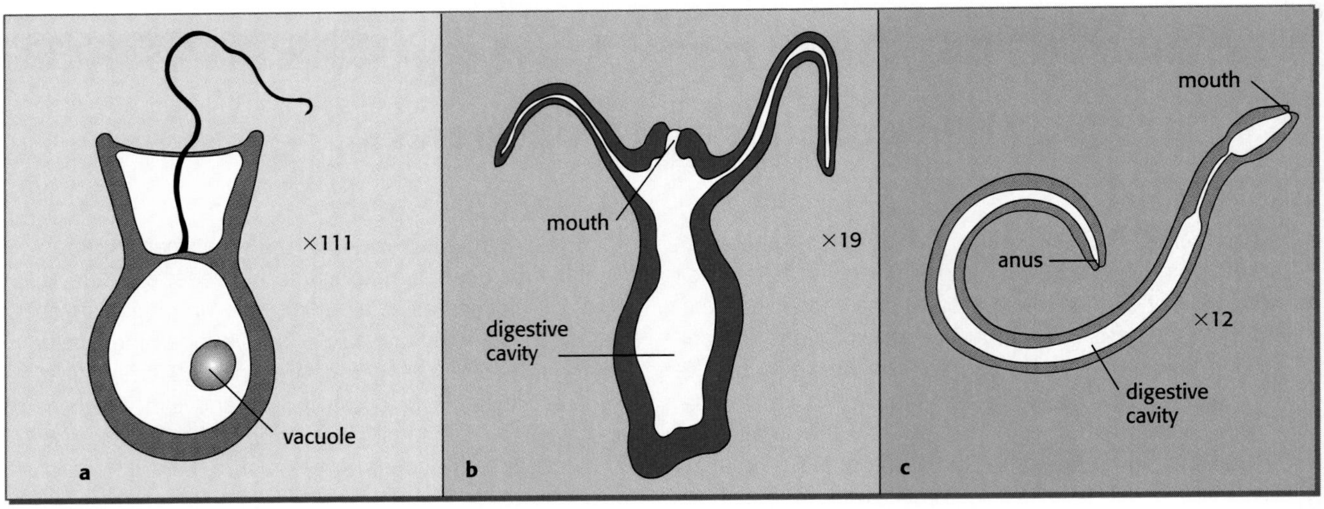

×111

vacuole

a

mouth

digestive
cavity

×19

b

mouth

anus

×12

digestive
cavity

c

Figure 14.22 Types of digestive cavities: (a) intracellular (vacuole) in a cell of a sponge; (b) extracellular with one opening (sac) in a hydra; (c) extracellular with two openings (tube) in a roundworm.

Ask students to suggest land animals that wait for their prey to come by.

Life Functions in Animals

14.8 Digestion May Be Intracellular or Extracellular

A variety of mechanisms for getting food have evolved in animals. Sponges demonstrate a rather primitive method. Individual cells, each with a single flagellum, keep a current of water moving through the sponge. When a food particle comes by, the cell may engulf it and draw it into a vacuole, just as a paramecium does. The process of taking food particles into a body cavity is called ingestion.

In contrast to sponges, most aquatic and terrestrial animals actively pursue their food and have means to capture it. Some predators poison their prey. For example, the tentacles of some cnidarians have stinging capsules. Within each capsule is coiled a long, hollow barbed thread. When a food organism brushes one of the tentacles, the thread shoots out with such force that it pierces the body of the victim. A paralyzing poison is injected into the prey, which then is drawn into the cnidarian's body cavity by the tentacles.

More complex food-getting structures also have evolved. Among vertebrates, jaws, beaks, and teeth aid food getting. Special adaptations of these structures often are present. A snake can unhook its lower jaw from its upper jaw and move the two independently, permitting it to swallow prey larger than its own head.

Once ingested, food enters a digestive cavity inside the animal's body. Figure 14.22 shows three types of digestive cavities: a vacuole within an individual cell, a sac, and a tube.

Most nutrient particles that animals ingest are too large to pass through plasma membranes. Even the microscopic particles that sponges take in are much too large to pass directly into cells. Food ingested by animals must be broken down into small molecules. The physical and chemical processes of this breakdown are known collectively as **digestion.** Digestion may be classified as either intracellular or extracellular, according to where it occurs.

The breakdown of large pieces of food into smaller ones is known as physical digestion. Most mammals have teeth that cut or grind food into smaller pieces. For others, movements of the digestive cavity itself break down food. The gizzard of a bird, for example, is a specialized part of the stomach in which food is ground up (see Figure 14.23). Some birds swallow sand and small pebbles that aid the grinding.

Figure 14.23 A bird swallows food without chewing and stores it temporarily in the crop. Sand and small pebbles in the gizzard aid in grinding food.

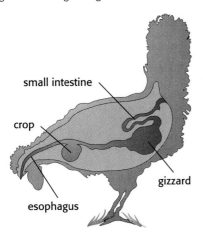

small intestine

crop

esophagus

gizzard

Physical digestion, whether in the mouth or the stomach, increases the surface area of food, so that the chemical part of digestion can take place more quickly. At this point, enzymes take over the task of breaking down food.

In sponges, and to some extent in cnidarians and flatworms, chemical digestion takes place in a vacuole inside a cell. This process is called intracellular digestion. The vacuole is formed when a food particle is surrounded by an infolding part of the plasma membrane. After the vacuole breaks free, it fuses with a lysosome, which contains digestive enzymes but is not digested by them. Chemical digestion occurs within this new structure. In this way the cell itself avoids being digested. Small molecules produced by chemical digestion pass out of the vacuole into the cell's cytoplasm, making the nutrients available to the cell.

In most animals, digestion is extracellular, meaning that it takes place in a digestive cavity into which enzymes are secreted. In the digestive sac of a cnidarian, for example, some of the cells lining the cavity secrete enzymes. Likewise, in the simple digestive tube of a roundworm, digestive
2 enzymes are produced by cells in the lining of the tube. In more complex animals, however, digestive enzymes are secreted by tissues specialized as glands. The enzymes may be specialized for a particular type of food. Some herbivorous vertebrates, such as cows, have a special chamber (the rumen), which contains cellulose-digesting microorganisms that enable the animal to utilize the cellulose fibers. Most of the fiber in the rumen is digested by numerous anaerobic prokaryotes—up to one billion per milliliter of rumen contents. The prokaryotes benefit from the constant environmental conditions and food source and the cow obtains much of its nutrition from prokaryote products.

The molecular products of digestion are absorbed into the animal's cells. Often the food molecules diffuse into an animal's blood and are carried to cells throughout the body. The rate of food absorption depends on the surface area of the digestive cavity lining—the greater the surface area, the faster the rate of absorption. In complex animals, the lining has many folds, an adaptation that increases surface area.

14.9 Animals Have Several Ways of Obtaining and Transporting Materials

To derive energy from cellular respiration, which requires oxygen, all animal cells must be able to obtain oxygen. Cells also must be able to get rid of carbon dioxide, the waste product of respiration.

In small animals, such as the hydra, gas exchange can occur entirely through the body wall (Figure 14.24a) because the surface area is large

Cellulose digestion in horses and rabbits occurs in projections of the large intestine. The horse uses very little of this digested material, but rabbits produce coecal pellets that they promptly re-eat. They also produce fecal pellets that they do not eat.

Figure 14.24   (a) Gas exchange in a hydra. (b) In fishes, gills are thin filaments supported by bony structures and richly supplied with blood vessels. Each filament is made of disks that contain numerous capillaries. Water flows past these disks in directions opposite the flow of blood through the capillaries. A covering over the gills, called the operculum, protects the delicate filaments.

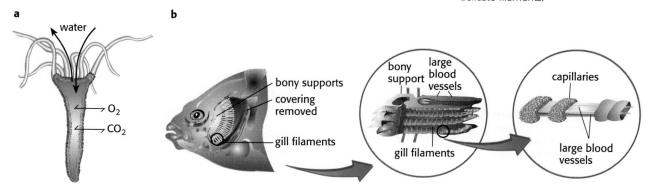

Ask students whether this water loss or gain is a problem for aquatic animals. They should consider the gills of fish living in fresh water. The question should cause some students to recall the problems of water diffusion discussed in Chapter 5.

Ask students if they know of any terrestrial vertebrate that normally has only one lung. Most snakes have only one lung.

enough to allow sufficient amounts of oxygen to enter the organism through diffusion. As body size increases, however, the ratio of surface area to volume becomes smaller. Large animals usually do not have enough body surface to allow sufficient gas exchange. Instead, they have organs that increase the surface area through which respiration can occur. In aquatic animals, these organs are usually gills (Figure 14.24b), which often have a feathery or plate like structure. Dissolved oxygen diffuses in and carbon dioxide diffuses out through the gill's cells. Gills are remarkably similar in a wide variety of animals.

Terrestrial animals lose a great deal of water through their respiratory organs because water can diffuse through any surface that also permits diffusion of respiratory gases. Also, cell surfaces must be kept moist so that gases can diffuse through them. For this reason, terrestrial animals that breathe through gills, such as sow bugs, or take in oxygen through their skin, such as frogs and earthworms, must live in places where the air is moist and evaporation is slow.

Many terrestrial animals live where the air is dry. These animals need an extensive surface *inside* the body where air can be kept moist. There are two principal ways of meeting this requirement. Insects and some other arthropods have a complicated system of air tubes that carry oxygen to all parts of the body. Body movements move air through the tubes. Air-breathing vertebrates have lungs, which are divided into a large number of tiny air sacs. Respiratory gases diffuse through the enormous moist surface area of these sacs.

In a single-celled organism, movement of the cytosol provides an adequate system for transporting materials. In some multicellular organisms such as sponges, cnidarians, and free-living flatworms, almost every cell has part of its surface exposed to the environment. Each cell can obtain its own oxygen and get rid of its own wastes. Although every cell may not take in food directly, no cell is very far from others that do. Diffusion and active transport are sufficient to move substances into and out of cells.

In roundworms, transport is complex. These organisms have a fluid-filled body cavity surrounding the digestive tube. As the roundworm wriggles, the fluid within the body cavity is squeezed about from one place to another. In this way substances dissolved in the fluid are carried to and from the body cells.

Most animals that have body fluid have a circulatory system, either open or closed. Both types of circulatory systems include a pump (the heart) and a series of tubes through which nutrients can be delivered and waste products removed from the cells. In arthropods and most mollusks, a heart pumps blood through blood vessels that empty into body spaces. The blood moves through these spaces, eventually emptying into another set of vessels that carry it back to the heart. This incomplete system is called an open circulatory system.

The diagram of the circulatory system of an annelid is somewhat misleading because the connections between vessels in the dorsal and ventral body walls cannot be shown in a longitudinal section.

Annelids, on the other hand, have a closed circulatory system. In a closed system, blood always flows within vessels. An earthworm has five pairs of hearts and a complex set of finely branched vessels. These vessels empty into a large dorsal vessel, which returns the blood to the hearts. Valves in the dorsal vessel keep the blood flowing in only one direction through the system. Vertebrates, including humans, also have closed circulatory systems. Figure 14.25 illustrates transport in several animals.

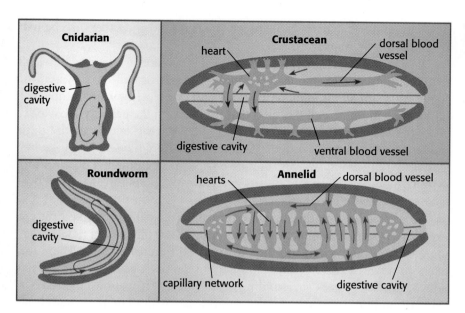

Figure 14.25 Diagrams of fluid transport in four invertebrate animals.

14.10 Excretory Systems Maintain Water and Chemical Balance

Wastes are removed from cells by a process called **excretion** (ek SKREE shun). What are wastes? Some products of metabolism are poisonous—ammonia, for example, which is formed in the breakdown of proteins. Other substances, however, are toxic only if large amounts accumulate. Sodium chloride is a normal cell substance that is toxic in large amounts. Ordinarily this salt is constantly diffusing into cells from the fluid surrounding them, and so the normal proportion of sodium chloride in cells can be maintained only by continuous excretion. In general, any substance can be called "waste" if an organism has too much of it.

In small aquatic animals, wastes may simply diffuse through plasma membranes to the outside environment. Sponges and cnidarians also can excrete wastes directly through their body surfaces, because all their cells contact the water environment or are very close to it. Most animals, however, have special systems for ridding their bodies of wastes.

Animal cells constantly produce water through cellular respiration. What happens to that water depends on the type of environment in which an animal lives. For example, the cells of a jellyfish contain a particular mixture of substances in water, whereas seawater contains a different mixture of substances in water. Normally the concentration of water molecules inside the jellyfish equals the concentration outside it. As fast as metabolic water is produced, it diffuses into the environment. This occurs in many other marine invertebrates as well.

Fresh water, on the other hand, contains few dissolved substances, and so the concentration of water molecules is high. In the cells of a freshwater planarian, the concentration of dissolved substances is high, and the water concentration is low. Therefore, water is always diffusing *into* a planarian, and metabolic water is constantly added to this excess. If water was not somehow expelled from the cells, eventually the cells would burst. Planarians and other freshwater animals excrete water by active transport.

Terrestrial animals have the opposite problem. As a human, for example, you take in a great deal of fresh water, and your cells also produce water by metabolism. As a land animal, however, your greatest danger is of drying out. Your kidneys reabsorb water that would otherwise be lost from the body with wastes.

Within the body, animals maintain water balance in a variety of ways. In planarians, for example, this function is performed by a system of flame cells spread throughout the body and connected by small tubes, or tubules. Each flame cell has cilia that project into the cell cavity. Wastes enter the flame cells from the tissues. The waving action of the cilia moves fluid and wastes out into the tubules. Many tubules join together and eventually empty wastes into the environment through a pore. In this system, there are many flame cells and many pores. Figure 14.26a shows the excretory system of a planarian.

Most annelids, mollusks, and crustaceans have a somewhat different excretory system. The functional unit for these organisms is a tubule leading from the body cavity to the outside (see Figure 14.26b). There are many of these tubules. Wastes from within the body cavity, and from the network of capillaries surrounding each tubule, diffuse into the tubules and are emptied outside. At the same time, useful materials may be absorbed back into the blood or body fluid from the tubules. Among vertebrates, a somewhat similar system of tubules is found in the kidneys, which are described in Chapter 16.

The excretory organs evolved chiefly as water-regulating devices. In most animals, however, they also regulate the excretion of wastes, especially those that contain nitrogen. Amino acids contain nitrogen and are used by cells to build proteins, but an animal may take in more amino acids than it can use. Some proteins are continually broken down to amino acids. Unlike carbohydrates and fats, amino acids and proteins cannot be stored in large amounts. As a result, there is usually a surplus of amino acids. This surplus cannot be used for energy until the amino group ($-NH_2$) is removed from each amino acid.

In vertebrates, amino groups are removed mainly in the liver. As they are removed, ammonia (NH_3) is formed. Ammonia is toxic but also soluble. If a large supply of water is available, the ammonia can be carried out of the body in solution. For example, in freshwater fishes, nitrogenous wastes may be excreted through the gills as ammonia.

In other vertebrates, the ammonia is converted to other forms, and the kidneys are the main route for excretion. In birds, reptiles, and insects, amino groups are excreted as uric (YOOR ik) acid. Uric acid is almost insoluble and

Figure 14.26   (a) The excretory system of a planarian. A flame cell and an excretory pore are shown enlarged. Wastes and excess water removed by flame cells are eliminated through pores in the body wall. (b) Excretory tubes in the earthworm. The near side of the body, including the digestive tube, is cut away. Most of the segments of the body have a pair of excretory tubes, one on either side. Capillaries from the circulatory system surround the excretory tubes.

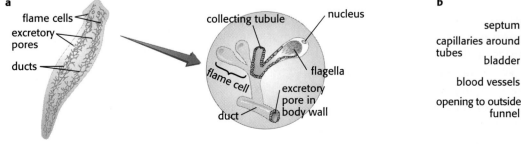

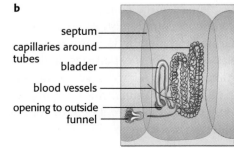

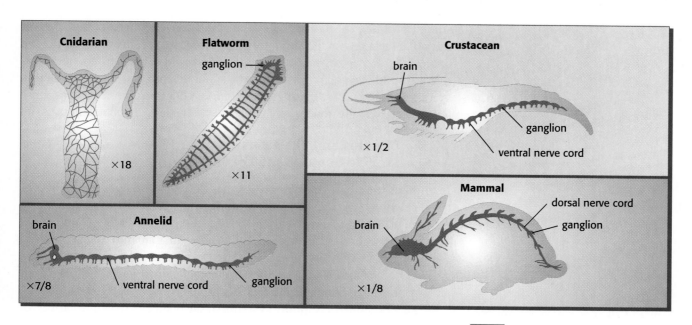

Figure 14.27   Nervous systems in five diverse types of animals.

is excreted with only a small loss of water. In most adult amphibians and mammals, amino groups are converted to a waste called urea (yoo REE uh). Unlike uric acid, urea is soluble. It diffuses into the blood, from which it is removed by the kidneys and then excreted in solution in the urine.

14.11 Nervous Systems Control Responses to Stimuli and Coordination of Functions

For the most part, animals can make rapid adjustments in response to changes in their environment. Even though sponges have no true nervous systems, they are able to adjust to environmental changes, or **stimuli** (STIM yoo ly), such as changes in temperature or chemical concentration. In more complex organisms, nerve cells, alone or in nervous systems, make these adjustments possible. Cnidarians have nerve cells, some of which are specialized to receive only certain stimuli. The nerve cells are connected in a network that permits simple coordinated responses.

Flatworms have nerve networks, but they also have a centralized system of two nerve cords that extend along the length of the body. At the anterior end is a large mass of nerve tissue, known as a ganglion (GANG glee un), where nerve impulses are exchanged. Flatworms also have cells specialized for receiving a variety of stimuli. For example, the eyespots of planarians cannot form images, but they can detect the direction and intensity of light.

Annelids have well-developed nervous systems. A main nerve cord extends along the ventral side of the body. Many ganglia occur along the nerve cord. A large ganglion, sometimes called a brain, is found at the anterior end of the body. Although earthworms have no obvious sense organs, they can detect many stimuli. Other annelids have specialized sense organs, including eyes. Mollusks and arthropods also have nervous systems, basically of the annelid type.

A dorsal, tubular nerve cord, usually called a spinal cord, is a distinctive characteristic of the chordate phylum. In vertebrates, an anterior enlargement of this nerve cord—the brain—dominates the rest of the nervous system. Figure 14.27 compares the nervous systems of several animals.

The ability to receive and react to stimuli from the environment is one of the basic characteristics of living things, although it is developed to different degrees in different organisms. Most animals detect different types of

The earlike flaps at the anterior end of a planarian contain chemical receptors by which the animal detects food—a type of sense of smell.

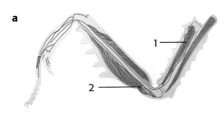

a

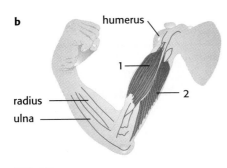

b

humerus

radius

ulna

1

2

Figure 14.28 Comparison of exoskeletal structure (a) with endoskeletal structure (b).

Review reproduction in plants if necessary for student understanding.

stimuli by using specialized receptors at the nerve endings. In humans, for example, receptors in the skin of the fingertips are sensitive to pressure but not to light. Receptors in the eyes are light sensitive, but they are not sensitive to sound waves. In most animals, some receptors are concentrated in special organs, such as eyes, ears, and nose. Other receptors, such as those sensitive to pressure, touch, cold, and heat, are widely distributed in the skin.

14.12 Muscles and Skeletons Provide Support and Locomotion

In animals, motion usually involves muscle cells. Sponges are an exception. They have no muscle cells, although individual cells are capable of movement. Sponge larvae swim by means of cilia, just as many protists do.

Cnidarians have specialized cells that allow them to move their tentacles. Flatworms have cells organized into definite muscle tissues, but locomotion still is accomplished largely by cilia. In all other animal phyla, muscle tissues are organized into bundles (muscles) controlled by nervous systems.

The skeletons or support systems of animals consist of a variety of mineral compounds. A sponge's skeleton is made of small, rigid parts scattered through the soft, living tissues. The stony skeletons built up by corals support the animals and protect them from predators. A mollusk's skeleton is its shell, which is protective.

Support and protection are the functions of arthropod and chordate skeletons. In these phyla, skeletons also aid in movement, especially in locomotion. Arthropod skeletons are external, with the muscles attached to the inner surfaces, as shown in Figure 14.28a. These **exoskeletons** (EK soh SKEL eh tunz) are composed of chitin, a tough, flexible material. In many arthropods, calcium compounds in combination with the chitin make the exoskeleton hard and strong. Thin, flexible chitin joints permit bending. Although exoskeletons provide good support, they do not allow for growth. Thus, arthropods periodically shed their exoskeletons and produce new, larger ones.

Vertebrates have **endoskeletons** (EN doh SKEL eh tunz) inside their bodies (see Figure 14.28b). Muscles are attached at points along the outside of the endoskeleton. The skeletons may consist of either cartilage or bone, both of which contain cells that secrete new material. Thus the skeleton can grow as the animal does.

14.13 Animals Are Adapted for Reproductive Success

Reproduction is primarily sexual in the animal kingdom. Even in species that ordinarily reproduce asexually, such as many of the cnidarians, sexual reproduction does take place. The alternation of haploid with diploid generations present in all plants is unknown in animals, although some animals have haploid and diploid groups. Cnidarians alternate a sexual generation with an asexual generation, but both are diploid. Male and female gametes are different in animals, and meiosis occurs just before or during gamete formation.

In most adult animals, an individual has either ovaries or testes, not both. A difference in reproductive organs usually is accompanied by differences in appearance (see Figure 14.29), behavior, or other characteristics. In some animals, an individual has both ovaries and testes. This is true of many annelids, most flatworms, and some crustaceans and mollusks. Among vertebrates, this type of individual occasionally occurs but is not considered normal.

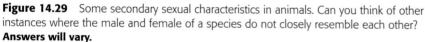

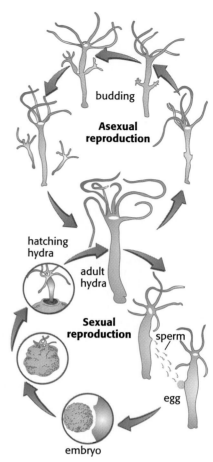

Figure 14.29 Some secondary sexual characteristics in animals. Can you think of other instances where the male and female of a species do not closely resemble each other? **Answers will vary.**

Figure 14.30 A hydra can reproduce either asexually or sexually.

Cnidarians reproduce both sexually and asexually. Hydras, for example, reproduce asexually by budding through much of the year. Under certain conditions, however, they produce ovaries and testes—an adaptation that increases the chances of fertilization in sessile or slow-moving animals. Eggs in the ovary are fertilized by sperm shed into the water. Figure 14.30 shows both types of reproduction in a hydra.

Flatworms have a complex reproductive system. Most individuals have both ovaries and testes, but they do not fertilize their own eggs. Instead, two worms exchange sperm. Free-living forms also reproduce asexually by fragmentation, and even a small piece can regenerate into an entire worm.

Parasitic forms such as *Schistosoma* (shis toh SOH muh) have complex life cycles involving many larval forms and two or more hosts, as you can see in Figure 14.31. Parasitic forms produce huge numbers of eggs, hatching an adaptation that increases the chances for survival of the species.

In roundworms, the sexes are separate, and asexual reproduction does not occur. The females may release as many as 200 000 eggs a day, and the egg cases and developing embryos can survive for many years before resuming growth. As with flatworms, parasitic roundworms have complex life cycles with several hosts.

Earthworms also have both ovaries and testes, but the ova are not fertilized by sperm from the same individual. When two earthworms mate, sperm from one individual are deposited in a special sac of the other. Likewise, sperm from the second worm are deposited in the first. As ova move from the ovary, they pass the sperm-storage sac. Sperm are released, and the ova are fertilized.

In high-school laboratories, hydras seldom produce sexual structures. The stimulus for sexual reproduction seems to be relative concentrations of oxygen and carbon dioxide that seldom occur naturally. You can obtain prepared slides of hydra bearing gonads and of budding hydras.

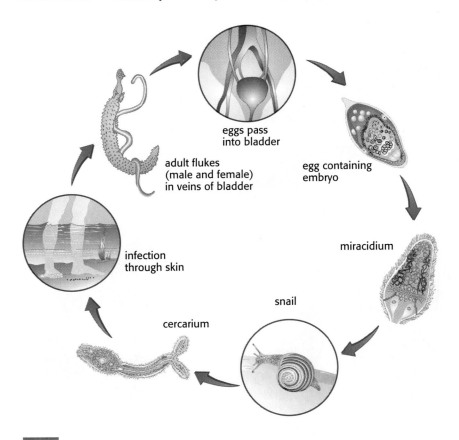

Figure 14.31   Life cycle of *Schistosoma haematobium*. Adult flukes in the veins of a human host produce fertilized eggs that are expelled with the urine. In the water, the larval stages, or miracidia, hatch from the egg and infect snails, the intermediate hosts. Further changes take place in the snail, and immature worms, or cercaria, emerge. The cercaria penetrate the skin of a human, migrate into the small blood vessels, and are carried by the circulatory system to small veins in the bladder, where they mature. Adult worms may live 30 years and produce thousands of eggs each day. This parasite is widespread in many parts of the world and causes serious disease.

In fishes and amphibians, fertilization occurs externally, and the zygote develops in the water. Fertilization is internal in all other verte-   10 brates. The zygote may develop in an egg laid by the female, as in birds and most reptiles, or within the body of the female, as in mammals.

CONCEPT REVIEW

1. How do the various food getting devices of animals illustrate structural diversity?
2. In what ways do digestive cavities differ among animals?
3. Why does a very small aquatic animal require no breathing system?
4. How do gills and lungs function in gas exchange?
5. What types of animals can survive without circulatory systems?
6. Distinguish between open and closed circulatory systems.
7. What are the basic functions of excretory systems?
8. Compare the nervous system of a vertebrate with that of a flatworm.
9. How do skeletons differ in vertebrates and arthropods?
10. Under what circumstances is internal fertilization necessary?

INVESTIGATION 14.1 Diversity in Animals: A Comparative Study

Animals are similar in many ways but differ in other ways. In this investigation, you will sharpen your observation skills by studying several different organisms. You also will examine the relationship between structure and function through observing the adaptations of different organisms to their environments.

Materials (per team of 6)
Station 1
3 stereomicroscopes or 10x hand lenses
3 Syracuse watch glasses
3 dropping pipets
3 compound microscopes
3 No. 2 paint brushes

3 prepared slides of longitudinal sections of hydra
5 to 10 live hydras
daphnia culture

Station 2
6 stereomicroscopes or 1OX hand lenses
3 Syracuse watch glasses

5 to 10 small pieces of raw liver
3 No. 2 paint brushes

Station 3
3 1OX hand lenses
moistened paper towels
6 boxes containing damp soil

3 prepared slides of cross sections of earthworms
5 to 10 live earthworms

Station 4
3 No. 2 paint brushes
1 small bowl containing lettuce and fruit pieces

3 shallow boxes or pans
3 land hermit crabs

Station 5
3 4-qt battery jars
aquarium

3 live frogs

Procedure

1. Review the guidelines in Appendix I concerning the proper care and handling of laboratory animals. Follow these guidelines carefully.

2. Make an enlarged copy of Table 14.1 (or use the table provided by your teacher). It should extend across two facing pages in your data book. Each of the 13 spaces should allow for several lines of writing.

3. In the *Characteristics* column, copy the italicized words from each of the following questions. The 13th space is for any other observations you make. All the specimens of one animal species and the materials and equipment needed for observing them are arranged at one station. Each team will have a turn at each station. Record only your observations, not what you have read or heard about the organism.

    a. What is the *habitat* of the animal? Does it live in water, on land, or both?

    b. Is *body symmetry* radial or bilateral?

    c. Does the animal have a *skeleton?* If it does, is it an endoskeleton or an exoskeleton?

    d. Is the animal's body *segmented* or is it *unsegmented?*

    e. Which type of *digestive cavity* does the animal have, a sac or a tube?

    f. Does it have *aired appendages?*

    g. How does the animal *obtain oxygen?* Through lungs, gills, skin, or a combination of these?

    h. Are any *sense organs* visible? If so, what types and where?

    i. How does the animal *move* from one place to another?

    j. Does it make any types of *movement* while it remains more or less in one spot?

k. How does the animal *capture* and take in *food?*

l. How does it *react* when touched lightly with a small brush?

Station 1 Observing Hydras

4. Place a single hydra in a small watch glass with some of the same water in which it has been living. Wait until the animal attaches itself to the dish and extends its tentacles, then slowly add a few drops of a daphnia culture with a dropping pipet.

5. Touch the hydra gently with the brush and observe its reactions.

6. Examine a prepared slide of a longitudinal section of a hydra under a compound microscope. Try to determine the presence or absence of a skeleton and of a digestive system.

Table 14.1

Characteristics	Hydra	Planarian	Earthworm	Hermit Crab	Frog
1.					
2.					
3.					
13.					

Station 2 Observing Planarians

7. Place 1 or 2 planarians in a watch glass containing pond or aquarium water. Add a small piece of fresh raw liver. Observe using a stereomicroscope or hand lens.

8. Use a compound microscope to examine cross sections of a planarian. Examine whole mounts with a stereomicroscope. Determine the presence or absence of a skeleton and a digestive system.

Station 3 Observing Earthworms

9. Pick up a live earthworm and hold it gently between your thumb and forefinger. Observe its movements. Do any regions on the body surface feel rough? If so, examine them with a hand lens.

10. Place a worm on a damp towel. Watch it crawl until you determine its anterior and posterior ends. Use a hand lens to see how the ends differ. Describe.

11. Place an earthworm on the loose soil and observe its movements as it burrows.

12. Use a compound microscope to examine cross sections of the body under low power and high power. Does it have a skeleton?

Station 4 Observing Hermit Crabs

13. Observe the movements of the appendages and the pattern of locomotion of a live land hermit crab. Observe the antennae. Touch them gently with the brush. Note the animal's reaction.

Keep your fingers away from the crab's pincers.

14. Place a small piece of food from the food dish in with the hermit crab. Observe how the hermit crab eats.

Station 5 Observing Frogs

15. Your teacher will show you how to catch and hold a frog so as not to injure it. Place a frog in a battery jar and study its breathing movements while it is not moving about. To do this, observe from the side, with your eyes at the level of the animal.

> **Do not rub your eyes while handling live frogs. Wash your hands immediately after handling.**

16. Remove the frog from the jar and place it in an aquarium. Observe its movements. How do they compare with the movements of a frog hopping and moving about on a laboratory table?

17. If a hungry frog is available, your teacher may be able to show you how it captures food.

18. Wash your hands thoroughly before leaving the laboratory.

Discussion

Review what you have learned about each of the organisms in Table 14.1. By reading across the table, you should be able to compare and contrast the characteristics of the five animals you have studied. For each animal, select five functions it performs as part of its way of life and describe how its structure enables it to perform these functions.

INVESTIGATION 14.2   Temperature and Circulation

Circulation in a multicellular animal is affected by many environmental factors. You might ask, for example, how increasing temperature would affect the rate at which an animal's heart beats. This investigation examines the effects of environmental temperature on the pulse rate of the dorsal blood vessel of an earthworm.

Construct a hypothesis about the effects of various temperatures on the pulse rate of a worm's dorsal blood vessel. Record this hypothesis in your data book.

Materials (per team of 3)

50-mL beaker of tap water
2 damp paper towels
500-mL beaker
ice
nonmercury thermometer

warm water (not more than 40°C)
10X hand lens
watch or clock with a second hand
clean, chemical-free dissecting tray
large earthworm

Procedure

1. Review the guidelines in Appendix I for the proper care and handling of laboratory animals. Follow these guidelines carefully.

2. Place an earthworm in the pan, dorsal side up. The dorsal side is much darker and rounder than the ventral side.

3. Moisten the earthworm with a few drops of water. Place a damp piece of paper towel gently over the head region so the worm will wiggle and move very little. Touch the worm as little as possible.

4. Locate the dorsal blood vessel running directly down the center of the dorsal surface. Find the part where you can most easily see the actual pulsing of the vessel. (It may be helpful to use a 10X hand lens.) Make a trial count of the number of beats in 1 minute. You will expose the worm to five different temperatures from cold (ice water) to warm but not hot. Do not expose the worm to temperatures higher than 40°C—it may die.

5. Add just enough ice water to the tray to submerge your worm (about 1 cm).

6. Lay the thermometer in the pan so the bulb is submerged.

7. Wait 3 minutes to allow the worm to adjust to the temperature.

8. Read the temperature of the water and immediately begin counting the pulse rate for 1 minute. Record the pulse rate per minute and the temperature in your data book.

Discussion

1. **Graphs should show that within the experimental range, the greater the temperature, the faster the pulse beat.**
2. **So there will be enough time for the new temperature to cause physiological changes in the worm. The effects will not be immediate.**
3. **There is insufficient oxygen in the water to allow the worm to breathe through its skin.**
4. **This will stabilize the physiology of the worm. Jumping from high to low temperatures may cause effects that are inconsistent or not due to the immediately preceding temperature.**
5. **Patterns will vary, but most likely the beat rate will increase with temperature. This relationship may be close to linear within this temperature range.**
6. **Responses will vary.**
7. **There are many possibilities here, including changes in water temperature during the counting, insufficient lag time for the worm to respond to each new temperature, unrepresentative beat rates caused by the worm being out of its natural habitat, and possible overuse of the worm as a subject.**
8. **Responses will vary considerably.**

9. Pour off the water and let the worm breathe for 2 minutes.
10. Add water that is about 10°C to the tray, as in Step 5.
11. Repeat Steps 6–8 with the 10°C water.
12. Repeat Steps 5–8 with water at or close to 20°C, 30°C, and 40°C.
13. When you are finished collecting data, return the worm to the place designated by your teacher. These worms will be retired to a nice compost pile.
14. Wash your hands thoroughly before leaving the laboratory.

Discussion

1. In your data book, prepare a graph from your data that shows the effect of temperature on the pulse rate per minute. Plot the independent variable on the horizontal axis.
2. Why is it important to wait a few minutes after adding water to the pan before counting the pulse rate?
3. Why must the water be poured off to allow the worm to breathe?
4. What would be the experimental advantage of starting either with the coldest or warmest temperatures and working gradually in the other direction instead of skipping around to different temperatures?
5. Describe any patterns revealed by the curve on your graph. Do the data support or refute your hypothesis? Explain.
6. To how many other types of organisms can you generalize your data?
7. What are some flaws or sources of error in the experimental procedure used here? Be very critical.
8. If you were to test your hypothesis again and had no limit on equipment or materials, how would you design the new experiment?

SUMMARY

All types of animal forms and functions have resulted from their heterotrophic way of life. Evolutionary development favored motility, which improved animals' ability to obtain food. Motility, in turn, called for streamlining and led to the evolution of bilateral symmetry and a head where sensory organs and nerve cells are concentrated. Larger animals have more choice of food sources but also more cells, requiring more coordination and greater specialization. Systems evolved to handle basic functions such as digestion, gas exchange, transport, excretion, and reproduction. Evolution, through natural selection, has resulted in a variety of adaptations that enable animals to solve the problems of life that different environments impose on them.

APPLICATIONS

1. Why do plants lack heads?
2. What is the advantage of having organs and of having a division of labor? In other words, what advantages do animals with organs have over those without?
3. What are the advantages and disadvantages of having an exoskeleton?
4. Many insects undergo metamorphosis. What are the advantages of having such different body forms within the same life cycle?
5. The importance of surface area to the function of a structure or system has been emphasized in this chapter. How is a greater surface area important to digestion? To respiration? Can you think of any other examples where surface area plays an important role?
6. Birds and mammals internally regulate their body temperature. In what ways is this adaptation an advantage?

7. What features of arthropods and vertebrates enabled them to adapt to terrestrial life?

8. One of the major principles of biology is that an organism's structure is related to its function. Choose one organism from any phylum and explain how it illustrates this principle.

9. Another of the major principles of biology is that an organism and its environment are complementary—the environment shapes the organism and vice versa. Choose one organism from any phylum and explain how that organism illustrates this principle.

PROBLEMS

1. Humans are omnivores. As part of their diet, they eat other animals to obtain their nutrients and energy. Make a list of the types of animals people around the world eat—arranging the food animals by phyla and classes.

2. In 1990, the 17-year cicadas emerged all over the Midwest after a long developmental period. Investigate these unusual organisms and report on their development. What other organisms show unusual developmental processes?

3. One of the major limitations to life on land is the lack of water. What adaptations do desert dwelling animals have that allow them to tolerate the hot and dry conditions?

4. Learn more about animals in their native habitats. To find out more about birds, contact a local Audubon Society chapter. Join members in one of their Christmas bird counts in your area.

5. Take a field trip to your local zoo or aquarium. Compare the food, behavior, artificial habitats, and the form and structure of different animals there.

6. List the principal life functions of animals, then compare the structures by which each is accomplished in a sponge, a planarian, an insect, and a mammal. (Because all these animals exist in large numbers, we must conclude that despite the diversity of their structures they all successfully carry on their life functions).

7. Make a list of animals from a number of phyla and arrange it in order of the care given the young—from least care to most care. List the same animals in order of numbers of young produced—from most to least. Explain any relationships you find between the two lists.

Windsurfing. How many different organs must work together here? Must they think about all of these organs to make them work? How can a complex multicellular organism, such as a human, coordinate so many different parts of the body?

Many organs, including the brain, heart, lungs, muscles, nerves, eyes, semicircular canals (balance), and skin, all are working together. Most of the actions are involuntary. Some actions take place automatically because of training, but some concentration is necessary. All of these parts are delicately coordinated by the nervous and endocrine systems.

Section Four

FUNCTIONING ORGANISMS IN THE BIOSPHERE

◆ Organisms vary greatly in size. A redwood tree may be 100 m tall and a blue whale may weigh 150 metric tons. A bacterium, on the other hand, may be only 0.4μm wide and weigh a small fraction of a milligram. No matter how different they appear on the outside, all organisms must accomplish the same basic tasks, such as acquiring food and eliminating wastes, to remain alive. The mechanisms and processes they use, however, may vary. Section Four examines how two different types of organisms go about the day-to-day business of life. ◆

The Human Animal: Food and Energy

15

A typical combination pizza. What probably are the most abundant nutrients in this meal? The least abundant? How do you think two slices of pizza compare in fat content to a chicken thigh and leg? To two beef tacos?

The most abundant nutrients are protein, iron, fat, and sodium; the least abundant are vitamin A, B complex vitamins, vitamin C, and calcium. The two chicken pieces total 10 g of fat (210 kcals), the thigh has 6 g of fat (120 kcal); 2 slices of pizza contain approximately 8 g of fat (290 kcals); 2 beef tacos contain 17g of fat (320 kcals).

◆ Humans consume an amazing variety of foods. Your lunch may include a hamburger and fries or several pieces of pizza and a soft drink. Later in the afternoon, you may add a candy bar, an ice cream cone, or popcorn and more soft drinks to your digestive tract. How can your body use the protein in hamburger and the potato starch in french fries to grow and provide your muscles with energy? How do the plant and animal tissues you eat become human cells and tissues?

More and more these days, processed and convenience foods are replacing foods rich in vitamins and nutrients. How does what you eat affect your health? Can you control your diet and still enjoy dessert? This chapter examines how the human digestive system operates and the importance of maintaining a complete and well-balanced diet. ◆

Ingestion and Digestion

15.1 Specific Actions and Reactions Take Place Along the Digestive Tract

Ingestion—the process of taking food into the entrance of the digestive tract—begins with the first bite of a hamburger. In humans, the entrance to the digestive tract is bounded by teeth, tongue, and palate. Once food has entered the mouth, the chewing action of the teeth begins the mechanical breakdown of food. This grinding action breaks the large pieces of hamburger and bun into smaller pieces, much like chopping a log into kindling to help start a fire.

As you chew, your highly muscular tongue keeps the food in contact with the teeth, moves it to mix it with saliva, and manipulates it into position for swallowing. Saliva, a watery secretion containing digestive enzymes, begins chemical digestion. Taste buds (Figure 15.1), sensory nerve endings in the tongue, tell you of the food's flavor. The tongue then moves the moistened and ground hamburger to the back of the mouth, where it is swallowed.

Connecting the mouth to the stomach is the **esophagus,** a tube made of two layers of muscles—an outer layer running lengthwise and an inner circular layer. Together, these muscles move the hamburger (now referred to as a food **bolus**) to the stomach by a process known as peristalsis (per ih STAWL sis), as shown in Figure 15.2. Contraction of the circular muscle

MAJOR CONCEPTS

◆ **Multicellular organisms must break food into particles small enough to enter the cell.**
◆ **Digestion in humans is both mechanical and chemical.**
◆ **Food and water are absorbed in the intestines.**
◆ **Cellular respiration provides the cell with an immediately available source of energy.**
◆ **The best human nutrition is a varied diet of natural high-fiber carbohydrates, unsaturated fats, and proteins.**
◆ **You are what you eat.**
◆ **Nutritional disorders can be treated or prevented altogether.**

KNOWLEDGE CHECK

◆ **Why is digestion necessary?**
◆ **What is nutrition?**
◆ **How do "junk" foods differ from other foods?**
◆ **Why is what you eat important to you?**
◆ **How does your body use the foods you eat?**

Guidepost
How is food prepared for absorption?

The secretion of saliva is under autonomic control. This is why the thought of food can cause salivation.

Figure 15.1 A cross section of a taste bud.

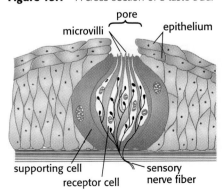

pore
microvilli
epithelium
supporting cell
receptor cell
sensory nerve fiber

Figure 15.2 Peristalsis in the wall of the esophagus forces food through the esophagus and into the stomach.

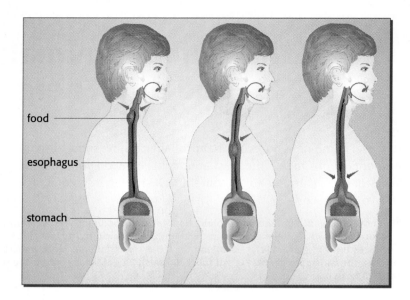

immediately above the swallowed food pushes the food down as the muscle in front of the food relaxes. Peristalsis proceeds in a wavelike sequence down the length of the esophagus.

Normally the food's journey is a one-way passage through the digestive tract, or gut, which may be described as a tube that begins at the mouth and ends where the undigested material exits the body. Figure 15.3 shows the human digestive tract and summarizes the activities that take place in it. As the food moves along approximately 9 m of digestive tract, it goes through various sections that function as a disassembly line. The **stomach,** which lies between the esophagus and the intestine, is an enlargement of the gut. A circular muscle, or sphincter (SFINK ter), at either end of the stomach closes the opening. Often thought of as a storage organ to be packed with food, the stomach actually holds food only long enough to prepare it for entrance into the small intestine.

A gastroscope, an illuminated tube of optic fibers with a lens system, makes possible direct viewing of the interior of the stomach in an anesthetized patient.

15.2 Chemical Digestion Reduces Large Food Molecules to Smaller, Absorbable Molecules

In the stomach, the food bolus is churned and diluted, the proper pH is established, and enzymatic action continues. The kneading and churning action of the stomach muscles continues the mechanical breakdown of food. This motion also helps mix in a watery fluid called gastric juice, which is secreted into the stomach by glands in the stomach wall. Gastric juice contains digestive enzymes and hydrochloric acid, which aids in the breakdown of proteins into polypeptides. The digestive enzymes that catalyze the breakdown require a low (acidic) pH to function effectively. The hydrochloric acid in the gastric juice provides an appropriate environment for enzymatic action.

Pepsin begins the digestion of proteins in the stomach and has an optimal functioning pH of 2.

Churning combined with the action of gastric juice and digestive enzymes changes the contents of the stomach into a thick liquid. The partially digested food is held in the stomach by the contraction of the pyloric valve, the sphincter between the stomach and small intestine. The pyloric valve relaxes irregularly, allowing the partially digested food to spurt into the small intestine. It takes about four hours to empty the stomach of food. **3**

When the lower esophageal, or cardiac, sphincter does not close adequately, food reenters the esophagus after mixing with gastric juices. The high acid content of the (backward flowing) gastric contents causes a burning sensation, known as heartburn, which may be relieved with antacid preparations.

Chemical digestion is completed and absorption of food molecules occurs within the **small intestine,** a tube approximately 6 m long.

2 Absorption occurs when the molecules move through the intestinal walls and enter the bloodstream. The blood then carries the molecules to the cells, where the food finally becomes a part of the body.

Food molecules go through their final chemical breakdown in the small intestine. Three types of molecules are used in all living cells as building materials and as sources of energy—carbohydrates, proteins, and fats (lipids). Although the food you eat must supply these molecules, your body cannot use the molecules directly from the food. Just as an old brick building can be demolished and the bricks used to construct a patio or a fireplace, so the human body tears food molecules apart and forms useful building blocks. Of all the molecules that make up a food, only the small molecules that make up carbohydrates, proteins, and lipids are absorbed and used to synthesize the specific molecules that make up the human body. Carbohydrates are strings of single sugars; proteins are composed of amino acids; and simple fats are made up of a glycerol molecule and three fatty acids.

Once the large molecules enter the small intestine, digestive enzymes begin to split them apart. Enzymes or digestive juices enter the upper part

The stomach wall absorbs some water, certain drugs, and alcohol.
Individuals who cannot tolerate milk and milk products may have a deficiency of the enzyme lactase, which is necessary to digest the milk sugar lactose. A lactase deficiency is found in 30 to 70% of Black and Asian adults and their children. Intake of milk by these individuals may cause gas buildup and diarrhea. Only 15% of Caucasians develop these symptoms; the difference may be related to the dependency of their culture on milk and milk products.

Figure 15.3   The human digestive system is a continuous tube with highly specialized organs and tissues along its length. Mechanical and chemical digestion are followed by absorption.

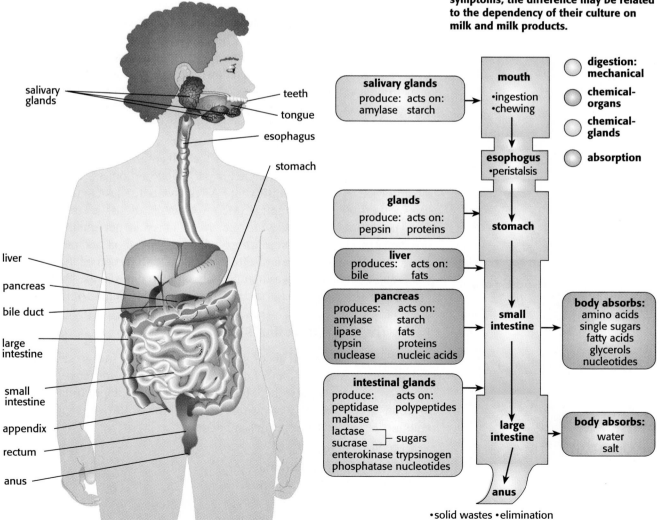

of the small intestine from the pancreas, the liver, and the intestine itself. The digestive juices all have a high (basic) pH, which helps to neutralize the hydrochloric acid of the soupy mixture that enters from the stomach. This is important because the enzymes of the small intestine require a neutral environment.

The pancreas delivers pancreatic (pan kree AT ik) juice through a duct leading into the small intestine. Pancreatic juice contains several enzymes that catalyze digestion of polypeptide, fats, and carbohydrates. See Figure 15.3. Intestinal juice, secreted by glands in the wall of the small intestine, contains other enzymes. These enzymes function in the digestion of dipeptides and double sugars.

Fats must be pretreated before they can be digested. This function is performed by bile, a substance secreted by the liver and stored in the gall bladder. Bile **emulsifies** the fat droplets (breaks them into fine particles) that pancreatic enzymes then convert to fatty acids and glycerol.

15.3   Absorption Takes Place in the Small Intestine

Almost all absorption of nutrients occurs in the lower part of the small intestine. The process of absorbing food molecules is much like wiping up a spill—the larger the sponge, the more that is absorbed. The small intestine is like a large sponge—if every wrinkle and bulge of the human small intestine were flattened, the total surface area would be about 300 m^2, about half the size of a basketball court. This enormous surface area is possible because of the presence of folds in the intestinal wall that are covered by tiny bulges known as villi (see Figure 15.4). These villi are covered by cells that have surface membranes folded into fine projections called microvilli. As proteins and carbohydrates are broken down, amino acids and single sugars are produced. These end products are absorbed into the blood by diffusion or active transport through the **microvilli** covering walls of the small intestine.

Fatty acids with 12 or fewer carbons are absorbed directly into the blood capillaries and carried to the liver for processing. Fatty acids with more than 12 carbons are absorbed by the cells of the intestinal wall, where they recombine with glycerol and are packaged in droplets. These droplets are absorbed by the **lymphatic** (lim FAT ik) **system,** a system of vessels that transports fluid called lymph (LIMF) from the body tissues to the blood.

The absorption processes are selective, but they are not infallible. There is no control mechanism to prevent absorption of excess food, and many harmful molecules can masquerade as useful molecules and be absorbed. For example, DDT, an insecticide that was used heavily on food crops in the United States until 1972, is absorbed and stored in fatty tissue and in human milk.

15.4   The Large Intestine Absorbs Water

Normally, digestion and absorption are completed in four to seven hours. Substances left in the small intestine then pass into the large intestine, where more water is absorbed. In a diet consisting of approximately 800 g of food and 1200 mL of water per day, the following approximate amounts of digestive secretions are added:

> 1500 mL saliva
> 2000 mL gastric secretions
> 1500 mL intestinal secretions
> 500 mL bile
> 1500 mL pancreatic secretions.

Approximately 2 to 3 L of clear yellow intestinal juice is secreted each day. It has a pH of 7.6 and contains some mucus, enzymes, and water.

The small intestine secretes hormones that help to regulate the production of bile, pancreatic juice, and intestinal juice.

Refer students to the discussion of DDT in Chapter 3.

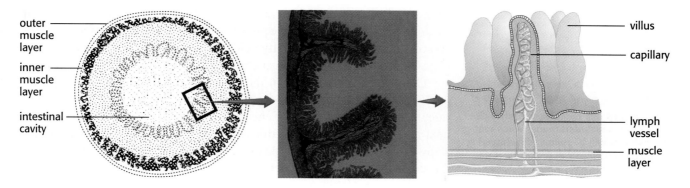

Figure 15.4   A cross section of the small intestine. Intestinal villi are shown enlarged in the micrograph and drawing. Digested foodstuffs enter the blood and the lymph through the villi.

3   Of this total—approximately 9000 mL including food and water—8500 mL are absorbed in the small intestine and 350 mL in the large intestine. The remaining 100 mL of water and 50 mL of solids are expelled from the body as solid wastes.

No further absorption of food substances occurs in the large intestine. Undigested foods, indigestible substances, mucus, dead cells from the digestive tube lining, and bacteria are concentrated by water removal to form waste matter, or feces (FEE seez). If the fecal matter remains in the large intestine too long, almost all the water is reabsorbed and very dry feces are passed. The feces exit the body through a sphincter muscle called the **anus** (AY nus).

What happens to the 8850 mL of absorbed solids and liquids? The next section traces the fate of molecules that are intended as fuel for the energy producing processes.

> Have students compare their typical diets with that of a younger child and an older adult. Have them suggest reasons for the differences.

CONCEPT REVIEW

1. Distinguish between mechanical and chemical digestion.
2. Distinguish among ingestion, digestion, and absorption.
3. Trace the movement of a hamburger and bun through the digestive system, indicating what happens in each part.
4. Discuss the digestion of carbohydrates, including where it takes place and the substances involved.
5. What is the role of the small intestine in the digestion of proteins and fats?

Cellular Respiration

15.5 Cellular Respiration Releases Energy from Food Molecules

Cells require a continuous supply of energy to maintain order, build organic molecules, grow, and carry on all their other activities. Because living things

> **Guidepost**
>
> How do cells obtain energy from food molecules?

can use only chemical energy, the energy for these activities must come from food. As a result of the digestive processes described in the previous sections, amino acids, single sugars, fatty acids, and glycerol are delivered to your cells. How do your cells obtain energy from the molecules?

Cells release energy from the basic molecules of food through the process of cellular respiration—a type of controlled burning. When something is burned, a great deal of energy is liberated all at once in the form of heat and light. Cells release the same amount of energy gradually in a stepwise series of reactions that are controlled by enzymes. As a result of **2** this controlled release, the energy present in these basic molecules is freed in small amounts that can be captured and stored in molecules of ATP. It might be said that the energy stored in food is like money in the bank, whereas energy stored in ATP is like "small change." It is readily available for use in cellular work.

Cellular respiration in humans generally is aerobic; that is, it requires oxygen, which is transported along with small food molecules in the blood. The oxygen and small food molecules move across the plasma membranes of every cell by diffusion, passive transport, or active transport and can be used by the cell in a variety of ways. The primary food molecule utilized in cellular respiration is glucose, a single sugar that has the formula $C_6H_{12}O_6$. During respiration, the energy stored in a glucose mole- **3** cule is released slowly as the molecule is broken down. Its gradual breakdown yields six carbon dioxide molecules. The overall reaction may be described in the hydrogens following equation:

$$C_6H_{12}O_6 \; + \; 6\,O_2 \xrightarrow{\text{enzymes}} 6\,CO_2 \; + \; 6\,H_2O \; + \; \text{energy}$$
(glucose) (oxygen)   (carbon   (water)
                              dioxide)

This equation summarizes a complex series of chemical reactions that involve many steps. Figure 15.5 provides a simplified diagram of this series of reactions.

The steps of cellular respiration can be divided into three stages. In the first stage, known as **glycolysis** (gly **KAWL** uh sis), glucose is split into two 3-carbon molecules and some of the released energy is transferred to ATP. Glycolysis takes place in the cytosol. The second stage is the **Krebs cycle.** During the Krebs cycle each of the 3-carbon glucose fragments is disassembled and six carbon dioxide molecules are formed. Hydrogen atoms also are released. Special molecules carry the hydrogen atoms to the third stage, the **electron transport system.** In this system, the relatively large amount of energy present in a glucose molecule is stored in several smaller amounts in ATP. Each hydrogen atom is separated into an electron and a proton and transferred through many small steps to oxygen, eventually forming water. The Krebs cycle and the electron transport system operate in the mitochondria (see Figure 15.6).

Stress that energy cannot be created nor destroyed, only transferred in chemical reactions.

15.6 Glycolysis begins the Energy-Yielding Process

As a single glucose molecule proceeds through the reactions of glycolysis, three important things happen. First, the carbon chain is broken into two pieces. Second, some ATP is formed. Third, hydrogen atoms from glucose become available for use in the electron transport system. Figure 15.7 diagrams the steps of glycolysis. Refer to the figure as you read the description that follows.

You may wish to review cellular respiration in Sections 4.5 and 4.6, Chapter 4.

Like switches in a complex electrical circuit enzymes may determine which reaction pathway will be followed by lowering the activation energy of a specific pathway *Glycolysis: glyco,* sugar; *lysis,* to split or break down.

Figure 15.5 Cellular respiration occurs in three stages. Glycolysis occurs in the cytosol; the Krebs cycle and the electron transport system take place in the mitochondrion.

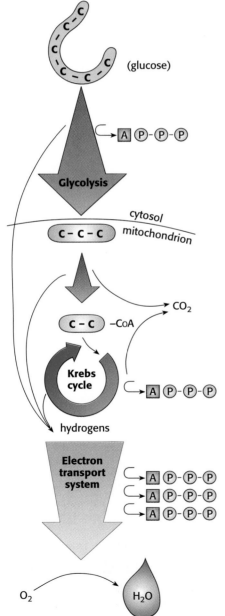

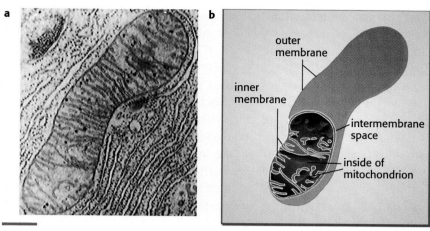

Figure 15.6   (a) Electron micrograph of a mitochondrion in the pancreas of a bat. (b) A three-dimensional drawing of the mitochondrion shown in (a). Note the projections formed by the inner membrane. Embedded in these projections are all the enzymes of the electron transport system and ATP formation. Some enzymes of the Krebs cycle are on the projections; most are in the inside area of the mitochondrion. Such organization makes possible the efficient function of this organelle.

The glucose molecule is a relatively stable molecule that contains a fairly large amount of chemical energy. To start the energy-releasing process, some energy must be expended, just as energy (in the form of a match) is required to make a fuel burn. The starting energy is provided by

Figure 15.7   In glycolysis, glucose breaks down in small steps to two 3-carbon molecules of pyruvic acid. ATP and NADH form in these reactions, which involve many rearrangements.

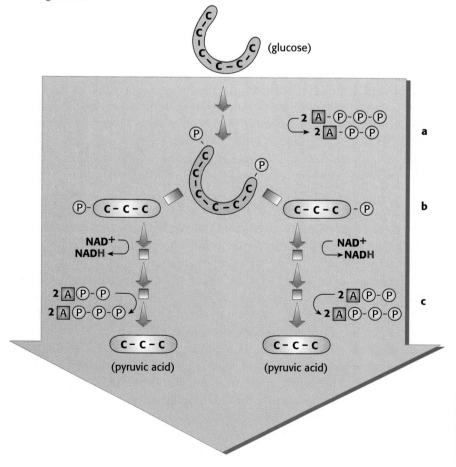

Use the analogy of a boulder on a hill; it is stable until a force is applied to start it rolling down. The potential energy in the boulder cannot be released without some activation energy.

the enzymatic transfer of phosphates from two ATP molecules, converting them to ADP, as shown at (a). The resulting molecule enters the next reaction (b) in which the 6-carbon chain is split into two 3-carbon molecules. In the remaining reactions of glycolysis (c), the 3-carbon molecules are rearranged to form two molecules of pyruvic acid. No further energy is expended in these rearrangements, each of which is catalyzed by a specific enzyme. Instead, these reactions yield energy.

Four ATP molecules are produced in glycolysis, which is a net gain of two ATP molecules. In addition, the hydrogen atoms released during the reactions combine with a hydrogen-carrier molecule called **NAD⁺,** or nicotinamide adenine dinucleotide (nik uh TEE nuh myd AD uh neen DY NOO klee oh tyd). A carrier molecule transfers substances from one reaction to another, much as a mail carrier transfers letters from the post office to a person's home. The molecule that is formed when NAD⁺ picks up a hydrogen molecule is NADH. NADH then transfers hydrogen atoms (electrons and protons) from the reactions of glycolysis to the electron transport system.

The end products of glycolysis are two molecules of pyruvic acid, two ATP molecules, and two NADH molecules. Most of the chemical energy in the original glucose molecule is retained in the pyruvic acid molecules, which are transported into the mitochondrion. There, a complex enzyme system converts each pyruvic acid molecule into a molecule of acetic acid, a 2-carbon compound. Each acetic acid molecule combines with a carrier molecule, **CoA** (coenzyme A), forming acetyl CoA. CoA delivers the acetyl group to the Krebs cycle, where respiration continues. Figure 15.8 shows these reactions, which also release two molecules of carbon dioxide and two hydrogen atoms that combine with NAD⁺, forming NADH.

15.7   The Krebs Cycle Completes the Breakdown of Glucose

Figure 15.9 illustrates the second stage of cellular respiration, the Krebs cycle. Study the diagram as you read the description of this stage, which takes place in the mitochondrion.

The acetyl CoA formed after glycolysis enters the cycle at (a). At this point, an enzyme combines the acetyl group with a 4-carbon compound (oxaloacetic acid) and releases CoA. The 6-carbon compound that results is citric acid. Citric acid then undergoes many rearrangements, each catalyzed by a specific enzyme. These rearrangements release the remaining two carbons of glucose as carbon dioxide (b). During one of the rearrangements, (c), enough energy is released to combine one phosphate with ADP, forming a molecule of ATP. Many hydrogen atoms also are released in the Krebs cycle. These are picked up by NAD⁺ or another carrier (d) and transferred to the electron transport system. Eventually oxaloacetic acid is regenerated, allowing the cycle to continue.

The Krebs cycle completes the breakdown of the carbon chain that makes up glucose. As a result of glycolysis and the Krebs cycle, one 6-carbon glucose molecule has been broken down to six 1-carbon molecules: the carbon dioxide molecules you exhale. These reactions often are called the carbon pathway. The direct energy gain has been two ATP molecules produced in glycolysis and another two ATP molecules formed during two turns of the Krebs cycle (one turn for each molecule of pyruvic acid formed from one glucose molecule in glycolysis). Most important, both processes have released hydrogen atoms that are transported to the electron transport system by carrier molecules. The complete aerobic respiration of 1 molecule of glucose to 6 molecules of carbon dioxide can

Figure 15.8 The conversion of pyruvic acid to acetyl CoA in a mitochondrion. This step prepares the pyruvic acid formed during glycolysis for entrance into the Krebs cycle.

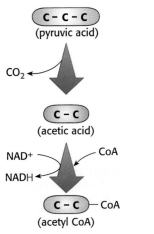

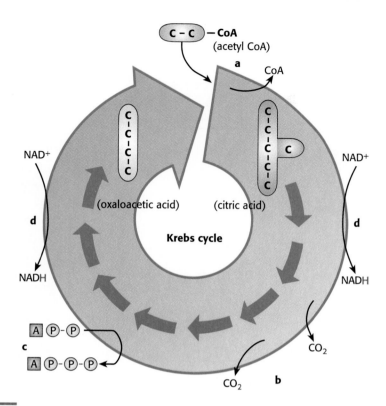

Figure 15.9 The reactions of the Krebs cycle complete the breakdown of glucose to carbon dioxide. In addition, ATP and NADH form in these reactions. Two molecules of acetyl CoA enter the cycle for each molecule of glucose.

produce up to 38 molecules of ATP. Stage three of respiration, the electron transport system, requires oxygen to produce most of that ATP.

15.8 The Electron Transport System Packages Energy from Glucose as ATP

Each electron transport system consists of a series of electron carriers embedded in the inner membrane of the mitochondria. During the final stage of respiration, illustrated in Figure 15.10 on page 382, hydrogen atoms are accepted by the electron transport system and separated into electrons and protons. The electron carriers transfer the electrons step by step through the system to a terminal carrier. The terminal carrier combines the electrons with protons and molecular oxygen, forming water. Only this final step of respiration uses oxygen.

As the electrons move from one carrier to the next (central part of the arrow in Figure 15.10), they release energy. Some of the energy is used to transport protons across the inner mitochondrial membrane. The protons become concentrated in the space between the two membranes (right side of the arrow), and then they diffuse through an enzyme complex in the inner membrane, where ATP is synthesized from ADP and phosphate. In this way, much of the energy from the electrons is transferred to ATP.

A maximum of 34 molecules of ATP can be formed in the electron transport system for each molecule of glucose. With the 4 ATP molecules formed during glycolysis and the Krebs cycle, a total of 38 ATP molecules can be formed in aerobic respiration. That represents approximately 44 percent of the energy available in a molecule of glucose, a remarkable efficiency. In comparison, an automobile engine converts only about 25 percent of the energy in gasoline into a usable form.

In anaerobic organisms and in muscles operating under anaerobic conditions, glycolysis is the main source of energy.

Sir Hans Krebs (1900–1981) of Oxford University, England, was the biochemist who first elucidated this cycle, also known as the citric-acid cycle or the tricarboxylic-acid cycle.

Active transport of protons produces a proton concentration gradient; protons then diffuse down their gradient, passing through the enzyme complex.

Stress that this discussion is about *aerobic* cellular respiration.

Summary of ATP production (approximate maximal values) per molecule yield:
NADH = 3ATP
FADH₂ = 2 ATP
Acetyl CoA = 12 ATP
Glucose = 38 ATP
Palmitate = 129 ATP (16-carbon fatty acid).

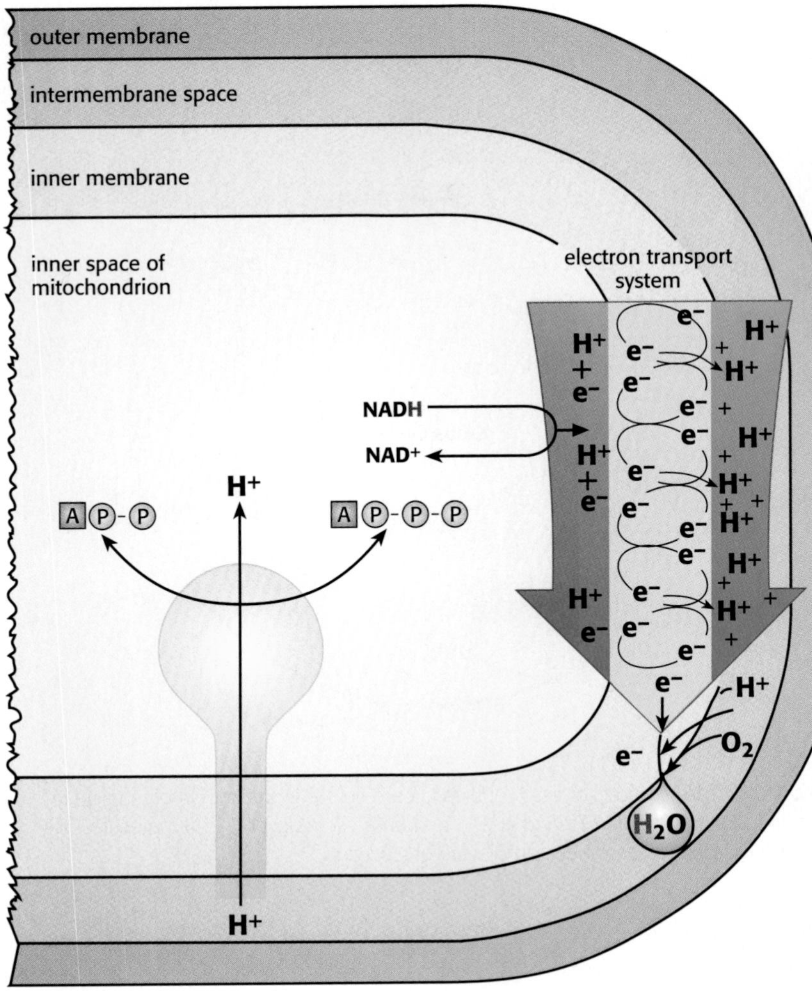

Figure 15.10 The electron transport system consists of a series of enzymes in the inner mitochondrial membrane through which electrons are transferred. Energy released from electron flow is used to build up a high concentration of protons in the intermembrane space. As protons move down their concentration gradient, they pass through an enzyme complex in the membrane that synthesizes ATP. Most of the cell's ATP is made in this way.

15.9 The Krebs Cycle Plays a Central Role in Cellular Metabolism

Both fat and protein molecules may be converted to glucose in the liver.

Carbohydrates, fats, and proteins all may be used as sources of energy in the process of respiration. Breakdown products of all three eventually enter the Krebs cycle. Carbohydrates can enter the carbon pathway as glucose or as any 3-carbon sugar. Fats first are separated into glycerol and fatty acids. Glycerol is converted into one of the 3-carbon intermediates of glycolysis before entering the carbon pathway. Fatty acids first are broken down to acetic acid and form acetyl CoA in the mitochondria. The acetyl group then is decomposed in the Krebs cycle and gives off two molecules of carbon dioxide.

When proteins are used in respiration, they first are decomposed to 5 amino acids. The amino group then is removed, and the remaining carbon skeletons may be broken down to the same 2-, 3-, 4-, and 5-carbon molecules formed in glycolysis and the Krebs cycle. These compounds then enter the carbon pathway at various points and are decomposed by the Krebs cycle, giving off carbon dioxide.

As Figure 15.11 shows, the Krebs cycle plays a central role in the carbon pathway. The intermediate compounds of the Krebs cycle and glycolysis also provide carbon skeletons that serve as building blocks for the
synthesis of other compounds. These biosynthesis (by oh SIN thuh sis)
6 reactions provide the cell with enzymes and all other materials needed for
maintenance, cell repair, and growth.

15.10 Energy Releasing Processes Are Essential to Life

Without the presence of oxygen, the entire process of cellular respiration
stops because all the electron carrier molecules become "loaded" and cannot transfer their loads. In fact, as the level of oxygen decreases, the
process slows down. Oxygen is the final acceptor of electrons. If enough
oxygen is not present, the electron flow stops, and the electron transport
system backs up like a plugged drain. NADH is unable to transfer the
hydrogen it is carrying, and therefore, it is unable to pick up more hydrogens. As a result, the Krebs cycle also stops.

When muscles are severely taxed, as they are during strenuous exercise, for example, the lungs and circulatory system may not be able to
meet the oxygen demands of the muscle cells. If this condition occurs, the
cells carry out **fermentation** (fer men TAY shun). Glycolysis continues,
producing two ATP molecules, pyruvic acid, and NADH. The pyruvic
acid does not proceed to the Krebs cycle, however. Instead, two hydrogen
atoms are transferred from NADH to the pyruvic acid. This converts the

**Stress the importance of regeneration of
NAD⁺. An analogy would be that of the
log jam that stops movement in a river.**

**There is great awareness today of the
importance of aerobic exercise. This
might be a good place to explain that
such exercise increases the ability of the
lungs and circulatory system to supply
oxygen to the muscles and, hence,
increases the length of time muscles can
continue to contract.**

Figure 15.11   Respiratory and biosynthesis
pathways. The Krebs cycle plays a central
role in both.

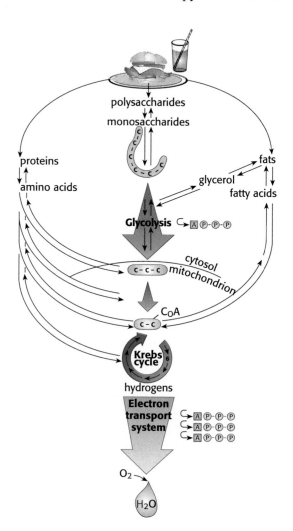

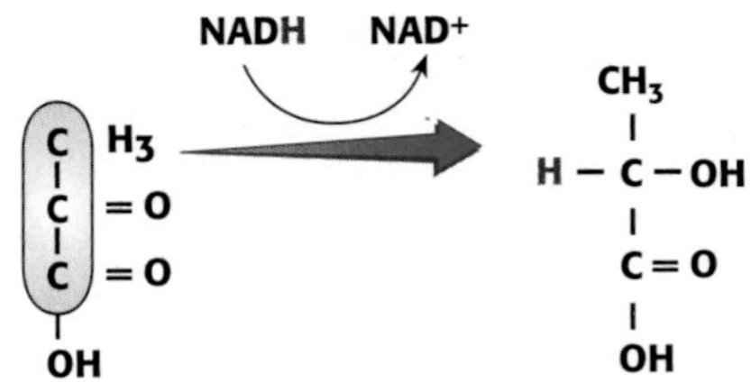

NADH   NAD+

pyruvic acid (from glycolysis)   lactic acid

Figure 15.12 Conversion of pyruvic acid to lactic acid in muscles. This reaction, in which NADH also is converted to NAD+, enables glycolysis to continue, supplying energy in the form of ATP to the muscle cells.

pyruvic acid to the 3-carbon molecule lactic acid (Figure 15.12). Thus NAD+ is freed to pick up more hydrogen, and glycolysis can continue, although it provides only limited amounts of ATP. The lactic acid accumulates in the muscles, eventually reducing the ability of muscle fibers to contract and causing muscle fatigue.

To illustrate the vital nature of these processes, consider what happens when they are stopped—for example, by the action of cyanide, one of the fastest acting poisons. Cyanide combines chemically with the final carrier in the electron transport system and blocks the production of ATP by the mitochondria. Without ATP, cellular activities in nerve cells and other cells cannot continue, and a person becomes unconscious within three to six minutes.

The food you had for lunch may serve one of two purposes. First, it can provide usable energy for life functions. Second, the breakdown of the food can provide molecules to be reassembled into larger molecules that build or repair cells in the body. This second process, known as biosynthesis, is dependent on the presence of appropriate building materials. These include the carbon skeletons formed in the Krebs cycle as well as various elements and vitamins that must be obtained in the diet. The next section examines some of the requirements for biosynthesis and some of the ways that today's environment may make it difficult to obtain these requirements.

CONCEPT REVIEW

1. Why can food be considered a fuel?
2. Compare the release of energy in burning and the release of energy in cells.
3. What role does glucose play in cellular respiration?
4. How is oxygen involved in cellular respiration?
5. How are fats and proteins used in cellular respiration?
6. In what ways does the Krebs cycle play a central role in the process of cellular metabolism?

Guidepost

What nutrients do humans need for optimal growth and health?

Historic evidence indicates that as recently as 1910 the typical American diet included lots of potatoes and whole grains such as corn, rye, and barley. Very little red meat or sugar was consumed.

Evidence has been found in pollen and bone content of garbage middens in caves, preserved coprolites, and tooth wear patterns. For more information, see S. B. Eaton and M. Konner, "Paleolithic Nutrition," *The New England Journal of Medicine* (31 January 1985): 283–88.

Nutrition

15.11   Dietary Habits Affect Health

During the approximately 2 million years the human genus has existed, the human digestive system has evolved in accordance with a dietary selection that remained fairly unchanged until this century. From the available evidence, early diets were high in fiber and vegetable matter and low in fat, sugar, and salt. During the last 50 years, however, diets have become high in fat, refined sugar, and salt and low in fiber. During these 50 years, there also has been a rise in disabilities and illnesses that may be linked to new dietary habits.

The function of the digestive system is to break down large molecules and to absorb the resulting smaller molecules that can meet the needs of the body's cells. The digestive system can use only the materials supplied to it, however. If you decided to have a hamburger, fries, and a soft drink for lunch, will this meal provide the necessary nutrients for your body to continue growing and functioning? Research has indicated that eating too much processed food, eating too much saturated fat (animal fats, tropical oils, and cottonseed oil), not eating enough complex carbohydrates (starches), and eating more kcals than are needed or expended may damage the human body.

Although Americans now consume about 3 percent fewer kcals than they did in 1910, they get so little exercise that two out of every five are overweight. Furthermore, because their intake of refined sugars and animal fats has increased, some overweight individuals actually are malnourished due to the absence of essential nutrients in their diets. The average American takes in more food than his or her body requires but even so does not ingest the proper amounts of basic and essential nutrients.

Does your diet supply you with all the nutrients you need? Look at Table 15.1, an analysis of a fast-food lunch. When he ate that lunch, a male between the ages of 15 and 18 years would have consumed 34.5 percent of his daily recommended kcal allowance. A female in the same age group, however, would have taken in 46 percent of her kcals for the day. These kcal recommendations are based on activities for the average high-school student and include walking to school and some daily physical activity, such as physical education class. A student on one of the athletic teams who works out regularly would have an increased allowance. A less active student would need to decrease the number of kcals to avoid gaining weight.

Of the carbohydrates in the fast-food lunch, only 28 g, or 23 percent, are complex carbohydrates. The 92 g of refined carbohydrates represent

Processed foods are foods that have been chemically or physically treated. Processing is used to prevent spoilage, maintain nutritional quality, maintain or enhance flavor or appearance, and make food convenient to use.

In the dietary sense, **complex carbohydrates** are the polysaccharides—starch, glycogen, and plant fibers.

Have students locate and compare weight charts based on frame size, age, or sex. Compare these to determine if there is a consensus on an "ideal" weight.

Discuss with the students how much of their diet is determined by (1) habit or family pattern, (2) peer relations and pressure, or (3) lack of understanding of good nutritional practices.

Table 15.1 Analysis of a Fast Food Lunch

	Kcals	Protein (grams)	Carbohydrates (grams)	Fats (grams)
Hamburger	606	29	51	32
Fries	215	3	28	10
Soft drink	45	0	41	0
Totals:	966	32	120	42

Age	Sex	Recommended Daily Kcal Intake	Kcals in This Meal		Percentage of Total Kcals
15–18	Male	2800	966	=	34.5
15–18	Female	2100	966	=	46.0

Nutrient (energy content)	Grams × Kcal/g	=	Total Kcals	Actual Percentage of Daily Kcal Intake	Recommended Percentage of Daily Kcal Intake
Protein (4 kcal/g)	32 × 4	=	128	4.6 (male) 6.1 (female)	12 12
Carbohydrates (3.8 kcal/g)	120 × 3.8	=	456	17.1 (male) 22.9 (female)	58* 58*
Fats (9.1 kcal/g)	42 × 9.1	=	382	13.5 (male) 18.0 (female)	30** 30**

*Only 10% of the total carbohydrate intake should be refined carbohydrates—77% of the carbohydrate in this lunch is refined.
**Only 10% of the total fat intake should be saturated fats—all of the fat in this lunch is saturated.

77 percent of the total carbohydrate intake, well above the recommended 10 percent. All 378 kcals of fats eaten probably are saturated fats. Current recommendations state that only 10 percent of the total daily intake of fats should be saturated fats.

In 1988, the first *Surgeon General's Report on Nutrition and Health* singled out saturated fat as the number one dietary problem in the United States. Five of the 10 leading causes of death in the United States—cardiovascular disease, certain types of cancer, stroke, diabetes mellitus, and atherosclerosis—are diseases in which the diet plays a role. According to the report, less fat and more fiber in the diet should reduce the risk of contracting many of these conditions.

The report recommends that most people should:

1. Reduce their intake of fat, especially saturated fat, by eating leaner meats and fish, and more whole grains, vegetables, fruits, and low-fat dairy products.
2. Get sufficient exercise to burn off the calories they consume.
3. Eat more whole grains and dried beans to increase the amount of fiber in their diet.
4. Limit their intake of sodium and sugar.
5. Limit their consumption of alcohol; pregnant women should not drink at all.

Many questions about diet and nutrition remain to be answered through research. Until more information becomes available, the most reasonable recommendation is to eat a variety of foods in moderation and to reduce intake of processed and convenience foods.

15.12 Lipids Are Concentrated Sources of Food Energy

Your body needs several types of nutrients. Carbohydrates and fats are the primary sources of energy. Proteins supply the body with building materials and enzymes. They also may be used to provide energy. Vitamins and several elements are essential substances that form cofactors in cellular respiration and other cellular activities. As cofactors, they enhance the activity of many enzymes, and they help to release the energy of carbohydrates, fats, and proteins and to maintain normal body function. The tables entitled *Elements Important in Humans* and *Vitamins Important in Humans* in Appendix 3 detail the major functions of the vitamins and essential elements. Figure 15.13 shows several foods that are high in vitamins.

Lipids, or fats, are the most concentrated source of food energy available. They release approximately 9 kcals for every gram of fat. Proteins and carbohydrates release only 4 kcals per gram. Lipids aid in the metabolism of cholesterol, a lipid associated with animal products and linked to heart disease, and they aid in the absorption of vitamins A, D, E, and K. They are used structurally in cellular membranes and in the production of hormones and hormone like substances. Lipids also supply essential fatty acids. Essential nutrients, such as fatty acids, are those the body cannot produce at all or cannot produce in adequate amounts and, therefore, must be obtained from ingested foods.

Although fats are important nutritionally, most Americans today consume far too many and the wrong type. Two types of fats are of particular concern: saturated fats, which include animal fats such as lard and butter, and unsaturated fats, which include most vegetable oils (except cottonseed oil and tropical oils such as palm oil and coconut oil). Fats containing fatty

One recent study using identical twins indicates that there are individual differences in what the body does with unneeded calories. Some people store the calories as fat, others as muscle. Most people fall somewhere in between. The study also provides more evidence that obesity is at least partly an inherited disorder.

Figure 15.13 Examples of foods high in vitamins.

2

3

Pioneers

The Limits of Human Performance

The Human Systems Center (HSC) is located at Brooks Air Force Base, San Antonio, Texas. HSC is the center for human-centered research and development for the United States Air Force. Its mission is to protect and enhance human capabilities and improve the interaction between humans and machines. HSC's origins go back to January 19, 1918, when the Medical Research Laboratory was formed at Hazelhurst Field, New York. In 1922, this laboratory was moved to Brooks Field, San Antonio, which was a center for beginning flight training. In 1959 in response to advances in space technology, this organization became a part of the Aerospace Medical Center. This Center represented an initial step for combining aerospace medical research, education, and clinical medicine. In 1961, the Center was assigned to a new organization, the Aerospace Medical Division. This was the forerunner of the Human Systems Center, so-named since 1992. A new organization called the Armstrong Laboratory was then founded under HSC. It is a world-class center in science and technology for protecting human crew members in all aerospace operations.

The Armstrong Laboratory (AL) has an interesting and exciting mission. It plans and conducts exploratory and advanced research all focused on maintaining, practicing, and enhancing human capabilities. The laboratory works with individuals, teams, and entire crews that travel into the aerospace environment. AL conducts programs in aerospace medicine, disease prevention and health services, crew systems, human resources, and occupational and environmental health. Highlighting the human as the most important element in aerospace systems, AL sponsors and carries out research and development in the fields of biodynamics (man-machine systems), biocommunications, toxic hazards, biological effects of radiation, human engineering, crew protection and life support, and training and simulator systems.

Within the Armstrong Laboratory, the Crew Systems Directorate is conducting some truly pioneering work. Its scientists research human physical, physiological, and behavioral characteristics and stress tolerance in order to understand and predict crew exposure limits in aerospace. They develop the best crew station equipment and protective countermeasures. Some specific pioneering efforts are in sustained acceleration, helmet mounted systems that extend human senses, escape systems, life support, high altitude protection, and protection against spatial disorientation in flight.

The Human Systems Center provides technical and instructional help to laboratories located in many countries. It maintains a vast aeromedical database, information resources and laboratory facilities. It's Hubertus Strughold Memorial Library is one of the largest aeromedical libraries in the world.

The Human Centrifuge operated by the Armstrong Laboratory is a simulator used to explore the reaction of humans to sustained acceleration.

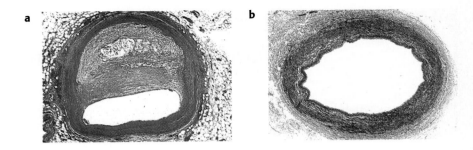

Figure 15.14   Atherosclerosis occurs as a fatty deposit on the walls of an artery (a). Note the small opening that remains. A healthy artery with large opening is shown in (b).

acids that do not have double carbon bonds (two carbon atoms sharing two chemical bonds) are known as saturated fats. Saturated fats tend to be solid when left at room temperature. Fats that contain one or more double carbon bonds in their fatty acids tend to be liquid at room temperature. These fats are known as monounsaturated or polyunsaturated fats, depending on the number of double carbon bonds present. (Review Figure 4.17.)

Reducing fat intake, particularly the intake of cholesterol and saturated fat, is important for individuals because of the apparent link between high blood cholesterol levels and cardiovascular disease (disease of the heart and blood vessels). During the past 30 years, scientific research has indicated that a high dietary intake of both cholesterol and saturated fat tends to raise blood cholesterol level in individuals. On the other hand, a diet that is rich in polyunsaturated fat may lower blood cholesterol levels for some people. In the disease atherosclerosis (ATH uh roh skleh roh sis), excess cholesterol is deposited on the inner lining of the arteries. The bulky fat streaks are called plaques. An accumulation of plaques, like that shown in Figure 15.14, can inhibit the flow of blood and increase blood pressure. Eventually a clot may form, obstruct the artery, and cause a heart attack or stroke. Table 15.2 lists the factors that increase the risk of heart

Table 15.2 Some Risk Factors Associated with Heart Disease

Risk Factors	Physiological Result	End Result
eating & drinking too much	overweight	
not exercising enough		
high total fat consumption	elevated blood cholesterol	higher risk of heart disease
high saturated fat consumption		
low polyunsat: sat. fat ratio		
high cholesterol consumption		
no exercise	elevated blood pressure	
overweight		
high fat diet	accelerates the atherosclerotic process	
diabetes		
smoking		

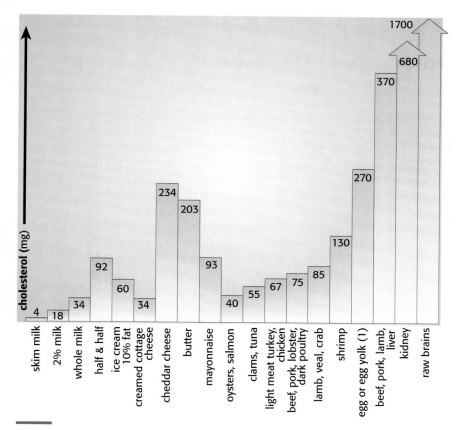

Figure 15.15   Cholesterol content of various foods, per 250 mL (1 cup) serving.

disease. Data show that half of all the deaths in the United States are caused by this type of disease.

In the last 30 years, autopsies of individuals between the ages of 18 and 30 have revealed significant development of cholesterol deposits in the arteries. A recent study has shown fatty streaks in arteries of children as young as two, indicating that atherosclerosis may begin in childhood. As a result of these studies, scientists think it is important to adopt moderate-fat and moderate-cholesterol diets early in life, before life-style patterns have been too firmly established. Figure 15.15 shows the cholesterol content of various foods.

Some of this research dates to the Korean conflict, when a systematic study of Americans killed revealed what then were considered "advanced" stages of atherosclerosis.

15.13   Carbohydrates Provide Energy, Nutrients, and Fiber

Fruits, vegetables, and whole grains, such as those shown in Figure 15.16, are important because they provide the essential elements and vitamins necessary for good health, as well as carbohydrates, the body's source of energy. Refined sugars, which contain no other nutrients, provide only kcals—sometimes called "empty" kcals. Reports indicate that three of every five kcals ingested by Americans come from fats or added sugars. In the American diet, processed and convenience foods, which have a high refined sugar content, are replacing more nutritious foods that supply vitamins and essential elements. Refined and processed sugars are added to almost every type of processed and convenience food. The food industry has many terms for the sugars and sweeteners it uses, such as sucrose, raw sugar, turbinado sugar, brown sugar, total invert sugar, corn syrups, honey, fructose, levulose, dextrose, lactose, and others. All of these, however, can

Biology Today

Cardiovascular Diseases

Cardiovascular diseases—diseases of the heart and blood vessels—are the leading cause of death in the United States. About one person in five has some form of cardiovascular disease. Heart attacks are the main cause of cardiovascular deaths and account for about one-fifth of all deaths. A heart attack occurs when some area of the heart muscle does not receive enough blood. Sometimes a blood clot forms in the blood vessels of the heart, causing a heart attack. The clot blocks the passage of blood through those vessels, so blood does not reach a portion of the heart muscle. Heart attacks also may result if a vessel is blocked by deposits of materials. If the damaged area of the heart is fairly small, recovery can occur in the manner illustrated in the figure. Sometimes the amount of damage is slight and thus difficult to detect. Because they reveal abnormalities in the timing of the heart contractions, electrocardiograms are useful for evaluating the overall condition of the heart. Abnormalities in the timing of heart contractions may be associated with damaged heart tissue.

Strokes are caused by an interference with the blood supply to the brain. They often occur when a blocked vessel bursts in the brain. Strokes also may be associated with a blood clot or a coagulation of other materials in a blood vessel. The overall effects of a stroke depend on how severe the brain damage is and where the stroke occurs in the brain.

Atherosclerosis, which often is a factor in heart attacks and strokes, is a buildup of deposits within the arteries.

The accumulation of fatty materials and deposits of cholesterol or other cellular debris within the arteries can impair the proper function of the arteries. When this condition is severe, the arteries can no longer expand and contract properly, and the blood moves through them with difficulty. The accumulation of cholesterol is thought to be the prime contributor to atherosclerosis. Diets low in cholesterol frequently are prescribed for this condition.

Arteriosclerosis, or hardening of the arteries, occurs when calcium is deposited in arterial walls. This usually occurs when atherosclerosis is severe. When atherosclerosis and arteriosclerosis occur, the flow of blood through the arteries is restricted, and the arteries cannot expand enough to handle the volume of blood pumped out by the heart. Thus, the heart must work harder.

At least three factors are linked to cardiovascular disease: smoking, high blood pressure, and high blood cholesterol. The current advice for reducing cardiovascular risk is simple: Do not smoke; have your blood pressure checked and regulated, if necessary; and follow a diet that lowers the levels of cholesterol in your blood. Finally, try to be aware of new information as it becomes available. Research on nutrition and cardiovascular disease is very active, and new data often result in changed dietary and activity recommendations. Look for valid reports based on controlled experiments and avoid popularized claims of quick fixes and miracle remedies.

thrombus

scar tissue

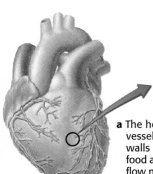

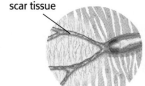

a The hearts own system of blood vessels supplies the muscular walls of this powerful pump with food and oxygen. Sometimes blood flow may be blocked (circled area). The muscles are damaged by lack of oxygen and foods.

b With smaller blood vessels taking over from the blocked artery, scar tissue begins to form in the dead muscle cells.

c If the heart attack is not too severe, the scar tissue will shrink over several months—eventually normal heart function may be established.

Events in a heart attack and recovery.

Figure 15.16   A variety of high-fiber foods.

contribute significantly to obesity and to the national $10 billion-a-year dental bill for cavities. Because of the current labeling laws, sometimes it is impossible for consumers to determine how much refined or processed sugar they actually consume.

Fruits, vegetables, and whole grains also are sources of dietary fiber, or roughage, as well as vitamins, essential elements, and carbohydrates. Dietary fiber consists of cellulose and other carbohydrates that are abundant in plants and in unprocessed plant foods. Humans lack the enzymes necessary to digest these substances. Instead, fiber absorbs water rather
5 like a sponge, causing bulkier feces that move more quickly and easily through the large intestine.

The current emphasis on fiber intake began with some observations made by a number of British physicians. They noted the reduced incidence of diseases such as cancer of the colon and atherosclerosis in certain African rural villages. The diet of these villagers is low in processed food and high in fiber. When these villagers moved to urban areas and began eating the low fiber, high fat diet of Western cultures, they suffered the same diseases as longtime urban dwellers. The physicians proposed the hypothesis that a high fiber diet would protect against the diseases of Western cultures. This hypothesis is still being tested, and caution is advised in adding large amounts of fiber to the diet. In large amounts, bran can interfere with absorption of calcium and other essential elements.

15.14   Protein Provides the Body Framework

Protein is needed daily for the repair and maintenance of body tissues, as well as for normal growth and development. Protein also supplies the amino acids needed to make the type of protein required by the body. Figure 15.17 shows some foods rich in protein. Humans can synthesize only 8 to 10 of the 20 amino acids they need. The others—called essential amino acids—must be obtained from foods. Eggs, milk, fish, soybeans, cheese, meat, and poultry provide complete proteins—proteins that contain all the essential amino acids in the proportions needed by the body.

Have students research the process of refining sugar. Contrast this with the processing of raw milk, that is, pasteurization, homogenization, and the addition of vitamin D. Students also may find it interesting to read labels to try to determine the sugar content of foods.

Figure 15.17 Foods rich in protein come from a variety of sources.

Although grains, nuts, and seeds are good sources of protein, they usually do not provide all the essential amino acids or the amounts or proportions. It is possible, however, to obtain complete proteins by combining these foods so that a missing amino acid in one is provided in the other. Grains and legumes (beans, peas, and peanuts), and legumes and nuts are examples of such combinations. Vegetarian diets should contain a balance of soybean products, leafy dark green vegetables, and some dairy products to supply the necessary balance of amino acids.

Most Americans exceed the protein requirement of 0.5 to 0.8 per kilogram of bodyweight per day. Contrary to popular claims, consuming excess protein does not result in "muscle building." Instead, the protein is respired to provide energy or converted to fat and stored.

15.15   Eating Disorders Are Widespread

Dieting has become almost epidemic in the United States, but in a growing number of cases, dieting becomes an obsession that leads to a potentially fatal eating disorder called anorexia nervosa (an oh REX ee uh ner VOH suh). A person suffering from anorexia literally starves. Bulimia (buh LEE mee uh), another eating disorder, also is increasing in the United States. A bulimic individual indulges in eating binges, usually involving large quantities of carbohydrate-rich "junk" food, and then induces vomiting or uses laxatives in an effort to purge the body of the food and thus not gain weight.

Since 1980, these disorders have increased at an unparalleled rate. For **6** reasons that are poorly understood, 90 to 95 percent of the individuals suffering from eating disorders are female. Researchers now estimate that 5 to 10 percent of the adolescent girls and young women in the United States suffer from eating disorders. As many as one in five female high-school and college students may suffer from bulimia.

People suffering from either of these diseases usually have an irrational fear of being fat, have low self-esteem, and are unable to see themselves as they actually are. Even in advanced cases, where the individuals may weigh only half of their normal weight, they still perceive themselves as fat. The disease can result in the cessation of the menstrual cycle, skin

Have students in groups of males and females develop their ideal image for the opposite sex. Discuss how this image is affected by TV and commercials.

rashes and dry skin, loss of hair and nail quality, dental cavities, and gum diseases. In its extreme form, affected individuals may die from damage to vital organs, heart failure, rupture of the esophagus, or other causes.

Although anorexia and bulimia are believed to be psychiatric disorders, scientists also are searching for physiological explanations for these conditions. Changes in the levels of certain hormones are involved, but whether these changes cause the conditions or result from the conditions has not been established. The mortality rate of these disorders is perhaps the highest of any condition classified as a psychiatric disorder. In treating these conditions, it is important to recognize the shame often associated with the condition as a problem. Professional help should be sought promptly. Both a psychiatrist or psychologist and a physician should be involved in treatment, and the family as a whole must play a strong, supportive role.

CONCEPT REVIEW

1. What are the three major causes of dietary damage to the human body?
2. Describe the role of each of the basic food molecules in normal functioning of the human body.
3. Distinguish between saturated and unsaturated fats. How are these differences important in nutrition?
4. What health problems may be related to excess fat intake?
5. Explain the importance of dietary fiber.
6. What problems are associated with taking in too many or too few kcals?

INVESTIGATION 15.1   Food Energy

All foods contain energy, but the amount of energy varies greatly from one food to another. This investigation uses a calorimeter (Figure 15.18) to measure the amount of energy, in calories, in some foods. Recall that a calorie is the amount of heat required to raise the temperature of 1 g (1 mL) of water 1°C. Caloric values of foods in diet charts are given in kilocalories (1000 calories) or kcals. Table 15.3 lists accepted caloric values for some common foods.

Using the LEAP-System™ with a thermistor, or a thermometer if the LEAP-System is unavailable, you can measure the change in temperature (ΔT) of a known volume of water. The temperature change is caused by the absorption of the heat given off by burning a known mass of food. Based on the change in temperature, you can calculate the amount of energy in the food. Instructions for use of the LEAP-System are provided in the LEAP-System BSCS Biology Lab Pac 2.

This investigation develops the idea that energy is contained in common food substances and that it can be measured. You may wish to discuss the idea that the energy released from foods is associated with a corresponding chemical change in the food itself.

The energy released in this experiment is rapid and uncontrolled. Point out that such an undirected release of energy within an organism would be harmful. Organisms control energy release through a series of enzymatically regulated steps. Teams of 3 allow for clearly defined roles.

Time: One class period; allow two periods if students have not previously used the LEAP-System™.

Materials (per team of 3)

3 pairs of safety goggles	balance
3 lab aprons	forceps
100-mL graduated cylinder	tin can with cutout air and viewing holes
250-mL Erlenmeyer flask	cork with sample holder
Apple lle, Apple IIGS, Macintosh, or	kitchen matche
IBM computer	20-cm X 30-cm piece of extra heavy
LEAP-System™	aluminum foil
printer	2 pot holders
thermistor	small container of water
extension cable for thermistor	3 whole peanuts
15-cm nonroll, nonmerucy thermometer	3 walnut halves
(optional)	

Table 15.3 Energy in Kcals of Common Foods Burned in a Highly Efficient Calorimeter

Food	Kcal/100 g
Almonds	640
Bacon	626
Brazil nuts	695
Bread, white	261
Butter	733
Cashews	609
Cornflakes	359
Dried coconut	570
Dried lima beans	341
Kidney beans	350
Lard, shortenings	900
Mayonnaise	720
Peanut butter	619
Pretzels	362
Puffed rice	363
Roasted peanuts	600
Salad oil	900
Walnuts, black	672
Walnuts, English	702

SAFETY

See the safety remarks for Investigation 2.1 when working with the LEAP-System equipment. Demonstrate how to assemble and disassemble the calorimeter and the best technique for igniting the samples. Tape a small piece of sandpaper to each lab table and distribute a container with 10 kitchen

Put on your safety goggles and lab apron. Tie back long hair and roll up long, loose sleeves.

Procedure

1. Decide who in your team will be the experimenter, who will be the recorder/computer operator, and who will be the safety monitor to assure correct safety procedure is followed.

2. Copy Table 15.4 in your data book or tape in the table your teacher provides.

3. Using the balance, determine the mass to the nearest 0.1 g of each peanut and each walnut half. Record each mass in the table.

4. Obtain a 250-mL flask, a tin can, a cork with sample holder, and a piece of extra heavy aluminum foil. This equipment may be used to make a calorimeter like the one shown in Figure 15.18. Practice assembling and disassembling the equipment.

5. With the calorimeter disassembled, measure 100 mL of tap water and pour it into the flask.

6. Bend the thermistor wire at a right angle about 3 cm from the tip of the sensing element. Insert the thermistor into the flask. The sensing element should be submerged in the top of the water and not touching the glass. Bend the thermistor wire at the top of the flask to hold the sensing element in position. If not using the thermistor, set the thermometer in the flask.

7. Measure the temperature of the water and record in the table.

8. Place a peanut in the wire holder anchored in the cork. Then, place the cork on the piece of aluminum foil.

9. Carefully set fire to the peanut. This may require several matches. Discard burned matches in the container of water.

Matches are flammable solids. In case of burns, call your teacher immediately; place burned area under *cold* running water.

Table 15.4 Energy Content of Walnut and Peanut Samples

| | Mass of sample | Temperature of Water (C)° | | | Food Energy | | |
		Before burning	After burning	Change	Calories	Kcal	Kcal per gram
Walnut sample 1							
Walnut sample 2							
Walnut sample 3							
Average							
Peanut sample 1							
Peanut sample 2							
Peanut sample 3							
Average							

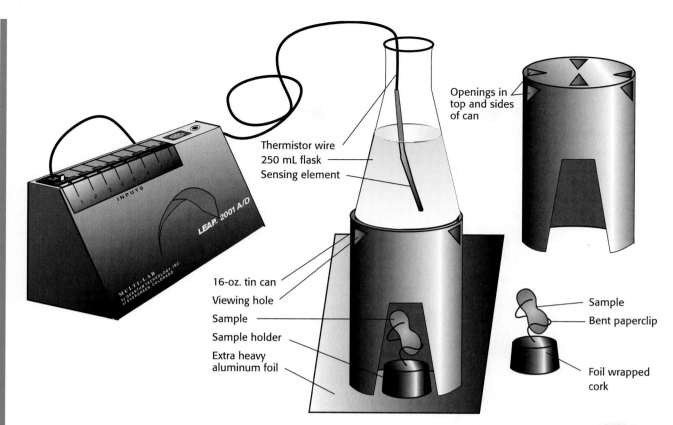

Figure 15.18   The calorimeter setup.

10. Place the tin can over the burning sample with the viewing hole facing you. Place the flask of water on top of the tin can.

11. Begin recording data. Continue recording, even when the peanut has burned completely, until the water temperature begins to decrease. (It will continue to rise after the sample has burned out as the water absorbs heat from the tin can.) Record the maximum water temperature. If using a thermometer, take readings as soon as the sample has burned out and then at 30-second intervals.

12. Allow the calorimeter to cool about 2 minutes before disassembling.

> **The flask and tin can will be hot; use pot holders to handle and place on the aluminum foil *only*. The sample holder also will be hot; use forceps to remove burned sample. In case of burns, call your teacher immediately; place burned area under *cold* running water.**

13. Repeat Steps 7 through 12 until you have data for three samples each of peanut and walnut. Change the water in the flask each time.

14. Wash your hands thoroughly before leaving the laboratory.

Discussion

1. Run strip charts to obtain graphs of the data. Determine the average change in temperature for each sample. Calculate the number of kcals produced per gram. To do this, multiply the increase in water temperature (average change) by 100 (the number of mL of water used). This step will give you the number of calories. To convert to kcals, divide by 1000. To calculate kcals produced per gram of food, divide this number by the number of grams of food burned. Enter all data in the table.

2. How do your data (adjusted for 100 g) compare with the values for 100 g of the same or similar food in Table 15.3 ? (The kcals listed in most diet charts

matches to each team; replenish matches as needed. Remind students to be cautious when using matches and handling hot objects and to place burned matches in the container of water. Be sure students place extra heavy aluminum foil under the calorimeter while samples are burned. Instruct safety monitors to throw water from the container on any flames that appear out of control. Warn students not to eat any foods in the laboratory.

Because this experiment involves an open flame, extra precautions are called for. Before students begin, be sure that:

CAUTION   No flammable liquids or vapors are present in the lab.
A recently tested ABC-type fire extinguisher is at hand.

Lab tables are clear of all materials except those required.

The piece of aluminum foil is in place under each calorimeter.

Textbooks and data books are at least 30 cm away from the piece of aluminum foil.

All students are wearing goggles and lab aprons and have tied back long hair and rolled up long sleeves.

All students are familiar with correct safety procedures.

DISCUSSION

1. The following sample data for peanuts may help students with the calculations.

$$T_2 = 37°$$

$$T_1 = 20°$$

$$\Delta T = 17°$$

$$17 \times 10 \text{ mL } H_2O = 170 \text{ calories}$$

$$170 \text{ cal} \div 1000 = 0.170 \text{ kcal}$$

$$0.170 \text{ kcal} \div 0.2\text{g} = 0.85 \text{ kcal/g}$$

$$\frac{0.85 \text{ kcal}}{g} \times \frac{100}{100} = \frac{85 \text{ kcal}}{100 \text{ g}}$$

2. **Comparisons will depend on student techniques and results. The values will be lower than the listed values for the food due to the inefficiency of the calorimeter. Results that are 20 to 40 percent of listed values are not uncommon.**
3. **Incomplete combustion, faulty observations and data recording, imperfect design of the calorimeter, or heat lost from the tin can and not absorbed into the water may account for the differences.**
4. **The results would be close to the values in charts of the caloric contents of foods.**
5. **The food with the higher kcal/100 g would be the best energy source.**
6. **It is possible that some foods may not be metabolized as efficiently as others even though they contain more kcal/unit.**
7. **All the energy can be traced to the sun through photosynthesis.**

This simulation allows students to experiment with some of the conditions that influence running a marathon and to develop and test hypotheses on race strategies. Students also can gain an understanding of how feedback mechanisms interact to maintain homeostasis. Teams of 3 control a single runner; up to 3 teams can "run" simultaneously on one LEAP-System. Teams can rotate on the LEAP-System; that is while one set of teams is planning strategy, another set can "run" the marathon, or you can organize the class as an elimination tournament. Logistics depend on your class size and equipment.

Time: **Three class periods. The first day, allows 15 to 20 minutes to explain the bars on the screen and what the paddles control. Then, let the students experiment with one runner, using trial and error to discover strategies that can produce a good running time. The second day, allow about 30 minutes to discuss the homeostatic mechanisms involved in running a marathon and to help the students develop "if/then" hypotheses before they go to the computer to test their hypotheses. Use the third day for a wrap-up discussion and further computer time, if necessary.**

are per 100 g, per ounce, per cup, or per serving. To compare your results, you may need to convert to common units.)

3. How do you account for any differences?
4. If the same amount of each food were completely burned in the cells of the human body, would you expect the energy release to be greater or less than your results? Than published results?
5. Which of the two foods you tested seems to be the better energy source?
6. Why might some foods with fewer kcals be better energy sources than other foods with more kcals?
7. What was the original source of energy in all the foods tested?

For Further Investigation

Determine the kcals in snack foods such as potato chips, corn chips, and cheese snacks.

INVESTIGATION 15.2   The Marathon

A marathon covers a distance of 42 195 meters. Athletes train for months or years to build up enough endurance to run this grueling race. Some of the many factors involved in running a marathon are diet, conditioning, race strategy (when to run fast or slow), and weather. Internal body conditions, such as body fluid levels, levels of glycogen in the liver, levels of glucose and lactic acid in the blood, heart rate, and breathing rate, are extremely important. These internal conditions are regulated by homeostatic mechanisms and vary from one person to another.

This investigation uses the LEAP-System™ to provide a computer simulation of a marathon. The simulation considers only a few of the factors involved. Figure 15.19 shows what appears on the computer screen. The marathon starts and ends in the "stadium" and is "run" on a track around the perimeter of the screen. The distance to the finish line is the same for all lanes of the track. Using the dials, you can control two factors for each runner—the rate of travel and the level of blood glucose. How fast a runner can go (rate of travel) depends on how much glucose and lactic acid are in the blood. The level of lactic acid is related to the rate of travel, and the level of glucose depends on the rate of travel and the amount of glycogen stored in the liver. Your task is to decide on a race strategy, based on regulating the rate of travel and blood glucose levels, that will result in the best possible marathon time. A good time is 2:40; an excellent time is 2:32. If you run the race in less than 2:30, please share your strategy with BSCS. Instructions for use of the LEAP-System are provided in the LEAP-System BSCS Biology Lab Pac 2.

Materials (per team of 3)
Apple IIe, Apple IIGS, Macintosh, or IBM computer
LEAP-System™ 2 dials

SAFETY
See the safety remarks accompanying investigation 2.1 for working with the LEAP-System equipment.

Procedure
PART A Trial Run

1. Decide which team member will control the runner's rate of travel (dial 1); which one will control the level of blood glucose (dial 2); and which one will watch the lactic acid and glycogen bars and advise the dial operators. The three of you must work together to achieve a good race time while keeping the glycogen, glucose, and lactic acid under control.
2. *For this trial run, the teams at each computer should work with one runner only.* Use the LEAP-System "run" procedure to select **MARTHNC1.SPU** for one runner.
3. Follow the prompt on the screen to start the marathon. Experiment with different levels of the variables to determine the effect on the time of the race.

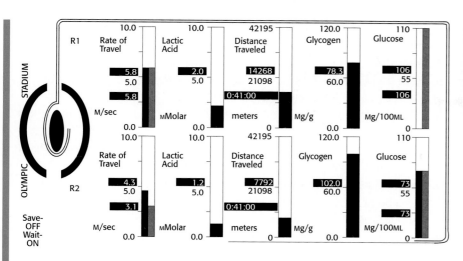

Figure 15.19 Appearance of the computer screen for two marathon runners, one across the top, the other across the bottom. From left to right the bars show the rate of travel (requested, a solid bar; attained, a hatched bar), lactic acid, distance traveled, glycogen, and glucose (requested in solid; attained in hatching).

You will need to cooperate with your partners to regulate the optimum levels for blood glucose and rate of travel. (Normal blood glucose range is 80 to 120 mL/100 mL blood.) Try several combinations of settings to see what happens as the variables change. Record these settings in your data book.

4. Use the LEAP-System "reinitialize" procedure to restart the race. Repeat as often as time permits, recording the settings for each run.

PART B Planning the Race

5. As a class, review the relationships that exist between glucose, glycogen, lactic acid, and exercise.

6. Decide on several race strategies that account for the level of blood glucose and the rate of travel at the beginning of the race, at various points in the race, and for the final kilometer. What assumptions are you making in your strategy? If, for example, you sprint the first 400 meters, what are you assuming is the effect of the sprint? Record your assumptions in your data book.

7. In your data book, write "if/then" hypotheses that include the strategies and assumptions from Step 6. The "if" part of each hypothesis should contain the conditions specified in your strategies. The "then" part of each hypothesis should be your predicted outcome, assuming your conditions are correct. For example, "If lactic acid rises to 8.0 mM, then rate of travel will slow to 3.5 m/sec."

PART C The Marathon

8. Start the race as in Step 2, but select **MARTHNC2.SPU** for two runners or **MARTHNC3.SPU** for three runners. Follow your strategies carefully. Record any changes in strategy made during the race. Record the results of the race in your data book. Include your finishing time if you finished, how far you ran, and so forth.

9. Based on the results in Step 8, decide with your partners on a strategy that will improve your runner's time in the marathon. Develop a hypothesis about how your runner's time could be improved and by how much. Record your hypothesis in your data book.

10. Use your new strategies and rerun the race.

Discussion

1. Compare how your strategies turned out. Were the prediction parts of your hypotheses correct? What changes did you make in the second trip?

2. Compare your strategies with those of other teams in the class. How were the strategies similar? How were they different?

DISCUSSION
1. **Answers will vary.**
2. **Answers will vary.**

3. **Results will vary, but usually a short sprint (not exceeding 7 m/sec) in the beginning followed by a steady pace at or near 4.8 m/sec will yield the best results.**

4. **The faster the rate of travel, the faster the lactic acid levels build up and the faster the store of glycogen is used up.**

5. **Glycogen is a carbohydrate storage compound formed in the liver. As blood glucose levels drop, glycogen is converted to glucose in the liver and released into the blood. It is the energy reservoir.**

6. **Falling glycogen levels have no direct effect on the runner until the glycogen is so depleted that it can no longer supply glucose. When the necessary glucose levels no longer can be supplied, the runner's performance is negatively affected.**

7. **A 3-carbon compound formed in muscles during anaerobic conditions; increasing levels of lactic acid decrease rate of travel.**

8. **A sugar used to produce ATP during cellular respiration in muscle cells; from glycogen stores.**

9. **Cooperation should lead to improved performances.**

10. **Depletion of glycogen levels and rise in lactic acid levels.**

11. **To increase the glycogen reservoir.**

12. **Body fluids, heart rate, breathing rate, cardiac output.**

The main objective of this investigation is to make students aware of the risk factors of cardiovascular disease. Most high-school students are young enough to make modifications in their life-style to minimize their chances of acquiring cardiovascular disease. Students should work in teams of 2 for Part A and individually for Part B.
   Time: **One class period.**

---

3. Describe the strategy that produced the best running time among all the runners. Tell why it was better than another strategy that was used.

4. What effect did the rate of travel have on your runner's performance?

5. What is glycogen? Where is it formed? How and when is it utilized? Why is glycogen supply important for a marathon runner?

6. What effect did the level of glycogen have on your runner's marathon performance?

7. What is lactic acid and when is it formed? What effect does an increasing level of lactic acid have on a marathon runner's performance?

8. What is glucose? How is it utilized in exercise? How is the supply replenished?

9. Describe how cooperation between the two controlled variables influences the outcome of the marathon.

10. Marathon runners frequently "hit the wall" late in a race and have difficulty continuing. Based on your experience with this simulation, what might produce the effect of "the wall"?

11. Why is carbohydrate loading important before a long race such as a marathon?

12. What important factors for a marathon runner were ignored in this simulation?

INVESTIGATION 15.3   Assessing Risk for Cardiovascular Disease

Several factors are known to contribute to cardiovascular disease, a leading cause of death in the United States. We can control some of these factors but not others. In this investigation, you will assess your chances of acquiring cardiovascular disease. Once you know your chances, if you wish, you can modify your life-style to reduce the risks of cardiovascular problems.

Materials (per team of 2)
centimeter tape measure   sphygmomanometer (optional)
bathroom scale   calculator

Procedure
PART A Calculating Your Ideal Weight

1. Work with your partner to make the measurements. Measure your wrists in centimeters at their *smallest* circumferences. Add both measurements and divide them by 2 to get an average.

$$\frac{\text{left wrist} + \text{right wrist}}{2} = \text{average}$$

2. Measure your forearms and calves at their *largest* circumferences. Measure your ankles over the ankle bones. Calculate an average for each as in Step 1.

3. Add the four averages and divide by 17.07 for males or 16.89 for females.

$$\frac{\text{wrist} + \text{forearm} + \text{calves} + \text{ankles}}{17.07 \text{ or } 16.89} = \text{quotient}$$

4. Square the quotient.

$$\text{quotient} \times \text{quotient} = \underline{\hspace{3em}}$$

5. Measure your height (in centimeters) without shoes.

6. Multiply your answer in Step 4 by your height.

$$\text{height} \times \text{quotient}^2 = \underline{\hspace{3em}}$$

7. Multiply the answer in Step 6 by 0.0111 to obtain your "ideal" mass in kilograms. To convert this value to weight in pounds, multiply by 2.2.

> height x quotient² x 0.0111 x 2.2 = ideal weight in pounds.

8. Determine your real weight by weighing yourself on a bathroom scale. Compare your real weight with your ideal weight. Is your real weight over, under, or equal to your ideal weight?

9. Find the difference between your real weight and ideal weight.

PART B A Self-Check

10. Take the self-check below. If a sphygmomanometer is not available, omit item 6.

Self-Check

Directions: Read each item and score the appropriate number in the space to the right of the number.

1. _____ Statistics show that males are more likely to suffer heart attacks than females. If you are (a) male, score 1 point, or (b) female, score 0.

2. _____ Heredity can influence your chances of heart disease. If one or more of your parents, grandparents, or siblings have suffered a heart attack, you may have an inherited tendency toward this condition. If one or more of your parents, grandparents, or siblings have suffered a heart attack (a) before age 60, score l 2 points, (b) after age 60, score 6 points, and (c) neither (a) nor (b), score 0.

3. _____ A person with diabetes is likely to build up fatty deposits in the arteries. If you have (a) diabetes and are now taking insulin or pills, score 10 points, (b) diabetes in your immediate family, score 5 points, or (c) no diabetes or can control it with diet, score 0.

4. _____ Smoking has been shown to contribute to cardiovascular disease. Substances inhaled during smoking damage the lining of blood vessels. If you (a) smoke 2 packs or more per day, score 10 points, (b) smoke 1-2 packs per day or quit less than a year ago, score 6 points, (c) smoke less than 1 pack per day or quit 1-10 years ago, score 3 points, (d) never smoked, score 0.

5. _____ High amounts of cholesterol in the diet can clog or narrow the arteries. This places added stress on the heart and arteries. If you eat (a) 1 serving of red meat per day, more than 7 eggs per week, use butter, whole milk, and cheese daily, score 8 points, (b) red meat 5-6 times a week, 4-7 eggs per week, use margarine, low-fat dairy products, and some cheese, score 4 points, (c) poultry, fish and little or no red meat, 3 or fewer eggs per week, use margarine and skim milk, score 0.

6. _____ High blood pressure increases the heart's work and places wear and tear on blood vessels. If you are able to measure your blood pressure, score this item. Otherwise, omit score. If your blood pressure at rest measures (a) 160/105, score 8 points, (b) between 160/105 and 140/90, score 4 points, (c) less than 140/90, score 0.

7. _____ Overweight people run a higher risk of heart disease than those not overweight. (Refer to the calculation of the difference between your ideal and real weight in Step 9, Part A.) If you are (a) 25 pounds overweight, score 4 points, (b) 10-25 pounds overweight, score 2 points, (c) less than 10 pounds overweight, score 0.

8. _____ Aerobic exercise is that type of activity that temporarily increases your heart rate, stimulates sweat production, and causes deep breathing (jogging, bicycling, swimming, and other similar activities). If you engage in aerobic exercise (a) less than once per week, score 2 points, (b) 1-2 times per week, score 1 point, (c) 3 or more times per week, score 0.

_____ Total Score

DISCUSSION

1. **Answers will vary depending on each student's self-check.**
2. **Risk factors that a person cannot change by modifications in life-style include: sex (males are statistically at a higher risk); possible inherited predisposition toward cardiovascular disease; and diabetes (in most young people, an inherited disease). It is important to note, however, that although factors of sex and inherited predisposition cannot be eliminated, they can be monitored and the risk of each one can be minimized. In some cases, regulating an individual's diet can minimize the risks of these factors.**
3. **Modifications include:**
   a. **Quitting smoking or reducing the number of cigarettes consumed each day.**
   b. **Reducing the amount of cholesterol in the daily diet.**
   c. **Continually monitoring blood pressure. If a person has high blood pressure, the person should seek proper medical assistance. High blood pressure can be controlled by diet medication, or by a combination of these two factors.**
   d. **Reducing or controlling weight problems. If a person is overweight, he or she should seek proper medical advice and assistance to remedy the problem.**
   e. **Developing a realistic plan of regular aerobic exercise. Medical advice should be sought to prevent possible adverse effects posed by an improper exercise program.**

FOR FURTHER INVESTIGATION

Students can use the Kcal Tally in Appendix 3. Another source of a calorie chart is the *Comprehensive List of Foods* (Rosemont, IL: National Dairy Council, 1980). The National Dairy Council has a variety of inexpensive or free materials related to nutrition and health.

SUMMARY

Digestion is the process by which food is broken down into small molecules. The process begins with mechanical breakdown in the mouth and continues in the various compartments of the digestive tract. Most of the chemical breakdown occurs in the upper part of the small intestine in the presence of specific digestive enzymes. The end products of digestion are absorbed from the small intestine and delivered to the cells by the blood and lymph. Inside the cell, the food molecules may be stored, respired to obtain usable energy, or used to synthesize other necessary molecules.

11. After completing the self-check, use the self-check score interpretation to determine your risk of cardiovascular disease.

Self-Check Score Interpretation	
Total Score	Risk of Heart Attack, Stroke, or Cardiovascular Disease
26 pts. or more	high risk
25–14 pts.	medium risk
13 pts. or less	no risk

Use the following information if item 6 (blood pressure analysis) was omitted.

Self-Check Score Interpretation	
Total Score	Risk of Heart Attack, Stroke, or Cardiovascular Disease
22 pts. or more	high risk
21–12 pts.	medium risk
11 pts. or less	low risk

Discussion

In Part A, you calculated your ideal weight. This weight is only an approximation of what your body weight should be. It is normal for your real weight to vary 1 to 10 pounds from your ideal weight. Part B in this investigation allowed you to assess your chances of acquiring cardiovascular disease. The information obtained in Part B is only a guide, but it can be useful in examining your life-style. Using this information, answer the following questions:

1. Based on your self-check score, do you appear to have a high, medium, or low risk of cardiovascular disease?
2. What aspects of your life that could increase your chances of cardiovascular disease are you incapable of changing?
3. What modifications, if any, can you make in your life-style to reduce your chances of having cardiovascular disease?

For Further Investigation

Does your daily diet have the proper number of kcals to maintain your ideal weight? To find out, keep a record of everything you eat for five days. After obtaining an average of your kcal intake for the five days, determine whether your average kcal intake is appropriate for your weight. A quick calculation to determine your approximate kcal needs per day follows:

$$\text{ideal weight} \times 15 = \text{kcal needs per day.}$$

In cellular respiration, the energy stored in food molecules is converted to ATP. The reactions of respiration take place in the cytosol and mitochondria and, in addition to providing usable energy in the form of ATP, provide carbon skeletons for biosynthesis. Carbohydrates, especially glucose, are the major source of energy in respiration, but fats and proteins also can be respired. These nutrients, as well as vitamins and essential elements, which are required in small amounts, must be supplied in the diet. Increased intake of processed foods, fats, and refined sugars in the United States and other industrialized nations appears to be related to major health problems such as obesity and heart disease.

APPLICATIONS

1. How is chemical digestion related to the various types of chemical syntheses carried on by the cells?
2. Research the various types of teeth in humans. How do these different teeth help in the digestion of the food you eat?
3. What changes occur in the pH of the human digestive system as food passes through it?
4. Based on comparative length, why does it make sense that most food is digested and absorbed in the small intestine and not in the large intestine?
5. Why are feces semisolid although digested food in the small intestine is semiliquid?
6. Compare the major events of glycolysis and the Krebs cycle with the electron transport system. Which reaction yields more energy?
7. Explain the central role of the Krebs cycle in both cellular respiration and cellular synthesis reactions.
8. If brain death usually occurs in a person who is without oxygen for 5 minutes, how is it possible for a child to survive for 20 minutes after falling into cold water under the ice of a lake?
9. Why do you lose weight rather than remaining the same weight if you do not eat?
10. A quotation from the Bible says that "all flesh is grass." In terms of the food you eat, what does this mean?
11. Why do we need complete proteins in our diet?
12. In what ways are vitamins and essential elements important to health?
13. Cellulose (fiber) and starch are both carbohydrates, yet humans can digest starch but not cellulose. Why is this, and why are people supposed to eat fiber?

PROBLEMS

1. Write a play to illustrate how sugar is completely broken down in cellular respiration. Have members of the class take different roles in the play and then videotape the play.
2. What are some of the factors in today's society that might cause a high-school student to become anorexic or bulimic?
3. People living in different parts of the world have very different diets. Most of the population in Africa, Asia, and Australia cannot drink milk. Investigate why this is so.
4. Suppose, because of the success of the artificial kidney, a biomedical engineer wished to design an artificial digestive system. What major functions would you have to build into the artificial digestive system?
5. What percentage of the calories you eat are fat? It may be difficult for you to measure what you eat, but try to keep a diary of the foods you eat that contain fat. Divide this list into fat or oils from plants and fat from animals. If the labels on processed foods provide nutrition information, record the percentage of fat in the various foods.
6. Many diet plants actually can be harmful. What is the strangest diet plan you have heard about? What is the evidence that is presented to potential dieters about its ability to help a person lose weight?
7. What did our ancestors eat? Use anthropology books to learn about the diet of hunter-gathers of today and of many thousands of years ago.
8. Compare the diet and health of ethnic people in their native land and after they move to the United States. For instance, Japanese and Japanese-Americans used to eat different foods, and the incidence of diet-related health problems in the two countries was different. Today, many Japanese people have adopted a Western diet, and as a consequence, their health has declined. Read about this problem.

The Human Animal: Maintenance of Internal Environment

The clotting of human blood, showing both red blood cells and the fibrin network (×5000). Use a metric ruler to measure the size of one of the red blood cells. Use the ratio of the magnification to this measurement to determine the *actual* size of this red blood cell. Check your answer with a classmate. Approximately how large is the widest opening in the fibrin? Develop a hypothesis based on these observations that explains how blood clots.

The measured length of the red blood cell in the illustration is about 4 cm or 40 000 μm. Because the red blood cell is enlarged 10 000 times, its actual size must be about 8 sm. The largest opening in the fibrin is only 5 to 6 μm. This network does not allow the red blood cell to pass through; thus, it prevents the blood from moving or flowing out of the body.

◆ Once a meal is eaten, digested, and absorbed through the cells of the digestive tract, the food molecules must be distributed throughout the body. The absorbed food molecules must be delivered, and they must be controlled carefully so the supply of food suits the needs of the cells. Excesses of certain nutrients can be as harmful as shortages. The balance of supply and demand is part of the homeostasis of the body.

This chapter examines how cells are supplied with the building materials and energy provided by food, as well as oxygen and other necessary substances, and how wastes are removed. The chapter also considers the role of the immune system in defending the body from invasion and how the various systems interact to regulate the internal environment of the body. ◆

MAJOR CONCEPTS

- ◆ Coordinated organ systems transport body materials.
- ◆ Human blood is a watery plasma of cells and chemicals.
- ◆ Red blood cells, white blood cells, and platelets have specific functions.
- ◆ The immune system protects against invasion of foreign materials.
- ◆ Kidneys filter wastes and recover needed chemicals.
- ◆ Several mechanisms maintain a constant body temperature.

KNOWLEDGE CHECK

- ◆ What is a circulatory system, and what is its function?
- ◆ What is the function of an immune system?
- ◆ Why do animals need a gas exchange system?
- ◆ What is the function of an excretory system?
- ◆ What is meant by homeostasis?
- ◆ How are the circulatory system, the respiratory system, and the excretory system related?

Point out that the ultimate destination of food is the working cell.

Circulation

16.1 The Heart Pumps Blood Through a Series of Tubes

Humans, like most animals that have body fluid, have a system of tubes or vessels that circulate fluid and transport materials throughout the body. These tubes or vessels make up the circulatory system, which carries raw materials to the cells and removes wastes from the cells' environment. Human circulation occurs in a closed system. A single, muscular heart with four chambers pumps blood through three types of blood vessels. **Arteries** have thick, muscular walls. They carry blood away from the heart. **Veins** have thin walls with little muscle. They carry blood toward the heart and have valves that prevent backflow. **Capillaries** (KAP ih layr eez) are thin-walled, narrow tubes that connect arteries and veins. Figure 16.1a shows the human circulatory system.

In the seventeenth century, William Harvey, a British physician, showed through a remarkable series of experiments that blood leaves a vertebrate's heart through arteries and returns to the heart through veins. In 1628, Harvey published a paper on the subject. Because microscopes

Guidepost
How does the circulatory system distribute materials to all functioning cells?

Each of the four major circuits—pulmonary, hepatic portal, renal, and systemic—involves a set of arteries, capillaries, and returning veins.

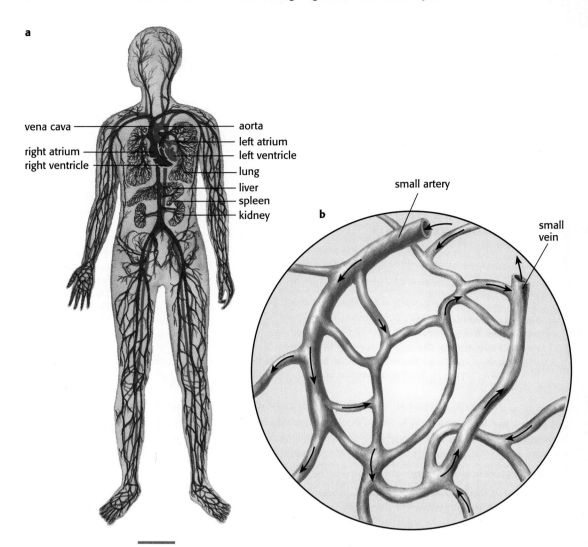

a

vena cava
right atrium
right ventricle

aorta
left atrium
left ventricle
lung
liver
spleen
kidney

b

small artery

small vein

Figure 16.1 (a) A simplified drawing of the human circulatory system. Oxygen-rich blood is shown in red; oxygen-poor blood is shown in blue. (b) A typical capillary bed.

were not yet widely used, Harvey never actually saw blood passing from arteries to veins. Later in the century, however, other scientists were able to observe capillaries, like those shown in Figure 16.1b, connecting the arteries and veins. These observations provided further evidence for Harvey's ideas about the circulation of the blood.

In humans, as in all mammals and birds, the heart is separated into right and left sides. Each side of the heart is connected to its own set of veins and arteries. The right side of the heart receives oxygen-poor blood from almost all parts of the body and sends it to the lungs. The left side receives the newly oxygenated blood from the lungs and sends it to all parts of the body. Each side of the heart has two chambers. The top chamber, or **atrium** (AY tree um), forms the upper part of the heart and receives incoming blood from the tissues of the body. The thin-walled atrium bulges with this blood, and as the lower heart muscle relaxes, the blood flows into a second chamber, the thick, muscular **ventricle** (VEN trih kul). Figure 16.2 shows these structures clearly.

On each side, the atrium and the ventricle are separated by one way valves, flaps of tissue similar to those shown in Figure 16.2a. The structure of these valves prevents the blood from flowing back into the atrium when the ventricle contracts. Ventricles, atria, and associated structures of

2

a

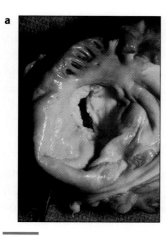

b

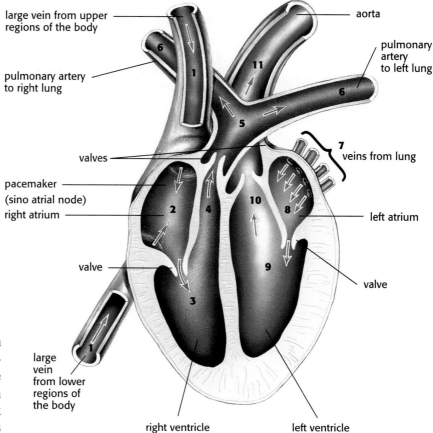

large vein from upper regions of the body

pulmonary artery to right lung

aorta

pulmonary artery to left lung

7 veins from lung

valves

pacemaker (sino atrial node)

right atrium

left atrium

valve

valve

large vein from lower regions of the body

right ventricle

left ventricle

Figure 16.2 (a) Aortic valve of the human heart. (b) A drawing of a cross section through a human heart and the blood vessels leading to and from it. Trace the flow of blood into and out of the heart by following the numbers and arrows.

the heart are shown in cross section in Figure 16.2b. As the ventricle contracts, the muscular wall squeezes the blood, forcing shut the valves between the two chambers and preventing back flow into the atrium. As the blood is pumped out of the heart, pressure forces open the valves at the entrance to the artery, and blood flows

3 through the vessel away from the heart. When the ventricles relax between heartbeats, back pressure from the blood forces the valves shut, thus blocking the artery and preventing blood from flowing back toward the heart. Figure 16.3 summarizes the pumping action of the heart. The contraction of muscles surrounding the veins helps to push blood through the body. Valves in the veins prevent backflow, so the blood flows only toward

3 the heart (see Figure 16.3d).

Atrioventricular valves (called tricuspid on the right side and bicuspid or mitral on the left) point into the ventricle. When intraventricular pressure rises (ventricular systole), the atrioventricular valves are driven closed.

Figure 16.3 The cardiac cycle. Blood enters the atria (a), which contract, forcing blood into the ventricles. The atria relax and fill (b), and the ventricles contract (c), forcing blood into the pulmonary artery and the aorta. Then, the ventricles relax and open the atria contract, repeating the cycle. Movement of blood in veins (d) is brought about by pressure from adjacent muscles. Compression forces blood in both directions, but valves keep it from flowing backward and away from the heart.

a   b   c

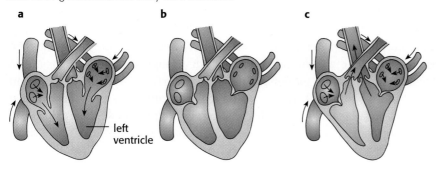

left ventricle

d   valve closed

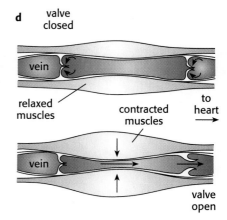

vein

relaxed muscles

contracted muscles

to heart →

vein

valve open

Water has a viscosity of 1.0 compared to blood viscosity of 4.5 to 5.5. Blood has a pH range from 7.35 to 7.45 and a salt (NaCl) concentration of 0.85 to 0.90%.

Albumin constitutes 55% of the plasma proteins and is largely responsible for blood viscosity. Fibrinogen functions in blood clotting and constitutes about 7% of the plasma proteins. Immunoglobulins include the antibodies.

If students do not know the difference between a solution and a suspension, this is a good time to make the difference clear.

Lack of either sufficient hemoglobin or red blood cells causes anemia. Ask students to consider why an anemic individual is less active than a normal one. Activity requires an energy source, usually cellular respiration. Cellular respiration, in turn, requires oxygen. Oxygen is delivered through oxyhemoglobin. Thus, a deficiency of hemoglobin leads to a decline in activity.

Figure 16.4 Human blood smear showing red blood cells, two large white blood cells, and a platelet (the small particle in the right half of the photo) (X500).

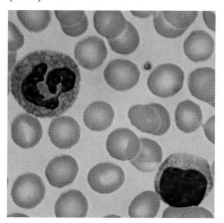

Capillary walls are only one cell thick. As blood flows through the capillaries, some substances move from the blood through the thin walls of the capillaries into the body tissues. Other substances move from the tissues into the blood. This exchange of materials occurs by diffusion because substances in the tissues and substances in the blood are present in unequal concentrations.

16.2   Plasma Contains Blood Cells and Other Substances

Blood is composed of cells and other substances suspended in **plasma** **4** (PLAZ muh), a clear straw-colored liquid that is 90 percent water and 10 percent dissolved substances. Most of the dissolved substances are proteins called plasma proteins. The other dissolved substances include essential elements, absorbed food molecules (such as single sugars, amino acids, and fatty acids), respiratory gasses (oxygen and carbon dioxide), waste products, and regulatory substances (hormones and enzymes). Some of the plasma proteins play important roles in homeostatic functions, such as blood clotting and maintenance of blood osmotic pressure. Other proteins are essential products of the immune system, which protects the body against invasion by organism or foreign particles.

Red blood cells are specialized for the transport of oxygen. As a red blood cell matures, it loses its nucleus and other cell structures and becomes filled with **hemoglobin** (HEE moh gloh bin), an iron-containing pigment that combines readily with oxygen. Because a human red blood cell lacks a nucleus and because it undergoes a great deal of wear and tear, it lives only 110 to 120 days. After this period of time, the red blood cells are removed from circulation and destroyed in the liver and spleen. Iron from the hemoglobin is salvaged in the liver and used by bone marrow cells to make new red blood cells.

White blood cells, shown in Figure 16.4, play a major role in defending the body against invading pathogens. Although there are several types of white blood cells that differ in size and function, all white blood cells contain a nucleus. Because they lack hemoglobin, white blood cells are colorless. White blood cells can move about like amoebas, slipping through the thin walls of capillaries and wandering among cells and tissues. Some white blood cells engulf pathogens in the same way as an amoeba engulfs food. Others synthesize antibodies, complex proteins that react with pathogens and other foreign substances.

16.3   Clotting Is an Interaction Between Platelets and Plasma Proteins

Normally, when you have a cut or scrape, the blood at the skin surface hardens, or clots. Blood clotting comes about through a complex sequence of events that involve some 30 proteins called clotting factors, as well as small cell fragments called **platelets** (PLAYT lets). Clotting begins when plasma and platelets come in contact with a rough surface, such as a torn tissue. The platelets gather on the rough surface, become sticky, and attract more platelets, forming a plug that partially seals the wound. They also release substances that act with clotting factors in the plasma to begin a chain of chemical reactions (see Figure 16.5). As a result of these reactions, a substance called prothrombin (proh THROM bin) activator is formed. In the presence of calcium ions, the prothrombin activator catalyzes the conversion of prothrombin, a plasma protein, to thrombin. Acting as an enzyme, thrombin converts the soluble plasma protein

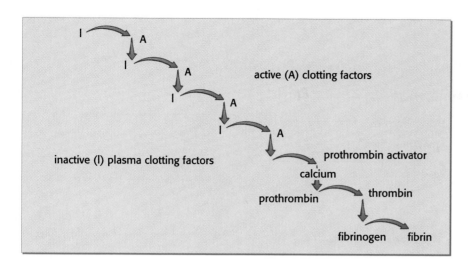

Figure 16.5   A blood clot is the result of a cascade of enzymatic reactions that end when the soluble protein fibrinogen is converted to insoluble fibrin strands. The fibrin strands provide a network in which platelets are trapped, forming a blood clot (see the chapter opener).

fibrinogen (fy BRIN oh jen) to its insoluble form, fibrin (FY brin). The fibrin forms a network of threads that trap platelets and other materials and form the clot (see the chapter opening photograph).

5   Sometimes a clot may form when there is no injury. For example, a clot may form within a blood vessel, blocking circulation. If the clot blocks one of the arteries supplying blood to the heart, blood flow to that area of the heart is cut off, resulting in a heart attack. A clot blocking an artery in the brain results in one type of stroke. Strokes usually result in paralysis of part of the body and may result in death. (See Chapter 15, Biology Today.)

Scientists believe that clotting within blood vessels is a continuous process that is combatted by normal clot-preventing and clot-dissolving processes.

16.4   Some Cells and Materials Move Between Blood Vessels and Tissue

Normally the water and nutrients in plasma pass through capillary walls to and from body tissues. White blood cells also move freely between blood vessels and tissues. Red blood cells do not leave the blood vessels, and only small amounts of plasma proteins are found in tissues. Thus, the fluid that bathes tissue cells is lower in protein content than the blood plasma. This difference in concentration is important in regulating the exchange of materials, illustrated in Figure 16.6, between the blood and the cells.

The maintenance of this internal environment is essential to the survival of the cells and plays a role in homeostasis.

Figure 16.6   The exchange of materials at a capillary. What do you think is happening at points 1, 2, and 3?   **Diffusion**

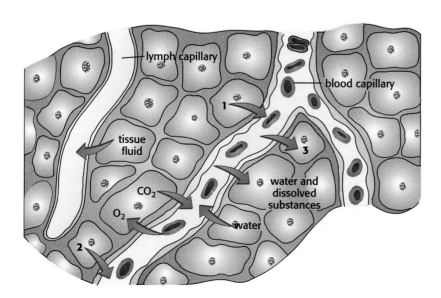

Figure 16.7   The human immune system. In addition to the lymphatic system, the bone marrow, thymus, tonsils, and spleen are important components of this system.

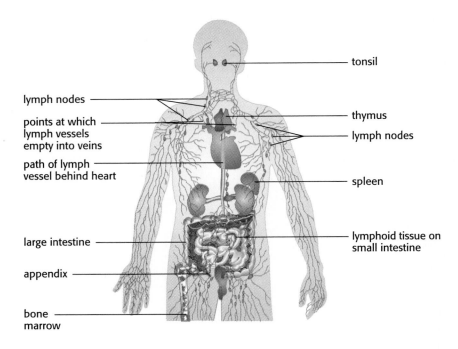

tonsil

lymph nodes

points at which lymph vessels empty into veins

thymus

lymph nodes

path of lymph vessel behind heart

spleen

lymphoid tissue on small intestine

large intestine

appendix

bone marrow

Some of the fluid that bathes the tissues may ooze back into the blood capillaries, but most of it collects in a different set of vessels. This tissue fluid is called lymph, and the tubes that carry it are called lymph vessels. **6** Muscle contractions caused by movement squeeze the vessels and move the lymph along. In the walls of the small intestine, lymph vessels in the villi absorb fats. Many of the metabolic wastes of cells also pass into the lymph. For these reasons, lymph has a higher fat content and a higher waste content than does blood.

The vessels of the lymphatic system divide into tiny, twisted passages that make up the lymph nodes. Lymph flows slowly through these nodes. Pathogens and other foreign materials that have entered the body can be engulfed by white blood cells in the lymph nodes. The lymphatic system, shown in Figure 16.7, is a part of the body's defense system.

Eventually all lymph vessels join, forming a duct in the region of the left shoulder. The duct empties into a vein, and the fluids that moved into the tissues at the capillaries return to the blood before it enters the heart.

CONCEPT REVIEW

1. Compare and contrast arteries, veins, and capillaries.
2. Explain how a four-chambered heart functions in circulation.
3. How is blood kept flowing in one direction?
4. Compare and contrast plasma and whole blood.
5. Why is it good to have a long chain of events leading to clot formation?
6. Describe the role of the lymphatic system in circulation.

Immunity

16.5   The Body Has Several Defenses Against Foreign Invaders

Guidepost

How does the immune system protect the body from invasion?

Immunity is the capacity of the human body to resist most pathogens that might damage tissues and organs. The body has two types of immunity against foreign substances. The first type is nonspecific protection. This type involves intact skin and mucous membranes and the inflammatory process. The second type is the immune system's specific immune response.

The layers of skin covering the body and the layers of mucous membranes lining the digestive and respiratory tracts provide a protective barrier against invasion. Mucous membranes are protected by secretions of mucus. Some membranes consist of ciliated cells that sweep foreign objects away. Other surfaces are washed by fluids such as saliva or tears that contain substances active against pathogens.

If the barrier formed by the skin or mucous membranes is broken, substances in the circulating blood and lymph initiate an inflammatory response. During an inflammatory response, injured cells release a chemical called histamine and other substances. These cause nearby capillaries to swell and become "leaky." Various types of white blood cells then pass through the capillary walls and gather at the site of injury, engulfing pathogens that may have entered through the cut.

The viscous nature of mucus allows it to trap many pathogens that attempt to enter the respiratory and digestive tracts.

16.6   Protection by the Immune System Is Specific

In contrast to the nonspecific defenses, the response of the immune system to foreign invaders is very specific. The immune system includes the structures shown in Figure 16.7 and a variety of interacting white blood cells.

The primary cells of the immune system are small white blood cells known as **lymphocytes** (LIM foh syts). Lymphocytes are specialized for the recognition of foreign substances. Like other blood cells, they develop in the bone marrow. Some lymphocytes also mature in the bone marrow. These types of lymphocytes are called **B cells** (**b**one-marrow-derived lymphocytes). Their function is to produce antibodies that act against foreign substances in the body. Other lymphocytes migrate to the thymus gland, where they mature into **T cells** (**t**hymus-derived lymphocytes). There are at least two types of T cells. Killer T cells attack infected host cells directly. Helper T cells interact with B cells to make antibodies. In addition, amoeba like white blood cells called **macrophages** (MAK roh fayj ez) act like processing plants that engulf and disassemble complex materials such as pathogens. Figure 16.8 shows the cells involved in the immune response.

Lymphocytes are activated by antigens—substances the body recognizes as foreign. Antigens (*anti*body *gen*erators) usually are large proteins or carbohydrates that make up the cell walls or other parts of viruses, fungi, bacteria, or other pathogens. Other large molecules or parts of molecules also can serve as antigens. Surface receptors on the lymphocytes bind to the antigen that has a shape that best fits the receptor. The fit between

B cells differentiate into plasma cells, which produce antibodies that move to the site of invasion. This is called humoral immunity. T cells migrate from the blood to all tissues of the body. Thus, their action is known as cellular immunity.

B cells come in thousands of different types. Each type is capable of combining with one specific antigen. For more information, see S. Tonegawa, "The Molecules of the Immune System," *Scientific American* (October 1985):122–31.

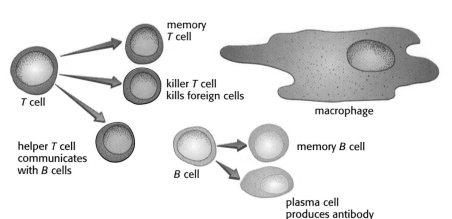

memory
T cell

T cell

killer T cell
kills foreign cells

helper T cell
communicates
with B cells

B cell

macrophage

memory B cell

plasma cell
produces antibody

Figure 16.8   Major cells of the immune system. *B* cells, when stimulated by an antigen, differentiate and become plasma cells that secrete antibodies. *T* cells are of two types. Killer *T* cells attack infected cells and invading pathogens by secreting toxic chemicals. Helper *T* cells secrete lymphokines, helper molecules that enhance the immune response. Both *B* cells and *T* cells give rise to long-lived memory cells. Macrophages ingest pathogens and display their antigens to *T* cells, triggering the specific immune response.

receptor and antigen is like a lock and key. Each lymphocyte may have several thousand receptors on its surface, but all the receptors are identical, so each cell can recognize only one foreign shape. To be able to recognize all potential antigens, humans have millions of different lymphocytes. To start **3** an antibody response, macrophages, *B* cells, and helper *T* cells must cooperate, as shown in Figure 16.9. When a particular pathogen enters the body, it may bind to a *B* cell surface receptor that fits its shape, as illustrated in (a). At the same time, a macrophage may engulf another pathogen of the same type and disassemble it, which is illustrated in (b). After the pathogen is disassembled, antigens from the pathogen appear in receptors on the plasma membrane of the macrophage (c). A helper *T* cell that binds to the antigen in these receptors (d) becomes activated.

The activated *T* cell begins to produce helper molecules known as lymphokines (LIM foh kynz). Lymphokines stimulate pathogen-bound *B* cells to enlarge and divide rapidly (e). This action gives rise to a clone of identical plasma cells, which secrete thousands of antibodies into the bloodstream (f). These antibodies are the same as the surface receptors on the original *B* cell. In the bloodstream, the antibodies bind to the antigens that stimulated their production.

When antibodies have bound to the antigens on a pathogen, it becomes easier for macrophages to engulf and destroy the pathogen.

These receptors on the macrophage are class I or class II MHC (major histocompatibility complex) receptors, which are unique for each organism. MHC receptors are present on every nucleated cell in an organism and enable the immune system to recognize "self."

Figure 16.9   Production of antibodies in the specific immune response.

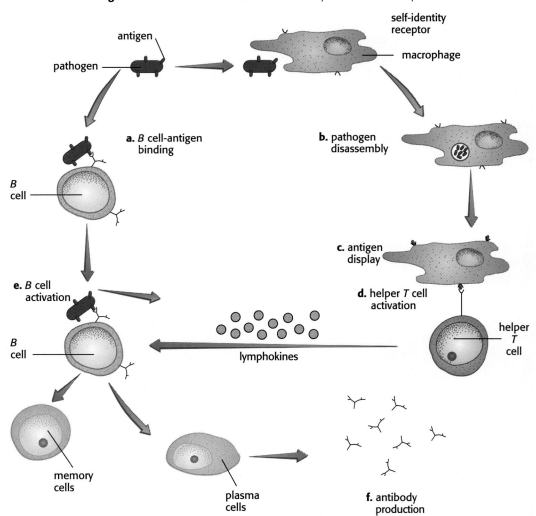

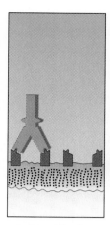

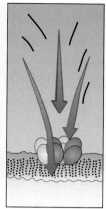
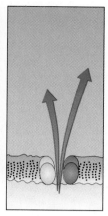

1. An antibody binds to antigen on the surface of a pathogen.

2. Another antibody binds, sending chemical signals that begin the process of activating complement factors.

3. The first activated complement factor binds to the antibodies and activates other complement factors.

4. Actvated complement factors gather in sequence on the surface, forming a pore in the membrane.

5. Fluid rushes in through the pore, . . .

6. . . . causing the pathogen to burst. Macrophages then dispose of the pathogen.

Figure 16.10   Activation of the complement system. When two antibody molecules bind to adjacent sites on a pathogen, they can activate the first factor in the complement system, which, in turn, activates other complement factors. Activated complement factors aggregate within the plasma membrane of a pathogen, forming a small pore. Fluid rushes in through the pore, causing the pathogen to burst.

Antibodies bound to antigens on a pathogen also can activate the **complement system**, a complex group of blood proteins. Once activated, the complement system can form holes in the plasma membrane of the pathogen, causing it to burst, as shown in Figure 16.10.

Lymphokines also stimulate the division of helper *T* cells, which produce still more lymphokines and ensure the continued stimulation of *B* cells and the continued secretion of antibodies. The invading antigen bearing pathogens usually are destroyed within 7 to 14 days by the overwhelming numbers of antibodies released into the bloodstream. As the numbers of antigens decline, the stimulus that began the antibody response is removed from the bloodstream, and the process gradually comes to a halt. The antibody response, however, also includes the production of long-lived *B* and *T* cells known as memory cells. These memory cells remain in the circulation for many years—even a lifetime—and protect against repeated invasions by the same pathogen.

Killer *T* cells act against the cells of the body that are infected by a virus or other pathogen. Although antibodies cannot act on a virus inside a cell, as the virus multiplies, viral antigens appear in the self-identity receptors on the cell's surface. Killer *T* cells then bind tightly to these antigens and form pores in the cell's plasma membrane (see Figure 16.11). As a result, the cells burst, exposing the viruses to antibody action.

The first time a pathogen enters the body, it takes 7 to 14 days for the production of mature plasma cells that then can secrete antibodies at the maximum rate of several thousand molecules per cell per hour. During this time, the antigen-bearing pathogens (for example, measles viruses) also are multiplying, producing the symptoms of disease. The second time measles viruses invade the body, however, memory cells immediately begin large-scale production of antibodies, and the infection is overtaken

For additional information, see H. M. Grey, A. Sette, and S. Buus, "How *T* Cells See Antigens," *Scientific American* (November 1989) 56–64, and K. A. Smith, "Interleukin-2," *Scientific American* (March 1990) 50–57.

Figure 16.11 A killer *T* cell (a) destroys a tumor cell by secreting a protein that makes holes in the tumor cell's membrane. As a result, the tumor cell becomes leaky, takes in water, and bursts. Scanning electron micrographs, X5000 and X9200.

before any symptoms appear. Thus we say the individual is immune to measles. Figure 16.12 shows the different immune responses to primary and secondary infections.

This immunity, brought about by the rapid response of memory cells, is the basis for vaccination against many infectious diseases. Vaccines are prepared from weakened, killed, or closely related or modified pathogens. When injected into the body, the vaccine stimulates production of plasma and memory cells, usually without producing disease symptoms. The immunity is specific. Immunity to measles does not provide immunity to other diseases such as chicken pox or mumps. Because there are so many similar, but different, viruses that can cause diseases such as the common cold and influenza, immunity does not develop. For the same reasons,

Figure 16.12 Response of the immune system to antigens. At the first exposure to an antigen, it takes 7 to 14 days to build a large clone of antibody-producing plasma cells. After the second exposure, however, the clone of memory cells immediately begins large-scale production of antibodies and also multiplies rapidly, further increasing antibody production.

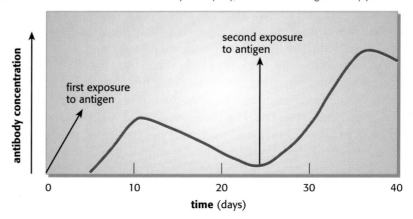

Biology Today

Monoclonal Antibodies

Antibodies are highly specific proteins produced by the immune system to help fight infection. Each type of antibody is effective against only one antigen. In 1975, biologists developed a technique using two different types of cells to produce a given antibody artificially.

Antibodies are produced by *B* cells, but *B* cells do not grow well in culture. They tend to die off after a few generations. Cancer cells, however, will grow in culture almost indefinitely. The technique for producing antibodies fuses *B* cells with cancer cells. First, mice are injected with a specific antigen—for example, the virus that causes hepatitis. Certain *B* cells in the mice have surface antibodies complementary to the virus antigens; those *B* cells multiply. Then, the *B* cells are collected and fused in culture with a special type of cancer cell called a myeloma cell. The resulting hybrids, known as hybridomas, are grown in a special culture that eliminates both the parent *B* cells and the myeloma cells. Only the hybridomas remain, and these cells possess the characteristics of both parent cells—they produce the specific antibody of the parent *B* cell, and they have the longevity of the parent myeloma cell. These genetically identical, or cloned, hybridomas are grown in culture to produce quantities of the desired antibody. The resulting product is called a monoclonal antibody.

This method of making monoclonal antibodies has several drawbacks. It usually takes several months, and it is difficult to make human monoclonal antibodies. In addition, this method, which is meant to yield large quantities of antibodies, will sometimes yield only one, or at most a few, of the desired antibodies.

A new technique being developed may solve these problems. It is comparatively fast, taking only a few days to get specific antibody clones, and it can be used to make human monoclonal antibodies. It also avoids the use of mammalian myeloma cells and uses instead the eubacterium *Escherichia coli,* which can be grown in enormous quantities.

Essentially, this genetic engineering technique calls for inserting into *E. coli* those genes from an animal cell that code for antibody production. The engineered eubacteria then would produce the complete range of antibodies, instead of just a specific one, and researchers would screen the eubacteria for the desired antibody.

The technique, however, has not been perfected. One problem is the structure of the antibody itself. An individual antibody consists of four proteins arranged into a Y-shaped molecule. The arms of the Y form the sites at which an antigen binds to the antibody. Initially, the engineered *E. coli* was able to produce only a portion of the antibody molecule—the antigen-binding sites at the arms of the Y. Some of these fragments, however, can bind antigen almost as well as a whole antibody molecule can. As long as the fragments bind antigens specifically, they can be used in research. As the technique is improved, researchers expect to insert longer gene segments into *E. coli,* making it possible to produce full-sized antibodies that can help fight infection.

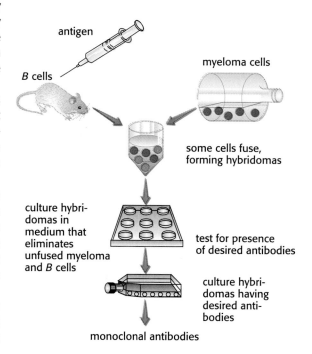

antigen

myeloma cells

B cells

some cells fuse, forming hybridomas

culture hybridomas in medium that eliminates unfused myeloma and *B* cells

test for presence of desired antibodies

culture hybridomas having desired antibodies

monoclonal antibodies

Flu shots offer protection against only the currently prominent flu-causing virus.

For a good coverage of autoimmune diseases and prospects for treatment, see J. Rennie, "The Body Against Itself," *Scientific American* (December 1990): 106–15.

Anaphylaxis is a localized reaction, such as hay fever, asthma, eczema, or hives. Anaphylactic shock, however, is a life-threatening, generalized effect on the circulatory and respiratory systems, often resulting in drastic drops in cardiac output and arterial pressure.

attempts to produce vaccines have not been successful. Figure 16.13 summarizes the steps involved in the specific immune response.

16.7 Many Problems Can Arise with the Immune System

The primary role of the immune system is to distinguish "self" from "nonself." In the developing embryo, the immune system learns to recognize substances that are present as "self." Later in life, this recognition may break down, and the immune system may produce antibodies or T cells that act against the individual's body cells. Disorders resulting from the action of the immune system against self are known as autoimmune diseases. In one **autoimmune** disease, antibodies are formed that interfere with the nerve stimulation of muscles, resulting in muscular weakness. In another, antibodies are produced against the individual's own DNA, thus disrupting many body functions. An autoimmune response appears to be involved in some other disorders, including rheumatoid arthritis and multiple sclerosis.

Allergies are apparently a poorly adapted response of the immune system to foods or substances commonly present in the environment, such as pollen and dust. Sensitive individuals produce a class of antibodies that combine with specialized cells found in the skin and membranes of the eye, nose, mouth, respiratory tract, and intestines. These cells release histamine and other substances that cause the congestion,

Figure 16.13   Interactions in the specific immune response.

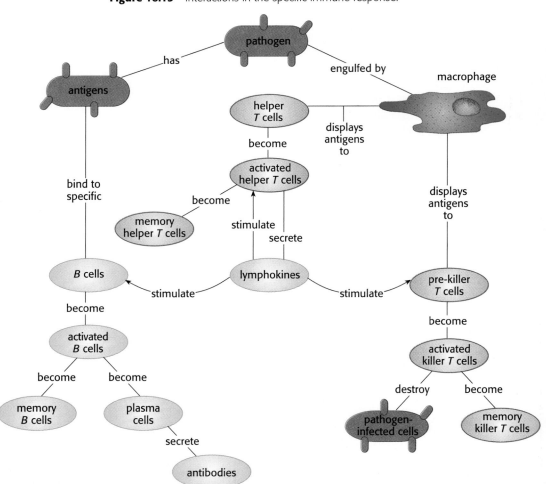

Table 16.1 ABO Blood Group Antigens and Antibodies

Blood Group	Antigen on Red Blood Cell	Antibody in Plasma
0	none	anti A, anti B
A	A	anti B
B	B	anti A
AB	A, B	none

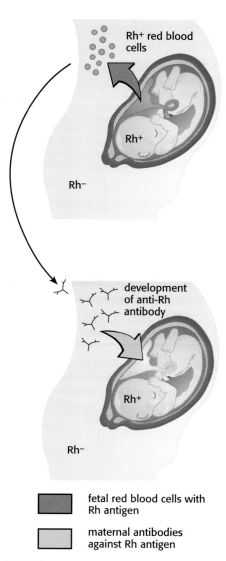

fetal red blood cells with Rh antigen

maternal antibodies against Rh antigen

Figure 16.14   The events leading to Rh disease. Fetal red blood cells may enter the maternal blood during childbirth, stimulating formation of Rh antibodies in the mother's plasma. These antibodies may pass into the fetal blood during a subsequent pregnancy, destroying the fetal red blood cells.

sneezing, and itching typical of hay fever or cramps and diarrhea in the case of food allergies. Treatment involves the use of antihistamines, which suppress some of the allergic symptoms but also may cause side effects such as drowsiness.

Some of the blood group systems present in humans can cause problems under certain circumstances. The blood types that arise from the ABO blood group system are not always compatible. This incompatibility comes about because antibodies in the plasma of one blood type (for example, type A) react with antigens on the red blood cells of another (for example, type B). The reaction causes the red blood cells to clump, or agglutinate (uh GLOO tin ayt). Because of this clumping, transfusing blood of one type into blood of another type can cause a fatal reaction.

Unlike other antibodies, which are produced only in response to an antigen, ABO blood group antibodies develop spontaneously and normally are present in the blood. Each blood group is characterized by the presence of a particular antigen and a different antibody, as shown in Table 16.1.

Rh factors are antigens sometimes found on the surface of red blood cells that can cause problems in transfusions and during pregnancy. Rh positive individuals have Rh antigens on their red blood cells; Rh negative individuals do not. Normally, antibodies to Rh antigens are not present in plasma. Antibodies may develop, however, if Rh positive blood is transfused into an Rh negative individual or if an Rh negative woman has an Rh positive baby. There usually is no problem for the fetus in the first pregnancy, unless the woman has been sensitized previously by a transfusion of Rh positive blood. During birth, however, red blood cells from the infant may enter the mother's blood, stimulating formation of Rh antibodies in her plasma. During a later pregnancy, these antibodies may pass into the fetal blood, causing the destruction of fetal red blood cells (see Figure 16.14). These problems are prevented if an Rh negative mother is injected with Rh antibodies within 72 hours of delivery of an Rh positive child. The Rh antibodies destroy the fetal Rh positive red cells in the mother's body, preventing her immune system from forming antibodies.

The immune system also is responsible for organ transplant rejections. Rejection is brought about by killer *T* cells. These cells identify the transplanted organ as a foreign antigen. A few days after an organ transplant, they invade the tissue and begin to destroy it. To prevent this reaction, physicians use drugs to suppress the immune response. Unfortunately the suppression of the immune response leaves the transplant recipient vulnerable to infection, the major cause of death among organ transplant recipients. Efforts to control the immune response and to improve matching of donor and recipient tissues are active areas of research.

Immunosuppressive drugs are given with organ transplants to reduce possibility of rejection. These must be controlled carefully so the immune system is not depressed to such a point that the body has no protection against infection.

CONCEPT REVIEW

1. Distinguish between specific and nonspecific immunity.
2. Define the term *antigen* and provide several examples of antigens.
3. Describe the steps involved in antibody production.
4. Why does the second exposure to a pathogen usually not produce disease?
5. In what way are allergies examples of an inappropriate immune response?
6. Describe some problems that can arise when the immune system malfunctions.

Gas Exchange and Excretion

16.8   Cellular Respiration and Gas Exchange Are Not the Same

Guidepost

What role do the respiratory and excretory systems play in maintenance of homeostasis?

In mammals, control of breathing movements is a function of the concentration of CO₂ in the blood.

Frequently, cellular respiration is confused with the process that exchanges vital gases with the environment. Cellular respiration is primarily a process of energy release. Because cellular respiration uses oxygen and releases carbon dioxide, gas exchange also is involved. As carbon dioxide accumulates, it produces acidic conditions that are poisonous to cells and must be removed. Respiration is the overall process of exchanging oxygen and carbon dioxide with the environment and the blood. The process is accomplished by the respiratory system. Respiration may be studied in three stages. The first stage is the process of breathing, which involves the movement of air in and out of the lungs. The second stage is the exchange of gases between the internal surface of the lungs and the blood. The third stage is the exchange of gases between the blood and the tissue cells.

The respiratory organs function to exchange gas molecules between the inner surface of the lungs and the blood. Breathing moves the air that has been inhaled by the mouth and nose through a series of air tubes into the lungs. Movement of the air is accomplished by the action of two groups of muscles. The first group makes up the **diaphragm** (DY uh fram), a muscular wall that separates the body cavity into two parts. The second groups the rib muscles. Working together, the diaphragm and rib muscles act to change the size of the chest cavity, as shown in Figure 16.15. When you inhale,

Figure 16.15   Movements of breathing in humans.

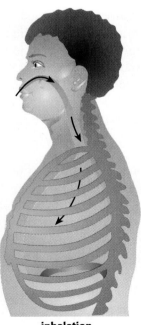

inhalation

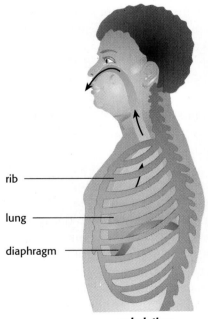

rib
lung
diaphragm

exhalation

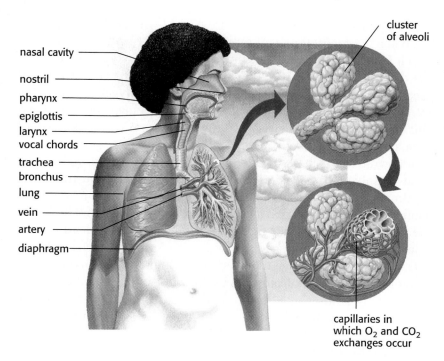

cluster of alveoli

nasal cavity
nostril
pharynx
epiglottis
larynx
vocal chords
trachea
bronchus
lung
vein
artery
diaphragm

capillaries in which O_2 and CO_2 exchanges occur

Figure 16.16   The parts of the human respiratory system involved in breathing. The lung has been cut away to expose the branching system of bronchial tubes. Part of the lung has been enlarged to show alveoli and their relation to capillaries. Millions of alveoli in each lung give the tissue a sponglike appearance.

your diaphragm moves down, your ribs move up and out, and the cavity enlarges. When the chest cavity expands, the pressure within the chest falls. Because the pressure within the chest cavity is lower than the atmospheric pressure outside, air rushes in. When you exhale, the chest cavity is reduced, the internal pressure becomes greater than the atmospheric pressure, and air is forced out. The rhythmic increase and decrease in the volume of the chest cavity is the mechanical pump that drives air in and out of the lungs.

Atmospheric air is dry, sometimes cold, and often dirty. The air you breathe passes through your nose and down the windpipe, or **trachea** (TRAY kee uh), into passageways that enter the lungs and end in air sacs, or **alveoli** (al VEE oh ly). The air is moistened, warmed, and cleaned by cells lining these air passageways. Alveoli are grapelike clusters of cavities formed by single-layered sheets of cells. Each lung has millions of these cavities, and the walls of these cavities are covered with a network of capillaries. Figure 16.16 illustrates the human respiratory system and details of the alveoli. Oxygen diffuses into the bloodstream through the alveolar walls. Because the many alveoli provide an enormous amount of surface area, the lung can exchange a large volume of gases in a short time. If all the alveoli of the human lung were spread out flat, the surface would cover an area of about 70m²—the size of about five parking spaces.

Some diseases affect the normal functioning of the respiratory system. In people with emphysema (em fuh SEE muh), for example, the alveolar walls lose elasticity, making it difficult for the lungs to remove air low in oxygen and high in carbon dioxide. As a result, less fresh air, which contains more oxygen and less carbon dioxide, can be brought in for air-blood gas exchange. Emphysema is associated with smoking or living in areas of high air pollution. The condition becomes progressively worse, and sufferers may become completely incapacitated.

There are approximately 300 million alveoli in the human being.

16.9   Respiratory Gases Are Transported in the Blood

Once oxygen has diffused from the alveoli into the blood, it must be transported to the tissues. Because oxygen does not dissolve readily in plasma, **3** special oxygen-transporting molecules are required. In humans and all other vertebrates, the red blood cells are packed with hemoglobin molecules. When the concentration of oxygen is relatively high, as in the alveoli, each hemoglobin molecule combines with four oxygen molecules. When the red blood cells reach the tissues, where the concentration of oxygen is relatively low, the hemoglobin gives up its oxygen. Hemoglobin enables blood to carry about 60 times more oxygen than could be transported by plasma alone. Figure 16.17 shows the steps in the transport of oxygen and carbon dioxide.

Carbon dioxide is more soluble than oxygen in the blood, and a small amount dissolves in plasma. The hemoglobin in the blood transports **3** approximately 25 percent of the carbon dioxide in the human body. The remaining 75 percent is carried in the blood as bicarbonate ion (HCO_3). To create this ion, carbon dioxide first reacts with water in the blood to form carbonic acid. The carbonic acid then ionizes to form hydrogen and bicarbonate ions. As with oxygen transport, the amount of carbon dioxide in the alveoli and the tissues determines the direction of the reaction. Carbon dioxide usually diffuses into the lungs because it is more concentrated in the blood than in the alveoli.

16.10   The Kidneys Are Major Homeostatic Organs

The body's internal chemical environment must be regulated at all times. **4** Regulation involves excreting cellular wastes, controlling concentrations of ions and other substances, and maintaining water balance. Carbon dioxide and nitrogen compounds are the chief cellular wastes. In humans, the lungs excrete carbon dioxide. The kidneys excrete nitrogen compounds, regulate ion concentrations, and maintain water balance.

Figure 16.18 shows the human urinary system. The kidneys are dark red, bean-shaped organs about 10 cm long, located on each side of the back wall of the body cavity, just above waist level. Each kidney contains about one million working units or **nephrons** (NEF rahnz). A nephron is a long, **5** coiled tubule. One end of a nephron opens into a duct that collects urine. The other end forms a cup that encloses a mass of capillaries. Other capillaries from a network that closely surrounds the entire nepron, as shown in Figure 16.19. This intimate relationship between the blood and the kidney is essential to kidney function.

Carbon dioxide combines with water to form carbonic acid which ionizes readily.

$$CO_2 + H_2O \longrightarrow H_2CO_3 \longrightarrow H^+ + HCO_3^-$$

The removal of nitrogenous wastes is complex and probably a new topic to most students. Take some class time to go over this function. Stress that blood flow through the kidneys is controlled by hormones that keep the water content of the body in balance.

The kidneys are the principal organs by which balance of water, glucose, salts, and many other substances that occur in living bodies is maintained within more or less narrow ranges of tolerance. They do not merely excrete: they also selectively reabsorb.

Students may be familiar with the term *Bowman's capsule,* an older name for glomerular capsule.

Figure 16.17   Oxygen and carbon dioxide transport. Oxygen from the alveolus enters red blood cells and combines with hemoglobin (Hb), forming oxyhemoglobin (HbO_2). The oxygen is carried in this form to body cells where it is released (see bottom of illustration) Carbon dioxide diffuses blood from body cells into the red blood cells. There, it combines with water to form carbonic acid (H_2CO_3), which ionizes to bicarbonate ions (HCO_3) and hydrogen ions (H^+). The bicarbonate ions diffuse into the plasma. The hydrogen ions combine with hemoglobin (H-Hb). Some of the carbon dioxide combines directly with hemoglobin ($HbCO_2$). In the lung capillary, the reverse reactions occur. Bicarbonate ions from the plasma are converted to carbon dioxide and diffuse into the alveolus.

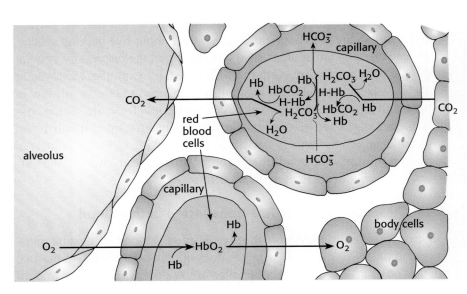

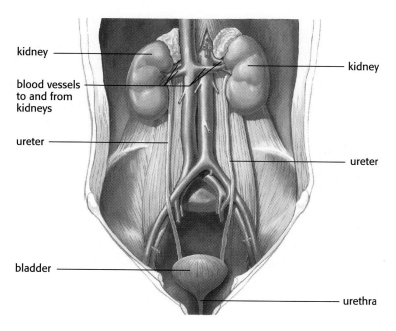

Figure 16.18 The human urinary system.

All the collecting ducts of the nephrons empty into the **ureter** (YOOR ee ter), a large tube that leads from each kidney to the urinary bladder. Urine is stored in the bladder until it is discharged to the outside through another tube called the **urethra** (yoo REE thruh). In males, the urethra also carries semen out of the body.

The ball of capillaries is called a **glomerulus** (glah MER yoo lus) and the cup of the nephron is called the glomerular capsule. The wall of a nephron is just one cell thick and is in direct contact with a capillary wall, also a single cell thick. Often the tubule and capillary walls appear merged into one undivided structure.

16.11 Nephrons Filter the Blood

Three processes are involved in the function of the nephrons: filtration, secretion, and reabsorption. Figure 16.20 summarizes these processes. Filtration occurs in the glomerulus of the nephron, where the fluid portion of the blood is forced into the glomerular capsule. Blood cells and most of the plasma proteins are retained in the glomerulus. The filtrate in the

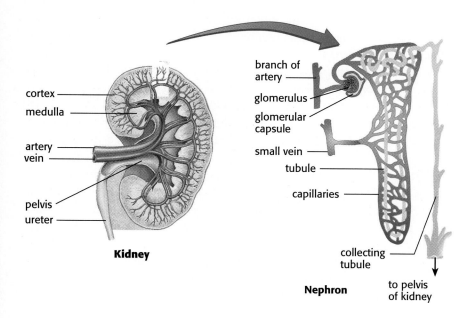

Figure 16.19 A section through the human kidney and an enlarged view of one nephron with its surrounding capillaries.

Glomerular blood pressure may be 60 mm Hg as compared with 30 mm Hg in other capillaries. This increased pressure aids in the absorption of material from the blood.

Without reabsorption a human would have to drink almost a bathtub full of water each day.

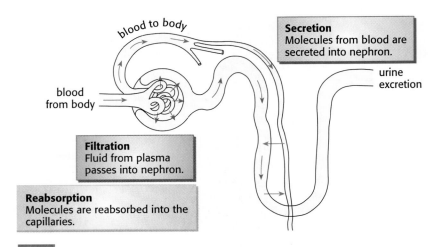

Figure 16.20   Major processes in the formation of urine.

glomerular capsule contains nitrogenous wastes, ions, and much of the blood's water content. Approximately 150 to 180 L of fluid enter the nephrons each day, yet only about 1.5 L of urine are eliminated from the bladder. More than 99 percent of the fluid arriving in the glomerular capsule is returned to the blood.

Secretion and reabsorption take place along the nephron tubule. As **6** the filtrate from the glomerulus moves through the nephron, the cells of the tubular wall selectively remove substances that are left in the plasma after filtration has taken place. The cells then secrete these substances into the filtrate. Substances such as penicillin may be removed from the blood in this manner.

Kidney stones may be formed from accumulation of uric acid crystals or calcium deposits and may block the passage of urine. Stones are more common in arid regions of the world, where water is scarce and sufficient fluids are not obtained to "wash out" the crystals.

Glycosuria is a malfunction of the tubular carrier mechanism that causes glucose to appear in the urine even though blood sugar level is normal.

Cells of the tubular walls also reabsorb useful substances from the filtrate and return these substances to the blood. Glucose, amino acids, essential ions such as sodium and potassium, and most of the water are conserved in this manner. Secretion and reabsorption require active transport for this exchange of materials.

The nephrons maintain appropriate blood-sugar levels by the process of glucose reabsorption. If the blood-sugar concentration rises too high, not all of the glucose can be reabsorbed, and some of it is excreted in the urine. If the blood-sugar level is low or normal, almost all the glucose in the nephrons is reabsorbed into the blood, and no glucose is present in the urine.

Water-salt balance also is regulated in the nephron and the nephron's collecting duct. Here, about 99 percent of the water that has left the capillaries and entered the nephron is reabsorbed into the blood. The mechanism is complex and involves the maintenance of a high salt concentration in the tissues surrounding the nephron and its collecting duct. As a result, water leaves the collecting duct by osmosis. The reabsorption of water is controlled by hormones, and the final concentration of the urine occurs in the collecting duct.

Drinking seawater is not merely a matter involving salt excretion. It also involves water diffusion from the lumen of the alimentary canal into the wall.

Although the kidneys can regulate blood-sugar levels, remove nitrogenous wastes from the blood, regulate water-salt balance, and excrete excess salt within limits, they cannot excrete high concentrations of salt. In fact, perspiration and the kidneys working together cannot remove high concentrations. People shipwrecked on the ocean can easily perish from dehydration by drinking the salt water. Instead of replenishing the body's water supply, they will lose more water than before drinking as the kidneys attempt to excrete the salt.

Pioneers

From the Depths to the Heights

As the center for aerospace medical education, the USAF School of Aerospace Medicine (USAFSAM) is the major provider of educational programs involving aviation, space, and environmental medicine for the Air Force, the Department of Defense, and for aerospace-faring nations world-wide. The programs conducted at the USAFSAM cover all levels from beginning aerospace and environmental specialty technicians through graduate medical education in aerospace medicine.

While pioneering educational activity is carried on in every department at the USAFSAM, the Aerospace Physiology Department operates at the cutting edge of new methods and technologies. The Aerospace Physiology Department trains US Air Force, US Army, US Navy and international aerospace educators, scientists, and technicians in the physiological stresses and human factors associated with modern aviation and prepares them to teach that knowledge to both military and civilian flying communities. Aerospace Physiology faculty members develop, plan, and teach courses in high altitude physiology and aircraft mishap investigation. They also teach courses in hyperbaric medical treatment. Hyperbaric medicine involves the use of oxygen under greater-than-normal pressure to treat decompression sickness (including divers' bends), aeroembolism (bubbles of air traveling in the bloodstream), carbon monoxide poisoning, surgical wound healing, and various bacterial and fungal diseases.

Hyperbaric medicine makes use of hyperbaric (compression) chambers for creating pressures similar to those encountered in SCUBA and deep-sea diving. Altitude physiology makes use of hypobaric (altitude) chambers to simulate the biological effects of altitudes ranging from sea level to near-space.

The Aerospace Physiology Department, in conjunction with the Armstrong Laboratory (see Pioneers, Chapter 15) conducts research focusing on spatial disorientation (the ability of a pilot to stay aware of where the earth's surface is) in flying. Cooperating with each other, the two agencies have installed a multi-million dollar Advanced Spatial Disorientation Demonstrator (ASDD) to conduct studies on this serious problem. The ASDD is fully computerized and is capable of simulating virtually any spatial disorientation problem that can be encountered in the aerospace environment from light aircraft flying to space operations.

The Aerospace Physiology Department trains over 3,000 students annually, maintains state-of-the-science facilities and simulation devices, and develops course materials supporting true pioneering technologies. The department also provides consultation service to numerous colleges and universities including many in foreign countries. Faculty members, all experts in human performance factors, also provide their expertise to aircraft mishap investigation boards.

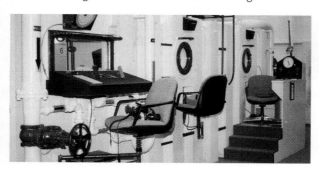

a. Altitude Chamber (outside)

b. Advanced Spatial Disorientation Demonstrator

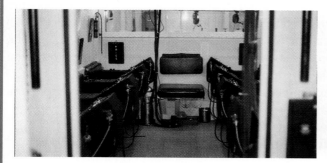

c. Altitude Chamber (inside)

d. Hyperbaric (compression) Chamber

The kidneys have still other functions, such as pH regulation, but blood sugar regulation, water-salt balance, and nitrogenous waste excretion are their major functions. If the kidneys fail, vital or poisonous blood components are not maintained at normal concentrations, homeostasis fails, and death follows.

CONCEPT REVIEW

1. Distinguish between respiration and cellular respiration.
2. Describe the mechanism that operates to move air into and out of the human lungs.
3. Describe how oxygen and carbon dioxide are carried in the blood.
4. List the major regulatory functions of the kidneys.
5. Describe the structure of the nephron.
6. Explain how filtration, secretion, and reabsorption contribute to formation of urine.

Temperature Regulation

16.12 The Rate of Chemical Reactions Is Influenced by Temperature

Guidepost

How do humans maintain a constant internal body temperature?

Metabolism is the sum of all the chemical reactions that occur within a cell or an organism. Like other chemical reactions, these enzyme-controlled reactions are influenced by changes in temperature. Within limits, they slow down at low temperatures and speed up at high temperatures. Therefore, maintaining a constant internal temperature allows for more   1 efficient chemical processes to take place. It also permits an animal to be active when environmental temperatures are low.

In humans and other animals that are able to maintain a constant internal temperature, the temperature of the skin and the tissues just beneath it may fluctuate. Internal body temperature in most individuals, however, changes little in the course of a day. On waking up, a human's temperature is usually about 36.2° C, and it increases to perhaps 37.6° C by late afternoon, averaging about 37° C. It is important to maintain the body's temperature within these narrow limits.

Death results if body temperature rises to between 44.4 and 45.5° C or drops to between 21.1 and 23.9° C.

To accomplish this regulation of internal temperature, the body must balance the amount of heat it produces with the amount it loses. Therefore, when   2 you become overheated, during strenuous exercise for example, the extra heat produced must be dispelled from the body. Otherwise, the body temperature could increase to the point at which enzyme functions would be impaired.

One of the by-products of cellular respiration is heat. It is this heat, produced by the breakdown of foods, that maintains the body's temperature. Because muscles and organs are the most active tissues in the body, they carry on more respiration and thus produce more heat than any other body tissues. The activity of the muscles determines the rate of heat production. When you sleep, there is little muscular movement, and heat production decreases. As you exercise or shiver, your active muscles produce more heat.

It is believed that high body temperatures inhibit the growth of certain bacteria and viruses.

16.13 Major Heat Loss Occurs Through Evaporation and Radiation

Most of the body's heat loss, 80 percent or more, occurs through the skin. The remainder takes place through the normal function of the respiratory, digestive, and excretory systems.

Evaporation at average temperatures and normal rates of activity is only one of the methods of cooling the body. At high temperatures or during vigorous activity, however, it is the only method by which heat can be lost through the skin. How can you explain that the same temperature on two different days may feel hotter on one of those days? At a temperature of 38° C (100° F) and a humidity of 90 percent, a person would feel hotter than at the same temperature with a humidity of 45 percent. That is because evaporation takes place more rapidly in a drier environment, thus cooling your skin more efficiently.

Radiated heat is transferred from one object to another even when the two objects do not touch. You can feel the radiant heat of the sun without actually being in contact with the sun's surface. The skin gains or loses heat through radiation, depending on the environment. If the environment is cool, body heat is lost through radiation. Conversely, in a hot environment heat is gained by radiation from hot surfaces (see Figure 16.21).

Figure 16.21 Lizard absorbing radiated heat. Animals such as this do not have internal mechanisms that regulate body temperature.

16.14   Internal Temperature Is Controlled by the Brain

Heat loss or gain also may be affected by the amount and type of clothing worn, the ingestion of hot or cold food or liquid, or even by standing in a breeze. Internal body temperature can remain constant only if the rate of heat loss is the same as the rate of heat production, as shown in Figure 16.22. Receptors that respond to temperature changes are necessary to keep body temperature constant. In an area of the brain called the hypothalamus,

Sunstroke or heatstroke occurs when temperature and relative humidity are high, making it difficult to lose heat by radiation or evaporation. Internal temperature may reach 43.5° C, at which point brain cells may be affected or destroyed.

Figure 16.22 How heat-dissipating and heat-conserving mechanisms operate to maintain normal body temperature.

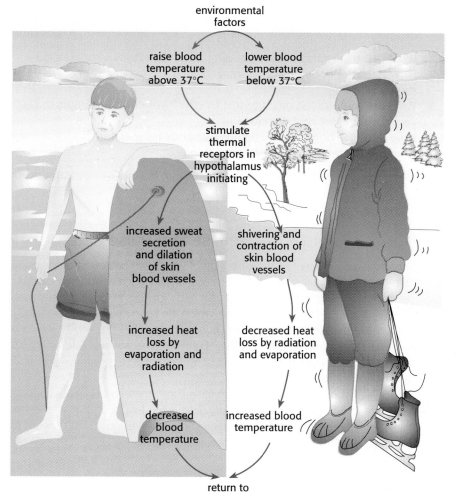

In heat exhaustion or prostration there is normal or slightly lower body temperature accompanied by profuse perspiration. Salt loss and heart problems may cause dizziness, cramps, vomiting, and fainting.

Although loss of water through perspiration depends on relative humidity, in general, the evaporation of 1 g of water removes 0.58 kcal of heat. In heavy exercise, evaporation of 4 L per hour can remove 2000 kcals of heat.

there are receptors that detect changes in blood temperature. Much as a thermostat controls a furnace, air conditioner, or water heater, the receptors in the hypothalamus act as a thermostat to regulate body temperatures.   5

The body thermostat is set at 37° C. When the temperature of the blood increases by as little as 0.01° C, the cells of the hypothalamus send messages to the sweat glands and blood vessels of the skin. In response to   4 these messages, the sweat glands increase their rate of secretion. The blood vessels in the skin begin to become larger, or dilate, allowing for greater blood flow to, and heat loss from, the body surface. If body temperature continues to increase, this process will accelerate the rate of sweating and dilation, causing more heat loss from the skin.

In a cold environment, the reverse of this process occurs. When the   3 temperature of the blood reaching the hypothalamus drops, messages are sent that slow production of sweat and that narrow, or constrict, the blood vessels in the skin. If this reaction does not raise the temperature sufficiently, a second mechanism is initiated. Shivering and skeletal muscle contractions cause an acceleration of cellular respiration and heat production. When you are very cold you use more oxygen and tire more quickly.

The deliberate use of hypothermia (lowering of body temperature) often is called for in open-heart surgery and other prolonged surgical procedures. The mechanisms involved in survival of extremely low body temperatures, however, are still poorly understood.

CONCEPT REVIEW

1. What is the advantage of maintaining a constant body temperature?
2. How does the body prevent damage from overheating as you are vigorously exercising?
3. What are some of the mechanisms by which your body attempts to keep you warm in very cold weather?
4. How does the human body regulate its body temperature?
5. How is body temperature related to homeostasis?

This investigation gives students a chance to design and carry out their own experiments. You many wish to have them write out their experimental designs before the lab period so you can discuss them. Students can work in pairs, but each student should do the experiment.
    Time: **One class period.**

SAFETY

See the safety remarks for Investigation 2.1 when working with the LEAP-System equipment.

Relate the problems of unicellular organisms to those of large multicellular ones in the distribution of nutrients and the elimination of wastes.

INVESTIGATION 16.1   Exercise and Pulse Rate

You probably have experienced the sensation of your heart beating strongly inside your chest when you participated in a physical activity such as running, aerobic exercise, or athletic competition. The heartbeat rate increases in response to signals from the nervous and endocrine systems, which are monitoring the entire body. Pulse rate, as measured by using a pulse probe with the LEAP-System™, or by taking the pulse rate manually at the wrist if a LEAP-System is not available, is a measure of how fast the heart is beating. The LEAP-System BSCS Biology Lab Pac 2 includes instructions for use of the system.

> Develop a hypothesis about how exercise (both mild and vigorous) affects heart rate and recovery time. Recovery time is the length of time it takes for the heartbeat to return to the normal resting rate. Design an experiment to test your hypothesis using the materials listed below.

Materials (per team of 2)

Apple lle, Apple IIGS, Macintosh,
    or IBM computer
LEAP-System™

pulse probe
watch (or clock) with second hand
    (optional)

Figure 16.23   Taking a pulse with (a) the pulse probe and the LEAP-system or by (b) the wrist method.

Procedure

Use the LEAP-System with a pulse probe (see Figure 16.23a) to carry out the experiment you designed. In your data book, record the pulse rate from the screen. Organize your data in a table, and construct graphs at the end of the experiment to display the data. If a LEAP-System is not available, use the wrist method (see Figure 16.23b). Count the pulse directly, organize your data in a table, and construct graphs to display the data.

Discussion

1. What was the control in your experiment?
2. What label did you put on the *y* axis of your graph? On the *x* axis?
3. What is the influence of exercise on heartbeat rate?
4. Are there any differences in heartbeat rates and recovery times for the different types of exercise among the students in your class? Explain.
5. Are there any differences in heartbeat rates and recovery times between males and females in your class?
6. Note any other conclusions you can make based on your data.

DISCUSSION
1. **The resting pulse rate.**
2. ***Y* axis should be pulse rate; *x* axis should be time.**
3. **More exercise produces higher pulse rates and longer recovery times.**
4. **There should be variation among the students, especially if you have a large class.**
5. **Results will vary, but there should be some differences.**
6. **Responses will vary.**

INVESTIGATION 16.2   The Alpine Slide Mystery

In this investigation you will develop hypotheses that might explain the severe allergic reaction experienced by several people after riding on a summertime amusement ride called an alpine slide.

Materials (per team of 3)
none

Procedure

1. Read the following information:

   During the two-week period from 22 July to 8 August 1982, five young men vacationing in Vermont experienced severe allergic reactions. Certain areas their skin developed slightly raised patches that were inflamed and itched. They had symptoms like those of hay fever in the nose and eyes. They also had some difficulty breathing and blood pressure in the low range.

   The doctors who treated these young men recognized that they were reacting in a highly sensitive manner to the presence of some foreign substance in their bodies. They treated them with injections of antihistamines and epinephrine, drugs that prevent the production of the agents that cause the reactions. All five young men recovered.

   The doctors were puzzled by these medical events. As they checked the histories of the five young men, they found a common experience. All their allergic reactions occurred within a 3- to 20-minute period after each one of the five sustained some skin abrasions while riding on an alpine slide.

Allergic reactions are a common result of hypersensitivity to a foreign antigen. This investigation, based on an actual case, asks students to develop hypotheses that explain severe allergic reactions. Teams of 3 or 4 allow for small-group interaction.
   ***Time:* One class period.**

PROCEDURE

2. The obvious relationship is between the abrasion and the reaction.

3. This new information indicates that not all people who suffered abrasions suffered severe allergic reactions. Therefore, students should be able to conclude that abrasions alone do not cause the reactions.

4. Students should come up with several possibilities, including the possibility the reactions could be related to the victims all being male. This is not the case, but no data are provided for females.

5. Pollen, animal dander, dust, insect stings, certain types of foods, penicillin or other drugs. Students will probably have some experience with one or more of these problems, but many are not seasonal.

6. The data suggest a seasonal pattern of the reactions. Many students should be able to relate hay fever symptoms to those seen in the young men and point to the seasonal nature of hay fever, although you may need to help some groups focus on pollen as an agent. Ask questions that help students eliminate the least likely possibilities.

7. Remind students that an association is not proof. Ultimately, students should see that these two major factors must be demonstrated: (1) the presence of pollen grains in the area of the slide, and (2) sensitivity to pollen of that type on the part of the five patients. In the actual case, factor 2 was determined by skin tests. Of the four patients tested, all were positive on the tests for grass and ragweed pollen. For factor 1, researchers used glass slides coated with silicone grease and affixed them to the front and the rear of five sleds to collect pollen grains as they went down the slide. More pollen grains were found on the rear of each sled than on the front. The range was from 4 grains (front of sled, slow speed) to 1600 grains (rear of sled, high speed). Have students speculate on why the number of pollen grains was greater on the rear than on the front. Researchers concluded that the severe allergic reactions were caused by pollen grains introduced into allergic persons through breaks in the skin as a result of the abrasions.

DISCUSSION

1. Answers may vary slightly, but all should state that pollen introduced through the abrasions caused the allergic reactions.

These slides are common in ski areas, where they are built on the ski runs. They generate income for the ski areas during the spring, summer, and fall. The slide consists of a free-running sled that glides down a mountainside on an asbestos-cement track. The abrasion injuries occurred as a result of falling off moving sleds.

A check of 27 similar alpine slides in the United States revealed 24 similar cases of severe allergic reactions, 22 of which followed skin abrasions. The evidence seemed to indicate a relationship between allergic reactions and skin abrasions sustained while riding the alpine slide.

2. What do you think might have caused the severe allergic reactions in these people? Make a list of possibilities.

3. Consider the following additional information:

Data from the Alpine Slide in Vermont	
Rides per day	approximately 1400
Rate of injury (abrasions or friction burns)	1–2% (14–18) per day
Rate of allergic reaction	0.0002%

Do you think that abrasions alone are sufficient to cause severe allergic reactions? Explain.

4. If abrasions alone are not the cause, what other factors might play a role in the apparent interaction between skin abrasions and severe allergic reactions? Discuss possible environmental factors and list them in order of likelihood.

5. Now consider the following historical data on the occurrence of severe allergic reactions from other alpine slide operators across the United States.

Time Period	Rate of Allergic Reaction
23 May–27 June (1976–1981)	0 cases
28 June–4 August (1976–1981)	1.71 cases/10^5 rides
5 August–6 November (1976–1981)	0 cases

Consider the symtoms that describe severe allergic reactions: itching and inflamed skin, watery and reddened nose and eyes, difficulty in breathing. What factors cause allergies in people you know? Are any of these allergies seasonal?

6. Look at the alpine slide environment shown in Figure 16.24. Describe what you see. Do you see any factors that should be included in the list of possible causes of the severe reactions?

Based on the list your team has developed, what might be the most likely cause of the reactions? Begin by eliminating proposed causes that do not vary with the calendar, and focus on those that consider the entire alpine slide environment.

7. Suppose you suspected that a skin abrasion alone could not cause the allergic reaction, that some other factor also was necessary. How could you demonstrate convincingly that the young men had to have an abrasion and had to be exposed to another factor for the allergic reaction to occur? In other words, how do you show an association between the allergic reactions in the young men, the abrasions they received, and any other factors?

Discussion

1. Compare your hypotheses about the causes of the allergic reactions with those of the other teams. Are they similar?

2. Compare the ways in which the different teams proposed to demonstrate the association between the allergic reactions and the factors that caused them. Are they similar?

Figure 16.24   An alpine slide.

3. If you were the operator of the alpine slide and knew that skin abrasions during certain seasons could lead to severe allergic reactions, what steps could you take to protect your customers?

INVESTIGATION 16.3   Exercise and Carbon Dioxide Production

When you exercise, you breathe faster than when you are at rest. Does exercise also affect the concentration of carbon dioxide in the air you exhale?

You can test for the presence of carbon dioxide by bubbling your breath through water. Carbon dioxide in your breath combines with water to form carbonic acid ($CO_2 + H_2O \rightarrow H_2CO_3$), which lowers the pH of the solution. Using a pH probe and the LEAP-System™, or a pH meter or wide-range pH paper if the LEAP-System is not available, you can measure the amount of carbon dioxide in the water by detecting a change in pH. The LEAP-System BSCS Biology Lab Pac 2 includes instructions for use of the system.

Read the procedure for the investigation and complete the statement: If carbon dioxide is a waste product of cellular respiration, then. . . .

Materials (per team of 2)

100-mL graduated cylinder
2 250-mL flasks
Apple IIe, Apple IIGS, Macintosh, or IBM
    computer
LEAP-System™
printer

2 wrapped drinking straws
2 pH probes
pH meter or wide-range pH paper
    (optional)
forceps (optional)
watch or clock with a second hand

Procedure

Each member of the team should do the investigation. One person should be responsible for the computer while the other keeps track of his or her own time breathing into the flask. Then switch responsibilities.

1. In your data book, prepare a table similar to Table 16.2 on page 428.
2. Label two flasks *A* and *B*.
3. Place 100 mL of tap water in each flask.
4. Record the initial pH in both flasks by using two pH probes with the LEAP-System (Figure 16.25), a pH meter, or by dipping small strips of pH paper into the water and comparing the color to a standard color chart.

2. Answers should note the same requirements: pollen must be in the area, and the patients must be sensitive to pollen. Answers for determining sensitivity to pollen should be basically the same: skin tests. Answers for determining the presence of pollen in the slide area may vary greatly. There probably will be many good and creative suggestions.

3. You might post warnings at the slide to alert persons who are allergic to pollens. Personnel at the slides could be trained in the treatment of severe allergic reactions. You might reduce the pollen in the air at the slide sites by frequent cutting of the grasses. Adjacent grassy areas could be moistened to keep pollen from flying onto the slide.

This investigation gives students a graphic demonstration of the relationship between cellular respiration and carbon dioxide production. Teams of 2 are best but your own facilities and equipment will dictate how you handle the logistics.

Time: Students should be able to collect the data in one class period.

SAFETY

See the safety remarks for Investigation 2.1 when working with the LEAP-System equipment.

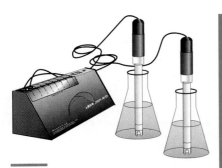

Figure 16.25   LEAP-System setup for measuring carbon dioxide production.

Table 16.2 pH Readings

Experimental Condition	pH Readings			
	Flask A		Flask B	
	Initial	*Final*	*Initial*	*Final*
At rest				
Mild exercise				
Vigorous exercise				

5. Put a drinking straw in the water in Flask B. Sit quietly for 1 minute and then gently blow air from your lungs through the straw into the water for 1 minute. Do not suck any material through the straw. Swirl the flask.

6. Determine the pH in both flasks and record the readings in the table.

7. Discard the water in both flasks. Rinse the flasks and the straw.

8. Repeat Steps 3–7, but this time, before blowing through the straw, exercise for 1 minute by walking in place.

9. Repeat Steps 3–7, but this time, exercise more vigorously for 1 minute (run in place, dance, or do jumping jacks). Discard the straw.

10. Wash your hands thoroughly before leaving the laboratory.

DISCUSSION

1. **Flask B. Carbon dioxide was added. Flask A. No Carbon dioxide was added from the breath.**
2. **The change should be used to adjust the experimental graphs.**
3. **The more vigorous the exercise, the more carbon dioxide is produced.**
4. **Physical conditioning of the students is one factor; others may be mentioned.**
5. **Normal variation among students in a class is one source of difference; experimental error is another.**
6. **Cellular respiration.**
7. **Control.**

Discussion

1. Run stripcharts to obtain graphs of the data. Observe the graphs carefully. In which flask was the pH decrease greatest? Why? In which flask was the pH decrease smallest? Why?

2. If the pH changed in Flask A, how would you use that information to explain the results in Flask B?

3. How is the production of carbon dioxide related to activity?

4. What factors other than exercise might play a part in determining the amount of carbon dioxide given off?

5. Compare your results with the results of other students in your class. Try to explain any differences.

6. What is the source of carbon dioxide in your breath?

7. What was the purpose of Flask A in this experiment?

This investigation is based on A *BSCS Classic Inquiry— The Kidney and Homeostasis,* developed by Media Design Associates, Inc., Box 3189, Boulder, CO 80301-3189. A worthwhile option would be to use the *Classic Inquiry* in place of this investigation. Besides doing what the investigation is designed to do, it will introduce students to a new type of learning strategy, and it will give you an opportunity to present this material in an interesting manner. Teams of 3 can use cooperative learning followed by a summarizing discussion in the entire class. Alteratively, the questions can be assigned as homework, followed by class discussion.
   ***Time:* One to 1 1/2 class periods for small group work and class discussion.**

INVESTIGATION 16.4   The Kidney and Homeostasis

The cells of the human body are surrounded by a liquid that is remarkably constant in its properties. The continuous regulation of the many dissolved compounds and ions in this internal environment is referred to as homeostasis.

The kidneys play an important role in homeostasis by regulating blood composition and by regulating the levels of many important chemicals and ions. The production of urine and its elimination from the body are critical functions of the kidneys and the urinary system.

Materials (per team of 3)
none

PART A Blood Versus Urine
Procedure

1. The relationship of structure and function in the kidney is illustrated in Figure 16.22 at the top of this page. Use this illustration and the data in Table 16.3 to answer the discussion questions.

Table 16.3 Comparison of Materials in Blood and Urine

	% in Blood as It Enters Kidney	% in Urine as It Leaves Kidney
Water	91.5	96.0
Protein	7.0	0.0
Glucose	0.1	0.0
Sodium	0.33	0.29
Potassium	0.02	0.24
Urea	0.03	2.7

Discussion

1. What do the data for water indicate?
2. Protein molecules are not normally found in the urine. Give some possible reasons for this.
3. The information for glucose is similar to that for protein. Can you explain these data?
4. Based on what the sodium data indicate, what do you think may happen to the sodium content in the urine of a person who increased his or her intake of sodium chloride?
5. How do the data for potassium differ from those for sodium?
6. How would you interpret the data for urea?
7. Summarize the functions that take place between blood and urine and identify the structures where these functions occur.

PART B Filtration, Reabsorption, and Secretion
Procedure

2. The micropuncture method was used in a second study of the six materials presented in Table 16.4. Under a microscope, a very fine pipet was used to withdraw samples of fluid at four points along the nephron. Study Table 16.4, which shows the data that were collected using this technique. Use the data to answer the discussion questions.

Discussion

1. Which function, secretion or reabsorption, is greatest as far as the movement of water in the kidney is concerned?
2. Proteins are involved in which of the three kidney functions?

DISCUSSION

1. Urine is a very dilute fluid that has a greater percentage of water than does blood.
2. Protein molecules remain in the blood because they are too large to pass through me glomerular membrane.
3. Glucose molecules are much smaller than protein molecules and pass into the nephron. Since these molecules are not found in the urine, they must be reabsorbed. Ask some leading questions to get students to develop these ideas.
4. If sodium intake is increased, the percentage of sodium in the urine is increased. This is an example of homeostasis in action.
5. A very small quantity of potassium is found in the blood. The percentage increases markedly in the urine.
6. The percentage of urea in the blood is very low compared with the percentage in the urine. One obvious function of the kidney is to remove urea from the blood.
7. Some materials in the blood are filtered from the glomerulus into the glomerular capsule of the nephron. Some of the materials are reabsorbed from the tubule back into the blood in the capillary. The materials remaining in the tubule become urine and are excreted as a waste product.

DISCUSSION

1. Of every 30 molecules that are filtered, all but one are reabsorbed. That one is secreted. The reabsorption function is greatest.
2. None of them—proteins do not normally enter the nephron.

Table 16.4 Comparison of Materials at Four Points Along a Nephron

	In Blood Entering Glomerulus	In Tubule from Glomerulus*	In the Urine	In Blood Leaving Nephron**
Water	100	30	1	99
Protein	100	0	0	100
Glucose	100	20	0	100
Sodium	100	30	1	99
Potassium	100	23	12	88

*The numbers in this column represent proportions, not actual numbers of molecules. For every 100 molecules of water in the blood, 30 are found in the tubule.

**The numbers in this column were obtained by subtracting the proportionate number of molecules of the substance in the urine from the proportionate number of molecules of the substance originally in the blood (100).

3. **No protein was found in the tubule, but 20 percent of the glucose molecules were. Glucose molecules are smaller than protein molecules and pass from the glomeruli into me tubules.**

4. **The kidney might be diseased. Too much glucose might be eaten for normal body needs. Glucose balance in the blood might be upset because of a lack of insulin.**

5. **An optimum quantity of glucose is essential for proper functioning of body cells. The kidney functions as a homeostatic organ by excreting the excess glucose molecules.**

6. **Going against a concentration gradient means active transport.**

7. **It can become toxic.**

8. **The discussion should summarize the basic points developed in this investigation. The following points from the chapter also should be developed during the discussion:**

   a. **Substances are selectively removed from the blood as they flow through the kidney.**

   b. **As blood flows through the glomerulus, water and many other molecules and ions are filtered in the nephron. The concentrations of these materials are changed in the nephron tubule through the processes of diffusion, selective reabsorption, and secretion.**

   c. **The kidney has both excretory and regulatory functions. By allowing the passage of certain substances and retaining others in optimum concentrations, the kidney plays a major role in maintaining the composition of body fluids.**

3. Compare the protein data with the glucose data. What is the difference? Explain the difference.

4. In some samples, glucose is found in the urine. What do you think might cause this condition?

5. Why are excess glucose molecules in the blood excreted?

6. The data tell us that the concentration of sodium in the blood is greater than in the urine, yet most of the sodium ions in the urine move back into the blood. What process makes this movement possible?

7. Urea is a by-product of amino acid metabolism. Next to water, urea is the most abundant material found in urine. If urea were allowed to accumulate in the blood, what might happen?

8. Homeostasis is the maintenance of a relatively stable internal environment in an organism. Discuss your ideas about how the kidney functions as a homeostatic organ.

SUMMARY

For its cells to function effectively, the body requires a constant internal environment. Homeostasis, the maintenance of this stable environment, is achieved by cooperative action of all body systems. The heart and blood vessels that make up the circulatory system are specialized to deliver raw materials to the cells of the body and to remove wastes from the cells' environment. The immune system serves to protect the body from foreign substances and pathogens through both nonspecific and specific types of protection. The respiratory system supplies the body with oxygen, and it aids the excretory system in eliminating waste substances. In addition, the excretory system plays a central role in regulating the concentration of many essential substances in the blood. A constant internal temperature enables chemical reactions to proceed at optimum rates. All of these systems are coordinated by the nervous and endocrine systems, the subject of the next chapter

APPLICATIONS

1. Describe the route followed by a molecule of oxygen as it moves from the air in the external environment to a mitochondrion in a muscle cell.

2. A temporary reddening of the skin surface sometimes called a "flush" and sometimes a "blush." The first term often is used in cases of fever. The second is used to describe a reaction to some situation in the external environment. Do you think the body mechanism is the same in both cases? If it is the same, how does it operate? If it is not the same, what are the differences?

3. Using chapters 15 and 16, trace the pathways of the hamburger, fries, and soft drink you might have had for lunch. Describe what happens to the food molecules from the time they are absorbed in

the intestine until their remains are excreted in the urine or eliminated through the anus.

4. Distinguish among excretion, secretion, and elimination. Explain which system(s) of the body are involved in each.

5. Construct a diagram or concept map showing the relationship between the circulatory and the respiratory systems.

6. What is the effect of the presence of respiratory pigment such as hemoglobin on the transport of oxygen?

7. How are breathing and gas transport affected by climbing to high altitudes?

8. If the human immune system has a long-term memory against foreign invaders, why is it neces-

sary for people to get vaccinated each year in order to protect themselves from contracting influenza?

9. How is the body's allergic reaction to pollen similar to how the body develops immunity to a disease?

10. Suppose you needed an organ transplant. Who would be the best choice as the donor of the organ—your spouse, one of your parents, an identical twin, or your fraternal twin? Who would be the worst choice as the donor, and why?

11. Some people used to be called "universal blood donors," and some used to be called "universal blood recipients." What was meant by this, and why are these terms no longer used?

 PROBLEMS

1. In cases of kidney failure, the patient must undergo periodic hemodialysis on a kidney machine. how does this machine function? Compare it with the functions of the human kidney.

2. Why do surgeons place their patients in hypothermia before procedures such as open heart surgery? What are the advantages?

3. Because the heart is often irreparably damaged by disease, much work has been done on designing artificial hearts. What problems must be overcome in order to design an effective artificial heart?

4. Under what circumstances would you apply cardiopulmonary resuscitation (CPR)? What are the precautions you must observe in exhaled air ventilation and external cardiac compression?

5. Take a field trip to a local heart clinic and meet with some of the technicians who operate the equipment. An opportunity to view such high-tech marvels as the ultrasound machine with Doppler image capabilities that produce an echocardiogram is an experience to remember.

6. Ask a representative of the American Lung Association to visit your class and to bring pictures, films, and slides of diseased lungs. Ask him or her to speak about the causes and prevention of lung disorders.

7. Your book has mentioned the importance of surface area to the functioning of structures in the body, find the references in this chapter and the previous chapter. Find references in other chapters to the relationship between surface area and function in various organisms. Why is surface area so important, and what types of structures often serve to increase it?

8. In a complete physical examination by your physician, samples of urine and blood are analyzed. What type of information about your health can be determined by analyzing these fluids?

9. Thermography is a technique used to detect diseases by checking the temperature patterns in the surface of a person's skin. Find out more about this technique, what diseases might be discovered by its use, and why the process works.

10. Hank Gathers was a very talented young college basketball player at Loyola Marymount College in California. He had a heart condition that he knew about, and still he decided to play. During a basketball game his heart began to beat arrhythmical, and he died. What was this heart condition, and what are other life threatening heart conditions? If you loved to participate in some activity, yet knew this activity was potentially life-threatening, would you continue? Make a list of reasons for and against.

The Human Animal: Coordination

This gymnast is performing a neuromotor task that requires a high level of concentration, training, and coordination. Name three sensory organs she is using to perform this task. What other organs is she using? Describe at least four separate functions the brain is carrying out. How is the gymnast able to do so many things at once?

Her eyes see the beam and position of her hands on the beam. Her ears provide tonal feedback, while touch receptors in her skin provide information about the pressure of her hands on the beam. The premotor and motor areas of the brain coordinate the arm, hand, and leg muscles. The hearing and vision centers of the brain interpret what is heard and seen, and the sensory area functions in the adjustment of touch and pressure. The brain also recalls from memory what was previously learned about this gymnastic skill and allows for a higher-level interpretation and judgment as to necessary adjustments. The cerebellum coordinates actions of the voluntary muscles. She is able to perform because all these organs and systems are coordinated by the nervous system.

◆ In performing any activity involving movement, the body must integrate several systems. A skateboarder's execution of a half-pipe, a musician's playing of a violin concerto, or an actor's speaking of the lines in a play all depend on the function and integration of the skeletal, muscular, nervous, and endocrine systems. Specialized cells in the nervous system transmit information from muscles, organs, and sensory receptors to the brain. The brain uses the information to determine the nature of electrical signals and chemical messages to be sent to muscles, organs, or glands. This chapter examines some of the mechanisms that make this integration possible and that maintain homeostasis in the body. ◆

MAJOR CONCEPTS

◆ Muscles contract in response to nerve impulses.
◆ Muscle contraction allows movement in only one direction.
◆ Nerve impulses travel by electrochemical mechanisms.
◆ The nervous system is highly specialized and intricately controlled.
◆ Endocrine glands secrete regulatory chemicals into the bloodstream.
◆ Emotional and physiological stress are related.

KNOWLEDGE CHECK

◆ How do bones, joints, and muscles allow for movement?
◆ How does the nervous system work?
◆ What are hormones, and how do they affect the body?
◆ What is stress, and what causes it?
◆ How do psychoactive drugs affect the nervous system?

Human Movement

17.1 Muscle Contraction Depends on Energy from ATP

Humans and other vertebrate animals have three types of muscles, as shown in Figure 17.1. Each of these three types has a different structure and function. Skeletal muscle moves the bones of the skeleton in response to conscious control. Because it is controlled by the individual, it is called voluntary muscle. Each skeletal muscle is composed of muscle tissue, which is made up of numerous individual cells acting together. Skeletal muscle is used in making adjustments to the external environment, such as raising your arm to swat an annoying mosquito. Figure 17.2 shows a skeletal muscle at successively greater magnifications. As shown in (a), the muscle consists of individual fibers that run its entire length. Each muscle fiber, the equivalent of a muscle cell, is made up of many parallel myofibrils (b), and each myofibril is made up of protein organized into thick and thin filaments (c). The thick filaments are composed of a protein called myosin. The thin filaments are made of another protein called actin. The thick and thin filaments are arranged in an overlapping pattern that gives skeletal muscle its characteristic striated appearance under the microscope. In muscle contraction, portions of the thick filaments attach to the thin filaments and pull the thin filaments toward each other.

Guidepost

How do muscles, bones, and joints enable humans to move?

Students may benefit from examining slides of each type of muscle.

433

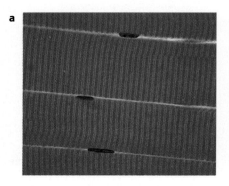
a

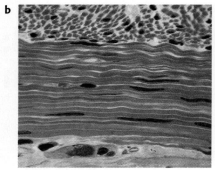
b

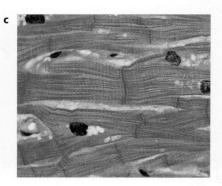
c

Figure 17.1 Types of muscle tissue. (a) Skeletal muscle cells fuse end to end, forming a long fiber with characteristic striations. (b) Smooth muscle is made of long, spindle-shaped cells arrayed in parallel. (c) Cardiac muscle cells are organized into interconnecting chains that form a lattice.

Muscle contraction helps maintain posture, accomplishes movement, and produces heat.

Smooth muscle is found in the walls of blood vessels, the digestive tract, and portions of the respiratory tract. Organized into sheets of cells, smooth muscle functions involuntarily in the operation of body organs and the regulation of the internal environment. In some tissues, the muscle cells contract only when they are stimulated by a nerve impulse or hormone, and then all of the cells contract as a unit. The smooth muscles in blood vessels operate in this fashion. In other body organs, such as the wall of the intestine, the individual cells of smooth muscle may contract spontaneously, creating a slow, steady contraction of the tissue.

Figure 17.2 The structure of skeletal muscle. See text for details.

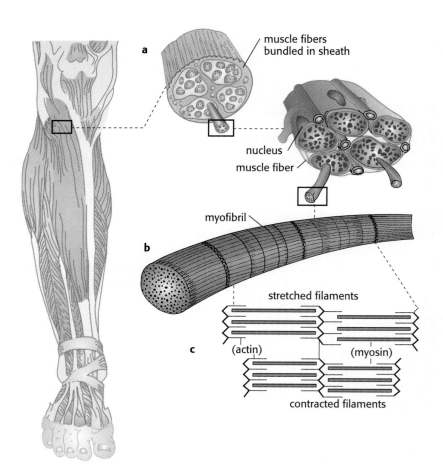

a

muscle fibers bundled in sheath

nucleus
muscle fiber

myofibril

b

stretched filaments

(actin)   (myosin)

c

contracted filaments

Cardiac (KARD ee ak) muscle is found only in the heart, and its contraction is involuntary. Cardiac muscle is striated. It is composed of chains of single cells, each with its own nucleus. These chains are organized into branches that interconnect and form a lattice. This latticelike structure is crucial to how heart muscle contracts. Heart contraction begins when
1 channels across the plasma membrane of the muscle cells open, admitting ions that change the electrical charge of the membrane. When two cardiac muscle fibers touch one another, their membranes make an electrical junction. As a result, the change in electrical charge of one fiber initiates a wave of contraction throughout the heart, with the change in electrical charge passing rapidly from one junction to the next. Thus, a mass of cardiac muscle tends to contract all at once, rather than a bit at a time.

Muscle contraction, whether voluntary or involuntary, is initiated by a nerve impulse. The nerve impulse causes the nerve endings on the muscle (Figure 17.3) to secrete chemicals that stimulate the muscle to contract. Muscle contraction also requires the presence of calcium ions and energy in the form of ATP (first discussed in Section 4.6). The supply of ATP in a single muscle fiber, however, can sustain full contraction for only about one second. For a muscle fiber to continue contracting the ATP must be regenerated rapidly. This is made possible by using three energy sources. The first is creatine phosphate (KREE uh tin FOS fayt), a substance in the muscle cell that can pass its energy to ADP. Creatine phosphate provides *immediate*
2 *diate* regeneration of ATP. The second energy source is the ATP formed during normal cellular respiration, and the third is glycogen, the form in which glucose is stored in the liver and in muscles (discussed in Section 4.8). In the muscle cells, the glycogen is broken down to glucose, which then is respired to produce more ATP. In the liver, glycogen is broken down to glucose, which is released into the blood in the presence of certain hormones. When an athlete is mentally preparing for exertion, the brain triggers the release of these hormones, which prepare the body for action.

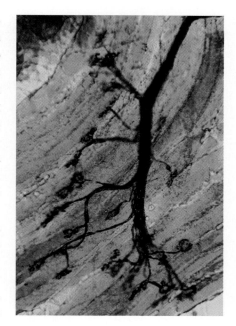

Figure 17.3   Nerve endings on skeletal muscle fibers (X60). Nerve impulses stimulate the nerve endings to secrete their chemical messengers, initiating the events that result in contraction of muscle fibers.

17.2   Interaction of Bones, Joints, and Muscles Produces Movement

Vertebrate movement is accomplished through the coordinated interaction of the bones of an animal's skeleton, the joints between the bones, and the contraction of muscles that span the joints. Figure 17.4 on page 436 shows the human skeletal and muscular systems. The skeleton, consisting of numerous bones and cartilage, is the internal framework of the body of a vertebrate. Most bones develop as cartilage, which gradually is replaced by bone through growth and further development. The hardness of bone results, in part, from deposits of calcium and magnesium compounds. In certain places of the body, for example in a portion of the rib cage and at the ends of bones, the cartilage is not replaced by bone. Retention of cartilage in these areas permits greater flexibility.

Joints are the places where two bones come together and body movement can occur. The structural features of the bones at the joint determine the degree of movement possible. Cords of connective tissue called ligaments hold together the bones that move at a joint. A joint capsule that surrounds the joint and encloses a lubricating joint fluid provides further support.

Muscles are attached to bones by flexible cords of connective tissue called tendons. Tendons span joints so that, when a muscle contracts, the force is applied at the joint and movement occurs. A muscle attached to a bone is like a rope attached to a wagon. You can pull a wagon with a rope,

Suggest that students locate on themselves through palpation, each of the structures identified in Figure 17.4.

Figure 17.4 Human skeletal and muscular systems.

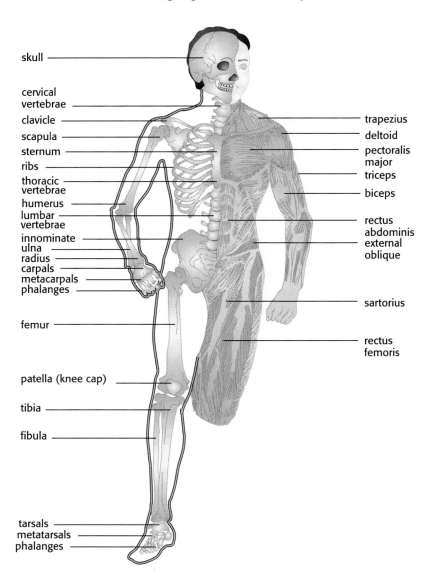

skull

cervical vertebrae
clavicle
scapula
sternum
ribs
thoracic vertebrae
humerus
lumbar vertebrae
innominate
ulna
radius
carpals
metacarpals
phalanges

femur

patella (knee cap)

tibia

fibula

tarsals
metatarsals
phalanges

trapezius
deltoid
pectoralis major
triceps
biceps
rectus abdominis
external oblique
sartorius
rectus femoris

but you cannot push it. To move the wagon back to its original position, you must attach a rope to the other end and pull it. Muscles act in the same way—in opposing pairs (see Figure 17.5). To touch your hand to your **3** shoulder, you contract one set of muscles, producing the flexed position. When you return your hand and arm to the extended position, you contract another set of muscles and relax the first set. These two sets of muscles, called flexors and extensors respectively, work in opposition to each other. All your skeletal movements are performed by contraction and relaxation of opposing sets of muscles.

Figure 17.5 A flexor-extensor muscle pair in the human arm. Flexors bend a limb at a skeletal joint; extensors straighten the limb again.

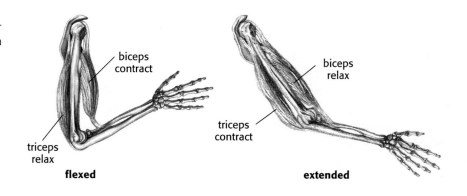

biceps contract

triceps relax

flexed

biceps relax

triceps contract

extended

17.3 Cardiovascular Fitness Depends on Regular Exercise

Even when an animal appears to be at rest, its muscles are not completely relaxed. The muscles of any healthy organism are always in a state of partial contraction called muscle tone. This produces the firmness that can be felt even in relaxed muscles. Today, physical fitness is a household word and a large business. For many, fitness means losing weight, whereas for others it means building muscles or improving muscle tone. The primary benefit of the effort to be physically fit, however, is to improve cardiovascular fitness—
4 the ability of the lungs, heart, and blood vessels to deliver oxygen to the cells.

The American Heart Association recommends exercising 20 to 30 minutes, three times a week, as one way of preventing coronary problems. The heart, like any muscle, gets stronger when it is exercised. The average person's heart pumps approximately 5 L of blood per minute. During intense exercise the amount of blood pumped may rise to between 15 and 30 L per minute. At this rate, a marathon runner's heart can deliver as much as 4000 L of blood to the body during a 43-km race—enough to fill a 12-gallon gas tank 80 times.

Cardiovascular fitness is important for everyone's health. The performance of an Olympic-caliber athlete, however, requires a great deal more than ordinary fitness. Centers such as the U.S. Olympic Training Center use the latest in modern electronic technology to help athletes improve their performances. Researchers also predict that within a decade or two, women may equal or surpass men in marathon events because of their more efficient method of utilizing stored energy.

The normal heart can increase is output by four or five times under stress conditions and by six or seven times in endurance athletes. Cardiac reserve is the maximum percentage that cardiac output can increase above normal. In a normal young adult, cardiac reserve may be 400% and in the trained athlete as high as 600%.

CONCEPT REVIEW

1. Describe the structure and action of the different types of muscle tissues.
2. How is the supply of ATP in muscle regenerated?
3. Explain how the contraction of muscles moves bones at joints.
4. How does cardiovascular fitness differ from physical fitness?

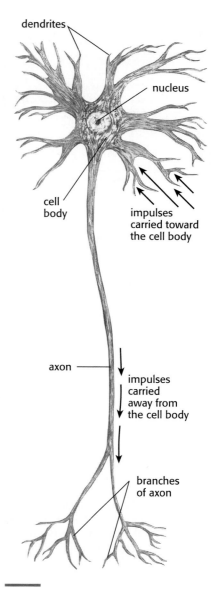

Figure 17.6   In this typical neuron, note that the axon always carries impulses away from the cell body.

The Nervous System

17.4 Neurons Transmit Nerve Impulses

An animal's rapid adjustments to the environment are brought about primarily by the nervous system. These adjustments or movements are initiated by an internal or external signal known as a **stimulus** (STIM yoo lus). The information is carried through the nervous system by **neurons** (NYOO rahnz), or nerve cells (see Figure 17.6). Each neuron has an enlarged region, the cell body, that contains the nucleus and other organelles. Most cell bodies are located in the brain or spinal cord. Two types of nerve fibers, dendrites, an axon, extend from the cell body. In humans, these fibers may be almost one meter long. Dendrites receive stimuli and conduct impulses *toward* the cell body. The axon carries impulses *away* from the cell body. Nerves are bundles of nerve fibers.

1   Neurons transmit **nerve impulses.** A stimulus occurring at the end of a nerve fiber starts a process of chemical and electrical change that travels

Guidepost

How does the brain coordinate body movements and determine behavior?

Neurotransmitters are in vesicles at the end of the axon. Norepinephrine acts in the sympathetic division of the autonomic nervous system; acetylcholine acts in muscles, the brain, and other organs.

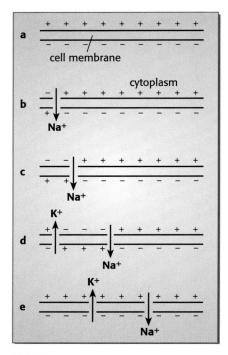

cell membrane

cytoplasm

Figure 17.7 The plasma membrane of a polarized neuron is more positive on the outside than on the inside. A stimulus changes membrane permeability, Na+ rushes in, and the membrane is locally depolarized (b). K+ diffuses out (c), repolarizing the membrane. The impulse moves as a wave of such changes in electrical charges (d, e, f).

The difference in electrical charge on the outside and the inside of the membrane is the electrical potential. This is a form of potential energy, much like a rock ready to roll down a hill.

A nerve impulse is both electrical and chemical in nature. The depolarization and repolarization comprise an action potential and take place in less than a millisecond.

In the development of many poisons and pesticides, the focus is on the synaptic sites and the functioning of the neurotransmitters in the target organisms. Have students speculate on why this is a popular approach.

Be certain students understand that they actually do not see with their eyes or hear with their ears but rather that these are specialized receptor organs. Perception is a function of the brain. Is it possible to have a sense of touch in paralyzed arm?

like a wave over the length of a neuron. This wave of change is the nerve impulse. The chemical reactions that occur during the nerve impulse require oxygen and energy and produce carbon dioxide and heat. Electrical changes also take place. The plasma membrane of a resting neuron (one not transmitting an impulse) is electrically more positive on the outside than on the inside, as shown in Figure 17.7a. This difference in electrical charge is due to the differential permeability of the membrane to sodium and potassium ions and to proteins.

Stimulation of a neuron causes a small segment of the plasma membrane to become more permeable to sodium ions. As positively charged sodium ions rush into the neuron, the electrical charge of the membrane suddenly changes (see Figure 17.7b). This local change is the nerve impulse, and it starts a similar change in the next small segment. In this manner, the impulse moves from one end of the neuron to the other in a wave of local chemical and electrical changes as sodium and potassium ions move across the membrane (Figure 17.7c, d, e, f). 2

One neuron is separated from another neuron by a small space called a **synapse** (SIN aps). When an impulse reaches the end of the axon of one neuron, it causes the release of a chemical substance known as a neurotransmitter (NYOOR oh trans MIT er). The neurotransmitter diffuses across the synapse to the membrane of the next neuron, where it causes a new nerve impulse to be initiated. Because transmission across a synapse takes place in only one direction, neurons conduct impulses in only one direction, from the axon of one neuron to the dendrites or cell body of the next neuron. Neurotransmitters also cause muscles to contract and certain glands to secrete hormones.

Nerve impulses may travel at speeds greater than 100 m per second in large fibers. The impulses are of short duration—usually a few ten thousandths of a second. The neuron, however, cannot carry another impulse until it has had time to recover. To restore the resting state, sodium ions must be pumped back outside the plasma membrane by means of active transport. Only after recovery can the neuron respond to a new stimulus.

In humans, the three types of neurons shown in Figure 17.8 can be distinguished by their functions. Sensory neurons receive information

Figure 17.8 Types of neurons.

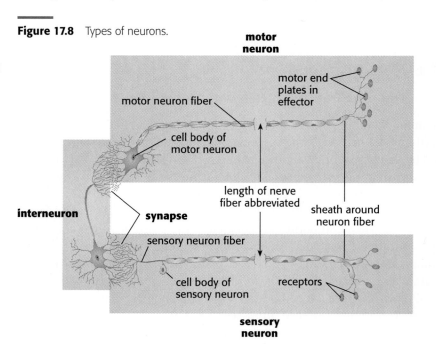

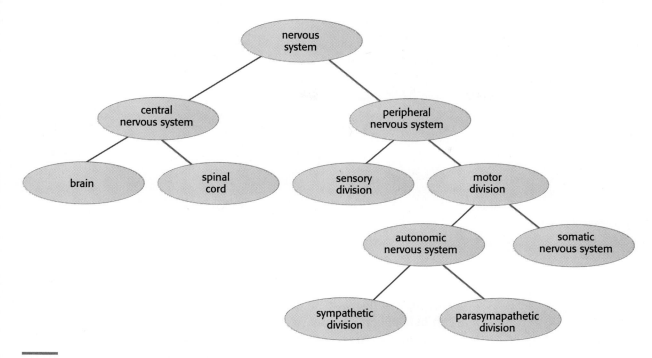

Figure 17.9   Organization of the human nervous system.

from the internal and external environments by means of specialized nerve endings called receptors. (Receptors also may be specialized neurons such as those in the retina of the eye that react to light.) They transmit impulses toward the brain or spinal cord. Interneurons transmit impulses from one neuron to another. Motor neurons transmit impulses that originate in the brain or spinal cord to an effector—a muscle or a gland. Effectors respond to an impulse with movement or secretion of some chemical.

17.5   The Central Nervous System Coordinates Body Functions

Nervous systems have two basic functions: to control responses to the external environment and to coordinate the functions of internal organs and thereby maintain homeostasis. Because the vertebrate nervous system is complex, it is convenient to divide it into parts that differ in function. The two main parts are the central nervous system, or CNS, and the peripheral nervous system, or PNS. Composed of the brain and spinal cord, the CNS is specialized to receive and process incoming information and to determine and initiate the appropriate responses. Communication between the CNS and the rest of the body is provided by the PNS, which consists of the sensory and motor neurons. Figure 17.9 shows the relationships between the parts of the nervous system.

Two major activities are carried out by the spinal cord: integration of simple responses to certain types of stimuli and transfer of information to and from the brain. Spinal integration normally is in the form of a reflex—an automatic, involuntary response to a stimulus, such as withdrawal from a pinprick (see Figure 17.10). Most reflexes are fairly complex and some times involve many interneurons between the sensory and motor neurons.

Personality, consciousness, learning, and memory are all functions of the **cerebrum** (seh REE brum), in which the senses of seeing, hearing, tasting, smelling, and feeling also are experienced. The cerebrum is made up of several lobes and is divided into a left and a right hemisphere, each having different functions. Verbal and analytical skills such as reading,

Figure 17.10 A reflex is an automatic response to a stimulus that occurs before the stimulus is received and interpreted by the brain, thus reducing the reaction time to a potentially life threatening situation. Getting jabbed with a pin causes an impulse to travel up a sensory neuron to the spinal cord. The impulse travels through an interneuron to a motor neuron and on to the arm muscles. The arm immediately jerks away from the stimulus. At the same time, the stimulus is traveling through the same interneuron to the brain, where it is received and interpreted as pain.

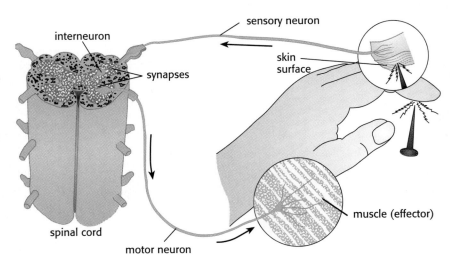

Have students research some problems associated with the nervous system, such as cerebral palsy, meningitis, multiple sclerosis, and the effects of strokes.

speech, writing, mathematics, and logic generally are associated with the left hemisphere, whereas spatial relationships and artistic skills such as music and art are associated with the right hemisphere. The right hemisphere controls functions on the left side of the body and, conversely, the left hemisphere controls functions on the right side of the body. Figure 17.11 illustrates the structures and functional areas of the human brain.

Located under the cerebrum, the **cerebellum** (ser eh BEL um) coordinates the contractions of skeletal muscle. Maintenance of posture and muscle tone as well as coordination of movements such as walking and running are under the influence of this part of the brain. The **medulla** (meh DUL lah) regulates automatic, unconscious activities, such as the heartbeat rate, contractions of the blood vessels, and the depth and rate of breathing. Other functions include the control of coughing, sneezing, swallowing, and vomiting. The medulla keeps several of the body's vital functions operating properly without any conscious intervention.

Figure 17.11 The major parts of the primary human brain are associated with specific senses and abilities.

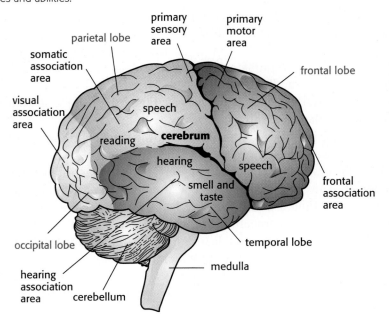

Biology Today

The Human Eye and Vision

All the input from sensory neurons to the central nervous system arrives in the same form—as nerve impulses. To the brain, every arriving nerve impulse is identical to every other one. The information the brain derives from sensory input is based on the frequency with which these impulses arrive and on the neuron that transmits the input. If the optic nerve, for example, is stimulated, the impulses generated are perceived as a flash of light. If the auditory nerve is stimulated, the stimulation is perceived as a noise. Most receptors are either chemical or mechanical, except for the receptors in the eye. Vision, the perception of light, is dependent on receptors that can detect the electromagnetic energy present in light.

The human eye is sensitive to almost 7 million shades of color and about 10 million variations of light intensity. What makes this sensitivity possible? Light waves that enter the eye through the cornea (shown in a) are bent toward the lens. The iris (the colored part of the eye) consists of smooth muscles that can contract and relax involuntarily, changing the size of the pupil and thus regulating the amount of light that enters the eye. The waves pass through the lens and are focused on the retina,

which has several different layers of tissue (b). Specialized sensory cells called rods and cones receive stimuli resulting from the incoming light waves (c). Rods are responsible for peripheral vision (the edges of the field of vision), and they respond to dim light for black-and-white vision. Cones are responsible for daylight color vision. Each cone contains pigments that are most sensitive to wavelengths corresponding to one of the colors, red, green, or blue.

The structure of the pigments in both rods and cones is altered by light, producing a change in the electrical charge of the cells. This change in charge is a signal to neighboring neurons of the presence of light. Signals converge and diverge among neurons of the retina. They may spread out and inhibit activity in nearby cells. As a result, the frequencies at which signals are transmitted to the brain change in response to changes in the size, location, and intensity of light stimuli. The neurons of the retina eventually relay the signal to the optic nerve fibers. From there, the signals to a part of the brain called the thalamus and then on to visual processing centers in the cerebrum.

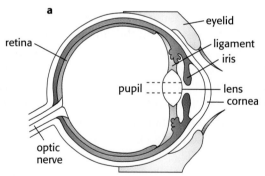
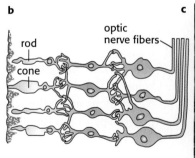
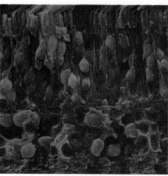

A cross-section diagram of the human eye (a); layers of tissue in the retina (b); rods and cones (c).

Figure 17.12 The autonomic nervous system is composed of two divisions that act like a brake and an accelerator. The parasympathetic division generally acts to maintain normal body function. The sympathetic division acts to prepare the body for emergency. The two divisions have opposing effects on a variety of organs and body functions.

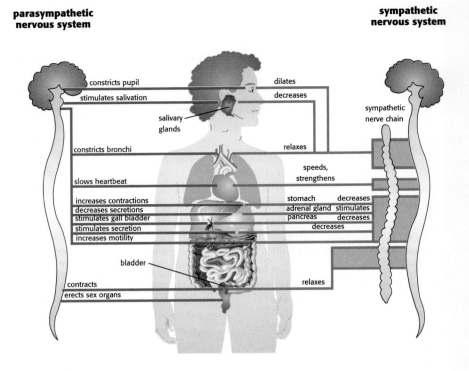

17.6   The Peripheral Nervous System Transmits Information

The peripheral nervous system (PNS) includes all the nerve pathways through which the central nervous system communicates with the rest of the body. Sensory neurons carry signals from the external and internal environments *to* the CNS. Motor neurons, by contrast, carry signals *from* the CNS to muscles and glands. The motor division of the PNS can be further subdivided into the somatic and the autonomic nervous systems (as diagrammed in Figure 17.9). Motor neurons in the somatic nervous system carry signals to skeletal muscles. The somatic nervous system often is considered "voluntary" because the skeletal muscles are subject to conscious control (although much skeletal movement is the result of reflexes). Motor neurons in the autonomic, or "involuntary," nervous system act to coordinate organ function and help to maintain homeostasis.

The autonomic nervous system itself is made up of two divisions—the sympathetic and the parasympathetic. These two divisions generally have opposite effects on an organ, acting much like an accelerator and a brake pedal. For example, the sympathetic division speeds up heart rate in emergencies, and the parasympathetic division slows it down again. For the most part, the sympathetic division serves as a stimulator and plays an essential role in preparing the body to respond to emergency situations. Figure 17.12 shows some of these contrasting actions.

The autonomic nervous system helps to control blood pressure, stomach movements and secretions, urine formation and excretion, body temperature, and many other functions necessary to maintenance of a steady state. It also plays a major role in emotions. The functions of the autonomic nervous system, in turn, are controlled by the hypothalamus, a part of the brain located under the cerebrum. In addition to receiving and sending nerve impulses, the hypothalamus has specialized cells that produce hormones, and thus it serves to coordinate the nervous system with the endocrine system.

4

1. What is the relationship between a stimulus and a nerve impulse?
2. Briefly describe the electrical and chemical changes involved in a nerve impulse.
3. How does the central nervous system operate to coordinate body functions?
4. How does the autonomic nervous system function in maintenance of homeostasis?

The Endocrine System

17.7   Hormones Are Chemical Messengers

Even before you were born, the endocrine (EN doh krin), or ductless, glands and the hormones they secrete were functioning in the homeostatic control of your body. The amount of growth you experience, the regulation of your blood sugar levels, the regulation of the calcium level in your bones and blood, and the rate at which you metabolize food are only a few examples of the endocrine glands at work. Endocrine glands are capable of sensing the level of specific hormones in the blood and responding with an increase or decrease in their own secretions. This network of glands, illustrated in Figure 17.13, makes up the endocrine system, which acts with the nervous system to coordinate body functions. Because these systems work together so closely, they sometimes are referred to as the neuroendocrine system. Endocrine response, however, is slower and longer lasting than that of the nervous system.

Although hormones are secreted directly into the bloodstream and carried all through the body, they produce their effects only on specific target organs, such as the ovaries or testes of the reproductive system discussed in Chapter 6. This specific action is possible because the cells of the target

Guidepost

How are the nervous and endocrine systems coordinated?

The first endocrine hormones to be recognized and studied were the male sex hormones. Early practices of castrating cattle to make them fatter and more docile go back to earliest agricultural ages. Eunuchs (castrated human males) lack normal sex drives and were used as harem guards or, because they retained their childhood soprano voices, as church choir singers.

Figure 17.13   The endocrine system. The dotted lines show organs such as stomach and kidneys.

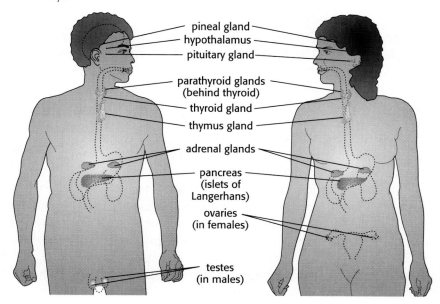

pineal gland
hypothalamus
pituitary gland
parathyroid glands (behind thyroid)
thyroid gland
thymus gland
adrenal glands
pancreas (islets of Langerhans)
ovaries (in females)
testes (in males)

organ have receptors that fit the shape of a particular hormone. When molecules of the hormone reach the cells of the target organ, they bind to the **1** receptors, much the way an antigen binds to an antibody. This binding begins a series of events that cause the target organ to take particular actions.

A hormone can be effective for as short a time as two minutes or as long a time as several hours. Hormones constantly are being removed from the blood by the target organs, excreted by the kidneys, or broken down to simpler compounds inside the liver. The effect of a hormone on the cells of its target organ depends to a large degree on the level of the hormone concentration in the blood. Feedback systems maintain the concentration of most hormones within a predetermined range. A feedback system operates as a cycle in which the last step affects the first step. Each endocrine gland continuously receives information, or feedback, through the blood in the form of chemical **2** signals about the cellular process its hormones affect. If conditions change, the gland can act on the information and adjust its rate of secretion.

17.8 Several Hormones Act Together to Control Blood Glucose Levels

Control of blood glucose demonstrates how a negative feedback system works. Endocrine tissues embedded in the pancreas produce two hormones: **insulin** (IN suh lin) and **glucagon** (GLOO kuh gahn). These are the primary regulators of glucose levels in the blood, although several other hormones are involved in different ways. Low levels of blood glucose provide feedback to the pancreas, which causes its cells to secrete glucagon. Glucagon increases the conversion of glycogen to glucose in the muscles and liver, thus increasing blood glucose levels. The feedback is negative because the release of glucagon reverses the low level of blood glucose, raising it toward an average level. Insulin has the opposite effect of glucagon and is secreted in response to high levels of blood glucose. Insulin enhances the uptake of glucose by muscle and liver cells, thus removing glucose from the blood. Insulin also is regulated by a negative feedback system. The release of **3** insulin reverses the high level of blood glucose, lowering it toward an average level. The negative feedback systems that control the production of insulin and glucagon, as well as other negative feedback systems that regulate most endocrine glands, thus work to maintain homeostasis.

Insulin and glucagon have opposite effects in the homeostatic control of the level of blood glucose. After a meal, glucose levels rise as carbohydrates are digested. As a result, insulin secretions increase and glucagon secretions decrease. Conversely, during fasting, insulin secretions decrease and glucagon secretions increase. Figure 17.14 summarizes these blood glucose control mechanisms.

Maintenance of appropriate glucose levels is crucial. In the absence of glucagon, blood glucose levels may drop too low, resulting in hypoglycemia (hy poh gly SEE mee uh), or low blood sugar. Hypoglycemia is dangerous because brain cells must obtain a constant supply of glucose from the blood, and if the supply is cut off, the person may lose consciousness.

In the absence of insulin, blood glucose levels tend to remain elevated, causing changes in many body systems. Complications resulting from the high blood glucose levels may include circulatory disorders and vision problems that can lead to blindness. In addition, the kidneys are unable to reabsorb all the glucose and, as a result of increased glucose in the collecting ducts, less water is removed and returned to the blood. Both glucose in the urine and thirst are symptoms of insulin deficiency. The lack of

In liver cells, insulin stimulates phosphorylation of glucose, so concentration gradients favor diffusion of glucose into the cells. In muscle cells, insulin enhances facilitated diffusion of glucose via a carrier.

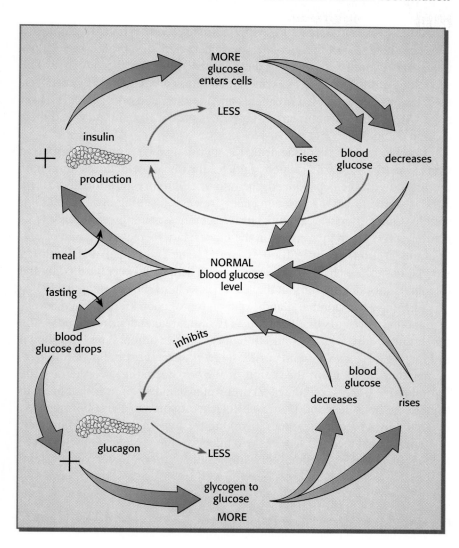

Figure 17.14   Insulin and glucagon function together to maintain a fairly stable level of blood glucose.

insulin also prevents glucose from entering body cells, which then metabolize fats and proteins instead. Products of fat and protein metabolism enter the blood and lower the pH, interfering with many body functions. At high levels, these products can cause coma and death.

The disease diabetes mellitus (MEL luh tus) results from insufficient insulin secretion. There are two major forms of diabetes mellitus, both apparently with a hereditary basis. In type I diabetes, the immune system attacks and destroys the insulin-secreting cells of the pancreas; type I diabetics are therefore insulin dependent. The causes of type II diabetes are not well understood, but type II often can be controlled by diet and drugs that stimulate the pancreatic cells to secrete insulin.

17.9   The Hypothalamus and Pituitary Control Other Endocrine Glands

Chief among the endocrine glands is the pituitary, which secretes hormones that control the rate at which several other endocrine glands function. The pituitary, in turn, is controlled by hormones from the hypothalamus. Some of these hormones pass through blood vessels into the pituitary, causing the pituitary to release its own hormones. Other hypothalamic hormones pass through secretory neurons and are stored in the pituitary for later release.
4 Pituitary hormones control the thyroid and the adrenal glands and, through the reproductive organs, regulate reproductive activities such as the

Synthesis of various hormones in humans has been shown to follow circadian rhythms: corticosteroids peak between 4:00 AM and 8:00 AM, growth hormone levels rise about an hour after falling asleep, and testosterone peaks about 9:00 AM. People are more likely to be born or die between 3:00 AM and 4:00 AM. This also is the time of lowest body temperatures.

menstrual cycle and ovulation, sperm production, embryonic development, and the production of the secondary sex characteristics. Figure 17.15 illustrates the relationship between the hypothalamus and the pituitary. The secretions of the hypothalamus and the pituitary are regulated by feedback mechanisms among these two glands and their target tissues. Table 17.1 on page 448 presents a summary of hormones produced in humans and their roles in coordination and homeostasis.

Located at the base of the neck, the thyroid gland secretes the hormone **thyroxine** (thy ROK sin), which regulates the rate of cellular respiration. If too little thyroxine is produced, the rate of cellular respiration decreases. Food is stored rather than used in energy release, resulting in an increase in weight. Other effects are sluggishness, sleepiness, and lowered body temperature. Too much thyroxine, on the other hand, increases the rate of cellular respiration, so little food is stored. The individual usually loses weight, has excess energy, and is nervous. The thyroid is controlled by the hypothalamus and pituitary in the feedback system shown in Figure 17.16.

There are two adrenal glands, one on top of each kidney. Each adrenal gland has two parts that function as two separate endocrine glands—the outer, larger adrenal cortex and the inner, smaller adrenal medulla shown in Figure 17.17. The adrenal cortex secretes several different steroid hormones that supplement the action of the sex hormones secreted by the reproductive organs, regulate salt and water balance, fight inflammation, or help the body respond to stress. Release of some of these hormones is regulated by a feedback system involving the hypothalamus and pituitary.

The adrenal medulla secretes two major hormones in response to nerve signals from the sympathetic division of the autonomic nervous system.

Figure 17.15 Connections between the hypothalamus and the two lobes of the pituitary. Under the influence of hormones from the hypothalamus, the anterior pituitary secretes six hormones, which affect the target organs shown. The hypothalamus also secretes two hormones that are stored in the posterior pituitary and released at a later time, affecting the target organs shown.

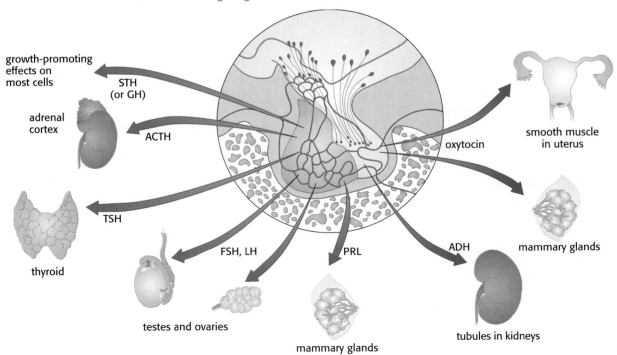

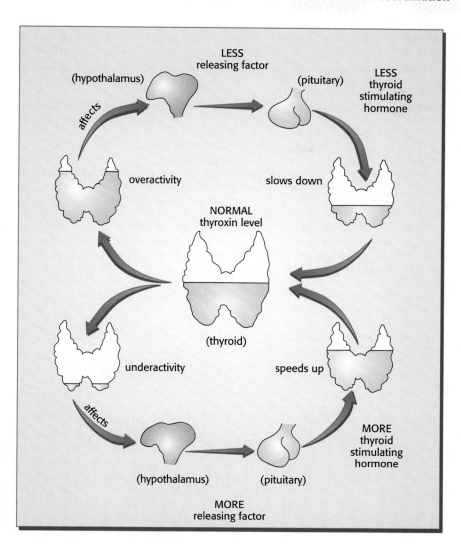

LESS
releasing factor

(hypothalamus)

(pituitary)

LESS
thyroid
stimulating
hormone

affects

overactivity

slows down

NORMAL
thyroxin level

(thyroid)

underactivity

speeds up

affects

MORE
thyroid
stimulating
hormone

(hypothalamus)

(pituitary)

MORE
releasing factor

Figure 17.16   The feedback mechanism between the pituitary, hypothalamus, and thyroid glands. This mechanism controls the level of thyroxine in the blood.

The hormones, **epinephrine** (ep ih NEF rin, also known as adrenalin) and **norepinephrine,** help increase mental alertness, raise blood pressure, speed up heart rate, dilate coronary blood vessels, increase respiratory rate, and speed up the rate of metabolism. These effects—often called the fight-or-flight response—supplement the immediate response of the sympathetic division and allow the body to react to threats or emergencies. Epinephrine also can enhance an athlete's performance by stimulating the breakdown of glycogen, thus providing additional energy sources to the cells.

Figure 17.17   The structure of the adrenal gland, showing the cortex and the medulla.

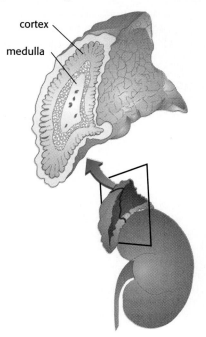

cortex

medulla

CONCEPT REVIEW

1. If hormones are transported throughout the body in the circulatory system, how are they able to act on a specific target organ?
2. Describe how a simple feedback system operates.
3. Compare the functions of insulin and glucagon in the control of blood glucose.
4. Describe the interrelationships of the hypothalamus and the pituitary.
5. Describe the feedback loop that controls secretion of thyroxine by the thyroid.
6. How does the adrenal gland function in the fight-or-flight response that enables the body to respond to emergency situations?

Table 17.1 Origins and Effects of Hormones

Gland/Organ	Hormone	Target	Principal Action
Adrenal cortex (outer portion)	aldosterone	kidneys	affects water and salt balance
	androgen, estrogen	testes, ovaries	supplement action of sex hormones
	cortisol	general	increases glucose, protein, and fat metabolism; reduces inflammation; combats stress
Adrenal medulla (inner portion)	epinephrine norepinephrine	general (many regions and organs)	increase heart rate and blood pressure; activate fight-or-flight response
Heart	atrial natriuretic factor (ANF)	blood vessels, kidneys, adrenal glands, regulatory areas of the brain	regulates blood pressure and volume; excretion of water, sodium, and potassium by the kidneys
Hypothalamus	corticotropin-releasing hormone (CRH)	anterior pituitary	stimulates secretion of ACTH
	gonadotropin-releasing hormone (GnRH)	anterior pituitary	stimulates secretion of FSH and LH
	prolactin-inhibiting hormone (PIH)	anterior pituitary	inhibits prolactin secretion
	somatostatin	anterior pituitary	inhibits secretion of growth hormone
	thyrotropin-releasing hormone (TRH)	anterior pituitary	stimulates secretion of TSH
Hypothalamus (via posterior lobe of pituitary)	antidiuretic hormone (ADH, vasopressin)	kidney	controls water excretion
	oxytocin	breasts, uterus	stimulates release of milk; contraction of smooth muscle in childbirth
Pancreas (endocrine tissues)	glucagon	liver	stimulates breakdown of glycogen to glucose
	insulin	muscles, liver cells	lowers blood sugar level; increases storage of glycogen
Parathyroid	parathyroid hormone (parathormone)	intestine, bone	stimulates release of calcium from bone; decreases excretion of calcium by kidney and increases absorption by intestine

Stress, Drugs, and the Human Body

17.10   Stress Is the Body's Response to Change

The physiological changes that characterize the fight-or-flight response affect the entire body and are known collectively as stress. Changes in the environment that cause stress are called stressors. Stressors may be physical, such as stepping in front of an oncoming car, exposure to extreme heat or cold, infections, injuries, prolonged heavy exercise, lack of sleep, or loud sounds. Stressors also may be emotional, for example fears about real or imagined dangers, personal losses, anger at a boss or teacher, or unpleasant social interactions (or lack of social interactions). Even pleasant stimuli, such as feelings of joy or happiness, can act as stressors.

Table 17.1 (continued)

Gland/Organ	Hormone	Target	Principal Action
Pineal	melatonin	unknown perhaps hypothalamus and pituitary	in humans- regulation of circadian rhythms (day/night cycles)
Pituitary (anterior lobe)	adrenocorticotropic hormone (ACTH)	adrenal cortex	secretes steroid hormones
	follicle stimulating hormone (FSH)	ovarian follicles, testes	stimulates follicle; estrogen production; spermatogenesis
	growth hormone (GH)	general	stimulates bone and muscle growth, amino acid transport, and breakdown of fatty acids
	luteinizing hormone (LH)	mature ovarian follicle interstitial cells of testes	stimulates ovulation in females, sperm and testosterone production in males
	prolactin	breasts	stimulates milk production and secretion
	thyroid stimulating hormone (TSH)	thyroid	secretes thyroxine
Reproductive organs ovaries	estrogen	general	development of secondary sex characteristics; bone growth; sex drive
	progesterone	uterus (lining)	maintenance of uterus during pregnancy
testes	testosterone	general	development of secondary sex characteristics; bone growth; sex drive
Thymus	thymosin	lymphatic system	possibly stimulates development of lymphatic system
Thyroid	calcitonin	intestine, kidney, bone	inhibits release of calcium from bone; decreases excretion of calcium by kidney and increases absorption by intestine
	thyroxine	general	stimulates and maintains metabolic activities

To understand what happens to the body during a stress reaction, imagine a stressful situation in your life. First, you experience a change in your environment that you consciously or unconsciously decide is a stressor. Then, your brain sends signals to the hypothalamus, which activates stress responses in the adrenal gland and other body tissues. These responses help your body prepare for vigorous muscular activity—to fight or flee. If your stressor is like most in today's society, however, neither action is appropriate, and the net result to the body is like stepping on the accelerator of a car while the brake is on. A lot of fuel is burned, but you do not get anywhere.

Long-term stress lasting days or weeks can cause considerable damage to the body. People under constant stress have increased levels of adrenal hormones in their blood and urine. One of these hormones (cortisol)

You may wish to discuss the term *psychosomatic* at this time. Point out that the American Heart Association has recognized stress as a risk factor in the development of heart disease.

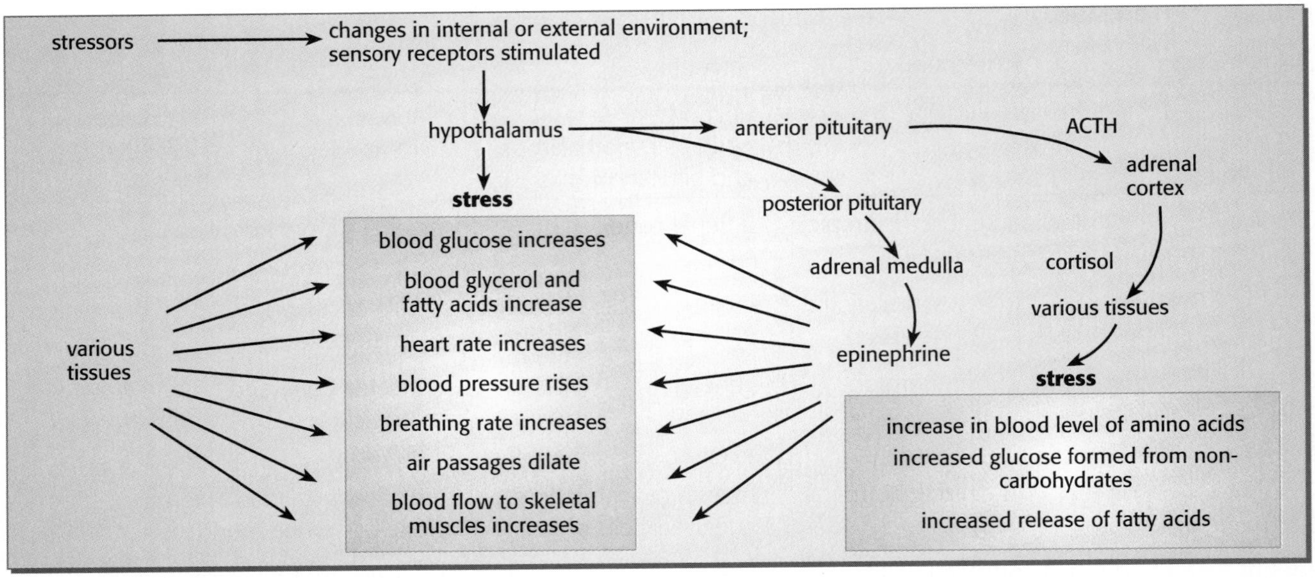

Figure 17.18 Changes in the internal or external environment lead to physiological changes in the body that prepare it for a fight-or-flight response. These changes are initiated by impulses arising from the hypothalamus that directly affect body tissues and by impulses directed toward the adrenal medulla that cause it to release epinephrine, which affects the same target tissues.

depresses the activity of the immune system, resulting in a lowered resistance to infectious diseases. Long-term stress is related to higher rates of high blood pressure, atherosclerosis, and ulcers in people who are relatively young. Psychological problems such as insomnia, anxiety, and depression also are associated with prolonged stress. Figure 17.18 summarizes the effects of stressors on the human body. Stress management offers ways reduce stressors in one's life and get rid of the extra energy the body builds under stress. It is healthy to learn how to manage long-term stress and reduce its effects on the body. Table 17.2 suggests ways to manage stress.

Table 17.2 Recommendations for Managing Stress

1. Do not take on too many responsibilities. The most common stressor is trying do too much with the time and mental and physical resources you have.
2. Balance your time. Make sure you have time to work and to play. Learn how to work effectively but also teach yourself how to relax—through relaxation techniques, music, hobbies, and/or exercise. The word *recreation* is just that—a "re-creation" of energies for your mind and body. It can stop the stress response.
3. Get enough rest, sleep, and exercise. Allow your body and mind to renew.
4. Work off the energy created by stress. Take a break and blow off steam by exercising regularly. Do something you enjoy, such as playing tennis, biking, walking, swimming, or playing basketball.
5. Avoid self-medication. Drugs, including alcohol and tobacco, might seem to produce a *temporary* reduction of stress, but drugs do not cure the causes, and they have harmful side effects. In addition, they can be addictive.
6. Talk to someone you trust about the stressors in your life. Your parents, a friend or a school counselor might help you see how you could avoid some stressors.
7. Check with your doctor or school nurse if you suspect you have a stress-related medical problem.

Physician and Physiologist

Dr. Sarah A. Nunneley was born in New York City and grew up in Washington, D.C. and Minneapolis, Minnesota. Her father is an attorney specializing in aviation law and her mother is a handweaver of national repute. Dr. Nunneley attended Mount Holyoke College where she majored in zoology and graduated Magna Cum Laude. She earned her medical degree at the University of Minnesota and interned at King's County Hospital in Brooklyn. She then studied Aerospace Medicine for three years at Ohio State University and the Lovelace Foundation. In 1972 she become the first woman to be board certified in Aerospace Medicine.

Dr. Nunneley's career reflects a lifelong interest in space flight and human response to environmental stress. She served for six months as flight surgeon at NASA's Flight Research Center in southern California, then joined the physiology faculty at the State University of New York, Buffalo, where she worked on problems in respiratory physiology, exercise, and acceleration. In 1975 she moved to the USAF School of Aerospace in San Antonio, Texas, where she heads up a group of scientists whose work focuses on the effects of heat and cold stress, acceleration, and exercise. She is the author or co-author of more than 30 scientific papers and has presented her work at national conferences and at international meetings in Europe and Australia. In 1979, Dr. Nunneley was Program Chairperson for the 50th Annual Meeting of the Aerospace Medical Association. She currently serves the School of Aerospace Medicine as a consultant on human experimentation and the problems of female flight crew members.

Dr. Nunneley is a Fellow and Past President of the Aerospace Medical Association, a member of the American Physiological Society, American Association for the Advancement of Science, Association for Women in Science, Sigma Xi Scientific Research Society, Undersea Medical Society, and the International Academy of Aviation and Space Medicine. She is also a Fellow of the American College of Sports Medicine. As a pioneer in her field of aerospace physiology and medicine, she has been the recipient of numerous awards including the Amelia Earhart Fellowship of Zonta International and the Julian E. Ward Memorial award for research excellence presented by the Aerospace Medical Association.

In her spare time, Dr. Nunneley jogs, cares for two horses and competes with them in combined training events (dressage, endurance, and jumping). She enjoys travel and is an avid reader of history, historical fiction, and science fiction.

Dr. Sarah A. Nunneley "Physician and Physiologist"

Have the students discuss their perception of the lack of limits on the number of prescriptions doctors write and of the effect of commercials on the use/abuse of over-the-counter drugs.

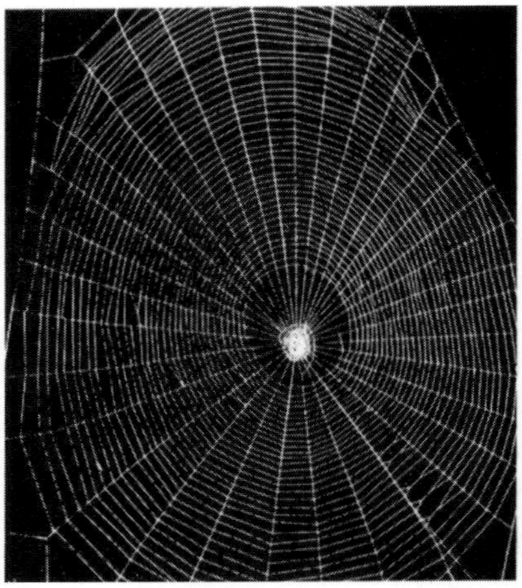

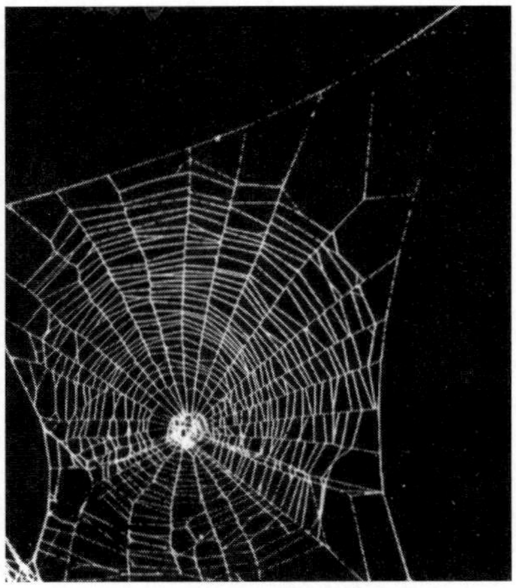

17.11 Psychoactive Drugs Affect the Central Nervous System

For some people, the effort to manage stress is difficult and confusing. They seek to escape stress and to find a short cut to confidence and satisfaction through the use of chemical substances that affect the nervous system. The functions of the central nervous system can be modified in a variety of ways, particularly by psychoactive (sy koh AK tiv), or mind-altering, drugs. Drugs are nonnutritive substances that change the way the body or brain functions. Used properly, certain drugs are important in maintaining health. Drugs include medicines, such as antibiotics and aspirin, that are used to cure or treat diseases or to relieve minor discomforts. Psychoactive drugs which include tranquilizers, opiates, barbiturates, and alcohol, may be prescribed to relieve severe pain, aid sleep, and calm nervousness. Because these substances work on the brain, they can change the way a person thinks, feels, or acts. Many of them are used illegally, without a prescription, for their pleasurable side effects.

Any drug can be abused, that is, used in a way that causes personal or
social problems. For example, overuse of aspirin can lead to severe 2
abdominal bleeding. Consumption of too much caffeine in coffee, tea,
coca, or cola drinks can lead to irritability and sleeplessness. Abuse of
psychoactive drugs, in addition to causing a variety of physical effects,
can lead to more serious problems. Many of the psychoactive drugs induce
tolerance, so that larger and larger amounts are required to produce the
same effect. Most also are addictive, producing physical or psychological
dependence. In physical dependence, the body becomes unable to function
normally without the drug. Withdrawal may produce symptoms ranging
from discomfort to convulsions, which can be fatal. In psychological 3
dependence, the user exhibits an intense craving for the drug and its
effects. Sources, Uses, and Effects of Some Psychoactive Drugs in
Appendix 3 summarizes information about the more commonly used psychoactive drugs.

Figure 17.19 Many of the complex behavior patterns of animals can be disturbed by chemical substances. A female spider builds webs in the characteristic regular form (a). Several hours after feeding on sugar water containing a small dose of amphetamine, the spider built the unusually small, irregular web (b). Although we cannot compare directly the behavior patterns of invertebrate animals and humans, experiments such as this provide important information about the effects of chemical substances on brain function.

a

b

Psychoactive drugs can be classified as depressants, stimulants, or hallucinogens. Depressants reduce the activity of the central nervous system, some of them by blocking nerve transmission. Alcohol, the most widely used and abused depressant, releases inhibitions and may produce feelings of happiness. Tranquilizers dull the emotions and reduce anxiety by calming and relaxing the user, but overuse can lead to psychological or physical dependence. Opiates (opium, morphine, heroin) are used to relieve pain and relax muscles; they also produce effects similar to tranquilizers. Opiates induce tolerance and are physically addictive. Barbiturates dull the mind and are used to induce sleep and relax tensions. Like opiates, they induce tolerance and are addictive. Barbiturates are especially dangerous when used with alcohol, opiates, anesthetics, or tranquilizers.

Stimulants increase activity of the central nervous system and usually
induce a feeling of well-being. Examples include caffeine, the nicotine in 4
tobacco, amphetamines, cocaine, and crack, a form of cocaine. Stimulants
increase alertness and reduce appetite. Cocaine and crack, in addition, create intense feelings of happiness and can block pain. All stimulants can
induce tolerance and produce powerful psychological and physical dependence. Nicotine and crack are two of the most addictive drugs known.
Figure 17.19 illustrates the distorting effect of amphetamines on the building of a spider's web.

Many drugs act synergistically, so that a combination produces greater or different effects than those of the same drugs taken separately.

Hallucinogens are drugs that alter thinking, self-awareness, emotion, space-time perception and the perception of incoming sensations. Marijuana, the most widely used of all hallucinogens, contains substances that affect transmission of nerve impulses in the brain. Long-term use may lead to loss of motivation and interest in daily life. LSD blocks nerve impulses and distorts sensory perceptions, producing hallucinations and mood swings. Peyote and psilocybin are naturally occurring hallucinogenic substances that have been used widely in religious ceremonies. The effects of all hallucinogens are related to the expectations of the user, the setting in which use takes place, and the potency of the drug.

4

CONCEPT REVIEW

1. How does long-term stress affect the body, and what can you do to reduce its effects?
2. What is meant by drug abuse?
3. Contrast physical and psychological dependence.
4. Compare and contrast the general effects of depressants, stimulants, and hallucinogens.

INVESTIGATION 17.1   Muscles and Muscle Fatigue

How do muscles move bones? Do muscles always work efficiently and at the same rate, or do they suffer from fatigue? In this investigation, you will observe the relationship between muscle movement, exercise, and muscle fatigue.

Materials (per team of 2)
watch or clock with a second hand
sheet of thin cardboard
scissors
string

brass brad
tape
metric ruler

PART A Muscles and Movement
Procedure

1. Using the materials provided and Figure 17.5 as a guide, construct a working model of the upper and lower arm that shows the attachment of the biceps and triceps. Use cardboard for bones, string for muscles, tape for tendons, and a brad for the elbow. (Hint: When cutting the string for the biceps and triceps muscles, be sure you have enough to compensate for the movement of the joint.)

2. Place the model on the table. Grasp the biceps string just below the upper arm attachment site. Gently pull the string. What do you observe? Release the string, but do not reposition the lower arm.

3. Now grasp the triceps string just below its upper attachment site and gently pull. What do you observe? How does this movement differ from that in Step 2?
   The only *direct* source of energy is ATP, some of which may be stored in myosin filaments.

Discussion

1. Many muscles in the body work in pairs. Some of these muscles are called flexors and extensors. How do such muscle pairs result in movement?

2. Is the biceps a flexor or an extensor? Is the triceps a flexor or an extensor?

3. What other joints of the body are moved by such muscle pairs?

4. Would movement be possible without flexors? Without extensors? Explain.

This investigation allows students to discover how muscles work in pairs to accomplish movement and acquaints them with muscle fatigue. Teams of 2 work best.
   Time: One class period.

PART A PROCEDURE

1. Figure T17.1 is one possible model that can be used as a guide. As shown in the figure, the biceps attaches lower down on the upper arm than does the triceps. The biceps also attaches toward the middle of the lower arm, whereas the triceps attaches on the underside of the elbow. Students probably will make the length of the tendons variable, but the string should be fastened so that it rides on or very near the thin edge of the cardboard. The strings should be cut long to allow for the movement of the joint. Help students position the tape fastening the distal end of the extensor so that it does not interfere with the movement of the elbow.
2. The joint flexes at the elbow.
3. The arm extends out straight. The biceps flexes the joint, and the triceps extends the joint.

DISCUSSION

1. The muscles work in opposing pairs. The contraction of one muscle bends, or flexes, a joint, contraction of the other straightens, or extends, the joint.
2. Biceps is a flexor; triceps is an extensor.
3. Answers will vary but may include knees, ankles, shoulders, and hips.
4. Skeletal movement would be impossible without opposing muscle pairs. There would be no force applied to the bones to make them move.

Figure T17.1   A possible model of the human arm.

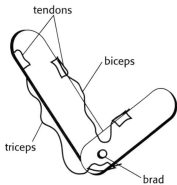

5. Movement is accomplished in the same manner, but the muscles are attached to the inside of the exoskeleton.

PART B PROCEDURE

4. **Help students get properly positioned. Be sure they grip their upper arms so they can feel movement in both flexor and extensor.**
5. **Both students should feel their flexors tighten more than the extensors.**
6. **Students who are right-handed probably will notice their left hand tires more quickly than the right. Conversely, left-handed students should notice their right hand tires more quickly than the left.**

DISCUSSION

1. **Answers will vary, but after a rest period fewer contractions may be possible.**
2. **Muscles in the hand and forearm tired.**
3. **Right-handed people usually have greater muscle development in their right hand and arm than in the left hand and arm. The opposite is true for left-handed people.**

This investigation supplements the text by allowing students to explore certain aspects of the senses of touch and sight and of the effects of the senses on response time. Parts A and C require teams of 2; students should perform part B individually.
Time: **One class period.**

5. How do animals with external skeletons, such as lobsters and beetles, move?

PART B Exercise and Muscle Fatigue Procedure

4. Rest your forearm on the table, palm up, and grasp your upper arm with your other hand. Have your partner rest his or her forearm on the table with the hand over your hand and wrist and grasp his or her own upper arm with the free hand.
5. Try to bend your arm up as your partner exerts pressure to keep your hand on the table. Describe what happens in your arm.
6. Switch roles with your partner and repeat Steps 4 and 5. Record any differences.
7. Make a tight fist with one hand, and then extend the fingers as far as you can. Do this as fast as possible for 30 seconds. Your partner should keep track of the elapsed time on a watch or clock with a second hand and the number of times you make a fist. Record the data. Do you feel any muscle fatigue? Where?
8. Rest one minute and repeat Step 7.
9. Repeat Steps 7 and 8 using your left hand. Compare the performance of your left hand with that of your right hand. Is there any difference?
10. Change roles with your partner and repeat Steps 7–9.

Discussion

1. How many times were you able to make a fist the first time? After the rest period?
2. What differences did you notice between when you started and when you finished? What might have caused those differences?
3. Account for any performance differences between your right and left hands.

INVESTIGATION 17.2   Sensory Receptors and Response to Stimuli

How do we know about the world in which we live? We receive information through receptors. The activities of these receptors, coordinated by the brain, are called the senses: touch, sight, hearing, smell, and taste. Sensory receptors respond to stimuli. In this investigation, you will test some aspects of the senses of touch and sight.

 In areas of the skin that are especially sensitive, are the touch receptors more numerous or simply more sensitive? Read the procedure for Part A, and then construct a hypothesis about the relationship between the sense of touch and the number and/or sensitivity of the touch receptors.

Materials (per team of 2)
centimeter ruler
probe or blunt dissecting needle

colored pencils (red, blue)
meter stick

PART A Touch
Procedure

1. In the center of a 4-cm square sheet of notebook paper, trace with a ruler a 1-cm square. Cut out the 1-cm square to form a hole in the larger square 1 cm² in area. You will use this paper template to find the number of touch receptors in 1 cm² of skin surface.
2. In your data book, trace six or seven of these templates in which to record data.
3. Work in pairs, with one partner applying the stimulus while the other reports the sensation felt.
4. With the subject's eyes closed, place the template on the underside of the subject's wrist. Using the probe, touch the skin *very lightly,* trying not to depress the skin surface. The subject should respond by saying yes whenever he or she feels the touch.
5. In one of the templates in your data book, map the subject's touch receptors,

using a red dot to represent a touch receptor (a yes response) and a blue dot to represent an area touched without sensation.

6. Try three different areas on the wrist. Record responses in the same way as in Step 5.

7. Proceed in a similar manner to locate touch receptors on the upper arm, back of the neck, and the fingertips. Record your data as before.

Discussion

1. What was the average number of touch receptors per cm_2 of wrist surface? What was the number on the upper arm surface? On the back of the neck surface? On the fingertip surface?

2. Do your data support or refute the hypothesis you developed before performing Part A?

3. Suggest other ways that you could test your hypothesis.

4. Which area had the most touch receptors? The fewest? Construct a hypothesis that accounts for the differences in receptor frequency in these various areas.

5. What is the relationship, if any, between hairs on the skin and touch receptors?

PART B Sight
Procedure

8. On a sheet of paper, place a cross (+) and a circle (O) 6.5 cm apart. Make the cross and circle about the size of the ring holes in your notebook paper. Enclose these marks within a 5-cm × 10-cm rectangle.

9. Working individually, hold the paper about 60 cm away from your eyes and close one eye. Slowly bring the paper closer while keeping your open eye fixed on the cross. In your data book, record your observations.

10. Draw a line from the cross almost to the circle, interrupting it in the vicinity of the circle and continuing it again on the other side of the circle.

11. Repeat Step 9. Record your observations.

Discussion

1. What happens to the circle and the cross as you bring the paper nearer your eye (Step 9)?

2. What happens to the line in the vicinity of the circle (Step 11)?

3. What do these experiments indicate about the sense of sight?

PART C Response to Stimuli
Procedure

12. The seated subject should rest an arm on the desk so the hand is extended past the edge of the desk. The last three fingers of the extended hand should be curled into a fist, leaving the thumb and forefinger free, about 3-4 cm apart, and parallel to the ground.

13. The experimenter should stand and hold a meter stick so its lowest number is between the thumb and forefinger of the subject.

14. Without any warning, the experimenter will drop the stick. The subject will try to catch it with just the thumb and forefinger of the extended hand.

15. Record the point on the meter stick at which the subject caught it.

16. Repeat the test four more times, recording this distance each time. Determine the average for the five trials.

17. Repeat the exercise while the subject counts by threes to 99. Average the time (distance on the meter stick) it takes the subject to catch the meter stick while counting.

18. Repeat the procedure while the subject's eyes are closed. The experimenter will snap his or her fingers to signal when the stick is released. Repeat four more times and calculate the average.

DISCUSSION

1. Variable, depending on student data.
2. Depends on student hypothesis.
3. Depends on student hypothesis.
4. The palmar surface of the fingers normally will have the most receptors (about 100 per cm_2) The back of the neck probably will have the fewest. The fingers are undoubtedly the structures of the body used most frequently in tactile reception and correlated with this is the presence of sensory receptors. Genetics and evolution are, of course, the determiners.
5. Most hairs seems to be associated with a touch receptor. The movement of the hair itself evokes the sensation of touch. These data support the fact that hairs is a tactile organ, as with cat's whiskers.

DISCUSSION

1. The circle disappears.
2. The line appears to be continuous, and the circle disappears.
3. The eyes are not equally discriminating at all distances and in all relationships.

DISCUSSION PART C

1. **The dropping of the stick is sensed by the eye. The message is relayed to the brain, which sends a message to the hand muscles, which contract and catch the meter stick.**
2. **Principally those of the eye.**
3. **Increases the time. The brain is distracted.**
4. **Much shorter with the eyes open.**
5. **Faster response rate when eyes rather than ears are used to detect the signal.**

This simulation allows students to experiment with some of the conditions that influence a long-distance bike trip and to develop and test hypotheses on trip strategies. Because one of the bike riders is an insulin-dependent diabetic, students can gain an understanding of the role of insulin in blood sugar regulation and of how a diabetic can live a normal life. Teams of 7 can control 2 riders. A team size of 4 is more desirable and may be achieved by staggering activities, but logistics depend on your class size and equipment. The simulation will accommodate 3 riders, but then the interactions become very complex.

Time: Three class periods. The first day, allow from 15 to 20 minutes to explain the bars on the screen and how to make pretrip selections from the menu. Then let the students at each computer experiment with one rider, using trial and error to discover strategies that can result in successful completion of the bike trip. The second day, allow about 30 minutes to discuss the homeostatic mechanisms involved in participating in a long distance bike trip and to help the students develop if/then hypotheses before they go to the computer to test their hypotheses. Use the third day for a wrap-up discussion and, if necessary, further computer time.

SAFETY

See the safety remarks accompanying Investigation 2.1 for working with the LEAP-System equipment.

Figure 17.20 Appearance of the computer screen for the bike trip with two riders, one across the top, the other across the bottom. From left to right, the bars show grade percentage, glucose demand, blood glucose, fluid demand, and body fluid level. The box to the left of the bars shows the gear, which is controlled by the dial.

Discussion

1. In the first test, describe the events in your body that enable you to respond and catch the meter stick.
2. What receptors are involved?
3. In the second test, what effect does thinking have on the response time?
4. Compare the response time in the first test with that in the third test. In which situation was response time shortest?
5. What can you conclude about the responses to different signals?

INVESTIGATION 17.3   A Bike Trip

During summer, an all-day bike trip can be fun, but it needs planning. In this computer simulation, planning is especially necessary for rider 1, an insulin-dependent diabetic. Among the factors to be considered are conditioning, diet, pretrip meals, breathing rate, heart rate, and levels of liver glycogen, blood glucose, body fluid, and blood lactic acid. These internal conditions are regulated by homeostatic mechanisms and vary from one person to another. This simulation considers only a few of these factors.

Figure 17.20 shows the conditions of the trips as they appear on the computer screen. The bike trip starts and ends at home; midway is a forest park where you can stop as long as you like. The temperature will rise to 95°F by midafternoon, and the trip covers hills as well as level stretches. The left bar on the screen shows the grade of the road at each point of the trip. Uphill is a positive percentage, with a maximum grade of 6 percent. Downhill is a negative percentage; a flat road shows as 0 percent. You control the speed of travel by using the dial to change the bicycle gears. This simulation assumes a constant rate of pedaling. The bikes can cover more distance in tenth gear but climb hills more easily in lower gears. You must watch the demand and level of both glucose and fluid. If either level drops too low, you will "crash." If one rider crashes, the other(s) must stop and the trip is over.

Cooperation is essential as you develop strategies and test your hypotheses about how to complete the bike trip, beginning with the pretrip selections—insulin dose for the diabetic rider, breakfast for all riders, and the food and drinks to pack in the saddlebags. Your task is to cooperate with your team in planning the best strategy for the trip. This is not a race—the goal is to complete the trip and return home.

Materials (per team of 7)

Apple IIe, Apple IIGS, Macintosh, or
   IBM computer

LEAP-System™
2 dials

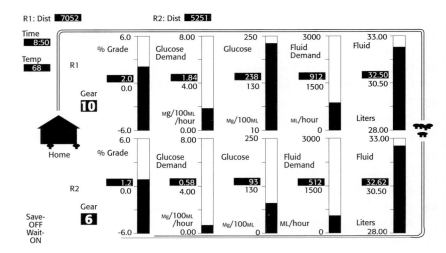

Table 17.3 Pretrip Selection Record

Menu: Biketrip			Rider Startup Information		
Trip No. 1 2			*Trip No.*		
Saddle bag weight allowed (grams)			R1 Insulin dose:		
Current saddle bag weight (grams)			R1 Breakfast:		
			R2 Breakfast:		
			R3 Breakfast:		
Meals/Drinks			**Snacks**		
Trip No.			*Trip No.*		
Peanut butter & jelly sandwich (120g)			Hard boiled egg (60g)		
Ham & cheese sandwich (9g)			Granola bar (30g)		
Tuna fish sandwich (90g)			Chocolate bar (45g)		
Pizza slice (120g)			Potato chips (30g)		
Hot dog in bun (90g)			Carrot sticks (30g)		
Water (250g)			Apple (120g)		
Sport drink (250g)			Orange (120g)		
Orange juice (250g)			Peanuts (30g)		
Milk (240g)			M & M's (45g)		
Diet soda pop (360g)			Canned pudding (120g)		
Soda pop (360g)			Cheese (30g)		
Vegetable juice (180g)					

PART A Practice Trip

Procedure

1. In each team, a group of three controls each rider, and one team member is the keyboard operator. Decide which group will control rider 1 (diabetic) and which rider 2. Each group includes one gear shift dial operator and two advisers—one to watch glucose demand and level, and one to watch fluid demand and level. Designate one of the advisers to be a recorder also. The dial operator changes gears according to the grade and input from the advisers. Each group must cooperate to enable its rider to complete the bicycle trip. Each group also must cooperate with the other rider's group, because if either rider crashes, both riders stop and you must start again with another strategy.

2. In your data book, prepare a pretrip selection record like Table 17.3 and a trip selection record like Table 17.4, or tape in the tables your teacher provides.

Table 17.4 Trip Selection Record

Trip No. _____

Rider No.	Stop No.	Distance Covered	Computer Time	Meals/Drinks/ Snacks Taken	Consequences/ Conclusions

Figure 17.21   Menu screen for the bike trip.

```
┌──────────────────────────────────┐   ┌──────────────────────────────────┐
│ Menu: BIKETRIP                    │   │     RIDER STARTUP INFORMATION     │
│ Saddle Bag Weight Allowed (grams) 5000│ │ R1 Insulin Dose:    Low         │
│ Current Saddle Bag Weight (grams) 1935│ │ R1 Breakfast:       Bacon and Eggs│
│                                   │   │ R2 Breakfast:       Captain Krunch│
│                                   │   │ R3 Breakfast:       Not going today│
└──────────────────────────────────┘   └──────────────────────────────────┘
```

MEALS/DRINKS			SNACKS	
Peanut Butter Jelly Sandwich (120g)	2		Hard Boiled Egg (60g)	0
Ham & Cheese Sandwich (90g)	0		Granola Bar (30g)	1
Tuna Fish Sandwich (90g)	0		Chocolate Bar (45g)	0
Pizza Slice (120g)	0		Potato Chips (30g)	0
Hot Dog in Bun (90g)	0		Carrot Sticks (30g)	0
Water (250g)	2		Apple (120g)	1
Gatorade (250g)	2		Orange (120g)	0
Orange Juice (250g)	2		Peanuts (30g)	0
Milk (240g)	0		M & M's (45g)	1
Diet Soda Pop (360g)	0		Canned Pudding (120g)	0
Soda Pop (360g)	0		Cheese (30g)	0
Vegetable Juice [V8] (180g)	0			

F10-Exit Menu   ↑ ←→ ↓   **Select**   **RET-Enter Box**

3. *For this practice trip, the team should work with the diabetic rider only.* Use the LEAP-System "run" procedure to select **BIKETRP1.SPU** for one rider. The menu shown in Figure 17.21 will appear on the screen.

4. Follow the prompts at the bottom of the screen to make pretrip decisions about insulin dose, breakfast, and what to pack in the saddlebags. Each rider is allowed 2500g. As you make selections, the computer displays total grams in the BIKETRIP box. If you overload the saddlebags, the computer will beep. Adjust the contents as needed. Record all selections in Table 17.3.

5. Follow the prompt on the screen to start the bike trip. To stop for a snack or a meal, press your rider number. A window will appear showing the items and the quantity of each remaining in the saddlebags. Follow the prompts to select items for your rider. Use Table 17.4 to record for each stop the rider number, stop number, distance covered, time from the screen, the snack/drink taken by the rider, and the resulting changes in glucose and/or fluid levels.

6. Watch the changing levels of glucose and fluid on the screen as the bike trip proceeds. Use the LEAP-System "reinitialize" procedure to restart the trip. The menu will reappear.

7. Experiment with different doses of insulin, with different breakfasts, and with different selections of food and drinks for the saddlebags. In your data book, record what problems you encounter and what steps you can take to overcome them.

8. Repeat the trip as often as time permits, recording the selections for each run.

PART B Planning the Trip

9. As a class, discuss the relationships between glucose, insulin, and exercise, and the effects of exercise on body fluid and fluid demand.

10. Decide on several trip strategies that account for all the variables you encountered in your practice trips. This could include not only what to eat for breakfast and what foods to take along but when to stop for food and/or drink. What assumptions are you making in your strategy? Record these assumptions in your data book. For example, if you stop for orange juice, what are you assuming about the effect of the orange juice on your ability to continue?

11. In your data book, write if/then hypotheses that include the strategies and assumptions from Step 10. The "if" part of each hypothesis should contain the conditions specified in the strategies. The "then" part of each hypothesis should contain the predicted outcome, assuming your conditions are met. For example, "If breakfast is skipped, then the rider will need to stop for a snack at 9:00 AM."

PART C The Trip

12. Start the trip as in Step 3, but select **BIKETRP2.5PU** for two riders or **BIKETRP3.SPU** for three riders. (All riders share the same saddlebags.) Follow your strategies carefully. Record any changes in strategies made during the trip. Record snacks/meals for all riders in the trip selection record.

13. Revise your strategies, if necessary. Develop a new hypothesis about how the trip will be completed, including insulin dosage, breakfast, and saddlebag contents. Record your hypothesis in your data book.

14. Use your new strategies and rerun the bike trip.

Discussion

1. Compare the results of your strategies. Were the prediction parts of your hypotheses correct? What changes did you make in the second trip?

2. Compare your strategies with those of other teams in the class. How were they similar? How were they different? Which strategies resulted in successful completion of the bike trip?

3. Describe the trip strategy that produced the most successful result in the bike trip. Tell why it was better than another strategy used.

4. What effect did the insulin dose have on rider 1's performance in the trip?

5. What effect did the food you selected for breakfast have on the outcome of the trip?

6. What effect did the type of food and drink you packed in the saddlebags have on the outcome of the trip? What effect did the time at which you stopped to eat or drink have?

7. Describe how cooperation between riders influences the outcome of the trip.

8. What effect does temperature have on the outcome of the trip? What body functions are influenced by temperature?

9. Now that you know the effects of selecting the proper food and drink for a bike trip, describe how what you eat and drink can be important in your everyday activities.

10. How does the diet of an insulin-dependent diabetic differ from the diet of another bike rider on the trip?

11. Write a short paragraph describing what you have learned about the type of activities in which an insulin-dependent diabetic can participate and what limitations diabetes can place on the person's life.

SUMMARY

The nervous and endocrine systems work in concert to bring about movement and homeostasis of internal systems. Muscle contractions are initiated by nerve impulses triggered by external and internal signals; bones, joints, and muscles work together to bring about the desired movements. A nerve impulse is a wave of chemical and electrical changes that travels over the length of a neuron, crossing the synapse between neurons by means of specialized neurotransmitters. Interpretation and coordination of nerve impulses resides in the central nervous system, which receives information about the external and internal environments through sensory neurons and sends messages about the action to be taken through motor neurons to muscles and endocrine glands. Most integration occurs in the brain, which is the seat of conscious activities such as speech, learning, memory, and the senses, as well as unconscious activities such as breathing and heartbeat rates. In addition to carrying information to and from the brain, the spinal cord integrates reflexes such as withdrawal from a pinprick.

DISCUSSION

1. Answers will vary, but improved strategies should result in completion of the trip.

2. Answers will vary, but good strategies will include a low insulin dose for rider 1, high carbohydrate meals and snacks, plenty of liquids, and timely stops to replenish fluid and glucose levels.

3. Answers will vary.

4. No insulin allows blood glucose to rise too high (because the glucose cannot enter the muscle cells); high blood glucose in turn increases fluid demand. A medium or high dose in combination with exercise causes blood glucose to drop too low (because too much glucose enters the muscle cells), and the rider will crash unless timely snacks are eaten to replenish the blood glucose.

5. Foods with high carbohydrate values will yield better results than those with low carbohydrate values.

6. Foods and drinks with high carbohydrate values will yield the best results. Responses to the question about timing of stops will vary, but usually the stops will be necessary before the glucose and/or fluid levels drop significantly, because of the time necessary for glucose to get from the digestive tract into the bloodstream (see Table T17.1 in Teaching Strategies by Chapter).

7. Unless rider groups cooperate, it is unlikely they can complete the trip; close cooperation should lead to improved performance.

8. The higher the temperature, the greater the fluid demand. Heart rate, breathing rate, metabolic rate.

9. Answers will vary, but there should be a noticeable emphasis on proper diet.

10. With a proper insulin dose, the diabetic rider's diet would be very similar to that of the nondiabetic rider. Both should be high in carbohydrates to provide energy for the trip.

11. Responses will vary, but it should be clear to the students that insulin-dependent diabetics can lead perfectly normal lives if they adhere to their diet and insulin administration.

FOR FURTHER INVESTIGATION

For a more thorough analysis of various strategies, save the data and run stripcharts. The marks on the stripchart indicate when snacks/drinks were taken and can help you understand the relationship between glucose and fluid levels and timing of snacks/drinks.

The hypothalamus integrates the network of endocrine glands that constitutes the endocrine system, which secretes hormones directly into the bloodstream. The hypothalamus secretes hormones that control the secretions of the pituitary and other endocrine glands through negative feedback mechanisms. Based on messages from sensory receptors throughout the body, the hypothalamus influences a variety of body functions and is especially important in the fight-or-flight response. Stress is the sum of the physiological changes that occur in response to stressors, which may be physical and/or emotional. Psychoactive drugs, which affect the central nervous system, cannot cure the causes of stress. Long periods of sustained stress do major harm to the body, so learning to manage stress is an important skill.

APPLICATIONS

1. When and how would food eaten 24 hours before exertion affect the amount of energy available for the muscles? What would be the effect of food eaten 4 hours before?

2. Explain the action of the muscles of the neck as you bring your left ear to your left shoulder and then move the right ear to the right shoulder.

3. Do all organisms need muscles to move? Explain your answer.

4. Why do muscles get tired and cramped? Relate physical activity to the physiology of the muscle.

5. A thermostat in your house works on a feedback system. Explain how this feedback system works, and how it resembles several systems in your body.

6. How do the structures of the eye relate to color blindness?

7. Do you get more muscle fibers when you pump iron, or do your muscle fibers just get bigger?

8. Beginning with a stimulus and ending with a movement, place the following in order: synapse, effector, receptor, sensory neuron, motor neuron, interneuron, muscle.

9. Drinking too much coffee, tea, or cola may cause the heart to skip a beat or muscles to twitch involuntarily. What is the explanation?

10. Describe how a pituitary tumor in a teenager could affect the onset of sexual maturity.

11. Select one endocrine gland and describe how a negative feedback loop works with the secretion of that gland and the hypothalamus and pituitary gland.

12. Do you have the same kind of stressors your parents have? How are they similar, and how are they different?

13. How does being able to express your feelings to someone help relieve stress symptoms?

14. Explain how having a hobby outside of an occupation can help reduce the effects of long term stress.

15. What would be the effects of a drug that heightens perception but depresses the function of the hypothalamus? How would it affect the performance of a high-wire walker, a musician, or a basketball player?

PROBLEMS

1. Make a working model of an arm (different from the one you made in Investigation 17.1) showing the relationships between bones, muscles, and joints.

2. Compare the skeletons of several different types of animals, including birds, four-legged animals,

and snakes. Can you determine where the muscles were attached to the bones or how the bones helped each animal move?

3. Make a graph showing the fastest recorded times in the marathon through history and another graph

of the fastest recorded times in the 100-meter dash. What is the trend in times? Do you think there are any limits to how fast people can run?

4. Research any one physical fitness fad, such as jogging, racketball, or aerobics, and determine its relative value for cardiovascular fitness.

5. Invite a speaker from Al-anon or Alateen to speak to your class about alcohol dependence.

6. What are some different methods to help people to stop smoking? Why do you think some methods are successful and others are not? Develop a program to be used in your school involving teachers, students, staff, and administrators to help people quit smoking. Challenge each other to stop.

7. Biofeedback has been used to control functions such as blood pressure and pulse. Research the validity of claims to control autonomic or subconscious functions.

8. In 1986, two young, strong, apparently well conditioned athletes died from heart attacks caused by cocaine. What might be the physiological mechanism?

9. Study the use of hormones in the raising of beef cattle, poultry, and dairy animals, as well as in the training of athletes.

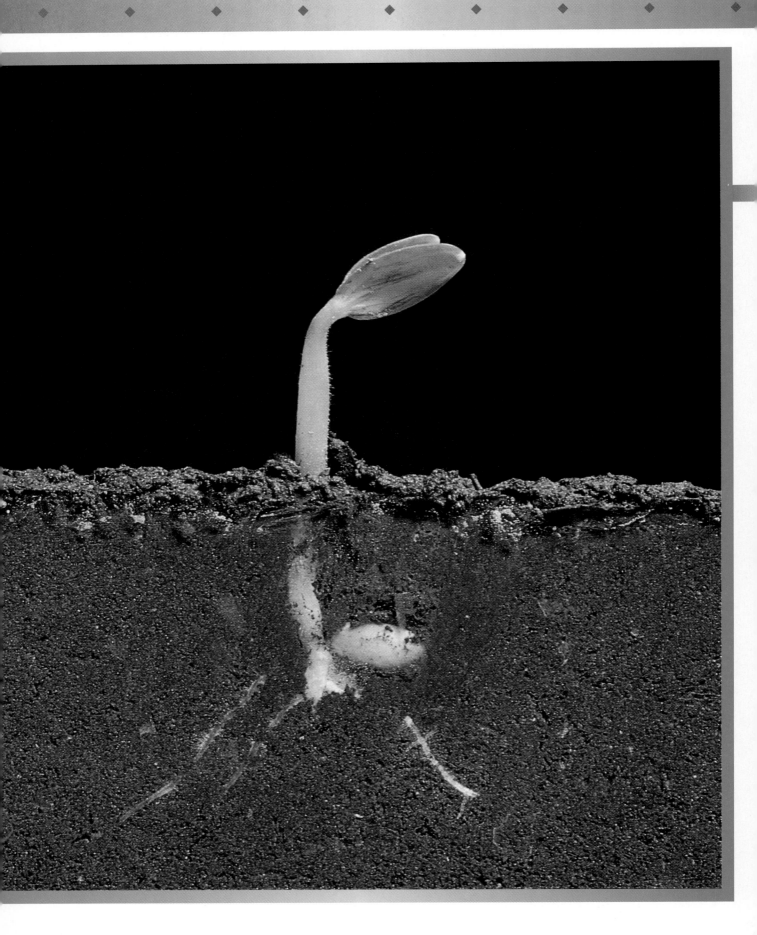

The Flowering Plant: Form and Function

How would you design an experiment to learn if this young cantaloupe seedling is responding to a source of light? What other forces are acting on the growth of this plant?

The leaves appear to be growing toward the sunlight, and roots appear to be growing toward the pull of gravity. An experiment in which the direction of the light and gravitational pull are varied would be interesting, as would attempts to make the plant grow upside down. The osmotic force of water against the cell walls makes the cells rigid.

◆ Most plants live in two different environments at the same time—air and soil. Leaves and stems usually are above ground, where they are surrounded by air and where light, carbon dioxide, and oxygen are available. Roots are in the soil, which supplies water and nutrients. Each part of the plant is adapted for its environment and for a specific function in the life of the entire plant. Most leaves carry out photosynthesis, stems support the leaves and transport materials to other plant parts, and roots anchor the plant and absorb water and soil nutrients. The plant uses sugars produced during photosynthesis to build its structure.

How is the plant's structure produced, and how does each part function? How are the activities of all the parts coordinated as a whole? How are the various parts adapted for their environments? This and the next chapter begin to answer these questions for flowering plants, the most familiar and important plants for humans. ◆

MAJOR CONCEPTS
- ◆ The basic organs of a vascular plant are leaves, stems, and roots.
- ◆ The structure of plant cells influences their function.
- ◆ Water loss from the leaf is controlled by guard cells.
- ◆ Leaves, stems, and roots can have a variety of functions.
- ◆ Movements of materials throughout the plant is due to a combination of forces.
- ◆ Meristematic tissue is responsible for vascular plant growth.

KNOWLEDGE CHECK
- ◆ How does leaf structure support leaf function?
- ◆ What prevents excessive water loss in plants?
- ◆ What are the functions of stems and roots?
- ◆ How does a plant grow?

Leaf Structure

18.1 Leaf Structure Allows for Movement of Carbon Dioxide

Usually the main function of a leaf is to produce food through photosynthesis. How is its structure adapted for that function? The leaf of a flowering plant consists of an enlarged flat portion, the blade, connected to the plant stem by a petiole (PET ee ohl). The blade may be in one piece or divided into separate leaflets (a compound leaf), as shown in Figure 18.1. The large surface area of the blade permits maximum exposure to light.

Figure 18.2 illustrates the cellular structure of a leaf blade, which is made of several different tissues. Covering both the upper and the lower surfaces of the leaf is a transparent layer of cells called the epidermis.

> **Guidepost**
>
> How are the parts of a leaf adapted for photosynthesis?

Ask students to name plants with compound leaves. Examples are clover, locust, ash, hickory.

a

b

Figure 18.1 Simple leaves (a) and a compound leaf (b). How does leaf structure enhance a plant's ability to carry out photosynthesis? **The broad, flat surface provides maximum exposure to the sun.**

463

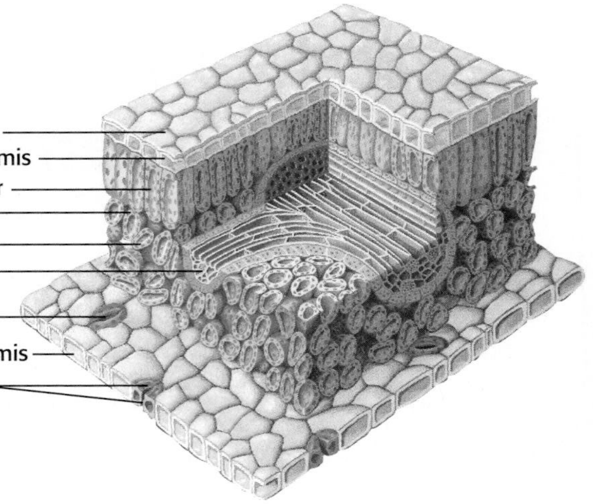

Figure 18.2 shows internal structures of a leaf with labels:
- waxy cuticle
- upper epidermis
- mesophyll
  - palisade layer
  - spongy layer
- air space
- vein
- stomate
- lower epidermis
- guard cells

Figure 18.2 A portion of a leaf blade showing internal structures. Colors are diagrammatic only.

Ask students to construct hypotheses about the functions of the two types of mesophyll cells. Their hypotheses should be based on the generalization that structures reflect the functions they perform. The generalization is related to the concept of adaptation stressed in Section Three, Diversity and Adaptation in the Biosphere, and throughout this chapter.

Figure 18.3 Scanning electron micrograph of a leaf cross section. Note that air can move freely throughout the interior of the leaf.

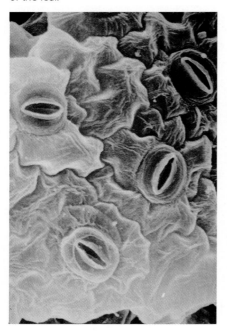

Epidermal cells secrete a waxy substance that forms a waterproof coating, the cuticle. Photosynthesis occurs in the **mesophyll** (MEZ oh fil) cells, which contain numerous chloroplasts. There are two types of mesophyll cells. In the upper, palisade layer, the mesophyll cells are elongated and packed together tightly. This space-saving arrangement exposes a maximum number of cells to light in a minimum amount of space. In the lower, spongy layer, the cells are rounder and packed loosely with many empty spaces between them. These empty spaces, which are clearly visible in the scanning electron micrograph in Figure 18.3, allow gases to move around freely and diffuse into all the mesophyll cells.   1

Substances enter and leave the leaf by two different routes: veins and stomates. Veins, which are continuations of the vascular tissues of the stem and root, supply the leaf cells with water and nutrients. Gases move into and out of a leaf by diffusion through millions of stomates (slitlike openings) in the leaf surfaces (see Figure 18.2). Because many of the stomates open into the spongy mesophyll's air spaces, the carbon dioxide used in photosynthesis and the oxygen used in respiration can reach all the cells of the leaf.

The concentration of carbon dioxide in the air is relatively constant. Within the leaf, however, the concentration varies. Cellular respiration goes on at all times in leaf cells, producing carbon dioxide and water as by-products. During the day, mesophyll cells use the carbon dioxide in photosynthesis. As a result, the concentration of carbon dioxide becomes lower in the leaf than it is in the air, and carbon dioxide diffuses in through   2
the stomates. At night, photosynthesis does not occur. As a result of continuous respiration, the level of carbon dioxide builds up in the leaf. Eventually, it becomes higher than the level of carbon dioxide in the air and diffuses out.

18.2   Guard Cells Regulate the Rate of Water Loss

Although water molecules are present in the air as water vapor, they are never as abundant in the air as they are in the leaf. Thus, the plant loses water as it diffuses into the air through the stomates. This water loss is called **transpiration** (trans pih RAY shun).

When plant cells have adequate supplies of water, the water exerts a pressure, known as turgor (TER ger) pressure, against the cell walls.   3

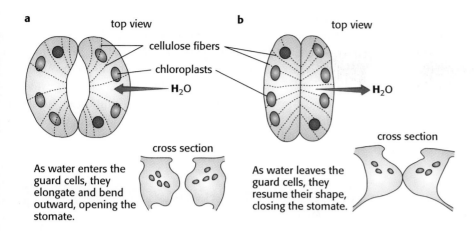

Figure 18.4   Guard cells control the opening of stomates. Because they are attached to each other at both ends and encircled by rigid cellulose fibers, the guard cells elongate and bend outward when they take up water.

Turgor pressure supports the stem and leaves. If more water is lost from a plant by transpiration than is replaced through the roots, the cells lose turgor pressure. As a result, the stem and leaves no longer are held upright and the plant wilts. If water is not replaced, the plant cells, and the plant, will die.

How can a plant conserve water and still allow carbon dioxide to enter? Each stomate is surrounded by a pair of specialized cells called guard cells. When the guard cells fill with water, they bend outward, as shown in Figure 18.4a. As they bend, the stomate opens and carbon dioxide can diffuse into the leaf. When a plant loses more water than it 4 can replace, the turgor pressure of the guard cells decreases. The guard cells no longer bend outward and as a result, the stomates close (Figure 18.4b). In this manner, the plan is able to reduce the loss of water under dry conditions. During these times however, little photosynthesis takes place.

18.3   Some Leaves Have Functions Other Than Photosynthesis

Although the major function of most leaves is photosynthesis, many leaves photosynthesize little, if at all. These leaves have other functions that are adapted to the environmental conditions in which the plants live. For example, the spines on spurge (Figure 18.5) and on cacti are modified leaves that are hard nonphotosynthetic, and often dangerous to touch. 5 Because water loss occurs mostly from leaves, plants with smaller leaves (and, therefore, less surface area) lose less water than do plants with larger leaves. In the dry habitats in which cacti are found, spines reduce water loss, and they help to protect the cactus from herbivores. In cacti, photosynthesis takes place in the stem rather than the spines.

Other plants in dry habitats have modified leaves that store water. These plants, known as succulents (SUK yoo lents) have tissues that are so juicy that water can be squeezed from them. After a rainstorm, the roots of succulents quickly absorb more water than the plant can use immediately. The water is stored in the leaves of succulents (and in the stems of cacti) until the plant needs it. A waxy cuticle so thick it can be scraped off with a fingernail also helps the leaves to hold water. **Some spines are modified stems, such as those of the hawthorn, *Crataegus,* or honey locust, *Gleditsia*.**

Ask students to consider the effects of temperature, wind, and rain on the rate of transpiration.

Bending of guard is thought to be related to the concentration of carbon dioxide dissolved in the cells. Respiration and photosynthesis both play an important role in the carbon dioxide supply. When carbon dioxide accumulates, the pH drops, favoring the formation of starch from soluble sugar, reducing turgor pressure, and resulting in narrowing of the stomate. When a leaf is illuminated, photosynthesis uses up carbon dioxide and the pH rises, favoring hydrolysis of starch. Water then enters the guard cells, increasing turgor pressure, and the stomate widens.

Ask students to explain the low rate of photosynthesis when stomates are closed.

Figure 18.5   Members of the spurge family have spines that actually are modified leaves. How might this modification be adaptive? **The reduced surface area reduces water loss in a dry environment.**

Plants living in dry environments have many adaptations that reduce transpiration. Among them are shedding leaves in dry seasons; thickened cuticle; gray color, which reduces heating by sunlight; and hairs that reduce passage of air across the surface.

Although bogs are watery habitats with soil that often lacks nitrogen, most bogs are populated with great numbers of plants—including insectivorous plants. These plants also have leaves adapted for their environment. Venus flytrap, the sundew, and the pitcher plant all have modified leaves that help attract and capture insects. The insects are digested within the leaf, and their nutrients, including nitrogen-containing compounds, are absorbed by the plant. In this way, insectivorous plants can live in the nitrogen-poor soils of bogs.

6

CONCEPT REVIEW

1. How is the arrangement of the two types of mesophyll cells in the leaf important to photosynthesis?
2. Explain why carbon dioxide moves into a leaf during the day.
3. How does turgor pressure provide support for a leaf?
4. How do guard cells function to prevent water loss through the stomates?
5. What advantage do small leaves have over large leaves in a dry habitat?
6. How do leaves of insectivorous plants help them live in nitrogen-poor soils?

Stems and Conduction

18.4   Stems Support Leaves and Conduct Materials

Guidepost

How are stems adapted for conduction and support?

Some plants are almost stemless. The leaves of dandelions, for example, spread out on the ground. They appear to grow directly from the top of the root.

Provide a number of specimens of dormant woody stems for comparison with the illustration. Collect the twigs before they begin developing and store them in a refrigerator. Bud packing is visible in the end bud of horse chestnut when opened with a dissecting needle.

The plant stem connects the leaves and roots and supports the leaves and reproductive organs in the light and air. Stems also carry water and nutrients from the roots to the leaves and sugars from the leaves to nonphotosynthetic parts of the plant. Air diffuses into stems through lenticels (LENT ih selz), small openings on the stem's surface, thus supplying oxygen to stem cells for cellular respiration. Stems may be either herbaceous or woody. Herbaceous stems are rather soft and rely on turgor pressure for support. Woody stems, on the other hand, have cells with thick, stiff walls that provide support.

Although stems differ greatly in structure, all have buds. A bud is a miniature shoot with a short stem, small immature leaves, and sometimes flowers. Buds on the bare twigs of deciduous trees are quite conspicuous, as shown in Figure 18.6. Growth from a terminal (tip) bud lengthens the twig. Growth from a lateral (side) bud starts a new branch. During the winter, a bud usually is covered with tough, protective scales that are modified leaves. Bud scales fall off when growth resumes in the spring, leaving a ring of scars where they were attached. Each ring of scars indicates where growth began for that year. Because these bud scales are formed only in the fall, the age of a branch can be determined from the number of scars.

1

Some plants have stems modified for storage. White potatoes are underground stems modified for storage; their "eyes" are buds. (Sweet potatoes, which do not have buds, are roots modified for storage.) An onion bulb is a small stem with modified leaves attached. The upper part of each leaf is green and produces sugars through photosynthesis. The bottom part of each leaf is white, lies underground, and stores the sugars in the form of starch. Figure 18.7 illustrates different types of underground stems. Most stems, however, grow above ground.

2

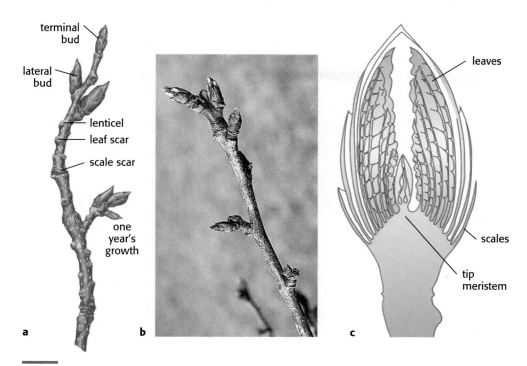

Figure 18.6 A diagram (a) and photograph (b) of a dormant woody twig, and a diagram of a longitudinal section of a terminal bud (c).

Figure 18.8 on page 468 shows the microscopic structure of a young wood dicot stem in three dimensions. The stem is composed of several tissues. Around the outside is the epidermis, a single, protective layer of cells that will be replaced by bark as the stem grows older. Just inside the epidermis is a layer of thin-walled cortex cells that carry on photosynthesis or function as storage areas, depending on the stem. Next to the cortex are the vascular tissues. **Phloem** (FLOH em) is the vascular tissue that transports sap containing dissolved sugars and amino acids from the leaves throughout the plant. **Xylem** (ZY lem) is the vascular tissue that helps

3 support the plant and transports water and dissolved nutrients from the roots throughout the plant. Figure 18.9 on page 467 illustrates these two tissues. At the center of a young woody stem there may be a group of loosely collected cells called the pith. The pith stores food in young stems and usually is destroyed as the stem grows older.

Figure 18.7   Structural adaptations of underground stems. What do you think is the principal function of each type? **Food storage and reproduction.**

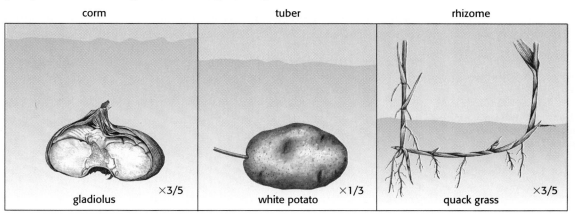

Figure 18.8 Diagram of a young woody stem showing the microscopic structures. Colors are diagrammatic only.
The color coding of phloem and xylem is as in Figures 18.2 and 18.13, but the cortex is green here because the stem cortex cells may contain chlorophyll and be photosynthetic.

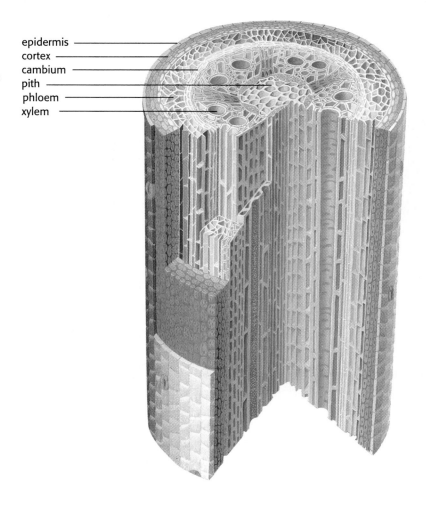

epidermis
cortex
cambium
pith
phloem
xylem

Phloem is composed of tubes formed from sieve cells. As each sieve cell develops, many small holes form in both end walls. The cytoplasm of one cell connects with the cytoplasm of adjacent cells through these holes, making a continuous cell-to-cell system of sieve tubes. As a sieve cell matures, its nucleus disintegrates. Located beside each sieve cell is a smaller, companion cell that has a nucleus. Companion cells appear to regulate the cell activity of sieve cells. Figure 18.10a illustrates the structure of phloem cells.

Lying between the phloem and xylem cells is the vascular **cambium** (KAM bee um). The cells of this tissue divide by mitosis and form other plant tissues. When a cambium cell divides into two new cells, one of these cells eventually becomes either a conducting or a supporting cell. The other cell remains a part of the cambium tissue and divides again. Phloem cells are formed continually at the outer surface of the cambium;

Figure 18.9 Stained longitudinal section of xylem tissue (a) and phloem tissue (b). How do these tissues help a plant survive on land? **Xylem transports water from the soil throughout the plant; phloem transports sugars to all nonphotosynthetic parts of the plant.**

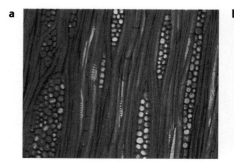

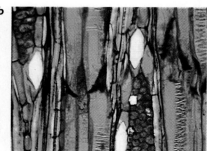

xylem cells are formed at its inner surface. The cambium also forms fiber cells, which are interspersed with the vascular xylem and phloem and strengthen the stem.

Eventually, most of a woody stem consists of xylem. Wood, which is found in older stems, is made completely of xylem cells. Xylem usually is composed of two types of cells: tracheids (TRAY kee idz) and vessels. Tracheids are long cells with thick walls that grow to their full size and then die. The walls of these dead cells contain thin areas called pits. Pits in one tracheid are lined up with those in adjacent tracheids, allowing water and dissolved substances to pass easily from one to another. Vessels are made of long thick-walled cells joined end to end to form miniature pipes that extend throughout the stem. The end walls of vessel cells disintegrate, and they, like tracheids, die. Both tracheids and vessels are arranged to form a continuous conducting system from the root, up through the stem, and out into the veins of the leaf. Figure 18.10b illustrates both types of xylem cells. You will be able to observe these structures in Investigation 18.2.

18.5 Pressure Within Phloem Cells Helps Move Sugars Down a Plant

Sugars move from the leaves to roots and other nonphotosynthetic parts of the plant through the phloem cells. Phloem cells contain living cytoplasm through which water and dissolved sugars must pass. The rate at which the fluid moves through phloem is thousands of times faster than diffusion alone could account for. What is the mechanism of phloem transport?

The best explanation of the movement of sugars through the phloem is the **pressure-flow** hypothesis. According to this hypothesis, water and dissolved sugars flow through the sieve tubes from an area of higher pressure to an area of lower pressure, as shown in Figure 18.11. Sugars secreted by mesophyll cells in the leaf move by active transport into the phloem cells. The resulting high concentration of sugar causes water to diffuse into phloem cells, increasing the turgor pressure there. This area of higher pressure forces the sugar-water solution to move into the next phloem cell. In this manner, sugars are moved from cell to cell—from higher-pressure area to lower-pressure area—until they reach a cell where they will be used or stored. There, sugars are removed from the phloem by active transport. At this time, water also leaves the phloem cell. Thus, the flow of sugar and water from the leaves to the root can continue. The entire process is dependent on an uptake of water by phloem cells in the leaves and a loss of water by phloem cells in the storage areas or in areas of rapid growth.

18.6 Transpiration Pulls Water Up Through the Xylem of a Plant

Water must travel more than 100 m to reach the top of the tallest trees. Many experiments have shown that water from roots rises through root xylem to stem xylem to the xylem in the veins of leaves. This movement is against the force of gravity. Imagine you are standing with a long soda straw on top of a building three stories tall. The straw reaches into a bottle of root beer on the ground. No matter how hard you try, you cannot suck the root beer up from the bottle, not even with a vacuum pump. How do water and materials dissolved in the water move to the top of a tree?

Water is thought to move by a process known as **cohesion-tension,** illustrated in Figure 18.12 on page 470. Under certain conditions, water can be pulled up a narrow tube if the water is in a continuous column. One

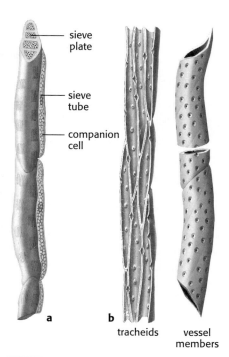

siee
plate

sieve
tube

companion
cell

a   b

tracheids   vessel
members

Figure 18.10 Cells from vascular tissues of flowering plants. Phloem cells (a) and xylem cells (b). What adaptations do these cells demonstrate? **Perforations (xylem) and sieve plates (phloem) in the end walls allow for formation of continuous tubes.**

Figure 18.11 The pressure-flow hypothesis, based on fluid pressure, for transport of food and water in phloem.

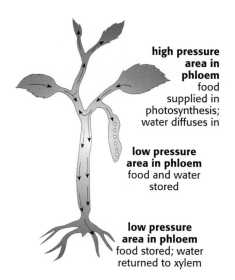

high pressure area in phloem
food supplied in photosynthesis; water diffuses in

low pressure area in phloem
food and water stored

low pressure area in phloem
food stored; water returned to xylem

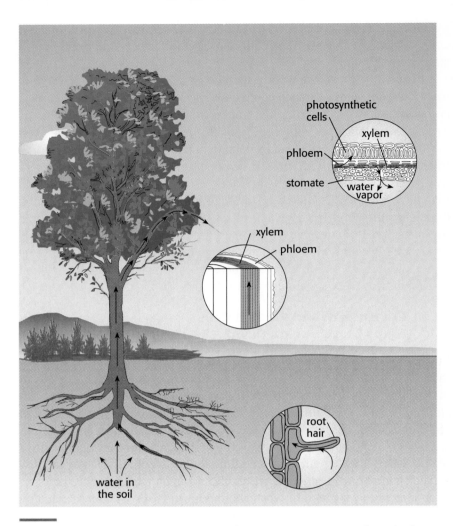

Figure 18.12 The principle water-conducting structures in a tree. Trace the path of water from the root hairs to the leaves.

To demonstrate cohesion, have students place several drops of water on a glass microscope slide, place a second slide on top, and try to pull the slides straight apart.

As water evaporates from the leaves, the concentration of dissolved materials in the mesophyll cells increases. Water molecules from the xylem then diffuse into these cells.

condition is that the tube must have a very small diameter. A second condition is that the tube must be made of a material to which water molecules will adhere. These two conditions exist within the xylem cells of plants. In addition, water molecules exhibit cohesion—they are attracted to adjoining water molecules, thus forming an unbroken chain of water molecules in each xylem tube. Tension is negative pressure that results from the pulling action of water molecules as they move out of the leaves during transpiration. The pull is transmitted through the column, and water is pulled from the roots up the plant to the leaves. Transpiration is the driving force of water transport, creating the negative pressure that keeps water moving from the soil into the roots and up through the stems into each leaf.

CONCEPT REVIEW

1. How are buds involved in stem growth?
2. In what way are leaves of succulents and leaves of onions alike?
3. Compare and contrast the structure and function of xylem and phloem.
4. How does the pressure-flow hypothesis explain the movement of sugars in the phloem?
5. Describe the roles of cohesion and transpiration in water movement.

Biology Today

City Forester

Ron Cousar is the City Forester for the City of Colorado Springs; a job that consists of caring for and maintaining for the estimated 70 million dollar urban forest resource. This includes trees along streets, as well as in neighborhood and regional parks. In addition, Ron also assists citizens with specific care concerns of trees on their private properties.

Ron has pursued and obtained a Bachelor's degree in Natural Resource Conservation, Urban Forestry and Parks Management, and has spent 15 years managing the urban forests for three major cities. During his tenure in Colorado Springs, Ron has implemented new legislation to regulate tree service licensing for those businesses that operate in the city, in addition to structuring an Urban Forestry Plan and providing citizens with an outlined tree planting matrix and tree care booklet. Ron has also initiated the Significant Tree Program. This program actually provides a vehicle to acknowledge and nominate those unique and historically significant trees in the community. City agencies such as Planning, Zoning, and Engineering are kept abreast of the trees that qualify under this

programs criteria in order to ensure tree preservation tactics are included in future planning schemes.

The care of trees involves both maintenance and disease control. Recently pruned trees are more vigorous and less susceptible than untrimmed trees to damage from wind and snow loads. It is also necessary to ensure that the trees do not obstruct street signs or interfere with power lines. About 80,000 trees along the streets of Colorado Springs, and 20,000 trees in the parks are individually maintained. If diseased trees are found (and pesticide spraying is ineffective) the trees (both living and dead) must be cut down and buried. Any trees that die naturally are examined for signs of wildlife nesting. If such signs are discovered and the tree is not a hazard to the life of the property, it is tagged as a "wildlife tree" and permitted to remain in place.

Every tree on city property is recorded on a computer. Information about the type of tree—its height, diameter, condition, and value—is readily available. The computer also stores information regarding the history of the tree and any treatment it has received.

Under Ron's supervision are fifteen full-time and several seasonal employees who comprise the Forestry Division. As the fifth Urban Forester for Colorado Springs, Ron is continuing the dream of the City's founder, General William Jackson Palmer, by providing tree-lined streets as this magnificent city continues to grow. Ron understands that the urban forest requires care and maintenance if we are to continue to be recipients of the precious benefits it offers, such as shade, oxygen production, air filtration, increased property value, and aesthetic value.

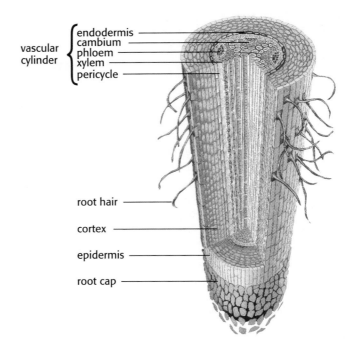

Figure 18.13 The terminal portion of a root. Colors are diagrammatic only.

vascular cylinder { endodermis
cambium
phloem
xylem
pericycle

root hair

cortex

epidermis

root cap

Some of the labeled structures in Figure 18.13 are mentioned briefly or not at all in the text. Students with questions about these structures may be referred to more advanced books. You may want to introduce *pericycle* and *endodermis* if you explain the origin of secondary roots.

Students may suggest that roots also function as reproductive structures. This is true of stems and leaves as well.

Guidepost

How are roots adapted for absorption of water and nutrients?

Ask students to what ecological conditions these two types of root systems are adapted. Judge their responses on reasonableness, rather than on correspondence to information students may not have. In general, taproots provide better access to water when surface layers of the soil dry out in temporary droughts. Some desert plants such as cacti, however, have wide-spreading, shallow, fibrous roots systems. These quickly absorb water from erratic rains that often fail to penetrate the soil. The water then is stored in the plant.

Roots and Absorption

18.7   Roots Anchor a Plant and Absorb Materials

Roots anchor plants and help absorb water and nutrients. They also conduct materials and serve as storage organs. The vascular tissues of many roots form a cylinder in the center that is surrounded by the cortex, which may form part of the bark around an older, woody root. Sometimes food in the form of starch or sugars is stored in the cortex. Carrots, sugar beets, radishes, and turnips are examples of storage roots. Figure 18.13 shows the microscopic structure of the tip of a root.

In some species of plants, many branching roots come from the bottom of the plant. This is known as a fibrous root system, characteristic of such plants as corn, beans, and clover. Fibrous root systems, which are relatively close to the soil surface, hold soil in place and thus reduce erosion. In other species, the plant is anchored by a long, tapering root with slender, short, side branches. This is a taproot system, found in such plants as dandelions, mesquite, and carrots. Because their roots penetrate deep into the soil, plants with taproots can use water sources unavailable to plants with fibrous root systems. Figure 18.14 shows a variety of roots.

Figure 18.14 Roots are of two basic types, both of which may be adapted for storage.

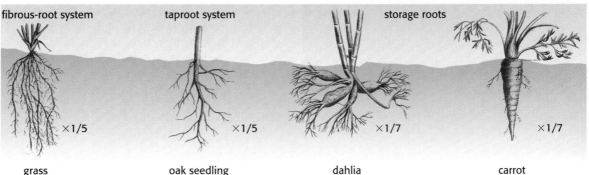

fibrous-root system   taproot system   storage roots

×1/5   ×1/5   ×1/7   ×1/7

grass   oak seedling   dahlia   carrot

Most water absorption occurs through **root hairs,** which penetrate the spaces between soil particles and absorb through their huge surface area the water and nutrients required by a plant. Each root hair is a thin-walled extension of a single epidermal cell. In Figure 18.15 the root hairs appear as a fuzzy white zone just behind the tip of the seedling root. Because of its root hairs, a single rye plant only 60 cm tall is estimated to have a root system with a total length of 480 km and a total surface area of more than 600 m²—twice that of a tennis court.

If a plant is pulled out of the soil roughly, most of its root system may remain behind. Plants often wilt after being transplanted because most of the root hairs have been pulled off during the transplant, and without them the plant cannot absorb enough water to prevent wilting.

2   The cell walls of root hairs absorb water from the soil, much as paper towels soak up liquid spills. Water diffuses into the inner part of the root hair cell from the cell walls. Diffusing from cell to cell, the water eventually makes its way to a xylem cell. Here, the water begins its upward journey to the leaves of the plant, pulled by transpiration, as described in Section 18.6.

Figure 18.15 A radish seedling showing root hairs. How are root hairs adaptive?

By providing an enormous surface area, they increase water absorption.

Have students suggest how the structure of the root hairs and the other cells of the root fits the function each performs.

18.8   Many Nutrients Move into a Plant by Active Transport

In addition to water, a plant requires a variety of elements for growth and maintenance. These nutrients enter the roots as ions dissolved in the soil water. Roots take in nitrogen, for example, in the form of ammonium or nitrate ions. Elements that a plant needs in relatively large amounts are called macronutrients. The most important of these macronutrients are nitrogen, phosphorus, and potassium. These three ingredients are identi-

3   fied by numbers such as 10-20-10 on commercial fertilizers, indicating percentage of the nutrient by weight. Elements such as zinc and copper, which are required in extremely small amounts, are called micronutrients. The essential elements have been identified by experiments in which plants were grown in solutions lacking a particular element (see Figure 18.16). These experiments have shown how the elements are absorbed and used by the plant, and what effect is produced when a particular element is missing. Table 18.1 summarizes that information.

Nutrients usually are more concentrated in root hairs than in soil water. Absorption of these nutrients involves active transport, which requires deficient plants. If root cells are deprived of oxygen, and thus

4   are unable to respire, the absorption of nutrients slows. Absorbed nutrients move deeper into the root by active transport. Eventually, nutrients reach the xylem cells, through which they are conducted to the stem and leaves.

Figure 18.16 Tomato plants showing nutrient deficiency. From left: potassium, phosphorus, and nitrogen deficient plants. The plant on the right received all essential elements.

CONCEPT REVIEW

1. Compare and contrast the structures of fibrous root and taproot systems.
2. Explain the importance of surface area in the role played by root hairs.
3. Distinguish between macronutrients and micronutrients.
4. What is the evidence that nutrients move into and through the root by active transport?

Table 18.1 Elements Required by Plants

Element	Function	Deficiency
Macronutrients		
Calcium	Component of cell walls; enzyme cofactor; involved in cell membrane permeability; encourages root development.	Characterized by death of the growing points.
Magnesium	Part of the chlorophyll molecule; activator of many enzymes.	Development of pale, sickly foliage, an unhealthy condition known as chlorosis.
Nitrogen	Component of amino acids, chlorophyll, coenzymes, and proteins. An excess causes vigorous vegetative growth and suppresses food storage and fruit and seed development.	Early symptom is yellowing of leaves, followed by a stunting in the growth of all parts of the plant.
Phosphorus	Promotes root growth and hastens maturity, particularly of cereal grains.	Underdeveloped root systems; all parts of plant stunted.
Potassium	Enzyme activator; production of chlorophyll.	Pale, sickly foliage.
Sulfur	Component of some amino acids.	Chlorosis; poor root system.
Micronutrients		
Boron	Pollen germination; regulation of carbohydrate metabolism.	Darker color, abnormal growth; malformations.
Chlorine	Evolution of oxygen during photosynthesis.	Small leaves; slow growth.
Copper	Component of some enzymes.	Lowered protein synthesis.
Iron	Needed for chlorophyll production.	Chlorosis, appearing first in youngest leaves.
Manganese	Activates Krebs cycle enzymes.	Mottled chlorosis.
Molybdenum	Part of enzymes for nitrate reduction.	Severe stunting of older leaves.
Zinc	Component of some enzymes; internode elongation.	Small leaves; short internodes.

Guidepost
What are the characteristics of growth in plants?

Cotyledons may have a variety of functions, depending on the plant. In many dicots, such as beans and peas, the cotyledons absorb the endosperm and take over its role of food storage. In other dicots, such as castor beans, the cotyledons remain thin but may produce enzymes that help transfer food from the endosperm to the germinating seedling. Some cotyledons are food storage organs in their own right. In grains, the endosperm persists in the seed (accounting for the usefulness of grains as food), and the cotyledon serves as a transfer agent. Some cotyledons become photosynthetic; others do not.

Plant Growth

18.9   A Seed Protects and Nourishes the Embryo Inside

A seed contains the offspring of a parent plant—the embryo—and stored food. The embryo is protected within the seed by a seed coat. Some seed coats are extremely thick, making digestion difficult for seed eaters and thus increasing the likehood that the seed will pass through the digestive system unharmed. Other seed coats contain chemicals that inhibit germination, or sprouting, until appropriate growing conditions occur.

When most seeds germinate, the first structure to appear is the embryonic root. The root turns downward into the soil, anchoring the seedling and absorbing water and nutrients. The stem appears next, and in the case of many dicot seeds, it carries two green cotyledons up with it. Each cotyledon contains energy and nutrients for the seedling. As the seedling grows it uses these stored reserves. After it has grown true leaves that can produce food through photosynthesis, the cotyledons fall off. The seedling continues to grow, and its cells differentiate into all the tissues of a mature plant. Figure 18.17 shows these growth stages for a bean.

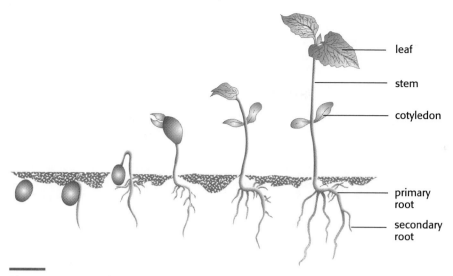

leaf

stem

cotyledon

primary
root

secondary
root

Figure 18.17   As a bean seedling develops from a seed, it grows both above and below the cotyledons.

18.10   Primary Growth Increases the Length of a Plant

Although a plant may be mature, it continues to increase in size throughout its life. A plant grows in length and in diameter, or girth. Growth in length is called primary growth because it produces primary tissues.
3 Young plants contain only primary tissues, which always are herbaceous, or nonwoody. Growth in plant diameter is called secondary growth. Secondary tissues often are woody.

In plants, cells that have differentiated usually do not duplicate. Instead, new cells are formed by special tissues called **meristems** (MER ih stemz) that are found throughout the plant. Meristems form new cells
2 through mitosis and cell division, which continue as long as the plant lives. Meristems are located at the tip of stems and roots, in buds, and between the xylem and phloem along the length of the stem, as shown in Figure 18.18.

The meristem responsible for growth in the length of the stem is located just behind the root cap, a layer of protective cells covering the root tip. Just behind the meristem is an area in which the newly formed cells grow longer. Behind this zone of elongation is a stem. Thus, the meristem gives rise to all stem tissues, including the cambium that produces new xylem and phloem cells.

The meristem responsible for root elongation is located just behind the root cap, a layer of protective cells covering the root tip. Part of this meristem forms the root cap cells, which are rubbed off as the root pushes through the soil. The root meristem also forms cells that differentiate into all the specialized root tissues. The meristems in the root and stem tips increase the length, or primary growth, of the plant Thus, a plant grows from its top upward and its bottom downward. Figure 18.19 compares the structures of a stem tip and a root tip.

In most leaves, all the cells differentiate into the leaf tissues (including xylem, phloem, and mesophyll) at an early stage of leaf formation. In late summer, deciduous woody plants develop tiny, fully formed leaves inside their buds. In spring, the new leaves expand mostly by the enlargement of the small cells. Leaves of grasses and some other plants are an exception; they grow continuously from a meristem at the base of the leaf. These plants evolved where they were constantly grazed by large herbivores. The

One of the differences between plants and animals is a plant's capacity for continuous growth—a capacity that resides in the various meristematic tissues.

This interaction of genes and environment is especially evident in plant development.

Figure 18.18   Location of plant meristems. Tip meristems bring about primary growth; cambium is responsible for secondary growth. Meristems in buds produce cells that differentiate into all the tissues of leaves, branches, and flowers.

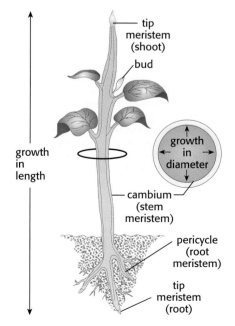

tip
meristem
(shoot)

bud

growth
in
diameter

growth
in
length

cambium
(stem
meristem)

pericycle
(root
meristem)

tip
meristem
(root)

Figure 18.19 A longitudinal section of a stem tip (a) and a root tip (b). Notice the correspondence between the zones in the stem and the root.

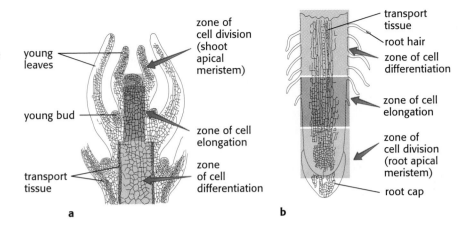

a

b

Enlargement of leaf cells takes place entirely through absorption of water, so the process is very rapid.

meristem at the leaf base is an adaptation to grazing—a leaf can grow even after most of the blade has been eaten.

18.11   Secondary Growth Increases the Diameter of a Plant

Vascular cambium is the meristem tissue that is responsible for secondary growth. Located between the xylem and phloem in the stem and root, vascular cambium produces new xylem and phloem cells throughout the life of a plant.

Ray cells, thin-walled living cells in the xylem and phloem, allow for horizontal transport in large stems. Pits in the walls of ray cells communicate with pores in both xylem and phloem. Ray cells supply water and nutrients to the phloem and sugars to the cambium cells.

Collect sample slices of small tree trunks. Pass out a variety of them and ask students to report what they can "read" from them. Rings indicate an alternation of weather favorable and unfavorable for growth. Such seasonal alternations are characteristic not only of temperate climes but of many tropical climates as well. A lack of rings indicates essentially no change in weather throughout the year, a characteristic of tropical rain forests.

Phloem cells, which are much more fragile than xylem cells, are crushed each year as the stem increases in diameter. The tough xylem cells, on the other hand, remain in the stem, forming one ring of cells each year. Figure 18.20 shows annual growth rings in a woody stem. Occurring predominantly in trees that grow in temperate zones, these annual rings can be used to determine the age of a tree as well as to reveal the growing conditions during a particular year. An extremely narrow ring could indicate bad growing conditions during the year the ring was formed. There could have been a drought or an infestation of leaf-eating caterpillars. In either case, the tree was unable to grow very much, and therefore the ring was small. The annual rings of xylem cells are the wood of the tree.

The cells in the central core of the stem usually fill with resins after a few years, forming heartwood. Heartwood is impermeable to water and thus is inactive in transport. Water is transported in the outer regions of xylem that form sapwood.

In older stems, the cambium remains as a boundary between the central core of wood (xylem) and the phloem and bark. Bark, which replaces the epidermis during the first few years of a tree's life, is composed of tough, dead cells that protect the thin layers of living phloem and cambium underneath. Because the walls of cambium cells are thin and easily broken, bark usually can be peeled easily from a tree trunk.

Cork cambium arises from outer cortical cells in many woody stems and gives rise to cork cells, forming a layer of cork beneath the epidermis. Deposits of fatty suberin in the walls make the cells almost impermeable to water and gases, providing protection against water loss.

CONCEPT REVIEW

1. Describe the role of the cotyledons in a bean seed.
2. What are the functions of cambium and other meristems?
3. How does primary growth differ from secondary growth?
4. What can annual rings tell about the life of a tree?

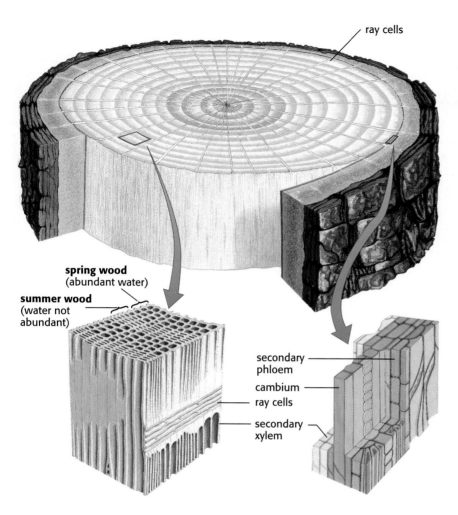

ray cells

spring wood
(abundant water)

summer wood
(water not
abundant)

secondary
phloem

cambium

ray cells

secondary
xylem

Figure 18.20   Section of a woody stem.
How old was this tree? What separates bark
from wood? **The tree was 12 years old
and there are 4 rings of heartwood. The
enlarged section shows that cambium
separates bark from wood.**

INVESTIGATION 18.1   Water and Turgor Pressure

Some plants resist the forces of gravity, wind, rain, and snow because of the rigidity
of the plant body. In herbaceous plants, however, many cell walls are thin and structurally weak. Their support depends on the turgor pressure the contents of individual cells exert against their walls. This turgor pressure varies as the volume of the
cell contents changes when water is taken in or lost. In these plants, maintaining
position is directly related to maintaining shape, which depends on turgor pressure.

   After reading the procedure, construct a hypothesis that predicts the change
for each potato core.

Materials (per team of 3)
3 pairs of safety goggles
3 lab aprons
50-mL graduated cylinder
3 25-mm × 200-mm test tubes
15-cm metric ruler
balance
glass-marking pencil
cork borer (5-to10-mm diameter)
scalpel

dissecting needle
test-tube rack
aluminum foil or plastic wrap
paper towels
10% sucrose solution
20% sucrose solution
distilled water
white potato

Put on your safety goggles and lab apron.

**Most of your students probably have
noticed wilted plants with bent-over
stems and drooping leaves. This
investigation enables students to
investigate the role of water in
maintaining the shape of nonwoody
plant structures. It is related to
Investigation 5.2 to which you may refer
students either by way of introduction or
during subsequent discussion. Teams of
3 allow each student to work with a
potato core.**
   Time: **Two class periods for the
student activity and another 20-30
minutes to discuss results.**

SAFETY

**Caution students to use care when
pushing the cork borer through the
potato so they do not cut themselves.**

**Provide small corks for
the dissecting needle
tips and store them and
the scalpels in a tray
when not in use.**

Table 18.2 Potato Core Data

Measurements	Core A (distilled)			Core B (10% sucrose)			Core C (20% sucrose)		
	1st day	2nd day	gain or loss (+ or–)	1st day	2nd day	difference (+ or –)	1st day	2nd day	difference (+ or–)
Length (mm)									
Diameter(mm)									
Mass (g)									
Rigidity (+,++,+++,++++)									
Volume (mL)									

Procedure

1. Label the test tubes *A, B*, and *C.*

2. Using a cork borer, cut three cores from a potato. With the scalpel, trim each core so it is at least 30 mm long. Make all cores as nearly the same length as possible. Place one core in each labeled test tube.

  Cork borers and scalpels are sharp, handle with care.

3. Copy Table 18.2 in your data book, or tape in the table provided by your teacher.

4. Measure the length and diameter of each core to the nearest millimeter and record the measurements in the table.

5. Determine the mass of each core to the nearest 0.1 or 0.01g. Record the data in your table.

PROCEDURE

6. **Some students may feel differences, although there should not be any. Accept these observations.**

6. Determine the rigidity of each core by holding it at each end between your fingertip, and gently bending it. Record the rigidity, using + + + + for very rigid, + + + for moderately rigid, + + for slightly rigid, and + for soft.

7. Measure the volume of each core by the following method. Pour water into the graduated cylinder until it is about half full. Hold the cylinder at eye level and read the line on the level with the lower part of the curved surface of the water, as in Figure 18.21a. This curved liquid surface is called the meniscus (men IS kus). Record this exact amount of water.

   Holding the core by a dissecting needle, sink it just under the water's surface and record the new water level. The difference between the two water levels represents the volume of the core in milliliters. Record the volume of each core in your table.

  Dissecting needles are sharp, handle with care.

8. **A 10% sucrose solution is 90% water; 20% sucrose is 80% water.**

8. Pour distilled water into test tube A until the core is covered. Pour 10% sucrose solution into test tube B until the core is covered, and pour 20% sucrose solution into test tube C in the same way. Cover each test tube with foil or plastic wrap, and store them in a test-tube rack until your next class period. What is the water concentration of the 10% sucrose solution? of the 20% solution?

9. Wash your hands thoroughly before leaving the laboratory.

10. The next day, remove the cores from the test tubes, blot them dry with paper towels, and repeat Steps 4–7.

11. Wash your hands thoroughly before leaving the laboratory.

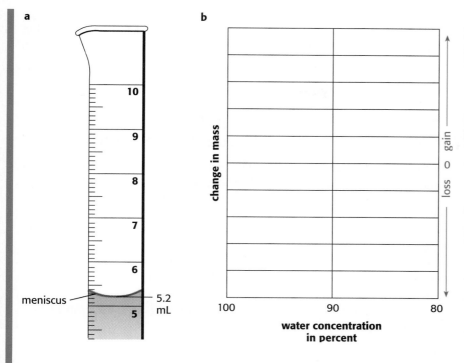

a

b

change in mass

gain
0
loss

meniscus ─   5.2
                                             mL

100   90   80

**water concentration
in percent**

Figure 18.21 (a) Read water volume from the bottom of the meniscus. (b) Units of measure for mass changes can be entered on the left and a graph plotted for changes in mass of potato cores at different water concentrations.

DISCUSSION

1. **Differences depend on student data. In distilled water, cores gain mass. In 90% water, most cores show little change. In 80% water, cores lose mass.**
2. **Graphs will vary depending on data collected by each team.**
3. **Answers will vary depending on graphs. Students should predict the water concentration that corresponds to the point on their plotted line where the mass change is zero.**
4. **Prepare sucrose solutions in a range of concentrations near the predicted point. Then experiment with potato cores, following the procedure used in this investigation.**
5. **When volume increases, mass also increases; when volume decreases, mass decreases.**
6. **Cores in 20% sucrose (80% water) usually show the greatest change in rigidity, cores in 10% sucrose (90% water) the least.**
7. **Increase in rigidity, a result of increase in cell volume, should be accompanied by increase in core mass, and a decrease in rigidity by decrease in core mass.**
8. **In the cores that increased in rigidity, the cells have taken in water from the sucrose solution; the volume of their cell contents has increased, and the cell contents are exerting greater pressure against the cell walls. The reverse is true for cores that decreased in rigidity.**
9. **Depends on hypotheses and data collected. Student hypotheses should relate the rigidity of potato tissue to diffusion of water into and out of cells in the tissue.**
10. **Determination of rigidity is subjective; cores may not have been thoroughly dried before mass was determined; measurements of length and diameter may vary; some potatoes may have been older than others, affecting their reactions.**

FOR FURTHER INVESTIGATION

1. **With 30% sucrose solution, there should be some decease in rigidity but little change in mass or volume as compared with data from cores placed in 20% sucrose. At 30% concentration, protoplasts so contract that little or no pressure is exerted on cell walls, so further decrease in**

Discussion

Review the discussion of diffusion in Investigation 5.2 and Section 5.7. Plasma membranes of potato cells are permeable to water and impermeable to sucrose molecules. The thin cell walls are permeable to both. The walls of these cells can be stretched or contracted only to a limited extent. Their shape when contracted, however, can be changed readily.

1. Complete the data table by calculating the differences, if any, between the initial and final measurements for each core. Use a + to indicate increase and a – to indicate decrease. What is the relationship between the concentration of water and the change in mass of the potato cores?
2. In your data book, prepare a graph such as the one in Figure 18.21b and plot the changes in mass of the three cores.
3. Predict the water concentration at which a potato core would not change in mass.
4. Describe an experiment that would test your prediction.
5. How does the change in volume compare with the change in mass for each core?
6. Which of the cores showed the greatest change in rigidity? Which showed the least change?
7. How does the change in rigidity compare with the change in mass for each core?
8. Explain the differences in rigidity among the cores placed in various concentrations of sucrose.
9. Do your data support the hypothesis you constructed at the beginning of this investigation? Explain.
10. What sources of error or other variables may have affected your results?

For Further Investigation

1. Using the same experimental approach, investigate the effects of 30% and 40% sucrose solutions. Compare the data obtained with the data derived from your original experiment.

core length is small. In 40% sucrose, rigidity and volume show little further change, but some increase in mass occurs as core tissue is infiltrated with the solution.

2. **This enables students to see contraction of cytoplasm to a degree where it pulls away from cell walls (plasmolysis). If leaves with plasmolyzed cells are mounted immediately in tap water, recovery from plasmolysis can usually be observed.**

This investigation focuses on structural adaptations and on the complementarity of structures and function. Do not obscure the major purpose with detailed considerations of structure. Teams of 2 are best.
    Time: **One to two class periods; time can be saved if you set up several stations with the materials to be studied and have students circulate among the stations.**

PROCEDURE PART A

1. **Greatest possible surface area exposed to the light.**

2. **Different patterns on different plants. Few plants have leaves that block the sun from other leaves.**

4. **Disk-shaped. Palisade and spongy cells. Chloroplasts are usually more abundant in palisade cells.**

5. **Because of their thick walls, they support the leaf.**

6. **Cuticle keeps the leaf from drying out, ensuring enough water for photosynthesis.**

7. **Cells are arranged irregularly with many large spaces between them.**

8. **Because the stomates open into the spaces between the cells of the spongy layer, gases can circulate freely in this layer, facilitating gas exchange.**

PART B

9. **Green, Photosynthesis. No, because mature woody stems lose their photosynthetic ability.**

2. Investigate the effects on *Elodea* leaves of sucrose solutions of the same range of concentrations as used in this investigation. Mount detached leaves in the sucrose solution on a microscope slide, and add a cover slip. Locate several cells under low power and then observe under high power for 5 minutes.

INVESTIGATION 18.2   Leaves, Stems, and Roots: Structural Adaptations

The leaves, stems, and roots of flowering plants have structural adaptations that enable them to perform their specific functions efficiently. In this investigation, you will examine those plant structures and attempt to deduce the functions for which they are adapted.

Materials (per team of 2)

microscope slide
cover slip
dropping pipet
compound microscope
stereomicroscope
forceps
colored pencils (red, blue, brown, green)
prepared cross section of a leaf
prepared cross section of an herbaceous
    dicot stem

prepared cross section of a root
variety of mature plants
young bean plant
radish seedlings germinating in a petri
    dish
grass seedlings germinating on water
carrot

Procedure

PART A Leaf Structure and Function

1. Look at the various plants and the different types of leaves. What might be the advantage of a flat, thin leaf blade to the photosynthetic capacity of a plant?

2. How are the leaves arranged on the stems of different plants? Relate this arrangement to the photosynthetic capacity of the plant.

3. Examine a prepared cross section of a leaf under the high power of your microscope. Compare your slide with Figure 18.2 to become familiar with the various regions of the leaf. Consider its structures in relation to the functions of light absorption, water supply, and carbon dioxide absorption.

4. Locate chloroplasts in the leaf cells. What shape are they? Which leaf cells contain the chloroplasts? Among those cells, what difference in abundance of chloroplasts can you see?

5. Locate a cross section of a small vein in the center tissues of the leaf. The vein is surrounded by a sheath of cells that also are photosynthetic. Notice some empty cells with thick walls in the upper part of the sectioned vein. What might be the function of these cells?

6. Examine the covering layers of the leaf—the upper epidermis and the lower epidermis. These single layers of cells are covered by cuticle, which may have been removed when the slide was made. How might the cuticle affect the efficiency of photosynthesis?

7. Locate a stomate and its guard cells. Locate the spongy tissue just above the lower epidermis. How does the structure of the spongy tissue differ from that of the other leaf tissues?

8. Note how the stomates are located in relation to the spongy tissue. What does that relationship suggest about the function of the spongy tissue?

PART B Stem Structure and Function

9. Observe the stem of the bean plant. Notice how the leaves are attached to the stem. The site of attachment is called the node (see Figure 18.22). What

color is the stem? Judging from its color, what might be the function of an herbaceous stem? Would this be a function of all stems? Why or why not?

10. Using low power, study a cross section through the internode of a stem. Compare your slide with Figure 18.8 to become familiar with the regions of the stem. Like a leaf, an herbaceous stem is covered by an epidermis. How many cells thick is the epidermis?

11. The most noticeable feature of the cross section is a ring of vascular bundles. Each of these vascular bundles is usually wedge-shaped, with the narrow end of the wedge directed toward the center of the stem. The center of the stem is made up of large, thin-walled cells. This is the pith. Between the outer edges of the vascular bundles and the epidermis is the cortex. Draw a circle about 15 cm in diameter. In the circle, outline the vascular bundles as they appear in your slide. Do not draw in any of the cells.

12. Using high power, look in the cortex and vascular bundles for thick-walled, empty cells. Show their location in red on your diagram. Suggest a possible function for these cells.

13. The leaves need water for photosynthesis. Water is absorbed by the roots and is conducted through the xylem of the stem to the leaves. Color the xylem in your drawing blue. In what structures of the stem did you look for xylem cells? What other functions might these cells have? What relation do these cells have to those previously noted in your stem drawing?

14. Materials manufactured in the leaves are transported to other parts of the plant through the phloem. Color the phloem in the diagram brown. Where in the stem is the phloem tissue? Was phloem also present in the leaf you observed earlier?

15. Food in the form of starch is stored in stems. Using high power, carefully examine the stem tissues to find cells that store food. Color the storage cells in the diagram green. How is the presence of stored food indicated in these tissues? What region of the stem is devoted primarily to storage?

PART C Root Structure and Function

16. What part of a plant absorbs water?

17. Recall your study of the absorbing surfaces of animal digestive tracts (Chapter 15). What would you expect one of the characteristics of water-absorbing structures of a plant to be?

18. Some roots store more food than others. Examine the root system of the grass plant and the carrot. Which probably contains more stored food?

19. Examine the radish seedlings in the petri dish. (Do not remove the cover from the dish.) Using the stereomicroscope, look through the cover at the fuzzy mass of tiny structures around the root. These structures are called root hairs. On what part of the root are the root hairs longest? What probably happens to the root hairs if a plant is pulled out of the soil?

20. With forceps, pick up a young grass seedling. Mount the seedling in a drop of water on a clean glass slide, without a cover slip. Using the low power of your compound microscope, find the root tip. (You can see that the root differs from the young shoot, which is green.) Look along the root for root hairs. Is a root hair made up of one or many cells? What relationship do root hairs have to the epidermis of the young root? (To answer this question, you may have to put a cover slip on the specimen and examine it with high power.) Are root hair cells living or dead? What important process is involved in the absorption of water by root hair cells?

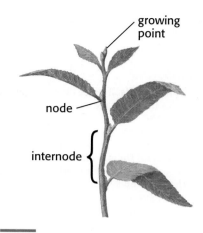

Figure 18.22 A leafy stem.

10. **One cell thick.**
12. **Because of their thick walls they must be strong and may provide good support.**
13. **In the vascular bundles. Support. They are the same; red indicates supporting tissue (xylem); blue indicates upward conducing tissue (xylem).**
14. **In the vascular bundle, just outside the xylem. Yes, in the vein.**
15. **By the presence of starch grains. Large cells in the cortex.**

PART C

16. **Roots.**

17. **A large amount of surface area.**

18. **The carrot.**

19. **On the older part of the root, away from the tip. They are broken off.**

20. **One cell. Extension of individual cells in the epidermis. Living. Diffusion.**

 Forceps are sharp; handle with care.

21. Between the radiating arms of the xylem.
22. Cortex. Starch.

DISCUSSION

1. Palisade cells. It places them as close to the source of light as possible.
2. Flat leaf surface and location of chloroplasts in palisade layer adapt leaf for maximum light absorption for photosynthesis. Thick-walled xylem cells are adapted for support and transport. Loosely arranged spongy cells adjacent to stomates adapt the leaf for efficient gas exchange.
3. Photosynthesis; support; transport of food and water; food storage.
4. Chloroplast-containing cortex cells; thick walled xylem cells; xylem and phloem cells; cortex cells.
5. Opposite function. Stem and leaf epidermis have cuticle that prevents water loss. Root epidermis has root hairs that absorb water.
6. Xylem of root is solid central strand; in stem, it is divided into many strands.
7. Storage.
8. Photosynthesis.
9. Root hair, epidermal cell, root cortex, root xylem, stem xylem, leaf xylem, space between spongy cells, stomate.

Parts A and B of this investigation allow students to observe some of the stages in germination. Part C focuses on enzymes that facilitate germination.

Time: One to one and one-half periods should be sufficient for observations and discussion.

SAFETY

Scalpels should be stored in a tray or similar container when not in use. If the investigation extends for a second day, remind students to put on their protective equipment and to wash up each day.

21. Observe with low power the prepared slide of a cross section of a mature root. Locate the epidermis, cortex, xylem, and phloem. Draw a circle to represent the root cross section. Color the location of the xylem blue on your diagram. Color the location of the phloem brown. Where is the phloem in relation to the xylem?
22. Examine the cells and tissues of the root with high power, looking for stored food. Color the region of food storage on your diagram green. What region of the root contains the greatest amount of stored food? In what form is the food stored in the root?
23. Wash your hands thoroughly before leaving the laboratory.

Discussion

1. According to your observations of chloroplast location in the leaf, which are the main photosynthetic cells? How might the location of chloroplasts in these cells be advantageous in light absorption?
2. How are the leaf structures you observed adapted for leaf functions?
3. According to your observations, what are the major functions of the stem?
4. What stem structures are adapted for each of those functions?
5. How does the function of root epidermis compare with the function of stem and leaf epidermis?
6. How does the position of xylem in the root compare with its position in the stem?
7. What function of stems and roots is lacking in most leaves?
8. What function of leaves and young stems is lacking in roots?
9. What structures would a molecule of water pass through from the soil to the air just outside a leaf? List them in order.

INVESTIGATION 18.3   Seeds and Seedlings

A seed is a packaged plant. Within it is a complete set of instructions for growing a plant such as a maple tree. The seed contains everything needed to establish the mature plant. How does the change from seed to mature (adult) plant occur? What are the functions of the structures of the seed and the growing plant? What changing characteristics can you observe?

➡ After reading the procedures, construct a hypothesis that predicts how petri dishes C and D will differ.

Materials (per team of 2)

2 pairs of safety goggles
2 lab aprons
10X hand lens
scalpel
petri dish with starch agar—dish A
petri dish with plain agar—dish B
petri dish with starch agar and germinating corn grains—dish C

petri dish with starch agar and boiled corn grains—dish D
Lugol's iodine solution in dropping bottle
soaked germinating corn grains
soaked bean seeds
bean seeds germinated 1, 2, 3, and 10 days

Put on your safety goggles and a lab apron.	

PART A The Seed
Procedure

1. Examine some of the external features of a bean seed. Notice that the seed is covered by a tough, leathery coat. Look along one edge of the seed and

find a scar. This scar marks the place where the seed was attached to the pod.

2. Remove the seed coat and examine the two fleshy halves, the cotyledons, which are part of the embryo.

3. Using a scalpel, cut a little sliver from one of the cotyledons.

 Scalpels are sharp; handle with care.

Test the sliver with a drop of Lugol's iodine solution. Record the results in your data book.

 Lugol's iodine solution is poisonous if ingested, irritating to skin and eyes, and can stain clothing. Should a spill or splash occur, call your teacher immediately; flush the area with water 15 minutes; rinse mouth with water.

4. Separate the two cotyledons and find the little plant attached to one end of one of the cotyledons. Use a hand lens to examine the plant. You will see that this part of the embryo has two miniature leaves and a root. The small leaves and a tiny tip make up the epicotyl (*epi* = above; above the cotyledon) of the embryo. The root portion is the hypocotyl (*hypo* = below; below the cotyledon).

Discussion

1. What do you think is the function of the seed coat?
2. What would you guess is the function of a connection between a parent plant and a developing seed?
3. What would you deduce is the primary function of the cotyledons?
4. What was the original source of the substance of the cotyledons?

PART B The Seedling
Procedure

5. Examine bean seedlings that are 1, 2, and 3 days old.
6. Compare the 10-day-old seedling with the 3-day-old seedling.

Discussion

1. What part of the plant becomes established first?
2. What part of the embryo gives rise to the root of the bean plant?
3. Where are the first leaves of the 3-day-old seedling?
4. What part of the embryo produces the first leaves?
5. What has happened to the cotyledons in the l0-day-old seedling?
6. Where is the seed coat in this plant?
7. Which part or parts of the embryo developed into the stem?
8. How are the first two tiny leaves arranged on the stem?

PART C Corn Seeds
Procedure

7. Cut a soaked, germinated corn grain lengthwise with the scalpel.
8. Test the cut surfaces with a few drops of Lugol's iodine solution. Record the results.
9. The starch agar and plain agar in petri dishes A and B were tested with Lugol's iodine solution as a demonstration. Observe the dishes. Record the results in your data book.

DISCUSSION

1. **Protects the embryo from injury and from drying out.**
2. **Parent plant supplies nutrients to developing seed.**
3. **To supply food to the growing plant during early development.**
4. **The photosynthetic activity of the parent plant.**

DISCUSSION

1. **The root.**
2. **The hypocotyl.**
3. **Hidden between the two halves of the cotyledon.**
4. **The epicotyl.**
5. **The cotyledons on most seedlings have dropped off or shriveled.**
6. **The seed coat was lost very early in development.**
7. **The epicotyl and the upper part of the hypotcotyl.**
8. **Opposite one another.**

DISCUSSION

1. Starch.
2. Protein, sugar, and fats.
3. **Petri dish B gives a positive test (blue) for starch.**
4. **The areas under the germinating corn grains did not give a positive test for starch, whereas the areas under the boiled corn grains did. An enzyme formed by the tissues of the germinating corn grains diffused into the agar and digested the starch. Thus, these areas remained unchanged after the iodine test. There are no clear areas under the boiled corn grains because the starch-digesting enzyme was inactivated by boiling; hence, the starch was not broken down. There are other equally defensible hypotheses—for example, that the corn grains absorbed the starch.**
5. **The result of starch digestion, a sugar.**

10. On petri dish C, 2 or 3 corn grains have started to germinate on starch agar. Each grain was cut lengthwise, and the cut surfaces were placed on the starch agar for about 2 days. Petri dish D contains starch agar and boiled corn grains. Remove the corn grains from both dishes.
11. Cover the surface of the starch agar in petri dishes C and D with Lugol's iodine solution.
12. After a few seconds, pour off the excess.
13. Wash your hands thoroughly before leaving the laboratory.

Discussion

1. When you tested the cut surface of corn grains, what food was present?
2. What other foods might be present in the corn that were not demonstrated by the test?
3. What difference did you observe in the demonstration tests of petri dishes A and B?
4. What difference did you observe when you tested petri dishes C and D? Suggest hypotheses that might account for what you observed.
5. What kind of food would you expect to find in the areas where the germinating corn grains were?

This investigation demonstrates that root growth occurs largely in the meristerm tissue near the tip.

Time: **Two and one-half class periods: one period on Day 1 to set up, one-half period on Day 2 to observe petri dish 1 and discuss the questions, and one period on Day 3 to observe petri dishes 2 and 3, analyze the data, and discuss the questions.**

INVESTIGATION 18.4   Root Growth

In this investigation, you will observe the growth patterns of corn seedlings just a few days after they have germinated. Are all parts of the root involved in growth? Do all parts of the root grow equally? If not, which part grows more quickly? Why?

> After reading the procedures, construct a hypothesis that predicts the answers to these questions.

Materials (per team of 3)

3 petri dishes
6 pieces of paper towel to fit petri dishes
glass-marking pencil
waterproof ink
scalpel

8 cardboard tags
15-cm metric ruler
3 flat toothpicks
distilled water
12 germinated corn seeds

PART A Setting Up
Procedure
Day 1

1. Select your seedlings. Observe the roots of one of them. Do you expect all parts of a root to grow equally? If not, which part would you expect to grow faster? Why? Label the petri dishes 1, 2, and 3 and add your team's symbol. Place a piece of paper towel in the bottom of each petri dish and add distilled water to dampen the paper. Pour off any excess water.
2. Select four seedlings. Using a toothpick and ink, carefully mark the shortest root with a straight, very narrow line exactly 2 mm from the tip. (You do not have to mark the entire circumference of the root.) Be careful not to crush or damage the root. Draw as many lines as possible, 2 mm apart, behind this first mark. Repeat this procedure for the other three roots. All of the roots should have the same number of marks.
3. Measure the distance from the tip to the last mark. This is the initial root length.
4. On the moist paper towel in petri dish 1, carefully place the four marked seedlings so the markings are visible.

5. Mark the remaining eight seedlings as follows: using a toothpick and ink, place a dot 5 mm from the tip of each root. Be sure to handle the roots carefully, and do not let them dry out.

6. Using a scalpel, cut off 1 mm of the tip of two corn seedling roots; cut off 3 mm of the roots of two others; cut off 5 mm of two others. Leave the remaining two seedlings untreated.

 Scalpels are sharp; handle with care.

 SAFETY
Scalpels should be stored in a tray or similar container when not in use.

7. Label each seedling as to the amount of root cut off, using a small cardboard tag tied to each seedling. Place four of these seedlings in petri dish 2 and the other four in petri dish 3.

8. Cover the seedlings in all petri dishes with a piece of paper towel. Add water to moisten them. Press down lightly to ensure that the paper is firmly placed in the dishes and pour off any excess water. Place the dishes away from direct sunlight and heat. Do not let the paper dry out during the experiment.

9. Wash your hands thoroughly before leaving the laboratory.

Day 2

10. After 24 hours, uncover the seedlings you prepared in dish 1. Examine each seedling. Measure the distance between the tip of the root and the last mark. Measure the distances between each of the lines in order, from the tip to the base. Record your measurements.

11. Note the appearance of all the lines. Are all the lines as clear as they were yesterday?

12. Discard these seedlings, and clean or discard the petri dishes.

13. Wash your hands thoroughly before leaving the laboratory.

Discussion

1. Add together the lengths of all four roots, from tip to last mark. Divide by 4. Subtract from this length the average initial root length in Step 3. What is the average amount of growth for each seedling?

2. In your roots, did growth occur at the tip, at the base, or all along the root?

3. How much growth occurred between the tip of the root and the first mark? Between the last two marks?

4. How do you explain the smears or wider marks at or near the tip of the roots?

5. On the basis of these results, what do you predict will be the results of cutting off the root tips?

PART B Comparing Results
Procedure
Day 3

14. After two days, examine the roots in petri dishes 2 and 3. Measure and record the distance from the original 5-mm mark to the tip of the root for each seedling.

15. Discard the seedlings and clean or discard the dishes.

16. Wash your hands thoroughly before leaving the laboratory.

Discussion

1. Add the measurements for the two seedlings that were not cut and determine the average. Has there been any growth in these two seedlings?

2. Determine the average amount of growth for the roots cut at 1 mm, 3 mm, and 5 mm.

3. Prepare a bar graph of the class data showing the amount of growth of roots cut at 1 mm, 3 mm, and 5 mm. Compare with the growth of uncut roots.

DISCUSSION
PART A

1. **Depends on student data.**

2,3. **Measuring the increase in length of a root alone does not tell where the growth actually is taking place. Growth does not occur uniformly over the whole length of the root; the greatest amount occurs just back of the root tip. Students probably will find that no detectable growth has occurred in their roots 4 to 5 mm from the tip.**

4. **The wider marks at or near the tips of the roots are due to growth of the root itself. The difference in the width of the marks between the start and end of the investigation indicates the actual amount of growth that has occurred at the mark.**

5. **Encourage your students to predict what will happen. Any reasoned prediction is acceptable.**

PART B

1-3. **Students may need some assistance in preparing a bar graph. Have them plot the amount of growth on the vertical axis and the four separate initial root lengths on the horizontal axis. Table T18.1 on page 486 illustrates some typical results.**

Table T18.1: Sample Data

Roots	Length (mm) Beyond the 5mm Mark
Not cut (control)	42
1 mm cut	20
3 mm cut	6
5 mm cut	3

4. **The result should indicate that the root tip is extremely important to continued growth of the root.**

5. **Because the control roots grew so well compared with the roots that were cut, we can conclude that the root tip is indispensable to growth of the root. The fact that the roots with excised tips continued to grow demonstrates that, although the greatest amount of growth in length occurs just behind the tip, some growth does occur (at decreasing rates) in parts of the root away from the tip.**

4. How important is the tip for growth of the root?

5. What information do these observations give you that the earlier observations did not?

For Further Investigation

1. Is the growth pattern of peas or beans the same as that of corn? Germinate some of these seeds and conduct the same experiment you did on corn.

2. Design and carry out an experiment to measure the growth rate of a leaf of a common houseplant.

SUMMARY

Each part of a flowering plant is adapted for a specific function. The leaf uses carbon dioxide, water, and light in photosynthesis. The mesophyll of the leaf is organized so that light and carbon dioxide can reach all the cells. Water loss through transpiration is reduced by the action of guard cells surrounding the stomates. The arrangement of the vascular tissues in stems provides support as well as transport of materials to all parts of the plant. Roots anchor the plant in the soil, and their thousands of root hairs are specialized to absorb water and nutrients. Water moves through the root cells until it reaches the xylem, where it is pulled up the root and stem by the force of transpiration. Pressure created by the sugar-water solution in phloem cells moves sugars to the nonphotosynthesizing parts of the plant. Leaves, stems, and roots may be modified for a variety of other functions.

A flowering plant begins its life as an embryo within a seed. The seed germinates, and the seedling develops characteristic plant structures. The plant grows in length and girth, forming new cells in meristems throughout its body. Growth depends on the production of sugars in photosynthesis and is regulated by the coordination of external and internal factors. These processes are discussed in the next chapter.

APPLICATIONS

1. Water lily plants are rooted in mud at the bottom of ponds, but their leaves float on the water surface. How might the cellular structure of the roots, stems, and leaves of water lilies differ from that of terrestrial vascular plants?

2. In what ways do the usual environments of a plant root and a plant shoot differ?

3. What are the principal substances that pass into and out of plants through the leaves? What forces are involved with this movement?

4. During the growing season farmers spend much time cultivating their crops—loosening the soil between plants. What advantage does this have for the crop plants?

5. How would you distinguish a root, a leaf, and a stem?

6. Explain how root hairs illustrate the biological principle of the complementarily of structure and function.

7. What part of the young tree becomes a knot in the wood of an older tree? How can you tell?

8. Living plants used to be taken out of a hospital room at night for fear they would remove oxygen from the room and thus harm the patient. How would you explain this? Should people worry about this?

9. How can grasses grow back when you cut off the tips of the plants each time you mow your lawn?

10. Are root hairs more important in helping to anchor a plant or in helping to absorb water and nutrients? Explain.

11. Could wood cells be involved in active transport? Explain.

12. If you want to have *crisp,* sliced vegetables for dinner, why is it a good idea to keep the vegetables in fresh water?

13. Compare an animal's fertilized egg to a plant seed. In what ways are they similar?

14. Would you expect to find root hairs along the "zone of elongation" of a root? Explain.

15. Would you expect roots to have a cuticle? Explain your answer.

PROBLEMS

1. Investigate how growth in plants differs from growth in animals.

2. Ten years ago a farmer built a fence 1.5 m high and attached one end of it to a tree that was 7 m high. Now the tree has grown to a height of 14 m. How far above the ground is the attached end of the fence now? Explain.

3. In a middle-latitude biome, a pine and an apple tree are growing side by side. Compare the amounts of water lost by these two trees and their growth rates throughout the year.

4. The following questions concern lateral growth in woody stems:

   a. Within a given biome, how would the annual ring formed in a wet year differ from one formed in a dry year?

   b. Sometimes two rings are formed in one year. How might this happen?

   c. What is the science of dendrochronology and how is it used?

   d. Would you expect to find annual rings in the bark of the tree? Why or why not?

5. Much of the grain of wood is the result of the pattern of annual rings. Examine different pieces of furniture and see if you can determine the direction in which the wood was cut. Can you clearly see annual rings?

6. Make a collection, or take photographs, or make drawings of leaves from a variety of plants, noting the type of habitat and environmental conditions in which they grow. (Make sure none of the leaves are poisonous or dangerous to touch.) What relationships can you find between (a) the size, shape, color, and coverings of leaves, and (b) where they grow?

7. Create a work of art based on the tissues of plants. Many tissues produce intricate, visually appealing patterns when seen in cross section.

8. Using any materials you like, make a working model of a stomate surrounded by a pair of guard cells. make the guard cells control the size of the stomate.

9. Investigate the growth of roots by growing a seedling in a narrow box with two plates of glass as the sides (similar to an ant farm). Fill the box with soil, and observe and record the growth of the root system.

10. Using a tree increment borer, take a sample of wood from a local tree. Can you determine the age of the tree, using annual rings as an indicator? What years were good growth years and what years were bad growth years for the tree? Can you link these years to any environmental factors in your area?

The Flowering Plant: Maintenance and Coordination

What do you think is the composition of the bubbles given off by this aquatic alga? What does the plant need to produce these bubbles?

If the cells are photosynthesizing, the bubbles probably are oxygen. A simple test for oxygen would determine this. These plant cells need water as a source of oxygen and light as a source of energy to split the water and release the oxygen.

◆ Like other multicellular organisms, plants are complex systems made of millions of cells that require energy for their activities. Energy for cellular work is made available through the processes of photosynthesis and cellular respiration. During photosynthesis, green plants convert light energy into chemical energy as they form ATP. The chemical energy in ATP can be used directly for cellular activities in the plant or can be stored in sugars. Photosynthesis also produces the oxygen gas necessary for cellular respiration. During cellular respiration the energy stored in sugars is transferred to ATP. Thus, photosynthesis maintains not only the plants themselves but also most of the consumers of the biosphere.

Each plant cell contributes to the life of the entire plant. The activities of all the individual cells must be coordinated for the plant to grow, mature, flower, and form fruit. That coordination is provided by a variety of plant growth substances called hormones. This chapter describes some of the internal activity of plants, how that activity is coordinated by the diverse effects of various plant hormones, and how plants respond to environmental stimuli. ◆

MAJOR CONCEPTS

- ◆ The structure of the plant is well adapted to carrying out photosynthesis.
- ◆ Light reactions convert solar energy to chemical energy.
- ◆ The Calvin cycle of photosynthesis combines hydrogen with carbon dioxide to produce carbohydrates.
- ◆ Environmental factors affect the rate of photosynthesis.
- ◆ Alternative ways of incorporating carbon dioxide have evolved in some plants.
- ◆ Plant hormones promote growth, survival, and fruiting.
- ◆ Tropisms are plant responses to environmental stimuli.

KNOWLEDGE CHECK

- ◆ What are the materials and products of photosynthesis?
- ◆ How does the environment affect photosynthesis?
- ◆ What are plant hormones, and how do they affect plant growth and development?
- ◆ How do plants respond to environmental stimuli?

You may wish to have students review Section 4.4 before beginning this chapter.

Photosynthesis

19.1 Photosynthesis Takes Place in Chloroplasts

Photosynthesis is a series of reactions in which plants use the sun's energy to synthesize complex, energy-rich molecules from smaller, simpler molecules. In eukaryotic cells, all of these reactions take place in the structures known as chloroplasts. Even when they are removed from a cell in a laboratory, chloroplasts can carry on the entire process of photosynthesis by themselves.

Electron micrographs such as the one in Figure 19.1 reveal the internal structure of chloroplasts. They show a highly organized array of internal membranes called thylakoids (THY luh koyds). Thylakoids may form stacks of flattened, disk-shaped structures called grana (GRAY nuh; singular: granum). Chlorophyll, other pigments, and enzymes are embedded in the thylakoids. Surrounding the thylakoids is the stroma (STROH muh), a colorless substance that contains other enzymes as well as DNA, RNA, and ribosomes.

Guidepost
How is the sun's energy changed into chemical energy in photosynthesis?

Students may be struck by the similarity in structure between chloroplasts and mitochondria. Although the physiology is different and chloroplasts are much larger than mitochondria, the stepwise changing of one compound into another requires an extensive and intimate interaction of enzymes. A layered construction appears to suit this purpose.

The similarity of chloroplast DNA to bacterial DNA is evidence for the bacterial origin of chloroplasts.

Figure 19.1   Electron micrograph of a chloroplast in a leaf of *Zea mays* (× 24 000). The darker areas are stacks of thylakoids called grana; the drawing shows the structure of a granum enlarged still more.

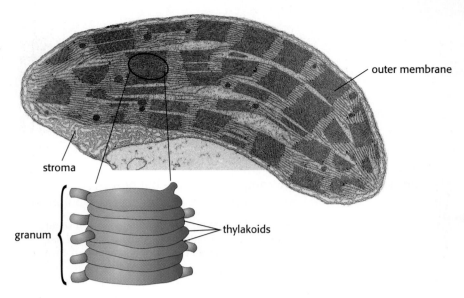

outer membrane

stroma

granum

thylakoids

19.2   Photosynthesis Requires Light

The meaning of *visible* may be different for different organisms but generally refers to the range of colored light from violet through blue, green, yellow, orange, and red. Bees see ultraviolet light, invisible to humans, and certain other differences of vision exist for other animals.

Most life on earth depends on the continual release of light energy from the sun. Of the sunlight that reaches our planet, only about 1 percent is involved in photosynthesis. The rest is absorbed or reflected by clouds or dust in the earth's atmosphere, or by the bare ground and bodies of water. Visible light, which is only a small fraction of the energy coming from the sun, consists of a spectrum of colors, each with a different wavelength and energy content, as shown in Figure 19.2. When light strikes an object it may be transmitted, absorbed, or reflected. In plant cells, several different pigments can absorb light energy. Each pigment is a chemical compound that absorbs only certain wavelengths of light and reflects or transmits all others. Green plants, for example, appear green because chlorophyll absorbs most wavelengths of visible light except green, which it reflects.

Chlorophylls are produced inside chloroplasts in the presence of light. Plants grown in the dark, or grass under an object on the lawn, show the lack of chlorophyll production in the absence of light.

Photosynthesis depends on the green pigment chlorophyll, which occurs in several forms. Chlorophyll *a* apparently is present in all photosynthetic plants; other forms may be present in different plants in different combinations. Figure 19.3 shows an absorption spectrum—a simple graph that displays the percentage of light absorbed by a pigment at each wave length—for chlorophylls *a* and *b*. Note that the chlorophylls absorb much of the light in the blue/violet and orange/red wavelengths and very little or none

Figure 19.2   Radiations from the sun form a continuous series. The range of radiations organisms can detect with their eyes— visible light—is roughly the same range plants use. Shorter wavelengths (blue light) are more energetic than longer wavelengths (red light). Leaves are green because chlorophyll absorbs red and blue wavelengths and only a little green.

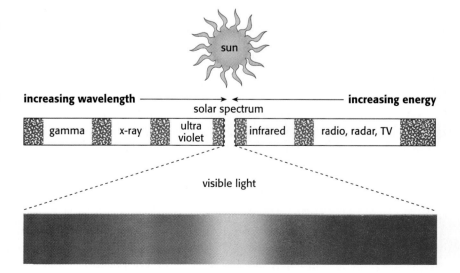

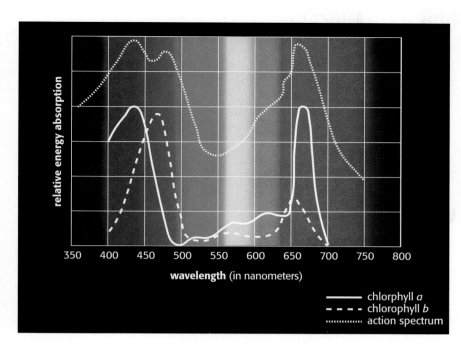

relative energy absorption

350   400   450   500   550   600   650   700   750   800

wavelength (in nanometers)

——— chlorphyll *a*
– – – – chlorophyll *b*
············· action spectrum

Figure 19.3 The upper curve shows the action spectrum for photosynthesis. The lower curves shows absorption spectra for chlorophylls *a* and *b*. It is clear from the graph that both the chlorophylls absorb the wavelengths of light used in photosynthesis. Which wavelengths do these chlorophylls absorb most? Least?
Chlorophylls absorb most light in the blue and red ends of the spectrum, and least in the green wavelengths. Ask students how well a plant would grow in green light only.

in the green/yellow wavelengths. Plants can utilize only the energy from absorbed wavelengths. The action spectrum at the top of Figure 19.3 shows this clearly. An action spectrum measures the rate of photosynthesis at cer-
2 tain wavelengths of light. In the wavelengths that are strongly absorbed, the rate of photosynthesis is high. In green light, the rate of photosynthesis is much lower because the pigment chlorophyll reflects those wavelengths.

In addition to chlorophylls, plants contain other pigments such as
1 those visible in the coleus leaves shown in Figure 19.4. Some of these accessory pigments work with chlorophyll to trap and absorb additional wavelengths of light. Their absorbed light energy is transferred to chlorophyll *a* before being used in photosynthesis. Accessory pigments are responsible for fall coloration of leaves. The pigments become more visible as the chlorophyll content of the leaves declines.

The classic equation for photosynthesis showing formation of glucose is not accurate and leads to misconceptions. A 3-carbon sugar-phosphate, not glucose, is the main sugar that leaves the Calvin cycle. Sucrose, not glucose, is the form of sugar transported within the plant. Sucrose and starch, not glucose, are the end products of photosynthesis. See R. D. Storey, "Textbook Errors and Misconceptions in Biology: Photosynthesis,"*ABT* (May 1989): 271–74.

Figure 19.4 These coleus leaves contain visible accessory pigments.

19.3   Photosynthesis Involves Many Interdependent Reactions

Three major events occur in photosynthesis: (1) absorption of light energy, (2) conversion of light energy into chemical energy, and (3) storage of chemical energy in sugars. The reactions by which these events occur may be considered in two distinct but interdependent groups, shown in Figure 19.5 on page 492.

In the first group, the **light reactions,** light energy is absorbed and converted into chemical energy as short-lived, energy-rich molecules are formed. These molecules then are used to make 3-carbon sugars from carbon dioxide in the second group, the series of reactions known as the **Calvin cycle.** In this cycle, chemical energy is stored in the sugars, and new carbon is incorporated into the plant for future growth.

The overall reactions of photosynthesis may be summarized as follows:

$$3CO_2 \;+\; 3H_2O \;\underset{\text{chlorophyll}}{\overset{\text{light energy}}{\rightleftharpoons}}\; C_3H_6O_3 \;+\; 3O_2$$

(carbon dioxide)   (water)   (3-carbon sugar)   (oxygen gas)

This equation indicates only the major raw materials and some of the end products. It does not show the many chemical steps and other substances involved. The next sections explain those steps in more detail.

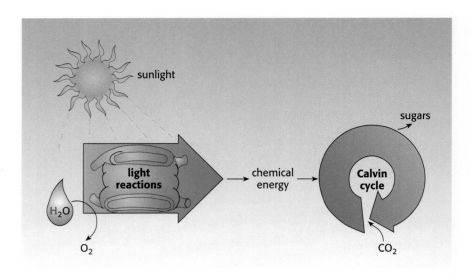

19.4   Oxygen Gas Is a By-Product of the Light Reactions

The light reactions of photosynthesis convert the energy in visible light into the chemical energy that powers sugar production. In these reactions, chlorophyll and other pigments in the thylakoid absorb light energy; water molecules are split into hydrogen and oxygen, releasing oxygen gas to the atmosphere; and light energy is stored as chemical energy. Figure 19.6 illustrates these reactions.

The pigments and other molecules involved in the light reactions are embedded in the thylakoid membranes and organized in a way that ensures their efficient operation. Two structurally distinct groups of molecules, called photosystem (PS) I and photosystem (PS) II, work in harmony to absorb light energy. The photosystems are connected by electron carriers that form an electron transport system. Acting together, the chlorophylls and accessory pigments absorb, concentrate, and transfer light energy to a special chlorophyll *a* molecule in each photosystem.

The absorption of light energy sets up a flow of electrons. In each photosystem, the special chlorophyll *a* molecule loses electrons, which are

Students may say that animals use respiration to obtain energy but that green plants use photosynthesis. In fact, plants, through photosynthesis, build up food reserves, which then are broken down by respiration to release energy for the plant. (The students might be partly correct, however, because the light reactions produce more ATP than is required for the Calvin cycle, and the ATP can be used to synthesize amino acids and lipids in the chloroplast.)

Figure 19.6 The light reactions of photosynthesis take place in the thylakoid of a chloroplast. Absorption of light energy sets up a flow of electrons from water to photosystem II to photosystem I to NADP+. As electrons flow along the electron transport system, protons are transported to the inside of the thylakoid. ATP forms as protons diffuse through an enzyme complex.

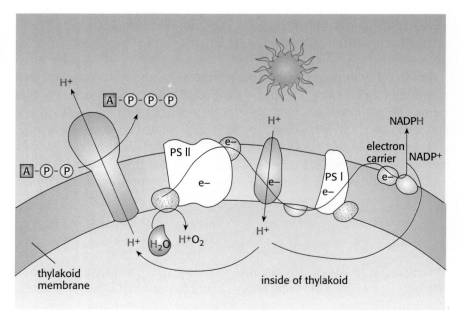

captured by electron carriers. The electrons lost by photosystem I are replaced by electrons lost from photosystem II. Electrons lost by photosystem II are replaced by electrons removed from water. As a result, water is separated into oxygen, protons, and electrons:

$$H_2O \longrightarrow 2H^+ + 2e^- + \tfrac{1}{2}O_2$$

The oxygen is given off as oxygen gas.

As electrons flow from carrier to carrier along the electron transport system, some of the energy they originally captured from the sun powers the active transport of protons across the thylakoid membrane. A relatively high concentration of protons accumulates inside the thylakoid. The protons then diffuse through an enzyme complex in the thylakoid membrane, releasing energy that powers ATP synthesis. When this occurs, light energy has been converted into chemical energy. At the end of the electron flow, the electrons, along with protons from water, combine with a hydrogen carrier called **NADP⁺,** nicotinamide adenine dinucleotide phosphate (nik uh TEE nuh mid AD uh neen DY NOO klee oh tyd FOS fayt). NADP⁺ is very similar to NAD⁺, a hydrogen carrier in cellular respiration. When electrons and protons combine with NADP⁺, NADPH is formed.

In the light reactions, the first two events of photosynthesis occur: absorption of light energy and the conversion of light energy into chemical energy. Three products are formed. One is oxygen gas, which is released to the atmosphere through the stomates of the leaf. The other two are the energy-rich molecules ATP and NADPH, both of which are used in the Calvin cycle to convert carbon dioxide into sugars.

19.5   Sugars Are Formed in the Calvin Cycle

The reactions of the Calvin cycle do not involve the absorption of light energy. They do, however, require the products of the light reactions. In addition, some of the enzymes involved in the reactions are activated in the light or deactivated in the dark. The Calvin cycle is a series of reactions in which carbon dioxide is combined with the hydrogen split from water in the light reactions. Energy for these reactions is provided by the ATP and NADPH formed in the light reactions.

Figure 19.7 is a simplified diagram of the reactions of the Calvin cycle, which take place in the stroma of the chloroplasts. As shown in Figure 19.7a, the cycle begins when a molecule of carbon dioxide combines with a 5-carbon molecule (RuBP, or ribulose bisphosphate), forming a 6-carbon molecule that immediately splits into two 3-carbon acid molecules. Each 3-carbon acid accepts hydrogen from NADPH and phosphate groups from ATP, shown in (b), forming a 3-carbon sugar phosphate. This molecule is the first sugar formed from carbon dioxide in the Calvin cycle. For each three molecules of carbon dioxide that enter the Calvin cycle, six molecules of 3-carbon sugar are formed. Of these, five molecules undergo many transformations in the Calvin cycle, thus regenerating the original 5-carbon molecules with which the cycle began, shown in (c). Without these molecules, the cycle would stop. The sixth molecule of 3-carbon sugar leaves the Calvin cycle.

Several things can happen to the 3-carbon sugar that leaves the Calvin cycle. In the stroma of the chloroplast, the sugar can be converted into starch and stored as starch grains, or it can be used to synthesize amino acids, proteins, lipids, chlorophyll, enzymes, and other compounds needed in the plant cell. Much of the 3-carbon sugar is exported to the cytosol,

---

Each special chlorophyll *a* molecule, in association with particular proteins and other membrane components, forms a reaction center and thus is made special by its chemical environment.

For details of the water-splitting process, see Govindjee and W. J. Colemen, "How Plants Make Oxygen," *Scientific American* (February 1990): 50–58.

You may wish to describe the classic experiments in which isotopes were used to determine that the oxygen given off in photosynthesis comes from water, not from carbon dioxide.

The mechanisms of ATP synthesis is similar in photosynthesis and cellular respiration; it depends on the buildup of a proton gradient.

The Calvin cycle often has been called the *dark reactions,* a misleading term because the cycle cannot operate in the dark.

The 3-carbon sugar-phosphates glyceraldehyde-phosphate and dilhydroxyacetone-phosphate are the Calvin cycle intermediates thought to exit the chloroplast as products of photosynthesis. These same molecules also are central intermediates in glycolysis.

Figure 19.7 The Calvin cycle converts carbon dioxide into sugars. Enzymes in the stroma of the chloroplast catalyze each step of the cycle. The reactions require ATP and NADH from the light reactions.

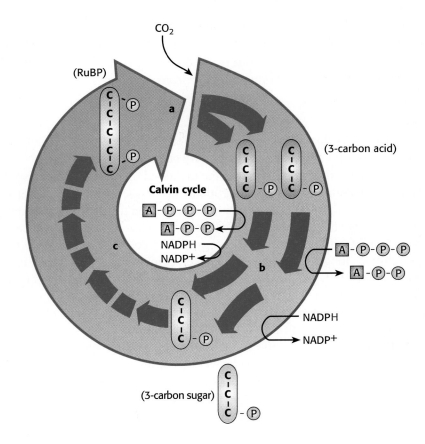

where it can be converted into sucrose, the major form in which sugar is transported between plant cells. The 3-carbon sugar also may enter the pathway of glycolysis and cellular respiration, providing energy for all **5** plant cells. Thus, in the Calvin cycle, chemical energy is stored in sugars as well as in other compounds synthesized by the plant cells, as shown in Figure 19.8.

Figure 19.8 Plants use the products of photosynthesis to supply energy and carbon skeletons for biosynthesis and other energy-requiring processes.

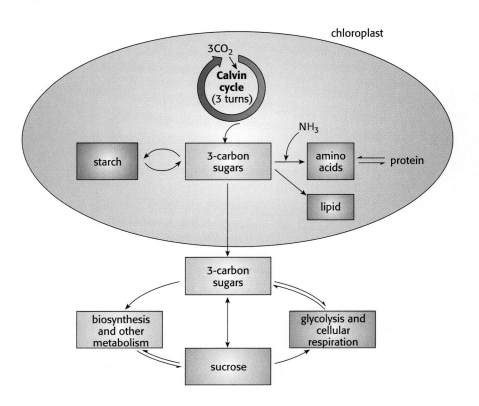

The Calvin cycle and the Krebs cycle (see Chapter 15) are similar in involving many rearrangements of carbon chains. Both produce carbon skeletons for use in biosynthesis reactions. Carbon dioxide is used in the Calvin cycle and released in the Krebs cycle. Much ATP is used in the Calvin cycle, and a little is formed in the Krebs cycle. NADPH is used in the Calvin cycle, and NADH is produced in the Krebs cycle.

The importance of the Krebs and Calvin cycles in the production of carbon skeletons for biosynthesis reactions often is overlooked and, therefore, has been stressed here and in Section 15.9.

CONCEPT REVIEW

1. Describe the roles of the chlorophylls and accessory pigments in photosynthesis. What is the special role of chlorophyll *a*?
2. How is light used in photosynthesis?
3. Describe the roles of photosystems I and II.
4. What are the products of the light reactions, and how are they used?
5. What are the products of the Calvin cycle, and how are they used?

Photosynthesis and the Environment

19.6 Environmental Factors Affect the Rate of Photosynthesis

Environmental conditions profoundly affect photosynthesis and, therefore, productivity in an ecosystem. Four environmental factors of particular importance are light intensity, temperature, and the concentrations of carbon dioxide and oxygen. The graph in Figure 19.9a illustrates the effects of light intensity on photosynthesis. Note that before the light reaches the intensity of full sunlight, the rate of photosynthesis levels off. (Rate is the activity per unit of time.) At that point, the light reactions are saturated with light energy and are proceeding as fast as possible, so a further increase in light intensity will not result in an increased rate of photosynthesis.

Figure 19.9b illustrates the effects of temperature on the rate of photosynthesis. In general, higher temperatures speed up, and lower temperatures slow down, the rate of chemical reactions. How, then, can you explain the decrease in rate above 30°C? Recall what you have learned about the effects of temperature on enzymes and other proteins.

This discussion refers to C-3 plants only. C-4 and CAM plants respond quite differently to the same environmental factors.

Guidepost

What adaptations enable plants to overcome rate-limiting environment factors.

Other environmental factors, such as humidity and soil nutrition, affect photosynthesis significantly although indirectly.

Shade plants grown and thrive in less than full light intensity. Sun plants require high light intensity and do not grow well in less than saturating light.

As a chemical reaction, the rate of photosynthesis increases with temperature. At higher temperatures, activity peaks and then falls as heat denatures enzymes and proteins and, in whole plants, as stomates close.

Figure 19.9   (a) As light intensity increases, the rate of photosynthesis increases, but only up to the point that is called the light saturation point (indicated by the brown arrow). (b) As the temperature increases, the rate of photosynthesis also increases up to a point where the rate levels off and then begins to decline. What are the optimum temperatures for photosynthesis? Data are generalized to show trends.   **Between 20°C and 30°C.**

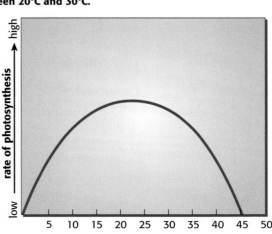

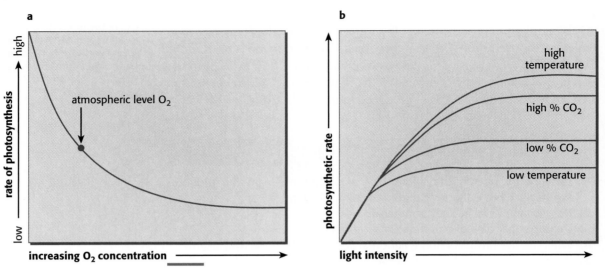

Figure 19.10   (a) Increasing concentrations of oxygen inhibit the rate of photosynthesis in most crop plants at atmospheric levels of carbon dioxide (0.035%). (b) Environmental factors have interacting effects on the rate of photosynthesis, and any one can be a limiting factor if present in less than optimum amounts.

The concentrations of carbon dioxide and of oxygen influence the rate of photosynthesis in opposite ways. An increased level of carbon dioxide concentration increases the rate of photosynthesis to a maximum point, after which the rate levels off. (This is similar to the effect of increased light intensity.) An increased oxygen concentration, on the other hand, slows the rate, as shown in Figure 19.10a. The present atmospheric concentrations of oxygen (about 20 percent) may slow the rate of photosynthesis by as much as 40 percent in some plants.

Because these four environmental factors interact, their combined effects must be considered. Any one of the factors, if present at less than optimal levels, may limit the rate of the reaction. For example, at optimal temperature and carbon dioxide concentration, the rate of photosynthesis may be limited by the light intensity. If temperature and light intensity are optimal, then carbon dioxide concentration may become the limiting factor. **2** And if light intensity and carbon dioxide concentration are optimal, the temperature may control the overall rate of photosynthesis. Figure 19.10b summarizes these interactions, which explain a great deal about the distribution of plants. For example, roses do not grow at the maximum rate in cold climates even in full sunlight, because the low temperature limits their growth. Cactus plants do not grow in forests because they require large amounts of light and can survive with the small amounts of water and the extremes of temperature common to deserts.

19.7   Photorespiration Inhibits Photosynthesis

Photorespiration is a light-dependent process involving the uptake of oxygen and release of carbon dioxide that occurs along with photosynthesis in all plants. Because it prevents incorporation of carbon dioxide in the Calvin cycle and results in the loss of previously incorporated carbon dioxide, photorespiration is counterproductive to photosynthesis. Figure 19.11 compares the two processes. As described earlier in Section 19.5, each molecule of carbon dioxide that enters the Calvin cycle first combines with a **3** 5-carbon molecule. This reaction, catalyzed by an enzyme called rubisco

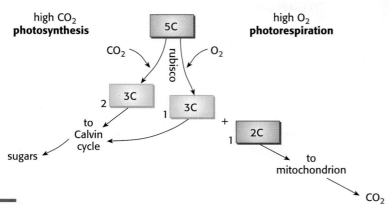

Figure 19.11 Photorespiration is the light-dependent uptake of oxygen and release of carbon dioxide in plants. It occurs simultaneously with photosynthesis and results in the loss of previously incorporated carbon dioxide. The enzyme rubisco, which can react with either carbon dioxide or oxygen, is involved in both of these processes. High levels of oxygen favor photorespiration; high levels of carbon dioxide favor photosynthesis.

Between 20% and 40% of the carbon fixed in photosynthesis may be reoxidized into carbon dioxide in photorespiration.

(roo BIS koh), forms two 3-carbon acids. Rubisco, however, can react with oxygen as well as with carbon dioxide. When rubisco reacts with oxygen, the rate of photosynthesis slows down. No new carbon dioxide is incorporated, and the 5-carbon molecule (containing previously incorporated carbon dioxide) is converted into carbon dioxide. Because there are thousands of molecules of rubisco and the 5-carbon compound in each chloroplast, photorespiration and photosynthesis can occur simultaneously in a plant. Relatively high levels of carbon dioxide favor photosynthesis. Relatively high levels of oxygen favor photorespiration.

Why would such an apparently damaging process have evolved in green plants? Evolution favors workable systems that develop from already existing systems, resulting in adequate but not optimum processes such as photosynthesis. When the Calvin cycle evolved there was little oxygen in the air, and the reaction of oxygen and rubisco was not significant. Millions of years of photosynthesis, however, have built up the atmospheric oxygen levels. Photorespiration has become a problem—one that may have a high cost because reduced plant growth reduces food production.

19.8 Special Adaptations Reduce Photorespiration

Hot, dry weather leads to high transpiration rates. In turn, high transpiration rates cause plants to lose valuable water through the stomates. In many plants, stomates are closed partially much of the time, thus avoiding excessive water loss. When stomates are closed, however, carbon dioxide levels in the leaves may drop so low that photorespiration is favored over photosynthesis.

Most plants use the Calvin cycle for the initial steps that incorporate carbon dioxide into organic material. These plants are called C-3 plants because the first stable molecule formed is a 3-carbon acid. In certain plants such as sugar cane, corn, crabgrass, and others that can tolerate hot conditions, the Calvin cycle is preceded by reactions that first incorporate carbon dioxide in a 4-carbon acid. These plants, known as C-4 plants, have two distinct types of photosynthetic cells. Surrounding each vein is a layer of tightly packed cells called bundle-sheath cells (see Figure 19.12a). Around the bundle sheath are mesophyll cells, which also are in contact with the air spaces in the leaf.

In C-3 plants, the bundle sheath, if present, is much less distinct and does not differ metabolically from mesophyll.

Rubisco is present only in the bundle-sheath cells, so the Calvin cycle occurs only in those cells. The carbon dioxide-fixing enzyme in the mesophyll cells does not bind oxygen. The rapid carbon dioxide fixation in mesophyll cells keeps the carbon dioxide concentration low in the cells, enhancing diffusion of carbon dioxide from air spaces and the atmosphere even through partially closed stomates.

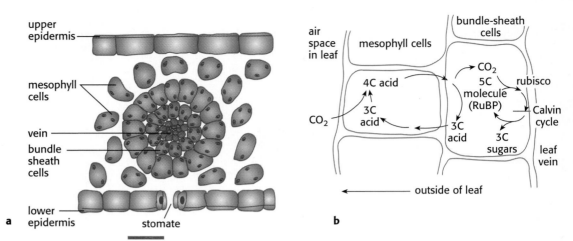

Figure 19.12   (a) Specialized leaf anatomy in a C-4 plant. Note the tightly packed layers of bundle-sheath and mesophyll cells surrounding the vein. (b) In C-4 plants, carbon dioxide first is incorporated into a 4-carbon acid in the outside mesophyll cells. The 4-carbon acid then is transported to the bundle-sheath cells, where carbon dioxide is released to the Calvin cycle.

C-4 plants utilize a two-step pathway involving two separate and different enzyme systems for carbon dioxide incorporation. The first system operates in the mesophyll cells, where carbon dioxide combines with a 3-carbon acid. The resulting 4-carbon acid is rearranged and then transport- **4** ed to the bundle-sheath cells where the second system operates. In the bundle sheath cells, the carbon dioxide is released from the 4-carbon acid and enters the Calvin cycle, eventually forming a 3-carbon sugar, as shown in Figure 19.12b.

The system in the mesophyll cells efficiently delivers carbon dioxide to the bundle-sheath cells. Thus, a higher concentration of carbon dioxide exists in the bundle-sheath cells than in the atmosphere—a state that favors photosynthesis and inhibits photorespiration. Even at high temperatures and at high light intensities (conditions that favor photorespiration), the high concentration of carbon dioxide around rubisco in C-4 plants overcomes photorespiration. Thus, C-4 plants can function at maximum efficiency in high temperatures and light intensities while the stomates are partially closed, avoiding excessive water loss. As a result, C-4 plants grow more rapidly than C-3 plants in warm climates and may be almost twice as efficient as C-3 plants in converting light energy to sugars under conditions of high temperature, high light intensity, and/or low water.

Photorespiration is a serious problem for many major food crops, such as soybeans, wheat, and rice, which are C-3 plants.

Another photosynthetic specialization was discovered in desert plants such as cactus and jade plants. Called CAM, for crassulacean (kras yoo LAY see un) acid metabolism, it is an important adaptation in arid-land ecosystems or very dry conditions. CAM plants, in contrast to C-3 and C-4 plants, can open their stomates at night and incorporate carbon dioxide **6** into organic acids in the cell vacuoles. During the day, with high temperatures and low humidity, the stomates close, preventing transpiration that could kill the plant. The organic acids release carbon dioxide that diffuses into the chloroplasts and feeds a normal Calvin cycle powered by the light reactions of photosynthesis. CAM photosynthesis, however, is not very efficient, and although desert plants that use this system can survive intense heat, they usually grow very slowly.

CAM also occurs in other plants such as Spanish moss.

1. Describe the influence of increasing temperature on the rate of photosynthesis.
2. How do various environmental factors interact to affect the rate of photosynthesis?
3. What is photorespiration, and how does it affect photosynthesis?
4. Describe the two-step pathway of carbon dioxide incorporation in C-4 plants.
5. Construct a graph that compares the effect of oxygen on C-3 and C-4 photosynthesis.   **See Figure T19.1.**
6. What is unique about CAM photosynthesis?

Figure T19.1. Effects of increasing oxygen concentration on C-3 and C-4 plants.

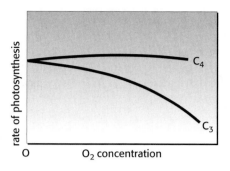

Hormonal and Environmental Control of Plant Growth

19.9   Plant Hormones Interact to Regulate Plant Processes

As a plant grows, it increases in size, but it also differentiates, forming a variety of cells, tissues, and organs. Although its genes determine the basic form of a plant, the environment can modify its development. Growth-regulating substances (plant hormones) that are produced by the plant play critical roles in plant responses to changing seasons and other environmental factors.

Guidepost

How are hormones involved in plant growth and development, and what role does the environment play?

At the present time, botanists recognize five main groups of plant hormones that have some effect on plant growth. Their production is under genetic control, but environmental factors have a part in determining when
1 the hormones are produced, how much of each hormone is produced, and which cells are sensitive to these hormones. Plant hormones are produced and are effective in extremely small concentrations, and they may interact with each other to produce their effects. The hormones may cause different effects in different parts of the plant, at different ages of the plant, or in different concentrations. Each hormone apparently has a primary effect on each affected cell type, producing a variety of responses. The response depends on poorly understood structural and physiological conditions in the affected cells. Because these conditions are not well understood, it is difficult to determine precisely how plant hormones operate.

19.10   Plant Hormones Have Diverse Effects

Auxins (AWK sins) are produced in actively growing regions such as the tip ends of shoots, from which they move by active transport to other parts of the plant. Auxins can cause cells in the growing regions of the plant to elongate, but the effects depend on a number of factors, including the concentration of the auxins. Laboratory experiments indicate that at extremely low concentrations auxins may promote elongation of roots, but at higher concentrations they may inhibit elongation (see Figure 19.13).
2 Applied to cuttings of stems or leaves, auxins can stimulate the formation
3 of roots; applied to certain flowers, they can promote fruit development without the need for pollination or fertilization. Such fruits lack seeds, and seedless tomatoes and cucumbers have been produced this way.

IAA appears to be the only important natural auxin, although other auxins or auxinlike substances are produced in plants.

Auxins appear to increase the plasticity of the cell wall, allowing it to stretch but not to recoil, so the stretching is irreversible.

Auxins are commercially available as rooting solutions and other treatments.

Figure 19.13 The concentration of applied auxin has opposite effects on cell elongation in roots and stems.

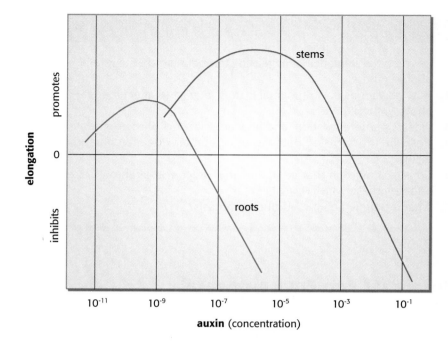

Students can research the literature to discover how 2,4,5-T and dioxin work and the source and effects of dioxin.

A synthetic auxin called 2,4-D is an herbicide used to kill dicot weeds. At the concentrations usually used, 2,4-D does not affect members of the grass family (monocots), so it can be used to control weeds such as dandelions in lawns and grain fields. 2,4-D and a similar herbicide, 2,4,5-T, were components of "Agent Orange" that was used during the Vietnam conflict to defoliate trees. Dioxin, a by-product that is produced when 2,4,5-T is made, often contaminates the 2,4,5-T and is known to be extremely toxic to humans. Dioxin is considered to be an environmental hazard all over the world. Agent Orange is no longer made and 2,4,5-T was banned by the United States Environmental Protection Agency in 1979 for use in the United States.

Gibberellins were named after the fungus in which they first were found, *Gibberella fujikuroi.*

Stem elongation in mature trees and shrubs is influenced by **gibberellins** (jib uh REL ens), which also are synthesized in the tips of stems and roots. Application of gibberellins can cause certain plants to grow faster, dwarf plants to grow to the height of normal varieties, and fruits to grow larger (see Figure 19.14). Gibberellins also can stimulate the production of enzymes in the endosperm of grains. The enzymes break down the stored nutrients, which are used by the embryo as it develops.

The embryo releases gibberellins after the seed has soaked up water when dormancy is broken.

Figure 19.14 Grapes with and without applied gibberellic acid.

Cytokinins, which are universal in the plant kingdom, were discovered by scientists trying to find chemicals that would promote the growth and development of plant cells in tissue cultures. They were named cytokinins because they promoted cytokinesis, or cell division.

The delicate balance between auxins and ethylene triggers leaf fall in deciduous trees. As auxin production is reduced, the production of ethylene is increased, starting the breakdown of stem tissue by enzymes that digest cell walls.

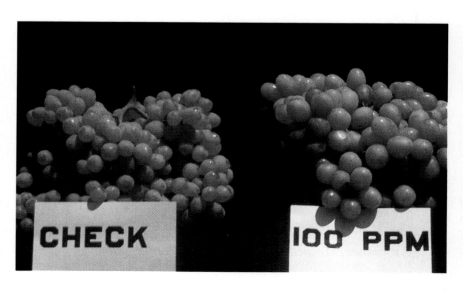

Biology Today

Development of New Plants

The use of plant hormones has enabled plant physiologists to devise new growing methods. In tissue culture, single cells or groups of cells are placed in sterile containers with nutrients, water, and (sometimes) hormones to stimulate growth and development. At equal concentrations of auxins and cytokinins, many cells in a culture continue to divide, forming an undifferentiated tissue called a callus. By adjusting the relative concentrations of auxins, cytokinins, and other growth substances, the callus cells can be induced to form roots or buds and, in some cases, to grow into a complete plant, as shown in the figure. Because one plant can contribute many cells, it is possible to develop many clones, or genetically identical plants. Tissue culture has been used to clone many different plants, but few plants have developed into adults. The cloning of animal cells is more difficult, and the cloning of human cells is not a present reality.

Cloning has an important advantage in the development of new plant varieties. Because there is much variation in most plant populations, there may be just one plant out of thousands that has a particular combination of desirable characteristics. It may be a new variety of rose or an unusually tasty carrot. If this one plant is crossed with another to get seeds, there is no guarantee the offspring will have the same characteristics as those of the desired parent. Can you explain why? Cloning, however, produces duplicate copies of any plant, and the cloned plants that

do survive then can contribute to the culturing of many more clones. In this way an entire field of identical plants can be developed.

Tissue culture is just one of many ways scientists use to develop new and better crops. Plant and animal breeders use selective breeding, allowing only those individuals with the "best" characteristics to breed. Most of the offspring of the individuals allowed to breed are expected to have some of the characteristics of the two parents.

Another technique used to develop new and better crops is protoplast fusion, in which cells from two different plants are placed in a special liquid in the same container. Enzymes are used to digest away the cell walls, leaving intact the living contents, or protoplasts. The protoplasts may fuse, develop a cell wall, and begin to grow, forming a hybrid cell.

Usually, individuals from two different species cannot form viable offspring. For example, it has not been possible to produce a hybrid offspring between a potato and a tomato by cross-fertilization. A hybrid cell of these two plants can be formed using protoplast fusion, however. Results such as these have raised hopes that protoplast fusion could lead to the combining of extremely different plants into "superplants" or "supercrops" that might, for example, be more drought resistant or require smaller amounts of nitrogen than present varieties.

Interaction of auxins and cytokinins in plant tissue culture. The culture media were supplemented with various levels of the two hormones.

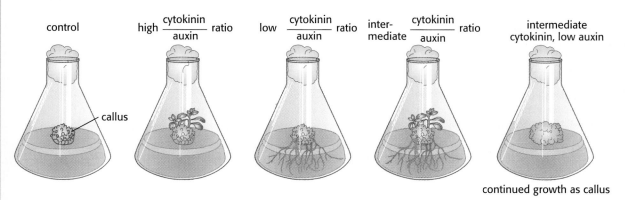

Cytokinins (syt oh KY ninz) are plant hormones that promote cell division and organ formation, at least in tissue culture. They are found mostly in plant parts with actively dividing cells, such as root tips, germinating seeds, and fruits. Cytokinins appear to be necessary for stem and root growth and to promote chloroplast development and chlorophyll synthesis. Cytokinins and auxins may interact closely to regulate the total growth pattern of the plant. **3**

Abscisic (ab SIS ik) **acid** is thought to be synthesized primarily in mature green leaves, root caps, and fruits. Apparently it is involved in the dormancy of buds and seeds, enabling them to survive extremes of temperature and dryness. Abscisic acid also affects stomate closing and is sometimes called the stress hormone because it may help protect the plant against water loss during unfavorable environmental conditions.

A simple gas produced by ripening fruits and other plant parts, **ethylene** (ETH ih leen) promotes fruit ripening and aging of tissues. Through interaction with auxins, cytokinins, and abscisic acid, ethylene apparently plays a major role in causing leaves, flowers, and fruits to drop from a plant. Ethylene appears to counteract many effects of auxins; for example, it inhibits flowering and the elongation of stems and roots. Hundreds of years ago, Chinese farmers unknowingly used ethylene when they burned incense (which produces ethylene) to ripen fruit stored in ripening rooms. Today, many farmers pick delicate fruits such as tomatoes when they are still green and hard and thus less susceptible to damage in shipping. Then, prior to sale, the tomatoes are treated with ethylene to speed and time their ripening.

Another example of controlled ripening of fruit is the "cold storage" of apples. Apples can be stored at a low temperature in an atmosphere low in oxygen (1–3%) and high in CO_2, with charcoal filters to absorb the ethylene given off by the fruit. These conditions successfully inhibit apple ripening by reducing the effects of ethylene. Oxygen is necessary for ethylene action and CO_2 is a competitive inhibitor of ethylene. Thus, apples are available year round. Figure 19.15 summarizes the actions of the five groups of plant hormones.

19.11   Plants Respond to Environmental Stimuli

Proper plant growth and development depend, in part, on **tropisms** (TROH piz umz)—movements of plants that occur in response to a stimulus. These movements are generally very small and take place so slowly that usually it is necessary to use time-lapse photography to see them. Most plant movements occur in response to changes in the environment. For example, when any part of the leaf of the sensitive plant, *Mimosa pudica,* is touched, the leaflets droop together suddenly, as shown in Figure 19.16. This movement results from a sudden loss of water from certain cells at the base of the leaves and leaflets. The significance of the movement is not completely clear, but the drooping may help prevent further water loss.

Many plant movements seem to result from differences in growth among cells of a tissue. You probably have observed that most plants can grow toward light (Figure 19.17 on page 503). Charles Darwin studied this response, called **phototropism** (foh toh TROH piz um), in which cells on the lighted side of the shoot can stop growing, while those on the shaded side can continue to elongate. That differential growth seems to cause the plant to bend toward the light, but what causes the differential growth? **4**

On the underside of the base of each petiole are enlarged cells that ordinarily are turgid. Stimuli of various types—sudden contact, rapid temperature change, and electricity—cause release of some diffusible chemical that stimulates these cells to lose liquid, reducing their turgor until the cell walls collapse and leaf folds. Although folding is almost instantaneous, recovery may take some minutes. *Mimosa pudica* is easy to grow; seeds are available from biological supply companies.

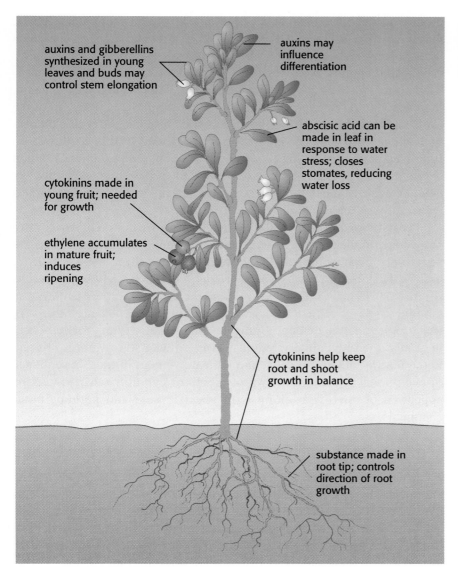

Figure 19.15 Some possible roles and interactions of hormones among the various parts of a plant.

auxins and gibberellins synthesized in young leaves and buds may control stem elongation

auxins may influence differentiation

abscisic acid can be made in leaf in response to water stress; closes stomates, reducing water loss

cytokinins made in young fruit; needed for growth

ethylene accumulates in mature fruit; induces ripening

cytokinins help keep root and shoot growth in balance

substance made in root tip; controls direction of root growth

Experiments designed to answer that question have been carried out on oat coleoptiles (koh lee OP tilz). A coleoptile is the protective sheath around monocot seedlings. In uniform lighting or in the dark, oat seedlings grow straight upward. With one-sided lighting, bending occurs. Evidence from the experiments indicated that the tip of a growing shoot

Figure 19.16 Mimosa plant before (a) and after (b) being touched.

a

b

Figure 19.17 Seedlings bending toward light.

The term *geotropism* may be more familiar. The response is to the acceleration force of gravity; therefore the term *gravitropism* is more accurate. For more information about the mechanism, see M. L. Evans, R. Moore, and K. H. Hasenstein, "How Roots Respond to Gravity," *Scientific American* (December 1986): 112–19.

produces a chemical that causes bending. Darwin's experiments showed that if the tip was covered, the shoot would not bend. Other researchers found a substance produced in the tip when it was exposed to light. This chemical, if applied to only one side of the shoot, could cause the shoot to bend even in the dark. Eventually the chemical was identified as an auxin. It appeared to researchers that the auxin was synthesized in the tip of the shoot and then transported to the shaded side, where it stimulated elongation of the cells. That hypothesis now is being questioned, however. Improved analytical techniques fail to show any redistribution of auxin within the shoot before bending occurs. At the present time it is not clear how, or even whether, light, auxin, and perhaps other plant hormones interact to produce bending.

Gravitropism (grav ih TROH piz um) is growth movement toward or away from the earth's gravitational pull. Stems and flower stalks are negatively gravitropic, that is, they grow in a direction away from gravity. If stems are placed in a horizontal position, they bend upward (Figure 19.18), although this response can be variable. Auxins are thought to play a critical role, as are calcium ions. Roots are positively **4** gravitropic, and the evidence indicates that sensitivity to gravity occurs in or near the root cap. There, plastids filled with starch grains migrate to the bottom of certain cells, and their movement initiates a sequence of events that leads to downward growth or curvature. Again, auxins appear to be involved, along with abscisic acid and perhaps other chemicals.

19.12 Many Plants Respond to the Length of Day and Night

The changing seasons are important environmental cues for all terrestrial organisms. Birds migrate, insects produce dormant eggs, animals hibernate, and deciduous trees lose their leaves as winter approaches. These activities are triggered by changes in day length. Organisms have mechanisms that allow them to respond to the relative amount of light and dark in a 24-hour period. This response is called **photoperiodism** (foh toh PIH ree ud iz um).

Many plants show the effects of photoperiodism in flowering. For example, spring flowering plants, such as daffodils, can reproduce only when day length exceeds a certain number of hours. Thus, they are called long-day plants. Fall-flowering plants, such as asters, can reproduce only when day length is shorter than a certain number of hours and therefore are called short-day plants. Many plants flower whenever **5** they become mature. These are day-neutral plants, although tempera-

Figure 19.18 When an upright plant is turned on its side, new roots grow downward and new stem growth is upward, resulting in a bent stem. What type of gravitropism do these reactions demonstrate? **Positive gravitropism in roots and negative gravitropism in stems.**

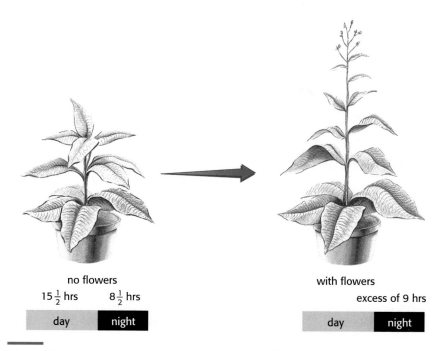

no flowers

$15\frac{1}{2}$ hrs $8\frac{1}{2}$ hrs

| day | night |

with flowers

excess of 9 hrs

| day | night |

Figure 19.19   An experiment illustrating the effect of different photoperiods on the flowering of Maryland mammoth tobacco.

ture may play an important role in their flowering. Florists use their knowledge of photoperiod to "force" short-day plants such as chrysanthemums to bloom year-round.

In fact, it is the length of the night rather than the length of the day that controls flowering and other photoperiodic responses. Experiments have demonstrated that a brief period of darkness during the daytime portion of the photoperiod has no effect on flowering. Even a few minutes of light during the dark period, however, prevents flowering, indicating that photoperiodic responses depend on a critical, uninterrupted night length. Long-day plants flower when the night is *shorter* than a critical length; short-day plants flower when the night is *longer* than a critical length (see Figure 19.19). Certain pigments that can be converted from one form to another are involved in the plant's ability to measure the length of darkness in a photoperiod. The conversion acts as a switching mechanism that may control many events in the life of the plant, such as flowering, seed germination, and chlorophyll formation.

A brief flash of red light during a long period of darkness causes a long-day plant to respond as though the night were short and to flower during periods of short day length.

CONCEPT REVIEW

1. What characteristics of plant hormones have made them especially difficult to study?
2. Compare and contrast the effects of applied auxins and gibberellins on plant growth.
3. What distinguishes the action of cytokinins from that of auxins?
4. How do phototropism and gravitropism affect plant growth?
5. What effect does photoperiodism have on flowering in various plants?

This investigation allows students to relate the number and distribution of stomates with the rate of photosynthesis. Teams of 2 are best for Part A; teams of 5 for Part B.

Time: Three and one-half class periods: one period for part A, one period for Day 0 of Part B, 15 minutes each on Days 1 and 3, and one period for Day 4 and discussion of the results.

INVESTIGATION 19.1 Gas Exchange and Photosynthesis

Following the principles of diffusion, carbon dioxide normally would flow into a leaf during the day, and oxygen would flow out as photosynthesis proceeds. In this investigation, you will disrupt the normal gas exchange process and observe the effects.

After reading the procedure for Part B, construct a hypothesis that predicts which leaf treatment will affect photosynthesis the most.

Materials Part A (per team of 2)

microscope slide	forceps
cover slip	scalpel
dropping pipet	fresh leaves, several types
compound microscope	

Materials Part B (per team of 5)

5 pairs of safety goggles	forceps
5 lab aprons	scissors
400-mL beaker	sheet of white paper
4 petri dish halves	paper towels
3 swab applicators	isopropyl alcohol (99%) in screw-cap jar
plastic bag labeled *waste*	Lugol's iodine solution in screw-cap jar
500-mL beaker of water	Histoclear™ in small corked bottle
petroleum jelly	2 potted plants

PART A PROCEDURE

4. **Some students may need help in computing field-of-view areas. Refer them to investigation A2.2 for calculating field-of-view diameter. The number depends on the type of leaf used and student observation.**

PART A Number of Stomates
Procedure

1. While holding the lower surface of a leaf upward, tear the leaf at an angle. The tearing action should peel off a portion of the lower epidermis. It will appear as a narrow, colorless edge extending beyond the green part of the leaf.

2. Lay the epidermis on a slide and use a scalpel to cut off a small piece. Add a drop of water and a cover slip. Do not allow the epidermis to dry out.

> **CAUTION: Scalpels are sharp; handle with care.**

3. Using the low power of your microscope, locate some stomates. Then switch to high power. Make a sketch to show the shape of a stomate, its guard cells, and a few adjacent cells in the epidermis.

4. Count the number of stomates in the high-power field for 10 different areas of the piece of lower epidermis. Calculate the average number of stomates per mm^2 of leaf surface. (Refer to Investigation A2.2 (Appendix 2) to calculate the diameter of the high-power field. Use this figure to calculate the area of the leaf observed under the microscope.)

5. In the same manner, count the stomates on the upper epidermis of the same leaf. Examine as many other types of leaves as possible. Compare the number of stomates per mm^2 on the upper and lower surfaces of each type of leaf.

6. Wash your hands thoroughly before leaving the laboratory.

DISCUSSION

1. **Depends on leaves used. Generally, there are more stomates on the lower than on the upper surface of a leaf.**

2. **Probably. Explanations may be based on the discussion in Sections 18.1 and 18.2, but all must be inconclusive.**

3. **Many counts musts be made and the counts averaged. Take care that counts are made on leaves of two plants that are comparable in (a) age of the plant and leaf and (b) position of leaves with respect to sun and shade.**

Discussion

1. How did the number of stomates per mm^2 compare in different areas of the piece of lower leaf epidermis? How did the number compare in different areas of the piece of upper leaf epidermis?

2. Did the stomates vary in the amount they were open? How can you explain your findings?

3. What would you do to assure a reliable comparison of the number of stomates per mm^2 for two species of plants?

4. What do your data suggest about the distribution of stomates in leaves of your species of plant? What assumption must you make in drawing this conclusion?

PART B Gas Exchange and Photosynthesis

Put on your safety goggles and lab apron.

Procedure
Day 0

7. Appoint one member of your team as safety monitor to be sure the correct safety procedures are followed.

8. Three days before this investigation, one plant was placed in the dark. A second plant of the same species was placed in sunlight. Remove one leaf from each plant. Use the scissors to cut a small notch in the margin of the one placed in sunlight.

9. Using forceps, drop the leaves into a beaker of *hot* (60°C) tap water.

10. When they are limp, use forceps to transfer the leaves to a screw-cap jar about half full of isopropyl alcohol. Label the jar with your team symbol and store it overnight as directed by your teacher.

 Isopropyl alcohol is flammable and toxic. Do not expose the liquid, or its vapors to heat or flame. Do not ingest; avoid skin/eye contact. In case of spills, flood the area with water and *then* call your teacher.

11. Select four similar leaves on the plant that has been kept in the dark, but do not remove them from the plant. Using a fingertip, apply a *thin* film of petroleum jelly to the upper surface of one leaf. Check to be sure the entire surface is covered. (A layer of petroleum jelly, although transparent, is a highly effective barrier across which many gases cannot pass.) Cut one notch in the leaf's margin.

12. Apply a thin film to the lower surface of a second leaf and cut two notches in its margin.

13. Apply a thin film to both upper and lower surfaces of a third leaf and cut three notches in its margin.

14. Do not apply petroleum jelly to the fourth leaf, but cut four notches in its margin. Place the plant in sunlight.

15. Wash your hands thoroughly before leaving the laboratory.

Day 1

16. Obtain your jar of leaf-containing alcohol from Day 0. Using forceps, *carefully* remove the leaves from the alcohol and place them in a beaker of room-temperature water. (The alcohol extracts chlorophyll from the leaves but also removes most of their water, making them brittle.) Recap the jar of alcohol and return it to your teacher.

17. When the leaves have softened, place them in a screw-cap jar about half full of Lugol's iodine solution.

  Lugol's iodine solution is poisonous if ingested, irritating to skin and eyes, and can stain clothing. Should a spill or splash occur, call your teacher immediately; flush the area with water for 15 minutes; rinse mouth with water if any solution entered mouth.

4. Distribution depends on the type of leaf used and student observations. The particular leaf examined is typical of the entire plant.

 SAFETY

Before beginning Part B, be sure all students are familiar with correct safety procedures and determine who in each team is the safety monitor. Remind students to put on their safety goggles and lab aprons each day.

Limit the amount of isopropyl alcohol in the laboratory by placing it in same (4-oz) screw-cap jars. Use the smallest amount adequate for the leaves, but no more than 60 mL per jar. Before the experiment, be sure there are no ignition sources in the laboratory and have on hand a recently tested ABC-type fire extinguisher. (Alcohol burns with a blue, almost invisible flame.)

 Store the student's leaf containing jars in a cool, dark, place overnight. Prepare a set of jars for each class; the alcohol can be reused on Day 3. To discard, empty the jars into a 2-L beaker, run water from the tap into the beaker for 10 minutes, and then run water down the drain for an additional 10 minutes to dilute the alcohol. In case of a spill, safety monitors should be instructed to throw water on the spill to dilute the alcohol, then call you. You (*not* the students) can wipe up the spill with paper towel.

Use 60 mL or less of Lugol's iodine solution per screw-cap jar; one set of jars is adequate for all classes, and the solution can be saved at the end of the experiment. Use protective gloves (nitrate rubber is recommended) and paper towels to wipe up spills of Lugol's solution.

Histoclear™ is a low-toxic solvent substitute for xylene. Remove the droppers from dropping bottles, place about 10 mL in each bottle, and add a cork. In case of spills, absorb onto paper towels and place the towels in a fume hood to allow the Histoclear to evaporate. Wearing protective gloves (nitrile rubber), moisten a paper towel with 99% isopropyl alcohol and rub the

 area to remove any reside. Place the paper towel in a fume hood to allow the alcohol to evaporate.

18. After several minutes, use forceps to remove both leaves, rinse them in a beaker of water, and spread them out in open petri dishes of water placed on a sheet of white paper. Record the color of each leaf. Recap the jar of Lugol's iodine solution and return it to your teacher.

19. Wash your hands thoroughly before leaving the laboratory.

Day 3

20. Remove from the plant the four notched leaves prepared on Day 0 and place them on paper towels. To remove the petroleum jelly, dip a swab applicator in the Histoclear™ and *gently* rub it over the surface of the film once or twice. Then blot gently with a paper towel to remove any residue of petroleum jelly. Discard the swab applicator and the paper towel in the waste bag.

> **WARNING: Histoclear is a combustible liquid. Do not expose to heat or flame. Do not ingest; avoid skin/eye contact. Should a spill or splash occur, call your teacher immediately; wash skin area with soap and water.**

21. Repeat Steps 9 and 10.

22. Wash your hands thoroughly before leaving the laboratory.

Day 4

23. Repeat Steps 16–18. Compare the color reactions of the four leaves and record your observations.

24. Wash your hands thoroughly before leaving the laboratory.

Discussion

1. What was the purpose of the iodine tests on Day 1?

2. If you use these tests as an indication of photosynthetic activity, what are you assuming?

3. What is the purpose of the leaf marked with four notches?

4. In which of the leaves coated with petroleum jelly did photosynthetic activity appear to have been greatest? Least? Do your data support your hypothesis?

DISCUSSION

1. **Control—to establish a condition in plants in the light (photosynthesizing plants) and in the dark (nonphotosynthesizing plants).**
2. **Starch accumulation in leaves usually is regarded as an indication of photosynthesis and lack of starch as an indication of lack of photosynthesis.**
3. **This leaf serves as a control.**
4. **The best support of the hypothesis comes from results that show no starch in either the two-notched or the three-notched leaves and much starch in both the one-notched and four-notched leaves.**

FOR FURTHER INVESTIGATION

Compare the number of stomates and their locations on leaves of two different species. Select one species that usually grows in full sunlight and one that grows in the shade.

This investigation allows students to compare the rate of photosynthesis under a variety of conditions and makes more meaningful the discussions in section 19.6. Teams of 4 work well with the LEAP-System™, with two teams per 8-input system.

Time: **Three or more class periods: one to set up the apparatus and collect data for the basic photosynthetic rate; one for each additional variable tested; one for class discussion of all results. Time can be reduced by having each team test a different variable, but all teams should perform Part B. Data and graphs should be combined and discussed as a class.**

INVESTIGATION 19.2   Photosynthetic Rate

There are several ways to measure the rate of photosynthesis. In this investigation, you will use *Elodea* and pH probes with the LEAP-System™, or narrow-range pH paper. The rate of photosynthesis is determined indirectly by measuring the amount of carbon dioxide removed from water by *Elodea*. Carbon dioxide is added by bubbling breath into the water, which absorbs the carbon dioxide. The carbon dioxide combines with water to produce carbonic acid ($H_2O + CO_2 \rightleftharpoons H_2CO_3$), a reversible reaction. As carbon dioxide is used by elodea in photosynthesis, less acid is present and the pH increases. The LEAP-System BSCS Biology Lab Pac 2 includes instructions for use of the system.

> After reading the procedure, construct hypotheses that predict the effects of light intensity and light color on the rate of photosynthesis.

Materials (per team of 4)

2-L beaker	250-mL flask
100-mL beaker or small jar	2 25-mm × 200-mm test tubes
2 1-mL pipets (optional)	1 thermistor
2 wrapped drinking straws	narrow-range pH paper (optional)
lamp with 100-watt spotlight	nonmercury thermometer (optional)

Apple lle, Apple IIGS, Macintosh,
    or IBM computer
LEAP-System™
printer
2 pH probes

red, blue, and green cellophane
distilled water
tap water at 25°C and 10°C
2 15-cm sprigs of young *Elodea*

Procedure
PART A Equipment Assembly

1. In your data book, prepare a table similar to Table 19.1, or tape in the table your teacher provides.

2. Put 125 mL of distilled water in the flask. Blow through a straw into the water for 2 minutes. Be careful not to suck any liquid through the straw. Discard the straw after use as your teacher directs. What is the purpose of blowing into the water in the flask?

3. Place two sprigs of *Elodea,* cut end up, into one of the test tubes (experimental tube).

4. Fill both test tubes 3/4 full with the water you blew into. What is the purpose of the second test tube?

5. Insert the pH probe from input 1 into the experimental test tube and the pH probe from input 2 into the other test tube, as shown in Figure 19.9.

6. Stand the test tubes in a 2-L beaker and add 25°C tap water to the beaker until it is about 2/3 full.

7. Insert the thermistor (or a thermometer) into the water in the 2-L beaker Determine the temperature in the beaker and maintain this temperature throughout the experiment. (Use the small beaker to add ice water and/or to remove water from the 2-L beaker.)

8. Let the entire assembly stand for about 5 minutes to permit the temperature to become uniform throughout the system. Note the initial pH in the test tubes and record both readings in the data table. (If the LEAP-System is not available, use the 1-mL pipets to transfer a drop of water halfway down each test tube to a piece of narrow-range pH paper and read the pH from the comparison chart on the strip).

PART B Basic Photosynthetic Rate

9. Place the lamp with the 100-watt spotlight 30 cm from the beaker and illuminate the two test tubes.

10. Take pH and temperature readings for 30 minutes. In the data table, record the current readings every 5 minutes.

11. Wash your hands thoroughly before leaving the laboratory.

PROCEDURE

  2. To add carbon dioxide to the water.
  4. It is the control.

Table 19.1 Rate of Photosynthesis

	Experimental Condition																	
	30 cm			10 cm			50 cm			Red			Blue			Green		
Reading #	pH 1	pH 2	°C	pH 1	pH 2	°C	pH 1	pH 2	°C	pH 1	pH 2	°C	pH 1	pH 2	°C	pH 1	pH 2	°C
1																		
2																		
3																		
4																		
5																		
6																		

Figure 19.20 The LEAP-System setup.

DISCUSSION

1. **In water, the CO$_2$ from your breath forms carbonic acid (H$_2$CO$_3$), which dissociates into H$^+$ and HCO$_3$, lowering the pH.**
2. **Photosynthesis removes CO$_2$, which reduces the H$^+$ concentration, raising the pH.**
3. **Temperature and pH.**
4. **Depends on student data. (With healthy, active *Elodea*, pH change at 30 cm may be about 0.5; at 10 cm, about 1.5.)**
5. **Any increase or decrease in pH without *Elodea* should be subtracted from or added to the results with *Elodea*.**
6. **Horizontal is time, vertical is pH.**
7. **The higher the light intensity, the higher the rate of photosynthesis for *Elodea*. Depends on student hypotheses.**
8. **Red and blue should produce higher rates than green. Depends on student hypotheses.**

This investigation offers a simple way to demonstrate gravitropism and phototropism. Appropriate hypotheses might concern expected effects of gravity (Part A), or color of light (Part B). Be sure all students observe both parts. Teams of 4 work well, each team performing Part A or Part B.

Time: One half to one period to prepare the setups; 5 minutes daily for making and recording observations, one half period for discussion.

Figure 19.21 Petri dish setup.

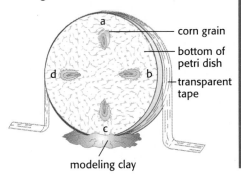

- corn grain
- bottom of petri dish
- transparent tape

modeling clay

PART C Effects of Light Intensity on Photosynthetic Rate

12. Repeat Steps 2–10 with the light source first 10 cm and then 50 cm away from the test-tube assembly.

PART D Effects of Light Color on Photosynthetic Rate

13. Repeat Steps 2–10 using first red, then blue, and then green cellophane over the beaker.
14. Wash your hands thoroughly before leaving the laboratory.

Discussion

1. What chemical change occurred in the water you blew into?
2. What happens during photosynthesis that causes the pH to increase in the test tubes?
3. What two environmental factors are being controlled by the test tube without *Elodea*?
4. How much did the pH change during the 30-minute period for each condition tested?
5. How would you use any change in pH in the test tube without *Elodea* to correct the data for the experimental test tube?
6. Run stripcharts to produce graphs for each part of the investigation. What variables are on the horizontal and vertical axes of the graphs? (If using pH paper, prepare graphs showing pH every 5 minutes under the various conditions being tested.)
7. Use the change in pH that occurs at the three light intensities you tested to determine the effect of light intensity on the rate of photosynthesis. Do your data support your hypothesis?
8. Use the change in pH with the three colors (red, blue, and green) to determine the effect of light color on the rate of photosynthesis. Do your data support the hypothesis that you constructed?

INVESTIGATION 19.3   Tropisms

This investigation allows you to observe tropisms. Each team will conduct only one part of the investigation. Observe the other part so you can discuss all the results as a class.

 After reading the procedures, construct two hypotheses—one predicting how the germinating corn grains will be affected and the other predicting the patterns of growth of the radish seeds.

Materials Part A (per team of 4)

petri dish	heavy blotting paper
scissors	transparent tape
glass-marking pencil	modeling clay
nonabsorbent cotton	4 soaked corn grains

Materials Part B

4 flowerpots, about 8 cm in diameter	scissors
4 cardboard boxes, at least 5 cm higher than the flowerpots	transparent tape
modeling clay	40 radish seeds
soil	red, blue, and clear cellophane

PART A Orientation of Shoots and Roots in Germinating Corn
Procedure

1. Place 4 soaked corn grains in the bottom half of a petri dish. Arrange them cotyledon side down, as shown in Figure 19.21.

2. Fill the spaces between the corn grains with wads of nonabsorbent cotton to a depth slightly greater than the thickness of the grains.

3. Cut a piece of blotting paper slightly larger than the bottom of the petri dish, wet it thoroughly, and fit it snugly over the grains and the cotton.

4. Hold the dish on its edge and observe the grains through the bottom. If they do not stay in place, repack with more cotton.

5. When the grains are secure in the dish, seal the two halves of the petri dish together with cellophane tape.

6. Use modeling clay to support the dish on edge, as shown in Figure 19.21, and place it in dim light.

7. Rotate the petri dish until one of the corn grains is at the top. With the glass-marking pencil, write an *A* on the petri dish beside the topmost grain. Then, proceeding clockwise, label the other grains *B, C,* and *D.* Also label the petri dish with your team symbol.

8. When the grains begin to germinate, make sketches every day for 5 days, showing the direction in which the root and the shoot grow from each grain.

9. Wash your hands thoroughly before leaving the laboratory.

Discussion

1. From which end of the corn grains did the roots grow? From which end did the shoots grow?

2. Did the roots eventually turn in one direction? If so, what direction?

3. Did the shoots eventually turn in one direction? If so, what direction?

4. To what stimulus did the roots and shoots seem to be responding?

5. In each case, were the responses positive (toward the stimulus) or negative (away from the stimulus)?

6. Why was it important to have the seeds pointed in four different directions?

PART B Orientation of Radish Seedlings

Procedure

10. Turn the four cardboard boxes upside down. Number them *1* to *4.* Label each box with your team symbol.

11. Cut a rectangular hole in one side of boxes 1, 2, and 3. (Use the dimensions shown in Figure 19.22.) Do not cut a hole in box 4.

12. Tape a piece of red cellophane over the hole in box 1, blue cellophane over the hole in box 2, and clear cellophane over the hole in box 3.

13. Number four flowerpots *1* to *4.* Label each with your team symbol. Fill the pots to 1 cm below the top with soil.

14. In each pot, plant 10 radish seeds about 0.5 cm deep and 2 cm apart. Press the soil box down firmly over the seeds and water the soil. Place the pots in a location that receives strong light but not direct sunlight.

15. Cover each pot with the box bearing its number. Turn the boxes so the sides with holes face the light.

16. Once a day remove the boxes and water the soil. (Do not move the pots; replace the boxes in the same position.)

17. When most of the radish seedlings have been above ground for 2 or 3 days, record the direction of stem growth in each pot—upright, curved slightly, or curved greatly. If curved, record in what direction with respect to the hole in the box.

Discussion

1. In which pot were the radish stems most nearly upright?

2. In which pot were the radish stems most curved? In what direction were they curved? Were the stems curved in any of the other pots? Which ones and in what direction were they curved?

Figure 19.22 Flowerpot setup

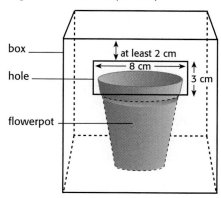

PART B

1. Box 4.
2. **The stems in box 3 (one-sided illumination with white light) should have stems curved most strongly in one direction; box 1 (red light) should show little or no bending; box 2 (blue light) should show bending toward the source of light, but not as much as in box 3. (Light intensity is a factor here; different colors of cellophane reduce intensity differently.)**
3. **Light.**
4. **See answer 2.**

3. To what stimulus do you think the radish stems responded?
4. What effect, if any, did the red and blue cellophane have on the direction of the radish stem growth?

SUMMARY

In the light reactions of photosynthesis, light energy is absorbed by a variety of plant pigments, primarily chlorophyll. A series of reactions follows in which energy-rich molecules are formed as the light energy is converted into chemical energy. In addition to these molecules, oxygen gas is released into the atmosphere as a by-product. In the Calvin cycle, the energy rich molecules are used to make 3-carbon sugars from carbon dioxide. During cellular respiration the energy from these sugars is released. Environmental factors such as temperature, oxygen and carbon dioxide concentration, and light intensity interact to affect the rate of photosynthesis. Photorespiration, a light-dependent process that results in the uptake of oxygen and release of carbon dioxide, occurs simultaneously with photosynthesis, resulting in loss of previously incorporated carbon dioxide and reducing photosynthetic efficiency. C-4 photosynthesis is a special adaptation that reduces photorespiration in certain warm-climate plants; CAM photosynthesis is an adaptation to hot, dry climates.

The growth and development of plants result from the interaction of plant hormones and environmental factors such as light and gravity. Auxins, gibberellins, cytokinins, abscisic acid, and ethylene are the groups of natural plant hormones. Some of these have been produced synthetically by humans and are used in research and agriculture.

APPLICATIONS

1. In what ways are photosynthesis and cellular respiration opposite sets of reactions?
2. Trace the production of ATP molecules in both respiration and photosynthesis.
3. What roles do photosynthesis and respiration play in the cycling of carbon?
4. How are plant hormones used to bring about differentiation and growth of a new plant from a callus?
5. Tomatoes are fruits. Many tomatoes are picked when they are green (unripe) and packaged with plastic wrap around them. What advantage does this packaging offer?
6. How much food do plants get from the soil? Explain your answer.
7. Energy can be changed from one form to another, but it cannot be created or destroyed. Give several examples from this chapter, or any part of the text, in which energy is changed from one form to another.
8. In the overall equation for photosynthesis, why is chlorophyll written under the arrow instead of to the left or right of the arrow?
9. Trace the flow of energy in photosynthesis, beginning with the sun and ending with a sugar molecule.

10. Different species of grasses grow all over the United States. Some are C-3 species and some are C-4. Would you expect C-4 species to make up a greater percentage of all the grasses in the northern part of the United States or in the southern part? Explain.
11. Describe possible hormonal and environmental stimuli that might cause deciduous trees to lose their leaves in the fall.
12. Some plants use changes in day length to help them determine the change of seasons. Why is photoperiodism a better measure of the oncoming winter than is falling temperature?
13. Why do you think visible light is the only radiation from the sun that triggers photosynthesis? Consider the characteristics of the wavelengths that are much shorter and much longer than those of visible light.
14. Potatoes grow very well during the summer in Idaho but not very well during the summer in Texas. Why? Consider the difference in temperature and its effect on photosynthesis and photorespiration.

PROBLEMS

1. Investigate some of the practical uses of auxins and synthetic growth substances as they apply to various types of farming.

2. Many botanists are searching for plants that synthesize fuel-related carbon compounds. Part of our energy needs could be met by growing such plants commercially. What other types of useful compounds do plants produce? Report your findings to the class.

3. Experiments have shown that when photosynthesizing vascular plants are grown in an atmosphere without oxygen they take up carbon dioxide at 1.5 times the rate at which they do in natural atmosphere, which is about 20 per cent oxygen. What does this indicate about the relationship between photosynthesis and photorespiration? How might this relationship affect the composition of the earth's atmosphere?

4. Plant development appears to be much more closely linked to environmental cues than does animal development. Explain how this adaptation would enhance survival of plants.

5. Write a science fiction short story in which the humans are genetically engineered to contain some of the genes of plants. What type of organisms would they be, and what advantages might they have?

6. Design and conduct an experiment that tests the effect of different colors of lights on the growth and development of plants. Make sure that the amount of light reaching each plant is the same—just different colors. What other factors must you control?

7. Different plant hormones may be obtained easily from your grocery store or local nursery. For example, cytokinin, along with many nutrients, may be found in coconut milk from fresh coconuts. Ethylene, a gas, is produced by ripening fruits. Auxin may be obtained in commercial root growth stimulators. Design an experiment to test the effect of one of these hormones on the growth and development of a plant.

8. Starch can be detected by bleaching the chlorophyll from a leaf and then staining it with an iodine solution. No sugar, and thus no starch, will be produced in those parts of the leaf that do not undergo photosynthesis. Designs can be made in leaves by first cutting a pattern, smaller than a leaf, out of a piece of aluminum foil and using it to cover a large green leaf of a geranium or other plant with big, sturdy blades. Leave the foil on the leaf, and leave the plant in the sun for several days. At the end of this period, remove the foil, bleach the leaf, and stain it with iodine. Can you see the design in the leaf?

9. For those of you with a solid chemistry background, look in any college biochemistry text for a complete explanation of photosynthesis, cellular respiration, and the relationship between these two processes. Make a wall chart showing a complete set of reactions relating these processes.

One of the observable patterns in the biosphere is the profound effects that organisms have on the physical environment. This coral reef in Australia illustrates their effects. Assuming that the island is approximately 10 km in length, estimate the magnitude of the effect these organisms have had on this environment. How could organisms as small as corals have been responsible for this effect?

The diameter of the island is about ⅕ of the diameter of the coral reef. If the radius of the island is approximately 10 km, the area of the entire reef would be πr^2, or $(10)^2 \times 3 = 300$ km². The buildup of exoskeletons as the corals grow produces the reef.

Section Five

PATTERNS IN THE BIOSPHERE

◆ Ecologists study living things as parts of interrelated patterns. Individuals are members of populations, and populations are members of communities and ecosystems. Ecosystems do not remain the same forever but change continuously. Catastrophes such as floods or fires and even the organisms themselves may change the character of entire ecosystems. The ecosystems of the earth today are the result of interacting organisms and patterns of the earth's past. The study of ecosystems—how and why they change—including how humans shape the changes, helps us predict future environments and, in a general way, the future of our world. ◆

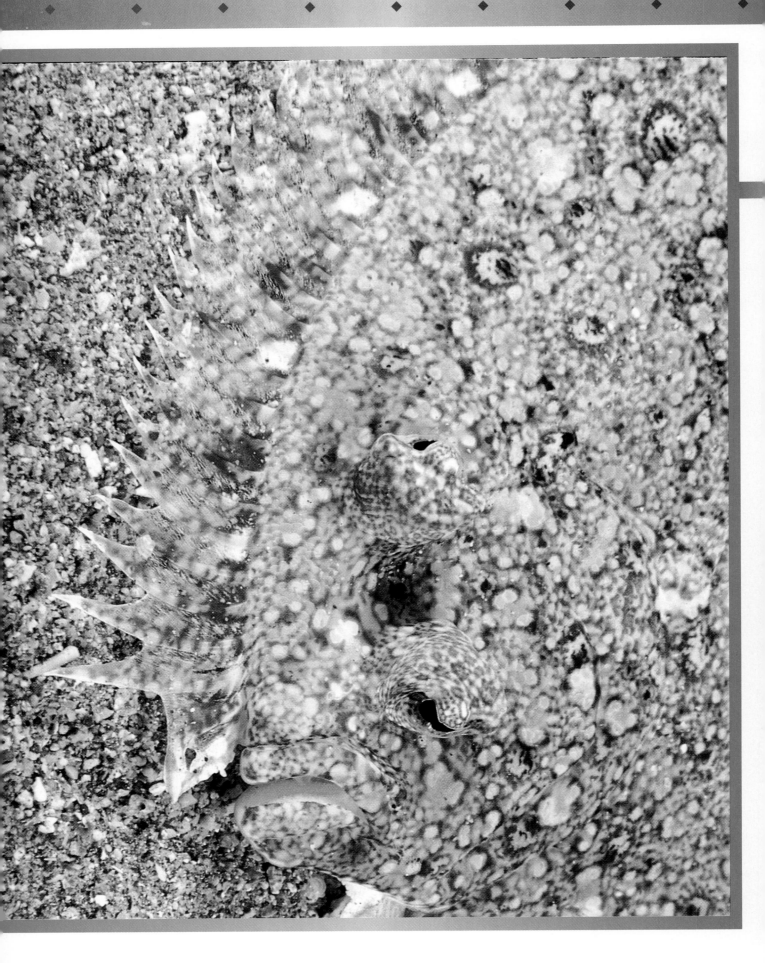

Behavior, Selection, and Survival

20

Can you find the animal in this photo? What do you see that would affect its chances of being selected by a predator? How would this affect its ability to pass its genetic information to the next generation? **There is a peacock flounder against the sea floor. It resembles the background closely and probably would go unnoticed by a predator. A survivor would have a good chance of reproducing.**

◆ All living organisms are *doing* something continually. They respond to stimuli from their external environment, and they respond to stimuli within their bodies. The sum of all the activities of an organism is called behavior. An organism's behavior is influenced by other organisms of the same type, by other organisms it uses as food, and by other organisms that use *it* as food. Behavior also includes all of an organism's responses to such things as the time of year, the time of day and night, changes in the amount of light, changes in temperature, and every other characteristic of the ecosystem in which the organism lives. The behavior of an organism must be adaptive if it is to survive and reproduce.

Organisms are what they are because the genes that they share with their ancestors allowed those ancestors to survive and reproduce. Environmental selection may act in favor of individuals with certain genes, but it removes other individuals (and their genes) from the gene pool. If the surviving individuals in a population are adapted to the environment, their genes will be retained. If the environment changes, however, the individuals and their genes may be removed. This chapter discusses how behavior and different types of selection affect an organism's ability to survive and reproduce. ◆

MAJOR CONCEPTS

- ◆ Behavior is the sum of all the activities of an organism in response to its environment.
- ◆ Innate behavior is genetically driven; learned behavior requires experience.
- ◆ Appropriate behavior improves the survival chances of the organism and its species.
- ◆ Sexual selection results in traits that allow individuals to compete more successfully for mates.
- ◆ Humans have major effects on the behavior of organisms and their ability to survive.

KNOWLEDGE CHECK

- ◆ What is behavior and how does it affect survival?
- ◆ What is sexual selection?
- ◆ How does the environment affect an organism?
- ◆ What is tolerance?
- ◆ How do humans affect the behavior of other animals?

Behavior and Selection

20.1 Behavior Usually Falls into Two Categories

In the broadest sense, behavior is considered to be either **innate** or **learned.** Innate, or inborn, behavior is not dependent on individual experience. Sometimes it is a pattern, such as crying or sucking, that is ready to function when the organism is born. Dogs that have spent their entire lives indoors will attempt to bury a bone by scratching the carpet as if they were digging in the soil. The same dogs will turn around several times before lying down, although there is no grass to trample or insects to chase out of the bed. These behavior patterns are innate and characterize the species. Behavior is considered to be innate if it occurs in response to a particular stimulus the first time an individual is exposed to that stimulus.

Environmental influences may trigger an innate behavior, but they do not explain it. The kitten in Figure 20.1 is standing on a checkerboard surface apparently looking down a drop, or cliff. Because a plate of glass covers it, the drop is only visual, not actual. The kitten will not venture out onto the glass, however, even though it has had no experience with heights

Guidepost

How does an organism's behavior affect its survival?

517

Figure 20.1 This Siamese kitten has never fallen or been dropped. It has had no experience with high places. The drop you see here is only visual; a plate of glass gives the kitten firm underfooting. Yet the kitten hesitates. Such experiments are used to help determine when behavior is innate and when it is learned.

This type of learning is called imprinting.

Seagulls exhibit similar behavior. Discuss how imprinting behavior could enhance survival.

This type of learning is called habituation.

Ask students how habituation would enhance survival and how it is helpful to humans. Habituation makes an individual more attentive to strange and potentially harmful stimuli.

Ask students to think of behaviors that animals *learn* from their environment; ask them how they know the behavior is learned.

or falling. Its reaction is an innate behavior in response to the visual stimulus. Such a complex innate behavior is called an instinct.

Learned behavior, on the other hand, consists of all those responses an organism develops as a direct result of specific environmental experiences. Most animals can learn to behave in new ways. A series of experiments with the southern toad shows how experience may change an innate behavior. The toad is a natural predator of the robber fly, which looks very much like a bumblebee. A toad may capture these robber flies until it mistakenly captures a real bumblebee, and the bee stings it on the tongue. After recovering from this experience, the toad does not attempt to eat either robberflies or real bumblebees, although it continues to eat other insects. The toad has changed its innate behavior.

One intriguing type of learning is shown in Figure 20.2. Once newly hatched goslings or ducklings are strong enough to walk, they will follow their mother away from the nest, and soon they will not follow any other animal. Through the experience of seeing and following a moving object, they learn something—to follow only their mother for weeks afterward. If the eggs are hatched away from the mother, however, the young will follow the first moving object they see, even a human. This type of learning is special because there is a sensitive period in which the learning occurs, and it requires little or no practice.

A simple type of learned behavior occurs when an animal is exposed to a stimulus over and over again. This repetitive exposure may cause the animal gradually to lose its response to the stimulus. For example, a dog responds to the sound of a car driving past by turning its head. If cars continue to drive by, the dog may learn to disregard the sound. Such a reaction can be thought of as learning how *not* to respond to unimportant stimuli.

A more complex type of learning occurs when the patterns of an innate behavior are changed. Such conditioned responses have been developed in many types of animals. One of the most interesting experiments was performed on planarians, as shown in Figure 20.3. First, the biologists shone a strong light on a planarian. Several seconds later, they administered a mild electric shock. The planarian's normal response to the light was to stretch, and its normal response to the shock was to contract or turn its head. This sequence, in which the light was followed by the shock, was

Figure 20.2 Many species of birds, such as these common mergansers, *Mergus merganser,* will follow the first moving object they see.

Biology Today
Camouflage and Mimicry

One evolutionary result of predator-prey relationships is the ability of many animals to "hide" in the open. Camouflage is an adaptation in form, patterning, color, or behavior that enables an organism to blend with its surroundings and thus better escape detection by predators. The canyon tree frog pictured in (a) exhibits camouflage. Camouflage, of course, is not exclusive to prey animals. Predators that rely on stealth, such as polar bears and tigers, often blend well with their background.

Another evolutionary consequence of predation is the existence of many prey species that are bad-tasting, poisonous, or able to harm their attackers in some fashion. Often, toxic prey species are colorful or have bold markings that warn potential predators to stay away. Inexperienced predators might attack a black-and-white striped skunk, a black and yellow wasp, or a poison arrow frog—but probably only once. As a result of their experience, these predators would quickly learn to associate the colors and patterns of the skunk, wasp, or frog with intestinal upsets. During the course of their evolution, some prey species without any such defenses have come to resemble distasteful ones that have bold colors or markings. If the unprotected animals are relatively few in number compared with the species they resemble, they too will be avoided by predators. The resemblance of an edible species to a relatively inedible species is one form of mimicry.

Many of the best examples of this type of mimicry occur among butterflies and moths. The larvae eat the plants and retain the poisonous molecules in their own bodies. The mimic butterfly larvae feed on a variety of different plants, but not those protected by toxic chemicals. One well-known mimic is the viceroy butterfly, *Limentis archippus* shown in (b). This butterfly resembles the poisonous monarch butterfly (c). Its larvae feed on willows and cottonwoods, and neither the larvae nor the adults are distasteful to birds. Interestingly, the larvae of the viceroy are camouflaged on the leaves of their host plants, where they resemble bird droppings. The distasteful larvae of the monarch butterfly, however, are very conspicuous.

If predators are to be deceived, mimic and model not only must look alike but also must act alike. For example, the members of several families of insects that resemble wasps behave surprisingly like the wasps they mimic. Mimics also must spend most of their time in the same habitats as do their models. If they did not, predators would discover that they are quite edible.

a

b

c

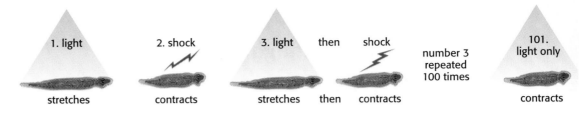

| 1. light | 2. shock | 3. light   then   shock | number 3 repeated 100 times | 101. light only |

stretches   contracts   stretches   then   contracts   contracts

Figure 20.3 A conditioned reflex experiment with a planarian. As a result of the conditioning, the planarian eventually responds to light as it would to electric shock.

repeated about 100 times. Eventually the planarian's response to the light alone was to contract or turn its head, the same as if the electric shock also had been given.

A somewhat higher level of learning occurs when an animal learns a behavior through trial and error. An animal faced with two or more possible responses eventually may learn to choose only the one for which a reward is given. Pigeons, for example, can be taught to peck only at a button of a specific shape or color if the correct choice is rewarded with food.

Learning gives an animal flexibility to adapt to a changing environment by acquiring new behavior patterns. Learning is adaptive when an animal, faced with unpredictable problems, develops new responses that prove helpful in a particular situation. Innate responses also are adaptive. For example, the kitten that avoids the cliff edge also avoids potentially fatal falls.

Most adaptive behavior has both innate and learned components. Consider the development of flight in birds. Flight is a largely innate behavior. Obviously, a young bird must be able to manage it fairly well on its first attempt, or it will crash to the ground. Young birds, however, do not fly as well as adults (see Figure 20.4). Even with flight, then, the innate pattern can be improved by learning through practice.

Communication also can be innate or learned. The prairie dog, for example, has an innate warning call that it voices when it sees a predator. Bees have a complex set of communication signals known as the bee dance. Through this dance, illustrated in Figure 20.5, a bee can communicate the location of food to other bees in the hive. Humans, on the other hand, must learn their vocal communication.

Discuss innate behaviors in humans. Discuss learned behaviors in humans and the long period of time necessary for this education.

You may wish to use the term *pheromones* in discussing chemical communication. Distinguish between hormones, which work *within* an individual, and pheromones, which work *between* individuals. Pheromones, instead of dangerous insecticides, now are being used in insect traps.

Figure 20.4 Although birds are born with the innate ability to fly, some maneuvers must be learned. How might this young bald eagle, *Haliaeetus leucocephalus*, benefit from practice? **Landings would be easier.**

a

b

a A foraging honeybee returns to the hive after finding a nectar source. Other bees soon leave the hive and fly to the same source. How does the first bee inform the other bees about the food source?

Figure 20.5 How bees communicate.

b The returning bee gives nectar to the other bees, then begins a dance on the verticle face of the hive.

c If the dance is a round dance, other bees search for plants nearby.

d If the dance is a figure 8 with wagging of the abdomen, bees search at a distance.

Communication may confer a number of advantages on the animal, but its ultimate function is to increase the animal's reproductive success. Communication can do this directly, as when the male elk gives its special call, called a bugle, to advertise for a mate, or indirectly, as when the prairie dog warns its offspring and other group members of danger. Communication, then, helps animals to survive and reproduce. Some communication is purely visual, such as the colorful breeding plumage of many male birds. Other types of communication include chemical signals, such as scents, and vocal communication, such as a cat's meow.

Although humans possess the most complex brain and language, they may not be alone in the ability to learn new methods of communication or to pass on knowledge. A chimp named Washoe (Figure 20.6), who was taught some American Sign Language as an infant, has shown her own

Using signals or displays, animals communicate with members of their own species or even different species. Ask students to speculate on the value of such communication.

You may wish to discuss the structure of many insect societies, in which there is a clear division of labor. Most animal societies do not have such clear-cut partitioning.

Figure 20.6 Chimp using sign language to communicate.

Figure 20.7   This swamp crayfish, *Procambarus*, has assumed a defensive posture that makes the animal appear larger and more threatening.

Ask students why animals might use rituals to determine the winner of a contest. Violent combat could reduce the reproductive ability of the winner as well as the loser; safety of the young may depend on other members of the group; the loser may eventually become the dominant animal.

adopted infant the sign for chair and has signed *food* to him and then molded his hands with hers into the sign for food. Not all researchers agree on the meaning of these actions, but data from these and other studies have suggested that new modes of communication can be learned by some non-human animals.

No matter what method of communication is used—vocal, visual, or chemical—they all are used to communicate with other animals. The other animals may be members of the same species, predators, or other animals within the same area. Communication, then, aids the interactions animals have with one another.

20.2   Organisms Interact with One Another

Any type of interaction between two or more animals, usually of the same species, is called **social behavior.** Mating behavior is a social behavior that contributes to the reproductive success of the interacting individuals. Other social interactions are competitive rather than cooperative, usually as a result of limited resources.

Because members of a population live together and use the same resources, there is strong potential for conflict. A contest of some sort usually determines which competitor gains access to something they both need, such as food or mates. Often, the encounter involves a test of strength or, more commonly, threat displays that make individuals appear large or fierce (see Figure 20.7). Usually some type of submissive or appeasement display by the contestant who surrenders or gives up the contest follows these displays. Much of this type of threatening behavior includes **ritual,** the use of symbolic activity, with the result that little or no real harm is done to either animal.

Dogs and wolves show aggression by baring their teeth, erecting their ears, tail, and fur, standing upright, and looking directly at their opponent—all actions that make the animal look large and threatening. The eventual loser, on the other hand, sleeks its fur, tucks its tail, and looks away, as shown in Figure 20.8. This appeasement display inhibits any further aggression. Although in some cases animals injure each other, most

Figure 20.8   Social behavior in wolves. The dominant male of the pack called the alpha male, reasserts his dominance over his brother who, in turn, assumes a submissive or appeasing posture. The interaction is ritualized—a fight that does not result in injuries.

Figure 20.9 Dominance relationships in white leghorn hens. The hen in the upper left in her first contact has been dominated by another hen. In the foreground, the lowest hen in the flock is pecked by the alpha hen. Three other hens, intermediate in the pecking order, rushed to the food when the top hen left.

often they end the contest as soon as the ultimate winner is clearly established. Appropriate behavior at the appropriate time is, therefore, adaptive both for the winner and the loser.

20.3   Many Animal Societies Are Hierarchical

If several chickens that are unfamiliar with one another are grouped together, they respond by pecking at each other and skirmishing. Eventually, they establish a clear "pecking order"—a more or less linear **dominance hierarchy.** The top-ranked hen drives off all others, often by mere threats rather than by actual pecking. The second-ranked hen similarly subdues all others except the top-ranked hen, and so on down the line to the lowest animal (see Figure 20.9).

In confrontations between animals, the larger, stronger, and more mature animal usually wins. Sex can make a difference: In some societies, males dominate females, whereas in others females dominate males. Animals with young tend to win out over animals without young. Animals that are ill tend to lose, and animals on winning streaks tend to continue to win. In a typical wolf pack, there is a dominance hierarchy among the females. The top female controls the mating of the others. When food is abundant, she mates and allows other females to do so also; when food is scarce, she allows less mating, sometimes restricting it to herself.

Although a social hierarchy may have survival value for the society, as in the case of a pecking order that reduces energy loss by bringing the activities of the group into harmony, the hierarchy really results from individuals striving for dominance within the group. The behavior that creates such hierarchies is best understood in terms of individual reproductive success. Those individuals closest to the top of the hierarchy stand the best chance of mating and producing offspring.

20.4   Many Animals Require a Territory

Although many animals live in groups, other species live alone or in pairs. Many of these more solitary animals expend much energy to keep other members of the same species out of their territory.

Figure 20.10 These gannets require very little space for individual nesting sites.

Dominance hierarchies and territoriality result from individual striving for enhanced reproductive success. Do not allow students to think of these mechanisms in terms of group selection.

Every living organism requires a certain amount of space in which to live, find food, and reproduce. The living space that an animal defends is called a territory, and the behavior of an animal defending its living area is called **territoriality.** Territories typically are used for feeding, mating, rearing young, or combinations of these activities. The size of a territory varies with the species, the use it makes of the territory, and the available resources. Song sparrows, for example, may have territories of about 3000 m², in which they live more or less permanently. Gulls and other seabirds, on the other hand, mate and nest in territories of only a few square meters or less, as Figure 20.10 illustrates.

Territories are established and defended through threat displays and challenges. An individual that has gained a territory may be difficult to dislodge. Ownership is announced continually—one function of most bird songs. Other animals, such as dogs and cats, use scent marks or frequent patrols that warn potential challengers.

Although dominance hierarchies and territories are mechanisms that primarily affect the reproductive ability of the individual, they do offer some incidental advantages to the population as a whole because they tend to stabilize density. For example, if resources were distributed evenly among all members of a population, the "fair share" that each member received might not be enough to sustain any one of them, particularly if a bad year reduced food supplies. Dominance and territoriality help ensure that at least some individuals receive an adequate amount of all resources.

20.5 Sexual Selection Plays an Important Role in Many Populations

If competition and the use of threatening displays are common among animals, how do the males and females manage to get close enough to mate? In many animals, a complex courtship interaction, unique to that species, occurs before the individuals mate. Each step of the courtship must be completed correctly and in the proper sequence to assure one partner that the other is not a threat and is the appropriate age and sex. Courtship often consists of a series of small behaviors. The correct performance of one behavior on the part of one partner leads the other partner to perform an answering behavior (Figure 20.11). In many cases, the sequences probably **3** evolved from threat displays and submissive gestures used in competitive interactions. The use of a submissive posture would allow one partner to remain close to the other.

Reemphasize that the ultimate goal of an individual is to reproduce. Behaviors that enhance the ability of an animal to attract a mate and produce offspring improve the individual's reproductive success.

Following the correct courtship behavior does not, however, guarantee reproductive success. A female may not pay much attention to a courting male, or she may be distracted by several males all courting her at the same time. This competition for mates and the possibility that the female will, for some reason, select one particular male from the group result in **sexual selection.**

Figure 20.11 The correct performance on the part of one of the blue-footed boobies, *Sula nebouxi,* signals the other to perform an answering display.

What factors influence a female's choice of a mate? Many animals that have intricate courtship rituals also have well-developed secondary sexual characteristics, such as antlers in male deer or colorful plumage in male birds. In most vertebrates, the male is showier than the female. Sometimes, the males with the most impressive masculine features are the most attractive to females, but many of these characteristics do not appear to be adaptive in the sense of promoting survival. For example,

showy feathers probably do not help male birds better cope with their environment and may even make them more noticeable to predators. If such showy feathers do not increase the individual's chance for survival, and may even decrease the chance, what function can they possibly serve? Particularly showy feathers displayed in a particularly flamboyant manner may attract the attention of the female. If such a display gives the male an advantage in gaining a mate, it improves his reproductive success.

Consider the example of the sage grouse of the central United States. During the breeding season, males gather in groups on small patches of prairie. Each male defends his territory, where he engages in a truly astonishing display. With tail feathers upright and splayed and neck pouches inflated, the male projects booming calls over the prairie while he prances about like a wind-up toy, as shown in Figure 20.12. Female grouse observing these males eventually select a partner from among the many competitors. Many females may choose the same male, greatly enhancing his reproductive success.

Among certain other species, males give their partners more than genes. In the case of the hang fly, the female's choice clearly benefits the female. Males of this species capture and kill a smaller fly, which they present to a female. The size of the prey may vary, but females prefer males that present them with larger prey. Moreover, the female will allow mating only as long as the food holds out.

20.6   Some Animals Display Cooperative Behavior

Many animals behave in ways that increase their ability to reproduce, but what about animals that give up all chances to reproduce in order to help feed another's young? In some species it is common for individuals to reduce their own reproductive success through cooperative behaviors. For example, Florida scrub jays (Figure 20.13) often flock in groups of about a dozen, but usually no more than one or two pairs mate and nest at a time. A single male and female will build a nest and produce a clutch of eggs. After the eggs hatch, the entire flock helps feed the young. Although the parents usually contribute most of the food, nonparents still contribute about 50 percent.

How can such cooperative behavior evolve if it reduces the reproductive success of the nonbreeding individuals? Consider this: Relatives share genes because they share a common ancestor. Brothers and sisters share 50 percent of their genes, the same as do parents and offspring. Which would be more effective in shaping the gene pool of a species—helping save the lives of two siblings or having one offspring? What, then, might be the composition of a flock of these cooperative scrub jays? Young jays are not easily accepted into new flocks; they tend to stay with their original flock. Therefore, a flock usually consists of closely related birds. Breeding is restricted, but nonbreeders still contribute to the survival of copies of their genes by caring for the young of the breeding pairs. The experience gained by the nonbreeders ultimately may be beneficial to the species when they have young of their own. At least among nonhuman animals, there probably is no such thing as truly unselfish behavior. Any behavior that appears to benefit another animal actually enhances the survival of the individual's own genes, directly or indirectly. If it did not, it would be selected against.

Figure 20.12   A male sage grouse, *Centrocerus unophasianus,* in display.

Be sure students understand such behaviors are not purposeful. The animals are not aware of the goal of reproductive success.

The alternate term, *altruistic behaviors,* is not widely accepted.

Ask students how the environment may affect cooperative behavior. The behavior of the Florida scrub jays may result from lack of appropriate breeding habitat. Therefore, the number of breeding pairs is restricted, and related individuals help out.

Ask students to name some other examples of cooperative behavior in humans and nonhuman animals and to speculate on the adaptive value of such behaviors. Many bird species, such as brown jays, Canada geese, and penguins, exhibit cooperative breeding behavior, which sometimes takes the form of "babysitting" the young of others. Deer also often babysit. In other cases, the young may be cared for by grandmothers, or other more experienced females (and even males).

Figure 20.13   Florida scrub jays, *Aphelocoma coerulescens,* exhibit cooperative behavior.

The Environment and Survival

20.7 Animals Adjust to Environmental Changes in Several Ways

Guidepost

How does the environment influence the survival of an organism?

Contrast hibernation with estivation—the ability of some animals in hot, dry conditions to enter a dormant state until reactivated by cooler temperatures and moisture.

Environments vary according to time as well as place. The temperature of the surface of a sand dune in Death Valley, for example, may climb to over 58° C in the day and drop to below 0° C at night. Individual organisms can respond to such drastic changes in their environment through changes in their behavior, physiology, and appearance or form.

The quickest response of many animals to an unfavorable change in the environment is to move to a new location, sometimes just a few feet away. Rainbow trout, for example, can sense the heat and lower oxygen levels of the upper zone in a lake and move to deeper water that is cooler and has more oxygen. Other animals are able to modify their immediate environment by cooperative social behavior. Honeybees, for example, cool the inside of their hives on hot days by collectively beating their wings. Through such behavior, the bees can keep the hive at a set temperature, plus or minus one or two degrees.

In general, physiological responses change more slowly than behavioral reactions. If you moved from Galveston, Texas, which is at sea level, to Santa Fe, New Mexico (about 2135 m above sea level), one physiological response would be an increase in your red blood cell count, but this reaction would take several weeks to develop. Some physiological responses are faster, such as the contraction of blood vessels close to the skin surface when you go outside on a cold day. This contraction helps reduce heat loss.

Many small- and medium-sized animals in northern regions tolerate low temperature and food scarcity by entering a state of prolonged dormancy, called **hibernation** (Figure 20.14). True hibernators, such as ground squirrels, jumping mice, and marmots, prepare for hibernation by building up large amounts of body fat. Entry into hibernation is gradual. After a series of "test drops" during which body temperature decreases a few degrees and then returns to normal, the animals cool to within a degree or so of surrounding air temperature. Metabolism decreases to a fraction of normal. Some mammals, such as bears, raccoons, badgers, and opossums enter a state of prolonged sleep in winter with little or no decrease in body temperature. This is not true hibernation.

Usually organisms either can regulate their internal conditions or conform to the prevailing external environmental conditions. Whether an animal regulates its internal environment or conforms to the external environment largely depends on the stability of the environment in which it normally lives. For example, brine shrimp live in waters with high but

Figure 20.14 Golden mantle ground squirrels, *Citellus lateralis,* are true hibernators.

variable salt concentrations and have the ability to maintain stable internal salt concentrations. Marine organisms that live where the salinity of the water is very stable have no such regulatory ability. If they are placed in water with a different salt concentration, they will gain or lose water to conform to that particular external environment.

1 Organisms may react to some change in the environment with growth responses that change their form or internal anatomy. One example is the arrowleaf plant, which can grow on land or in water. The leaves of submerged plants are flexible and lack a waxy cuticle. Thus, they are able to absorb nutrients from the surrounding water. Arrowleaf plants that live on land, however, have extensive root systems and more rigid leaves covered with a thick cuticle that retards water loss. In general, plants are much more capable of this type of response to environmental change than are animals.

Make clear the distinction between *adaptability* and *adaptation*.

20.8 Organisms Vary in Their Tolerance

How do environmental factors result in the selection of certain organisms for survival? Specific environmental factors set limits for growth, which affect selection. One important environmental limiting factor is temperature. If a household geranium is left outdoors during a Minnesota winter, it will die. A blacksnake left in a shadeless cage and exposed to the August sun of Texas also will die. Household geraniums cannot tolerate long periods of freezing temperatures, and blacksnakes cannot tolerate long periods of high temperatures. If an organism cannot tolerate the highest or lowest temperature in an area, it will die or move elsewhere, if it can. Thus, **tolerance** is the ability of an individual or a species to withstand particular environmental conditions. If an individual cannot tolerate the conditions of the environment, it may not survive to reproduce.

Household geraniums, *Pelargonium,* are native to South Africa.

All organisms function most efficiently under certain environmental conditions. We can determine the tolerance limits of a species for one environmental factor by varying that factor and measuring some aspect of the organism's performance. The resulting curves typically are bell-shaped, with the tails of the curve representing the limits of the organism's tolerance for a particular environmental variable and the values in between representing the range within which the organism can live, as shown in **2** Figure 20.15. The conditions most favorable for growth and reproduction

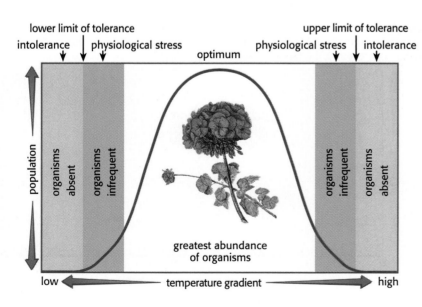

Figure 20.15 A typical bell-shaped curve showing the tolerance range of a given organism, in this case a geranium.

for that species are called the optimum conditions. Tolerance limits may seem clear-cut, but complications can arise. The duration of a condition is important, especially in determining maximum and minimum limits. For example, geraniums can withstand short periods of freezing temperature but not long ones. The limiting factor of temperature selected those geraniums that can grow in moderate climates.

Within any population, individuals vary genetically for some characteristics, including tolerance. Within a species, also, populations of organisms vary in their tolerance for certain conditions. Populations that come from different parts of a species' geographic range differ in their tolerance to environmental factors. A population of jellyfish, *Aurelia aurita* (Figure 20.16a), which lives off the coast of Maine, can tolerate temperatures between 5° C and 18° C. For the *Aurelia* population off the southern coast of Florida, however, the optimum varies between 28° C and 30° C. Both the northern and the southern jellyfish have been selected for survival in their particular environments. Figure 20.16b shows the tolerance ranges for each jellyfish population. Such variation in a population arises by mutation or genetic recombination in some individuals. Not all individuals in a population are equally suited to the same environmental conditions. Those that are best suited will survive and out-reproduce the others.

The interaction of several variables also may affect an organism's tolerance. For example, several variables interact to determine the geographic range of the field sparrow. Field sparrows can survive northern winter temperatures in the United States if they have enough food for energy. Sparrows need more food in the winter because they lose more heat from their bodies than in summer. They can hunt for food only during the day, and winter days are short. Is it the temperature, the food supply, or the length of day that sets the northern winter boundaries for field sparrows? It could be any one or a combination of these factors.

Abiotic conditions within a species' tolerance limits may allow it to survive in a particular ecosystem, but these conditions do not guarantee that the species will survive there. The species may fail because of conditions in the biotic environment. It may have to compete with other organisms or be affected by them in other ways. For example, the geographic range of the koalas of Australia is determined not by abiotic conditions that directly affect them, such as temperature or rainfall, but by the availability of

Ask students how they would determine the importance of each of these factors. Ask them to design experiments testing each.

Figure 20.16 *Aurelia aurita* (a) and the tolerance ranges for two populations of *A. aurita*, one in Florida and one in Maine. Even though they are the same species, they no longer can live in water that is the same temperature (b).

a

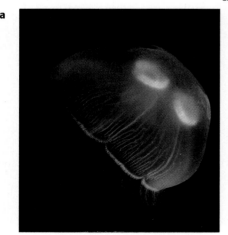

b

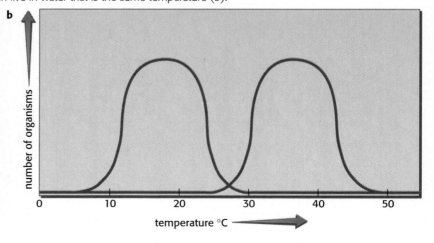

eucalyptus leaves. Only the leaves of about a dozen species are acceptable. Koalas (Figure 20.17) certainly can tolerate the abiotic conditions of other areas of Australia, but because eucalyptus trees are not found there, the koalas would not survive. Similarly, if the eucalyptus trees in the current range were destroyed to create farmland, the koalas would be unable to survive even though the abiotic conditions remained unchanged.

20.9 Human Activity Affects the Behavior and Survival of Other Animals

The human population is growing rapidly and increasing its demand for food, housing, water, and other necessities of life. In the course of obtaining these necessities, humans affect other organisms in many, usually negative, ways. Changing the environment and destroying habitats have the greatest effect on other animals. For example, building houses on a tract of wooded land means that the resident animals either must move to another area, adapt to life among humans, or must die. Of course, most of the plants must give way to houses and lawns.

If the animals move and join another population, competition among the members of that population for food, territory, mates, and other resources probably will increase. The needs of the increased population also may outstrip the area's carrying capacity, leading to famine, increased disease, and higher mortality. If the new houses totally cover a unique habitat, the animals that lived there may well become extinct.

A few animals have found that living close to humans confers some benefits. The bighorn sheep in Figure 20.18 have found that the grass in lawns is more abundant, more succulent, and probably easier to obtain than wild grasses. Elk often eat the apples from trees in backyards, and raccoons are common nighttime garbage can visitors. Often, people who move into areas that were totally undeveloped a short time ago have difficulty with some of the wildlife. It is not uncommon for great horned owls (or other large birds of prey) or coyotes to dine on small dogs and cats. In fact, the human population has moved into so many different places and modified so much of the environment that we often forget our place in the web of life (see Figure 20.19).

Figure 20.17 Koalas, *Phascolarctos cinereus* are Australian marsupials that feed on the leaves and bark of eucalyptus trees.

Raccoons pose a threat to suburban songbird populations because they feed on their eggs in suburban woodlands.

Figure 20.19 A grizzly reminder of how humans fit into the food chain. People often forget that humans are a part of the web of life. Just because humans are present in an area does not guarantee that the other animals will behave in what we consider to be an appropriate manner.

Figure 20.18 These desert bighorn sheep, *Ovis candensis,* have adapted to life near humans. What might be some of the advantages and disadvantages? **Answers will vary, but should include mention of increased (and more constant) food supply, but also increased death rate from vehicles and dogs.**

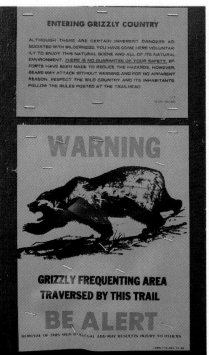

Figure 20.20   Rock doves, *Columba ivia,* in Times Square, New York City. These birds have adapted quite completely to urban living and are rarely found away from humans.

In many areas, urban wildlife has created great problems. Ask students to list local examples. In most cases, these animals live off human refuse.

Ask students to name some other examples of human interference and human solutions to such problems. Use local examples if possible.

Figure 20.21   (a) Manatees, *Trichechus manatus,* are threatened by the loss of suitable habitat and food due to dredging and other human activities. (b) Even in protected areas encounters with motorboat propellers often prove fatal to the manatees.

Sometimes the animals have adapted so well to life with humans that they rarely are found away from cities and towns. Pigeons, or rock doves, such as those shown in Figure 20.20, are a good example. Other birds, such as sparrows and finches, are found in the greatest numbers near human occupied areas. Raccoons, squirrels, skunks, pigeons, finches, and sparrows are typical representatives of urban wildlife—wildlife that has successfully adapted to life among humans.

Humans can sometimes obtain the necessities of life without totally disrupting native wildlife. In an effort to minimize harmful effects, some new housing developments hire ecologists to perform animal and plant surveys of an area. If the types of wildlife and their movements are known, provisions can be made that will encourage the animals to remain in the area. For example, greenbelts and parks can provide deer with the open areas they need to move from one place to another as well as with a rich source of food.

Most often, however, human activities result in great damage to native animal populations. The greatest threat to most wild species is the destruction or degradation of their habitats, as in the case of the Florida manatee, **5** shown in Figure 20.21. If a parking lot is built over a swamp, for example, nothing will help the turtles, frogs, snakes, and other animals that live there. The disruption of natural communities threatens wild species by destroying migration routes, breeding areas, and food sources. One study found that two-thirds of the original wildlife habitat in tropical Africa and in much of southeast Asia has been lost or severely degraded. The tropical rain forests in Central and South America are being destroyed at an alarming rate. In the United States, forests have been reduced by one third and wet lands by 50 percent.

Many rare and threatened species live in vulnerable, specialized habitats, such as small islands or even single trees. Human alteration of these fragile ecosystems often creates habitat "patches" that are too small to support the number of individuals needed to sustain a population. A great deal of careful planning and cooperation on a global scale is required to help wild populations survive human activity and interference.

CONCEPT REVIEW

1. How are physiological responses to environmental change different from changes in form or appearance?

a

b

2. What is meant by an organism's range of tolerance, and how is it important?

3. Is the range of tolerance the same for all populations of a species? Explain.

4. What factors, other than abiotic factors, may affect the survival of an organism?

5. How can humans affect natural communities and the animals that live there?

INVESTIGATION 20.1   Conditioning

Conditioning is one type of learned response to a stimulus. In this investigation, you will have an opportunity to experience conditioning firsthand.

Materials (per team of 1)
paper and pen

Procedure

1. In your data book, prepare two tables similar to Table 20.1 and Table 20.2, or use the tables provided by your teacher.

2. Your teacher will read a paragraph about behavior to you. Each time the word *the* is read, make a slash mark (/) on your paper. The object is to record the number of times *the* is used in this paragraph (paragraph A).

Table 20.1 Conditioning Tally A

	Number of Words								
	38	39	40	41	42	43	44	45	46
Paragraph A									

Table 20.2 Conditioning Tally B

	Number of Words								
	6	7	8	9	10	11	12	13	14
Paragraph B									

3. Do not correct mistakes. You will make the data inaccurate if you do. If you make a mark when the word *the* has not been read, you must leave it. You cannot skip the next *the* to make up for your mistake.

4. After your teacher has read the paragraph, count the number of marks you have made and record the total on your paper.

5. Your teacher will read a second paragraph (paragraph B). As before, make a mark on your paper each time the word *the* is read.

6. After your teacher has finished reading, total your marks for paragraph B.

7. Enter your totals for each paragraph in the tables your teacher has drawn on the chalkboard.

This investigation is designed to condition the students to respond to a stimulus other than the one to which they are instructed to respond. After completing this investigation, students should be able to describe condition and how some unusual behavior can be produced by the use of substitute stimuli. Students should work individually.

Time: One class period.

Procedure

1. Copy Tables 20.1 and 20.2 on the board. Blackline masters of these tables are in the Teacher's Resource Book.

2. Read paragraph A to the students:

The study of behavior in the animal kingdom can be difficult. The animals do not respond the way we expect. The experimenter may attempt the experiment often, but the animals may ignore the stimulus and thus not exhibit the expected response. The animals also may respond to stimuli the experimenter did not allow for. The experimental situation may be the best that the experimenter can devise, but the animals may not appreciate the time, the effort, the materials and expense, and the skill the student has devoted to the study. Thus, the experimenter must have the ability to change the situation to meet the requirements of the experimental animal. Also, the experimenter has to fit the observation into the available time in the day. One of the best approaches to the study of the social behavior of animals is the field study—the study of the organism in the natural environment of the organism

3. The word *the* is accompanied by the bell 38 times in paragraph A. Other words are accompanied by the bell 8 times. Thus, any total above 38 represents conditioning. A tally of 46 marks indicates maximum conditioning, which probably will not occur.

4. This next paragraph, paragraph B, should be read in the same way as paragraph A. As before, sound the bell simultaneously with or just before pronouncing the boldfaced words.

In the conditioning process, it is necessary to present the normal stimulus almost simultaneously with the substitute stimulus. Otherwise very little conditioning will occur. The simultaneous presentation of stimulus and substitute stimulus is called reinforcement **The natural stimulus reinforces the substitute stimulus. Presentation of the substitute stimulus without reinforcement, over a period of time, can lead to extinction of the conditioning. The animal no longer responds to the substitute stimulus by itself. Thus, at extinction, the conditioned response has been unlearned.**

In paragraph B, the word *the* **is accompanied by the bell only 6 times and only in the early sentences of the paragraph. Other words are accompanied by the bell 17 times, and the last 13 of these follow the final conditioning** *the,* **although not the final use of** *the* **paragraph. When the conditioned stimulus ceases to accompany the official or instructed stimulus, the conditioned response is partly unlearned. Of the 11** *thes* **in paragraph B, 6 are boldfaced. A score of 11 on this paragraph indicates that the conditioned response to the bell has been unlearned. However, the conditioning may carry over from the first part of the passage. Thus, some students may have more than 11 marks. Some may make fewer marks, because 5** *thes* **are not accompanied by the bell.**

5. **Circle the correct number of times** *the* **was accompanied by the bell— 38 for paragraph A and 6 for paragraph B.**

Discussion

1. **Answers will vary.**
2. **The sound of the word** *the.*
3. **The sound of the bell.**
4. **As paragraph A was read, the sound of the word** *the* **was a stimulus. Because a bell was sounded each time** *the* **was pronounced, marks were made at the sound of the bell regardless of what word was pronounced. The sound of the bell was a substitute stimulus for the sound of the word** *the.*
5. **When the word** *the* **was not accompanied by the bell and when the bell was sounded with words other than** *the,* **the conditioning started to become unlearned.**
6. **Answers will vary but should include the idea of conditioning, that is, of the transfer of a response from one type of stimulus to another. For example, a dog may sit up on its haunches when the stimulus of food is held over its head. If the spoken word** *sit* **is always used simultaneously with the presentation of food, in time the dog will respond to the word** *sit* **alone when no food is present.**

8. Enter the class results from the tables on the board in your data book tables. Construct a bar graph of the class results. What was the average number of marks for each paragraph?

Discussion

1. Look at the numbers your teacher has circled on the tables. They represent the actual number of times the word *the* was read for each paragraph. How do these numbers compare with the numbers you and your classmates recorded?
2. What stimulus caused you to respond when the first part of paragraph A was read?
3. What stimulus caused you to respond when the last part of paragraph A was read?
4. In the context of this investigation, what is a substitute stimulus?
5. Using what you discovered when paragraph B was read, how do you think a conditioned response can be unlearned?
6. Apply the principles of conditioned responses to describe the procedure for teaching a pet to do a trick.

INVESTIGATION 20.2   A Field Study of Animal Behavior

The study of an organism's behavior in an ecosystem is called ethology. This behavior includes the organism's responses to the abiotic environment, to other species, and to other members of its own species. You can learn a great deal about an organism by observing it under natural conditions in the field, where its behavior is likely to be typical of the species. In this investigation, you will develop a systematic field study of an organism in its natural habitat and then relate its behaviors to its basic needs. Be careful not to attribute human values and emotions to the organism.

Materials (per person)
data book
pen

Procedure

1. Select a nondomesticated animal that is available for observation. Some animals that would make good subjects are insects and spiders; birds such as robins, blue jays, pigeons, sparrows, ducks, and geese; mammals such as deer, squirrels, chipmunks, muskrats, and mice; snails, slugs, and fish.
2. Most of these organisms live in fields, ponds, parks, or forests. Even an aquarium, though not the same as a natural habitat, provides a suitable environment for observation. The best times to observe are just before sunset and just after sunrise. Many animals feed at these times. You do not need to study the same individual each time because the behaviors you observe should be typical of the species.
3. Some good behaviors on which to focus include:
   a. Orientation to external stimuli, such as wind, sun, and moisture.
   b. Communication with other members of the species or with other species.
   c. Feeding.
   d. Courtship and mating.
   e. Interaction with other members of the species.

f. Interaction with members of other species, including protective measures for itself or its young.

g. Reactions to the presence of other species, including humans.

4. Once you have selected an organism to study, devise a procedure and submit it to your teacher for approval. Your plan should include:

a. The scientific and common names of your organism.

b. Where and when you will make your observations. (If possible, the organism should be observed several different times.)

c. The question you plan to investigate.

d. A hypothesis related to your question.

5. Conduct your observations. Your results will reflect natural behaviors more accurately if your subject is not aware of your presence.

6 Record your observations accurately and comprehensively in your data book. You will use your data to prepare your report.

Discussion

1. Write a report relating the behaviors you observed to the organism's basic needs. Remember not to attribute human motives to the animal. A good report should include the following:

a. Title, your name, and date.

b. The procedure approved by your teacher.

c. The actual procedure you followed if it differed from your proposal.

d. Your question and hypothesis.

e. The data you collected. Organize your notes by quantifying as much data as possible (how many, how much, how often, and so on), using tables, charts, or graphs.

f. Conclusions: interpretations or explanations of the behaviors observed in light of the basic needs of the animal.

g. Evaluation of your hypothesis: Did the data support or refute your hypothesis?

h. Recommendations for further study: What you would suggest if someone else were going to do the same study, or what you would do differently if you did the study again?

This investigation permits students to study animal behavior under natural conditions. Even students who live in an urban environment can complete this investigation successfully, probably much to their own surprise. Students should work individually.

Time: One-half to one class period to discuss with students individually their study plans. Observations must extend through several days but will be conducted outside of class.

Safety

Warn students not to invade an animal's territory and to stay away from any animal exhibiting any unusual behavior, which may indicate the animal is sick. Remind students that although squirrels and other rodents can become tame, they are know reservoirs of infectious diseases. Animals caught in the wild should *never* be brought into the laboratory for observation.

Discussion

Use the guidelines in the student discussion section to evaluate their written reports. Have students give oral presentations to the class.

For Further Investigation

Students could investigate "handedness" in squirrels and "footedness" in birds. Some preference for either right or left has been observed in many animal species.

INVESTIGATION 20.3   Environmental Tolerance

The seeds of some desert plants will not germinate (sprout) until rain washes out the chemicals in the seeds that inhibit germination. Other seeds must pass through the digestive tracts of animals before germinating. Some wheat seeds will not germinate until they have been exposed to low temperatures for a certain period. In this investigation, you will examine the tolerance of some seeds to certain environmental factors.

Before beginning work, read through the procedure and develop hypotheses that predict which seeds will germinate best in which environment(s). Record your hypotheses in your data book.

This investigation explores the effects of light and temperature on seed germination, highlighting and the role of tolerance in the survival and distribution of organisms. Teams of 5 allow each student to study the effects of one environment.

Time: One class period to set up the experiment, 10 minutes of observation time each day for 10 days, and one-half period on the final day for discussion.

Materials (per team of 5)

5 pairs of safety goggles	2 shoe boxes (per class)
5 lab aprons	forceps
5 pairs of gloves	glass-marking pencil
4 50-mL beakers	refrigerator
2 nonmercury thermometers (per class),	incubator
−10° C to +110° C	500 mL distilled water

Table 20.3 Seed Germination

Dish No. _____ Environment _____			
Type of Seed	Number Germinated		
	Day 1	Day 2	Day 10
Tomato			
Radish			
Vetch			
Lettuce			

5 100-mm × 15-mm petri dishes
5 clear plastic bags with twist ties
2 100-mm diameter paper towel disks
20 100-mm × 15-mm cardboard strips

150 mL fungicide (dilute household bleach)
50 seeds each of radish, vetch, tomato, and lettuce

Put on your safety goggles, lab apron, and gloves.

Procedure

1. In your data book, prepare a table similar to Table 20.3, or tape in the table provided by your teacher.

2. Label the beakers *Tomato, Radish, Vetch,* and *Lettuce.* In each beaker, place 40 seeds of each of the types named. Add enough fungicide to cover the seeds. Soak them for the period of time recommended by your teacher.

Household bleach is an irritant and can damage clothing. Avoid skin/eye contact; do not ingest. Should a spill or splash occur, call your teacher immediately; flush the area with water for 15 minutes; rinse mouth with water.

3. Place four paper towel disks in each petri dish and moisten thoroughly with distilled water. Use the cardboard dividers to divide each dish into four sections, as shown in Figure 20.22.

4. On the bottom of each dish, label one section *Tomato,* one *Lettuce,* one *Vetch,* and one *Radish.*

5. Pour the fungicide solution from the beakers and rinse the seeds with water. (Fungicide can be poured down the drain.)

6. Using the forceps, place 10 tomato seeds in the section of each dish labeled *Tomato.* Repeat with the other three types of seeds in the corresponding three sections of each dish.

7. Label the dishes with your team symbol and number them from 1 to 5. Place each dish in a clear plastic bag and close the bag securely with a twist tie.

8. Place each dish in a different environment, as follows:

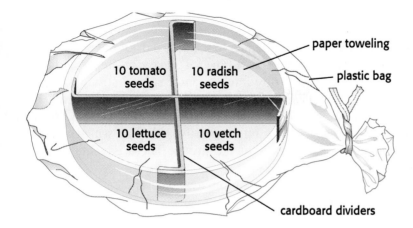

Figure 20.22 Experimental setup.

   a. Dish 1: continuous light and cold—a refrigerator with a light (10° C to 12° C).

   b. Dish 2: continuous dark and cold—a light-tight box in a refrigerator (10° C to 12° C).

   c. Dish 3: continuous light and warm—an incubator with a light (30° C to 32° C).

   d. Dish 4: continuous dark and warm—a light-tight box in an incubator (30° C to 32° C).

   e. Dish 5: variable temperature and light—on a windowsill.

  9. Wash your hands thoroughly before leaving the laboratory.

10. Each day, count the number of germinated seeds. Record the counts in your data book.

11. On the tenth day, take a final reading and combine the data of all teams.

12. Each day, wash your hands thoroughly before leaving the laboratory.

Discussion

 1. Why should the data from all teams be combined?

 2. In which environment did the greatest percentage of tomato seeds germinate? Radish seeds? Vetch seeds? Lettuce seeds?

 3. Did any seeds of one species germinate fastest in one environment but have the greatest number of germinated seeds in another environment? If so, which species and in which environments?

 4. Which species has the greatest tolerance for continuous light?

 5. Which species has the greatest tolerance for low temperature?

 6. Which, if any, species germinated similarly in all the experimental environments?

 7. How do your results compare with your hypotheses? What do the data suggest about the ability of species to establish themselves in different environments?

 8. Which factor might give a species a greater advantage—rapid germination or germination of a large percentage of seeds? Why?

 9. On the basis of your experimental results, describe ecosystems in which each species you studied in this investigation might have an advantage.

For Further Investigation

From a grocery store, obtain seeds of plants that grow in a variety of climates, such as avocados, dates, grapefruits, oranges, pomegranates, lentils, and many types of beans. Test these for germination in experimental environments. In some cases, you may have to lengthen the time allowed for germination.

Discussion

 1. To get a larger sample size.

2-3. Answers may vary, but in general, tomato seeds germinate most rapidly in complete darkness at about 26° C; lettuce seeds, especially the Great Lakes variety, germinate best in light; vetch seeds require a cool environment; and radish seeds germinate well under a wide range of conditions.

4-5. Students may have difficulty separating the influence of individual factors from combinations of factors. Lend assistance when necessary.

 6. Answers may vary.

 7. Insist that the answers be consistent with the data.

8-9. Both of these questions are speculative. Emphasize that, in scientific reasoning, the evidence must support the conclusion. Of course, in some cases there may be reason to believe that the evidence itself is misleading or unreliable.

SUMMARY

An organism's behavior is adapted to the demands of the environment in which it lives. There are two basic types of behavior—innate and learned. Both types are influenced by internal and external stimuli, but innate behavior is less dependent on environmental experience. Learned behavior consists of the responses an animal develops as a result of environmental experience. Some species exhibit cooperative behavior, which confers certain advantages on members of the group. These advantages include increased reproductive success.

Population growth usually is limited by environmental factors, which result in the selection of individuals for survival or death. Thus, limiting factors determine what the organisms' characteristics are. Through a long period, the interacting populations in an area are shaped by their biotic and abiotic environments.

Human population growth results in the destruction of habitat, which forces animals to move or to modify their behavior in order to survive. All too often the animals do not survive human interference. Worldwide, a great deal of natural habitat is being destroyed at an alarming rate. Careful planning and global cooperation are needed to help wild populations survive the impact of humans.

APPLICATIONS

1. Predators actually can benefit the population of their prey as a whole, even though certain individuals will be killed. Explain the beneficial role of predators.
2. What disadvantages do you see in cooperative breeding behavior?
3. Do plants show behavior? Explain.
4. The text states that "organisms usually either can regulate their internal conditions or conform to the prevailing environmental conditions." Some people would argue that humans do neither of these things. Explain.
5. Why is it sometimes difficult to distinguish between innate and learned behaviors?
6. How would you plan an experiment to test learning in baby chicks? What useful background information might be obtained from a commercial hatchery?
7. How would you design an experiment to test whether cooperative breeding behavior in the Florida scrub jay is innate, learned, or a mixture of both?
8. The destruction of natural habitat by humans also may reduce the natural predators of a certain species. For example, mountain lions eat deer. If the habitat of mountain lions is destroyed, the cats no longer can live in the area, but the deer might remain. Explain how this might affect the deer population, the habitat, and the other animal populations in the area.
9. How is hibernation similar to sleeping? How are they different?

PROBLEMS

1. Learn how to communicate in a nonvocal system, such as signing with your hands or waving two semaphore flags.

2. Pheromones are commonly used to control insect pests. Go to your library or contact an extension agent to find information on the use of pheromones to fight certain insects.

3. Many languages are "foreign" to our ears, such as the clicking language of the !Kung bushmen in Africa and the whistling language of the Canary Islands. Listen to audiotapes of any foreign language. What do all languages have in common?

4. Investigate your community and the use of its land resources. Use the module by BSCS, *Investigating the Human Environment: Land Use* (Dubuque, IA: Kendall/Hunt, 1984), to help you study a local land use issue. Identify a local land use problem, investigate it, produce suggestions, and report these to the appropriate authorities.

5. Report on the research attempts to train chimpanzees and gorillas to use sign language to communicate with humans. How successful do you think these programs are?

6. Observe how people behave toward one another in a park, shopping center, or bank. Write a report detailing what you observed.

7. Research an extinct species in the library. Write an essay on how the behavior of that species might have influenced its ability to survive. Possible topics could be the passenger pigeon or the Carolina parakeet.

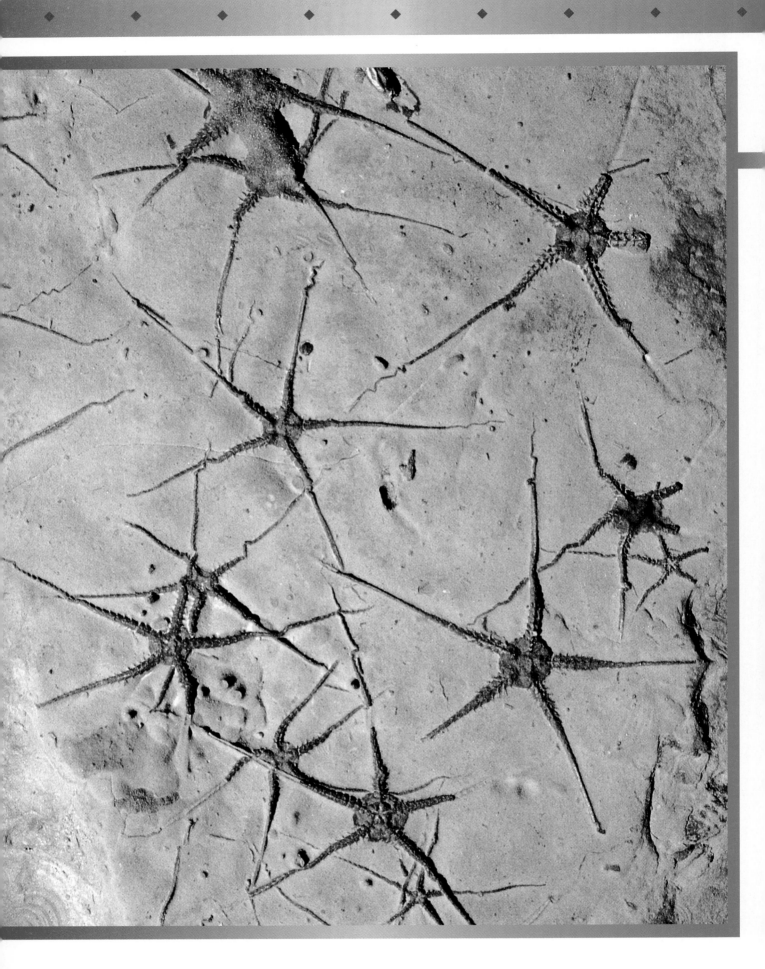

Ecosystems of the Past

This is a brittle star fossil from the Jurassic Period (165 million years ago). What living organisms does it resemble? Based on these observations, what appears to be the pattern of evolution of this group of animals?

Not all organisms exhibit major changes through time. This life form has not changed much at all in 165 million years, probably because changes were not needed to remain successful.

◆ Ecologists study the interrelationships of organisms and their environments. The size, shape, and other characteristics of organisms may change quickly or may persist for hundreds of years. Selection as a result of environmental factors may promote one population's survival but extinguish another population. If the same area could be viewed once every million years, few populations would be seen to have remained the same as their ancestors of a million years before. The entire pattern of the ecosystem would have changed, or evolved. This chapter discusses how scientists study earth's history and what their studies have revealed about past ecosystems. ◆

Reconstructing the Past

21.1 Several Techniques Are Used to Date Fossils

When organisms of the distant past died, most decayed and disappeared. Here and there, however, skeletons, teeth, and other decay-resistant structures became fossils. Sometimes, parts of organisms were replaced by minerals, or petrified. At other times a mold, or cast, of an organism was formed in mud that hardened. Figure 21.1 shows a variety of fossils.

Even as fossils were forming they were covered with sand or mud that hid them from view. In some cases millions of years passed and many layers of sediment were added. The layers, or **strata** (STRAYT uh), turned to rock. Finally, some event, such as the erosion of the Grand Canyon or the building of a highway (Figure 21.2), exposed some of the fossils so scientists could study them and learn about the past.

Fossils in layers of rock are the only remains of past ecosystems. These layers are studied by paleontologists (pay lee un TOL uh jists)—scientists who use fossils to reconstruct the history of life. These scientists have pieced together information from many parts of the world and have put most of the known rock strata in chronological sequence. The *Geologic Time Scale* in Appendix 3 is based on the ages of the rock strata. The relative ages of fossils can be determined by their position in the strata.

How old are the strata? The most reliable method for dating rocks depends on the presence of radioactive elements such as potassium-40 and uranium-235. The atomic nuclei of radioactive elements are unstable and emit radiation. These radioactive elements break down to stable elements at a constant rate known as the half-life, the time it takes for half of the element to break down.

MAJOR CONCEPTS

◆ Radioactive dating is used to estimate the age of an object.
◆ Fossils give clues to characteristics of ancient organisms.
◆ The layers of the earth reveal the ordering of life in the past.
◆ The earth's crust floats, slowly changing the configuration of the continents.
◆ Some major events help explain the evolution of life.
◆ Life has existed for about 3.5 billion of the earth's 5 billion years.
◆ Evidence exists that the genus Homo emerged in Africa and has evolved for about 2 million years.

KNOWLEDGE CHECK

◆ How have ecosystems changed through time?
◆ How do scientists date and interpret fossils?
◆ What geologic theory explains the distribution of animals?
◆ What evidence do anthropologists use to reconstruct the evolution of modern humans?

Guidepost
What evidence do researchers use to construct the past?

Some objects dug from the earth show mere chance resemblances to fossil organisms. Often it takes an expert to distinguish true fossils, and experts may disagree.

Ask students to list conditions that would favor the preservation of organisms as fossils. Conditions that retard decay are a principal requirement for fossilization. Among these are coolness, acidity, and an anaerobic environment such as may occur deep in mud at the bottom of a lake or sea.

a

b
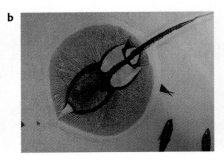

c
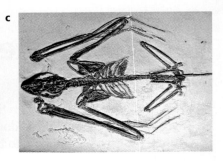

Figure 21.1   A variety of fossils: (a) *Pecopteris,* a fossil fern; (b) *Heliobatus radions,* a stingray; and (c) a fossil bat.

The potassium-40 dating method, which usually is used for inorganic material, indicates the age of volcanic rocks. Potassium-40 isotopes decay into argon gas (argon-40). In many types of rocks, the argon gas can accumulate as the potassium decays. In uncontaminated samples, the ratio of potassium-40 to argon-40 indicates how long it has been since the argon began to accumulate. Potassium-40 has a relatively long half-life of 1.3 billion years. Because of this long half-life, the potassium-40 technique is not accurate for dating deposits younger than 400 000 or 500 000 years.

The most important method for dating young fossils, or other organic material, measures the decay of radiocarbon, carbon-14. Figure 21.3 outlines the principles of carbon-14 decay. In the upper atmosphere, bombardment by cosmic rays transforms nitrogen into carbon-14, an unstable form of carbon. The carbon-14 filters down into the lower atmosphere where living things absorb it. When the organism dies however, it absorbs no new carbon, and the carbon-14 slowly decays into nitrogen-14. Assuming the production of carbon-14 in the atmosphere is constant, the proportion of carbon-14 to nitrogen-14 in once-living material indicates how long it has been since the organism died. However, carbon-14 dating has limitations. First, there is little carbon-14 in fossils and other organic material. Second, the half-life of carbon-14 is only 5730 years. If a specimen is 57 300 years old, 10 half-lives have elapsed, leaving so little of the original carbon-14 that it is hard to measure. (Currently there is no reliable radiometric technique for dating fossils between 500 000 and 50 000 years old.)

Paleontologists using radiometric dating methods have determined the age of the oldest fossils to be approximately 3.5 billion years. Despite uncertainties in radiometric dating techniques, if two or more methods yield similar data, the age usually is accepted as reliable. Such data have enabled paleontologists to construct a fairly accurate time scale.

21.2   Fossils Provide Information About Ancient Life

Inferences about ancient ecosystems are based on studies of ancient rock layers and the fossils they contain. Early paleontologists were puzzled by fossils that did not resemble any living forms. The theory of evolution by natural selection, however, made it possible to link fossils and living forms and to use fossils as clues to the history of life.

Nothing about the study of fossils is easy, however. The soft parts of organisms usually decay, and even when all the hard parts are preserved, fossils are seldom found in perfect condition. They usually must be reassembled and placed in proper relation to each other. To do this accurately, paleontologists must know the anatomy of modern organisms.

Figure 21.2   The stratification of rock layers in this mountainside was exposed when a highway was built. Fossils of different ages are found in different layers.

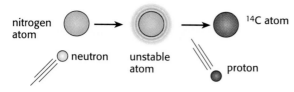

a   Nitrogen atom becomes carbon-14 atom in the atmosphere.

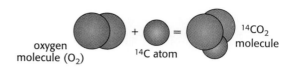

b   An O_2 molecule combines with a carbon-14 atom to form carbon-14 dioxide.

c   Living organisms absorb ^{14}C.

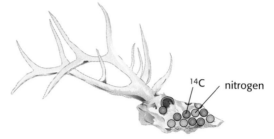

d   When the organism dies, the ^{14}C atoms begin to disintegrate.

Then, they must determine the placement of structures, such as muscles or leaves, that were not preserved. Finally, the paleontologists make a life-size reconstruction of the plant or animal based on its fossils. The colors used in these models are inferred from the colors of modern organisms. Different researchers may interpret the evidence differently. The farther they move from the basic evidence—the fossils themselves—the greater is the possibility of differing interpretations, as shown in Figure 21.4.

In reconstructing the evolution of species, paleontologists need to determine the relationships between modern organisms and their fossilized relatives. Today, relationships between fossils themselves can be studied as well. The DNA and protein of some fossils preserved in ice and mud can be compared with the DNA and protein of other fossils. Such biochemical studies, however, are a relatively recent development. Observable structural characteristics still are used widely for determining relationships and classification.

3   Consider the problem of classifying a fossil jaw that has both mammalian and reptilian characteristics. The fossil is from the early Mesozoic era, about 225 million years ago. Mammals were beginning to appear at that time. If mammals and reptiles shared a common ancestor, as scientists think then there once must have been animals with a combination of mammalian and reptilian characteristics. The jaw could belong to one of these ancestral animals. Figure 21.5 shows one artist's interpretation of the evolution of mammals from bony fishes.

Figure 21.3   A certain amount of carbon-14 exists in the atmosphere, and plants incorporate it into their tissues as carbon dioxide. Animals absorb the carbon-14 by eating plants. When an organism dies, the carbon-14 disintegrates at a known rate. It then is possible to measure the amount of carbon-14 remaining and relate that to the age of the organism.

Provide artistic students with the skull of a modern animal and ask them to visualize and sketch the whole animal. Compare the drawings produced and then show a picture of the species of animal from which the skull came.

The therapsids of the Triassic period and the icteridosaurians of the Jurassic period are examples of such classification problems.

Figure 21.4   An artist's reconstruction of a stegosaurus, a late Mesozoic reptile (a), and another artist's reconstruction (b).

a

b

Figure 21.5 Artist's conception of the animals along the main line of evolutionary descent from primitive bony fishes to mammals. *Eusthenopteron,* a primitive lobe-finned fish; *Ichthyostega,* a large primitive amphibian; *Varnapos,* a primitive mammal-like reptile; *Probelesodon,* a more advanced mammal-like reptile showing distinct mammalian characteristics, and *Megazostrodon,* the small, shrewlike first mammal.

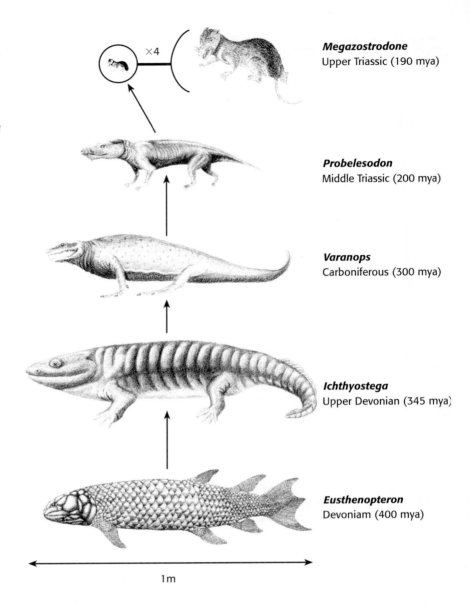

Megazostrodone
Upper Triassic (190 mya)

Probelesodon
Middle Triassic (200 mya)

Varanops
Carboniferous (300 mya)

Ichthyostega
Upper Devonian (345 mya)

Eusthenopteron
Devoniam (400 mya)

1m

Figure 21.6 A fossil trilobite. Its actual size was about that of a paper clip.

21.3   The Fossil Record Reveals Both Change and Stability

No one can examine a large collection of fossils without being impressed by the evidence of change, especially if the fossils of one type are arranged in chronological order. Many species have died out and left no descendants; others have changed so much through many generations that the successive forms are given different species names.

The fossil record shows that organisms in one geological period are descended from ancestral forms present in earlier periods, though not necessarily directly. For example, although mammals are thought to have evolved more recently than reptiles, the reptiles of the Mesozoic era did not give rise to mammals in the Cenozoic. Rather, the ancestors of the Mesozoic reptiles apparently also were the ancestors of the mammals. **4**

At least 130 000 species of animals are known only from their fossils. In the Cambrian period, trilobites (Figure 21.6) were varied and numerous, but they disappeared suddenly from the geological record. There have been many large-scale extinctions. Evidence suggests the trilobites perished in one of these catastrophes. Other groups, including the dinosaurs, died out in other extinctions. What caused such extinctions? Many **5**

a

b

hypotheses have been proposed, but testing them is difficult because we are dealing with the past. Although we may never learn what caused the extinction of trilobites and dinosaurs, new evidence continues to be found.

Just as organisms have changed, whole ecosystems also have changed. Change in ecosystems depends on the changes in a particular environment. It also depends on the adaptability of the associated organisms. A characteristic that adapts an organism to one environment might be a hindrance in a different environment. Similarly, a characteristic that has no adaptive value in one situation may prove to be adaptive if conditions change.

Although the fossil record offers abundant evidence of change, it also reveals great stability. For example, twentieth-century oceans contain brachiopods (BRAY kee oh podz), animals with hinged shells, similar to those from the Paleozoic seas, as shown in Figure 21.7. Ocean environments are usually more constant than land environments. Thus, many adaptations that allowed early forms to survive apparently are still useful.

Figure 21.7   (a) Present-day brachiopods, and (b) a fossil cast of a brachiopod.

Refer students to the discussion of adaptive radiation in Chapter 9.

Have students research geoecological changes in your region. Most states issue publications on historical geology.

Marine organisms probably have changed less than land or freshwater ones. Ask students why this might be. Marine environments are more stable, both short-and long-term, than terrestrial or freshwater ones. Because changes in organisms can be linked to environmental changes, they could be expected less frequently in marine environments.

CONCEPT REVIEW

1. How are fossils formed, and how are they used?
2. How can the age of a fossil be determined?
3. How would a paleontologist decide whether a particular fossil represented a reptile or a mammal?
4. How does the fossil record support both the idea of change and the idea of stability in living things?
5. What problem does the extinction of large groups of organisms—such as the dinosaurs—present to paleontologists?

Evolution and Plate Tectonics

21.4   Evolution Has Occurred on Moving Plates

Reconstructing the history of earth and the evolution of organisms from fossils is complicated by certain geological events. Those events have been involved in cases like the ones shown in Figure 21.8. *Nothofagus* (noth oh FAY gus) may be found in New Guinea, New Zealand, Australia, and South America. Fossils of *Glossopteris* (glah SOP ter is), a genus of Paleozoic seed ferns, show a similar pattern of distribution except that

Guidepost

What steps led to the broad array of organisms found on earth today?

Figure 21.8 Distribution map of extinct *Glossopteris* ferns and living *Nothofagus* trees.

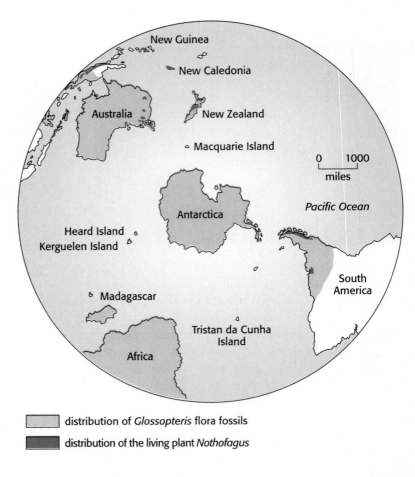

distribution of *Glossopteris* flora fossils

distribution of the living plant *Nothofagus*

For a description of the scientist and his work, see A. Hallam, "Alfred Wegener and the Hypothesis of Continental Drift," *Scientific American* (February 1975): 88–97.

they occur also in India and South Africa. Other Paleozoic and early Mesozoic terrestrial organisms have similar patterns of distribution. The areas in which these organisms are known to exist are separated by thousands of miles of ocean. How did the plants and their fossils get where they are now?

An answer to that and similar puzzling questions was proposed by a German scientist, Alfred Wegener, who in 1912 suggested that the present continents, including South America and Africa, had once been joined in a supercontinent he called Pangaea (pan GEE uh). He suggested further that all the continents have not always been where they are now, but rather that these big land masses might be slowly "floating" over the earth's hot liquid interior. Although Wegener's hypothesis was not well accepted by most of the geologists of his time, during the last 20 years additional data have been gathered that support his idea. We now know that about 200 million years ago all the continents were together in one huge land mass that eventually split up. Wegener's idea is now called continental drift.

The continental drift theory is now part of a broader theory of **plate tectonics** (tek TON iks), illustrated in Figure 21.9. According to the broader theory, land masses are the highest parts of huge floating plates that make up the earth's crust. New material from the earth's mantle is added to the plates along ridges on the ocean bottom. The mid-Atlantic ridge is one example; other ridges exist in the Pacific Ocean. The addition of new material to the earth's crust forces the plates to move slowly away from the ridges. At other locations, the plates are grinding into one another and forcing crustal materials into the mantle through deep ocean trenches. Many important geological phenomena, such as mountain building, volcanic activity, and earthquakes, happen at the boundaries of plates. The San Andreas

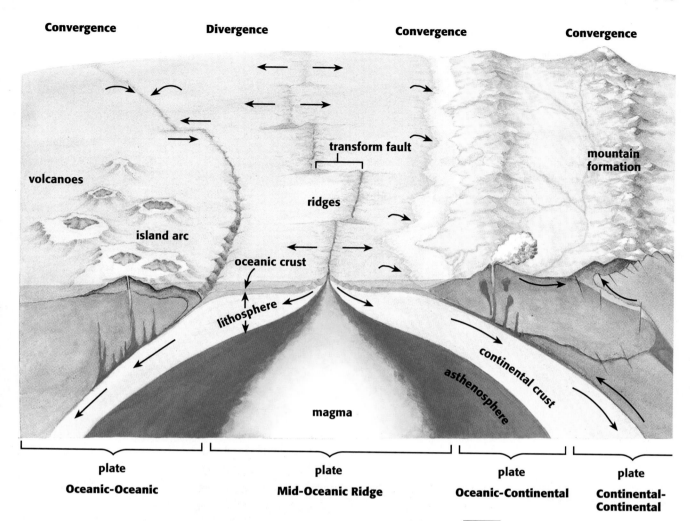

Convergence   **Divergence**   **Convergence**   **Convergence**

transform fault

mountain formation

volcanoes

ridges

island arc

oceanic crust

lithosphere

continental crust

asthenosphere

magma

plate	plate	plate	plate
Oceanic-Oceanic	**Mid-Oceanic Ridge**	**Oceanic-Continental**	**Continental-Continental**

Figure 21.9   Diagram illustrating the mechanism of continental drift. Magma, or molten rock from the earth's interior, forces its way through the thin oceanic crust and forms new crust. This new crust moves away from the mid-oceanic ridge as the process continues. Where the heavy oceanic crust meets the lighter continental coast it is subducted, or forced to slide beneath the continental crust, resulting in volcanoes and island arcs.

fault in California is part of a border where two plates slide past each other. Mount St. Helens in Washington state is near another plate boundary.

21.5   Plate Tectonics Explains the Distribution of Species

The history of the earth helps to explain the current geographic distribution of species, as shown in Figure 21.10. For example, the emergence of islands, such as the Galapagos, creates new environments for founding populations, and adaptive radiation fills many of the available niches with new species. On a global scale, the drifting of continents is the major fac-
3 tor in the spatial distribution of life. Although these plate movements rearrange geography all the time, two such past events have been especial-
4 ly significant in the distribution of organisms. About 250 million years ago, plate movements brought all the land masses together to form Pangaea. Species that had been evolving in isolation now were in competition. The total amount of shoreline was reduced, and there is some evidence that the ocean basins became deeper, draining most of the shallow coastal seas. Then as now, most marine species inhabited the shallow waters, and the formation of Pangaea destroyed much of that habitat. Changing ocean currents also would have affected land and sea life.

The other major event occurred about 180 million years ago. Pangaea began to break up, and this breakup served as a giant isolating
4 mechanism. As the continents drifted apart, each became a distinct

CAMBRIAN EARTH

Four large (and two small) continents had been formed and were moving toward each other. Sea levels were high. Vermont mountains were formed.

DEVONIAN EARTH

Laurentia and Baltica had collided to form one continent. Sea levels were low. In some regions there were periods of drought.

CARBONIFEROUS EARTH

Because of high humidity, swamps and forests covered much of the land.

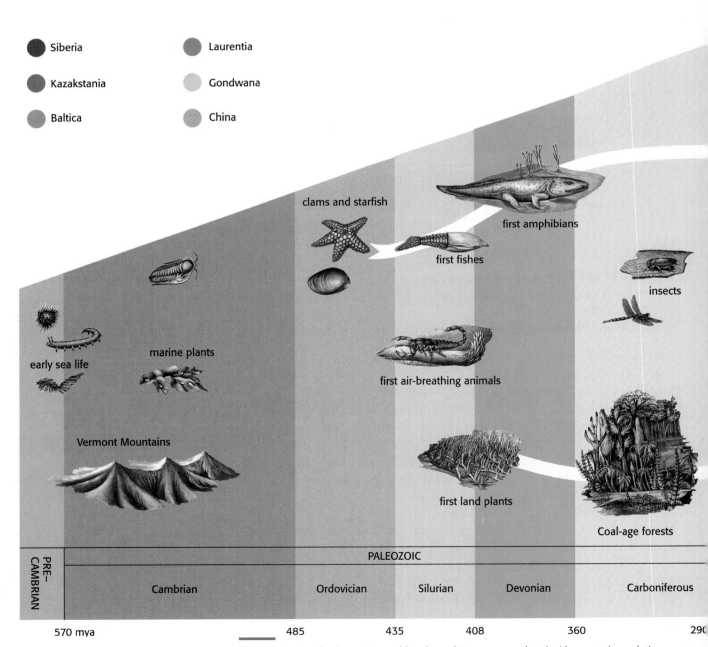

Siberia

Kazakstania

Baltica

Laurentia

Gondwana

China

Figure 21.10   The formation and breakup of Pangaea correlated with events in evolution.

TRIASSIC EARTH

Pangaea had been formed by the four major continents. Sea levels were low and there were many deserts. Appalachian mountains were formed.

CRETACEOUS EARTH

North America and Europe began splitting apart and South America and Africa separated. The sea level was high. Formation of the Rocky Mountains began.

PLEISTOCENE EARTH

During each of four "ice ages" the sea was very low. Warm periods followed periods of glaciation. The continents had drifted almost to their positions of today.

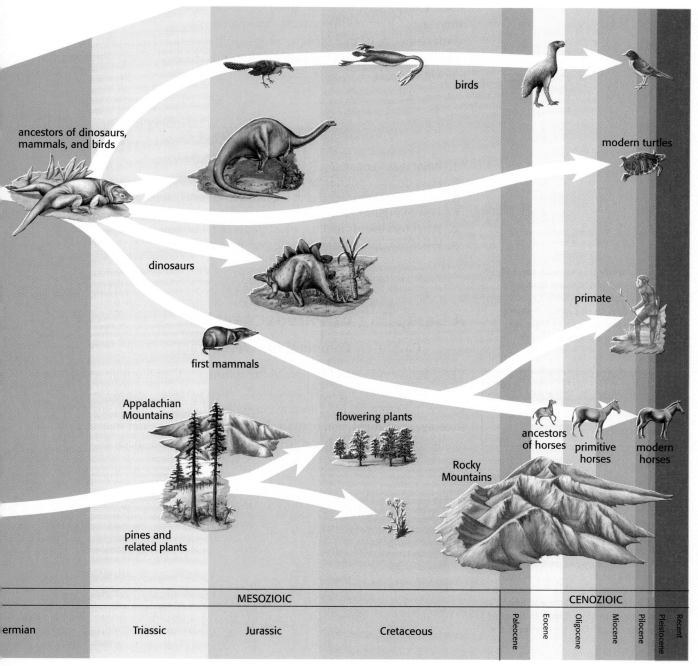

birds

ancestors of dinosaurs, mammals, and birds

modern turtles

dinosaurs

primate

first mammals

Appalachian Mountains

flowering plants

Rocky Mountains

ancestors of horses

primitive horses

modern horses

pines and related plants

	MESOZIOIC			CENOZIOIC					
ermian	Triassic	Jurassic	Cretaceous	Paleocene	Eocene	Oligocene	Miocene	Pliocene	Pleistocene / Recent
245	208	144		66.4	57.8	36.6	23.7	5.3	1.6

evolutionary theater, with the organisms on each diverging from those on the other continents. The pattern of continental separation solves the puzzle of geographical distribution of such animals as marsupials and placental mammals.

Australia provides a good example of the effects of continental drift. Until about 35 million years ago, Australia was attached to a much warmer Antarctica, as was South America. Marsupials, best represented in Australia and South America today, appear to have migrated overland between these continents via Antarctica. As the two continents moved farther away from Antarctica, the land bridges that served migrating animals were lost. From then on, as Australia moved steadily northward, its peculiar plants and animals evolved in isolation.

A general principle of paleontology is, "The present is the key to the 5 past." Conversely, as in the study of discontinuous distribution, the past often sheds light on the present. The patterns found in ecosystems today result from events that occurred in past ecosystems. The next section presents a broad overview of some of these paleoecosystems—the ecosystems of the past.

CONCEPT REVIEW

1. What was Pangaea?
2. How do the continental plates move?
3. How does continental drift explain some cases of discontinuous distribution of organisms?
4. What two events had the most impact on the distribution of species?
5. What principle guides the interpretation of paleontologic evidence?

As students study this section have them refer to *A Catalog of Living Things* to relate the organisms in the paleoecosytems to the types that are living today—particularly in terms of higher classification levels.

Guidepost

What organisms dominated various ecosystems in the past?

A Sampling of Paleoecosystems

21.6 Cambrian Communities Were Aquatic

There is no evidence of terrestrial life during the Cambrian period, around 570 million years ago, but marine ecosystems were well developed, as shown in Figure 21.11. There were shallow- and deep-water organisms— floating, swimming, and bottom-dwelling types. The chief marine producers then, as now, probably were microscopic plankton species.   1

Figure 21.11 An artist's reconstruction of life in a Cambrian sea. How many types of organisms can you identify? **At the left, a jelly-fish floats in front of an alga; beneath are brachiopods and a trilobite; next to them stands a sea cucumber, with a large annelid beside it. Farther to the right is another trilobite and above it another type of large anthrop in front of a sponge. In the middle background are mollusks.**

Figure 21.12   Reconstruction of a forest in the Carboniferous period.

There are no chordates in the Cambrian fossil record. Otherwise, the major animal phyla known today also were represented in the Cambrian period. There were sponges and cnidarians, and the most abundant animals were marine brachiopods. Although many of the brachiopods were similar to species living today, the arthropods were so different that none of them can be placed in modern arthropod classes. Among Cambrian arthropods, the trilobites left the most abundant fossils.

21.7   Land Plants and Animals Appeared in the Devonian

After the Cambrian period, the seas receded a little but remained widespread. During this time, microscopic plants thrived, and the earliest fishes flourished. In fact, the Devonian period often is called the Age of Fishes. Many fishes were covered with armorlike plates. Sharks and the ancestors of modern bony fishes evolved then also.

Land plants multiplied both in number and variety during the Devonian period. Spore-bearing ferns, seed-bearing ferns, and horsetails developed. Trees, including the ancestors of cone-bearing pines and firs, first appeared. Soon after, insects crawled around and flew above them.

Some fishes had adaptations (sometimes called preadaptations) for life on land, such as lungs, internal nostrils, and skeletons that better supported their paired fins. In some, lobe-shaped fins became modified into stumpy, leglike paddles. With the appearance of these fishes during the Devonian period, the vertebrate invasion of land had begun.

21.8   Giant Forests Grew During the Carboniferous

By the Carboniferous (coal age) period, about 360 million years ago, the earliest forests fringed the shallow seas that still covered most of North America. Unlike those in present-day forests, these trees were mostly huge relatives of present-day horsetails, club mosses, and ferns (Figure 21.12). Strata of the late Carboniferous period contain fossils of conifers and related plants, but the flowering plants of today were absent. Plant life of the Carboniferous suggests a warm and humid climate, like that in today's tropical forests. There probably was little or no seasonal change.

Invertebrates were numerous. Some predatory insects resembled modern dragonflies. One—with a wingspread of almost 75 cm—was the largest insect ever. Cockroaches abounded, some species nearly 10 cm long.

The only large animals were the true amphibians, which probably evolved from the lobe-finned fishes of the Devonian period. Most of the many types of amphibians had four legs but could not really stand. Instead, they resembled modern salamanders and snakes, and they preyed on fishes in the streams and ponds. In the seas, trilobites remained, but these would disappear from the fossil record around the end of the Paleozoic era.

21.9   Dinosaurs Dominated Triassic Ecosystems

Following the Carboniferous period, Pangaea was formed. Most of the seas covering North America drained away, and by the Triassic period, about 115 million years later, large areas were dry or mountainous. By then the amphibians had some formidable competitors.

Evidence of the Triassic period (about 245 million years ago) is found in the Connecticut Valley of New England, where certain rocks contain the large, three-toed footprints seen in Figure 21.13. Characteristics of the rocks suggest that a slow, winding stream flowed through a dry valley, carrying materials from highlands on both sides to a broad, flat plain, where it deposited them. Narrow bands of coal-like rocks mark the presence of small ponds. Almost all the plant fossils are from such rocks. Aquatic insects and several types of fishes lived in the ponds. Fossils of lungfish (Figure 21.14), which could obtain oxygen from the air, suggest that many ponds sometimes dried up, just as do ponds where modern lungfish live.

Dinosaurs and other large reptiles were abundant. Lizardlike creatures scurried about. The structure of their leg bones suggests they moved quickly, and their small, sharp teeth indicate they ate insects. Several species of slender dinosaurs about 2.5 m high roamed the mud flats, leaving the threetoed tracks that first called attention to this paleoecosystem.

Figure 21.13   The fossil track in the photo could have been made by a bipedal dinosaur such as the one on the left.

× 1/23

21.10   Large Mammals Roamed North America in the Cenozoic

The Cenozoic era, represented in Figure 21.15, often is called the Age of Mammals because mammals have become so widespread in the last
5 60 million years. Though the first small, ratlike mammals lived at the same time as the dinosaurs, larger mammals did not evolve until the dinosaurs died out. Whatever killed the dinosaurs spared some ancestral mammals.

In what is now downtown Los Angeles, molasseslike pools of asphalt still entrap unwary small animals, although today the La Brea tar pits are enclosed with fences that protect most animals. The huge animals that once roamed the area often bogged down in similar pools of asphalt and died. Many rabbits, rodents, shrews, and perching birds, all similar to modern species, also were trapped in the asphalt. Their similarity to modern forms suggests that the climate of that ecosystem was probably like the warm, comfortable climate of southern California today. Bishop pines and cypresses grew there, along with juniper and coast live oak trees.

Western horses and other herbivores—peccaries, tapirs, antelope, bison, camels, and deer—grazed the area. Ground sloths ate yucca and other plants. Perhaps the most impressive remains are those of mastodons and mammoths. Some of the mammoths were more than 4 m high at the shoulders. Small reptiles and amphibians also lived in the La Brea ecosystem, eating such insects as flies and crickets. There were small carnivorous mammals. The most striking carnivores, though, were saber-toothed cats and dire wolves. (Dire wolves were similar to modern timber wolves.) Figure 21.16 shows some of these animals.

Is it surprising to find no humans here, considering the relatively recent age of this ecosystem? Most evidence indicates that the first humans to reach this continent probably crossed a bridge of land or ice that connected North America with Asia. They may have been hunters who followed the mammoths and eventually brought about their extinction.

Figure 21.14   An Australian lungfish.

Figure 21.16   Cenozoic animals. From top to bottom: *Diatryma; Smilodon,* a saber-toothed cat; and a ground sloth.

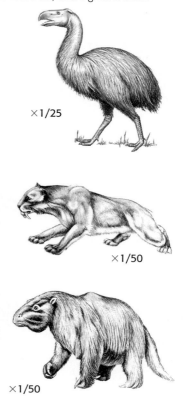

Figure 21.15   A Cenozoic landscape (Pliocene epoch, about 3.4 million years ago).

CONCEPT REVIEW

1. Compare the Cambrian environment with that of the Carboniferous. What types of organisms lived in each?
2. What preadaptations of some fishes may have led to the evolution of amphibians during the Devonian period?
3. Compare and contrast a forest of the Carboniferous time with a forest of today.
4. Compare and contrast the organisms of the Carboniferous period with those of the Triassic.
5. State the general trend in characteristics of animals from the Cambrian period into the Cenozoic era.

The Emergence of Humans

21.11   Humans Are Classified in the Order Primates

Guidepost
How do humans differ from other primates?

To convince students that each eye views a scene from a different angle, have them view an object in the classroom first with one eye then with the other.

Humans, monkeys, and apes are similar in the details of their anatomy. They are, therefore, grouped together in the order Primates, along with such animals as lemurs and tarsiers (Figure 21.17).

Most primates are arboreal and possess structures and behaviors that relate to the arboreal way of life. The digits of the primate, which have nails rather than claws, are well developed and give the animal a powerful grasp. In addition, the digits are sensitive and can easily tell if a surface is crumbly or slippery. The eyes of primates are directed forward instead of to the side, as in most mammals. Both eyes view the same object from

Figure 21.17 Although primates share many characteristics, there is also great diversity within the group. Identify as many similarities and differences as you can.

bush baby
tarsier
×1/10
squirrel monkey
gibbon
gorilla

slightly different angles, thus allowing the brain to perceive the object in three dimensions and to make accurate judgments of distance. The brains of most primates are exceptionally large compared with those of all other mammals except whales, dolphins, and porpoises. Unlike many mammals, primates have a sense of color.

Bearing one offspring at a time is the rule among primates, although twins are not unusual. An active arboreal animal cannot carry many offspring. To feed these young, a female primate has only two mammary glands. Young primates, unlike young horses or jackrabbits, receive maternal care for a long time. Unlike other organisms, they do not learn all they need to know by trial and error but are taught by their parents. Primates tend to be omnivorous. They gather food in social groups and have well-established patterns of communication that range from elaborate body language to complex vocal signals and calls.

21.12 **Hominids Are Primates That Walk Upright**

The first humans on this continent migrated here from Asia, but the human species did not arise in Asia. Early Asians probably came from Africa, through the Mideast. Fossil evidence and early tool sites suggest that it was in Africa that **hominids** (HOM ih nidz)—primates that walked upright—developed.

The first undisputed hominids appear between 4 million and 3.75 million years ago in Tanzania and Ethiopia. Scientists uncovered the fossilized footprints of hominids that were bipedal—capable of walking on two legs. These fossils have been firmly dated to 3.75 million years ago. In addition, almost half of a complete skeleton of an individual called "Lucy" (Figure 21.18) as well as other fossil finds have been less firmly dated at between 3 and 4 million years ago. Anthropologists (scientists who study humans) generally agree that these finds represent a distinct group of hominids classified as **australopithecines** (AH strahl oh PITH uh seenz). Different forms of australopithecines have been found in East and South Africa and have been dated to different times. There is considerable doubt about the relationship of these groups to one another as well as to modern humans. Anthropologists, however, are convinced that all australopithecines were bipedal, although they may not have walked exactly as modern humans do.

Although these populations shared some traits with humans, other characteristics show how different they actually were. The length of their arms in proportion to the length of their legs and torso suggests that they were more apelike in appearance. Their cheek teeth were comparatively large and had a thick coating of enamel, possibly an adaptation to eating large quantities of fruit, seeds, tubers, roots, and pods—some of which may have been quite tough. There is no concrete evidence that they used or made tools.

At some point between 2.5 and 2 million years ago, African hominids underwent adaptive radiation. By the later date, there were at least two, possibly three, and perhaps even more species of hominids. In eastern Africa, a group of large and robust australopithecines lived at the same time as a larger-brained hominid called *Homo habilis*. Both hominids are shown in Figure 21.19. Many anthropologists believe *Homo habilis* to be the first member of our own genus *Homo,* but there is still much debate about the subject.

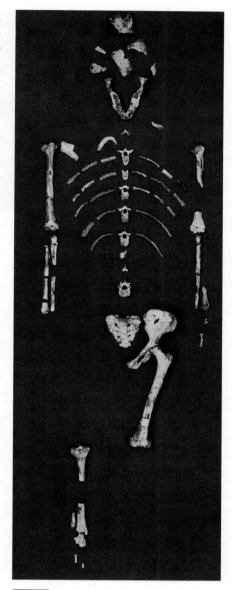

Figure 21.18 Lucy is the most complete skeleton of an australopithecine. She stood only slightly more than a meter tall and lived about 3.2 million years ago.

a
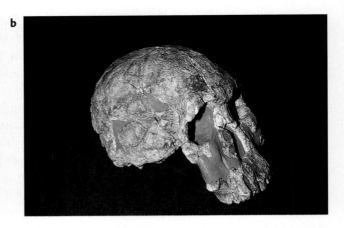
b

Figure 21.19   (a) An *Australopithecus boisei* skull, and (b) *Homo habilis* skull 1470, named for its index number at the National Museum of Kenya, East Africa. The skull is at least 1.8 million years old. Skull fragments of other individuals have been found, but skull 1470 is the most nearly complete.

21.13   The First Humans Appeared in Africa

By about 1.6 million years ago, *Homo habilis* seemingly disappeared and was replaced by an even larger-brained hominid, *Homo erectus*. *Homo erectus*, the first widely distributed hominid species, first appeared in Africa. By one million years ago, the species was present in southeastern and eastern Asia and survived in that area until about 300 000 years ago. In that span of time, the fossil record shows *Homo erectus* to have remained relatively unchanged (Figure 21.20).

Homo erectus resembled later species of *Homo* (excepting modern humans) in both body size and robustness. There is evidence that *Homo erectus* groups made large, symmetrically flaked stone tools, called hand axes, of the sort illustrated in Figure 21.21, and that some populations used fire. Toward the end of the *Homo erectus* occupation, there is fossil evidence of another group of hominids, *Homo sapiens neanderthalensis* (nee an der tal EN sis), shown in Figure 21.22. The Neanderthals, a subspecies of *Homo sapiens*, were much more robust than *Homo erectus*, or even modern humans, and the muscle attachments on their bones show that they

Anatomically modern fossil humans once were called Cro-Magnon people. This name has fallen from favor because it implies they originated where fossils were found at Cro-Magnon in France.

Figure 21.20   Two views of a reconstruction of a *Homo erectus* skull. The original fossil skull fragments were unearthed near Peking, China. Further quantities of *Homo erectus* fossils have since been discovered in China, Java, East Africa, Europe, and the Middle East.

a
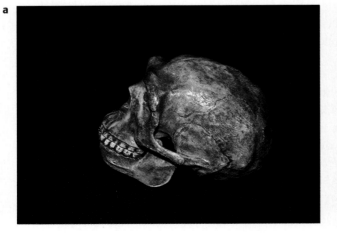
b

Biology Today

Biological Anthropologist

Mike Hoffman is a biological anthropologist who teaches at a liberal arts college. His major interest is in the study of the human skeleton, both prehistoric and modern.

As a child Mike was always picking up bones, and he loved to visit the dinosaur and fossil hominid displays in museums. During a family vacation, he visited Mesa Verde National Park and saw the artifacts and skeletal remains of prehistoric Southwestern Indians in the park's museum. These made a lasting impression on him.

Although he was a biology major in college, Mike took a number of anthropology courses, mostly in archaeology and biological anthropology. Through his class work, he began to see how a background in biology could be useful for understanding some of the questions asked by anthropologists. As his senior year in medical school began, he found he really was more interested in being an anthropologist than in practicing medicine. Following graduation from medical school, he started graduate studies in biological anthropology.

After completing his Ph.D. in anthropology, Mike taught for several years at universities in Arizona and California. While in California he became interested in some of the diseases found in prehistoric Native Americans of the region. Since then he has maintained a professional interest in human skeletal studies.

Today Mike teaches a variety of courses in biological anthropology, including human evolution, biological variation and adaptation, prehistory of the American Southwest, and his real love—bones. His current research on prehistoric skeletons involves material from California and from an archaeological site near Mesa Verde in Colorado. One of his major interests is in questions about the patterns of diseases in prehistoric populations, at least those diseases that are reflected in the human skeleton. Why does a population living in a particular environment at a particular time have certain diseases? Are the differences in these patterns related to the age of the individual, to male and female differences, to dietary differences, or to different activities? Diseases commonly seen in the human skeleton include arthritis, tooth cavities and abscesses, infectious diseases, and occasional tumors.

Mike's knowledge of human skeletons also allows him to be helpful to police and coroners who often call on him to help identify human skeletons or remains. He also works closely with dentists in the identification process. Using a variety of techniques, he can tell the age at death, sex, stature, ethnic group, right-versus left-handedness, general body build, and other distinguishing characteristics of individuals. This type of work is known as forensic anthropology.

In Mike's own words, "Make no bones about it, this is a very interesting profession. It also happens to be great fun."

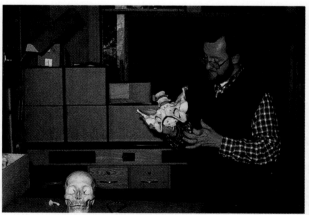

Figure 21.21 Stone tools made by *Homo erectus.*

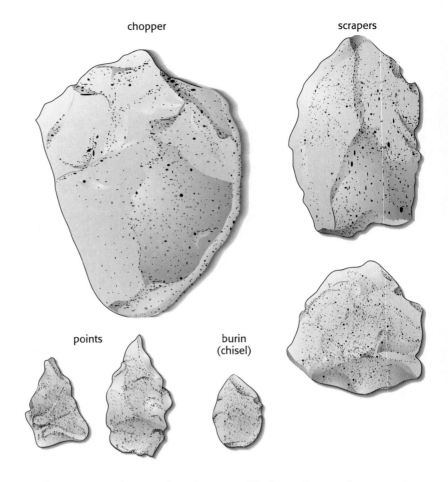

chopper   scrapers

points   burin
                (chisel)

were far stronger than modern humans. Their teeth were larger and more worn, probably from being used as tools, such as for chewing hides to soften them. The Neanderthals buried their dead and perhaps wore jewelry.

The Neanderthals spread out over much of Europe and southwestern Asia. Fossil evidence for them disappears about 30 000 years ago. Some anthropologists believe that many Neanderthals evolved into modern humans. Others insist that all Neanderthals became extinct and were replaced by modern humans who evolved from unknown genetic stock.

A case may be made, however, for the evolution of modern humans from a group of hominids that lived at the same time as the Neanderthals. These anatomically modern hominids, *Homo sapiens sapiens,* also exhibit a great deal of skeletal variation, as shown in Figure 21.23. The European fossil record suggests that these anatomically modern humans **4** replaced the Neanderthals in a rather complicated manner and may have intermixed with the Neanderthal population. Evidence from southwestern Asia suggests that modern people had replaced the Neanderthals there by 40 000 years ago. Evolutionary events in Africa may have led to the emergence of these modern humans, whose subsequent intrusion into Europe may have led to the demise of the Neanderthals.

Certainly, Neanderthals and the first anatomically modern humans were quite different, and great variations also existed within each population. At least three distinct modern human cultures existed in Europe at the same time. Other cultures of humans probably inhabited every continent except Antarctica. The age of modern humans began with much of the genetic variation found in human populations today. Among present-day human populations, however, gene pools have changed significantly because of increased travel and colonization by groups from other cultures.

Figure 21.22 A fossil Neanderthal skull. The disappearance of these widely distributed people has never been satisfactorily explained. They survived as recently as 40 000 to 50 000 years ago.

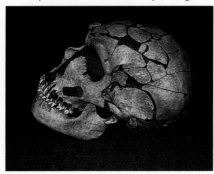

The study of genetics cannot provide a complete understanding of the history of human evolution, but geneticists and biochemists recently have found a new means of gathering data. They can analyze the structure of the DNA of human genes and of proteins made under the control of these genes, and they can compare these structures to DNA and protein structures of other organisms. The comparisons can yield clues about human evolution. Unfortunately, most proteins and DNA do not survive in fossils. Otherwise, biochemical tests of kinship between extinct species and the one living species of *Homo* might yield even more information about human evolution. Thus, much of our knowledge must be derived from the interpretation of fossils and other artifacts. Investigation 21.3 deals with the interpretation of evidence from two archaeological sites.

Certainly, the people of 30 000 to 35 000 years ago were *Homo sapiens*. They were cave dwellers, taller than their ancestors. Many were taller than most people today. Anthropologists also note important behavioral

4 changes based on the archaeological record. These changes include the use of superior bone and stone tools; the use and control of fire; the use of clothing; shifts in hunting patterns, settlement patterns, population size, and ecological range; and the development of art and of ritual activity. All these characteristics point to the emergence of a species possessing modern behavioral capabilities from a *Homo erectus* line. Figure 21.24 shows one hypothesis about the course of human evolution.

If the people who lived as recently as 30 000 to 40 000 years ago can leave behind questions that anthropologists cannot answer, it is easy to see that questions about still older populations always may exist. The origins of human speech, for example, have not been mentioned because fossil evidence has not helped directly in discovering them. Yet the development of meaningful speech must have had a dramatic effect on the evolution of humans from that point onward.

Have students speculate on the evolution of human speech.

CONCEPT REVIEW

1. What structural characteristics do primates share? How are humans different from other primates?

Figure 21.23   Two anatomically modern human skulls, possibly representing two somewhat genetically different populations. These people, like all people today are classified as *Homo sapiens*.

a
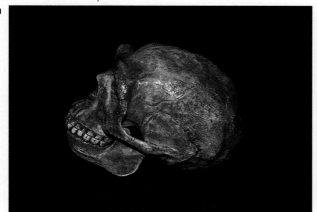

b
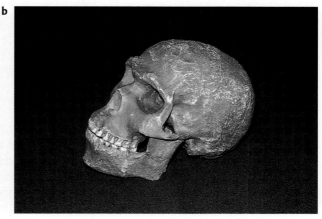

Figure 21.24 One hypothesis of human evolution. It is impossible to define the exact phylogeny of humans because numerous theories are being presented and many debates are taking place. Only recently, new discoveries have thrown doubt on previously accepted theories.

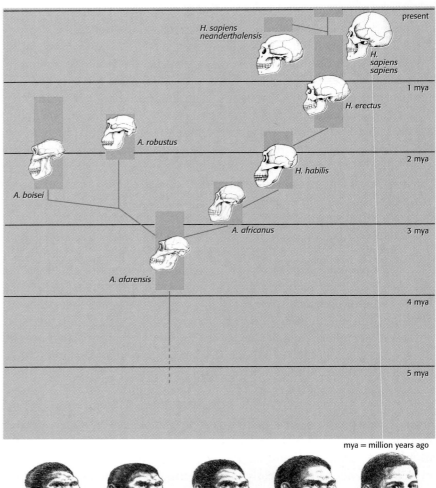

mya = million years ago

Australopithecus   *Homo habilis*   *Homo erectus*   *H. sapiens neanderthalensis*   *H. sapiens sapiens*

2. What is the best accepted current hypothesis on the evolution of humans? How do the Neanderthals fit in?
3. In human evolution, did brain expansion or upright posture probably occur first?
4. Distinguish between living modern humans and anatomically modern humans. How are these groups different from *Homo erectus?*

INVESTIGATION 21.1   Evolution and Time

In this investigation, students construct a scale model of earth's history to gain a sense of the enormous expanse of geologic time. Team sizes are variable. *Time:* Two class periods.

Life has been evolving for more than 3.5 billion years, a time span that is difficult to comprehend. Our limited experience with time involves tens and hundreds of years, not millions or billions. To understand evolution, we can represent time concretely, convert it into a linear dimension, and scale it down to a more manageable size. Time lines can be used to represent the breadth of time and the changes that have occurred during the history of earth and life. In this investigation, you will construct a scale model of major events in earth's history—a concrete time line—to gain a sense of the vast amounts of time during which evolution occurs.

Materials (per team)

Team 1

15-m tape measure
1 each black and red broad,
    permanent, felt-tip markers

150-cm tape measure
100 m of rope clothesline

Team 2

3 copies of the *Geologic Time Scale*
    (see Appendix 3)
5 dark-colored crayons or broad
    felt-tip markers
15-cm metric ruler

25 sheets of construction paper
    (one color)
19 clothespins
150-cm tape measure
pocket calculator

Team 3

3 copies of *Major Biological Events*
    (see Appendix 3)
5 dark-colored crayons
    or broad felt-tip markers
15-cm metric ruler

25 sheets of construction paper
    (one color)
14 clothespins
150-cm tape measure
pocket calculator

Team 4

3 copies of *Major Biological Events*
    (see Appendix 3)
5 dark-colored crayons
    or broad felt-tip markers
15-cm metric ruler

25 sheets of construction paper
    (one color)
20 clothespins
150-cm tape measure
pocket calculator

Procedure

Day 1

1. Your teacher will assign you to one of four teams for this activity. Team 1 will measure the area to be used, mark a clothesline at 1-m intervals, and help the other teams position their labels of the major events. Team 2 will prepare labels for the geologic time scale, Team 3 will prepare labels for the major geological events, and Team 4 will prepare labels for the major biological events.

2. **Team 1:** Use the 15-m tape measure to determine the length of the area, designated by your teacher, in which you will establish your scale model of geologic time. Use a 150-cm tape measure to mark the clothesline at 1-m intervals. Use the red marker for every 5 m and the black marker for the other marks.

    Teams 2, 3, and 4: Fold the construction paper in half crosswise. Each team should use a different color of construction paper. Label both sides of each sheet with one of the major events, using the underlined words on your team's table. Write large enough so the labels can be seen at a distance.

3. Select the sheets of construction paper with the following labels: *Rocky Mountains, Mississippi River, Ice Age, Life, Land Plants, Dinosaurs,* and *Modern Humans.* As a class, decide how many millions of years ago you think each of these events occurred. With the help of the measuring team, place the labels along the clothesline at the points that indicate the times you have chosen. Mark these times in your data book and compare them with the information in your text.

4. **Teams 2, 3, and 4:** The length of the measured clothesline represents the length of time since the earth was formed, and the far end represents the origin of the earth 4600 million (4.6 billion) years ago. If the length of your model is 100 m, then 1 million years is represented by a distance of 0.0217 m or 2.17 cm. Thus, an event that occurred 1 million years after the earth formed would be marked 2.17 cm from the *far* end (the past) of the time line; an event that occurred 1 million years ago would be marked 2.17 cm from the *near* end (the present) of the time line. Use the following equation to calculate the scaling factor for your model, based on information from the measuring team. (The example is based on a 100-m model.)

Procedure

1. **For a class of 30, use 4 students for the measuring team, 6 for the geological events team, and 10 each for the other two teams. Adjust team sizes according to your class size.**

2. **The area selected should be as close to 100 m as possible—any convenient indoor or outdoor area will do (gymnasium, hallway, football field, parking lot). A 100-yd football field is 91.44 m.**

5. Review the tables with the students and make sure they understand how to use the scaling factor. See Appendix 3, *Geologic Time Scale, Major Geological Events,* and *Major Biological Events.*

Discussion

1. This will depend on the initial placement of the labels. Because many students have a distorted view of time, it is important to have as long a time line as possible. Ancient events can be perceived visually to have occurred many, many years ago.
2. Students should observe that events are few and spread out at one end, and yet are numerous and condensed at the other.
3. Students should observe that events do not occur in strictly linear fashion, that one event does not necessarily lead to the next.
4. Large gaps appear in the time line for several reasons. First, evolutionary changes require a great deal of time. Second, the spacing of events on the time line depends on arbitrary factors, such as our perceptions of what constitutes an important event. For example, if bacteria were to construct a time line, they might show a cluster of important events between 3500 million years ago and 1500 million years ago with relatively few events since that time.
5. As constructed, the time line indicates that events occurred more rapidly when atmospheric oxygen (O_2) reached levels close to 20 percent (600 million years ago), or when animals appeared in the fossil record (800 million years ago).
6. Answers will vary. Emphasize that change will continue into the future and that it is very difficult to predict the direction or rate of change. Evolution, the formation and transformation of species, certainly will continue as long as life exists.

This investigation will help students understand what the anthropologist looks for when examining a skull or other fossil remains, and it will give them practice in graphing and data interpretation. Teams of 2 allow students to exchange ideas.
Time: One class period.

$$\frac{\text{length of model}}{\text{age of earth}} = \frac{\text{meters}}{\text{million years}} \; or \; \frac{100 \text{ m}}{4600 \text{ million years}} = \frac{0.0217 \text{ m (2.17 cm)}}{\text{million years}}$$

5. **Teams 2,3, and 4:** The scaling factor allows you to convert into meters the number of years that have elapsed since any past event. For example, using the knowledge that the first mammals appeared about 225 million years ago, you can determine the proper placement for that label in the following way:

$$\text{event} \times \text{scaling factor} = \text{label placement}$$
or
$$225 \text{ million years} \times 0.0217 \text{ m/million years} = 4.88 \text{ m}$$

The label for this event should be placed 4.88 m from the near end (the present) of the time line. Events that occurred more recently may be measured in centimeters. Refer to your team's table and use the scaling factor determined in Step 4 to calculate the distance for placing labels for each geologic time period and for each geological and biological event. Enter each distance on the table and write it on the lower right corner of the appropriate label.

Day 2

7. With the help of the measuring team, use the marked clothesline, metric tape measures, and rulers to place your team's labels the correct distance from the near end of the time line. Use a clothespin to clip a corner of each label to the clothesline so that it can be read from either side of the line.
8. Walk through your scale model of geologic time.

Discussion

1. How close to their actual occurrence were you able to place the events on the time line? What might account for the differences?
2. Look at the end of the time line near the formation of the earth. Compare it with the opposite end, near where modern humans appear. What differences do you see?
3. When you look at the time line, do the events seem to occur one after the other, or do several things appear to happen at about the same time?
4. Some portions of the time line include very few events. What are some possible reasons for this?
5. At what point do events start to occur at an increased rate?
6. Predict the changes you think might take place in the future. How might plants and animals change? How might humans change?

INVESTIGATION 21.2   Interpretation of Fossils

How do anthropologists learn about evolution? Fossil remains record the evolution of early humans, hominids, and other primates. Even though humans and other primates share many similar features, humans did not evolve from apes. Although primate brains have increased in both size and complexity during the course of evolution, more than an examination of braincase casts (or endocasts) is needed to help determine the relationships between modern humans and other primates and fossil hominids. Some of the most important features anthropologists use in comparing fossils include jaws and teeth: the cranium, or skull; the pelvis, or hips; and the femur, or thighbone.

This investigation simulates some of the comparisons anthropologists make when studying fossil hominid remains, including examination of the brain casts of several hominids and some skeletal measurements for a human, an early hominid, and a gorilla.

Materials (per team of 2)
protractor graph paper   metric ruler

PART A   Comparison of Endocasts
Procedure

1. Examine the five endocasts shown in Figure 21.25. What is the volume of each? Which species do you predict to have the largest brain volume compared with body weight?
2. Using the information in Table 21.1 and a sheet of graph paper, plot the brain volume to body mass ratio for each of the five species.
3. The area circled in color on the *Homo sapiens* endocast is Broca's area, an enlargement of the frontal lobe of the cerebrum. Broca's area sends signals to another part of the brain that controls the muscles of the face, jaw, tongue, and pharynx. If a person is injured in Broca's area, normal speech is impossible. (There are, however, two other areas of the brain that are important in language and speech.) In which species can you see Broca's area? What might the presence of Broca's area indicate?
4. Convert the ratios of brain mass to body mass shown in Table 21.2 to decimal factions and then draw a histogram (see Investigation 9.1) of the fractions.

Discussion

1. Based on your first graph, what would you say is the relationship between evolution in primates and the ratio of brain mass to body mass?
2. What major portion of the brain has become most noticeably enlarged during primate evolution?
3. Do you think the endocast of *Australopithecus* indicates that this hominid had a Broca's area in its brain?

Figure 21.25

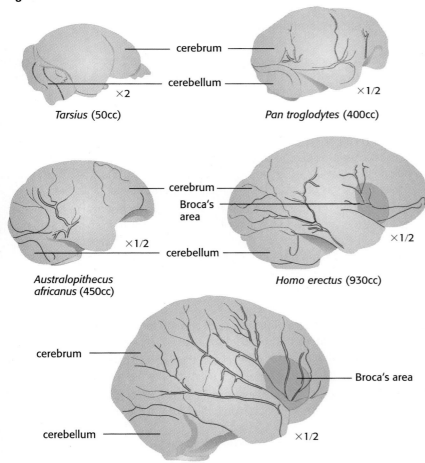

Tarsius (50cc)

Pan troglodytes (400cc)

Australopithecus africanus (450cc)

Homo erectus (930cc)

Homo sapiens sapiens (1400cc)

PART A Procedure

1. **Brain volumes are identified in the art. Students may predict that modern humans will have the highest ratio.**
3. **Broca's area is identified in *Homo erectus* and *Homo sapiens*. The presence of Broca's area might indicate that speech was possible for the species.**

Discussion

1. **The trend has been a rapid increase in the ration. See Figure T21.1.**
2. **The cerebrum has enlarged and covered much of the rest of the brain.**
3. **Some anthropologists think the *Australopithecus* endocasts do indicate a Broca's area; others disagree.**

Table 21.1 Body Weight of Selected Species

Species	Average Body Mass (grams)
Tarsius (tarsier)	900
Australopithecus	22 700
Homo erectus	41 300
Pan troglodytes (chimpanzee)	45 360
Homo sapiens	63 500

Table 21.2 Brain-Mass-to-Body-Mass Ratios

Mammal	Brain-Mass-To-Body-Mass Ratio
Tree shrew	1 : 40
Macaque	1 : 170
Blue whale	1 : 10 000
Human	1 : 45
Squirrel monkey	1 : 12
House mouse	1 : 40
Elephant	1 : 600
Porpoise	1 : 38
Gorilla	1 : 200

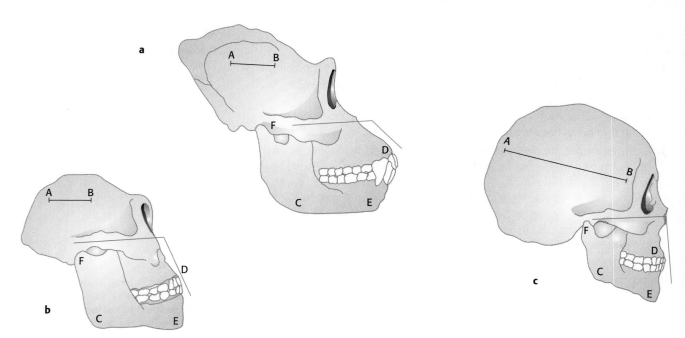

Figure 21.26

4. No, because at least two additional areas of the brain are involved (as well as the associated throat structures).

5. In some primates (e.g., the squirrel monkey) the ratio is greater than in humans. No; many other variables, including brain organization, must be considered.

PART B

7. The volume of sand (or shot) that will fill the cranium is normally used for determining cranial capacity. The technique of multiplying by 175 is suitable only for these diagrams.

8. Sometimes an index of facial area to cranial capacity is used as a measurement for comparison. Generally,

Figure T21.1

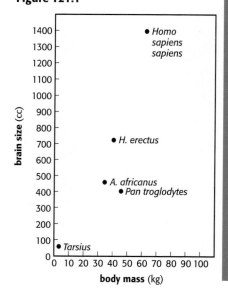

4. Does the presence or absence of Broca's area alone determine the language capabilities of a hominid? Explain.

5. How does your histogram affect your answer to Question 1? Are brain size and brain mass to body mass ratios reliable indicators of the course of primate evolution? Why or why not?

PART B   Skeletal Comparisons
Procedure

5. Examine the three primate skulls in Figure 21.26 and the drawings of lower jaws and the pelvic area in Figure 21.27. Imagine you are an anthropologist and these fossils have been placed before you for identification. Complete this hypothesis in your data book: "If the skulls, jaws, and pelvises are significantly different, then . . . "

6. In your data book, prepare a table similar to Table 21.3 or tape in the table your teacher provides. Your task is to determine which skull, jaw, and pelvis belong to a human, which belong to an early hominid, and which belong to a gorilla. Record all observations and measurements in your table.

7. Brain capacity: The straight line drawn on each skull represents the brain capacity for each primate. Measure in centimeters the distance from point A to point B for each skull. Multiply by 175 to approximate the brain capacity in cm^3.

8. Facial area: Measure from point C to point D and from point E to point F on each skull. Multiply these two measurements of each skull to determine the approximate area of the lower face. How might the size of facial area relate to evolution?

9. Facial projection: Use a protractor to determine the angle created by the colored lines. What does this measurement tell you?

10. Brow ridge: This is the bony ridge above the eye socket. Record the presence or absence of this feature for each skull. Also note the relative sizes.

11. Teeth: Study the drawings of jaws in Figure 21.27. Record the number of teeth and the number of each type of tooth. Also look at the relative sizes of the different types of teeth.

12. Pelvis: Study the drawings of the pelvis in Figure 21.27. Notice whether the flange, or lower portion, projects to the rear. Use your ruler to measure in millimeters the diameter of the pelvic opening at the widest point.

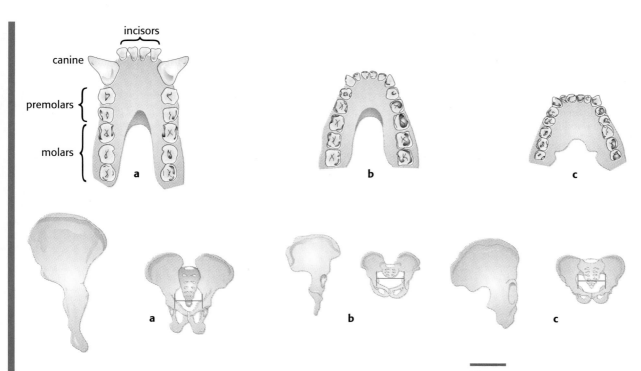

Figure 21.27

13. Read through these additional data:

   a. A large brain capacity is characteristic of humans.

   b. A smaller lower facial area is characteristic of humans.

   c. A facial projection of about 90° is characteristic of humans.

   d. Modern humans have lost most of the brow ridge.

   e. All primates have the same number of teeth and the same number of each type.

   f. Humans and other hominids have smaller canine teeth than do gorillas.

   g. A smaller pelvis, with a broad blade, a flange extending toward the rear, and a wider opening are characteristics of primates that walk on two feet.

   h. As brain size increased, the width of the pelvic opening increased to accommodate the birth of offspring with a larger head-to-body ratio.

there has been a reduction in the lower facial area of the hominid line. Nonhuman primates, however, have a greater lower facial area.

9. The degree to which the lower facial area protrudes. The degree of prognathism (protrusion) can be an indication of ancestry—human vs. nonhuman.

11. All skulls in this investigation have 16 teeth in half a jaw, and all have similar types of teeth. The sizes of the teeth vary—canines in gorillas are very large; molars in early hominids and gorillas also are very large.

12. Increased width of the pelvic opening can indicate bipedality,

Table 21.3 Comparison of Primate Characteristics

Trait	Fossil A	Fossil B	Fossil C
Brain capacity			
Facial area			
Jaw angle			
Brow ridge			
Number of teeth			
Number of each type of teeth			
Relative size of teeth			
Relative size of pelvis			
Direction of flange			
Width of pelvic opening			

Discussion

1. Specimens B and C. Number and type of teeth. Cranial capacities of A and B are somewhat similar; both have brow ridges and greater prognathism than C.
2. Specimen C is human, specimen B is an early hominid, and specimen A is a gorilla.
3. Increase width of the pelvic opening indicates increased cranial capacity and possible bipedal locomotion.
4. Answers may vary, but should include cranial capacity, prognathism, size of canine teeth, and the presence of brow ridges as the most useful characteristics. Least helpful characteristics are the number of type of teeth and size of pelvis. Additional measurements may include ration of cranial area to lower facial area, ration of width of dental arcade (canine to canine and second molar to second molar), and width of nasal opening. Access to a more complete skeleton would be helpful because more observations and measurements could be made.

This investigation allows students to analyze and interpret some data concerning human cultures of the past. The investigation is quite open-ended, encouraging hypothesis formulation, challenge, and defense. Divergent responses to the questions afford many chances for discussion and evaluation of creative ideas and opinions.

Archaeological inquiry is a valid approach in investigating past cultures. Archaeological evidence can be a strong support in laboratory investigations and can aid in the interpretation of laboratory findings. Teams of 3 are recommended for small-group interaction.

Time: One to 1½ periods; time can be cut if the reading is assigned before class or if Part C is completed as homework and followed by a class discussion.

PART A Procedure

1. High infant mortality; average life span less than that of today, few people lived beyond 50; teenagers may have been killed in battle.

Discussion

Only one or two of the most likely responses are given here. Students may suggest equally valid possibilities.

1. A beach. The sea has gone down since the time the Indians lived there.

Discussion

1. Compare the data you have assembled for each specimen. Which ones are hominid? Which traits are similar in all three primates? Which traits are similar in A and B?
2. Compare your data with the additional data. On the basis of your observations and measurements and the additional data, which fossil remains would you say are human, which are gorilla, and which are early hominid?
3. Why might an anthropologist be interested in the size of the pelvic opening?
4. Write a few paragraphs discussing the methods anthropologists use to determine human ancestry. Which of the characteristics you examined were most helpful? Which were least helpful? List any other additional observations or measurements that could be made for these specimens. Would having a more complete skeleton be helpful? Why?

INVESTIGATION 21.3   Archaeological Interpretation

Because we cannot travel back through time to see how people worked and lived thousands of years ago, we can never be sure that we understand the details of earlier cultures. Archaeologists search for clues among the remains of ancient peoples and civilizations. When they have found, dated, and studied the evidence, they formulate hypotheses to explain their findings. What emerges is a picture of life in a particular place hundreds or thousands of years ago. Although it is impossible to complete a detailed picture, many reasonable and logical conclusions can be drawn. Some findings, however, can be interpreted in a variety of ways, and archaeologists may disagree about which interpretation is correct.

This investigation requires the analysis of data found in archaeological digs. In nearly every case, several interpretations are possible. Think of as many interpretations as you can. Remember that your interpretations should account for all, not just some, of the data. In the final section of this investigation, you will predict how archaeologists far in the future might interpret today's culture and life-style.

Materials (per team of 3)
paper and pens

PART A   An Indian Cemetery in Newfoundland
Procedure

1. Read the following paragraphs:

   One scientist investigated an Indian burial site located near Port aux Choix, a small village on the west coast of Newfoundland (Figure 21.28 *left*). The burial ground appeared to have been used between 4000 and 5000 years ago. The graves were strips of very fine sand, often covered with boulders or slabs of rock. One strip of sand was almost 1.6 km long, varying from about 9 to 22 m in width. The sandy strip lay about 6 m above the high water mark of the ocean shore.

   The ages of the skeletons were estimated as shown in Table 21.4. What do these data suggest about the lives of the Indians?

   The teeth of many of the adults were worn, often so worn that the pulp was exposed. Some skeletons also had teeth missing, but there was no evidence of tooth decay. Figure 21.28 (*right* a, b, c, and d) shows some of the artifacts found in the graves. In addition, a great number of woodworking tools were found.

Discussion

1. What type of an area might the Newfoundland burial site have been thousands of years ago when Indians lived there (for example, forest, mountain, beach)?
2. What might account for the sandy strip being chosen as a prime burial site?
3. Boulders or slabs of rock covered many graves. Why might that practice have been common?

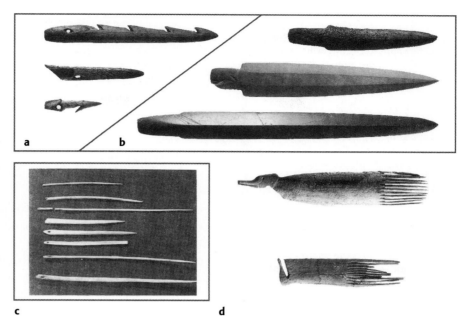

Figure 21.28

4. What do data on teeth suggest about the diet and life-style of the Indians?

5. What do you think the small objects shown in Figure 21.28a were used for? Explain your reasoning. What purpose might the holes have served?

6. What might the objects in Figure 21.28b have been used for? The objects in Figure 21.28c?

7. What modern implement do the objects in Figure 21.28d resemble? What might their function have been?

8. Considering the site location, what might the Indians have built from wood?

PART B   Ancient People of Greece
Procedure

2. Read the following paragraphs:

Stone Age people inhabited Greece long before the dawn of classical Greek civilization. Until recently, little was known of the culture of these people who lived between 5000 and 20 000 years ago. Excavations in and around a particular cave on the east coast of Greece began in 1967. Since then, archaeologists have uncovered much evidence to show the changes that occurred in the culture and life-styles of Stone Age Greeks living there.

The oldest remains (20 000 years old) in the cave primarily are the bones of a single species of horse and some tools made from flint (rock that forms sharp edges when it breaks). What might explain the presence of these horse bones mixed with flint tools?

Newer remains (10 000 years old) include bones of red deer, bison, horses, and a species of wild goat, as well as remains of wild plants such as vetch and lentils (pealike plants), shells of land snails, marine mollusks, and some small fish bones. These newer remains were left at the time when a great ice age had come to an end. In still later finds, dated about 7250 BC, very large fish bones were found—the fish might have been 100 kg or more. At about the same time, tools made of obsidian, a type of volcanic glass, were found. The nearest source of obsidian is 150 km away from the excavation site and across a body of water. The site also yielded the oldest complete human skeleton found in Greece. A male of about 25 years of age was buried in a shallow grave covered with stones. Certain bone abnormalities show that he may have suffered from malaria.

Remains dated about 6000 BC are dominated by the bones of goats and sheep found in and around the cave. These remains were of a different type

Table 21.4   Port au Choix Skeletal Ages

Age	Number of Skeletons
Newborn infants	12
Under 2 years	15
2–6 years old	2
6–18	15
18–21	1
21–50	36
50+	7

2. **Easy to dig in; a sacred site; near their settlement, thus yielding protection from enemies or wild animals that might destroy the graves.**

3. **Some religious significance; protection from weather or wild animals.**

4. **Either genetic stock or absence of foods such as honey and sugar could explain absence of tooth decay. Teeth may have been worn by chewing tough foods such as stems or roots. (Actually, experts think the wearing away of enamel was probably a result of chewing animal hides to soften the leather.)**

5. **Harpoon points (note barbs and hole for retrieval line).**

6. **Spear points for hunting animals. Needles for sewing.**

7. **Combs used for grooming, preparing furs, combing fibers such as wool or cotton. (Other artifacts suggest that hair grooming was most likely.)**

8. **Boats, because they lived near the sea; shafts for their weapons.**

PART B

2. **Because the tools are found along with bones, tools might have been used to kill the animals or prepare the meat.**

Discussion

1. **Environment became more diverse, supporting a greater variety of animals and plants. People took advantage of this diversity, even obtaining some food from the sea.**

2. **Large-scale fishing in the sea was practiced, along with development of tools to catch large prey.**

3. **The people could travel long distances at sea. Trade thus became possible. Such trade might explain the sudden appearance of obsidian. Obsidian might have a better cutting edge than other rock.**

4. **May have had low, marshy areas where mosquitoes could breed.**

5. **Indicate the beginning of animal domestication and agriculture. (Note that earlier remains indicate a hunter-gatherer life-style.)**

6. **The grave items might be related to high status in the community or a growing sense of personal property. The rarity might suggest no belief in afterlife or no particular value placed on personal property. It is possible that easily decomposed materials were placed in graves and have left no trace.**

Discussion

1-5. **Responses to questions will vary. Encourage diverse ideas and free interchange of points of view. If students express an interest, Question 5 can form the basis for a class project.**

from the animal remains found away from the cave. Evidence of wheat and barley seeds also was found. Among the tools were axes and millstones. Only a few of the human graves at the site contained objects such as tools or jewelry. A 40-year-old woman, who probably died about 4500 BC, was buried with some bone tools and obsidian blades. One infant was found near a marble vessel and a broken clay pot.

Discussion

1. What happened to the diet of the cave's inhabitants through the centuries? Explain your answer.

2. What does the evidence for the period dated 7250 BC suggest about a change in the life-style of the people?

3. How do you think these ancient people obtained the obsidian?

4. What does the skeleton with the bone abnormalities suggest about the environment at that time?

5. For the remains dated 6000 BC, what do the findings suggest about the life-style of the cave's inhabitants at that time?

6. What might be the significance of finding items buried along with people?

PART C   Excavating the Present
Procedure

3. Imagine that you are an archaeologist living in the year 4000 AD. A catastrophe has destroyed the written records of the twentieth century. You have only the remains that have been preserved through 2000 years as clues to what life was like in the 1980s or 1990s. Imagine, furthermore, that you are excavating a site that was a school 2000 years ago—your own school.

Discussion

1. What things do you think you would find? Make a list and then exchange it with your partners.

2. For each object, suggest several interpretations. (Remember, you know nothing about the time when the objects were left—the only clues are the fossils.)

3. Some materials such as stone and metal can last thousands of years. Other materials, such as wood, decay rapidly. What important things would be absent in the archaeological dig that once was your school? What ideas about our society might you miss because the clues were not preserved?

4. Consider Questions 1 through 3 again. This time, imagine a different type of site, such as a mall or business park, a residential street, a farm, or an industrial area. How would your findings and interpretations differ?

5. Suppose you wanted to build a time capsule to tell people in the far distant future about our lives today. What would you put in the capsule and why?

SUMMARY

Paleontologists interpret fossil evidence to reconstruct the organisms (and their interactions) of ancient ecosystems. The evolution from simple beginnings of the invertebrates, the flowering plants, the reptiles, the amphibians, and the mammals of today is apparent in the fossil record of those ecosystems. Evolutionary mechanisms, including the effects of natural selection of organisms occurring on continents that moved, account for the variety and distribution of today's plants and animals. Our own group, the hominids, is a subgroup of primates. Ancestors of both humans and apes radiated from early hominidlike primates, evolving into hominids and apes.

We share many characteristics with other primates but also have distinct differences. Chief among these differences are our upright posture, brain size and head shape, and speech.

APPLICATIONS

1. The ocean makes up about 2/3 of the surface of our planet, yet there are only approximately 19 000 species of algae, which are mostly restricted to water. There are, however, approximately 250 000 species of flowering plants, which are mostly restricted to land. What could account for the difference in numbers of species?

2. Birds and mammals are endothermic—they use metabolic energy to maintain a constant internal body temperature. Reptiles, fishes, and amphibians are ectothermic, that is, they use environmental energy and behavioral adaptations to regulate their body temperature. Were dinosaurs endothermic or ectothermic animals, and how would this affect their lives?

3. The formation and breakup of Pangaea greatly changed the amount of shoreline in the world. How do ocean currents affect life on land? What influence do these currents have on the climate of terrestrial habitats?

4. If you plotted all of the known volcanic eruptions in the world, what would this tell you about the location of the tectonic plates?

5. Some scientists suggest that the extinction of dinosaurs was the result of an ancient "nuclear winter." What is a nuclear winter, how could it have occurred *before* humans were on earth, and how might it have killed the dinosaurs?

6. Evidence suggests our ancient ancestors walked on four limbs instead of two. Standing upright may have some advantages for our survival, but how are the following problems related to our relatively recent stand on two feet: slipped disks in the lower back, flat feet, varicose veins in the legs, and hernias?

7. Why are tropical forests poor sources of fossil evidence? Consider the conditions for fossilization and the conditions for finding fossils.

8. If you found a fossil skeleton with a cranial capacity of 550 cm^3 and a foot with a high arch, would you classify it as a hominid? Explain.

9. Would a fossil skeleton with a cranial capacity of 600 cm^3, holding an axelike stone, be classified as a hominid? Explain.

PROBLEMS

1. Ask a geologist to organize a field trip to an interesting site in your area. Before going on the trip, obtain information about the geology from the nearest geological survey office.

2. What type of weapons and tools did our ancient ancestors use? Try to find drawings or photographs of ancient equipment, or visit a museum to see what was used. Compare these weapons and tools with those used by primitive tribes people alive today.

3. Go to the zoo and observe the behavior of the primates there. How easy is it to misinterpret what an animal does or to give an incorrect "reason" for a particular behavior?

4. In many museum displays, background paintings illustrate the environment and organisms of a particular time and place. Choose a prehistoric time and make a large painting showing interactions between the plants and animals living in your area at that time.

5. Using a large map of the world today, try to create the ancient continent of Pangaea. Cut out tracings of the continents and try to fit them together. Do any continents seem to fit together like pieces of a jigsaw puzzle?

6. Where and how do nonhuman primates live in the world today? What is the relationship between these wild animals and humans? Read about any of these primates and write a report.

7. Read S. J. Gould's book, *Wonderful Life* (New York: W. W. Norton, 1989), to learn about the incredible array of fossilized organisms found in Canada, many of which are extremely different from anything alive on earth today. Sketch these weird creatures and see what modern organisms they most resemble.

8. Construct a La Brea ecosystem food web.

9. Consider the causes of death in modern urban and suburban environments. Include factors that kill individuals before, during, and after the age of reproduction. Consider also factors that reduce health or impair development and factors that reduce fertility or the survival rate of offspring. Could these factors cause changes in characteristics of future human populations?

Biomes Around the World

Anza-Borrego Desert State Park, California. What types of vegetation do you see here? Based on the photo, what do you think might be some limiting factors for this biome? How might these limiting factors affect the types of plants and animals found here?

Century plants and beavertail cactus probably are the major vegetation types; grasses might grow in the small pockets of soil between the rocks. Water, available soil, and temperature probably are the most important limiting factors. Only plants and animals that are able to withstand dry conditions, high daytime temperatures, and low nighttime temperature would be able to live here. Plants that can establish themselves in limited amounts of soil would have a better chance.

◆ Across the face of the earth there are great differences in climate. Through billions of years, these climatic differences have resulted in the evolution of diverse ecosystems that harbor distinct groups of plants, animals, and microorganisms, even though they may be located at the same altitude or latitude. The small town of White River in Ontario, Canada, which lies about 20 km from the northeast shore of Lake Superior, is surrounded by a forest of spruce, fir, and some pine. A sign with a large thermometer, shown in Figure 22.1, proclaims that White River is the coldest town in Canada, with a low temperature of "72° below zero." A town better known than White River but at the same latitude—48° north—is Paris, France. What is the climate in Paris? Is it as cold? Do the same types of trees grow there as in White River? This chapter examines the major types of habitat, the organisms found in each, and the reasons for the differences between habitats. It also discusses some ways in which humans affect ecosystems. ◆

MAJOR CONCEPTS

- ◆ Distribution of solar energy determines world climatic patterns.
- ◆ Biomes are regions of the world with unique climatic characteristics and life forms.
- ◆ Temperature and precipitation determine the types of life present.
- ◆ Similar climates permit similar biomes.
- ◆ Many factors can cause communities to change.
- ◆ Humans have drastically altered many biomes around the world.

KNOWLEDGE CHECK

- ◆ What abiotic factors determine the character of biomes?
- ◆ What can cause ecosystems to change?
- ◆ How do human activities affect ecosystems?

Climate and Ecosystems

22.1 Abiotic Factors Influence the Character of Ecosystems

The earth is a patchwork of habitats that differ in abiotic factors such as temperature and the amount of precipitation and sunlight received. The interaction of these limiting factors determines the climate of the area, which in turn affects the distribution and diversity of living organisms.

In each major type of climate, a characteristic type of vegetation develops. For example, warm, arid climates—those with little rainfall—are associated with desert vegetation. Semiarid climates usually are covered with grassland. Moist climates support forests. Each type of plant life, in turn, supports a characteristic variety of animal life. The resulting ecological community of plants and animals is called a **biome.** (Ecosystems, in contrast, also include abiotic factors such as soils and precipitation.) Biomes extend over large natural areas, as shown in Figure 22.2.

Where does the energy that drives an ecosystem come from? Ecosystems are powered primarily by sunlight, or radiant energy, which is important for two reasons. First, radiant energy is the form of energy that most producers use to make food. All organisms in an ecosystem ultimately

Guidepost
What abiotic factors influence the climate of an ecosystem?

Figure 22.1 White River, Ontario.

569

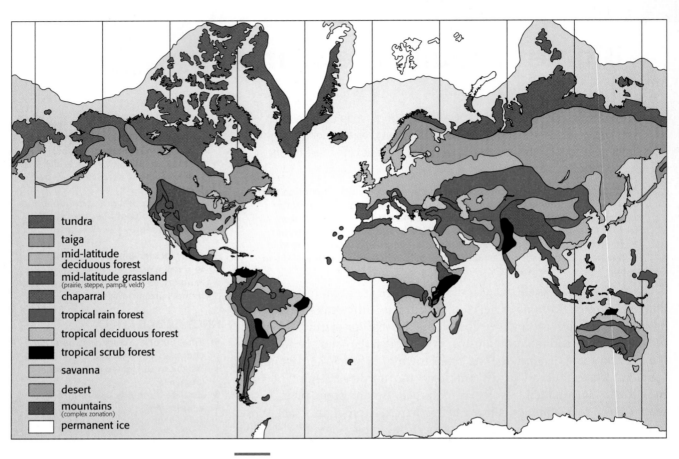

Figure 22.2 Major biomes of the earth. Many parts of the earth have not been thoroughly studied, and even where observations are plentiful, ecologists sometimes disagree about the correct interpretation.

Legend:
- tundra
- taiga
- mid-latitude deciduous forest
- mid-latitude grassland (prairie, steppe, pampa, veldt)
- chaparral
- tropical rain forest
- tropical deciduous forest
- tropical scrub forest
- savanna
- desert
- mountains (complex zonation)
- permanent ice

depend on food made by producers. Second, the temperature of an ecosystem is determined by the amount of radiant energy it receives. The energy is absorbed at the earth's surface and is radiated to the air as heat.

The position of the earth in relation to the sun affects the amount of radiant energy different parts of the earth receive. The earth is nearly spherical and is tilted with respect to its orbit around the sun. At the summer solstice, about June 21, the Northern Hemisphere is tilted 23.5° toward the sun. As a result, the Northern Hemisphere receives more solar **2** energy than the Southern Hemisphere, so it is summer north of the equator and winter south of the equator. At the winter solstice, December 21, the Southern Hemisphere receives more solar radiation than the Northern Hemisphere, and the seasons are reversed. Figure 22.3 illustrates the position of the earth relative to the sun at different times of the year. Other factors, however, also affect the distribution of solar energy.

22.2 Air Circulation and Ocean Currents Affect Climate

The circulation of air in the atmosphere is powered by solar energy. The movements of the air helps to distribute the heat that comes to the earth as radiant energy. At the same time, circulating air currents carry water vapor from the oceans over land surfaces, where it falls as rain.

Climates occur in broad belts that encircle the earth. The boundaries of these belts are disrupted by land masses and oceans. Climates are

Figure 22.3 The distribution of solar energy on the earth's surface. At which point does North America receive the most radiant energy? **The summer solstice.**

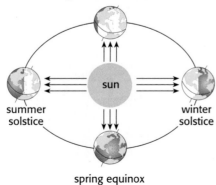

summer solstice

winter solstice

spring equinox

further modified by mountains, particularly when mountain ranges block
3 moisture-laden winds from the oceans (see Figure 22.4). When moisture-carrying wind from the ocean meets the mountains, the moisture condenses and is released as precipitation. As the air travels down the other side of the mountains, the leeward side, it warms and begins to collect moisture. As a result, the leeward sides of mountains are drier than the windward sides, and the vegetation often is very different. This phenomenon is called the rain-shadow effect.

Ocean currents also affect climate because they redistribute heat. For
3 example, the Gulf Stream (in the North Atlantic) swings away from North America near Cape Hatteras, North Carolina, and reaches Europe near the southern British Isles. Because of the Gulf Stream, western Europe is much warmer than is eastern North America at similar latitudes. As a general rule, the sides of continents in the temperate zones of the Northern Hemisphere are warmer and drier than the eastern sides. The opposite is true in the Southern Hemisphere. Figure 22.5 shows the major ocean currents.

It is easier to map the distribution of a particular climate factor than it is to map a climate as a whole because climatic factors overlap and interact with one another in complex ways. To simplify climate-mapping, ecologists frequently use climatograms that summarize monthly measurements of temperature and precipitation. Appendix 3 contains several climatograms for different areas.

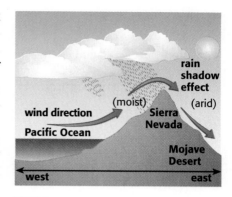

Figure 22.4 The rain-shadow effect. As air containing water vapor rises, it cools. Cool air holds less moisture than warm air; thus clouds form on the western side of the Sierra Nevada range. Rainfall tends to be heavy on this side. As the air descends on the eastern side, it picks up moisture from its surroundings rather than releasing it. This rain-shadow effect produces desert conditions on the eastern side of the Sierra Nevada range.

CONCEPT REVIEW

1. How does sunlight affect an ecosystem?
2. What causes the winter and summer solstices?
3. How do mountains and ocean currents affect precipitation and temperature?
4. What determines the characteristic plants and animals found in a biome?

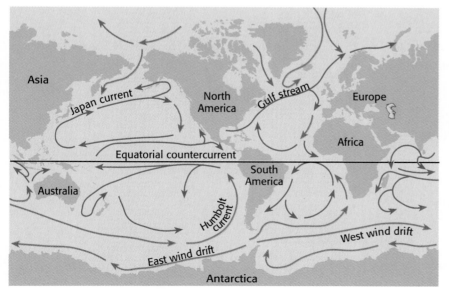

Figure 22.5 The major ocean currents have profound effects on climate. How does the Humboldt Current affect the western side of South America. What affects the eastern side of South America?
The Humboldt Current cools the western side of South America. The eastern side of South America is warmed by water from the equator.

Guidepost

How does the vegetation change from north to south?

Figure 22.6 Tundra in Alaska. Note the pond in the foreground and the lack of trees. What climatic factor is responsible for these features? **Permafrost.**

Leaves of most tundra plants are small. Many are hairy or have margins rolled inward, thus reducing evaporation of water from the leaf surface.

One of the concerns about the Alaskan oil pipeline was that it would interfere with the annual migrations of caribou. The pipeline was raised high enough above ground that the caribou can pass under it.

Biomes with Increasing Radiant Energy

22.3 Tundra Is Characterized by Low Vegetation

Beyond the northern limit of tree growth and at high altitudes is the tundra (TUN druh). Arctic tundra, shown in Figure 22.6, is determined by latitude and characterized by permafrost—continuously frozen ground that prevents plant roots from penetrating very far into the soil. Spring comes in mid-May; only plants near the sun-warmed ground can grow, flower, and produce seeds before the advent of fall in August. Most plant reproduction and growth occurs during the brief, warm summers, when daylight lasts for almost 24 hours. The tundra may receive less precipitation than many deserts, yet the combination of permafrost, low temperatures, and low evaporation leaves the soils saturated with water. These conditions further restrict the species that grow there. Dwarf perennial shrubs, sedges, grasses, mosses, and lichens are the dominant vegetation.

The change from summer to winter is rapid. Snowfall is light, only 25 to 30 cm per year, and high winds sweep open areas free of snow. There are few daylight hours, and above the Arctic Circle around the winter solstice there are three months of near darkness.

Alpine tundra (determined by altitude) is found above the tree line on high mountains. Because of the similarity in conditions, many plant and animal species found in the arctic tundra also are found in the alpine tundra. Alpine tundra can occur at any latitude if the elevation is high enough (see Figure 22.7). Day length in the alpine tundra is shorter in summer than it is in the arctic tundra, but longer in winter. Brief and intense periods of productivity occur daily in summer, but freezes may happen any night. Winter snow or gale winds result in short growing seasons.

Animals of the arctic tundra include year-round residents such as ptarmigan, arctic foxes, and snowshoe hares. During the summer, large flocks of migratory water birds raise their young, and caribou graze on grasses and reindeer moss. Figure 22.8 shows two typical tundra animals.

How do animals cope with the coming of winter? Many animals that stay in the tundra become dormant or hibernate. As winter approaches, they find shelter and their body processes slow down. The energy needed to keep them alive while they are hibernating is derived from body fat stored during the warmer months. Other animals withstand the cold by

Figure 22.7 Types of vegetation and animal life are affected both by altitude and latitude.

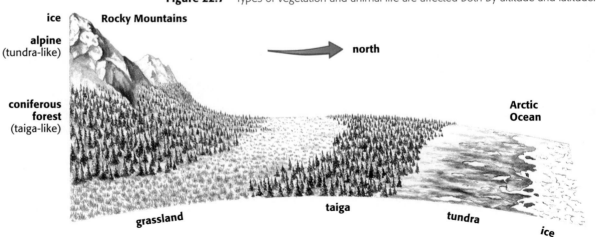

a

b

Figure 22.8   Animals of the arctic tundra: (a) snowy owl, *Nyctea scandiaca*; and (b) musk ox, *Ovibos moschatus*. How are each adapted to life in the tundra?   **The owl is camouflaged and has thick down to protect it from cold. The musk ox relies on its very large size, layers of fat, and thick coat of fur to protect it.**

living in burrows, having a large body size, or having good insulation that retains heat. Still other species, especially birds, are migratory and leave the area during the harshest part of the winter.

22.4   Taiga Is a Coniferous Forest

At the southern edges of the arctic tundra (or at a slightly lower elevation than alpine tundra) scattered groups of dwarf trees appear in sheltered places. Eventually the tundra gives way to the great coniferous forest or taiga (TY guh), shown in Figure 22.9. This forest extends in a broad zone across Europe, Asia, and North America, but there is no similar biome in the Southern Hemisphere. The taiga receives more radiant energy, both daily and annually, than does the tundra. The summer days are shorter but warmer, and the ground thaws completely. Winters are not as long as they are in the tundra, although more snow falls.

2

Taiga vegetation usually is dense, and the woody plants prevent much undergrowth. Because the conifers keep out the sunlight, there are mostly mosses, lichens, and a few shrubs growing near the ground. Most food production, therefore, takes place in the upper parts of trees. Conifers, such as spruce, fir, pine, and hemlock, grow in an interesting variety of combinations with deciduous trees, such as willow, alder, and aspen. Usually only the conifers are conspicuous. The soil tends to be acidic under the coniferous trees and supports few species of decomposers. Earthworms are uncommon, but large numbers of tiny arthropods live in the soil and decompose organic matter.

Taiga animals include deer and elk that browse the young leaves, and moose that eat aquatic vegetation (Figure 22.10). Smaller animals include squirrels, jays, hares, porcupines, and beavers. Predators of the taiga include grizzly bears, wolves, lynxes, and wolverines. What adaptations to cold might these predators have?

22.5   Midlatitude Forest Biomes Have Four Distinct Seasons

Midlatitude, or temperate, forests develop throughout the midlatitude regions where there is enough moisture to support the growth of large trees. This biome has four distinct seasons, as shown in Figure 21.11. Much of the farmland in the United States was once forest.

In summer, the daily supply of radiant energy is greater than it is in the tropics on all but a few days of the year. In winter, the earth's

Figure 22.9   The northern coniferous forests make up the taiga biome in which ponds, lakes, and bogs are abundant.

Taiga **is the Russian word for forest.**

Ask students to hypothesize why no taiga exists in the Southern Hemisphere.

Coniferous trees that are not evergreen include bald cypress, which does not occur in the taiga, and larch, which does.

In the taiga, decaying needles from the conifers produce an acidic soil that inhibits the germination and growth of annuals and other herbaceous plants.

Figure 22.10   Moose, *Alces alces*.

Winter   Spring   Summer   Fall

Figure 22.11   Seasons in midlatitude deciduous forest. These pictures show the same area. What effects do seasonal changes have on herbivores such as deer?

Taiga animals avoid the cold by sleeping for long periods, or they have great bulk and thick fur sufficient to keep them warm at −40° C.

Deer are mainly browsers; they eat twigs and buds all year but obtain most nourishment from leaves during summer. Snow hinders travel.

Ask students what effects such conditions might have on producers. High radiant energy, warmth, and humidity in combination favor rapid growth of producers during a relatively long summer. The cool, dry air masses are never sufficient to hinder growth.

Ask students to speculate on why reptiles are more abundant in temperate forests than in the taiga.

orientation to the sun changes, and the midlatitude forests receive less radiant energy. Thus, the total annual supply of radiant energy is much less in midlatitude forests than in the tropics. Temperature ranges from cold in the winter (−30° C) to hot in the summer (+35° C). Both rainfall and snowfall are abundant, with an average annual precipitation between 75 and 150 cm per year. Humidity tends to be high, growth and decomposition rates are high, and the soil is deep and rich.

The many species of deciduous trees include oaks, hickory, maple, beech, chestnut, and basswood. The tallest ones form a canopy—an upper layer of **3** leaves that catches the full sunlight. Enough sunlight penetrates this layer to provide energy for a lower layer of trees and a layer of shrubs beneath. On the forest floor, mosses and ferns receive the remaining faint light.

The diverse vegetation supports a large number of consumers. Squirrels collect nuts and berries from trees. Deer mice climb in the shrubs and search the ground for seeds. White-tailed deer browse on shrubs and the lower branches of trees. In summer, insects are abundant in the soil and in all layers **3** of the forest canopy. A variety of birds prey on these insects. For example, in some forests red-eyed vireos consume canopy insects, acadian flycatchers catch insects flying below the canopy, ovenbirds search out insects on the ground, and woodpeckers extract boring insects from the bark of trees (Figure 22.12). Foxes, skunks, and an occasional black bear also may be present. There are few large predators in most of this biome. Figure 22.13 shows some animals and plants found in a midlatitude deciduous forest.

In autumn, the leaves of deciduous trees cover the ground with a thick mass of organic matter. Many mammals become fat on the abundant nuts, acorns, and berries, and some store the food. Woodchucks form thick layers of fat and then hibernate in burrows. Reptiles, much more abundant here than in the taiga, also hibernate. Many insect-eating birds migrate to the tropics. Many mammals, such as squirrels, rest during the cold spells of winter. Winter birds, which are more abundant here than in the coniferous forest, eat seeds and fruits and search out dormant insects and insect eggs from the cracks in tree bark.

22.6   Tropical Rain Forests Have a Uniform Climate

Tropical rain forests are found on lowlands near the equator, where temperature and day length show little seasonal variation. The noon sun is almost directly overhead throughout the year, making the amount of radiant

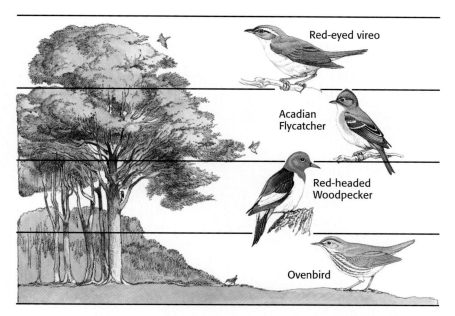

Red-eyed vireo

Acadian
Flycatcher

Red-headed
Woodpecker

Ovenbird

Figure 22.12   Resource partitioning in a deciduous forest. Red-eyed vireos in the canopy, flycatchers directly below, ovenbirds on the ground, and woodpeckers on the trunks. How does such partitioning affect these species? **It allows them to occupy the same area in a forest without directly competing with one another.**

energy high and fairly constant. Rain falls almost every day, and the humidity is always high. Average rainfall is about 200 cm a year. Rainfall, however, depends on elevation and winds that bring moisture.

Vegetation is dense. Figure 22.14 shows the typical vegetation layers in a tropical rain forest. Weaving through the branches are many woody vines. Because the tall dense canopy of foliage competes for light, little light reaches the ground. Beneath the taller trees are shorter trees that can tolerate shade. Beneath these are still others that are even more shade tolerant. Low light intensity limits the growth of vegetation on the forest floor and, hence, the animal species that live there.

Along the trunks and branches of the trees and the twisting stems of the vines are many epiphytes (EP ih fyts). These plants use the branches on which they perch for support but not for nourishment. Epiphytes have no contact with the ground, so they have special adaptations for obtaining water and nutrients. Some have roots that absorb moisture from the humid atmosphere the way blotting paper soaks up water. Many catch the

In rain forests, unlike most biomes, light rather than water is the greatest limiting factor.

Abundant rainfall and high humidity are necessary for the prolific growth of epiphytes.

Figure 22.13   Animals and plants of the deciduous forest: (a) white-tailed deer; (b) least chipmunk; (c) marsh marigold and may apple.

a

b

c

Figure 22.14 The layering effect in a tropical rain forest. Virtually no sunlight reaches the rain forest floor, and comparatively fewer organisms are able to live on the ground than in the trees.

Intense competition results in specialized organisms that fill the multitude of microniches.

Unknown thousands of species remain to be described.

Figure 22.15 Arboreal rain forest animals include (a) the scarlet macaw, *Ara macao*, and (b) the squirrel monkey, *Ateles*.

a

b

daily rain in special hollow leaves. Mosquitoes, water beetles, and other aquatic insects may live in such treetop puddles.

Ripe fruits drop to the forest floor and supply food for some ground dwellers. Although the trees are always green, leaves die and fall continuously for most of the year. In the warm, moist environment, huge numbers of insects, fungi, and bacteria attack this food supply rapidly. Organic remains, therefore, do not build up on the ground. Predators and parasites are abundant at all levels of the forest. Large herbivores, such as hoofed mammals, are rare or live only near riverbanks.

What environmental factors might be related to the rarity of large, hoofed mammals? On the forest floor there is little herbaceous vegetation for herbivores to eat, and the continuously dropping debris is decomposed **3** quickly. The shallow-rooted plants reabsorb the released nutrients through mycorrhizae. Because most of the nutrients are held in the vegetation, the soil is nutrient-poor.

All forests have some arboreal (ar BOHR ee uhl) animals—animals that live in the trees. In the tropical rain forest, most animals live in the canopy (Figure 22.15). In one study of rain forests, 90 percent of the birds were found to feed mostly in the canopy. This may not be surprising in regard to birds, but in the tropical rain forest, a large number of mammals, more than **4** 50 percent of the species, also are arboreal. Because there is more food in the canopy layers than on the ground, there is a greater diversity of animals to be found in the various canopy layers. Moreover, there are many tree snakes, tree lizards and tree frogs, as well as an untold number of arboreal insects.

The seasonless rain forests with a uniform climate cover only a small part of the tropics. Most tropical regions have seasons. Instead of being warm and cold, however, the seasons are wet and dry. Many woody plants **6** lose their leaves during the dry season. Tropical regions with uniform temperatures but wet and dry seasons produce tropical deciduous forests or tropical seasonal forests. Consult the map in Figure 22.2 to locate regions of tropical deciduous forest. The canopy is not as dense as that in the rain forest, and light filters all the way to the forest floor.

CONCEPT REVIEW

1. What adaptations of structure and function are found among tundra organisms?
2. What is the most noticeable difference between the tundra and taiga landscapes?
3. Compare the vegetation in a midlatitude deciduous forest and a tropical rain forest. Explain the differences.
4. Many arboreal animals live in tropical rain forests. What are the reasons for this?
5. How do epiphytes obtain water and nutrients?
6. What is the major limiting factor in the tropical rain forest? In the tropical deciduous forest?

Biomes with Decreasing Precipitation

22.7   There Are Three Distinct Types of Grassland

Along the latitudes of deciduous forest the precipitation decreases from east to west in North America. This lack of water, rather than of radiant energy, is the major factor limiting plant and animal life. Grasslands, such as the prairies of North America, typically receive 63 to 71 cm of annual precipitation, have low winter temperatures, and may be relatively high in elevation. Drought and periodic fires keep woody shrubs and trees from invading the environment, except along streams.

There are three distinct types of grasslands in North America: the tall grass prairie in the northern Midwest, the short-grass prairie that stretches eastward from the Rocky Mountains into Kansas, and the mixed-grass prairie in the central Great Plains. Figure 22.16 shows these typical grasslands. The geographic distribution of these grasses results mostly from the nearness of the Rocky Mountains and the rain shadow the mountains produce. Tall grass prairies receive more precipitation than do mixed-grass prairies, and mixed-grass prairies receive more than short-grass prairies. The boundaries of these grasslands are indistinct, and they share many plant and animal species.

The most conspicuous consumers are hoofed mammals. Once there were many bison and pronghorns in North American grasslands. Now

> **Guidepost**
>
> How does the vegetation change as the precipitation decreases?

Similar areas are found in other midlatitude regions of the world. Many names are give to midlatitude grassland: prairie and plains in North America, steppe in Asia, veldt in South Africa, pampas in Argentina.

Figure 22.16   Tall-grass prairie (a); mixed-grass prairie (b); short-grass-prairie (c). What is the most important climatic factor in determining each type of prairie?   **Precipitation.**

a

b

c

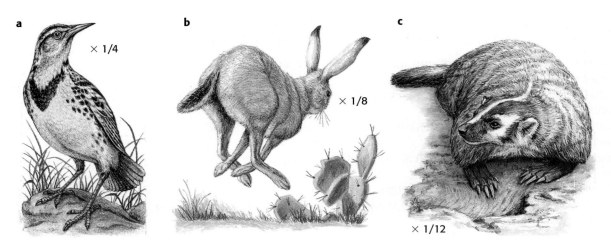

Figure 22.17 Animals of the grassland include (a) western meadowlarks, *Stumella neglecta*; (b) black-tailed jackrabbits, *Lepus californicus*; and (c) badger, *Taxidea taxus*.

nearly all of the bison and many of the pronghorns have been replaced by cattle and sheep. Less conspicuous herbivores are jackrabbits and ground squirrels. Many types of insects also feed on the vegetation. At times grasshopper populations reach large numbers, devouring the plants down to ground level.

Wolves and coyotes were once the chief large predators. Today, wolves have been exterminated almost everywhere in the contiguous 48 states, but coyotes survive in many places. Rattlesnakes and badgers are important predators of ground squirrels and prairie dogs. Many insect-eating birds, such as meadowlarks and mountain plovers, nest on the ground under the cactus on the short-grass prairie. Figure 22.17 shows some typical grassland animals.

What adaptations might be important for animals that live in such open areas? Because of the openness of the grasslands, speed and endurance are adaptations that enable many organisms to escape predators. Small mammals seek shelter below ground, and grassland birds and reptiles escape detection because of protectively colored skins.

22.8   Midlatitude Deserts Border the Grasslands

In North America, the western edge of the grassland is bordered by desert. The climatic situation is complicated by mountains, but between the western ranges, deserts occur from eastern Washington south into Mexico. There are two types of North American deserts. The northern part of the desert biome is the cool desert of the Great Basin and has an average temperature of about 10 to 12° C in the shade. In the southwestern or hot deserts, the temperature averages around 20 to 22° C. Death Valley in the Mojave Desert has reached 57° C in the shade. Figure 22.18 shows vegetation typical of both hot and cool deserts.

Latitude and temperature have profound effects on the formation of deserts. The amount of precipitation that results in a desert at the equator may support a fine grassland at higher latitudes. When precipitation does occur in a desert, it is likely to be heavy but brief and is often the result of thunderstorms or cloudbursts. Much of the water runs off instead of sinking into the soil. The rate of evaporation is always high compared with precipitation. For example, although Tucson, Arizona, receives about 20 cm of rain a year, it is so hot that the evaporation rate exceeds the equivalent of about 195 cm of rain. Loss of heat from the earth's surface is greatly slowed by water vapor in the air. Because desert air is dry, the heat that builds up in the soil during the day is quickly lost at night by radiation.

Students should now be able to make a generalization about the relationships between production of vegetation (in kg/hectare/year) and the abiotic environment. Production is a function of total annual radiant energy, given optimal water and nutrients.

a

b

Desert plants have several adaptations that enable them to survive the
3 harsh conditions. Most desert plants resist drought by having root systems that
spread out in all directions from the stems but extend only a short distance
into the ground. When it rains, these widespread, shallow roots soak up the
moisture rapidly. Cactus plants resist drought by storing water in the tissues of
their thick stems. The seeds of some plants are covered with a chemical that
prevents their germination during unfavorable conditions. The chemical is
washed away during a heavy cloudburst so that germination is possible. Some
plants have root systems that grow down near the water table, ensuring a continuous supply of water. Mesquite plants (Figure 22.19) can have taproots up
to 25 or 30 m long and have been seen growing down through the ceilings of
horizontal mine tunnels. Still other plants have small leaves with thick cuticles
that reduce water loss, and some even shed their leaves when it is very dry.
Some plants, such as the creosote bush, produce a substance that inhibits the
growth of other plants that otherwise would compete for space and water.

Seed-eating animals, such as ants, birds, and rodents, are common in
deserts. Like the desert plants, many of these animals are especially adapted to living where there are extreme temperatures and little available water.
Many desert animals live in burrows and are active only at night. Some
4 rodents wait out the hot summer days in their burrows in a state of sleep
called **estivation** (es tih VAY shun). Estivation is similar to hibernation, but
it is a response to heat and dryness rather than to cold. Others animals are

Figure 22.18   Hot desert in Arizona (a).
What characteristics do many of the plants
have in common? Cool desert (b). How
do these plants differ from those of the
hot desert?   **Plants of the hot desert have
numerous adaptations that enable them
to conserve water: shallow, spreading
root systems that quickly absorb the
infrequent but torrential rain; leaves
reduced in size or modified as thorns
or spines that reduce water loss;
water storage.**

**Thorns and spines on desert plants
probably are protection against
herbivores in a land where their food is
sparse. More clearly demonstrable is the
relationship between reduction in leaf
surface and retardation of water loss.**

Figure 22.19   How does mesquite escape
drought?   **By having long tap roots that
reach down to the water table.**

Figure 22.20   The kangaroo rat can survive without drinking any water. What adaptations might make this possible? **The kangaroo rat obtains almost 90% of its water as a product of carbohydrate metabolism; the remaining 10% comes from water present in food. Waste products are excreted as uric acid crystals.**

Guidepost

What is the effect of variable rainfall and uniform temperature on a biome?

light-colored, thereby reflecting light, and have adaptations that enable them to conserve water. The kangaroo rat, shown in Figure 22.20, which lives in the American Southwest, can survive without drinking any water.

CONCEPT REVIEW

1. Why are there grasslands in the same latitude as forests?
2. What two abiotic factors must be considered together in describing desert climate?
3. How are plants adapted to desert conditions?
4. In what way is estivation an adaptive behavior?

Biomes with Variable Precipitation

22.9   Savannas Are Tall Grasslands in Tropical Dry Areas

The savanna (suh VAN uh) is a tropical or subtropical grassland with a few scattered trees. Three distinct seasons characterize the savanna: cool and dry; hot and dry; and warm and wet. Most savanna soils are rather poor. Savannas cover wide areas of central South America, central and South Africa, much of Australia, and portions of southern Asia.

In Africa, the savanna is home to some of the world's largest herbivores, including elephants, giraffes, and rhinos. These animals, especially the young, are preyed on by lions, leopards, and cheetahs. Scavengers, such as hyenas and vultures, clean up the carcasses left by the predators. Often, leopards will place their kill in trees so other animals cannot steal the carcass. Large herds of zebras, wildebeest, and gazelles, as well as elephants, giraffes, and other animals, are found on the Serengeti Plains of the savanna (see Figure 22.21). At present these herds make up the largest concentration of large herbivores to be found on any continent.

Consider the character of the savanna. What adaptations might savanna dwelling animals have? Burrowing animals are common, and nest sites

Figure 22.21   Acacias, water bucks and impalas on the savanna in Kenya.

Figure 22.22 A chaparral in California.

and shelters often are on the ground. Animals in the savanna are most
2 active during the rainy season, and many species are nocturnal. In the winter, or dry season, when the aboveground vegetation is dry, many animals
are dormant or subsist on seeds and dead plant parts.

Ask students to what environmental conditions a plant with tough evergreen leaves might be adapted. They are structurally adapted to seasonal lack of water.

22.10   Chaparral Covers Dry Areas with Thin Soil

Midlatitude areas along coasts where cool ocean currents circulate off-
3 shore often are characterized by mild, rainy winters and long, hot, dry
summers. These areas, such as the California coast, are dominated by
chaparral (shap uh RAL), or brushland, communities composed of dense,
spiny shrubs with tough evergreen leaves, often coated with a waxy material. Chaparral vegetation also is found in the Mediterranean region, along
the coastlines of Chile, southwestern Africa, and southwestern Australia.
Plants from these regions are unrelated, but they resemble one another in
form and function and show the same type of adaptations. Annual plants
also are common in the chaparral during winter and early spring when
rainfall is most abundant. Figure 22.22 shows a typical chaparral.

The plants of the chaparral are adapted to and maintained by periodic,
fires. Many of the shrubs have deep and extensive root systems that permit
3 quick regeneration. The fires burn all the plant structures above the
ground, thus releasing the nutrients in them. The nutrients then are available for use by the new plant shoots. In addition, many chaparral species
produce seeds that germinate only after a hot fire. Deer, fruit-eating birds,
rodents, lizards, and snakes are characteristic chaparral animals.

CONCEPT REVIEW

1. Describe the savanna biome and the abiotic factors that influence it.
   How does the savanna differ from other grasslands?
2. How does the character of the savanna influence the animals that live
   there?
3 What is chaparral? How are chaparral plants adapted to the abiotic factors?

Have students try to find successional ecosystems in your biome. Even in urban areas some stages of secondary succession can be observed on recently bared patches of ground—especially those resulting from the often slow pace of urban renewal.

22.11   Ecosystems Are Not Permanent

Do the abiotic and biotic factors that determine the character of biomes always remain the same? Landscapes, mountains, and seascapes may seem eternal, but they are not. The great sheets of ice that once covered the northern parts of North America and Eurasia have melted. In many spots, they have left bare rock and soil. In all regions of the world, erosion constantly exposes bare rock. In many areas, fire is a natural factor. Floods, blowing sand, and volcanic eruptions also may destroy the vegetation of an area and leave bare ground or create new land.   2

Nuclear explosions also can destroy life. Before atmospheric tests of nuclear explosions were stopped in 1958, some biologists began a five-year study of the effects of earlier nuclear explosions at the Nevada test site. The explosions had removed the topsoil at ground zero and destroyed all the desert vegetation within a 1 km radius, as shown in Figure 22.23b. Farther out from ground zero, only a few hardy species (such as the creosote bush and the ground thorn) survived. A year after the explosion, some spring annuals (stickleaf, stork's bill, and desert pincushion) appeared. These grew from seeds carried in by wind or birds. Small plants began covering the area, as shown in Figure 22.23c.

As the plants died, their remains were added to the sandy soil. The plants changed the abiotic environment in other ways, also. Their roots held more moisture in the soil, which made the temperature suitable for the germination of seeds of other species. By 1961, there was a community of plants, including foxtail chess, wild buckwheat, and others. This was not the pre-explosion community, however. It was similar to the community that eventually appears in a grassland area previously swept by fire.

22.12   Communities Change Through Relatively Long Periods of Time

The process by which one type of community replaces another is called **succession.** The traditional idea of succession starts with some bare life- 1 less substrate, such as rocks. Lichens may grow first, forming small pockets of soil and breaking down the stone. Mosses then may colonize, eventually followed by ferns and the seedlings of flowering plants.

Figure 22.23   The aftereffects of atomic testing: (a) before testing at ground zero, showing typical vegetation; (b) after testing. The exposed rocks indicate extensive soil removal. Russian thistle (c) was the only species that grew within 0.5 km of ground zero in the first year. Its presence indicates the start of recovery.

a

b

c

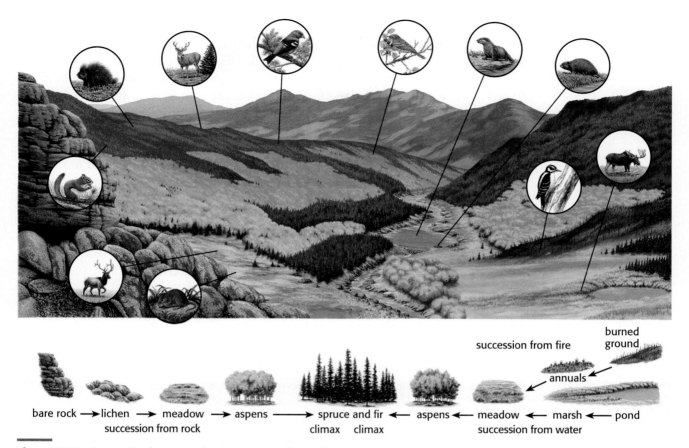

bare rock →lichen → meadow → aspens ——→ spruce and fir ← aspens ← meadow ← marsh ← pond
        succession from rock   climax   climax   succession from water

succession from fire burned ground

annuals

Figure 22.24 Successional stages and animals in a coniferous forest. Typical animals of the coniferous forest include (clockwise from the left) voles, elk, squirrels, porcupines, mule deer, crossbills, yellow-rumped warblers, otters, beavers, woodpeckers, and moose.

Through many thousands of years, or even longer periods, the rocks may be completely broken down, and the vegetation in an area where there once was a rock outcrop may be just like that of the surrounding forest or grassland. Figure 22.24 illustrates succession in a woodland. You can observe succession in your own region in a fairly short time by comparing pieces of land that have been abandoned for varying numbers of years.

3   The community that ends a succession, at least for a long period of time, is called the climax community. Unlike the other communities in the succession process, the climax community is not replaced by any other community unless a disturbance or a change in the climate occurs. If a single spruce dies of old age, the space that it occupied is too shaded by neighboring spruces for aspen to grow. A young spruce, tolerant of shade, probably will take its place. Thus, once established, the climax community is relatively permanent. Although other factors (such as soil) may play important roles, the climax community is determined mainly by the cli-

4   mate of the region. In some areas, for example, the climate is unsuitable for trees, and the climax community is dominated by grasses. With the

3   increasing realization that the climate keeps changing, that the process of succession often is very slow, and that the nature of a region's vegetation is greatly determined by human activities, many ecologists consider the concept of climax communities to be less useful than it once was.

Whatever causes succession, early stages are likely to be dominated by plants that are dispersed readily, grow rapidly, and have adaptations that favor colonization. Most colonizers are small annuals that grow close to the ground. They use most of their energy to produce seeds,

The reality of succession can be observed in a few weeks in a laboratory culture of microorganisms and in a few years in any abandoned agricultural area, but there is much controversy among ecologists about how it occurs. Ecology tests offer a variety of viewpoints. For an excellent demonstration of succession that can be set up easily in the laboratory, see H. K. Pigage, "The Winogradsky Column: A Miniature Pond Bottom, "*ABT* (April 1985): 239–240.

rather than to develop large root, stem, and leaf systems. Later stages of succession are likely to be dominated by longer-living species that are able to endure increased competition and increased shade.

In any community, the plants and the abiotic environment determine which animals can live there. When the former Nevada nuclear test site was invaded by plants again, animals soon followed. No forest or tropical animals appeared—only pocket mice, kangaroo rats, and other species that feed on desert plants. Through time, the climate in an area may change. If it grows colder or drier, some animals may die or move elsewhere. Other animals may find the area more hospitable than it was earlier. The selective effect on the animals depends on the adaptations they possess. The adaptations may be to climatic factors themselves or to certain plants that are, in turn, adapted to a certain climate.

22.13   Succession May Not Always Be Orderly

The traditional idea of succession holds that the diversity of species in an ecosystem increases as succession progresses. As a community matures, species that are better competitors usually replace the initial colonizers. Interactions between animals, such as predation, symbiosis, and competition, become more complex and extensive during successions, thus opening new niches and increasing the diversity of animals that can live in the community. According to this model, succession reaches climax when the web of interaction becomes so intricate that no additional species can fit into the community unless niches should become available through localized extinction.

Recently, many ecologists have challenged this traditional model of succession in which predictable and orderly changes result in a climax community. Most communities are in a continual state of flux, or change, and the types and number of species change throughout all the stages of succession, even at the so-called climax stage. The course of succession may vary depending on which species happen to colonize the area in the early stages and on the nature of the disturbance that causes further change. Disturbances such as fire, flood, hurricanes, and even unseasonable temperatures can prevent a community from ever reaching a state of equilibrium. When disturbances are severe and frequent, community membership may be restricted to those organisms that are good colonizers. If disturbances are mild and rare, then the species that are most competitive will make up the community. Species diversity, then, would be greatest when disturbances are intermediate in frequency and severity, because organisms typical of different successional stages are likely to be present.

The recovery of the Mount St. Helens area after the volcanic eruption in 1980 is an interesting study (see Figure 22.25). Scientists were divided on their predictions of how the recovery would progress. One group of scientists expected an orderly progression beginning with pioneer plants and eventually ending in a mature forest. Others thought recovery would begin as a matter of chance—a seed would arrive, perhaps blown by the wind, take root, and then play a role in influencing what other plants would grow nearby. To the surprise of both groups, recovery is happening in both ways. Seeds of alpine lupine, a pioneer plant, began to grow in areas that were nearly sterile. After they died, their stalks and dead leaves collected sand and dust particles, which allowed middle-stage plants to root. This supports the orderly succession idea. In contrast, individual trees, mostly alder, have become established out in the middle of very desolate areas without benefit of any pioneer plants.

Have students give several different examples of each model using different pioneer vegetation and animals and different types of disturbances as variables.

Discuss the recent Yellowstone fires and their effect on the vegetation and animals in the park. Discuss forest fire policy and its effect. A good reference is W. H. Romme and D. G. Despain, "The Yellowstone Fires," *Scientific American* **(November 1989): 37–46.**

Figure 22.25   An area of Mount St. Helens colonized by *Penstemon*.

Biology Today

Biodiversity in Danger

All organisms on earth, including humans, are dependent on each other. The interplay between biotic and abiotic factors makes possible the diversity of organisms that exists today.

Natural ecosystems have a vast amount of genetic diversity, yet researchers have strong evidence that ecosystems are in the opening stages of an extinction episode. There is a pronounced and sudden decline in the worldwide abundance and diversity of ecologically different groups of organisms, particularly those in tropical rain forests. Such a rate of extinction is the most extreme in the past 65 million years. Some researchers estimate that at the current rate more than one million species may become extinct by the end of this century.

The primary cause of this decline in diversity is not direct human exploitation but the habitat destruction that results from the expansion of human populations and activities. As the human population grows, it places greater demands on the ecosystem for space and goods. As a result, land is cleared for cultivation and living space, destroying the habitat of many organisms.

Organisms require appropriate habitats if they are to survive. Just as humans cannot live in an atmosphere with too little oxygen, trout cannot breed in water that is too warm or too acidic. Red-eyed vireos, scarlet tanagers, and other birds must have mature tropical forest in which to spend winters. Black-footed ferrets (see figure) require prairie that still supports prairie dogs, their major source of food.

Although many organisms provide humans with crops, domesticated animals, industrial products, and medicines, the roles that microorganisms, plants, and animals play within the natural ecosystem are more important. All these organisms exchange gases with their environments and thus help to maintain the mix of gases in the atmosphere. Changes in that mix, such as increases in carbon dioxide, nitrogen oxides, and methane, may lead to rapid climatic change and agricultural disaster.

The extent to which humans have devastated the diversity of organisms and the effect this devastation will have on the ecosystems of the world are as yet undetermined. Currently, however, procedures to alleviate some of the problems are more widely practiced than ever before. Some of theses procedures require a change in life-style. People are learning to recycle aluminum, plastic, paper, and glass, and to refuse to purchase items intended to be used only once and then thrown away. Other procedures involve new techniques to perpetuate diversity, such as embryo transplants between an endangered species and a domestic species (see Chapter 6, Biology Today). In the months and years to come, researchers will continue to develop new techniques and products to help alleviate the human pressure on natural resources and to create a "sustainable-earth" economy.

A black-footed ferret. Attempts have been made to breed these endangered animals in captivity so they can be reintroduced in the wild.

Human Influence on Biomes

22.14   The Need for Land Has Changed Some Biomes

Guidepost

How are human activities changing biomes?

A flight from Cleveland, Ohio, to Nashville, Tennessee, crosses the mid latitude deciduous forest, yet only a few traces of forest can be seen. A flight from Chicago, Illinois, to Lincoln, Nebraska, crosses the eastern portion of North American grassland, but here, too, only traces of the original biome remain. In fact, in these two flights, the landscapes below appear remarkably similar. In both cases the climate remains an important factor, but the present landscape has been shaped by humans, as shown in Figure 22.26. The changes have been caused by the need for food and homes. In Ohio, for example, only about 2 percent of the original deciduous forest remains, and the land now supports farms and cities instead. This is the inevitable consequence of a growing population.

Biomes modified for human use are simplified. For example, when grasslands are plowed, thousands of interrelated plant and animal species in that ecosystem may be replaced with a greatly simplified single-crop ecosystem, or with structures such as highways and parking lots. Modern agriculture is based on the practice of deliberately keeping ecosystems in the early stages of succession, which are highly productive but vulnerable because of the diminished diversity of species. For example, destruction of some insects by insecticides can lead to the failure of crops that depend on insect pollination. Extermination of the enemies of insect pests, such as

Figure 22.26   The rich, flat plains of the Midwest became the corn and wheat belt of the United States.

some birds and beneficial insects, can lead to severe pest outbreaks. The extermination of soil organisms by fungicides and herbicides can destroy the fertility of the soil. Because of such problems, there is an increased interest in "organic" or natural farming techniques that maintain productivity but are much less destructive in their effects on the soil and the ecosystem.

22.15 Tropical Forests May Disappear

The demands of a growing population have placed the tropical deciduous forests (Figure 22.27) and tropical rain forests of the world in jeopardy. These tropical rain forests, which account for two-thirds of all tropical forests, are home to more than 50 percent of all species on earth. Already, more than half the original area of tropical rain forest has been cleared for timber, cattle grazing, fuelwood, and farming. If clearing and degradation continue at the current rate, almost all tropical forests will be gone or severely damaged in about 30 years.

The present destruction and degradation of tropical forests is considered one of the world's most serious environmental and resource problems. Why? Two-thirds of the plant and animal species and as much as 80 percent of the plant nutrients in a tropical rain forest are in the canopy. Hence, removing all or most of the trees in a forest destroys the habitat and food supply of most of its plant and animal life. Tropical deforestation also exposes the nutrient-poor soil to drying by the hot sun and to erosion by torrential rains. Ecologists warn that the loss of these diverse biomes could cause the premature extinction of as many as one million plant and animal species by the beginning of the next century.

From 50 to 80 percent of the moisture in the air over tropical forests comes from trees through transpiration. If large areas of these forests are cleared, average annual precipitation would decrease and the climate would become hotter and drier. Any rain that fell would run off the bare soil very rapidly instead of being absorbed and slowly released by vegetation. Eventually these changes could convert a tropical forest into a sparse grassland or even a desert. The carbon dioxide emitted by deforestation and burning also would affect global climate, food production, and sea levels by warming the earth.

Deforestation also will have a serious economic impact. Tropical forests supply a variety of food products including coffee, cocoa, spices, nuts, and fruits. They also provide the raw materials for latexes, gums, resins, dyes, waxes, and essential oils. These substances are used in the manufacture of ice cream, toothpaste, shampoo, deodorant, perfume, varnish, tires, tennis rackets, jogging shoes, and many other products. In addition, the raw materials of one-fourth of the prescription drugs used, such as vincristine, come from plants that grow in tropical rain forests. The loss of tropical forest would severely impact the manufacture of these products and medicines. In addition, plants that could be vitally important in the development of new drugs and medicines would be gone.

22.16 Desertification Is Spreading Rapidly

Presently, about 35 percent of the earth's surface is classified as arid or semi arid desert. In the drier parts of the world, desert areas are increasing at an alarming rate from a combination of natural processes and human activities. The conversion of rangeland, rain-fed cropland, or irrigated cropland to desertlike land is called **desertification** (Figure 22.28). A grassland that once would support a large herd of livestock becomes

Figure 22.27   Tropical deciduous forest. Many food products and medicines are made from plants that grow in such forests.

Figure 22.28   Crops being covered by sand near Portales, New Mexico.

unable to support any at all. Natural desertification occurs near the edges of existing deserts and is caused by the dehydration of the top soil layers during periods of prolonged drought and evaporation. Drought is the absence of sufficient precipitation to make the plants grow. The few plants that remain soon are eaten by the livestock. With the ground cover gone, the winds begin moving the soil of the grazing lands into dunes, and in some places the sands of the deserts encroach on the margins of the grasslands. This is happening in the sub-Saharan region called the Sahel and has brought great misery and death to those nomads who live there (see Chapter 2, Biology Today). It is estimated that hundreds of millions of hectares of land that border deserts are at risk of becoming deserts by the year 2000. Desertification also is occurring in the Midwest, the region that currently produces the bulk of the United States grains. Experts predict that unless soil erosion is controlled another dust bowl may result.

Natural desertification, however is accelerated by such short-sighted practices as overgrazing. Overgrazing is a result of keeping too many head of livestock on too small an area of land. Poor soil and water resource management and deforestation also can speed up desertification. It is estimated that worldwide, an area 12 times the size of Texas has become desert during the past 50 years. Despite attempts to halt or sharply reduce the spread of desertification, each year an estimated 6 million hectares (15 million acres)—an area the size of West Virginia—of new desert are formed, mostly in Africa.

22.17   Acid Deposition Has Many Harmful Effects

When electric power plants and industrial plants burn coal and oil, they emit large amounts of sulfur dioxide, particulate matter, and nitrogen oxides. As these emissions are transported by winds, they form secondary pollutants such as nitrogen dioxide, nitric acid vapor, and droplets containing solutions of sulfuric acid, sulfate, and nitrate salts. These chemicals fall to the earth's surface as acid rain or snow, or as gases, fog, dew, and solid particles. Gases in this mixture can be absorbed directly by leaves. The combination of dry and wet deposition and absorption of acids and acid-forming compounds is called acid deposition, or acid rain. Normal rain has a pH of 5.6. The pH of acid rain ranges from 5.5 to as low as 2.4.

What happens to forests that are watered with acid rain? The soils supporting the forests no longer can buffer or neutralize the acid precipitation, which then leaches nutrients out of the soil. This deprives the trees of the elements they require for growth, and they begin to weaken and die. The waxy cuticle on leaf surfaces can be destroyed directly by the acid rain. Needles and leaves then turn yellow and drop prematurely, and the trees eventually die (Figure 22.29). Whole forest ecosystems are threatened by acid rain; the forests of the eastern United States, Canada, and Europe show the greatest destruction so far.

Acid deposition has other harmful effects. It damages statues, buildings, and car finishes. It can kill fish, aquatic plants, and microorganisms in lakes and streams, at the same time increasing populations of *Giardia*, a parasite associated with a severe gastrointestinal disease. It also can affect the ability of trout and salmon to reproduce. The acid stunts the growth of crops, such as tomatoes, soybeans, spinach, carrots, and cotton, and it leaches toxic metals from water pipes into the drinking water itself.

Figure 22.29 Trees exposed to acid rain eventually die. How does their death affect the atmosphere? **Less oxygen is produced, and less CO$_2$ is absorbed.**

CONCEPT REVIEW

1. What has happened to the deciduous forests of the Midwest?
2. How does the cutting of a tropical rain forest affect the soil under the trees?
3. What factors, human and natural, increase the size of deserts?
4. What causes acid deposition, and how is it harmful?

INVESTIGATION 22.1   Climatograms

Climatograms, such as those in Appendix 3, show monthly variations in only two climatic factors, precipitation and temperature. Other factors also affect climate, but a climatogram gives a rough idea of the climate in a particular area. By daily observation you can associate the climate with the biome of your own locality. This investigation allows you to study the worldwide relationships between climates and biomes. Refer to the pictures and descriptions of biomes in this chapter to help you visualize the relationships between biotic and abiotic factors in some of the earth's major biomes.

Materials (per team of 1)
1–3 sheets of graph paper or climatogram forms

Procedures

1. Construct climatograms from the data in Table 22.1. These four, plus the six in Appendix 3, represent the major biomes on the earth.
2. Using the data in Table 22.2, construct the climatograms assigned by your teacher.
3. Obtain monthly averages of precipitation and temperature from the weather station closest to your school. These data may be expressed as inches of precipitation and degrees Fahrenheit. If so, convert the data to centimeters and degrees Celsius using the information in Figure A1.2 in Appendix 3. From these data, construct a climatogram for your area.

Discussion

1. Compare your local climatogram with the four from Step I and the six in Appendix 3. Which one does it most closely resemble?
2. What similarities are there between the two?
3. What differences are there?

This investigation allows students to examine the relationships between precipitation, temperature, and the biome. This point is worth discussing in class. Even students who are familiar with the mechanics of graph making often do not know how to select a form suitable for a given set of data. Students should work alone.
Time: One class period.

Procedure

1-2. After some of the climatograms in the student's book have been studied, the graph making phase of this investigation may be assigned as homework. Each student should construct these four climatograms.
2. The number of climatograms assigned to each student is a matter of choice.

Discussion

For each of the 10 identified biomes, discuss the possible relationships between the climatic data and the characteristic features of the biome. Focus attention on the relation of climatic factors to biota. The discussion of known climatogram-biome relationships serves as a background for developing hypotheses about the biomes represented by the data in Table 22.2. Point out the limitation of the data the students are working with. Students should realize that a climatogram does not summarize precipitation and temperature for an entire biome.

1-5. The major biomes of the United States, excluding Hawaii, are in the set of 10 climatograms. Some local climatic data however, may depart from the data of the representative station.
6-8. Answers will vary.
9. Following are the stations from which the data in Table 22.2 were obtained.
   a. Washington, DC (midlatitude deciduous forest)

Table 22.1   Group 1

T = temperature (in degrees Celsius) P = precipitation (in centimeters)											
J	F	M	A	M	J	J	A	S	O	N	D
a. Tropical Deciduous Forest: Cuiaba, Brazil											
T 27.2	27.2	27.2	26.7	25.6	23.9	24.4	25.6	27.8	27.8	27.8	27.2
P 24.9	21.1	21.1	10.2	5.3	0.8	0.5	2.8	5.1	11.4	15.0	20.6
b. Chaparral: Santa Monica, California											
T 11.7	11.7	12.8	14.4	15.6	17.2	18.9	18.3	18.3	16.7	14.4	12.8
P 8.9	7.6	7.4	1.3	1.3	0.0	0.0	0.0	0.3	1.5	3.6	5.8
c. Savanna: Moshi, Tanzania											
T 23.2	23.2	22.2	21.2	19.8	18.4	17.9	18.4	19.8	21.4	22.0	22.4
P 3.6	6.1	9.2	40.1	30.2	5.1	5.1	2.5	2.0	3.0	8.1	6.4
d. Tropical Desert: Aden, Aden											
T 24.6	25.1	26.4	28.5	30.6	31.9	31.1	30.3	31.1	28.8	26.5	25.1
P 0.8	0.5	1.3	0.5	0.3	0.3	0.0	0.3	0.3	0.3	0.3	0.3

Table 22.2 Group 2

		J	F	M	A	M	J	J	A	S	O	N	D
a.	T	1.1	1.7	6.1	12.2	17.8	22.2	25.0	23.3	20.0	13.9	7.8	2.2
	P	8.1	7.6	8.9	8.4	9.2	9.9	11.2	10.2	7.9	7.9	6.4	7.9
b.	T	10.6	11.1	12.2	14.4	15.6	19.4	21.1	21.7	20.0	16.7	13.9	11.1
	P	9.1	8.9	8.6	6.6	5.1	2.0	0.5	0.5	3.6	8.4	10.9	10.4
c.	T	25.6	25.6	24.4	25.0	24.4	23.3	23.3	24.4	24.4	25.0	25.6	25.6
	P	25.8	24.9	31.0	16.5	25.4	18.8	16.8	11.7	22.1	18.3	21.3	29.2
d.	T	12.8	15.0	18.3	21.1	25.0	29.4	32.8	32.2	28.9	22.2	16.1	13.3
	P	1.0	1.3	1.0	0.3	0.0	0.0	0.3	1.3	0.5	0.5	0.8	1.0
e.	T	−3.9	−2.2	1.7	8.9	15.0	20.0	22.8	21.7	16.7	11.1	5.0	-0.6
	P	2.3	1.8	2.8	2.8	3.2	5.8	5.3	3.0	3.6	2.8	4.1	3.3
f.	T	19.4	18.9	18.3	16.1	15.0	13.3	12.8	13.3	14.4	15.0	16.7	17.8
	P	0.0	0.0	1.5	0.5	8.9	14.7	12.2	8.1	2.0	1.0	0.3	0.8
9.	T	−22.2	−22.8	−21.1	−14.4	−3.9	1.7	5.0	5.0	1.1	−3.9	−10.0	−17.2
	P	1.0	1.3	1.8	1.5	1.5	1.3	2.3	2.8	2.8	2.8	2.8	1.3
h.	T	11.7	12.8	17.2	20.6	23.9	27.2	28.3	28.3	26.1	21.1	16.1	12.2
	P	3.6	4.1	4.6	6.9	8.1	6.9	6.4	6.6	8.9	5.1	5.6	4.6
l.	T	23.3	22.2	19.4	15.6	11.7	8.3	8.3	9.4	12.2	15.1	18.9	21.7
	P	5.1	5.6	6.6	5.6	2.8	0.9	2.5	4.1	5.8	5.8	5.1	5.3
j.	T	17.2	18.9	21.1	22.8	23.3	22.2	21.1	21.1	20.6	19.4	18.9	17.2
	P	0.3	0.5	1.5	3.6	8.6	9.2	9.4	11.4	10.9	5.3	0.8	0.3
k.	T	−20.0	−18.9	−12.2	−2.2	5.6	12.2	16.1	15.0	10.6	3.9	−5.6	−15.0
	P	3.3	2.3	2.8	2.5	4.6	5.6	6.1	8.4	7.4	4.6	2.8	2.8
l.	T	−0.6	2.2	5.0	10.0	13.3	18.3	23.3	22.2	16.1	10.6	4.4	0.0
	P	1.5	1.3	1.3	1.0	1.5	0.8	0.3	0.5	0.8	1.0	0.8	1.5

b. Lisbon, Portugal (chaparral)
c. Iquitos, Peru (tropical rain forest)
d. Yuma, Arizona (midlatitude desert)
e. Odessa, USSR (midlatitude grassland)
f. Valparaiso, Chile (chaparral)
g. Upernavik, Greenland (tundra)
h. San Antonio, Texas (midlatitude grassland)
l. Bahia Blanca, Argentina (midlatitude grassland)
j. Oaxaca, Mexico (tropical deciduous forest)
k. Moose Factory, Ontario, Canada (taiga)
l. Fallon, Nevada (midlatitude desert)

4. Consider the biotic characteristics of your local area. What characteristics of the local climate would be important factors in determining these biotic characteristics?

5. Does your local climatogram *exactly* match any of the original 10 climatograms?

6. Explain how any differences between your local climatogram and the one most nearly matching it might affect the biotic characteristics of your biome.

7. Compare each unidentified climatogram from Table 22.2 with the 10 identified climatograms. Label each graph with the name of the biome that you think the climatogram represents.

8. Using the climatic information shown by each climatogram constructed from Table 22.2, describe the biotic characteristics of each corresponding biome.

9. Check the location of each climatogram from the information your teacher gives you. How often were you correct in your reasoning?

This investigation allows students to study the living world around them. After they discover what organisms are found near their school, they can study how the physical environment affects these organisms and how humans affect this

INVESTIGATION 22.2   Studying a Piece of the Biosphere

The biosphere is composed of many parts, including the empty lots, lawns, fields, forests, and ponds near you. Wherever there is enough water, plants will grow. Wherever plants grow, there will be animals as well. Sometimes the plants and animals may not be immediately apparent, but once you begin this in-depth study of a piece of the biosphere, you will discover many different organisms and begin to learn about how they interact.

Materials (per team of 6)

2 metal coat hangers, or pieces of stiff wire, for small quadrats	wire screen of approximately 1-mm mesh curved into a bowl shape
4 wooden stakes and string for large quadrats	6 data books or clipboards and paper plastic gloves
meterstick or metric tape	2 30-cm pieces of PVC pipe (½-1 inch in diameter)
trowel or small-bladed garden shovel	4 corks to fit pipe
white-enameled pan or large sheet of white paper	50-cm wooden dowel that fits inside pipe
collection containers such as jars and paper or plastic bags	4 sterile nutrient agar plates
6 pairs of forceps	field guides

Procedure

PART A   Field Study Preparations

Different schools have different opportunities for outdoor studies. Your class and your teacher must work out procedures to fit the piece of biosphere that is most convenient to your school setting.

1. As a class, select a study area such as a local park, forest, or grassland. Choose an area that is convenient to get to but seldom disturbed by other people. The more natural your area of study, the greater the diversity of organisms you will find. A forest is complex and provides opportunities to collect abundant data, but it is difficult to picture as a whole. A prairie is almost as complex and is somewhat easier to study. Cultivated areas, such as cornfields and pastures, are relatively simple to study.

   Cities also contain pieces of the biosphere suitable for study. Many schools have lawns with trees and shrubs. You can study vacant lots and spaces between buildings. Even cracks in pavement, gutters, and the area around trees often contain a surprising number of organisms.

2. Make and record observations of the plants and animals in your study area. Generate a list of questions about the plants and animals. Do the same type of plants grow in all parts of your study area? Why are some plants more abundant in one part of the area than in another? Do you see any animals, and if so, what do they appear to be doing? What do you think is the relationship between the plants and the animals in your study area?

   Choose a question you would like to study. The discussion questions should give you some ideas. With your teacher, determine whether your entire class will study the same question or whether each team will study a different question in the same area. Decide what type of data to collect to answer your question and how large an area you need to study. Your teacher can direct you to other investigations where you will find techniques for studying your piece of the biosphere.

   If teams are studying different aspects of the area, one study site should be selected. For example, one team could study the plants at the site, one team the insects, and another team the other animals. Other teams might want to study the physical characteristics of the area, such as temperature, soil moisture, or light intensity. In forests, study sites for different plants should be of different sizes. In any habitat, one or several teams can study the organisms that live in the soil.

   To study organisms in the soil, take soil cores of equal length from several different parts of your sampling area. Push a length of pipe into soft soil approximately 25 cm deep. Gently twist the pipe in the ground, and then pull it up with the soil core inside. Plug up the ends of the pipe with the corks, and take all pipes back to the lab for study. In the lab, carefully push the soil core out of the pipe with a wooden dowel. Count the number of earthworms, pill

piece of the biosphere. If carefully planned, this investigation can be one of the most rewarding experiences of the year. Teams of 6 allow team members to work in pairs.

Time: The amount of time you devote to this investigation depends on the study area. If the study area is part of the school yard, three or four class periods should suffice. Allow at least one class period for preparation and one period for discussion. If at all possible, try to devote at least half a day to the actual field study. If students need to construct equipment, allow additional class time. In some cases, students may be able to conduct Part B during their free time.

Suggested Overall Procedure

1. Locate a study site—the more natural the better. Obtain permission from the owner to conduct a long-term study and, if possible, to collect a few specimens and soil samples.

2. Collect all the field equipment necessary. This includes soil sampling materials, quadrats, and any equipment needed to study the physical environment.

3. Make preliminary observations of the study site before you take students there. Generate a list of possible questions that could be studied by different teams.

4. Take your class to the site and allow them to make observations of plants, animals, and any interactions they find. One class period could be devoted to this important stage.

5. Back in class, divide your class into teams, and have each team select a different question to study. (All teams may study the same question, but if each team differs in its goal, your class will be able to focus on may parts of the web of life.) Discuss with them the plan of action and the form in which you want their report. One possibility is for each team to make a large poster on butcher paper with illustrations and data presented in an easy-to-read manner. Each team can give an oral presentation to the entire class at the end of the project.

6. Make certain each team has a plan of action before they go back out to the filed even though their activities may change as they begin to work on their projects. Allocate materials and equipment based on these initial plans.

7. Spend one, two, or more days observing and collecting data in the field. Return to class and allow your students to work with any materials they have collected or to analyze their data. Reserve another day or two to allow them to complete their

collection of information.

8. Allow enough time for students to complete their study and to prepare their presentations.

PART A Procedure

1. **Try to locate a study area near the school so repeated visits can be made—for this first study and for comparative studies later on. If the resources in the immediate vicinity of the school are poor, a single excursion to a desirable site at a greater distance may be preferable. In no school is this investigation impossible. In urban situations, vacant lots, the area around a billboard, even cracks in cement and asphalt contain plants, insects, and nematodes. Mice, rats, cats, and pigeons also are common.**

2. **If you have students who are camera buffs, you may want to use their skills and equipment. Close-up and microphotography are excellent techniques for recording data. Encourage students with artistic skills to keep a record through pictures and those with writing skills to write about their observations. Visit the selected area—with a committee of students, if possible—to determine its possibilities for observation. On the basis of your visit and the suggested procedures, the class can make detailed plans and select team leaders.**

3. **The plans must fit the personnel. Each student must have something to do. Each must know what his or her responsibility is and how to carry it out. Written instructions are essential. Work with individual teams to devise forms that are appropriate for recording their field data. Go through a dry run with team leaders to check out directions and data forms. This is a good time to link this investigation with others in the text, for example Investigations 3.1, Abiotic Environment: A Comparative Study; 20.2 A Field Study of Animal Behavior; and 20.3, Environment Tolerance.**

4. **If possible, the investigation should be a comparative study. For example, the border of a woods may be compared with the interior, a grazed pasture with a mowed meadow; a well-trodden section of the school campus with a less-disturbed section. Consider comparing studies by classes in successive seasons or years. This gives firsthand meaning to the concept of change, ecological**

bugs, or other larger organisms you see, noting at what depth in the core they are found. Sprinkle a small sample of soil from different depths of the core onto a sterile plate of nutrient agar—one sample for each plate. Cover the plates, tape closed, and incubate them at room temperature for two days. *Without opening the plates,* observe and count the number of bacterial and fungal colonies that grow.

To study the plants, use quadrats—square or rectangular samples of the study areas. For trees, a quadrat of 10×10 m is common; for shrubs, 4×4 m is recommended. Where there are few, if any, trees or woody species, use quadrats made of stiff wire that are 0.5 m on each side, or unwind two metal coat hangers and reshape them into a quadrat. Note the size of the area inside the quadrat, and use this same size quadrat throughout the study.

3. Select a team leader to divide tasks among the team members. Decide what materials your team needs to gather the data, and how and where to obtain them. Draw up a form on which data can be quickly recorded. It may be easier to handle sheets of paper on a clipboard than to take data books into the field. Paste the sheets into your data book when you return to the laboratory. NOTE: Before you collect any specimens to take back to class, observe these important rules. First, never collect anything unless you have consulted your teacher. Second, never collect anything unless you have obtained permission from the owner of the property. Third, never collect any organism that is poisonous, harmful, or rare and endangered—make sure you know what you are touching. Finally, if you do collect organisms for the classroom, collect as few specimens as possible. These should be added to a permanent collection of plants and invertebrates that can be used as a reference by future generations of students.

4. Decide how your team will collect and identify the data. Either identify organisms only by general group names, such as trees, spiders, grasses, beetles, or turtles—or else collect specimens. Your teacher will help you decide the best method. If you collect specimens, assign each specimen a letter (A, B, C, and so on), and whenever you need to refer to that species of organism, refer to its letter. After you return to the classroom, you may be able to identify your specimen by looking it up or showing it to an expert. Your teacher can show you how to make permanent specimen displays of your plants by pressing them between pieces of paper, drying them, and mounting them on pieces of cardboard.

Some plants you may find in your study area are:

a. Tree—tall, woody plant with a single stem (trunk).
b. Shrub—woody plant that branches at or near the ground and lacks a trunk.
c. Sapling—young tree with a trunk 1 to 5 cm in diameter.
d. Herb—nonwoody plant that dies back at least to ground level in winter.
e. Tree seedling—very young tree with a trunk less than 1 cm in diameter.

5. If you decide to collect animal data, it should be done after the studies of plants have been completed, possibly at a later date. Work in pairs, with one person searching and the other recording. Turn over stones, logs, and other cover to find animals, being sure to return these sheltering objects to their original position. Examine plants, especially the flowers. Record notes of the types, numbers, and activities of animals you find. If possible, collect specimens of insects of unknown types for identification. In thick forests, it may be difficult to catch flying insects, but you can beat the shrubs and saplings with a stout insect net, holding the open end up. In open fields, sweep your net through the plants. Try to identify the plants on which you are making your sweeps. If you are able to identify the insect, you can release it. Otherwise, you may want to preserve it according to directions from your teacher and take it back to class for further identification.

6. During your study you may uncover toads or snakes. To study reptiles, birds, and other large animals, it may be necessary for you to cover larger areas than

the ones described and observe the areas over longer periods of time. Do not collect these animals. Birds may not be present or active when your class is collecting data—around daybreak and dusk are usually the best times to observe them. Look for animal tracks and animal droppings, which also can be identified. In cities you are likely to see birds, rats, mice, dogs, and cats. Look for signs of humans, too.

PART B   Conducting the Field Study

7. If your team is studying a small area, toss the quadrat on the ground to determine a site. If not, use stakes and a metric measuring tape to define a quadrat.

8. List all the different types of plants you find within your quadrat. If you cannot name the plants, describe them well enough so that you will be able to identify them in your other quadrats, or collect samples using the system you developed in Step 4. If you collect samples, be sure to wear plastic gloves.

> **Do not collect any plants you cannot identify as harmless.**

9. For each species on your list, estimate its cover. This is the percentage of ground within the quadrat covered by the leaves and stems of all the plants of that type. Suppose there are three different types of plants, as in Figure 22.30. Estimate, to the nearest 10 percent, the amount of cover for each type of plant, as well as the amount of bare area. The greater the amount of cover for a plant species, the greater is its importance within the plant community.

10. To increase accuracy, at least two members of your team should estimate the cover of each plant species. Average the two estimates, and record this number. If the numbers are extremely different, reevaluate the cover. If some tall plants over shadow smaller plants within the quadrat, your total estimated cover may be greater, than 100 percent when you finish.

11. Complete at least three different quadrat samples within the study area. Calculate an average cover for each plant species you find. To do this, add up the percentages for each species and divide by the number of quadrats measured (see Table 22.3).

12. Examine a sample of the surface soil and/or organic litter. Insert a trowel or small garden shovel into the ground to a depth of approximately 10 cm, and remove a cubic sample of approximately 10 cm × 10 cm × 10 cm. Using the wire screen that has been shaped into a bowl, sift the contents onto a white-enameled pan or a large sheet of white paper. Pick through the remaining litter with forceps. Organisms present probably will include a variety of insects (especially beetles), sow or pill bugs, millipedes and centipedes, spiders or other arachnids, worms, and fungi.

13. When your field work is completed, share your team data with the rest of the class by combining data from each team in a large table drawn on the

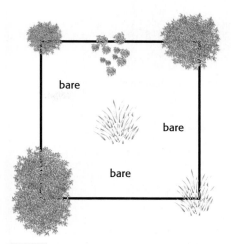

Figure 22.30   Vegetation coverage in a quadrat.

succession, and long-term studies. Population densities of several selected species may be obtained by counting individuals on measured quadrats. Choice of species depends on the habitats being compared. For example, in comparing a much-trodden land with an out-of-the-way one, count dandelions and plantains rather than grass. The mean cover in one habitat then can be compared with that in another. Let students decide subjectively whether the differences are significant.

4–5. Several of the specific procedures suggest collecting specimens. Caution students not to collect any plant or animal specimens they cannot identify as harmless. If you are working in a public park, obtain permission to collect leaves and other specimens. Be sure you and your students do not violate preservation laws. If possible, allow students to collect something; plants and insects probably are most acceptable for this purpose. Collecting is *not* the aim of this investigation—see that student enthusiasm for it does not interfere with the study of the relationships among organisms. If possible, develop a permanent collection of local plants by pressing them and mounting them on stiff paper. These collections can be displayed throughout the investigation and will allow students to identify many of their specimens by sight.

Table 22.3   Percentage of Cover Within Sample Quadrats

Species	Sample 1	Sample 2	Sample 3	Average Cover
Bluegrass	30	20	70	40.0
Dandelions	20	20	10	16.7
Ragweed	40	30	20	30.0
Bare ground	10	30	10	16.7

Discussion

1-14. **Answers will vary.**

chalkboard. Answer as many of the discussion questions as possible. If different types of areas were included, make comparisons.

Discussion

1. What producers are in the area? Identify as many of the plants as you can, but answer in general terms—trees, shrubs, and so on—if you cannot name the producers.

2. The plant species with the greatest amount of cover is said to be dominant within the study area. What effect does the dominant species you found have on the rest of the study area?

3. Are the producers evenly distributed throughout the entire study area? What do you think may be responsible for the pattern of plant growth you find?

4. Do you have any evidence of seasonal changes in the types and numbers of producers?

5. Are there different groups of producers? If so, which group produces the most food in the area?

6. Are there layers of producers? If so, what relationships can you find among the layers?

7. Does the area produce all its own food, or is food carried in from beyond its boundaries? What evidence do you have for your answer?

8. What consumers are in the community? Answer in general terms or with names of the organisms you have identified.

9. Which consumers are herbivores and which are carnivores? What evidence supports your answer?

10. What relationships can you find between the numbers of a particular herbivore and the numbers of a carnivore that eats it?

11. Using the information you have, construct an energy-flow diagram for the area.

12. What is the evidence that one type of organism affects another in ways other than those involving food relationships?

13. What biome type best characterizes the area you studied? Give reasons for your choice.

14. An investigation such as this should raise more questions than it answers. In studying the data, part of your job is to look for questions that need answering. Write several questions concerning the organisms in your area. Although your data from this study are incomplete, you have taken a biologist's first step in describing a study area.

This investigation gives students a model of ecosystem change through time. It also lets them practice visualizing change and diagramming food webs. Teams of 3 allow small-group interaction.
   Time: **One class period.**

INVESTIGATION 22.3   Long-Term Changes in an Ecosystem

Through many years, one biotic or abiotic factor in an ecosystem may change. This change causes a second factor to change, which, in turn, changes a third factor. Eventually, the entire ecosystem may be quite different from what it was earlier. In this investigation, you will trace the effects of a gradual climatic change on a model ecosystem. The model is much simplified from a real situation.

Materials (per team of 3)
none

Procedure

1. Read the following information, and diagram the food web in the ecosystem described.

    An oak grows next to a shallow stream in a grassy area. There are no shrubs here. Grasshoppers and blacktail jackrabbits feed on the little bluestem grass.

Table 22.4 Ecosystem Changes

Round	Years	Average Annual Precipitation	Average Annual Temperature	Changes in Organisms
1	200			
2	400			
etc. (to 50)				

Duckweed grows in the water, providing some of the food for wood ducks. These birds also feed on grasshoppers, acorns, and grass. The ducks are a major food source for red wolves.

The average annual precipitation is 84 cm. The average annual temperature is 10° C.

2. In your data book, prepare a table similar to Table 22.4.

3. Working as a team, fill in the table using this additional information:

To the west of the grassy area is a desert. There, glossy snakes eat Merriam pocket mice. These mice must burrow under shrubs or cacti to hide from the snakes. The mice, therefore, cannot live in a grassy area. In the desert, they burrow under mesquite shrubs and eat the seeds. Grasshoppers and blacktail jackrabbits eat the mesquite leaves.

The climate is changing in the grassy area. For the next 10 000 years (50 rounds on the table), the average annual precipitation will decrease by 1.25 cm every 200 years, and the average annual temperature will increase by 0.3° C every 200 years.

An average annual temperature of more than 20° C or an average annual precipitation of less than 34 cm will cause the stream to begin drying up and the oak trees to die.

If the average annual temperature reaches 23° C, or the average annual precipitation is less than 24 cm, the water table will be too low for the grass.

In filling out the table, assume that the organisms in the desert area are trying to disperse to the grassy areas at all times and that they will invade it whenever conditions allow.

4. After filling out the table, diagram the food web for the same area at the end of 10 000 years.

Discussion

1. When did the first changes in the ecosystem occur? What were they?
2. What subsequent changes occurred, and in which years?
3. What similarities or differences are there between the initial and final food webs?
4. What is a barrier to dispersal for Merriam pocket mice?
5. What limited the population of red wolves?
6. What climatic factor was responsible for the other changes?
7. Why does the use of average annual values for temperature and precipitation make this model very artificial?

For Further Investigation

1. Create a computer program to determine when critical points are reached and what the outcomes are.
2. Substitute different pioneer organisms and determine the effects on the ecosystem and other organisms. For example, what would happen if creosote colonized in the new area instead of mesquite?
3. Create a new model using your local biome and nearby biomes.

Discussion

1. **At 6800 years, the temperature rises above 20° C. Oaks die; duckweed is disappearing.**
2. **At 8200 years, the precipitation falls to 34 cm. Duckweed, ducks and wolves are gone. At 8800 years, temperatures reach 23° C and grass disappears. At 10 000 years, the precipitation falls below 23 cm. Grass is replaced by mesquite; mice and snakes move in.**
3. **The grasshoppers and jackrabbits are the only organisms common to both the initial and final food webs.**
4. **Lack of mesquite.**
5. **Number of ducks.**
6. **A decrease in precipitation was solely responsible for the other changes.**
7. **Maximum and minimum levels would be much more realistic.**

This investigation intro-
duces students to the
effects of a natural
catastrophe on different
biomes. Different biomes
do not respond in the same manner to
the same disturbance. Teams of 3 allow
for small-group interaction.
   Time: One class period.

PART A Procedure

1. **Both are increasing.**
2. **Access to deep supplies of moisture during long droughts.**
3. **Mesquite brush with less grass.**
5. **Mesquite.**

Figure 22.31

INVESTIGATION 22.4   Effects of Fire on Biomes

Fire is an important ecological factor in terrestrial ecosystems. Some fires start from natural causes, but many are caused by humans—deliberately or accidentally. No matter how they begin, fires have many effects on the organisms in their paths. The most easily observed effects are on vegetation.

This investigation considers three different North American biomes and the changes that fire might cause in each.

Materials (per team of 3)
none

Procedure
PART A   Fire in a Midlatitude Grassland

1. Figure 22.31 depicts a series of events in the southern part of the North American grassland. Two types of populations are involved: grasses of various species and mesquite shrubs. Study Figure 22.31a and 22.31b. What is happening to the sizes of the populations?

2. Roots of mesquite have been found in mine shafts many meters below the surface of the soil. What competitive advantage might this type of root growth give mesquite over grasses?

3. If occasional droughts strike the area (as actually happens), what type of community do you think might result?

4. Refer to Figure 22.31c and 22.31d. In these figures, plant parts shown in light color represent unharmed tissue. Do both types of plants survive fires?

5. In which type of plant has more growing tissue been killed?

a. grasses and mesquite bush   **b.** same area 10 years later

c. fire   **d.** 2 years after fire

6. Grasses usually reach maturity and produce seeds in 1 to 2 years; mesquite usually requires 4 to 10 years. Which type of plant has lost more growing time?

7. In Figure 22.31d, which type of plant occupies the most land?

8. What might you expect this area to look like 4 or 5 years after a fire?

9. Make a generalization about the effect of fire on this community. Describe the probable landscape if fires did not occur at all.

10. How would the landscape look if fires occurred every few years?

11. What seems to be necessary for maintaining grassland in this region?

PART B   Fire in a Forest of the Great Lakes Region

Around the Great Lakes of North America is a region of transition between the coniferous and midlatitude deciduous forests. In many places much of the forest consists of pines (Figure 22.32a). Early in the settlement of the region, Europeans brought about a great change in the landscape (Figure 22.32b)

12. What was this change?

13. Fires apparently had been rare in this region, but following the change in landscape, they became more frequent. What might have brought about the increase in the number of fires?

14. If a fire does not occur, what might the area shown in Figure 22.32b look like in later years?

15. Study Figures 22.32b, 22.32c, and 22.32d, which depict jack pine. What characteristic of jack pine gives that species a competitive advantage when there is a fire? (Hint: Compare the cones in Figure 22.32b and 22.32c.)

16. Describe the probable appearance of the area shown in Figure 22.32d 5 or 6 years later.

6. Mesquite.
7. Grass.
8. Somewhat like (d) but with the regrowth of the mesquite as in (a).
9. Mesquite brush with less grass.
10. Dominantly grassland.
11. Fire.

PART B

12. Lumbering.
13. More people; brush piles; further clearing by fire in attempts at agriculture.
14. Eventually, much like (a). Deciduous woody plants would be favored at first because they spring up as sprouts from stumps.
15. Small fires cause the cones of jack pine to open; otherwise, seeds may be held in the cones many years.
16. An area dominated by young jack pine.
17. Jack pine would became rare, and other types of trees would take its place.

Figure 22.32

a. before settlement

b. after settlement

c. fire

d. 3 years after fire

18. **Dominated by jack pine.**

PART C

19. **Fire does not kill the roots of the young longleaf pines.**
20. **Fire usually kills young deciduous plants.**
21. **Longleaf pines.**
22. **Ground fires kill the lower branches but not the tops or the trunks.**
23. **In the absence of fires, deciduous trees grow faster than the young longleaf pines and eventually shade them out.**
24. **Fire.**

Figure 22.33

17. Jack pines produce cones in 8 to 10 years but do not live to a very great age. Their seedlings do not thrive in shade. Suppose no fires occur for 200 years. What changes in appearance might take place in this area during that period?
18. Suppose fires occur about once every 20 years. What might the area look like at the end of 200 years?

PART C   Fire in a Forest of the Southeastern United States

In the southeastern United States are great forests in which longleaf pine is almost the only large tree. Occasionally there may be seedlings and saplings of deciduous trees. Between 3 and 7 years of age the longleaf pines look somewhat like a clump of grass (Figure 22.33a). While in this stage, the young trees develop deep roots in which food is stored. (See the cutaway, lower corner of Figure 22.33a.)

19. What is the effect of fire on young longleaf pines?
20. What is the effect on the deciduous shrubs and saplings?
21. Which plants have a competitive advantage after a fire?
22. After the grass stage, longleaf pines grow rapidly in height and develop a thick, fire-resistant bark. What is the effect of ground fires at this stage in the development of the pines (Figure 22.33c)?
23. Which plants have a competitive advantage when fires do not occur (Figure 22.28d)?
24. What factor seems to maintain a forest of longleaf pines within the deciduous forest biome?

a. longleaf pines, deciduous saplings and shrubs

b. fire in same area

c. fire in mature pine forest

d. pine forest—no fire for 5 years

Discussion

1. If you were interested in raising cattle in a midlatitude grassland, would occasional fires be an advantage or a disadvantage? Why?

2. Jack pine is not as valuable a lumber tree as other trees of the Great Lakes region. If you were a landowner in that region, would fire be an advantage or a disadvantage? Why?

3. If you were interested in maintaining a longleaf pine forest to obtain turpentine, would ground fires be an advantage or a disadvantage? Why?

4. Suppose you wanted bobwhites (game birds that nest on the ground) in your turpentine forest. What effect might this have on your management of the forest?

5. What things must ecologists know before deciding whether to recommend fire as a method of management to a landowner?

SUMMARY

In each type of climate, a characteristic type of vegetation develops. That vegetation supports a characteristic variety of animal life. The tundra biome of the Northern Hemisphere slowly blends with the taiga in a southerly direction. In the middle latitudes are the deciduous forest biome in the east, the grasslands to the west, and the desert biome in the southwest.

The tropical rain forest biome, located just north and south of the equator, receives abundant rainfall and has warm temperatures all year long. The tropics also support a deciduous forest, but unlike the rain forest, this biome has wet and dry seasons. Still within the tropics is the savanna biome, a grassland that supports a few trees and the world's largest concentration of hoofed animals. Southern California and other scattered areas support evergreen shrubs that make up the chaparral biome. Because of changes in altitude, the mountains have a number of biomes that form bands of vegetation ranging from grasslands at the base to alpine tundra on the top of the tallest peaks.

No ecosystem is permanent. Natural and human-made catastrophes can change the character of an ecosystem. As a result, a succession of communities may occur, ending, at least temporarily, in a relatively stable climax community. All ecosystems around the world are undergoing changes as a consequence of human activities. The cutting of the tropical rain forests, the loss of grasslands to desertification, and the acid rain problem in North America and Europe are just a few examples.

Discussion

1. Advantage, because they maintain grass.

2. Disadvantage, because if favors growth of jack pine.

3. Advantage, because they favor longleaf pine in competition with deciduous trees.

4. Burning would have to be done at a season when bobwhites are not nesting.

5. First, the landowner must decide what is wanted. In any case, it is impossible to predict the effects of fire without prior empirical evidence because the variables are numerous. For example, periodic fires in certain tall-grass prairies tend to increase their productivity as rangeland for cattle; fires in the mixed-grass prairies tend to have the opposite effect. Some ecologist suggest that the depth of mulch is the important variable in this case.

   To some degree, fire is used as a management tool in all regions. Ecologist do not always agree on the role fires should play in land management. Frequent fires prevent the growth of understory in a forest. After a long period without fire, the understory provides a route by which fire can reach the crowns of trees. The controlled burning of forests at frequent intervals is sometimes suggested as a means of preventing disastrous forest fires. On the other hand, allowing natural fires to burn many keep the forests in as natural a state as possible. These issues are currently being debated as regards the Yellowstone fires.

APPLICATIONS

1. Explain why the four seasons are more pronounced in the midlatitudes than in the tropics.
2. What biomes would you encounter if you were to travel from Washington, DC, to San Francisco?
3. Give several examples of changes in ecosystems caused by human activities.
4. Describe what you think would happen if all humans left the following places: a farm in Nebraska; a sidewalk in New York; a swimming pool in Seattle, Washington; a landscaped park in Las Vegas, Nevada.
5. Make a list of terrestrial organisms that humans have brought into your locality. Divide the list into two parts; organisms that survive because of human activities, and organisms that would survive without human help. Give reasons for your decision on each organism.
6. Why do you think biomes are characterized by the types of plants that grow there rather than the types of animals that live there?
7. During the winter in the United States, the earth is closer to the sun than at any other time of the year. How do you explain that this is the coldest season of the year?
8. What do you think would happen if all of the major ocean currents reversed direction? Speculate as to the changes in climate we would see in the United States.
9. In which biomes would you expect to find the most productive farms? Why?
10. In which area would you expect the food webs to be more complex—the tundra or the tropical rain forest? Explain.
11. Would you expect the tree line to be lower or higher in the mountains of Mexico than in the mountains of Montana? Explain.
12. Why might a large animal have an easier time keeping warm than a small animal?
13. Is there such a thing as a "good" forest fire?
14. Compare and contrast hibernation and estivation. How are they adaptive?

PROBLEMS

1. Investigate your area and determine the biome in which you live.
2. The prairie dog once was abundant in the prairie ecosystem. What is the role of the prairie dog in that ecosystem? How are prairie dog populations being reduced? How will other species be affected by a reduction in numbers of prairie dogs?
3. The search for oil in the Arctic has ecologists concerned about damage to the tundra. What happens to the tundra when heavy equipment is driven over it? What might be the effects on migrating herbivores such as the caribou?
4. Find out what different countries are doing to protect their particular ecosystems. Some things that could be explored are: national parks, forest preserves, wildlife preserves, and city parks.
5. Investigate the ecological recovery from the 1988 Yellowstone fires.
6. Before Europeans settled the grasslands of Australia, kangaroos were the ecological equivalent of the bison in North American grasslands. These animals had similar niches in the community structure of their regions—they were the largest grazing herbivores. What are the ecological equivalents of these animals in most of the Australian and North American grasslands today? What were the ecological equivalents of these animals in the steppes of Asia, the pampas of Argentina, and the veldt of South Africa before these regions were highly modified by humans? What are the ecological equivalents of these animals in the tundra?

   In the desert of South Africa, what are the ecological equivalents of the cacti of North American deserts? In the tropical forests of the Old World, what are the ecological equivalents of the hummingbirds of the New World tropical forests?

7. Contact an environmental group and ask about their programs to help preserve parts of biomes around the world.

8. Start a collection of photographs and slides illustrating different biomes around the world. Display groups of these photographs around your classroom, or create a slide show about biomes for your class.

9. Arguably, the tropical rain forest is the most important biome in the world. Estimates are that 20.24 hectares/minute are being destroyed. What can you, your school, and your community do to help save this endangered biome?

10. Choose any biome and compare the plants and animals in that biome in different parts of the world. Are the species the same all around the earth?

11. Movies and television shows that are filmed "on location" use the natural vegetation as a backdrop. The next time you view a movie or TV program, check to see in what type of biome the filming was done. Is the vegetation consistent with the location identified in the script?

12. How has your community's ecosystem changed through time? Visit a local museum or library to find out what types of plants grew in your community I000 years ago, 100 years ago, or even 10 years ago. What types of plants replaced the native plants?

13. Use BSCS, *Land Use* (Dubuque, IA: Kendall/ Hunt, 1984) to study a land use issue in your community. What is the problem, what are alternative solutions, and how would you manage the problem?

In memory of Rich Buzzelli

Aquatic Ecosystems

How deep is the water in this pond? What organisms can you think of that might be able to live here? Where do you think this pond is found? What are some limiting factors in this particular biome?

Numerous microscopic algae and other protists probably live here, as well as some small insects. Fish probably do not live here. This is a shallow alpine pond. Temperature, nutrients, and oxygen content probably are the biggest limiting factors.

MAJOR CONCEPTS

- Ponds and lakes differ physically and biologically.
- Unique properties of water cause mixing of lake waters and of ocean waters.
- Brooks, streams, and rivers generally are progressively larger, slower, and less oxygenated.
- Water circulates through the bio-sphere in long and short cycles.
- Most ocean life exists where photo-synthesis can occur.
- The intertidal zone is an environment of extremes.
- Wetlands ecosystems are vital to the world's ecology.
- We must be very careful about what we allow to run into the oceans.

KNOWLEDGE CHECK

- What are some differences between lakes and ponds?
- How do some streams differ from rivers?
- What are wetlands? Why are they important ecosystems?
- How can pollution affect inland waters and the oceans?

◆ The earth, when viewed from outer space, is blue with water. All this water makes up the hydrosphere (HY droh sfir), just as all the air makes up the atmosphere. The hydrosphere is a vast heat reservoir that absorbs, stores, and circulates the heat that results when sunlight strikes the earth. The hydrosphere also is a reservoir of chemical elements and compounds that are continuously dissolved in water that eventually drains into the oceans. The oceans contain 97 percent of all water found on earth and form the great interconnecting water system that surrounds the continents, covering 70 percent of the face of the earth.

Many inland waters are found on the surface of the land. Inland-water ecosystems are classified either as standing waters or flowing waters, although the boundary between these classifications often is not distinct. Standing waters range in size from roadside puddles to the Caspian Sea, which occupies more than 350 000 km². These waters also range from very shallow to very deep, from clear to muddy, and from fresh to salty. A river is an example of flowing water, yet in some places a river may have such a slow current it is difficult to see any movement at the surface.

Although humans are terrestrial organisms, they always have lived close to a source of fresh water. Throughout the world, villages, towns, and cities have developed close to bodies of water, which have supplied food, transportation, and a waste removal system. This chapter examines the major aquatic ecosystems and the impact human activities have on them. ◆

Standing Fresh Waters

23.1 Ponds Are Shallow Enough for Rooted Plants

A pond is an example of standing water, yet many ponds are fed by springs or brooks, and many have outlets. Ponds usually are defined as small, completely enclosed bodies of water that are shallow enough so that light penetrates to the bottom. Rooted plants can grow throughout a pond, and temperatures from top to bottom are relatively uniform during the warm months of the year. Whether ponds are natural or artificial, their characteristics are much the same. Figure 23.1 shows a cross section of a typical pond in the northeastern United States.

In studying any ecosystem, an ecologist first looks for its source of energy. In a pond, the most important producers are not the most obvious

Guidepost
What are the characteristics of ponds and lakes?

Figure 23.1 Cross section through the edge of a natural pond in the northeastern United States. Where is the boundary of the ecosystems?

The boundary is indistinct.

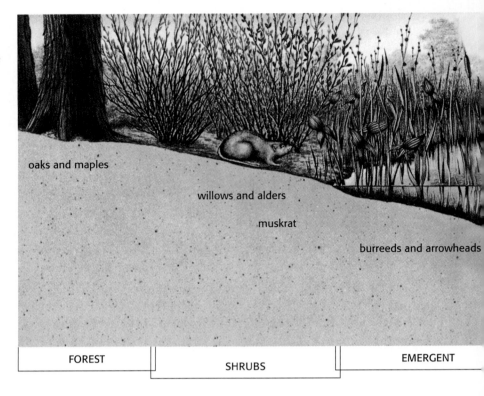

oaks and maples

willows and alders

muskrat

burreeds and arrowheads

| FOREST | SHRUBS | EMERGENT |

Figure 23.2 Plankton.

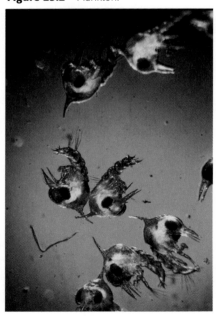

ones. The large rooted plants, such as cattails and rushes, are conspicuous, but they produce little of the ecosystem's food. Indeed, in some ponds—especially artificial ones—rooted plants may be scarce or absent. Many plants live within the water itself—some floating on the surface, some submerged. These plants may become so numerous that they hamper a swimmer, yet even these are not the most important pond producers.

Light shining into a pond reveals moving specks that look like dust in a beam of sunlight. These specks, shown in Figure 23.2, are living organisms called **plankton** (PLANK ton), which may belong to any of the five kingdoms. Few of these organisms actually can swim, but they are carried about by the currents in the water. Plankton make up for their small size by their incredible numbers; research shows that most food production in all parts of the hydrosphere—not just in ponds—is the result of photosynthesis by plankton.

The plankton producers collectively are called phytoplankton (FYT oh PLANK ton). Different types of phytoplankton vary in their abundance from one body of water to another. Diatoms usually are the most numerous, though not the most conspicuous. Green algae may become so abundant in late summer that the entire surface of a pond appears green.

Microscopic herbivores, called zooplankton (ZOH oh PLANK ton), include a variety of protists and also tiny animals such as rotifers and small crustaceans. Zooplankton are not the only pond herbivores, however. Many fishes, such as minnows, and bottom dwellers, such as mussels, also are herbivores. Other pond fishes are carnivores; the larger fishes eat the smaller ones. Fishes of all sizes eat aquatic insects, many of which are carnivores themselves. Thus, pond food webs have numerous links. Most of the macroscopic carnivores stay near the edge of a pond, where floating plants provide a hiding place. Larger predatory fishes swim throughout the entire pond.

Layers of organic matter—mostly dead organisms and waste products—build up on pond bottoms. Decomposers such as tubifex worms

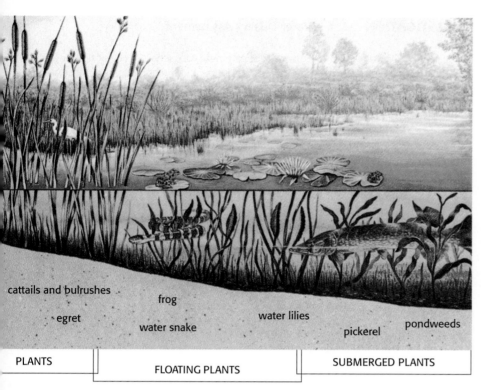

cattails and bulrushes

frog

egret

water snake

water lilies

pickerel

pondweeds

PLANTS | FLOATING PLANTS | SUBMERGED PLANTS

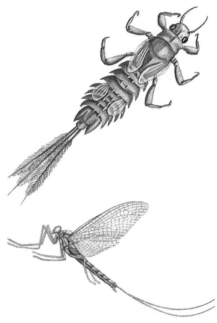

Figure 23.3   The mayfly lives its entire life in and near the water. Nymph (top) and adult (bottom).

(annelids) reduce the dead organic matter to smaller bits and pieces, which are further broken down by other decomposers such as fungi and bacteria.

3   It often is difficult to determine the boundaries of an ecosystem. In a pond ecosystem, the surface of the water is not a distinct boundary. Insects, usually are flying above and around ponds. Many of them hatch from eggs laid in the water and spend most of their lives there. A mayfly's entire life, for example, is spent in the water except for the last few hours, when the adults mate in the air. The females then lay eggs in the water, and the immature mayflies grow there for a year or more. Figure 23.3 shows both the adult and immature forms of a mayfly. Frogs and many fish live primarily on these and other insects. Water snakes eat fish and frogs. Many species of birds and some mammals feed on fish, crayfish, and mussels. These consumers, although they spend a great deal of their time on the land, actually are a part of pond ecosystems. Their supply of energy can be traced back to the phytoplankton.

The growing season is short; the average water temperature is cold in the summer; there is little shore vegetation to contribute nutrients.

23.2   Lakes Are Larger and Deeper than Ponds

Lakes, such as the one illustrated in Figure 23.4, are larger than ponds and
1   have some deep areas that remain dark. Most of the world's lakes were formed by the action of glaciers and, through the process of succession, eventually will become terrestrial ecosystems, as shown in Figure 23.5. Phytoplankton are the major producers in lakes, as they are in ponds. The phytoplankton of a lake, however, can exist only near the water's surface, where there is enough light for photosynthesis. In deep lakes, therefore, all the food is produced in the upper part of the water.

The existence of aquatic consumers is limited not by light but by the amount of oxygen in the water. Water at the surface of a lake constantly receives dissolved oxygen from photosynthetic organisms and from the air. This oxygen diffuses slowly downward from the surface, but most of it

Figure 23.4   A mountain lake. What are the environmental conditions here, and what are their effects?

Figure 23.5 An example of lake succession. At what stage is the lake apt to be nutrient-rich? **Probably by c.**

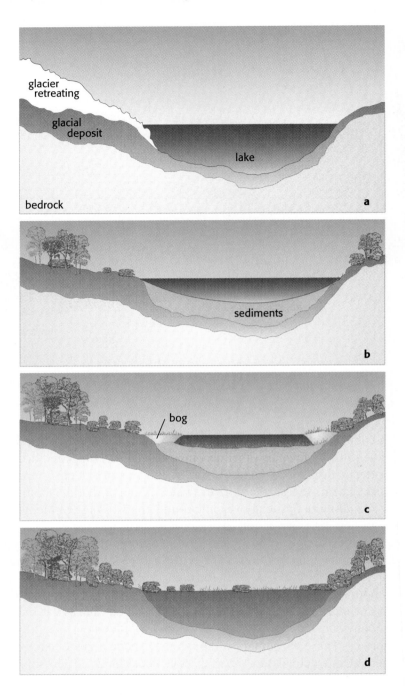

Wave action, which is an important agent of mixing in oceans, is of minor importance in any but the largest lakes.

Other substances, such as DDT and herbicides—many containing mercury—accumulate temporarily even in lakes that have outlets. These substances may poison animals or make fish unsuitable for human consumption.

Ask students what factors might affect the amount of water that sinks into the ground. Answers might include the character of the precipitation (snow versus rain), rate of fall, slope of land, amount of vegetation, and amount of humus in the soil.

is used up by zooplankton before it diffuses very far. The spread of oxygen into deep water thus depends on other factors.

During the summer, the sun heats the lake surface and the water expands, becoming lighter or less dense than water at the lake bottom. This expansion forms a warm water layer that floats on the colder water below. Separating these layers is a zone, called the thermocline (THER moh klyn), across which both temperature and oxygen concentration drop sharply. The thermocline prevents vertical mixing of the upper and lower layers of water, resulting in important differences between these layers. The concentration of dissolved oxygen remains high in the upper layer because of photosynthesis but decreases in the lower layer as oxygen is used up by consumers, thus restricting the activities of aerobic decomposers. As a result, the dead bodies of organisms that fall from above are only partly decomposed, and the nutrients released through decomposition remain on the bottom of the lake.

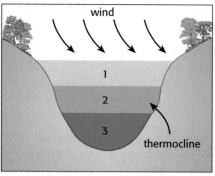
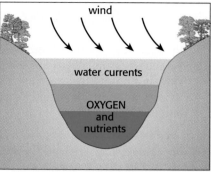
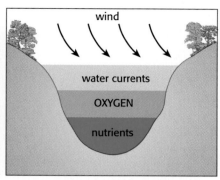

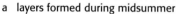
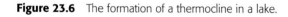

a   layers formed during midsummer

b   mixing of water currents during fall and early spring in a northern lake

c   a northern lake in late spring and early summer

Figure 23.6   The formation of a thermocline in a lake.

In the fall, as the air temperature drops, the surface water becomes colder and more dense. The heavier surface water sinks, forcing deep water to the top. Wind striking the surface helps mix the water and distribute oxygen throughout the lake. During winter, the surface of the lake may freeze, and a warmer layer remains at the bottom. The oxygen concentration declines but anaerobic decomposers continue to release nutrients from dead organic matter. In the spring, rising temperatures again produce movement of the waters, which mixes oxygen and nutrients throughout the lake. Figure 23.6 illustrates this mixing of water layers at different times of the year.

If a lake has no outlet, nutrients and minerals washed in from the surrounding land become concentrated in the water because only water molecules can evaporate from the surface. The water may become saturated with nutrients and minerals, the most abundant of which is usually sodium chloride. The buildup of salt and other compounds in a lake, called salinization (sal ih nih ZAY shun), takes place faster in arid regions where evaporation is more rapid, resulting in an environment unfavorable to most organisms, such as that shown in Figure 23.7.

Some of the water that falls on land runs directly into lakes, ponds, and streams; some evaporates. Most of the water, however, soaks into the ground. This groundwater reappears in springs, from which the course to the ocean may be short or long. This is the way water cycles in the biosphere, as shown in Figure 23.8.

Figure 23.7   Mono Lake in California is an example of the salinization of a lake. Salt accumulation is responsible for the large tufa formations on the right. The water is so saline that only brine shrimp live in it.

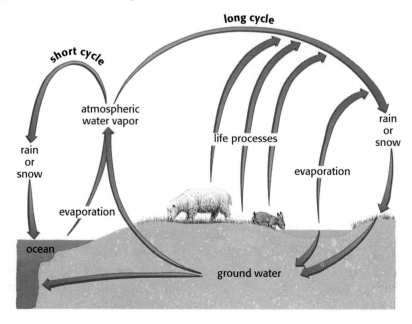

Figure 23.8   The water cycle. Water is necessary to life in many ways. Land plants absorb water from the soil, and land animals drink water or obtain it from their foods. Water constantly bathes organisms that live in ponds, lakes, rivers, and the oceans. Water also evaporates from the surface of all bodies of water and from land. Plants and animals also lose water, but through different processes. All the water comes from the nonliving environment. What is the difference between the long and the short cycles? **The long cycle would be eliminated, except for groundwater to the oceans, if living organisms were absent. The water cycle would revert to the short cycle only. Notice also that the carbon cycle virtually would be eliminated by the absence of living organisms.**

Hot-water brooks flow for some distance from hot springs and geysers, notably in Yellowstone Park, New Zealand, and Iceland.

Flowing Waters

23.3   Stream Size and Flow Affect Aquatic Organisms

Most springs give rise to small brooks. Such flowing water usually is cool, although some hot springs do exist. Tumbling through rapids and falls, the water traps many air bubbles from which oxygen easily dissolves. Because cool water can hold relatively large amounts of gases in solution, the water in brooks usually contains much oxygen.   **1**

Plankton generally are absent in the swift-flowing water of brooks (Figure 23.9) because they are swept away. Producers, such as cyanobacteria, diatoms, and water mosses, grow on the stones at the edge or on the bottom of the brook. These producers provide some food and much shelter for aquatic insects. The insects, in turn, are food for small fish. Most of the food supply in a brook ecosystem comes from the land around it. Small insects fall in the brook and are eaten by stream inhabitants. Every rain washes in dead organic matter. Anything not used immediately is washed downstream, and very little food is left for decomposers.   **3**

          2

Brooks meet and form larger streams with wider beds. The water in streams usually moves more slowly, and solid substances carried along by the swift brooks are deposited as sediments. Bits of organic matter that accumulate among the sediments provide food for decomposers. As the stream widens, the relative amount of water surface shaded by trees and shrubs along its banks decreases, and direct sunlight reaches most of the water surface. Some phytoplankton organisms live in this slower-moving water, but many still are carried downstream. Rooted plants grow in the sediments of stream bottoms, although they may be uprooted and washed away during floods.

Streams support a larger number of consumers than do brooks because they contain more producers although the diversity in a stream is more a result of decreased flow and increased habitat. Mussels, snails, crayfish, and many immature insects inhabit the bottom. Larger consumers, such as turtles and fish, are dependent on these bottom dwellers. Figure 23.10 illustrates some typical stream consumers.

Many streams come together to form a river. As a river approaches the sea, it usually moves more slowly and deposits larger amounts of sediments. Thus, instead of eroding the land, a river often builds up land near its mouth, as does the Mississippi River, shown in Figure 23.11. Riverbanks actually may become higher than the land behind them. During floods a river often breaks through these natural levees. The water left behind is slow to drain away and forms a swamp where many rooted and floating plants grow. Subsequent floods may sweep fruits, seeds, and other plant parts back into the river and thus contribute to the food supply of the river ecosystem. Phytoplankton, which grow well in the unshaded, slow-moving   **3**
swamp water, supply much of the food for consumers in large rivers.   **1**

The current is running rapidly, which prevents plankton from accumulating, and the water is clear, which indicates little suspended organic matter. Also, there are no rooted plants or evidence of submerged ones, though the latter might not be apparent.

A brook ecosystem might be compared with a small town; it has its gardens, but most of its food comes from elsewhere.

The terminology for flowering waters of various sizes is confusing. Brook, stream, and river are used here to indicate size (roughly, volume).

Many ancient riverside civilizations were agricultural. Ask students how river ecosystems contributed agriculture. Consider the Nile Valley.

Figure 23.9 A mountain brook. What evidence can you see that this aquatic ecosystem has few producers?

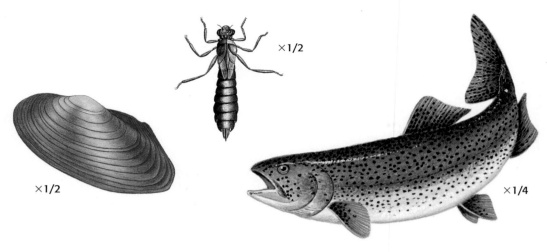

Figure 23.10   Stream consumers (left to right): freshwater mussel, dragonfly nymph, and rainbow trout.

Consumers in rivers are varied and numerous. Zooplankton are food for larger consumers. Mollusks, crustaceans, and fish often grow very large. Crocodiles are common in tropical rivers. Many birds and mammals obtain their food from rivers. Humans also have long taken advantage of the abundant food in rivers.

23.4   Flowing Waters May Serve as a Laboratory

Not all springs give rise to brooks. Some springs produce so much water that large streams flow from them. Many of the short rivers of northwest Florida, for example, originate in this manner. These short Florida rivers are excellent outdoor laboratories. Because these rivers have areas in which the water volume, current, chemical composition, and temperature are stable, ecologists have used several of them in studies of ecological productivity.

The productivity of any community depends on the photosynthetic rate of its producers. To determine productivity indirectly, biologists first

Figure 23.11   Part of the Mississippi delta. What major types of ecosystems can you see here?   **Swamp or wetland, river, salt marsh, estuarine, and ocean are among possible answers.**

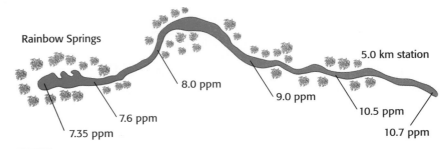

Figure 23.12   Diagram of Rainbow Springs, Florida, and the river that flows from them. Ecologists measured the amounts of oxygen in the springs and in the river water at several places downstream. Such measurements are expressed as parts per million (ppm). One ppm = 0.0001 percent.

This section may be confusing to some students. Attempt to get them all to realize that a phenomenon can be understood indirectly. Productivity here refers to the amount of organic substance produced per unit of time.

Point out that a stream has depth. Why, then, was surface area rather than volume used in calculating productivity? Productivity is concerned with energy, not matter. The energy in food comes from the sun. The amount of sunlight available depends on the amount of water surface exposed to sunlight. Water depth may affect the proportion of sunlight used, but it cannot affect the amount of solar energy available.

measure the amount of oxygen produced during a given time. Because **4** photosynthetic organisms give off oxygen in proportion to the amount of organic substance they produce, biologists can use this figure to calculate the amount of organic substance produced by photosynthesis during the same period.

Any increase in the amount of oxygen in the spring water as it flows downstream comes largely from photosynthesis. Figure 23.12 shows the oxygen content measured at given places along the river flowing from Rainbow Springs. The biologists also measured the amount of water that flowed past each point every day. Based on this information, they calculated the total productivity of the water between its source in the springs and the point of measurement. This productivity is expressed as grams of organic substance produced per square meter of stream surface per day. The investigators found an average productivity of 17.5 g/m²/day.

In further studies the investigators estimated the **biomass**—the total mass of all individuals of a species in a given area—for each major species in the community. Figure 23.13 shows the result of biomass measurements in the Silver River, another short Florida river. Such measurements help ecologists understand the interrelationships in an ecosystem.

Note that Figure 23.13 is a pyramid of biomass whereas Figure 3.10 is a pyramid of energy; the latter is not a mere generalization of the former. Energy always decreases along a trophic chain, but biomass might not appear to do so when generations of producers turn over more rapidly than those of consumers.

Figure 23.13   Average biomass measurements from Silver River, Florida. Largemouth bass are top carnivores. Compare this figure with Figure 3.10.

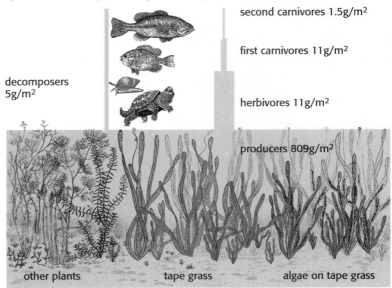

CONCEPT REVIEW

1. How do abiotic conditions in a brook differ from those in a river?
2. What is the source of energy for organisms in a brook?
3. Why is plankton more abundant in a river than in a brook?
4. How do biologists measure productivity?
5. What usually happens to biomass between one link in a food chain and the next?

Ocean Ecosystems

23.5 Many Abiotic Factors Influence the Ocean

Ocean environments, such as the one shown in Figure 23.14, differ in many ways from inland-water environments, but the principal difference lies in the chemical composition of the water itself. Seawater is about 3.5 percent dissolved substances, most of which are salts. Because sodium chloride accounts for more than 75 percent of the substances present, the content of dissolved substances in seawater is referred to as salinity (suh LIN ih tee). Salinity varies somewhat at different depths and in different places on the surface. Table 23.1 shows the average content of various elements in sea water.

Evidence from rocks and fossils shows that oceans have had a high salinity for hundreds of millions of years. Dissolved substances constantly washed into oceans from the land appear to have had only a slight effect on the composition of the water. Ocean water represents a steady state. Substances are added to it continually, but substances also are removed continually by marine organisms and remain in the accumulated skeletons of these organisms for long periods of geologic time. Ocean water is the environment of marine organisms, but it is also a product of their activities. The hydrosphere, like the atmosphere, undoubtedly would have a much different composition if life were absent.

Figure 23.14   Drake Passage between Tierra del Fuego and the Antarctic peninsula.

Guidepost

What factors affect ocean ecosystems?

Table 23.1 Average Content of Different Elements in Seawater*

Element	Seawater (Part per thousand)
Chlorine	18.98
Sodium	10.56
Magnesium	1.27
Sulfur	.88
Calcium	.40
Potassium	.38
Bromine	.065
Carbon	.028
Strontium	.013
Silicon	.003
Fluorine	.001
Aluminum	.0005
Phosphorus	.0001
Iodine	.00005

* These elements occur as parts of compounds.

Exceptions such as the Great Salt Lake contain more salts than seawater does.

The shells of most sea animals are made primarily of calcium carbonate. Ask students how the existence of sea animals affects the supply of calcium ions in seawater. The supply is affected as marine organisms remove both calcium carbonate and silica from the reservoir of dissolved substance in the sea.

Depending on students' earth science background (and the availability of a globe), you may want to discuss the relationship between differential interception of solar radiation and the powering of ocean current systems.

Rapid changes in the atmosphere result in great differences in the climates of terrestrial ecosystems. Such rapid changes do not occur in the hydrosphere. Temperature changes between day and night and between 5 the seasons are very small at the ocean surface. Although the temperature of the surface water at the equator is very different from that of antarctic and arctic waters, those differences diminish with increasing depth. At great depths the oceans have uniform temperature.

The amount of radiant energy available to photosynthetic organisms is greatest at the water's surface. The depth to which light penetrates into the water depends on several factors, including the angle at which light strikes 6 the water's surface and whether the surface is smooth or broken by waves. Another factor is the rate at which the water absorbs the light. As light passes through the water, the longer wavelengths are absorbed most rapidly. Red and yellow disappear first; blue and violet penetrate farthest (Figure 23.15). The rate at which water absorbs light also depends on the clarity of the water. The cloudier the water, the faster the rate of light absorption.

Ocean currents distribute chemicals that are useful to organisms. Currents also affect water temperatures and salinities at any given place in the ocean, and they affect climate (see Chapter 22). Currents, in turn, are affected by the world pattern of winds and by the earth's rotation.

23.6   Productivity Is Limited in the Open Ocean

The chief producers of the open ocean are diatoms, other microscopic algae, and certain flagellates. The zooplankton depend directly on these phytoplankton. Many food chains are based on these producers and include large consumers such as tuna, sharks, whales, and oceanic birds such as albatrosses. The oceans vary greatly in productivity. One study of phytoplankton

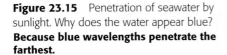

Figure 23.15   Penetration of seawater by sunlight. Why does the water appear blue? **Because blue wavelengths penetrate the farthest.**

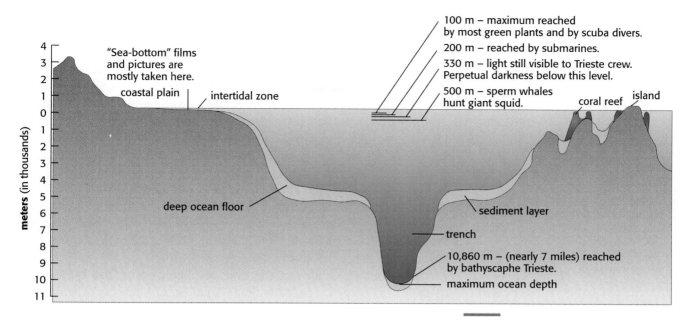

population density, conducted at a laboratory in Plymouth, England, found that the ocean water there contains, at the very least, 4.5 million phytoplankton organisms in each liter of water. Biologists do not have enough data to give accurate density averages for the ocean as a whole.

A limiting factor in the ocean, as on land, is the availability of chemical elements, especially phosphorus and nitrogen. Phytoplankton constantly exhaust the supply of these elements, and their continued growth depends on constant resupply. Elements may be added from sediments washed from the land or from water welling up from below. Upwelling
1 may result from seasonal changes or from offshore winds moving surface water and deep water rising to take its place. Because of upwelling, ocean areas at higher latitudes and close to the land—such as the North Sea and the Grand Banks of Newfoundland—are very productive. These areas of upwelling are small, however, compared with the entire ocean. Fifty percent of all fish and shellfish harvested from the ocean are taken from 0.1 percent of its surface. In contrast, ocean regions far from land, especially those in the tropics, have very low productivity.

Regardless of the availability of nutrients, phytoplankton need light. All of the photosynthesis in the open ocean occurs in the upper layer of
1 water, where light penetrates beneath the surface. This layer is very thin compared with the total depth of ocean water, as shown in Figure 23.16. Despite the deep penetration of blue light (550 m near Bermuda), phytoplankton are largely limited to the upper 100 m of water.

23.7   Ocean Depths Support Some Forms of Life

For a long time biologists thought life could not exist in the dark, cold ocean depths because of the tremendous pressure of the water. The first clear evidence that this idea was wrong came in 1858, when one of the telegraph cables lying on the bottom of the Mediterranean Sea broke and was hauled up for repair. It was encrusted with bottom-dwelling animals, mostly sponges, some of which had grown at depths as great as 2000 m.
2 Water pressure has no ill effect on organisms unless they contain spaces filled with air or other gases. Because deep-sea organisms lack such spaces, pressure is exerted equally from both inside and outside the organism, thus preventing damage.

Figure 23.16 Diagram of an ocean in cross section. What is the most productive zone? **The intertidal zone and the areas around coral reefs, to a maximum of 100 m.**

Ask students how upwelling relates to the location of the great ocean fisheries. Most fisheries are located near oceanic areas rich in phytoplankton, the basis of marine food chains.

When brought to the surface, fish caught at great depths often look as if they have exploded. Gases within the bodies of these fish are adjusted to the great external pressure of the ocean depths. When they are hauled to the surface, the pressure of their internal gases is no longer balanced; the gases expand and may rupture membranes within the fish.

The ocean depths are special ecosystems that require unusual adaptations in the organisms that survive there. The depths are cold, dark, and quiet, and food is scarce. Life depends on organic substances that settle from the water above. The adaptations of organisms to the pressures of the deep make their ascent to upper levels fatal. Thus, the consumers in the ocean depths form one of the most isolated communities of the biosphere.

In the eternal darkness of the ocean depths, most animals are either black or dark red and have very sensitive eyes. Many are bioluminescent (by oh loo mih NES ent)—they produce their own light. One species of fish dangles a special luminescent organ in front of its mouth that apparently lures unwary prey close enough to be caught and eaten. Deep-sea shrimp escape some predators by emitting clouds of a luminescent secretion they give off when disturbed. Patterns of light on an animal's body also may help organisms recognize one another in the dark. Figure 23.17 shows some fishes that live in the deep ocean.

Nearly all life on the earth depends on radiant energy that is converted to chemical energy by photosynthetic organisms. A few organisms, however, are not dependent on photosynthesis or photosynthetic organisms for their chemical energy. A type of ecosystem based on chemosynthesis was discovered by researchers aboard the submersible *Alvin* in 1977 at the Galapagos Rift, a boundary between tectonic plates. Seawater seeps into the vents created by volcanic rocks erupting along this boundary, becomes heated, and rises again. As a result, oases of warmth are created in the normally near-freezing waters. More important, the water reacts chemically,

Figure 23.17   Some fishes of the deep ocean and the depths at which they have been caught. All are small. They are shown here close to actual size.

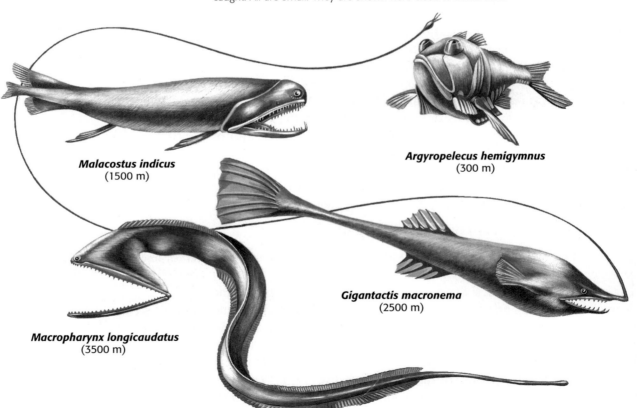

Malacostus indicus
(1500 m)

Argyropelecus hemigymnus
(300 m)

Gigantactis macronema
(2500 m)

Macropharynx longicaudatus
(3500 m)

under extreme heat and pressure, with the rocks deep in the earth's crust. The crucial reaction appears to be the conversion of sulfate SO_4^{2-} in the seawater to sulfite (S^{2-}). Heat supplies the energy necessary for the reaction. Chemosynthetic bacteria obtain energy by converting the sulfite back to sulfate. These bacteria use the energy to extract carbon from carbon dioxide in the seawater and incorporate it in organic compounds, thereby becoming the primary producers in the ecosystem.

About a dozen such vent ecosystems have been identified, each with a somewhat different animal community. Among the most spectacular inhabitants are clams measuring some 20 cm in diameter. Other permanent inhabitants include crabs and mussels. The clams and mussels feed on the chemosynthetic bacteria and are, in turn, fed on by the crustaceans. One oasis, called the Garden of Eden, is dominated by huge tube worms, shown in Figure 23.18. Apparently these worms, which lack mouths and digestive tracts, utilize carbon compounds synthesized by chemosynthetic bacteria living within them.

23.8 Coastal Waters and Coral Reefs Support a Diversity of Life

With few exceptions, oceans are relatively shallow near the continents. These bands of shallow water, which average less than 200 m in depth, are widest at the mouths of large rivers and along areas of broad lowlands. Along mountainous coasts, as in California, shallow waters may be almost absent.

In shallow waters some light reaches the bottom. Where there is plenty of light, a luxuriant growth of algae may be found. In middle and higher latitudes, the brown algae are most common and conspicuous. Among the brown algae are kelps (Figure 23.19), which may reach lengths of 35 m or more.

The physical characteristics of the ocean bottom—sand, mud, or rock—determine the types of organisms that live there. Plants usually are not abundant on unstable sandy bottoms, but many types of animals burrow into the sand, especially crustaceans, mollusks, and annelid worms. Muddy bottoms have even more burrowers, and most of the species are unlike those adapted to sand. Sea cucumbers, clams, and some types of crabs are common in the mud.

Because they are shallow and close to land, coastal waters offer more opportunity for human use than does the open ocean. Much of the marine food supply of humans comes from the coastal waters. Underwater farming of kelp, fish, crustaceans, and mollusks is a future possibility. It is unlikely, however, that the ocean can supply all the food humans will need in the future. Except in coastal waters (including coral reefs), the ocean is very limited in some of its dissolved nutrients. Rich fertile soils may contain 0.5 percent nitrogen, an important nutrient for growth, whereas the richest ocean water, in comparison, contains only about 0.00005 percent nitrogen.

Among the most productive and diverse ecosystems are coral reefs (Figure 22.20), found only in coastal tropical waters where there is adequate light and oxygen. The reefs are large formations of calcium carbonate laid down by individual coral polyps, which harbor symbiotic green algae in their tissues. These algae carry out most of the photosynthesis of the reef community, and they also assist in the cycling of chemicals such as nitrogen. Because of the diversity of life forms that coral reefs support, they have been called the tropical rain forests of the sea.

Figure 23.18 Among the organisms that live in vent ecosystems are giant tube worms, *Riftia pachyptila*, many more than a meter long.

Ask students why plants and algae are scarce on ocean bottoms. Most plants cling by roots and larger algae cling by holdfasts, so they require a stable substrate.

Figure 23.19 Channel Islands kelp bed.

Figure 23.20 A Caribbean coral reef. What organisms can you identify? **Answers will vary.**

Figure 23.21 The ocean shore. What is the effect of wave action?

Wave action produces slow erosion, even of a rocky shore.

Figure 23.22 A tide pool community including sea stars, *Piaster ocuraceus* rockweed, and barnacles. What conditions must these organisms be able to tolerate?

Low tide exposes them to higher temperatures and decreased oxygen supplies; if the surf is strong, they must be able to tolerate repeated pounding.

23.9   The Intertidal Zone Is a Difficult Place to Live

Everywhere along the margins of the oceans you can see the effects of waves and tides. In the Bay of Fundy (Nova Scotia) the maximum vertical change between high and low tides is 15.4 m. At the other extreme, the average tidal difference in the Mediterranean is only 0.35 m. Wave action, too, varies depending on the place and time. Some shores, such as those of Maine, are pounded by heavy surf, as shown in Figure 23.21. Others, especially in small, protected bays, may be no more exposed to wave action than the shores of small lakes. High and low tides each occur twice a day. The zone between high and low tides—the intertidal zone—is a difficult environment for life. Twice a day intertidal organisms are submerged in salt water and then exposed to air and the hot sun or freezing wind. Between these times the intertidal zone is pounded by advancing or retreating surf.

On rocky coasts, life in the intertidal region is surprisingly abundant, as shown in Figure 23.22. In cold water, different species of brown algae, protected from the drying sun by a jelly-like coating, cling to the rocks. These tangled algae provide protection and support for other algae, protists, and many animals. In addition, barnacles, chitons, and snails cling firmly to rocks or seaweeds, closing up tightly during the periods when they are exposed to air.

On sandy coasts, life in the intertidal zone is limited to organisms that can burrow in the sand or skitter over it. There are no attached producers, but phytoplankton may be present when the tide is in. The burrowers, such as small crustaceans, eat food particles brought by the high tides. Shorebirds, which prey on the sand burrowers or forage in the debris left behind by the retreating tide, are a link with land ecosystems. Land crabs, such as those in Figure 23.23, release their young in the water. Sea turtles crawl out on the beach and bury their eggs. Terrestrial and marine ecosystems merge in the intertidal zone.

Wherever ocean meets land there is constant change. In only a few years sediments deposited at the mouths of rivers stretch the land into the ocean. Elsewhere you can observe the ocean pushing back the land as it carves out sandy beaches with its wave action. Such changes have occurred throughout most of the earth's history.

Figure 23.23   A ghost crab, *Ocypode puadrata*, showing defensive posture.

CONCEPT REVIEW

1. Why do some parts of the oceans have a higher productivity than others?
2. Describe the environment of the ocean depths and the organisms that live there.
3. How do coastal ecosystems differ from those of the open ocean?
4. How do marine organisms affect the amount of dissolved substances in ocean water?
5. Compare variations of temperature in the oceans with those in the atmosphere.
6. What factors determine the amount and color of light found beneath the water surface?

Human Influences on Aquatic Ecosystems

23.10 Wetland Ecosystems Are Threatened

Wetlands, including marshes, swamps, shallow ponds, and coastal areas near estuaries, are flooded with fresh or salt water all or part of the year. Among the most productive of all ecosystems, wetlands form the base of many aquatic and terrestrial food chains.

Inland wetlands provide freshwater habitats for a diversity of fish, waterfowl, and other organisms, including a number of threatened and endangered species such as the whooping crane. Wetlands near rivers help control erosion by storing water during periods of high runoff and by regulating stream flow. Because wetlands trap stream sediments and absorb, dilute, and degrade many toxic pollutants, they improve water quality. If overburdened, however, they can accumulate too many toxins and be destroyed. Wetlands also play a significant role in the cycling of carbon, nitrogen, and sulfur. Important crops, such as cranberries, blueberries, and rice, are grown in wetlands. Such areas also are used for recreation—especially hunting. Figure 23.24 shows a typical inland wetland.

Coastal wetlands, such as the one in Figure 23.25, are found near estuaries where rivers join the oceans. These areas supply food and serve as spawning and nursery grounds for many species of marine fish and shell

Guidepost
How do human activities affect aquatic ecosystems?

Figure 23.24 Swamps and marshes such as this cypress swamp in North Carolina, are important inland wetlands.

Figure 23.25 Low tide in a coastal estuary. What conditions must organisms tolerate? **Repeated wetting and drying; varying salinity.**

Estuaries are among the most productive areas on the earth. Tidal salt and brackish marshes produce four times the plant biomass of a good cornfield—without fertilizers. If your school is near an estuary, pursue this topic further.

fish, including salmon, shrimp, oysters, clams, and haddock. Coastal wet lands also are important breeding grounds and habitat for waterfowl and other wildlife, and they help protect coastal areas by absorbing the impact of damaging waves caused by storms and hurricanes. Like inland wetlands, coastal areas dilute and filter out large amounts of water-borne pollutants.   2

Because the importance of wetlands may go unnoticed, these areas often are dredged or filled in and used as cropland, garbage dumps, and sites for urban and industrial development. So far, about 56 percent of coastal and inland wetland acreage in the contiguous 48 states—an area about four times the size of Ohio—has been destroyed. Most of this loss is   1 due to drainage and clearing for agriculture, and the remainder has been lost to real estate development and highways. Although concerned groups are making attempts to slow the rate of wetland loss, only a small fraction (8 percent) of inland wetlands is under federal protection.

Although the drainage of standing waters continues, the surface area of inland waters has actually increased during the last 50 years as a result of the construction of dams across running waters. Dams create waterfalls   1 that turn electric generators, and most dams, especially those in the western states, also form reservoirs that store floodwaters for irrigation and for supplying water to cities. Major changes in the environment, such as   3 reservoirs and controlled stream flow, bring about major changes in communities within that environment. A large dam may block the passage of fish, such as salmon, that swim up rivers to lay their eggs in headwater streams. On the other hand, dams may greatly increase the amount of habitat favorable for other species, such as pike.

23.11   Sewage and Industrial Wastes Affect Aquatic Ecosystems

Although fresh water is a renewable resource, it can become so contaminated by human activities that it is no longer safe for many purposes and is harmful to living organisms. Water pollution is a local, regional, and global problem connected with how we use the land. Some of the fertilizers and pesticides humans apply to croplands run off into nearby surface waters or leach into groundwaters. Poor land use accelerates erosion, and wastes produced on land often are dumped into the oceans. In

Biology Today

Plant Ecologist

Joy Zedler is a plant ecologist and the director of the Pacific Estuarine Research Laboratory (PERL) in San Diego, California. PERL operates two field research facilities that conduct a wide range of applied ecological studies involving wetlands, including research projects related to wastewater management. Two such projects currently are under way in Southern California. One project is aimed at determining at what level and for how long sensitive estuarine organisms, such as halibut, arrow gobies, and other native fishes and invertebrates, can tolerate reduced salinity of their water. Historically, these species have experienced population reductions when freshwater inflows have occurred following sewage spills. The second project uses wetlands to improve the quality of waste water discharged to coastal areas. The wastewater is discharged in pulses, thus reducing the amount of time that the inflow of fresh water has to affect coastal waters.

Joy's personal research interest include salt marsh ecology, the structure and functioning of coastal wetland ecosystems, the effects of rare events on estuarine ecosystems, the interactions of native and exotic species in coastal wetlands, and the use of scientific information in the management of coastal habitats. Currently, Joy and her associates are researching the restoration of salt marsh ecosystems. In addition to studying how well-restored salt marshes function compared with natural salt marshes, they also are developing new methods to accelerate the restoration process. By understanding what factors act to slow ecosystem development in human-made wetlands, they are hoping to correct similar problems in new restoration sites.

Joy grew up on a farm where she always felt a close relationship with the land—a factor that greatly influenced her decision to become an ecologist. She attributes her interest in figuring out how to restore wetlands to the "fix-it" ethic practiced on the farm by her parents. As a high-school student in rural South Dakota, she was expected to get a secretarial job, marry, and raise children. Joy rebelled against that stereotype and pursued her education in botany and plant ecology. Along the way, however, her secretarial skills helped her earn money for college. She did get married, and her children now work on wetland experiments for their high-school science fair projects.

Joy's interest in biology, and eventually in wetlands, is the consequence of various opportunities and many teachers who encouraged her. In Joy's words. "Very little of my career was planned, but I have always followed leads and tried to find a match between what I can do and what someone needs to have done. You have to find the link between skills and available funds, and then be willing to grow with the opportunities that develop. I doubt if it's very different in any other field."

Figure 23.26   Diagram of sewage pollution in a stream. Organisms are not drawn to scale and distances are decreased greatly. At the left, sewage is attacked by decomposers and the oxygen supply decreases. Aerobic organisms die. Further downstream the amount of sewage decreases and the oxygen supply increases. The phytoplankton and aerobic consumers increase in number.

large, rapidly flowing rivers, relatively small amounts of pollutants are diluted quickly to low concentrations, and the supply of dissolved oxygen needed for aquatic life and the decomposition of organic wastes is renewed quickly. One common pollutant, sewage, provides nutrients for the growth and reproduction of aquatic producers, as shown in Figure 23.26. If the amount of sewage is large relative to the volume of river water, the growth of phytoplankton is great. Phytoplankton have short life spans, and the organisms that decompose the phytoplankton use up the available oxygen. Anaerobic conditions result, a situation that is deadly for all aerobic organisms. As the water flows farther downriver, the sewage is diluted by tributary streams, and eventually decomposers use up the sewage materials.

Every day, rivers also receive enormous amounts of natural sediment runoff and surface runoff from urban and agricultural uses of land. Factories and mills contribute alkalis, chromium, lead, and mercury ions—all of which are poisonous to some living things, particularly to the organisms that decompose organic wastes. Industrial processes, especially the generation of electricity, often result in the discharge of hot water into rivers. Such thermal pollution abruptly changes the abiotic environment and makes it intolerable for many aquatic organisms.

Initially the mercury ions are harmless, but microorganisms incorporate them and change them into mercury compounds that enter food chains and poison other organisms. An additional pollution problem is the magnification of pesticides described in Chapter 3.

The warmer water also may bring about a change in the types of species that can inhabit the area. All of these pollutants also affect the health of the human populations that depend on the water.

3 In contrast to rivers, lakes and reservoirs act as natural traps that col-
4 lect nutrients, suspended solids, and toxic chemicals—such as those that result from acid rain—in their bottom sediments. Many pollutants, such as pesticides, fertilizers, and sewage, contain high levels of plant nutrients, specifically nitrates and phosphates. The accumulation of these nutrients in lakes and other standing bodies of water encourages the growth of photosynthetic organisms, including algae. These organisms produce great quantities of oxygen during the day, but at night cellular respiration greatly reduces the amount of oxygen. As the organisms die and organic material accumulates, the metabolism of the decomposers uses up the available oxygen. A body of water, such as that shown in Figure 23.27, that is nutrient-rich yet oxygen-poor is **eutrophic** (yoo TROH fik; from the Greek *eutrophos*, for well nourished). Although eutrophication is a natural process, it is accelerated by the intrusion of pollutants into the water. Eutrophic conditions may make it impossible for some organisms to survive. For example, eutrophication of Lake Erie resulting from pollution completely wiped out commercially important fish species such as blue pike, lake trout, and whitefish in the 1960s. Tighter regulations on the dumping of wastes into the lake have enabled many fish populations to rebound, although some fish and invertebrate populations have not recovered.

Eventually other organisms suffocate.

Figure 23.27   A highly eutrophic pond. How does the fast growth of the algae affect other aquatic organisms?

Figure 23.28 An oil-soaked grebe in west San Francisco Bay.

23.12   The Oceans Receive the Major Part of Human Waste

Oceans are the ultimate dump for much of the wastes produced by humans. In addition to natural runoff, oceans receive agricultural and urban runoff, atmospheric fallout, garbage and untreated sewage from ships, and accidental oil spills from tankers and offshore oil drilling platforms. Barges and ships also dump industrial wastes, sludge from sewage plants, and other wastes.

Oceans can dilute, disperse, and degrade large amounts of some organic pollutants, especially in deep-water areas. Other pollutants, however, do not readily degrade. What happens to these pollutants when they reach the ocean? Some, such as pesticides and mercury, enter the food chains and can eventually end up in the seafood we eat. The people **5** of Minamata, Japan, learned firsthand what mercury pollution can do. They ate fish with high concentrations of methyl mercury in their tissues, and before long there were several thousand people, many of them children, who were left crippled as a result of damage to the nervous system.

Oil spills are a constant threat to ocean ecosystems. After an oil spill, **6** hydrocarbons such as benzene and toluene are the primary killers of fish and shellfish. Other chemicals remain on the surface and form floating tarlike globs that adhere to marine birds, sea otters, seals, sand, rocks, and almost all objects they encounter (Figure 23.28). This oily coating destroys the animal's natural insulation and buoyancy, and many animals drown or die from exposure due to loss of body heat. These tarlike globs gradually are degraded by naturally occurring bacteria. Heavy oil components that sink to the ocean floor probably have the greatest impact on marine ecosystems. These components can kill bottom-dwelling organisms, such as crabs, oysters, mussels, and clams, or else can make them unsuitable for human consumption.

Everything that is dumped into a river or stream eventually reaches the ocean, and many of those pollutants will end up in the food chain. Most pollutants, however, stay in coastal waters, the most productive parts of the ocean. They could cause the destruction of important species of fish that many people depend on for food.

23.13   Some Practices Can Reduce Water Pollution

Several measures can reduce water pollution, both directly and indirectly. The indirect pollution of water by fertilizer runoff and leaching can be controlled in several ways. Farmers should avoid using excessive amounts of fertilizer and should use no fertilizers on land with steep slopes. Contour plowing, such as that shown in Figure 23.29, can control erosion and thus the amount of sediment- and nutrient-rich runoff that reaches lakes and streams and that leaches into groundwater. Periodically planting crops that restore the nitrogen content of the soil can reduce the need for fertilizers. The need for pesticides can be reduced by employing biological pest control methods, such as using spiders to reduce the numbers of insect pests. Already more than 300 biological pest control projects have been successful worldwide.

Directly controlling water pollution involves improving the current methods of waste treatment and disposal. When sewage reaches a treatment plant, it can undergo up to three levels of purification, depending on the type of plant and degree of purity desired. Once the desired level of purity is achieved, the wastewater is disinfected before being released into the environment. Primary treatment of wastewater screens out debris, such as sticks and leaves, and the suspended solids are allowed to settle out. Secondary treatment uses aerobic bacteria to remove degradable organic wastes. The wastewater then goes into a tank where most suspended solids and microorganisms settle out. The resulting sludge is removed and disposed of by burning, dumping in the ocean or landfill, or using it as fertilizer. The third level of treatment uses specialized chemical and physical processes that lower the quantity of specific pollutants remaining after primary and secondary treatment. For example, wastewater from secondary treatment plants can be purified further by allowing it to flow slowly through long canals filled with water hyacinths or bulrushes. These plants remove toxic organic chemicals and metal compounds not removed by conventional treatment. Ensuring that all wastewater receives secondary treatment and is disposed of properly would greatly reduce water pollution.

As individuals, we can reduce water pollution by safely disposing of products that contain harmful chemicals, by using less water, and by using low-phosphate or nonphosphate detergents and cleansers. Inorganic fertilizers, pesticides, detergents, bleaches, and other chemicals should be used only when necessary and then in the smallest effective amounts.

Figure 23.29   How does contour plowing reduce water pollution?   **By controlling the amount of runoff that reaches surface water and groundwater supplies.**

1. What benefits do humans get from draining inland water? From damming flowing waters?
2. How are wetlands important ecosystems?
3. How may draining and damming waters be harmful to human interests?
4. Compare the effects of sewage dumped in a river with the effects of sewage dumped in a lake.
5. How did the people of Minamata, Japan, get mercury poisoning?
6. How do oil spills affect ocean ecosystems?

This investigation introduces students to the use of biological indicator species in determining the type and degree of pollution in a body of water. Teams of 2 work best.

Time: Two to three class periods: one for Parts A and B, one or two for Parts C and D.

Figure 23.30 (a) Microscopic view of 1 mm² graph paper at ×100, field diameter = 2 mm. (b) The area of a strip = the length of the cover slip (22 mm) × diameter of the microscope field.

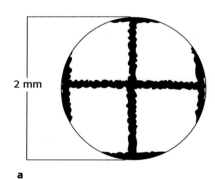

2 mm

a

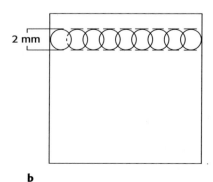

2 mm

b

INVESTIGATION 23.1 Water Quality Assessment

How clean is the water in a lake, pond, or stream near your school? Technical tests can determine exact types and amounts of water pollution, but a simple survey of algae and cyanobacteria can be a reliable index of water quality. Clean water usually has many different species of algae and cyanobacteria, and no single species predominates. Polluted water has fewer species and high numbers of each. This investigation consists of three parts: learning to identify key genera of algae and cyanobacteria; testing known samples of clean, intermediate, and polluted water; and testing a local water source.

Materials (per team of 2)

3 microscope slides
3 22-mm × 22-mm cover slips
dropping pipet
1 L beaker
plankton net
compound microscope
protist/cyanobacteria key
2-cm² piece of graph paper with 1-mm squares

4 flat toothpicks
pocket calculator (optional)
Detain™ in squeeze bottle
water sample with algae and cyanobacteria
water samples—clean, intermediate, and polluted
A Catalog of Living Things

Procedure
PART A Field of View Calculations

1. Place the 2 cm² piece of graph paper on a microscope slide and add a cover slip. Examine the slide with the low power (100X) of your microscope. Align the squares so one edge of a square just touches the bottom edge of the field of view. Count the number of squares from the bottom to the top of the field (see Figure 23.30a). The number of squares equals the diameter of the field of view in millimeters. In your data book, record the results and the low-power magnification.
2. Repeat Step 1 with the high power (400X). (Do not use the oil immersion lens.)
3. You will count individual organisms on a slide made from your water samples. To do so, you will move the slide across the microscope stage, the width of one field of view at a time, as illustrated in Figure 23.30b. The area of the strip of fields of view equals the diameter of the field of view multiplied by the width of the cover slip (22 mm). Practice moving the slide one field of view at a time, going from one side of the cover slip to the other.

PART B Algae/Cyanobacteria Identification

4. Place a drop of Detain™ (protist slowing agent) on a clean microscope slide. Using a dropping pipet, add one drop (0.1 mL) of a sample, and use a toothpick to mix the two drops. Carefully add a cover slip.

5. Practice locating and identifying different organisms using the protist/cyanobacteria key and *A Catalog of Living Things.* Locate organisms under low power of the microscope and then use high power to make a positive identification.

6. Move the slide, one field of view at a time, going from one side of the cover slip to the other. In each strip, count the numbers of individuals for each genus you identify. Ask your teacher for help if you cannot identify some of the organisms.

7. Wash your hands thoroughly before leaving the laboratory.

PART C   Identification of Water Samples

8. Table 23.2 lists the pollution index of each genus used to determine water quality with the Palmer method. You will use this information to determine the pollution index of several samples of water.

9. For each of the three water samples, you will count the number of organisms in each genus in three strips of fields of view. Then you will compute the pollution index for each sample in order to decide which sample is clean, which is intermediate, and which is polluted. In your data book, prepare four worksheets with the headings in Table 23.3 or tape in the worksheets your teacher provides. Use the worksheets for your calculations.

10. Put one drop of Detain™ and one drop (0.1 mL) of Sample 1 on a slide. Mix with a clean toothpick. Add cover slip. In your worksheet, note the sample number.

11. Count and record the number of individuals from each genus present in a strip of fields of view as you move the slide from one side to the other. Count *units* (colonies or filaments are counted as one unit). Large filaments and colonies only partly lying in the strip should be counted as fractions. For example, if a filament of *Stigeoclonium* is only half in your field of view, record 0.5 instead of 1.

12. Repeat Steps 10 and 11 two more times for a total of three strips.

13. The calculations for the Palmer score assume the sample from which you extracted the organisms was 1 L (1000 mL) of water that was filtered and concentrated in 100 mL of water. Study the example in Table 23.3. The first calculation (column F) determines the number of individuals from each genus in your 0.1-mL concentrated sample. The second calculation (column I) calculates the number of units per mL in the original 1-L sample. *You must obtain at least 50 units per mL for the 1-L sample or you cannot enter a Palmer score for that genus.* If you are having difficulty with the calculations, ask your teacher for help.

14. After obtaining the final count for each genus with a frequency of at least 50 units per mL, look up the pollution index for that genus and enter it in your worksheet. (Only genera included in the Palmer list are scored as indicator organisms.) A total score (the sum of the Palmer scores for all species in a sample) of 20 or more for a sample is evidence for high organic pollution. Scores between 15 and 19 indicate moderate pollution, and scores less than 15 indicate absence of pollution.

Table 23.2 Algae/Cyanobacteria Genera Pollution Index (Palmer 1969)

Genera	Pollution Index
Anacystis	1
Ankistrodesmus	2
Chlamydomonas	4
Chlorella	3
Closterium	1
Cyclotella	1
Euglena	5
Gomphonema	1
Lepocinclis	1
Navicula	3
Nitzschia	3
Oscillatoria	5
Pandorina	1
Phacus	2
Phormidium	1
Scenedesmus	4
Stigeoclonium	2
Synedra	2

Table 23.3 Worksheet for Determining Water Quality

Genus	Number of Cells or Units Strip 1	Strip 2	Strip 3	A Total	B Area of Sample	C Strip Area	D Number of Strips	E Volume of Sample	F $\frac{A \times B}{C \times D \times E}$	G Sample Volume	H Water Filtered	I $\frac{F \times G}{H}$	Palmer Score
(Example) Oscill	﷼﷼﷼ ﷼﷼ ﷼﷼	﷼﷼ ﷼﷼ 0.2	﷼﷼ ﷼﷼ III	38.2	484 mm²	44 mm²	3	0.1 mL	1400/mL	100mL	1000mL	140/mL	5

15. Repeat Steps 10 through 13 for Samples 2 and 3.

16. Check with your teacher to determine whether your total Palmer scores correctly identify the water quality of the three samples.

PART D   Investigation of a Local Water Source

17. Using water from a local pond or lake, filter 1 L through a plankton net to trap the organisms. Rinse the organisms from the net into the 1-L beaker and add clean water to total 100 mL. Repeat Steps 10 through 13 to determine the pollution index.

18. Wash your hands thoroughly before leaving the laboratory.

Discussion

1. Which genera can tolerate the highest levels of organic pollution?

2. If a genus has a Palmer score of 1, what does that indicate?

3. If your local water source has moderate or high pollution, what might be the source of pollution?

4. How can algae and cyanobacteria be helpful in polluted water?

5. How can algae and cyanobacteria be harmful in polluted water?

INVESTIGATION 23.2   A FIeld Study of a Stream

The surface of bodies of water usually hides the diversity and activity of life below. Animals may live in the stream or just come to the stream to use its resources. Dead plants and animals that fall into the running water are used as food by stream dwellers. Underwater, an entire community of plants, animals, and other organisms interact with one another and their environment. This investigation examines the types of organisms that inhabit a local stream, and the physical environment in which they are found.

Materials (per team of 5)

2 10X hand lenses	100-m measuring tape
nonmercury thermometer	2 laminated black-and-white cards
3 pairs of forceps	5 pairs of thick cotton gloves
stopwatch or wrist watch with second hand	5 pairs of waders, hip boots, or old sneakers
6 30-cm wooden or metal stakes	1-m × 0.75-m screened frame
1-m × 0.75-m screenless frame	keys for aquatic organisms
plankton net (optional)	data books or clipboards
2 large trays filled with stream water	lemon

PART A   Stream Flow and Physical Characteristics
Procedure

1. Mark a 50-m section of a stream with the six stakes, 10 m apart. At each of the six stakes, measure the width of the stream and its depth midway between the two banks. Determine the average depth and width.

> **Do not walk into the stream with bare feet or sandals. Wear waders, hip boots, or an old pair of sneakers.**

2. Determine the area of the stream by multiplying the average depth by the average width.

3. To estimate the velocity of the water, toss the lemon into the water upstream from the first stake. Measure the time it takes for the lemon to travel from the first stake to the last. Repeat this step twice and average your results. Remove the stakes.

Discussion

1. *Euglena* and *Oscillatoria,* followed by *Chlamydomonas* and *Scenedesmus,* then *Chlorella.*

2. Only slightly tolerant to pollution.

3. Responses will vary, depending on the local situation.

4. By adding oxygen to the water and absorbing pollutants.

5. When algae and cyanobacteria cannot carry on photosynthesis, they die. Decomposers use up oxygen in the water as they break down the algae and cyanobacteria.

This investigation allows students to examine stream ecosystems firsthand and to develop hypotheses that explain the abundance and distribution of different aquatic organisms. Teams of 5 are recommended.

Time: If the stream is near the school, one class period for the measurements and calculations, and one or two additional periods for the observations. If the stream is located away from the school, one-half day should be sufficient.

Safety

Make sure no part of the stream is too deep for any student to wade across or swift enough to knock anyone down. Caution students not to handle any of the organisms with their bare hands.

4. Determine the rate of flow of water through this section of the stream by multiplying the average velocity by the average area.

5. Determine the water temperature at several locations and record these data.

6. Make note of other physical characteristics of the stream, such as whether the bottom is rocky or sandy.

Discussion

1. What is the average width of the stream? What is its average depth?

2. What is the area of the stream?

3. What is the average time it takes for the lemon to travel 50 m?

4. What is the volume of the water that passes by any stake per minute? (Biologists record these data in cubic feet.) What is the rate of flow?

PART B Stream Organisms
Procedure

7. Put on your gloves. Place the screenless frame on a flat part of the stream bottom so the entire frame rests on the streambed. Within the frame, estimate the area that is covered by algae, leaf litter, submerged plants, emergent plants, and the area that is bare. Refer to Investigation 22.2 for information about estimating ground cover.

8. Place the screened frame upright immediately downstream from the open frame so the two frames touch and the water will pass through the screen, as shown in Figure 23.31. If a plankton net is available, hold it a little downstream from the screened frame to collect smaller organisms.

9. Disturb the substrate within the screenless frame, turning over rocks and stirring up gravel. Dislodge any organisms found on the rocks or underneath them. Allow these organisms to flow downstream to be caught on the screen. One team member should hold the screenless frame on the streambed, one should hold the screened frame upright, one should hold the plankton net, and one should turn over rocks.

10. The fifth team member should examine rocks carefully for any invertebrates clinging to them. Place rocks and gravel into one tray on the bank for further examination.

11. Carefully lift the screened frame with its collected material. Try not to let any material be washed away before observing it. Using forceps, transfer any organisms on the screen into the trays of fresh stream water on the bank. Also transfer any organisms in the tray containing the gravel and rocks.

Discussion

1.–3. Answers depend on student data.

4. To calculate the volume of water passing by any stake: 50 m ÷ average number of seconds × cross-sectional area in m². Rate of flow is ft³/second, or volume in ft³ times velocity.

PART B Procedure

11. Collecting organisms from within a frame of known area lets students compare the number and type of organisms in different parts of the stream.

Figure 23.31   Screen setup.

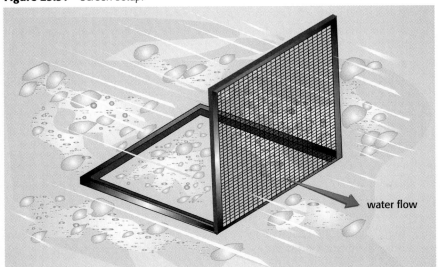

water flow

12. Collectors filter the water for small particles of plant material. This group includes larval blackflies and net-spinning caddisflies. Some collectors use particles that have become trapped on the slimy surfaces of rocks. Included in this group are many mayflies and midges. In parts of streams exposed to direct sunlight algae may be abundant on the rocks. Here, grazers may dominate. These organisms have modified mouthparts adapted for shearing and scraping algae from rock surfaces. This group includes snails, mayflies, and caddisflies. Predators are present at low densities. Common predators are dragonflies, damselflies, some stoneflies, and several true flies; predators also include several types of vertebrates. Shredders use large leaf particles that have fallen into the water. They derive nutrition from the microbes living on the surface of the leaves. Larval stoneflies, caddisflies, and true flies are common shredders.

Discussion

1–3. **Answers depend on student data.**

Discussion

1–2. **Hypotheses and students depend on student data.**

> ✋ **Do not touch any of the organisms with your bare hands. Be careful not to damage the organisms by squeezing them too tightly with the forceps. Forceps are sharp; handle with care.**

12. Identify the organisms and divide them into groups based on what they eat. Your teacher may have you develop your own classification system for the organisms you collect. One suggested grouping is the following:

   a. Collectors—animals that live on decomposing organic matter.
   b. Grazers—herbivores that scrape algae off rocks.
   c. Predators—animals that eat part or all of other animals.
   d. Shredders—herbivores that chew on plant material, or mine or bore through leaves.

13. When you have finished viewing and counting organisms, return them to the stream.

14. Place a drop or two of the water from the jar at the end of the plankton net on the laminated card and examine this sample using the hand lens. Identify the organisms present.

Discussion

1. What types of organisms did you collect?
2. Compare your results with those of other teams in your class.
3. Determine the number of organisms collected per m².

PART C   Hypothesis Forming and Testing
Procedure

15. After making observations of the stream and its surroundings, make a list of factors that might affect life in the stream. Does the rate of water flowing downstream make a difference to stream life? For example, do some organisms prefer fast-moving water? Are there many plants growing along the banks or in the water, and what type are they? How important to life in the stream is the amount of rock, gravel, or sand on the bottom? How do physical conditions above the water affect conditions in the water? What abiotic and biotic factors may affect life in the stream? Which of these factors might be related to the types of organisms you collected from the stream?

Discussion

1. What were the most common organisms you found? Are they found in the same abundance throughout the stream? If not, why not? What factor do you think is important in controlling the population size of the most common organism? Develop a hypothesis about the types of organisms you found and one factor that might affect their lives and their abundance within the stream. How would you design a study to test your hypothesis?

2. Each team should develop a different hypothesis to study. Return to the stream to conduct your investigation and report your findings to the entire class.

For Further Investigation

Colonization studies can yield additional information about microorganisms in the stream ecosystem. Place a commercial sponge in the streambed. After about two weeks, remove the sponge, place it in a jar, and return it to the lab for examination. Suspend glass microscope slides in the stream, remove them two weeks later, and return them to the lab for staining and examination.

INVESTIGATION 23.3   Effects of Salinity on Aquatic Organisms

This investigation examines the idea of tolerance, discussed in Chapter 20. Tolerance to variations in concentration of dissolved substances is an important characteristic of all aquatic organisms. In a more general way, the investigation is concerned with osmoregulation. Students should be encouraged to relate the observations in this investigation to their work with osmosis and diffusion in Chapter 5. Teams of 2 are recommended.
Time: **One class period**

If you were to move freshwater organisms into the ocean or ocean organisms into fresh water, they probably would die very quickly. Although salmon migrate from the open ocean into freshwater rivers in order to reproduce, they first spend several days in water with decreasing salinity. In this way, their bodies gradually adjust to the lower salt concentrations.

Small freshwater ponds become salty as the summer sun evaporates the water. Small aquatic organisms that cannot tolerate the change either die or go into a dormant stage.

This investigation examines the tolerances of some freshwater organisms to various concentrations of salt solutions.

Review the procedure and construct hypotheses predicting the reaction of your assigned organisms to changes in salinity.

Materials (per team of 2)

2 microscope slides	aged tap water
2 cover slips	living specimens of small aquatic
dropping pipet	organisms in aged tap water and in
compound microscope	1%, 3%, and 5% NaCl solutions
paper towel	*Elodea* or filamentous algae
5% NaCl solution in dropping bottle	

Procedure
PART A   Plants and Salinity

1. Remove an *Elodea* leaf from near the tip of the plant and place it upside down or a microscope slide. Add a drop of aged tap water and a cover slip.

2. Use the low power of your microscope to find a cell you can observe clearly. Describe the cell contents. Does the cell seem to be completely filled?

3. Add a few drops of 5% NaCl solution to one side of the cover slip. Draw the solution under the cover slip by applying a small piece of paper towel to the opposite side of the cover slip. Describe what you observe.

4. Rinse the dropping pipet. Determine if the process is reversible by drawing aged tap water under the cover slip. Record any changes you observe.

Procedure

4. The process can be reversed by flushing away the 5% NaCl with aged tap water.

PART B   Animals and Salinity

5. On a clean slide, place a drop of the freshwater sample assigned by your teacher. Add a cover slip.

6. Use the low power (100X) of your microscope to observe the organisms for a few minutes to determine their normal appearance and actions.

7. Prepare a slide of the organisms in 1% NaCl solution. Observe the organisms and record any difference in movements and shape from the freshwater sample.

8. Repeat Step 7 with the organisms in 3% and 5% NaCl solutions. Which of these solutions is most like ocean water?

9. Replace the 5% NaCl solution with water as you did in Step 4. The organisms may be drawn up by the paper towel. If any organisms remain on the slide, record any changes in movement or shape you observe.

8. The 3%. (Ask students to express the percentages as parts per thousand.)
9. Motile organisms tend to be drawn up by the paper towel when solutions are changed by the method used with *Elodea* or filamentous algae.

Discussion

1. Did all individuals of the species react in the same way? If not, what differences were noted?

2. Did the type of reaction differ with different NaCl solutions? Try to explain any differences.

3. Compare observations of teams that worked on different types of organisms. Which type was most tolerant to changes in NaCl concentration? Which was least tolerant?

Discussion

All the questions depend on the actual observations made. Much variation is to be expected. Have students check each other's observations to help eliminate observational error. In addition, students will carry out the procedure differently— especially with respect to timing—and tolerance among individual organisms will vary. For microscopic freshwater organisms, individual variation may be greater than variation between species. However, if *Artemia* is used it should provide marked contrast to most other species in its ability to tolerate even a 5% salt solution.

4. What type of aquatic habitat do you think each type of organism normally inhabits?
5. Which of your observations support your hypothesis?
6. Do any of your observations weaken your hypothesis?
7. Taking all your observations into account, draw a conclusion that relates your hypothesis to the experiment.

This investigation allows students to observe the effects of detergents and fertilizers on eutrophication. You can increase the complexity (and the level of student interest) by adding more variables to the design of the experiment. Include different temperatures, different pH levels, different amounts of fertilizer (with varying amounts of phosphate), and different low-and high-phosphate detergents. Assign different variables or multiple sets of test tubes to each team, depending on the amount of equipment and space available. While waiting for the cultures to reproduce, place the test tubes in a window that receives indirect sunlight or under a grow light. Teams of 3 work well for the experiment as written, but larger teams are appropriate if each team tests more variables.

Time: One class period for teams to set up their experiments and take the initial readings; a few minutes at the beginning of each class period to take readings for the next one to two weeks.

Safety

See the safety remarks for Investigation 2.1 when working with the LEAP-System equipment.

INVESTIGATION 23.4   Eutrophication

Many ponds and lakes suddenly turn green at various times of the year, fish die, and the water smells bad. This process, known as eutrophication, occurs frequently around areas where fertilizers and/or detergents enter streams that drain into the ponds or lakes. Are the fertilizers and/or detergents responsible for this "greening" effect? Using the LEAP-System™ with a turbidity probe, or density indicator strips if a LEAP-System is not available, you can test the effects of various substances on the growth of euglena, a common pondwater inhabitant. The LEAP-System BSCS Biology Lab Pac 2 includes instructions for use of the system.

After reading the procedure, construct hypotheses that predict the effects of the test substances on the growth of euglena.

Materials (per team of 3)

3 pairs of safety goggles
3 lab aprons
3 18-mm X 150-mm test tubes
dropping pipet
Apple IIe, Apple IIGS, Macintosh, or IBM computer
LEAP-System™
turbidity probe

3 density indicator strips (optional)
3 stoppers to fit test tubes
test-tube rack
glass-marking pencil
aged tap water
aged fertilizer water
aged detergent water
Euglena culture

Put on your safety goggles and lab apron.

Procedure

1. Label the three test tubes *T, D,* and *F.* Fill each of the test tubes about 3/4 full of the appropriate aged water—tap water in the T tube, detergent water in the D tube, and fertilizer water in the F tube. Place all three test tubes in the test-tube rack. (If you are using density indicator strips, be sure to affix a strip in each t tube before filling with water.)

 The fertilizer water may contain chemicals that are irritating to skin. Flush spills or splashes with water.

Procedure

2. **To obtain base-line readings before the experiment starts.**

2. Use the turbidity probe to take readings of each test tube for about 10 seconds. Record the readings in your data book. (Or, record initial readings from the density indicator strips.) What is the purpose of this reading?
3. Thoroughly mix the euglena culture and transfer one full dropping pipet of culture into each of the three labeled test tubes.

4. Securely cork each test tube and invert several times to distribute the *Euglena* evenly. Remove the cork and immediately take turbidity readings as in Step 2. Record the readings in your data book. What is the purpose of this reading?

5. Label the test tubes with your team symbol and store them, uncorked, in the location specified by your teacher.

6. Repeat Step 4 each school day for one or two weeks, or for the time specified by your teacher.

7. Wash your hands thoroughly before leaving the laboratory.

Discussion

1. Account for any differences in turbidity readings between test tubes T, D, and F.

2. Describe any visual differences between the three test tubes.

3. What is the purpose of the test tube labeled *T*?

4. Prepare a graph of the readings you recorded for each test tube. Label the *Y* axis *percent turbidity* and the *X* axis *time*. Summarize the effects of detergent and fertilizer on the growth of euglena.

5. Describe what happens in eutrophication.

6. Why would the ponds on golf courses be especially susceptible to the process of eutrophication?

7. Would farm ponds be susceptible to eutrophication? Explain.

SUMMARY

Humans always have depended on various sources of water to meet their needs. Inland or freshwater ecosystems are classified as standing or flowing waters. Ponds are shallow and smaller than lakes, but phytoplankton are the primary producers in both. Plankton are usually absent in brooks and streams but may be present in slow-moving rivers. Rivers carry sediments and pollutants into the sea. Drainage, dam building, and pollution may have adverse effects on the quantity and quality of water and on the organisms that use it.

The oceans cover about 70 percent of the earth's surface. The concentration of salt and other substances in the oceans remains in a steady state through the action of marine organisms. Temperature, the amount of light available to photosynthetic organisms, and currents also affect ocean waters. The four major zones in the ocean blend together, but each one is different in terms of physical factors and biota. All of the zones are affected by human activities, making ocean pollution one of our most demanding challenges.

Discussion

4. To obtain base-line readings after the euglena have been added and before the experiment starts.

1. *Euglena* are growing more rapidly in the fertilizer and/or detergent water. Results will vary, depending on type of fertilizer or detergent used and other variables included in the investigation.

2. Both fertilizer and detergent water should produce a green or dark green color.

3. Control.

4. Both detergent and fertilizer should enhance the growth and reproduction of euglena.

5. Phosphates enable some *Euglena* to reproduce rapidly, which fills the water with the organisms. Students may infer that large numbers of euglena use up all available oxygen when photosynthesis is not occurring. This is what kills fish.

6. Fertilizers used on golf courses to keep the grass green are washed into ponds by runoff from sprinkler water, which drains into the ponds and enhances eutrophication.

7. Farm ponds can be subject to eutrophization if fertilizers are washed from the soils and drain into the ponds.

APPLICATIONS

1. Why does the concentration of oxygen in a small pond differ from night to day?
2. How would a cloudy, windy day affect the productivity of phytoplankton in a lake?
3. Is a layer of ice on the surface of a lake harmful to fish? Explain.
4. When ponds or lakes have no inlet or outlet, how do plants and animals find their way into these bodies of water?
5. Compare the sources of oxygen in a brook, a slow-moving river, a pond, and a lake.
6. Many rivers overflow their banks almost every year. The Amazon River is one such ecosystem in which fish, during the flood season, are able to swim among the trees growing near the banks of the river. What effect might these fish have on the trees, and what effect might the trees have on the fish?
7. Are decomposers likely to be more active on the bottom of a pond or of a lake? Explain.
8. Why is the deep sea blue? What makes the Red Sea red? What other descriptive names are appropriate, or inappropriate, for bodies of water with which you are familiar?
9. Some people have called the open ocean (the part away from the coastline) a desert. Why?
10. Why is a poisonous substance in a pond—such as a mercury compound—likely to be more concentrated in the flesh of a bass than in the flesh of a ciliate?
11. A program designed to improve the fishing was introduced in a midwestern pond. First, a fish poison was used to kill all the many small fish that were in the pond. Then the pond was restocked with game fish. Instead of large game fish, the new population contained many stunted individuals. Explain.
12. If you collect 167 snails of a given species from five plots totaling 5 m² and find their total mass is 534 g, what would be the biomass of the snail population?
13. What do you think happens to the household cleansers that you and your family wash down the sink? Where do they go, and how much do you throw away per year? Also, what happens to the unused household cleansers and other chemicals that you throw in your garbage?
14. A common saying about the problem of pollution used to be, "Dilution is the solution to pollution." The idea was that people could get rid of wastes simply by dumping them into the ocean or flushing them down the sink with a lot of water. What do you think is the problem with this philosophy?

PROBLEMS

1. Current is one of the most important limiting factors in a stream ecosystem. Stream organisms must be adapted for maintaining a constant position. Cite at least five specific organisms showing five different adaptations to stream currents.
2. Differences in the physical characteristics of water and air are important in understanding the contrast between aquatic and terrestrial environments. Consider such questions as these:
   a. What land organisms have the most streamlined bodies? With what form of locomotion do you associate this streamlining?
   b. What water organisms have the most streamlined bodies? What niches in aquatic ecosystems do these organisms occupy?
   c. Why do most plankton organisms have little or no streamlining?
   d. How does locomotion by walking on the bottom of the ocean differ from locomotion by walking on land?
3. The St. Lawrence River in North America is about 1197 km long. Find out what changes have been made in the river in the last 25 years and what these changes have done to the river's ecosystem.
4. Investigate the effects of the El Nino current.
5. Estuaries, such as Chesapeake Bay and San Francisco Bay, represent a special type of aquatic environment. Find out what characteristics distinguish estuarine environments from marine and inland-water environments and how these affect aquatic life.
6. Investigate the various types of adaptations that allow organisms to survive in the intertidal zone.

7. Though the seas are immense, marine waters can become polluted. Investigate the types of oceanic pollution and their effects on marine ecosystems.

8. What does the damming of a river do to the terrestrial habitats upstream from the dam? What does the damming do to the aquatic and wetland habitats downstream? Study a dam that has been built in your state, or a projected dam to be built, and investigate what impact it had (or may have) on other ecosystems related to the river.

9. What are organizations such as "Ducks Unlimited" doing to help preserve wetlands in the United States? Ask a spokesperson from the group to inform your class.

10. What were the environmental effects of the oil spilled along the coast of Alaska in 1989 and into the Persian Gulf in 1991? What were the predicted effects, and how many of these predictions were realized?

11. What are the most important game fish in the streams and lakes of your area? What do they eat, and where do they live? Create a food web relating the game fish of your area to the food supply, or draw a profile of a lake showing the types of habitats in which you might expect to find different fish species

12. One of the great freshwater ecosystems in the world is the Everglades, found at the tip of Florida. Unfortunately, this community is in great danger of being completely destroyed. What types of plants and animals live in the Everglades, and what makes them unusual? What threatens the existence of the Everglades?

13. Create an intertidal ecosystem in an aquarium. A real challenge is to maintain such organisms as sea anemones, sea cucumbers, and sea urchins in your classroom. Before any organisms are ordered, make certain you have all of the necessary equipment and materials to maintain a healthy environment for them.

Managing Human-Affected Ecosystems

24

What has been planted and what is happening to it? This area was once part of a prairie. As humans modified this ecosystem, many changes took place that require about twice the energy needed for a natural ecosystem. Where does this added energy come from? What might be the benefits of the extra expenditure of energy? What are some of the other costs associated with maintaining this artificial ecosystem? For how many years in the future do you think humans can afford to sustain such an ecosystem? What are some alternatives for the future?

Corn is being harvested. Humans supply the energy in the form of work, fertilizers, and gasoline or diesel fuel to run the machines. In return for this expenditure, humans eat the diet they prefer rather than the products of a natural prairie. Other costs are lowered diversity (the monoculture may not survive as well as a natural ecosystem with more diversity); erosion due to lessened resistance to climatic factors, such as rain, snow, and heat; and the need to add artificial poisons to kill predators and competing plants. The future is conjecture, but as energy costs increase it may ultimately be too expensive to support these human modified ecosystems in the numbers that now exist.

◆ For thousands of years, humans have lived apart from their natural environment. We have dominated the land and other organisms by building houses and roads, cutting trees and clearing natural vegetation for cropland and cities, damming rivers, and polluting the air and water. The consequences of our actions now are all too clear.

Although the outlook is not as bright as we might hope, there still is time to change the course of events. By passing and enforcing tougher laws, developing more energy-efficient practices, and changing our attitudes and personal habits, we can help ensure that our species, and countless others, will continue for generations to come.

Previous chapters have examined the principles of biology and ecology and have shown how human activities can sometimes disrupt natural ecosystems. This chapter summarizes the relationships between human culture, technology, and resources, how this interaction affects ecosystems, and what we can do to lessen our effect on the earth and help manage it in a responsible fashion. ◆

Human Culture and Technology

24.1 Human Culture Has Affected Our Role in the Ecosystem

Early hominids lived in a comparatively simple fashion. If they ate meat, it probably was scavenged from animals killed by other predators. Their main sources of food most likely were berries, fruits, nuts, and roots, which they gathered. Until about 25 000 years ago, humans and their ancestors existed as a stable part of the equilibrium between plants, animals, and other organisms. The size of human groups probably fluctuated according to the amount of resources in the immediate environment—in good years the populations grew, in bad years the populations declined. Figure 24.1 illustrates such a self-regulating population control model.

MAJOR CONCEPTS

- ◆ Our ability to modify our environment causes environmental problems.
- ◆ Nonconservative agricultural practices are a cause of ecological decay.
- ◆ Technology causes depletion of resources and the accumulation of wastes in the environment.
- ◆ Human population growth must be checked.
- ◆ Coordinated international efforts are required for humans to coexist with the environment.
- ◆ Sustainable agriculture must be understood and practiced.
- ◆ There is hope if we change some of our ways.

KNOWLEDGE CHECK

- ◆ What effect did the Agricultural Revolution have on the human population?
- ◆ How did urban areas develop?
- ◆ How has technology affected human evolution?
- ◆ What is sustainable agriculture?
- ◆ What can individuals do to conserve resources and energy?

Guidepost

How has the relationship of humans to our environment changed?

There is evidence that through time, the cranial capacity of hominids has increased from about 400 cm^3 some 5 million years ago to between 1000 and 2000 cm^3 today. This relatively rapid increase in brain size resulted in a tremendously complex brain, which allowed hominids to respond to environmental stresses in a totally new fashion. Previously, if a group was biologically suited to its environment, it survived; if it was not well suited, the group either adapted to meet the conditions of its environment, moved somewhere else more suitable, or became extinct. A complex brain, however, allowed our ancestors to devise other means of coping with environmental stressors, such as using furs to keep warm in colder climates. These responses are cultural not biological, adaptations. Human culture consists of those ideas and behaviors perpetuated through teaching and learning. Probably the most important factor for humans is the use of symbolic language, which provides a means of remembering and transmitting culture.

New ideas for coping with problems and changes required new tools. The earliest technological developments probably came slowly. First using naturally occurring stone flakes as knives, our ancestors learned to make their own simple stone tools. Over time, they improved tools to serve different purposes. The rate of technological change eventually quickened, and cultural developments came more rapidly.

24.2   Agriculture Had Wide-Reaching Effects

Eventually humans learned to change their environment in a fundamental way. One of the most significant changes in human history began about 10 000 years ago when groups of people began to switch from hunting wild animals and gathering other foods to domesticating—taming, herding, and breeding—wild game. They also began domesticating selected wild food plants, consciously planting and harvesting them. Many crop species first were domesticated in areas called centers of origin, including Southeast Asia, the Middle East, Northeast Africa, and northern South America. Most of the plant foods we eat originally came from these centers of origin.

Some plant cultivation may have begun in tropical forest areas. To prepare for planting, early agriculturists cleared small patches of forest by cutting down trees and other vegetation. They allowed the cut vegetation to dry and then burned it. The resulting ashes added plant nutrients to the normally nutrient-poor soils of the tropical forest. This type of agriculture, shown in Figure 24.2, is called slash-and-burn agriculture. After a plot had been cultivated for several years, few if any crops could be grown because of the depleted condition of the soil. When crop yields dropped, growers shifted to a new area of forest and cleared a new plot for planting.

These early growers practiced subsistence agriculture, growing only enough food to feed their families. Because their dependence on human labor and crude stone tools allowed them to cultivate only small plots, they had relatively little impact on their environment.

Large-scale agriculture began about 7000 years ago with the invention of the plow, pulled by domesticated animals and steered by the farmer, as illustrated in Figure 24.3. Animal-drawn plows allowed farmers to break up fertile grassland soils, which could not be cultivated before because of the thick, extensive root systems of grasses. In some arid regions, farmers further increased crop yields by diverting water into hand-dug ditches and canals to irrigate crops. With animal-and irrigation-based agriculture, families usually grew enough crops to survive, and sometimes they had enough left over for sale or for storage.

The gradual shift from hunting and gathering to farming had several major effects. The human population began to increase because of the larger, more stable food supply. Humans cleared increasingly larger areas of land and began to control and shape the surface of the earth to suit their needs. The formation of cities—urbanization—began because a small number of farmers could produce enough food to feed their families as well as a surplus that could be traded to other people. Specialized occupations and long-distance trade developed as people not involved in farming learned new crafts, such as weaving, toolmaking, and pottery. These goods could be exchanged for food. Small villages grew into towns and cities, which served as centers for trade, government, and religion.

About 5500 years ago, the interdependence of farmers and city-dwellers led to the gradual development of a number of agriculture-based urban societies near early agricultural settlements. The growing populations of these emerging civilizations needed more food and more wood for building and for fuel. To meet these needs, more forests were cleared and grasslands were plowed. Such massive land clearing destroyed and degraded many habitats. Poor management of the cleared areas led to increased deforestation, soil erosion, and overgrazing. The gradual degradation of the soil, water, forests, grazing land, and wildlife probably was a factor in the downfall of many great civilizations in the Middle East, North Africa, and the Mediterranean.

Figure 24.2   Slash-and-burn agriculture in the Brazilian rain forest.

The discovery that animals could be put to use apparently never was fully developed in ancient America. Animal muscle was used only for carrying loads. Llamas were used extensively in South America; dogs (travois and sled) in North America. Otherwise, American Indians fully domesticated only turkeys and guinea pigs. Some mammals may have been casually adopted at times.

Figure 24.3   Animal-powered agriculture. Preparing rice fields with water buffalo in Java, Indonesia. Compare its effects on the biosphere with the effects of the types of agriculture shown in Figure 24.4.

Figure 24.4   Industrialized agriculture.

24.3   The Industrial Revolution Increased the Demand for Resources

The next major cultural change, the Industrial Revolution, spread to the United States from England in the 1800s. Goods formerly produced on a small scale by hand now were being produced on a much larger scale by machines. Eventually, locomotives, cars, trucks, tractors, and ships powered by fossil fuels replaced horse-drawn wagons, plows grain harvesters, and wind-driven ships.

Within a few decades, agriculture-based urban societies became early industrial-based urban societies. The growth of industries increased the flow of raw materials into the industrial centers. The increased demand for resources caused the environment of the nonurban areas supplying these resources to be further degraded. Industrialization also produced greater amounts of smoke, ash, garbage, and other wastes in urban areas.

Farm machines powered by fossil fuels, the use of commercial fertil- **2** izers, and new plant-breeding techniques greatly increased the amount of food that could be grown on each acre of land. Fewer people were needed to raise food and, thus, the number of people migrating from rural to urban areas increased. The development of more efficient machines and mass production techniques led to the advanced industrialized societies of today, such as Japan, Germany, and the United States.

The combination of industrialized agriculture (Figure 24.4), increased mining, and urbanization has accelerated the degradation of potentially renewable topsoil, forests and grasslands, and wildlife populations—the same problems that contributed to the downfall of earlier civilizations.

24.4   Human Culture and Technology Have Created Many Problems

Today, the world's population approaches 6 billion. Most of these people will live in urban areas, such as that shown in Figure 24.5, which are ill-equipped to supply them with the necessities of life. Meeting the demands of the growing population has increased economic activity worldwide but also has increased pollution levels, depleted many of the world's resources, and degraded a large portion of the world's farmlands, rangelands, forests, and other nonurban areas. Industrialized countries have assumed a throw-

Figure 24.5   An urban slum. Every day, hundreds of people leave rural areas for cities. What effect does this immigration have on the city?   **Cities cannot provide adequate shelter, water, or services, and there are not enough jobs. Existing problems become compounded.**

away life-style in which they consume the bulk of the world's energy and goods and create the bulk of the world's waste and pollution.

Burning fossil fuels increases the emission of carbon dioxide, sulfur, nitrogen oxide, and other harmful gases. Acid rain, damage to the ozone layer, and other problems created by this pollution now are damaging public health and harming forests, fish, and crops. Increased carbon dioxide levels could cause the climate to warm, thus disrupting agriculture and fisheries and affecting weather patterns. Fewer plant and animal species 3 can survive as favorable habitats are either degraded or destroyed.

Toxic metals and synthetic chemicals from industrial wastes, fertilizers, and pesticides pollute many rivers and lakes and leach into groundwater supplies. Oil spills threaten ocean and coastal habitats. Excessive amounts of sediment from erosion destroy fish-spawning areas. Nutrients 4 entering water supplies with runoff can cause eutrophication, thus threatening fish and other aquatic organisms. Housing developments and industrial developments, along with environmental pollution, degrade and destroy important wetland ecosystems.

The human species still is changing over time, and invisible and immeasurable alterations slowly are taking place. The agents of selection may no longer be climate and disease, but factors such as atmospheric nitrogen oxide and emotional stressors. Humans as a species still depend on their culture to help them adapt and survive, a process called cultural adaptation. This reliance, however, creates a problem: Each new cultural adaptation alters the environment to which we are adapting. New adaptations increase 5 the population, and greater population densities require further adaptations. New adaptations deplete world resources, and resource depletion requires more adaptations. Figure 24.6 shows this feedback cycle. What is vital now is a concerted effort to break out of the cycle and bring about a new stability in population size, in technology, and in the consumption of resources.

Have students investigate the proportion of the world's energy and other resources consumed by the United States. Have them compare these numbers with the proportion of the world's population that lives in the United States.

24.5   All Environmental Issues Are Connected to Population Growth

Environmental degradation occurs slowly, and the early stages are not always obvious. In fact, individuals who stand to profit from exploiting the environment often deny that their projects will degrade it. As the

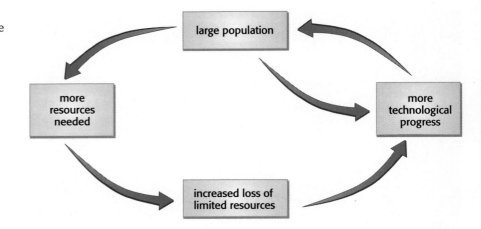

Figure 24.6   The positive feedback cycle of technology, population growth, and resource use. How can the cycle be broken?   **Decreasing population and decreasing resource demand.**

population continues to increase, there is more pressure to take more resources from the limited biotic and abiotic environment. To what extent are we justified in satisfying our needs at the expense of future generations?

Between 1958 and 1962 an estimated 30 million people died from famine in China. China has since made tremendous efforts to feed its people and bring its population growth under control. Between 1972 and 1985, China managed to lower its birthrate from 5.7 children per woman to 2.1 children, and today many experts believe that China's population actually will decline after peaking early in the twenty-first century. To accomplish the drop, China established an extensive and strict population control program, which has since been modified. The program encourages couples to postpone marriage and provides easy and free access for married couples to sterilization, contraception, and abortion. If a couple already has two children, one of the parents is encouraged to be sterilized. Couples who sign and adhere to an agreement to have no more than one child receive economic rewards such as salary bonuses, extra food, larger pensions, better housing, and free medical care and schooling for their child. Although this population control program greatly restricts individual freedom, the unchecked growth of the population would have far worse consequences for the entire population.

Simply reducing the number of children each woman has, however, does not immediately slow population growth. Any country with a large number of people below age 15 has a powerful built-in momentum toward increasing population size unless death rates rise sharply (see Figure 24.7). Because the number of women who can have children increases greatly as the large number of young people reach their reproductive years, the number of births rises even if women have only one or two children. The effects of family planning practices such as having only one or two children are not felt for an entire generation—until the children themselves reach reproductive age.

Most other countries cannot or do not want to employ the major features of China's population program. The best lesson other countries, including the United States, can learn from China's experience is to curb population growth now through family planning and economic incentives and not wait until the choice is between mass starvation and strict coercive measures.

Ask students what effect an increased average life span has on a population. The principal effect is on population structure. Other factors remaining the same, a longer life span increases the proportion of people in older age groups, resulting in problems of economic support. Also, any decrease in the death rate increases the rate of population growth, if natality remains constant.

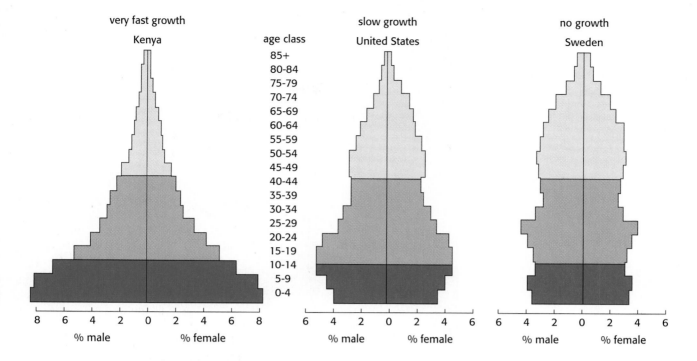

Figure 24.7 Population age structure diagrams for countries with rapid, slow, and zero population growth rates. What does each shaded portion represent? **Darkly shaded portions represent reproductive years (0–14), medium portions represent reproductive years (15–44), and light portions represent postreproductive years (45–58+).**

CONCEPT REVIEW

1. Explain the relationship between culture and technology.
2. What effects did the invention of agriculture have on the human population? How has it affected the environment?
3. What is the relationship between growing population numbers, technology, and resource depletion?
4. How has technology led to increased levels of pollution?
5. Describe the feedback cycle involving population size, resources, and technology.

Change and the Future

24.6   International Cooperation Is Necessary to Solve Modern Problems

One nation alone cannot curb worldwide pollution or ensure the preservation of endangered species. Just as ecological problems fail to stop at state boundaries, so they fail to stop at national boundaries. Air pollution drifts into other countries, and water pollution flows into the oceans.

In April 1986, a nuclear power plant in Chernobyl, in the western part of the Soviet Union, exploded and caught fire (Figure 24.8). The carbon rods used to control the rate of nuclear fission burned for several days, and many deaths and injuries were reported. The effects of the accident were not restricted by national boundaries, however. Large clouds of radioactive particles, released by the fire in the reactor, were carried by wind to Scandinavia and countries in eastern Europe—over 2000 km away.

Soviet and Western medical experts estimate that 5000 to 100 000 people in the Soviet Union and the rest of Europe will die prematurely during the next 70 years from cancer caused by exposure to the radiation

Guidepost

What measures are needed to alleviate the current environmental crisis?

Figure 24.8 The Chernobyl nuclear power plant. This photo was taken May 9, 1986.

Ask students to trace what happens to the chemicals in the rain after it has fallen. They should make the connection with both chemical cycling and the water cycle.

released at Chernobyl. Thousands of others will be afflicted with tumors, cataracts, and sterility. The death toll would have been much higher if the accident had occurred during the day, when people were outdoors, and if the wind had been blowing toward Kiev and its 2.4 million people.

Another problem that has global environmental implications is acid rain, which is carried by winds across national boundaries, as shown in Figure 24.9. Mexican copper smelters produce acid rain that falls on the United States, and U.S. industries produce acid rain that falls on Canada. All three countries must cooperate to solve this problem.

24.7 Technology Can Help Solve Some Problems

As humans we are an integral part of nature, yet we have attempted to "conquer" nature. Rather than trying to conquer nature, we must learn to coexist with it. We have an obligation to our successors and to the offspring of all species to preserve the environment in which we live. The time for correcting our past mistakes is running out. In many parts of the world, biomes have been changed permanently to deserts or urban wastelands, but

Biology Today

Bioremediation

Bioremediation is a process in which certain eubacteria are used to degrade hazardous substances found in the environment. Recently, bioremediation has been used to treat oil spills. Research on this process has been conducted on several oil spills, including the *Exxon Valdez* disaster in Prince William Sound, Alaska. How are eubacteria used treating oil spills?

Biodegradation is a natural process in which erubacteria break down, or consume petroleum and release carbon dioxide. These eubacteria are sometimes called hydrocarbon-oxidizing eubacteria because they destroy petroleum molecules by adding oxygen to them. The newly oxygenated molecules then are further broken down by the eubacteria until only end products remain.

Hydrocarbon-degrading eubacteria occur naturally, although their numbers vary from one environment to another. In Prince William Sound and nearby areas, the community of such degraders contains hundreds of different organisms that act together to degrade the petroleum and its products.

Eubacteria that degrade hydrocarbon require three components to grow: hydrocarbons, oxygen and nutrients. Oxygen is plentiful on the surface layers of the beaches in Prince William Sound because of their contact with the air. Normally, the population of hydrocarbon-degrading eubacteria is limited by the availability of hydrocarbons. Before the *Exxon Valdez* spill, these eubacteria accounted for less than 0.1 percent of the naturally occurring eubacteria found on the beaches. After the oil spill, the availability of hydrocarbons was unlimited, and the hydrocarbon-degrading eubacteria accounted for more than 10 percent of all eubacteria on the beaches. Any further growth in the population of these eubacteria is limited only by the availability of nutrients such as nitrogen and phosphorus.

One way to increase the population of these naturally occurring eubacteria is to add the nutrients as fertilizers. The application of fertilizers to beaches is difficult, however, because tidal and action quickly wash away water-soluble nutrients. In addition, introducing nitrogen and phosphorus into sheltered bays and estuaries might cause algal blooms and eutrophication. To minimize these problems, researchers have developed several special methods of applying fertilizers. One method uses fertilizers that stick to or dissolve in the oil, another uses granules that release the fertilizers over several days or weeks, and a third uses an automatic spray system to apply fertilizers at low tide.

In May 1989, the Environmental Protection Agency and Exxon agreed to test the different fertilizer application strategies on an island in Prince William Sound. After application of the fertilizer developed to stick to the oil, the beaches showed a perceptible improvement within 10 days (see Chapter 11 opening photo). The beaches were monitored during and after the test applications to determine any adverse effects. To date, none have been recorded. Apparently, the numbers of eubacteria return to normal levels after the fertilizer has been used or washed away. Researchers continue to be concerned, however, about the effects of introducing high levels of fertilizers into marine ecosystems.

Ward's Natural Science Establishment offers an oil-degrading microbes kit (catalog #85M3503) that allows students to see microbes grow and degrade petroleum products.

A worker spraying a section of beach with fertilizers to promote the growth of oil-degrading bacteria.

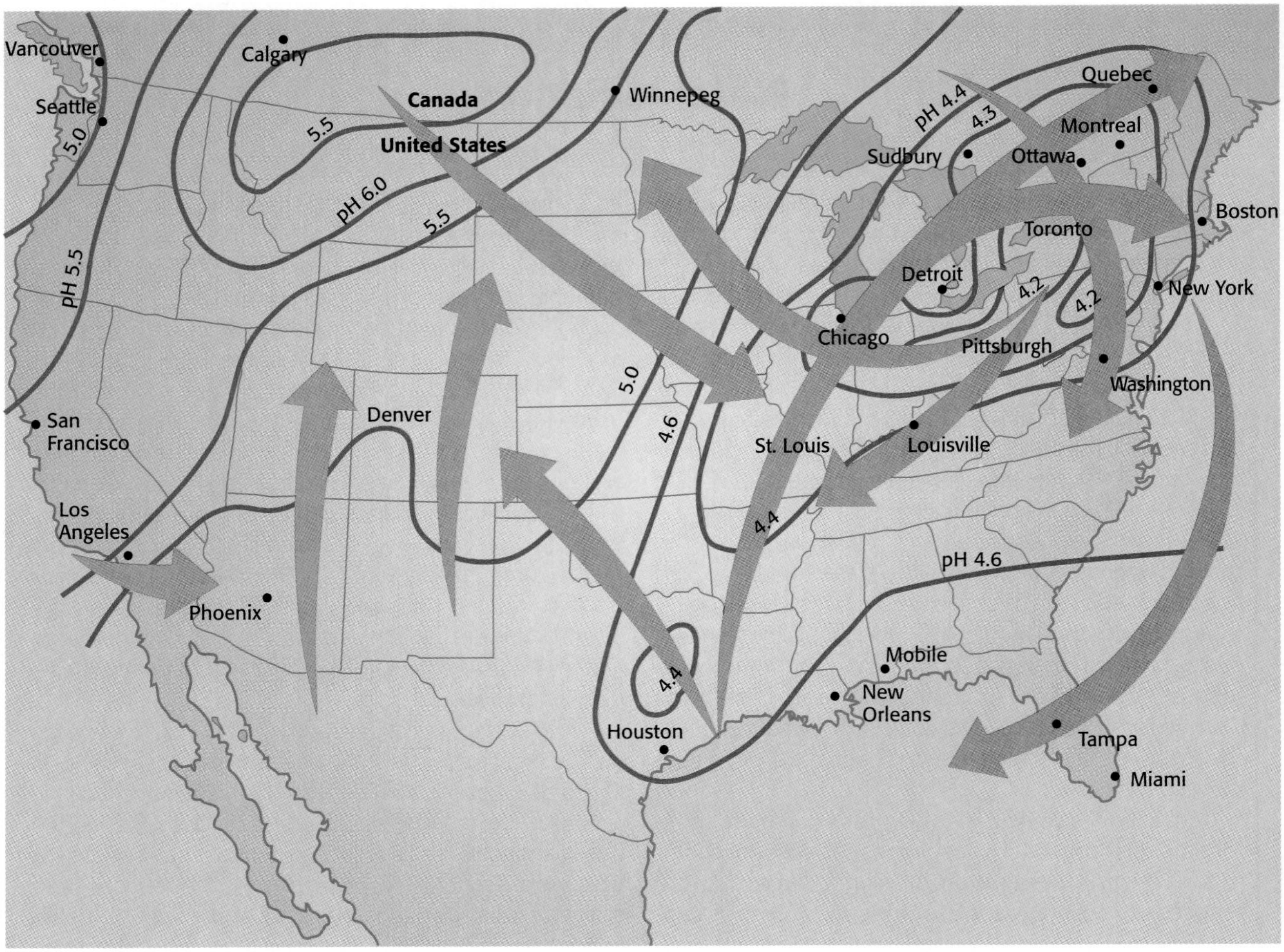

Figure 24.9 Direction traveled by pollutants responsible for acid rain, and the average acidity of rainfall. Recall that low pH figures indicate high acidity. What other factors should be taken into account when considering damage done by acid rain? **The most crucial factor is the ground's capacity to neutralize or buffer the acid rain. In the northeastern United States, rain with high acidity falls on soils with poor buffering capacity; environmental damage results. Alkaline soils of the plains can neutralize the rain.**

there still is some wilderness left, there still is farmland that can be restored, and there still are countless species that share this planet with us. The technology that helped create the current situation can be used to develop solutions to some, if not all, of the problems we now face. It is not too late to take steps to reverse the environmental downward spiral.

Population growth can be controlled by decreasing the fertility rate of women through intensive family planning and increased access to inexpensive or free contraceptives. New birth control devices can prevent pregnancy for up to five years, and other methods, such as male contraceptives currently are under development. Figure 24.10 shows three population predictions for different fertility rates in the United States.

Air pollution can be controlled by regulating population growth, reducing the unnecessary waste of metals, paper, and other resources, increasing energy efficiency, reducing energy use, switching from coal to natural gas, and switching from fossil fuels to energy from the sun, wind and flowing water or nuclear power.

In the past 25 years, the United States has come a long way toward increased energy efficiency. The typical house built today is 35 percent more energy efficient than one built before 1973, and improvements still can be made. Cost-effective conservation measures currently available would reduce the amount of energy, consumed in buildings by 30 to 50 percent. New car mileage now averages about 27 miles per gallon of gasoline—nearly double that of new cars in 1973. New automobile designs with cleaner-burning engines, improved lubricants, electronic emissions

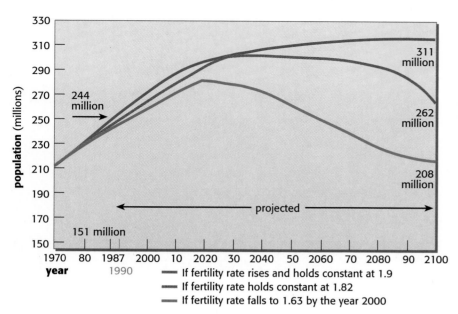

Figure 24.10 Projections of changes in the population size of the United States for different fertility rates.

controls, improved tires, and improved aerodynamics would result in still more energy-efficient cars. Increasing automobile efficiency, however, will help reduce energy use only so far. Alternative modes of transportation also are necessary.

The Clean Air Act of 1990 sets standards for the maximum allowable levels of several key air pollutants. As a result of earlier air pollution control laws, levels of most major air pollutants in the United States either have decreased or have not risen significantly. Much more needs to be done, however, to reduce levels of ozone-destroying chemicals and greenhouse gases. The 1990 Clean Air Act limits certain emissions that cause acid rain, calls for a reduction of toxic air pollutants and tailpipe emissions, and addresses the phasing out of ozone-destroying chemicals by the year 2000. Some experts believe restrictions should be even tighter and the use of some chemicals should be banned altogether.

Have students investigate what measures are being taken in your community to alleviate air and water pollution.

Water pollution can be decreased by decreasing air pollution, by improving erosion controls, farming techniques, and human waste treatment and disposal techniques, and by strictly controlling the disposal of industrial wastes. Runoff from feedlots and barnyards can be collected in detention ponds. The nutrient-rich water from these ponds can be applied to croplands or forests as fertilizer.

Biotechnology will continue to aid us in the future, although bioethical questions surely will follow. Investigation 24.2 examines some of the bioethical issues involved in environmental decisions.

24.8 Sustainable Agriculture Is Important

One way to reduce world hunger and the harmful environmental effects of industrialized agriculture is to develop a variety of sustainable agricultural systems that produce adequate amounts of high-quality food but conserve resources. Such systems combine appropriate aspects of existing agricultural systems with new agricultural techniques to take advantage of local climates, soils, and resources (Figure 24.11). Sustainable agricultural systems do not require large inputs of fossil fuels, and they conserve topsoil, build new topsoil with organic fertilizers, conserve irrigation water, and control pests with little, if any, use of pesticides.

For more information about this topic, see J.P. Reganold, R.I. Papendick, and J.F. Parr, "Sustainable Agriculture," *Scientific American* (June 1990): 112–120.

Figure 24.11 Crops that are plowed under instead of being harvested for sale, as shown here, enrich the soil and improve future crop productivity. Many of these crops contribute nitrogen to soil, thus reducing the need for synthetic nitrogen fertilizers.

Have students investigate the basics of crop rotation and develop additional crop rotation systems.

The primary focus of sustainable agricultural systems is on the longterm condition of the soil, rather than on short-term agricultural productivity. Instead of producing a single type of crop, sustainable systems produce a diverse mix of fruit, vegetable, and fuelwood crops and use organic fertilizers made from animal and crop wastes. Special fast-growing trees planted along with the food crops supply fuelwood and add nitrogen to the soil. Whenever possible, locally available alternatives to fossils fuels—such as wind energy or solar power—are used.

A central component of sustainable agricultural systems is crop rotation—the planned succession of various crops on one field. Rotation provides better weed and insect control, improves nutrient cycling, and thus improves crop yields. A typical seven-season rotation might call for three seasons of planting alfalfa and plowing it back into the soil, followed by four seasons of harvested crops—one of wheat, then one of soybeans, another of wheat, and finally one of oats. The cycle then would start over. The wheat grown the first season would remove some of the nitrogen produced by the alfalfa (a legume). Soybeans, which are nitrogen-fixing legumes also, restore some nitrogen to the soil. Oats are grown at the end of the cycle because they demand fewer nutrients than does wheat. Figure 24.12 illustrates a crop rotation system.

A second important component of sustainable agricultural systems is the regular addition of crop wastes, manures, and other organic materials to the soil. Organic matter improves soil structure, increases its ability to store water, and enhances fertility. Soil in good condition is easier to till, and it allows seedlings to emerge and root more easily. Water readily seeps into rich soil, thus reducing surface runoff and erosion. Organic materials also provide food for earthworms and other soil organisms, which in turn improve the condition of the soil.

In industrialized countries such as the United States, a shift from industrialized to sustainable agriculture is difficult to accomplish. Already, however, 50 000 to 100 000 of the 650 000 full-time farmers in the United States have shifted partly or completely away from conventional industrialized agriculture to forms of sustainable agriculture. Some experts believe that with governmental support of research, subsidies and tax breaks, and training programs for farmers, the shift from industrialized agriculture to sustainable agriculture could be accomplished in 10 to 20 years.

Figure 24.12   Growing combinations of crops can make farms more sustainable by maintaining the productivity of the soil and reducing the farm's reliance on a single crop. On the farm shown here, the parallel strips of land have been planted on the contour of the terrain with oats (yellow), corn, and alfalfa (both green). Within each strip, the crops are rotated on a four-year cycle. Corn is grown for one year and then replaced by an oat crop, which also is grown for one year. The oat crop is then replaced by alfalfa, which is grown for two years.

24.9   Many Measures Require Ethical Decisions

Each environmental issue has unique, local aspects, in addition to biotic, abiotic, and social aspects. Decisions about specific problems cannot be made from a distance or on a broad basis only. They must, in part, be made by those who are close to the situation and who are themselves affected.

Some people do not believe we should send food to countries suffering from famine unless they agree to control their growing populations, thus helping to prevent future famines. In contrast, some people oppose abortion and feel we should not give any financial aid to countries that promote the use of abortion for population control. Do we have the right to interfere with other countries' ethical decisions when those decisions affect the global environment?

There are similar difficult bioethical decisions to be made in other 5 areas. Does a childless couple have the right to use fertility drugs that may produce multiple births? Do we have the right to use medical technology to correct potentially fatal genetic disorders? Do we have the right to use other primates as sources of organs for transplants or as subjects for research that may be painful? Do industrialized countries have the right to maintain high consumption life-styles at the expense of other, less-developed countries? Do we have the right to destroy the habitats of other animals for our own purposes (Figure 24.13)? These are but a few of the issues we need to address if we are to be successful in turning around the current course of events.

Ask students to name other issues that involve ethical decisions. Ask them who should make these decisions and why.

24.10   Managing the Earth Effectively Requires Personal Commitment

Most large-scale measures to combat the global population and pollution crisis require legislation and worldwide cooperation. There are, however, choices that individuals can make now that also will make a difference. To move from causing these environmental problems to correcting them, we must each ask ourselves what type of world we want, what type of world we can get, and what changes we are willing to make to get it.

Technology cannot cure *all* the ailments humans have created, but if we become ecologically informed, choose a simpler life-style by reducing

Stress the importance of a good outlook. Pessimism only ensures that nothing is done and that the situation worsens. A healthy, positive attitude that recognizes the problems and the difficulties involved in solving them will ensure that measures are taken to help.

Figure 24.13 The power of humans to shape our environment may lead to continuing usefulness, or this power may be directed toward waste, depletion, and ruin. Consider the choices shown here and in the world around you.

Ask students to cite additional energy- and water-saving measures. The Teacher's Resource Book contains a detailed list of ways to save water, energy, and money.

our consumption of resources and our production of wastes, and limit the number of children we have, there is hope for the future.

For example, you can improve the energy efficiency of your home in numerous ways. Simple things include replacing incandescent bulbs with fluorescent bulbs, replacing the weather stripping around the doors, and turning down the thermostat on the furnace and water heater. Lessen household water use: Use low-flow shower heads and replace the washers on all leaky faucets. Do not let the water run the entire time you brush your teeth or shave. Turn off lights when you leave the room, and do not leave on outdoor lights all night.

Further measures include riding a bicycle, walking, or using public transportation instead of driving. If you have to drive, use a car that gets 6 good gas mileage and carpool whenever possible. In the store, do not buy overpackaged and overprocessed products. Eat lower on the food chain by consuming more fruit, vegetables, and whole grains, and less meat and dairy products. Recycle paper, aluminum, steel, glass, and plastic. Try to distinguish between needs and wants before you purchase any item.

Each of these small acts constitutes a personal economic and political decision that, when combined with those of many others, can lead to largescale changes with profound effects on the earth. Remember that all organisms on the earth, including ourselves, depend on many types of energy flows and chemical cycles that are affected by human activity. These life-sustaining processes and cycles are a beautiful and diverse web of interrelationships—the web of life.

CONCEPT REVIEW

1. Explain why countries must cooperate to solve environmental problems. Give examples.

2. How can technology help solve some of the environmental problems humans have created?

3. How does sustainable agriculture differ from present agricultural practices?

4. What measures can be taken to control population growth?

5. How can ethics affect environmental decisions? Give examples.

6. What steps can individuals take to help correct environmental problems?

INVESTIGATION 24.1 Views of the Earth from Afar

This investigation acquaints students with some of the technology used to study current environmental problems. Students should work alone.
Time: One class period.

Aerial photos give the "big picture" of events in an area. Infrared film and computer imaging disclose events invisible to the naked eye. Infrared film is sensitive to heat. Consequently, objects giving off heat appear dark in black-and-white photos or show up as a different color in color photos. Some color prints show warmer areas as green; others show them as red. In either case, the contrast with cooler areas is striking.

This investigation uses infrared photos and other modern technology to examine a current environmental problem.

Materials (per team of 1)

Procedure

1. Examine the data in Table 24.1, which were obtained from studies of "old air" in tightly sealed containers. These data were collected in Hawaii, and similar data have been collected in other areas. Between 1958 and 1990, by what percentage did the proportion of carbon dioxide in Hawaii's air change and in which direction did it change? If this trend is typical of the industrial United States, by what percentage has the atmospheric level of carbon dioxide increased in the last 100 years?

2. Study the aerial photo of the Vermont forest (Figure 24.14). In this photo, healthy trees appear green and damaged trees appear red. In areas west of this forest, industries are producing acid rain and other forms of air pollution.

3. Study the maps shown in Figure 24.15, which were produced by a computer from satellite data. What color has been used for your area for January? For July? What do these colors indicate about temperatures?

Procedure

1. The level has increased by 12.06 percent in 32 years. If this increase is typical, it has increased by at least 36 percent in 100 years.

3. Depends on locality. Redder areas are hotter.

Figure 24.14

Table 24.1 Change in Atmospheric CO_2 in Hawaii

Year	Average Annual CO_2 Level (ppm)	Year	Average Annual CO_2 Level (ppm)
1958	315	1976	333
1960	317	1978	335
1962	318	1980	337
1964	319	1982	340
1966	321	1984	343
1968	322	1986	346
1970	324	1988	350
1972	326	1990	353
1974	311		

MEAN SURFACE TEMPERATURE FOR JANUARY 1979
USING HIRS 2 AND MSU DATA

Degrees Kelvin

CHAHINE   SUSSKIND
JPL   GSFC
(1984)

243   253   263   273   283   293   303   313

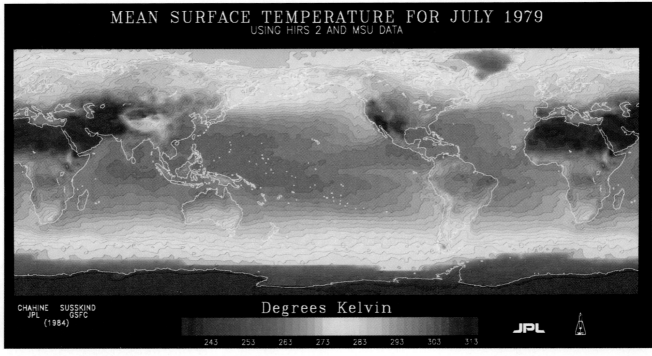

MEAN SURFACE TEMPERATURE FOR JULY 1979
USING HIRS 2 AND MSU DATA

Degrees Kelvin

CHAHINE   SUSSKIND
JPL   GSFC
(1984)

243   253   263   273   283   293   303   313

Figure 24.15

Discussion

1. Increase.
2. By causing deforestation, acid rain lowers the use of carbon dioxide by plants; the gas then increases in concentration in the atmosphere. Carbon-containing air pollutants also increase carbon dioxide levels.
3. A positive correlation. Although less sunlight may penetrate polluted air containing particular matter, thus lowering the earth's temperature, the gradual rise in the earth's

Discussion

1. What is the general trend in atmospheric concentration of carbon dioxide from 1958 to 1990?

2 How might this trend be related to acid rain? To air pollution in general?

3. During the last 100 years, measurements taken around the world show that the mean global temperature has risen 0.4° C. What is the relationship between the level of atmospheric carbon dioxide and temperature?

4. If the trend represented in Table 24.1 continues, what colors might be used in a map of your area 100 years from now? How might the food web in your area change as a result of a change in temperature?

5. What cause-and-effect relationship, if any, is demonstrated here?

5. None. The correlation is interesting and supports the hypothesis that a rise in carbon dioxide has brought about the rise in temperature, but more research is needed. Emphasize to students that correlation does not equal causation, although it may help to support a hypothesis.

INVESTIGATION 24.2   Decisions About Land Use

Land is a finite resource necessary to support human life. Because many interests compete for the limited available land, the problems involved in deciding how land will be used are complex. Land use decisions involve not only biology and technology but also political, economic, legal, and attitudinal problems. The solutions often require tradeoffs—compromises of values or life-styles in exchange for a portion of the available land.

Before a community land use decision is made, the public should be given the opportunity to express opinions and offer information or insights. This investigation simulates a real-life situation and asks you to evaluate different land use proposals in a fictitious case.

Materials (per team of 10)

set of role cards   map

Procedure

1. Read the following background information for a situation involving proposed land uses:

Winterville is a growing city, located somewhere in the United States, that has just annexed 2 square miles of land. Although Winterville owns the land, it may be sold or used for any purpose the city wishes.

Ten different groups are interested in what happens to this land. Speaking for some of these interest groups are a farmer, a college student, a single parent, a retired citizen, and an industrialist. Other groups are represented by a laborer, an environmentalist, a construction worker, a developer, and a city manager.

Figure 24.16 is a topographic map of the annexed land. The lake on the annexed land is unpolluted and is one source of irrigation and drinking water

temperature since 1860 indicates to many scientists that carbon dioxide has produced a green house effect that has raised the temperature.

4. The color might be slightly redder if the mean temperature rose another 0.4°C. Local conditions might be more severe.

This investigation asks students to decide a hypothetical land use issue. Students first express individual concerns and then, as a team, decide on the final use of the land. The activity asks students not only to draw from their pool of information about land and its uses but also to think about the dynamics of land use policy making and ethics. Many students will be exposed to ideas that differ from their own.

Time; **Two class periods.**

Figure 24.16

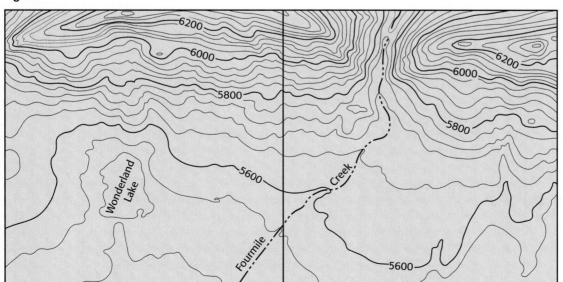

Table 24.2 Age Structure

Age	Number
Under 5	14 000
5–9	12 000
10–14	10 000
15–19	11 000
20–24	10 000
25–34	13 000
35–44	8000
46–65	15 000
Over 65	7000

Procedure

2. **Each student should have a role card. If this is not possible, have two or three students work together to play each role. Have each of the 10 students pick a card at random; and explain that it represents an individual from the city of Winterville whose role they must assume. Each individual has certain concerns and interests, and each has a different use in mind for the land the city has annexed. Be certain each student takes particular notice of the proposed land use stated on the card.**
3. **Give the students at least 15 minutes to think of all the positive consequences and arguments they can for their proposed land use. Point out that although their personal opinions are important, they must defend the proposed land use on their card, even if they do not agree with it.**

for the city. The present sewage treatment plant and sewer lines of Winterville are working at their top capacity. The garbage dump is only large enough to handle the garbage produced by the current population.

The population of Winterville is 100 000, and the growth rate during the last 10 years was 10 percent. Projections for the future show a similar growth rate for the next 10 years. The birthrate is 20 births per 1000 people per year. Table 24.2 shows the present age structure of the community.

Average per capita income per household for the city is $25 000, and there are approximately 40 000 jobs in the city. The unemployment rate is 6 percent; that is, 6000 people of working age need jobs. Table 24.3 shows the income distribution of the city.

There are 18 elementary schools, each of which can accommodate up to 700 students. Each of the 6 junior high schools can serve 1000 students. The 3 high schools have room for 3000 students each.

2. In your assigned team, select a role card at random and read about the individual described on your card. Take particular notice of the proposed land use this individual wants. From this person's point of view, think of arguments and consequences that will persuade the others in your team to support your cause. The total amount of land needed for each use is shown on the role card, but the total area of all the proposed land uses is much greater than the 1280 acres available for development (see Table 24.4). This means some of the proposed land uses cannot be accommodated or will have to be reduced greatly in size.
3. List the five most important positive consequences of the land use you have proposed. Some relevant ideas you may want to consider are listed in Table 24.5. Remember, you have assumed the role of the person on your card and are speaking from that position.
4. Discuss the concerns of each member of your team, one at a time. Record in your data book each person's occupation, the use of the land he or she proposes, and the five positive consequences he or she claims will result.
5. After each person has presented his or her proposal, evaluate each of the five arguments, using the values analysis scale shown in Figure 24.17. Assign only positive scores. Do not evaluate your own arguments.
6. Total the five positive scores for each proposal and compare your evaluations with those of the other team members.
7. Form a small group with other members of your team who share similar opinions and values concerning the use of the annexed land. If you feel you have nothing in common with the other citizens, continue to act as an individual.

Table 24.3 Family Income Distribution

Annual Income	Percent of Population
Less than $10 000	10
$10 000 to 15 000	11
$15 000 to 20 000	12
$25 000 to 30 000	27
$30 000 to 35 000	12
More than $35 000	6

Table 24.4 Proposed Land Uses

Proposed Land Use	Amount of Land Needed (acres)
Shoe manufacturing plant	40
Low-cost housing	40
Dump and sewage treatment plant	10
Luxury homes	640
Park	320
Shopping center	40
Open space	1280
School	20
Houses and condominiums	240
Farm	240
Total	2870
Total land available	1280

Table 24.5 Land Use Considerations

Will the Proposed Land Use Have an Effect on Individuals?
1. Will it change individual life-styles? Will it cause people to live in different ways? Will it allow them to do something that they want to do? Will it force people to do something that they do not want to do, or to stop doing something they currently do?
2. Will it change individual financial situations? How will the proposed land use be financed? Will it affect the cost of goods and services? Will it affect taxes? Will the action affect people's jobs—create new jobs or eliminate current jobs?
3. Will it change the health of individuals? Will the land use have a foreseeable positive or negative effect on the physical or mental health of individuals in the area? For example, will it disrupt the social structure of a neighborhood?

What Are the Consequences for the Whole Community?
1. Will the action conflict with the land's economic (agricultural, mining, industrial) use or recreation use? Will the action conflict with the land's aesthetic value?
2. Is the land especially well-suited for one use and less appropriate for other uses?
3. Can the local government provide necessary services: water, sewers, electrical and gas connections, waste disposal, transportation, police and fire protection, and schools?
4. What about conflicting interests? Will the action satisfy the wants or needs of only a small part of the community's population?

How Will Various Parts of the Natural Environment Be Affected?
1. Does the area have unique geological, biological, archeological, or historical features that should be preserved?
2. What effect will the action have on vegetation, wildlife and wildlife habitats?
3. What effect will the action have on temperature, climate, or air quality?
4. What effect will the action have on water quality?
5. Are there natural hazards in the area that should be considered? These may include floods, types of soil, erosion, landslides or snow avalanches, sinking or settling of the earth, earthquakes, and high winds or other dangers due to extreme weather.

8. In your small groups, discuss the negative consequences of all the land proposals your group is against. List the five most serious negative consequences for each proposal your group is against.
9. Record the negative consequences and evaluate them as you did in Step 6, but assign only negative scores. Total the negative scores for each land use.
10. Add your negative evaluation score to your positive evaluation score for each land use. Compare your results with those of the others in your team.
11. As a team, decide which of the proposals to accept. You may eliminate any of the land uses you wish, or you may keep all of them. Remember, however, that the total area of land available for use is limited and that your final decision must be made as a team.

8. Remind students to remain in their assigned roles.

Figure 24.17

−10	−9	−8	−7	−6	−5	−4	−3	−2	−1	0	+1	+2	+3	+4	+5	+6	+7	+8	+9	+10
very negative				negative				slightly negative		don't know		slightly positive			positive			very positive		

12. Remind students to keep the topography of the land in mind. If they select a farm, for example, it probably should not be placed in the steep hills. Or, if they select a garbage dump and a school, these probably should not be located next to one another. Each student can develop a map using the land uses the team has decided to accept.

Discussion

1-4. Answers will vary depending on the outcomes of the teams and each student's personal opinions.

12. Using a large copy of Figure 24.16, identify where each land use will be located. Keep the following information in mind as you locate each item: (a) the area required for each land use (the area given includes roads that must be built for access); (b) the location of each use in relation to the physical aspects of the land; and (c) the location of each use in relation to the other uses.

Discussion

1. How does your team's decision on land use compare with the other teams'?

2. Compare your completed map with those of the other teams. How do they differ? Do you agree with the placement of land uses on other maps?

3. Which land use map do you think is best? Why?

4. What consequences were different? Did you overlook any consequences that others included? Did you include consequences that others did not think of?

5. As a class, develop a master plan for the land based on input from all the teams.

For Further Investigation

Invite someone from the city planning department to talk to your class about how land use decisions are made in your community.

SUMMARY

The earliest humans changed the biosphere little, living within it much like other animals. With the development of culture and technology, however, our ancestors began adapting environments to make them more hospitable to themselves. The advent of agriculture and industry greatly increased the size of the human population and the impact of humans on their surroundings.

Today, the exploding population and the subsequent demand for resources have created severe environmental problems that threaten life. Changes in governmental policies, industrial practices, and life-styles must be implemented soon to slow the environmental decline. Shifting agricultural practices away from industrialized systems to sustainable systems will provide food without degrading the soil and depleting resources. Using our knowledge of ecology and the environment, we can help manage the earth in a responsible manner. Humans are a small but powerful part of the web of life. We have the chance to help determine the future. By learning about ecology and caring about the earth and the living things on it we can help bring about the rebirth of our planet.

APPLICATIONS

1. What are the major disadvantages of the life of a hunter-gatherer?

2. Many hunter-gatherers were semi-nomadic people—they settled in one place for a while and then moved to another. Why do you think hunter-gatherers moved from one location to another within a broad territory?

3. Do you think agriculture developed in one part of the world and then spread to other parts, or do you think it developed independently in many parts of the world? What evidence might support your position?

4. Why is the use of symbolic language by humans considered to be so important?

5. Very few plants and animals have been domesticated by humans compared with the great number of species in the wild. What do you think are important characteristics of plants and animals that have been domesticated?

6. Does the United States have a population problem? Support your answer.

7. In what ways have automobiles contributed to the human impact on the biosphere?

8. A population map of the United States would show large areas that are barely inhabited. If someone argued that the map showed there is no population problem here, what arguments might you present for another conclusion?

9. How does annual crop rotation—the alternation of crops such as corn with leguminous plants such as soybeans aid—the soil?

10. Do you think the United States should send food to starving children in other countries? Support your answer.

11. Certain Native American people believe that no human has a right to "own" land and to do with it what he or she wants. Rather, everyone has a right to use the land, and people merely are caretakers or stewards who have the responsibility of preserving the land for use by future generations. How does this philosophy differ from that of most Americans? Is this an important or realistic philosophy for America? Why or why not?

12. If Americans were to eat more corn muffins and less corn-fed beef, what would be the effect of such a change on corn consumption and production?

PROBLEMS

1. Suppose you are among the first colonizers on the moon. You must build a habitat, large enough for 20 people, in which you can live for several months. What environmental factors should you take into consideration in your building design, and what would your habitat look like?

2. Make a list of the plants you eat during a week. Find out from where the ancestors of these food plants originally came. How many of them are from what is now the United States?

3. Demographers are concerned not only with the total numbers of persons but also with the *structure* of populations—the relative numbers of individuals of various types. For example two populations of the same size may have different proportions of males and females. Or, two populations of the same size may have different proportions of children and adults. Such data often provide useful information about a population.

   a. The total populations of the United States and Sri Lanka are quite different. More important, the United States' population has a smaller proportion of children than that of Sri Lanka. What hypotheses about the two countries can you suggest on the strength of this information?

   b. We can divide the population of the United States into three age groups: (1) persons under 20, most of whom are not selfsupporting; (2) persons 20 to 65, most of whom are working; (3) persons over 65 most of whom are retired. In recent years the first and third groups have been increasing more rapidly than the second group. What hypotheses can you suggest to explain this? Can you see a future economic problem in this situation?

   c. In human females, reproduction occurs mostly between the ages of 15 and 45. Suppose this age group increases more slowly than the age group over 45 but the number of children per female remains the same. What will happen to the birthrate in the population, expressed as births per 1000 of population?

   d. The average age at which a female has her first child is higher in nation A than in nation B. The average age at death is about the same in both nations. From this information, make a guess about the rate of population growth in the two countries. What additional information would make your guess more reliable?

4. If enough nuclear power plants were built, they could provide all of the energy the United States needs. The U.S. public, however, has been reluctant to support the building of such plants. Is there a place for nuclear energy in the United States of the future? What are the advantages and disadvantages of nuclear power plants?

5. Suppose the United States was not able to import any natural resources from other countries. What effect would this have on our economy? What products would not be made, and what activities would you have to stop?

6. Research what changes have taken place during the last 50 years in the farming industry, and report to the class.

7. What can you do to help reduce the amount of resources you consume and the amount of wastes you produce each year? What can you, your family, your school, your city, and your state do? Make a list of suggestions, and carry out as many as possible.

8. The term *ecology* has become a household word, but it is often used as if it were a synonym for *environment*. Sometimes it is used merely as a vague indication of something good. How would you explain the scientific meaning of the word *ecology* to a person who has never studied biology?

9. Identify a local environmental problem in your community. Research the biotic, abiotic, and social aspects of the problem.

General Procedures
for the Laboratory

The laboratory is a scientist's workshop—the place where ideas are tested. In the laboratory portion of this course, you will see evidence that supports major biological concepts. To pursue your investigations effectively, you need to learn certain basic techniques, including safe laboratory practices, record keeping, report writing, and measurement. The information on the following pages will help you acquire these techniques.

Laboratory Safety

The laboratory has the potential to be either a safe place or a dangerous place. The difference depends on your knowledge of and adherence to safe laboratory practices. It is important that you read the information here and learn how to recognize and avoid potentially hazardous situations. Basic rules for working safely in the laboratory include the following:

1. Be prepared. Study the assigned investigation before you come to class. Resolve any questions about the procedures before you begin to work.

2. Be organized. Arrange the materials needed for the investigation in an orderly fashion.

3. Maintain a clean, open work area, free of everything except those materials necessary for the assigned investigation. Store books, backpacks, and purses out of the way. Keep laboratory materials away from the edge of the work surface.

4. Tie back long hair and remove dangling jewelry. Roll up long sleeves and tuck long neckties into your shirt. Do not wear loose-fitting sleeves or open-toed shoes in the laboratory.

5. Wear safety goggles and a lab apron whenever working with chemicals, hot liquids, lab burners, hot plates, or apparatus that could break or shatter. Wear protective gloves when working with preserved specimens, toxic and corrosive chemicals, or when otherwise directed to do so.

6. Never wear contact lenses while conducting any experiment involving chemicals. If you must wear them (by a physician's order), inform your teacher *prior* to conducting any experiment involving chemicals.

7. Never use direct or reflected sunlight to illuminate your microscope or any other optical device. Direct or reflected sunlight can cause serious damage to your retina.

8. Keep your hands away from the sharp or pointed ends of equipment such as scalpels, dissecting needles, or scissors.

9. Observe all cautions in the procedural steps of the investigation. **CAUTION, WARNING,** and **DANGER** are signal words used in the text and on labeled chemicals or reagents that tell you about the potential for harm and/or injury. They remind you to observe specific safety practices. ***Always read and follow these statements.*** They are meant to help keep you and your fellow students safe.

   CAUTION statements advise you that the material or procedure has *some potential risk* of harm or injury if directions are not followed.

   WARNING statements advise you that the material or procedure has a *moderate risk* of harm or injury if directions are not followed.

   DANGER statements advise you that the material or procedure has a *high risk* of harm or injury if directions are not followed.

If an ophthalmologist approves a student's use of contact lenses, have the student wear cup-type safety goggles. Supervise carefully.

safety goggles
Safety goggles are for eye protection. Wear goggles whenever you see this symbol. If you wear glasses, be sure the goggles fit comfortably over them. In case of splashes into the eye, flush the eye (including under the lid) at an eyewash station for 15 to 20 minutes. If you wear contact lenses, remove them *immediately* and flush the eye as directed. Call your teacher.

reactive
These chemicals are capable of reacting with any other substance, including water, and can cause a violent reaction. *Do not* mix a reactive chemical with any other substance, including water, unless directed to do so by your teacher. Wear your safety goggles, lab apron, and protective gloves.

lab apron
A lab apron is intended to protect your clothing. Whenever you see this symbol, put on your apron and tie it securely behind you. If you spill any substance on your clothing, call your teacher.

corrosive
A corrosive substance injures or destroys body tissue on contact by direct chemical action. When handling any corrosive substance, wear safety goggles, lab apron, and protective gloves. In case of contact with a corrosive material, *immediately* flush the affected area with water and call your teacher.

gloves
Wear gloves when you see this symbol or whenever your teacher directs you to do so. Wear them when using *any* chemical or reagent solution. Do not wear your gloves for an extended period of time.

flammable
A flammable substance is any material capable of igniting under certain conditions. Do not bring flammable materials into contact with open flames or near heat sources unless instructed to do so by your teacher. Remember that flammable liquids give off vapors that can be ignited by a nearby heat source. Should a fire occur, *do not* attempt to extinguish it yourself. Call your teacher. Wear safety goggles, lab apron, and protective gloves whenever handling a flammable substance.

sharp object
Sharp objects can cause injury, either a cut or a puncture. Handle all sharp objects with caution and use them only as your teacher instructs you. *Do not* use them for any purpose other than the intended one. If you do get a cut or puncture, call your teacher and get first aid.

poison
Poisons can cause injury by direct action within a body system through direct contact (skin), inhalation, ingestion, or penetration. *Always* wear safety goggles, lab apron, and protective gloves when handling any material with this label. Before handling any poison, inform your teacher if you have any preexisting injuries to your skin. In case of contact, call your teacher *immediately.*

irritant
An irritant is any substance that, on contact, can cause reddening of living tissue. Wear safety goggles, lab apron, and protective gloves when handling any irritating chemical. In case of contact, flush the affected area with soap and water for at least 15 minutes and call your teacher. Remove contaminated clothing.

biohazard
Any biological substance that can cause infection through exposure is a biohazard. Before handling any material so labeled, review your teacher's specific instructions. *Do not* handle in any manner other than as instructed. Wear safety goggles, lab apron, and protective gloves. Any contact with a biohazard should be reported to your teacher *immediately.*

Figure A1.1

10. Become familiar with the caution symbols identified in Figure A1.1.

11. Never put anything in your mouth and never touch or taste substances in the laboratory unless specifically instructed to do so by your teacher.

12. Never smell substances in the laboratory without specific instructions. Even then, do not inhale fumes directly; wave the air above the substance toward your nose and sniff carefully.

13. Never eat, drink, chew gum, or apply cosmetics in the laboratory. Do not store food or beverages in the lab area.

14. Know the location of all safety equipment and learn how to use each piece of equipment.

15. If you witness an unsafe incident, an accident, or a chemical spill, report it to your teacher immediately.

16. Use materials only from containers labeled with the name of the chemical and the precautions to be used. Become familiar with the safety precautions for each chemical by reading the label before use.

17. When diluting acid with water, ***always add acid to water.***

18. Never return unused chemicals to the stock bottles. Do not put any object into a chemical bottle, except the dropper with which it may be equipped.

19. Clean up thoroughly. Dispose of chemicals and wash used glassware and instruments according to the teacher's instructions. Clean tables and sinks. Put away all equipment and supplies. Make sure all water, gas jets, burners, and electrical appliances are turned off. Return all laboratory materials and equipment to their proper places.

20. Wash your hands thoroughly after handling any living organisms or hazardous materials and before leaving the laboratory.

21. Never perform unauthorized experiments. Do only those experiments assigned by your teacher.

22. Never work alone in the laboratory, and never work without the teacher's supervision.

23. Approach laboratory work with maturity. Never run, push, or engage in horseplay or practical jokes of any type in the laboratory. Use laboratory materials and equipment only as directed.

In addition to observing these general safety precautions, you need to know about some specific categories of safety. Before you do any laboratory work, familiarize yourself with the following precautions.

Heat

1. Use only the source of heat specified in the investigation.

2. Never allow flammable materials, such as alcohol, near a flame or any other source of ignition.

3. When heating a substance in a test tube, point the mouth of the tube away from other students and yourself.

4. Never leave a lighted lab burner, hot plate, or any other hot object unattended.

5. Never reach over an exposed flame or other heat source.

6. Use tongs, test-tube clamps, insulated gloves, or potholders to handle hot equipment.

Glassware

1. Never use cracked or chipped glassware.

2. Use caution and proper equipment when handling hot glassware; remember that hot glass looks the same as cool glass.

3. Make sure glassware is clean before you use it and clean when you store it.

4. When putting glass tubing into a rubber stopper, moisten the tubing and the stopper. When putting glass tubing into or removing it from a rubber stopper, protect your hands with heavy cloth. Never force or twist the tubing.

5. Broken glassware should be swept up immediately (never picked up with your fingers) and discarded in a special, labeled container for broken glass.

Electrical Equipment and Other Apparatus

1. Before you begin any work, learn how to use each piece of apparatus safely and correctly in order to obtain accurate scientific information.
2. Never use equipment with frayed insulation or with loose or broken wires.
3. Make sure the area under and around electrical equipment is dry and free of flammable materials. Never touch electrical equipment with wet hands.
4. Turn off all power switches before plugging an appliance into an outlet. Never jerk wires from outlets or pull appliance plugs out by the wire.

Living and Preserved Specimens

1. Specimens for dissection should be properly mounted and supported. Do not cut a specimen while holding it in your hand.
2. Wash down your work surface with a disinfectant solution both before and after using live microorganisms.
3. Always wash your hands with soap and water after working with live or preserved specimens.
4. Animals must be cared for humanely. General rules are as follows:
    a. Always follow carefully the teacher's instructions concerning the care of laboratory animals.
    b. Provide a suitable escape-proof container in a location where the animal will not be constantly disturbed.
    c. Keep the container clean. Cages of small birds and mammals should be cleaned daily. Provide proper ventilation, light, and temperature.
    d. Provide water at all times.
    e. Feed regularly, depending on the animal's needs.
    f. Treat laboratory animals gently and with kindness in all situations.
    g. If you are responsible for the regular care of any animals, be sure to make arrangements for weekends, holidays, and vacations.
    h. When animals must be disposed of or released, your teacher will provide a suitable method.
5. Many plants or plant parts are poisonous. Work only with the plants specified by your teacher. Never put any plant or plant parts in your mouth.
6. Handle plants carefully and gently. Most plants must have light, soil, and water, although requirements differ.

Accident Procedures

1. Report *all* incidents, accidents, and injuries, and all breakage and spills, no matter how minor, to the teacher.
2. If a chemical spills on your skin or clothing, wash it off immediately with plenty of water and notify the teacher.
3. If a chemical gets into your eyes or on your face, wash immediately at the eyewash fountain with plenty of water. Wash for at least 15 minutes, flushing the eyes, including under each eyelid. Have a classmate notify the teacher.
4. If a chemical spills on the floor or work surface, do not clean it up yourself. Notify the teacher immediately.
5. If a thermometer breaks, do not touch the broken pieces with your bare hands. Notify the teacher immediately.
6. Smother small fires with a wet towel. Use a blanket or the safety shower to extinguish clothing fires. Always notify the teacher.
7. All cuts and abrasions received in the laboratory, no matter how small, should be reported to your teacher.

Record Keeping

Science deals with verifiable observations. No one—not even the original observer—can trust the accuracy of a confusing, indefinite, or incomplete observation. Scientific record keeping requires clear and accurate records made **at the time of observation.**

The best method of keeping records is to jot them down in a data book. It should be a hardcover book, permanently bound (not loose-leaf), and preferably with square-grid (graph-paper type) pages.

Keep records in diary form, recording the date first. Keep observations of two or more investigations separate. Data recorded in words should be brief but to the point. Complete sentences are not necessary, but single words are seldom descriptive enough to represent accurately what you have observed.

You may choose to sketch your observations. A drawing often records an observation more easily, completely, and accurately than words can. Your sketches need not be works of art. Their success depends on your ability to observe, not on your artistic talent. Keep the drawings simple, use a hard pencil, and include clearly written labels.

Data recorded numerically as counts or measurements should include the units in which the measurements are made. Often numerical data are most easily recorded in the form of a table.

Do not record your data on other papers to be copied into the data book later. Doing so might increase neatness, but it will decrease accuracy. Your data book is **your** record, regardless of the blots and stains that are a normal circumstance of field and laboratory work.

You will do much of your laboratory work as a member of a team. Your data book, therefore, will sometimes contain data contributed by other team members. Keep track of the source of observations by circling (or recording in a different color) the data reported by others.

Writing Laboratory Reports

Discoveries become a part of science only when they are known to others. Communication, therefore, is important. In writing, scientists must express themselves so clearly that a person can repeat their procedures exactly. The reader must know what material was used (in biology this includes the species of organism) and must be able to understand every detail of the work. Scientific reports frequently are written in a standard form as follows:

1. Title
2. Introduction: a statement of how the problem arose, often including a summary of past work
3. Materials and equipment: usually a list of all equipment, chemicals, specimens, and other materials
4. Procedure: a complete and exact account of the methods used in gathering data
5. Results: data obtained from the procedure, often in the form of tables and graphs
6. Discussion: a section that demonstrates the relationship between the data and the purpose of the work
7. Conclusion: a summary of the meaning of the results, often suggesting further work that might be done
8. References: published scientific reports and papers that you have mentioned specifically

Your teacher will tell you what form your laboratory reports should take for this course. Part of your work may include written answers to the discussion questions at the end of each investigation. In any event, the material in your data book will be the basis for your reports.

Measurement

Measurement in science is made using the *Systeme Internationale d 'Unites* (International System of Units), more commonly referred to as "SI." A modification of the older metric system, SI was used first in France and is now the common system of measurement throughout the world.

Among the basic units of SI measurement are the meter (length), the kilogram (mass), the kelvin (temperature), and the second (time). All other SI units are derived from these four. Units of temperature that you will use in this course are degrees Celsius, which are equal to kelvins.

SI units for volume are based on meters cubed. In addition, you will use units based on the liter for measuring volumes of liquids. Although not officially part of SI, liter measure is used commonly. Like meters cubed, the liter is metric ($1 \, L = 0.001 \, m^3$).

Some of the SI units derived from the basic units for length, mass, volume, and temperature follow. Note especially those units described as most common in your laboratory work.

Length

1 kilometer (km) = 1000 meters
1 hectometer (hm) = 100 meters
1 dekameter (dkm) = 10 meters
1 meter (m)
1 decimeter (dm) = 0.1 meter
1 centimeter (cm) = 0.01 meter
1 millimeter (mm) = 0.001 meter
1 micrometer (pm) = 0.000001 meter
1 nanometer (nm) = 0.000000001 meter

Measurements under microscopes often are made in micrometers. Still smaller measurements, as for wavelengths of light used by plants in photosynthesis, are made in nanometers. The units of length you will use most frequently in the laboratory are centimeters (cm).

Units of area are derived from units of length by multiplication. One hectometer squared is one measure that is often used for ecological studies; it commonly is called a hectare and equals 10 000 m^2. Measurements of area made in the laboratory most frequently will be in centimeters squared (cm^2).

Units of volume also are derived from units of length by multiplication. One meter cubed (m^3) is the standard, but it is too large for practical use in the laboratory. Centimeters cubed (cm^3) are the more common units vou will see. For the conditions you will encounter in these laboratory investigations, 1 cm^3 measures a volume equal to 1 mL.

Mass

1 kilogram (kg) = 1000 grams
1 hectogram (hg) = 100 grams
1 dekagram (dkg) = 10 grams
1 gram (g)
1 decigram (dg) = 0.1 gram
1 centigram (cg) = 0.01 gram
1 milligram (mg) = 0.001 gram
1 microgram (ug) = 0.000001 gram
1 nanogram (ng) = 0.000000001 gram

Measurements of mass in your biology laboratory usually will be made in kilograms, grams, centigrams, and milligrams.

Another type of measurement you will encounter in this course is molarity (labeled with the letter *M*). Molarity measures the concentration of a dissolved substance in a solution. A high molarity indicates a high concentration. The solutions you will use in the investigations frequently will be identified by their molarity.

Volume

1 kiloliter (kL) = 1000 liters
1 hectoliter (hL) = 100 liters
1 dekaliter (dkL) = 10 liters
1 liter (L)
1 deciliter (dL) = 0.1 liter
1 centiliter (cL) = 0.01 liter
1 milliliter (mL) = 0.001 liter

Your volume measurements in the laboratory usually will be made in glassware marked for milliliters and liters.

Temperature

On the Celsius scale, 0° C is the freezing point of water, and 100° C is the boiling point of water. (Atmospheric pressure affects both of these temperatures.) Figure A1.2 illustrates the Celsius scale alongside the Fahrenheit scale, which still is used in the United States. On the Fahrenheit scale, 32° F is the freezing point of water, and 212° F is the boiling point of water. Figure A1.2 is useful in converting from one scale to the other.

If you wish to learn more about SI measure, write to the U.S. Department of Commerce, National Institute of Standards and Technology (NIST), Washington, DC 20234.

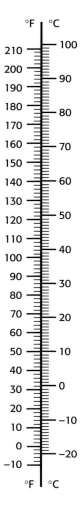

Figure A1.2 A comparison of Fahrenheit and Celsius temperature scales.

Appendix Two
Supplementary Investigations

INVESTIGATION A2.1 Chemical Safety

All chemicals are hazardous in some way. A hazardous chemical is defined as a substance that is likely to cause injury. Chemicals can be placed in four hazard categories: flammable/combustible, toxic, corrosive, and reactive. Table A2.1 summarizes their characteristics. In the investigations for this course, every effort is made to minimize the use of dangerous materials. Many "less hazardous" chemicals can cause injury, however, if not handled properly. This investigation will help you become aware of the four types of chemical hazards and of how you can reduce the risk of injury when using chemicals. Be sure to review the safety rules in Appendix One before you work with any chemical.

These activities are designed to increase student awareness of hazards and safety measures. In combination with the safety guidelines in Appendix 1 and in *Guidelines for Laboratory Safety,* **these materials should help ensure safe laboratory experiences for you and your students. Teams of 2 and 3 are best for the experimental procedures.**
   Time: **Two or three class periods. Allow adequate time for discussion of all aspects of this investigation.**

Materials (per team of 2)
2 pairs of safety goggles
2 lab aprons
2 pairs of plastic gloves

Safety

Demonstrate all safety guidelines as you work through this investigation with your students, particularly with the demonstration in Part C.

PART A
none

PART B
3 microscope slides
3 cover slips
dropping pipet
3 flat toothpicks
compound microscope

concentrated biological stain in
   dropping bottle
dilute biological stain in dropping bottle
Detain™ in squeeze bottle
paramecium culture

PART C
2 glass dropping pipets
2 100-mL beakers
white paper

100 mL red food dye
scissors

PART D
100-mL glass beaker
spatula

baking soda (sodium bicarbonate)
vinegar (acetic acid)

PART A   Flammable/Combustible Substances
Procedure
Flammable/combustible substances are solids, liquids, or gases that will burn or that will sustain burning. The process of burning involves three interrelated components: fuel (any substance capable of burning), oxidizer (often air or a specific chemical), and ignition source (a spark, flame, or heat). The three components are represented in the diagram of a fire triangle in Figure A2.1. For burning to occur, all three components (sides) of the fire triangle must be present. To control fire hazard, one must remove, or otherwise make inaccessible, at least one side of the fire triangle. Flammable chemicals should not be used in the presence of ignition sources such as lab burners, hot plates, and sparks from electrical equipment or static electricity. Containers of flammables should be closed when not in use. Sufficient ventilation in the laboratory will help to keep the

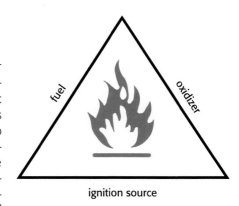

Figure A2.1   The fire triangle.

665

Table A2.1 Categories of Hazardous Materials

Type of Material	Categories
Flammable/Combustible Substance Common Lab Hazard 	**Flammable Liquid**—Liquid having a flashpoint* less than 100° F (37.8° C) **Combustible Liquid**—Flashpoint* equal to or greater than 100° F (37.8° C) but less than 200° F. *Flashpoint—the lowest temperature at which a liquid gives off vapor that forms an ignitable mixture with air near the surface. **Flammable Solid**—Causes fire through friction, absorption of moisture, or spontaneous chemical change. Ignites readily; burns vigorously. **Flammable Gas**—Forms a flammable mixture with air at ambient temperature and pressure.
Reactive Chemical Violent reaction under certain ambient or induced conditions. Spontaneous generation of great heat, light, and flammable and nonflammable gases or toxicants. 	**Acid-Sensitive**—Reacts with acids or acid fumes to generate heat, flammable or explosive gases, or toxicants. **Water-Sensitive**—Reacts with moisture to generate heat and/or flammable or explosive gases. **Oxidizer**—Promotes combustion in other materials through release of oxygen or other gases.
Corrosive Chemical Injures body tissue and corrodes metal by direct chemical action. 	**Corrosive Liquid, Solid, Gas**—Causes visible destruction or irreversible alterations in living tissue. **Irritant**—Causes reversible inflammation in living tissue. **Sensitizer**—Causes allergic reaction in normal tissue of a substantial number of individuals after repeated exposure.
Toxic Chemical Injures by direct action with body systems when tolerable limits are exceeded. Exposure routes: direct contact, inhalation, ingestion, and penetration. 	**Inhalation**—Toxic gases may pass rapidly into capillary beds of lungs and be carried to all parts of the body via circulatory system. **Ingestion**—Toxics may damage tissues of mouth, throat, and gastrointestinal tract; produce systemic poisoning if absorbed through these tissues. **Skin/Eye Contact**—Hair follicles, sweat glands, as well as cuts and abrasions are the main portals of entry. Eyes are acutely sensitive to chemical irritants as well as corrosives **Injection/Penetration**—Exposure to toxics by injection seldom occurs; but wounds by broken glass or metal are frequent entry points for injected chemicals.

Information courtesy of Ward's Natural Science Establishment.

concentration of flammable vapors to a minimum. Wearing safety goggles, lab aprons, and protective gloves are important precautionary measures when using flammable/combustible materials.

1. Use your knowledge to draw fire triangles with the appropriate side removed to represent the following control measures.

   a. Isolate flammable materials from reactive chemicals.

   b. Do not store more than 600 mL of flammable material in glass in the laboratory.

   c. Transfer flammable liquids in a working fume hood whenever possible.

   d. Eliminate sources of ignition.

   e. Store flammable substances in a cool area, at a maximum temperature of 27° C (80° F).

   f. Use special absorbent materials to reduce vapor pressure when wiping up spills.

   g. Transport flammable liquids in metal or other protective containers.

2. Use your knowledge to circle the side of the fire triangle that is most vulnerable to these common safety problems.

   a. Improperly stored glass containers of flammable solvents.

   b. A lack of adequate grounding to prevent sparking (by static electricity) generated by flowing liquids.

   c. Flammable liquids mixed in a waste disposal can with other chemicals.

   d. Open flames on a laboratory bench near flammable liquids.

   e. Transporting glass containers containing flammable liquids on top of a cart.

   f. Flammable substances stored on a shelf with acids and other reactive chemicals.

Discussion

1. What fire control measures are used at gas stations when fuel is pumped?

2. Why do firefighters apply foam when a flammable material is spilled?

PART B   Toxic Substances

Put on safety goggles, lab apron and plastic gloves.

Procedure

Most of the chemicals in a laboratory are toxic, or poisonous to life. The degree of toxicity depends on the properties of the substance, its concentration, the type of exposure, and other variables. The effects of a toxic substance can range from minor discomfort to serious illness or death. Exposure to toxic substances can occur through ingestion, skin contact, or through inhaling toxic vapors. Wearing safety goggles, lab apron, and protective gloves are important precautions when using toxic chemicals. A clean work area, prompt spill cleanup, and good ventilation also are important.

3. Prepare a wet mount with one drop each of paramecium culture, concentrated biological stain, and Detain™ (protist slowing agent). Mix with a toothpick and label the slide *A*.

4. Repeat Step 3, but this time use dilute biological stain. Label this slide *B*.

5. Repeat Step 3, but do not add any stain. Label this slide *C*.

6. Observe each of the three slides for 15 to 20 minutes under various magnifications of a compound microscope.

7. Wash your hands thoroughly before leaving the laboratory.

PART A
Procedure

1. a. ∠   2. a. ◿
   b. ◿   b. ◿
   c. ◿   c. ◿
   d. ∧   d. ◿
   e. ∧   e. ◿
   f. ◿   f. ◿
   g. ◿

Discussion

1. **Answers will vary. Shutting off the car's engine; no smoking; the presence of a canopy or pump-level spray system that is activated to reduce vapors should a significant spill occur, thus preventing ignition.**

2. **Such commercial foam materials lower the vapor pressure, thus preventing ignition.**

Preparations

Prepare a concentrated stain for Part B as a 2% alcoholic solution of brilliant cresyl blue; dilute 1 : 9 in distilled water for the diluted preparation.

Discussion

1. Control.
2. The concentrated stain. (The ten-fold dilution used on slide B should not injure the cells.)
3. Signs of cell toxicity are marked reduction in contractile vacuole and ciliary activity coupled with eventual cell death. (You may wish to speculate with students whether the stain or the alcohol used to dissolve it is toxic to paramecia. Brilliant cresyl blue also is soluble in water, and additional investigations can be conducted to demonstrate that at high enough concentrations, both the stain and the alcohol can be toxic.)

SAFETY

Preparations

For Part C, wear safety goggles, face shield, lab apron, and nitrile rubber gloves to set up the demonstration of corrosive solids. Do not allow students to come into contact with any of the corrosive solids used. All students should wear safety goggles, lab apron, and protective gloves.

Materials

- ◆ **5 plates of plain agar numbered *1* through *5***
- ◆ **plastic forceps**
- ◆ **sodium hydroxide crystals or pellets (plate 1) labeled *DANGER: Corrosive solid***
- ◆ **potassium hydroxide crystals or pellets (plate 2) labeled *DANGER: Corrosive solid***
- ◆ **phenol crystals (plate 3) labeled *DANGER: Corrosive solid; Poison***
- ◆ **sodium chloride—lumps of rock salt (plate 4)**
- ◆ **sugar cubes (plate 5)**

Use plastic forceps to remove small quantities (1–3 pellets or small lumps) of each of the test solids and place them on top of the agar in separate, numbered petri dishes. Cover and seal the plates and record the start time of the experiment.

    At the conclusion of this demonstration, add 1*M* HCl, drop by drop, to each of the plates containing corrosive bases until neutrality is reached according to litmus test. Discard plates only after this step.

Discussion

1. What is the purpose of the third slide?
2. Which concentration of stain is toxic to paramecia?
3. What observations would yield criteria to measure cell toxicity of the stain?

PART C   Corrosive Substances

Put on your safety goggles, lab apron and plastic gloves.	

Procedure

Corrosive chemicals are solids, liquids, or gases that by direct chemical action either destroy living tissue or cause permanent change in the tissue. Corrosive substances can destroy eye and respiratory-tract tissues, causing impaired sight or permanent blindness, severe disfigurement, permanent severe breathing difficulties, and even death. Safety goggles, lab apron, and protective gloves should be worn when handling corrosive chemicals to prevent contact with the skin or eyes. Splashes on the skin or in the eyes should be washed off immediately while a classmate notifies the teacher.

I Corrosive solids

Corrosive solids may appear relatively harmless because they can be removed more easily than liquids and because they may not cause an immediate destructive effect. The effect depends largely on their solubility in skin moisture and even more rapid solubility in the moisture present in the respiratory and intestinal tracts.

    Caustic alkalies (sodium or potassium hydroxides) are perhaps the greatest potential hazard because of their wide use in general science laboratories. The hazards of corrosive solids include the following:

    a. Solutions of corrosive solids are readily absorbed through the skin.
    b. Caustic alkalies and other corrosive solids may produce delayed reactions rather than immediate, painful reactions.
    c. Many corrosive solids dust easily, thus increasing the hazard through other exposure routes (inhalation).
    d. Molten corrosive solids greatly increase the threat of exposure by acting as liquid corrosives.

    The most serious hazard associated with corrosives is from materials in the gaseous state. In this state, corrosives are rapidly absorbed into the body by dissolving skin moisture and by inhalation. Many corrosives give off dangerous vapors either by themselves or during spills or chemical reactions.

    8. Your teacher has set up a demonstration that shows the action of corrosive solids. Agar is used to represent body tissue. Observe the effects of each substance on the agar during the next 30 to 45 minutes.

II Inorganic acids

Everyone knows that acids can cause burns, but few people realize the extent to which they can damage body tissues. For example, sulfuric acid can destroy skin and eye tissues and dissolve teeth, and even the vapor can destroy lung and other respiratory tissues.

    Every student has heard the admonishment, "Do as you oughta . . . add acid to water." The following investigation demonstrates why this practice is critical to chemical safety.

    9. For illustration purposes, the red food coloring will represent hydrochloric acid. Cut and position 7-cm wide white paper strips around the top edge of two beakers. Fill one beaker with red dye, the other with tap water. Fill the beakers to the same height, about 2 cm from the lip.

10. Use a dropping pipet to add 5 to 10 drops of water to the beaker containing the "acid" (red dye). Hold the dropping pipet 15 to 20 cm above the beaker—normal dispensing height.

11. Use a dropping pipet to add 5 to 10 drops of "acid" (red dye) to the beaker containing the water. Hold the dropping pipet at normal dispensing height when adding the red dye.

12. Remove the white paper from around each beaker and check carefully for signs of splashing.

Discussion

1. Which solid(s) had the greatest corrosive effect? Cite evidence for your observation.

2. Which solid(s) had the least corrosive effect?

3. What factor present in the agar is responsible for initiating the corrosive effect?

4. Are some substances noncorrosive? On what observed characteristics is your answer based?

5. Based on your observations, what can you conclude about those solids that are corrosive?

6. What can you conclude from the evidence (the color of the splash marks) about acid dilution?

PART D   Reactive Substances

> **Put on your safety goggles, lab apron, and plastic gloves.**
>

Procedure

Reactive chemicals (see Table A2.1) promote violent reactions under certain conditions. A chemical may explode spontaneously or when mechanically disturbed. Reactive chemicals also include those that react rapidly when mixed with another chemical, releasing a large amount of energy. Keep chemicals separate from each other unless they are being combined according to specific instructions in an investigation. Heed any other cautions your teacher may give you. Always wear safety goggles, lab apron, and protective gloves when handling reactive chemicals.

13. Pour a small amount of vinegar in the glass beaker.

> **Vinegar is an irritant. Avoid skin/eye contact; do not ingest. In case of spills or splashes, call your teacher immediately; flush with water 15 minutes; rinse mouth with water.**

14. Use the spatula to scoop up a pinch of baking soda and add it to the vinegar. Record what happens.

15. Wash your hands thoroughly before leaving the laboratory.

Discussion

1. What reaction category did this chemical reaction demonstrate?

2. Would you expect this reaction to proceed faster if sulfuric acid were used? Explain.

Discussion

1. Phenol is the most corrosive of the three corrosive solids investigated. The hazy area penetrated deep into the agar. (In agar, tissue damage is mimicked by a roughened surface in conjunction with a white, hazy area that spreads and deepens through time.)

2. Sodium hydroxide and potassium hydroxide are less corrosive than phenol—the hazy area did not penetrate deep into the agar.

3. Water. (All substances liquefy on contact with the agar, which, like tissue, is mostly water. It is skin moisture that initiates the corrosive effects of the solid. As observed, these effects usually are not immediate, but progress slowly and become evident during a 30-minute period.)

4. Sugar and salt are noncorrosive. No roughened surface or hazy area formed on the agar.

5. A corrosive material causes tissue damage. Not all are fast acting; as the solid liquefies, it penetrates the tissue. Damage is initiated by moisture.

6. This model demonstrates that little dilution occurs when water is added directly to an acid. Numerous red splash marks indicate a high acid content. When water is added to acid in this manner, serious injury could result. Conversely, when acid is added to water, great dilution capability exists and splashes have a low acid content.

Discussion

1. Baking soda, or sodium bicarbonate, is in the acid-sensitive reaction category; that is, it can react with an acid (in this case vinegar) to generate heat and gases, among other products.

2. Inorganic acids such as sulfuric acid are extremely strong acids compared with vinegar (a solution that usually contains about 4–8% acetic acid). If baking soda were sprinkled onto a quantity of sulfuric acid, the intensity of the reaction would be much greater, with more heat and toxic gas generated at a much faster rate.

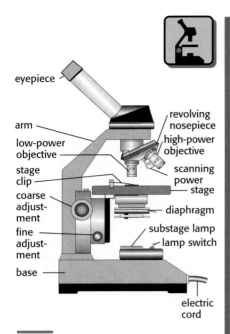

eyepiece

arm

revolving nosepiece

low-power objective

high-power objective

stage clip

scanning power stage

coarse adjustment

diaphragm

fine adjustment

substage lamp

lamp switch

base

electric cord

Figure A2.2 Parts of a compound microscope.

Many of the investigations in this course involve microscopic examination of materials. It is to the student's advantage to learn to use a microscope efficiently at the beginning of the school year. Even students who have used a microscope before will find it useful to take part in this investigation along with inexperienced students.

One student per microscope is preferable. A limited number of microscopes may make it necessary to work in teams of 2 or even 3 students per microscope.

Time: Students who have had previous experience with the compound microscope may be able to complete the exercise within one class period. Students who have not used a microscope before may require two periods.

Table A2.2 Microscopic Observations

Object Being Viewed	Observations and Comments
Letter *o*	
Letter *c*	
Etc.	
mm ruler	

INVESTIGATION A2.2   The Compound Microscope

The human eye cannot distinguish objects much smaller than 0.1 mm in size. The microscope is a biological tool that extends vision and allows observation of much smaller objects. The most commonly used compound microscope (Figure A2.2) is monocular (one eyepiece). Light reaches the eye after being passed through the objects to be examined. In this investigation, you will learn how to use and care for a microscope.

Materials (per person or team of 2)

3 cover slips
3 microscope slides
dropping pipet
compound microscope

scissors
transparent metric ruler
lens paper
newspaper

Procedure

PART A   Care of the Microscope

1. The microscope is a precision instrument that requires proper care. Always carry the microscope with both hands, one hand under its base, the other on its arm.

2. When setting the microscope on a table, keep it away from the edge. If a lamp is attached to the microscope, keep its wire out of the way. Keep everything not needed for microscope studies off your lab table.

3. Avoid tilting the microscope when using temporary slides made with water.

4. The lenses of the microscope cost almost as much as all the other parts put together. Never clean lenses with anything other than the lens paper designed for this task.

5. Before putting the microscope away, always return it to the low-power setting. The high-power objective reaches too near the stage to be left in place safely.

PART B   Setting Up the Microscope

6. Rotate the low-power objective into place if it is not already there. When you change from one objective to another you will hear the objective click into position.

7. Move the mirror so that even illumination is obtained through the opening in the stage, or turn on the substage lamp. Most microscopes are equipped with a diaphragm for regulating light. Some materials are best viewed in dim light, others in bright light.

> **DANGER: Never use a microscope mirror to capture direct sunlight when illuminating objects viewed under a microscope. The mirror concentrates light rays, which can permanently damage the retina of the eye. Always use indirect light.**

8. Make sure the lenses are dry and free of fingerprints and debris. Wipe lenses with lens paper only.

PART C   Using the Microscope

9. In your data book, prepare a data table similar to Table A2.2.

10. Cut a lowercase letter *o* from a piece of newspaper. Place it right side up on a clean slide. With a dropping pipet, place one drop of water on the letter. This type of slide is called a wet mount.

11. Wait until the paper is soaked before adding a cover slip. Hold the cover slip at about a 45° angle to the slide and then slowly lower it. Figure A2.3 shows these first steps.

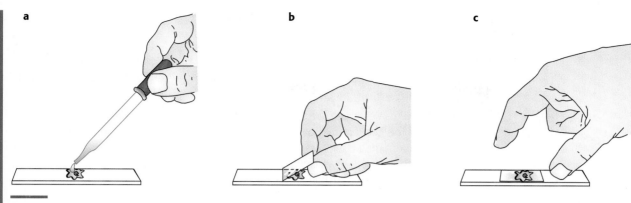

Figure A2.3 Preparing a wet mount with a microscope slide and cover slip.

12. Place the slide on the microscope stage and clamp it down. Move the slide so the letter is in the middle of the hole in the stage. Use the coarse-adjustment knob to lower the low-power objective to the lowest position.

13. Look through the eyepiece and use the coarse-adjustment knob to raise the objective slowly, until the letter *o* is in view. Use the fine-adjustment knob to sharpen the focus. Position the diaphragm for the best light. Compare the way the letter looks through the microscope with the way it looks to the naked eye.

14. To determine how greatly magnified the view is, multiply the number inscribed on the eyepiece by the number on the objective being used. For example, eyepiece (10X) X objective (10X) = total (100X).

15. Follow the same procedure with a lowercase *c*. In your data book, describe how the letter appears when viewed through a microscope.

16. Make a wet mount of the letter *e* or the letter *r*. Describe how the letter appears when viewed through the microscope. What new information (not revealed by the letter *c*) is revealed by the *e* or *r?*

17. Look through the eyepiece at the letter as you use your thumbs and forefingers to move the slide slowly *away* from you. Which way does your view of the letter move? Move the slide to the right. Which way does the image move?

18. Make a pencil sketch of the letter as you see it under the microscope. Label the changes in image and movement that occur under the microscope.

19. Make a wet mount of two different-colored hairs, one light and one dark. Cross one hair over the other. Position the slide so the hairs cross in the center of the field. Sketch the hairs under low power; then go to Part D.

PART D   Using High Power

20. With the crossed hairs centered under low power, adjust the diaphragm for the best light.

21. Turn the high-power objective into viewing position. *Do not* change the focus.

22. Sharpen the focus with the *fine-adjustment knob only. Do not focus under high power with the coarse-adjustment knob.*

23. Readjust the diaphragm to get the best light. If you are not successful in finding the object under high power the first time, return to Step 20 and repeat the whole procedure carefully.

24. Using the fine-adjustment knob, focus on the hairs at the point where they cross. Can you see both hairs sharply at the same focus level? How can you use the fine adjustment knob to determine which hair is crossed over the other? Sketch the hairs under high power.

Safety

Warn students never to point any optical device at the sun or its direct rays. They never should use the mirror to capture direct sunlight when illuminating objects viewed under a microscope. The mirror concentrates light rays, which can permanently damage the retina of the eye. Only indirect light should be used. Southern exposure is best, if possible.

PART E   Measuring with a Microscope

25. Because objects examined with a microscope are usually small, biologists use units of length smaller than centimeters or millimeters for microscopic measurement. One such unit is the micrometer, which is one thousandth of a millimeter. The symbol for micrometer is μm, the Greek letter μ (called mu) followed by m.

26. You can estimate the size of a microscopic object by comparing it with the size of the circular field of view. To determine the size of the field, place a transparent metric ruler on the stage. Use the low-power objective to obtain a clear image of the divisions on the ruler. Carefully move the ruler until its marked edge passes through the exact center of the field of view. Now, count the number of divisions that you can see in the field of view. The marks on the ruler will appear quite wide; 1 mm is the distance from the center of one mark to the center of the next. Record the diameter, in millimeters, of the low-power field of your microscope.

27. Remove the ruler and replace it with the wet mount of the letter e. (If the mount has dried, lift the cover slip and add water.) Using low power, compare the height of the letter with the diameter of the field of view. Estimate as accurately as possible the actual height of the letter in millimeters.

Discussion

1. Summarize the differences between an image viewed through a microscope and the same image viewed with the naked eye.

2. When viewing an object through the high-power objective, not all of the object may be in focus. Explain.

3. What is the relationship between magnification and the diameter of the field of view?

4. What is the diameter in micrometers of the low-power field of view of your microscope?

5. Calculate the diameter in micrometers of the high-power field. Use the following equations:

$$\frac{\text{magnification number of high-power objective}}{\text{magnification number of low-power objective}} = A$$

$$\frac{\text{diameter of low-power field of view}}{A} = \text{diameter of high-power field of view}$$

For example, if the magnification of your low-power objective is 12X and that of your high-power is 48X, A = 4. If the diameter of the low-power field of view is 1600 μm, the diameter of the high-power field of view is $\frac{1600}{4}$, or 400 μm.

Discussion

Many questions are asked in the procedures. Students should record their observations and make notes pertinent to them.

1. **Images are larger. Type will be grainier and show a dot pattern. Objects are inverted and reversed. They appear to move in the opposite direction of actual motion of the slide.**

2. **The high-power objective can focus on only a portion of the depth of the object at any one time.**

3. **As the magnification increases, the diameter of the field of view decreases.**

4. **Answers depend on the microscope used.**

5. **Answers depend on the microscope used. 1 mm = 1000 μm.**

This investigation allows students to increase their skills with the microscope. Teams of 2 are satisfactory. but allow students to work alone if you have enough microscopes.
Time: **One class period.**

INVESTIGATION A2.3   The Compound Microscope: Biological Material

This investigation provides practice in using a compound microscope.

Materials (per team of 2)

2 pairs of safety goggles
2 lab aprons
microscope slide
cover slip
dropping pipet
100-mL beaker containing water

compound microscope
lens paper
paper towel
Lugol's iodine solution in dropping bottle
yeast culture
small piece of white potato

Put on your safety goggles and lab apron.

Remind students to put on their safety goggles and lab aprons for steps 3 and 8.

Procedure

For setting up the microscope and preparing wet mounts, follow the directions in Investigation A2.2.

1. Place a *small* piece of potato in the center of a clean slide. Place the slide on your laboratory table and carefully press the potato with a finger until some juice is forced out. Distribute the juice evenly over the center of the slide by moving the piece of potato in a circle. Discard the potato. Add a drop of water and a cover slip. Avoid getting air bubbles in the wet mount.

2. Examine the wet mount under low power. To increase contrast between the starch grains and the water surrounding them, decrease the size of the opening in the substage diaphragm. Move the slide on the stage until you locate a field containing well-separated grains. Center several grains in the field and switch to high power. Describe the shape of an individual starch grain. Describe any internal structure you can see in these grains.

3. Turn again to low power. Stain the grains by placing a small drop of Lugol's iodine solution on the slide at one side of the cover slip, as shown in Figure A2.4. Tear off a small piece of paper towel. Place the torn edge in contact with the water at the opposite edge of the cover slip. As water is absorbed by the paper towel at one edge, the Lugol's iodine solution will be drawn under the cover slip at the opposite edge. Continue until the Lugol's solution covers half the space under the cover slip. The Lugol's solution will continue to spread slowly throughout the wet mount.

> **!** **Lugol's iodine solution is poisonous if ingested, irritating to skin and eyes, and can stain clothing. In case of spills or splashes, call your teacher immediately; flush area with water 15 minutes; rinse mouth with water.**

4. Examine various regions of the wet mount to observe the effects of different concentrations of Lugol's iodine solution on starch grains. Examine under low power and then high power. What changes occur in the starch grains exposed to relatively high concentrations of Lugol's iodine solution? What differences do you see between these grains and others exposed to lower concentrations of Lugol's solution? Describe any internal structures you can see in the stained grains. Using the method described in Investigation A2.2, estimate the size (in micrometers) of the larger starch grains.

5. Remove the slide, lift off the cover slip, and wash and dry them. Carefully wipe off any liquid from the microscope stage.

Procedure

Follow the laboratory work with a brief class discussion. There is no need to achieve consensus on all observations, but emphasize the possible reasons for differences in observations.

2. Detection of the layered structure of starch grains depends on lighting. Many students will not see it. Does it exist, or are students who report it seeing something that is not there? This is a good opportunity for a brief discussion of artifacts in microscopy.

4. Differences depend on a gradient of stain concentration and easily may be missed as the stain spreads. The layered structure is more evident in stained starch grains but also could be an artifact.

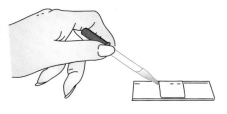

Figure A2.4 Putting liquid under a cover slip.

6. If starch grains turn up in the yeast preparations students have not cleaned the slides carefully.

7. Students may have difficulty recognizing that the presence of smaller organisms attached to larger ones indicates budding. It is unusual to see much structure in unstained yeasts.

8. Nuclei and possibly vacuoles may be seen in stained yeasts.

Discussion

1. Stains make transparent structures visible.

2. The stains may kill living organisms or introduce artifacts.

6. Place one drop of yeast culture on the clean slide and add a cover slip. Examine first under low power and then under high power. Describe the shape of the yeast organisms.

7. Study the arrangement of small groups of these organisms. From your observations, what can you conclude about how new yeast organisms develop? Sketch any internal structures you see.

8. Using Lugol's iodine solution, stain the yeast as you stained the starch grains. Compare the effects of the Lugol's solution on the yeast organisms with its effects on starch grains. Describe any structures that were not visible in the unstained yeast organisms.

9. Using the method from Investigation A2.2, estimate the size (in micrometers) of an average yeast organism.

10. Wash your hands thoroughly before leaving the laboratory.

Discussion

1. Based on your observations, what do you think is the purpose of adding stain to specimens to be examined with a compound microscope?

2. What problems may arise as a result of adding stain to a specimen?

Appendix Three
Tables

Table of Relative Humidities

Dry-bulb Temperature	Differences Between Dry-bulb and Wet-bulb Temperatures																																		
(°C)	0.5	1.0	1.5	2.0	2.5	3.0	3.5	4.0	4.5	5.0	5.5	6.0	6.5	7.0	7.5	8.0	8.5	9.0	9.5	10.0	10.5	11.0	11.5	12.0	12.5	13.0	13.5	14.0	14.5	15.0	16.0	17.0	18.0	19.0	20.0
10	94	88	82	77	71	66	60	55	50	44	39	34	29	24	20	15	10	6																	
11	94	89	83	78	72	67	61	56	51	46	41	36	32	27	22	18	13	9	5																
12	94	89	83	78	73	68	63	58	53	48	43	39	34	29	25	21	16	12	8																
13	95	89	83	79	74	69	64	59	54	50	45	41	36	32	28	23	19	15	11	7															
14	95	90	84	79	75	70	65	60	56	51	47	42	38	34	30	26	22	18	14	10	6														
15	95	90	85	80	75	71	66	61	57	53	48	44	40	36	32	27	24	20	16	13	9	6													
16	95	90	85	81	76	71	67	63	58	54	50	46	42	38	34	30	26	23	19	15	12	8	5												
17	95	90	86	81	76	72	68	64	60	55	51	47	43	40	36	32	28	25	21	18	14	11	8												
18	95	91	86	82	77	73	69	65	61	57	53	49	45	41	38	34	30	27	23	20	17	14	10	7											
19	95	91	87	82	78	74	70	65	62	58	54	50	46	43	39	36	32	29	26	22	19	16	13	10	7										
20	96	91	87	83	78	74	70	66	62	59	55	51	48	44	41	37	34	31	28	24	21	18	15	12	9	6									
21	96	91	87	83	79	75	71	67	64	60	56	53	49	46	42	39	36	33	29	26	23	20	17	14	12	9	6								
22	96	92	87	83	80	76	72	68	64	61	57	54	50	47	44	40	37	34	31	28	25	22	19	16	14	11	8	6							
23	96	92	88	84	80	76	72	69	65	62	58	55	52	49	45	42	39	36	33	30	27	24	21	19	16	13	11	8	6						
24	96	92	88	84	80	77	73	69	66	62	59	56	53	50	47	43	40	37	34	31	29	26	23	20	18	15	13	10	8	5					
25	96	92	88	84	81	77	74	70	67	63	60	57	54	51	48	45	42	39	36	33	30	28	25	22	20	17	15	12	10	8					
26	96	92	88	85	81	78	74	71	67	64	61	58	55	52	49	46	43	40	37	34	32	29	26	24	21	19	16	14	12	10					
27	96	92	89	85	82	78	75	71	68	65	61	58	56	53	49	46	44	41	38	36	33	31	28	26	23	21	19	16	14	12	5				
28	96	93	89	85	82	78	75	72	68	65	62	59	56	53	50	48	45	42	40	37	34	32	29	27	25	22	20	18	16	13	7	5			
29	96	93	89	85	82	79	76	72	69	66	63	60	57	54	51	48	46	43	41	38	36	33	31	29	26	24	22	19	17	15	9	7	5		
30	96	93	89	86	83	79	76	73	70	66	63	61	58	55	52	50	47	44	42	39	37	35	32	30	28	25	23	21	18	17	10	9			
31	96	93	90	86	83	80	77	73	70	67	64	61	59	56	53	51	48	45	43	40	38	36	33	31	29	27	25	22	20	18	12	11	5		
32	96	93	90	86	83	80	77	74	70	67	64	62	59	56	54	52	49	46	44	42	39	37	35	32	30	28	26	24	22	20	14	12	7		
33	97	93	90	87	84	80	78	74	71	68	65	63	60	57	55	53	50	47	45	43	40	38	36	33	31	29	27	25	23	21	15	14	9	5	
34	97	93	90	87	84	81	78	75	72	69	66	63	61	58	56	54	51	48	46	44	42	39	37	35	33	31	29	27	25	23	17	15	11	7	5
35	97	94	90	87	84	81	78	75	72	69	67	64	61	59	56	54	51	49	47	44	42	40	38	36	34	32	30	28	26	24	19	17	12	8	7
36	97	94	91	87	84	81	79	76	73	70	67	64	62	59	57	54	52	50	48	45	43	41	39	37	35	33	31	29	27	25	20	18	14	10	8
37	97	94	91	87	84	82	79	76	73	70	68	65	63	60	58	55	53	51	48	46	44	42	40	38	36	34	32	30	28	26	21	20	15	11	10
38	97	94	91	88	84	82	79	76	74	71	68	66	64	61	59	56	54	51	49	47	45	43	41	39	37	35	33	31	29	27	23	21	17	13	11
39	97	94	91	88	85	82	79	77	74	71	69	66	64	61	59	57	54	52	50	48	46	44	42	39	38	36	34	32	30	28	24	22	18	14	12
40	97	94	91	88	85	82	80	77	74	72	69	67	64	62	59	57	54	53	51	48	47	44	42	40	38	36	35	33	31	29	26	23	20	16	14

Periodic Table of the Elements

Key:
- 14 — Atomic number
- Si — Symbol
- Silicon — Name
- 28.0855 — Atomic mass

TRANSITION METALS

1	2	3	4	5	6	7	8	9	10	11	12	13	14	15	16	17	18
1 H Hydrogen 1.007																	2 He Helium 4.0026
3 Li Lithium 6.941	4 Be Beryllium 9.012											5 B Boron 10.81	6 C Carbon 12.0111	7 N Nitrogen 14.0067	8 O Oxygen 15.9994	9 F Flourine 18.998	10 Ne Neon 20.179
11 Na Sodium 22.98977	12 Mg Magnesium 24.305											13 Al Aluminum 26.9815	14 Si Silicon 28.0855	15 P Phosphorus 30.973	16 S Sulfur 32.06	17 Cl Chlorine 35.453	18 Ar Argon 39.948
19 K Potassium 39.098	20 Ca Calcium 40.08	21 Sc Scandium 44.955	22 Ti Titanium 47.88	23 V Vanadium 50.9415	24 Cr Chromium 51.996	25 Mn Manganese 54.938	26 Fe Iron 55.847	27 Co Cobalt 58.933	28 Ni Nickel 58.69	29 Cu Copper 63.546	30 Zn Zinc 65.39	31 Ga Gallium 69.72	32 Ge Germanium 72.59	33 As Arsenic 74.92	34 Se Selenium 78.96	35 Br Bromine 79.904	36 Kr Krypton 83.80
37 Rb Rubidium 85.467	38 Sr Strontium 87.62	39 Y Yttrium 88.905	40 Zr Zirconium 91.224	41 Nb Niobium 92.906	42 Mo Molybdenum 95.94	43 Tc Technetium (98)	44 Ru Ruthenium 101.07	45 Rh Rhodium 102.906	46 Pd Palladium 106.42	47 Ag Silver 107.868	48 Cd Cadmium 112.41	49 In Indium 114.82	50 Sn Tin 118.71	51 Sb Antimony 121.75	52 Te Tellurium 127.60	53 I Iodine 126.905	54 Xe Xenon 131.29
55 Cs Cesium 132.905	56 Ba Barium 137.3	57 La Lanthanum 138.906	72 Hf Hafnium 178.49	73 Ta Tantalum 180.948	74 W Tungsten 183.85	75 Re Rhenium 186.207	76 Os Osmium 190.2	77 Ir Iridium 192.22	78 Pt Platinum 195.08	79 Au Gold 196.967	80 Hg Mercury 200.59	81 Tl Thallium 204.383	82 Pb Lead 207.2	83 Bi Bismuth 208.980	84 Po Polonium (209)	84 At Astatine (210)	86 Rn Radon (222)
87 Fr Francium (223)	88 Ra Radium (226.0)	89 Ac Actinium (227.028)	104 (261)	105 (262)	106 (263)	107 (262)	108	109 (266)									

INNER TRANSITION METALS

Lanthanide series

58 Ce Cerium 140.12	59 Pr Praseodymium 140.908	60 Nd Neodymium 144.24	61 Pm Promethium (145)	62 Sm Samarium 150.36	63 Eu Europium 151.96	64 Gd Gadolinium 157.25	65 Tb Terbium 158.925	66 Dy Dysprosium 162.50	67 Ho Holmium 164.930	68 Er Erbium 167.26	69 Tm Thulium 168.934	70 Yb Ytterbium 173.04	71 Lu Lutetium 174.96

Actinide series

90 Th Thorium 232.038	91 Pa Protactinium 231.036	92 U Uranium 238.029	93 Np Neptunium (244)	94 Pu Plutonium (244)	95 Am Americium (243)	96 Cm Curium (247)	97 Bk Berkelium (247)	98 Cf Californium (251)	99 Es Einsteinium (252)	100 Fm Fermium (257)	101 Md Mendelevium (258)	102 No Nobelium (259)	103 Lr Lawrencium (260)

Protist/Cyanobacteria Key

Identification may be made either by comparing the observed cell to the line drawings or by using the key. Start at number 1, comparing the observed cell to each of the characteristics listed (1a and 1b). Advance to the number corresponding to the characteristic you have chosen (2 or 17). Proceed according to the key until it terminates in the name of an organism.

la
White or colorless
.... 2

lb
Colored
.... 17

2a
Slow-creeping or floats without apparent motion
.... 3

2b
Exhibits other motion
.... 8

3a
Slow-creeping
.... 5

3b
Appears to float without apparent motion
.... 4

4a
Cell large, spherical in shape; with central core and radiating spines (axopods)

Actinosphaerium

(200–1000 μm)

4b
Cell small, spherical in shape with radiating spines; with no central core

Actinophrys

(40–50 μm)

5a
Cell possesses a test or shell sometimes composed of small grains of sand
.... 6

5b
Cell naked, has no shell
.... 7

6a
Shell flattened; without attached or embedded material; pale brown in color; with colorless pseudopods ("false feet")

Arecella

(45–100 μm, shell diameter)

6b
Shell dome-shaped; with attached particles, usually of sand; with colorless pseudopods

Difflugia

(60–580 μm × 40–240 μm shell)

7a
Small cell; creeps along using pseudopods; changes shape constantly, usually by extending a single pseudopod. Cell with a single, large nucleus

Amoeba

(to 600 μm)

7b
Large cell; creeps using pseudopods; changes shape constantly, usually by extending many pseudopods. Cell with many (1000+) nuclei

Chaos

(150–450 μm)

8a
Cell has many hairlike structures (cilia) used for movement
.... 9

8b
Cell moves by one or more whiplike flagella
.... 15

9a
Cell body uniformly covered with cilia
.... 10

9b
Cell body not uniformly covered with cilia; instead has either bands or tufts of cilia or distinct pointlike projections (fused groups of cilia called cirri)
.... 16

10a
Cell body elongate, either cylindrical or flattened
.... 11

10b
Cell body other shape
.... 12

11a
Large cell, visible to naked eye, with elongate, flattened body with blunt ends; swims slowly. Cell can contract to 1/4 body size when stimulated.

Spirostomum

(1–3 mm)

11b
Small cell with cylindrical (cigar-shaped) body having rounded ends; swims rapidly in corkscrew fashion

Paramecium

(200–300 μm)

12a
Cell body oval in shape
.... 13

12b
Body other shape
.... 14

13a
Large oval cell, visible to naked eye, with large round mouth at anterior end

Bursaria

(0.5–1 mm)

13b
Small oval cell, mouth on side; rapid swimmer

Colpidium

(50–70 μm)

14a
Large cell with an elongate body having a pointed tip and a pronounced long neck; mouth at base of neck

Dileptus

(250–500 μm)

14b
Small cell with pear-shaped body; mouth at top of cell

Tetrahymena

(40–60 μm)

15a
Small, elongate colorless cell with single long flagellum. Posterior end of cell rounded. Cell highly plastic when

Protist/Cyanobacteria Key   (continued)

stationary; often appears to vibrate when in motion

Parenema

(27–70 µm)

15b
Small, elongate colorless cell; with two visible flagella. Posterior end of cell narrowed

Chilomonas

(20–40 µm)

16a
Cell on stalk
. . . . 42

16b
Cell not on stalk
. . . . 44

17a
Cell green, golden brown, or blue-green in color
. . . . 18

17b
Color not green
. . . . 47

18a
Cells contain chloroplasts
. . . . 19

18b
Cells without chloroplasts, blue-green pigment seemingly distributed and diffused throughout the cell or nearly so (Cyanobacteria)
. . . . 33

19a
Cells arranged in filaments (threads) or branching filaments
. . . . 30

19b
Cells not arranged in filaments
. . . . 20

20a
Single cell or a group of cells that show no apparent motion; with distinct glasslike walls etched with grooves or holes forming delicate patterns
. . . . 21

20b
Single cell or group of cells with or without motion; without distinct glasslike walls or delicate etchings
. . . . 25

21a
Single cell
. . . . 22

21b
Cells in groups or clusters
. . . . 23

22a
Cell cigar-shaped, or cigar-shaped with end knobs.

Navicula

(200 µm)

22b
Cell bean pod-shaped

Hantzschia

(900 µm)

23a
Cell rectangular in shape
. . . . 24

23b
Cells barrel-shaped, arranged in a filament with clear center plates

Melosira

(14–25 µm, cell)

24a
Rectangular-shaped cells arranged in a zig-zag pattern

Tabellaria

(50 µm, cell)

24b
Rectangular-shaped cells arranged in ribbons

Fragilaria

(100 µm, cell)

25a
Single cell (sometimes may occur in clusters)
. . . . 26

25b
Groups of cells arranged in a definite colony
. . . . 39

26a
Cell with flagella
. . . . 27

26b
Cell usually single; no flagella (sometimes may occur in clusters)
. . . . 37

27a
Cell colored golden-brown; differentiated into two distinct cones, each composed of plates. Cell has two small flagella in an area between cones

Peridinium

(50–65 µm)

27b
Cell green-colored without cones made of plates
. . . . 28

28a
Cell with single flagellum
. . . . 29

28b
Small, pear-shaped cell with two flagella and a horseshoe-shaped green chloroplast

Chlamydomonas

(20 µm)

29a
Green-colored cell flattened and oval in shape. Cell has conspicuous eyespot with a long, conspicuous, single flagellum. Cell sometimes is twisted

Phacus

(40–170 µm)

29b
Small green-colored cell, cylindrical in shape with a pointed end; with a long, single flagellum

Euglena

(35–55 µm)

30a
Cells arranged in a single filament (thread); cells mostly longer than wide
. . . . 31

30b
Cells arranged in a branching filament
. . . . 32

31a
Individual cells in filament each with two star-shaped chloroplasts

Zygnema

(40 µm, diameter)

31b
Individual cells in filament each with many chloroplasts that are arranged in a definite spiral

Spirogyra

(50-60 µm, diameter)

Protist/Cyanobacteria Key   (continued)

32a
Visible to naked eye. Branching
filaments appear as tufts. Each
filament branch with a thornlike tip.
Each cell with a single chloroplast
usually at the side

Stigeoclonium
_____
(12–22 μm, diameter)

32b
Visible to naked eye. Branching
filaments, each with a rounded tip.
Each cell contains numerous netlike
chloroplasts

Cladophora
_____
(75–100 μm, diameter)

33a
Cells arranged in a filament (thread)
. . . . 49

33b
Cells not arranged in a filament
. . . . 34

34a
Oval-shaped cells arranged in pairs as
a flat, platelike colony

Merismopedia
_____
(10 μm, diameter)

34b
Cells arranged in a round, globular
colony
. . . . 35

35a
Group of oval-shaped cells arranged
within concentric rings of mucilage

Gloeocapsa
_____
(1–2 μm, cell diameter)

35b
Cells not arranged in concentric rings
of mucilage
. . . . 36

36a
Numerous tiny round cells crowded
together within a colony inside a
mucilage matrix. Cells appear
brownish black or purplish

Microcystis
_____
(2–4 μm, cell diameter)

36b
Spherical-shaped cells arranged in
pairs and evenly distributed
throughout a mucilage matrix

Anacystis
_____
(2–4 μm, cell diameter)

37a
Cells green-colored and spherical in
shape. Sometimes may occur in
clusters

Chlorella
_____
(7–10 μm, diameter)

37b
Cells curved or needlelike
. . . . 38

38a
Cells curved or crescent shaped;
green in color

Closterium
_____
30–70 μm

38b
Cells needlelike, slightly curved

Ankistrodesmus
_____
(25–100+ μm)

39a
Colony flat and disc-shaped; usually
containing 16 cells

Gonium
_____
(5–15 μm, cell diameter)

39b
Colony spherical in shape
. . . . 40

40a
Colony contains 32 cells or less
. . . . 41

40b
more than 32 cells
Colony visible to naked eye; contains
more than 32 cells (hundreds)

Volvox
_____
(2.5–6 μm, cell; 300–500 μm, colony)

41a
Colony contains 32 cells

Eudorina
_____
10–20 μm, cell; to 200 μm, colony)

41b
Colony contains 16 cells

Pandorina
_____
(10–20 μm, cell; to 220 μm, colony)

42a
Stalk not branched; cells contract (stalk
appears to contract like a spring)

Vorticella
_____
(50–150 μm)

42b
Stalk branched; cells may or may not
contract
. . . . 43

43a
Cells contract independently (stalk
appears to contract like a spring)

Carchesium
_____
(100–250 μm, cells; 3 mm, colony)

43b
Cells do not contract (lack myoneme)

Epistylis
_____
(50–150 μm, cells; 3 μm, colony)

44a
Cell with tufts or bands of cilia
. . . . 46

44b
Cell with distinct pointlike projections
(cirri)
. . . . 45

45a
Cell with oval body, flattened on
ventral side, convex on dorsal side
with 4 posterior cirri. Travels by
walking on cirri

Euplotes
_____
(80–200 μm)

45b
Cell with oval body, flattened on
ventral side, convex on dorsal side
with 3 posterior cirri. Travels by
walking on cirri

Stylonychia
_____
(100–300 μm)

Protist/Cyanobacteria Key   (continued)

46a
Cell body barrel-shaped with dark colored plates. Tuft of cilia around the mouth; with 8 posterior spines (four at and four near the rear)

Coleps

(80–110 µm)

46b
Cell oval-shaped with two distinct ciliary bands—one anterior and one in the middle of the body. Swims with a spiral motion.

Didinium

(80–200 µm)

47a
Single cell with cilia

. . . . 48

47b
Red-colored tufted mass visible to naked eye. Colony composed of branched filaments with each filament having numerous bead-shaped cells

Batrachospermum

(5–10 cm, plant mass)

48a
Pink or rose-colored ciliate

Blepharisma

(130–200 µm)

48b
Dark bluish-green ciliate; trumpet shaped

Stentor

(1–2 mm extended)

49a
Flat, rectangular-shaped cells arranged in a filament or thread

. . . . 51

49b
Round or barrel-shaped cells arranged in a filament or chain

. . . . 50

50a
Single straight chain of beadlike cells; looks like a necklace

Anabaena

(3–11 µm, diameter)

50b
Numerous coiled filaments or chains of beadlike cells which may comprise a visible globular (ball-shaped) mass looking like a hollow egg or bead

Nostoc

(5–7 µm, cell; 5–10 mm, colony)

51a
Filaments inside a clear covering or sheath; cells at ends of each filament rounded

Phormidium

(4–6 µm, diameter)

51b
Filament not inside a clear sheath; cells at ends of filament tapered or rounded

Oscillatoria

(5–6 µm, diameter)

Elements Important in Humans

Name	Food Sources	Function	Deficiency	Excess
Major Elements				
Calcium (Ca)	Dairy products, green vegetables, eggs, nuts, dried legumes, broccoli, canned salmon, hard water	Development of bones and teeth, muscle contraction, blood clotting, nerve impulse transmission, and enzyme activation	Osteoporosis, stunted growth, poor quality bones and teeth, rickets, convulsions	Rare—excess blood calcium, loss of appetite, muscle weakness, fever
Chlorine (Cl)	Table salt, most water supplies, soy sauce	Acid-base balance, hydrochloric acid in stomach, bone and connective tissue growth	Rare—metabolic alkalosis, constipation, failure to gain weight (in infants)	Vomiting; elemental chlorine is a poison used for chemical warfare
Magnesium (Mg)	Whole grains, liver, kidneys, milk, nuts, green leafy vegetables	Component of chlorophyll; bones and teeth, coenzyme in carbohydrate and protein metabolism	Infertility, menstrual disorders	Loss of reflexes, drowsiness, coma, death
Phosphorus (P)	Soybeans, dairy foods, egg yolk, meat, whole grains, shrimp, spinach, peas, beef, green vegetables	Development of bones and teeth, energy metabolism, pH balance, cold adaptation	Rare—bone fractures, disorders of red blood cells, metabolic problems, irritability, weakness	As phosphorus increases, calcium decreases; muscle spasm, jaw erosion
Potassium (K)	Whole grains, meat, bananas, vegetables, cantaloupe, milk, peanut butter	Body water and pH balance, nerve and muscle activity, insulin release, glycogen and protein synthesis	Muscle and nerve weakness, poor digestion	Rare—abnormalities in heartbeat or stoppage, muscle weakness, mental confusion, cold and pale skin
Sodium (Na)	Table salt, dairy foods, eggs, baking soda and powder, meat, vegetables	Body water and pH balance, nerve and muscle activity, glucose absorptive	Weakness, muscle cramps, diarrhea, dehydration, nausea	High blood pressure, edema, kidney disease
Sulphur (S)	Dairy products, nuts, legumes, garlic, onion, egg yolk	Component of some amino acids, enzyme activator, blood clotting	Related to intake and deficiency of sulfur-containing amino acids	Excess sulfur-containing amino acid intake leads to poor growth
Trace Elements				
Cobalt (Co)	Common in foods, meat, milk, fish	Component of Vitamin B12, essential for red blood cell formation	Rare	Dermatitis, excessive production of red corpuscles
Copper (Cu)	Liver, meat, seafood, whole grains, legumes, nuts	Production of hemoglobin, bone formation, component of electron carriers, melanin synthesis	Anemia, bone and connective tissue disorders, scurvylike conditions, an early death	Toxic concentrations in liver and eyes of persons with genetic inability to metabolize

Elements Important in Humans   (continued)

Name	Food Sources	Function	Deficiency	Excess
Fluoride (F)	Most water supplies, seafood	Prevents bacterial tooth decay, aids in hardening and stability of bones and teeth	Osteoporosis, tooth decay, bone weakness	Mottling and brown spots on teeth, deformed teeth and bones
Iodine (I)	Seafood, iodized salt, dairy products	Component of thyroid hormone, which controls cellular respiration	Inadequate synthesis of thyroid goiter (enlarged thyroid), cretinism, fatigue	Antithyroid compounds
Iron (Fe)	Liver, meat, eggs, spinach, enriched bread and cereals, asparagus, raisins	Component of hemoglobin (oxygen and electron transport system); immune system	Iron-deficiency anemia, chronic fatigue, weakness, poor digestion	Accidental poisoning of children, cirrhosis of liver
Manganese (Mn)	Liver, kidneys, legumes, cereals, tea, coffee, green vegetables	Ions necessary in protein and carbohydrate metabolism, krebs cycle	Infertility, menstrual problems	Brain and nervous system disorders
Molybdenum (Mo)	Organ meats, milk, whole grains, leafy vegetables, legumes	Enzyme component, uric acid excretion	Edema, lethargy, disorientation, and coma	Weight loss, growth retardation, and changes in connective tissue
Zinc (Zn)	Liver, sea foods, common foods, nuts, chicken, carrots, peas	Essential enzyme component, necessary for normal senses of taste and smell, disease resistance	Slow sexual development, loss of appetite, retarded growth, loss of taste and smell, lower disease resistance	Nausea, bloating, and cramps, depresses copper absorption

Kcal Tally

Meat
140   Beef pot roast, lean, 2 thin slices
85   Ham, boiled, 1 slice
245   Hamburger patty, plain
170   Hot dog, plain
115   Meatloaf, 1 slice
150   Porkchop, 1 lean

Poultry
215   Chicken breast, fried, 1/2
185   Chicken, broiled, 1/4
150   Turkey, light, 1 slice
175   Turkey, dark, 1 slice

Fish
50   Fish stick, breaded, 1
230   Halibut, broiled, 1 steak
195   Ocean perch, breaded, fried, 1 piece
215   Tuna, in oil, 1/2 cup

Dairy
100   Butter or margarine, 1 tbs
59   Cheese, American, 1 slice
60   Cottage cheese, 1/4 cup
150   Milk, whole, 1 cup

Main Dishes
345   Bean tostada with cheese
460   Beef potpie, 1 individual
180   Beef stew, 1 cup
95   Chicken chow mein, 1 cup
365   Chicken with noodles, 1 cup
545   Chicken potpie, individual
170   Chili con carne with beans, canned, 1/2 cup
365   Enchiladas, 2
380   Lasagna, 8 ounces
240   Macaroni and cheese, 1/2 cup
153   Pizza, cheese, 1/8 of a pie
157   Pizza, sausage, 1/8 of a pie
260   Spaghetti w/meat sauce, 3/4 cup

Salads
285   Chef's salad (lettuce, ham, cheese, dressing)
50   Coleslaw, 1/2 cup
245   Potato salad, 1/2 cup
180   Tuna salad, 1/2 cup
185   Waldorf salad, 1/2 cup

Sandwiches
280   Bacon, lettuce, and tomato
560   Big Mac™

360   Bologna
330   Cheese
545   Cheeseburger
410   Fish fillet, fried
445   Hamburger
280   Hot dog, 1 bun
310   Roast beef
360   Salami

Soups
100   Beef, 1 cup
100   Chicken noodle, 1 cup
140   Split pea, 1 cup
90   Tomato, 1 cup
80   Vegetable, 1 cup

Vegetables
165   Beans, baked, with pork, 1 cup
15   Beans, green, snap, wax, or yellow, 1/2 cup
50   Broccoli, 1 cup
35   Beets, diced, 1/2 cup
30   Carrots, 1 medium, raw
15   Cauliflower, 1/2 cup
70   Corn, 1 medium ear
210   Corn, creamed, 1 cup
5   Lettuce, 4 small leaves
38   Onion, raw, 1/2 cup
60   Peas, green, 1/2 cup
15   Pepper, green, 1 medium
145   Potato, baked
115   Potato chips, 10 medium
155   Potatoes, French-fried, 10
135   Potatoes, mashed, 1/2 cup

Desserts
330   Apple pie, 1 slice
103   Brownie, frosted, with nuts
420   Chocolate cake, fudge icing, 2" wedge
120   Devil's-food cupcake with chocolate icing
80   Gelatin, plain, 1/2 cup
135   Ice cream, vanilla 1/2 cup
300   Lemon meringue pie, 1 slice
265   Pumpkin pie, 1 slice

Breakfast Foods
95   Bacon, 2 strips
145   Bran flakes, with raisins, 1 cup
60   Bread, whole wheat, 1 slice
76   Bread, white, 1 slice
85   Cornflakes, 1 cup
135   Doughnut, cake

85   Egg, fried in butter
80   Egg, poached or hard cooked
95   Egg, scrambled (milk added)
110   Muffin, blueberry
110   Oatmeal, 3/4 cup
60   Pancake, 4" cake
170   Pork sausage, 2" patty
95   Pork sausage, 1 link
200   Toaster pastry
210   Waffle

Fruits
80   Apple, raw, 1 medium
380   Avocado, raw, 1 whole
100   Banana
5   Cucumber, 6 thin slices
75   Grapefruit, l/2
70   Orange
40   Peach
100   Pear
30   Strawberries, 1/2 cup
30   Tomato, 1 medium

Drinks
117   Apple juice, 1 cup
175   Cocoa (all milk), 3/4 cup
0   Coffee or tea, unsweetened
145   Cola, 12 ounces
1   Cola, diet, 12 ounces
340   Eggnog, 1 cup
75   Grape juice, 1/2 cup
100   Grapefruit juice, unsweetened, 1 cup
210   Milk, whole, chocolate, 1 cup
135   Orange juice, 1 cup
350   Shake, chocolate, 8 ounces

Candies/Cookies/Snacks
145   Candy bar, chocolate, 1 ounce
200   Chocolate chip cookies, 4
100   Chocolate kisses, 5
200   Vanilla cookie, sandwich type, 4
90   Marshmallow
840   Peanuts, roasted, 1 cup
23   Popcorn, plain, 1 cup
41   Popcorn, oil and salt, 1 cup
50   Pretzels, 3-ring, 10
810   Sunflower seeds, dry, hulled, 1 cup

Vitamins Important in Humans

Name	Food Sources	Function	Deficiency Symptoms
A, retinol	Liver, green and yellow vegetables, fruits, egg yolks, butter	Formation of visual pigments especially of epithelial cells; cell	Night blindness, flaky skin, lowered resistance to infection, growth stunting, faulty reproduction
D, calciferol	Fish oils, liver, action of sunlight on lipids in skin, fortified milk, butter, eggs	Increases calcium absorption from gut; important in bone and tooth formation	Rickets—defective bone formation
E, tocopherol	Oils, whole grains, liver, mayonnaise, margarine	Protects red blood cells, plasma membranes, and vitamin A from destruction; important in muscle maintenance; DNA and RNA synthesis	Fragility of red blood cells, muscle wasting, sterility
K, menadione	Synthesis by intestinal bacteria; green and yellow vegetables	Synthesis of clotting factors by liver	Internal hemorrhaging (deficiency can be caused by oral antibiotics, which kill intestinal bacteria)
B_1, thiamine	Whole grains, legumes, nuts, liver, heart; kidney, pork, macaroni, wheat germ	Carbohydrate metabolism, nerve transmission, RNA synthesis	Beriberi, loss of appetite, indigestion, fatigue, nerve irritability, heart failure, depression, poor coordination
B_2, riboflavin	Liver, kidney, heart, yeast, milk, eggs, whole grains, broccoli, almonds, cottage cheese, yogurt, macaroni	Forms part of electron carrier in electron transport system, production of FAD, activates B_6 and folic acid	Sore mouth and tongue, cracks at corners of mouth, eye irritation, scaly skin, growth retardation
Pantothenic acid	Yeast, liver, eggs, wheat germ, bran, peanuts, peas, fish, whole grain cereals	Part of Coenzyme A; essential for energy release and biosynthesis; stimulates antibodies and intestinal absorption	Fatigue, headache, sleep disturbances, nausea, muscle cramps, loss of antibody production, irritability, vomiting
B_3, niacin	Yeast, liver, kidney, heart, meat, fish, poultry, legumes, nuts, whole grains, eggs, milk	Coenzyme in energy metabolism; part of NAD^+ and NADP	Pellagra, skin lesions, digestive problems, nerve disorders, diarrhea, headaches, fatigue
B_6, pyridoxine	Whole grains, potatoes, fish, poultry, red meats, legumes, seeds	Coenzyme for amino acid and fatty acid metabolism; synthesis of brain chemicals, antibodies, red blood cells, DNA; glucose tolerance	Skin disorders, sore mouth and, tongue, nerve disorders, anemia, weight loss, impaired antibody response, convulsive seizures
Biotin (vitamin B)	Cauliflower, liver, kidney, yeast, egg yolks, whole grains, fish, legumes, nuts, meats, dairy products; synthesis by intestinal bacteria	Fatty acid, amino acid, and protein synthesis; energy release from glucose, insulin activity	Skin disorders, appetite loss, depression, sleeplessness, muscle pain; elevated blood cholesterol and glucose
Folacin (folic acid)	Liver, yeast, leafy vegetables, asparagus, salmon	Nucleic acid synthesis, amino acid metabolism, new cell growth	Failure of red blood cells to mature, anemia, intestinal disturbances and diarrhea
B_{12} (cyanocobalamin)	Liver and organ meats, muscle meats, fish, eggs, shellfish, milk; synthesis by intestinal bacteria	Nucleic acid synthesis; maintenance of nervous tissue; glucose metabolism	Pernicious anemia, fatigue, irritability, loss of appetite, headaches
C, ascorbic acid	Citrus fruits, tomatoes, green leafy vegetables, sweet peppers, broccoli, cauliflower	Essential to formation of collagen and intercellular substance; protects against infection; maintains strength of blood vessels; increases iron absorption from gut; important in muscle maintenance; stress tolerance	Scurvy, failure to form connective tissue, bleeding, anemia, slow wound healing, joint pain, premature wrinkling and aging

Sources, Uses, and Effects of Some Psychoactive Drugs

Psychoactive drugs affect body and nervous functions and may change behavior. This table lists some of the drugs most commonly abused in the United States today. Effects vary a great deal from person to person; only a few are described.

Name	Source and Use	Effects
Depressants		
Alcohol (in wine, beer, ale, spirits, liquors, mixed drinks, etc.)	Synthesized or produced naturally by fermentation of fruits, vegetables, or grains.	Affects brain in proportion to amount in bloodstream. Small dose produces euphoria, drowsiness, dizziness, flushing, and release of inhibitions and tensions. Slurred speech, staggering, and double vision may result from large dose. Impairs driving in any dose and exaggerates effects of other drugs. Liver, heart, and brain damage, and vitamin deficiencies can occur with excessive use. Use during pregnancy may result in underweight babies with physical, mental and behavioral abnormalities. May produce tolerance and addiction. Most common of all forms of drug abuse today.
Tranquilizers—Valium, Librium, Miltown	Synthetic chemicals. Used to treat anxiety, nervousness, tension, and muscle spasm.	Produce mildly impaired coordination and balance, reduced alertness and emotional reactions, loss of inhibition. May produce sleep disturbances or depression. Use during pregnancy may cause congenital malformations. Use with alcohol or while driving is dangerous. Tolerance-inducing and addictive.
Opiates—opium, morphine, codeine, heroin; also called dope, horse, smack, skag	Seed capsules of the opium poppy *(Papaver somniferum)* produce a milky juice. From this juice, many drugs can be prepared, including morphine and codeine. Heroin, is formed by chemical synthesis from morphine. Taken by smoking, by mouth, or by injection. Used in anti-diarrhea and anti-cough medication and as pain killers.	Injection produces surge of pleasure, state of gratification, and stupor. Sensations of pain, hunger, and sexual desire are lost. Physical effects include nausea, vomiting, insensitivity to pain, constipation, sweating, slowed breathing. Long-term effects include severe constipation, moodiness, menstrual irregularities, and liver and brain damage. Use with alcohol very dangerous. Tolerance inducing and highly addictive.
Barbiturates—pentobarbital, secobarbital; also called goof-balls, downers, yellow jackets, red devils, etc.	Many compounds, developed to suppress pain, treat sleeplessness, anxiety, tension, high blood pressure, convulsions, and to induce anesthesia.	Effects similar to alcohol: release from tension and inhibition, followed by interference with mental function, depression, and hostility. Overdoses cause unconsciousness, coma, and death. Extremely dangerous in combination with alcohol. Long-term effects include liver damage and chronic intoxication. Addictive.
Stimulants		
Caffeine	White, bitter, crystalline substance, found in coffee beans, tea leaves, coca leaves, and kola nuts. Ingredient in many pain relievers, cold remedies, and beverages.	Increases metabolic rate, blood pressure, urination, and body temperature; shortens sleep, decreases appetite, and impairs coordination of movement. Long-term or heavy use can cause insomnia, anxiety, depression, or birth problems. Regular consumption of 4 cups of coffee a day can produce a form of physical dependence.

Sources, Uses, and Effects of Some Psychoactive Drugs   (continued)

Name	Source and Use	Effects
Tobacco	Shredded, dried leaves of tobacco plant. Contains tar and nicotine and cancer-causing substances. Smoke contains carbon monoxide. Generates 2000 different chemical compounds. Tobacco is smoked or chewed.	Increases heart and breathing rates and blood pressure; decreases appetite. Nicotine enters bloodstream; tars accumulate in lungs. Carbon monoxide interferes with or oxygen transport. Long-term effects include blockage of blood vessels, respiratory disease, cancers of lung, mouth, or throat, stomach ulcers, and reduced immunity. Chewing tobacco increases the risk of oral cancer by 40 times. Smoking during pregnancy increases risk of miscarriage, low birth weight, and developmental problems. Causes psychological and physical dependence.
Amphetamines— benzadrine, dexedrine, methedrine, also called speed, splash, uppers, bennies, dexies, crystal	Synthetic drugs taken by mouth or injected into veins.	Increases alertness and energy, feeling of well-being, heart and breathing rates, and blood pressure. Restlessness and hostility common. Very large doses can produce irregular heartbeat, hallucinations, or heart failure. Long-term use can produce mental illness, kidney damage, or lung problems. Produce powerful psychological and physical dependence.
Cocaine—also called coke, snow, nose candy, crack	Comes from the leaves of the coca plant. Sniffed, ingested, injected, or smoked as freebase.	Effects similar to amphetamines but of shorter duration. Toxic; causes heart irregularities that can lead to death, lung damage, hallucinations, muscle spasms, and convulsions. Use produces restlessness, insomnia, delusions, and weight loss. Produces powerful psychological and physical dependence.

Hallucinogens

Name	Source and Use	Effects
Cannabis—hashish and marijuana; also called pot, grass, weed, reefer, and joint	Flowers, leaves, oils, or resin of *Cannabis sativa*. Smoked in pipe or hand-rolled cigarette, or ingested by mouth.	Impairs concentration, short-term memory, coordination, and motor skills. Enhances sensory perception and feelings of relaxation. Distorts space-time sense and induces withdrawal, fearfulness, anxiety, depression, or hallucinations. Long-term effects include loss of motivation, interest, memory, and concentration. The smoke can cause chronic lung disease and lung cancer. Psychological dependence common.
LSD—lysergic acid diethylamide or acid; mescaline and psilocybin, also called mind-expanding drugs	LSD is an artificial chemical compound developed in the 1940s. Mescaline is a natural product of peyote (a cactus) long used by American Indians in religious practices. Psilocybin is formed by a mushroom. Usually taken by mouth.	Vivid psychic effects, including hallucinations. Sensations intensified and distorted. Extreme mood swings, including joy, inspiration, depression, anxiety, terror, and aggression. Long-term effects of LSD include decreased motivation and interest, prolonged depression, and anxiety. Fetal abnormalities may result from use during pregnancy.

Geologic Time Scale

Eon	Era	Period	Epoch	Beginning Time (Millions of Years Ago)	Distance From Near End
		Quaternary	Holocene	10 000 years ago to present	.2 mm
			Pleistocene	1.6	3.5 cm
			Pliocene	5.3	.12 m
			Miocene	23.7	.51 m
			Oligocene	36.6	.80 m
			Eocene	57.8	1.25 m
	Cenozoic	Tertiary	Paleocene	66.4	1.44 m
		Cretaceous		144	3.12 m
		Jurassic		208	4.51 m
	Mesozoic	Triassic		245	5.32 m
		Permian		286	6.21 m
		Carboniferous		360	7.81 m
		Devonian		408	8.85 m
		Silurian		438	9.50 m
		Ordovician		505	11.0 m
Phanerozoic	Paleozoic	Cambrian		570	12.4 m
Proterozoic				2500	54.2 m
Archaean				3800	82.5 m
Hadean				4600	100 m

Adapted from the Geological Society of America, "A Decade of North American Geology," compiled in 1983.

Major Geological Events

First evidence of . . .	Millions of Years Ago	Distance From Near End
Pleistocene ice age begins	1.6	3.5 cm
land bridge between North and South America	5.7	.12 m
Antarctic ice cap	24	.52 m
Mississippi River	35	.76 m
separation of Antarctica and Australia	50	1.1 m
collision of India with Asia and formation of the Himalayan Mountains	55	1.2 m
formation of the present Rocky Mountains	70	1.5 m
breakup of Pangaea and formation of the Atlantic Ocean	165	3.6 m
formation of the super continent Pangaea	225	4.9 m
extensive coal deposits	350	7.6 m
free oxygen (O_2) nears present levels (1–20 percent)	600	13 m
free oxygen (O_2) building up in the atmosphere	2500	54 m
an atmosphere and hydrosphere	3800	83 m
the earth as a solid planet	4600	100 m

Major Biological Events

First evidence of . . .	Millions of Years Ago	Distance From Near End
modern humans (*Homo sapiens sapiens*)	.04	9 mm
"Lucy" (*Australopithecus afarensis*)	3.2	6.9 cm
members of the human family (hominids)	4.0	8.6 cm
monkeys	35	.76 m
primates	65	1.4 m
flowers	140	3.0 m
birds	150	3.2 m
mammals	225	4.9 m
dinosaurs	235	5.1 m
reptiles	300	6.5 m
seeds	350	7.6 m
amphibians	360	7.8 m
land animals	400	8.7 m
land plants	430	9.3 m
vertebrates (jawless fishes)	520	11 m
animals with hard shells (marine invertebrates)	590	13 m
animals (soft-bodied marine invertebrates)	680	15 m
multicellular organisms (algae)	1000	22 m
cells with a nucleus (eukaryotes)	1400	30 m
life (bacteria)	3500	76 m

Climatograms

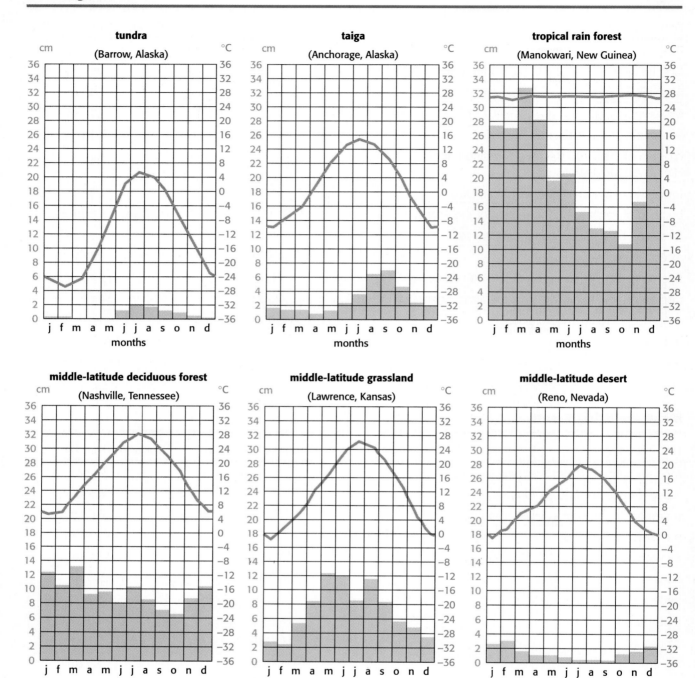

Role Cards

Industrialist
Position: wants city to grow
Land use: shoe factory
Area: 40 acres
Cost to city: none
Income to city: $10,000/acre; property taxes, 750 jobs
Concerns: low property and income taxes; availability of work force; source of water for cooling purposes; good transportation (railroad, highway, or river) for bringing in raw materials and distributing manufacturing goods; price and availability of raw materials

Retired Citizen
Position: wants a nice, affordable home
Land use: low-rent housing
Area: 40 acres (400 units, 1200 people)
Cost to city: 20 percent of building costs (80% from federal government)
Income to city: increased utility use
Concerns: economical housing with low maintenance costs; shopping nearby; safety of neighborhood; availability of social programs; medical facilities nearby; quiet neighborhood; mass transportation; libraries nearby; restrictions of a fixed income

Developer
Position: wants city to grow
Land use: shopping center
Area: 40 acres
Cost to city: extension of services
Income to city: property taxes and sales tax from sale or leasing of land
Concerns: property and business taxes; availability of retail leasing sites; promotion of business and residential development; expansion of commercial zoning; good roads to area; public transportation to area; population size

College Student
Position: interested in environment and recreation
Land use: park with recreational facilities
Area: 320 acres
Cost to city: landscaping, construction, maintenance
Income to city: minor, such as entrance fee to swimming pool
Concerns: environmental concerns; using park for jogging and swimming; strong controls and restrictions on land development; politics; availability of part-time jobs

Single Parent
Position: wants good education for child
Land use: elementary school
Area: 20 acres
Cost to city: cost of buildings and maintenance
Income to city: none
Concerns: does not want crowded schools; good transportation to school; safety of neighborhood; jobs for teachers and other personnel; no increase in taxes

Construction Worker
Position: wants job security
Land use: middle income homes and condominiums
Area: 240 acres (720 homes, 2800 people)
Cost to city: extension of services to these homes
Income to city: $1000/acre paid by developer; property taxes
Concerns: home to live in; property taxes; promotion of community growth and expansion; construction (building) codes

City Manager
Position: wants growth of city carefully managed
Land use: landfill site and sewage treatment plant
Area: 10 acres total (5 acres each)
Cost to city: building and maintenance
Income to city: dumping fees, increased rate for sewage service
Concerns: all aspects of development; services for all people (water, sewers, electricity, street maintenance); safety and health of population; police and fire protection; increased income to city from property and sales taxes

Environmentalist
Position: wants to preserve land in its natural condition
Land use: open space
Area: 1280 acres
Cost to city: maintenance
Income to city: none
Concerns: city needs open space; organisms living on the land now are worth saving; keep pollution under control; strong controls and restrictions on land development; political activist

Farmer
Position: wants to grow cash crops (fruits and vegetables) for market
Land use: farm
Area: 240 acres
Cost to city: none
Income to city: $1000/acre; annual property tax
Concerns: topography of land: length of growing season and climate; type of soil; amount of water available for irrigation; market for crops; not very interested in politics; wants to limit expansion of city; needs good roads into city to transport produce

Laborer
Position: wants a job
Land use: low-middle income houses and apartments
Area: 640 acres (100 homes, 400 people; 5 apartment complexes, 800 people)
Cost to city: extension of services to these homes
Income to city: $20 000/acre paid by developer; property taxes
Concerns: apolitical; interested in meeting immediate needs of feeding and housing family; not interested in environmental issues.

Appendix Four
A Catalog of Living Things

This appendix shows one way taxonomists arrange the major groups of living organisms. It does not take into account the many extinct groups known only from fossils.

In general the classification is not carried below class level. In some cases, however, examples are given at the family level. In two groups—insects and mammals—a more detailed classification at the order level is given. To show how complicated classification can become at lower levels, the primate order of mammals is carried to the family level.

In examples, common names of groups are used wherever appropriate, as, for example, phylum Chlorophyta: green algae. The illustrated organisms are identified by common names if they have them, or by names of genera.

A Catalog of Living Prokaryotes

Prokaryotic; lack membrane-enclosed organelles, circular DNA without protein, plasmids usually present. Autotrophic (photosynthetic or chemosynthetic) or heterotrophic; some parasitic; important decomposers; essential to many chemical cycles. Prevalent in all environments. Minute (usually 1 to 5 μm); single cells, chains, colonies, or slender branched filaments; most have rigid cell walls that lack cellulose. Locomotion, if present, by unique flagella or gliding. Great diversity in metabolic pathways; many anaerobic; photosynthetic species may use H_2, H_2S, or other substances. Reproduction by fission; exchange of genetic material occurs in some species. More than 5000 species described, two major groups; classification to phyla difficult.

Archaebacteria

Biochemically unique; composition of cell walls and plasma membranes differs from that of eubacteria; transfer RNA and ribosomal RNA also unique. May represent the first organisms; adapted to extreme conditions. Three major groups: methanogens, thermoacidophiles, extreme halophiles.

X7500

X5833

methanogens

X4000

halophiles

Eubacteria

Three major groups based on characteristics of cell wall: wall-less, gram-positive, and gram-negative. Mycoplasmas lack a cell wall; have extremely simple cell structure. Many gram-positive eubacteria produce lactic acid; others are source of antibiotics. Gram-negative eubacteria include photosynthetic and chemosynthetic types; many are anaerobic.

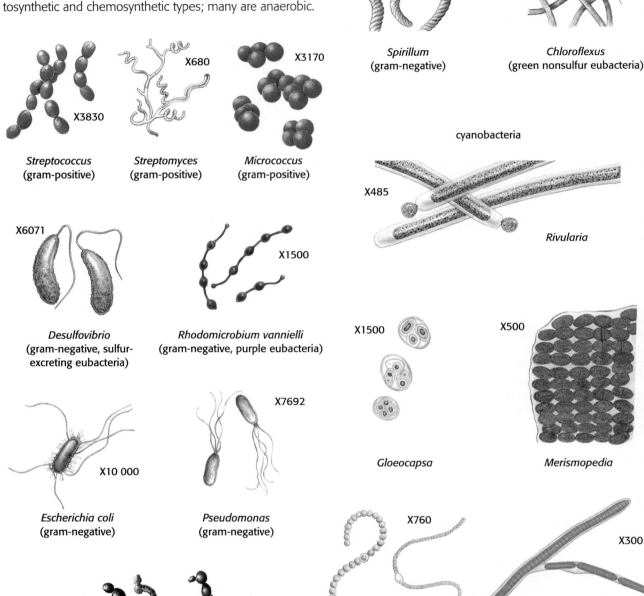

X790

Spirillum
(gram-negative)

X1980

Chloroflexus
(green nonsulfur eubacteria)

X3830

X680

X3170

Streptococcus
(gram-positive)

Streptomyces
(gram-positive)

Micrococcus
(gram-positive)

X6071

X1500

Desulfovibrio
(gram-negative, sulfur-excreting eubacteria)

Rhodomicrobium vannielli
(gram-negative, purple eubacteria)

cyanobacteria

X485

Rivularia

X1500

X500

Gloeocapsa

Merismopedia

X10 000

X7692

Escherichia coli
(gram-negative)

Pseudomonas
(gram-negative)

X760

X300

X8000

Nostoc

Tolypothrix

Mycoplasma

A Catalog of Living Protists

Eukaryotic, with membrane-enclosed nuclei and organelles; DNA organized into chromosomes complexed with protein. Autotrophic or heterotrophic; many can switch from one form to the other; many parasitic. Aquatic; marine, freshwater, or in watery tissues of other organisms. Size ranges from microscopic to 100 m; unicellular, colonial, or multicellular. Great variability in cell organization, chromosome structure, mitosis, meiosis, and life cycles. Reproduction asexual, by fission; or sexual, by conjugation or fertilization; often involving complex life cycles. Algae, protozoa, slime molds, slime nets, water molds, and others; 27 phyla; more than 100 000 species.

Green Algae (Phylum Chlorophyta) [Green *chloros,* green; *phyton,* plant]
Autotrophic; chlorophyll seldom masked by other pigments; food usually stored as starch. Microscopic to macroscopic; unicellular, filamentous, ribbons, sheets, tubes, or irregular masses. Probably ancestral to plants. Reproduction asexual and sexual. About 7000 species.

Diatoms (Phylum Bacillariophyta) [Latin *bacillus,* little stick; Greek *phyton,* plant]
Autotrophic; chlorophyll usually masked by yellow pigments; food stored as oil. Mostly microscopic; unicellular, colonial, or multicellular; shells (tests) made of two parts impregnated with silica. Reproduction asexual and sexual. As many as 10 000 species.

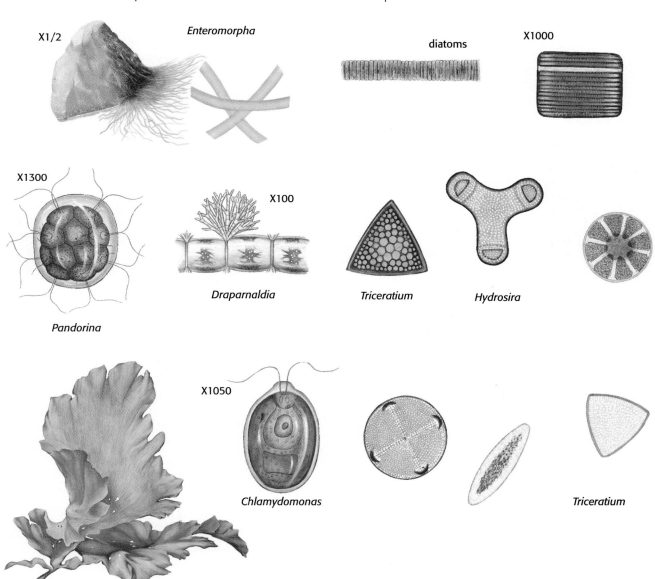

X1/2   *Enteromorpha*

diatoms   X1000

X1300

X100

Draparnaldia   *Triceratium*   *Hydrosira*

Pandorina

X1050

Chlamydomonas   *Triceratium*

X1   *Ulva*

Brown Algae (Phylum Phaeophyta) [Greek *phaios,*
brown; *phyton,* plant]
Autotrophic; chlorophyll usually masked by brownish pigments; food stored as carbohydrates, but not as starch.
Almost all marine and macroscopic (up to 100 m); many
with specialized structures. Reproduction mostly sexual.
About 1500 species.

Red Algae (Phylum Rhodophyta) [Greek: *rhodos,* red;
phyton, plant]
Autotrophic; chlorophyll usually masked by red pigments;
food stored as carbohydrates but not as starch. Almost all
marine and macroscopic. Reproduction sexual, with complex
life cycles; reproductive cells nonmotile. About 4000
species.

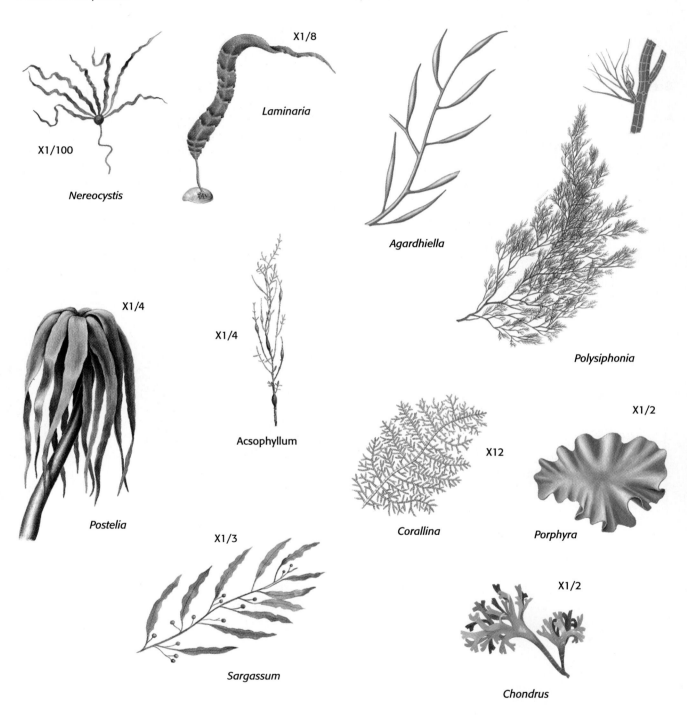

X1/8

Laminaria

X1/100

Nereocystis

Agardhiella

X1/4

X1/4

Polysiphonia

Postelia

Acsophyllum

X12

X1/2

Corallina

Porphyra

X1/3

X1/2

Sargassum

Chondrus

Flagellates (several phyla) [Latin *flagellatus,* whipped]
Heterotrophic, autotrophic, or both; many form symbiotic
relationships; many parasitic species. Locomotion by whiplike flagella; microscopic or almost so; unicellular or colonial.
Reproduction asexual. About 2000 species.

Sarcodines (several phyla) [Greek *sarx,* flesh; *eidos,* form]
Heterotrophic; some parasitic species. Locomotion by
pseudopods; microscopic or almost so; unicellular; many
with intricate skeletal structures. Reproduction asexual. About
8000 species.

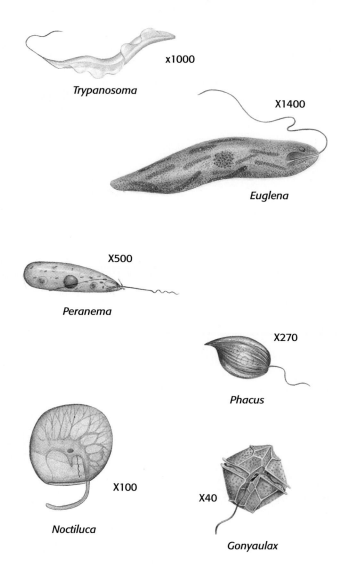

x1000

Trypanosoma

X1400

Euglena

X500

Peranema

X270

Phacus

X100

Noctiluca

X40

Gonyaulax

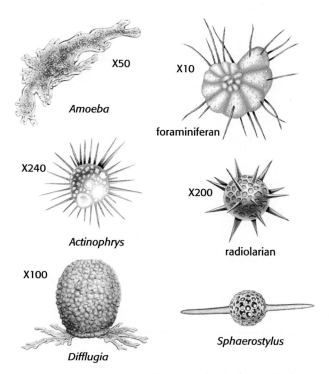

X50

Amoeba

X10

foraminiferan

X240

Actinophrys

X200

radiolarian

X100

Difflugia

Sphaerostylus

Sporozoans (Phylum Apicomplexa) [Latin *apex,* summit;
complexus, an embrace]
Parasites of animals, with complex life cycles often involving
several species of hosts. Usually nonmotile, but pseudopods
or flagella may occur in certain stages of some species;
microscopic. Reproduction sexual, with alternation of generations. About 2000 species.

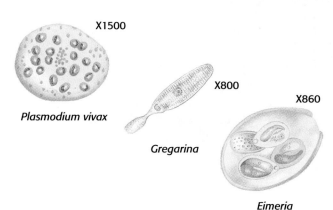

X1500

Plasmodium vivax

X800

Gregarina

X860

Eimeria

Ciliates (Phylum Ciliophora) [Latin *cilium,* eyelash; Greek *phorein,* to bear]
Heterotrophic. Locomotion by cilia; microscopic or almost so; unicellular; cells with macro- and micronuclei. Reproduction asexual, by fission or budding, or sexual by conjugation. About 5000 species.

Slime Molds (Phylum Myxomycota) [Greek *myxa,* mucas; *mykes,* fungus]
Heterotrophic (saprotrophic). Amoebold, multinucleate stage engulfs decaying vegetation in damp habitats. Sexual reproduction by spores formed in sporangia. About 450 species.

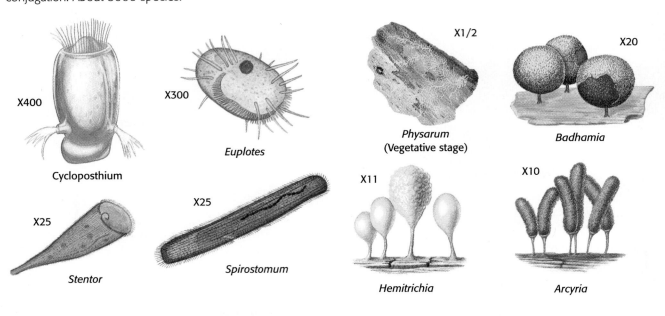

X400 Cycloposthium

X300 *Euplotes*

X25 *Stentor*

X25 *Spirostomum*

X1/2 *Physarum* (Vegetative stage)

X20 *Badhamia*

X11 *Hemitrichia*

X10 *Arcyria*

A Catalog of Living Fungi

Eukaryotic; heterotrophic (saprotrophic), many parasitic; important decomposers; many are sources of antibiotics. Mostly terrestrial; a few marine species. Nonmotile; mostly multicellular; structure primarily a system of threadlike cells or hyphae. Reproduction by asexual and sexual spores that develop directly into hyphae. Five classes; more than 100 000 species.

Conjugating Fungi (Phylum Zygomycota) [Greek *zygon,* pair; *mykes,* fungus]
Hyphae usually not divided by cross walls. Reproduction asexual, by spores, and sexual, by conjugation and formation of thick-walled zygospores. About 600 species.

Sac Fungi (Phylum Ascomycota) [Greek *askos,* bladder; *mykes,* fungus]
Hyphae divided by cross walls; a few unicellular species. Reproduction asexual, by conidiospores, and sexual, by conjugation and formation of a saclike structure, the ascus, in which sexual spores (ascospores) form by meiosis. Often form lichen partnerships with green algae or cyanobacteria. About 25 000 species.

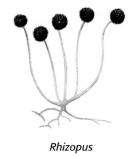

Rhizopus

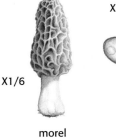

X1/6 morel

X3200 yeast

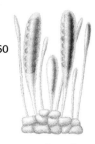

X60 apple scab

Club Fungi (Phylum Basidiomycota) [Greek *basidion,* small base; *mykes,* fungus]
Hyphae divided by cross walls. Reproduction sexual, by spores produced on the surface of a clublike structure, the basidium. Many form mycorrhizae. About 25 000 species.

Imperfect Fungi (Phylum Deuteromycota) [Greek *deuteros,* second; *mykes,* fungus]
Hyphae divided by cross walls. Reproduction asexual, by conidiospores; sexual reproduction stages unknown. Taxonomic grouping of convenience rather than of relationship. Includes members of the genus Penicillium. About 25 000 species.

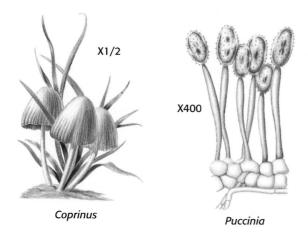

X1/2

X400

Coprinus

Puccinia

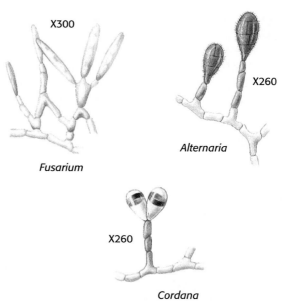

X300

X260

Fusarium

Alternaria

X260

Cordana

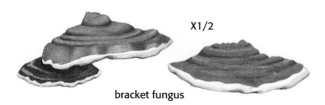

X1/2

bracket fungus

Lichens (Phylum Mycophycophyta) [Greek *mykes,* fungus; *phykos,* alga; *phyton,* plant]
Symbiotic relationship between a fungus and an alga or cyanobacterium; extremely slow-growing. Reproduction sexual, by formation of an ascus or basiduium, depending on the fungal partner. About 25 000 species.

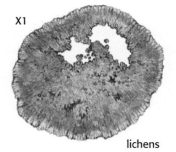

X1

lichens

A Catalog of Living Plants

Eukaryotic, autotrophic; nonmotile; multicellular. All have cellulose-containing cell walls and chloroplasts containing chlorophylls *a* and *b* plus other pigments. Mostly terrestrial; some aquatic. Reproduction sexual, with alternation of generations; in some cases asexual, by vegetative means; develop from a multicellular embryo that does not have a blastula stage. Nine divisions; more than 500 000 species.

Bryophytes (Division Bryophyta) [Greek *bryon,* moss; *phyton,* plant]
Small (less than 40 cm tall); mostly terrestrial but require moist habitats, lacks vascular (conducting) tissue. Well-developed alternation of generations; haploid gametophyte predominant; sporophyte more or less dependent on it. About 24 000 species.

Class Hepaticae (liverworts) [Greek *hepatikos,* liverlike (from the shape of the leaves)]
Gametophytes flat, often simple, branching masses of green tissue, sometimes with leaflike structures. About 8500 species.

X1

Marchantia

Class Anthocerotae (Hornworts) [Greek *anthos,* flower; *keras,* horn]
Gametophytes similar to those of liverworts; sporophytes live longer and are capable of continuous growth. About 50 species.

X1

Anthoceros

Class Musci (true mosses) [Latin *mucus,* moss]
Plants usually erect (not flat), with leaflike structures arranged in radial symmetry around a stalk. Gametophytes develop from algalike masses of green threads. About 15 000 species.

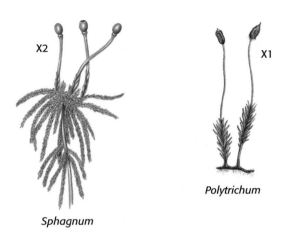

X2

X1

Sphagnum

Polytrichum

Whisk Ferns (Division Psilophyta) [Greek *psilo,* bare; *phyton,* plant]
Vascular; forking stems produce spore cases at the tips of short branches; no roots. Oldest fossil land plants. Alternation of generations with conspicuous sporophyte. Three species.

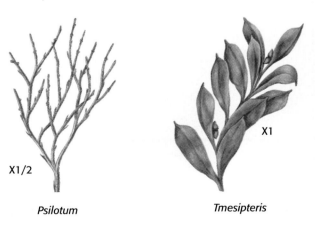

X1/2

X1

Psilotum

Tmesipteris

Club Mosses (Division Lycophyta) [Greek *lycos*, wolf; *phyton*, plant (so named because the roots of a lycopod were thought to resemble a wolf's claw)]

Low-growing evergreen vascular plants with roots, horizontal stems, and small leaves. Spore cases borne in various ways, usually on modified leaves grouped to form structures something like cones. About 1000 species.

X2/3

Lycopodium

Horsetails (Division Sphenophyta) [Greek *sphen*, wedge (from the shape of the leaves); *phyton*, plant]

Vascular plants with roots and jointed hollow stems containing silica; small scalelike leaves arranged in a circle around each stem joint. Spores borne in conelike structures. About 40 species.

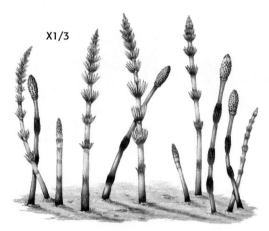

X1/3

Equisetum

Ferns (Division Pterophyta) [Greek *pteron*, feather; *phyton*, plant]

Vascular plants with roots, stems, and leaves; compound leaves grow from horizontal underground stems. Sporophyte generation dominant, but gametophytes independent; free-swimming sperm cells. About 12 000 species.

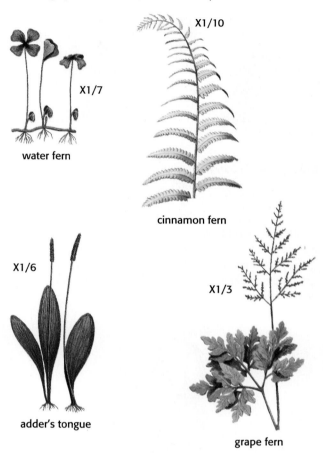

X1/10

X1/7

water fern

cinnamon fern

X1/6

adder's tongue

X1/3

grape fern

X1/4

climbing fern

Cycads (Division Cycadophyta) [Greek *kyos,* a palm; *phyton,* plant]
Tropical and subtropical vascular plants with palmlike or fern-like compound leaves; unbranched stems. Naked seeds borne in cones. About 100 species.

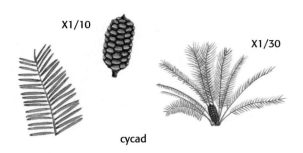

cycad

Ginkgoes (Division Ginkgophyta) [Japanese *ginkyo,* silver apricot; Greek *phyton,* plant]
Ancient division of vascular woody plants with fan-shaped leaves; cultivated worldwide. Male trees produce pollen that is carried by wind to female trees. Pollen that lands on a fleshy ovule is pulled through its opening; a sperm is released and fertilizes the egg nucleus. The embryo develops within a naked seed (not enclosed in an ovary wall). One species.

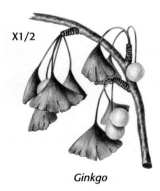

Ginkgo

Gnetophytes (Division Gnetophyta) [Latin *gnetum* from Moluccan Malay *ganemu,* a gnetophyte species found on the island of Ternate; Greek *phyton,* plant]
Vascular plants with naked seeds borne in cones; some characteristics like those of flowering plants. About 70 species.

Welwitschia

Conifers (Division Coniferophyta) [Latin *conus,* cone; *ferre,* to bear; Greek *phyton,* plant]
Vascular plants; most have needle-shaped leaves, many are evergreen. Gametophytes microscopic, within tissues of sporophytes; naked seeds borne on scales of a cone. About 700 species.

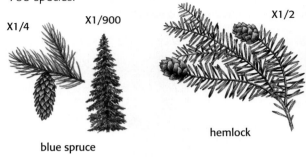

blue spruce

hemlock

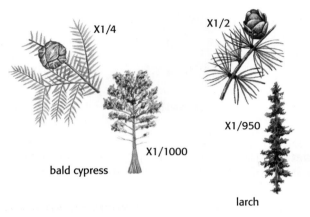

bald cypress

larch

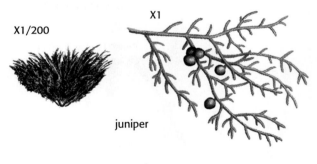

juniper

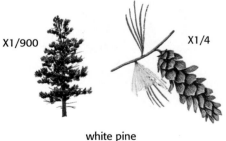

white pine

Flowering Plants (Division Anthophyta) [Greek *anthos*, flower; *phyton*, plant]
Vascular plants with seeds enclosed in a fruit; sperm cells in pollen tubes. Gametophytes microscopic, within tissues of sporophytes. About 230 000 species. (There are more than 300 families in the division Anthophyta. A few of the common families are listed for the two classes. Orders are omitted.)

Monocots (Class Monocotyledoneae) [Greek *monos*, one; *cotyledon*, a cavity]
Seed with one cotyledon; leaves with parallel veins; flower parts usually in threes or multiples of three; mostly herbaceous. About 34 000 species.

Family Alismataceae (Water Plantain Family) Aquatic or marsh plants; flowers radially symmetrical; 3 sepals, 3 petals, 6 to many stamens, 6 to many carpels, ovary superior (located *above* the attachment of sepals, petals, and stamens).

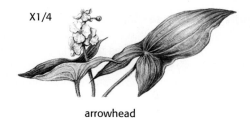

X1/4

arrowhead

Family Amaryllidaceae (Amaryllis Family)
Flowers radially symmetrical, 3 sepals, 3 petals, 6 stamens, one carpel, ovary inferior (located *below* the attachments of sepals, petals, and stamens).

1/3

narcissus

1/100

century plant

Family Commelinaceae (Spiderwort Family) Flowers radially or somewhat bilaterally symmetrical; 3 sepals, 3 petals, 3 or 6 stamens, one carpel, ovary superior.

X1/4

spiderwort

Family Cyperaceae (Sedge Family) Stems usually solid; leaves sheath the stem. Flowers radially symmetrical; no sepals or petals (but scalelike structures present), 1 to 3 stamens, one carpel, ovary superior.

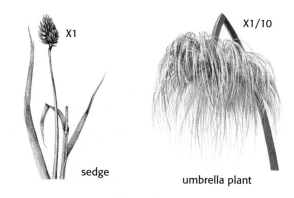

X1

X1/10

sedge

umbrella plant

Family Iridaceae (Iris Family) Flowers radially or somewhat bilaterally symmetrical; 3 sepals, 3 petals, 3 stamens, one carpel, ovary inferior.

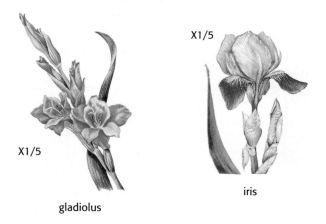

X1/5

X1/5

gladiolus

iris

Family Liliaceae (Lily Family)

Flowers radially symmetrical; 3 sepals, 3 petals (sepals and petals often colored alike, thus appearing to be 6 petals), 3 or 6 stamens, one carpel, ovary superior.

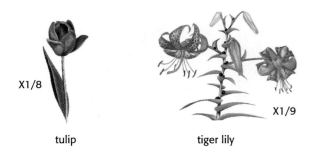

X1/8

tulip

X1/9

tiger lily

Family Orchidaceae (Orchid Family)

Flowers bilaterally symmetrical; 3 sepals, 3 petals (united), 1 or 2 stamens, one carpel, ovary inferior.

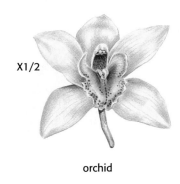

X1/2

orchid

Family Poaceae (Grass Family)

Stems usually hollow; leaves sheath the stem. Flowers radially symmetrical; no sepals or petals (but scalelike structures present), 1 to 6 stamens, one carpel, ovary superior.

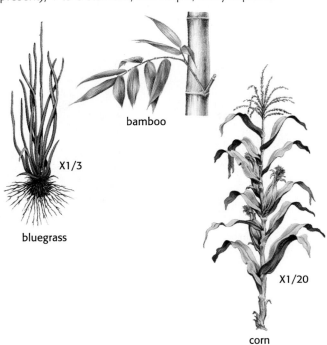

bamboo

X1/3

bluegrass

X1/20

corn

Dicots (Class Dicotyledoneae) [Greek *dis*, two; *cotyledon*, a cavity]

Seed with two cotyledons; leaf veins form a network; flower parts usually in four or fives or multiples of these numbers. About 166 000 species.

Family Apiaceae (Parsley Family)

Herbaceous; small, usually numerous, flowers in an umbel (umbrella-shape); flowers radially symmetrical; 5 small sepals, 5 petals, 5 stamens, one carpel, ovary inferior.

X1/2

Queen Anne's Lace

Family Asteraceae (Sunflower Family)

Herbaceous; small flowers in dense groups forming a head that appears to be a single, large flower. Individual flowers radially or bilaterally symmetrical; sepals reduced to bristles or scales, 5 petals (united), 5 stamens, one carpel, ovary inferior.

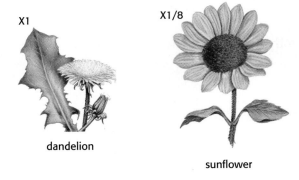

X1

dandelion

X1/8

sunflower

Family Brassicaceae (Mustard Family)

Herbaceous; flowers radially symmetrical; 4 sepals, 4 petals arranged in the form of a cross, 2 sets of stamens (4 long and 2 short), one carpel, ovary superior. Often having a turniplike or cabbage like odor.

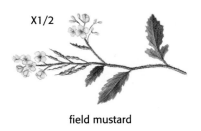

X1/2

field mustard

Family Caprifoliaceae (Honeysuckle Family)

Woody shrubs or vines; flowers radially or bilaterally symmetrical; 4 or 5 sepals, 4 or 5 petals (united), 4 or 5 stamens, one carpel, ovary inferior.

twinflower

snowberry

Family Fabaceae (Pea Family)

Woody or herbaceous; flowers bilaterally symmetrical; 5 sepals, 5 petals, 10 stamens, one carpel, ovary superior. Legumes.

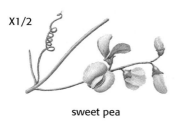

sweet pea

Family Fagaceae (Oak Family)

Trees and shrubs; flowers radially symmetrical, 4 to 7 sepals, no petals, few to many stamens, one carpel, ovary inferior; carpels and stamens in separate flowers.

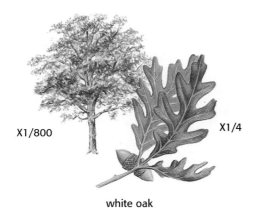

white oak

Family Lamiaceae (Mint Family)

Herbaceous; stems usually square in cross section; flowers bilaterally symmetrical; 5 sepals (united), 5 petals (united), 2 or 4 stamens, one carpel, ovary superior.

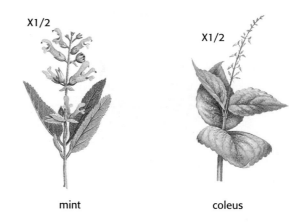

mint

coleus

Family Polemoniaceae (Phlox Family)

Herbaceous; flowers radially symmetrical; 5 sepals, (united), 5 petals (united), 5 stamens, one carpel, ovary superior.

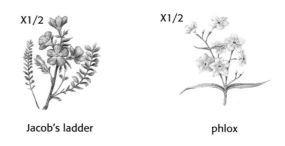

Jacob's ladder

phlox

Family Ranunculaceae (Buttercup Family)

Herbaceous; flowers radially symmetrical; few to many sepals and petals, many stamens and carpels, ovary superior.

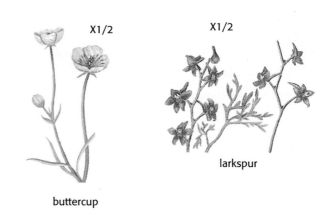

buttercup

larkspur

Family Rosaceae (Rose Family)

Herbaceous or woody; flowers radially symmetrical; 5 sepals, 5 petals, numerous stamens, one to many carpels, ovary either superior or less inferior.

Family Scrophulariaceae (Figwort Family) Herbaceous; flowers bilaterally symmetrical; 5 sepals, 5 petals, (2 forming an upper lip and 3 forming a lower lip), 4 stamens (in 2 unlike pairs), one carpel, ovary superior.

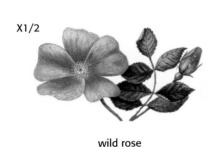

X1/2

wild rose

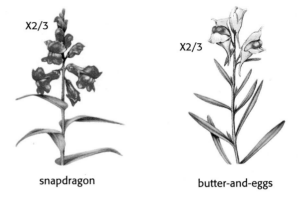

X2/3

snapdragon

X2/3

butter-and-eggs

A Catalog of Living Animals

Eukaryotic, heterotrophic; most are motile at some stage in the life cycle; multicellular. Reproduction usually sexual, with organisms developing from an embryo that has a blastula stage. 33 phyla; more than one million species.

Sponges (Phylum Porifera) [Latin *porus*, pore; *ferre*, to bear]

Free-living mostly marine. Sessile—adults attached to solid object. Asymmetrical; two cell layers; body with pores connected to an internal canal system. Reproduction asexual and sexual. About 10 000 species.

Cnidarians (Phylum Cnidaria) [Greek *knide*, nettle]

Free-living; mostly marine. Motile stages. Radially symmetrical; two cell layers with jellylike material between; digestive sac; mouth surrounded by stinging tentacles; nerve network. Reproduction asexual and sexual. About 9500 species.

Class Hydrozoa (hydras) [Greek *hydra*, water serpent; *zoon*, animal]

Single individuals or colonies; digestive sac undivided; simple sense organs. About 3100 species.

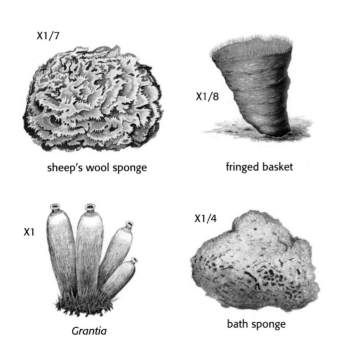

X1/7

sheep's wool sponge

X1/8

fringed basket

X1

Grantia

X1/4

bath sponge

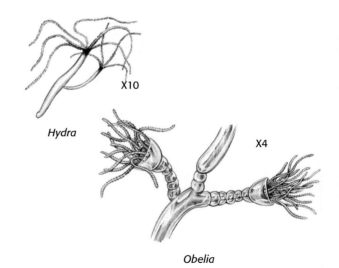

X10

Hydra

X4

Obelia

Class Scyphozoa (jellyfish) [Greek *skyphos,* cup; *zoon,* animal]

Single individuals that float or swim; a few species have attached stages in the life cycle. Digestive sac divided; rather complex sense organs. About 200 species.

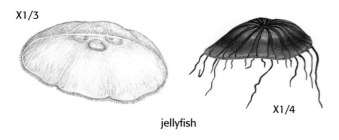

X1/3

X1/4

jellyfish

Class Anthozoa (corals and sea anemones) [Greek *anthos,* flower; *zoon,* animal]

Single individuals or massive colonies; no floating or swimming stages. Often produce limy skeletons; digestive sac divided. About 6200 species.

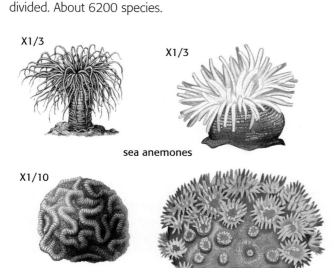

X1/3

X1/3

sea anemones

X1/10

brain coral

X1

coral

Comb Jellies (Phylum Ctenophora) [Greek *kteis,* comb; *pherein,* to bear]

Free-living; marine. Floating or free-swimming by means of eight bands of paddlelike comb plates; somewhat resemble jellyfish without stinging cells. Reproduction sexual. About 90 species.

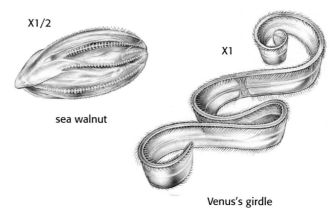

X1/2

X1

sea walnut

Venus's girdle

Mesozoans (Phylum Mesozoa) [Greek *mesos,* middle; *zoon,* animal]

Parasitic in flatworms, mollusks, and annelids. Small, wormlike, bilaterally symmetrical; two cell layers; lack organs and systems except for ovaries and testes. About 45 species.

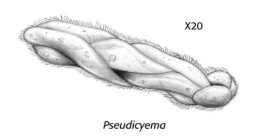

X20

Pseudicyema

Flatworms (Phylum Platyhelminthes) [Greek *platys,* flat; *helmins,* worm]
Free-living or parasitic; motile. Bilaterally symmetrical; flat; three cell layers; branched or unbranched digestive sac or no digestive cavity; excretion by flame cells. Reproduction asexual and sexual. About 15 000 species.

Class Turbellaria (planarians) [Latin *turba,* disturbance (so named because the cilia cause tiny currents in the water)]
Free-living; usually with external cilia. Mostly marine, but some freshwater or terrestrial species.

X8

planarian

Class Trematoda (flukes) [Greek *trematodes,* with holes]
Parasitic; no external cilia. Usually possess hooks and suckers; digestive system present.

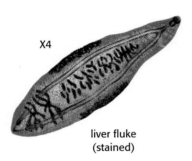

X4

liver fluke
(stained)

Class Cestoda (tape worms) [Greek *kestos,* girdle]
Internal parasites of animals; complex life cycles involving alternate hosts; no digestive system.

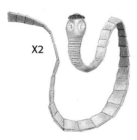

X2

tapeworm

Ribbon Worms (Phylum Nemertina) [Greek *Nemertes,* a sea nymph]
Free-living; mostly marine. Bilaterally symmetrical; flat; unsegmented; digestive tube with two openings; circulatory system. About 900 species.

X1/2

Cerebratulus

Gastrotrichs (Phylum Gastrotricha) [Greek *gaster,* stomach; *thrix,* hair]
Free-living; freshwater or marine. Bilaterally symmetrical; flat; unsegmented; spiny bristly, or scaly. Reproduction sexual. About 400 species.

X22

Chaetonotus

Rotifers (Phylum Rotifera) [Latin *rota,* wheel; *ferre,* to bear]
Mostly free-living; some symbiotic or parasitic; freshwater or marine. Mostly motile; microscopic. Bilaterally symmetrical; body cavity; digestive tube with two openings; cilic form a wheel around mouth; no circulatory system. Reproduction asexual and sexual. About 2000 species.

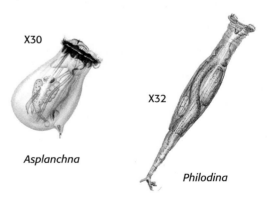

X30

Asplanchna

X32

Philodina

Kinorynchs (Phylum Kinorhyncha) [Greek *kinein,* to move; *rhynchos,* beak]
Free-living; marine; minute. Spiny, retractable head; pull themselves along by rings of recurved spines. About 150 species.

Echinoderella

Spiny-headed Worms (Phylum Acanthocephala) [Greek *akantha,* spine; *kephale,* head]
Young parasitic in arthropods; adults parasitic in intestine of vertebrates; no digestive system. Reproduction sexual. More than 600 species.

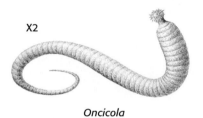

Oncicola

Roundworms (Phylum Nematoda) [Greek, *nema,* thread]
Free-living or parasitic, especially on roots of plants; about 30 species parasitic in humans; many species are decomposers; motile. Bilaterally symmetrical; cylindrical; digestive tube with two openings; no circulatory system. Reproduction sexual. More than 80 000 species.

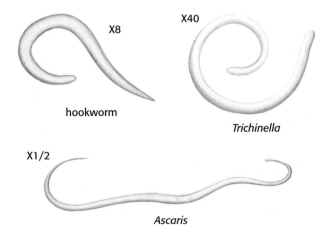

hookworm

Trichinella

Ascaris

Horsehair Worms (Phylum Nematomorpha) [Greek *nema,* thread; *morphe,* form]
Young are parasitic in arthropods; adults free-living, with much reduced digestive tubes. About 240 species.

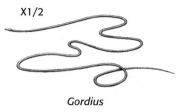

Gordius

Bryozoans (Phylum Ectoprocta) [Greek *ektos,* outside; *proktos,* anus]
Free-living; mostly marine, forming attached colonies. U-shaped digestive tube; mouth encircled by a crown of tentacles. Reproduction sexual. About 5000 species.

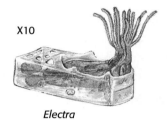

Electra

Phoronids (Phylum Phoronida) [Greek *pherein,* to bear; Latin *nidus,* nest]
Free-living; marine. Live in tubes in mud; U-shaped digestive tube; ring of tentacles forms a food-gathering organ; closed circulatory system with hemoglobin-containing red blood cells. Reproduction sexual. About 15 species.

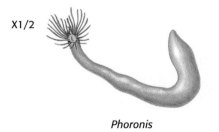

Phoronis

Brachiopods (Phylum Brachiopoda) [Latin *brachion,* arm; Greek *pous,* foot]
Free-living; attached or burrowing; marine. Bilaterally symmetrical; dorsal-ventral shells enclosing tentacle-bearing arms; open circulatory system. Reproduction sexual; sexes separate. About 335 species.

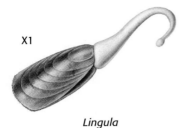

X1

Lingula

Mollusks (Phylum Mollusca) [Latin *molluscus,* soft]
Free-living; marine, freshwater, or terrestrial; sessile or motile. Bilaterally symmetrical or asymmetrical; no segmentation; true body cavity; well-developed digestive, circulatory, and nervous systems. Soft-bodied, with external or internal shell lined with a fold of tissue, the mantle, which secretes the hard, limy shell. Reproduction sexual. About 110 000 species.

Class Monoplacophora [Greek *monos,* one; *plax,* tablet, flat plate; *phora,* bearing]
Marine; found in deep ocean trenches. Single shell with a curved apex, broad, flattened foot. Ten species.

X2

Neopilina

Class Polyplacophora (chitons) [Greek *polys,* many; *plax,* plate; *phora,* bearing]
Marine; shell composed of eight overlapping plates (exposed or hidden); no distinct head.

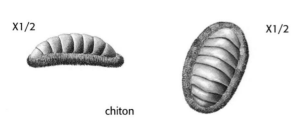

X1/2   X1/2

chiton

Class Scaphoda (tusk shells) [Greek *skaphe,* boat; *pous,* foot]
Marine; shells form a tapering tube; food-catching tentacles on head.

X1/2

tuskshell

Class Gastropoda (slugs and snails) [Greek *gaster,* stomach; *pous,* foot]
Marine, freshwater, or terrestrial; shell (if present) coiled; head usually distinct.

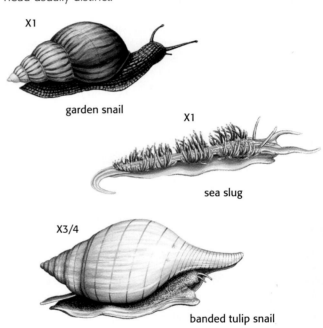

X1

garden snail

X1

sea slug

X3/4

banded tulip snail

Class Bivalvia (bivalves) [Latin *bi,* two; *valva,* folding door]
Marine or freshwater; some attached; others burrow in mud or sand. Shells in two parts, hinged.

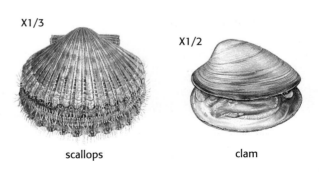

X1/3

X1/2

scallops   clam

Class Cephalopoda (octopus and squid) [Greek *kephale,* head; *pous,* foot]
Marine; small, internal shell. In a few cases shell is external; coiled, and internally divided. Several tentacles on head; locomotion by jet of water.

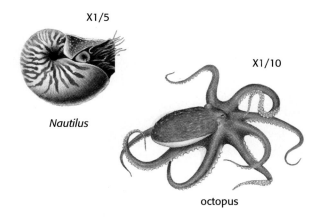

X1/5

Nautilus

X1/10

octopus

Priapulids (Phylum Priapulida) [Latin *priapos,* phallus]
Free-living; marine; burrowing; mouth with curved spines that grasp prey; body covered with rings of cuticle. Reproduction sexual. About 5 species.

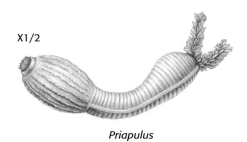

X1/2

Priapulus

Segmented Worms (Phylum Annelida) [Latin *annellus,* little ring]
Free-living or parasitic; marine, freshwater, or terrestrial; motile. Bilaterally symmetrical; body internally and externally segmented; appendages either not jointed or tracking; main nerve cord ventral. Reproduction asexual and sexual. About 8800 species.

Class Oligochaeta (earthworms) [Greek *oligos,* few; *chaite,* hair]
Mostly freshwater or terrestrial. Appendages small or lacking.

X1

earthworm

Class Hirudinea (leeches) [Latin *hirudo,* leech]
Parasitic, with suction disks at each end; flat; appendages lacking.

X1

leech

Class Polychaeta [Greek *polys,* many; *chaite,* hair]
Mostly marine; burrowers or tube-builders; usually with paddlelike appendages on each body segment.

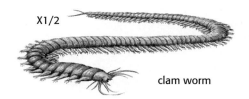

X1/2

clam worm

Velvet Worms (Phylum Onychophora) [Greek *onyx,* claw; *pherein,* to bear]
Free-living; terrestrial; tropical; require high humidity. Wormlike, unsegmented; paired legs; one pair short antennae. Combine many annelid and arthropod characteristics. About 80 species.

X1/2

Peripatus

Arthropods (Phylum Arthropoda) [Greek *arthron,* joint; *pous,* foot]
Free-living; important pollinators; many transmit parasites; marine, freshwater, or terrestrial; motile. Bilaterally symmetrical; segmented body fused to two or three parts; jointed appendages; joined exoskeleton; ventral nerve cord. Reproduction sexual, metamorphosis common in life cycle. About 800 000 described species.

Subphylum Chelicerata [Greek, *chele,* claw]
First pair of appendages modified to grasp food; four pairs of legs; no antennae; two body parts (head-thorax and abdomen); no wings.

Class Merostomata (horseshoe crabs) [Greek *meros,* thigh; *stoma,* mouth]
Aquatic, compound lateral eyes; gas exchange through gills.

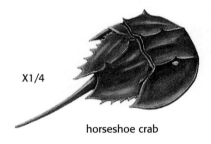

X1/4

horseshoe crab

Class Arachnida (spiders, ticks, and mites) [Greek *arachne,* spider]
Terrestrial; simple eyes; gas exchange through trachea, lungs, or gill books.

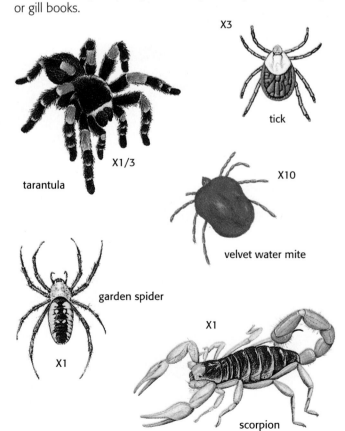

X3

tick

X1/3

tarantula

X10

velvet water mite

garden spider

X1

X1

scorpion

Subphylum Crustacea [Latin *crusta,* shell]
Mostly aquatic; compound eyes; gas exchange through gills; two pairs of antennae; jaws; three distinct body parts.

Class Branchiopoda (fairy shrimp, brine shrimp, water fleas) [Greek *branchia,* gills; *pous,* foot]
Reduce first antennae; gas exchange through gills on flattened, leaflike legs.

X15

X16

Daphnia

brine shrimp

Class Copepoda (cyclops) [Greek *kope,* oar; *pous,* foot]
Free-living and parasitic. Major component of plankton; important food item for whales, sharks, many fish.

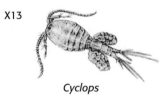

X13

Cyclops

Class Cirripedia (barnacles) [Latin *cirrus,* curl of hair; *pes,* foot]
Free-living filter feeders or parasites; marine; sessile as adults; frequently foul ship bottoms in such large numbers that ship speed is reduced 30-40 percent.

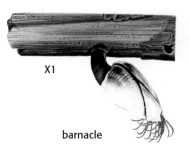

X1

barnacle

X1/2

rock barnacle

Class Malaostraca (sow bugs, crabs, lobsters) [Greek *malakos,* soft; *ostrakon,* shell]
Mostly marine; sow bugs (isopods) terrestrial in moist soil litter under stones; decapods (lobsters, crabs, true shrimp) exist in great variety of forms; important food source.

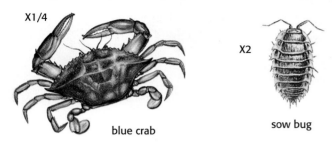

X1/4

blue crab

X2

sow bug

X1/3

shrimp

Subphylum Uniramia [Latin *unus,* one; *ramus,* branch]
Three distinct body parts; one pair of antennae.

Class Diplopoda (millipedes) [Greek *diploos,* double; *pous,* foot]
Important forest decomposers. Many-segmented, cylindrical body; one pair of short antennae; two pairs of legs per segment; gas exchange through trachea.

X1/2

millipede

Class Chilopoda (centipedes) [Greek *cheilos,* lip; *pous,* foot]
Predators, mostly of insects. Many-segmented, flattened body; one pair of long antennae; one pair of legs per segment; gas exchange through trachea.

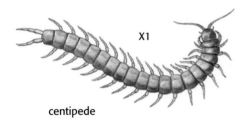

X1

centipede

Class Insecta (insects) [Latin *insectus,* cut into (from the segmented bodies)]
Three body parts (head, thorax, and abdomen); three pairs of legs on thorax; generally one or two pairs of wings; gas exchange through trachea. Perhaps 10 million species. (The following are the more common orders in the class Insecta.)

Order Thysanura [Greek *thysanos,* tassel; *oura,* tail]
Small; wingless; soft scales on the body; three long bristles at posterior end.

X4

silverfish

Order Ephemeroptera [Greek *ephemeros,* temporary (literally, lasting but a day); *pteron,* wing]
Two pairs of transparent wings, hind wings smaller; two or three long tails; immature forms aquatic.

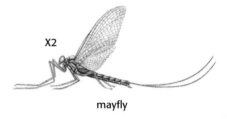

X2

mayfly

Order Odonata [Greek *odous,* tooth; *ata,* characterized by]
Two similar pairs of long wings; antennae short; abdomen long and slender; immature forms aquatic.

X1

damselfly

Order Orthoptera [Greek *orthos,* straight; *pteron,* wing]
Front wings leathery; hind wings folded, fanlike; chewing mouth parts; terrestrial.

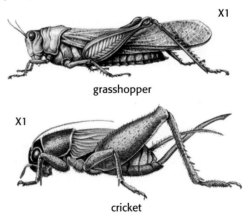

X1

grasshopper

X1

cricket

Order Isoptera [Greek *isos,* equal; *pteron,* wing]
Four wings alike in size, with many fine veins, or wings lacking; chewing mouth parts; social.

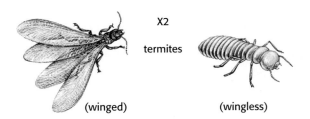

X2
termites

(winged)   (wingless)

Order Anoplura [Greek *anoplos,* unarmed; *oura,* tail]
Wingless, flat body; sucking or piercing mouth parts; parasitic on mammals.

X15

louse

Order Hemiptera [Greek *hemi,* half; *pteron,* wing]
Front wings thick at the base, thin at the tips; hind wings thin; jointed beak on front of head.

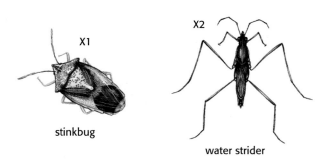

X1

stinkbug

X2

water strider

Order Homoptera [Greek *homos,* same; *pteron,* wing]
Two pairs of wings arched above the body, or wings lacking; jointed sucking beak at the base of head.

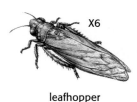

X6

leafhopper

X5

aphid

Order Coleoptera [Greek *koleos,* sheath; *pteron,* wing]
Front wings form a hard sheath; hind wings folded beneath them; chewing mouth parts; young usually wormlike (grubs).

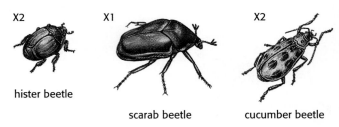

X2

X1

X2

hister beetle

scarab beetle   cucumber beetle

Order Lepidoptera [Greek *lepidos,* scale; *pteron,* wing]
Two pairs of wings covered with soft scales; coiled sucking mouth parts in adults; young wormlike (caterpillars).

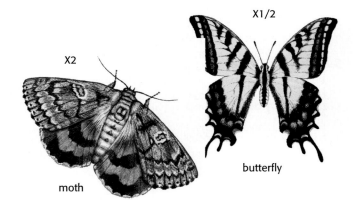

X1/2

X2

butterfly

moth

Order Diptera [Greek *dis,* two; *pteron,* wing]
One pair of wings; hind wings reduced to small rods; antennae short; sucking mouth parts; young wormlike and either terrestrial (maggots) or aquatic.

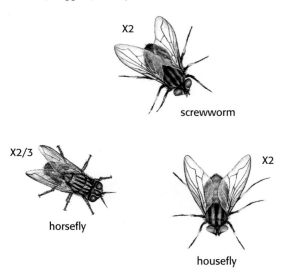

X2

screwworm

X2/3

X2

horsefly

housefly

Order Hymenoptera [Greek *hymen,* membrane; *pteron,* wing]

Front wings much larger than hind wings, with the two pairs hooked together, or wings lacking, chewing or sucking mouth parts; young wormlike; many social species.

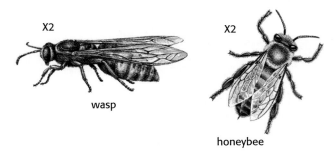

X2

X2

wasp

honeybee

Echinoderms (Phylum Echinodermata) [Greek *echinos,* sea urchin; *derma,* skin]

Free-living; marine; feeding, locomotion, and gas exchange by tube feet. Adults radially symmetrical; unsegmented; no head or brain. Internal, limy skeleton, usually with many projecting spines. Reproduction asexual and sexual; larvae bilaterally symmetrical; chordatelike. About 6000 species.

Class Crinoidea (crinoids) [Greek *krinon,* lily; *eidos,* appearance]

Many highly branched arms; attached (at least when young).

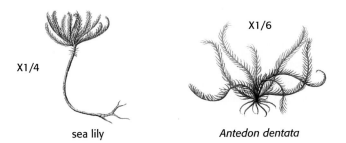

X1/4

X1/6

sea lily

Antedon dentata

Class Asteroidea (sea stars) [Greek *aster,* star; *eidos,* form]

Usually five arms, joined to the body at broad bases.

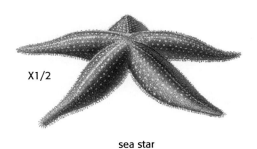

X1/2

sea star

Class Ophiuroidea (brittle stars) [Greek *phis,* snake; *oura,* tail; *eidos,* form]

Usually five long, slim arms (sometimes branched), clearly distinguished from the body; pharyngeal gill slits, at least in embryo. Reproduction sexual; sexes separate. About 45 000 species.

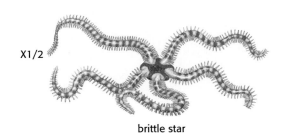

X1/2

brittle star

Class Echinoidea (sand dollars) [Greek *echinos,* sea urchin; *eidos,* form]

No arms; spherical or disk-shaped; long spines or short, hairlike projections from body; skeleton of interlocking plates.

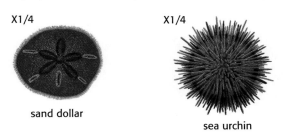

X1/4

X1/4

sand dollar

sea urchin

Class Holothuroidea (sea cucumbers) [Greek *holothourion,* sea cucumber; *eidos,* form]

No arms; somewhat cylindrical; tentacles around mouth; no spines, skeleton consists of particles embedded in the leathery skin.

X1/4

sea cucumber

Arrowworms (Phylum Chaetognatha) [Greek *chaite,* hair; *gnathos,* jaw]
Predators; grasp prey with moveable hooks; marine; free swimming or floating. Bilaterally symmetrical; straight digestive tube. Reproduction sexual; hermaphroditic. More than 100 species.

X1

Sagitta

Acorn Worms (Phylum Hemichordata) [Greek *hemi,* half; Latin *chorda,* cord]
Predators, marine. Bilaterally symmetrical; unsegmented; wormlike, conspicuous proboscis used for burrowing in mud and sand; dorsal nerve cord and pharyngeal slits; notochord doubtfully present. Reproduction asexual, by budding, or sexual; sexes separate. About 90 species.

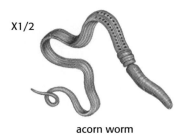

X1/2

acorn worm

Chordates (Phylum Chordata) [Latin *chorda,* cord]
Marine, freshwater, or terrestrial. Bilaterally symmetrical; usually with paired appendages; some segmentation, especially in arrangement of muscles and nerves. Hollow dorsal nerve tube; cartilaginous notochord that may be lost or replaced by vertebral column during development.

Subphylum Urochordata (tunicates) [Greek *oura,* tail; Latin, *chorda,* cord]
Marine, larvae free-swimming; adults usually attached. Notochord disappears during development; no brain.

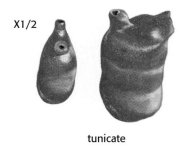

X1/2

tunicate

Subphylum Cephalochordata (lancelets) [Greek *kephale,* head; Latin *chorda,* cord]
Marine; free-swimming; translucent; well-developed hollow dorsal nerve cord, notochord, and pharyngeal slits in adults.

X3/4

lancelet

Subphylum Vertebrata (vertebrates) [Latin *vertebra,* joint]
Notochord replaced by a backbone of vertebrae during development; brain protected by cartilage or bone; usually with paired appendages.

Class Agnatha (jawless fishes) [Greek *a,* without; *gnathos,* jaw]
Cartilaginous skeleton; no true jaws or paired appendages; two-chambered heart; gas exchange through gills.

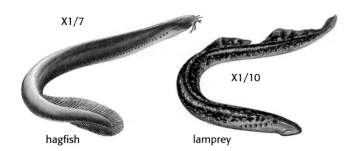

X1/7

X1/10

hagfish   lamprey

Class Chondrichthyes (cartilaginous fishes) [Greek *chondros,* cartliage; *ichthyes,* fish]
Cartilaginous skeleton; gill slits externally visible; jaws; paired fins; two-chambered heart; gas exchange through gills.

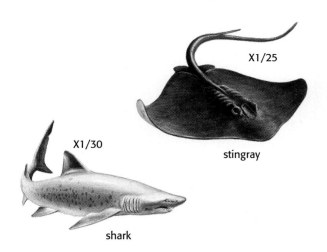

X1/25

stingray

X1/30

shark

Class Osteichthyes (bony fishes) [Greek *osteon,* bone; *ichthyes,* fish]

Bony skeleton; gill slits covered; jaws; paired fins; scales; two-chambered heart; gas exchange through gills.

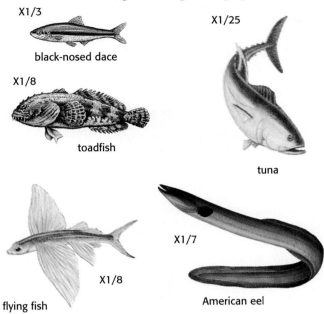

X1/3
black-nosed dace

X1/8
toadfish

X1/25
tuna

flying fish
X1/8

X1/7
American eel

Class Amphibia (amphibians) [Greek *amphis,* both; *bios,* life]

Bony skeleton; moist glandular skin; most with pairs of limbs; three-chambered heart; gas exchange through skin or lungs in adult; through gills in aquatic larvae. About 2000 species.

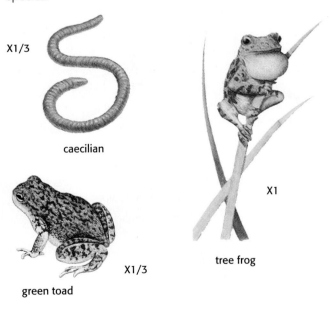

X1/3
caecilian

X1
tree frog

X1/3
green toad

X1/2   tiger salamander

Class Reptilia (reptiles) [Latin *repere,* to creep]

Bony skeleton; dry, scaly skin; most with two pairs of limbs; leathery-shelled eggs; incompletely divided four-chambered heart; gas exchange through lungs. About 5000 species.

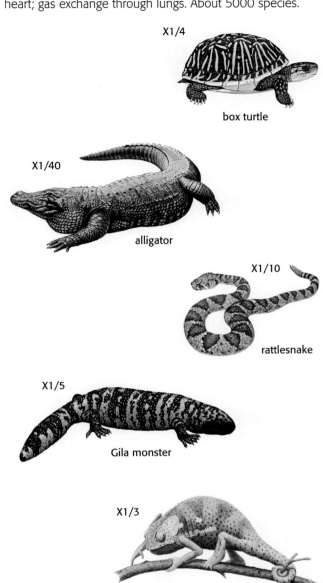

X1/4
box turtle

X1/40
alligator

X1/10
rattlesnake

X1/5
Gila monster

X1/3
chameleon

X1/2
six-lined race runner

Class Aves (birds) [Latin *avis*, bird]
Bony skeleton; scales modified as wings; hard-shelled eggs; four-chambered heart; endothermic; gas exchange through lungs. Nearly 9000 species.

X1/14
turkey vulture

X1/8
quetzal

X1/36
albatross

X1/3
broad-tailed hummingbird

X1/7
mourning dove

robin
X1/5

X1/16
barred owl

X1/24
king penguin

X1/27
brown pelican

X1/3
lazuli bunting

X1/21
tree sparrow

X1/25
flamingo

Class Mammalia (mammals) [Latin *mamma*, breast]
Bony skeleton; scales modified as hairs; mammary glands in females secrete milk; four-chambered heart; endothermic; gas exchange through lungs; teeth usually of four well-defined types (incisors, canines, premolars, molar). About 4500 species.

Order Monotremata [Greek *monos*, single; *trema*, hole]
Egg-laying; mammary glands without nipples. 5 species

X1/8
platypus

Order Marsupialia [Greek *marsypos*, pouch]
Young born in undeveloped state and transferred to a pouch, where they remain tightly attached to the nipples. About 250 species.

1/16
opossum

X1/43
great red kangaroo

X1/16
koala

Order Insectivora [Latin *insectum*, insect; *vorare*, to eat]
Numerous teeth of all four mammalian types, none highly specialized. About 400 species.

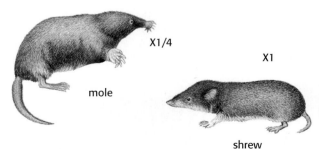

X1/4
mole

X1
shrew

Order Chiroptera [Greek *cheir,* hand; *pteron,* wing]
Web of skin between fingers and between front limbs and hind limbs allows flight. About 900 species.

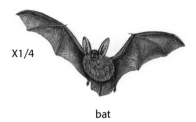

X1/4

bat

Order Carnivora [Latin *carnis,* flesh; *vorare,* to eat]
Incisors small; canines large; premolars adapted for shearing; claws usually sharp. About 280 species.

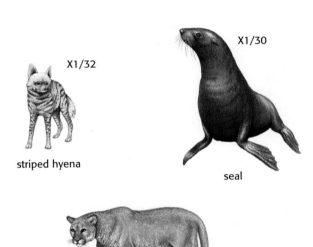

X1/32

X1/30

striped hyena

seal

X1/32

cougar

Order Tubulidentata [Latin *tubulus,* tube; *dens,* tooth]
Teeth few in adults but numerous in embroys; toes ending in nails that are intermediate between claws and hoofs. One species.

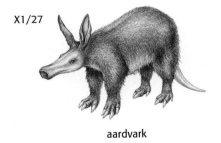

X1/27

aardvark

Order Rodentia [Latin *rodere,* to gnaw]
Chisel-like incisors that grow continually from the roots; no canines; broad molars. About 1700 species.

X1/10

X1/10

woodchuck

squirrel

Order Pholidota [Greek *pholis,* horny scale]
No teeth; body encased in scales formed from modified hairs. 8 species.

X1/12

pangolin

Order Lagomorpha [Greek *lagos,* hare; *morphe,* form]
Harelike mammals; teeth similar to those of rodents, but with four incisors instead of two; tail very short. About 60 species.

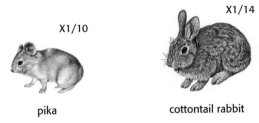

X1/10

X1/14

pika

cottontail rabbit

X1/20

porcupine

Order Edentata [Latin *edentatus,* toothless]
No front teeth; molars in some species. About 30 species.

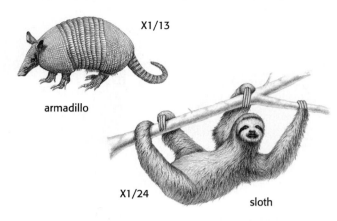

X1/13

armadillo

X1/24

sloth

Order Cetacea [Latin *cetos,* whale]
Marine; front limbs modified as flippers; hind limbs absent; no hair on adults; eyes small; head very large. About 80 species.

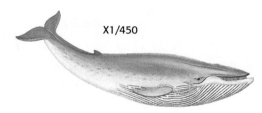

X1/450

blue whale

Order Proboscidea [Greek *pro,* before; *baskein,* to feed]
Herbivorous; upper incisors modified as tusks; molars produced two to four at a time as older ones wear out; nose and upper lip modified as a trunk. 2 species.

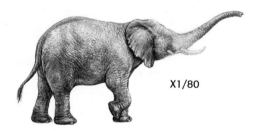

X1/80

elephant

Order Sirenia [Greek *seiren,* sea nymph]
Herbivorous; aquatic; no hind limbs; broad, flat tail expanded as a fin; few hairs. 5 species.

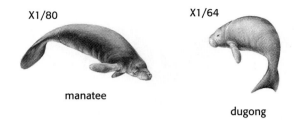

X1/80

manatee

X1/64

dugong

Order Perissodactyla [Greek *perissos,* odd; *daktylos,* toe]

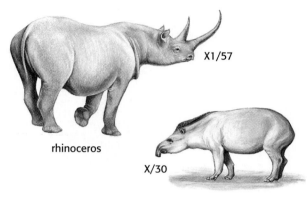

X1/57

rhinoceros

X/30

American tapir

Order Artiodactyla [Greek *artios,* even; *daktylos,* toe]
Herbivorous most with complex stomachs; even-toed, hoofed mammals; often have horns or antlers. About 170 species.

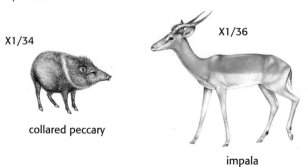

X1/34

collared peccary

X1/36

impala

X1/55

hippopotamus

Order Primates [Latin *prima,* first]

Eyes usually directed forward; nails usually present instead of claws; teeth much like those of insectivores. About 200 species.

X1/30

chimpanzee

X1/40

gorilla

X1/24

baboon

X1/10

red howler

X1/8

marmoset

X1/32

human

To show how complex classification can become, a complete classification of the order Primates follows.

Suborder Prosimii
Infraorder Lemuriformes
Superfamily Tupaioidea
Family Tupaiidea: tree shrews
Superfamily Lemuroidea
Family Lemuridae: lemurs
Family Indriidae: indris
Superfamily Daubentonioidea
Family Daubentoniidae: aye-ayes
Infraorder Loirisformes
Family Lorisidae: Lorises, pottos, galagos
Infraorder Tarsiiformes
Family Tarsiidae: tarsiers
Suborder Anthropoidea
Superfamily Ceboidea
Family Cebidae: New World monkeys
Family Callithricidae: marmosets
Superfamily Cercopithecoidea
Family Cercopithecidae: Old World monkeys, baboons
Superfamily Hominoidea
Family Pongidae: apes
Family Hominidae: humans

Glossary

A

abiotic (AY by OT ik) **factor:** a physical or nonliving component of an ecosystem. p. 27

abscisic (ab SIS ik) **acid:** a plant hormone that protects a plant in an unfavorable environment by promoting dormancy in buds and seeds and closing of stomates. p. 502

absorption: the movement of water and of substances dissolved in water into a cell, tissue, or organism. p. 375

absorption spectrum: the characteristic array of wavelengths (colors) of light that a particular substance absorbs. p. 491

acid deposition: The falling of acids and acid-forming compounds from the atmosphere on the earth's surface; commonly known as acid rain. p. 588

acidic: having a pH of less than 7, reflecting more dissolved hydrogen ions than hydroxide ions. p. 74

acquired characteristics: characteristics acquired during the life of an organism; once thought to be heritable. p. 206

actin: a protein in a muscle fiber that, together with myosin, is responsible for muscle contraction and relaxation. p. 433

action spectrum: a representation of the rate of an activity, especially photosynthesis, under different wavelengths of light at a given light intensity. p. 491

active site: the specific portion of an enzyme that attaches to the substrate through weak chemical bonds. p. 82

active transport: the movement of a substance across a biological membrane against its concentration gradient with the help of energy input and specific transport proteins. p. 110

adaptation: in natural selection, a hereditary characteristic of some organisms in a population that improves their chances for survival and reproduction in their environment compared with the chances of other organisms in the population. p. 206

adaptive radiation: the emergence of numerous species from a common ancestor introduced to an environment presenting a diversity of conditions. p. 212

adenosine diphosphate: see ADP.

adenosine triphosphate: see ATP.

ADP: adenosine diphosphate (uh DEN oh seen dy FOS fayt); the compound that remains when a phosphate group is transferred from ATP to a cell reaction site requiring energy input. p. 78

adrenal cortex: the outer portion of the adrenal gland. p. 446

adrenal medulla: the central portion of the adrenal gland that secretes epinephrine. p. 446

aerobic (eh ROH bik)**:** occurring or living in the presence of free or dissolved oxygen. p. 249

agglutinate (uh GLOO tin ayt)**:** to unite in a mass; clump. p. 415

aggression: forceful or hostile behavior. p. 522

algae: unicellular or multicellular photosynthetic organisms lacking multicellular sex organs. p. 288

alkaline: basic; having a pH greater than 7, reflecting more dissolved hydroxide ions than hydrogen ions. p. 73

allele (al LEEL)**:** one of two or more possible forms of a gene, each affecting the hereditary trait somewhat differently. p. 166

alpine tundra: a biome roughly similar to arctic tundra that occurs above the timberline on mountains. p. 572

alternation of generations: a reproductive cycle in which a haploid (n) phase, the gametophyte, gives rise to gametes that, after fusion to form a zygote, produce a diploid ($2n$) phase, the sporophyte; spores from the sporophyte give rise to new gametophytes. p. 316

alveoli (al VEE oh ly)**:** air sacs in a lung. p. 417

amino (uh MEEN oh) **acid:** an organic compound composed of a central carbon atom to which are bonded a hydrogen atom, an amino group ($-NH_2$), an acid group ($-COOH$), and one of a variety of other atoms or groups of atoms; amino acids are the building blocks of polypeptides and proteins. p. 81

ammonia: (1) a toxic nitrogenous waste excreted primarily by aquatic organisms, p. 267; (2) a gas in the earth's early atmosphere. p. 360

amniocentesis (AM nee oh sen TEEsus)**:** a technique used to detect genetic abnormalities in a fetus through the presence of certain chemicals or defects in fetal cells in the amniotic fluid; the fluid is obtained through a needle inserted into the amniotic sac. p. 150

amnion: a sac or membrane, filled with fluid, that encloses the embryo of a reptile, bird, or mammal. p. 150

amniotic fluid: the liquid that bathes an embryo or fetus. p. 150

amphibian: any of the various ectothermic, smooth skinned vertebrate organisms, belonging to the class Amphibia, that characteristically hatch as aquatic larvae with gills and that develop into adult forms that use lungs for gas exchange. p. 352

anaerobic (an eh ROH bik)**:** occurring or living in conditions without free oxygen. p. 248

anaphase: the stage in mitosis in which chromosomes on the spindle separate and are pulled toward opposite ends of the cell. p. 115

Animalia (an ih MAYL yuh)**:** the animal kingdom. p. 238

Annelida (an NEL ih duh)**:** the phylum that includes segmented worms, such as earthworms and leeches. p. 349

annelids (AN ih lidz)**:** worms belonging to the phylum Annelida. p. 349

anorexia nervosa (an oh REX ee uh ner VOH suh)**:** a condition characterized by abnormal loss of appetite and induced self-starvation to reduce body weight. p. 392

anterior: situated toward the front; having a head end. p. 344

anther: the enlarged end of a stamen, inside which pollen grains with male gametes form in a flower. p. 318

Anthophyta (an THOF ih tuh)**:** the division containing the flowering plants. p. 328

anthropologist (an thruh PAH loh jist)**:** a scientist who studies humans—human evolution, variability, and both past and present cultures and behavior. p. 555

antibody: a blood protein produced in response to an antigen, with which it combines specifically. p. 270

antigen: any material usually a protein that is recognized as foreign and elicits an immune response. p. 409

anus (AY nus)**:** the outlet of the digestive tube. p. 377

Apicomplexa (ap ih kum PLEX suh)**:** the phylum containing the sporozoans. p. 292

appendage (uh PEN dij)**:** a structure attached to the main part of a body; in animals, a tentacle, a leg, a flipper, a wing, etc. p. 344

aquatic (uh KWAH tik)**:** describing a water environment or organisms that live in water. p. 344

arboreal: adapted for living in or around trees, as are monkeys and some apes. p. 552

archaebacteria (AR kee bak TIR ee uh)**:** the more ancient of the two major lineages of prokaryotes represented today by a few groups of bacteria inhabiting extreme environments. p. 263

archaeology: the systematic study of the human past; locating and interpreting the cultural products of prehistoric humans. p. 564

arctic tundra: biome at the northernmost limits of plant growth, where plant forms are limited to low, shrubby, or matlike vegetation. p. 572

artery: a vessel that transports blood away from the heart. p. 403

arthropod: any of the numerous invertebrate organisms of the phylum Arthopoda. p. 349

Arthropoda (ar THRAP uh duh)**:** the phylum that includes insects, crustaceans, arachnids, and others. p. 350

artificial selection: the selective breeding of domesticated plants and animals to encourage the occurrence of desirable traits. p. 205

Ascomycota (AS koh my KOH tuh)**:** the phylum containing the sac fungi. p. 297

ascus (AS kus)**:** a reproductive structure formed by sac fungi when two hyphae conjugate. p. 155

asexual reproduction: any method of reproduction that requires only one parent or one parent cell. p. 127

atherosclerosis (ATH uh roh skleh roh sis)**:** a chronic cardiovascular disease in which plaques develop on inner walls of arteries, restricting the flow of blood. p. 389

atom: the smallest particle of an element; in turn, an atom is made of smaller particles that do not separately have the properties of the element. p. 12

ATP: adenosine triphosphate (uh DEN oh seen try FOS fayt); a compound that has three phosphate groups and is used by cells to store energy. p. 78

atrium (AY tree um; plural atna)**:** a chamber of the heart that receives blood from the veins. p. 404

australopithecine (AH strahl oh PITH uh seen)**:** any of the earliest known species of hominids that walked erect and had humanlike teeth but whose skull, jaw, and brain capacity were more apelike; may include several species. p. 553

autoimmune: a response in which antibodies are produced against some of the body's own cells. p. 414

autonomic nervous system: a division of the nervous system that controls involuntary activities of the body such as blood pressure, body temperature, and other functions necessary to the maintenance of homeostasis. p. 442

autosome: a chromosome that is not directly involved in determining sex. p. 178

autotroph (AWT oh trohf)**:** an organism able to make and store food, using sunlight or another nonliving energy source. p. 238

auxin (AWK sin)**:** a plant hormone produced in an actively growing region of a plant and transported to another part of the plant, where it produces a growth effect. p. 499

axon: a structure that extends out from a neuron and conducts impulses away from the cell body p. 437

B

Bacillariophyta (buh SIL er ee OFF ih tuh)**:** the phylum containing the diatoms. p. 288

basic: alkaline; having a pH greater than 7, reflecting more dissolved hydroxide ions than hydrogen ions. p. 72

basidia (buh SID ee uh)**:** specialized reproductive cells of basidiomycetes, often club-shaped, in which nuclear fusion and meiosis occur. p. 298

Basidiomycota (buh SID ee oh my KOH tuh)**:** the phylum containing the club fungi. p. 298

B cell: a type of lymphocyte that develops in the bone marrow and later produces antibodies. p. 409

bilateral symmetry (by LAT er ul SIM eh tree)**:** a body having two corresponding or complementary halves. p. 343

bile: a secretion of the liver, stored in the gall bladder, and released through a duct to the small intestine; breaks large fat droplets into smaller ones that enzymes can act upon more efficiently p. 375

binomial nomenclature (by NOH mee ul NOH men klay chur)**:** the two-word naming system used in systematics or taxonomy. p. 234

biocide: a poisonous substance produced and used to kill forms of life considered to be pests to humans or that spread diseases. p. 56

biodiversity: the diversity of different species and the genetic variability among individuals within each species. p. 58

biology: the study of living organisms and life processes. p. 6

biomass: the dry weight of organic matter composing a group of organisms in a particular habitat. p. 610

biome: the distinctive plant cover and the rest of the community of organisms associated with a particular physical environment; often the biome is named for its plant cover. p. 569

biosphere (BY oh sfir)**:** the outer portion of the earth—air, water, and soil—where life is found. p. 13

biosynthesis (by oh SIN thuh sis)**:** the process of putting together or building up the large molecules characteristic of a particular type of cell or tissue. p. 383

biotic (by OT ik) **factor:** an organism or its remains in an ecosystem. p. 27

bipedal: capable of walking erect on the hind limbs, freeing the hands for other uses. p. 553

blade: the broad, expanded part of a leaf. p. 463

blastocyst: the mammalian embryonic stage that corresponds to the blastula of other animals. p. 148

blastula (BLAS choo luh)**:** an animal embryo after the cleavage stage, when a pattern of cell movements toward the outside of the ball of cells results in a fluid-filled cavity inside. p. 143

brachiopods (BRAY kee oh podz)**:** marine invertebrates with dorsal and ventral shells and a pair of tentacled structures on either side of the mouth. p. 543

Bryophyta (bry OFF ih tuh)**:** the division containing the bryophytes. p. 320

bryophytes (BRY oh fyts)**:** the mosses, liverworts, and hornworts; a group of nonvascular plants that inhabit land but lack many of the terrestrial adaptations of vascular plants. p. 314

bulimia (buh LEE mee uh)**:** an abnormal craving for food beyond the body's needs; frequently expressed as gorging followed by forced vomiting. p. 392

bundle sheath cells: tightly packed cells surrounding veins in the leaves of C-4 plants that receive a rearranged 4-carbon acid in the first stage of C-4 photosynthesis; within the bundle sheath cells, carbon dioxide is released and reincorporated by an enzyme into PGA. p. 497

C

callus: a mass of dividing, undifferentiated cells at the cut end of a shoot or in a tissue culture, from which adventitious roots develop. p. 501

calorie: the amount of heat required to raise the temperature of one gram of water 1° C. p. 35

Calvin cycle: the cycle (named for its discoverer) by which carbon dioxide is incorporated in sugars during photosynthesis; uses chemical energy previously converted from light energy. p. 491

CAM: crassulacean (kras yoo LAY see un) acid metabolism; an adaptation for photosynthesis in arid conditions, in which carbon dioxide entering open stomates at night is converted into organic acids, which release carbon dioxide during the day when the stomates are less open. p. 498

cambium (KAM bee um)**:** a layer of meristem tissue in the stems and roots of plants that produces all growth in diameter, including new xylem and phloem cells. p. 468

cancer: malignancy arising from cells that are characterized by profound abnormalities in the plasma membrane and in the cytosol, and by abnormal growth and division. p. 152

capillary (KAP ih layr ee)**:** a microscopic blood vessel that penetrates the tissues and that has walls consisting of a single layer of cells that allows exchange between the blood and tissue fluids. p. 403

carbohydrate (kar boh HY drayt)**:** an organic compound made of carbon, hydrogen, and oxygen, with the hydrogen and oxygen atoms in a 2 : 1 ratio; examples are sugars, starches, glycogen, and cellulose. p. 79

carbon cycle: the biogeochemical cycle in which carbon compounds made by some organisms are digested and decomposed by others, releasing the carbon in small inorganic molecules that can be used again by more organisms to synthesize carbon compounds. p. 87

carbon-14 dating: a technique to date fossils that uses the known disintegration rate of radioactive carbon; the amount of carbon-14 remaining in fossils indicates their age. p. 540

cardiac (KARD ee ak) **muscle:** a specialized type of muscle tissue found only in the heart. p. 435

cardiovascular disease: disease of the heart and/or blood vessels, such as atherosclerosis. p. 388

cardiovascular fitness: the relative state of health or fitness of the lungs, heart, and blood vessels, especially as regards their ability to deliver oxygen to the cells. p. 437

carnivore (KAR nih vor): any organism that eats animals; a meat-eater, as opposed to a plant-eater, or herbivore. p. 50

carpel (KAR puhl): the female reproductive organ of a flower, consisting of the stigma, style, and ovary. p. 317

carrying capacity: the maximum population size that can be supported by the available resources of a given area. p. 28

cartilage: a tough, elastic connective tissue that makes up the skeleton of cartilaginous fishes, but that in other vertebrates is replaced mostly by bone as the animal matures. p. 352

catalyst (KAT uh list): a chemical that promotes a reaction between other chemicals and may take part in the reaction but emerges in its original form. p. 73

cell cycle: an ordered sequence of events in the life of a dividing cell, composed of the M, G_1, S, and G_2 phases. p. 111

cell theory: the theory that organisms are composed of cells and their products and that these cells all are derived from preexisting cells. p. 99–100

cellular respiration (SEL yoo ler res pih RAY shun): the series of chemical reactions by which a living cell breaks down food molecules and obtains energy from them. p. 76

cellulose (SEL yoo lohs): a carbohydrate found in cells walls. p. 80

cell wall: a nonliving covering around the plasma membrane of certain cells, as in plants, many algae, and some prokaryotes; in plants the cell wall is constructed of cellulose and other materials. p. 105

central nervous system: the brain and spinal cord in vertebrates. p. 439

centriole (SEN tree ohl): one of two structures in animal cells, composed of cylinders of nine triplet microtubules in a ring; centrioles help organize microtubule assembly during cell division. p. 104, 107

centromere (SEN troh meer): the specialized region of a chromosome that holds two replicated chromosomal strands together and that attaches to the spindle in mitosis. p. 130

cerebellum (ser eh BEL um): the part of the brain in vertebrates that is associated with regulating muscular coordination, balance, and similar functions. p. 440

cerebrum (seh REE brum): the largest portion of the brain in humans and many other animals; controls the higher mental functions such as learning. p. 439

chancre (SHAN ker): an ulcer located at the initial point of entry of a pathogen; a dull red, hard lesion that is the first manifestation of syphilis. p. 276

chaparral (shap uh RAL): a scrubland biome of dense, spiny evergreen shrubs found at midlatitudes along the coast where cold ocean currents circulate offshore; characterized by mild, rainy winters and long, hot, dry summers. p. 581

chelicerates (chih LIS er ayts): arthropods possessing paired appendages near the mouth that are modified for grasping. p. 350

chemical bond: the attraction between two atoms resulting from the sharing or transfer of outer electrons from one atom to another. p. 72

chemical energy: energy stored in the structure of molecules, particularly organic molecules. p. 10

chemosynthesis (KEE moh SIN thih sis): a pathway that uses energy from the oxidation of inorganic substances to drive the formation of organic molecules. p. 268

chitin (KYT in): a hard organic substance secreted by insects and certain other invertebrates as the supporting material in their exoskeletons. p. 296

chlorophyll (KLOR uh fil): the green pigments of plants and many microorganisms; converts light energy (via changes involving electrons) to chemical energy that is used in biological reactions. p. 76

Chlorophyta (kloh ROF it uh): the phylum containing the green algae. p. 288

chloroplast (KLOR oh plast): an organelle found only in plants and photosynthetic protists; contains chlorophyll that absorbs light energy used to drive photosynthesis. p. 105, 107

cholesterol (koh LES ter ol): a lipid that is associated particularly with animal plasma membranes and is linked to deposits in blood vessels and to corresponding disorders of the heart. p. 80

Chordata (kohr DAH tuh): the phylum containing the chordates. p. 351

chordates: a diverse phylum of animals that possess a notocord, a dorsal hollow nerve cord, a tail, and pharyngeal gill slits at some stage of the life cycle. p. 351

chorion (KOR ee on): an embryonic membrane that surrounds all the other embryonic membranes in reptiles, birds, and mammals. p. 148

chorionic villi sampling (CVS): a procedure by which a small piece of membrane surrounding a fetus is removed and analyzed to detect genetic defects. p. 149

chromatid (KROH muh tid): one of two strands of a replicated chromosome before their separation during mitosis or meiosis. p. 130

chromosome (KROH moh sohm): a long, threadlike group of genes found in the nucleus of all eukaryotic cells and most visible during mitosis and meiosis, chromosomes consist of DNA and protein. p. 106

chromosome theory of heredity: the ideas that both eggs and sperm make equal genetic contributions to a new organism, that genes are located in the nucleus of both gametes, and that chromosomes contain the genes. p. 170

cilia (SIL ee uh): short, hairlike cell appendages specialized for locomotion and formed from a core of nine outer doublet microtubules and two inner single microtubules. p. 293

ciliates: unicellular, heterotrophic organisms that move by means of cilia. p. 293

Ciliophora (sil ee OFF er uh): the phylum containing the ciliates. p. 293

class: the third largest grouping, after kingdom and phylum or division, in the biological classification system. p. 232

cleavage (KLEE vaj): the process of cell division in animal cells, characterized by rapid cell divisions without growth that occur during early embryonic development and that convert the zygote into a ball of cells. p. 143

climatogram: a graph of monthly measurements of temperature and precipitation for a given area during a year. p. 571

clone: a lineage of genetically identical individuals. p. 501

closed circulatory system: a type of internal transport in which blood is confined to vessels. p. 358

closed population: a population in which there is no immigration or emigration; a completely isolated population. p. 38

club fungi: fungi belonging to the phylum Basidiomycota and possessing basidia, on which spores are produced. p. 298

Cnidaria: the phylum containing cnidarians. p. 345

cnidarians (nih DAR ee unz): radially symmetrical, stinging-tentacled animals that possess nerve networks and digestive sacs. p. 345

CoA: an enzyme present in all cells and necessary for cellular respiration and fatty acid metabolism. p. 380

coacervate (koh AS er vayt): a cluster of proteinlike substances held together in a small droplet within a surrounding liquid; used as a model for a precell to investigate the formation of the first life on the earth. p. 246

codominance: a condition in which both alleles in a heterozygous organism are expressed. p. 175

codon: the basic unit of the genetic code; a sequence of three adjacent nucleotides in DNA or mRNA. p. 172

coevolution: the evolution of two different species interacting with each other and reciprocally influencing each other's adaptations. p. 215

cohesion-tension hypothesis: the hypothesis that water is raised upward in a column in the xylem of plants by the force holding the water molecules together and by the continuing evaporation of water from the top of the column, in leaves. p. 469

communication: the exchange of thoughts, messages, or information by vocalizations, signals, or writing. p. 520

community: all the organisms that inhabit a particular area. p. 8

companion cell: a specialized elongated cell in the phloem of flowering plants that is associated with a sievetube cell. p. 468

competition: interaction between members of the same population or of two or more populations to obtain a mutually required resource in limited supply p. 52

complement system: a complex group of serum proteins that can destroy antigens. p. 411

complete protein: a protein that contains all the essential amino acids in the proportions required by humans. p. 392

compound: a substance formed by chemical bonds between atoms of two or more different elements. p. 12

concentration gradient: a difference in the concentration of certain molecules over a distance. p. 118

conditioning: training that modifies a response so that it becomes associated with a stimulus different from the stimulus that originally caused it. p. 531

Coniferophyta: the division containing conifers. p. 325

conjugation: the process by which sexual exchange and reproduction take place in many microscopic organisms, principally protists and fungi. p. 263

consumer: a heterotroph; an organism that feeds on other organisms or on their organic wastes. p. 8, 50

contact inhibition: the phenomenon observed in normal animal cells that causes them to stop dividing when they come in contact with one another. p. 153

continental drift: change in distance between continents and in their relative positions, caused by the spreading of new crustal material in sea-floor and by land rifts. p. 544

control: the norm (unchanged subject or group) in an experiment. p. 20

convergent evolution: the independent development of similarity between species as a result of their having similar ecological roles and selection pressures. p. 215

cooperative behavior: behavior that reduces the reproductive fitness of the performing individual and increases the fitness of the recipient. p. 525

cornea (KOR nee uh): the transparent outer layer of the vertebrate eye. p. 146

corpus luteum (KOR pus LOOT ee um): the structure that forms from the tissues of a ruptured ovarian follicle and secretes female hormones. p. 137

cortex: (1) the outer layers of the adrenal gland, p. 472; (2) a layer of cells under the epidermis or bark of some plant stems and roots. p. 467

cotyledons (kot ih LEE dunz): the single (monocot) or double (dicot) seed leaves of a flowering plant embryo. p. 329

C-4 plants: plants with a photosynthetic pathway that incorporates carbon dioxide into 4-carbon compounds before beginning the Calvin cycle. p. 498

cranium: the skull (without the jaw). p. 560

creatine phosphate (KREE uh tin FOS fayt)**:** an energy storage compound used by muscle cells of vertebrates to replenish ATP supplies; replenished by the breakdown of glycogen. p. 435

crossing-over: during prophase I of meiosis, the breakage and exchange of corresponding segments of chromosome pairs at one or more sites along their length, resulting in genetic recombination. p. 130

cultural adaption: an adaptation to new pressures or situations resulting from cultural innovation. p. 639

culture: a system of learned behaviors, symbols, customs, beliefs, institutions, artifacts, and technology characteristic of a group and transmitted by its members to their offspring. p. 635

cuticle: in plants, a noncellular, waxy outer layer covering certain leaves and fruits. p. 314

cyanobacteria (SY an oh bak TIR ee uh)**:** the blue-green bacteria, which carry on oxygen-producing photosynthesis much like plants, but without membrane-enclosed chloroplasts isolating their chlorophyll. p. 267

cytokinesis (syt oh kih NEE sus)**:** the division of the cytoplasm of a cell after nuclear division. p. 112, 115

cytokinin (syt oh KY nin)**:** a plant hormone that promotes cell division, stem and root growth, chlorophyll synthesis, and chloroplast development. p. 502

cytoplasm (SYT oh plaz um)**:** the entire contents of the cell, except the nucleus, bounded by the plasma membrane. p. 103

cytoskeleton: a network of microtubules, microfilaments, and intermediate filaments that run throughout the cytoplasm and serve a variety of mechanical and transport functions. p. 104, 107

cytosol (SYT oh sol)**:** the gelatinlike portion of the cytoplasm that bathes the organelles of the cell. p. 105

D

data (singular, *datum*)**:** observations and experimental evidence bearing on a biological question or problem. p. 14

decomposer: an organism that lives on decaying organic material, from which it obtains energy and its own raw materials for life. p. 8

decomposition (de kom poh ZISH un)**:** the process of taking molecules apart; heat and chemicals are the chief agents. p. 73

deforestation: the removal of trees from a forested area without adequate replanting. p. 587

dendrite: a structure that extends out from a neuron and transmits impulses toward the cell body p. 437

denitrifying (DEE ny trih fy ing) **bacteria:** bacteria that break down nitrogen compounds in the soil and release nitrogen gas to the air. p. 269

density: the number of individuals in a population in proportion to the size of their environment or living space. p. 28

dental caries: cavities; tooth decay caused, in part, by lactic acid bacteria. p. 272

dental plaque: a film composed of eubacteria in a sugar matrix on the surface of teeth. p. 272

deoxyribose (dee OK sih RY bohs)**:** a sugar used in the structure of DNA; it contains one less oxygen atom than does ribose. p. 83

depressant: a drug that slows the functioning of the central nervous system. p. 452

desert: a biome characterized by lack of precipitation and by extreme temperature variation; there are both hot and cold deserts. p. 578

desertification: the conversion of rangeland or cropland to desertlike land with a drop in agricultural productivity; usually caused by a combination of overgrazing, soil erosion, prolonged drought, and climate change. p. 587

development: (1) cell division, growth, and differentiation of cells from embryonic layers into all the tissues and organs of the body; (2) later changes with age, including reproductive maturity with its effects on appearance and body function. p. 143.

diabetes mellitus (MEL luh tus)**:** a disease resulting from insufficient insulin secretion by the pancreas or impairment of insulin receptors on body cells; characterized by abnormal absorption and use of glucose. p. 445

diaphragm (DY uh fram)**:** (1) the sheet of muscle that separates the chest and abdominal cavity in mammals and, along with rib muscles, is important in breathing, p. 416; (2) a caplike rubber device inserted in the vagina and used as a contraceptive. p. 149

diatom (DY uh tom)**:** any of a large group of algae with intricate, patterned, silica-containing shells made in two halves; unusually shaped pores create the patterns. p. 288

dicot (DY kot)**:** flowering plant that has two embryonic seed leaves, or cotyledons. p. 331

differentially permeable: open to passage or penetration by some substances, but not all, depending on molecule size, charge, and other factors. p. 118

differentiation: specialization, as when developing cells become ordered into certain tissues and organs. p. 143

diffusion (dih FYOO zhun)**:** the movement of a substance down its concentration gradient from a more concentrated area to a less concentrated area. p. 108, 118

digestion: the process by which larger food molecules are broken down into smaller molecules that can be absorbed. p. 356

dinoflagellate: any of the numerous, chiefly marine, protozoans characteristically possessing two flagella; one of the major types of plankton. p. 291

dipeptide (DY PEP tyd)**:** two amino acid molecules bonded to one another; the dipeptide may be the start of a chain for a protein or a product of digestion of a protein and a polypeptide. p. 31

diploid (DIP loyd)**:** a cell containing both members of every chromosome pair characteristic of a species (2*n*). p. 129

dispersal: (1) the spreading of organisms from a place of concentration, p. 39; (2) the scattering of spores and seeds that promotes spreading of nonmotile organisms. p. 31

divergent evolution: evolutionary change away from the ancestral type, with selection favoring newly arising adaptations. p. 215

division: the second largest grouping, after kingdom, in the biological classification system for plants. p. 233

DNA: deoxyribonucleic (dee OK sih ry boh noo KLEE ik) acid; the hereditary material of most organisms; DNA makes up the genes; these nucleic acids contain deoxyribose, a phosphate group, and one of four bases. p. 83

dominance hierarchy: a linear "pecking order" of animals in which position dictates characteristic social behaviors. p. 523

dominant: a trait that is visible in a heterozygous organism. p. 165

dormant: inactive but alive or viable. p. 302

dorsal: in animals, situated toward the top or back side. p. 344

E

ecology: the study of living and nonliving components of the environment and of the interactions that affect biological species. p. 6

ecosystem (EE koh sis tum)**:** a biological community in its abiotic environment. p. 49

ectoderm (EK toh derm)**:** the outer layer of cells in the gastrula stage of an animal embryo. p. 144

effector: a muscle or gland activated by nerve impulses or hormones. p. 439

electron (ee LEK trahn)**:** a negatively charged particle that occurs in varying numbers in clouds surrounding the nuclei of atoms. p. 71

electron transport system: the process in which electrons are transferred from one carrier molecule to another in photosynthesis and in cellular respiration; results in storage of some of the energy in ATP molecules. p. 378

element: a substance composed of atoms that are chemically identical—alike in their proton and electron numbers. p. 12

embryo: an organism in its earliest stages of development. p. 133

emigration (em uh GRAY shun)**:** departure by migration; out-migration. p. 27

emphysema (em fuh SEE muh)**:** a lung condition in which the air sacs dilate and the walls atrophy, resulting in labored breathing and susceptibility to infection. p. 417

endocast: a plaster or plastic cast of the inside surface, or the contents, of a covering or enclosing structure; examples include endocasts of a skull's braincase, a plant's spore case, or a marine microorganism's glasslike shell. p. 560

endocrine (EN doh krin) **gland:** a ductless gland that secretes one or more hormones into the bloodstream. p. 443

endocrine system: the system of glands that secrete their products from their cells directly into the blood. p. 443

endoderm (EN doh derm)**:** an inner layer of cells, as in an embryo. p. 144

endoplasmic reticulum (en doh PLAZ mik reh TIK yoo lum)**:** an extensive membranous network in eukaryotic cells composed of ribosome-studded (rough) and ribosomefree (smooth) regions. p. 105, 106

endoskeleton (EN doh SKEL eh tun)**:** a hard skeleton buried in the soft tissues of an animal, such as the spicules of sponges and bony skeletons of vertebrates. p. 362

endosperm (EN doh sperm)**:** a nutrient-rich structure, formed by the union of a sperm cell with two polar nuclei during double fertilization, that provides nourishment to the developing embryo in seeds of flowering plants. p. 318 .

endospore (EN doh spor)**:** a thick-walled spore of a particular type, like that produced by the anaerobic bacterium that causes botulism. p. 263

energy: the active component of all atoms; the ability to do work or to cause change. p. 74

energy pyramid: a graphic representation of the energy available for use by producers and consumers as levels of a pyramid: at the bottom, with the greatest amount of available energy, are the producers, followed by primary consumers, secondary consumers, and top-level consumers; only about 10 percent of the energy available at the preceding level is available for use at the next higher level. p. 55

environment (en VY run ment)**:** everything living and nonliving in an organism's surroundings, including light, temperature, air, soil, water, and other organisms. p. 27

enzyme (EN zym)**:** a protein or part-protein molecule made by an organism and used as a catalyst in a specific biochemical reaction. p. 74

enzyme-substrate complex: an enzyme molecule together with the molecules on which it acts, correctly arranged at the active site of the enzyme. p. 82

epinephrine (ep ih NEF rin)**:** an adrenal hormone, also called adrenaline, that speeds up heart rate and raises blood sugar level and blood pressure, the "fight-or flight" hormone that is secreted during a sudden fright or emergency. p. 447

epiphyte (EP ih fyt)**:** a plant that takes its moisture and nutrients from the air and from rainfall and that usually grows on a branch of another plant; not a parasite. p. 575

esophagus: the tubular portion of the digestive tract that leads from the pharynx to the stomach. p. 373

essential nutrient: a nutrient that an organism cannot synthesize, or cannot synthesize in the quantities it requires; plants obtain these nutrients from the soil, animals from food they ingest. p. 386

estivation (es tih VAY shun)**:** a physiological state, characterized by decreased metabolism and inactivity, that permits survival during long periods of elevated temperature and diminished water supplies. p. 579

estrogen (ES troh jen)**:** a hormone that stimulates the development of female secondary sexual characteristics. p. 137

ethylene (ETH ih leen)**:** a gaseous plant hormone that promotes fruit ripening while inhibiting further plant growth in roots and stems. p. 502

eubacteria (YOO bak TIR ee uh)**:** the bacterial group including the cyanobacteria but not the archaebacteria; sometimes called the "true bacteria"; they differ from archaebacteria in their ribosomal RNA, tRNA, and in other ways. p. 263

euglenoids: a small group of freshwater flagellates similar to the organisms of the genus *Euglena;* many contain chloroplasts, others absorb organic substances or ingest prey; all lack rigid cell walls. p. 291

eukaryote (yoo KAIR ee oht)**:** an organism whose cells have a membrane-enclosed nucleus and organelles; a protist, fungus, plant, or animal. p. 102

eutrophic (yoo TROH fik)**:** characterizing a body of water that is nutrient-rich, leading to population explosions of photosynthetic organisms and then of decomposers, which deplete the dissolved oxygen and cause fish and other aquatic organisms to die. p. 621

excretion (ek SKREE shun)**:** the elimination of wastes, especially by-products of body metabolism, by organisms. p. 359

exon (EKS on)**:** a segment of DNA that is transcribed into RNA and translated into protein, specifying the amino acid sequence of a polypeptide; characteristic of eukaryotes and some prokaryotes. p. 184

exoskeleton (EK soh SKEL eh tun)**:** a hard encasement deposited on the surface of an animal, such as the shell of a mollusk, that provides protection and points of attachment for muscles. p. 362

extensor: a muscle that extends a limb or skeletal part. p. 436

extinct: no longer surviving as a species. p. 59

F

F$_1$: the first filial, or hybrid, offspring in a genetic cross-fertilization. p. 165

F2: offspring resulting from interbreeding of the hybrid F$_1$ generation. p. 165

family: the fifth largest grouping after kingdom, phylum or division, class, and order, in the biological classification system; a group of related genera. p. 232

famine (FAM in)**:** severe shortage of food, causing wide spread hunger and starvation within a population. p. 35

feces (FEE seez)**:** the waste material expelled from the digestive tract. p. 377

feedback system: a relationship in which one activity of an organism affects another, which in turn affects the fist, yielding a regulatory balance. p. 134

femur: the thighbone. p. 560

fermentation (fer men TAY shun)**:** the incomplete breakdown of food molecules, especially sugars, in the absence of oxygen. p. 383

fertilization: the union of an egg nucleus and a sperm nucleus. p. 129

fetus: a vertebrate embryo in later stages of development when it has attained the recognizable structural plan and features of its type. p. 150

fibrin (FY brin)**:** an insoluble, fibrous protein that forms a network of fibers around which a blood clot develops. p. 407

fibrinogen (fy BRIN oh jen)**:** a soluble blood protein that is changed into its insoluble form as fibrin during the blood clotting process. p. 407

fight-or-flight response: a response of the neuroendocrine system to stressors in which the body prepares to fight or flee; includes the secretion of epinephrine that speeds up heart rate and elevates blood pressure. p. 447

flagella (fluh JEL uh)**:** singular, flagellum; the long cellular appendages specialized for locomotion and formed from a core of nine outer doublet microtubules and two single inner microtubules; many protists and certain animal cells have flagella. p. 237

flame cell: in planarians, a cell with cilia in a network connected by tubules; absorbs fluid and wastes and moves fluid through the tubules to eliminate excess water. p. 360

flatworm: any member of the phylum Platyhelminthes. p. 347

flexor: a muscle that bends a joint; its action is opposite to that of an extensor. p. 436

follicle: in mammals, an ovarian sac from which an egg is released. p. 137

follicle-stimulating hormone (FSH): a substance secreted by the anterior lobe of the pituitary that stimulates the development of an ovarian follicle in a female or the production of sperm cells in a male. p. 137

food: a substance containing energy-rich organic compounds made by organisms and used as a source of energy and matter for life. p. 13

food chain: the transfer of food from one feeding level to another, beginning with producers. p. 8

food web: food chains in an ecosystem taken collectively, showing partial overlapping and competition for many food organisms. p. 8

foraminifera (FOR am ih NIF er uh)**:** any of the unicellular microorganisms of the order Foraminifera, characteristically possessing a calcareous shell with perforations through which numerous pseudopodia protrude. p. 292

fossil: a cast of an organism preserved in rock that formed where it died; the organism itself in ice, volcanic glass, or amber; or the skeleton preserved in deposits. p. 231

Fungi (FUN jy)**:** a kingdom of heterotrophic organisms that develop from spores; many are decomposers; some are parasites of other organisms. p. 240

G

G₁ (Growth 1)**:** the first growth phase of the cell cycle, starting just after offspring cells form. p. 115

G₂ (Growth 2)**:** the second growth phase of the cell cycle, beginning after DNA synthesis. p. 115

gamete (GAM eet)**:** a sex cell, either an egg cell or a sperm, formed by meiosis. p. 126

gametophyte (guh MEET oh fyt)**:** a plant of the gamete producing generation in a plant species that undergoes alternation of generations. p. 316

gastric juice: mixed secretions of the glands in the stomach wall; in humans, principally mucus, hydrochloric acid, and protein-fragmenting enzymes. p. 374

gastrula (GAS truh luh)**:** an early embryo at the stage when infolding of cells from the outside occurs. p. 144

gastrulation: the formation of the gastrula stage of an embryo; the stage at which the embryo acquires three layers. p. 144

gene: the fundamental physical unit of heredity, which transmits a set of specifications from one generation to the next; a segment of DNA that codes for a specific product. p. 106

gene flow: the loss or gain of alleles in a population due to the emigration or immigration of fertile individuals. p. 210

gene pool: the total aggregate of genes in a population at any one time. p. 209

genetic drift: changes in the gene pool of a small population due to chance. p. 210

genome: the total genetic content or complement of a haploid cell from any given species. p. 184

genotype (JEE noh typ)**:** the genetic makeup of an organism. p. 166

genus (JEE nus)**:** the next largest grouping after kingdom, phylum or division, class, order, and family in the biological classification system; a group of related species. p. 232

gibberellin (jib uh REL en)**:** a plant hormone that stimulates elongation of the stems, triggers the germination of seeds, and with auxin, stimulates fruit development. p. 500

gills: the respiratory organs chiefly of aquatic organisms such as fish; the chief excretory organs in many ocean fish. p. 358

gizzard: a muscular sac in the digestive system of birds, earthworms, and other animals, that mechanically changes food by grinding it against sand particles. p. 356

glomerulus (glah MER yoo lus)**:** a ball of capillaries surrounded by a capsule in the nephron and serving as the site of filtration in the kidneys. p. 419

glucagon (GLOO kuh gahn)**:** a pancreatic hormone that acts to raise the blood glucose level. p. 444

glycerol: a 3-carbon alcohol molecule that combines with fatty acids to form fats and oils. p. 80

glycogen: the chief carbohydrate used by animals for energy storage. p. 80

glycolysis (gly KAWL uh sis)**:** the initial breakdown of a carbohydrate, usually glucose, into smaller molecules at the beginning of cellular respiration. p. 378

Golgi (GOHL jee) **apparatus:** an organelle in eukaryotic cells consisting of stacked membranes that modifies and packages materials for export from the cell. p. 105, 106

grana (GRAY nuh)**:** stacks of thylakoids within a chloroplast. p. 489

gravitropism (grav ih TROH piz um)**:** a positive or negative response of a plant or animal to the acceleration force of gravity. p. 504

guard cells: a pair of cells that surround a stomate in a leaf's epidermis; turgor pressure in the guard cells regulates the opening and closing of the stomate. p. 465

gut: the alimentary canal, or portions thereof, especially the stomach and intestines. p. 394

H

habitat (HAB ih tat)**:** place where an organism lives; even in the same ecosystem, different organisms differ in their habitats. p. 51

half-life: in radioisotopes, the time required for half of a specified quantity to decay. p. 539

hallucinogen: a drug that causes the user to see, feel, or hear things that are not there. p. 453

halophile (HAL uh fyl)**:** an organism that requires a salty environment; usually refers to a type of archaebacteria. p. 265

haploid (HAP loyd)**:** a cell containing only one member (n) of each chromosome pair characteristic of a species. p. 129

hemoglobin (HEE moh gloh bin)**:** the pigment in red blood cells responsible for the transport of oxygen. p. 406

herbaceous (her BAY shus)**:** herblike; without woody tissues. p. 331

herbivore (HER bih vor)**:** a plant-eating consumer; one of the class of consumers most closely associated with producers. p. 50

heterotroph (HET eh roh trohf)**:** an organism that obtains carbon and all metabolic energy from organic molecules previously assembled by autotrophs; a consumer. p. 238

heterotroph hypothesis: the hypothesis that the first life forms used the supply of naturally occurring organic compounds for food. p. 246

heterozygous (HET er oh ZY gus)**:** having two different alleles for a given trait. p. 167

hibernation: a physiological state that permits survival during long periods of cold and diminished food, in which metabolism decreases, the heart and respiratory system slow down, and body temperature is maintained at a lower level than normal. p. 526

histamine: a substance released by injured cells that causes blood vessels to dilate during an inflammatory response. p. 409

homeostasis (hoh mee oh STAY sis)**:** the tendency for an organism, or a population of organisms, to remain relatively stable under the range of conditions to which it is subjected. p. 29

hominid (HOM ih nid)**:** a primate of the family Hominidae, which includes modern humans, earlier subspecies and australopithecines. p. 553

Homo erectus: an extinct tool-using human species that lived from 1.6 million to 300 000 years ago; the first undisputed human species. p. 554

Homo habilis: a fossil hominid, larger than an australopithecine, thought to be between 1.6 and 2 million years old. p. 554

homologies (hoh MOL uh jeez)**:** likeness in form, as a result of evolution from the same ancestors. p. 230

Homo sapiens neanderthalensis (nee an der tal EN sis)**:** a subspecies of modern humans that lived in Europe Africa and Asia from about 100 000 to 35 000 years ago; their relationship to modern humans is debated. p. 554

Homo sapiens sapiens: a subspecies of anatomically modern humans composed of present-day human groups. p. 556

homozygous (HOH moh ZY gus)**:** having two identical alleles for a given trait. p. 167

hormone: a substance, secreted by cells or glands, that has a regulatory effect on cells and organs elsewhere in the body; a chemical messenger. p. 134

host: an organism that serves as a habitat or living food source, or both, for another organism. p. 270

hybrid: having different alleles for a given trait, one inherited from each parent; heterozygous. p. 167

hyphae (HY fee)**:** threadlike growth of a fungus; in an irregular mass they compose the mycelium of many fungi; in an orderly and tightly packed arrangement they compose the body of a mushroom or bracket fungus. p. 294

hypoglycemia (hy poh gly SEE mee uh)**:** low blood sugar level. p. 444

hypothalamus: a specialized part of the base of the brain that in humans combines neuron and hormone activity; it links that autonomic nervous system with the endocrine system in regulating many body functions. p. 136

hypothesis (hy POTH uh sis)**:** a statement suggesting an explanation for an observation or an answer to a scientific problem. p. 16

I

immigration (im uh GRAY shun)**:** arrival by migration; in migration. p. 27

immunity: disease-resistance, usually specific for one disease or pathogen. p. 270

imperfect fungi: any of the fungi not belonging to other phyla, in which the method of reproduction, if known, is asexual; a taxonomic grouping of convenience. p. 299

incubation period: the time from the entry of an infection or its initiation within an organism up to the time of the first appearance of signs or symptoms. p. 276

industrialized agriculture: using large inputs of energy primarily derived from fossil fuels to produce large quantities of crops and livestock. p. 638

infectious disease: a disease caused by viruses or microorganisms that can be transmitted directly from an affected individual to a healthy individual. p. 271

ingestion: the process of taking a substance from the environment, usually food, into the body. p. 356

innate behavior: that is genetically determined, as in the organization of an ant society; also called instinctive behavior. p. 517

instinct: the capacity of an animal to complete a fairly complex, stereotyped response to a key stimulus without having prior experience. p. 518

insulin (IN suh lin)**:** a pancreatic hormone that promotes cell absorption and use of glucose; impairment of its secretion or its action results in diabetes. p. 444

interneuron: associative neuron; a neuron located between a sensory neuron and a motor neuron. p. 439

interphase (IN ter fayz)**:** a normal interval between successive cell divisions when the only evidence of future divisions is that chromosomes begin to be replicated; a cell at work, rather than a cell dividing. p. 111, 115

intertidal zone: the shallow zone of the ocean where land meets water; also called the littoral zone. p. 616

intestinal juice: secretions of the glands in the small intestinal wall containing enzymes that act on dipeptides and double sugars. p. 376

intron (IN tron): a segment of DNA that is transcribed into precursor mRNA but then removed before the mRNA leaves the nucleus. p. 184

ion (EYE on): an atom or molecule that has either gained or lost one or more electrons, giving it a positive or negative charge. p. 72

ionization: the conversion of a nonionic substance, such as water, into ions. p. 73

J

joint: a point of movement, or of fixed calcium deposits preventing movement, marking where two bones meet in the skeleton. p. 435

K

karyotype: a method of organizing the chromosomes of a cell in relation to number, size, and type. p. 178

kcal: kilocalorie; a measure of food energy equal to 1000 calories. p. 35

kingdom: the largest grouping in the biological classification system. p. 233

Krebs cycle: the cycle in cellular respiration that completes the breakdown of intermediate products of glycolysis, releasing energy; also a source of carbon skeletons for use in biosynthesis reactions. p. 378

L

larva (LAR vuh): an immature stage of development in offspring of many types of animals. p. 351

latent: present but not evident or active. p. 276

learned behavior: behavior developed as a result of experience. p. 517

lens: in the eye, a transparent tissue layer behind the iris that focuses light rays entering through the pupil to form an image on the retina. p. 146

lenticel (LENT ih sel): an opening in the surface of a plant stem through which air can diffuse. p. 466

life cycle: all the events that occur between the beginning of one generation and the beginning of the next generation. p. 125

ligaments: a cord or sheet of connective tissue by which two or more bones are bound together at a joint. p. 435

light reactions: the energy-capturing reactions in photosynthesis. p. 491

limiting factor: an environmental condition such as food, temperature, water, or sunlight that restricts the types of organisms and population numbers that an environment can support. p. 27

lipid (LIP id): a fat, oil, or fatlike compound that usually has fatty acids in its molecular structure; an important component of the plasma membrane. p. 79

luteinizing hormone (LH): a hormone secreted by the anterior lobe of the pituitary gland that controls the formation of the corpus luteum in females and the secretion of testosterone in males. p. 137

Lycophyta (ly KOF ih tuh): the division containing the club mosses. p. 321

lymph (LIMF): the fluid transported by the lymphatic vessels. p. 376

lymphatic (lim FAT ik) **system:** a system of vessels through which body lymph flows, eventually entering the bloodstream where the largest lymph duct joins a vein. p. 376

lymph node: a tiny, twisted portion of a lymph vessel in which white blood cells attack any pathogenic organisms in the lymph and engulf any foreign particles. p. 406

lymphocyte (LIM foh syt): a type of small white blood cell important in the immune response. p. 409

lymphokine (LIM foh kyn): any of a class of proteins by which the cells of the vertebrate immune system communicate with one another. p. 410

lymph vessel: a vessel in which lymph from the body tissues, or from villi in the intestine, flows until it enters the largest lymph duct, which empties into a vein. p. 409

lysosome (LY soh zohm): a cell vesicle that contains digestive enzymes. p. 104, 107

M

macronucleus: the larger of two types of nuclei in ciliates; one or more of the macronuclei may be present in each organism. p. 293

macronutrient: a nutrient required in large amounts by a plant or other organism. p. 473

macrophage (MAK roh fayj): a large white blood cell that ingests pathogens and dead cells. p. 409

mammary glands: the milk-producing organs in female mammals, consisting of clusters of alveoli with small ducts terminating in a nipple or teat. p. 354

marsupials: a group of mammals, such as koalas, kangaroos, and opossums, whose young complete their embryonic development inside a maternal pouch called a marsupium. p. 354

mating behavior: the behaviors, including courtship rituals and displays, that lead to mating; such behavior usually is specific for each species. p. 522

medulla (meh DUL lah): (1) the inner portion of an organ, such as the kidney, p. 419; (2) the most posterior region of the vertebrate brain, the hindbrain. p. 440

meiosis (my OH sis): two successive nuclear divisions (with corresponding cell divisions) that produce gametes (in animals) or sexual spores (in plants) having one-half of the genetic material of the original cell. p. 129

memory cell: B- or T-lymphocyte, produced in response to a primary immune response, that remains in the circulation and can respond rapidly if the same antigen is encountered in the future. p. 412

menopause (MEN oh pahz)**:** in human females, the period of cessation of menstruation, usually occurring between the ages of 45 and 50. p. 138

menstrual cycle: the female reproductive cycle that is characterized by regularly recurring changes in the uterine lining. p. 136

menstruation: periodic sloughing of the blood-enriched lining of the uterus when pregnancy does not occur. p. 136

meristem (MER ih stem)**:** plant cells at the growing tips of roots and stems and in buds and cambium that divide and produce new cells that can differentiate into various plant tissues. p. 475

mesoderm (MEZ oh derm)**:** in most animal embryos, a tissue layer between the ectoderm and endoderm that gives rise to muscle, to organs of circulation, reproduction, and excretion, to most of the internal skeleton (if present), and to connective tissue layers of gut and body covering. p. 144

mesophyll (MEZ oh fil)**:** the green leaf cells between the upper and lower epidermis of a leaf; the primary site of photosynthesis in leaves. p. 464

mesosome (MEZ oh sohm)**:** an infolding of the plasma membrane of a prokaryotic cell. p. 262

metabolism (meh TAB oh liz um)**:** the sum of all the chemical changes taking place in an organism. p. 108

metamorphosis (met uh MOR phuh sis)**:** in the life cycles of many animals, marked changes in body form and functions from the newly hatched young to the adults. p. 351

metaphase: the stage in mitosis in which replicated chromosomes move to the center of the spindle and become attached to it. p. 115

metastasize: to spread, as in the spread of cancer cells. p. 153

methanogen (meh THAN uh jen)**:** a methane-producing archaebacterium. p. 265

micronucleus: the smaller of two types of nuclei in ciliates; one or more micronuclei may be present in each organism. p. 293

micronutrient: a nutrient required in only small amounts by a plant or other organism. p. 473

microorganism: an organism too small to be seen with the unaided human eye. p. 8

microsphere: a cooling droplet from a hot water solution of polypeptides; the droplet forms its own double-layered boundary as it cools; used as a model for precells to study the formation of the first life on the earth. p. 246

microtubule: a hollow rod of protein found in the cytoplasm of all eukaryotic cells, making up part of the cytoskeleton and involved in cell contraction. p. 107

midlatitude forest: a biome located throughout midlatitude regions where there is sufficient moisture to support the growth of large trees, mostly of the broad-leaved deciduous type; also called temperate forest. p. 573

mitochondria (my toh KON dree uh)**:** the cell organelles in eukaryotic cells that carry on cellular respiration, releasing energy from food molecules and storing it in ATP. p. 105, 106

mitosis (my TOH sis)**:** the replication of the chromosomes and the production of two nuclei in one cell; usually followed by cytokinesis. p. 112, 115

mixed-grass prairie: the biome composing the central Great Plains that contains both short and tall grasses and is characterized by low precipitation, low winter temperatures, and relatively high elevation. p. 577

molecule (MOL uh kyool)**:** a particle consisting of two or more atoms of the same or different elements chemically bonded together. p. 12

Mollusca (muh LUS kuh)**:** the phylum containing mollusks. p. 349

mollusk: any invertebrate of the phylum Mollusca, typically having a hard shell that wholly or partly encloses a soft, unsegmented body; includes chitons, snails, bivalves, squid, octopuses, and similar animals. p. 349

monocot (MON oh kot)**:** a subdivision of flowering plants whose members have one embryonic seed leaf, or cotyledon. p. 331

monotremes: a group of egg-laying mammals, represented by the platypus and echidnas. p. 354

mortality (mor TAL ih tee)**:** death rate, measured as the proportion of deaths to total population over a given period; often expressed as number of deaths per 1000 or 10 000 individuals. p. 26

motile (MOH tuhl)**:** capable of movement from place to place, characteristic of most animals. p. 343

motor neuron: a specialized neuron that receives impulses from the central nervous system and transmits them to a muscle or gland. p. 439

multicellular: composed of many cells, as contrasted with a single-celled organism. p. 99

multifactorial inheritance: a pattern of inheritance in which characteristics are determined by at least several genes with a large number of environmental variables. p. 176

multiple alleles: the existence of several known alleles for a gene. p. 176

mutation (myoo TAY shun)**:** a chemical change in a gene, resulting in a new allele; or, a change in the portion of a chromosome that regulates the gene; in either case the change is hereditary. p. 168

mycelium (my SEE lee um)**:** the densely branched network of hyphae in a fungus. p. 296

mycoplasma (my koh PLAZ muh)**:** eubacteria that lack cell walls; probably the smallest organisms capable of independent growth. p. 266

mycorrhizae (my koh RY zee)**:** symbiotic associations of plant roots and fungi. p. 300

myosin: a protein that, together with actin, is responsible for muscular contraction and relaxation. p. 433

Myxomycota (mik soh my KOH tuh)**:** the phylum containing the slime molds. p. 293

N

NAD$^+$: nicotinamide adenine dinucleotide (nik uh TEE nuh myd AD uh neen DY NOO klee oh tyd); an electron and hydrogen carrier in cellular respiration. p. 380

NADP$^+$: nicotinamide adenine dinucleotide phosphate; a hydrogen-carrier molecule that forms NADPH in the light reactions of photosynthesis. p. 493

natality (nay TAL ih tee)**:** birthrate, as measured by the proportion of new individuals to total population over a given period; often expressed as new individuals per 1000 or 10 000 in the population. p. 27

natural selection: a mechanism of evolution whereby members of a population with the most successful adaptations to their environment are more likely to survive and reproduce than members with less successful adaptations. p. 206

nectar: a secretion, mainly a sugar solution, produced by small glands in the petals of many flowers. p. 329

Nematoda (nee muh TOH duh)**:** the phylum containing the unsegmented worms. p. 348

nephron (NEF rahn)**:** the functional unit of a kidney, consisting of a long, coiled tubule, one end of which forms a cup that encloses a mass of capillaries and the other end of which opens into a duct that collects urine; the entire nephron is surrounded by a network of capillaries. p. 418

nerve impulse: a wave of chemical and electrical changes that passes along a nerve fiber in response to a stimulus. p. 437

nerve: a bundle of nerve fibers; the cell bodies from which the fibers extend usually are located together at one end of the fiber. p. 437

nervous system: a coordinating mechanism in all multicellular animals, except sponges, that regulates internal body functions and responses to external stimuli; in vertebrates, it consists of the brain, spinal cord, nerves, ganglia, and parts of receptor and effector organs. p. 361

neural tube: the structure formed in the embryo that eventually gives rise to the brain and spinal cord. p. 144

neuron (NYOO rahn)**:** a nerve cell; a name usually reserved for nerve cells in animals that have a complex brain and specialized associative, motor, and sensory nerves. p. 437

neurotransmitter (NYOOR oh trans MIT er)**:** a chemical messenger, often similar to or identical with a hormone, that diffuses across the synapse and transmits a nerve impulse from one neuron to another. p. 438

neutral: a solution in which the concentrations of hydrogen and hydroxide ions are equal. p. 73

neutron (NOO trahn)**:** a particle carrying no electrical charge; found in the nuclei of all atoms except those of hydrogen. p. 71

niche (NITCH)**:** the total of an organism's utilization of the biotic and abiotic resources of its environment. p. 51

nitrifying (NY trih fy ing) **eubacteria:** bacteria that use ammonium ions to produce nitrite and nitrate ions. p. 269

nitrogen-fixing eubacteria: bacteria that take in free nitrogen from the atmosphere and use it to produce ammonia; subsequent reactions by other bacteria produce ammonium ions and nitrates, from which plants obtain their nitrogen. p. 268

nodule (NOD yool): a rounded growth of tissue that usually contains microorganisms or some other agent associated with the growth; in certain plants, nitrogen-fixing bacteria live in nodules on the plant roots. p. 269

norepinephrine: a hormone secreted by the adrenal medulla in response to nerve signals from the sympathetic division of the autonomic nervous system. p. 447

notochord (NOH toh kord)**:** in chordates, a flexible, dorsal, rodlike structure that extends the length of the body; in vertebrates it is replaced in later stages of development by the vertebrae. p. 352

nuclear envelope: the membrane in eukaryotes that encloses the genetic material, separating it from the cytoplasm. p. 106

nucleic (noo KLEE ik) **acid:** DNA or RNA; an organic compound composed of nucleotides and important in coding instructions for cell processes. p. 79

nucleoid (NOO klee oyd)**:** a region in a prokaryotic cell consisting of a concentrated mass of DNA. p. 262

nucleotide (NOO klee oh tyd)**:** a subunit or building block of DNA or RNA, chemically constructed of a 5-carbon sugar, a nitrogen base, and a phosphate group. p. 83

nucleus (NOO klee us)**:** in atoms, the central core, containing positively charged protons and (in all but hydrogen) electrically neutral neutrons, p. 71; in eukaryotic cells, the membranous organelle that houses the chromosomal DNA. p. 102, 104

nutrient: (1) a food substance usually digested or not requiring digestion in a form, that is interchangeable among organisms, p. 46; (2) certain chemicals used by plants. p. 473

O

omnivore (OM nih vor): a consumer organism that feeds partly as an herbivore and partly as a carnivore, eating plants, animals, and fungi. p. 50

oncogene: a gene found in viruses or as part of the normal genome that is crucial for triggering cancerous characteristics. p. 153

open circulatory system: a system in which the blood does not travel through the body completely enclosed in vessels; the return to the heart is usually through open spaces, or sinuses, as in the grasshopper. p. 358

open population: a population that gains members by immigration or loses them by emigration, or both. p. 30

opportunistic infections: infections that occur in organisms with weakened immune responses; normally these infections would be repelled by healthy immune systems. p. 277

organelle (or guh NEL): an organized structure within a cell, with a specific function; a chloroplast and a mitochondrion are examples. p. 102

organic compounds: compounds built of carbon combined with other elements. p. 71

osmosis (os MOH sis): the movement of water across a selectively permeable membrane. p. 109, 118

ovaries (singular, *ovary*): the primary reproductive organs of a female; egg-cell-producing organs. p. 133

oviduct: a tube leading from an ovary to the uterus. p. 136

ovulation: in vertebrates, the release of one or more eggs from an ovary p. 134

ovule (OHV vool): a structure that develops in the plant ovary and contains the female egg. p. 318

ovum: a female gamete, or egg. p. 128

P

P_1: the parental organisms in a genetic cross-fertilization. p. 165

paleontologist (pay lee un TOL uh jist): a specialist who studies extinct organisms through fossils. p. 539

palisade layer: the layer of tightly packed and column shaped mesophyll cells containing many chloroplasts. p. 464

pancreatic (pan kree AT ik) **juice:** fluid secreted by the pancreas. p. 376

Pangaea (pan GEE uh): in the Triassic period, the super continent formed by the four major continents; the breakup of Pangaea affected the distribution and evolution of organisms. p. 544

parallel evolution: the type of evolution that occurs when organisms from a single species become so different that they no longer can interbreed, yet both groups continue to evolve in a similar direction; the ostrich and the rhea are the result of parallel evolution. p. 215

parasitism (PAIR uh sih tiz um): an ecological niche in which one organism is the habitat and the food for another, which lives and feeds on the host organism, usually without killing it. p. 52

parasympathetic division: in vertebrates, one of the two divisions of the autonomic nervous system; stimulates resting activities, such as digestion, and restores the body to normal after emergencies. p. 442

passive transport: the diffusion of a substance across a biological membrane through a transport protein in the membrane. p. 110

pathogen: a disease-causing organism. p. 270

penicillin: any of several antibiotic compounds obtained from penicillium mold and used to prevent or treat a wide variety of diseases and infections. p. 208

penis: in vertebrates, the male organ through which sperm are passed to the female and through which nitrogenous wastes from the kidneys—in the form of urine—are discharged outside the body. p. 133

peptidoglycan (PEP tid oh GLY kan): a substance found only in eubacteria cell walls that consists of modified sugars cross-linked by short polypeptides. p. 263

peripheral nervous system: the sensory and motor neurons that connect the central nervous system to the rest of the body p. 442

peristalsis (per ih STAWL sis): rhythmic waves of contraction of smooth muscle which push food along the digestive tract. p. 373

permafrost: the sublayers of soil that remain frozen through the summer thaw in the tundra of northern latitudes. p. 572

petal: one of the leaflike, often brightly colored, structures within the ring of green sepals in a developing flower; in the mature flower the petals may form their own ring or fuse to form a cuplike or tubular structure. p. 317

petiole (PET ee ohl): the slender structure at the base of a leaf that attaches it to a plant stem. p. 317

pH scale: a scale from 0 to 14 reflecting the concentration of hydrogen ions; the lower numbers denote acidic conditions, 7 is neutral, and the upper numbers denote basic or alkaline conditions. p. 73

Phaeophyta (fay OFF it uh): the phylum containing the brown algae. p. 290

phenotype (FEE noh typ): the expression of a genotype in the appearance or function of an organism; the observed trait. p. 166

pheromone (FAYR uh mohn): a small chemical signal functioning in communication between animals and acting much like hormones to influence physiology and behavior. p. 155

phloem (FLOH em): a portion of the vascular system in plants, consisting of living cells arranged into elongated tubes that transport sugar and other organic nutrients throughout the plant. p. 467

photoperiodism (foh toh PIH ree ud iz um): a physiological response to day length, such as in flowering plants. p. 504

photorespiration: a metabolic pathway that uses oxygen, produces carbon dioxide, generates no ATP, and slows photosynthetic output; generally occurs on hot, dry, bright days, when stomates are nearly closed and the oxygen concentration in the leaf exceeds that of carbon dioxide. p. 496

photosynthesis: the process by which living cells that contain chlorophyll use light energy to make organic compounds from inorganic materials. p. 11

photosystems (PS) I and 11: the light-harvesting units in photosynthesis, located in the thylakoid membranes of the chloroplast. p. 492

phototropism (foh toh TROH piz um): movement or growth curvature toward light. p. 502

phylum (FY lum): the second largest grouping, after kingdom, in the biological classification system, for all organisms except plants, which are classified in divisions. p. 233

phytoplankton (FYT oh PLANK ton): very small aquatic organisms, many microscopic, that carry on photosynthesis. p. 604

pigment: any coloring matter or substance. p. 490

pith: cells at the center of the young stems of many plants in some plants the pith disappears as the stems age. p. 467

pituitary (pih TOO ih ter ee): endocrine gland attached to the base of the brain, consisting of an anterior and a posterior lobe. p. 137

placenta (pluh SEN tuh): a structure in the pregnant uterus for nourishing a fetus with the mother's blood supply, formed from the uterine lining and embryonic membranes. p. 148

plankton (PLANK ton): very small aquatic organisms, many microscopic, that usually float or feebly swim near the surface. p. 604

Plantae (PLAN tee): the plant kingdom, containing multicellular, autotrophic eukaryotes; all members have cellulose containing cell walls and chloroplasts; reproduction may be sexual or asexual. p. 238

plaque: an abnormal change that occurs on inner arterial walls when lipids, such as cholesterol, infiltrate a matrix of smooth muscle; plaques are a product of atherosclerosis. p. 388

plasma (PLAZ muh): the liquid portion of the blood in which the cells are suspended. p. 406

plasma cell: an antibody-producing cell that is formed as a result of the proliferation of sensitized B-lymphocytes. p. 4

plasma membrane: the membrane at the boundary of every cell that serves as a selective barrier to the passage of ions and molecules. p. 103, 104

plasmid: in prokaryotes, a small ring of DNA that carries accessory genes separate from those of the bacterial chromosome. p. 185

plasmodium (plaz MOHD ee um): the motile sheetlike stage of life formed by the fusion of many amoebalike slime molds. p. 293

platelet (PLAYT let): a small plate-shaped blood factor that contributes to blood clotting at the site of a wound; platelets release substances that begin formation of a network in which the platelets are caught, forming a clot. p. 406

plate tectonics: the theory and study of the great plates in the earth's crust and their movements, which produce earthquakes, sea-floor spreading, continental drift, and mountain building. p. 544

Platyhelminthes (plah tih hel MIN theez): the phylum containing bilaterally symmetrical, somewhat flattened worms. p. 347

pollen grain: a haploid spore produced by a flowering plant; it gives rise to sperm nuclei. p. 316

pollen tube: a tube that develops from a germinating pollen grain and penetrates the carpel until it reaches the ovary; sperm nuclei pass through the tube. p. 318

pollination: the placement of pollen onto the stigma of a carpel by wind or animal vectors, a prerequisite to fertilization. p. 318

polypeptide (POL ee PEP tyd): a long chain of chemically bonded amino acids. p. 81

population: a group of organisms of one species that live in the same place at the same time. p. 25

Porifera (poh RIH fer uh): the phylum containing sponges and their relatives. p. 345

posterior: situated toward the rear, or coming last; the tail end in most animals. p. 344

potassium-40 dating: a radiometric dating technique that involves measuring the relative amounts of potassium-40 to argon-40 in a volcanic rock sample; the ratio is compared to the known half-life of potassium-40, thus yielding an estimate of the age of the rocks. p. 540

predator-prey relationship: the relationship between organisms in which one, the predator, feeds on the other, the prey p. 52

pressure-flow: the hypothesis that food is transported through the phloem as a result of differences in pressure. p. 469

primary growth: growth in the length of plant roots and stems. p. 475

principle of independent assortment: the inheritance of alleles for one trait does not affect the inheritance of alleles for another trait. p. 175

principle of segregation: during meiosis, chromosome pairs separate into different gametes such that each of the two alleles for a given trait appears in a different gamete. p. 166

probability: the chance that any given event will occur. p. 164

producer: an autotroph; any organism that produces its own food using matter and energy from the nonliving world. p. 8

productivity: the amount of available solar energy converted to chemical energy by producers during any given period. p. 610

progesterone (proh JES tuh rohn)**:** a female hormone secreted by the corpus luteum of the ovary and by the placenta that acts to prepare and maintain the uterus for pregnancy and to prepare the breasts for lactation. p. 137

Prokaryotae (proh kayr ee OH tee)**:** the kingdom of bacteria; their cells are prokaryotic and lack membrane-enclosed organelles such as nuclei and mitochondria. p. 277

prokaryote (pro KAIR ee oht)**:** an organism whose cells do not have membrane-enclosed organellcs, such as nuclei, mitochondria, and chloroplasts; a bacterium. p. 102, 238

prophase: the stage in mitosis during which replicated strands of chromosomes condense to become thicker and shorter, the nuclear envelope begins to disappear, and a spindle forms. p. 115

protein (PROH teen)**:** an organic compound composed of one or more polypeptide chains of amino acids; most structural materials and enzymes in a cell are proteins. p. 79

prothrombin (proh THROM bin)**:** a plasma protein that functions in the formation of blood clots. p. 406

prothrombin activator: in the clotting process, a substance that catalyzes the conversion of prothrombin to thrombin. p. 406

Protista (pro TISt uh)**:** a kingdom of mostly aquatic organisms whose cells are eukaryotic, but which are mostly microscopic; exceptions in size include the larger seaweeds such as kelp. p. 240

proton (PRO tahn)**:** a particle bearing a positive electrical charge, found in the nuclei of all atoms. p. 71

pseudopod (sOO doh pod)**:** literally, a false foot; a term used to describe an amoebalike extension of a cell or unicellular organism; pseudopods are used in locomotion and obtaining food, or, as in white blood cells, in engulfing bacteria and foreign particles. p. 292

psychoactive (sy koh AK tiv) **drug:** a drug that affects the mind or mental processes. p. 452

Pterophyta (ter OFF ih tuh)**:** the division containing the ferns. p. 324

puberty: the stage of development in which the reproductive organs become functional. p. 134

punctuated equilibria: a theory of evolution positing spurts of relatively rapid change followed by long periods of stasis. p. 214

pure breeding: genetically pure; homozygous. p. 165

pyloric valve: the valve located between the stomach and the small intestine that regulates the passage of food from the stomach. p. 374

R

radial (RAYD ee ul) **symmetry:** correspondence in size, shape, and position of parts as though they all radiated equally from a center point, or from a center line or axis. p. 344

rate: change per unit of time; the amount of change measured over a period of time, divided by the length of time. p. 26

ray cell: any of the cells in the stems of woody plants that form a channel allowing the passage of material laterally back and forth between the xylem and the phloem. p. 476

receptor: a specialized sensory cell, as in the eye or the skin, that is sensitive to a particular type of stimulus. p. 439

recessive: a term used to describe an allele or trait that is masked by a dominant allele or trait. p. 165

recombinant DNA: DNA that incorporates parts of different parent DNA molecules, as formed by natural recombination mechanisms or by recombinant DNA technology. p. 184

reduction division: the first meiotic division, in which the chromosome number is reduced from diploid ($2n$) to haploid (n). p. 131

reflex: an involuntary reaction or response to a stimulus. p. 439

replication: the process of making a copy of the chromosome in a cell nucleus, and other genes in certain organelles outside the nucleus-particularly chloroplasts and mitochondria; the process is unlike duplication in that each gene and each chromosome in the double set is partly new but also includes part of the old gene or chromosome. p. 112

reproductive isolation: the state of a population or species in which successful mating outside the group is impossible because of anatomical, geographical, or behavioral differences. p. 201

reptile: any of the ectothermic, usually egg-laying vertebrates that belong to the class Reptilia; includes snakes, lizards, crocodiles, turtles. p. 353

resistance: relative immunity; the ability of a host organism to destroy a pathogen or prevent the disease symptoms it causes. p. 270

resource: in ecology, an environmental supply of one or more of an organism's requirements (light energy, food energy, water, oxygen or carbon dioxide, living space, protective cover, and so on); in a human society, a resource may be anything useful. p. 28

retina (REH tin uh)**:** the photosensitive layer of the vertebrate eye that contains several layers of neurons and light receptors (rods and cones); the retina receives the image formed by the lens and transmits it to the brain via the optic nerve. p. 146

retrovirus: certain RNA viruses that must make copies of DNA from their RNA before they can reproduce. p. 273

reverse transcriptase: an enzyme that transcribes RNA into DNA, found only in association with retroviruses. p. 273

rhizoid (RY zoid)**:** the threadlike structure that in nonvascular plants absorbs water and nutrients from the soil and helps hold the plants in place. p. 320

Rhodophyta (roh DOF it uh)**:** the phylum containing the red algae. p. 290

ribose (RY bohs)**:** a 5-carbon sugar found in RNA molecules. p. 83

ribosome (RY boh sohm)**:** a cell organelle constructed in the nucleus, consisting of two subunits and functioning as the site of protein synthesis in the cytoplasm. p. 105, 107

ritual: a detailed method of procedure that is faithfully or regularly followed. p. 522

RNA: ribonucleic (ry boh noo KLEE ik) acid; the hereditary material of certain viruses, and the material coded by the DNA of other cells to carry out specific genetic functions, for example, messenger RNA and transfer RNA. p. 83

root cap: a layer of protective cells that covers the growing tip of a plant root. p. 475

root hair: a hairlike extension of an epidermal cell of a plant root that absorbs water and nutrients. p. 473

roundworm: an unsegmented, cylindrical worm having a digestive tube with two openings; a nematode. p. 348

rubisco (roo BIS koh)**:** ribulose bisphosphate carboxylase (RY byoo lose BIS fos fayt kar BOX uh lays); an enzyme that catalyzes the initial incorporation of carbon dioxide in the Calvin cycle. p. 497

rumen (ROO men)**:** an enlargement of the digestive tract of many herbivorous mammals, in which microorganisms that can digest cellulose live. p. 265

S

sac fungi: fungi belonging to the phylum Ascomycota, in which sexual spores are formed in a saclike structure, the ascus. p. 297

sarcodines: microscopic, or almost microscopic, protists that move by using pseudopods; many produce intricate shells or skeletal structures; includes amoebas, foraminiferans, and radiolarians. p. 292

saturated fat: a fat containing fatty acids with no double carbon bonds; usually solid at room temperature. p. 336

savanna (suh VAN uh)**:** a tropical grassland biome with scattered individual trees and large herbivores; water is the major limiting factor. p. 580

scavenger (SKAV en jer)**:** a consumer organism that feeds on the dead carcasses of other consumer organisms it did not kill. p. 53

scrotum: a pouch of skin that encloses the testes. p. 133

scurvy: a disease caused by a deficiency of vitamin C, characterized by spongy, bleeding gums, bleeding under the skin, and extreme weakness. p. 271

secondary growth: growth in girth or diameter of plant and root. p. 475

secondary sexual characteristics: differences in appearance or behavior associated with male or female hormones; in many animal species, males are larger or more colorful. p. 524

seed: an embryo plant, along with food storage tissue (endosperm), both enclosed within protective coatings formed of tissues from an ovule in the parent plant. p. 319

seed coat: the tough, protective outer covering of a seed. p. 474

semen (SEE men)**:** the thick fluid in which sperm are transported in mammalian males. p. 138

sensory neuron: a neuron that receives impulses from a sensory organ or receptor and transmits them toward the central nervous system. p. 438

sepal (SEE pul)**:** one of the leaflike structures that enclose and protect a flower bud; in the mature flower they are on the underside, next to the stem; they often are green. p. 317

sessile (SES il)**:** not free to move about in the environment; sessile animals usually are attached by the base to an object in the environment. p. 344

sex chromosome: one of a pair of chromosomes that differentiates between and is partially responsible for determining the sexes. p. 178

sexual reproduction: reproduction involving the contribution of genetic material from two parents. p. 126

sexual selection: selection based on variation in secondary sexual characteristics, leading to the enhancement of individual reproductive fitness. p. 211, 524

short-grass prairie: a grassland biome that stretches from the central Great Plains to the Rocky Mountains; grasses usually are less than 0.5 m tall. p. 577

sieve cell: a type of phloem cell in plants that forms a column. p. 468

skeletal muscle: a type of muscle tissue found in muscles attached to skeletal parts and responsible for voluntary muscle movement. p. 433

slash-and-burn cultivation: cutting down trees and other vegetation in a patch of forest, leaving the cut material to dry, and then burning it; the ashes add plant nutrients to the nutrient-poor soils of most tropical forest areas. p. 636

slime molds: members of the phylum Myxomycota; slime molds form amoebalike colonies, are heterotrophic, and reproduce by spores. p. 293

small intestine: a digestive organ of vertebrates and some invertebrates; in vertebrates it is located between the stomach and the large intestine and is the organ in which the digestive processes are completed. p. 374

smooth muscle: the type of muscle tissue found in the walls of hollow internal organs. p. 434

social behavior: animal behavior that shows evidence of differing individual roles in the organization of a group and of cooperation or division of labor in tasks. p. 522

somatic nervous system: those nerves leading from the central nervous system to skeletal muscles. p. 442

species (SPEE sheez)**:** all individuals and populations of a particular type of organism, maintained by biological mechanisms that result in their breeding mostly with their type. p. 59

sperm cell: a male gamete, usually motile in swimming movements; its motility increases its chance of encountering and fertilizing an egg. p. 128

S phase (Synthesis)**:** the synthesis phase of the cell cycle during which DNA is replicated. p. 115

Sphenophyta (sfen OFF ih tuh)**:** the division containing horsetails. p. 322

sphincter (SFINK ter)**:** a circular muscle that functions to close an opening of a tubular structure. p. 374

spina bifida (SPY nuh BIH fih duh)**:** a birth defect resulting from the failure of the spinal cord and brain to close properly during development. p. 147

spinal cord: a complex band of neurons that runs through the spinal column of vertebrates to the brain. p. 439

spongy layer: a layer of round, chloroplast-containing mesophyll cells surrounded by air spaces. p. 464

sporangia (spoh RAN jee uh)**:** spore-producing structures formed in one stage of the life cycle of many organisms, but not animals. p. 293

spores: one-celled reproductive bodies that are usually resistant to harsh environmental conditions and may remain dormant, in a dry covering, for long periods; in some organisms, spores are asexual and may initiate the growth of a new organism under favorable conditions; in other organisms, spores are sexual and must unite with those of the other sex before producing a new organism. p. 155

sporophyte (SPOR oh fyt)**:** a plant of the spore-producing generation in a plant species that undergoes alternation of generations; in some species, the sporophyte is reduced to a dependent structure that grows from the gametophyte plant. p. 316

sporozoans: usually nonmotile, heterotrophic, parasitic protists of the phylum Apicomplexa, with complex life cycles. p. 292

stamen (STAY men)**:** the pollen-producing male reproductive organ of a flower, consisting of an anther and filament. p. 317

sterile: not capable of reproducing. p. 201

stigma: the tip of the carpel in a flower; it secretes a sticky substance that traps pollen. p. 317

stimulant: a drug that increases the activity of the central nervous system. p. 452

stimulus (STIM yoo lus)**:** a change or signal in the internal or external environment that causes an adjustment or reaction by an organism. p. 437

stomach: the digestive organ located between the esophagus and the small intestine. p. 374

stomate (STOH mayt)**:** the opening between two guard cells in the epidermis of a plant leaf; gases are exchanged with the air through stomates. p. 315

strata (STRAYT uh; singular, stratum)**:** layers, usually of deposited earth sediments carried by erosion; many strata become mineralized into rock layers. p. 539

stress: a physiological response to factors causing disruptive changes in the body's internal environment p. 448

stressor: a factor capable of stimulating a stress response. p. 448

stroma (STROH muh)**:** the colorless substance in a chloroplast surrounding the thylakoids; the enzymes of the Calvin cycle also are in the stroma. p. 489

stromatolite (stroh MAT uh lyt)**:** a rock made of banded domes of sediment in which are found the most ancient forms of life, fossil prokaryotes dating back as far as 3.5 billion years. p. 243

subsistence agriculture: producing only enough food to feed oneself and family members; in good years, enough may be left over to sell or store. p. 637

substrate (SUB strayt)**:** a molecule on which enzymes act. p. 82

succession: the replacement of one community by another in a progression to a climax community. p. 582

sustainable agriculture: a method of growing crops and raising livestock based on organic fertilizers, soil conservation, water conservation, biological pest control, and minimal use of nonrenewable energy sources. p. 645

symbiosis (sim by OH sis)**:** an ecological relationship between organisms of two different species that live together in direct contact. p. 52

symmetry: correspondence in form and arrangement of parts on opposite sides of a boundary. p. 343

sympathetic division: a division of the autonomic nervous system of vertebrates that functions in an alarm response; increases heart rate and dilates blood vessels while placing the body's everyday functions on hold; mobilizes the body for response to stressors, danger, or excitement. p. 442

synapse (SIN aps): an open junction between neurons, across which an impulse is transmitted by a chemical messenger, a neurotransmitter. p. 438

synthesis (SIN thih sis): the process of putting together or building up; applicable to ideas, chemical compounds, and so on. p. 72

T

taiga (TY guh): the coniferous or boreal forest biome, characterized by much snow, harsh winters, short summers, and evergreen trees. p. 573

tall-grass prairie: a grassland biome found at the western edge of the deciduous forest in the United States; grasses may grow 1.5 to 2 m tall. p. 577

target organ: a specific organ on which a hormone acts. p. 443–444

taxonomy (tak SAHN uh mee): the study of species and their classification by genus, family, order, class, phylum, and kingdom. p. 229

T cell: a lymphocyte that matures in the thymus stimulated by the presence of a particular antigen; it differentiates and divides, producing offspring cells (killer cells) that attack and kill the cells bearing the antigen. p. 409

telophase: the final stage in mitosis; two new cell nuclei are completed as nuclear envelopes form around the two clusters of chromosomes at opposite ends of the cell, and the cell itself divides. p. 115

tendon: a cordlike mass of white fibrous connective tissue that connects muscle to bone. p. 435

terrestrial (ter ES tree uhl): living on land. p. 344

territoriality: the behavior pattern in animals consisting of the occupation and defense of a territory. p. 524

testes: the primary reproductive organs of a male; sperm cell-producing organs. p. 133

thermoacidophiles (THER moh a SID uh fylz): archaebacteria requiring high temperatures and/or acidic conditions for life. p. 264

thermocline (THER moh klyn): a layer in a thermally stratified body of water that separates upper, oxygen-rich and nutrient-poor warm water from lower, oxygen-poor and nutrient-rich cold water. p. 606

threat display: a behavior in which an animal attempts to make itself appear larger or fiercer than it actually is in an attempt to discourage intruders. p. 522

thrombin: a blood protein that is important in the clotting process. p. 406

thylakoid (THY luh koyd): a flattened sac in a chloroplast; many of the thylakoids are arranged in stacks known as grana; the pigments and enzymes for the light reactions of photosynthesis are embedded in the sac membrane. p. 489

thyroid gland: an endocrine gland in the neck region of most vertebrates that controls the rate of cell metabolism in the body through one of its hormones, thyroxine. p. 446

thyroxine (thy ROK sin): a principal hormone of the thyroid gland; regulates the rate of cell metabolism. p. 446

tolerance: the ability to withstand or survive a particular environmental condition. p. 527

toxin: a substance produced by one organism that is poisonous to another. p. 272

trachea (TRAY kee uh): the windpipe of an air-breathing vertebrate, connecting the air passage in the throat with the lungs. p. 417

tracheid (TRAY kee id): a water-conducting and supportive element of xylem composed of long, thin cells with tapered ends and hardened walls. p. 469

transpiration (trans pih RAY shun): the loss of water to the atmosphere by a plant through the stomates in its leaves. p. 464

trimester: a period or term of three months; the gestation period of humans usually is divided into trimesters. p. 152

tropical deciduous forest: a tropical forest biome with wet and dry seasons and constant temperatures. p. 576

tropical rain forest: the most complex of all communities, located near the equator where rainfall is abundant and harboring more species of plants and animals than any other biome in the world; light is the major limiting factor. p. 574

tropism (TROH piz um): a change in the orientation of a plant, or part of a plant, in response to light, gravity, or other environmental factors. p. 502

tumor: a mass that forms within otherwise normal tissue, caused by the uncontrolled growth of a transformed cell. p. 153

turgor (TER ger) **pressure:** pressure exerted by plant cells against their cell walls whenever the plant is adequately supplied with water. p. 464

U

ultrasound: high frequency sound waves; a method used to determine some fetal abnormalities. p. 149

ultraviolet light: the range of radiation wavelengths just beyond violet in the visible spectrum, on the border of the X-ray region. p. 490

umbilical (um BIL ih kul) **cord:** in placental mammals, a tube connecting the embryo with the placenta. p. 148

unicellular: one-celled. p. 99

Uniramia (yoo nih RAY mee uh): the subphylum containing animals with three distinct body parts and one pair of antennae; includes millipedes, centipedes, and insects. p. 350

unsaturated fat: a fat containing fatty acids with one or more double-bonded carbon atoms; each double bond in the carbon chains reduces by one the number of hydrogen atoms that can be bonded to the carbons; unsaturated fats usually are liquid at room temperature. p. 386

uranium-235 dating: a radiometric dating technique in which the decay of uranium into lead isotopes is measured; used to date rocks 2 billion years old. p. 539

urban wildlife: wildlife living in urban and suburban areas, including raccoons, skunks, pigeons, songbirds, raptors, mice, rats, and similar animals. p. 530

urea (yoo REE uh)**:** a nonprotein nitrogenous substance produced as a result of protein metabolism. p. 360

ureter (YOOR ee ter)**:** a muscular tube that carries urine from the kidney to the urinary bladder. p. 419

urethra (yoo REE thruh)**:** the tube through which urine is carried from the bladder to the outside of the body in vertebrates. p. 419

uric (YOOR ik) **acid:** the insoluble precipitate of nitrogenous waste excreted by land snails, insects, birds, and some reptiles. p. 360

uterus: a hollow muscular organ, located in the female pelvis, in which a fetus develops. p. 136

V

vaccine: a substance that contains antigens and is used to stimulate the production of antibodies. p. 271

vacuole (VAK yoo ohl)**:** a membrane-enclosed structure in the cytoplasm of a cell or a unicellular organism; different tapes of vacuoles serve different functions. p. 105, 107

vagina: a tubular organ that leads from the uterus to opening of the female reproductive tract. p. 136

valve: a membrane or similar structure in an organ or passage, such as an artery or vein, that retards or prevents the return flow of a bodily fluid. p. 404

variable: a condition that varies from one organism to another (size, shape, color) or is subject to change for an individual organism (humidity, temperature, light intensity, fatigue). p. 20

variation: small differences among individuals within a population or species that provide the raw material for evolution. p. 202

vascular tissue: plant tissues specialized for the transport of water and nutrients; also for support. p. 314

vegetative reproduction: asexual reproduction by plants that also may reproduce sexually; examples include potato plants from "eyes" and grass plants from runners. p. 127

vein: a vessel that carries blood toward the heart. p. 403; (2) a continuation of the vascular tissues of the stem and root in a plant leaf. p. 464

ventral: in animals, situated toward the lower or belly side. p. 344

ventricle (VEN trih kul)**:** one of two chambers of the heart that pump blood out of the heart. p. 404

vertebra (VER teh bruh)**:** an articulated bone; the vertebrae make up the backbone, or spinal column, of vertebrates. p. 352

vertebrate: a chordate animal with a backbone; mammals, birds, reptiles, amphibians, and various classes of fishes are examples. p. 200

vesicle: a small, intracelluar membrane-enclosed sac in which various substances are transported or stored. p. 106

vessel: in plants, a type of water-conducting xylem cell. p. 469

villi (VIL eye)**:** fingerlike projections of the small intestine that increase surface area. p. 376; (2) fingerlike projections of the chorion that together with the uterine lining form the placenta. p 148

virulence (VIR yuh lents)**:** the relative ability of a pathogen to overcome body defenses and cause disease. p. 270

virus: a submicroscopic pathogen composed of a core of nucleic acid surrounded by a protein coat that can reproduce only inside a living cell. p. 282

W

wetland: land that stays flooded all or part of the year with fresh or salt water; includes coastal and inland wetlands. p. 617

X

X-linked trait: a trait determined by a gene carried on the X chromosome. p. 179

xylem (ZY lem)**:** conducting tissue that transports water and dissolved nutrients in vascular plants. p. 467

Z

zooplankton (ZOH oh PLANK ton)**:** very small, feebly swimming aquatic organisms that are herbivorous or carnivorous or both. p. 604

Zygomycota (ZY goh my KOH tuh)**:** the phylum containing conjugation fungi. p. 296

zygospore (ZY goh spor)**:** a zygote that forms a spore; produced in some fungi and many plants following the union of sexual cells or nuclei. p. 297

zygote (ZY goht)**:** the diploid product of the union of haploid gametes in conception; a fertilized egg. p. 129

Index

Illustrations and tables are referenced in italics.

A

Aardvark, *721*

Abiotic environment, 27; community and, 49; comparative study of, 62–64

Abiotic factors: ecosystems and, 569–70; effects of, *54;* ocean and, 611–12; population size and, 28–30

ABO blood types, 176, *176, 415*

Abortion, 135, 647; in China, 640

Abscisic acid, 502

Absorption: food, 375; roots and, 472–74; in small intestine, 376

Abstinence (sexual), 135

Acacia, *580*

Acadian flycatcher, 574

Acanthocephala, 711

A Catalog of Living Things, 695–793

Accidents, laboratory, 660

Accipiter cooperi, 142, 143

Acetabularia, 288

Acetyl CoA, 380; conversion to pyruvic acid, *380*

Acid: abscisic, 502; amino, 81, 382; ascorbic, *685;* carbonic, 418; citric, 380; diluting with water, 659; fatty, 376, 382; folic, *685;* inorganic, 668–69; lactic, *384;* oxaloacetic, 380; pantothenic, *685;* pyruvic, 383, *384*

Acid deposition, 588

Acidianus infernus, 264

Acidic solution, 73

Acid rain, 5, 35–36, 588; effects of, 621, 639; global implications of, 642; U.S. map of, *644;* seed germination and, 67–68; trees exposed to, *588*

Acorn worms, 718, *718*

Acquired characteristics, 206

Acquired Immune Deficiency Syndrome. *See* AIDS

Acsophyllum, 698

Actinophrys, 678, 699

Actinosphaerium, 678

Active site, 82, *82*

Active transport, 110, *110*

Activity, symbolic, 522

Adaptation, 206; in animals, 343, 345–54; cultural, 636, 639; diversity and, 227–369

Adaptive learning, 520

Adaptive radiation, 212; in mammals, *214*

Adder's tongue, *703*

Addiction, drug, 452

Adenine, *78,* 171, *171, 173*

Adenosine, *78*

Adenovirus, *273*

ADP, 380, 435

Adrenal cortex, 446; hormones originated in, *448*

Adrenal gland, 445–46; in endocrine system, *443;* structure of, *447*

Adrenalin, 447

Adrenal medulla, 446; hormones originated in, *448*

Aerobic bacteria, 249, *249*

Aerobic respiration, 378

Aerospace Physiology Department, 421

AFP. *See* Alphafetoprotein

Africa: famine in, 37; first humans in, 554–57

African elephant, 62

African sleeping sickness, 271, 291

Agar cubes, comparison of, *120*

Agardhiella, 698

Age of Fishes, 549. *See also* Devonian period

Age of Mammals, 551. *See also* Cenozoic period

Agent orange, 500

Age of Reptiles, *243*

Agglutinate, 415

Aging, 152

Agnatha, 718

Agriculture: animal-powered, *637;* effects of, 636–38; industrialized, *638;* slash-and-burn methods of, 36; sustainable, 645–47

AIDS, 276–77; genetic engineering and, 184; linking toxin to virus molecule of, *278;* retrovirus causing, 274; screening for, 282; treatment for, 278

Ailuropoda melanoleuca, 231

Air: in biosphere, 13; circulation, 570–71; pollution, 644. *See also* Pollution

Alanine, *172*

Albatross, 354, 612, *720*

Alces alces, 573

Alcohol, *686*

Alga, 52; aquatic, *488,* 489; brown, 698; green, 288–89, 604, 615, 697; identification of, 624–25; mixed, *288;* multicellular, 290, 313; in ocean, 612; as photosynthetic protists, 287–88; red, 698; single-celled, *241;* symbiotic, 345; unicellular green, *100*

Alismataceae, 705

Alkalies, caustic, 668

Alkaline, 73

Alleena mexicana, 347

Alleles: defined, 166; frequency in population, 210; multiple, 176

Allergies, 414

Alligators, *50,* 51, *59, 719*

Alphafetoprotein (AFP), 149

Alpine lupine, 584

Alpine slide mystery, 425–27

Alpine tundra, 572

Alternaria, 701

Alternation of generations, 316

Altitude: affects on vegetation and animal life, *572;* chamber, *421*

Altman, Sidney, 247

Alveoli, 417

Alvin, 614

Amanita, 298

Amanita muscaria, 298

Amaryllidaceae, 705

Ambystoma opacum, 352

American eel, *719*

American Heart Association, 437

American Sign Language, chimp's use of, 521, *521*

American silkworm, *351*

American tapir, *722*

Amino acids, 108; defined, 81; essential, 391; in Krebs cycle, 382; microspheres and, *246;* in polypeptide chains, *81;* tRNA and, 172

Ammonia: conversion to nitrite, *270;* nitrogen fixers and, 268; as volcanic gas, 244; as waste substance, 108, 359

Ammonium ions, 269

Amniocentesis, 149–50, *150*

Amnion, *148,* 150

Amniotic cavity, *148*

Amniotic fluid, 150

Amoeba, 242, 292, 292, 678, 699

Amoebic dysentery, 271

Acknowledgements   continued from page iv

Inv. 3.2 Adapted with permission from the National Association of Biology Teachers; **Inv. 3.3** Adapted with permission from the National Association of Biology Teachers; **Inv. 10.1** Adapted from Cleveland P. Hickman, Jr., Larry S. Roberts, and Frances M. Hickman, *Integrated Principles of Zoology*, 8th ed. (St. Louis: Times Mirror/ Mosby College Publishing, 1988): 118; **Inv. 17.1** Reprinted by permission from Peter DeDecker, *The Responsible Use of Animals in Biology Classrooms*, (Hastings High School, Reston, Virginia: National Association of Biology Teachers.).

Illustration Acknowledgements

All illustrations by BSCS except the following: **3.20** Adapted with permission from the National Association of Biology Teachers; **3.12** *From Living in the Environment*, 6th ed. by G. Tyler Miller, Jr., Copyright © 1990 by Wadsworth, Inc. Adapted with permission of the publisher; **4BT** *From Living in the Environment*, 6th ed. by G. Tyler Miller, Jr., Copyright © 1990 by Wadsworth, Inc. Adapted with permission of the publisher; **8BT** Cystic Fibrosis Foundation; **8.28** Cystic Fibrosis Foundation; **9.9 top** Adapted from illustration on page 30 of Life Nature Library, *Evolution* © 1962 by Time Inc ; **9.9 bottom** Adapted from page 142 of *Morphological Differentiation and Adaptation In The Galapagos Finches* by Robert I. Bowman, University of California Publication in Zoology Volume 58 (1961); **10.3** Dr. Walter M. Fitch; **11.6** C. R. Woese; **18.8c** From *Biology,* by Leland G. Johnson. Dubuque IA: Wm. C. Brown, Copyright © 1983. Reprinted by permission. **21.21** "Peking Man," by Wu Rukang and Lin Shenglong. Copyright © 1983 by Scientific American, Inc. All rights reserved; **22.14** From "The Canopy of the Tropical Rainforest," by Donald R. Perry. Copyright© 1984 by Scientific American, Inc. All rights reserved; **24.1** *From Living in the Environment*, 6th ed. by G. Tyler Miller, Jr., Copyright © 1990 by Wadsworth, Inc. Adapted with permission of the publisher; **24.10** *From Living in the Environment*, 6th ed. by G. Tyler Miller, Jr., Copyright © 1990 by Wadsworth, Inc. Adapted with permission of the publisher.

Photograph Acknowledgements

Unit 1 Opener JPL/TSADO/Tom Stack and Associates; **Chapter 1 Opener** Jeff Foott; **1.1 (a)** C. Allan Morgan; **1.1 (b)** Milton Rand/Tom Stack and Associates; **1.2** Carlye Calvin; **1.3** Courtesy of USDA-APHIS; **1.6** Carlye Calvin; **1.9 (a)** Carlye Calvin/BSCS; **1.9 (b)** Karl Aufderheide/Visuals Unlimited; **1.14** Mark Newman/Tom Stack and Associates; **Pioneers** BSCS; **Pioneers** Nancy L. Cushing/Visuals Unlimited; **1.15 (a)** Carlye Calvin; **1.15 (b–d)** Carlye Calvin/BSCS.

Chapter 2 Opener Kenneth D. Edwards; **2.1** C. Allan Morgan; **2.2** William J. Cairney; **2.3** Jackie Ott-Rodgers/BSCS; **2.8 (a)** Maurice Cariale/Carlye Calvin; **2.8 (b)** Kjell B. Sandved/Visuals Unlimited; **2.9** Dwight Kuhn; **2.12 (a)** Carlye Calvin/BSCS; **2.12 (b)** Carlye Calvin/Linda Ward/BSCS; **2.13** Gary Braasch; **BT** Courtesy of NASA; **2.14 (a)** Tim McCabe/Courtesy of Soil Conservation Service; **2.14 (b)** Tim McCabe/Courtesy of Soil Conservation Service.

Chapter 3 Opener Dwight Kuhn; **3.1** Carlye Calvin; **3.5** William J. Cairney; **3.6** Brian Parker/Tom Stack and Associates; **3.7** Jackie Ott-Rodgers/BSCS; **3.8** Mark Newman/Tom Stack and Associates; **3.9** Carlye Calvin; **3.11** Courtesy of USDA; **3.14** John D. Cunningham/Visuals Unlimited; **3.15** Brian Parker/Tom Stack and Associates; **3.16** Nancy L. Cushing/Visuals Unlimited; **3.17 (b)** Kerry T. Givens/Tom Stack and Associates; **3.18** G. Prance/Visuals Unlimited; **3.19** William J. Cairney; **BT** Courtesy of Bear Creek Nature Center.

Chapter 4 Opener Doug Sokell; **4.7** Bruce Gaylord/Visuals Unlimited.

Unit 2 Opener Joe McDonald/Visuals Unlimited; **Chapter 5 Opener** Ed Reschke; **5.1 (a)** Ed Reschke; **5.1 (b)** Biodisc, Inc. 1-800-453-3009; **5.1 (c&d)** Ed Reschke; **BT (a)** Linda Hall Library; **5.2 (a&b)** Ed Reschke; **5.3 (a)** Biodisc, Inc. 1-800-453-3009; **5.3 (b)** Carolyn I. Noble/Courtesy of Colorado College; **5.4 (a)** William J. Cairney; **5.4 (b)** Gerald Van Dyke/Visuals Unlimited; **5.5** Veronica Burmesiter/Visuals Unlimited; **5.7** K. G. Murti/Visuals Unlimited; **5.8** David M. Phillips/Visuals Unlimited; **5.9** K. G. Murti/Visuals Unlimited; **5.10 (a–c)** BSCS; **5.16 (a–i)** Andrew S. Bajer, Department of Biology, University of Oregon; **5.17** Werner Heim/BSCS; **5.19** David M. Phillips/Visuals Unlimited; **5.20** David M. Phillips/Visuals Unlimited.

Chapter 6 Opener David M. Phillips/Visuals Unlimited; **6.2** Ed Reschke; **6.3 (a)** Jackie Ott-Rodgers/BSCS; **6.3 (b)** Brian Parker/Tom Stack and Associates; **6.5** David M. Phillips/Visuals Unlimited; **6.6** R. DeGoursey/Visuals Unlimited; **6.11(c)** Ed Reschke; **6.14** BSCS.

Chapter 7 Opener C. Allan Morgan; **7.1 (a–f)** Carolina Biological Supply Company; **7.7** Carolina Biological Supply Company; **7.9** Courtesy of Carnegie Institution of Washington; **7.12 (a)** Courtesy of Carnegie Institution of Washington, Department of

Embryology Pairs Division; **7.12 (b)** Lennart Nilsson, A CHILD IS BORN, Dell Publishing Company; **7.12 (c–e)** Robert Rugh; **7.12 (f)** Lennart Nilsson, A CHILD IS BORN, Dell Publishing Company; **7.13** Courtesy of Dr. James W. Hanson, M.D., Professor of Pediatrics, The University of Iowa; **7.17** Ward's Natural Science Establishment, Inc.

Chapter 8 Opener Dave Watts/Tom Stack and Associates; **8.2** Courtesy of Dr. Orel Vitezlav, Brno, Czech Republik; **8.7** R. Calentine/Visuals Unlimited; **8.16** David C. Peakman, Cytogenetics Laboratory, Reproductive Genetic Center, Denver, Colorado; **8.17** David C. Peakman, Cytogenetics Laboratory, Reproductive Genetic Center, Denver, Colorado; **8.20 (a&b)** Douglas P. Haug/Courtesy of Down Syndrome Association, Colorado Springs, Colorado; **8.22** Courtesy of Carolyn Trunca, Ph.D., Department of Ob & Gyn, Genetics Division, School of Medicine, State University of New York at Stonybrook; **Pioneers** Will & Dent McIntyre/Photo Researchers, Inc.; **8.29** BSCS.

Chapter 9 Opener (a) David M. Dennis/Tom Stack and Associates; **Chapter 9 Opener (b)** John Shaw/Tom Stack and Associates; **9.1 (left, middle & right)** Biodisc, Inc. 1-800-453-3009; **9.3** Kaylene Silvester, Bucklin, Kansas, mules owned by Bill and Oneta Silvester, Champion, Nebraska; **9.4 (a–c)** W. D. Bransford/National Wildflower Research Center; **9.7** C. Allan Morgan; **9.8** The Bettman Archive; **9.11** Dr. L. M. Cook, Department of Environmental Biology, University of Manchester, Manchester, U.K.; **9.17 (a)** Jeff Foott/Tom Stack and Associates; **9.18 (b)** Mary Clay/Tom Stack and Associates; **BT (a&b)** Courtesy of Wildlife Safari, Winston, Oregon; **9.20 (a&b)** Walt Anderson/Visuals Unlimited; **9.21 (top left-top right-bottom left)** Carlye Calvin/BSCS; **9.21 (bottom right)** Gordon Uno/BSCS; **9.22 (left)** Carlye Calvin/BSCS; **9.22 (right)** Dan Suzio/Visuals Unlimited.

Unit 3 Opener C. Allan Morgan; **Chapter 10 Opener** David Fleetham/Tom Stack and Associates; **10.2** Barbara Von Hoffman/Tom Stack and Associates; **10.4** Martin G. Miller/Visuals Unlimited; **10.7** Jackie Ott-Rodgers/BSCS; **BT** Courtesy of The Cincinnati Zoo, Center for Reproduction of Endangered Wildlife; **10.11 (a)** B. Ormerod/Visuals Unlimited; **10.11 (b)** Biodisc, Inc. 1-800-453-3009; **10.11 (c)** Courtesy of J. G. Zeikus, Michigan Biotechnology Institute, Lansing, Michigan; **10.12 (a)** Doug Sokell; **10.12 (b)** Terry Donnelly/Tom Stack and Associates; **10.12 (c)** Carlye Calvin; **10.13 (a)** Carlye Calvin; **10.13 (b)** Denise Tackett/Tom Stack and Associates; **10.13 (c)** Stan Elems/Visuals Unlimited; **10.14 (a)** C. Allan Morgan; **10.14 (b)** Milton H. Tiernery/ Visuals Unlimited; **10.14 (c)** Kenneth D. Whitney/Visuals Unlimited; **10.15 (a)** Philip Sze/Visuals Unlimited; **10.15 (b)** T. E. Adams/Visuals Unlimited; **10.15 (c)** Kerry T. Givens/Tom Stack and Associates; **10.18 (a)** S. M. Awranik/BPS/University of California/Tom Stack and Associates; **10.18 (b)** BPS/Tom Stack and Associates; **10.21** Courtesy of Dr. Glen Mason, Indiana University Southeast; **10.23** Sidney Fox/Visuals Unlimited; **10.26** Science VU/Visuals Unlimited; **10.28** Doug Sokell/BSCS.

Chapter 11 Opener Mark Newman/Tom Stack and Associates; **Chapter 11 Opener Insert** Courtesy of United States Environmental Protection Agency; **11.3 (a–c)** Courtesy of Lynn Margulis, Department of Botany, University of Massachusetts; **11.4** BPS/Pankratz, Beakman, Gerhard/Tom Stack & Associates; **11.5** T. F. Anderson, Institute for Cancer Research, Philadelphia, Pennsylvannia; **11.7 (a–c)** Courtesy of K. O. Stetter, University of Regenburg, Munich, Federal Republic of Germany; **11.8** Courtesy of Lynn Margulis, Department of Botany, University of Massachusetts; **11.9 (a&b)** Courtesy of J. G. Zeikus, Department of Botany, University of Massachusetts; **11.11** David M. Phillips/Visuals Unlimited; **11.12 (a)** Dr. E. S. Boatman; **11.12 (b)** Courtesy of J. G. Zeikus, Department of Botany, University of Massachusetts; **11.12 (c&d)** Courtesy of J. G. Zeikus, Department of Bacteriology, University of Wisconsin; **11.13 (a)** T. E. Adams/Visuals Unlimited; **11.13 (b)** Ed Reschke; **11.13 (c)** T. E. Adams/Visuals Unlimited; **11.15 (a)** John D. Cunningham/Visuals Unlimited; **11.15 (b)** C. P. Vance/Visuals Unlimited; **11.16 (a&b)** Courtesy of Stan Watson, Woods Hole Oceanographic Institute, Woods Hole, Massachusetts; **11.17** Science VU/Visuals Unlimited; **11.18** Courtesy of Bill Jacobi, Department of Agriculture Colorado State University; **11.19** Thomas D. Brock/Science Tech Publishers; **11.20** David M. Phillips/Visuals Unlimited; **11.21 (a)** Science VU/Visuals Unlimited; **11.21 (b)** Hans Gerlderblon/Visuals Unlimited; **11.21 (c)** Cabisco/Visuals Unlimited; **11.22** Science VU/Visuals Unlimited; **11.24 (a)** E. Couture-Tosi, The Electron Microspy, Group of Biozentrum, Basel, Schweiz, Switerland; **11.24 (b)** J. V. deu Broek, The Electron Microspy, Group of Biozentrum, Basel, Switzerland; **11.26** A. M. Siegelman/Visuals Unlimited; **11.27** Science VU/Visuals Unlimited; **11.28** Science VU/Visuals Unlimited.

Chapter 12 Opener John Shaw/Tom Stack and Associates; **12.1** Dwight Kuhn; **12.2 (a)** Linda Sims/Visuals Unlimited; **12.2 (b)** T. E. Adams/Visuals Unlimited; **12.2 (c)** John D. Cunningham/Visuals

Unlimited; **12.4 (a)** Cabisco/Visuals Unlimited; **12.4 (b)** Veronika Burmeister/Visuals Unlimited; **12.5 (a&b)** Ed Reschke; **12.5 (c)** Ralph Robinson/Visuals Unlimited; **12.6 (a)** T. E. Adams/Visuals Unlimited; **12.6 (b)** David M. Phillips/Visuals Unlimited; **12.6 (c)** Dr. Ron Hathaway/Courtesy of The Colorado College; **12.7 (a)** M. Abbey/Visuals Unlimited; **12.7 (b)** T. E. Adams/Visuals Unlimited; **12.8 (a)** A. M. Siegelman/Visuals Unlimited; **12.8 (b)** Stan Elems/Visuals Unlimited; **12.9 (a)** T. E. Adams/Visuals Unlimited; **12.9 (b)** Karl Aufderheide/Visuals Unlimited; **12.10** Biodisc, Inc. 1-800-453-3009; **12.11 (a)** David M. Dennis/Visuals Unlimited; **BT** Dwight Kuhn; **12.12** James W. Richardson/Visuals Unlimited; **12.13** Jack Bostrack/Visuals Unlimited; **12.15 (a)** David Newman/Visuals Unlimited; **12.15 (b)** John Gerlach/Visuals Unlimited; **12.15 (c)** David Newman/Visuals Unlimited; **12.17 (a)** Dick Poe/Visuals Unlimited; **12.17 (b)** George Loun/Visuals Unlimited; **12.17 (c)** Carlye Calvin; **12.19 (a)** Carlye Calvin/BSCS; **12.19 (b)** John D. Cunningham/Visuals Unlimited; **12.21** Courtesy of David Preamer, Rutgers University; **12.22** Stanley Fleger/Visuals Unlimited; **12.23 (a&b)** Doug Sokell/BSCS; **12.24** John D. Cunningham/Visuals Unlimited.

Chapter 13 Opener Kerry T. Givens/Tom Stack and Associates; **13.1** Jack S. Grove/Tom Stack and Associates; **13.2** John D. Cunningham/Visuals Unlimited; **13.4** Carlye Calvin/BSCS; **13.5** E. J. Cable/Tom Stack and Associates; **13.6 (a)** Doug Sokell/BSCS; **13.6 (b)** William S. Ormerod/Visuals Unlimited; **13.8** Fred Hossler/Visuals Unlimited; **13.11 (a)** Biodisc, Inc. 1-800-453-3009; **13.11 (b)** John Cunningham/Visuals Unlimited; **13.15 (a)** Carlye Calvin; **13.15 (b)** John Shaw/Tom Stack and Associates; **13.15 (c)** Ken Davis/Tom Stack and Associates; **13.17** William S. Ormerod/Visuals Unlimited; **13.18 (a)** Greg Vaughn/Tom Stack and Associates; **13.18 (b)** Kevin Schafer/Tom Stack and Associates; **13.20 (a)** David M.Dennis/Tom Stack and Associates; **13.21 (a)** Hugh Spenser/BSCS; **13.21 (b)** Doug Sokell/BSCS; **13.22** George Loun/Visuals Unlimited; **13.23 (a)** Doug Sokell; **13.23 (b)** BSCS; **13.23 (c)** John Cunningham/Visuals Unlimited; **Pioneers** Jerry L. Grant/BSCS; **13.24 (a)** Doug Sokell/BSCS; **13.26 (a&b)** Doug Sokell/BSCS; **13.27 (a)** Carlye Calvin; **13.27 (b)** Carlye Calvin/BSCS; **13.27 (c&d)** Carlye Calvin; **13.28 (a&b)** Doug Sokell/BSCS; **13.31 (a)** Doug Sokell/BSCS; **31.32** Jackie Ott-Rogers; **BT** Doug Sokell/BSCS.

Chapter 14 Opener Jeff Foott/Tom Stack and Associates; **14.2 (a)** Bruce Russell/BioMEDIA Associates; **14.2 (b)** Carlye Calvin; **14.4 (a)** Brian Parker/Tom Stack and Associates; **14.4 (b)** Tom Stack/Tom Stack and Associates; **14.5 (b)** Denise Tackett/Tom Stack and Associates; **14.5 (c)** Mike Bacon/Tom Stack and Associates; **14.5 (d)** Jeff Foott; **14.6 (a)** C. Allan Morgan; **14.7 (a)** Arthur M. Siegleman/Visuals Unlimited; **14.7 (b)** R. Calentine/Visuals Unlimited; **14.9 (a)** David B. Fleetham/Tom Stack and Associates; **14.9 (b)** Bill Tronca/Tom Stack and Associates; **14.10 (a)** Carlye Calvin; **14.10 (b)** Brian Parker/Tom Stack and Associates; **14.12 (a)** Milton H. Tierney, Jr./Visuals Unlimited; **14.12 (b)** D. Newman/Visuals Unlimited; **14.12 (c)** Milton H. Tierney, Jr./Visuals Unlimited; **14.13 (a)** Larry Lapsky/Tom Stack and Associates; **14.13 (b)** Milton Rand/Tom Stack and Associates; **14.14 (a)** George D. Dodge & Dale R. Thompson/Tom Stack and Associates; **14.14 (b)** John Gerlach/Tom Stack and Associates; **14.14 (c)** David M. Dennis/Tom Stack and Associates; **14.16 (a)** David B. Fleetham/Tom Stack and Associates; **14.16 (b)** Tom Stack/Tom Stack and Associates; **14.17 (a)** Randy Morse/Tom Stack and Associates; **14.17 (b)** Brian Parker/Tom Stack and Associates; **14.18 (a)** D. G. Barker/Tom Stack and Associates; **14.18 (b)** Richard Thom/Tom Stack and Associates; **14.19 (a)** Kevin Schafer/Tom Stack and Associates; **14.19 (b)** David M. Dennis/Tom Stack and Associates; **14.20 (a&b)** Carlye Calvin; **14.21 (a)** Carlye Calvin; **14.21 (b)** David B. Fleetham/Tom Stack and Associates; **14.21 (c)** Carlye Calvin; **BT** Dave Watts/Tom Stack and Associates.

Unit 4 Opener Eric Sanford/Tom Stack and Associates; **Chapter 15 Opener** Carlye Calvin/BSCS; **15.4 (b)** Don Fawcett/Visuals Unlimited; **15.6 (b)** Courtesy of Keith R. Porter, University of Maryland; **15.13** Carlye Calvin/BSCS; **Pioneers** William J. Cairney/BSCS; **15.14 (a&b)** Biodisc, Inc. 1-800-453-3009; **15.16** Doug Sokell/BSCS; **15.17** Doug Sokell/BSCS.

Chapter 16 Opener Robert Caughey/Visuals Unlimited; **16.2 (a)** SIU/Visuals Unlimited **16.4** Ed Reschke; **16.11** Science VU/Boehringer Ingelheim/Visuals Unlimited; **Pioneers** William J. Cairney/BSCS; **16.21** Carlye Calvin; **16.23 (b)** Doug Sokell/BSCS; **16.24** Carlye Calvin.

Chapter 17 Opener Tom Stack/Tom Stack and Associates; **17.1 (a-c)** Ed Reschke; **17.3** John Cunningham/Visuals Unlimited; **BT** R. G. Kessel & C.Y. Shih/Visuals Unlimited; **Pioneers** William J. Cairney/BSCS; **17.19 (a&b)** Peter N. Witt/Courtesy of North Carolina Department of Mental Health.

Chapter Opener 18 Milton H. Tierney, Jr./Visuals Unlimited; **18.1 (a)** Carlye Calvin/BSCS; **18.1 (b)**

Gordon Uno/BSCS; **18.3** Gerald VanDyke/Visuals Unlimited; **18.5** Jackie Ott-Rodgers/BSCS; **18.6 (b)** Doug Sokell/BSCS; **18.9 (a)** Randy Moore/Visuals Unlimited; **18.9 (b)** George J. Wilder/Visuals Unlimited; **BT** Jan Chatlain Girard/BSCS; **18.15** Doug Sokell/BSCS; **18.16** Gordon Uno/BSCS.
Chapter 19 Opener Doug Sokell/Tom Stack and Associates; **19.1 (a)** Courtesy of Dr. Lewis K. Shumway, College of Eastern Utah, Blanding; **19.4** Carlye Calvin/BSCS; **19.14** Courtesy of Don Luvisis, University of California Cooperative Extension; **19.16 (a&b)** Carlye Calvin/BSCS; **19.17** David Newman/Visuals Unlimited.
Unit 5 Opener Manfred Gottschalk/Tom Stack and Associates; **Chapter 20 Opener** David B. Fleetham/Tom Stack and Associates; **20.1** Courtesy of William Vandivert; **20.2** Tom J. Ulrich/Visuals Unlimited; **BT (a)** C. Allan Morgan; **BT (b)** John Gerlach/Tom Stack and Associates; **BT (c)** Dan Kline/Visuals Unlimited; **20.4 (a&b)** Stephen J. Krasemann/DRK PHOTO; **20.6** H. S. Terrace/Anthro-Photo; **20.7** Joe and Carol McDonald/Tom Stack and Associates; **20.8** C. Allan Morgan; **20.10** Dick Poe/Visuals Unlimited; **20.11** Jeff Foott; **20.12** Jeff Foott; **20.13** Joe McDonald/Visuals Unlimited; **20.14** Jeff Foott; **20.16 (a)** Gerald and Buff Corsi/Tom Stack and Associates; **20.17** Brian Parker/Tom Stack and Associates; **20.18** Jeff Foott; **20.19** Jeff Foott; **20.20** C. Allan Morgan; **20.21(a&b)** Jeff Foott.
Chapter 21 Opener Ken Lucas/Visuals Unlimited; **21.1 (a)** Jeff Foott; **21.1 (b)** John Cancalosi/Tom Stack and Associates; **21.1 (c)** Jeff Foott/Tom Stack and Associates; **21.2** Carlye Calvin/BSCS; **21.4 (a)** Smithsonian Institution Negative No.28531; **21.4 (b)** Peabody Museum of Natural History Yale, University; **21.6** Carlye Calvin; **21.7 (a)** James R. McCullagh/Visuals Unlimited; **21.7 (b)** Ed Reschke; **21.11** Negative #318465 (Photo by John Bowen) Courtesy of Department of Library Services, American Museum of Natural History; **21.12** Trans. No. K10234 (Photo by Earth History Hall/AMNH) Courtesy of Department of Library Services, American Museum of Natural History; **21.13** Betty Seacrest/Carlye Calvin; **21.15** Smithsonian Institution; **21.18** Institute of Human Origins; **21.19 (a&b)** Doug Sokell/BSCS; **21.20 (a&b)** Doug Sokell/BSCS; **BT** Doug Sokell/BSCS; **21.22** Doug Sokell/BSCS; **21.23 (a&b)** Doug Sokell/BSCS.
Chapter 22 Opener Terry Donnelly/Tom Stack and Associates; **22.1** Courtesy of Paul McIver; **22.6** Richard Tolman/BSCS; **22.8 (a)** S. Maslowski/Visuals Unlimited; **22.8 (b)** Beth Davidow/Visuals Unlimited; **22.9** Kirtley-Perkins/Visuals Unlimited; **22.10** Beth Davidow/Visuals Unlimited; **22.11(a–d)** William J. Weber/Visuals Unlimited; **22.13 (a)** Tom Kitchin/Tom Stack and Associates; **22.13 (b)** Mary Clay/Tom Stack and Associates; **22.13 (c)** John Shaw/Tom Stack and Associates; **22.15 (a)** Kjell B. Sandved/Visuals Unlimited; **22.15 (b)** Roy Toft/Tom Stack and Associates; **22.16 (a&b)** Kevin Magee/Tom Stack and Associates; **22.16 (c)** BSCS; **22.18 (a)** Carlye Calvin; **22.18 (b)** Shattil/Rozinski/Tom Stack and Associates; **22.19** Doug Sokell/Tom Stack and Associates; **22.20** Jeff Foott; **22.21** Walt Anderson/Visuals Unlimited; **22.22** John D. Cunningham/Visuals Unlimited; **22.23 (a–c)** Courtesy of Dr. Lora Mangum Shields, Navajo Community College; **22.25** Gary Braasch; **BT** Jeff Foott/Tom Stack and Associates; **22.26** Larry Brock/Tom Stack and Associates; **22.27** Greg Vaughn/Tom Stack and Associates; **22.28** Larry Brock/Tom Stack and Associates; **22.29** John D. Cunningham/Visuals Unlimited.
Chapter 23 Opener Rich Buzzelli/Tom Stack and Associates; **23.2** William C. Jorgensen/Visuals Unlimited; **23.4** C. Allan Morgan; **23.7** Christine Case/Visuals Unlimited; **23.9** Terry Donnelly/Tom Stack and Associates; **23.11** TSA/Tom Stack and Associates; **23.14** C. Allan Morgan; **23.15** Brian Parker/Tom Stack and Associates; **23.18** Whol, D. Foster/Visuals Unlimited; **23.19** Cindy Garoutte/Tom Stack and Associates; **23.20** Brain Parker/Tom Stack and Associates; **23.21** T. Kitchin/Tom Stack and Associates; **23.22** Daniel W. Gotshall/Visuals Unlimited; **23.23** John D. Cunningham/Visuals Unlimited; **23.24** Bob Pool/Tom Stack and Associates; **23.25** Gary Braasch; **BT** Alan Decker; **23.27** Doug Sokell/Tom Stack and Associates; **23.28** Jeff Foott; **23.29** John Shaw/Tom Stack and Associates.
Chapter 24 Opener D.Cavagnaro/Visuals Unlimited; **24.2** G. Prance/Visuals Unlimited; **24.3** Gary Milburn/Tom Stack and Associates; **24.4** Inga Spence/Tom Stack and Associates; **24.5** Dick Poe/Visuals Unlimited; **24.8** The Bettman Archive; **BT** Courtesy of Exxon Research and Engineering Company; **24.11** Courtesy of USDA; **24.12** Courtesy of USDA; **24.13 (a)** William J. Cairney/BSCS; **24.13 (b)** Bob Wilson/BSCS; **24.13 (c)** Don W. Fawcett/Visuals Unlimited; **24.13 (d)** Terry Donnelly/Tom Stack and Associates; **24.14** Courtesy of M.Chahine, Jet Propulsion Laboratory; and J. Susskind, Goddard Space Flight Center, NASA; **24.15** Courtesy of M. Chahine, Jet Propulsion Laboratory, California Institute of Technology, under contract with NASA.